THE LIBRARY
ST. MARY'S COLLEGE OF MARYLAND
ST. MARY'S CITY, MARYLAND 20686

KETENES

KETENES

THOMAS T. TIDWELL
Department of Chemistry
University of Toronto
Toronto, Ontario

A Wiley-Interscience Publication
JOHN WILEY & SONS, INC.
New York / Chichester / Brisbane / Toronto / Singapore

This text is printed on acid-free paper.

Copyright ©1995 by John Wiley & Sons, Inc.

All rights reserved. Published simultaneously in Canada.

Reproduction or translation of any part of this work beyond that permitted by Section 107 or 108 of the 1976 United States Copyright Act without the permission of the copyright owner is unlawful. Requests for permission or further information should be addressed to the Permissions Department, John Wiley & Sons, Inc., 605 Third Avenue, New York, NY 10158-0012.

Library of Congress Cataloging in Publication Data:
Tidwell, Thomas T., 1939-
   Ketenes / Thomas T. Tidwell.
      p. cm.
   "A Wiley-Interscience publication."
   Includes index.
   ISBN 0-471-57580-1
   1. Ketenes. I. Title.
QD305.K2T53   1995
547'.436–dc20   95-7884

Printed in the United States of America

10 9 8 7 6 5 4 3 2 1

To Sarah, who makes it all worthwhile

CONTENTS

Preface	**xiii**
Introduction	**1**
References / 2	

1. STRUCTURE, BONDING, AND THERMOCHEMISTRY OF KETENES 4

    1.1   Theoretical Studies of Ketenes / 4
            1.1.1   Molecular and Electronic Structure and Energy / 4
            1.1.2   Theoretical Studies of Ketene Reactions / 6
            1.1.3   Substituent Effects on Ketenes / 13
            References / 18
    1.2   Molecular Structure Determinations / 20
            References / 31
    1.3   Thermochemistry of Ketenes / 31
            References / 32

2. SPECTROSCOPY AND PHYSICAL PROPERTIES OF KETENES 33

    2.1   Nuclear Magnetic Resonance Spectroscopy / 33
    2.2   Ultraviolet and Photoelectron Spectroscopy / 36
    2.3   Infrared Spectra / 38
    2.4   Dipole Moments / 44
    2.5   Mass Spectrometry and Gas Phase Ion Chemistry / 44
            References / 47

3. PREPARATION OF KETENES 52

    3.1   Ketenes from Ketene Dimers / 52

3.2 Ketenes from Carboxylic Acids and Their Derivatives / 53
    3.2.1 Ketenes from Carboxylic Anhydrides / 53
        References / 56
    3.2.2 Ketenes from Acyl Halides and Activated Acids / 57
        References / 66
    3.2.3 Ketenes from Esters / 67
        3.2.3.1 Ketenes from Ester Enolates / 67
        3.2.3.2 Ketenes by Ester Pyrolysis / 70
        3.2.3.3 Other Preparations of Ketenes from Esters / 73
        References / 73
    3.2.4 Ketenes by Dehalogenation of α-Halo Carboxylic Derivatives / 74
        References / 76

3.3 Ketenes from Diazo Ketones (Wolff Rearrangements) / 77
    3.3.1 Thermal Wolff Rearrangement / 82
    3.3.2 Photochemical Wolff Rearrangement / 84
        References / 96

3.4 Ketenes by Photochemical and Thermolytic Methods / 100
    3.4.1 Ketenes from Cyclobutanones and Cyclobutenones / 100
        References / 107
    3.4.2 Ketenes from Photolysis of Cycloalkanones and Enones / 108
        References / 118
    3.4.3 Ketenes from Cyclohexadienones and Other Cycloalkenones / 120
        References / 124
    3.4.4 Ketenes from Dioxinones / 125
        References / 128
    3.4.5 Ketenes by Thermolysis of Alkynyl Ethers / 129
        References / 130
    3.4.6 Ketenes from Other Thermolytic and Photochemical Routes / 130
        References / 137

3.5 Ketenes from Alkenylcarbene Metal Complexes / 138
    References / 144

3.6 Ketene Formation under Acidic Conditions / 145
    References / 145

3.7 Ketenes from Oxidation of Alkynes / 146
    References / 147

3.8 Other Routes to Ketenes / 147
    References / 149

4. TYPES OF KETENES 150

4.1 Carbon-Substituted Ketenes / 150
   4.1.1 Alkylketenes / 150
          References / 173
   4.1.2 Alkenylketenes / 176
          References / 191
   4.1.3 Alkynylketenes and Cyanoketenes / 193
          References / 202
   4.1.4 Arylketenes / 203
          References / 216
   4.1.5 Cyclopropylketenes / 218
          References / 226
   4.1.6 Acylketenes / 227
          References / 251
   4.1.7 Imidoylketenes / 254
          References / 259
   4.1.8 Cumulene-Substituted Ketenes / 259
          References / 262
   4.1.9 Ketenes with Charged, Radical, or Carbenic Side Chains / 262
          References / 268
   4.1.10 Fulvenones / 269
          4.1.10.1 Triafulvenones / 269
          4.1.10.2 Pentafulvenones / 270
          4.1.10.3 Heptafulvenones / 275
          4.1.10.4 Oxoquinone Methides and Related Species / 276
          References / 297
4.2 Nitrogen-Substituted Ketenes / 299
     References / 309
4.3 Oxygen-Substituted Ketenes / 310
     References / 320
4.4 Halogen-Substituted Ketenes / 321
   4.4.1 Fluoroketenes, Perfluoroalkylketenes, and Perfluoroarylketenes / 321
          References / 334
   4.4.2 Chlorine, Bromine, and Iodine-Substituted Ketenes / 336

References / 347
4.5 Silyl-, Germyl-, and Stannylketenes / 348
References / 366
4.6 Phosphorous- and Arsenic-Substituted Ketenes / 368
References / 379
4.7 Sulfur-Substituted Ketenes / 380
References / 387
4.8 Metal-Substituted Ketenes / 388
    4.8.1 Lithium Ketenes (Lithium Ynolates) and Ynols / 388
    References / 394
    4.8.2 Boron-Substituted Ketenes / 395
    References / 399
    4.8.3 Other Metal-Substituted Ketenes and Metal Ketenides / 399
    References / 402
    4.8.4 Metal-Complexed Ketenes / 402
    References / 405
4.9 Bisketenes / 405
References / 427
4.10 Ketenyl Radicals, Anions, and Cations / 429
References / 433
4.11 Cumulenones / 434
References / 442

5. REACTIONS OF KETENES   445

5.1 Oxidation and Reduction of Ketenes (Electron Transfer) / 445
References / 447
5.2 Photochemical Reactions / 448
References / 453
5.3 Thermolysis Reactions / 454
References / 458
5.4 Cycloaddition Reactions of Ketenes / 459
    5.4.1 Intermolecular [2 + 2] Cycloaddition / 460
        5.4.1.1 Dimerization of Ketenes / 460
        5.4.1.2 Cycloadditions with Alkenes and Dienes / 473
        5.4.1.3 Mechanism of Ketene [2 + 2] Cycloadditions with Alkenes / 486
        5.4.1.4 Cycloaddition of Ketenes with Nucleophilic Alkenes / 502
        5.4.1.5 Cycloaddition of Ketenes with Allenes / 511

             5.4.1.6   Cycloaddition of Ketenes with Alkynes / 514
             5.4.1.7   Cycloaddition of Ketenes with Imines / 518
             5.4.1.8   Cycloadditions of Ketenes with Other
                       Substrates / 527
             References / 529
      5.4.2 [3 + 2] Cycloaddition Reactions of Ketenes / 536
      5.4.3 [4 + 2] Cycloadditions of Ketenes / 544
             References / 553
      5.4.4 Intramolecular Cycloadditions of Ketenes / 554
             References / 563
      5.4.5 Intermolecular and Intramolecular Cycloaddition of
             Ketenes with Carbonyl Groups / 564
             References / 570
5.5   Nucleophilic Addition to Ketenes / 571
      5.5.1 Mechanisms / 571
             5.5.1.1   Theoretical Studies / 572
             5.5.1.2   Kinetics of Hydration: Neutral and Base
                       Reactions / 576
             5.5.1.3   Acid-Catalyzed Hydration / 585
             5.5.1.4   Alcoholysis and Aminolysis / 587
      5.5.2 Nucleophilic Additions to Ketenes: Preparative Aspects / 590
             5.5.2.1   Hydride Addition / 590
             5.5.2.2   Oxygen Nucleophiles / 592
             5.5.2.3   Nitrogen Nucleophiles / 597
             5.5.2.4   Carbon Nucleophiles / 604
             5.5.2.5   Other Nucleophiles / 611
             References / 613
      5.5.3 Wittig Reactions / 619
             References / 621
5.6   Electrophilic Addition to Ketenes / 622
      5.6.1 Protonation of Ketenes / 622
      5.6.2 Electrophilic Addition of Hydrogen Halides to Ketenes / 623
      5.6.3 Electrophilic Additions of Other Reagents / 624
             References / 628
      5.6.4 Oxygenation of Ketenes / 630
             References / 636
5.7   Radical Reactions of Ketenes / 636
      References / 639
5.8   Polymerization of Ketenes / 639

    References / 642
5.9 Stereoselectivity in Ketene Reactions / 642
    References / 651
5.10 Other Additions to Ketenes / 653
    5.10.1 Reaction with Diazomethanes / 653
    5.10.2 Reaction with Sulfur Dioxide / 655
    References / 655

Index 657

PREFACE

The inspiration for this text is Staudinger's classic book *Die Ketene,* which was published in 1912 and encompassed the entire field of ketene chemistry. This period was the first golden age of ketene chemistry, and a remarkable amount had already been learned following the identification of the first ketene in 1905.

Comprehensive coverage of the entire field of ketene chemistry would now require many authors and volumes, and such a work would be too costly for individuals to own. The previous work of this type, *The Chemistry of Ketenes, Allenes, and Cumulenes,* edited by Saul Patai, appeared in 1980, and the field has expanded far beyond what was known at that time. Important topics, such as cycloaddition, were largely neglected in that text because the planned chapters were never completed. Furthermore, the coverage in such a multiauthor text highlights the particular interests of the various authors, and so while many areas are covered in great depth, others are omitted altogether.

It is the ambitious goal of this volume to provide an up-to-date introduction to almost all aspects of ketene chemistry, so that those interested in the field can use this book as a starting point for work on any topic dealing with ketenes. By following the leading references provided, full details about specific subjects of interest can be found. To keep the book at a manageable length much of the discussion must be quite brief.

The *Chemical Abstracts* Registry File has been examined by means of a substructure search for all ketenes, namely those molecules containing the C=C=O grouping. These are listed in the tables in Chapter 4, grouped according to structural type, along with Registry Numbers, a brief note about the method of preparation, and one or two references. It is not practical to give all references to each ketene (there are more than 800 in *Chemical Abstracts* to diphenylketene), but it is hoped that sufficient information is given to lead one to any further information required.

Many ketenes which have been generated are not indexed in *Chemical Abstracts* because this journal does not cite reactive intermediates involved in chemical reactions, and many ketenes fall in this category. In many cases where such ketenes are implicated in the literature they have been included in the tables, but no Registry Numbers exist. Also, in several cases more than one Registry Number appears to have been assigned to the same structure, and all of these are included. In some cases Registry Numbers have been assigned to compounds which have not been reported in the literature. Several errors in the assignment of ketene structures during abstracting have been noted, as well as several ketene structural assignments which have subsequently been revised. Furthermore, most of the metal-complexed ketenes found in *Chemical Abstracts* were omitted from the tables, as the formulae are too complicated for simple representation.

Several areas of active investigation that have been given rather superficial treatment include metal-complexed ketenes, ketene photochemistry, thioketenes, and oxocarbons, such as carbon suboxide (C_3O_2). There is a large body of current literature dealing with each of these topics, and leading references are provided to assist the interested reader in accessing these topics. The survey of the literature includes selected references from throughout 1994 and even a few from 1995, the Year of the Pig in the Chinese calendar.

During the course of writing this volume many unanswered questions became apparent, and where possible these have been noted, with the hope of stimulating the readers to fill in the gaps. However, despite the advance of understanding the horizons of ketene chemistry are receding even more rapidly, and it is certain that when the Centennial of the Discovery of Ketenes is marked in 2005, the opportunities for research will be even greater than they are today.

The study of ketene chemistry has been marked from the beginning by scientific controversy, disagreement, and uncertainty. The structure of ketene itself was originally a matter of dispute, and the structure of the ketene dimer was a puzzle for nearly 50 years. Even today the mechanisms of ketene cycloadditions and ketene hydration are not agreed upon.

The attempt to cover such a diversified area has meant that the knowledge of the author has been severely taxed, and it is certain that errors, misinterpretations, and omissions have occurred. The author takes full responsibility for these faults and will be grateful to individuals who provide corrections, further information, or better interpretations.

Many individuals have shared results prior to publication, or have provided criticisms of various drafts of this manuscript or assistance with other aspects of the preparation of this book. These include Curt Wentrup, O. S. Tee, D. P. N. Satchell, Amitai Halevi, Herbert Mayr, Heinrich Zollinger, V. A. Nikolaev, Ernst Schaumann, Zvi Rappoport, John Baldwin, Ekkehard Lindner, Mircea Gheorghiu, T. S. Sorensen, and W. T. Brady. F. P. Cossio and K. N. Houk provided copies of published figures for reproduction here. In some cases differences of opinion remain, but the discussion of the points has improved the manuscript. Thanks are given to Alfred Bader for a helpful conversation and access to his private correspondence with Michael Carroll, which greatly assisted in the writing of Section 3.4.4 on dioxinones. Special thanks are given to Grace Baysinger and the other staff of the Chemistry Library at Stanford University for carrying out the *Chemical Abstracts* substructure search and for their hospitality while a significant part of the literature survey was carried out. I am particularly grateful to Pat Woodcock, who cheerfully typed the many versions of the manuscript, and created all of the structures using ChemDraw. The project would have been impossible without her.

The principal credit is owed to my research co-workers who made possible our own investigations of ketenes, and maintained my enthusiasm for the subject. These include Hani Seikaly, Lynn Baigrie, Leyi Gong, Regis Leung-Toung, Dachuan Zhao, Jihai Ma, Mike McAllister, Ronghua Liu, and especially Annette Allen. Collaborative studies with Professor A. J. Kresge have been very beneficial to our studies. Quite naturally the efforts of all these individuals have been given prominent coverage in the text, but this is not meant to slight the contributions of the many others who have helped make ketene chemistry the exciting field that it is today.

Thomas T. Tidwell

Toronto, Ontario
April, 1995

INTRODUCTION

The first ketene to be prepared and characterized was diphenylketene (**1**), made by Hermann Staudinger in 1905,[1] while he was an instructor (*Unterrichtsassistent*) at the University of Strassbourg in the Institute of J. Thiele. This preparation involved the reaction of α-chlorodiphenylacetyl chloride with zinc (equation 1), and as recounted by Staudinger in his scientific autobiography[2] was inspired by the preparation of the stable radical triphenylmethyl by Gomberg. Rather than the expected stable radical **2**, the ketene **1** was obtained instead. Dimethylketene (**3**)[3] and dibenzopentafulvenone (**4**)[4] were soon prepared by the same method.

$$Ph_2CClCOCl \xrightarrow{Zn} Ph_2C=C=O \qquad\qquad Ph_2\overset{\bullet}{C}COCl \qquad (1)$$
$$\qquad\qquad\qquad\qquad\qquad 1 \qquad\qquad\qquad\qquad 2$$

$(CH_3)_2C=C=O$

3

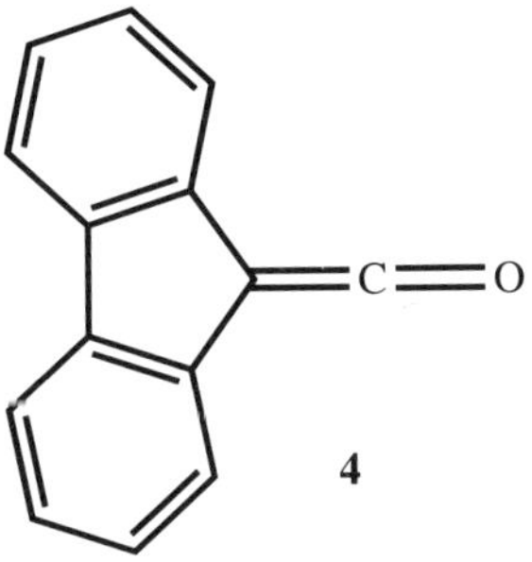

4

Shortly thereafter N. T. M. Wilsmore at University College, London, prepared the parent ketene $CH_2=C=O$ from the pyrolysis of acetic anhydride using a hot platinum wire.[5] Staudinger and Klever then prepared ketene from bromoacetyl bromide and zinc,[6] and engaged in a discussion with Wilsmore and Stewart as to whether ketene was correctly represented as $CH_2=C=O$ or $HC\equiv COH$, as was suggested by the latter authors.[7,8]

Staudinger then was successively "associate professor"[2] at the Technische Hochschule in Karlsruhe (1907–1912), Professor at the ETH in Zurich (1912–1926), and Professor at Freiburg (1926–1951). He summarized his early findings in a monograph in 1912,[9] and published further studies on ketenes through 1925. However he later devoted most of his attention to the study of macromolecules, and this work was recognized by the award of the Nobel Prize in Chemistry in 1953.

Wilsmore had been at the ETH in Zurich, and left there in 1903 for University College, London, where he subsequently became Assistant Professor. In 1912 he was appointed to the Chair of Chemistry at the University of Western Australia, and owing to the demands of setting up a new department, made no more contributions to ketene chemistry.[10]

Edgar Wedekind missed the opportunity to discover ketenes while at Tübingen in 1901, when he reacted $Ph_2CHCOCl$ with n-Pr_3N, noted the formation of n-Pr_3NHCl, and proposed the formation of the intermediate **5**, which was equivalent to $Ph_2C=C=O$ (equation 2).[11,12] However the species was not isolated, and this achievement was left to Staudinger in 1905.[1]

$$Ph_2CHCOCl \xrightarrow{n\text{-}Pr_3N} Ph_2\overset{|}{C}-\overset{|}{C}=O \ + \ n\text{-}Pr_3NHCl \qquad (2)$$

5

The chemical utility of ketenes was recognized quickly and these species became a popular topic of investigation. For many years the hydration of ketene was a major industrial process for the preparation of acetic acid. Many Nobel Prize winning chemists have taken a particular interest in ketenes, and besides Staudinger they include R. B. Woodward, E. J. Corey, R. Hoffmann, R. G. W. Norrish, K. Fukui, L. Ruzicka, O. Diels, G. Natta, and W. N. Lipscomb. Ketenes have been of major importance in organic synthesis, for example, in the formation of β-lactams leading to penicillins by [2 + 2] cycloadditions with imines (Section 5.4.1.7), the formation of prostaglandin precursors (Section 5.4.1.2), and syntheses of quinones (Section 5.4.4). The uses of ketenes are increasing and a bright future for ketene chemistry is assured.

Since Staudinger's book there has been a succession of extensive reviews on ketene chemistry,[13–17] and all of these are worthy of consultation for detailed treatment of specific topics.

References

1. Staudinger, H. *Chem. Ber.* **1905**, *38*, 1735–1739.
2. Staudinger, H. *From Organic Chemistry to Macromolecules;* Wiley: New York, 1970.

3. Staudinger, H.; Klever, H. W. *Chem. Ber.* **1906,** *39,* 968–971.
4. Staudinger, H. *Chem. Ber.* **1906,** *39,* 3062–3067.
5. Wilsmore, N. T. M. *J. Chem. Soc.* **1907,** *91,* 1938–1941.
6. Staudinger, H.; Klever, H. W. *Chem. Ber.* **1908,** *41,* 594–600.
7. Wilsmore, N. T. M.; Stewart, A. W. *Chem. Ber.* **1908,** *41,* 1025–1027.
8. Staudinger, H.; Klever, H. W. *Chem. Ber.* **1908,** *41,* 1516–1517.
9. Staudinger, H. *Die Ketene*; Verlag Enke: Stuttgart, 1912.
10. Obituary Notice of N. T. M. Wilsmore. *J. Chem. Soc.* **1941,** 59–60.
11. Wedekind, E. *Chem. Ber.* **1901,** *34,* 2070–2077.
12. Wedekind, E. *Liebigs Ann. Chem.* **1901,** *323,* 246–257.
13. Hanford, W. E.; Sauer, J. C. *Organic Reactions* **1946,** *3,* 108–140.
14. Lacey, R. N. In *The Chemistry of the Alkenes;* Patai, S., Ed.; Interscience: New York, 1964; pp. 1161–1227.
15. *Chemistry of Ketenes, Allenes, and Related Compounds;* Patai, S., Ed.; Wiley: New York, 1980.
16. Bormann, D. *Methoden der Organischen Chemie;* Theime Verlag: Stuttgart, 1968; Vol. 7, Part 4.
17. Schaumann, E.; Scheiblich, S. *Methoden der Organischen Chemie;* Theime Verlag: Stuttgart, 1993; Vol. E15, Part 3, Chaps 4, 6, 8.

CHAPTER 1

STRUCTURE, BONDING, AND THERMOCHEMISTRY OF KETENES

1.1 THEORETICAL STUDIES OF KETENES

Because of its small size, rigid geometry, and fascinating chemical properties, ketene has been a prominent subject of theoretical investigations. The fundamental theory of the electronic structure of ketene has been the subject of a comprehensive review,[1] and this work and the results of prior studies[2,3] in this field are not repeated here. The results of some recent theoretical studies dealing with the formation, structure, and reactions of ketenes are mentioned below, and relevant theoretical studies are also cited at various places in the text.

1.1.1 Molecular and Electronic Structure and Energy

The most distinguishing feature of the electronic structure of ketene is the presence of the highest occupied molecular orbital (HOMO) perpendicular to the ketene plane, and the lowest unoccupied molecular orbital (LUMO) in the ketene plane (Figure 1). These place substantial negative charge on oxygen and C_β and positive charge on C_α, and show that electrophiles are expected to attack ketene perpendicular to the molecular plane at the former positions, while nucleophiles will approach in the plane at the latter position.

There have been a number of calculations of the geometry and electronic states of ketene.[2-4] Recent calculations[4] of the vertical excitation energies of ketene are given in Table 1, and are in good agreement with the experimental values.

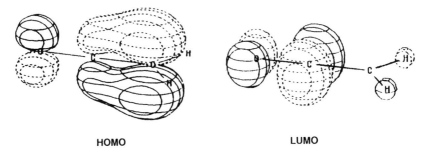

Figure 1. Frontier molecular orbitals of ketene (reproduced from ref. 16 by permission of the American Chemical Society).

TABLE 1. Vertical Excitation Energies (cm^{-1}) for Ketene (from $^1A_1$)^4

	$^3A_2$	$^1A_2$	$^3A_1$	$^1B_1$	$^2A_1$
Theoretical^a	29,800	30,900	44,700	46,600	56,100
Experimental^b	29,200	31,000	43,000	47,300	54,680

^{a}DZP + R−CISD.
^{b}Selected experimental values compiled in reference 4.

The photodissociation of ketene to methylene and CO (equation 1) is a classical problem in theoretical[5,6] and experimental chemistry, and is discussed more in Section 5.2. Recent ab initio calculations have considered dissociation from the in-plane bent $^3A''$ and the out-of-plane bent $^3A'$ structures, and the energy for dissociation by the former path is proposed to be 22.3 kcal/mol.[6]

$$CH_2=C=O \xrightarrow{h\nu} CH_2 + CO \qquad (1)$$

1

The stabilities and possible interconversions of the various C_2H_2O isomers has also been studied in great detail including ketene (**1**), hydroxyacetylene (**2**), oxirene (**3**), and formylmethylene (**4**).[7–10] The species and their interconversion are of interest in the formation of ketenes by the Wolff rearrangement (equation 2), and for the rearrangement of photoexcited ketene (equation 3) as revealed by isotopic labeling.[8–12] The dissociation of diazoacetaldehyde (**5**) to formylmethylene (**4**) has also been studied, and at the MP4SDTQ/6-31G*//6-31G* level was calculated to be exothermic by 42.3 kcal/mol.[13]

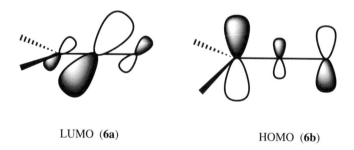

These studies suggest that excitation of **1** can lead to **4**, which is 84.3 kcal/mol above **1**, and that rearrangement occurs by conversion of **4** to **3**, which is 2 kcal/mol lower in energy.[10] The dissociation energy of ketene is calculated to be 86.1 kcal/mol.[10] Ethynol **2** is calculated to be 36.1 kcal/mol above ketene, and is thought to be formed by rearrangement of **4**.[7] These structures and energies are combined in Figure 2, although different levels of theory are involved, and improved and consistent values are needed.

1.1.2 Theoretical Studies of Ketene Reactions

The reactivity of ketenes, particularly in cycloaddition reactions, has been evaluated in terms of the in-plane LUMO (**6a**) of ketene and the perpendicular HOMO (**6b**).[14–16] Cycloadditions of ketene with $CH_2=CH_2$,[17–22] $HC\equiv CH$,[21] $CH_2=NH$,[16,22–24] $CH_2=O$,[24] and cyclopentadiene,[24a] and ketene dimerization[25–28] have been studied in some detail, and are considered in Chapter 5.

LUMO (**6a**)   HOMO (**6b**)

Cyclizations of ketenes with conjugated unsaturated groupings, as in the conversion of vinylketene to cyclobutenone (equation 4), have also been examined[29,30] and are discussed in Section 5.4.4.

1.1 THEORETICAL STUDIES OF KETENES   7

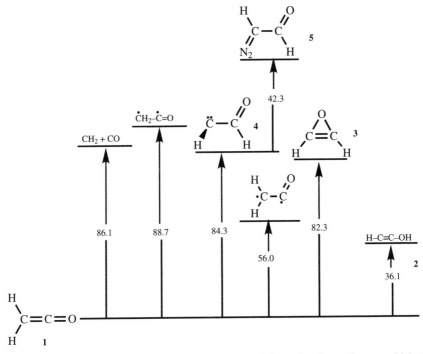

Figure 2. Isomerization and dissociation of ketene and formation from diazoacetaldehyde (**5**) (kcal/mol).[7,10,13]

$$\begin{array}{c}\text{H}\\ \diagdown\\ \text{C}=\text{C}=\text{O}\\ \diagup\\ \text{H}\quad\text{CH}_2\end{array}\quad\longrightarrow\quad\square\!=\!\text{O} \qquad (4)$$

The protonation of ketene and the structures and energies of the various protonated isomers have been of considerable interest.[31–40] Protonation of ketene (**1**) at C_β gives the acetyl cation (**7**), a process that in the gas phase is experimentally exothermic by 196.2 kcal/mol.[40] Protonation on oxygen gives **8**, which is calculated at the MP3/6-31G**//4-31G level to be 43.3 kcal/mol higher in energy.[38] Protonated isomers at C_α are the oxiranyl cation **9** and the planar formylmethyl cation **10**, which are 58.3 and 78.9 kcal/mol, respectively, higher in energy than **7**.[38] The conversion of **9** to **7** involves conversion to **10**, which forms **7** with no barrier (Figure 3).[38]

STRUCTURE, BONDING, AND THERMOCHEMISTRY OF KETENES

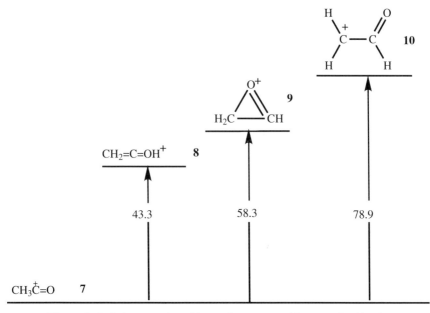

Figure 3. Relative energies of isomeric protonated ketenes (kcal/mol).

The effects of ketene substitution on the relative propensity for protonation at different positions as obtained in several different investigations[33,38,39] are given in Table 2. Variations in basis set do not appear to affect the relative ordering. Protonation at C_β (equation 5) is preferred, but the substitution of carbocation stabilizing groups at this position tends to enhance protonation at C_α, as this allows carbocation generation at C_β (equation 6).[33] Substitution at C_β by groups such as phenyl, which stabilize the double bond by conjugation and destabilize positive charge development at C_α, inhibit proton attack at the former carbon.[33] With vinylketene protonation at C_δ is favored (see also Section 5.5.1.3) (equation 7).

$$RCH=C=O \xrightarrow{H^+} R\overset{+}{C}H_2C=O \qquad (5)$$

$$\mathbf{7a}$$

$$RCH=C=O \xrightarrow{H^+} R\overset{+}{C}HCH=O \qquad (6)$$

$$CH_2=CHCH=C=O \xrightarrow{H^+} CH_3\overset{+}{C}H-CH=C=O \longleftrightarrow CH_3CH=CH-\overset{+}{C}=O \qquad (7)$$

The energies of some substituted acetyl cations have been calculated[32] and combined with the calculated energies for the corresponding substituted ketenes.[41]

TABLE 2. Relative Energies (kcal/mol) Calculated for Protonation of Ketenes

Ketene	RR^{1}CHC$\overset{+}{=}$O	RR^{1}C=C=OH$^+$	RR$^1\overset{+}{C}$—CH=O	Basis	Ref.
CH$_2$=C=O	0	43.3	78.9	MP3/6-31G**//4-31G	38
CH$_2$=C=O	0	37.6	76.1	3-21G//3-21G	33
(CH$_3$)$_2$C=C=O	0	34.5	39.4	3-21G//3-21G	39
(CH$_3$)$_2$C=C=O	0	40.4	38.7	CI/4-31G//3-21G	39
CH$_2$=CHCH=C=O	15.8	54.2	38.1	6-31G*//3-21G	33
CH$_2$=CHCH=C=O	0.0 (CH$_3\overset{+}{C}$HCH=C=O)				

These give the effect of substitution on the protonation of ketenes at C_β according to equation 5 (Table 3).

The relative proton affinities according to equation 5 are affected by the substituent effects on both the ketenes and on the acylium ions **7a**. Both ketenes and acylium ions are destabilized by electronegative groups, and the greatest effect is due to fluorine, which is the most destabilizing ketene substituent,[41] but the destabilizing effect of fluorine on the acylium ion is even greater, so that fluoroketene is even less basic than ketene. The energy changes parallel those that have been found in the analogous ethyl cations **11**, in which the geometries are fixed to prevent bridging,[42] as given in Table 3 for the isodesmic reaction of equation 8.

$$RCH_2CH_3 + CH_3CH_2^+ \longrightarrow RCH_2CH_2^+ + CH_3CH_3 \qquad (8)$$

11

Oxygen protonation of ketene was found to be less effective than for acetaldehyde, but more efficient than for formaldehyde, which appears explicable on the basis of the expected stabilities of the carbocationic structures formed.[36] However C_β-protonation of ketene is more favorable than O-protonation of acetaldehyde.[35]

$$CH_3\overset{+}{C}H-OH > CH_2=\overset{+}{C}-OH > \overset{+}{C}H_2-OH$$

Proton abstraction from ketene to form the ynolate anion has been considered,[37,43–45] and the linear geometry **12** with most of the negative charge in oxygen is favored over the bent geometry **13** with charge concentration on carbon (equation 9).[44]

TABLE 3. Calculated (6-31G*//6-31G*) Energies (Hartrees) of Protonation of Ketenes: RCH=C=O + H$^+$ → RCH$_2\overset{+}{C}$=O

R	E(RCH=C=O)^a	E(RCH$_2\overset{+}{C}$=O)^b	ΔE_{rel} (kcal/mol)	SE^c
H	−151.7247	−152.0593	0.0	0.0
CH$_3$	−190.7592	−191.1030	−5.8^d	−3.8
NH$_2$	−206.7382	−207.0834	−6.7	−5.5
OH	−226.5540	−226.8932	−2.9	2.6
F	−250.5422	−250.8698	4.4	14.5

^{a}Reference 41.
^{b}Reference 32.
^{c}SE for ethyl cations **11**, equation 8 (ref. 42).
^{d}Experimentally the value is −6.0 kcal/mol (ref. 40).

$$\begin{array}{c}\text{H}\\ \diagdown\\ \text{C}=\text{C}=\text{O}\\ \diagup\\ \text{H}\end{array} \xrightarrow{-\text{H}^+} \text{H-C}\equiv\text{C-O}^- \rightleftharpoons \begin{array}{c}\text{H}\\ \diagdown\\ \text{C}=\text{C}=\text{O}\\ \\ \phantom{\text{H}}\end{array}^{-} \quad (9)$$

$$ \textbf{12} \textbf{13}$$

The reaction of ketene with H⁻ and with MeO⁻/MeOH was calculated to occur with a lower barrier by proton abstraction from the ketene to give **14** in preference to nucleophilic attack at the carbonyl carbon to form **15**, although the energies of the two products were essentially the same (equation 10).[45] Deprotonation was also the path observed in ion cyclotron resonance experiments involving the gas phase reaction of ketene with alkoxides.[45]

$$\begin{array}{c}\text{H}\\ \diagdown\\ \text{C}=\text{C}=\text{O}\\ \diagup\\ \text{H}\end{array} \xrightarrow{\text{MeO}^- \text{---HOMe}} \text{H}-\text{C}\equiv\text{C}-\text{O}^-\begin{array}{c}\text{HOMe}\\ \\ \text{HOMe}\end{array} \quad \text{or}$$

$$ \textbf{14}$$

$$\begin{array}{c}\text{H}\\ \diagdown\\ \text{C}=\text{C}\\ \diagup \diagdown\\ \text{H} \text{OMe}\end{array}\begin{array}{c}\text{O}^-\\ \\ \\ \end{array} + \text{MeOH} \quad (10)$$

$$ \textbf{15}$$

The hydration of ketene by various numbers of H₂O molecules has been examined.[46–51] The most detailed calculations are of the reaction of ketene with water dimer, and in this case addition to the carbonyl group via the transition state **16** is preferred (equation 11).[46–50] The calculated energy change in this process does not agree well with the experimental value, and in solution further H₂O molecules are involved in the hydration. This process is discussed in more detail in Section 5.5.1.

$$\begin{array}{c}\text{H}\\ \diagdown\\ \text{C}=\text{C}=\text{O}\\ \diagup\\ \text{H}\end{array} \xrightarrow{(\text{H}_2\text{O})_2} \begin{array}{c}\text{H}\\ \diagdown\text{O}\cdots\text{H}\\ \text{C}=\text{C}\diagup\\ \diagup\diagdown\\ \text{H}\text{O}\cdots\text{O}\\ \text{H}\text{H}\text{H}\end{array} \longrightarrow \begin{array}{c}\text{H}\text{O}-\text{H}\\ \diagdown\diagup\\ \text{C}=\text{C}\\ \diagup\diagdown\\ \text{H}\text{O}-\text{H}\end{array} \quad (11)$$

$$ \textbf{16}$$

The hydration of hydroxyketene has also been studied computationally.[49] Studies of H₂O and CH₃OH addition to the C=C bond of ketenes[51,52] using rather low levels of theory have also appeared. Addition of H₂O to formylketene is calculated to occur through a cyclic 6-membered transition state (equation 12).[53] This process is discussed further in Section 4.1.6.

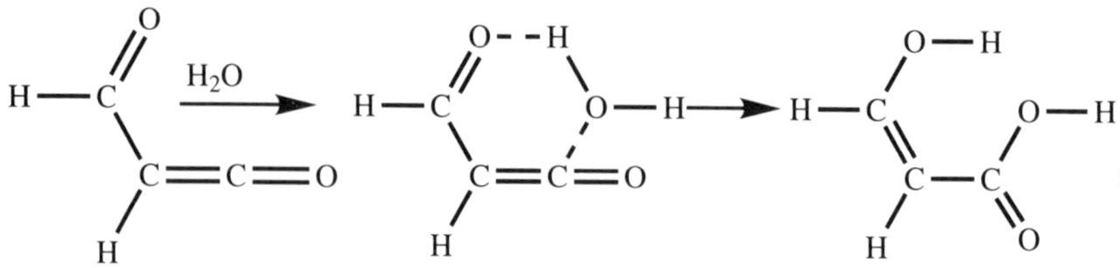

12 STRUCTURE, BONDING, AND THERMOCHEMISTRY OF KETENES

$$\text{(12)}$$

The reactions of ketene with LiH,[54] LiCH$_3$,[54] and LiOCH=CH$_2$[55] have also been studied computationally by ab initio methods. These reactions are calculated to occur by initial lithium complexation with oxygen, followed by in-plane nucleophilic attack on the ketene LUMO at C$_\alpha$ in **6a** and formation of the enolate product (equation 13). The calculations also predict the experimentally observed preferred attack on the least hindered face of the ketene, and the kinetic preference for O-acylation by ketene of LiOCH=CH$_2$, although the product of C-acylation is more stable (equation 14). These reactions are discussed in more detail in Section 5.5.1.1 as are structures calculated by AM1 for the addition of trimethylamine and pyridine to ketene.[56]

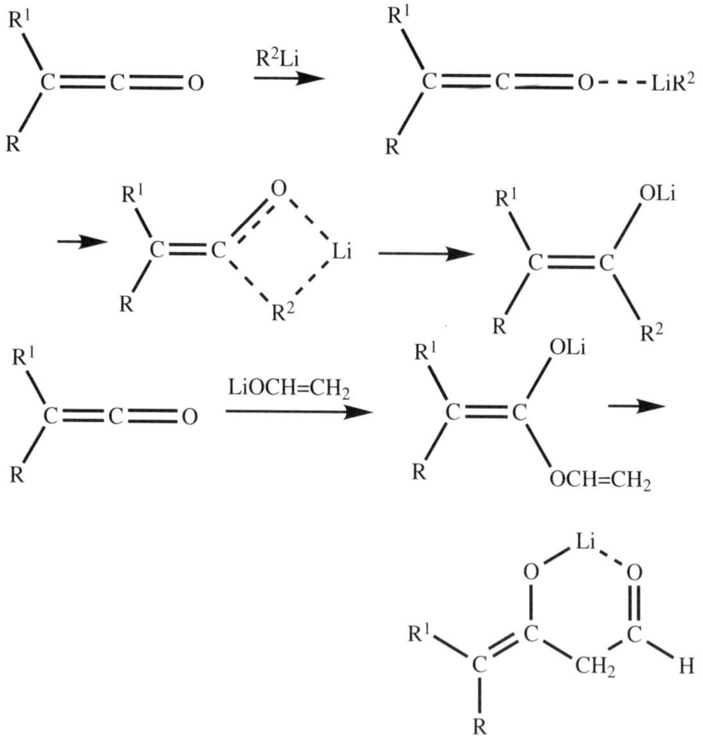

$$\text{(13)}$$

$$\text{(14)}$$

1.1.3 Substituent Effects on Ketenes

The energies of substituted ketenes and the corresponding alkenes were calculated at the HF/6-31G*//HF/6-31G* level for a large number of substituents, as given in Table 4.[41,57] Further calculations at the MP2/6-31G*//MP2/6-31G* level lead to similar conclusions.[57a]

Ketene stabilization energies (SE) were calculated for the isodesmic reaction of equation 15, and are included in Table 4.[41,57] The SE values were correlated by group electronegativity values χ_{BE}[58,59] by equation 16 with a correlation coefficient $r = 0.978$.[41] Thus ketenes are stabilized by electropositive substituents and destabilized by electronegative substituents.

$$RCH=C=O + CH_3CH=CH_2 \xrightarrow{\Delta E} CH_3CH=C=O + RCH=CH_2 \quad (15)$$

$$\Delta E = -15.6\chi_{BE} + 42.3 \quad (16)$$

The results did not differ much from those obtained at the 3-21G//3-21G level, lending confidence that calculations at even higher levels would not cause significant changes. In cases where experimental geometries, dipole moments, and energies were available for comparison, they were in reasonable agreement with the calculated values. Details of the calculated geometries, atomic charges, and dipole moments are given in Section 1.2.

The stabilization energies obtained from equation 15 for the BH_2 and $CH=O$ groups are 16.8 and 3.6 kcal/mol, and are larger than those of 12.5 and 1.7 kcal/mol, respectively, as predicted by equation 16. The extra stabilization due to these groups, besides that predicted to be due to electronegativity, is attributed to conjugative stabilization by these π-acceptor substituents, as shown for example in **17**.[41]

17

Ketenes substituted with the n-π donor groups OH and NH_2 adopt the nonplanar conformations **18** and **19**, in contrast to the planar conformations found for the corresponding alkenes. These nonplanar conformations **18** and **19** serve to minimize π-donation to ketenes and indicate repulsion between such groups and the electron-rich C_β.[41,57a]

TABLE 4. Calculated Energies (HF/6-31G*//HF/6-31G*) (Hartrees) for Ketenes RCH=C=O and Alkenes RCH=CH$_2$ and ΔE (kcal/mol) for the Isodesmic Reaction of Equation 15[41]

R	−E(RCH=C=O)	−E(RCH=CH$_2$)	ΔE (kcal/mol)	χ_{BE}[a]
H	151.7247	78.0317	3.3	2.20
Li	158.5935	84.8614	27.9	1.00
BeH	166.3753	92.6587	18.1	1.47
BH$_2$	177.0041	103.2897	16.8	1.93
CH$_3$	190.7592	117.0715	0.0	2.56
NH$_2$	206.7382	133.0620	−7.2	3.10
OH	226.5540	152.8889	−14.2	3.64
F	250.5422	176.8820	−17.2	4.00
Na	312.9798	239.2456	29.2	1.00
MgH	351.3195	277.5969	21.9	1.33
AlH$_2$	394.2252	320.5076	18.7	1.62
SiH$_3$	441.8176	368.1125	10.9	1.91
PH$_2$	493.0237	419.3259	6.3	2.17
SH	549.2317	475.5419	1.3	2.63
Cl	610.6094	536.9337	−7.5	3.05
CF$_3$	487.3439	413.6568	−0.4	2.68
c-Pr	267.6250	193.9402	−1.8	2.56
CH=CH$_2$[b]	228.6070	154.9197	−0.2	2.61
CH=O[c]	264.4573			2.60
CH=O[b]	264.4559	190.7624	3.6	2.60
Ph[d]	381.2742	307.5854	0.7	2.58
CO$_2$H	339.3488	265.6536	4.7	2.66
C≡CH	227.3959	153.7079	0.2	2.66
CN	243.4551	169.7680	−0.4	2.69
CH=C=O[c]	302.2832			2.58
CH=C=O[b]	302.2858	228.6070	−5.3[e]	2.58
![cyclobutenedione]	302.2968			
N=O	280.3647	206.6774	−0.3	3.06
NO$_2$[f]	335.1867	281.5041	−3.2	3.22
N≡C[g]	243.4125	169.7384	−8.5	3.30

[a]Group electronegativity from References 58 and 59, except Pauling electronegativity for H.
[b]Transoid.
[c]Cisoid.
[d]Reference 57.
[e]For the process (CH=C=O)$_2$ + (CH$_2$=CH)$_2$ → 2CH$_2$=CHCH=C=O.
[f]Reference 57a.
[g]Isocyano.

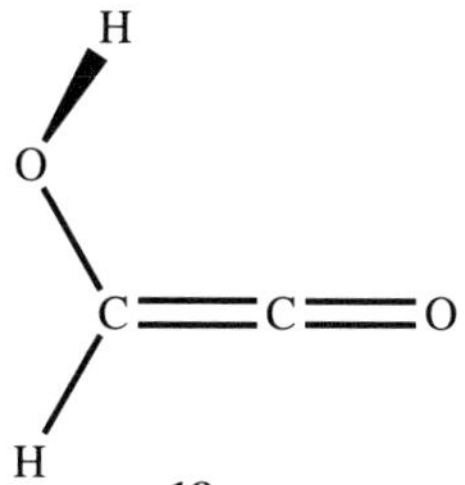

[Structures 18 and 19 shown: hydroxy- and amino-substituted ketenes]

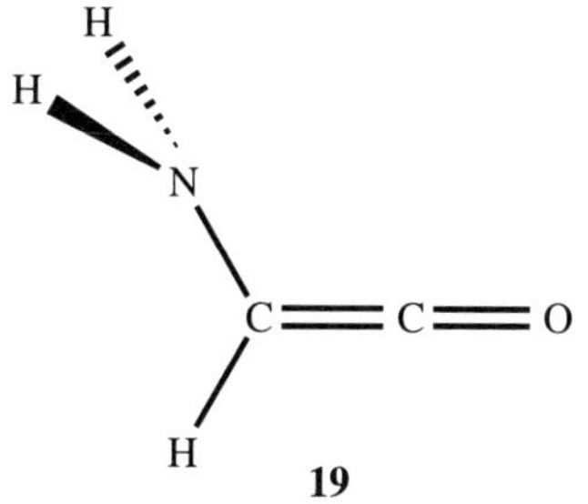

Studies of the fulvenenones **20–22** suggest that **20** and **22** have enhanced *antiaromatic* destabilization compared to the corresponding fulvenes, whereas **21** has enhanced aromatic stabilization.[60] These conclusions were based on isodesmic energy comparisons (Table 5), comparative bond lengths, and atomic charges.[60] These fulvenones are discussed further in Section 4.1.10.

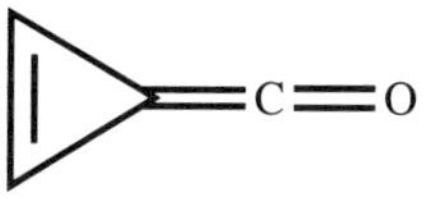

[Structures 20, 21, 22: cyclopropenylidene ketene, cyclopentadienylidene ketene, cycloheptatrienylidene ketene]

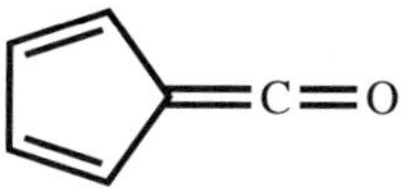

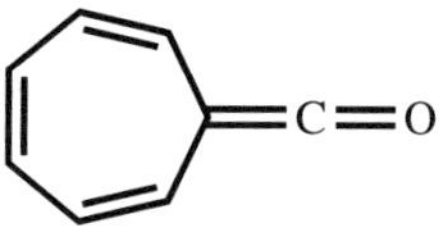

The cycloaddition reactions of phenylketene and diphenylketene were interpreted based on CNDO/2 studies of phenylketene **23** and the isomers **24** and **25**.[61]

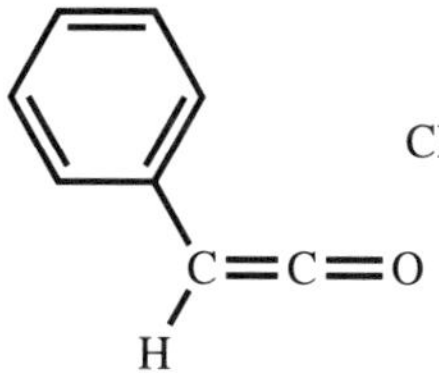

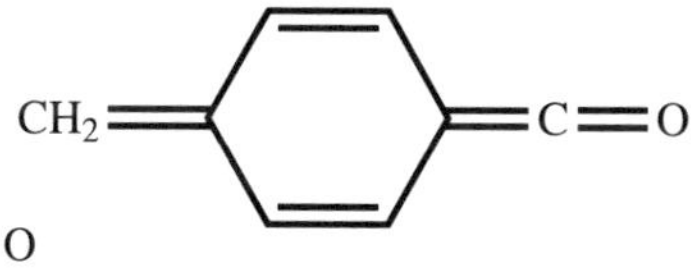

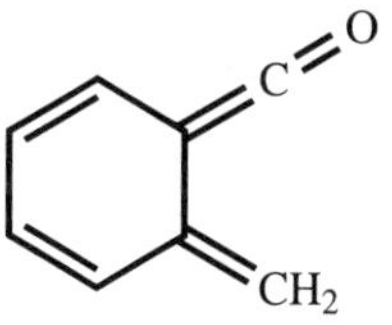

[Structures 23, 24, 25: phenylketene and its quinoid isomers]

Carbanions are known to be highly stabilized by substituents of third-period elements such as silicon, phosphorous, and sulfur, and the ketenes substituted with silicon and sulfur are remarkably stable. In the case of carbanions this stabilization is attributed partly to "polarizability," and this factor may play a role in ketene stabilization as well.

The use of natural valence coordinates in the calculation of the structure of ketene did not improve the agreement of the calculated geometry with experimental values.[62] Molecular mechanics calculations of ketene structures are also available.[63]

TABLE 5. Energies (Hartrees) (HF/6-31G*//HF/6-31G*) for Fulvenones and Fulvenes, and Stabilization Energies (ΔE, kcal/mol) for some Isodesmic Reactions of Fulvenones[60]

Fulvenone		Fulvene	ΔE (kcal/mol)
cyclopropenone −227.3303 + CH₃CH=CH₂ −117.0715	→	methylenecyclopropene −153.6698 + CH₃CH=C=O −190.7592	−17.1
cyclopropanone −228.5758 + CH₃CH=CH₂	→	methylenecyclopropane −154.8873 + CH₃CH=C=O	0.5
cyclobutenone −266.4275 + CH₃CH=CH₂	→	methylenecyclobutene −192.7481 + CH₃CH=C=O	−5.2
cyclobutanone −267.6264 + CH₃CH=CH₂	→	methylenecyclobutane −193.9424 + CH₃CH=C=O	−2.3
cyclopentadienone −304.3387 + CH₃CH=CH₂	→	fulvene −230.6444 + CH₃CH=C=O	4.1

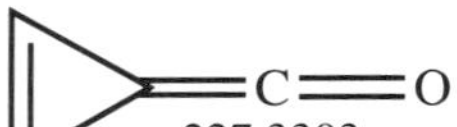

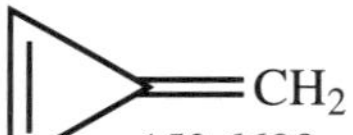

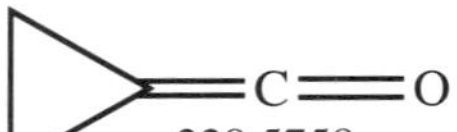

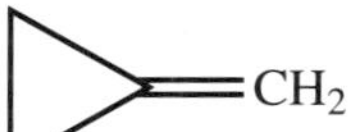

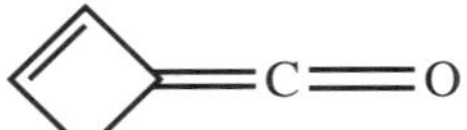

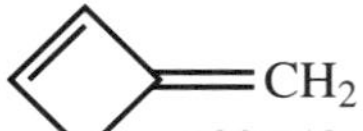

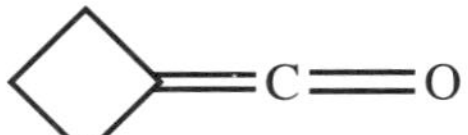

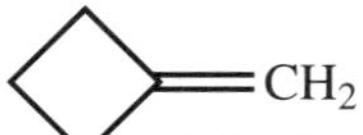

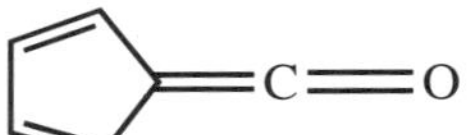

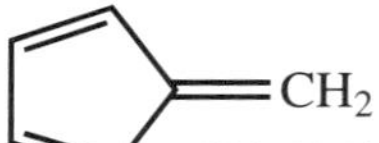

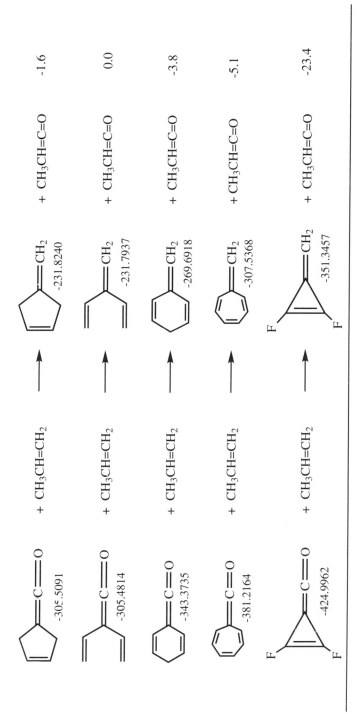

References

1. Dykstra, C. E.; Schaefer, H. F., III. In *The Chemistry of Ketenes, Allenes, and Related Compounds;* Patai, S., Ed.; Wiley: New York, 1980; pp. 1–44.
2. Del Bene, J. E. *J. Am. Chem. Soc.* **1972**, *94*, 3713–3718.
3. Dykstra, C. E.; Schaefer, H. F., III. *J. Am. Chem. Soc.* **1976**, *98*, 2689–2695.
4. Allen, W. D.; Schaefer, H. F., III. *J. Chem. Phys.* **1986**, *84*, 2212–2225.
5. Yamabe, S.; Morokuma, K. *J. Am. Chem. Soc.* **1978**, *100*, 7551–7556.
6. Allen, W. D.; Schaefer, H. F., III. *J. Chem. Phys.* **1988**, *89*, 329–344.
7. Tanaka, K.; Yoshimine, M. *J. Am. Chem. Soc.* **1980**, *102*, 7655–7662.
8. Bouma, W. J.; Nobes, R. H.; Radom, L.; Woodward, C. E. *J. Org. Chem.* **1982**, *47*, 1869–1875.
9. Dykstra, C. E. *J. Chem. Phys.* **1978**, *68*, 4244–4247.
10. Yoshimine, M. *J. Chem. Phys.* **1989**, *90*, 378–385.
11. Vacek, G.; Gailbraith, J. M., Yamaguchi, Y.; Schaefer, H. F., III; Nobes, R. H.; Scott, A. P.; Radom, L. *J. Phys. Chem.* **1994**, *98*, 8660–8665.
12. Scott, A. P.; Nobes, R. M.; Schaeffer, H. F., III; Radom, L. *J. Am. Chem. Soc.* **1994**, 10159–10164.
13. Wang, F.; Winnik, M. A.; Peterson, M. R.; Csizmadia, I. G. *J. Mol. Struct. (Theochem)* **1991**, *232*, 203–210.
14. Sonveaux, E.; Andre, J. M.; Dehalle, J.; Fripiat, J. G. *Bull. Soc. Chim. Belg.* **1985**, *94*, 831–847.
15. Houk, K. N.; Strozier, R. W.; Hall, J. A. *Tetrahedron Lett.* **1974**. 897–900.
16. Cossio, F. P.; Ugalde, J. M.; Lopez, X.; Lecea, B.; Palomo, C. *J. Am. Chem. Soc.* **1993**, *115*, 995–1004.
17. Bernardi, F.; Bottoni, A.; Olivucci, M.; Robb, M. A.; Schlegel, H. B.; Tonachini, G. *J. Am. Chem. Soc.* **1988**, *110*, 5993–5995.
18. Bernardi, F.; Bottoni, A.; Robb, M. A.; Venturini, A. *J. Am. Chem. Soc.* **1990**, *112*, 2106–2114.
19. Wang, X.; Houk, K. N. *J. Am. Chem. Soc.* **1990**, *112*, 1754–1756.
20. Burke, L. A. *J. Org. Chem.* **1985**, *50*, 3149–3155.
21. Fang, D.; Fu, X. *Beijing Shifan Daxue Xuebao, Ziran Kexueban,* **1991**, *27*, 69–74; *Chem. Abstr.* **1992**, *116*, 127846z.
22. Yamabe, S.; Minato, T.; Osamura, Y. *J. Chem. Soc., Chem. Commun.* **1993**, 450–453.
23. Sordo, J. A.; Gonzalez, J.; Sordo, T. L. *J. Am. Chem. Soc.* **1992**, *114*, 6249–6251.
24. Fang, D. C.; Fu, X. Y. *Chin. Chem. Lett.* **1992**, *3*, 367–368; *Chem. Abstr.* **1993**, *118*, 59108z.
24a. Jiang, J.; Fang, D.; Fu, X. *Chin. Chem. Lett.* **1992**, *3*, 713–714; *Chem. Abstr.* **1993**, *119*, 159390z.
25. Seidl, E. T.; Schaeffer, H. F., III. *J. Am. Chem. Soc.* **1990**, *112*, 1493–1499.
26. Seidl, E. T.; Schaeffer, H. F., III. *J. Phys. Chem.* **1992**, *96*, 657–661.
27. Seidl, E. T.; Schaeffer, H. F., III. *J. Am. Chem. Soc.* **1991**, *113*, 5195–5200.
28. Schaad, L. J.; Gutman, I.; Hess, B. A., Jr.; Hu, J. *J. Am. Chem. Soc.* **1991**, *113*, 5200–5203.

29. Nguyen, M. T.; Ha, T.; More O'Ferrall, R. A. *J. Org. Chem.* **1990**, *55*, 3251–3256.
30. Miller, R. D.; Theis, W.; Heilig, G.; Kirchmeyer, S. *J. Org. Chem.* **1991**, *56*, 1453–1463.
31. Pericas, M. A.; Serratosa, F.; Valenti, E.; Font-Altaba, M.; Solans, X. *J. Chem. Soc., Perkin Trans. 2* **1986**, 961–967.
32. Lien, M. H.; Hopkinson, A. C. *J. Org. Chem.* **1988**, *53*, 2150–2154.
33. Leung-Toung, R.; Peterson, M. R.; Tidwell, T. T.; Csizmadia, I. G. *J. Mol. Structure (Theochem)* **1989**, *183*, 319–330.
34. Wellington, C. A.; Khowaiter, S. H. *Tetrahedron* **1978**, *34*, 2183–2190.
35. Olivella, S.; Urpi, F.; Vilarrasa, J. *J. Comput. Chem.* **1984**, *5*, 230–236.
36. Hopkinson, A. C.; Csizmadia, I. G. *Can. J. Chem.* **1974**, *52*, 546–554.
37. Hopkinson, A. C. *J. Chem. Soc., Perkin Trans. 2*, **1973**, 795–797.
38. Nobes, R. H.; Bouma, W. J.; Radom, L. *J. Am. Chem. Soc.* **1983**, *105*, 309–314.
39. Bouchoux, G.; Hoppilliard, Y. *J. Phys. Chem.* **1988**, *92*, 5869–5874.
40. Armitage, M. A.; Higgins, M. J.; Lewars, E. G.; March, R. E. *J. Am. Chem. Soc.* **1980**, *102*, 5064–5068.
41. Gong, L.; McAllister, M. A.; Tidwell, T. T. *J. Am. Chem. Soc.* **1991**, *113*, 6021–6028.
42. White, J. C.; Cave, R. J.; Davidson, E. R. *J. Am. Chem. Soc.* **1988**, *110*, 6308–6314.
43. Hopkinson, A. C.; Lien, M. H.; Yates, K.; Mezey, P. G.; Csizmadia, I. G. *J. Chem. Phys.* **1977**, *67*, 517–523.
44. Smith, B. J.; Radom, L.; Kresge, A. J. *J. Am. Chem. Soc.* **1989**, *111*, 8297–8299.
45. Hayes, R. N.; Sheldon, J. C.; Bowie, J. H. *Aust. J. Chem.* **1985**, *35*, 355–362.
46. Nguyen, M. T.; Hegarty, A. F. *J. Am. Chem. Soc* **1984**, *106*, 1552–1557.
47. Skancke, P. N. *J. Phys. Chem.* **1992**, *96*, 8065–8069.
48. Andraos, J.; Kresge, A. J.; Peterson, M. R.; Csizmadia, I. G. *J. Mol. Struct. (Theochem)* **1991**, *232*, 155–177.
49. Andraos, J.; Kresge, A. J. *J. Mol. Struct. (Theochem)* **1991**, *233*, 165–184.
50. McAllister, M. A.; Kresge, A. J.; Csizmadia, I. G. *J. Mol. Struct. (Theochem)* **1992**, *258*, 399–400.
51. Lee, I.; Song, C. H.; Uhm, T. S. *J. Phys. Org. Chem.* **1988**, *1*, 83–90.
52. Choi, J. Y.; Lee, I. *Kich'o Kwahak Yonguso Nonmunjip* **1984**, *5*, 61–67; *Chem. Abstr.* **1984**, *101*, 54096h.
53. Allen, A. D.; McAllister, M. A.; Tidwell, T. T. *Tetrahedron Lett.* **1993**, *34*, 1095–1098.
54. Leung-Toung, R.; Schleyer, P. v. R.; Tidwell, T. T. unpublished results.
55. Leung-Toung, R.; Tidwell, T. T. *J. Am. Chem. Soc.* **1990**, *112*, 1042–1048.
56. Chelain, E.; Goumont, R.; Hamon, L.; Parlier, A.; Rudler, M.; Rudler, H.; Daran, J.; Vaissermann, J. *J. Am. Chem. Soc.* **1992**, *114*, 8088–8098.
57. Allen, A. D.; Andraos, J.; Kregse, A. J.; McAllister, M. A.; Tidwell, T. T. *J. Am. Chem. Soc.* **1992**, *114*, 1878–1879.
57a. McAllister, M. A.; Tidwell, T. T. *J. Org. Chem.* **1994**, *59*, 4506–4515.
58. Boyd, R. J.; Edgecombe, K. E. *J. Am. Chem. Soc.* **1988**, *110*, 4182–4186.
59. Boyd, R. J.; Boyd, S. L. *J. Am. Chem. Soc.* **1992**, *114*, 1652–1655.
60. McAllister, M. A.; Tidwell, T. T. *J. Am. Chem. Soc.* **1992**, *114*, 5362–5368.

61. Kuzuya, M.; Miyake, F.; Kamiya, K.; Okuda, T. *Tetrahedron Lett.* **1982**, *23*, 2593–2596.
62. Fogarasi, G.; Zhou, X.; Taylor, P. W.; Pulay, P. *J. Am. Chem. Soc.* **1992**, *114*, 8191–8201.
63. Stewart, E. L.; Bowen, J. P. *J. Comput. Chem.* **1992**, *13*, 1125–1137.

1.2 MOLECULAR STRUCTURE DETERMINATIONS

The molecular structures[1] of a number of ketenes have been determined experimentally, using microwave,[2–14] electron diffraction,[15] and X-ray techniques.[16–18] The reported geometrical parameters of simple ketenes are compiled in Table 1, together with dipole moments and comparative data obtained by molecular orbital calculations.[19–23]

Two recent microwave structures of ketene have appeared,[2,3] following earlier studies,[4–6] as well as studies of FCH=C=O,[7,8] ClCH=C=O,[9] CH_3CH=C=O,[10] and NCCH=C=O.[11] The microwave spectrum of vinylketene was measured, but only the dipole moment was deduced from the data.[12] The structure of $(CH_3)_2$C=C=O was obtained[13] from the microwave spectrum based on the assumption that the C=O bond length was the same as that of CH_3CH=C=O.[10] Electron diffraction was used to derive the molecular structure of CCl_2=C=O.[15]

There is rather good agreement between the experimental gas phase values and the theoretical values obtained at the HF/6-31G* or MP2/6-31G* levels.[20–22] These results provide confidence that values calculated for ketenes which have not been studied experimentally provide a good guide to the actual structures.

The structure of thioketene was derived from FTIR studies, as shown in **1**,[14] along with comparative data (parentheses) for ketene.[3]

$$
\begin{array}{c}
\text{H} \quad (1.316) \; (1.161) \\
(1.080) \; 1.085 \diagdown 1.356 \quad 1.556 \\
(122.0) \; 119.8 \diagup \text{C}=\!=\!\text{C}=\!=\!\text{S} \\
\text{H} \\
\mathbf{1}
\end{array}
$$

The X-ray crystal structures of dimesitylketene (**2**) and bis(3,5-dibromomesityl)ketene (**3**) have been reported as shown.[16] The structure of **2** shows C_1 symmetry with nonequivalent aryl groups in a propeller conformation, with dihedral angles of 48.8 and 56.8°. For **3** the molecule has C_2 symmetry with equivalent aryl groups and dihedral angles of 51.8°.

TABLE 1. Experimental and Calculated (HF/6-31G*//HF/6-31G*) (parentheses) Bond Distances (Å), Bond Angles (deg), and Dipole Moments of Ketenes $RR_1C_2=C_1=O$

R	R^1	C=O	$C_1=C_2$	C_2-R	C_2-R_1	OC_1C_2	RC_1C_2	RC_2R_1	R_1C_2Cl	μ (Debye)	Ref
H	H	1.1614	1.3165	1.0800	1.0800	180.0	119.01	121.98	119.01	1.41	3
		1.1626	1.3147	1.0905	1.0905	180.0	118.27	123.46	118.27		2
		(1.145)	(1.306)	(1.075)		(180.0)	(119.3)	(121.4)	(119.3)	(1.63)	20
		(1.1681)	(1.3166)	(1.0745)	(1.0745)	(180.0)	(119.38)	(121.25)	(119.38)		2^a
F	H	1.167	1.317	1.080	1.356	178.0	119.5	118.2	122.3	1.29	7,8
		(1.147)	(1.308)	(1.067)	(1.339)	(177.9)	(120.1)	(117.4)	(122.5)	(1.59)	20
		(1.181)	(1.323)	(1.074)	(1.363)	(178.1)	(120.0)	(117.7)	(122.3)		7^b
Cl	H	1.161	1.316	1.082	1.726	180.0	120.9	119.3	119.8	1.2	9
		(1.142)	(1.308)	(1.068)	(1.731)	(179.2)	(120.5)	(118.7)	(120.8)	(1.39)	20
CH_3	H	1.171	1.306	1.083^c	1.518	180.5	122.6^d	123.7	113.7	1.79	10
		(1.149)	(1.305)	(1.074)	(1.512)	(180.0)	(123.2)	(120.8)	(116.0)	(1.99)	20
CN	H									3.52	11
		(1.133)	(1.321)	(1.072)	(1.424)	(178.7)		(121.3)	(120.6)	(3.72)	20
$CH=CH_2$	H									0.97	12
										(1.20)	20
Cl	Cl	1.160	1.299	1.726	1.726	180.0	120.4	119.2	120.4		15
		(1.141)	(1.309)	(1.722)	(1.722)	(180.0)	(120.2)	(119.6)	(120.2)	(0.07)	18
CH_3	CH_3	1.171	1.300	1.514		180.0	120.6	118.8	120.6		13

^{a}MP3/6-31G**.
^{b}MP2/4-31G**.
^{c}For CH_3 C—H 1.083 (1.084), 1.11.
^{d}H—C—H = 109.9, 108.8; H—C—C_2 = 111.1 (111.4).

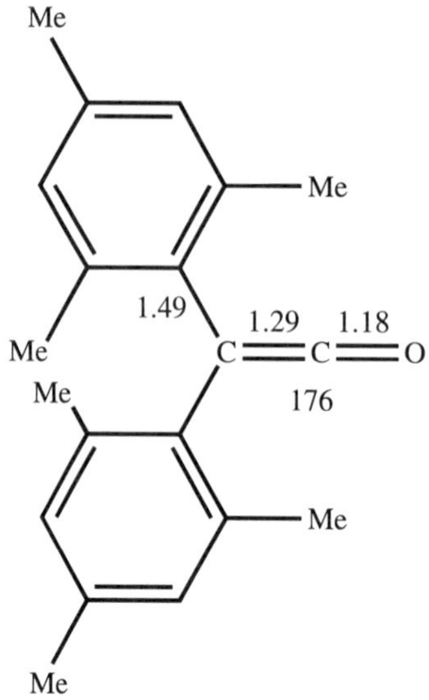

STRUCTURE, BONDING, AND THERMOCHEMISTRY OF KETENES

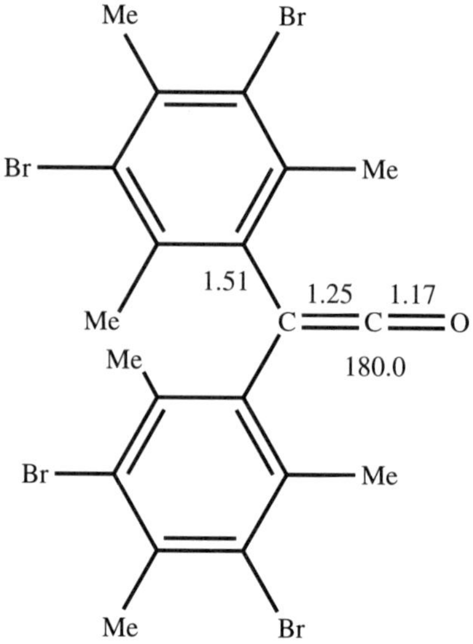

[Structural diagrams of compounds 2 and 3 with bond lengths and angles:]

Compound **2**: dimesityl ketene with bond lengths 1.49, 1.29, 1.18 and angle 176°

Compound **3**: bis(bromomesityl) ketene with bond lengths 1.51, 1.25, 1.17 and angle 180.0°

Dihedral angles: 48.8 (Ar$_1$)   51.8
                  56.8 (Ar$_2$)   51.8

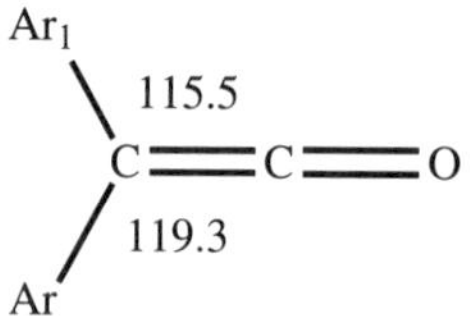

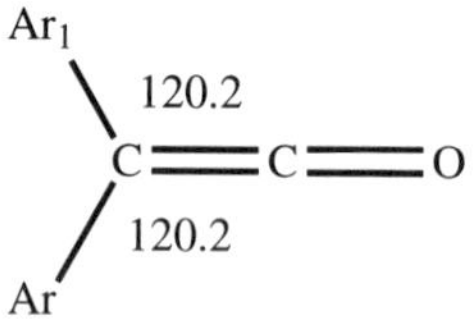

Angles at ketene carbon:
- Compound 2: Ar$_1$–C 115.5°, Ar–C 119.3°
- Compound 3: Ar$_1$–C 120.2°, Ar–C 120.2°

The structures of phenylketene and diphenylketene have been calculated by AM1 (with fixed bond distances and bond angles in the phenyl rings)[19] and the structure of the former has also been deduced by HF/6-31G*//HF/6-31G* ab initio calculation.[22] The former compound was calculated to be planar,[19,22] while the aryl rings in the latter were calculated to be conrotated by 30–40° from the ketene plane.[19] Comparative geometries are shown, with the ab initio results for phenylketene in parenthesis.

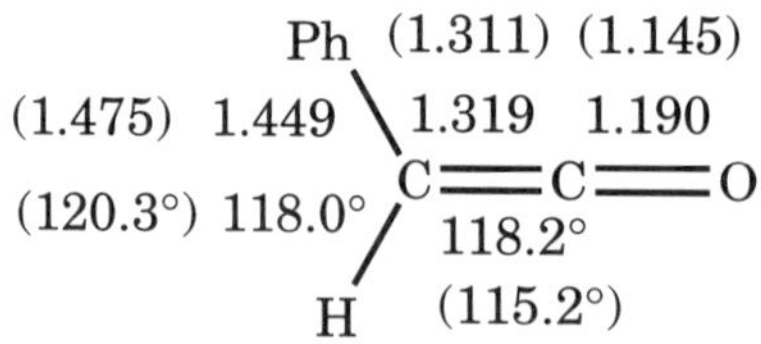

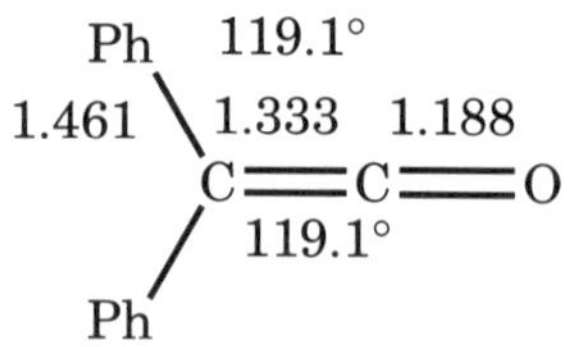

Phenylketene: Ph (1.311) (1.145), (1.475) 1.449, 1.319 1.190, (120.3°) 118.0°, 118.2°, H (115.2°)

Diphenylketene: Ph 119.1°, 1.461, 1.333 1.188, 119.1°, Ph

The X-ray crystal structures of the aliphatic ketene **4** and the corresponding thioketene **5** are summarized below.[17] Interestingly for structures **2–4**, obtained from X-ray data, the ketene C=C bonds of 1.25–1.29 Å are shorter than any of the microwave or electron diffraction based bond lengths obtained (Table 1), and the only calculated C=C bonds that are less than 1.30 Å are for the highly electropositive Li, Na, and MgH substituents (Table 2). The C=O bond lengths obtained by X-ray[16,17] range from 1.17 to 1.18 Å, and on average are slightly longer than the experimental gas phase values, which range from 1.160 to 1.171 Å (Table 1). Another ketene X-ray structure is noted on p. 230.

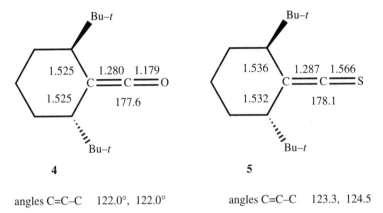

angles C=C–C 122.0°, 122.0°   angles C=C–C 123.3, 124.5

The X-ray structure of the chromium tricarbonyl complexed vinylketene **6** has been reported.[18] The Cr(CO)$_3$ ligand in this structure is far removed from the ketenyl group and appears not to affect the ketenyl structure which is, however, heavily substituted, and this crowding is at least partly responsible for the compressed SiCC angle of 111.8° in the ketenyl moiety. The C=C and C=O bond lengths of the ketene are quite close to those of **2–5**.

6

The molecular structures of a large number of ketenes have been calculated at the 6-31G* level,[20–22] and are presented in Table 2, and atomic charges and dipole moments are presented in Table 3. Similarly, the structures and atomic charges of fulvenones have been calculated[21] and are presented in Tables 4 and 5, respectively.

Molecular mechanics calculations of ketene structures have also been carried out and compared to the available experimental geometries.[23] Both MM2 and MM3 methods were used, and vibrational frequencies were also calculated.[23]

TABLE 2. Bond Distances (Å) and Bond Angles (deg) Calculated (HF/6-31G*//HF/6-31G*) for Ketenes

$$H_nM\diagdown_{C_2}=C_1=O_1 \diagup^{H_2}$$

(ref. 20)

MH_n	$C_1=O_1$	$C_1=C_2$	C_2-H_2	C_2-M	$M-H$	$O_1C_1C_2$	C_1C_2M	MC_2H_2	HMC_2
H	1.145	1.306	1.071			180.0	119.3	121.4	
Li	1.163	1.290	1.077	1.946		184.8	118.3	124.1	
BeH	1.144	1.309	1.079	1.659	1.332	182.1	122.1	123.2	179.4
BH_2	1.134	1.327	1.075	1.519	1.190	181.3	117.7	126.3	120.1
CH_3	1.149	1.305	1.074	1.512	1.084	180.0	123.3	120.8	110.9, 111.4
NH_2	1.145	1.313	1.076	1.423	1.000	178.9	118.6	124.4	111.9
OH	1.147	1.312	1.072	1.374	0.948	178.7	120.4	120.5	109.3[a]
F	1.147	1.308	1.067	1.339		177.9	120.1	117.4	
Na	1.171	1.282	1.074	2.273		185.6	113.8	125.7	
MgH	1.152	1.299	1.078	2.058	1.711	183.4	122.4	122.3	177.8
AlH_2	1.142	1.312	1.078	1.922	1.580	182.2	121.6	123.2	118.9, 118.3
SiH_3	1.143	1.308	1.077	1.860	1.475	181.1	122.0	122.5	108.3, 110.9
PH_2	1.142	1.309	1.073	1.825	1.402	181.3	123.4	119.6	100.8
SH	1.140	1.313	1.072	1.766	1.329	180.0	119.7	122.0	99.4[b]
Cl	1.142	1.308	1.068	1.731		179.2	120.8	118.7	
CF_3	1.137	1.312	1.071	1.484	1.324[c]	178.8	120.9	120.2	110.9[d]
c-Pr	1.149	1.305	1.074	1.494	1.076	179.8	123.3	120.8	120.3
$CH=CH_2$	1.144	1.312	1.074	1.467	1.077[e]	179.8	122.6	121.3	116.2[f]
CH=O	1.136	1.321	1.073	1.463	1.094[g]	180.4	120.7	121.2	115.6
Ph^h	1.145	1.311	1.074	1.475		179.6	124.5	120.3	
CO_2H	1.134	1.322	1.071	1.458	1.335[i]	178.6	121.3	120.2	124.8[j]
C≡CH	1.139	1.317	1.073	1.428	1.188[k]	179.0	121.3	121.8	179.1[l]
CN	1.133	1.321	1.072	1.424	1.137[m]	178.7	120.6	121.3	179.4[l]
$C=C=O^n$	1.148	1.307	1.073	1.476		180.7	123.2	121.1	

C=C=O^o	1.147	1.308		1.073	1.479		179.2	127.0	117.9
N=O			1.175	1.502^p	1.071	1.337^q	1.185^u	136.9^r	137.3^s	132.4^t
N≡C			1.131	1.330	1.072	1.400		179.1	115.9	123.2	114.4^v
			1.139	1.317	1.070	1.376	1.157^w	178.1	121.0	119.9	177.7^x

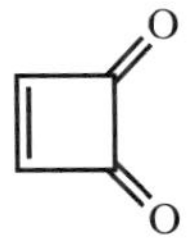

^{a}HOC$_2$C$_1$ (84.2°).
^{b}HSC$_2$C$_1$ (97.3).
^{c}CF (ave).
^{d}FCC (ave).
^{e}CH$_2$=CH (1.322).
^{f}CH$_2$CHC (124.5).
^{g}CH=O (1.190).
^{h}Ref. 22.
^{i}C—O, C=O (1.188), O—H (0.952).
^{j}O—CC, HOC (108.2).
^{k}C≡C, C≡C—H (1.057).
^{l}C≡C—C.
^{m}CN.
^{n}Transoid.
^{o}Ciscoid.
^{p}C—C.
^{q}C=C.
^{r}C—C—C.
^{s}O=C—CO.
^{t}C—C—H, H—C=C (133.4).
^{u}N=O.
^{v}O=N—C.
^{w}N—C.
^{x}CNC.

TABLE 3. Net Atomic Charges[a] on Ketenes $H_nMC_2H_2=C_1=O_1$ (HF/6-31G*//HF/6-31G*)[20,22]

MH_n	O_1	C_1	C_2	M	H_2	H_n	μ (Debye)
H	−0.44	0.57	−0.61	0.24	0.24		1.63
Li	−0.55	0.54	−0.69	0.51	0.18		7.39
Be	−0.45	0.53	−0.55	0.30	0.24	−0.07	1.47
B	−0.41	0.58	−0.52	0.19	0.24	−0.04	1.08
C	−0.45	0.53	−0.37	−0.48	0.23	0.18	1.99
N	−0.44	0.58	−0.23	−0.83	0.21	0.36	2.90
O	−0.43	0.53	−0.07	−0.70	0.22	0.46	2.33
F	−0.41	0.48	0.06	−0.36	0.23		1.59
Na	−0.58	0.49	−0.72	0.61	0.18		9.01
Mg	−0.49	0.56	−0.71	0.53	0.23	−0.12	2.03
Al	−0.44	0.60	−0.72	0.64	0.25	−0.17	1.36
Si	−0.43	0.57	−0.67	0.71	0.25	−0.14	1.39
P	−0.43	0.60	−0.66	0.27	0.26	−0.02	0.44
S	−0.42	0.61	−0.59	0.03	0.26	0.10	1.57
Cl	−0.41	0.60	−0.49	0.02	0.27		1.39
CF_3	−0.39	0.59	−0.55	1.16		−0.36[b]	1.88
c-Pr[c]	−0.45	0.50	−0.34	−0.16	0.23	0.20	2.04
$CH=CH_2$[d]	−0.43	0.52	−0.37	−0.10	0.24	0.20	1.20
CH=O	−0.39	0.57	−0.46	0.35	0.27	−0.16[e]	2.41
Ph	−0.43	0.52	−0.37	0.10	0.25		1.50
CO_2H	−0.39	0.61	−0.51	0.83	0.28		1.12
C≡CH	−0.41	0.58	−0.42	0.14[f]	0.27		1.01
CN	−0.38	0.61	−0.40	0.32	0.30		3.72
CH=C=O	−0.43	0.51	−0.33	−0.33	0.25		0.0
N=O	−0.49	0.62	−0.14	0.03	0.33		2.73
NC	−0.39	0.57	−0.13	−0.40	0.28	0.08[g]	2.74

[a] From Mulliken population analysis.
[b] F.
[c] C_4 (−0.36), H_4 (0.19).
[d] C_4 (−0.42), Z-H_4 (0.18), E-H_4 (0.17).
[e] O_3 (−0.51).
[f] C_4 (−0.46), H_4 (0.29).
[g] C_4.

TABLE 4. HF/6-31G* Optimized CC, CO, and CF Bond Distances (Å) and Bond Angles (deg) for Fulvenones and Fulvenes[21]

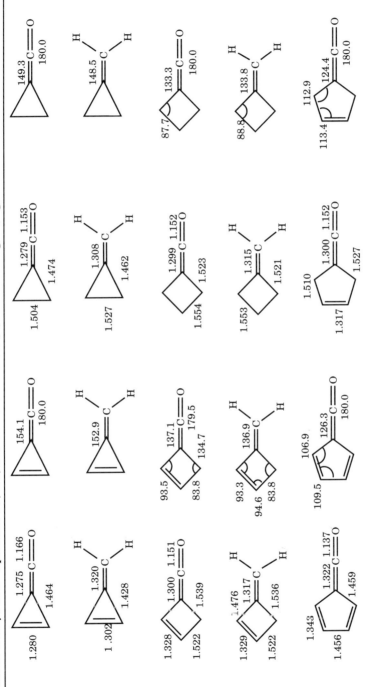

TABLE 4. (*Continued*)

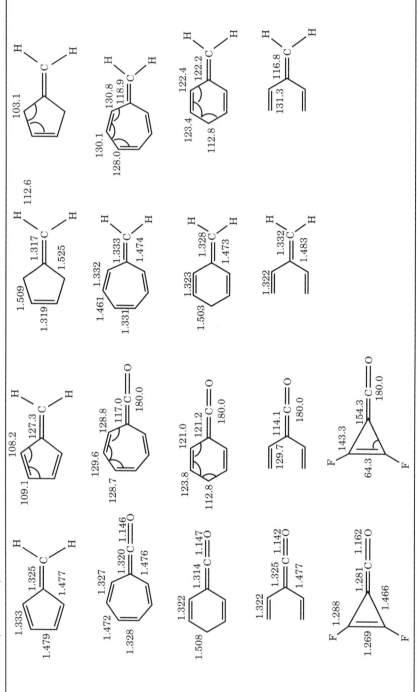

TABLE 5. Mulliken Charges (HF/6-31G//HF/6-31G*) for Fulvenones and Fulvenes[21]

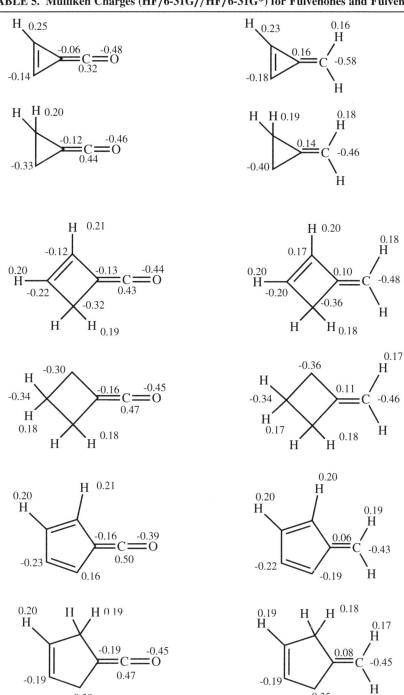

TABLE 5. *(Continued)*

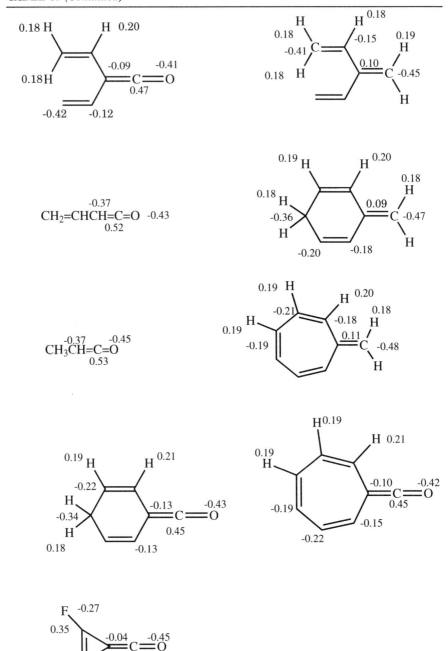

References

1. Runge, W. In *The Chemistry of Ketenes, Allenes, and Related Compounds;* Patai, S., Ed.; Wiley: New York, 1980; pp. 45–95.
2. Brown, R. D.; Godfrey, P. D.; McNaughton, D.; Pierlot, A. P.; Taylor, W. H. *J. Mol. Spec.* **1990**, *140*, 340–352.
3. Duncan, J. L.; Munro, B. *J. Mol. Struct.* **1987**, *161*, 311–319.
4. Johnson, H. R.; Strandberg, M. W. P. *J. Chem. Phys.* **1952**, *20*, 687–695.
5. Cox, A. P.; Thomas, L. F.; Sheridan, J. *Spectrochim. Acta* **1959**, *15*, 542–543.
6. Moore, C. B.; Pimentel, G. C. *J. Chem. Phys.* **1963**, *38*, 2816–2829.
7. Brown, R. D.; Godfrey, P. D.; Wiedenmann, K. H. *J. Mol. Spec.* **1989**, *136*, 241–249.
8. Brown, R. D.; Godfrey, P. D.; Kleibomer, B. *Chem. Phys.* **1986**, *105*, 301–305.
9. Gerry, M. C. L.; Lewis-Bevan, W.; Westwood, N. P. C. *J. Chem. Phys.* **1983**, *79*, 4655–4663.
10. Bak, B.; Christiansen, J. J.; Kunstmann, K.; Nygaard, L.; Rastrup-Andersen, J. *J. Chem. Phys.* **1966**, *45*, 883–887.
11. Hahn, M.; Bodenseh, H. In *Tenth Colloquium on High Resolution Molecular Spectroscopy;* Dijon, September 1987, paper M16. H. Bodenseh, private communication.
12. Brown, R. D.; Godfrey, P. D.; Woodruff, M. *Aust. J. Chem.* **1979**, *32*, 2103–2109.
13. Nair, K. P. R.; Rudolph, H. D.; Dreizler, H. *J. Mol. Spec.* **1973**, *48*, 571–591.
14. Duncan, J. L.; Jarman, C. N. *Struct. Chem.* **1990**, *1*, 195–199.
15. Rozsondai, B.; Tremmel, J.; Hargittai, I.; Khabashesku, V. N.; Kagramanov, N. D.; Nefedov, O. M. *J. Am. Chem. Soc.* **1989**, *111*, 2845–2849.
16. Biali, S. E.; Gozin, M.; Rappoport, Z. *J. Phys. Org. Chem.* **1989**, *2*, 271–280.
17. Schaumann, E.; Harto, S.; Adiwidjaja, G. *Chem. Ber.* **1979**, *112*, 2698–2708.
18. Schubert, U.; Dötz, K. H. *Crystal Struct. Commun.* **1979**, *8*, 989–994.
19. Sammynaiken, R.; Westwood, N. P. C. *J. Chem. Soc., Perkin Trans. 2*, **1989**, 1987–1992.
20. Gong, L.; McAllister, M. A.; Tidwell, T. T. *J. Am. Chem. Soc.* **1991**, *113*, 6021–6028.
21. McAllister, M. A.; Tidwell, T. T. *J. Am. Chem. Soc.* **1992**, *114*, 5362–5368.
22. McAllister, M. A.; Tidwell, T. T. *J. Org. Chem.* **1994**, *59*, 4506–4515.
23. Stewart, E. L.; Bowen, J. P. *J. Computational Chem.* **1992**, *13*, 1125–1137.

1.3 THERMOCHEMISTRY OF KETENES

The most recent heats of formation ΔH_f° that have been reported for ketene,[1] $CH_3CH=C=O$,[2,3] and $CH_2=CHCH=C=O$,[4] are -14.6, -20.6, and 4 kcal/mol, respectively. These data in combination with the respective ΔH_f° values of ethylene and propylene of 12.4 and 4.7 kcal/mol[5] lead to $\Delta\Delta H_f^\circ$ values for the isodesmic reaction of equation 1 of 1.8 and -3.1 kcal/mol for the H and $CH_2=CH$ groups, respectively.[6] The HF/6-31G*//6-31G* calculated energies for these molecules lead to calculated ΔE values for the same reaction of 3.3 and -0.2 kcal/mol, and the agreement between these values lends confidence to the calculated energy values of the ketenes.

$$RCH=C=O + CH_3CH=CH_2 \rightarrow CH_3CH=C=O + RCH=CH_2 \quad (1)$$

Previous measurement of ΔH_f° for ketene of -11.4[7] and -14.8[8] kcal/mol were derived from the reactions of ketene with NaOH and alcohols, respectively. Some estimated[9] values for ΔH_f° (kcal/mol) of some substituted ketenes are $CH_3CH=C=O$ (-25),[10] $(CH_3)_2C=C=O$ (-37), $PhCH=C=O$ (6), and $Ph_2C=C=O$ (29). A value for ΔH_f of 156.8 kcal/mol was measured for $CH_3CH=C=O^{+\bullet}$.[3]

References

1. Rosenstock, H. M.; Draxl, K.; Steiner, B. W.; Herron, J. T. *J. Phys. Chem. Ref. Data* **1977**, *6*, Suppl. 1.
2. Traeger, J. C. *Org. Mass Spectrom.* **1985**, *20*, 223–227.
3. Turecek, F.; Drinkwater, D. E.; McLafferty, F. W. *J. Am. Chem. Soc.* **1990**, *112*, 5892–5893.
4. Terlouw, J. K.; Burgers, P. C.; Holmes, J. L. *J. Am. Chem. Soc.* **1979**, *101*, 225–226.
5. Cox, J. D.; Pilcher, G. *Thermochemistry of Organic and Organometallic Compounds;* Academic: London, 1970.
6. Gong, L.; McAllister, M. A.; Tidwell, T. T. *J. Am. Chem. Soc.* **1991**, *113*, 6021–6028.
7. Nuttall, R. L.; Laufer, A. H.; Kilday, M. V. *J. Chem. Thermodyn.* **1971**, *3*, 167–174.
8. Rice, F. O.; Greenberg, J. *J. Am. Chem. Soc.* **1934**, *56*, 2268–2270.
9. Deming, R. L.; Wulff, C. A. In *The Chemistry of Ketenes, Allenes, and Related Compounds;* Patai, S., Ed.; Wiley: New York, pp. 154–164.
10. Lias, S. G.; Bartmess, J. E.; Liebman, J. F.; Holmes, J. L.; Levin, R. D.; Mallard, W. G. *J. Phys. Chem. Ref. Data* **1988**, *17*, Suppl. 1.

CHAPTER 2

SPECTROSCOPY AND PHYSICAL PROPERTIES OF KETENES

Earlier reviews have dealt with the spectroscopy of ketenes.[1,2] Some representative results are given here, but there is a significant opportunity for a systematic study in this area.

2.1 NUCLEAR MAGNETIC RESONANCE SPECTROSCOPY

The ^{13}C NMR spectra of a variety of ketenes are listed in Table 1.[3-14] The most distinctive feature of the ^{13}C NMR chemical shifts of ketenes is the very high field shift of C_β, with a value of $\delta 2.5$ for ketene itself.[4] This value reflects the high degree of negative charge at C_β as predicted by the resonance structure **A** (Section 1.1), and provides a powerful diagnostic tool for ketene identification.

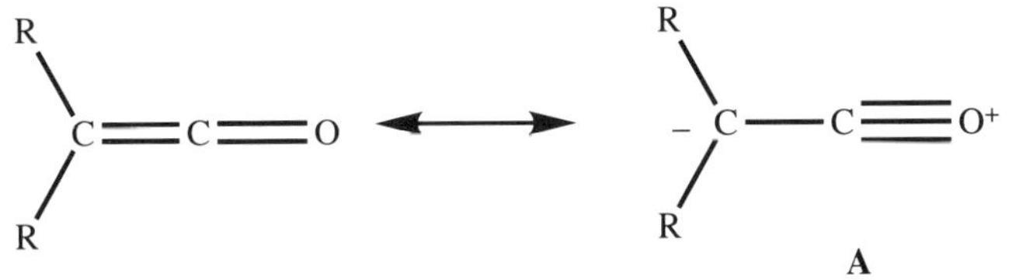

$$\underset{R}{\overset{R}{\diagdown}}C=C=O \longleftrightarrow \underset{R}{\overset{R}{\diagdown}}\overset{-}{C}-C\equiv O^+$$

A

There are large effects of substituents at C_β on the chemical shifts of this carbon, and qualitatively these effects appear to parallel the effects on alkenes, although no detailed comparisons have been made. These substituents also cause significant changes in the chemical shifts at C_α, although again these effects have not been analyzed in detail. It is striking that group IV substituents such as Si at C_β cause significant upfield shifts at C_α, consistent with σ-π hyperconjugative electron donation to C_α. However, as discussed in Section 4.5, other evidence for this effect is not strong. Large downfield shifts at C_β substituted by two Br or CF_3Se groups were observed (δ = 98.5 and 124.9, respectively) that were attributed to a special type of heavy atom effect.[5] A more detailed study of substituent effects on ketene ^{13}C NMR chemical shifts is needed to fully appreciate the factors that affect these shifts.

The ^{13}C NMR spectrum of CH_2=C=O in the solid state has been recorded, along with those of other linear and pseudolinear molecules.[15] For ketene the assignments of the shielding tensor axes were made using ab initio calculations by the IGLO version of the coupled Hartree–Fock method.[15]

TABLE 1. ^{13}C NMR Chemical Shifts of Ketenes RR^1C_β=C_α=O (δ ppm, TMS)[a]

R	R^1	C_α	C_β	R	R^1	C_α	C_β
H[b]	H	194.0	2.5	Me_3Si	H	179.2	−0.1
Me[b]	H	200.0	10.9	Et_3Si	H	179.2	−4.9
Et[b]	H	200.0	18.6	t-$BuMe_2Si$[h]	H	180.0	−4.6
Me[c]	Me	204.9	24.2	t-$BuPh_2Si$[h]	H	178.5	−3.6
Et[b]	Me	206.1	26.9	Me_3Si	Me_3Si	166.8	1.7
Ph[b]	Me	205.6	33.8	Me_2SiH[i]	Me_2SiH	—	−4.2
Ph[b]	Et	205.6	42.1	Me_3Ge	H	179.4	−4.8
Ph[b]	Ph	201.2	47.6	Et_3Ge	Me_3Ge	167.0	0.0
Cl[d]	H	166.2	70.1	Et_3Ge	Et_3Ge	165.7	−8.5
Me[d]	Cl	168.9	85.0	Me_3Sn	Me_3Sn	161.7	−13.9
Br[c]	Br	178.6	98.5	Et_3Sn	Et_3Sn	161.3	−20.5
CH=CH_2[e]	H	200.2	28.6	Me_3Si	Me_3Ge	166.2	0.5
CF_3S[f]	CF_3S	171.8	18.8	Me_3Si	Me_3Sn	164.4	−5.8
C_6F_5S[c]	C_6F_5S	172.6	13.6	Me_3Ge	Me_3Sn	164.6	−6.3
CF_3Se[g]	CF_3Se	169.3	124.9	t-$Bu_2P(S)$[j]	CH_3CO	170.2	83.0
CF_3O_2S[g]	CF_3O_2S	181.0	13.5	t-$Bu(MeO)PhSi$[k]	Ph	182.9	20.9
t-Bu[l]	CO_2Me	191.8	50.0	t-Bu[l]	t-BuCO	196.5	53.4
				1-Ad[m]	1-Ad	205.2	29.1

[a] Reference 3 unless noted.
[b] Reference 4.
[c] Reference 5.
[d] Reference 6.
[e] Reference 7.
[f] Reference 8.
[g] Reference 9.
[h] Reference 11.
[i] Reference 12.
[j] Reference 13.
[k] Reference 14.
[l] Reference 53.
[m] Reference 14a.

The ^{1}H NMR chemical shifts and some coupling constants are given in Table 2.[16–19] The high field position of the protons at C_β reflect the importance of the negative charge at this position as reflected in the resonance structure **A**. A linear correlation was found between ^{1}H and ^{13}C chemical shifts for $CH_2=C=O$, CH_2N_2, $CH_2=C=NPh$, and $CH_2=C=CH_2$.[20] Coupling constants $J_{^{13}C^1H}$ (Hz) for C_β are 171.5 $(CH_2=C=O)^4$ and 184.4 $(CHCl=C=O)$,[6] and $^2J_{^{13}C^1H}$ is 4.5 Hz for $CH_3CCl=C=O$.[6]

The ^{17}O chemical shifts (δ) of $R_2C=C=O$ are reported[5] (in parentheses) for the following groups R: CH_3 (329), Ph (340), Br (76), CF_3S (284), C_6F_5S (245), CF_3SO_2 (223), and CF_3Se (43). The ^{17}O shifts for $(CH_3)_2C=C=O$ and $Ph_2C=C=O$ are smaller than those for the sp^2 carbonyl oxygens of ketones, consistent with the triple bond character in the C—O bond shown in **A**.[5] The higher field shifts of the ^{17}O in the sulfur-substituted ketenes were attributed to polarization of the carbonyl group by the inductive effects of the electronegative substituents, while the very strong upfield shifts of the Br and CF_3Se substituted ketenes were attributed to a special type of heavy-atom effect.[5]

TABLE 2. ^{1}H NMR Chemical Shifts and Coupling Constants of Ketenes $RR^1C_\beta=C_\alpha=O^a$

R	R^1	$\delta(H^1)$	J (Hz)	Other δ, J(Hz)	Ref.
H	H	2.43^b	15.8 (H, H')		17
H	H	2.46			18,19
D	H	2.40^b	2.42 (H, D)		17
Me	H	2.67	7.4	1.56 (CH_3, J = 7.4)	18,19
Et	H	2.80	6.9	2.04 (CH_2), 1.06 (CH_3, J = 7.4)	18,19
i-Pr	H	2.82	5.7	2.49 (CH), 1.07 (CH_3, J = 6.7)	18,19
t-Bu	H	2.81		1.19 (t-Bu)	18,19
Me_3Si	H	1.65			11
Et_2HSi	H	1.65		3.0 (HSi—CH)	10
$(CD_3)_3Si$	H	1.70			10
$SiCl_3$	H	2.84			10
$CH_3CH=CH$	H	3.95, 4.10 (E, Z)	10		22
$CH_2=CH$	H	4.09	10	4.66(Z), 4.95(E), 6.06 (vinyl, J = 11.0, 18)	7,18,19
$CH_2=CH$	Me	1.76 (Me)		4.76, 4.77, 6.30 (vinyl, J = 11, 16.6)	18,19
Me	Me	1.58, 1.65^c			18,19

^{a}CDCl$_3$ unless noted.
^{b}85% in TMS.
^{c}CCl$_4$ (ref. 21).

2.2 ULTRAVIOLET AND PHOTOELECTRON SPECTROSCOPY

The UV spectrum of $CH_2=C=O$[23-25] shows a C=C $\pi \to \pi^*$ transition at 183 nm,[23] the C=O $\pi \to \pi^*$ transition at 215 ($\varepsilon = 80$),[24] and the n $\to \pi^*$ transition at 325 nm ($\varepsilon = 10$).[2,23a] Other available information on ketene UV spectra is compiled in Table 1.[23-30] Photochemical reactions are discussed in Section 5.2.

The effects of substituents on the UV spectra of ketenes have not been studied in a systematic way, and no consistent interpretation of how substituents affect these spectra has been put forward.

A compilation of photoelectron spectral data[31-39] on ketenes is given in Table 2. For these measurements CHCl=C=O, CHBr=C=O, MeCH=C=O, and $Me_2C=C=O$ were formed by pyrolysis of the acyl chlorides at 700–800 °C (equation 1).[35] In this procedure it was necessary to remove the liberated HHal by

TABLE 1. UV Spectra of Ketenes[a]

Ketene	λ_{max} [nm(ε)]	Ref.
$CH_2=C=O$	183, 215 (80), 325 (10)[b]	23–26
$t\text{-BuMe}_2\text{SiCH}=C=O$	206 sh (2,380), 292 (34.7)[c]	27
$Me_2C=C=O$	370	26
$Et_2C=C=O$	378 (13.3)[d]	26, 27
$t\text{-Bu}_2C=C=O$	225 (1,200), 360 (11.6)[c]	27
$t\text{-BuCH}=C=O$	360 (13.2)[d]	27
$t\text{-BuC}(CO_2Et)=C=O$	228	27
$Ph_2C=C=O$	350, 405; 399 (310)[e]	28, 28a
$(EtS)_2C=C=O$	288 (6,000)[f]	68
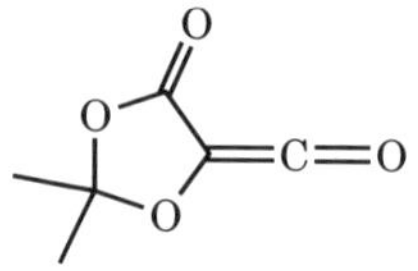 (dioxolanone-C=O structure)	~280[d]	29
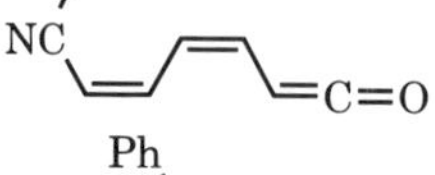 NC-CH=CH-C=O (Ph)	313 (15,500), 323 sh (12,500)[c]	30
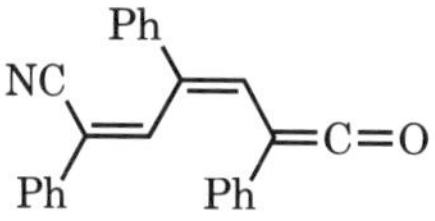 NC(Ph)C=C(Ph)-C=O	221 (14,000), 324 (11,500)[c]	30
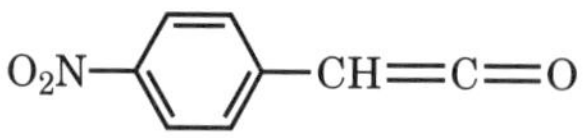 O_2N-C$_6$H$_4$-CH=C=O	495[g]	30a

[a] Gas phase unless noted.
[b] Values estimated from reference 25.
[c] Hexane.
[d] THF.
[e] Cyclohexane.
[f] MeOH.
[g] DMSO/H_2O.

2.2 ULTRAVIOLET AND PHOTOELECTRON SPECTROSCOPY

TABLE 2. Vertical Ionization Potentials (IP in eV) for Ketenes $RR^1C{=}C{=}O$ from Photoelectron Spectroscopy[a]

R	R^1	IP	Ref.
H	H	9.64 14.2 15.0 16.3	32
		9.63 13.84 14.60 16.08 16.7 18.2 24.3	33
CH_3	H	8.92 13.29 13.72 14.11 15.22 15.55 16.63 17.61	34
CH_3	H	8.95 13.3 (14.3) (15.2)	32
C_2H_5	H	8.80 (12.3) (12.8) (14.3)	32
Cl	H	9.24 12.13 (13.2) (14.6)	32
Cl	H	9.25 12.15 13.10 14.54 15.06 16.39 17.20 18.30	35
Br	H	9.14 11.37 12.24 14.0 14.82 15.95 17.09 18.03	35
Br	H	9.10 11.05	32
CN	H	10.07 12.44 12.73 13.07	32
Ph	H	8.06 9.31 10.39 11.77 12.36 14.5 15.62 16.34	36
$CH_2{=}CH$	H	8.29 10.23	37
MeS	H	8.24 10.85	38
CH_3	CH_3	8.45 (12.6) (13.8) (15.4) (16.1)	32
Cl	Cl	9.15 12.20 12.50 12.80	32
Cl	Cl	9.07 12.18 12.52 12.85 13.90 14.90 15.62 16.75	35
Br	Br	8.89 11.20 11.58 11.87 12.98 14.0 15.07 16.05	35
▷=C=O		8.78 10.7 11.3 12.7	32
⬠=C=O		8.39 8.98 (12.1) (12.7)	32
Ph	Ph	7.64 9.18 10.19 11.64 12.1 14.1 14.7 15.64	36
$CH_2{=}C(OMe)$	CO_2H	8.80 9.61 10.97	38
$CH_2{=}C(OMe)$	H	8.80 9.68 10.97 12.85	38
CF_3	CF_3	10.95 14.51 15.72	39

[a]Band maxima in parentheses where overlapping.

reaction with NH_3. However for $CCl_2{=}C{=}O$ and $CBr_2{=}C{=}O$, decarbonylation took place under these conditions and so the trihaloacyl halides were reacted with metallic zinc at 300–400 °C (equation 2).[35] This reaction had the further advantage that no HX was formed.

$$CH_2XCOX \xrightarrow{700\text{-}800\ °C} CHX{=}C{=}O \qquad (1)$$

$$CX_3COX \xrightarrow[Zn]{300\text{-}400\ °C} CX_2{=}C{=}O \qquad (2)$$

The photoelectron spectra of phenylketene and diphenylketene together with MNDO and AM1 calculations were interpreted in terms of a planar phenyl group for the former, while the phenyl groups in the latter were each conrotated by 30–40° out of the ketene plane.[36] The photoelectron ionization potentials of a variety of ketenes and bisketenes have been calculated by ab initio methods, and are in good agreement with available experimental data.[40]

2.3 INFRARED SPECTRA

The infrared spectrum of ketene has been thoroughly studied,[41-45] and the band assignments for $CH_2=C=O$, $CHD=C=O$, and $CD_2=C=O$ are given in Table 1.[42,44] The rotational constants have been assigned.[45] Several bands for some silylketenes have also been assigned.[10]

The strong C=O stretch between 2100 and 2200 cm^{-1} is the most characteristic IR band of ketenes and is frequently used for characterization purposes. However, as found by Moore and Pimentel,[46] the position of this band is dependent upon the medium, and for $CH_2=C=O$ it occurred at 2151, 2142, and 2133 cm^{-1} in the vapor, in an argon matrix, and in the solid, respectively.[46] No systematic study of the variation in the position of this band in different solvents has been reported.

Values for the C=O stretching frequency for many ketenes have been reported, and some of these are listed in Table 2.[47-84] These values were obtained under widely different conditions of temperature and medium, using different instruments, and therefore their utility for comparative purposes is limited.

A correlation of the IR frequencies of disubstituted ketenes has been reported[47] using a data set for five ketenes, $R_2C=C=O$ (R = H, Cl, CF_3, CN, CH_3), with argon matrix frequencies. The correlation obtained (equation 1) used the substituent parameters F for field (inductive) and R for resonance effects, respectively.

$$\nu \text{ (cm}^{-1}) = 24F + 30R + 2142 \tag{1}$$

This correlation indicates that the C=O stretching frequency ν is increased both by substituents that are inductive and resonance electron acceptors. This is

TABLE 1. Assigned IR Absorption Bands of Ketene (cm^{-1})

Mode	$CH_2=C=O$[44]	$CHD=C=O$[42]	$CD_2=C=O$[42]
C—H stretch	3071	2268	3115
C=O stretch	2152	2121	2150
CH_2 deformation	1388	1228	1293
C=C stretch	1118	890	1022
C—H stretch (anti sym)	3166	2375	2297
CH_2 rocking	977	798	815
CH_2 wagging	588	530	—
C=C=O bending	438	712	—
C=C=O bending (oopl)	528	450	515

TABLE 2. Carbonyl Stretching Frequencies of Ketenes (cm^{-1})^a

Ketene	ν(C=O)	Ref.	Ketene	ν(C=O)	Ref.
CH$_2$=C=O	2151^b	46	CCl$_2$=C=O	2160^b	62
MeCH=C=O	2130	58	(NC)$_2$C=C=O	2175^c	63
EtCH=C=O	2132^b	59	(CF$_3$)$_2$C=C=O	2196^c	64
Me$_2$C=C=O	2120	26	CH$_2$=CHCH=C=O	2130^b	65
PhCH=C=O	2250^b	36	MeCH=CHCH=C=O	2118	22
Ph$_2$C=C=O	2100	48	CH$_2$=CHC(SiMe$_3$)=C=O	2085	67
Ph$_2$C=C=O	2098^c	36			
(Me$_3$Si)$_2$C=C=O	2085	10	Me$_2$C=CHCMe=C=O	2097	66
Me$_3$SiCH=C=O	2112	10	(CF$_3$S)$_2$C=C=O	2139	8
(CD$_3$)$_3$SiCH=C=O	2122	10	(CF$_3$Se)$_2$C=C=O	2127	9
Et$_2$SiHCH=C=C	2115	10	(CF$_3$O$_2$S)$_2$C=C=O	2139	9
Cl$_3$SiCH=C=O	2165	10	(EtS)$_2$C=C=O	2160	68
t-BuMe$_2$SiCH=C=O	2110	11	(t-BuS)$_2$C=C=O	2170	68
t-BuPh$_2$SiCH=C=O	2110	11	(ArS)$_2$C=C=O	2190	68
t-BuCOCH=C=O	2140^b	52	F$_5$SCH=C=O	2177	69
EtO$_2$CCMe=C=O	2137^b	52	F$_5$SCMe= C=O	2149	70
MeO$_2$CCMe=C=O	2137^b	52	F$_5$SCCl=C=O	2175	70
CH$_3$COCMe=C=O	2121^b	52	F$_5$SCBr=C=O	2169	70
CH$_3$COCH=C=O	2137^b	52	MeOCH=C=O	2136	53
CHF=C=O	2142.2^c	60	EtOCH=C=O	2120	71
CHCl=C=O	2141.4^c	60	(C$_6$Cl$_5$)$_2$C=C=O	2130	72
CHCl=C=O	2150^b	35	(C$_6$F$_5$)$_2$C=C=O	2140	73
CCl$_2$=C=O	2158^c	60	ĊH=C=O	2023	74
CCl$_2$=C=O	2155^c	61	Ph$_3$P=C=C=O	2080	75
			CH$_2$=CHC(CO$_2$H)=C=O	2135^c	75a

TABLE 2. (*Continued*)

Ketene	$\nu(C=O)$	Ref.	Ketene	$\nu(C=O)$	Ref.
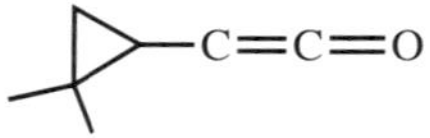 (2,2-dimethylcyclopropyl)-C=C=O	2110	76			
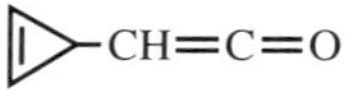 cyclopropyl-CH=C=O	2108^c	77	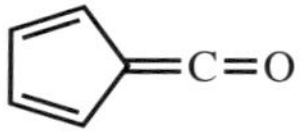 cyclopentadienylidene C=O	2130, 2133^c	81
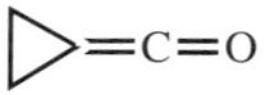 cyclopropenylidene =C=O	2125, 2145^c 2135, 2154	78 55	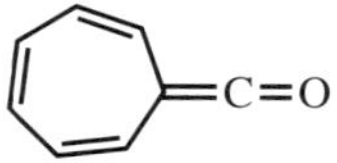 cycloheptatrienylidene C=O	2103^c	82
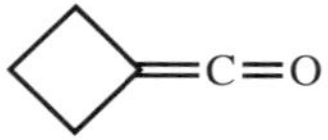 cyclobutylidene =C=O	2098, 2150	55	Ph(C=O)C(Ph)=C=O		
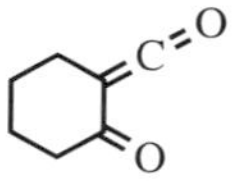 cyclohexanone-2-ylidene C=O	2124^c	54	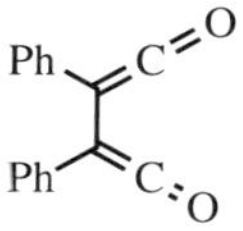	2100, 2112^c	83

Structure	IR (cm⁻¹)	Ref.	Structure	IR (cm⁻¹)	Ref.
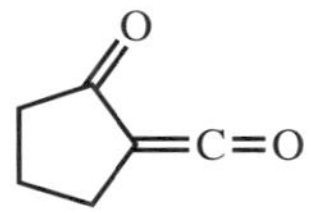 cyclopentanone-ylidene ketene	2133[c]	54	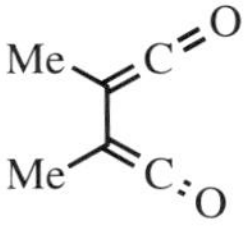 Me, Me-substituted bisketene	2096, 2117[c]	83
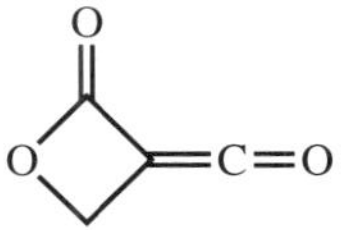 4-methyleneoxetan-2-one (ketene)	2154, 2176[c]	79	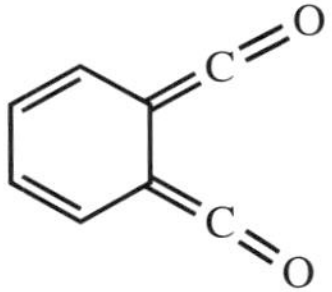 cyclohexadienylidene bisketene	2077, 2138	56
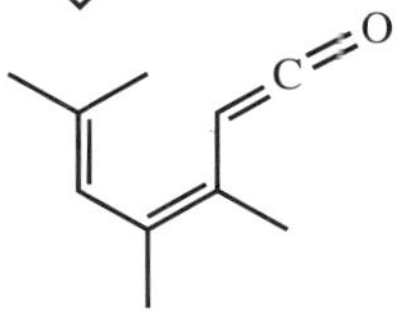 dimethyl-substituted dienyl bisketene	2100	80	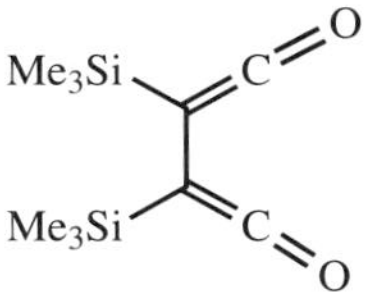 Me₃Si-substituted bisketene	2084	84

[a] In solution unless noted.
[b] Gas phase.
[c] Matrix or solid.

qualitatively reasonable based on the contribution of the resonance structure **A**, which would be favored by electron acceptors, and would lead to higher frequencies. The IR stretching frequencies in 1,2-dichloroethane of diarylketenes (4-XC$_6$H$_4$)$_2$C=C=O were 2096, 2100, and 2109 cm^{-1} for X = MeO, H, and O$_2$N, respectively,[48] and this trend is consistent with the hypothesis discussed above.

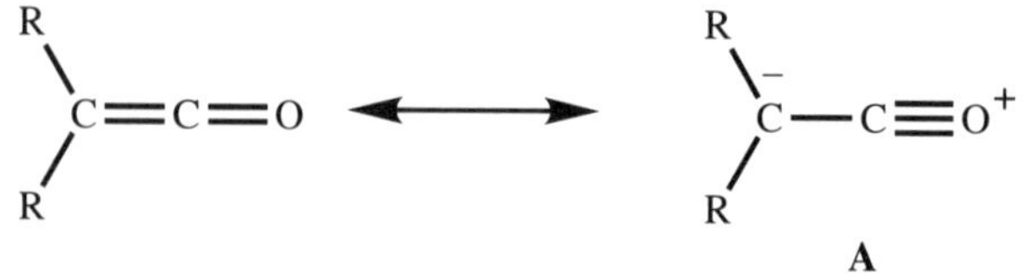

For a more thorough study, a large set of ketene stretching frequencies were obtained by ab initio molecular orbital calculations at the 6-31G*//6-31G* level and are presented in Table 3.[49] These calculated frequencies gave good agreement with available experimental data, and were correlated by equation 2, which had a modest correlation coefficient of 0.87, but showed no dependence on the resonance parameter R.[49] It was also found that the calculated IR intensities increased with increasing substituent electropositive character, and were also enhanced by π-acceptor substituents.[49]

$$\nu \text{ (cm}^{-1}) = 88F + 2120 \qquad (2)$$

The IR spectra of four different conformers of ketene **1**, labelled in the carbonyl carbon of the ketene with ^{13}C, have been measured in an Ar matrix at 8 K,[50] with bands at 2068, 2074, 2081, and 2089 cm^{-1}. In cyclohexane solution the absorption of **1** was observed at 2129 cm^{-1} in Table 4.[51]

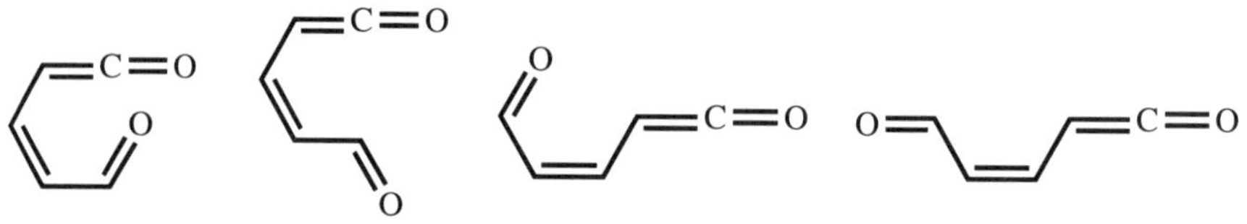

The IR spectra of various acylketenes have been measured in the gas phase (Table 2),[52] while the ketenes **2–7** formed by flash vacuum pyrolysis and isolated in Ar matrices displayed bonds assigned to the *s-Z* and *s-E* conformations as shown in Table 4.[53–57]

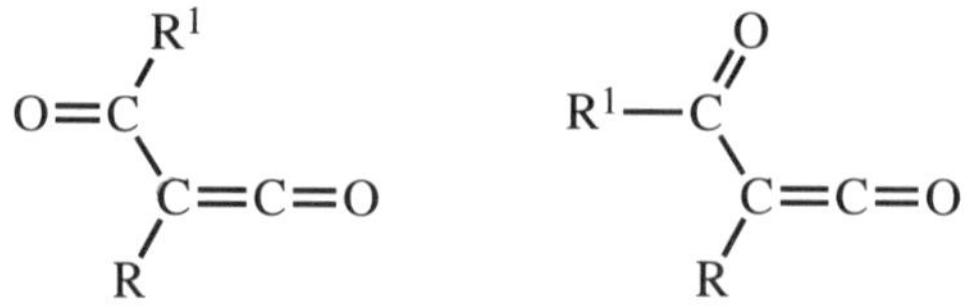

TABLE 3. Calculated (6-31G*//6-31G*) IR Stretching Frequencies (cm^{-1}) of Ketenes RCH=C=O and R$_2$C=C=O, and Fulvenones[49]

R					
H	2141	Na	2010	c-Pr	2130
Li	2023	MgH	2067	CH=CH$_2$	2125
BeH	2100	AlH$_2$	2097	C≡CH	2125
BH$_2$	2119	SiH$_3$	2112	CH=O	2142
CH$_3$	2132	PH$_2$	2121	CO$_2$H	2157
NH$_2$	2139	SH	2139	CN	2162
OH	2146	Cl	2159	CH=C=O	2109
F	2169	NC	2156	CH$_2$=C=CH	2126
N=O	2161	(CN)$_2$	2182	Ph	2119
F$_2$	2227	Me$_2$	2120	(NH$_2$)$_2$	2130
Cl$_2$	2172	CF$_3$	2170	NO$_2$	2186

Fulvenones

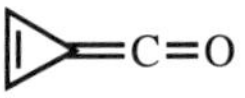

cyclopropenylidene=C=O	2192	cyclobutenylidene=C=O	2135	cyclopentadienylidene=C=O	2125
cyclopropylidene=C=O	2172	cyclobutylidene=C=O	2126	cyclopentenylidene=C=O	2125
F$_2$-cyclopropenylidene=C=O	2217	cyclohexadienylidene=C=O	2111	cyclopentadienylidene=C=O	2105
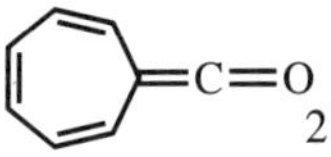 cycloheptatrienylidene=C=O	2103				

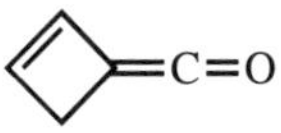

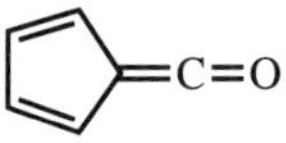

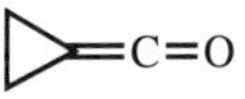

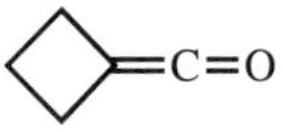

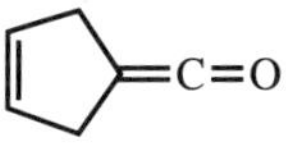

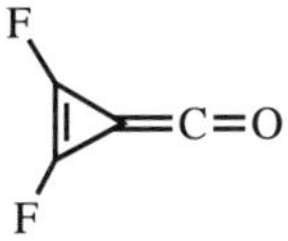

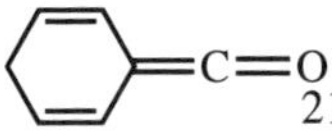

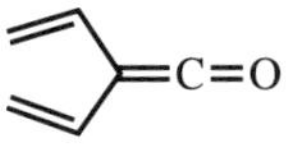

TABLE 4. Measured IR Stretching Frequencies (cm^{-1}) of Acylketenes R^{1}COCR=C=O in Ar Matrices[53–55]

			ν (cm^{-1})		
	R	R^1	s-E	s-Z	Ref.
2	H	OMe	2132	2143(s), 2147(m)	53
3	t-Bu	OMe	2129		53
4	H	t-Bu	2134	2142	53
5	t-Bu	t-Bu	2104		53
6	Me	Me	2123(s), 2129(s)	2131	56
7	Ph	H	2147, 2144	2134, 2132	57

2.4 DIPOLE MOMENTS

A comparison of experimental and calculated dipole moments of ketenes is presented in Section 1.2. It was found very early[85] that the dipole moment of ketene (1.45 D) was less than those of formaldehyde (2.27 D) and acrolein (CH_2=CHCH=O, 3.04 D). This result facilitated an understanding of the electronic structure of ketene, and was interpreted in terms of the resonance structures shown, which include a contribution with negative charge on C_β and positive charge on oxygen, with a consequent decrease in the net dipole moment directed toward oxygen. Walsh suggested an alternative explanation of the low dipole moment of ketene in terms of the sp hybridization on C_α, which would reduce the negative charge on oxygen.[86]

$$CH_2=C=O \longleftrightarrow \overset{+}{CH_2-C}-O^- \longleftrightarrow \overset{-}{CH_2}-C\equiv O^+$$
$$\mathbf{1}\mathbf{2}$$

The dipole moments of some ketenes in benzene gave the following values: ketene, 1.43; dimethylketene, 1.87; diphenylketene, 1.76; and mesitylphenylketene, 1.74.[87] These values were interpreted in terms of a major contribution of structures similar to **2** in all these ketenes, but with no major delocalization of negative charge into the aryl groups.[87] A dipole moment of 1.7 has been measured for $Me_3SiCH=C=O$.[40]

The reduced dipole moment of vinylketene (0.97 D) was interpreted in terms of further electron delocalization away from oxygen as in **3a**.[65] The calculated dipole moment of $H_2BCH=C=O$ is also quite low (1.08 D), supportive of the resonance structure **4a**.[88] Many other calculated dipole moments are given in Section 1.2.[88]

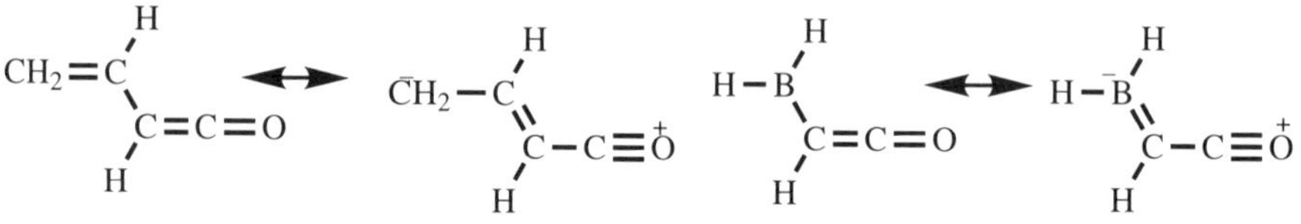

$$CH_2=C\overset{H}{\underset{H}{\diagdown}}C=C=O \longleftrightarrow CH_2-C\overset{H}{\underset{H}{\diagdown}}C-C\equiv\overset{+}{O} \qquad H-B\overset{H}{\underset{H}{\diagdown}}C=C=O \longleftrightarrow H-\bar{B}\overset{H}{\underset{H}{\diagdown}}C-C\equiv\overset{+}{O}$$

$$\mathbf{3} \qquad\qquad \mathbf{3a} \qquad\qquad \mathbf{4} \qquad\qquad \mathbf{4a}$$

2.5 MASS SPECTROMETRY AND GAS PHASE ION CHEMISTRY

Because of the tendency of many ketenes to dimerize and react in other ways there are only a few studies of the mass spectrometry of ketenes where the substrates themselves are introduced into the spectrometer.[89] However, an extensive series of alkylketenes were quantitatively generated from the acyl chlorides by the flash vacuum thermolysis at 650 °C with loss of HCl (equation 1), and their mass spectra recorded, as listed in Table 1.[90]

$$RR^1CHCOCl \xrightarrow[-HCl]{\Delta} RR^1C=C=O \qquad (1)$$

TABLE 1. Mass Spectrometry of Ketenes from Acyl Chlorides[a]

Ketene	Principal Fragment Ions
$CH_2=C=O$	42 (M^+, 100), 41 ($M^+ - 1$, 25)
$MeCH=C=O$	56 (M^+, 100), 55 ($M^+ - H$, 21), 29 ($MH^+ - CO$)
$EtCH=C=O$	70 (M^+, 100), 55 ($M^+ - CH_3$, 97), 41 ($M^+ - COH$, 48)
$n\text{-}PrCH=C=O$	84 (M^+, 34), 55 ($M^+ - C_2H_5$, 100)
$n\text{-}BuCH=C=O$	98 (M^+, 21), 55 ($M^+ - C_3H_7$, 100)
$n\text{-}PnCH=C=O$	112 (M^+, 25), 55 ($M^+ - C_4H_9$, 100)
$i\text{-}PrCH=C=O$	84 (M^+, 75), 69 ($M^+ - CH_3$, 80), 55 (M^+, $-C_2H_5$, 36), 41 ($C_3H_5^+$, 100)
$s\text{-}BuCH=C=O$	98 (M^+, 35), 69 ($M^+ - C_2H_5$, 100), 55 ($M^+ - C_3H_7$, 24)
$t\text{-}BuCH=C=O$	98 (M^+, 52), 83 ($M^+ - CH_3$, 100), 55 ($M^+ - C_3H_7$, 80)
$Me_2C=C=O$	70 (M^+, 100), 41 (M^+, C_2HO^+ or $C_3H_5^+$, 74)
$MeCEt=C=O$	84 (M^+, 50), 69 ($M^+ - CH_3$, 16), 55 ($M^+ - C_2H_5$, 53), 41 (100)
$n\text{-}PrCMe=C=O$	98 (M^+, 46), 69 ($M^+ - C_2H_5$, 52), 41 (100)
$Et_2C=C=O$	98 (M^+, 51), 83 ($M^+ - CH_3$, 19), 69 ($M^+ - Et$, 6), 55 ($M^+ - C_3H_7$, 100)

[a]Reference 90.

The main fragmentation pathway of alkylketenes is loss of an alkyl radical by α-cleavage followed by loss of CO. Thus β-cleavage of the radical ions to give vinylacylium ions **1**, as shown in equation 2, is a common fragmentation pathway, with preferential loss of the larger R group. The ions **1** then undergo decarbonylation (equation 2). The solution phase observation of vinylacylium ions **1** is discussed in Section 4.1.9.

$$\underset{R}{\overset{R^1}{>}}C=C=O \xrightarrow{-e^-} \underset{R}{\overset{R^1}{>}}\overset{\bullet}{C}-\overset{+}{C}=O \xrightarrow{-R\bullet} \underset{-C}{\overset{R^1}{>}}C-\overset{+}{C}=O \xrightarrow{-CO} R^1C\overset{+}{\equiv}C \qquad (2)$$

1

The 70-eV mass spectrum of vinylketene (**2**) gave a major parent ion for $C_4H_4O^{+\bullet}$ (67%), and other prominent ions at m/z (relative abundance) 42 (CH_2CO^+, 14), 40 ($C_3H_4^+$, 37), 39 ($C_3H_3^+$, 78), 38 ($C_3H_2^+$, 17), 37 (C_3H^+, 11), 28 ($C_2H_4^+$, 100), 27 ($C_2H_3^+$, 52), 26 ($C_2H_2^+$, 48), and 25 (C_2H^+, 9).[91] The vinylketene radical cation was also generated by mass spectral fragmentation of a number of other precursors, including Z-crotonic acid and 2-cyclohexenone (equation 3), furan, 3-butenoic acid, cyclopropanecarboxylic acid, and methylcrotonate.[92]

$$\underset{H}{\overset{H_2C}{>}}C=C\underset{H}{\overset{H}{>}}\underset{}{\overset{O}{\underset{}{\parallel}}} \xrightarrow{+\bullet\ -H_2O} CH_2=CH-CH=C=\overset{+\bullet}{O} \xleftarrow{+\bullet\ -C_2H_4} \left[\bigcirc = O\right]^{+\bullet} \qquad (3)$$

2

The mass spectrum of PhOCH$_2$CH=C=O (**3**), generated by dehydrochlorination of the acyl chloride, was observed, and gave the same ion **3**$^+$ formed by electron impact induced Wolff rearrangement in the mass spectrometer (equation 4).[93] The former process at 70 eV gave M$^+$ (10), M$^+$−CO (24), C$_6$H$_6$O$^+$ (67), and C$_3$H$_3$O$^+$(M$^+$−PhO, 100).[93]

$$\text{PhOCH}_2\text{COCHN}_2 \xrightarrow[-N_2, -e^-]{EI} \text{PhOCH}_2\text{COCH}^{+\bullet} \longrightarrow \text{PhOCH}_2\text{CH=C=O}^{+\bullet} \quad (4)$$

$$\mathbf{3^+}$$

Other studies include the kinetic energy release in the metastable fragmentation of ketene[94,95] and the photoionization mass spectrometry of ketene and CD$_2$=C=O.[96] The fragmentation patterns of some alkylketene radical cations formed from electron impact on cyclobutanones have also been observed.[97]

Ketene reacts with alkoxide ion alcohol pairs in the gas phase by proton removal to give a hydrogen-bonded ynolate ion (equation 5).[98]

$$\underset{H}{\overset{H}{>}}C=C=O \xrightarrow{ROH\text{---}^-OR} H-C\equiv C-O^-\text{---}HOR \quad (5)$$

The reaction of the ketene radical cation with NH$_3$ in a selected ion flow tube (SIFT) proceeds by the reaction channels in equations 6 and 7.[99,100]

$$CH_2=C=O^{+\bullet} + NH_3 \longrightarrow CH_2NH_3^{+\bullet} + CO \quad \Delta H = -28.7 \text{ kcal/mol} \quad (6)$$

$$CH_2=C=O^{+\bullet} + NH_3 \longrightarrow NH_4^+ + H-\overset{\bullet}{C}=C=O \quad \Delta H = -7.1 \text{ kcal/mol} \quad (7)$$

A number of ketene complexes with ionic species in the gas phase have been examined by mass spectrometry. These include the [acylium ketene] ion-neutral complex **4** formed from the metastable ion **5** (equation 8).[101] The structure for **4** shown is that calculated to be the most stable.

$$CH_3\overset{O}{\overset{\|}{C}}CH_2\overset{+}{C}=O \longrightarrow [CH_2=C=O\text{---}HCH_2\overset{+}{C}=O] \quad (8)$$

$$\mathbf{5} \qquad\qquad\qquad \mathbf{4}$$

The complex [CH$_2$=C=O$^{+\bullet}$/H$_2$O] has been examined,[102] and Me$_2\overset{+}{C}$CH$_2$CO$_2$H is proposed to rearrange to [Me$_2$C=O---H---CH$_2$=C=$\overset{+}{O}$].[103] The iminium/ketene complex as shown in equation 9 was proposed to form in the 70 eV mass spectral fragmentation of *n*-propyl acetamide.[104]

$$CH_3\overset{O}{\underset{\|}{C}}NHCH_2CH_2CH_3^{+\bullet} \xrightarrow{-C_2H_5^\bullet} CH_2=\overset{+}{N}H---H---CH_2=C=O \qquad (9)$$

The generation of $CH_2=C=O^{+\bullet}$ from fragmentation of the acetone radical cation $(CH_3)_2C=O^{+\bullet}$ has been examined.[105] The ion $^\bullet CH_2CH_2CH_2CO^+$ transferred $CH_2=C=O^{+\bullet}$ to acetone.[106] The ion $CH_2=C=O^{+\bullet}$ abstracts CH_2 from ketene to form the distonic ion $CH_2=C^+-OCH_2^\bullet$.[107]

In the electron impact fragmentation of methyl 3-hydroxybutyrate (**6**) it was proposed that the ion complex **7** of ketene formed and reacted to give the acylium ion as shown in equation 10.[108]

(10)

References

1. Munson, J. W. In *The Chemistry of Ketenes, Allenes, and Related Compounds;* Patai, S., Ed.; Wiley: New York, 1980; pp. 165–188.
2. Runge, W. *Prog. Phys. Org. Chem.* **1981**, *13*, 315–484.
3. Grishin, Yu. K.; Ponomarev, S. V.; Lebedev, S. A. *Zh. Org. Khim.* **1974**, *10*, 404–405; *Engl. Transl.* **1973**, *10*, 402.
4. Firl, J.; Runge, W. *Z. Naturforsch.* **1974**, *29B*, 393–398; *Angew. Chem., Int. Ed. Engl.* **1973**, *12*, 668–669.
5. Duddeck, H.; Praas, H.-W. *Magn. Reson. Chem.* **1993**, *31*, 182–184.
6. Lindner, E.; Steinwand, M.; Hoehne, S. *Angew Chem., Int. Ed. Engl.* **1982**, *21*, 355–356; *Chem. Ber.* **1982**, *115*, 2181–2191.
7. Trahanovsky, W. S.; Surber, B. W.; Wilkes, M. C.; Preckel, M. M. *J. Am. Chem. Soc.* **1982**, *104*, 6779–6780.
8. Haas, A.; Lieb, M.; Praas, H.-W. *J. Fluorine Chem.* **1989**, *44*, 329–337.
9. Haas, A.; Praas, H.-W. *Chem. Ber.* **1992**, *125*, 571–579.
10. Shchukovskaya, L. L.; Kol'tsov, A. A.; Lazarev, A. N.; Pal'chik, R. I. *Dokl. Akad. Nauk SSSR* **1968**, *179*, 892–895; *Engl. Transl.* **1968**, *179*, 318–320.
11. Valenti, E.; Pericas, M. A.; Serratosa, F. *J. Org. Chem.* **1990**, *55*, 395–397.
12. Groh, B. L.; Magrum, G. R.; Barton, T. J. *J. Am. Chem. Soc.* **1987**, *109*, 7568–7569.
13. Lukashev, N. V.; Fil'chikov, A. A.; Zhichkin, P. E.; Kazankova, M. A.; Beletskaya, I. P. *Zh. Obshch. Khim.* **1991**, *61*, 1014–1016; *Engl. Transl.* **1991**, *61*, 920–921.

14. Maas, G.; Gimmy, M.; Alt, M. *Organometallics* **1992**, *11*, 3813–3820.
14a. Reetz, M. T.; Schwellnus, K.; Hübner, F.; Massa, W.; Schmidt, R. E. *Chem. Ber.* **1983**, *116*, 3708–3724.
15. Beeler, A. J.; Orendt, A. M.; Grant, D. M.; Cutts, P. W.; Michl, J.; Zilm, K. W.; Downing, J. W.; Facelli, J. C.; Schindler, M. S.; Kutzelnigg, W. *J. Am. Chem. Soc.* **1984**, *106*, 7672–7676.
16. Yannoni, C. S.; Reisenauer, H. P.; Maier, G. *J. Am. Chem. Soc.* **1983**, *105*, 6181–6182.
17. Allred, E. L.; Grant, D. M.; Goodlett, W. *J. Am. Chem. Soc.* **1965**, *87*, 673–674.
18. Masters, A. P.; Sorensen, T. S.; Ziegler, T. *J. Org. Chem.* **1986**, *51*, 3558–3559.
19. Sorensen, T. S., personal communication.
20. Runge, W. *Z. Naturforsch.* **1977**, *32B*, 1296–1303.
21. Ogata, Y.; Adachi, K. *J. Org. Chem.* **1982**, *47*, 1182–1184.
22. Schiess, P.; Radimerski, P. *Helv. Chim. Acta* **1974**, *57*, 2583–2597.
23. Price, W. C.; Teegan, J. P.; Walsh, A. D. *J. Chem. Soc.* **1951**, 920–926.
23a. Kirmse, W. *Carbene Chemistry,* 2nd ed.; Academic: New York, 1971; p. 9.
24. Rabalais, J. W.; McDonald, J. M.; Scherr, V.; McGlynn, S. P. *Chem. Revs.* **1971**, *71*, 73–108.
25. Knox, K.; Norrish, R. G. W.; Porter, G. *J. Chem. Soc.* **1952**, 1477–1486.
26. Holroyd, R. A.; Blacet, F. E. *J. Am. Chem. Soc.* **1957**, *79*, 4830–4834.
27. Allen, A., University of Toronto, unpublished results.
28. Nadzhimutdinov, Sh.; Slovokhotova, N. A.; Kargin, V. A. *Zh. Fiz. Khim.* **1966**, *40*, 893–897; *Chem. Abstr.* **1966**, *65*, 4851h.
28a. Huisgen, R.; Fciler, L. A.; Otto, P. *Chem. Ber.* **1969**, *102*, 3444–3459.
29. Winnik, M. A.; Wang, F.; Nivaggioli, T.; Hruska, Z.; Fukumura, H.; Masuhara, H. *J. Am. Chem. Soc.* **1991**, *113*, 9702–9704.
30. Hobson, J. D.; Al Holly, M. M.; Malpass, J. R. *Chem. Commun.* **1968**, 764–766.
30a. Broxton, T. J., Duddy, N. W. *J. Org. Chem.* **1981**, *46*, 1186–1191.
31. Clouthier, D. J.; Moule, D. C. *Topics Curr. Chem.* **1989**, *150*, 167–247.
32. Bock, H.; Hirabayashi, T.; Mohmand, S. *Chem. Ber.* **1981**, *114*, 2595–2608.
33. Hall, D.; Maier, J. P.; Rosmus, P. *Chemical Physics* **1977**, *24*, 373–378.
34. Chong, D. P.; Westwood, N. P. C.; Langhoff, S. R. *J. Phys. Chem.* **1984**, *88*, 1479–1481.
35. Colbourne, D.; Westwood, N. P. C. *J. Chem. Soc., Perkin Trans. 2*, **1985**, 2049–2054.
36. Sammynaiken, R.; Westwood, N. P. C. *J. Chem. Soc., Perkin Trans. 2*, **1989**, 1987–1992.
37. Terlouw, J. K.; Burgers, P. C.; Holmes, J. L. *J. Am. Chem. Soc.* **1979**, *101*, 225–226.
38. Chuburu, F.; Lacombe, S.; Pfistser-Guillouzo, G.; Ben Cheik, A.; Chuche, J.; Pommelet, J. C. *J. Am. Chem. Soc.* **1991**, *113*, 1954–1960.
39. Gleiter, R.; Saalfrank, R. W.; Paul, W.; Cowan, D. O.; Eckert-Maksic *Chem. Ber.* **1983**, *116*, 2888–2895.
40. Werstiuk, N. H.; Ma, J.; McAllister, M. A.; Tidwell, T. T.; Zhao, D., *J. Chem. Soc., Farad. Trans.* **1994**, *90*, 3383–3390.
41. Drayton, L. G.; Thompson, H. W. *J. Chem. Soc.* **1948**, 1416–1419.

42. Arendale, W. F.; Fletcher, W. H. *J. Chem. Phys.* **1957**, *26*, 793–797.
43. Duncan, J. L.; Ferguson, A. M.; Harper, J.; Tonge, K. H. *J. Mol. Spec.* **1987**, *125*, 196–213.
44. Leszczynski, J.; Kwiatkowski, J. S. *Chem. Phys. Lett.* **1993**, *201*, 79–83.
45. Johns, J. W. C.; Nemes, L.; Yamada, K. M. T.; Wang, T. Y.; Domenech, J. L.; Santos, J.; Cancio, P.; Bermejo, D.; Ortigoso, J.; Escribano, R. *J. Mol. Spec.* **1992**, *156*, 501–503.
46. Moore, C. B.; Pimentel, G. C. *J. Chem. Phys.* **1963**, *38*, 2816–2829.
47. Gano, J. E.; Jacob, E. J. *Spectrochim. Acta* **1987**, *43A*, 1023–1025.
48. Melzer, A.; Jenney, E. F. *Tetrahedron Lett.* **1968**, 4503–4506.
49. McAllister, M. A.; Tidwell, T. T. *Can. J. Chem.*, **1994**, *72*, 882–887.
50. Huang, B. S.; Pong, R. G. S.; Laureni, J.; Krantz, A. *J. Am. Chem. Soc.* **1977**, *99*, 4154–4156.
51. Arnold, B. R.; Brown, C. E.; Lusztyk, J. *J. Am. Chem. Soc.* **1993**, *115*, 1576–1577.
52. Emerson, D. W.; Titus, R. L.; Gonzalez, R. M. *J. Org. Chem.* **1991**, *56*, 5301–5307.
53. Leung-Toung, R.; Wentrup, C. *Tetrahedron* **1992**, *48*, 7641–7654.
54. Freiermuth, B.; Wentup, C. *J. Org. Chem.* **1991**, *56*, 2286–2289.
55. Leung-Toung, R.; Wentrup, C. *J. Org. Chem.* **1992**, *57*, 4850–4858.
56. Mosandl, T.; Wentrup, C. *J. Org. Chem.* **1993**, *58*, 747–749.
57. Andreichkov, Yu. S.; Kollenz, G.; Kappe, C. O.; Leung-Toung, R.; Wentrup, C. *Acta Chem. Scand.* **1992**, *46*, 683–685.
58. McCarney, C. C.; Ward, R. S. *J. Chem. Soc, Perkin Trans. 1* **1975**, 1600–1603.
59. Coomber, J. W.; Pitts, J. N., Jr.; Schrock, R. R. *Chem. Commun.* **1968**, 190–191.
60. Davidovics, G.; Monnier, M.; Allouche, A. *Chem. Phys.*, **1991**, *150*, 395–403.
61. Torres, M.; Ribo, J.; Clement, A.; Strausz, O. P. *Nouv. J. Chim.* **1981**, *5*, 351–352.
62. Gerry, M. C. L.; Lewis-Bevan, W.; Westwood, N. P. C. *Can. J. Chem.* **1985**, *63*, 676–677.
63. Gano, J. E.; Jacobson, R. H.; Wettach, R. H. *Angew. Chem., Int. Ed. Engl.* **1983**, *22*, 165–166.
64. Mal'tsev, A. K.; Zuev, P. S.; Nefedov, O. M. *Izv. Akad. Nauk SSSR, Ser. Khim.* **1985**, 957–958; *Engl. Transl.* **1985**, 876.
65. Brown, R. D.; Godfrey, P. D.; Woodruff, M. *Aust. J. Chem.* **1979**, *32*, 2103–2109.
66. Mayr, H. *Angew. Chem., Int. Ed. Engl.* **1975**, *14*, 500–501.
67. Danheiser, R. L.; Sard, H. *J. Org. Chem.* **1980**, *45*, 4810–4812.
68. Inoue, S.; Hori, T. *Bull. Chem. Soc. Jpn.* **1983**, *56*, 171–174.
69. Krugerke, T.; Seppelt, K. *Chem. Ber.* **1988**, *121*, 1977–1981.
70. Bittner, J.; Seppelt, K. *Chem. Ber.* **1990**, *123*, 2187–2190.
71. Krantz, A. *J. Chem. Soc, Chem, Commun.* **1973**, 670–671.
72. O'Neill, P.; Hegarty, A. F. *J. Org. Chem.* **1992**, *57*, 4421–4426.
73. Lubenets, E. G.; Gerasimova, T. N.; Barkhash, V. A. *Zh. Org. Khim.* **1972**, *8*, 654; *Engl. Trans.* **1972**, *8*, 663.
74. Unfried, K. G.; Curl, R. F. *J. Mol. Spect.* **1991**, *150*, 86–98.
75. Bestmann, H. J.; Sandmeier, D. *Chem. Ber.* **1980**, *113*, 274–277.

75a. Wentrup, C.; Lorencak, P. *J. Am. Chem. Soc.* **1988**, *110*, 1880–1883.
76. Agosta, W. C.; Smith, A. B., III *J. Am. Chem. Soc.* **1971**, *93*, 5513–5520.
77. Maier, G.; Hoppe, M.; Lanz, K.; Reisenauer, H. P. *Tetrahedron Lett.* **1984**, *25*, 5645–5648.
78. Baxter, G. J.; Brown, R. F. C.; Eastwood, F. W.; Harrington, K. J. *Tetrahedron Lett.* **1975**, 4283–4284.
79. Chapman, O. L.; Miller, M. D.; Pitzenberger, S. M. *J. Am. Chem. Soc.* **1987**, *109*, 6867–6868.
80. Griffiths, J.; Hart, H. *J. Am. Chem. Soc.* **1968**, *90*, 5296–5298.
81. Baird, M. S.; Dunkin, I. R.; Hacker, N.; Poliakoff, M.; Turner, J. J. *J. Am. Chem. Soc.* **1981**, *103*, 5190–5195.
82. McMahon, R. J.; Chapman, O. L. *J. Am. Chem. Soc.* **1986**, *108*, 1713–1714.
83. Chapman, O. L.; McIntosh, C. L.; Barber, L. L. *Chem. Commun.* **1971**, 1162–1163.
84. Zhao, D.; Tidwell, T. T. *J. Am. Chem. Soc.* **1992**, *114*, 10980–10981.
85. Hannay, N. B.; Smyth, C. P. *J. Am. Chem. Soc.* **1946**, *68*, 1357–1360.
86. Walsh, A. D. *J. Am. Chem. Soc.* **1946**, *68*, 2408–2409.
87. Angyal, C. L.; Barclay, G. A.; Hukins, A. A.; LeFevre, R. J. W. *J. Chem. Soc.* **1951**, 2583–2588.
88. Gong, L.; McAllister, M. A.; Tidwell, T. T. *J. Am. Chem. Soc.* **1991**, *113*, 6021–6028.
89. Schwarz, H.; Köppel, C. In *The Chemistry of Ketenes, Allenes, and Related Compounds;* Patai, S., Ed.; Wiley: New York, 1980; pp. 189–222.
90. Maquestiau, A.; Flammang, R.; Pauwels, P. *Org. Mass Spec.* **1983**, *18*, 547–552.
91. Terlouw, J. K.; Burgers, P. C.; Holmes, J. L. *J. Am. Chem. Soc.* **1979**, *101*, 225–226.
92. Holmes, J. L.; Terlouw, P. C.; Vijfhuizen, P. C.; A'Campo, C. *Org. Mass Spectrom.* **1979**, *14*, 204–212.
93. Selva, A.; Ferrario, F.; Saba, A. *Org. Mass Spec.* **1987**, *22*, 189–196.
94. Lorquet, J. C. *Int. J. Mass. Spectrom. Ion Phys.* **1981**, *38*, 343–350.
95. Stockbauer, R. *Int. J. Mass Spectrom. Ion Phys.* **1977**, *25*, 401–410.
96. Vogt, J.; Williamson, A. D.; Beauchamp, J. L. *J. Am. Chem. Soc.* **1978**, *100*, 3478–3483.
97. Audier, H.; Conia, J.-M.; Fetizon, M.; Gore, J. *Bull. Soc. Chim. Fr.* **1967**, 787–791.
98. Hayes, R. N.; Sheldon, J. C.; Bowie, J. H. *Austr. J. Chem.* **1985**, *38*, 355–362.
99. Iraqi, M.; Lifshitz, C.; Reuben, B. G. *J. Phys. Chem.* **1991**, *95*, 7742–7746.
100. Drewello, T.; Heinrich, N.; Maas, W. P. M.; Nibbering, N. M. M.; Weiske, T.; Schwarz, H. *J. Am. Chem. Soc.* **1987**, *109*, 4810–4818.
101. Tortajada, J.; Berthomieu, D.; Morizur, J.-P.; Audier, H.-E. *J. Am. Chem. Soc.* **1992**, *114*, 10874–10880.
102. Heinrich, N.; Schwarz, H. *Int. J. Mass Spectrom. Ion Proc.* **1987**, *79*, 295–310.
103. Van Baar, B. L. M.; Terlouw, J. K.; Akkok, S.; Zummack, W.; Weiske, T.; Schwarz, H. *Chimia* **1988**, *42*, 226–229.
104. Burgers, P. C.; Kingsmill, C. A.; McGibbon, G. A.; Terlouw, J. K. *Org. Mass Spectrom.* **1992**, *27*, 398–405.

105. Heinrich, N.; Louage, F.; Lifshitz, C.; Schwarz, H. *J. Am. Chem. Soc.* **1988**, *110*, 8183–8192.
106. Stirk, K. M.; Kenttämaa, H. I. *Org. Mass Spectrom.* **1992**, *27*, 1153–1154.
107. Smith, R. L.; Franklin, R. L.; Stirk, K. M.; Kenttämaa, H. I. *J. Am. Chem. Soc.* **1993**, *115*, 10348–10355.
108. Morizur, J.-P.; Chapon, E.; Berthomieu, D. *Org. Mass Spectrom.* **1991**, *26*, 1119–1120.

CHAPTER 3

PREPARATION OF KETENES

There are many methods of ketene preparation, some of which almost defy classification, and several reviews of the methods are available.[1-5] These include considerable detail on the preparation of particular ketenes and should be consulted for experimental procedures. Nevertheless this is still an area in which there is ample room for improvements in methodology, and it remains a worthy goal to create procedures by which free ketenes of diverse structural type can be generated under mild conditions in environments where they can be utilized in structural, mechanistic, and synthetic studies. Many simple ketenes have still not been successfully prepared in circumstances where they can be directly characterized by spectroscopic means.

3.1 KETENES FROM KETENE DIMERS

In some cases ketene dimers provide a good source of ketenes, most particularly in the case of ketene itself (equation 1).[1,6] Dimethylketene may also be prepared by thermolysis of the symmetrical dimer (equation 2).[7,8]

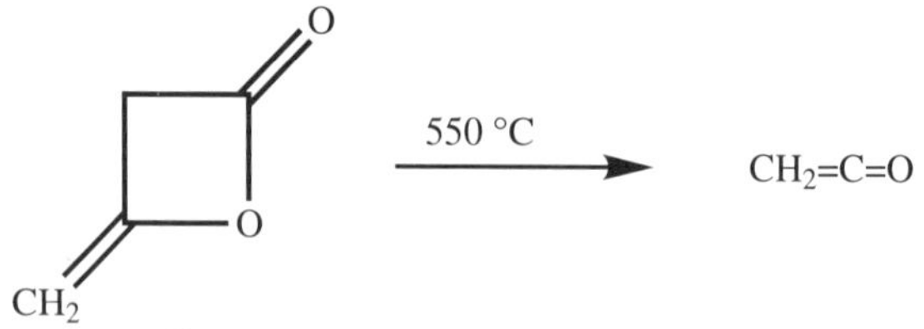

$$\text{1} \xrightarrow{550\ °C} CH_2=C=O \qquad (1)$$

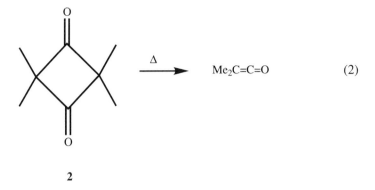

2

Photolyses of **2** and other tetraalkyl-1,3-cyclobutanediones give dialkylketenes in yields of 20–30%, along with products of decarbonylation.[7,8]

3.2 KETENES FROM CARBOXYLIC ACIDS AND THEIR DERIVATIVES

The direct preparation of ketene on an industrial scale is achieved by the pyrolysis of acetic acid.[9] The formation of $CH_2=C=O$, $CH_3CH=C=O$, and $Me_2C=C=O$ can be accomplished by reaction of the appropriate carboxylic acids over alkali metal-exchanged zeolites at 350° C.[10] The preparation of some very stable ketenes such as $t\text{-}Bu_2C=C=O$ and $Me_3SiCH=C=O$ has been carried out by the direct dehydration of the carboxylic acids using dicyclohexylcarbodiimide (equation 3).[11]

$$t\text{-}Bu_2CHCO_2H \xrightarrow[Et_3N]{DCC} t\text{-}Bu_2C=C=O \qquad (3)$$

In a few cases ketenes such as vinylketene[12] have been prepared directly for gas phase spectroscopic studies by high temperature pyrolysis of carboxylic acids, but this method is not generally useful for laboratory preparation. Usually activated forms of the acids are required for conversion to the ketene.

3.2.1 Ketones from Carboxylic Anhydrides

The industrial preparation of higher ketenes can be carried out by the pyrolysis of acid anhydrides.[13] Thus pyrolysis of cyclohexanecarboxylic anhydride gives ketene **3** (equation 4).[14] A laboratory scale procedure for preparation of ketene by acetic anhydride pyrolysis at 500–600° C (equation 5)[15] has been recommended as a superior method for producing laboratory amounts of ketene.[6] Methylketene is also available by this route.[16]

$$\text{Cyclohexyl-COC-(C}_6\text{H}_4\text{)} \xrightarrow{\Delta} \text{cyclohexylidene}=C=O \quad (4)$$

$$3$$

$$(\text{CH}_3\text{CO})_2\text{O} \xrightarrow{\Delta} \text{CH}_2=C=O + \text{CH}_3\text{CO}_2\text{H} \quad (5)$$

The original preparation of ketene devised by Wilsmore involved the pyrolysis of acetic anhydride, ethyl acetate, or acetone with a heated metal filament.[17] This method was developed as a standard laboratory procedure for ketene preparation using acetone and a "Hurd lamp."[18-20] On an industrial scale acetic anhydride is made from the reaction of acetic acid and ketene, and the ketene is prepared from pyrolysis of acetone.[21]

The photolyses of the mixed anhydrides **4–6** were found to proceed partly by Norrish type II cleavage yielding ketenes to the extent of 70.6, 39,8, and 47.4%, respectively, in competition with other photochemical processes.[22] The formation of the ketenes was confirmed by trapping with H_2O or CH_3OH, and the process of equation 6 was proposed to explain ketene formation.[22]

$$\text{RCO-O-CO-CHR}^1\text{R}^2 \xrightarrow{h\nu} \text{biradical} \xrightarrow{-\text{RCO}_2\text{H}} \text{R}^1\text{R}^2\text{C}=C=O$$

(6)

4 R, R^1 = Ph; R^2 = H

5 R, R^1, R^2 = Ph

6 R = Ph_3C; R^1R^2 = $(CH_2)_5$

The use of malonic acids for the preparation of ketenes by conversion to their polymeric anhydrides followed by pyrolysis was developed early in the history of ketene chemistry.[1-4] Formation of the anhydrides may be conveniently achieved by reaction with acetic anhydride or trifluoroacetic anhydride (TFAA).[23] This procedure is suggested[23] to involve initial formation of a mixed anhydride which is converted to the anhydride by pyridine. These anhydrides, which are polymeric, decarboxylate on heating (equation 7). Arylmalonic acids evidently rapidly form ketenes at room temperature on reaction with TFAA as evidenced by the appearance and gradual disappearance of IR bands in the 2100–2200 cm^{-1} region.[23]

$n-Bu_2C(CO_2H)_2 \xrightarrow{TFAA} n-Bu_2C(CO_2H)CO_2COCF_3 \xrightarrow{pyridine} (n-Bu_2C(CO)_2O)n$

7

$7 \xrightarrow[-CO_2]{140 \text{ to } 180°} n-Bu_2C=C=O$ (7)

Monomeric malonic anhydride (**8**) is obtained from ozonolysis of ketene dimer at −78 °C and was identified by its spectra (equation 8).[24] At −30 °C (**8**) undergoes decarboxylation to ketene which was observed directly by ^{13}C NMR and was trapped by aniline. Dimethylmalonic anhydride was obtained similarly,[24] and characterized by its Raman spectrum,[25] and it also undergoes decarboxylation, to give $(CH_3)_2C=C=O$.[24]

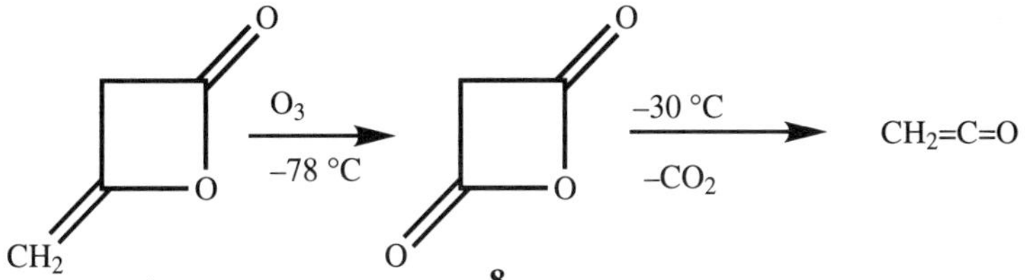

$\xrightarrow[-78 °C]{O_3} \qquad \xrightarrow[-CO_2]{-30 °C} CH_2=C=O$ (8)

1   **8**

Malonic acids are also converted by acetone and acid to acylals (Meldrum's acid derivatives) which give ketenes on thermolysis.[26–28] Thus reaction of dimethylmalonic acid with acetone in the presence of an acidic catalyst at 60 °C led to an exothermic reaction, giving the crystalline acylal **9** in high yield (equation 9).[28] Heating **9** at 150 °C in the presence of K_2CO_3 efficiently forms $Me_2C=C=O$ in high yield (equation 9), or the ketene can be effectively generated in situ in solution for reactions with alkenes, alcohols, or amines.[28] Meldrum's acid derivatives are also utilized for the preparation of cumulenones (equation 10).[29,30]

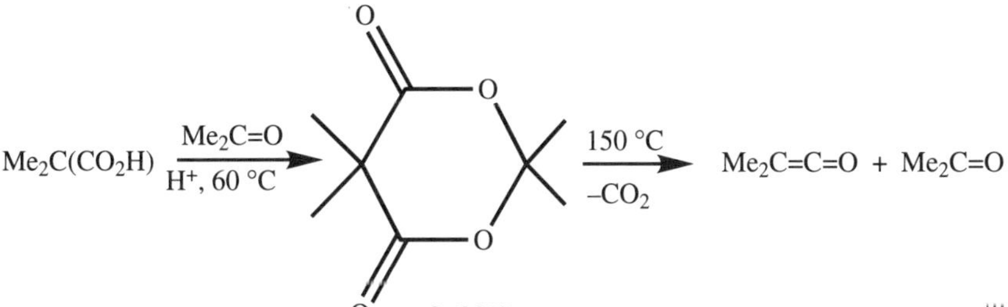

$Me_2C(CO_2H) \xrightarrow[H^+, 60 °C]{Me_2C=O} \text{ [structure]} \xrightarrow[-CO_2]{150 °C} Me_2C=C=O + Me_2C=O$

9 (80%) (9)

$9 \xrightarrow{495° C} CH_2=C=C=O$ (10)

References

1. Hanford, W. E.; Sauer, J. C. *Organic Reactions* **1946**, *3*, 108–140.
2. Lacey, R. N. In *The Chemistry of the Alkenes;* Patai, S., Ed.; Interscience: New York, 1964; pp. 1161–1227.
3. Ward, R. S. In *The Chemistry of Ketenes, Allenes, and Related Compounds;* Patai, S., Ed.; Wiley: New York, 1980; pp. 223–277.
4. Borrmann, D. *Methoden der Organischen Chemie;* Theime Verlag: Stuttgart, 1968; Vol. 7, Part 4.
5. Schaumann, E.; Scheiblich, S. *Methoden der Organischen Chemie;* Theime Verlag: Stuttgart, 1993; Vol. E15; pp. 2353–2530, 2818–2881, 2933–2957.
6. Andreades, S.; Carlson, H. D. *Organic Syntheses,* Coll. Vol. V, Baumgarten, H. E., Ed.; Wiley: New York, 1973; pp. 679–684.
7. Turro, N. J.; Leermakers, P. A.; Wilson, H. R.; Neckers, D. C.; Byers, G. W.; Vesley, G. F. *J. Am. Chem. Soc.* **1965**, *87*, 2613–2619.
8. Haller, I.; Srinivasan, R. *J. Am. Chem. Soc.* **1965**, *87*, 1144–1145.
9. Wagner, F. S. In *Kirk-Othmer Encyclopedia of Chemical Technology,* 3rd ed.; Wiley: New York, 1980; Vol. 1, p. 155.
10. Parker, L. M.; Bibby, D. M.; Miller, I. J. *J. Catal.* **1991**, *129*, 438–446.
11. Olah, G. A.; Wu, A.; Farooq, O. *Synthesis,* **1989**, 568.
12. Mohmand, S.; Hirabayashi, T.; Bock, H. *Chem. Ber.* **1981**, *114*, 2609–2621.
13. Hasek, R. H. In *Kirk-Othmer Encyclopedia of Chemical Technology,* 3rd ed.; Wiley: New York, 1980; Vol. 13, pp. 874–893.
14. Sioli, G.; Mattone, R.; Giuffre, L.; Trotta, R.; Tempesti, E. *Chim. Ind. (Milan)* **1971**, *53*, 133–139; *Chem. Abstr.* **1971**, *74*, 99490x.
15. Fisher, G. J.; MacLean, A. F.; Schnizer, A. W. *J. Org. Chem.* **1953**, *18*, 1055–1057. See note 13 in ref. 6.
16. Jenkins, A. D. *J. Chem. Soc.* **1952**, 2563–2568.
17. Wilsmore, N. T. M. *J. Chem. Soc.* **1907**, *91*, 1938–1948.
18. Hurd, C. D. *J. Am. Chem. Soc.* **1923**, *45*, 3095–3101.
19. Williams, J. W.; Hurd, C. D. *J. Org. Chem.* **1940**, *5*, 122–125.
20. Hurd, C. D. *Organic Syntheses,* Coll. Vol. I, Blatt, A. H., Ed.; Wiley, New York, 1941; pp. 330–334.
21. Cook, S. L. In *Acetic Acid and Its Derivatives;* Agreda, V. H.; Zoeller, J. R., Eds.; Dekker: New York, 1993; Chapter 9.
22. Penn, J. H.; Owens, W. H. *J. Am. Chem. Soc.* **1993**, *115*, 82–86.
23. Duckworth, A. C. *J. Org. Chem.* **1962**, *27*, 3146–3148.
24. Perrin, C. L.; Arrhenius, T. *J. Am. Chem. Soc.* **1978**, *100*, 5249–5251.
25. Perrin, C. L.; Magde, D.; Berens, S. J.; Roque, J. *J. Org. Chem.* **1980**, *45*, 1705–1706.
26. Davidson, D.; Bernhard, S. A. *J. Am. Chem. Soc.* **1948**, *70*, 3426–3428.
27. Hurd, C. D.; Hayao, S. *J. Am. Chem. Soc.* **1954**, *76*, 5563–5564.
28. Bestian, H.; Günther, D. *Angew. Chem., Int. Ed. Engl.* **1963**, *2*, 608–613.
29. Brown, R. F. C.; Eastwood, F. W.; McMullen, G. L. *J. Am. Chem. Soc.* **1976**, *98*, 7421–7422.

30. Brown, R. F. C.; Eastwood, F. W. In *The Chemistry of Ketenes, Allenes, and Related Compounds*; Patai, S., Ed.; Wiley: 1980; pp. 757–778.

3.2.2 Ketenes from Acyl Halides and Activated Acids

The most common method of activation of carboxylic acids is conversion to the acyl chlorides, which react with amines, usually Et_3N, in a widely used method for the preparation of ketenes that was first demonstrated by Staudinger.[1] In the case of stable ketenes these may be isolated from this reaction (equation 1),[2] or the ketenes may be trapped in situ with reactive alkenes such as cyclopentadiene (equation 2)[3] or vinyl ethers (equation 3),[4] or in intramolecular reactions.

$$t\text{-}Bu_2CHCOCl \xrightarrow{Et_3N} t\text{-}Bu_2C=C=O \quad (1)$$

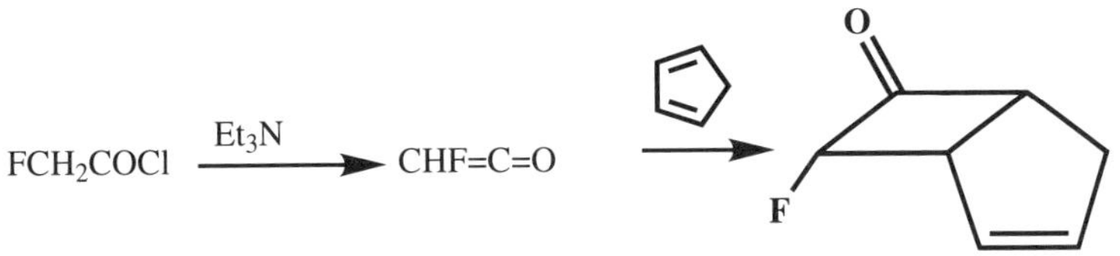

$$FCH_2COCl \xrightarrow{Et_3N} CHF=C=O \longrightarrow \quad (2)$$

$$Cl_2CHCOCl \xrightarrow{Et_3N} CCl_2=C=O \longrightarrow \quad (3)$$

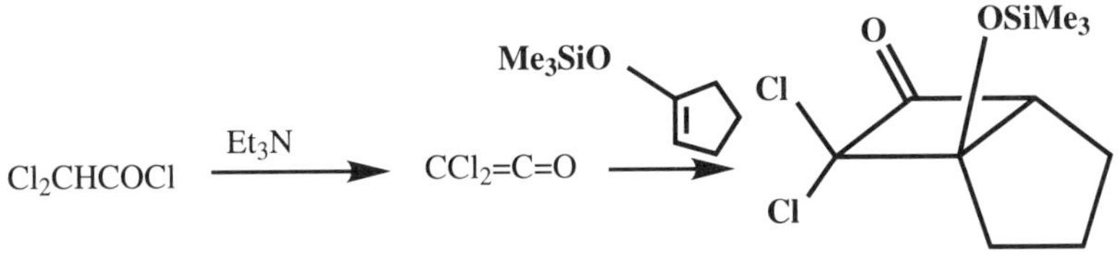

The reaction of isobutyryl chloride (**1**) with Et_3N was examined by ^1H NMR and IR at $-60\ °C$ and interpreted as showing the formation of the acylammonium salt **2**, with an IR band at $1810\ cm^{-1}$.[4] A peak at δ 1.4 in the ^1H NMR increased with time and was assigned to the formation of dimethylketene **3** (equation 4).[5] The formation of the acylammonium intermediate in ketene formation is consistent with previous studies.[6–8]

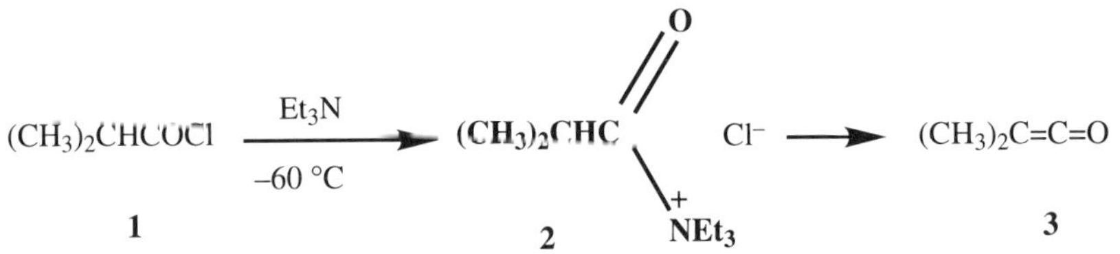

$$(CH_3)_2CHCOCl \xrightarrow[-60\ °C]{Et_3N} (CH_3)_2CHC(=O)\overset{+}{N}Et_3\ Cl^- \longrightarrow (CH_3)_2C=C=O \quad (4)$$

$$\quad\ \ \mathbf{1} \qquad\qquad\qquad\qquad \mathbf{2} \qquad\qquad\qquad\qquad \mathbf{3}$$

The reaction of α-chloropropionyl chloride (**4**) and other α-haloacyl halides with Et_3N is more complex.[5,9–11] The reactions involved are proposed to include formation of the acylammonium salt **5** and the ketene **6**, but enolization to form **7** which reacts with the starting material to give enol ester **8** also occurs. Even with stop flow NMR chloroketene **6** could not be directly observed.[10]

$$CH_3CHClCOCl \xrightarrow{Et_3N} CH_3CHClC(=O)NEt_3^+ \; Cl^- \rightleftharpoons CH_3CCl=C=O + Et_3NHCl \quad (5)$$

$$\mathbf{4} \qquad\qquad \mathbf{5} \qquad\qquad \mathbf{6}$$

$$CH_3CHClCOCl \xrightarrow{Et_3N} CH_3CCl=C(O^-)(Cl) \xrightarrow{\mathbf{4}} CH_3CCl=C(O_2CCHClCH_3)(Cl) \quad (6)$$

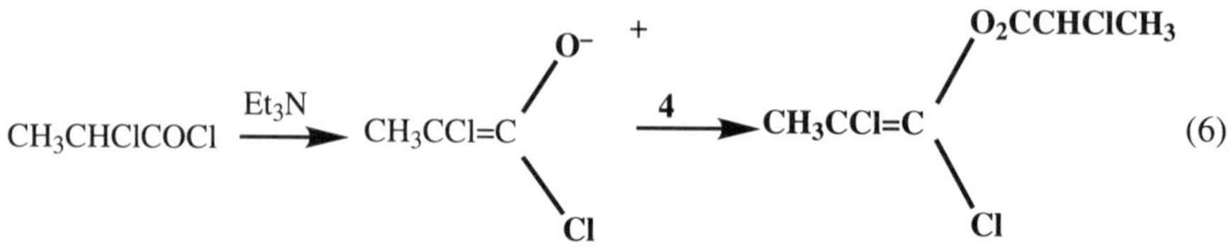

$$\mathbf{4} \qquad\qquad \mathbf{7} \qquad\qquad \mathbf{8}$$

Scheme 1

The reaction of $CHCl_2COCl$ with Et_3N to give $CCl_2=C=O$ and products analogous to **8** is discussed in Section 4.4.2. It has been proposed that the formation of ketenes from the acyl ammonium salts is reversible, because it is often difficult to drive these reactions to completion.[3] Attempts to directly observe fluoroketenes in similar reactions were also unsuccessful.[12] Both **6** and $CHCl=C=O$ were generated for direct observation by the route of equation 7.[13]

$$CH_3CCl_2COCl \xrightarrow{Me_3SiPPh_2} CH_3CCl_2COPPh_2 \xrightarrow{80\,°C} CH_3CCl=C=O \quad (7)$$

$$\qquad\qquad\qquad\qquad\qquad\qquad\qquad\qquad\qquad\qquad \mathbf{6}$$

In the reaction of acyl chlorides including CH_3COCl, $ClCH_2COCl$, $PhOCH_2COCl$, and $PhCH_2COCl$ with Et_3N in the presence of alcohols, formation of esters was shown by deuterium labeling experiments to involve competitive ketene formation along with direct alcoholysis.[14] In reactions in which ketenes are generated in situ and not observed directly but captured with trapping agents, there is always the question of whether free ketenes are actually formed in the reaction. This question is particularly prominent in the [2 + 2] cycloaddition of ketenes with imines to form β-lactams. This reaction is also discussed separately in Section 5.4.1.7, and it has been extensively studied because of the pharmaceutical interest in the β-lactam products.

The reaction of the acyl chloride **9** with diisopropylethylamine (DIPEA) gave a solution with a strong band in the IR at 2120 cm^{-1} assigned to the ketene **10**.[15] When **9**, DIPEA, and the imine **11** were monitored in an IR cell at −22 °C the rapid formation of the ketene **10** was observed, followed by slower formation of

3.2 KETENES FROM CARBOXYLIC ACIDS AND THEIR DERIVATIVES

β-lactams from cycloaddition of the ketene with the imine. The kinetics were fit by a scheme based on reactions 8 and 9, involving formation of ketene **10** which then reacted with the imine to form the β-lactam.[15]

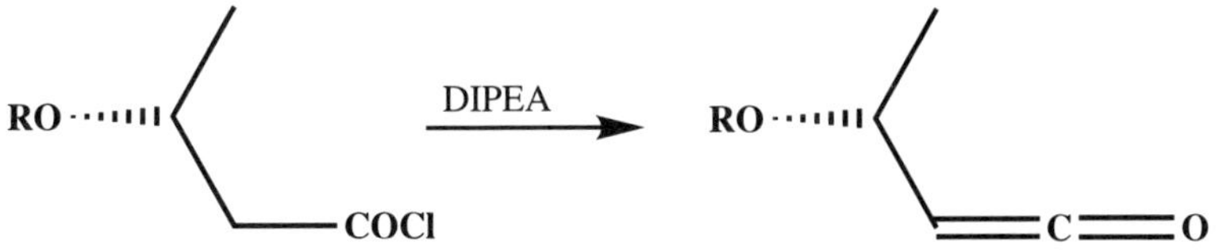

$$\text{RO}\cdots\text{CH(Me)CH}_2\text{COCl} \xrightarrow{\text{DIPEA}} \text{RO}\cdots\text{CH(Me)CH=C=O} \quad (8)$$

9 R = i-Pr$_3$Si

10

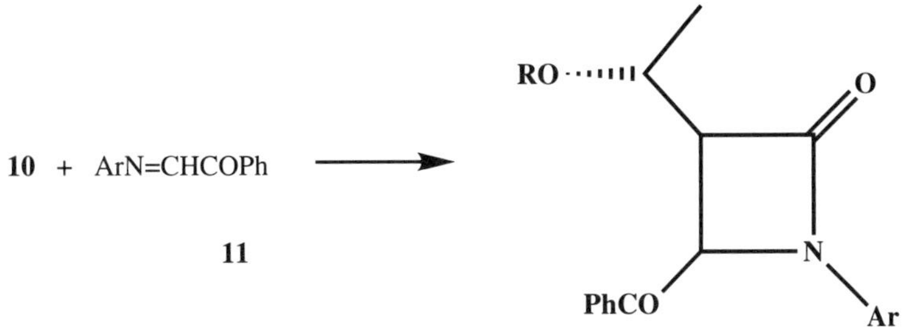

$$\mathbf{10} + \text{ArN=CHCOPh} \longrightarrow \text{[β-lactam 12]} \quad (9)$$

11

12

The reaction of 2-butenoyl chloride with tertiary amines to give vinylketene that is trapped by imine cycloaddition can be effectively carried out using a microwave oven and chlorobenzene solvent (equation 10).[16] Ketene generation from PhCH$_2$OCH$_2$COCl was carried out similarly.[17]

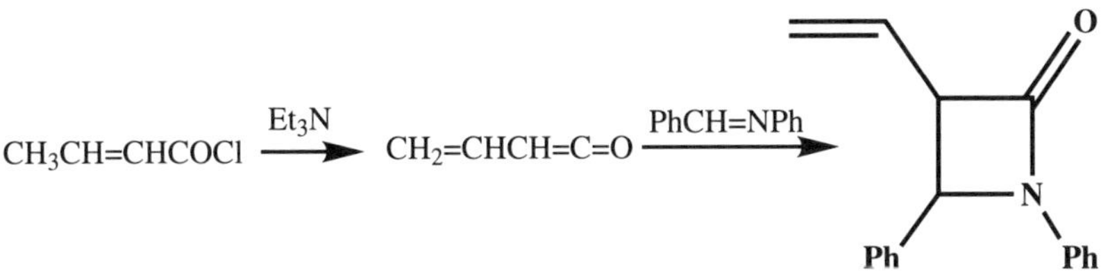

$$\text{CH}_3\text{CH=CHCOCl} \xrightarrow{\text{Et}_3\text{N}} \text{CH}_2\text{=CHCH=C=O} \xrightarrow{\text{PhCH=NPh}} \text{[β-lactam]} \quad (10)$$

The reaction of Me$_2$C=CHCOCl with Me$_3$N led to a solid product assigned an acyl ammonium structure, which on stirring at room temperature apparently gave the vinylketene which could be trapped with ethyl vinyl ether (equation 11).[18] Other examples of vinylketene intermediates prepared similarly are given in Section 4.1.2.

$$\text{Me}_2\text{C=CHCOCl} \xrightarrow{\text{Me}_3\text{N}} \text{Me}_2\text{C=CHCONMe}_3^+ \longrightarrow \text{CH}_2\text{=CMeCH=C=O} \quad (11)$$

The dehydrochlorination of acyl chlorides has also been effected by Ru(0)[19] and low-valent Pt compounds.[20]

60 PREPARATION OF KETENES

Other methods for the activation of carboxylic acids for ketene formation include the in situ formation of acyl tosylates which react with Et_3N to generate ketenes (equation 12).[21-23] The reaction of carboxylic acids under Perkin reaction conditions, sodium acetate in refluxing acetic anhydride (equations 13[24] and 14[25]), has also been proposed to involve ketene intermediates.[24-26] Ketoketenes generated in this way are suggested to undergo cyclization and decarboxylation to form cycloalkenes (equation 14).[26]

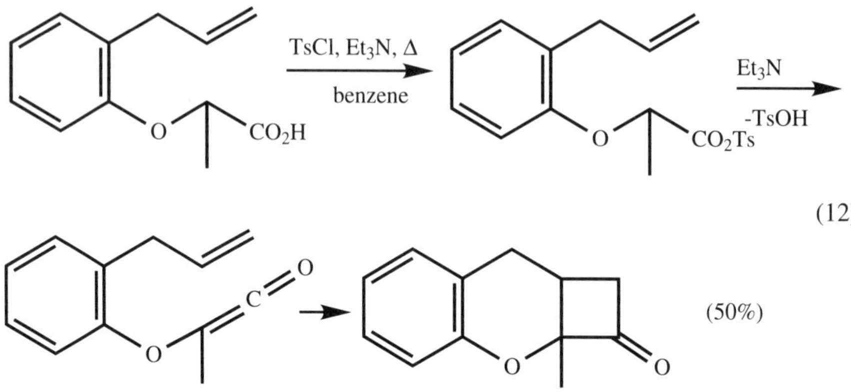

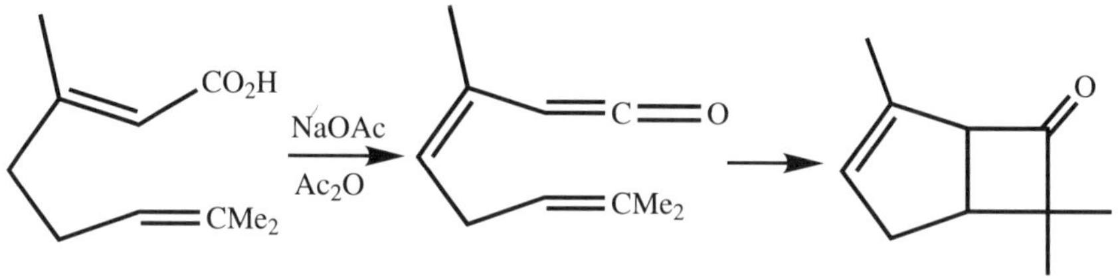

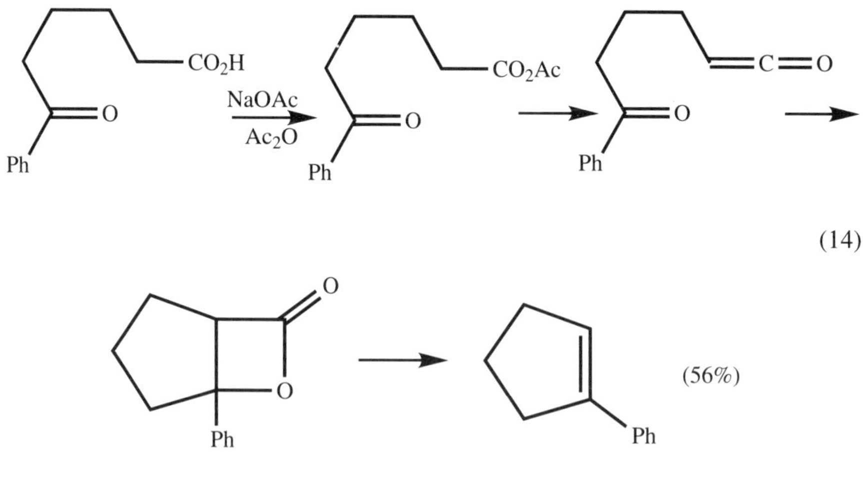

3.2 KETENES FROM CARBOXYLIC ACIDS AND THEIR DERIVATIVES

A variety of mixed sulfonic anhydrides of cyclohexanecarboxylic acid (**13**) were prepared and gave ketene **14** on pyrolysis (equation 15).[27] Reaction of cyclohexanecarboxylic anhydride (equation 16) with 100% H_2SO_4 gave **15**, suggested to result from electrophilic α-sulfonation of **14** (equation 16).[27]

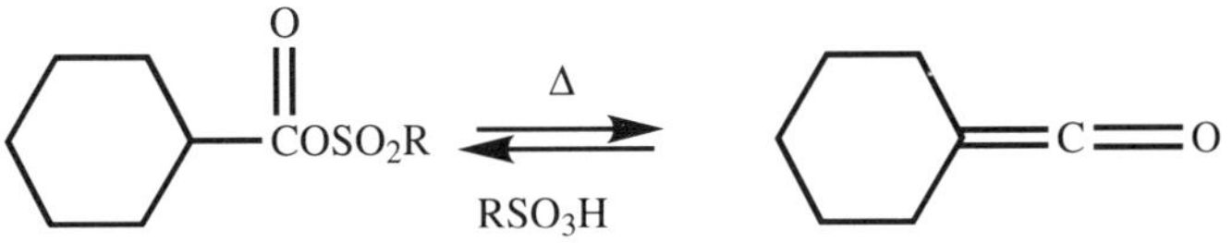

$$Cy\text{-}COSO_2R \underset{RSO_3H}{\overset{\Delta}{\rightleftharpoons}} Cy\text{=}C\text{=}O \qquad (15)$$

13   **14**

R = CH_3, 4-$CH_3C_6H_4$, Ph

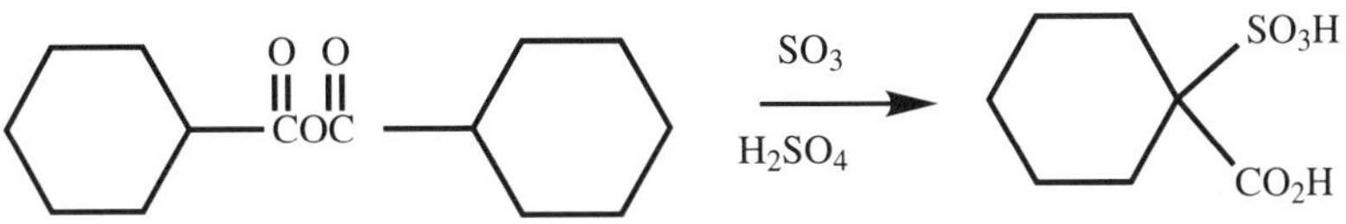

$$Cy\text{-}CO\text{-}O\text{-}CO\text{-}Cy \xrightarrow[H_2SO_4]{SO_3} Cy(SO_3H)(CO_2H) \qquad (16)$$

15

The mixed anhydrides of CH_3SO_3H with other carboxylic acids are reported to form ketenes on pyrolysis at 130° C[28] or on reaction with Et_3N at 0° C.[29]

The reaction of substituted acetic acids with trifluoroacetic anhydride or alkyl chloroformates in CH_2Cl_2 at room temperature to generate the mixed anhydrides followed by reaction with Et_3N and imines gave the β-lactams **16**.[30] The reactions may well involve ketenes as intermediates which react with the imines in [2 + 2] cycloadditions (equation 17).

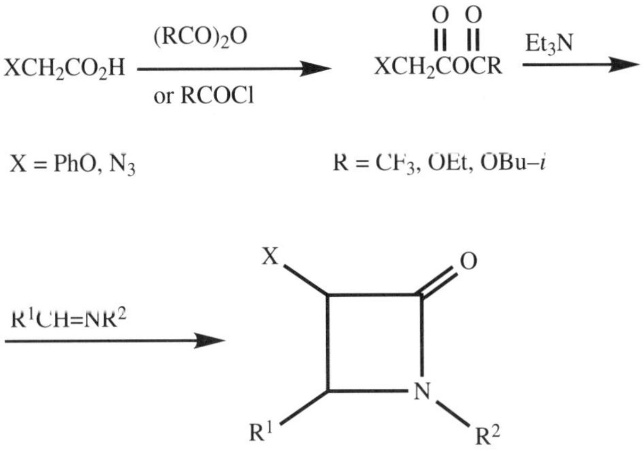

$$XCH_2CO_2H \xrightarrow[\text{or RCOCl}]{(RCO)_2O} XCH_2COCR(O)(O) \xrightarrow{Et_3N} [XCH\text{=}C\text{=}O] \qquad (17)$$

X = PhO, N_3 R = CF_3, OEt, OBu–i

$$\xrightarrow{R^1CH=NR^2} \text{β-lactam}$$

16

The activation of acetic acids bearing chiral substituents by the trifluoroacetic anhydride method leads to diastereomeric selection in the product β-lactams.[31] β-Lactams are also obtained by TsCl/Et$_3$N activation of carboxylic acids followed by reaction with imines,[32] and by the use of Mukaiyama's reagent **17**, in a process that may also involve ketene formation (equation 18).[33,34] The use of **17** in the preparation of ketenes for intramolecular trapping has also been reported.[35]

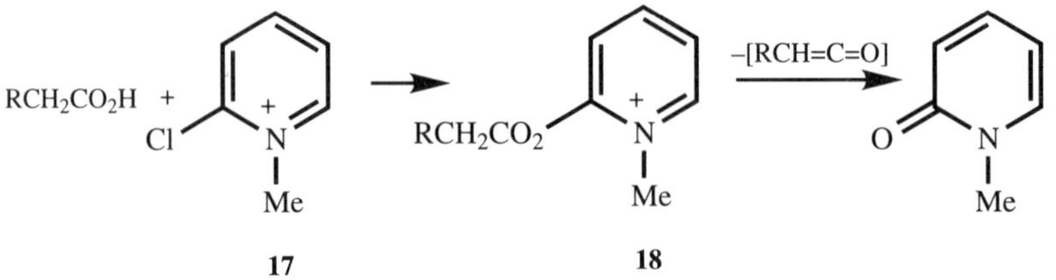

$$RCH_2CO_2H + \text{[17]} \longrightarrow RCH_2CO_2\text{-[Py-Me]}^+ \xrightarrow{-[RCH=C=O]} \text{[pyridone]} \quad (18)$$

17   **18**

A variety of other phosphorous containing[36–45] and other[46] activating reagents of carboxylic acids have been examined in reactions with imines to yield β-lactams. Examples of some of these reactions are shown in equations 19–26, along with the presumed activated carboxylic acids that may be involved.

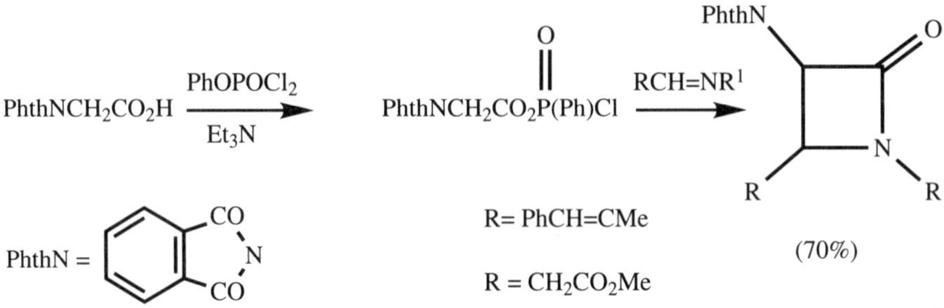

$$PhthNCH_2CO_2H \xrightarrow{PhOPOCl_2, Et_3N} PhthNCH_2CO_2P(Ph)Cl \xrightarrow{RCH=NR^1} \text{β-lactam} \quad (70\%)$$

PhthN = phthalimido

R = PhCH=CMe
R = CH$_2$CO$_2$Me

(19)[38]

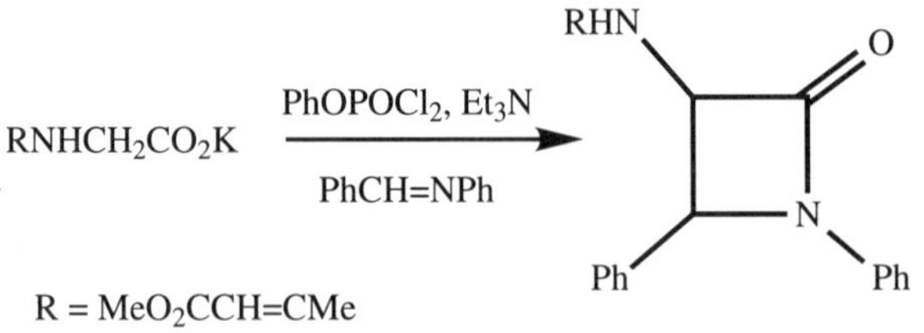

$$RNHCH_2CO_2K \xrightarrow[PhCH=NPh]{PhOPOCl_2, Et_3N} \text{β-lactam} \quad (20)^{39,40}$$

R = MeO$_2$CCH=CMe

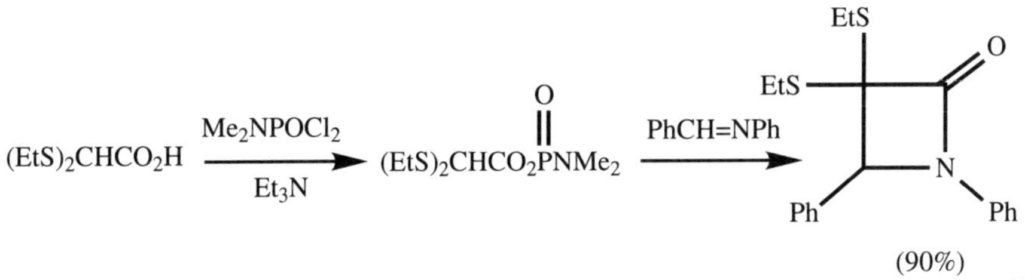

3.2 KETENES FROM CARBOXYLIC ACIDS AND THEIR DERIVATIVES

$(EtS)_2CHCO_2H$ $\xrightarrow{Me_2NPOCl_2, Et_3N}$ $(EtS)_2CHCO_2PNMe_2$ (with =O) $\xrightarrow{PhCH=NPh}$ β-lactam with EtS, EtS at C3; Ph at C4; N-Ph (90%) (21)[41]

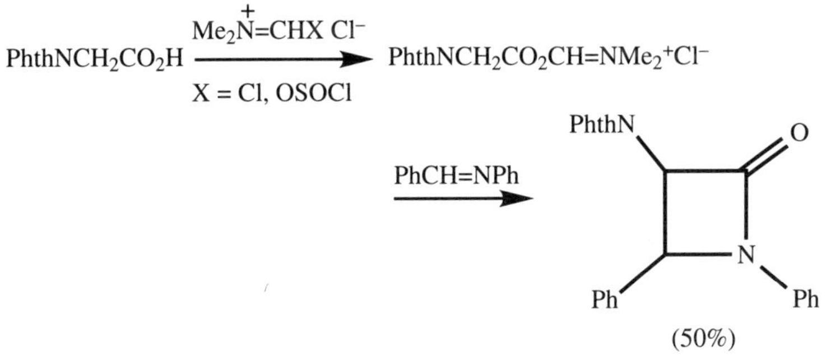

$PhthNCH_2CO_2H$ $\xrightarrow{Me_2\overset{+}{N}=CHX\ Cl^-,\ X = Cl,\ OSOCl}$ $PhthNCH_2CO_2CH=NMe_2{}^+Cl^-$ $\xrightarrow{PhCH=NPh}$ β-lactam with PhthN at C3, Ph at C4, N-Ph (50%) (22)[42]

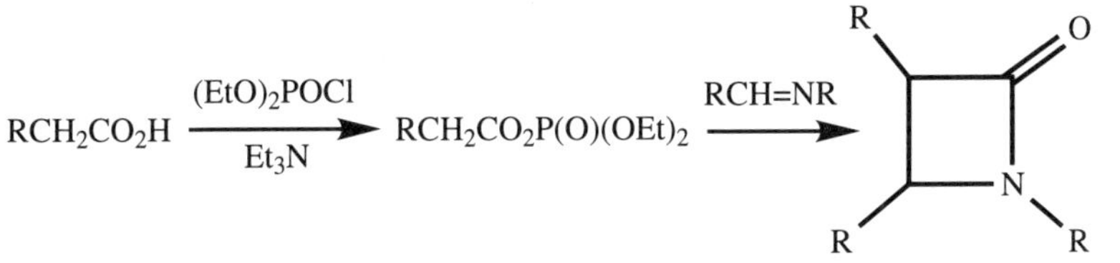

RCH_2CO_2H $\xrightarrow{(EtO)_2POCl, Et_3N}$ $RCH_2CO_2P(O)(OEt)_2$ $\xrightarrow{RCH=NR}$ β-lactam with R at C3, R at C4, N-R (23)[43]

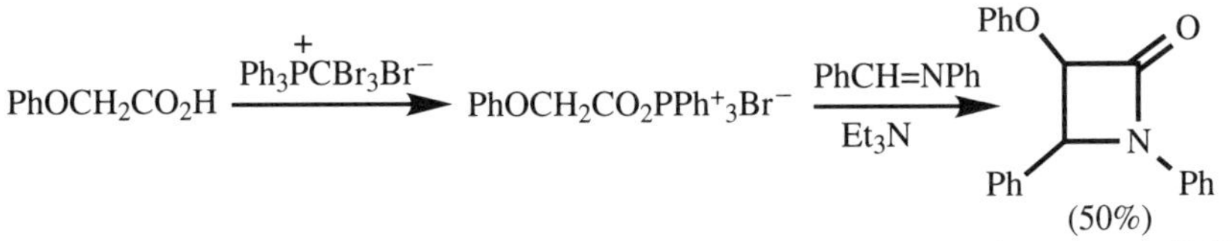

$PhOCH_2CO_2H$ $\xrightarrow{Ph_3\overset{+}{P}CBr_3Br^-}$ $PhOCH_2CO_2PPh_3{}^+Br^-$ $\xrightarrow[Et_3N]{PhCH=NPh}$ β-lactam with PhO at C3, Ph at C4, N-Ph (50%) (24)[44]

$CHCl_2CO_2H$ $\xrightarrow{Ph_3PBr_2}$ $CHCl_2CO_2PPh_3{}^+Br^-$ $\xrightarrow{PhCH=NPh}$ β-lactam with Cl, Cl at C3, Ph at C4, N-Ph (50%) (25)[45]

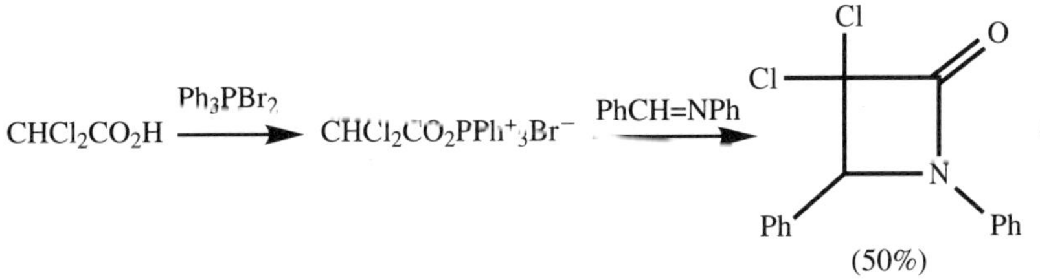

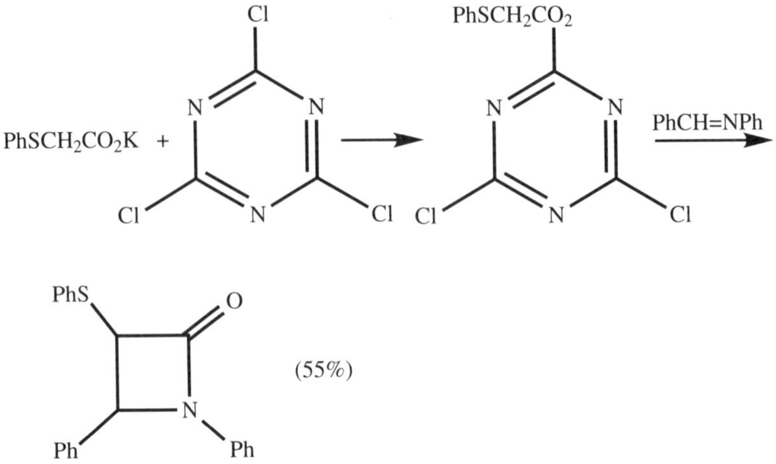

PhSCH$_2$CO$_2$K + [2,4-dichloro-6-chloro-1,3,5-triazine] ⟶ [triazine with PhSCH$_2$CO$_2$ and Cl substituents] $\xrightarrow{\text{PhCH=NPh}}$

(26)[46]

[β-lactam: PhS, Ph, N-Ph, C=O] (55%)

Some diverse reactions involving amide intermediates include the reaction of acid chlorides with saccharin followed by Et$_3$N and an imine (equation 27),[47] the reaction of chloroacetylhydrazines with base, which is proposed to involve ketene (equation 28),[48] and the reaction of N-trifluoroacetyl derivatives of penicillins and cephalosporin esters with DBN to give phenylketene elimination (equation 29).[49]

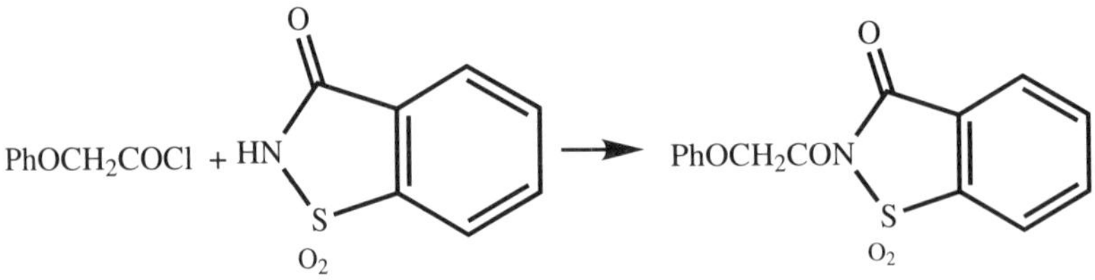

$$\text{PhOCH}_2\text{COCl} + \text{HN-saccharin} \longrightarrow \text{PhOCH}_2\text{CON-saccharin} \quad (27)^{47}$$

$$\text{ClCH}_2\text{CONHNH}_2 \xrightarrow{\text{OH}^-} \text{ClCH}_2\text{CONHNH}^- \xrightarrow[-\text{HN=NH}]{-\text{Cl}^-} \text{CH}_2\text{=C=O} \quad (28)^{48}$$

$$\text{PhCH}_2\text{CON(COCF}_3\text{)R} \xrightarrow{\text{DBN}} \text{PhCH=C=O} + (\text{RNHCOCF}_3) \quad (29)^{49}$$

The reaction of crotonyl chloride with enamines of 4-benzoylcyclohexanone (**19**) was proposed to proceed by initial N-acylation of the enamine followed by a [3.3] sigmatropic rearrangement forming ketenes **20** (equation 30).[50] These underwent intramolecular nucleophilic attack to form bicyclo[3.3.1] nonane derivatives, which cyclized to adamantanone **21** after hydrolysis (equation 31).[50] Hydrolysis of the reaction mixture at intermediate times gave the carboxylic acid **22**, supportive of the intermediacy of the ketenes **20** (equation 32).[50]

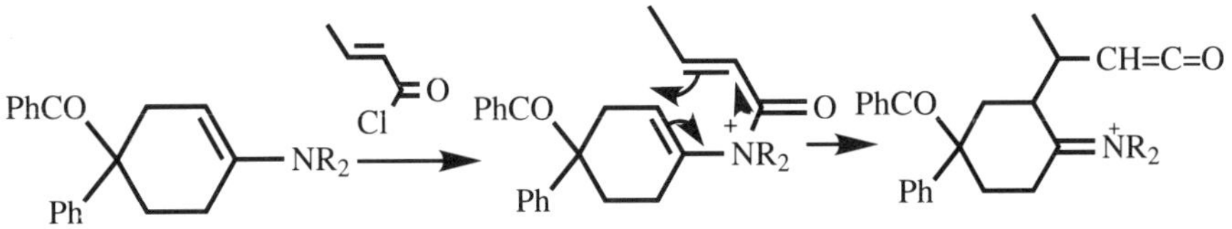

(30), (31), (32)

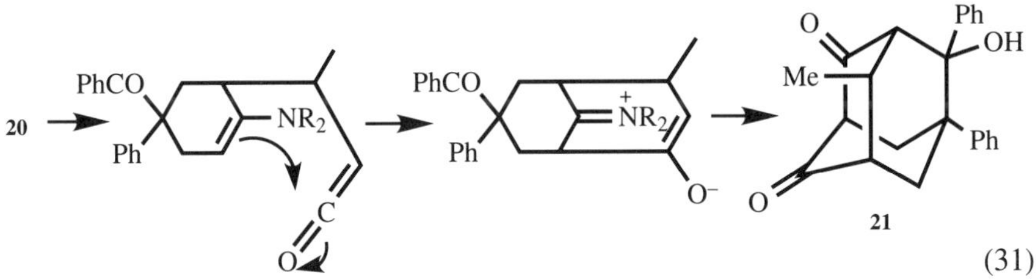

The reaction of acrylyl chloride with the morpholine enamine **23** was proposed to proceed similarly through the ketene **24** which after cyclization and hydrolysis gave **25** (equation 33).[51] However, no direct evidence for the intermediacy of **24** was presented and alternative pathways are conceivable. Some related reactions of ketenes with enamines are considered in Section 5.5.2.4.

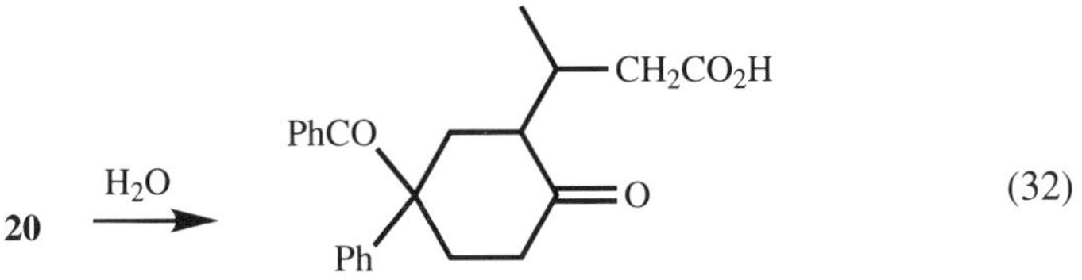

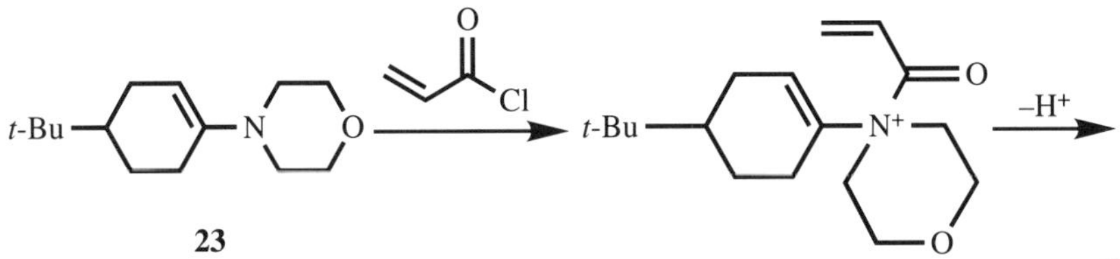

(33)

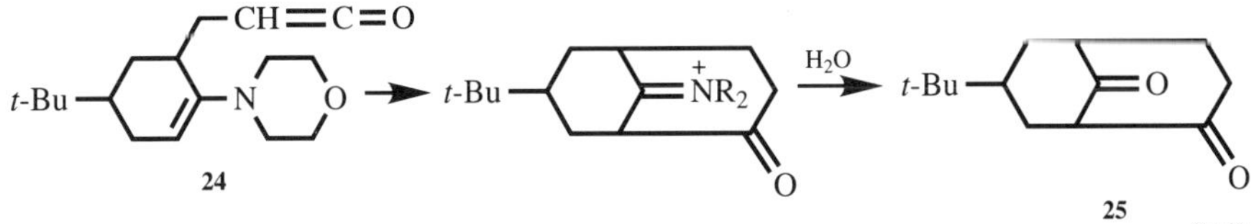

(34)

References

1. Staudinger, H. *Chem. Ber.* **1911**, *44*, 1619–1623.
2. Newman, M. S.; Arkell, A.; Fukunaga, T. *J. Am. Chem. Soc.* **1960**, *82*, 2498–2501.
3. Brady, W. T.; Ting, P. L. *J. Org. Chem.* **1976**, *41*, 2336–2339.
4. Brady, W. T.; Lloyd, R. M. *J. Org. Chem.* **1980**, *45*, 2025–2028.
5. Brady, W. T.; Scherubel, G. A. *J. Am. Chem. Soc.* **1973**, *95*, 7447–7449.
6. Adkins, H.; Thompson, Q. E. *J. Am. Chem. Soc.* **1949**, *71*, 2242–2244.
7. Walborsky, H. W. *J. Am. Chem. Soc.* **1952**, *74*, 4962–4963.
8. Cook, D. *Can. J. Chem.* **1962**, *40*, 2362–2368.
9. Brady, W. T.; Scherubel, G. A. *J. Org. Chem.* **1974**, *39*, 3790–3791.
10. Cocivera, M.; Effio, A. *J. Org. Chem.* **1980**, *45*, 415–420.
11. Giger, R.; Rey, M.; Dreiding, A. S. *Helv. Chim. Acta* **1968**, *51*, 1466–1469.
12. Dolbier, W. R.; Jr.; Lee, S. K.; Phanstiel, O., IV *Tetrahedron* **1991**, *47*, 2065–2072.
13. Lindner, E.; Steinwand, M.; Hoehne, S. *Angew. Chem., Int. Ed. Engl.* **1982**, *21*, 355–356; *Chem. Ber.* **1982**, *115*, 2181–2191.
14. Truce, W. E.; Bailey, P. S., Jr. *J. Org. Chem.* **1969**, *34*, 1341–1345.
15. Lynch, J. E.; Riseman, S. M.; Laswell, W. L.; Tschaen, D. M.; Volante, R. P.; Smith, G. B.; Shinkai, I. *J. Org. Chem.* **1989**, *54*, 3792–3796.
16. Bose, A. K.; Manhas, M. S.; Ghosh, M.; Shah, M.; Raju, V. S.; Bari, S. S.; Newaz, S. N.; Banik, B. K.; Chaudhary, A. G.; Barakat, K. J. *J. Org. Chem.* **1991**, *56*, 6968–6970.
17. Banik, B. K.; Manhas, M. S.; Kaluza, Z.; Barakat, K. J.; Bose, A. K. *Tetrahedron Lett.* **1992**, *33*, 3603–3606.
18. Payne, G. B. *J. Org. Chem.* **1966**, *31*, 718–721.
19. Singh, S.; Baird, M. C. *J. Organomet. Chem.* **1988**, *338*, 255–260.
20. Ishii, Y.; Kobayashi, Y.; Iwasaki, M.; Hidai, M. *J. Organomet. Chem.* **1991**, *405*, 133–140.
21. Brady, W. T.; Marchand, A. P.; Giang, Y. F.; Wu, A. *Synthesis* **1987**, 395–396.
22. Brady, W. T.; Giang, Y. F.; Marchand, A. P.; Wu, A. *J. Org. Chem.* **1987**, *52*, 3457–3461.
23. Corey, E. J.; Rao, K. S. *Tetrahedron Lett.* **1991**, *32*, 4623–4626.
24. Beereboom, J. J. *J. Am. Chem. Soc.* **1963**, *85*, 3525–3526.
25. Brady, W. T.; Gu, Y. *J. Heterocyclic Chem.* **1988**, *25*, 969–971.
25a. Marotta, E.; Medici, M.; Righi, P.; Rosini, G. *J. Org. Chem.* **1994**, *59*, 7529–7531.
26. Brady, W. T.; Gu, Y. *J. Org. Chem.* **1988**, *53*, 1353–1356.
27. Tempesti, E.; Giuffre, L.; Sioli, G.; Fornaroli, M.; Airoldi, G. *J. Chem. Soc., Perkin Trans. 1*, **1974**, 771–773.
28. Karger, M. H.; Mazur, Y. *J. Org. Chem.* **1971**, *36*, 528–531.
29. Karger, M. H.; Mazur, Y. *J. Org. Chem.* **1971**, *36*, 532–540.
30. Bose, A. K.; Kapur, J. C.; Sharma, S. D.; Manhas, M. S. *Tetrahedron Lett.* **1973**, 2319–2320.
31. Ikota, N.; Hanaki, A. *Heterocycles* **1984**, *22*, 2227–2230.
32. Miyake, M.; Tokutake, N.; Kirisawa, M. *Synthesis* **1983**, 833–835.

33. Amin, S. G.; Glazer, R. D.; Manhas, M. S. *Synthesis,* **1979,** 210–213.
34. George, G. I.; Mashava, P. M.; Guan, X. *Tetrahedron Lett.* **1991,** *32,* 581–584.
35. Funk, R. L.; Abelman, M. M.; Jellison, K. M. *Synlett* **1989,** 36–37.
36. Shridhar, D. R.; Ram, B.; Narayana, V. L. *Synthesis,* **1982,** 63–65.
37. Miyake, M.; Kirisawa, M.; Tokutake, N. *Synthesis,* **1982,** 1053.
38. Cossio, F. P.; Lecea, B.; Palomo, C. *J. Chem. Soc., Chem. Commun.* **1987,** 1743–1744.
39. Arrieta, A.; Lecea, B.; Cossio, F. P.; Palomo, C. *J. Org. Chem.* **1988,** *53,* 3784–3791.
40. Arrieta, A.; Cossio, F. P.; Palomo, C. *Tetrahedron* **1985,** *41,* 1703–1712.
41. Cossio, F. P.; Ganboa, I.; Garcia, J. M.; Lecea, B.; Palomo, C. *Tetrahedron Lett.* **1987,** *28,* 1945–1948.
42. Arrieta, A.; Lecea, B.; Palomo, C. *J. Chem. Soc., Perkin Trans. 1* **1987,** 845–850.
43. Manhas, M. S.; Lal, B.; Amin, S. G.; Bose, A. K. *Synth. Commun.* **1976,** *6,* 435–441.
44. Manhas, M. S.; Amin, S. G.; Ram, B.; Bose, A. K. *Synthesis* **1976,** 689–690.
45. Cossio, F. P.; Ganboa, I.; Palomo, C. *Tetrahedron Lett.* **1985,** *26,* 3041–3044.
46. van der Veen, J. M.; Bari, S. S.; Krishnan, L.; Manhas, M. S.; Bose, A. K. *J. Org. Chem.* **1989,** *54,* 5758–5762.
47. Miyake, M.; Tokutatake, N.; Kirisawa, M. *Synth. Commun.* **1984,** *14,* 353–362.
48. le Noble, W. J.; Chang, Y. *J. Am. Chem. Soc.* **1972,** *94,* 5402–5406.
49. Barrett, A. G. M. *J. Chem. Soc., Perkin Trans. 1* **1979,** 1629–1633.
50. Hickmott, P. W.; Ahmed, M. G.; Ahmed, S.A.; Wood, S.; Kapon, M. *J. Chem. Soc., Perkin Trans. 1,* **1985,** 2559–2571.
51. Hickmott, P. W. *S. Afr. J. Chem.* **1989,** *42,* 17–19.

3.2.3 Ketene from Esters

3.2.3.1 *Ketenes from Ester Enolates*

The reaction of bases with esters possessing acidic hydrogens adjacent to the carbonyl group and good alkoxy or aryloxy leaving groups can proceed by ketene formation, as shown in equation 1. Indeed this is a well-documented route for ester hydrolysis, termed the E1cB mechanism.[1-5]

$$RR^1CHCO_2R^2 \xrightleftharpoons{\text{base}} RR^1\bar{C}HCO_2R^2 \longrightarrow RR^1C{=}C{=}O + R^2O^- \quad (1)$$

This reaction has been proposed to occur for aryl ethyl malonates (**1**)[1], *o*-nitrophenyl cyanoacetate (**2**)[2] and possibly for *o*-nitrophenyl dimethylsulfonium acetate (**3**)[2], and for a series of aryl acetoacetate esters (**4**).[3] For the thio ester $EtO_2CCH_2COSCH_2CF_3$[3] a definite decision as to the extent of occurrence of the ketene pathway could not be made.[3] Linear correlations of log k are found for hydrolysis of aryl acetoacetates **4** with pK_a of the phenolate leaving group, giving strong evidence for the E1cB mechanism.[3] Bulky α-substituents on the aryl group were proposed to promote this process.[6,7] The E1cB reaction is also characterized by positive activation volumes, in contrast to the negative activation volumes for associative-type ester hydrolyses.[8,9]

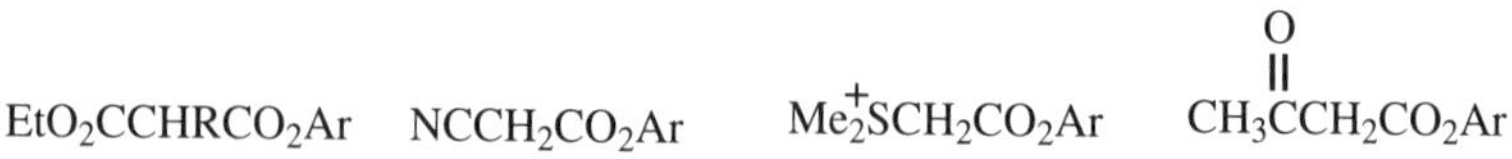

$EtO_2CCHRCO_2Ar$ $NCCH_2CO_2Ar$ $Me_2\overset{+}{S}CH_2CO_2Ar$ $CH_3\overset{\overset{O}{\|}}{C}CH_2CO_2Ar$

 1   2   3   4

The carbanion necessary for the occurrence of the E1cB mechanism can also be formed by Michael-type addition to aryl acrylates, as in equation 2.[10]

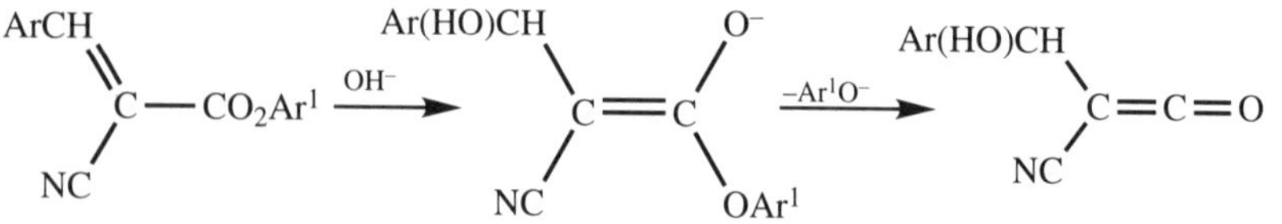

$$\underset{NC}{\overset{ArCH}{\diagdown}}C-CO_2Ar^1 \xrightarrow{OH^-} \underset{NC}{\overset{Ar(HO)CH}{\diagdown}}C=C\underset{OAr^1}{\overset{O^-}{\diagup}} \xrightarrow{-Ar^1O^-} \underset{NC}{\overset{Ar(HO)CH}{\diagdown}}C=C=O \quad (2)$$

Reactions of aryl 4-nitrophenylacetates **5** with OH⁻ in 80% DMSO/H$_2$O proceed through formation of carbanions **6**, observed by UV, which formed ketenes **7** as evidenced by the correlation with $\rho = 4.4$ of the rates of carbanion decay with the σ^- parameters of the aryl substituents.[11] The UV absorption ascribed to **7** was also observed with $\lambda_{max} = 495$ nm, and was used to observe the kinetics of the hydration of **7** in this solution.[11] Similar mechanisms have been proposed for aryl phenylacetate hydrolysis in 20% H$_2$O/CH$_3$CN.[12]

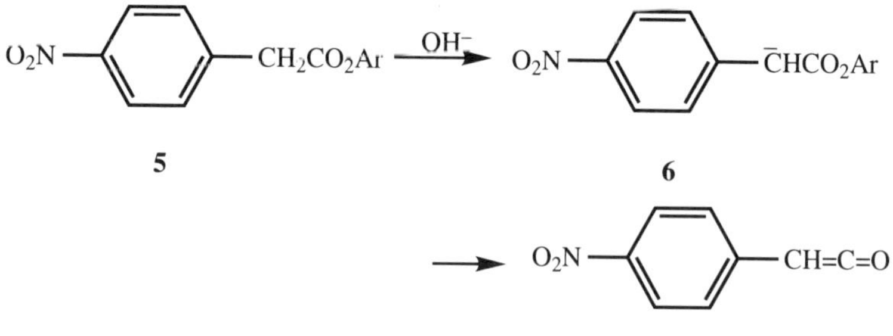

$$O_2N-C_6H_4-CH_2CO_2Ar \xrightarrow{OH^-} O_2N-C_6H_4-\bar{C}HCO_2Ar$$

           5   **6**   (3)

$$\longrightarrow O_2N-C_6H_4-CH=C=O$$

                               7

The hydrolysis of aryl 4-hydroxybenzoate esters in basic solution has been proposed to involve ketene products **8** (equation 4).[13] Reaction of the appropriate methyl ester in base has been suggested to form the charged ketene $Me_2S^+C_6H_4CH=C=O$.[14]

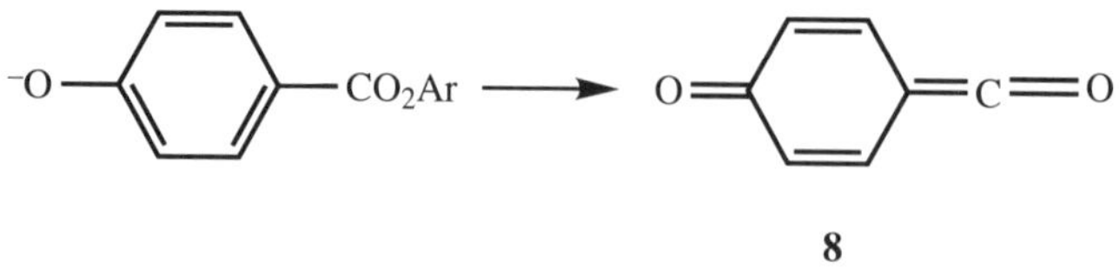

$$^-O-C_6H_4-CO_2Ar \longrightarrow O=C_6H_4=C=O \quad (4)$$

                                                8

Ketene formation from ester enolates is a major pathway in the gas phase.[15,16]

3.2 KETENES FROM CARBOXYLIC ACIDS AND THEIR DERIVATIVES

The reaction of 2,6-di-*tert*-butyl-4-tolyl (BHT) esters of carboxylic acids (**9**) with *n*-BuLi has been developed as a method for the in situ generation of ketenes; under these conditions the ketenes are captured by excess *n*-BuLi to form enolates (equation 5).[17,18] Attack of the organolithium reagent occurs from the least hindered side of the ketene. The product enolates can be reacted with Me_3SiCl to give silyl enol ethers, or with aldehydes to give aldols.

$$\underset{R}{\overset{R^1}{>}}CHCO_2Ar \xrightarrow{n\text{-BuLi}} \underset{R}{\overset{R^1}{>}}C=C=O \xrightarrow{n\text{-BuLi}} \underset{R}{\overset{R^1}{>}}C=\underset{R}{\overset{O^-}{<}}C \quad (5)$$

9 (Ar = 2,6-*t*-Bu$_2$-4-MeC$_6$H$_2$)

The formation of $(Me_3Si)_2C=C=O$ from a lithium ester enolate has been observed.[19,20] The reaction of esters with hindered Grignard reagents to give ester enolates which form ketenes had been suggested previously.[21]

The hydrolyses of aryl esters **10** of fluorene-9-carboxylic acid were proposed to occur by an E1cB mechanism for X = 4—NO_2, 4—CN, 4—Ac, 3—NO_2, 3—Cl, and 4—Cl leading to fluorenylideneketene (**11**) (equation 6), but for less stable leaving groups hydroxide attack on the ester carbonyl was favored.[22] For X = 3—O_2N the effects of neutral, ionic, and zwitterionic micelles on the reactivity were determined and interpreted in terms of electrostatic effects on the relative energies of the initial and transition states.[23]

$$\text{Fluorene-H, }CO_2C_6H_4X \rightleftharpoons \text{Fluorenyl}^- \text{-}CO_2C_6H_4X \xrightarrow{-XC_6H_4O^-} \text{Fluorenylidene}=C=O \quad (6)$$

10   **11**

Ketene formation from an ester enolate with a dienolate leaving group has been reported[24] to occur above -108 °C, and to be facilitated by the presence of HMPA (equation 7). The replacement of *t*-butoxy by methoxy by this mechanism has also been proposed.[25] Thermolysis of the lithium enolate **12** of *tert*-butyl acetate at 130° C gave ketene, which reacted with **12** to give **13** (equation 8).[26] Thiolate groups have also been shown to be leaving groups from acetothiolacetates (equation 9).[27]

$t\text{-BuO}_2\text{C(CH}_2)_6\text{CH}=\text{C} \begin{smallmatrix} \text{O}^- \\ \text{O(CH=CH)}_2\text{Pn-}n \end{smallmatrix} \xrightarrow[\text{HMPA}]{>-108\ °C} t\text{-BuO}_2\text{C(CH}_2)_6\text{CH=C=O}$ (7)

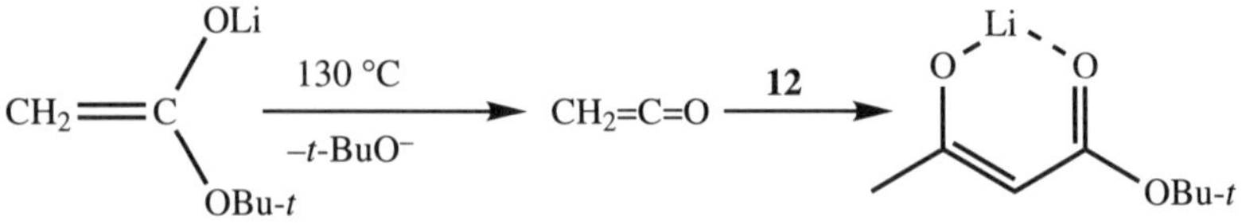

$$\underset{\textbf{12}}{\text{CH}_2=\text{C}\begin{smallmatrix}\text{OLi}\\\text{OBu-}t\end{smallmatrix}} \xrightarrow[-t\text{-BuO}^-]{130\ °C} \text{CH}_2\text{=C=O} \xrightarrow{\textbf{12}} \underset{\textbf{13}}{\text{[dioxinone with OBu-}t\text{]}} \quad (8)$$

$$\text{CH}_3\text{COCH}_2\text{COSR} \xrightarrow[-\text{RSH}]{\text{base}} \text{CH}_3\text{COCH=C=O} \quad (9)^{27}$$

3.2.3.2 Ketenes by Ester Pyrolysis
Pyrolysis of isopropenyl acetoacetate gives acetylketene **14** by the process shown in equation 10.[28] The kinetics of the formation of **14** were observed, and the presence of the ketene was confirmed by capture with ketones to give dioxinones **15** (equation 11) or by self reaction to give **16** (equation 12).[28] This reaction has been used for the preparation of hexadecylketene as a reactive intermediate (equation 13).[29]

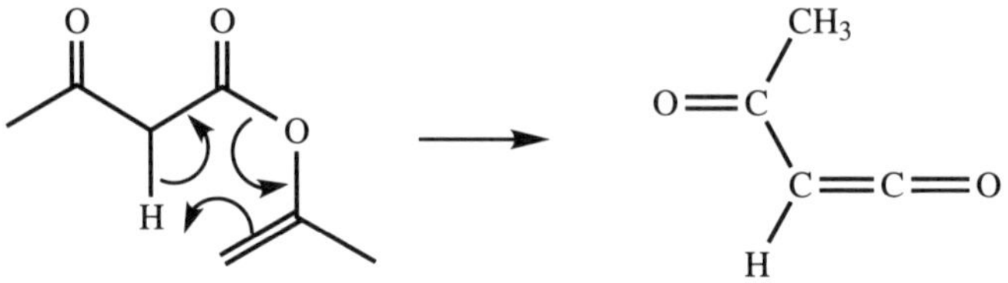

(10)

14

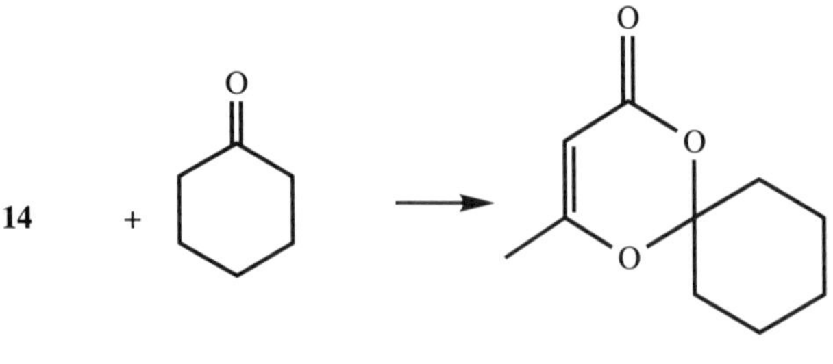

14 + cyclohexanone → **15** (11)

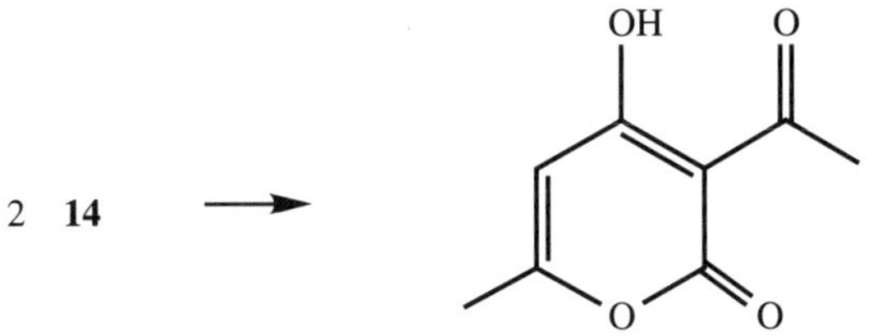

$$2 \ \mathbf{14} \longrightarrow \mathbf{16} \qquad (12)$$

$$n\text{-}C_{17}H_{35}CO_2C(CH_3)\text{=}CH_2 \xrightarrow{170\ °C} n\text{-}C_{16}H_{33}CH\text{=}C\text{=}O \qquad (13)$$

Evidence for the intramolecular pericyclic reaction of equation 10[28] comes from kinetic studies of the pyrolysis of a series of acetoacetate esters (Table 1).[30,31] The faster reaction of the isopropenyl ester supports the involvement of the process of equation 10, but other confirmatory evidence is desirable.

Two possible mechanisms proposed[32] for the formation of ketenes from the thermolysis of ethyl acetoacetic esters are shown in equations 14 and 15. This route to acylketenes is discussed in more detail in Section 4.1.6, and there is good evidence that the route of equation 15 involving assistance by the enolized acyl group is followed. Pyrolysis of the trimethylsilyl enol ether of dimethyl malonate gave carbomethoxyketene, and the route of equation 16 analogous to equation 15 was proposed.[33]

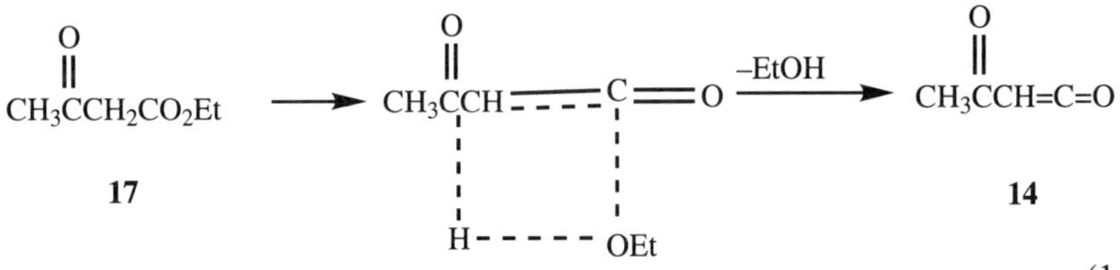

$$CH_3CCH_2CO_2Et \longrightarrow CH_3CCH\cdots C\text{=}O \xrightarrow{-EtOH} CH_3CCH\text{=}C\text{=}O$$

$$\mathbf{17} \qquad\qquad\qquad\qquad\qquad\qquad \mathbf{14}$$

$$(14)$$

TABLE 1. Relative Rates of Thermolysis of Esters $CH_3COCH_2CO_2R$ in n-BuOH at 91.7 °C[30]

R	k_{rel}
Me	1.0
Et	1.1
i-Pr	1.4
t-Bu	17
Isopropenyl[29]	7.3

17 14 (15)

$MeO_2CCH{=}C{=}O + Me_3SiOMe$ (16)

In a similar reaction the ethyl β–keto ester **18,** labelled with ^{13}C as indicated, undergoes alkoxyl group exchange on heating with methanol (equation 17).[34] This reaction was also interpreted[34] in terms of enolization followed by formation of ketene **19,** and then methanol addition. When the reaction was done in H_2O the major product was $CH_3CH_2COCO_2Et$, evidently resulting from decarboxylation of the acid formed by hydration of **19.** Synthetic applications of this reaction have been presented.[34,35]

18 **19**

(17)

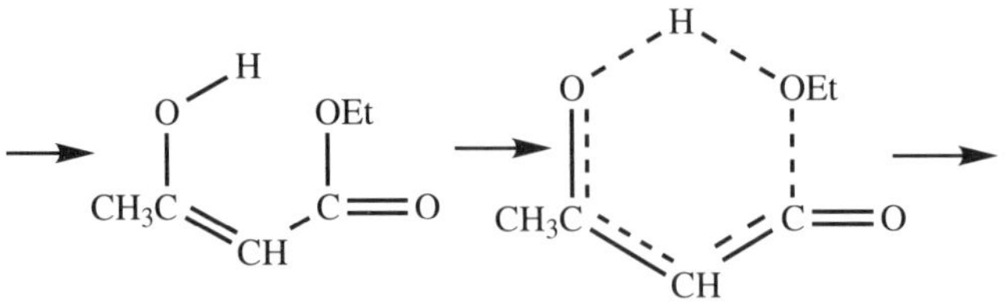

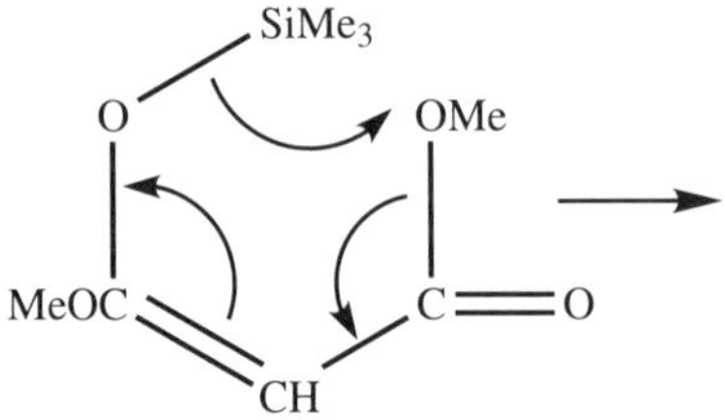

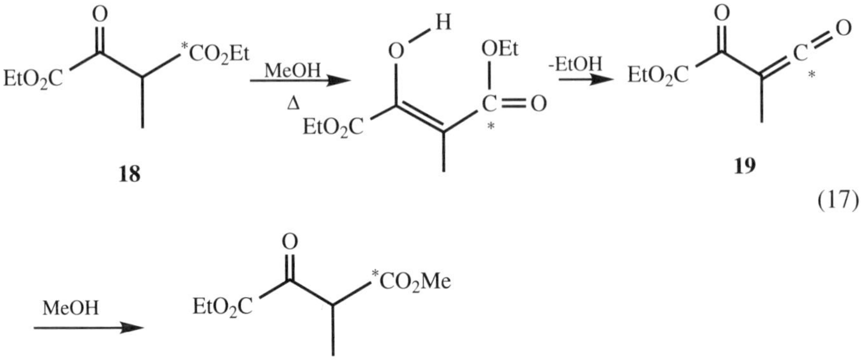

Pyrolysis of alkyltrimethylsilyl acetates above 300 °C gives ketene by a proposed "semiconcerted" mechanism shown in equation 18.[36]

$CH_2{=}C{=}O + Me_3SiOMe$ (18)[36]

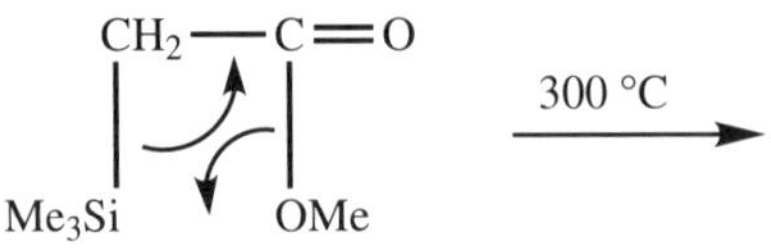

3.2.3.3 Other Preparations of Ketenes from Esters

It has been suggested that ynol phosphates **20** hydrolyze by way of ketenes.[37] Reaction of a bacterial phosphotriesterase with the alkynyl phosphate led to deactivation of the enzyme, and it was proposed that this was due to the ketene.[37]

$$n\text{-BuC}{\equiv}\text{COPO}_3\text{Et}_2 \xrightarrow[-(\text{EtO})_2\text{PO}_2\text{H}]{\text{H}_2\text{O}} n\text{-BuCH}{=}\text{C}{=}\text{O} \quad (19)$$

20

References

1. Holmquist, B.; Bruice, T. C. *J. Am. Chem. Soc.* **1969**, *91*, 2993–3002.
2. Holmquist, B.; Bruice, T. C. *J. Am. Chem. Soc.* **1969**, *91*, 3003–3009.
3. Pratt, R. F.; Bruice, T. C. *J. Am. Chem. Soc.* **1970**, *92*, 5956–5964.
4. Williams, A.; Douglas, K. T. *Chem. Rev.* **1975**, *75*, 627–649.
5. Satchell, D. P. N.; Satchell, R. S. In *The Chemistry of Functional Groups. Supplement B. The Chemistry of Acid Derivatives;* Patai, S., Ed.; Wiley: New York, 1992; Vol. 2, pp. 747–802.
6. Inoue, M.; Bruice, T. C. *J. Org. Chem.* **1983**, *48*, 3559–3561.
7. Inoue, M.; Bruice, T. C. *J. Org. Chem.* **1986**, *51*, 959–963.
8. Isaacs, N. S.; Najem, T. S. *Can. J. Chem.* **1986**, *64*, 1140–1144.
9. Isaacs, N. S.; Najem, T. S. *J. Chem. Soc. Perkin Trans. 2* **1988**, 557–562.
10. Inoue, M.; Bruice, T. C. *J. Am. Chem. Soc.* **1982**, *104*, 1644–1653.
11. Broxton, T. J.; Duddy, N. W. *J. Org. Chem.* **1981**, *46*, 1186–1191.
12. Cheong, D. Y.; Yoh, S. D.; Choi, J. H.; Shim, K. T. *J. Korean Chem. Soc.* **1992**, *36*, 446–452; *Chem. Abst.* **1992**, *117*, 89660v.
13. Cevasco, G.; Guanti, G.; Hopkins, A. R.; Thea, S.; Williams, A. *J. Org. Chem.* **1985**, *50*, 479–484; Cevasco, G.; Thea, S. *J. Org. Chem.* **1995**, *60*, 70–73.
14. Casanova, J., Jr.; Werner, N. D.; Kiefer, H. R. *J. Am. Chem. Soc.* **1967**, *89*, 2411–2416.
15. Froelicher, S. W.; Lee, R. E.; Squires, R. R.; Freiser, B. S. *Org. Mass Spectrom.* **1985**, *20*, 4–9.
16. Hayes, R. N.; Bowie, J. H. *J. Chem. Soc., Perkin Trans. 2*, **1986**, 1827–1831.
17. Häner, R.; Laube, T.; Seebach, D. *J. Am. Chem. Soc.* **1985**, *107*, 5396–5403.
18. Seebach, D.; Amstutz, R.; Laube, T.; Schweizer, W. B.; Dunitz, J. D. *J. Am. Chem. Soc.* **1985**, *107*, 5403–5409.
19. Sullivan, D. F.; Woodbury, R. P.; Rathke, M. W. *J. Org. Chem.* **1977**, *42*, 2038–2039.
20. Woodbury, R. P.; Long, N. R.; Rathke, M. W. *J. Org. Chem.* **1978**, *43*, 376.
21. Huet, F.; Emptoz, G.; Jubier, A. *Tetrahedron* **1973**, *29*, 479–485.
22. Alborz, M.; Douglas, K. T. *J. Chem. Soc., Perkin Trans. 2* **1982**, 331–339.
23. Correia, V. R.; Cuccovia, I. M.; Stelmo, M.; Chaimovich, H. *J. Am. Chem. Soc.* **1992**, *114*, 2144–2146.
24. Corey, E. J.; Wright, S. W. *J. Org. Chem.* **1990**, *55*, 1670–1673.

25. Remers, W. A.; Roth, R. H.; Weiss, M. J. *J. Org. Chem.* **1965**, *30*, 2910–2917.
26. Waldmüller, D.; Mayer, B.; Braun, M.; Hanuschik, A.; Krüger, C.; Guenot, P. *Chem. Ber.* **1992**, *125*, 2779–2782.
27. Douglas, K. T.; Yaggi, N. F. *J. Chem. Soc., Perkin Trans. 2*, **1980**, 1037–1044.
28. Clemens, R. J.; Witzeman, J. S. *J. Am. Chem. Soc.* **1989**, *111*, 2186–2193.
29. Rothman, E. S. *J. Am. Oil Chem. Soc.* **1968**, *45*, 189–193.
30. Witzeman, J. S. *Tetrahedron Lett.* **1990**, *31*, 1401–1404.
31. Witzeman, J. S.; Nottingham, W. D. *J. Org. Chem.* **1991**, *56*, 1713–1718.
32. Campbell, D. S.; Lawrie, C. W. *J. Chem. Soc., Chem. Commun.*, **1971**, 355–356; Berkowitz, W. F.; Ozorio, A. A. *J. Org. Chem.* **1971**, *36*, 3787–3792.
33. Jullien, J.; Pechine, J. M.; Perez, F. *Tetrahedron Lett.* **1983**, *24*, 5525–5526.
34. Emerson, D. W.; Titus, R. L.; Gonzalez, R. M. *J. Org. Chem.* **1990**, *55*, 3572–3576.
35. Emerson, D. W.; Titus, R. L.; Gonzalez, R. M. *J. Org. Chem.* **1991**, *56*, 5301–5307.
36. Taylor, R. *J. Chem. Soc., Chem. Commun.* **1987**, 741–742.
37. Blankenship, J. N.; Abu-Soud, H.; Francisco, W. A.; Raushel, F. M.; Fischer, D. R.; Stang, P. J. *J. Am. Chem. Soc.* **1991**, *113*, 8560–8561.

3.2.4 Ketenes by Dehalogenation of α-Halo Carboxylic Derivatives

The dehalogenation of α-bromodiphenylacetyl bromide by zinc was used by Staudinger in the first preparation of a ketene (equation 1),[1] and the procedure has been used widely since.[2-10] An updated procedure appeared in 1975.[9] This reaction can be carried out in various solvents such as ether, ethyl acetate, or THF. Transfer of the ketene formed can be effected by codistillation with these solvents,[9,10] or by syringe transfer in cases where the zinc salts are insoluble. The use of dimethoxyethane as a complexing agent for the zinc salts has been recommended.[3]

$$Ph_2CBrCOBr \xrightarrow{Zn} Ph_2C=C=O \qquad (1)$$

Activation of the zinc is needed for this reaction, and these methods have been reviewed.[11,12] Usually fairly simple activation procedures for the zinc are sufficient for insuring efficient reaction,[13] although for preparation of $CCl_2=C=O$ from CCl_3COCl the use of Zn(Cu) activated with $POCl_3$ is recommended.[4,8] This reaction is discussed further in Section 4.4.2. Simplified methods for zinc activation include heating under N_2 at 140–150 °C[7] and the use of ultrasound.[5,6] Both have proven to be very effective in the generation of $CCl_2=C=O$ from CCl_3COCl. Interfering processes include the cationic polymerization of alkenes used to react with the ketenes by the product Zn salts.[2] The addition of phosphorous oxychloride to the reaction mixture is suggested to reduce the reactivity of the product zinc salts by complexation.[8]

The metal anions $Mn(CO)_5^-$ [14,15] and $Cr(CO)_4NO^-$ [16] have been found to be extremely active in dehalogenation reactions. Thus reaction of γ-bromotigloyl chloride in $CHCl_3$ with a $CHCl_3$ solution of $Mn(CO)_5^-$ followed by codistillation gave a solution of methylvinylketene in 80% yield (equation 2).[14] Reaction of a mixture of **1** and **2** led to formation of 75% of the mixed dimer **3** along with 2–3% of the symmetrical dimers (equation 3).[16]

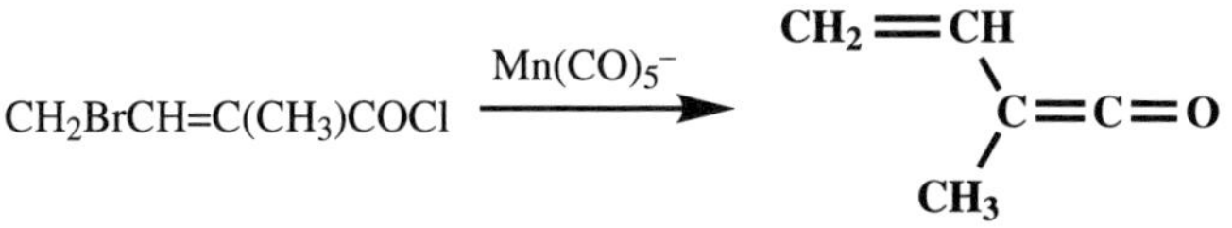

Use of the Zn–Ag couple has been utilized in preparation of cyclopropylideneketenes from α-bromoacyl chlorides (equation 4) and also from α-bromoacyl phosphates $R_2CBrCOPO_3Et_2$.[17,18]

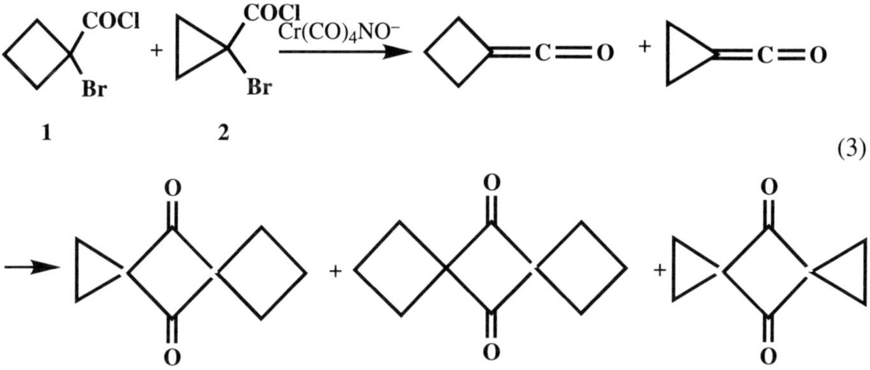

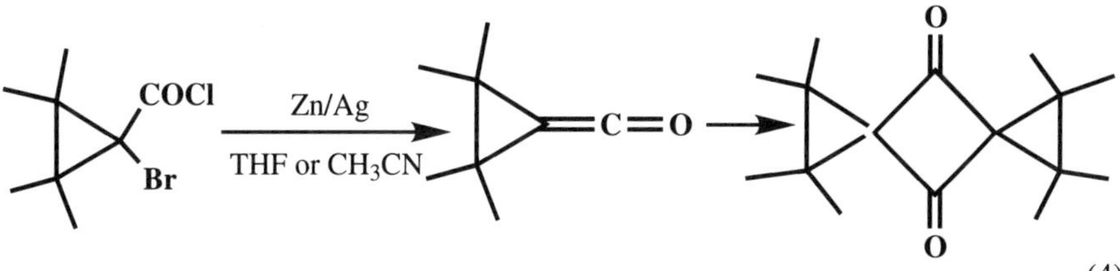

The ketene $CCl_2=C=O$ generated by Zn-dehalogenation of CCl_3COCl gives a cycloaddition reaction with silyl enol ethers[19] but this reaction is unsuccessful when the ketene is generated from the acyl halide.[20] It was suggested that the cycloaddition might have been catalyzed by zinc complexation with the ketene.

Triphenylphosphine reacts with bromodiphenylacetyl bromide at 20 °C to give diphenylketene in 82% yield after distillation (equation 5).[21] It was proposed that in the preparation of α-iodo acyl chlorides that the species may undergo dehalogenation either thermally or with chloride ion assistance to give ketenes (equation 6).[22] Reaction of **4** with Ph_3P led to the formation of dibromoketene, which was trapped by cyclopentadiene (equation 7).[23]

$$\text{Ph}_2\text{CBrCOBr} \xrightarrow[-\text{Ph}_3\text{PBr}_2]{\text{Ph}_3\text{P}} \text{Ph}_2\text{C=C=O} \qquad (5)$$

$$\text{Ph}_2\text{ClCOCl} \xrightarrow[\text{or Cl}^-]{\Delta} \text{Ph}_2\text{C=C=O} \qquad (6)$$

$$\underset{\mathbf{4}}{\text{CBr}_3\text{COSiMe}_3} \xrightarrow{\text{Ph}_3\text{P}} \text{CBr}_2\text{=C=O} \qquad (7)$$

References

1. Staudinger, H. *Chem. Ber.* **1905**, *38*, 1735–1739.
2. Brady, W. T. *Tetrahedron*, **1981**, *37*, 2949–2966.
3. Danheiser, R. L.; Savariar, S.; Cha, D. D. *Org. Synth.* **1989**, *68*, 32–40.
4. Depres, J.; Greene, A. E. *Org. Synth.* **1989**, *68*, 41–48.
5. Mehta, G.; Rao, H. S. P. *Synth. Commun.* **1985**, *15*, 991–1000.
6. Wulferding, A.; Wartchow, R.; Hoffmann, H. M. R. *Synlett.* **1992**, 476–479.
7. Stenstrøm, Y. *Synth. Commun.* **1992**, *22*, 2801–2810.
8. Hassner, A.; Krepski, L. R. *J. Org. Chem.* **1978**, *43*, 2879–2881.
9. McCarney, C. C.; Ward, R. S. *J. Chem. Soc., Perkin Trans. 1*, **1975**, 1600–1603.
10. Smith, C. W.; Norton, D. G. *Organic Syntheses*, Coll. Vol. IV; Rabjohn, N., Ed.; Wiley: New York, 1963, pp. 348–350.
11. Reike, R. *Acc. Chem. Res.* **1977**, *10*, 301–306.
12. Erdik, E. *Tetrahedron*, **1987**, *43*, 2203–2212.
13. Tsuda, K.; Ohki, E.; Nozoe, S. *J. Org. Chem.* **1963**, *28*, 783–785.
14. Masters, A. P.; Sorensen, T. S.; Ziegler, T. *J. Org. Chem.* **1986**, *51*, 3558–3559.
15. Masters, A. P.; Sorensen, T. S.; Tran, P. M. *Can. J. Chem.* **1987**, *65*, 1499–1502.
16. Masters, A. P.; Sorensen, T. S. *Tetrahedron Lett.* **1989**, *30*, 5869–5872.
17. Hoffmann, H. M. R.; Geschwinder, P. M.; Hollwege, H.; Walenta, A. *Helv. Chim. Acta* **1988**, *71*, 1930–1936.
18. Hoffmann, H. M. R.; Walenta, A.; Eggert, U.; Schomberg, D. *Angew. Chem., Int. Ed. Engl.* **1985**, *24*, 607.
19. Brady, W. T.; Lloyd, R. M. *J. Org. Chem.* **1979**, *44*, 2560–2564.
20. Brady, W. T.; Lloyd, R. M. *J. Org. Chem.* **1980**, *45*, 2025–2028.
21. Darling, S. D.; Kidwell, R. L. *J. Org. Chem.* **1968**, *33*, 3974–3975.
22. Harpp, D. N.; Bao, L. Q.; Black, C. J.; Gleason, J. G.; Smith, R. A. *J. Org. Chem.* **1975**, *40*, 3420–3427.
23. Okada, T.; Okawara, R. *Tetrahedron Lett.* **1971**, 2801–2802.

3.3 KETENES FROM DIAZO KETONES (WOLFF REARRANGEMENTS)

Diazo ketones react thermally, photochemically, or catalytically to form ketenes (equations 1). The ketenes are usually trapped by hydroxylic solvents R^2OH to give acids or esters, and since the discovery of this process by Wolff in 1902[1] this has been a useful synthetic procedure. There is a very extensive literature and numerous reviews[2-12] dealing with various aspects of this process, and only a brief summary and some recent developments will be considered here.

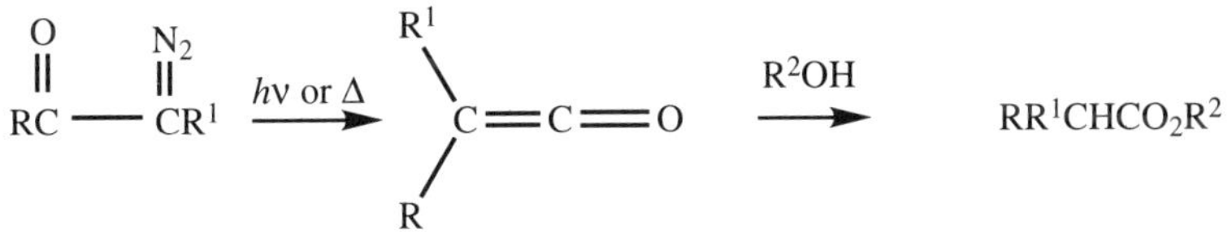

$$\underset{RC-CR^1}{\overset{O\;\;\;\;N_2}{\overset{\parallel\;\;\;\;\parallel}{}}} \xrightarrow{h\nu \text{ or } \Delta} \underset{R}{\overset{R^1}{}}C=C=O \xrightarrow{R^2OH} RR^1CHCO_2R^2 \quad (1)$$

α-Diazo carbonyl compounds are available from a number of routes,[3,5,8,13] and in addition to the Wolff rearrangement give a number of other reactions, including Rh(II) catalyzed cyclopropanation with alkenes and C—H and X—H bond insertion,[14] acid-induced reactions,[15] and intramolecular and intermolecular addition reactions.[16-18]

The electronic structure of diazo ketones may be represented by the resonance structures shown, in which electron donation by the diazo group to the carbonyl plays an important role. This phenomenon makes diazo ketones more stable than nonconjugated azo compounds, and also results in restricted rotation around the C—C bond between the carbonyl group and the diazo linkage.

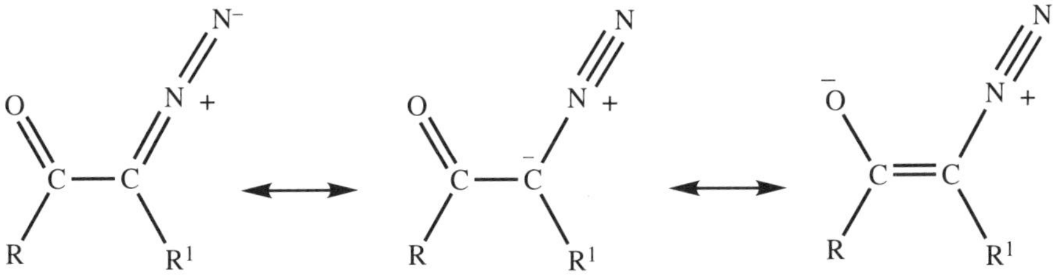

The effect of substituents on the structure and energy of diazo compounds and of diazirines have been studied by ab initio MO calculations.[19] These confirm the important role of conjugation in the stabilization of α-diazo ketones. The conformations of diazo ketones are important for their reactivity and have been studied by a variety of methods.[20]

The two conformers due to restricted rotation around the double bond (equation 2) in diazoketones may be observed by ^1H NMR, and for CH_3COCHN_2 there is a barrier for interconversion of the two conformers of 15.5 kcal/mol, with a 92/8 preference for the conformer with the oxygen and nitrogen *cis*.[21] For $HCOCHN_2$ the *cis/trans* ratio is 69/31.[21] If the coplanarity and the conjugation of the π-system is restricted by bulky groups (equation 3)[22,22a] or twisting of a ring (equation 4),[23] then these compounds are much more labile and nitrogen loss occurs at low temperature. However, in the former example migration of methyl is much more effective than for *tert*-butyl.[22]

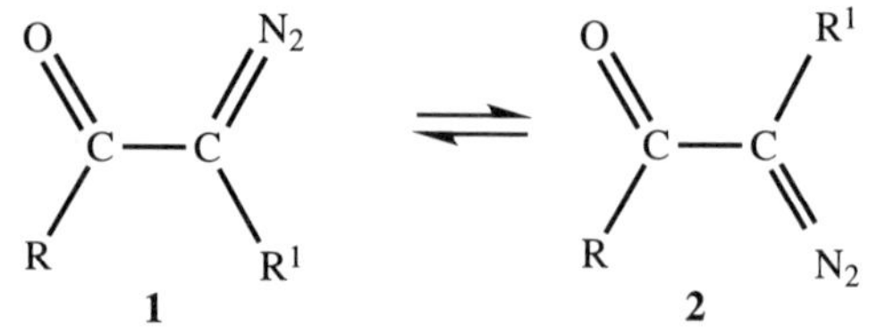

$$\underset{1}{\underset{R}{\overset{O}{\|}}{\overset{\|}{C}}-\underset{R^1}{\overset{N_2}{\overset{\|}{C}}}} \quad \rightleftharpoons \quad \underset{2}{\underset{R}{\overset{O}{\|}}{\overset{\|}{C}}-\underset{N_2}{\overset{R^1}{\overset{\|}{C}}}} \qquad (2)$$

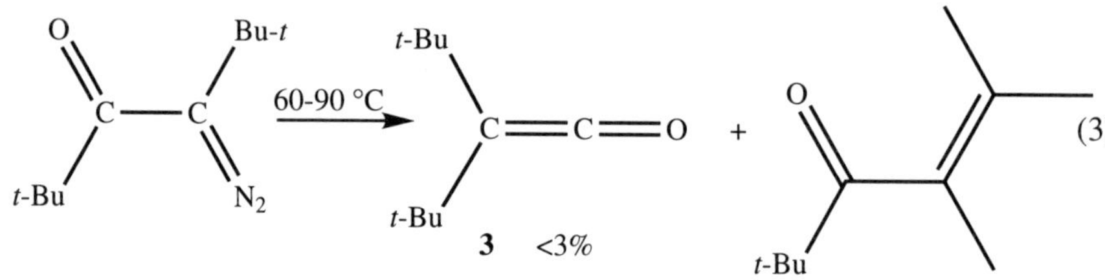

$$\underset{t\text{-Bu}}{\overset{O}{\|}}{\overset{\|}{C}}-\underset{N_2}{\overset{\text{Bu-}t}{\overset{\|}{C}}} \xrightarrow{60\text{-}90\ °C} \underset{t\text{-Bu}}{\overset{t\text{-Bu}}{C}}=C=O \;+\; \text{(ketone with }t\text{-Bu)} \qquad (3)$$

3 <3%

$$\xrightarrow{20\ °C} \qquad (4)$$

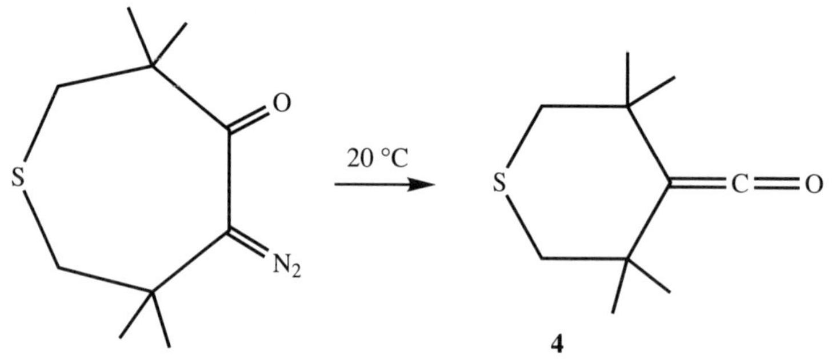

4

The general pathway of the Wolff rearrangement involves initial loss of N_2 to form a keto carbene, which can react either by migration of the group R^1 to form the ketene (equation 5), or by a variety of other processes. Some of the competing reactions of keto carbenes include the reaction with nucleophilic solvents such as CH_3OH (equation 6), the addition to C—C double bonds to form cyclopropanes (equation 7), and intramolecular insertion (equation 8). Reversible photoisomerization of diazoketone **5** to the diazirine **6** has also been observed,[24] and both **5** and **6** evidently also form the keto carbene **7**, which gives the ketene **8** as well as tricyclene (equation 9).[24] Rearrangements of keto carbenes occur which suggest the formation of oxirenes as transition states or intermediates (equation 10).[9,25,26] The reactions with other reagents are facilitated if these are present in the same molecule, and in such cases the Wolff rearrangement may be completely suppressed, as in the example of equation 11.[17]

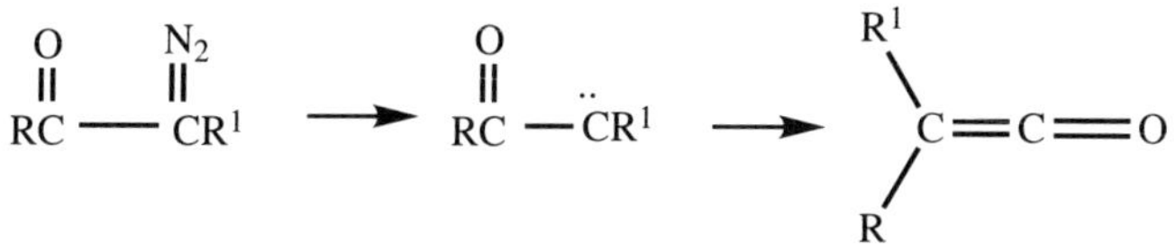

3.3 KETENES FROM DIAZO KETONES (WOLFF REARRANGEMENTS)

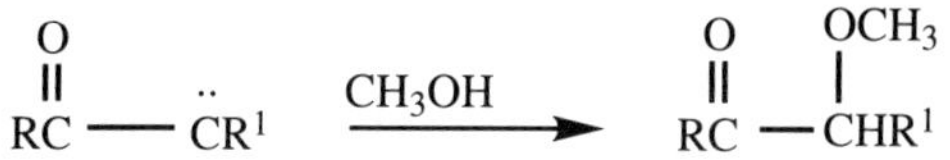

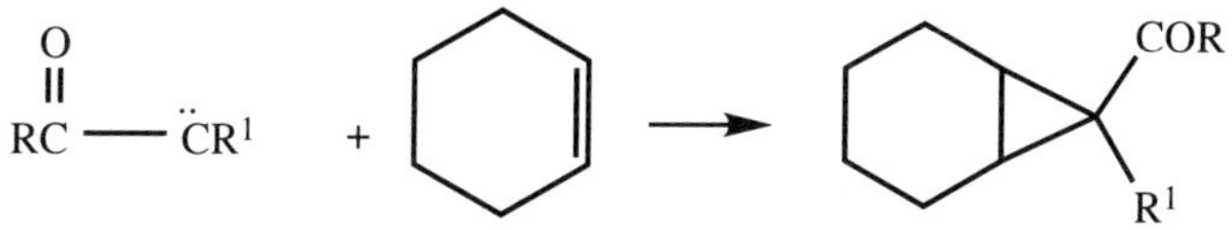

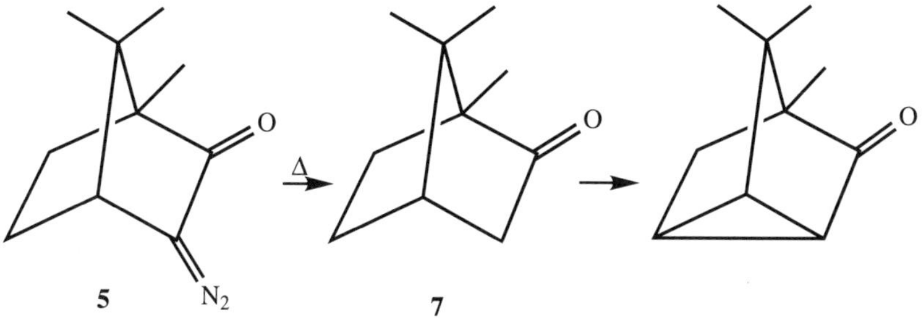

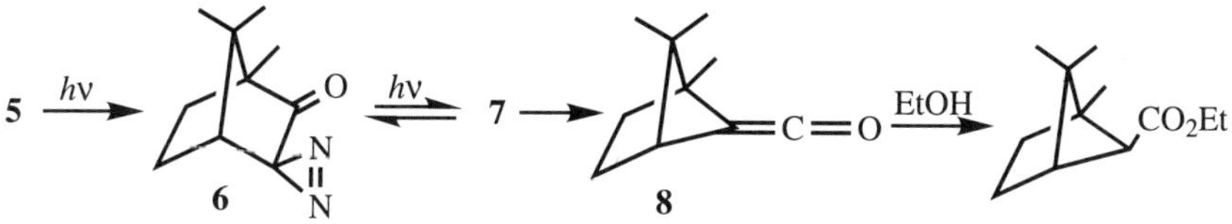

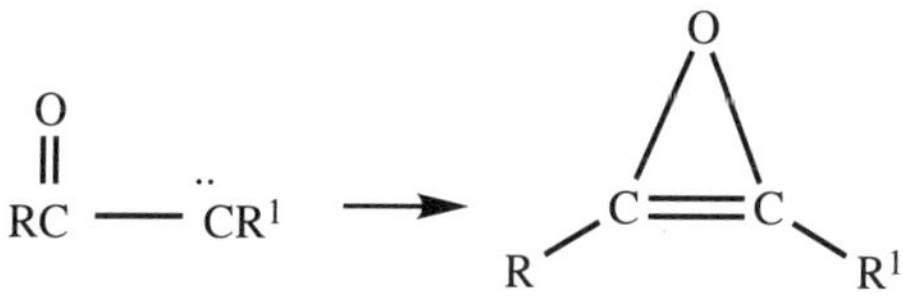

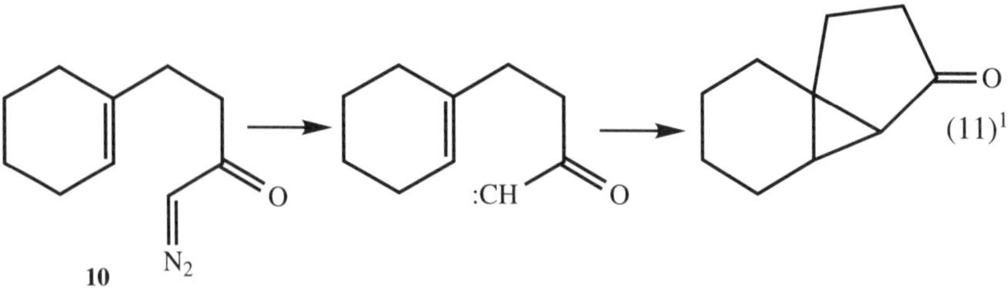

Photolysis of diazoacetaldehyde in an Ar matrix gave the parent ketene as identified by its IR spectrum (equation 12), and N_2CHCO_2Et similarly gave $EtOCH=C=O$.[27]

$$N_2CHCH=O \xrightarrow{h\nu} CH_2=C=O \qquad (12)$$

There has been a great deal of study devoted to the role of conformational factors[28,29] and whether the Wolff rearrangement is a concerted process or whether intermediate keto carbenes or oxirenes can be observed (equation 13).[30-36] The ESR spectrum of the ketocarbene PhCOČPh has been observed,[34] providing strong evidence for the involvement of such species in some Wolff rearrangements.

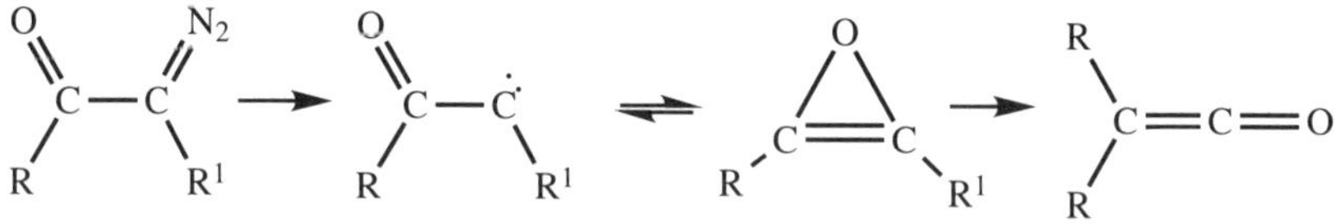

Thermolysis or photolysis of the diazo ketone **11** in dioxane containing 5% MeOH led to the ketene **12** by alkyl group migration, as evidenced by the formation of ester **13** in 83–88% yield (equation 14).[29] Thermolysis or photolysis of the crowded acylic diazoketone **14** proceeded by exclusive aryl group migration to the ketene **15**, which was captured as the ester **16** (equation 15).[29] The rates of thermal decomposition of **11, 15,** and other diazodiketones are correlated with the angle of twisting of the diazodiketone moiety.[29a]

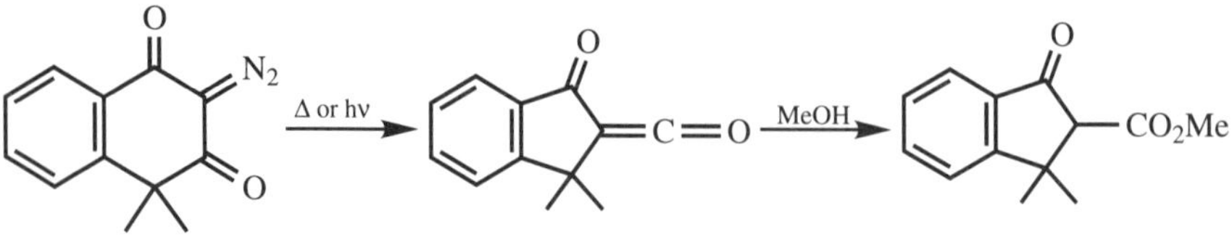

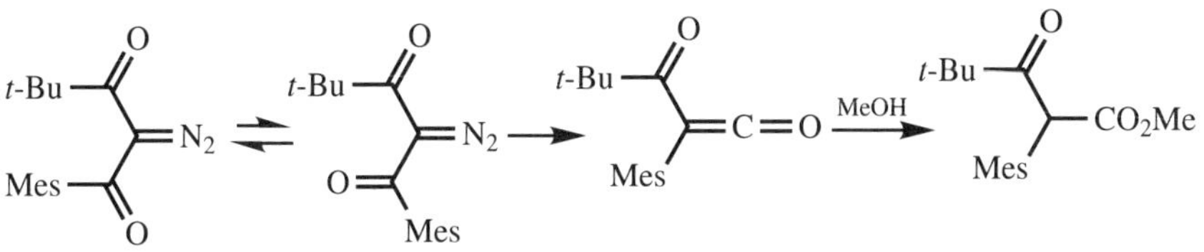

$$\text{(15)}$$

It was proposed that Wolff-rearrangement of **11** was a concerted process in which alkyl migration occurred during nitrogen loss. However **14** exists as a 2:1 mixture of the Z,Z and E,Z conformers, which are calculated to be twisted out of planarity by 88 and 81°, respectively.[29] Therefore concerted migration is impossible and a ketocarbene is formed which reacts by aryl migration as the only ketene forming pathway, along with 2–4% of C—H insertion by the ketocarbene.[29]

A number of acylketenes formed from Wolff rearrangements of 2-diazo-1,3-diones in Ar matrices were observed by IR, and on further irradiation the ketenes underwent decarbonylation to give further Wolff rearrangement to new ketenes (equation 16).[37]

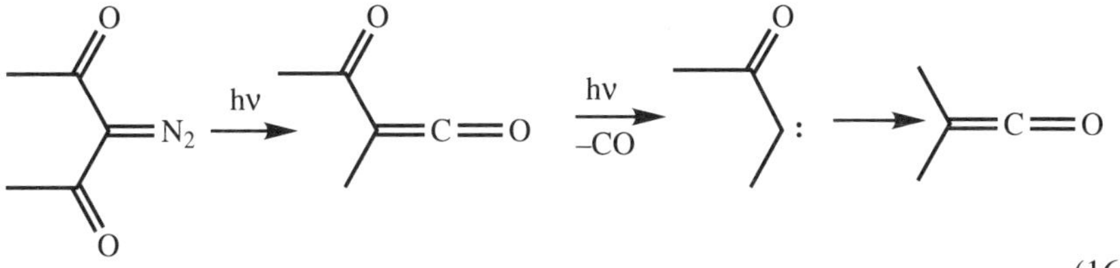

$$\text{(16)}$$

The Wolff rearrangement has been studied by ab initio MO calculations (Section 1.1.1).[38] The relative stabilities of ketocarbenes, oxirenes, and ketenes have also been calculated.[38] A "retro-Wolff rearrangement," namely the conversion of **17** to **18,** has been studied experimentally[39] and theoretically (equation 17).[40] The formation of a ketene analog of **18a** by a similar process has been studied theoretically.[40a] The extent of oxygen shift during the photochemical Wolff rearrangement of $PhCOCN_2Me$ and $MeCOCN_2Ph$ has been examined using ^{13}C labeling.[41]

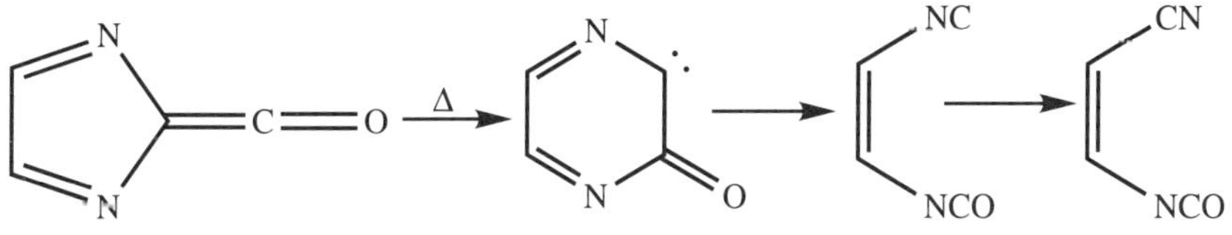

$$\text{(17)}$$

The Wolff rearrangement provides an extremely versatile method for the generation of ketenes of a wide variety of structural types, including alkyl, aryl, cycloalkylidene, and others, as illustrated in the examples below. When the ketenes are sufficiently stable they may be isolated from the reactions, as in the cases of diphenylketene (equation 18),[42] some large-ring cycloalkylideneketenes (equation 19),[43,44] and silylated ketenes (equation 20).[45] Reactive ketenes generated by flash photolysis of diazoketones have been studied by fast reaction techniques and the kinetics of their reactions measured, as discussed in Section 5.5.1.2.[46,47] Ketocarbenes have been generated in inert matrices at low temperature and their spectral properties examined.[9,30,32,48,49]

$$PhC(=O)-C(=N_2)Ph \xrightarrow{\Delta} Ph_2C=C=O \quad (18)$$
$$\mathbf{19}$$

20

(19)

$$Me_3SiC(=O)-C(=N_2)R \longrightarrow (Me_3Si)(R)C=C=O \quad (20)$$

3.3.1 Thermal Wolff Rearrangement

There is extensive evidence[4,5] that the thermal Wolff rearrangement proceeds as shown in equation 5 with initial loss of N_2 to form a ketocarbene which then rearranges to a ketene. Evidence against participation of the solvent or a nucleophile in the initial step is the lack of a rate dependence on these reagents, and the capture of the ketocarbene by various reagents, including nitriles and amines. Evidence against concerted migration with loss of N_2 is the lack of a dependence of the rate on the anticipated migratory ability of the groups (*vide infra*). There are accelerations observed due to acidic materials, possibly due to hydrogen bonding to the carbonyl group, and by polar solvents, indicating some ionic character of the transition state.

Thermolysis of α–diazoacetophenone **21** in dodecane led to the lactone **22** and the dimer **23**, which was believed to be formed by oxidation of **22**.[50] The formation of this product was proposed to occur by reaction of phenylketene with unreacted diazoketone, as shown in equation 22.[50] Evidence for this proposal was found in the observation that the kinetics of thermal decomposition of **21** were unimolecular, but were greatly accelerated by the addition of diphenylketene, leading to the product **24**.[50,51]

3.3 KETENES FROM DIAZO KETONES (WOLFF REARRANGEMENTS)

$$\text{PhCCHN}_2 \longrightarrow \text{PhCCH} \longrightarrow \text{PhCH=C=O} \quad (21)$$
21

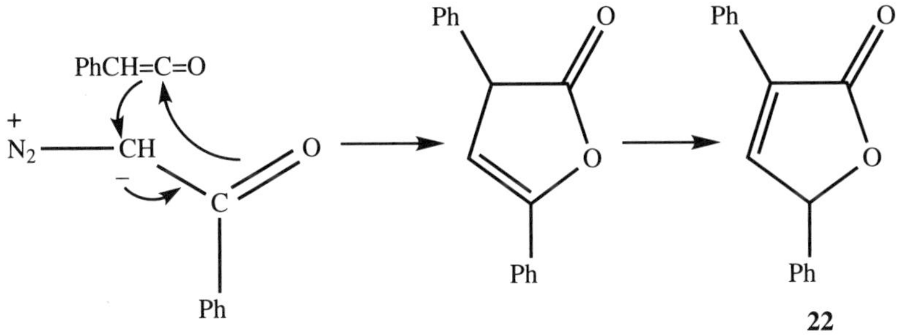

(22)

22

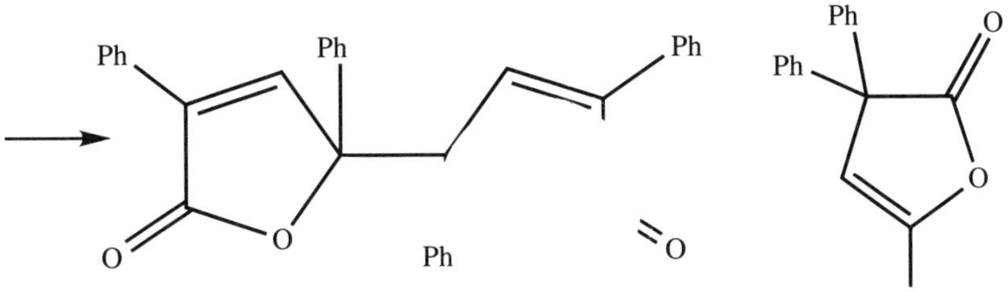

23　　**24**

There have been several kinetic studies of the thermal decomposition of diazoketones.[52-56] These show phenyl groups are more activating than acyl groups for decomposition.[52] The effect of substituents X and Y on the rate of thermal decomposition of **25** and **26** (Table 1) is correlated with the Hammett σ values with a slope of 0.75, whereas the correlation with Y groups gives a correlation with σ^+ with slope of -1.49.[55] For disubstituted substrates the total effects are larger.[56] Electron withdrawal by Y would stabilize the negative charge on the nitrogen substituted carbon, whereas electron donors X can conjugate with the carbonyl and stabilize the ground state.

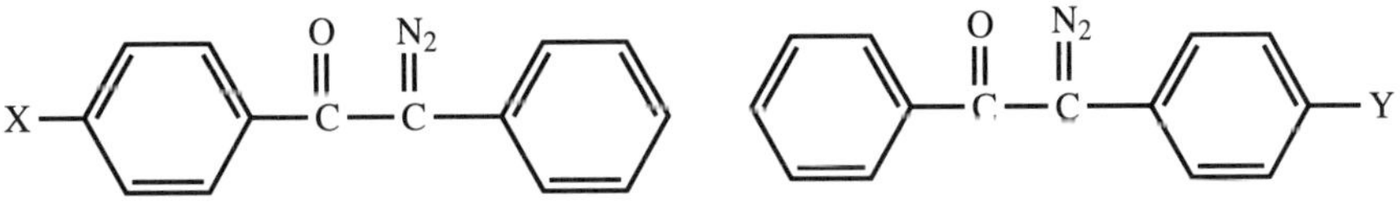

25　　**26**

Based on these results, rate-limiting loss of nitrogen followed by rearrangement to the ketene was proposed.[55,56] In other kinetic studies no effect of solvent

TABLE 1. Relative Rates of Decomposition of 25 and 26 in n-Butyl Ether

X	k_{rel} (70 °C)[55]	Y	k_{rel} (70°C)[55]
MeO	0.73	Me	3.04
Et	0.76	H	1.0
Me	0.80	F	1.92
H	1.0	Cl	0.74
F	1.29	Br	0.76
Cl	1.56	NO$_2$	0.071
Br	1.66		

$$X-\underset{}{\text{C}_6\text{H}_4}-\underset{\underset{\text{O}}{\|}}{\text{C}}-\underset{\underset{\text{N}_2}{\|}}{\text{C}}-\underset{}{\text{C}_6\text{H}_4}-Y \qquad \underset{\underset{\text{O}}{\|}}{\text{Ar}\text{C}}-\underset{\underset{\text{N}_2}{\|}}{\text{C}}-\underset{\underset{\text{O}}{\|}}{\text{C}\text{R}}$$

X	Y	k_{rel}[56a]	Ar	R	k_{rel} (101°C)[29a,b]
NO$_2$	MeO	220	4-Tol	Me	26.7
MeO	MeO	40	4-Tol	i-Pr	45.6
H	H	1.0	4-Tol	t-Bu	101
NO$_2$	NO$_2$	0.04	Mes	t-Bu	306
MeO	H	0.026			3.9
H	NO$_2$	0.022			

[a] In 1,2-dichloroethane.
[b] In dioxane.

nucleophilicity was observed, although in some cases acceleration by proton acids was detected.[53]

A variety of catalysts have been used to assist the thermal Wolff rearrangement. In diazo ketone reactions induced by Rh$_2$(OAc)$_4$ increased amounts of the insertion of ketocarbene intermediates into the OH bond of solvent MeOH was observed (equation 6).[29] Metal catalysts that have been used in thermal Wolff rearrangements include Ag$_2$O,[57] PhCO$_2$Ag,[58] and various Cu(II) salts.[59]

3.3.2 Photochemical Wolff Rearrangement

The mechanism of this process has been extensively discussed[2,3,5,8] and interpreted in terms of an initial excitation either by direct photolysis to a singlet or sensitized generation of the triplet state. Singlet state excited diazoketones in the s-Z conformation (27) can rearrange in a concerted fashion to ketenes (equation 24), whereas either singlet or triplet ketocarbenes can form from excited singlet or triplet state diazoketones, respectively. Wolff rearrangement of the singlet state carbene directly to the ketene should be facile while the triplet carbene would evidently need to

3.3 KETENES FROM DIAZO KETONES (WOLFF REARRANGEMENTS)

undergo intersystem crossing before rearrangement could occur, and characteristic carbene reactions such as addition and insertion should be more prominent.

$$\underset{RC-CR^1}{\overset{O\quad N_2}{\overset{\|\quad\|}{}}} \xrightarrow[\text{or sens.}]{h\nu} \left[\underset{RC-CR^1}{\overset{O\quad N_2}{\overset{\|\quad\|}{}}}\right]^* \longrightarrow \underset{RC-\ddot{C}R^1}{\overset{O}{\overset{\|}{}}} \quad (23)$$

$$\left[\underset{\underset{R}{}}{\overset{O}{\overset{\diagdown}{C}}}\underset{\underset{R^1}{}}{\overset{N_2}{\overset{\diagup}{C}}}\right]^{*1} \xrightarrow{-N_2} \underset{R}{\overset{R^1}{\diagdown}}C=C=O \quad (24)$$
$$\mathbf{27}$$

The formation of oxirenes along the reaction coordinate from the ketocarbenes has been demonstrated in a number of cases by the use of ^{13}C labeling which showed scrambling,[60–62] and by the observation of structural isomerism.[25,26,63,64] The latest theoretical study of the parent oxirene **9** (R = H) suggests that this species is an energy minimum,[32,65] but the transition state for rearrangement to the much lower energy ketene $CH_2=C=O$ could not be located, and the barrier is evidently quite low. Photolysis of hexafluoro-3-diazo-2-butanone permitted the direct observation of the ketene **29** at 10 K,[48,49] but an intermediate reported[48] to be the ketocarbene **28** was later shown to be the diazirine **31** which was isolable at room temperature.[49] The claimed observation of **30** could not be evaluated, as the reported[48] IR spectrum was not observed.[49]

$$\underset{CF_3C-CCF_3}{\overset{O\quad N_2}{\overset{\|\quad\|}{}}} \longrightarrow \underset{CF_3C-\ddot{C}CF_3}{\overset{O}{\overset{\|}{}}} \longrightarrow \underset{CF_3}{\overset{CF_3}{\diagdown}}C=C=O$$

$$\mathbf{28} \qquad\qquad \mathbf{29} \qquad (25)$$

30 **31**

Another report[30] of spectroscopic detection of an oxirene has been questioned,[47] and although there are other reports of observation of these species,[32] these should be critically examined. The intermediacy of an oxirene in formation of a fulvenone was excluded by ^{13}C labeling.[65a] Methyloxirene generated from the radical cation was proposed to form methylketene.[65b]

The expected dependence of the course of diazoketone photolysis on spin multiplicity is indeed observed. Thus direct photolysis of diazocyclohexanone in the presence of olefins and alcohols leads to capture of the ketene by the alcohol (equation 26) and no trapping of the ketocarbene by the olefin, whereas photosensitized reaction of diazocyclohexanone or diazoacetone in the presence of olefins and alcohols leads to capture of the carbene and not the ketene (equations 27 and 28).[66,67]

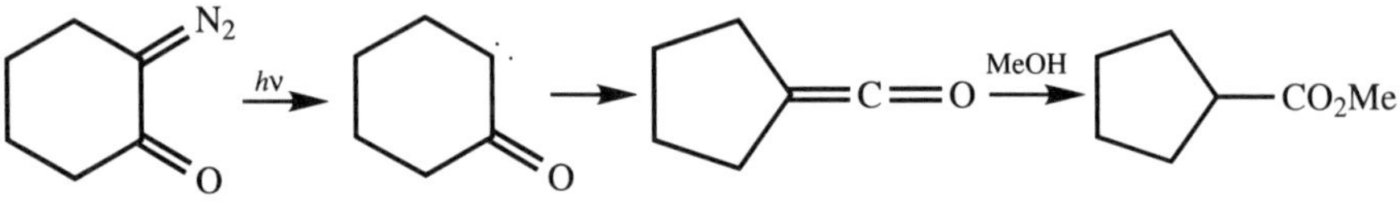

(26)

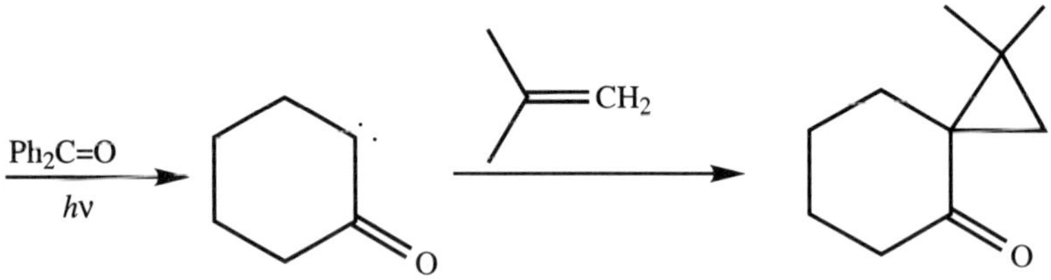

(27)

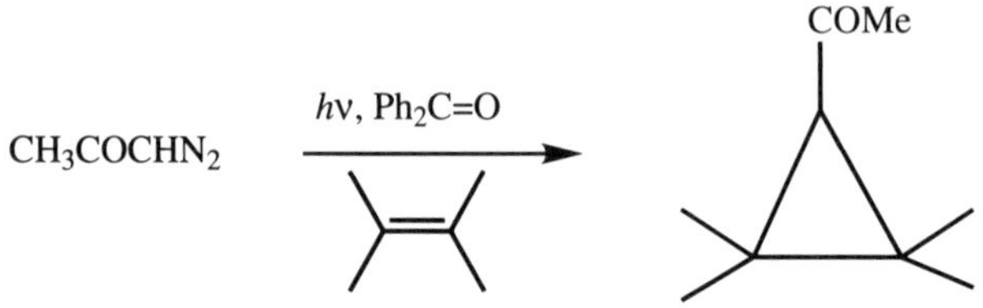

(28)

The nature of neighboring alkoxy groups affects the reactivity of ketocarbenes.[68] Photolysis of ethyl diazoacetate in alcohols led to the suggested reaction path in equation 29.[69,70] The formation of ester $ROCH_2CO_2R$ was evidence for the formation of **34**.[69] The formation of $EtOCH=C=O$ from **33** in an Ar matrix was confirmed by IR.[27]

$$N_2CHCO_2Et \xrightarrow{h\nu} N_2\overset{+}{C}H=C=O \longrightarrow EtOCH=C=O \xrightarrow{ROH} EtOCH_2CO_2R$$
$$\phantom{N_2CHCO_2Et \xrightarrow{h\nu}\ } {}^{-}OEt$$

33   **34**   **35**

(29)

3.3 KETENES FROM DIAZO KETONES (WOLFF REARRANGEMENTS) 87

Trapping of a ketocarbene by insertion into the O—H bond of water has been reported (equation 30).[71]

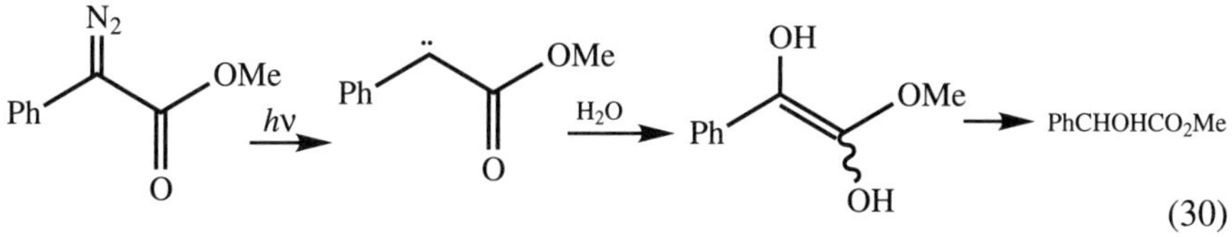

$$Ph-C(N_2)-C(=O)-OMe \xrightarrow{h\nu} Ph-\ddot{C}H-C(=O)-OMe \xrightarrow{H_2O} Ph-C(OH)=C(OMe)-OH \rightarrow PhCHOHCO_2Me \quad (30)$$

Photochemical ring contraction by Wolff rearrangement has led to a large number of cycloalkylideneketenes, many of which are listed in Section 4.1.1 (Table 1). Some representative examples, including large and small rings, polycyclic systems, and benzannelated derivatives, are shown in equations 31–34.[72–74] The strained ketenes **37** and **39** were evidently formed in very low yields.

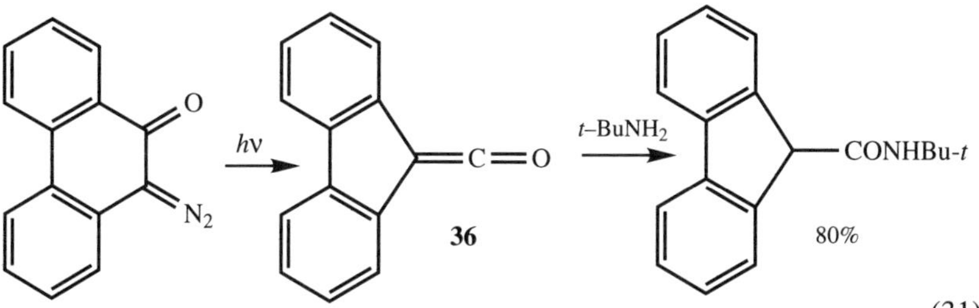

(phenanthrene-9(10H)-one diazo) $\xrightarrow{h\nu}$ fluorenylidene ketene **36** $\xrightarrow{t\text{-BuNH}_2}$ fluorene-CONHBu-t, 80% (31)[72]

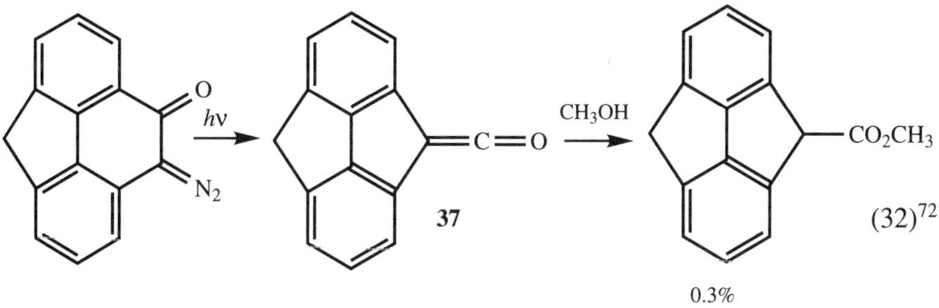

(acenaphthylene-fused diazo ketone) $\xrightarrow{h\nu}$ **37** C=O $\xrightarrow{CH_3OH}$ —CO$_2$CH$_3$, 0.3% (32)[72]

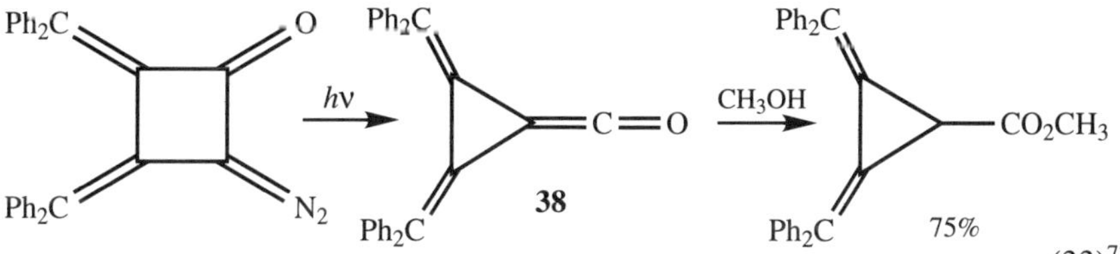

$Ph_2C=\square=CPh_2$ (with O and N$_2$) $\xrightarrow{h\nu}$ **38** $Ph_2C=C=C(CPh_2)=C=O$ $\xrightarrow{CH_3OH}$ $Ph_2C=C(CPh_2)$—CO$_2$CH$_3$, 75% (33)[73]

39 <2%

(34)[74]

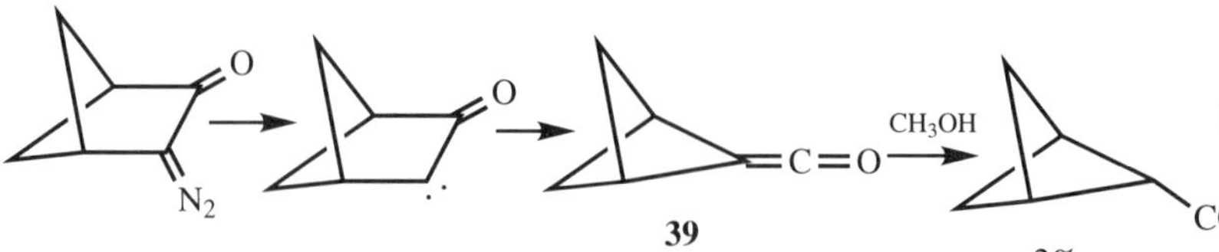

Photolysis of bis-1,3-diazoketones leads to typical products of Wolff rearrangements, and this reaction has been interpreted in terms of formation of α-(diazomethyl)ketenes (equation 35).[75,76] However photolysis of 1,3-bisdiazo-1,3-diphenylacetone (**42**) leads to isolation of diphenylcyclopropenone (**43**) (equation 36) and the formation of cyclopropenones by the route shown is an alternative for the cyclic bisdiazoketones as well.[77] Reaction of either stereoisomer of **44** led to **45** and **46** in similar ratios, presumably through the ketenes **47** and **48**, or by a cyclopropenone (equation 37).[76]

40 **41**

(35)

42 **43**

(36)

44 **47** **48** **45** (44%)

(37)

46 (16%)

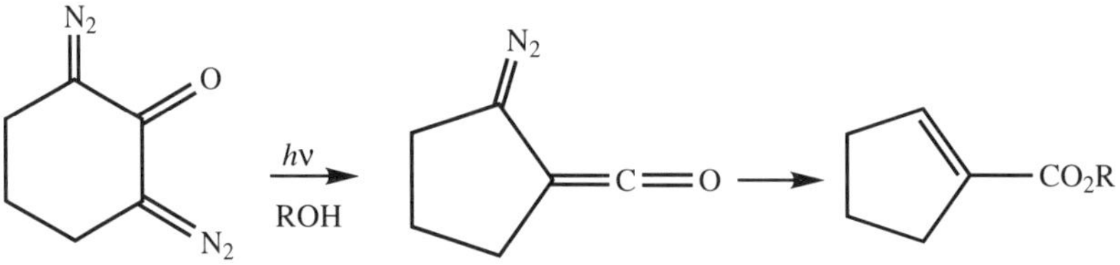

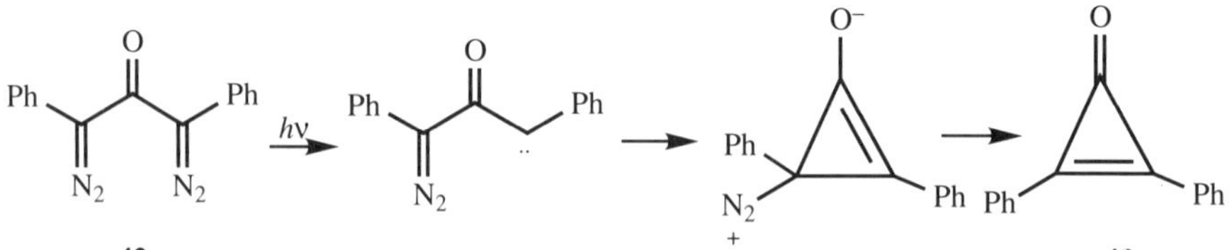

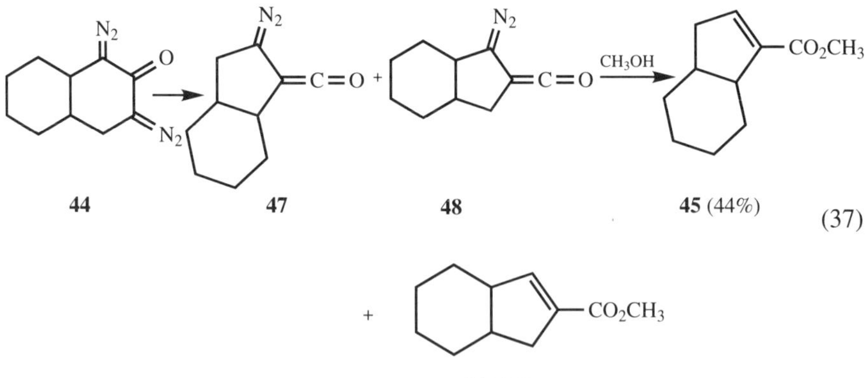

3.3 KETENES FROM DIAZO KETONES (WOLFF REARRANGEMENTS) 89

Photolysis of **40** in an Ar matrix at 8 K gave a species identified as **41** on the basis of its IR bands at 2118 and 2120 cm^{-1}.[78] Similarly, photolysis of **49** gave **50** with an IR band at 2106 cm^{-1}, and a broad UV–VIS band at 507 nm giving a red color (equation 38).[79] However, photolysis of **49** in fluid solutions led to capture of the intermediate ketocarbenes and no Wolff rearrangement.[79] It was proposed that conjugation in the diazoketone group in the intermediate ketocarbene resulting from N_2 loss by **49** inhibited Wolff rearrangement.[79]

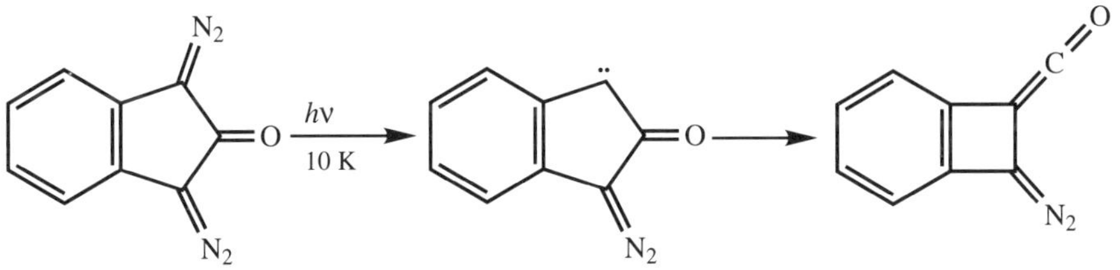

(38)

Ketenes generated by photo Wolff rearrangements have sometimes been trapped in cycloaddition reactions by alkenes.[80–82] Thus photolysis of ethyl diazoacetate (**33**) in E or Z-2-butene, or 2-methylpropene, gave 16–18% yields of the stereospecifically formed [2 + 2] cycloaddition product of the ketene **35** with the alkene (equations 29 and 39).[69,80] Products attributed to the addition or insertion of the intermediate carboethoxycarbene were also observed. Thermolysis or triplet sensitization of **33** led to either singlet or triplet carbene which did not form detectable ketene.[80] It appears that rearrangement of the carbene to EtOCH=C=O is relatively slow, thus allowing carbene reactions to compete rather effectively with ketene formation. Pyrolysis of **33** above 210 °C gave **35**, which could be trapped by alkynes to give cyclobutenones.[70]

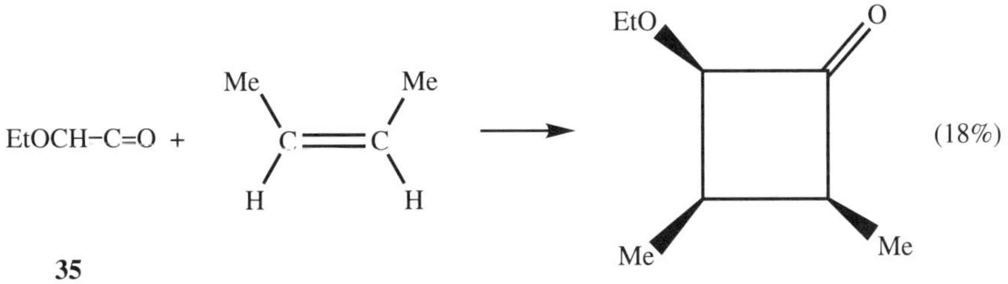

(39)

Photolysis **33** in MeOH leads to a comparable yield of EtOCH=C=O to that in EtOH, as indicated by the yield of ester (equation 40).[81] Phenyl diazoacetate gives a somewhat greater percentage of rearrangement (equation 41), and amino group rearrangement also occurs as indicated in equation 42.[81,82]

$$N_2CHCO_2Et \xrightarrow[MeOH]{h\nu} EtOCH_2CO_2Me + MeOCH_2CO_2Et \quad (40)$$
$$\mathbf{33} \qquad\qquad (20\text{--}25\%) \qquad (75\text{--}80\%)$$

$$N_2CHCO_2Ph \xrightarrow[MeOH]{h\nu} PhOCH_2CO_2Me + MeOCH_2CO_2Ph \quad (41)$$
$$(45\text{--}60\%) \qquad (40\text{--}55\%)$$

$$N_2CHCONHMe \xrightarrow[H_2O]{h\nu} MeNHCH_2CO_2H + MeNHCOCH_2OH \quad (42)$$
$$(30\%) \qquad\qquad (70\%)$$

Ketenes formed by Wolff rearrangements have also been trapped with imines (equation 43)[83] and azo compounds (equation 44).[84] However ketenes derived from $N_2CHCOCO_2Et$ or N_2CHCO_2Et were not trapped in this way. This is probably due to the inefficiency of the ketene formation and competition from other pathways.

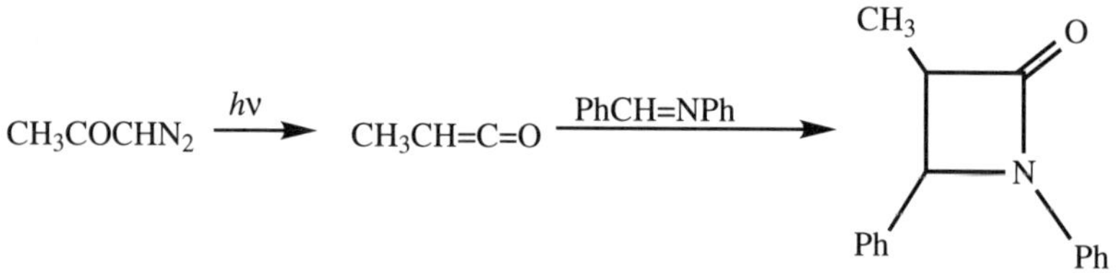

$$CH_3COCHN_2 \xrightarrow{h\nu} CH_3CH=C=O \xrightarrow{PhCH=NPh} \text{[β-lactam]} \quad (43)$$

$$PhCOCHN_2 \xrightarrow{h\nu} PhCH=C=O \xrightarrow{PhN=NPh} \text{[1,2-diazetidinone]} \quad (44)$$

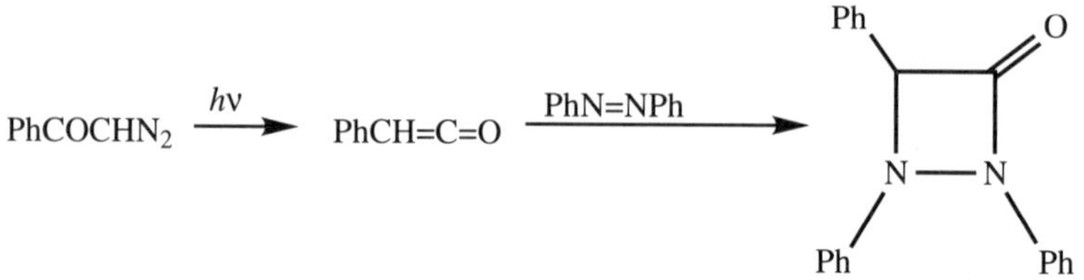

(Diacyl)azo compounds can react with two moles of ketene (equation 45).[84] This product could form by reaction of an initial zwitterion with a second mole of ketene as shown.

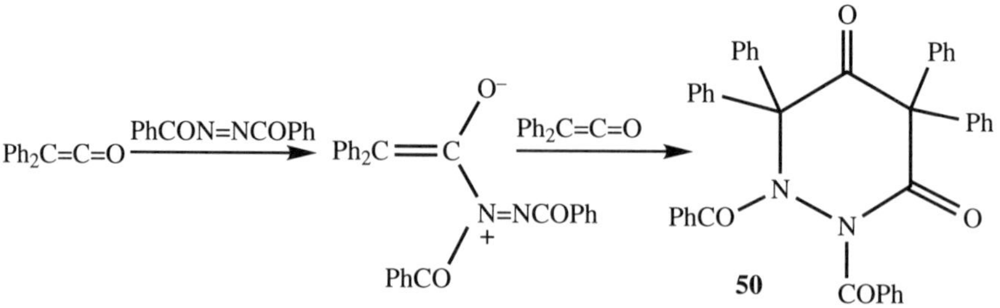

$$Ph_2C=C=O \xrightarrow{PhCON=NCOPh} Ph_2C=C(O^-)\text{-}N=NCOPh\ (+PhCO) \xrightarrow{Ph_2C=C=O} \mathbf{50} \quad (45)$$

3.3 KETENES FROM DIAZO KETONES (WOLFF REARRANGEMENTS) 91

Photolysis of methyl diazoacetate in benzene led to the tetramers **51** and **52**.[85] This was attributed[85] to formation of the cyclopropanone **53** which then formed the tetramers. Formation of **53** via the zwitterion **54** can be formulated as shown. Further examples of cyclopropanone formation from ketenes and diazo compounds are given in Section 5.10.1.

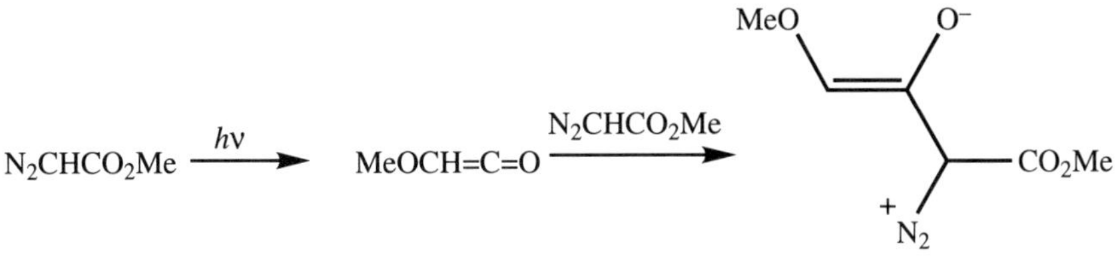

$$N_2CHCO_2Me \xrightarrow{h\nu} MeOCH=C=O \xrightarrow{N_2CHCO_2Me} \mathbf{54}$$

(46)

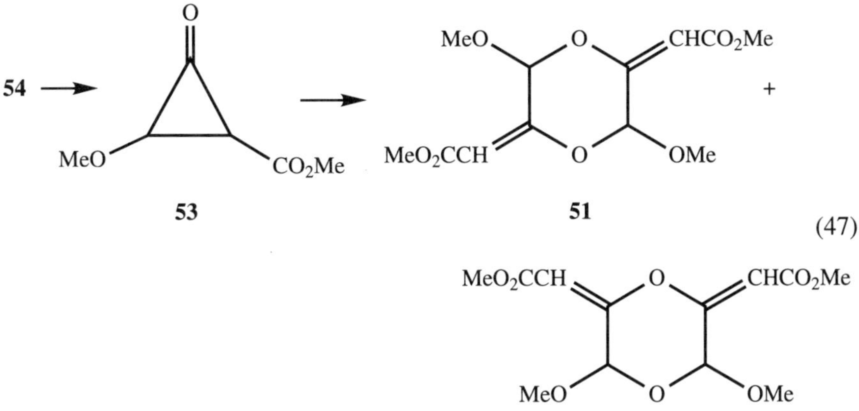

$$\mathbf{54} \longrightarrow \mathbf{53} \longrightarrow \mathbf{51} \; + \; \mathbf{52}$$

(47)

Pyrazoles may be formed upon diazoketone photolysis,[86] presumably by cyclization of zwitterions **55**.

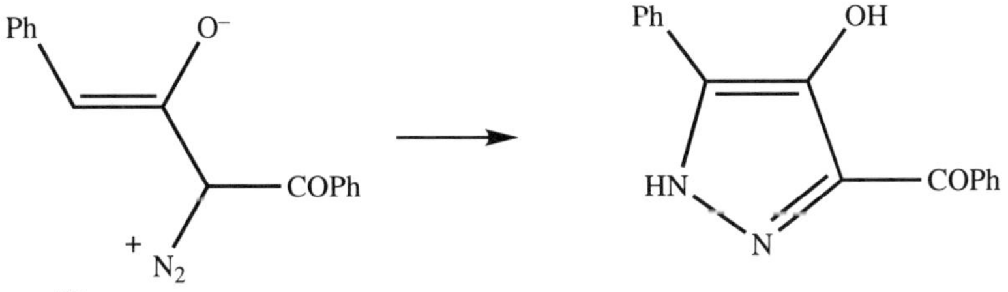

(48)

Many ketenes have presumably been generated in the Arndt–Eistert[6] systhesis and while not observed directly have been trapped by water or alcohols as the acid or ester, respectively.[87–94] One notable example is cubylketene **56** (equation 49).[87] Trapping with ammonium ion[95] or amines[96] to give amides has also been used.

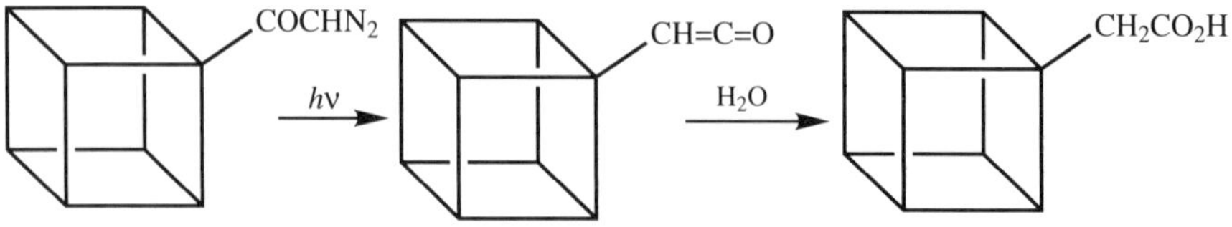

$$\text{Cubyl-COCHN}_2 \xrightarrow{h\nu} \text{Cubyl-CH=C=O} \xrightarrow{H_2O} \text{Cubyl-CH}_2\text{CO}_2\text{H}$$

56

(49)

A double Arndt–Eistert reaction on the bis(diazoketone) **57** was reported, but presumably involved sequential formation and trapping of the ketenes (equation 50).[97] Similar reactions were reported during studies directed to dodecahedrane synthesis.[98]

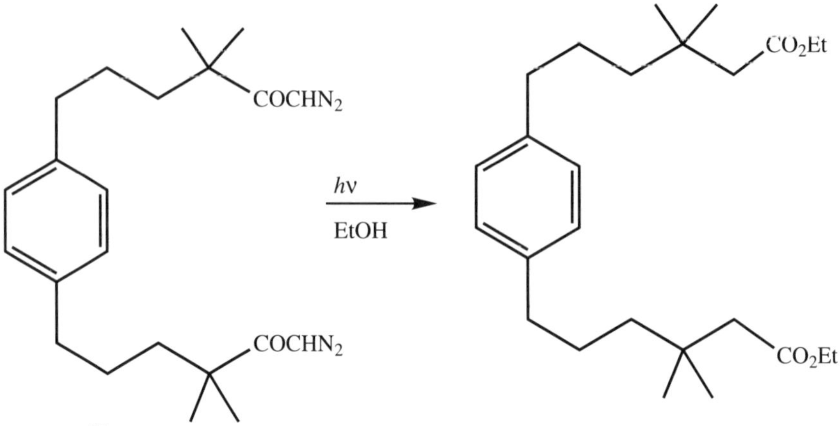

57 $\xrightarrow[\text{EtOH}]{h\nu}$ (bis-CO$_2$Et product)

(50)

Photolysis of the diacyldiazomethane **58** in a poly(methyl methacrylate) film led to the acylketene **59**, observed by a UV λ_{max} at about 280 nm.[99] The decay of the absorption was ascribed to hydration by adventitious H$_2$O to the acid (equation 51). Photolysis of **58** in the presence of H$_2$O or MeOH gave the acid and ester, respectively, as in equation 51, along with a minor amount of the diazirine **60**.[100]

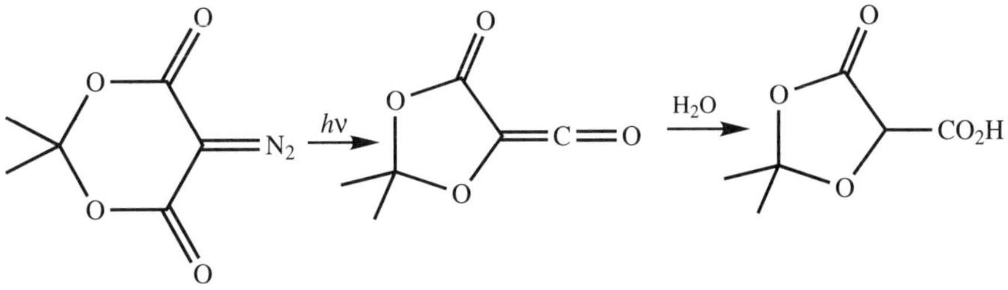

Ketene **59** reacted with 1,3,3-trimethylcyclohexene by [2 + 2] cycloaddition to give **61** (equation 52).[101] The structure of **61** was originally incorrectly misassigned as a cyclopropane formed from a ketocarbene.[102] Further photolysis of **59** to give further Wolff rearrangement to **62** has also been proposed (equation 53).[103]

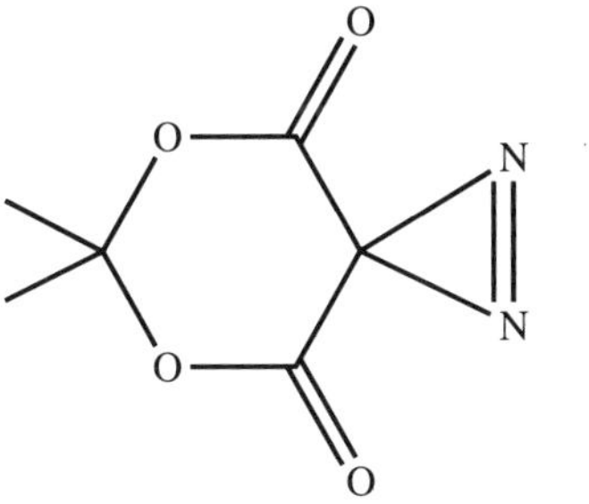

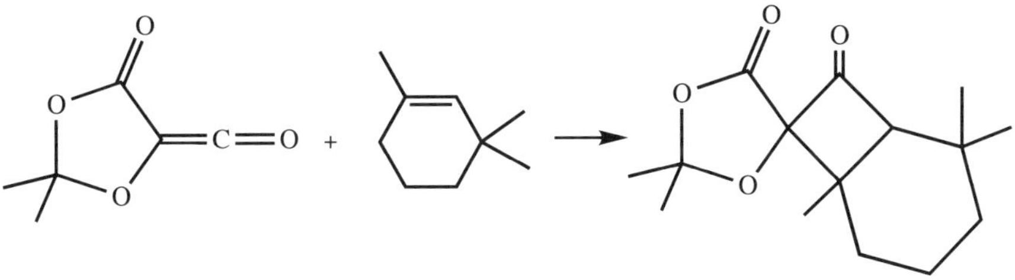

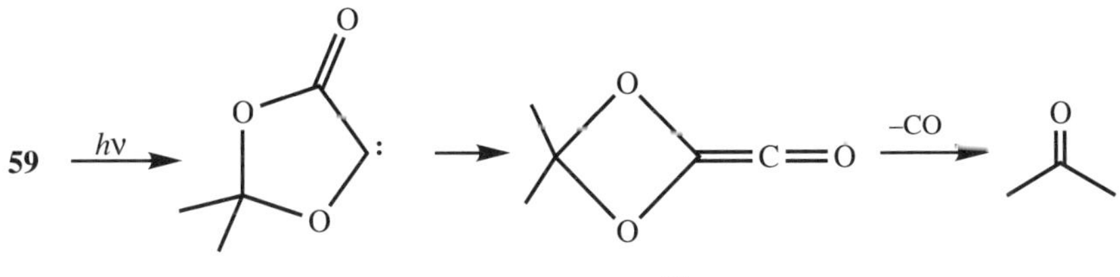

94 PREPARATION OF KETENES

Gas phase pyrolysis of the acyldiazo ketone **63** and addition of CH_3OH vapor gave capture of the ketene **64** as the ester **65**. Photochemical reaction of **63** in CH_3OH gave **66,** and this was proposed to occur by photochemical ring opening of **65** via a fulvenone intermediate.[104]

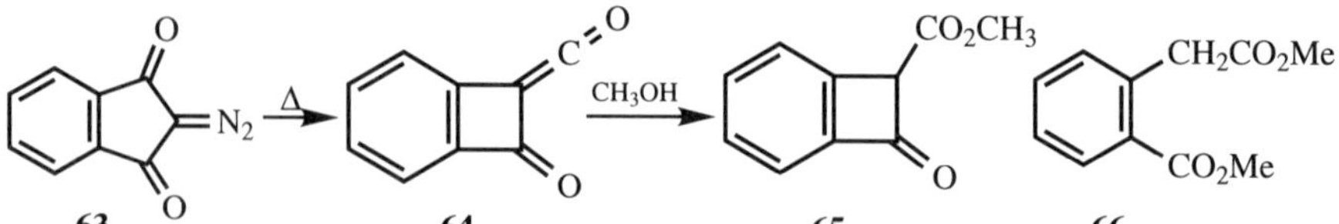

(54)

Photolysis of **67** gave the dimer **68** (equation 55).[105] Further examples of such acylketene-forming reactions are given in Section 4.1.6.[37]

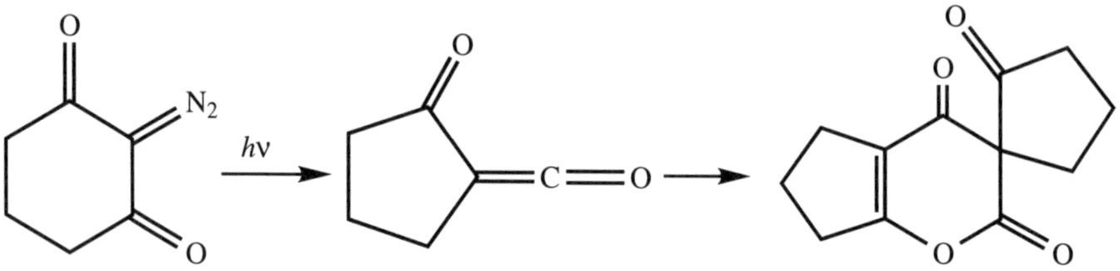

(55)

Migratory aptitudes of groups in the reaction of equation 56 were Me 1.0; Et, 0.96; Ph, 0.43; and *t*-Bu, 0.15.[106] Comparative migratory aptitudes of aryl and other groups in thermal and photochemical (direct and photosensitized) reactions, as well as the relative amount of capture of the presumed intermediate diacylcarbene, are given in Table 2.[107]

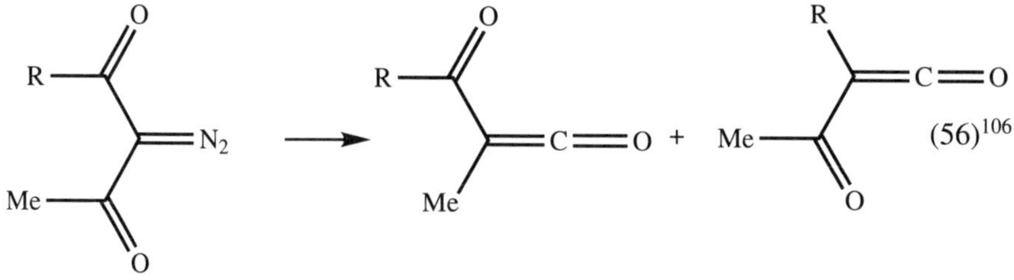

(56)[106]

The photolysis of 1,3-diacyldiazomethanes forming acylketenes has found applications in KrF excimer laser resists, and has been extensively examined for this purpose.[108]

Photolysis of the diazomalonate derivative **69** in CH_3OH gave a mixture of products **70–72**.[82] Similar products were obtained in EtOH or *i*-PrOH. The ketene **73** is a reasonable precursor to the ester **70**, but could not be trapped with styrene as the cyclobutanone, and so cyclization of the acyl carbene **75** to the α-lactam **74**

3.3 KETENES FROM DIAZO KETONES (WOLFF REARRANGEMENTS)

TABLE 2. Percentage Yields of Products from Diacyldiazomethanes[107]

$$\underset{\text{ArC-C-CR}}{\overset{O\ N_2\ O}{\underset{\|\ \|\ \|}{}}} \xrightarrow{\text{EtOH}} \text{ArCOCHRCO}_2\text{Et} + \text{RCOCHArCO}_2\text{Et} + \text{ArCOCHOEtCOR}$$

Ar	R				
Ph	H	Δ	>90.0	—	—
		hν	>90	—	—
Ph	Me	Δ	28	46.0	—
		hν	61	2.5	—
4-MeOC$_6$H$_4$	Me	Δ	18	56	7
		hν (direct)	40	13	23
		hν (sens.)	14.5	4.5	39
4-O$_2$NC$_6$H$_4$	Me	Δ	21	—	37
		hν	14.5	—	32

was suggested as a possibility. However, it is often difficult to trap ketenes from Wolff rearrangements in cycloadditions, and so formation of ketene **73** should not be discounted. Products **71** and **72** apparently form from the acyl carbene **75**.

$$\underset{\textbf{69}}{\overset{\text{EtO}_2\text{C}}{\underset{\text{PhHN}}{}}}\! \xrightarrow{h\nu} \underset{\textbf{75}}{\text{EtO}_2\text{C-C(:)-C(=O)-NHPh}} \longrightarrow \underset{\textbf{73}}{\text{EtO}_2\text{C, PhHN}\diagdown\text{C=C=O}} \longrightarrow \underset{\textbf{70}}{\text{EtO}_2\text{C-CHCO}_2\text{CH}_3\text{-NHPh}} \quad (57)$$

$$\textbf{75} \longrightarrow \underset{\textbf{71}}{\text{EtO}_2\text{CCH}_2\text{CONHPh}} + \underset{\textbf{72}}{\text{EtO}_2\text{CCH(OCH}_3)\text{CONHPh}} \quad (58)$$

$$\textbf{75} \longrightarrow \underset{\textbf{74}}{\underset{\text{PhN}}{\text{EtO}_2\text{C}\diagup\triangle\diagdown}\!\!=\!\!\text{O}} \quad (59)$$

The vinylogous Wolff rearrangement is a process that occurs for β,γ-unsaturated diazoketones in competition with normal Wolff rearrangement and has been proposed to occur by cyclization to a bicyclo[2.1.0]pentanone which opens to an unsaturated ketene (equations 60 and 61).[59,109-111] The same reaction has been reported when the intermediate has been generated by $Rh_2(OAc)_4$ catalyzed reaction of the diazoketone.[112]

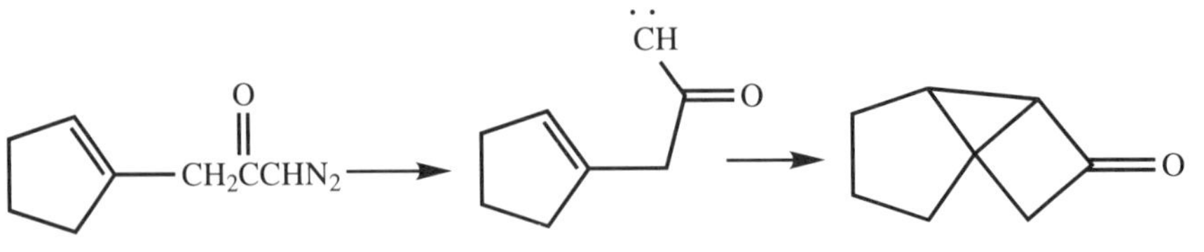

(60)

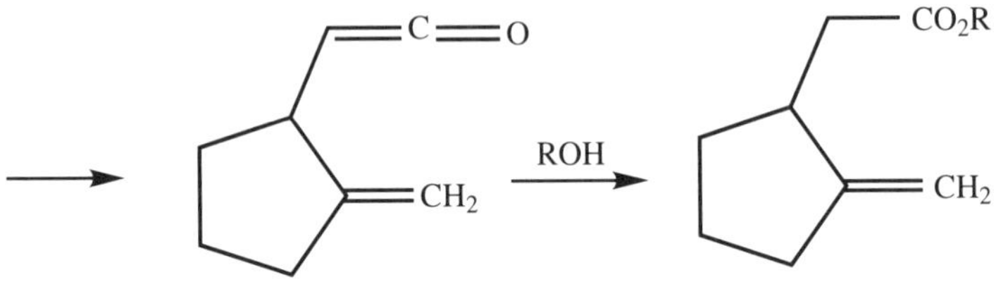

(61)

References

1. Wolff, L. *Liebigs Ann. Chem.* **1902**, *325*, 129–195.
2. Maas, G. *Methoden der Organischen Chemie,* 4th ed., Vol. 19b; *Carbene (Carbenoide);* M. Regitz, Ed., Thieme: Stuttgart, 1989; pp. 1231–1259.
3. Gill, G. B. In *Comprehensive Organic Synthesis,* Vol. 3; Pattenden, G., Ed.; Pergamon: Oxford, 1991; pp. 887–912.
4. Meier, H.; Zeller, K.-P. *Angew. Chem., Int. Ed. Engl.* **1975**, *14*, 32–43.
5. Regitz, M.; Maas, G. *Diazo Compounds;* Academic: Orlando, FL, 1986.
6. Eistert, B.; Regitz, M.; Heck, G.; Schwall, H. In *Methoden der Organischen Chemie,* Müller, E., ed., Vol. 10/4., Thieme: Stuttgart, 1968, pp. 473–893.
7. Bachmann, W. E.; Struve, W. S. *Organic Reactions* **1942**, *1*, 38–62.
8. Ando, W. In *The Chemistry of Diazonium and Diazo Groups;* Patai, S., Ed.; Wiley: New York, 1978; Vol. I, pp. 341–487.
9. Lewars, E. G. *Chem. Rev.* **1983**, *83*, 519–534.
10. Rodina, L. L.; Korobitsyna, I. K. *Russ. Chem. Rev.* **1967**, *36*, 260–272.
11. Chapman, O. L. *Pure Appl. Chem.* **1979**, *51*, 331–339.
12. Miranda, M. A.; Garcia, H. In *The Chemistry of Functional Groups. Supplement B: The Chemistry of Acid Derivatives;* Patai, S., Ed.; Wiley: New York, 1992, Vol. 2; pp. 1271–1394.
13. Danheiser, R. L.; Miller, R. F.; Brisbois, R. G.; Park, S. Z. *J. Org. Chem.* **1990**, *55*, 1959–1964.
14. Adams, J.; Spero, D. M. *Tetrahedron* **1991**, *47*, 1765–1808.
15. Smith, A. B., III; Dieter, R. K. *Tetrahedron,* **1981**, *37*, 2407–2439.
16. Burke, S. D.; Grieco, P. A. *Organic Reactions* **1979**, *26*, 361–475.

17. Padwa, A.; Krumpe, K. E.; *Tetrahedron* **1992**, *48*, 5385–5453.
18. Dave, V.; Warnhoff, E. W. *Organic Reactions* **1970**, *18*, 217–401.
19. McAllister, M. A.; Tidwell, T. T. *J. Org. Chem.* **1994**, *59*, 4506–4515.
20. Nikolaev, V. A.; Popik, V. V.; Korobitsyna, I. K. *Zh. Org. Khim.* **1991**, *27*, 505–521; *Engl. Transl.* **1991**, *27*, 437–450.
21. Kaplan, F.; Meloy, G. K. *J. Am. Chem. Soc.* **1966**, *88*, 950–956.
22. Newman, M. S.; Arkell, A. *J. Org. Chem.* **1959**, *24*, 385–387.
22a. Marfisi, C.; Verlaque, P.; Davidovics, G.; Pourcin, J.; Pizzala, L.; Aycard, J.-P.; Bodot, H. *J. Org. Chem.* **1983**, *48*, 533–537.
23. de Groot, A.; Boerma, J. A.; de Valk, J.; Wynberg, J. *J. Org. Chem.* **1968**, *33*, 4025–4029.
24. Rau, H.; Bokel, M. *J. Photochem. Photobiol. A* **1990**, *53*, 311–322.
25. Cormier, R. A.; Freeman, K. M.; Schnur, D. M. *Tetrahedron Lett.* **1977**, 2231–2234.
26. Cormier, R. A. *Tetrahedron Lett.* **1980**, *21*, 2021–2024.
27. Krantz, A. *J. Chem. Soc., Chem. Commun.* **1973**, 670–671.
28. Tomioka, H.; Okuno, H.; Izawa, Y. *J. Org. Chem.* **1980**, *45*, 5278–5283.
29. Nikolaev, V. A.; Popik, V. V. *Tetrahedron Lett.* **1992**, *33*, 4483–4486.
29a. Popic, V. V.; Nikoleav, V. A. *J. Chem. Soc., Perkin Trans. 2*, **1993**, 1791–1793.
30. Tanigaki, K.; Ebbesen, T. W. *J. Am. Chem. Soc.* **1987**, *109*, 5883–5884.
31. McMahon, R. J.; Chapman, O. L.; Hayes, R. A.; Hess, T. C.; Krimmer, H.-P. *J. Am. Chem. Soc.* **1985**, *107*, 7597–7606.
32. Bachmann, C.; N'Guessan, T. Y.; Debu, I.; Monnier, M.; Pourcin, J.; Aycard, J.-P.; Bodot, H. *J. Am. Chem. Soc.* **1990**, *112*, 7488–7497.
33. Maier, G.; Reisenauer, H. P.; Sayrac, T. *Chem. Ber.* **1982**, *115*, 2192–2201.
34. Murai, H.; Safarik, I.; Torres, M.; Strausz, O. P. *J. Am. Chem. Soc.* **1988**, *110*, 1025–1032.
35. Torres, M.; Ribo, J.; Clement, A.; Strausz, O. P. *Can. J. Chem.* **1983**, *61*, 996–998.
36. Torres, M.; Gosavi, R. K.; Lown, E. M.; Piotrkowski, E. J.; Kim, B.; Bourdelande, J. L.; Font, J.; Strausz, O. P. *Stud. Phys. Theor. Chem.* **1992**, 184–211; *Chem. Abstr.* **1993**, *118*, 123794z.
37. Leung-Toung, R.; Wentrup, C. *J. Org. Chem.* **1992**, *57*, 4850–4858.
38. Wang, F.; Winnik, M. A.; Peterson, M. R.; Csizmadia, I. G. *J. Mol. Struct. (Theochem)* **1991**, *232*, 203–210.
39. Bender, H.; Meutermans, W.; Qiao, G.; Sankar, I. V.; Wentrup, C., as cited in reference 40.
40. Nguyen, M. T.; Hajnal, M. R.; Ha, T.-K.; Vanquickenborne, L. G.; Wentrup, C. *J. Am. Chem. Soc.* **1992**, *114*, 4387–4390.
40a. Nguyen, M. T.; Hajnal, M. R.; Vanquickenborne, L. G. *J. Chem. Soc., Perkin Trans. 2* **1994**, 169–170.
41. Zeller, K.-P. *Angew. Chem., Int. Ed. Engl.* **1977**, *16*, 781–782.
42. Smith, L. I.; Hoehn, H. H. *Organic Syntheses* Col. Vol. III; Horning, E. C.; Ed.; Wiley, New York, 1955; pp. 356–358.
43. Concannon, P. W.; Ciabattoni, J. *J. Am. Chem. Soc.* **1973**, *95*, 3284–3289.
44. Regitz, M.; Rüter, J. *Chem. Ber.* **1969**, *102*, 3877–3890.
45. Maas, G.; Brückmann, R. *J. Org. Chem.* **1985**, *50*, 2801–2802.

46. Allen, A. D.; Andraos, J.; Kresge, A. J.; McAllister, M. A.; Tidwell, T. T. *J. Am. Chem. Soc.* **1992**, *114*, 1878–1879.
47. Barra, M.; Fisher, T. A.; Cernigliaro, G. J.; Sinta, R.; Scaiano, J. C. *J. Am. Chem. Soc.* **1992**, *114*, 2630–2634.
48. Torres, M.; Bourdelande, J. L.; Clement, A.; Strausz, O. P. *J. Am. Chem. Soc.* **1983**, *105*, 1698–1700.
49. Laganis, E. D.; Janik, D. S.; Curphey, T. J.; Lemal, D. M. *J. Am. Chem. Soc.* **1983**, *105*, 7457–7459.
50. Yates, P.; Clark, T. J. *Tetrahedron Lett.* **1961**, 435–439.
51. Huisgen, R.; Binsch, G.; Ghosez, L. *Chem. Ber.* **1964**, *97*, 2628–2639.
52. Regitz, M.; Bartz, W. *Chem. Ber.* **1970**, *103*, 1477–1485.
53. Bartz, W.; Regitz, M. *Chem. Ber.* **1970**, *103*, 1463–1476.
54. Regitz, M.; Förster, U.; unpublished results cited in reference 5, pp. 80, 81.
55. Jugelt, W.; Schmidt, D. *Tetrahedron* **1969**, *25*, 969–984.
56. Melzer, A.; Jenny, E. F. *Tetrahedron Lett.* **1968**, 4503–4506.
57. Kropf, H.; Reichwaldt, R. *J. Chem. Res. (S)*, **1992**, 284–285.
58. Lee, V.; Newman, M. S. *Organic Syntheses*, Coll. Vol. VI, Noland, W. E.; Ed.; Wiley, New York, 1988; pp. 613–615.
59. Smith, A. B. III; Toder, B. H.; Branca, S. J. *J. Am. Chem. Soc.* **1984**, *106*, 3995–4001.
60. Csizmadia, I. G.; Font, J.; Strausz, O. P. *J. Am. Chem. Soc.* **1968**, *90*, 7360–7361.
61. Thornton, D. E.; Gosavi, R. K.; Strausz, O. P. *J. Am. Chem. Soc.* **1970**, *92*, 1768–1769.
62. Zeller, K.-P. *Chem. Ber.* **1979**, *112*, 678–688.
63. Timm, U.; Keller, K.-P.; Meier, H. *Chem. Ber.* **1978**, *111*, 1549–1557.
64. Tomioka, H.; Okuno, H.; Kondo, S.; Izawa, Y. *J. Am. Chem. Soc.* **1980**, *102*, 7123–7125.
65. Fowler, J. E.; Galbraith, J. M.; Vacek, G.; Schaefer, H. F., III *J. Am. Chem. Soc.* **1994**, *116*, 9311–9319.
65a. Blocher, A.; Zeller, K.-P. *Chem. Ber.* **1994**, *127*, 551–555.
65b. Turecek, F.; Drinkwater, D. E.; McLafferty, F. W. *J. Am. Chem. Soc.* **1991**, *113*, 5958–5964.
66. Jones, M., Jr.; Ando, W. *J. Am. Chem. Soc.* **1968**, *90*, 2200–2201.
67. Padwa, A.; Layton, R. *Tetrahedron Lett.* **1965**, 2167–2170.
68. Tomioka, H.; Hirai, K.; Tabayashi, K.; Murata, S.; Izawa, Y.; Inagaki, S.; Okajima, T. *J. Am. Chem. Soc.* **1990**, *112*, 7692–7702.
69. DoMinh, T.; Strausz, O. P.; Gunning, H. E. *J. Am. Chem. Soc.* **1969**, *91*, 1261–1263.
70. Shapiro, E. A.; Dolgii, I. E.; Nefedov, O. M. *Izv. Akad. Nauk SSSR, Ser. Khim.* **1980**, 2096–2102; *Chem. Abstr.* **1981**, *94*, 83858a.
71. Chiang, Y.; Kresge, A. J.; Pruszynski, P.; Schepp, N. P.; Wirz, J. *Angew. Chem., Int. Ed. Engl.* **1991**, *30*, 1366–1368.
72. Trost, B. M.; Kinson, P. L. *J. Am. Chem. Soc.* **1975**, *97*, 2438–2449.
73. Ueda, K.; Toda, F. *Chem. Lett.* **1975**, 779–780.
74. Bond, F. T.; Ho, C. *J. Org. Chem.* **1976**, *41*, 1421–1425.
75. Kirmse, W. *Angew. Chem.* **1959**, *71*, 537–541.
76. Borch, R. F.; Fields, D. L. *J. Org. Chem.* **1969**, *34*, 1480–1483.
77. Trost, B. M.; Whitman, P. J. *J. Am. Chem. Soc.* **1974**, *96*, 7421–7429.

78. Chapman, O. L.; Gano, J.; West, P. R.; Regitz, M.; Maas, G. *J. Am. Chem. Soc.* **1981**, *103*, 7033–7036.
79. Murata, S.; Yamamoto, T.; Tomioka, H. *J. Am. Chem. Soc.*, **1993**, *115*, 4013–4023.
80. DoMinh, T.; Strausz, O. P. *J. Am. Chem. Soc.* **1970**, *92*, 1766–1768.
81. Chamovitch, H.; Vaughn, R. J.; Westheimer, F. H. *J. Am. Chem. Soc.* **1968**, *90*, 4088–4093.
82. Buu, N. T.; Edward, J. T. *Can. J. Chem.* **1972**, *50*, 3719–3728.
83. Kirmse, W.; Horner, L. *Chem. Ber.* **1956**, *89*, 2759–2765.
84. Horner, L.; Spiestchka, E. *Chem. Ber.* **1956**, *89*, 2765–2768.
85. Schenck, G. O.; Ritter, A. *Tetrahedron Lett.* **1968**, 3189–3190.
86. Takebayashi, M.; Ibata, T. *Bull Chem. Soc. Jpn.* **1968**, *41*, 1700–1707.
87. Eaton, P. E.; Yip, Y. C. *J. Am. Chem. Soc.* **1991**, *113*, 7692–7697.
88. Gibson, T.; Erman, W. F. *J. Org. Chem.* **1966**, *31*, 3028–3032.
89. Adams, J.; Frenette, R.; Bettey, M.; Chibante, F.; Springer, J. P. *J. Am. Chem. Soc.* **1987**, *109*, 5432–5437.
90. Rao, V. B.; George, C. F.; Wolff, S.; Agosta, W. C. *J. Am. Chem. Soc.* **1985**, *107*, 5732–5739.
91. Press, J. B.; Schechter, H. *J. Org. Chem.* **1975**, *40*, 2446–2458.
92. van Haard, P. M. M.; Thijs, L.; Zwanenburg, B. *Tetrahedron Lett.* **1975**, 803–806.
93. Sonawane, H. R.; Bellur, N. S.; Ahuja, J. R.; Kulkarni, D. G. *J. Org. Chem.* **1991**, *56*, 1434–1439.
94. Becker, D.; Harel, Z.; Birnbaum, D. *J. Chem. Soc., Chem. Commun.* **1975**, 377.
95. Eaton, P. E.; Jobe, P. G.; Reingold, I. D. *J. Am. Chem. Soc.* **1984**, *106*, 6437–6439.
96. Cossy, J.; Belotti, D.; Leblanc, C. *J. Org. Chem.* **1993**, *58*, 2351–2354.
97. Gleiter, R.; Krämer, R.; Irnagartinger, H.; Bissinger, C. *J. Org. Chem.* **1992**, *57*, 252–258.
98. Melder, J.-P.; Pinkos, R.; Fritz, H.; Wörth, J.; Prinzbach, H. *J. Am. Chem. Soc.* **1992**, *114*, 10213–10231.
99. Winnik, M. A.; Wang, F.; Nivaggioli, T.; Hruska, Z.; Fukumura, H.; Masuhara, H. *J. Am. Chem. Soc.* **1991**, *113*, 9702–9704.
100. Nikolaev, V. A.; Khimich, N. N.; Korobitsyna, I. K. *Khim. Geterotsikl. Soedin.* **1985**, 321–325; *Chem. Abstr.* **1985**, *103*, 87818q.
101. Stevens, R. V.; Bisacchi, G. S.; Goldsmith, L.; Strouse, C. E. *J. Org. Chem.* **1980**, *45*, 2708–2709.
102. Livinghouse, T.; Stevens, R. V. *J. Am. Chem. Soc.* **1978**, *100*, 6479–6482.
103. Kammula, S. L.; Tracer, H. L.; Shevlin, P. B.; Jones, M., Jr. *J. Org. Chem.* **1977**, *42*, 2931–2932.
104. Cava, M. P.; Spangler, R. J. *J. Am. Chem. Soc.* **1967**, *89*, 4550–4551.
105. Stetter, H.; Kiehs, K. *Chem. Ber.* **1965**, *98*, 1181–1187.
106. Nikolaev, V. A.; Kotok, S. D.; Korobitsyna, I. K. *Zh. Org. Khim.* **1974**, *10*, 1334–1335; *Engl. Transl.* **1974**, *10*, 1343–1344.
107. Zeller, K. P.; Meier, H.; Müller, E. *Tetrahedron* **1972**, *28*, 5831–5838.
108. Horiguchi, R.; Onishi, Y.; Hayase, S. *J. Electrochem. Soc.* **1990**, *137*, 3561–3568.
109. Smith, A. B., III; Toder, B. H.; Richmond, R. E.; Branca, S. J. *J. Am. Chem. Soc.* **1984**, *106*, 4001–4009.

110. Lokensgard, J. P.; O'Dea, J.; Hill, E. A. *J. Org. Chem.* **1974**, *39*, 3355–3357.
111. Wilds, A. L.; Von Trebra, R. L.; Woolsey, N. F. *J. Org. Chem.* **1969**, *34*, 2401–2406.
112. Motallebi, S.; Müller, P. *Chimia* **1992**, *46*, 119–122.

3.4 KETENES BY PHOTOCHEMICAL AND THERMOLYTIC METHODS

3.4.1 Ketenes from Cyclobutanones and Cyclobutenones

Ketenes are formed from cyclobutanones and cyclobutenones by routes that are often reversible, and this general topic has been recently reviewed.[1] Ketene is formed by the thermolysis[2] or photolysis[3] of cyclobutanone, while with 2,3-dimethylcyclobutanone the thermolysis reaction is highly stereoselective at 325 °C (equation 1).[4] With activating groups such as acyl (equation 2)[5] or vinyl[6] the reactions are much more facile, but the intermediacy of ketenes in these latter processes has not been established.[6] Strained cyclobutanones also cleave readily to ketenes (equation 3).[7] Other examples of these reactions have been reviewed,[8] including preparatively useful bicyclic examples.[9]

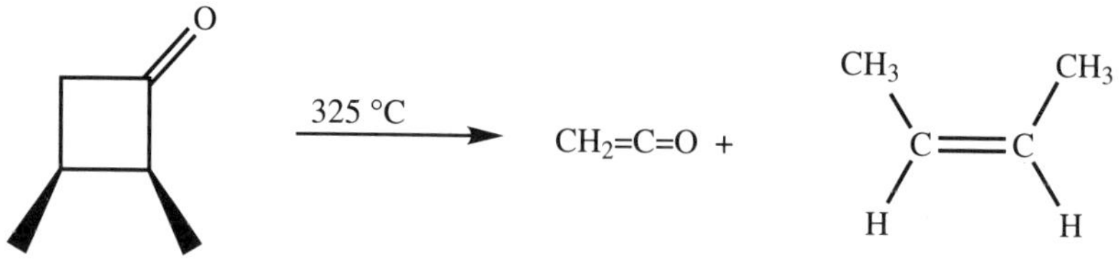

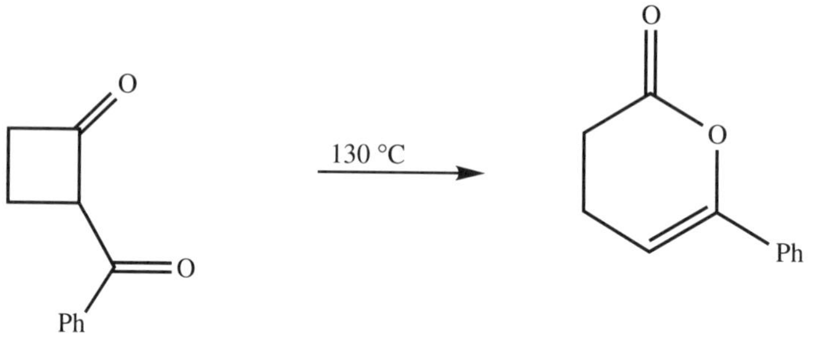

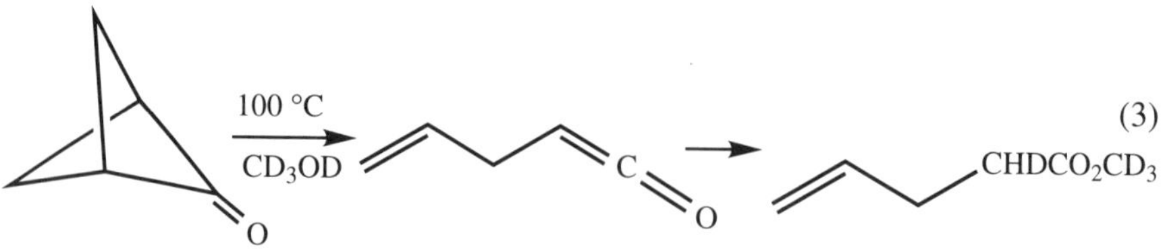

The kinetics of the decomposition of cyclobutanones including **4** in the gas phase from 192 to 285 °C have been studied.[10–15] These cyclobutanones cleave in two directions to give different pairs of products. The results were interpreted in terms of zwitterionic transition states **5** and **6** resembling $[\pi 2_s + \pi 2_a]$ transition states for ketene formation (equation 4). There is a large 11.5 kcal/mol stabilization of the transition state by the ethoxy group, and the results appear consistent with a stepwise process with **5** and **6** as reaction intermediates.

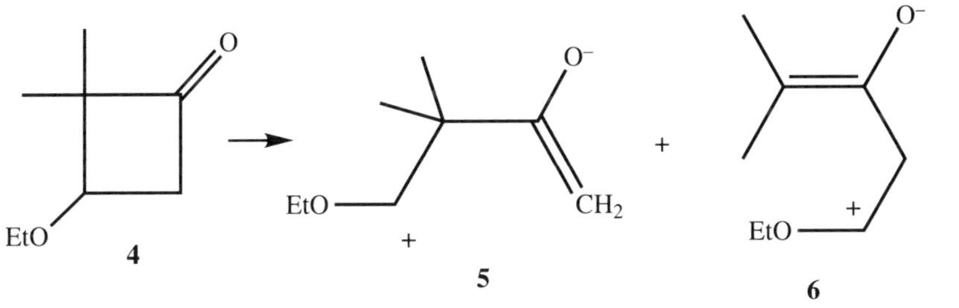

(4)

The thermal and photochemical cleavage of cyclobutanones has been proposed to proceed by different pathways, as illustrated by the products from 2,2,3-triphenylcyclclobutanone shown in equations 5 and 6.[16] The formation of isomeric biradicals was proposed for these reactions as shown,[16] but alternative paths involving zwitterionic intermediates analogous to **5** and **6** cannot be excluded.

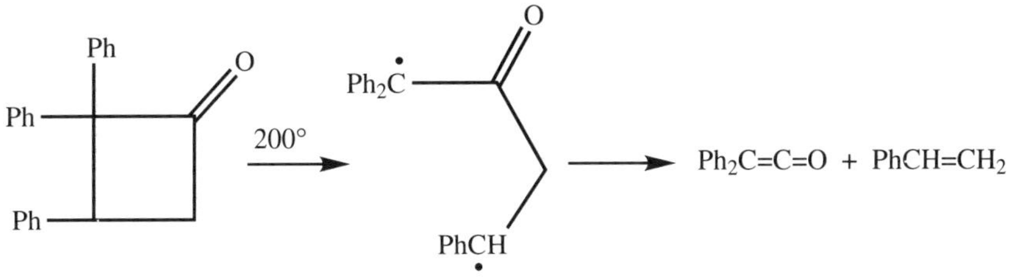

(5)

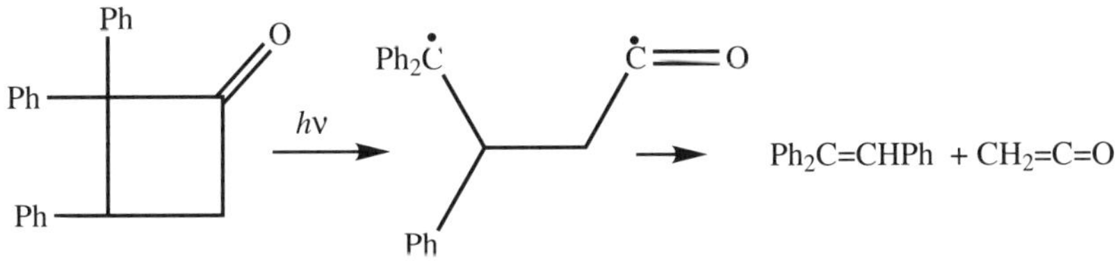

(6)

The thermal isomerization of the diphenylketene-tropone cycloadduct **7** has been proposed to occur by initial cleavage to a diradical intermediate in a process with substantial homolytic character (equation 7),[16] but a zwitterionic intermediate appears more plausible.

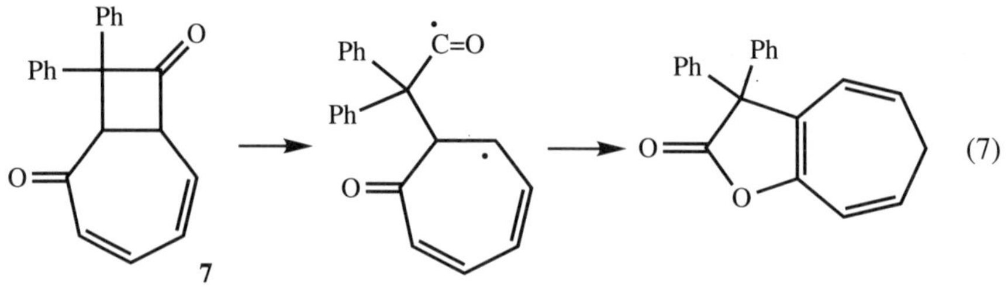

102 PREPARATION OF KETENES

(7)

In photolysis of bicyclo[3.2.0]heptan-6-one systems such as **8** not only are the products from trapping of ketenes observed, but also products from ring expansion (equation 8).[17]

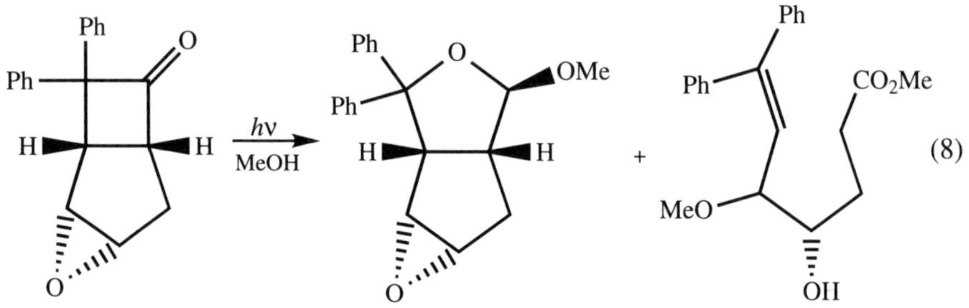

(8)

Arndt–Eistert reaction of **9** gave 36–38% of the normal products **10**, and 12–30% of products **11** (equation 9) suggested to arise from formation of **12** and thermal ring opening to ketene **13** (equation 10).[18] This reaction is an aromatic analogue of the vinylogous Wolff rearrangement discussed in Section 3.3.2.

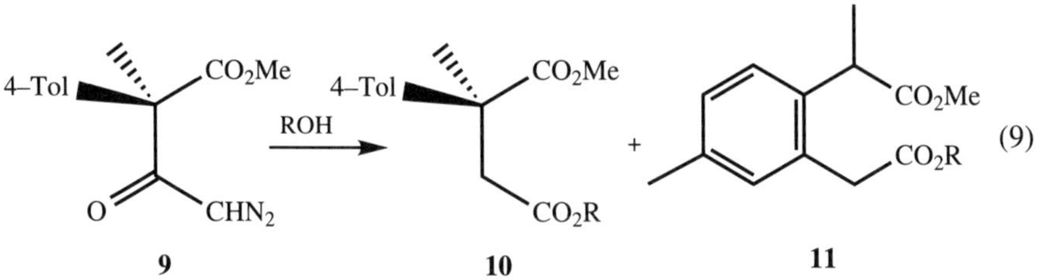

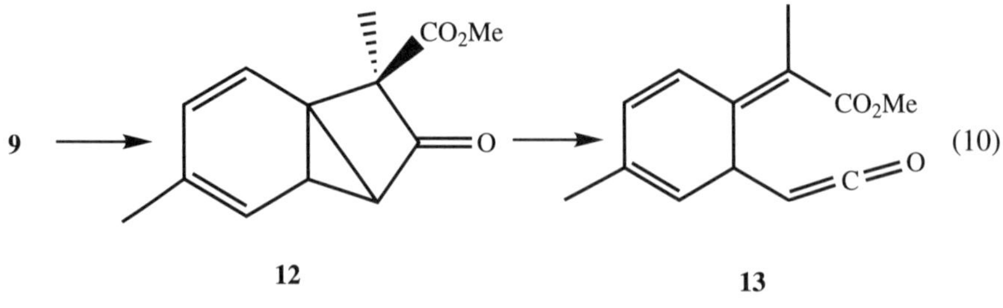

(9)

(10)

It was found by Jenny and Roberts in 1956[19] that optically active 2,4-dichloro-3-phenylcyclobutenone (**14**) undergoes racemization on heating in CHCl$_3$ at rates that can be monitored conveniently at 100 °C. It was established that this racemization occurred by reversible electrocyclic ring opening to the ketene **15**. When the reaction was conducted in ethanol the ketene was trapped as the ethyl ester **16** (equation 11). Another example of ring closure to a chlorinated cyclobutenone is shown in Section 4.1.2 (equation 8).

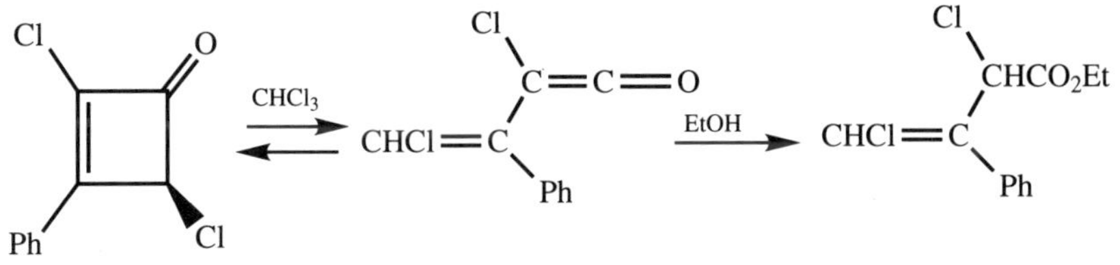

$$14 \underset{CHCl_3}{\rightleftharpoons} 15 \xrightarrow{EtOH} 16 \quad (11)$$

It was later found[20] that thermolysis of **14** gave Z-**16** with *cis* Cl and phenyl, while photolysis of **14** gave E-**16**. Other mechanistic studies of thermal and photochemical generation of ketenes from cyclobutenones have been reported.[21]

2-Methyl-4,4-diphenylcyclobutenone (**17**) forms esters **18** at 55 °C in alcoholic solvents, and the naphthol **19** in nonnucleophilic solvents (equations 12 and 13).[22,23] Both reactions are proposed to occur through rate-limiting formation of the ketene **20**, and the rather small dependence of the rate on solvent polarity, as measured by E_T values, with lower rates in more polar solvents (Table 1), suggests a slight decrease in the polarity of the substrate on going to the transition state.[23]

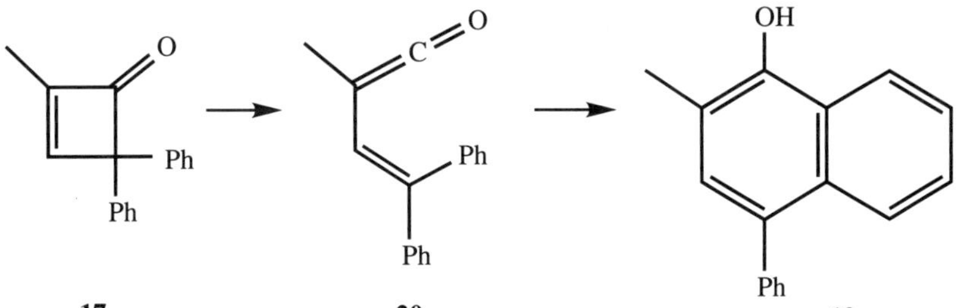

$$17 \rightarrow 20 \rightarrow 19 \quad (12)$$

$$20 \xrightarrow{ROH} Ph_2C=CHCHMeCO_2R \quad (13)$$

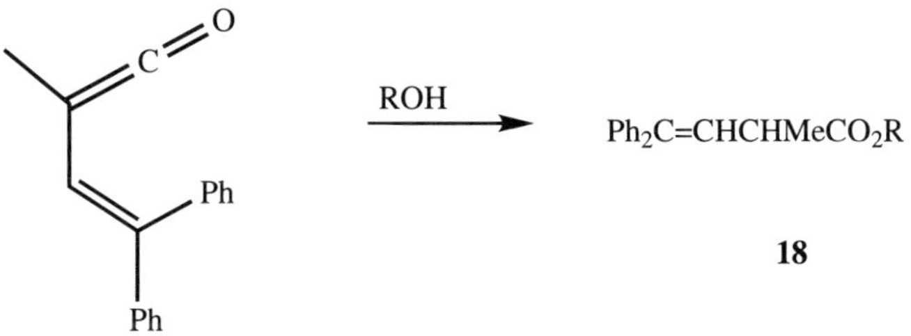

TABLE 1. Solvent Effects on Rates of Ring Opening of 17[22,23]

Solvent	E_T	$k \times 10^4, s^{-1}$ (55 °C)
Cyclohexane	31.2	5.36
Benzene	34.6	4.14
EtOAc	38.1	2.53
DMF	43.8	1.75
HOAc	51.1	1.83
EtOH	51.9	2.01
MeOH	55.5	1.78

There are major structural effects on the rate of cyclobutenone to vinylketene ring opening (equation 14), as shown (Table 2) along with some extrapolated rates for cyclobutenes.[22] These show strong rate accelerations by phenyls which can conjugate with the newly formed vinyl group, and some steric retarding effects. Interestingly the rates of ring opening of 1,3,3-trimethylcyclobutene and 1,3,3-trimethylcyclobutenone are estimated to be similar.[22] Thus the replacement of two hydrogens on cyclobutene by a carbonyl group evidently has little effect on the reactivity.[22]

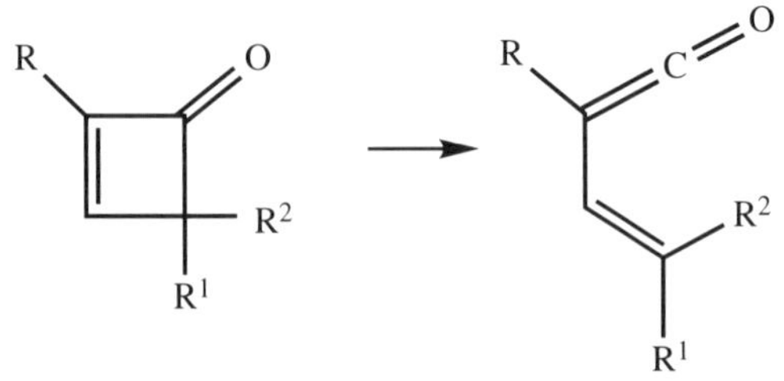

(14)

21

Many synthetic applications of ketenes formed by thermal cyclobutenone ring openings[24] have also been developed since it was discovered by Smith and Hoehn in 1939[25] that phenylcyclobutenones formed naphthols on heating. This reaction may be interpreted in terms of ring opening of the cyclobutenone **17** to form 4,4-diphenyl-2-methylvinylketene **20**, which undergoes intramolecular cycloaddition with the aryl group to give a naphthol (equation 12).[26] The thermolysis of cyclobutenones derived from cyclobutenediones has been particularly useful in the preparation of hydroquinones and quinones (equations 15 and 16).[27–43]

Heating of cyclobutenones in toluene in the presence of 1-triisopropylsilyloxyalkynes results in intermolecular [4 + 2] cycloaddition of the vinylketene intermediate leading to formation of resorcinols (equation 17).[44]

TABLE 2. Structural Effects on Rates of Ring Opening of 21[22]

R	R¹	R²	$k \times 10^6 (s^{-1})$ 80 °C
Me	Me	Me	1.1
Me	Me	Ph	90
Me	Et	Ph	33
Me	i-Pr	Ph	12.7
Me	Ph	Ph	2430
Et	Ph	Ph	3710
H	Ph	Ph	20400

$k \times 10^6$ (s⁻¹)	☐	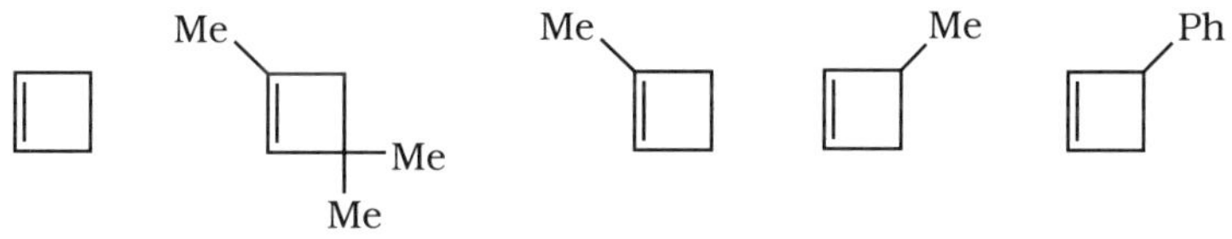 Me-cyclobutene with Me,Me	Me-cyclobutene	Me-cyclobutene	Ph-cyclobutene with Me
80 °C	0.093	1.4 (est)	0.012	1.0	200

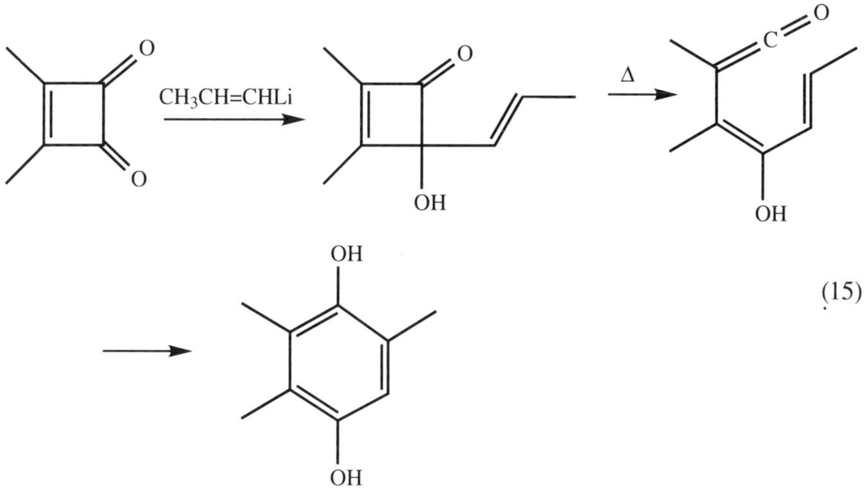

(15)

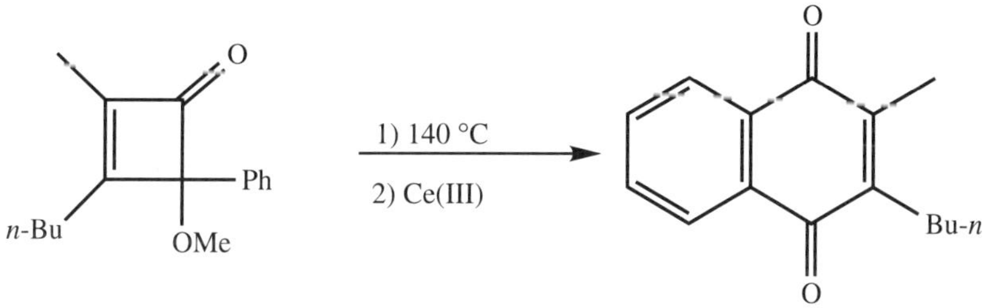

(16)

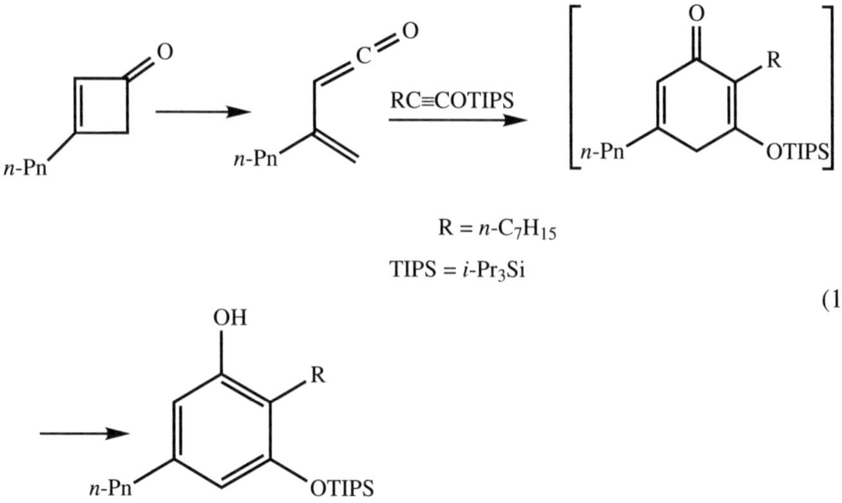

A novel route to substituted cyclobutenones involves photochemical Wolff rearrangement of aryl or vinyl diazoketones to give aryl or vinylketenes which react with alkoxyalkynes to give cyclobutenones, which spontaneously ring open and cyclize to phenols, as in the example of equation 18.[45]

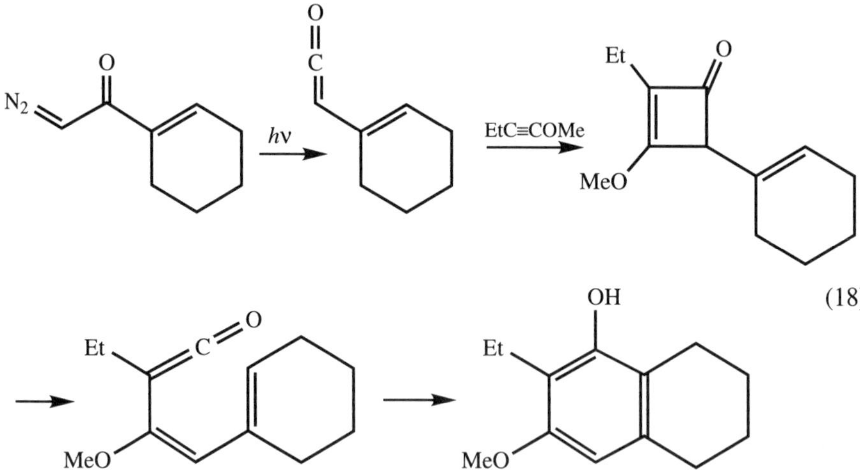

Rearrangement of the fused cyclobutanone **23** formed from homoketonization of 4-methyl-1-hydroxycubane **22** is reported to give the ketene **24** (equation 19).[46]

                                                                    (19)

References

1. Lee-Ruff, E. *Adv. Strain Org. Chem.* **1991**, *1*, 167–213.
2. Das, M. N.; Kern, F.; Coyle, T. D.; Walters, W. D. *J. Am. Chem. Soc.* **1954**, *76*, 6271–6274.
3. Benson, S. W.; Kistiakowsky, G. B. *J. Am. Chem. Soc.* **1942**, *64*, 80–86.
4. Carless, H. A. J.; Lee, E. K. C. *J. Am. Chem. Soc.* **1970**, *92*, 4482.
5. Huet, F.; Lechevallier, A.; Conia, J. M. *Chem. Lett.* **1981**, 1515–1519.
6. Huston, R.; Rey, M.; Dreiding, A. S. *Helv. Chim. Acta* **1982**, *65*, 451–461.
7. Sponsler, M. D.; Dougherty, D. A. *J. Org. Chem.* **1984**, *49*, 4978–4984.
8. Bellus, D.; Ernst, B. *Angew. Chem., Int. Ed. Engl.* **1988**, *27*, 797–827.
9. Rahman, S. S.; Wakefield, B. J.; Roberts, S. M.; Dowle, M. D. *J. Chem. Soc., Chem. Commun.* **1989**, 303–305.
10. Egger, K. W. *J. Am. Chem. Soc.* **1973**, *95*, 1745–1750.
11. Cocks, A. T.; Egger, K. W. *J. Chem. Soc., Perkin Trans. 2*, **1973**, 835–838.
12. Egger, K. W.; Cocks, A. T. *J. Chem. Soc., Perkin Trans. 2*, **1972**, 211–214.
13. Egger, K. W.; Cocks, A. T. *J. Chem. Soc., Perkin Trans. 2*, **1972**, 2014–2018.
14. McGee, T. H.; Schleifer, A. *J. Phys. Chem.* **1972**, *76*, 963–967.
15. Schiess, P.; Fünfschilling, P. *Tetrahedron Lett.* **1972**, 5191–5194.
16. Kende, A. S. *Tetrahedron Lett.* **1967**, 2661–2666.
17. Davies, H. G.; Rahman, S. S.; Roberts, S. M.; Wakefield, B. J.; Winders, J. A. *J. Chem. Soc., Perkin Trans. 1*, **1987**, 85–89.
18. Canet, J.-L.; Fadel, A.; Salaün, J. *J. Org. Chem.* **1992**, *57*, 3463–3473.
19. Jenny, E. F.; Roberts, J. D. *J. Am. Chem. Soc.* **1956**, *78*, 2005–2009.
20. Baldwin, J. E.; McDaniel, M. C. *J. Am. Chem. Soc.* **1968**, *90*, 6118–6124.
21. Chapman, O. L.; Lassila, J. D. *J. Am. Chem. Soc.* **1968**, *90*, 2449–2450.
22. Mayr, H.; Huisgen, R. *J. Chem. Soc., Chem. Commun.*, **1976**, 57–58.
23. Huisgen, R.; Mayr, H. *J. Chem. Soc., Chem. Commun.*, **1976**, 55–56.
24. Moore, H. W.; Yerxa, B. R. *Chemtracts: Org. Chem.* **1992**, *5*, 273–313.
25. Smith, L. I.; Hoehn, H. H. *J. Am. Chem. Soc.* **1939**, *61*, 2619–2624.
26. Mayr, H. *Angew. Chem. Int. Ed. Engl.* **1975**, *14*, 500–501.
27. Perri, S. T.; Moore, H. W. *J. Am. Chem. Soc.* **1990**, *112*, 1897–1905.
28. Xu, S. L.; Moore, H. W. *J. Org. Chem.* **1989**, *54*, 6018–6021.
29. Xu, S. L.; Moore, H. W. *J. Org. Chem.* **1992**, *57*, 326–338.
30. Xu, S. L.; Taing, M.; Moore, H. W. *J. Org. Chem.* **1991**, *56*, 6104–6109.

31. Xu, S. L.; Xia, H.; Moore, H. W. *J. Org. Chem.* **1991**, *56*, 6094–6103.
32. Heerding, J. M.; Moore, H. W. *J. Org. Chem.* **1991**, *56*, 4048–4050.
33. Chow, K.; Moore, H. W. *J. Org. Chem.* **1990**, *55*, 370–372.
34. Xia, H.; Moore, H. M. *J. Org. Chem.* **1992**, *57*, 3765–3766.
35. Gayo, L. M.; Winters, M. P.; Moore, H. W. *J. Org. Chem.* **1992**, *57*, 6896–6899.
36. Lee, K. H.; Moore, H. W. *Tetrahedron Lett.* **1993**, *34*, 235–238.
37. Karlsson, J. O.; Nguyen, N. V.; Foland, L. D.; Moore, H. W. *J. Am. Chem. Soc.* **1985**, *107*, 3392–3393.
38. Leibeskind, L. S.; Iyer, S.; Jewell, C. F., Jr. *J. Org. Chem.* **1986**, *51*, 3065–3067.
39. Liebeskind, L. S. *Tetrahedron* **1989**, *45*, 3053–3060.
40. Liebeskind, L. S.; Granberg, K. L.; Zhang, J. *J. Org. Chem.* **1992**, *57*, 4345–4352.
41. Karabelas, K.; Moore, H. W. *J. Am. Chem. Soc.* **1990**, *112*, 5372–5373.
42. Danheiser, R. L.; Gee, S. K. *J. Org. Chem.* **1984**, *49*, 1672–1674.
43. Pollart, D. J.; Moore, H. W. *J. Org. Chem.* **1989**, *54*, 5444–5448.
44. Kowalski, C. J.; Lal, G. S. *J. Am. Chem. Soc.* **1988**, *110*, 3693–3695.
45. Danheiser, R. L.; Brisbois, R. G.; Kowalczyk, J. J.; Miller, R. T. *J. Am. Chem. Soc.* **1990**, *112*, 3093–3100.
46. Hormann, R. E. Ph.D. Dissertation, University of Chicago, cited by Eaton, P. E.; Yip, Y. C. *J. Am. Chem. Soc.* **1991**, *113*, 7692–7697.

3.4.2 Ketenes from Photolysis of Cycloalkanones and Enones

The photolysis of a cycloalkanone can lead by Norrish Type I cleavage to a diradical which can then react intramolecularly by hydrogen atom transfer to give either a ketene **1** or an alkenylaldehyde, depending on the direction of the hydrogen transfer (equation 1).[1–3]

Deuterium labeling studies of methylcyclohexanones led to the conclusion that there was essentially free rotation in the intermediate biradical.[4] Cyclopentanones give very little ketene product, while the yields of ketenes relative to aldehydes are higher for cyclohexanones.[5–7] Formation of both of the products are suppressed by piperylene and so the α-cleavage of the ketones occurs from a triplet state.[5,6]

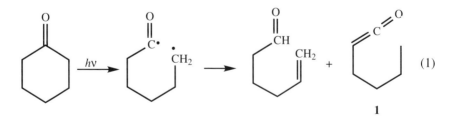

For 4-tert-butylcyclohexanone (**2**) photolysis in MeOH, little or no aldehyde was observed, and the ester from trapping of the ketene **4** was the major product. This was attributed to a conformational preference for hydrogen atom transfer by transition state **3** leading to ketene **4** (equation 2) as opposed to **5** leading to the aldehyde **6**,[6] in which there is an interaction of the *tert*-butyl group with the developing CH_2 (equation 3).[6]

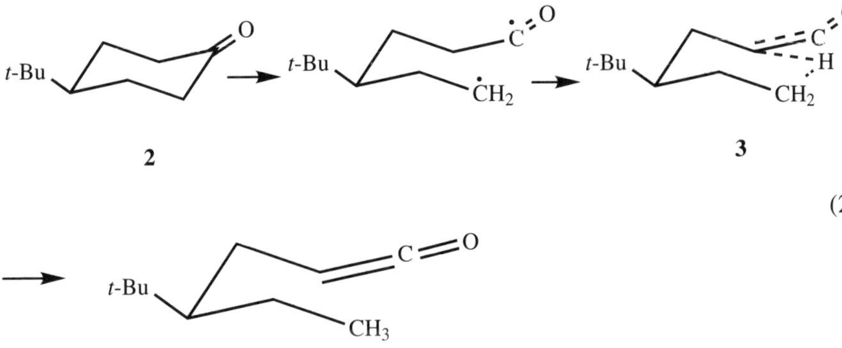

3.4 KETENES BY PHOTOCHEMICAL AND THERMOLYTIC METHODS

(2)

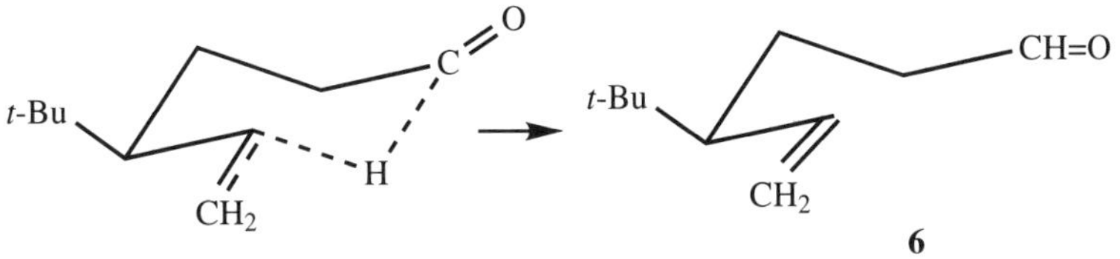

(3)

In other suitable cases ketenes are also the predominant products, as in the example of equation 4.[8] Steroidal ketones also may give major amounts of ketenes.[9] As shown, these reactions are often carried out in alcoholic solvents which convert the ketenes to esters. The ketenes formed may undergo further photoreaction to give decarbonylation products (equation 5).[10]

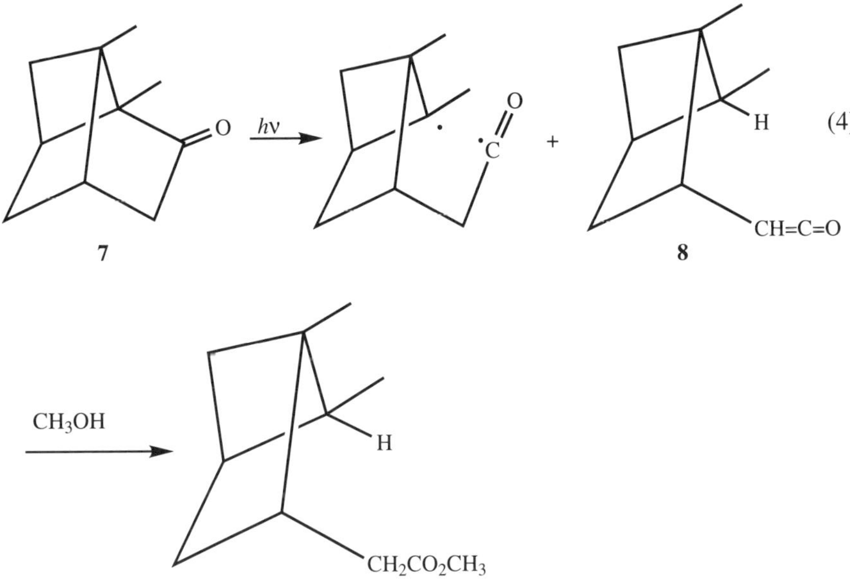

(4)

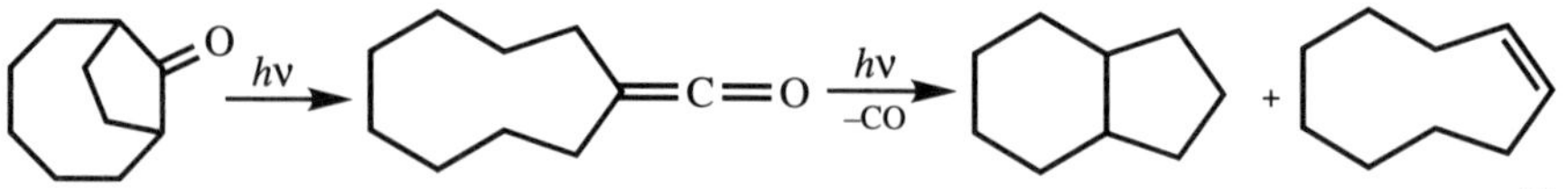

110 PREPARATION OF KETENES

(5)

The ^{13}C CIDNP signals of the acids derived from the ketenes formed from photolysis of 2-phenylcyclohexanone, cycloheptanone, and cyclooctanone gave evidence regarding spin interactions in the intermediate diradical (equation 6).[11]

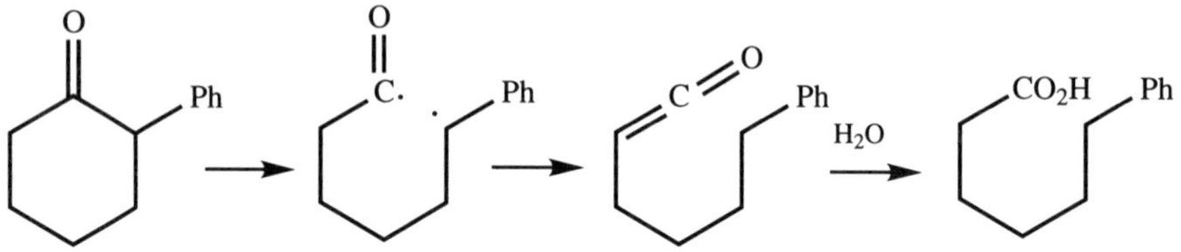

(6)

Cleavage of cyclohexanone by flash photolysis to n–butylketene (**9**) serves as a route to this ketene for kinetic studies of the hydration (equation 7).[12] Photolysis of the cyclohexanone **10** was proposed to proceed to the diradical **11** which gave the cyclooctenone **12**, enol, and ketene **13** with quantum yields $\phi = 0.43$, 0.06, and 0.08, respectively (equation 8).[13]

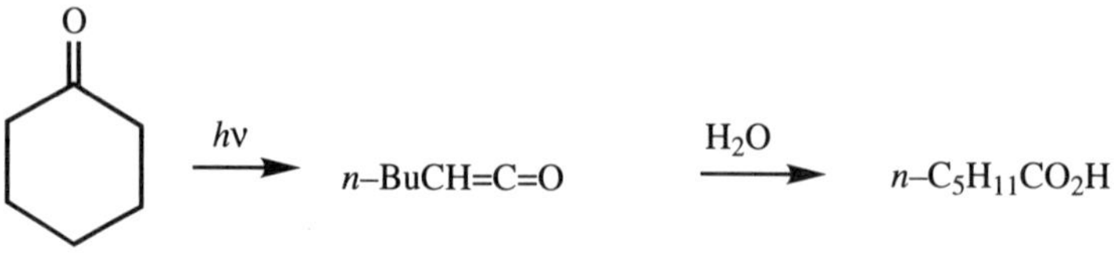

(7)

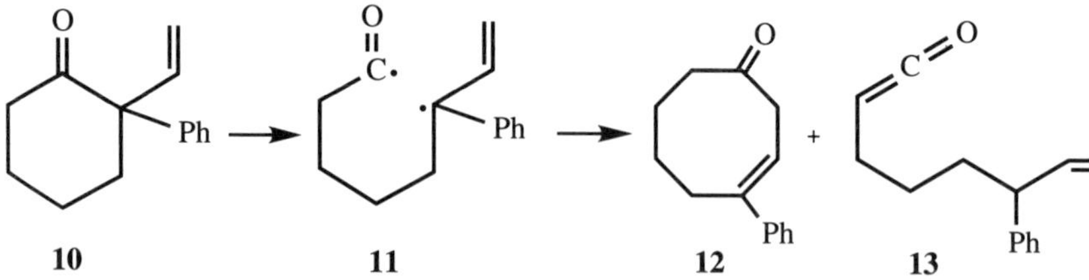

(8)

10 **11** **12** **13**

Photolysis of a cyclobutyl fused steroidal ketone **14** gave a rather stable ketene **15** (equation 9) whose IR absorption at 2096 cm^{-1} was observed in EtOAc at room temperature.[14] Addition of MeOH gave the corresponding ester. Further photolysis of **15** led to decarbonylation.

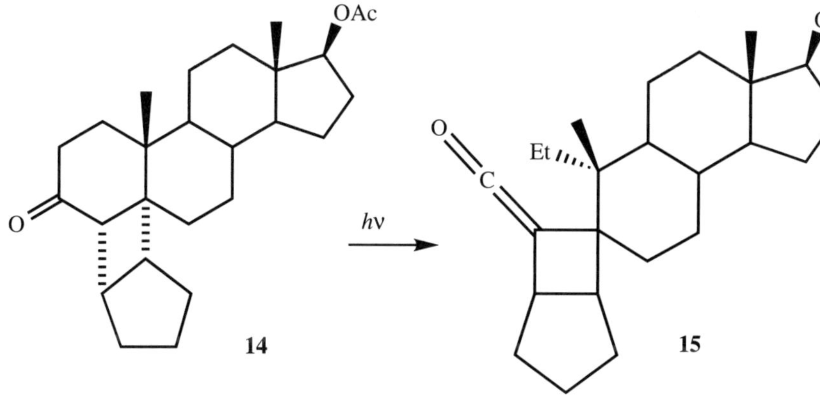

3.4 KETENES BY PHOTOCHEMICAL AND THERMOLYTIC METHODS

(9)

Photolysis of **16** in MeOH was proposed to involve the ketene **17**, which was captured by MeOH to give **18** in 73% yield (equation 10).[15]

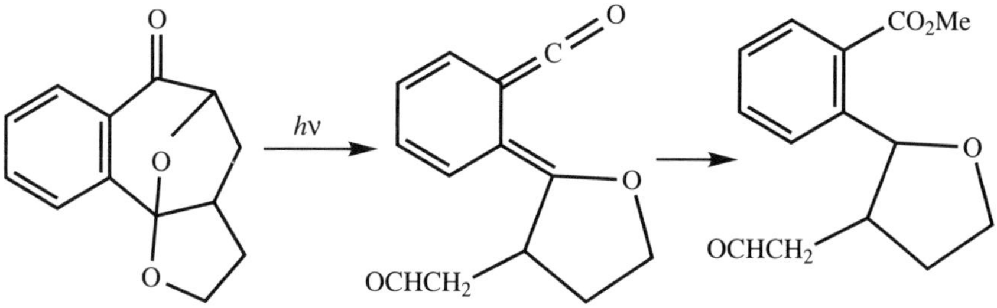

(10)

Initial Norrish Type I cleavage of the ketone **19** leads to the ketenylaldehyde **20**, directly observed by IR, by a multiple bond scission (equation 11).[16] The ketene **20** was also captured by MeOH.[16]

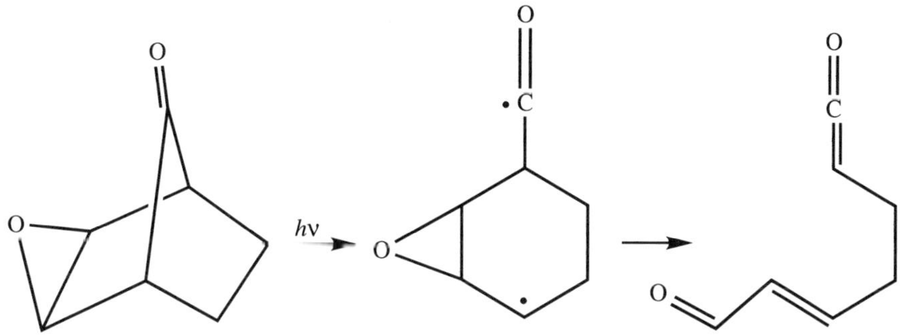

(11)

112 PREPARATION OF KETENES

Photolysis of bicyclo[2.2.1]hept-2-en-5-one (**21**) proceeded by rearrangement to the isomer **22**, presumably after initial Type I cleavage, and then to ketene and cyclopentadiene on further photolysis (equation 12).[17,18]

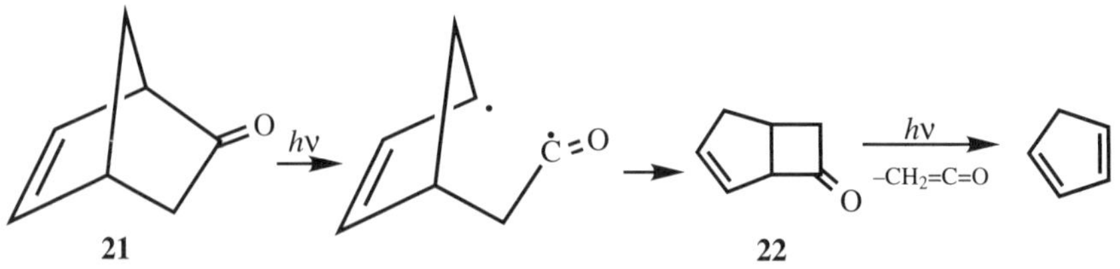

(12)

Photolysis of the dihydrofuranone **23** in methanol led to the ketene **24** as evidenced by the formation of **25** (equation 13).[19] The product **26** was also observed and its formation was ascribed to an intramolecular rearrangement of **24** (equation 14).[19]

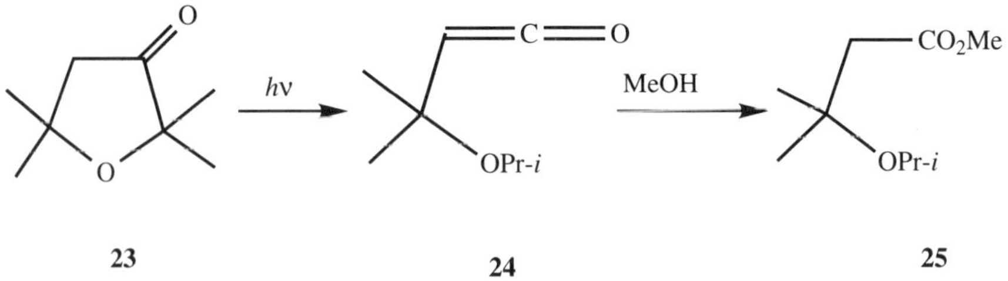

(13)

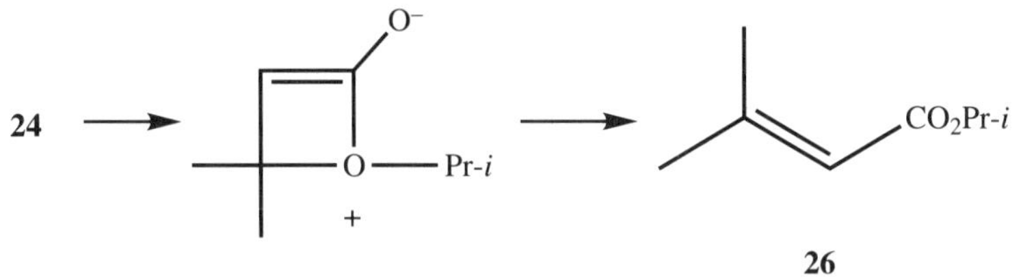

(14)

Photolysis of 4-aryl-2-cyclopentenones leads to ketenes **27** by di-π-methane rearrangements (equation 15).[20,21] The ketenes were observed by IR and trapped with alcohols.[20] It was suggested that the bicyclic ketones **28** might be intermediates in the process at low temperature, while the ketenes were the initial products at higher temperatures.[20] Bicyclo[2.1.0]pentan-2-ones such as **28**, which open to allylketenes such as **27**, are also formed in the vinylogous Wolff rearrangement (Section 3.3.2, equation 60).

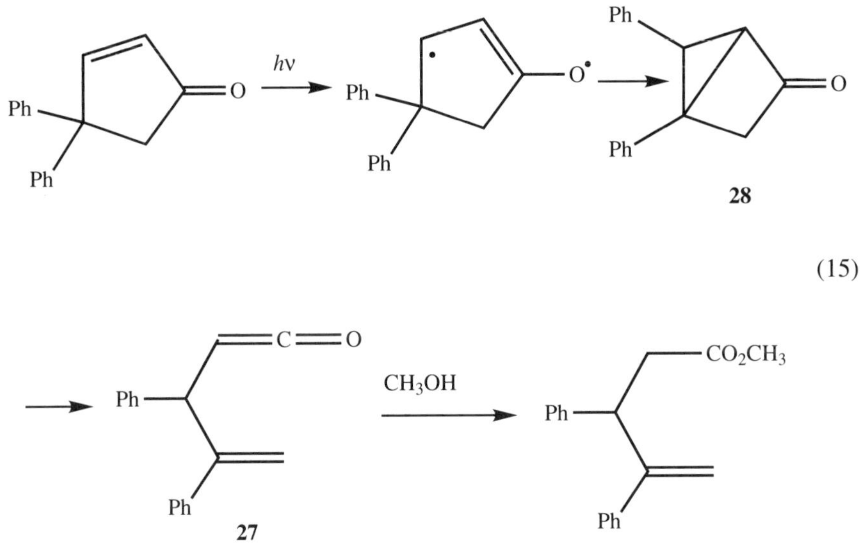

3.4 KETENES BY PHOTOCHEMICAL AND THERMOLYTIC METHODS

(15)

Photolysis of 5-substituted cyclopentenones can lead to cyclopropylketenes, as discussed in Section 4.1.5 (equation 16).[22]

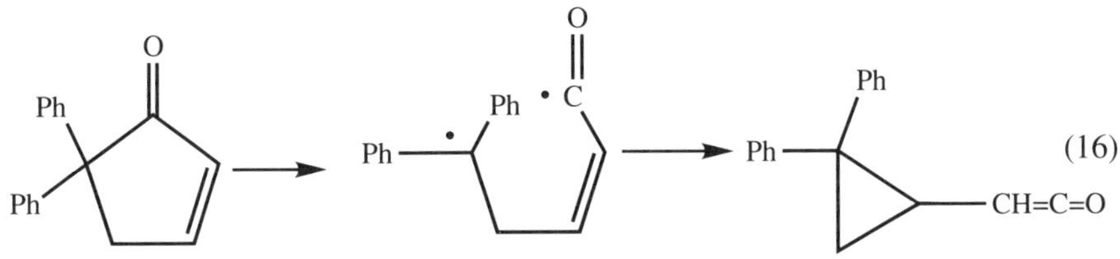

(16)

Photolysis of **29** results in hydrogen atom transfer to give a diradical which gives predominant closure to a cyclobutanone, and also formation of ketene **30**, directly observed by IR, and isolated in 25% yield as the methyl ester when the reaction was conducted in CH_3OH (equation 17).[23]

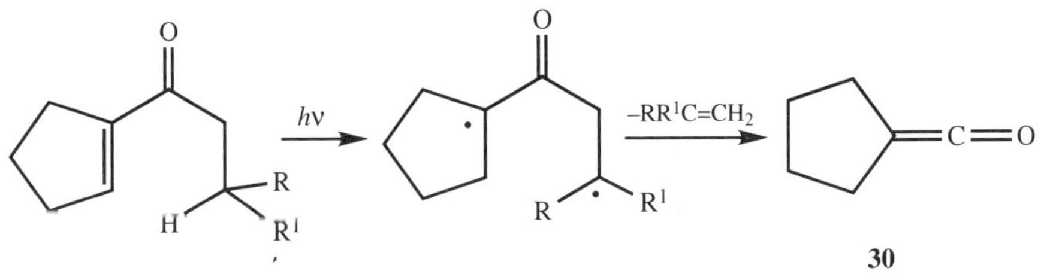

29 R = R^1 = CH_3; RR^1 = $(CH_2)_4$ (17)

Photolysis of (−)-verbenone (**31**) leads to ketene **33**, presumably by the diradical **32**. Ketene **33** is relatively long-lived, for when **31** was photolyzed in cyclohexane the IR spectrum of the crude reaction product showed a ketene band at 2120 cm^{-1}, and when CH_3OH was added to the product the ester **34** was formed.[24]

31 **32** **33**

(18)

33

(19)

34

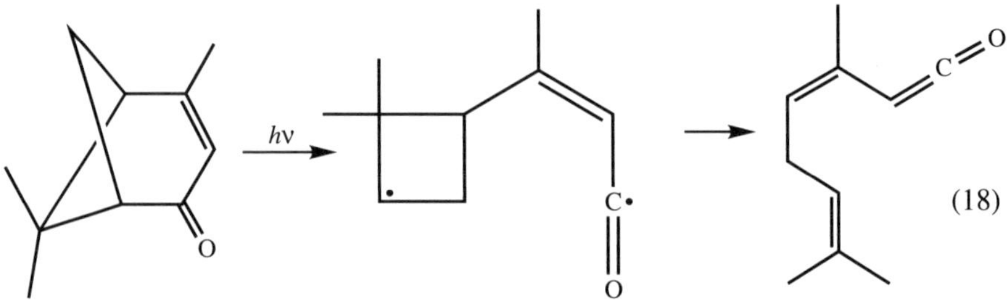

Ketenes are produced from photolyses of 1,5-hexadiene-3-ones by different routes. Photolysis of **35** in benzene–CH_3OH gave competitive 1,5-closure to a bicyclohexanone **36** in 43% yield (equation 20), as well as 1,6-closure to a diradical leading to ketene **37,** captured as the methyl ester in 27% yield (equation 21).[25,26]

(20)

35 **36**

35 **37**

(21)

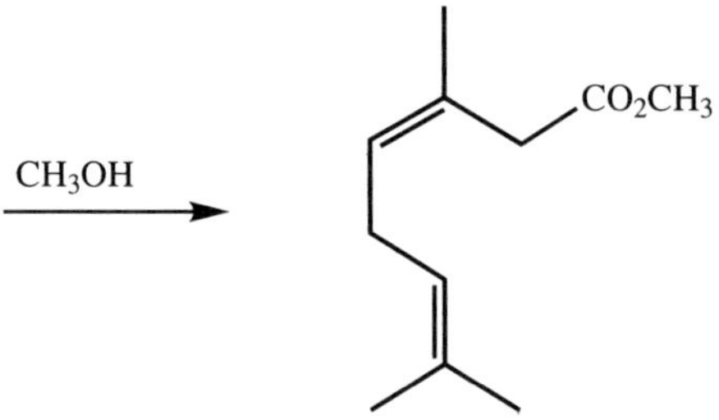

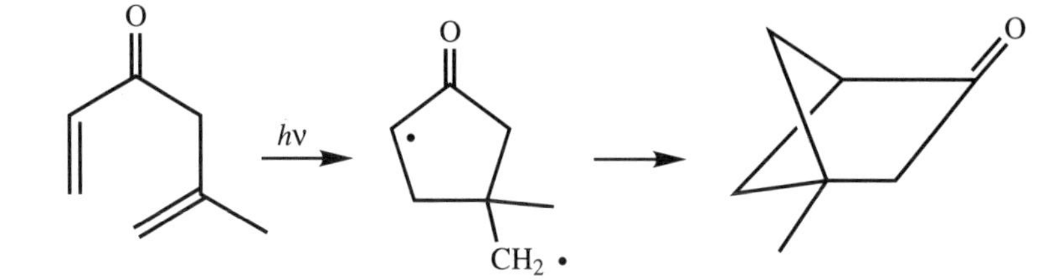

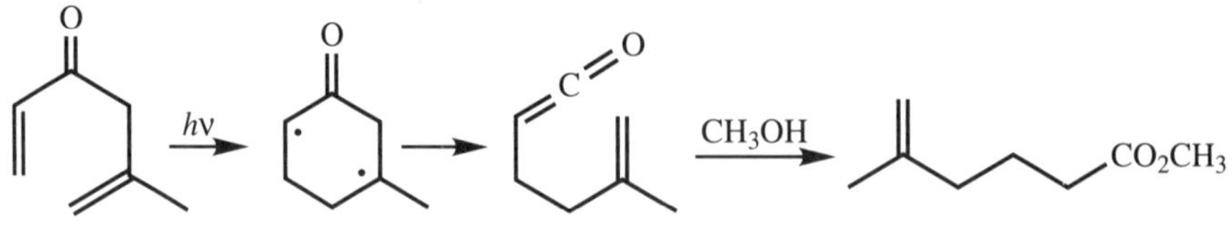

3.4 KETENES BY PHOTOCHEMICAL AND THERMOLYTIC METHODS

Further study by Agosta and co-workers[26] showed that **35-d$_2$** gave both the methyl esters **38** and **39**, and that the portion of the ester **39** varied from 10% at wavelengths above 340 nm, to 40–45% for light at 313 nm.[26] These results were interpreted as showing that the ketene **40** formed by the 1,6-rearrangement of equation 21 predominated at λ > 340 nm, whereas at 313 nm there was extensive α-cleavage and recombination via the ketenes **40** and **41**, leading to **38** and **39**, respectively (equations 22 and 23).

$$\text{35-d}_2 \xrightarrow{h\nu} \cdots \longrightarrow \mathbf{40} + \mathbf{41} \tag{22}$$

$$\mathbf{40} \xrightarrow{CH_3OH} \mathbf{38} \tag{}$$

$$\mathbf{41} \xrightarrow{CH_3OH} \mathbf{39} \tag{23}$$

Other examples of this wavelength dependent α-cleavage have been demonstrated by Dauben et al.,[27] who by triplet sensitization studies concluded that a predominant cause of the wavelength dependence was a competition between α-cleavage occurring in the singlet state and triplet cycloaddition.[27]

Photolysis of the enone **42** in benzene led to IR observation of a ketene band assigned to **43**, which was proposed to form as shown in equation 24.[28] When the reaction was conducted in 1/1 *t*-BuOMe/MeOH containing 0.4% HOAc, the ester **44** was formed as well as the indanone **45** ascribed to acid-induced cyclization of **43** (equations 25 and 26).[28] As discussed in Section 5.6.1, protonation of ketenes on carbon to give acylium ions also occurs, and this is a possible alternative route to **45**.

$$\underset{\mathbf{42}}{Ph_2C=CHCCHPh_2} \xrightarrow{h\nu} Ph_2C=CH\overset{\bullet}{C}=O + Ph_2\overset{\bullet}{C}H \longrightarrow \underset{\mathbf{43}}{Ph_2CHCPh_2CH=C=O} \tag{24}$$

$$\mathbf{43} \xrightarrow{MeOH} \underset{\mathbf{44}}{Ph_2CHCPh_2CH_2CO_2Me} \tag{25}$$

116 PREPARATION OF KETENES

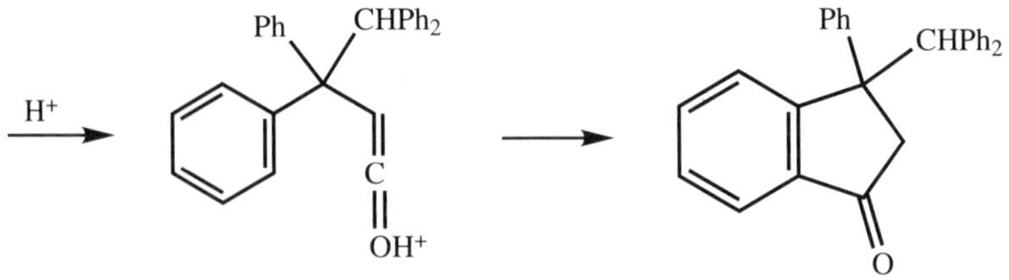

(26)

Photolysis of **46** gave the ketene **47**, which was not observed but was trapped with MeOH (equation 27).[29] Photolysis of **48** gave the ketene **49**, which was trapped with nucleophiles or as its dimer **50** (equation 28).[30] Pyrolysis of **49** gave **51** (equation 29).[30]

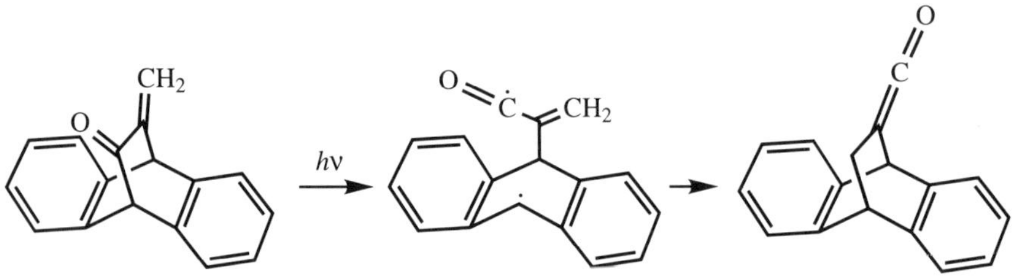

(27)

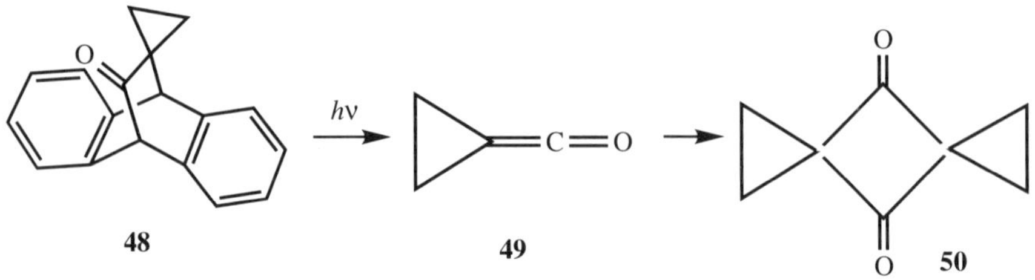

(28)

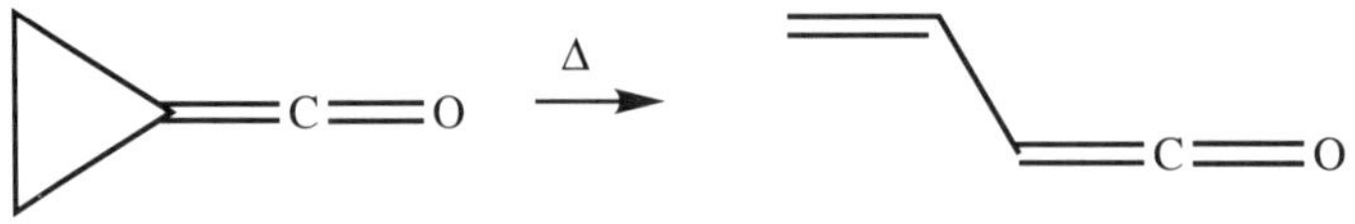

(29)

Photolysis of **52** and related bicyclo[2.2.2]octadienone derivatives gave dimethylketene (equation 30).[31]

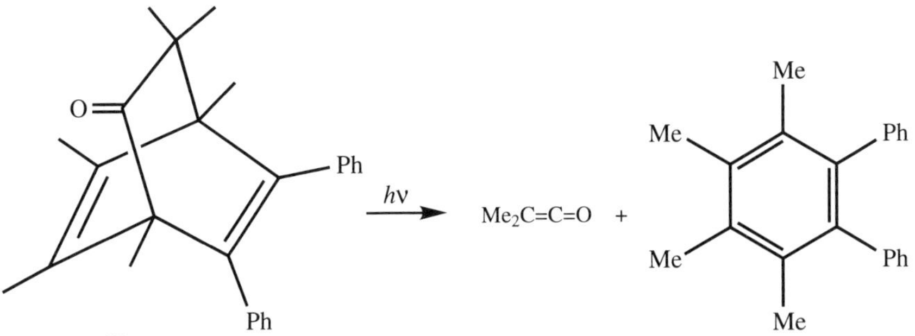

3.4 KETENES BY PHOTOCHEMICAL AND THERMOLYTIC METHODS

(30)

A multistep rearrangement process was proposed for the photolysis of **53** leading to ketene **54** (equation 31).[32]

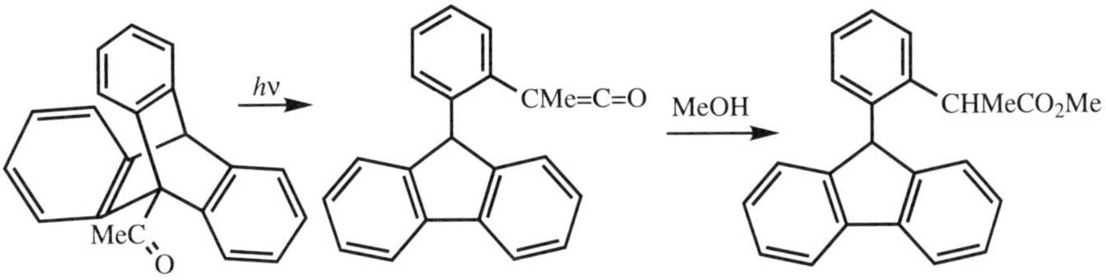

(31)

The photolysis of Z-dibenzoylethylene in alcohols is proposed to lead by the route shown to ketene **55** as identified by the capture product **56** (equation 32).[33–35]

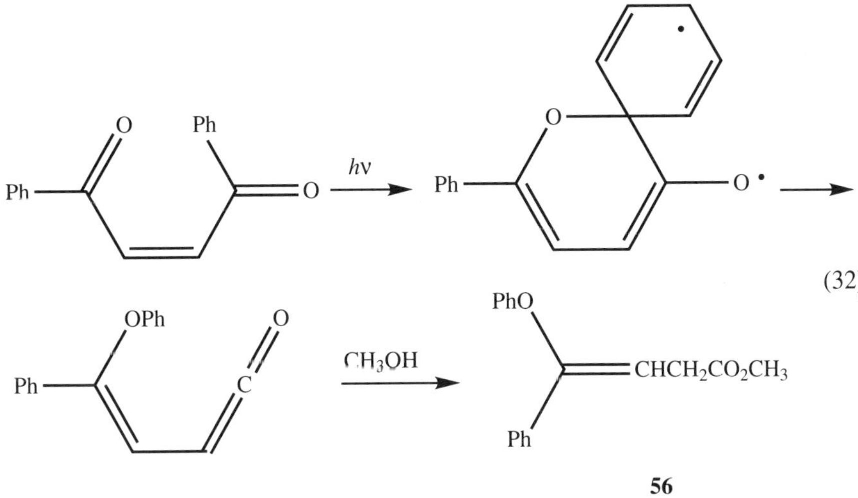

(32)

This reaction has been observed in other substituted dibenzoylethylenes,[35–38] and in tetrabenzoylethylene (equation 33).[39]

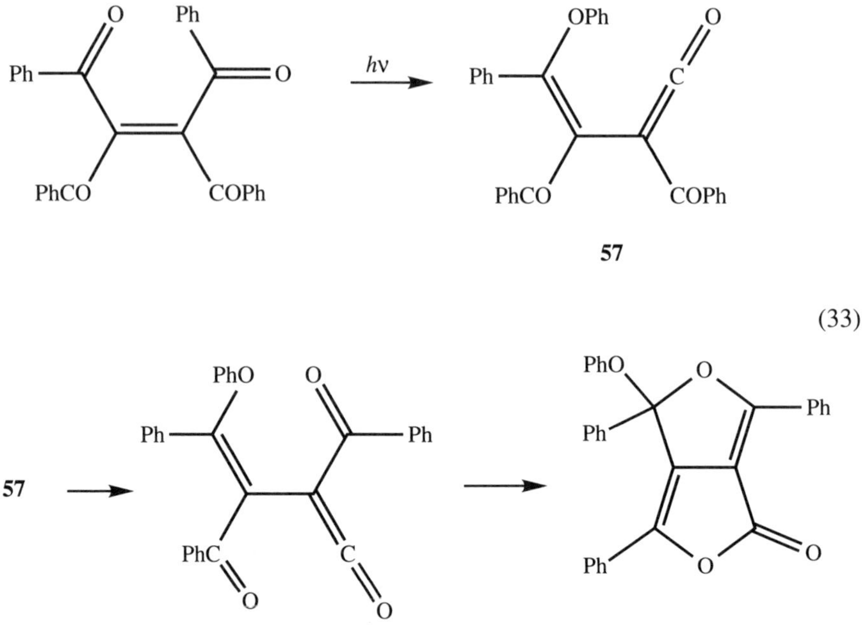

(33)

Photochemical cleavage of dihydrocoumarin (**58**) leads to the ketene **59**, as evidenced by the IR band at 2115 cm^{-1} at 80 K, and the formation of the acid or ester on reaction with H_2O or CH_3OD, respectively (equation 34).[40] Photochemical reactions of pyrones are considered in Section 3.4.6.

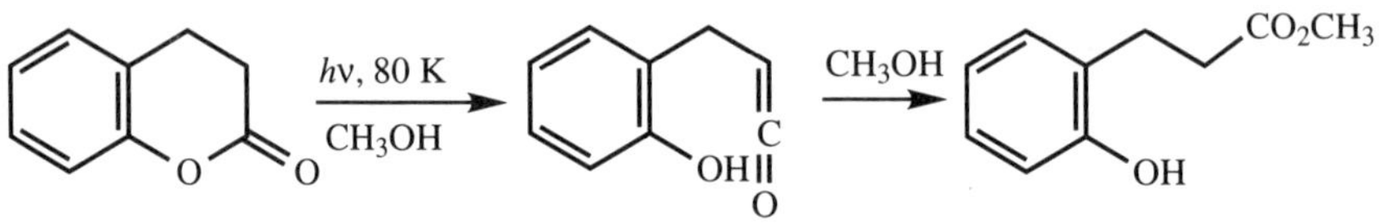

(34)

References

1. Wagner, P. J. In *Rearrangements in Ground and Excited States;* de Mayo, P., Ed.; Academic: New York, 1980; Vol. 3, pp. 381–444.
2. Yates, P., *Pure Appl. Chem.* **1968**, *16*, 93–113.
3. Weiss, D. S. *Org. Photochem.*, **1981**, *5*, 347–420.
4. Agosta, W. C.; Schreiber, W. L. *J. Am. Chem. Soc.* **1971**, *93*, 3947–3952.
5. Wagner, P. J.; Spoerke, R. W. *J. Am. Chem. Soc.* **1969**, *91*, 4437–4440.
6. Dalton, J. C.; Dawes, K.; Turro, N. J.; Weiss, D. S.; Barltrop, J. A.; Coyle, J. D. *J. Am. Chem. Soc.* **1971**, *93*, 7213–7221.

7. Coyle, J. D. *J. Chem. Soc. B*, **1971**, 1736–1740.
8. Meinwald, J.; Schneider, R. A.; Thomas, A. F. *J. Am. Chem. Soc.* **1967**, *89*, 70–73.
9. Quinkert, G.; Blanke, E.; Homburg, F. *Chem. Ber.* **1964**, *97*, 1799–1810.
10. Gutsche, C. D.; Baum, J. W. *J. Am. Chem. Soc.* **1968**, *90*, 5862–5867.
11. Doubleday, C., Jr. *Chem. Phys. Lett.* **1979**, *64*, 67–70.
12. Allen, A. D.; Chiang, Y.; Kresge, A. J.; Schepp, N. P.; Tidwell, T. T. *Can. J. Chem.* **1987**, *65*, 1719–1723.
13. Koppes, M. J. C. M.; Cerfontain, H. *Recl. Trav. Chim. Pays-Bas*, **1988**, *107*, 412–417.
14. Kaprinidis, N. A.; Woning, J.; Schuster, D. I.; Ghatlia, N. D. *J. Org. Chem.* **1992**, *57*, 755–757.
15. Padwa, A.; Carter, S. P.; Nimmesgern, H.; Stull, P. D. *J. Am. Chem. Soc.* **1988**, *110*, 2894–2900.
16. Ayral-Kaloustian, S.; Wolff, S.; Agosta, W. C. *J. Org. Chem.* **1978**, *43*, 3314–3319.
17. Schenk, G. O.; Steinmetz, R. *Chem. Ber.* **1963**, *96*, 520–525.
18. Schuster, D. I.; Axelrod, M.; Auerbach, J. *Tetrahedron Lett.* **1963**, 1911–1916.
19. Hagens, G.; Wasacz, J. P.; Joullie, M.; Yates, P. *J. Org. Chem.* **1970**, *35*, 3682–3685.
20. Zimmerman, H. E.; Little, R. D. *J. Am. Chem. Soc.* **1974**, *96*, 4623–4630.
21. Wolff, S.; Agosta, W. C. *J. Chem. Soc., Chem. Commun.* **1972**, 226–227.
22. Agosta, W. C.; Smith, A. B., III. *J. Am. Chem. Soc.* **1971**, *93*, 5513–5520.
23. Smith, A. B., III; Agosta, W. C. *J. Am. Chem. Soc.* **1973**, *95*, 1961–1968.
24. Erman, W. F. *J. Am. Chem. Soc.* **1967**, *89*, 3828–3841.
25. Gibson, T. W.; Erman, W. F. *J. Org. Chem.* **1972**, *37*, 1148–1154.
26. Matlin, A. R.; Wolff, S.; Agosta, W. C. *Tetrahedron Lett.* **1983**, *24*, 2961–2964.
27. Dauben, W. G.; Cogen, J. M.; Ganzer, G. A.; Behar, V. *J. Am. Chem. Soc.* **1991**, *113*, 5817–5824.
28. Adam, W.; Berkessel, A.; Peters, E.; Peters, K.; von Schnering, H. G. *J. Org. Chem.* **1985**, *50*, 2811–2813.
29. Hart, H.; Dean, D. L.; Buchanan, D. N. *J. Am. Chem. Soc.* **1973**, *95*, 6294–6301.
30. Ripoll, J. L. *Tetrahedron* **1977**, *33*, 389–391.
31. Murray, R. K., Jr.; Hart, H. *Tetrahedron Lett.* **1968**, 4995–4998.
32. Iwamura, H.; Yoshimura, K. *J. Am. Chem. Soc.* **1974**, *96*, 2652–2654.
33. Griffin, G. W.; O'Connell, E. J. *J. Am. Chem. Soc.* **1962**, *84*, 4148–4149.
34. Zimmerman, H. E.; Dürr, H. G. C.; Lewis, R. G.; Bram, S. *J. Am. Chem. Soc.* **1962**, *84*, 4149–4150.
35. Zimmerman, H. E.; Dürr, H. G.; Givens, R. S.; Lewis, R. G. *J. Am. Chem. Soc.* **1967**, *89*, 1863–1874.
36. Padwa, A.; Crumrine, D.; Shubber, A. *J. Am. Chem. Soc.* **1966**, *88*, 3064–3069.
37. Lahiri, S.; Dabral, V.; Chauhan, S. M. S.; Chakachery, E.; Kumar, C. V.; Scaiano, J. C.; George, M. V. *J. Org. Chem.* **1980**, *45*, 3782–3790.
38. Barik, R.; Bhattacharyya, K.; Das, P. K.; George, M. V. *J. Org. Chem.* **1986**, *51*, 3420–3428.
39. Rubin, M. B.; Sander, W. W. *Tetrahedron Lett.* **1987**, *28*, 5137–5140.
40. Chapman, O. L.; McIntosh, C. L. *J. Am. Chem. Soc.* **1969**, *91*, 4309–4310.

3.4.3 Ketenes from Cyclohexadienones and Other Cycloalkenones

Photolysis of 2,4-cyclohexadienones (**1**) gives ring opening to dienylketenes **2**.[1–4] These are proposed[1–4] to undergo photochemical isomerization to isomeric ketenes **3** (equation 1). In solvents containing H_2O or amines, the ketenes are captured by H_2O or RNH_2 to give acids or amides, respectively (equation 2).

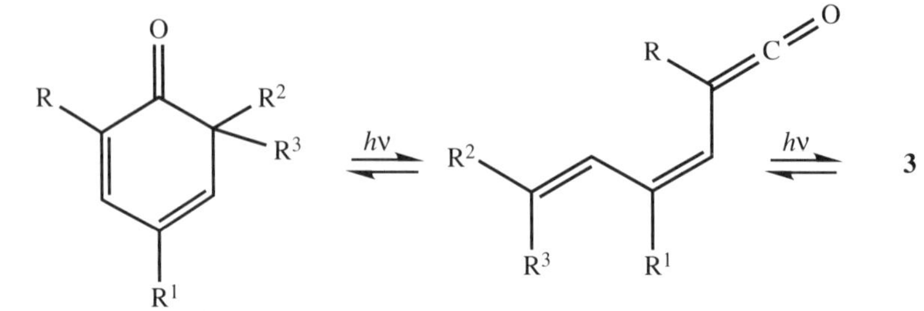

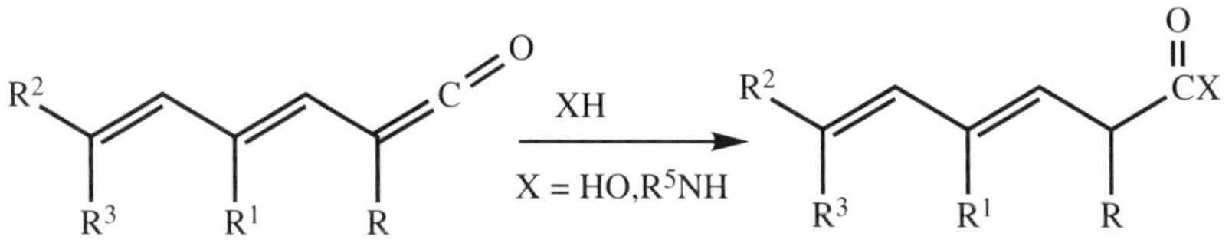

Photolysis of either **4**[5] or **5**[6] evidently gives the parent 2,4-cyclohexadienone **6** which undergoes further photolysis to the ketene **7** (equation 3), identified by its IR band at 2112 cm^{-1}, its UV spectrum, and its trapping product with EtOH. The UV spectrum of **6** was detected following flash photolysis.[6]

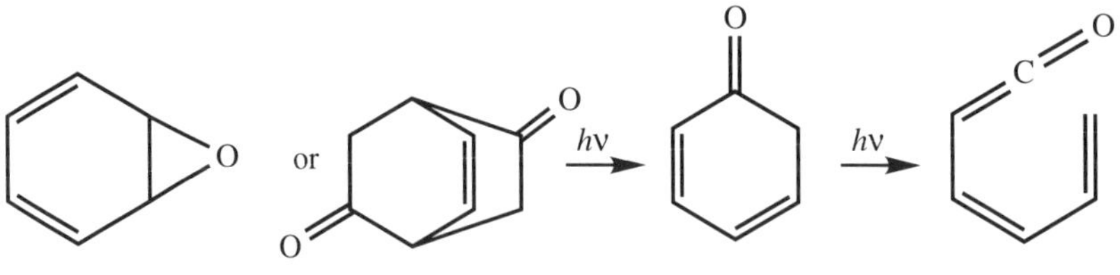

Photolysis of **8** as a neat glass cooled by liquid N_2 was proposed to give the *cis*-dienyl ketene **9** (equation 4) on the basis of the IR spectrum, which showed a strong ketene C=O stretch at 2118 cm^{-1} and bands between 1500 and 1700 cm^{-1}, characteristic of a *cis* diene.[7] Continued irradiation led to a new ketene band at 2145 cm^{-1}. Thermolysis of **8** gave no reaction at 205 °C.[8]

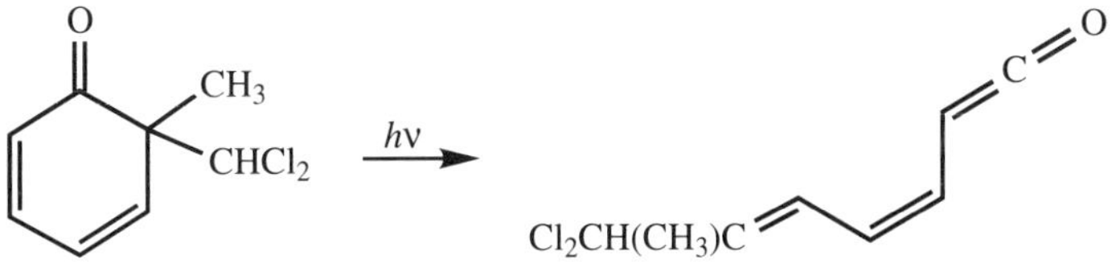

3.4 KETENES BY PHOTOCHEMICAL AND THERMOLYTIC METHODS

[Equation (4): Structure **8** (cyclohexadienone with CH₃ and CHCl₂ substituents) → hv → Structure **9** (dienylketene with Cl₂CH(CH₃)C= group)]

Photolysis of **10** in CH₃OD gave a single methyl ester, and this was proposed to form via the ketene **11**, which adds CH₃OD to form the observed product **12** (equation 5).[8]

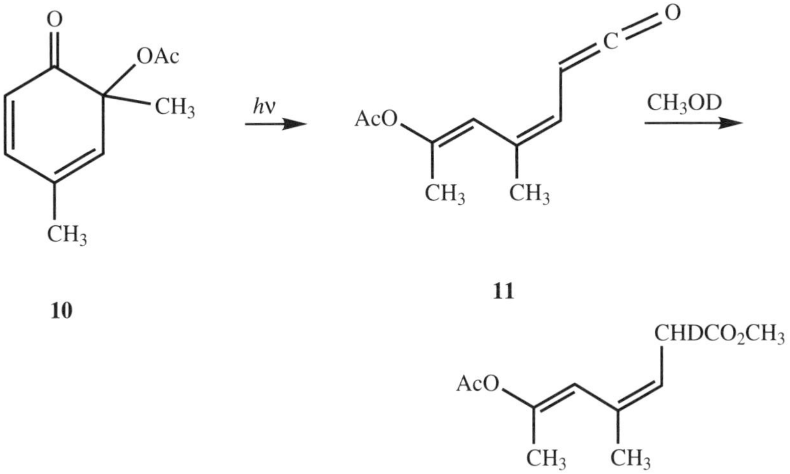

[Equation (5): Structure **10** (cyclohexadienone with OAc, CH₃, and CH₃ substituents) → hv → Structure **11** (dienylketene with AcO, CH₃, CH₃ groups) → CH₃OD → Structure **12** (with CHDCO₂CH₃, AcO, CH₃, CH₃)]

Perfluoro-2,4-cyclohexadienone was also proposed to form a ketene **13** on photolysis (equation 6), although this was not directly observed (see Section 4.4.1).[9]

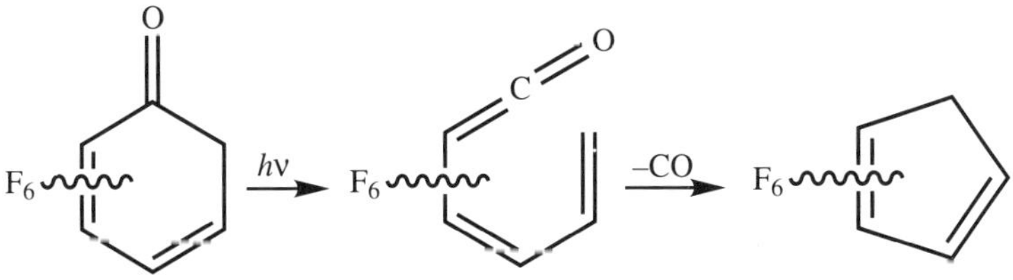

[Equation (6): F₆-cyclohexadienone → hv → F₆-dienylketene **13** → –CO → F₆-cyclopentadiene]

Cyclization of the dienylketenes to bicyclic ketones can occur, and these thermally regenerate the ketenes (equation 7).[10]

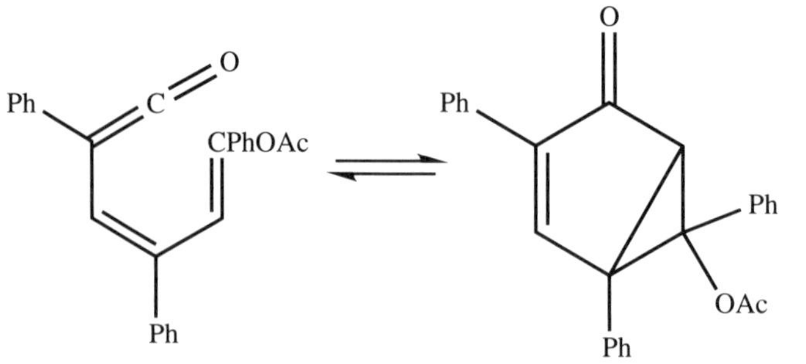

$$\text{(7)}$$

Photolysis of the azide **14** in hexane led to formation of the ketene[15] identified by its IR absorption at 2120 cm^{-1}, its UV spectrum, and its capture by nucleophilic solvents.[11] The half-life of **15** was 4 min at 22 °C, with formation of the dimer shown as the major product, evidently by way of the dienone **16**.[11,12]

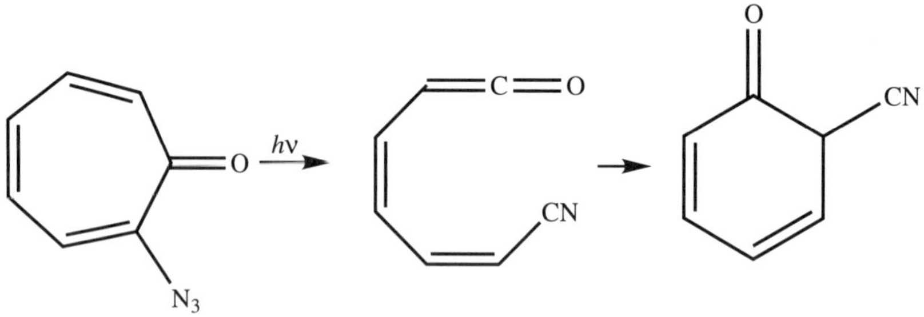

$$\text{(8)}$$

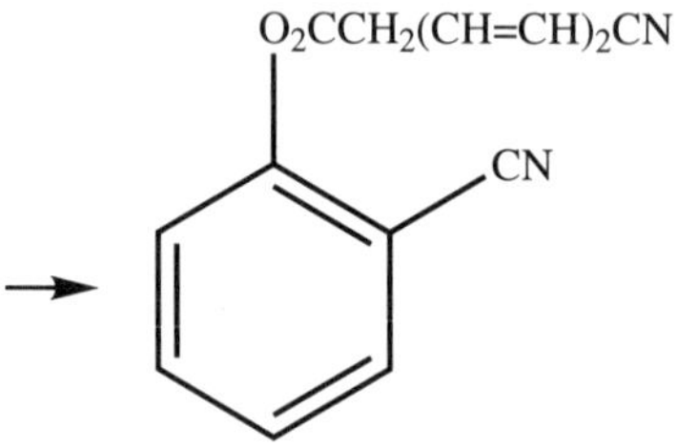

Photolysis of the trimethylazidotropone **17** at −50 to −77 °C gave the ketene **18** whose kinetics of conversion to **19** could be monitored.[11,12] Photolysis of **19** at −77 °C regenerated **18**. The cyclization rate of **18** was estimated to be 10^3 times greater than that of **15**, and generation of **18** in the presence of H$_2$O or CH$_3$OH led to formation of products of nucleophilic capture of **18**, together with dienone **19**.

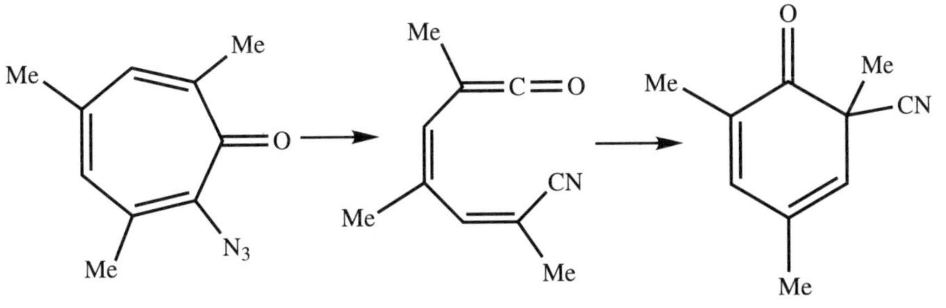

3.4 KETENES BY PHOTOCHEMICAL AND THERMOLYTIC METHODS

(9)

Photolysis of oxazinones **20**,[13] pyrimidinones **20a**,[13a] and pyrones **21**[14] also leads to ketenes in related processes (equations 10 and 11). The photochemical conversion of **22** to **23** was proposed to involve the unobserved ketene **24** (equation 12).[15]

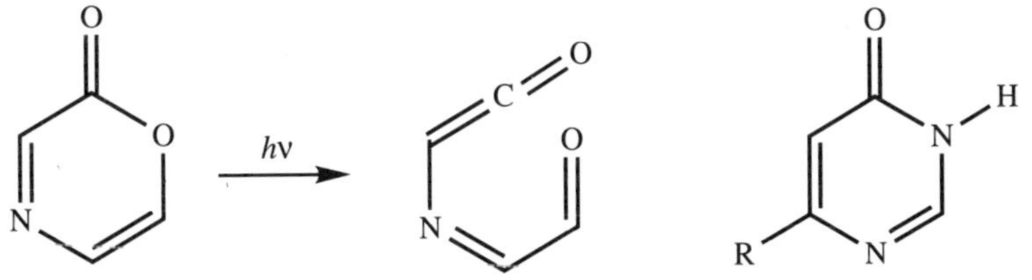

(10)

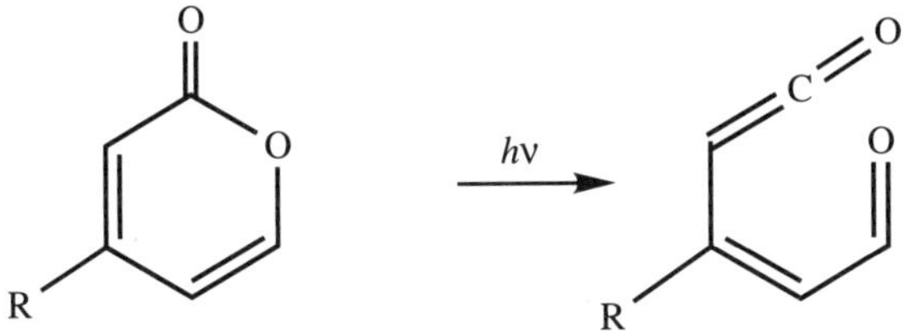

(11)

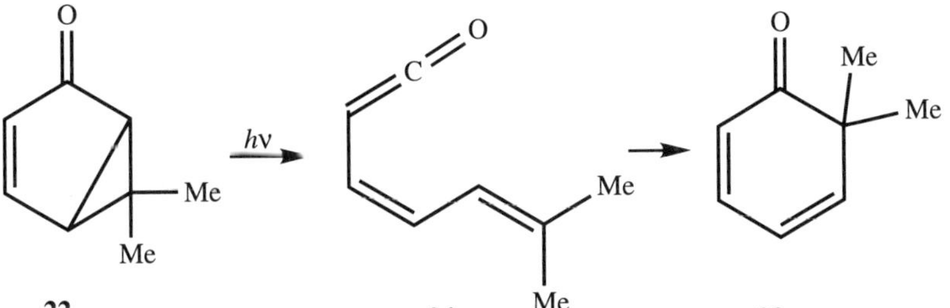

(12)

Photolysis of **25** gave a ketene (or diradical) **26** which led to **27** after decarboxylation (equation 13).[16]

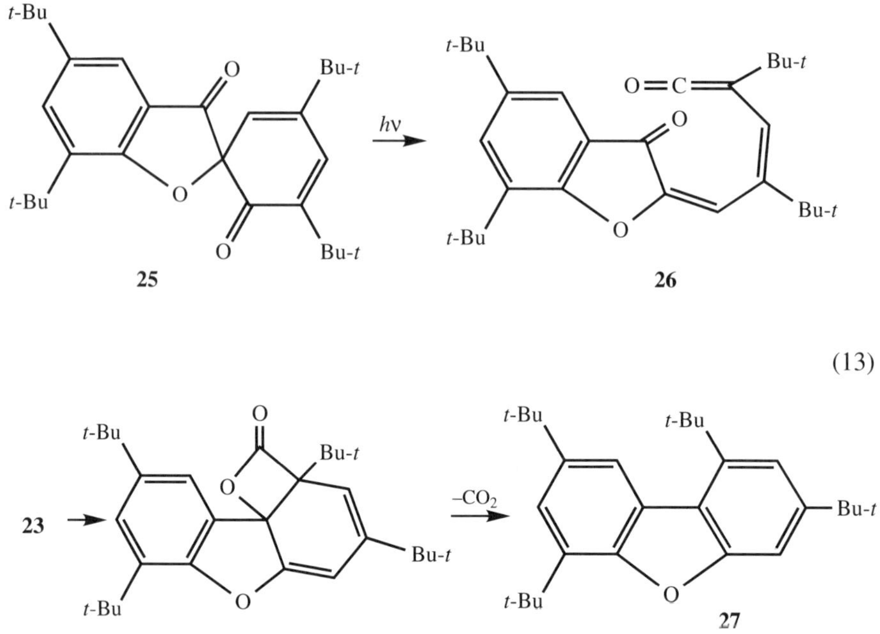

(13)

Photolysis of **28** gave **29** by a bond-cleavage recyclization route. Isolation of **29** followed by further photolysis was proposed to give ring opening to **30** and its stereoisomer, and these were captured by MeOH as the esters (equation 14).[17]

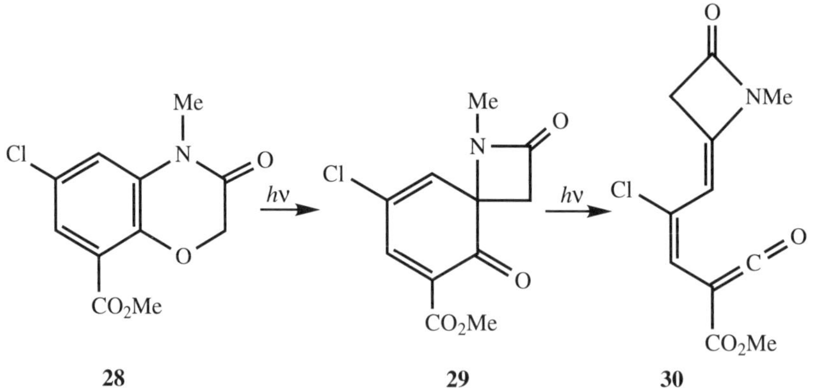

(14)

References

1. Quinkert, G.; Kleiner, E.; Freitag, B.; Glenneberg, J.; Billhardt, U.; Cech, F.; Schmieder, K. R.; Schudok, C.; Steinmetzer, H.; Bats, J. W.; Zimmermann, G.; Dürner, G.; Rehm, D.; Paulus, E. F. *Helv. Chim. Acta* **1986**, *69*, 469–535.

2. Quinkert, G. *Angew. Chem., Int. Ed. Engl.* **1965**, *4*, 211–222.
3. Barton, D. H. R.; Quinkert, G. *J. Chem. Soc.* **1960**, 1–9.
4. Waring, A. J. *Adv. Alicyclic Chem.* **1966**, *1*, 129–256.
5. Jerina, D. M.; Witkop, B.; McIntosh, C. L.; Chapman, O. L. *J. Am. Chem. Soc.*, **1974**, *96*, 5578–5580.
6. Capponi, M.; Gut, I.; Wirz, J. *Angew. Chem., Int. Ed. Engl.* **1986**, *25*, 344–345.
7. Chapman, O. L.; Lassila, J. D. *J. Am. Chem. Soc.* **1968**, *90*, 2449–2450.
8. Baldwin, J. E.; McDaniel, M. C. *J. Am. Chem. Soc.* **1968**, *90*, 6118–6124.
9. Soelch, R. R.; Mauer, G. W.; Lemal, D. M. *J. Org. Chem.* **1985**, *50*, 5845–5852.
10. Dannenberg, W.; Perst, H. *Liebigs Ann. Chem.* **1975**, 1873–1894.
11. Hobson, J. D.; Al Holly, M. M.; Malpass, J. R. *Chem. Commun.* **1968**, 764–766.
12. Hobson, J. D.; Malpass, J. R. *J. Chem. Soc. (C)* **1967**, 1645–1648.
13. Maier, G.; Schäfer, U. *Liebigs Ann. Chem.* **1980**, 798–813.
13a. Lapinski, L.; Nowak, M.-J.; Les, A.; Adamowicz, L. *J. Am. Chem. Soc.* **1994**, *116*, 1461–1467.
14. Pirkle, W. H.; Seto, H.; Turner, W. V. *J. Am. Chem. Soc.* **1970**, 92, 6984–6985.
15. Swenton, J. S.; Saurborn, E.; Srinivasan, R.; Sonntag, F. I. *J. Am. Chem. Soc.* **1968**, *90*, 2990–2991.
16. Izuoka, A.; Miya, S.; Sugawara, T. *Tetrahedron Lett.* **1988**, *29*, 5673–5676.
17. Marubayashi, N.; Ogawa, T.; Kuroita, T.; Hamasaki, T.; Ueda, I. *Tetrahedron Lett.* **1992**, *33*, 4585–4588.

3.4.4 Ketenes from Dioxinones

The dioxinone **1** is formed from the reaction of diketene and acetone, but even though the preparation of diketene in acetone was used industrially on a large scale, the formation of **1** was not recognized for many years. The structure of **1** was established in a collaboration by Michael F. Carroll of the A. Boake, Roberts and Company, Ltd., and Alfred Bader, founder of the Aldrich Chemical Company, using UV and IR analysis, and later NMR.[1,2] Compound **1** was one of the early offerings of the Aldrich Co., but initial sales were very slow.

1

Dioxinones cleave on thermolysis or photolysis[3] with formation of acylketenes and a ketone. When **1** was pyrolyzed at 180–240 °C and the product trapped in an Ar matrix at 5–12 K the IR spectrum showed the presence of acetylketene (**2**) and acetone (equation 1).[4] Distinct ketene absorptions at 2142 and 2135 cm^{-1}

could be resolved, and it is possible that these are due to the presence of both the s-cis and s-trans rotamers of **2,** although this was not established. On the basis of MNDO calculations it was proposed that the s-trans form depicted is more stable.[4] As discussed in Section 4.1.6, this question has also been addressed by ab initio calculations, and for the parent formylketene the s-cis form is more stable.

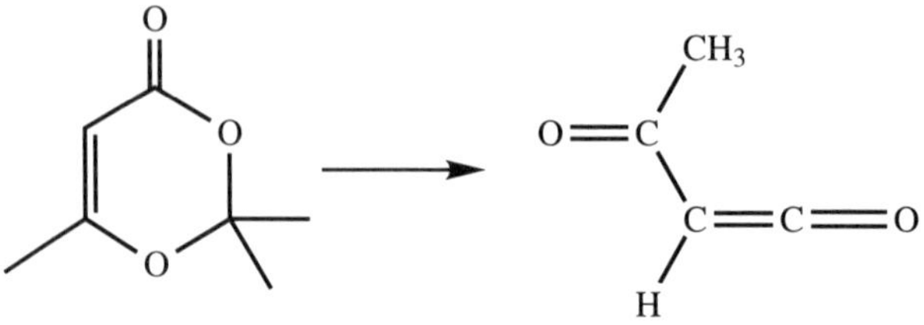

$$\mathbf{1} \longrightarrow \mathbf{2} + (CH_3)_2C=O \quad (1)$$

Thermolyses of a variety of dioxinones to give acylketenes have been examined. The acylketenes have been trapped intramolecularly by hydroxy groups to give ketolactones (equation 2),[5,6,15,16] intermolecularly by nucleophiles[7-11] such as alcohols (equation 3),[4] amines,[4,12] and thiols,[7] by dienophiles, including enol ethers,[12] and by dimerization (equation 4).[4] 5-Trifluoromethyl-1,3-dioxin-4-ones give trifluoromethyl acyl ketenes (equation 5).[14] 5-Halodioxinones react similarly to form α-haloacylketenes which are trapped with alcohols to give esters, or with dimethylcyanamide to give oxazinones (equation 6).[17,18] The chloroketene formed more readily than the fluoro analogue,[17] consistent with the known low stability of fluoroketenes.

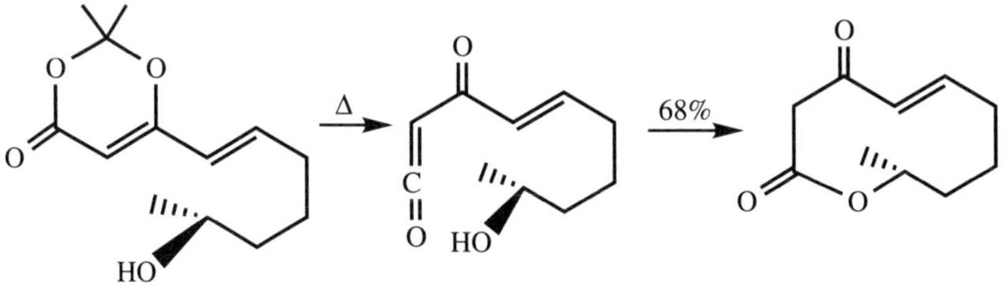

$$(2)^5$$

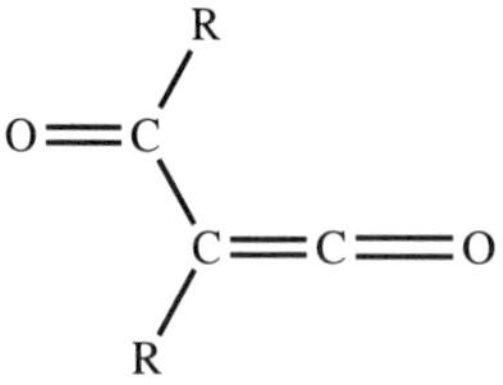

$$\text{R}^1\text{OH} \longrightarrow RCOCHRCO_2R^1 \quad (3)$$

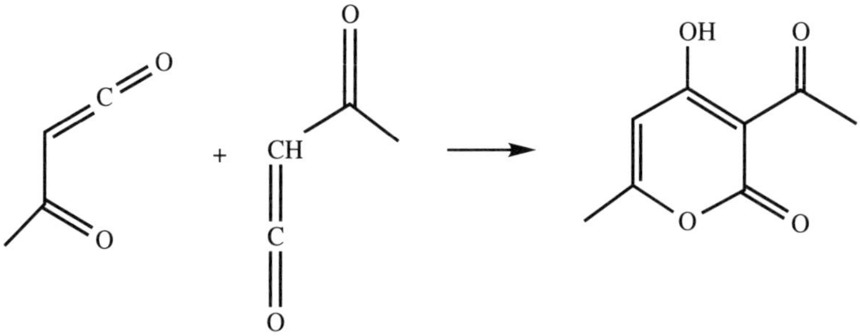

The bisdioxinone **3** was prepared from the reaction of oxalyl chloride, ketene, and acetone (equation 7) and on heating with alcohols gave the corresponding diesters.[19,20] This reaction can formally be written as proceeding through the bisketene intermediate **4** (equation 8),[19] but it is more likely that the ketenyl moieties are formed and esterified sequentially.

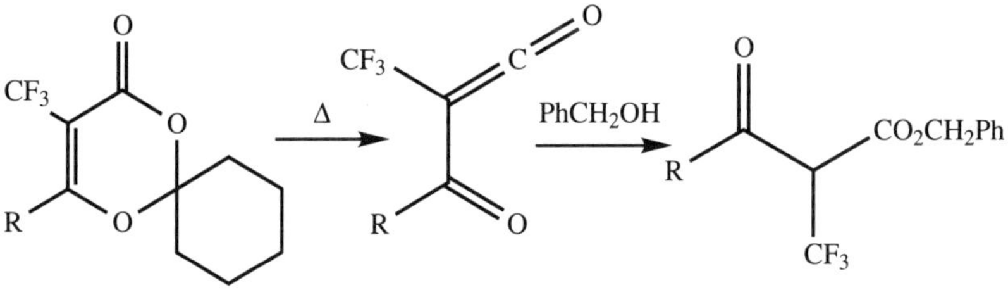

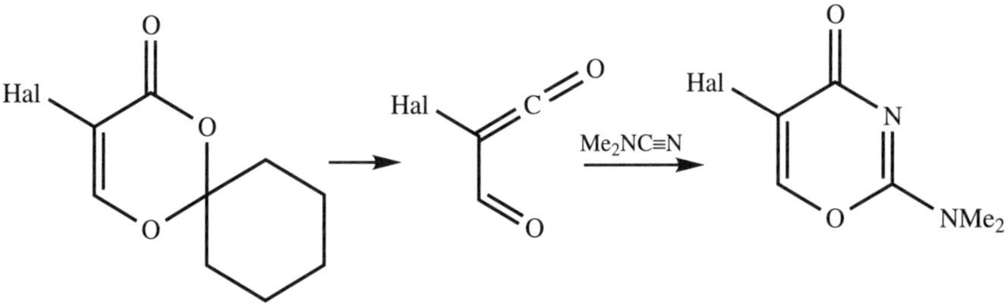

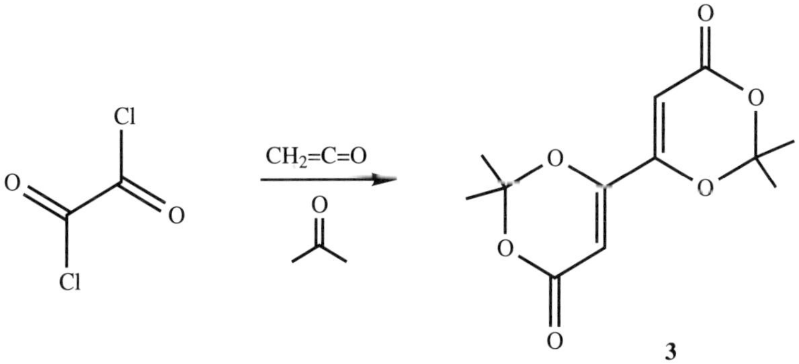

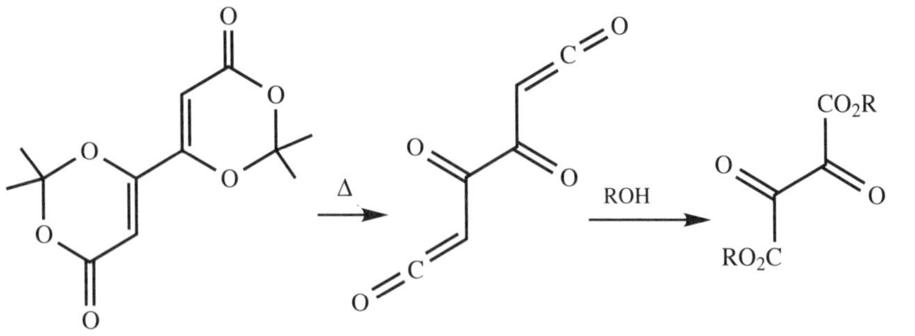

(8)

References

1. Carroll, M. F.; Bader, A. R. *J. Am. Chem. Soc.* **1952**, *74*, 6305; **1953**, *75*, 5400–5402.
2. Bader, A. R.; Gutowsky, H. S.; Heeschen, J. P. *J. Org. Chem.* **1956**, *21*, 821–822.
3. Kaneko, C.; Sato, M.; Sakaki, J.; Abe, Y. *J. Heterocyl. Chem.* **1990**, *27*, 25–31.
4. Clemens, R. J.; Witzeman, J. S. *J. Am. Chem. Soc.* **1989**, *111*, 2186–2193.
5. Boeckman, R. K., Jr.; Pruitt, J. R. *J. Am. Chem. Soc.* **1989**, *111*, 8286–8288.
6. Petasis, N. A.; Patane, M. A. *J. Chem. Soc., Chem. Commun.* **1990**, 836–837.
7. Sakaki, J.; Kobayashi, S.; Sato, M.; Kaneko, C. *Chem. Pharm. Bull.* **1990**, *38*, 2262–2264.
8. Clemens, R. J.; Hyatt, J. A. *J. Org. Chem.* **1985**, *50*, 2431–2435.
9. Hyatt, J. A.; Feldman, P. L.; Clemens, R. J. *J. Org. Chem.* **1984**, *49*, 5105–5108.
10. Sato, M.; Kanuma, N.; Kato, T. *Chem. Pharm. Bull.* **1982**, *30*, 1315–1321.
11. Sato, M.; Ogasawara, H.; Yoshizumi, E.; Kato, T. *Chem. Pharm. Bull.* **1983**, *31*, 1902–1909.
12. Sato, M.; Ogasawa, H.; Komatsu, S.; Kato, T. *Chem. Pharm. Bull.* **1984**, *32*, 3848–3856.
13. Coleman, R. S.; Grant, E. B. *Tetrahedron Lett.* **1990**, *31*, 3677–3680.
14. Iwaoka, T.; Sato, M.; Kaneko, C. *J. Chem. Soc, Chem. Commun.* **1991**, 1241–1242.
15. Sato, M.; Sakaki, J.; Sugita, Y.; Yasuda, S.; Sakoda, H.; Kaneko, C. *Tetrahedron* **1991**, *47*, 5689–5708.
16. Sakaki, J.; Sugita, Y.; Sato, M.; Kaneko, C. *Tetrahedron* **1991**, *47*, 6197–6214.
17. Sato, M.; Kaneko, C.; Iwaoka, T.; Kobayahsi, Y.; Iida, T. *J. Chem. Soc., Chem. Commun.* **1991**, 699–700.
18. Iwaoka, T.; Murohashi, T.; Sato, M.; Kaneko, C. *Synthesis* **1992**, 977–981.
19. Stachel, H. D. *Arch. Pharm.* **1962**, *295*, 735–744. *Chem. Abstr.* **1963**, *59*, 1582d.
20. Stachel, H. D. *Angew. Chem.* **1957**, *69*, 507.

3.4.5 Ketenes by Thermolysis of Alkynyl Ethers

Thermolysis of alkynyl ethers provides a mild and specific route to a variety of ketenes,[1-8] as in equations 1[2] and 2.[3]

$$CH_3CC(=O)C{\equiv}C-O-CH_2-CH_2-H \xrightarrow{-CH_2=CH_2} CH_3C(=O)CH=C=O \quad (1)$$

$$CH_2=CHC{\equiv}COEt \xrightarrow[\Delta]{-CH_2=CH_2} CH_2=CHCH=C=O \quad (2)$$

For alkynyl ethers RC≡COR[1] it was found that the rate of thermolysis increased in the order R[1] = t − Bu > i − Pr > Et, as predicted by a semiempirical molecular orbital study of the reaction.[9] *tert*-Butyl alkynyl ethers *t*-BuOC≡CR undergo ketene formation at temperatures as low as 40 °C,[10,11] and provide a route to ketene itself.[10] When the ketene is thermally stable, it may be prepared and isolated from the alkyne, as in the preparation of Me₃SiCH=C=O from Me₃SiC≡COBu-*t*.[11] Other ketenes were trapped by reaction with nucleophiles, or in [2 + 2] cycloadditions. The main limitations on this process are the availability of the desired 2-substituted 1-alkoxyalkynes and their stability under the conditions needed to generate the corresponding ketenes.[13] Reaction of the alkynyl ethers with the product ketene to give cyclobutenones is a common occurrence.[12] New preparative methods for such alkynes make them more available as synthetic reagents.[13] Alkynyl ether pyrolysis has been particularly useful in the preparation of silyl-substituted ketenes (Section 4.5) and phosphorous-substituted ketenes (Section 4.6). Ketenes generated from alkynyl ether pyrolysis are also useful in intramolecular cyclization to give lactones.[14]

Irradiation of the alkynyl ether **1** in methanol gave 32% yield of the ester **2**, evidently formed through the ketene **3** (equation 3),[15] along with 30–40% cyclohexanol. The corresponding reaction of the alkynyl ether of (+)-2-octanol gave 40% racemization in formation of the rearranged ester, so a mechanism involving the intermediacy of a short-lived radical pair **4** appears likely (equation 3), although some concerted 1,3-migration is also possible.[15]

$$HC\equiv C-O-\underset{1}{\bigcirc} \xrightarrow[MeOH]{h\nu} \left[HC\equiv C-O^{\bullet} + \underset{4}{\bigcirc^{\bullet}}\right] \longrightarrow$$

$$\underset{3}{\bigcirc-CH=C=O} \xrightarrow{MeOH} \underset{2}{\bigcirc-CH_2CO_2Me} \quad (32\%)$$

(3)

References

1. Nieuwenhuis, J.; Arens, J. F. *Recl. Trav. Chim. Pays-Bas,* **1958**, *77,* 761–768.
2. Hyatt, J. A.; Feldman, P. L.; Clemens, R. J. *J. Org. Chem.* **1984**, *49,* 5105–5108.
3. Terlouw, J. K.; Burgers, P. C.; Holmes, J. L. *J. Am. Chem. Soc.* **1979**, *101,* 225–226.
4. Brandsma, L.; Bos, H. J. T.; Arens, J. F. In *Chemistry of Acetylenes;* Viehe, H. G.; Ed.; Marcel Dekker: New York, 1969; pp. 808–810.
5. Ficini, J. *Bull. Soc. Chim. Fr.* **1954**, 1367–1371.
6. van Daalen, J. J.; Kraak, A.; Arens, J. F. *Recl. Trav. Chim. Pays-Bas* **1961**, *80,* 810–818.
7. Pericás, M. A.; Serratosa, F. *Tetrahedron Lett.* **1977**, 4437–4438.
8. Pericas, M. A.; Serratosa, F.; Valentí, E. *Synthesis,* **1985**, 328.
9. Mayano, A.; Pericas, M. A.; Serratosa, F.; Valenti, E. *J. Org. Chem.* **1987**, *52,* 5532–5538.
10. Valentí, E.; Pericas, M. A.; Serratosa, F.; Mana, D. *J. Chem. Res. (S)* **1990**, 118–119.
11. Valentí, E.; Pericas, M. A.; Serratosa, F. *J. Org. Chem.* **1990**, *55,* 395–397.
12. Serratosa, F. *Acc. Chem. Res.* **1983**, *16,* 170–176.
13. Pericas, M. A.; Serratosa, F.; Valentí, E. *Tetrahedron* **1987**, *43,* 2311–2316.
14. Liang, L.; Ramaseshan, M.; MaGee, D. I. *Tetrahedron* **1993**, *49,* 2159–2168.
15. Smith, B. A.; Callinan, A. J.; Swenton, J. S. *Tetrahedron Lett.* **1994**, *35,* 2283–2286.

3.4.6 Ketenes from Other Thermolytic and Photochemical Routes

Many of the reactions that proceed by bond reorganizations, rearrangements, or cleavages to yield ketenes may be initiated either photochemically or thermally. Both types of reaction are listed here, although only one or the other means of inducing ketene formation may have actually been demonstrated in particular examples.

(–)-Chrysanthenone (**1**) undergoes racemization on heating to 81 °C, and this result is interpreted[1–4] in terms of the facile reversible formation of the ketene **2** from **1** (equation 1).[1-4] Confirmatory evidence for the formation of the ketene **2** is provided by the capture of the ketene as the methyl ester **3** on heating **1** to reflux in CH_3OH at 65 °C (equation 1). However, direct reaction of **1** with CH_3OH under these conditions cannot be discounted.

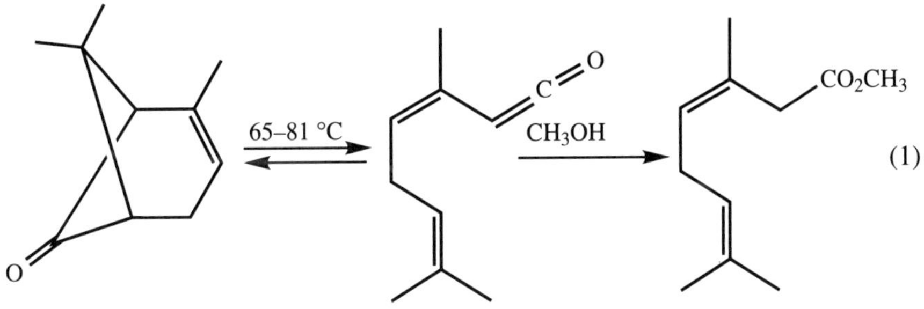

$$(-)\text{-}\mathbf{1} \underset{}{\overset{65-81\,°C}{\rightleftharpoons}} \mathbf{2} \xrightarrow{CH_3OH} \mathbf{3} \qquad (1)$$

The tricyclic ketone **4** rearranged above 50 °C to the ketene **5**, which was identified by the products of trapping (equation 2).[5]

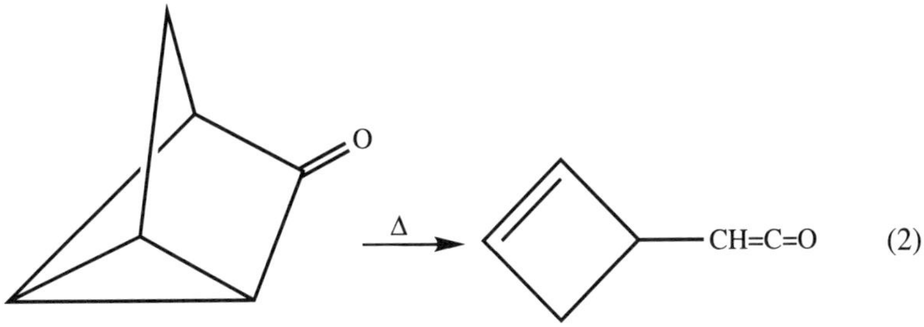

$$\mathbf{4} \xrightarrow{\Delta} \mathbf{5} \qquad (2)$$

The diketone **6** showed a band in the IR at 2300 cm^{-1} attributed to ketene **7** present in thermal equilibrium with **6** at room temperature (equation 3).[6] On heating to 180 °C in CCl_4 **6** was converted mainly to **8**, while in CH_3OH isomers **9** and **10** were formed at room temperature. These products were assigned as rearrangement products from the ketene. The thermal cleavage of intermediate bicyclo[2.1.0]pentan-2-ones such as **6** to allylketenes has been invoked in the vinylogous Wolff rearrangement (Section 3.3.2) and in other reactions.

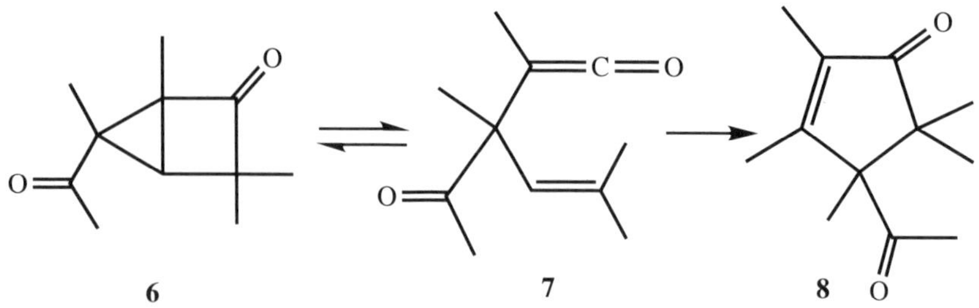

Pyrolysis of 3,4-epoxycyclopentene at 490 °C and trapping of the products at −80 °C in CS_2 led to a 90% yield of the E and Z-1-propenyl ketenes (**11**), as identified by their IR and ^{1}H NMR spectra, and by their conversion to esters and anilides on reaction with alcohols and aniline, respectively.[7] The formation of Z-**11** was proposed to occur by a [1,5] hydrogen shift, while E-**11** was formed by thermal isomerization of Z-**11**.[7]

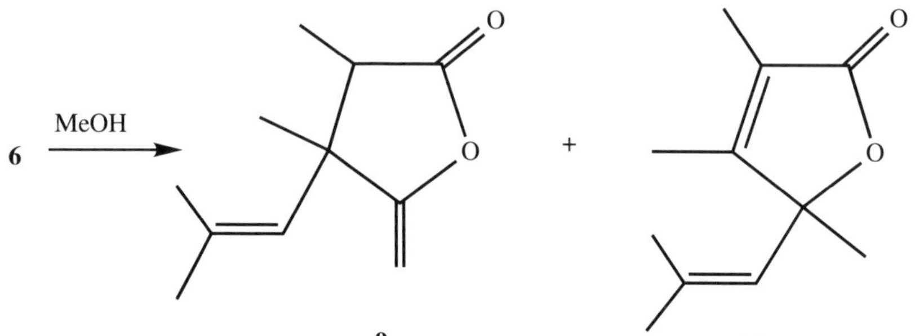

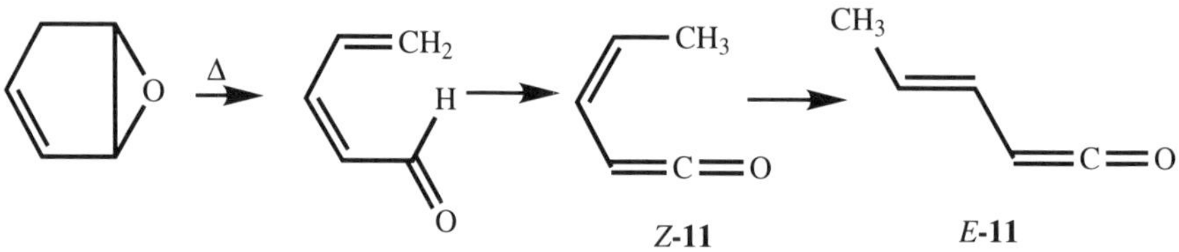

Photolysis of 2-pyrone (**12**) in a CH_2Cl_2–THF glass at 90 K gave rise to an IR absorption at 2128 cm^{-1} attributed to the ketene **13,** which could be trapped with MeOH.[8] Thermolysis at 550–625 °C of 5-methyl and 5-bromo 2-pyrones gave the 3-isomers,[8] while 3-carboethoxypyrone isomerized to the 5-isomer,[9] and 5-^{2}H-2-pyrone gave a mixture of the 3- and 5-isomers[8] (equation 4). It was proposed that these pyrones give ring openings to isomeric ketenes which interconvert by hydrogen shifts to form the more stable products.[8–11] These reactions are related to the formation of ketenes from cyclohexadienones, as discussed in Section 3.4.3.

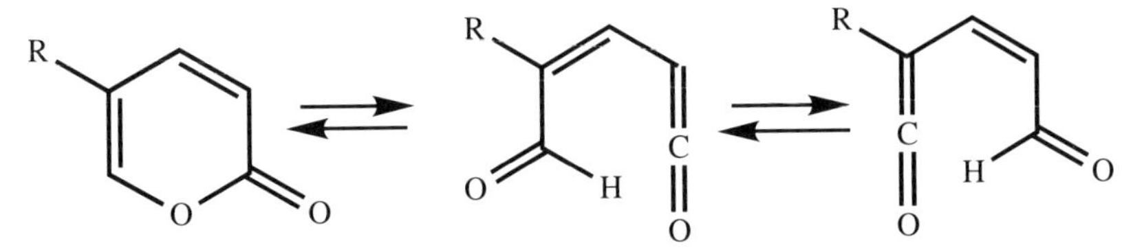

R = H (**12**), Me, Br, CO$_2$Et, ^{2}H   **13** (R = H)

(4)

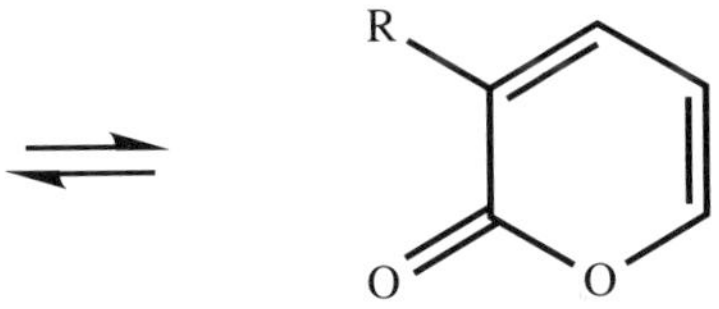

The conversion of **12** to **13** has also been effected by laser flash photolysis,[12] and the lifetime of **13** was only 2.9 μs, as measured by TRIR of the band at 2129 cm^{-1}.

Additions to cyclopropenone **14** have been proposed to lead to ketenes **15** by thermal ring openings (equation 5).[13] These ketenes then isomerize to the observed cyclopentenones **16**. Related reactions are discussed in Section 4.1.9.

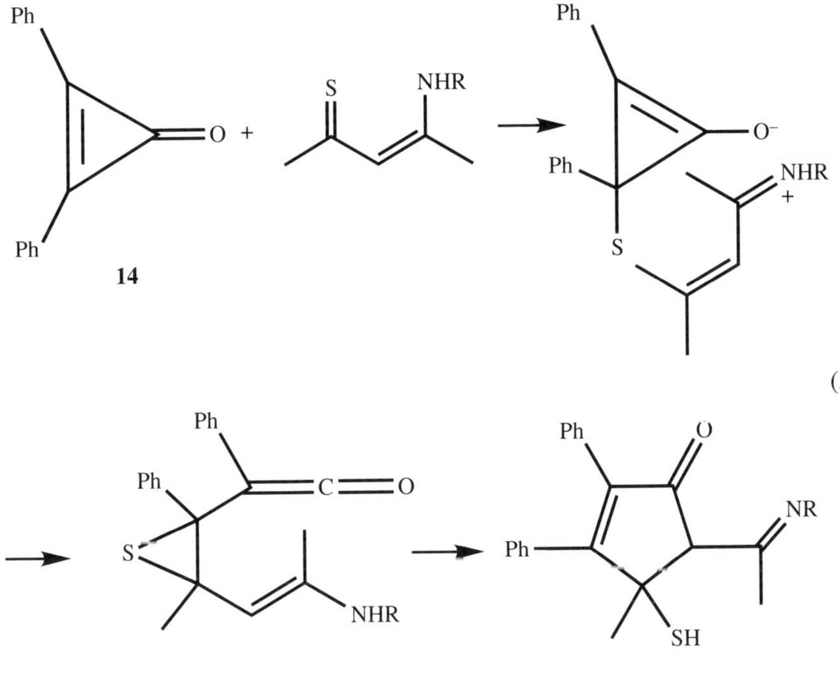

(5)

134   PREPARATION OF KETENES

Photolysis of the cyclopropenyl aldehyde **17** in benzene gave the naphthol **18**, while reaction in MeOH gave **19**.[14] These reactions implicate the vinylketene **20**, whose formation was proposed to occur from a Wolff-type rearrangement from the vinyl carbene **21** (equations 6 and 7).[14]

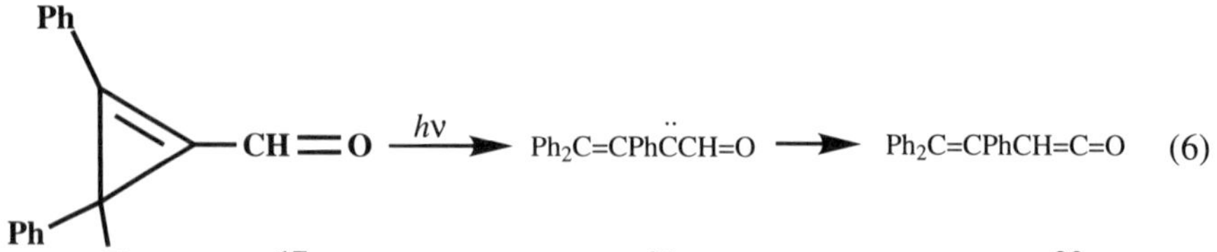

$$\text{17} \xrightarrow{h\nu} \text{Ph}_2\text{C=CPhCCH=O} \longrightarrow \text{Ph}_2\text{C=CPhCH=C=O} \quad (6)$$
$$\quad\quad\quad\quad\quad\quad\quad\quad\quad\quad\quad \text{21} \quad\quad\quad\quad\quad\quad\quad \text{20}$$

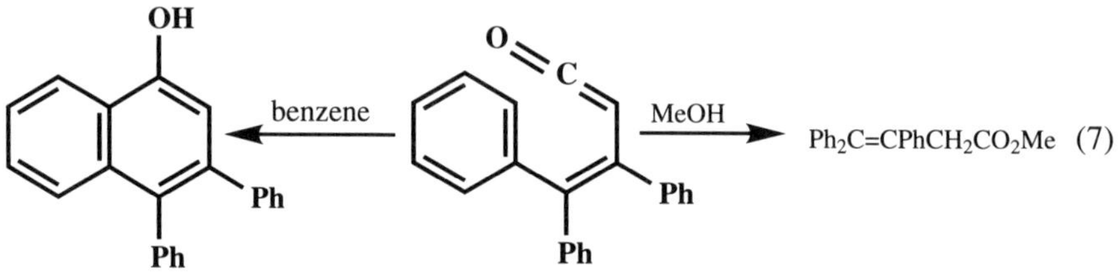

18 ← benzene — **20** — MeOH → $\text{Ph}_2\text{C=CPhCH}_2\text{CO}_2\text{Me}$ (7)
                                                                19

Photolysis of *E*-crotonaldehyde in the gas phase led to ethylketene among other products, as detected by the IR absorption at 2132 cm^{-1} (equation 8).[15,16] The formation of ethylketene was attributed to a unimolecular process involving vibrationally excited crotonaldehyde in the first excited singlet state.[16]

$$\text{CH}_3\text{CH=CHCH=O} \xrightarrow{h\nu} \text{CH}_3\text{CH=CHCH=O}^* \longrightarrow \text{CH}_3\text{CH}_2\text{CH=C=O} \quad (8)^{15}$$

The 1,3,4-oxadiazin-6-one **23** on reaction with cycloalkenes leads to adducts such as the crystalline **24** (equation 9) which itself reacted at 24 °C in CHCl$_3$ to form the long-lived ketene **25**, which is in equilibrium with the less stable isomer **26** (equation 10).[17-20] A related reaction is shown in Section 5.4.5 (equation 15).[21]

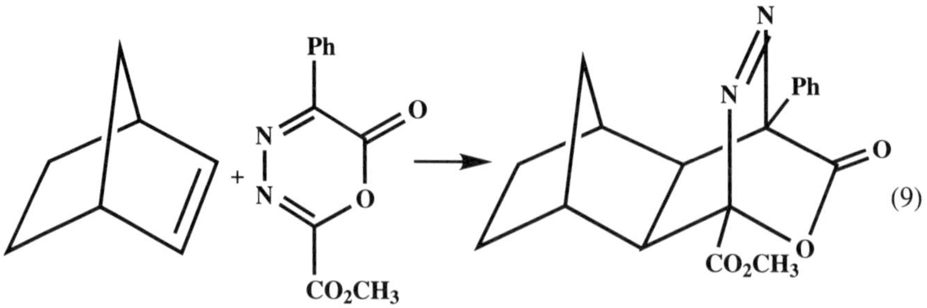

(9)

             23   **24**

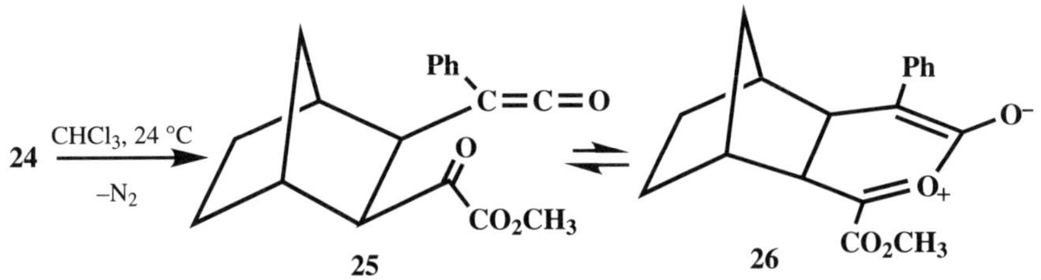

3.4 KETENES BY PHOTOCHEMICAL AND THERMOLYTIC METHODS

$$24 \xrightarrow[-N_2]{\text{CHCl}_3,\ 24\ °C} \quad 25 \rightleftharpoons 26 \tag{10}$$

Pyrolysis of the furandione **27** led to dibenzoylketene (**28**), as evidenced by the IR band at 2140 cm^{-1} at -196 °C.[22,23] At high temperature this band disappeared, and the products **29** and **30** were isolated, presumably via the dimer **31**. Other furandiones also give ketenes on pyrolysis.[24]

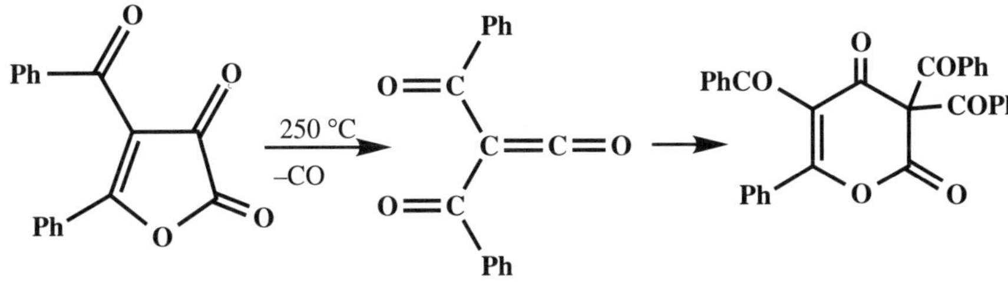

$$27 \xrightarrow[-CO]{250\ °C} 28 \rightarrow 31 \tag{11}$$

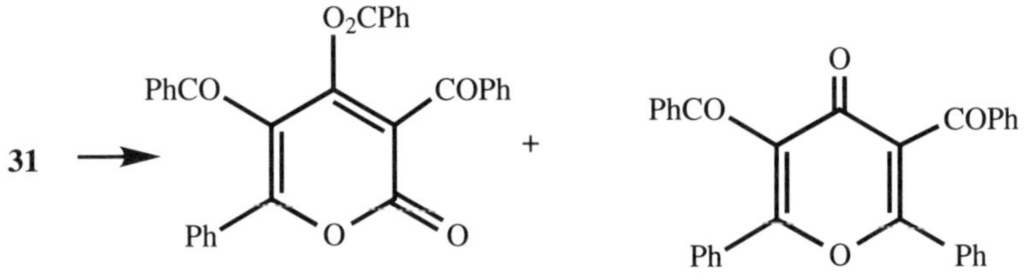

$$31 \rightarrow 29 + 30 \tag{12}$$

Pyrolysis of **32** at 500 °C was proposed to lead to the thioketoketene **33**, which cyclized to **34**. The latter product was not stable but was identified by its IR and NMR spectra at -40 °C.[22]

32 33 34

(13)

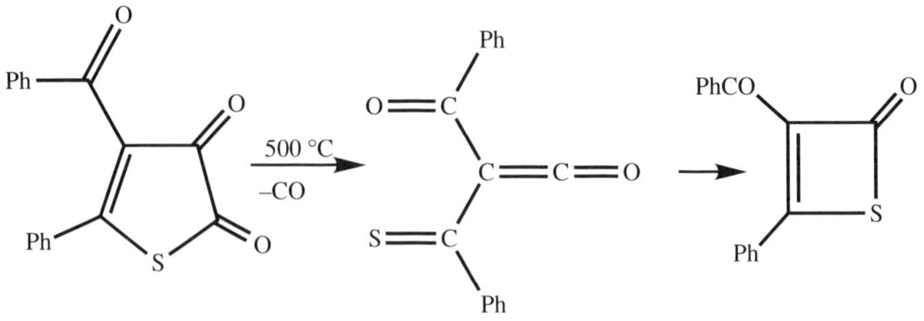

Thermolysis at 400–600 °C of 2,3-dihydropyrrole-2,3-diones **35** or Meldum's acid derivatives gave imidoylketenes **36** which could be trapped in an Ar matrix at 18 K and identified by their IR band at 2132 cm^{-1}.[25] Trapping at 77 K gave ketenimines **37,** and it was proposed that **36** and **37** were interconverting (equation 14).[25] Phenyl migration was also observed in the example of equation 15.[26]

35 36 37

(14)

(15)

2113 cm^{-1} Ad = 1-adamantyl

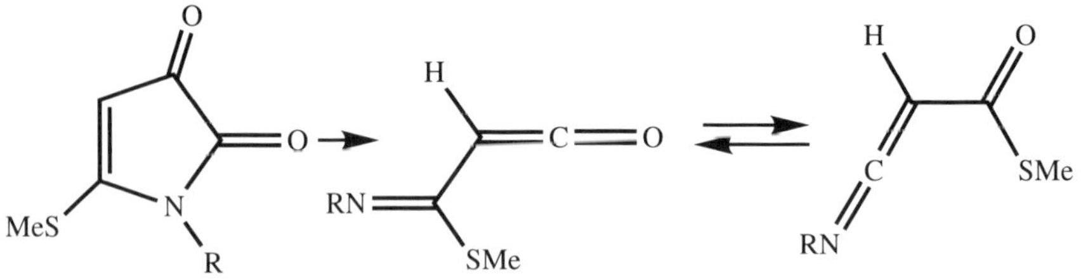

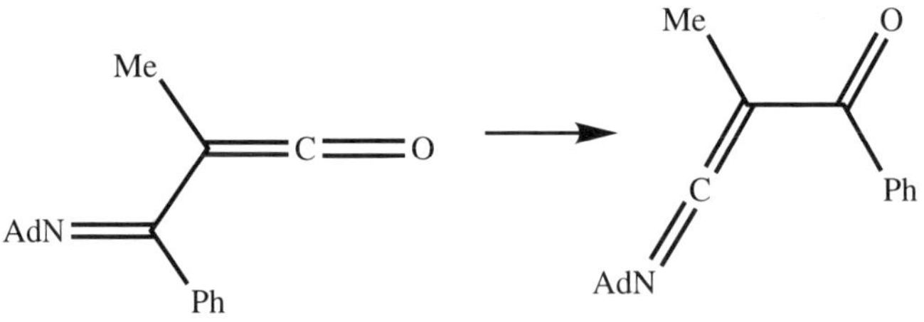

Pyrolysis of trialkylsilyl acetals gives ketenes, and as shown in equation 16 ^{18}O labeling implicates a 4-center transition state.[27]

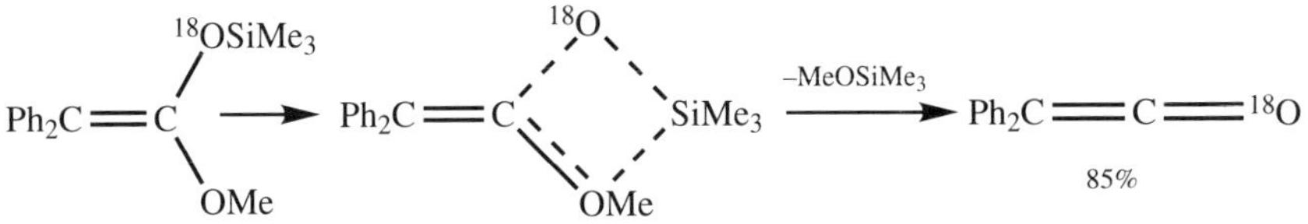

3.4 KETENES BY PHOTOCHEMICAL AND THERMOLYTIC METHODS

$$Ph_2C=C(^{18}OSiMe_3)(OMe) \longrightarrow Ph_2C=C\langle\substack{^{18}O\\OMe}\rangle SiMe_3 \xrightarrow{-MeOSiMe_3} Ph_2C=C=^{18}O \quad 85\%$$

(16)

The thermolysis of azidoquinones is a convenient route to cyanoketenes that is discussed in Section 4.1.3.

Pyrolysis of dioxolene **38** at 530 °C gave diphenylketene in 52% yield by capture with aniline.[28] This reaction may involve an oxirene intermediate as shown (equation 17).

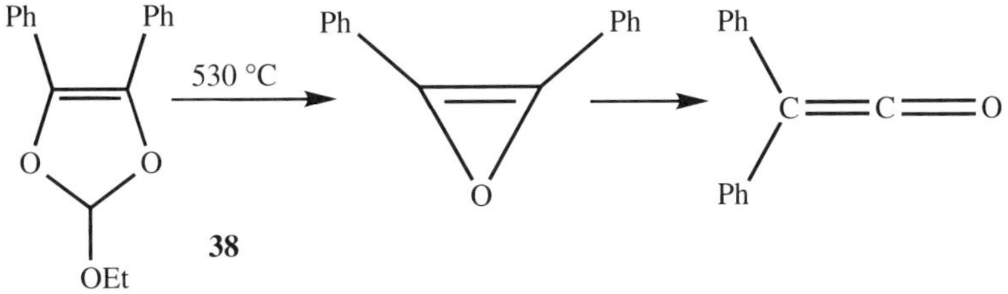

$$\mathbf{38} \xrightarrow{530\,°C} [\text{Ph}_2\text{-oxirene}] \longrightarrow Ph_2C=C=O \quad (17)$$

The addition of carbon monoxide to carbenes is another thermal route to ketenes, as in the example of equation 18.[29] Other examples are cited in Section 3.5. The reported[30] formation of a stable ketene by this route could not be replicated.[31]

$$(c\text{-Pr})_2CN_2 \xrightarrow[6\,K]{h\nu} (c\text{-Pr})_2C: \xrightarrow{CO} (c\text{-Pr})_2C=C=O \quad (18)$$

References

1. Erman, W. F. *J. Am. Chem. Soc.* **1967**, *89*, 3828–3841.
2. Erman, W. F. *J. Am. Chem. Soc.* **1969**, *91*, 779–780.
3. Erman, W. F.; Wenkert, E.; Jeffs; P. W. *J. Org. Chem.* **1969**, *34*, 2196–2203.
4. Gibson, T. W.; Erman, W. T. *J. Org. Chem.* **1972**, *37*, 1148–1154.
5. Ho, C.; Bond, F. T. *J. Am. Chem. Soc.* **1974**, *96*, 7355–7356.
6. Hart, H.; Shih, E. *J. Org. Chem.* **1976**, *41*, 3377–3381.
7. Schiess, P.; Radimerski, P. *Angew. Chem., Int. Ed. Engl.* **1972**, *11*, 288–289; *Helv. Chim. Acta* **1974**, *57*, 2583–2597.
8. Pirkle, W. H.; Seto, H.; Turner, W. V. *J. Am. Chem. Soc.* **1970**, *92*, 6984–6985.
9. Huang, B.-S.; Pong, R. G. S.; Laureni, J.; Krantz, A. *J. Am. Chem. Soc.* **1977**, *99*, 4154–4156.

10. Chapman, O. L.; McIntosh, C. L.; Pacansky, J. *J. Am. Chem. Soc.* **1973**, *95*, 244–246.
11. Engle, C. R.; DeKrassny, A. F.; Belanger, A.; Dionne, E. *Can. J. Chem.* **1973**, *51*, 3263–3271.
12. Arnold, B. R.; Brown, C. E.; Lusztyk, J. *J. Am. Chem. Soc.* **1993**, *115*, 1576–1577.
13. Kascheres, A.; Kascheres, C.; Braga, A. C. H. *J. Org. Chem.* **1993**, *58*, 1702–1703.
14. Schrader, L.; Hartmann, W. *Tetrahedron Lett.* **1973**, 3995–3996.
15. Coomber, J. W.; Pitts, J. N., Jr.; Schrock, R. R. *Chem. Commun.* **1968**, 190–191.
16. Coomber, J. W.; Pitts, J. N., Jr. *J. Am. Chem. Soc.* **1969**, *91*, 4955–4960.
17. Hegmann, J.; Christl, M. *Tetrahedron Lett.* **1987**, *28*, 6429–6432.
18. Christl, M.; Hegmann, J.; Reuchlein, H. *Tetrahedron Lett.* **1987**, *28*, 6433–6436.
19. Christl, M.; Lanzendörfer, U.; Grötsch, M. M.; Hegmann, J. *Angew. Chem., Int. Ed. Engl.* **1985**, *24*, 886–888.
20. Christl, M.; Lanzendörfer, U.; Grötsch, M. M.; Ditterich, E.; Hegmann, J. *Chem. Ber.* **1990**, *123*, 2031–2037.
21. Rubin, M. B.; Sander, W. W. *Tetrahedron Lett.* **1987**, *28*, 5137–5140.
22. Wentrup, C.; Winter, H.-W.; Gross, G.; Netsch, K.-P.; Kollenz, G.; Ott, W.; Biedermann, A. G. *Angew. Chem., Int. Ed. Engl.* **1984**, *23*, 800–801.
23. Ziegler, E.; Kollenz, G.; Ott, W. *Synthesis* **1973**, 679–680.
24. Kolesnikova, O. N.; Livantsova, L. I.; Shurov, S. N.; Zaitseva, G. S.; Andreichikov, Yu. S. *Zh. Obshch. Khim. Engl. Trans.* **1990**, *60*, 467–468.
25. Kappe, C. O.; Kollenz, G.; Leung-Toung, R.; Wentrup, C. *J. Chem. Soc., Chem. Commun.* **1992**, 487–488.
26. Kappe, C. O.; Kollenz, G.; Netsch, K.; Leung-Toung, R.; Wentrup, C. *J. Chem. Soc., Chem. Commun.* **1992**, 488–490.
27. Kuo, Y.; Chen, F.; Ainsworth, C. *J. Am. Chem. Soc.* **1971**, *93*, 4604–4605.
28. Moss, G. I.; Crank, G.; Eastwood, F. W. *Chem. Commun.* **1970**, 206.
29. Ammann, J. R.; Subramanian, R.; Sheridan, R. S. *J. Am. Chem. Soc.* **1992**, *114*, 7592–7594.
30. Lyashchuk, S. N.; Skrypnik, Y. G. *Tetrahedron Lett.* **1994**, *35*, 5271–5274.
31. Dixon, D. A.; Arduengo, A. J., III; Dobbs, K. D.; Khasnis, D. V. *Tetrahedron Lett.* **1995**, *36*, 645–648.

3.5 KETENES FROM ALKENYLCARBENE METAL COMPLEXES

The reaction of chromium carbonyl complexes of carbenes with alkynes leading to the formation of a naphthol complex was reported by Dötz in 1975 (equation 1).[1] It was suggested[2] that this reaction proceeds via a complexed alkenylketene **1** as shown, as supported by the reaction kinetics[2] and the isolation of such complexed products.[2] However, in many other examples the respective roles of complexed ketenes, free ketenes, and other possible intermediates in these reactions have not been established.

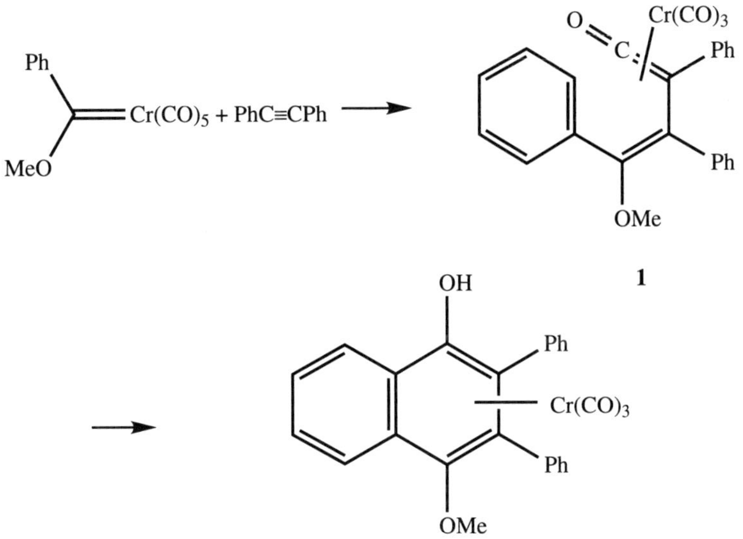

3.5 KETENES FROM ALKENYLCARBENE METAL COMPLEXES

(1)

An advantage of the use of metal complexed ketenes over free ketenes in synthetic procedures is that dimerization and other undesirable ketene reactions are supressed.[3-5]

Reaction of the cobalt complex **2** with 3-hexyne at 25 °C gave the vinylketene complex **3** whose structure was determined by X-ray crystallography.[6] Oxidation of **3** gave the furanone **4** (equation 2).[6] In other reactions chromium, molybdenum, and tungsten complexes have also been utilized.[7-13]

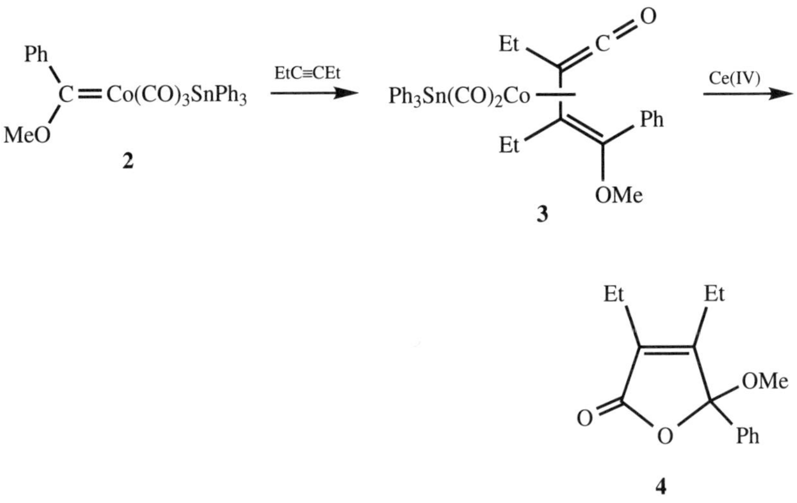

(2)

These processes have been extensively studied and widely reviewed,[4,5,8,14–21] and a simplified reaction scheme has been presented.[22] Thus on photolysis chromium carbene complexes undergo direct carbon monoxide insertion to give ketene complexes, which can be trapped by cycloaddition with alkenes (usually with the same selectivity as ketenes from other routes)[23,23a] or by nucleophilic addition (equation 3) or in suitable cases undergo intramolecular cyclization (equation 4).[22] Alternatively thermolysis of alkenyl chromium carbene complexes in the presence of alkynes proceeds by displacement by the alkyne of CO, followed by alkyne metathesis and CO insertion (equation 5).[22]

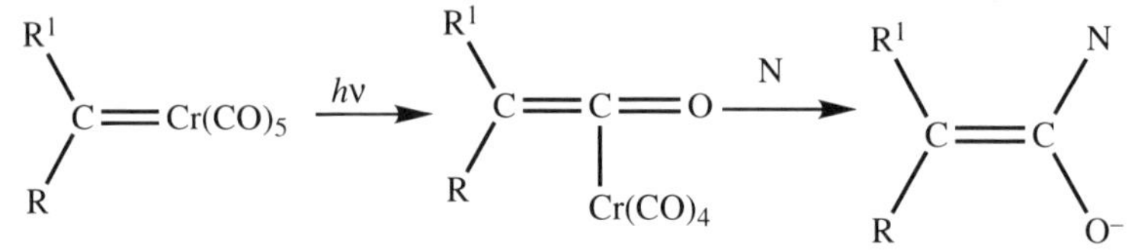

N = nucleophile

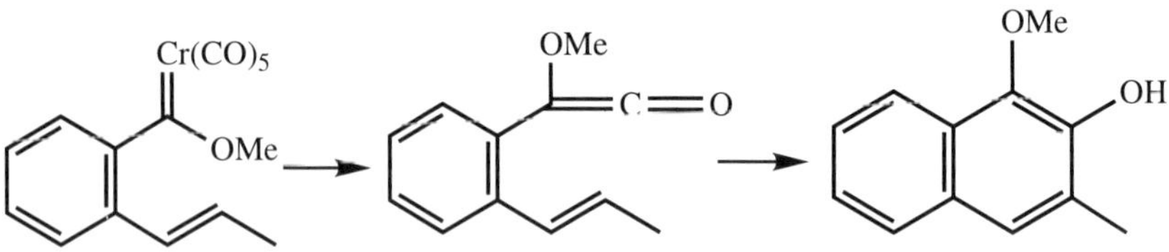

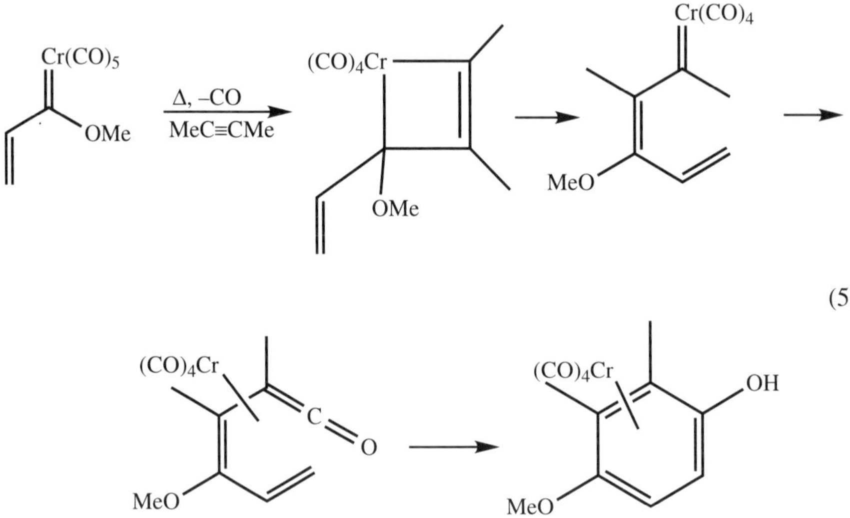

3.5 KETENES FROM ALKENYLCARBENE METAL COMPLEXES

Many parallels with the chemistry of ketenes prepared by different routes have been noted, and in some cases different reactivity is observed for free ketenes compared to when metals are present. Many other types of metal complexes of ketenes are noted in Section 4.8.4, but a detailed discussion of these species is beyond the scope of this section.

Nucleophilic reactions of metal–ketene complexes have also been widely studied, and in one example comparison of the reactivity of ketenes generated from photolysis of the (oxazolidinone)carbene complex **5** and the ketene generated by the triethylamine reaction of the acid chloride **6** led to the conclusion that the ketene generated from the chromium derivative was complexed with chromium and gave different product diastereoselectivity.[10] Related complexes **7** have also been examined.[24,25]

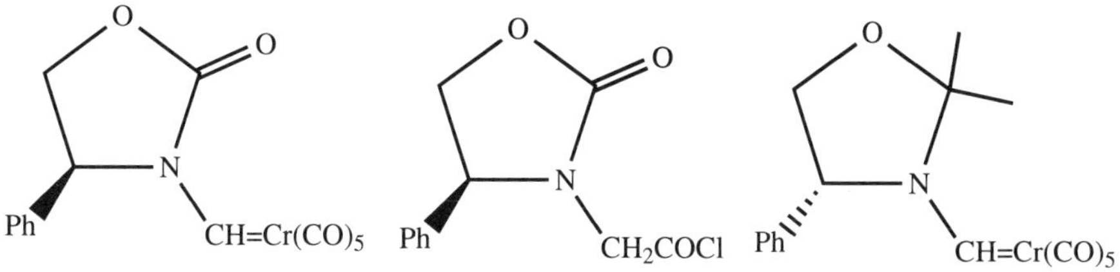

The reaction of ^{13}C-enriched chromium hexacarbonyl with methyllithium provides the ^{13}C-labeled chromium carbene complexes **9** and **10** (equation 6).[26] On photolysis these provide doubly ^{13}C-labeled ketene complexes such as **11**, which were used to prepare doubly labeled amino acid derivatives such as **12** (equation 7), as well as ketene derived cyclobutenones, β-lactams, azapenams, and β-keto enoates.[26]

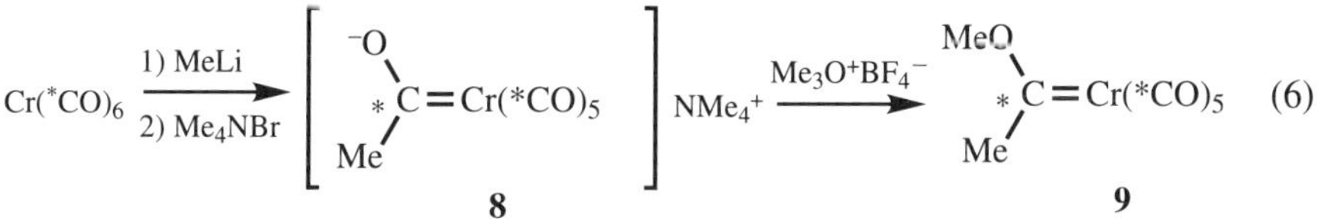

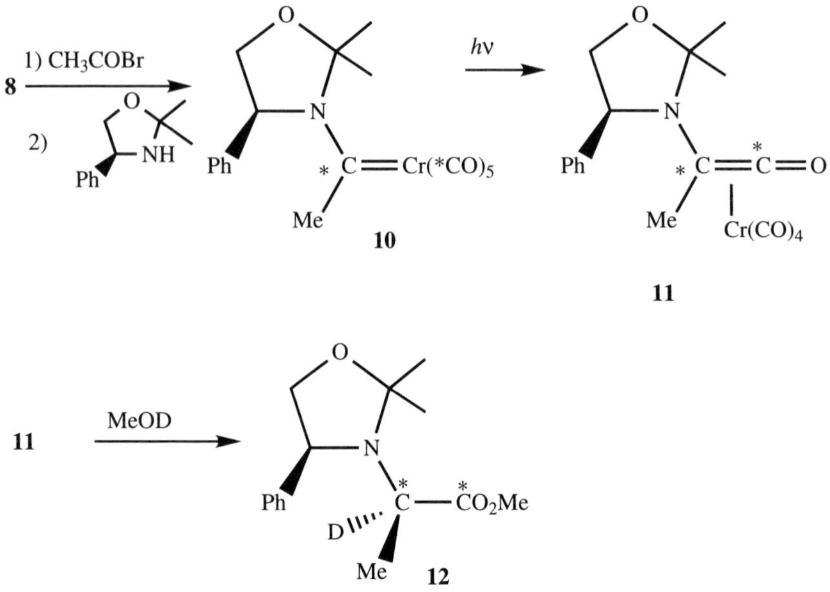

In a reaction which tested the competitive intramolecular reactivity of a metal complexed carbene with an arene and an OH group the carbene complex **13** gave **14** as the sole nonpolar product after oxidation, showing greater reactivity for the arene cyclization (equation 8).[27]

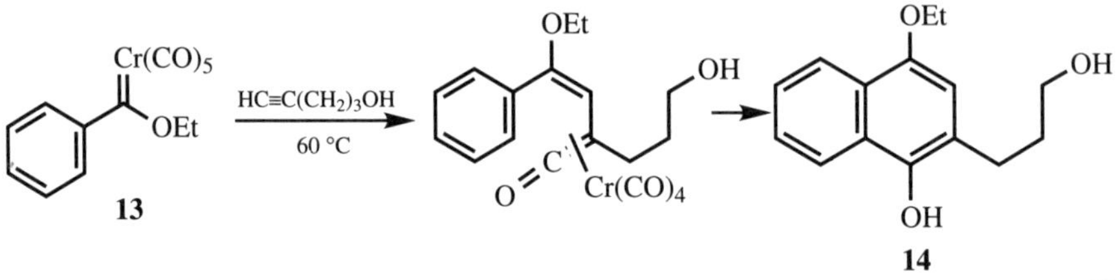

(8)[15]

Alkyl chromium carbene complexes react with alkynes to give cyclopentenones, and a mechanism was proposed which involved metal hydride intermediates, and possibly a vinylketene metal complex **15** (equation 9),[28] although this was not proven. A free vinylketene precursor to the cyclopentenone was excluded. Furans were obtained from similar complexes.[28a]

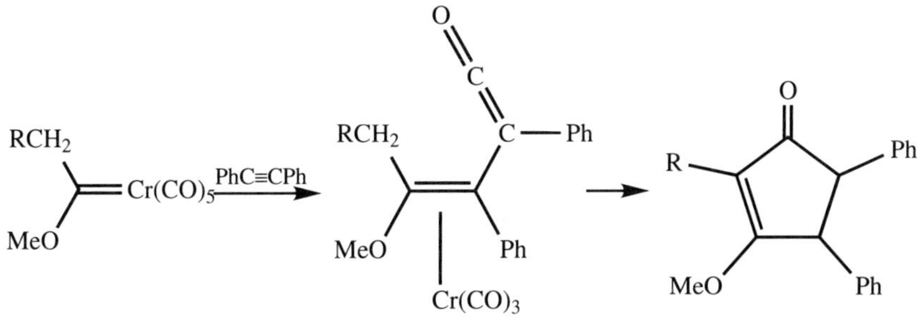

3.5 KETENES FROM ALKENYLCARBENE METAL COMPLEXES

(equation 9, showing structure **15** as intermediate)

The reaction of cyclopropylcarbene–chromium complexes and alkynes provides a general synthesis of cyclopentenones. Cyclopropylvinylketene complexes such as **16** were possible intermediates in this process but were not required by the evidence (equation 10).[29]

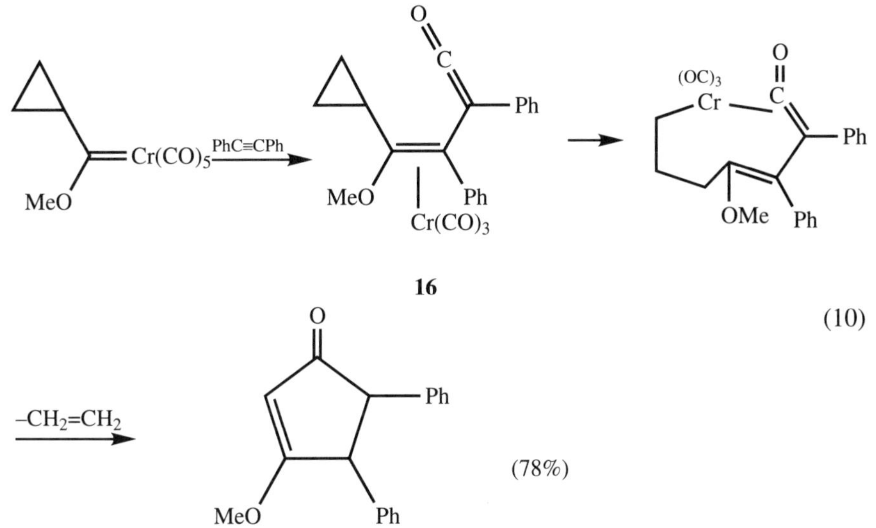

(equation 10, showing intermediate **16**, with $-CH_2=CH_2$ loss, giving product in 78%)

An aminovinylketene complex generated similarly cyclized to an ylide **17**, whose structure was determined by X-ray (equation 11).[30]

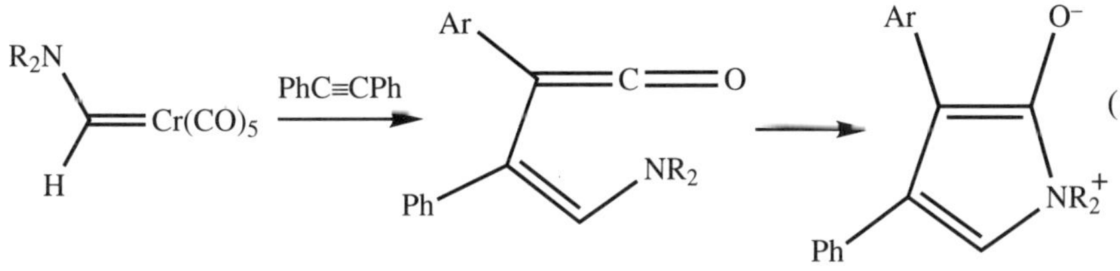

(equation 11, with $Ar = PhCr(CO)_3$, $R_2 = (CH_2)_5$, structure **17**)

References

1. Dötz, K. H. *Angew. Chem., Int. Ed. Engl.* **1975**, *14*, 644–645.
2. Fischer, H.; Múhlemeier, J.; Märkl, R.; Dötz, K. H. *Chem. Ber.* **1982**, *115*, 1355–1362.
3. Dötz, K. H.; Schäfer, T. O.; Harms, K. *Synthesis,* **1992**, 146–150.
4. Dötz, K. H. *New J. Chem.* **1990**, *14*, 433–445.
5. Dötz, K. H. In *Advances in Metal Carbene Chemistry;* Schubert, U., Ed.; Kluwer Academic Publishers: Boston, 1989; pp. 199–210.
6. Wulff, W. D.; Gilbertson, S. R.; Springer, J. P. *J. Am. Chem. Soc.* **1986**, *108*, 520–522.
7. Wulff, W. D.; Yang, D. *J. Am. Chem. Soc.* **1984**, *106*, 7565–7567.
8. Wulff, W. D.; Tang, P. C.; Chan, K.-S.; McCallum, J. S.; Yang, D. C.; Gilbertson, S. R. *Tetrahedron* **1985**, *41*, 5813–5832.
9. Bauta, W. E.; Wulff, W. D.; Pavkovic, S. F.; Zaluzec, E. J. *J. Org. Chem.* **1989**, *54*, 3249–3252.
10. Brandvold, T. A.; Wulff, W. D.; Rheingold, A. L. *J. Am Chem. Soc.* **1990**, *112*, 1645–1647.
11. Bos, M. E.; Wulff, W. D.; Miller, R. A.; Chamberlin, S.; Brandvold, T. A. *J. Am. Chem. Soc.* **1991**, *113*, 9293–9319.
12. Brandvold, T. A.; Wulff, W. D.; Rheingold, A. L. *J. Am. Chem Soc.* **1991**, *113*, 5459–5461.
13. Bao, J.; Dragisich, V.; Wenglowsky, S.; Wulff, W. D. *J. Am. Chem. Soc.* **1991**, *113*, 9873–9875.
14. Wulff, W. D. *Adv. Metal-Org. Chem.* **1988**, *1*, 209–393.
15. Wulff, W. D. In *Comprehensive Organic Synthesis;* Trost, B. M.; Fleming, I. Eds.; Pergamon Press: New York, 1991; Vol. 5, pp. 1065–1113.
16. Chan, K. S.; Peterson, G. A.; Brandvold, T. A.; Faron, K. L.; Challener, C. A.; Hyldahl, C.; Wulff, W. D. *J. Organomet. Chem.* **1987**, *334*, 9–56.
16a. Chamberlin, S.; Wulff, W. D.; Bax, B. *Tetrahedron* **1993**, *49*, 5531–5547.
17. Hegedus, L. S.; Montgomery, J.; Narukawa, Y.; Snustad, D. C. *J. Am. Chem. Soc.* **1991**, *113*, 5784–5791.
18. Thompson, D. K.; Suzuki, N.; Hegedus, L. S.; Satoh, Y. *J. Org. Chem.* **1992**, *57*, 1461–1467.
19. Sestrick, M. R.; Miller, M.; Hegedus, L. S. *J. Am. Chem. Soc.* **1992**, *114*, 4079–4088.
20. Betschart, C.; Hegedus, L. S. *J. Am. Chem. Soc.* **1992**, *114*, 5010–5017.
21. Hegedus, L. S. In *Advances in Metal Carbene Chemistry;* Schubert, U., Ed.; Klumer Academic Publishers: Boston, 1989; pp. 233–246.
22. Merlic, C. A.; Xu, D. *J. Am. Chem. Soc.* **1991**, *113*, 7418–7420.
23. Köbbing, S.; Mattay, J. *Tetrahedron Lett.* **1992**, *33*, 927–930.
23a. Köbbing, S.; Mattay, J. *Chem. Ber.* **1993**, *126*, 1849–1858.
24. Narukawa, Y.; Juneau, K. N.; Snustad, D.; Miller, D. B.; Hegedus, L. S. *J. Org. Chem.* **1992**, *57*, 5453–5462.
25. Schwindt, M. A.; Miller, J. R.; Hegedus, L. S. *J. Organomet. Chem.* **1991**, *413*, 143–153.
26. Lastra, E.; Hegedus, L. S. *J. Am. Chem. Soc.* **1993**, *115*, 87–90.

27. King, J. D.; Quayle, P. *Tetrahedron Lett.* **1991**, *32*, 7759–7762.
28. Challener, C. A.; Wulff, W. D.; Anderson, B. A.; Chamberlin, S.; Faron, K. L.; Kim, O. K.; Murray, C. K.; Xu, Y.; Yang, D. C.; Darling, S. D. *J. Am. Chem. Soc.* **1993**, *115*, 1359–1376.
28a. O'Connor, J. M.; Ji, H.; Rheingold, A. L. *J. Am. Chem. Soc.* **1993**, *115*, 9846–9847.
29. Tumer, S. U.; Herndon, J. W.; McMullen, L. A. *J. Am. Chem. Soc.* **1992**, *114*, 8394–8404.
30. Chelain, E.; Goumont, R.; Hamon, L.; Parlier, A.; Rudler, M.; Rudler, H.; Daran, J.; Vaissermann, J. *J. Am. Chem. Soc.* **1992**, *114*, 8088–8098.

3.6 KETENE FORMATION UNDER ACIDIC CONDITIONS

The reaction of methylmalonic acid or its dimethyl ester in SbF_5–FSO_3H gave rise to an ^1H NMR signal at δ 3.05 and an IR band at 2170 cm^{-1}. These signals were attributed to formation of the ketenyl acylium ion shown in equation 1,[1] but confirmatory experiments are needed to verify this assignment.

$$CH_3CH(CO_2H)_2 \xrightarrow{SbF_5-FSO_3H} CH_3C\begin{matrix}\nearrow C^+=O \\ \searrow C=O\end{matrix} \longleftrightarrow CH_3C\begin{matrix}\nearrow C=O \\ \searrow C^+=O\end{matrix} \quad (1)$$

Ketenes are reactive towards acids and so their preparation under acidic conditions is problematic. Thus at one time it was suggested that ketenes are intermediates in the halogenation of aliphatic carboxylic acids in strong acids.[2,3] However a more recent study of this reaction has instead supported that the enol of monacyl sulfate and not the ketene is the decisive intermediate in this halogenation.[4]

The reaction of acyl halides with Lewis acids leads to polyketenes, and a process involving acylium ions that are interconverting with ketenes was proposed.[5] This reaction is discussed further in Section 5.6. Some diverse cationic ketenyl species are mentioned in Sections 4.1.9 and 4.10.

References

1. Conrow, K.; Morris, D. L. *J. Chem. Soc., Chem. Commun.* **1973**, 5–6.
2. Ogata, Y.; Harada, T.; Sugimoto, T. *Can. J. Chem.* **1977**, *55*, 1268–1272.
3. Ogata, Y.; Sugimoto, T. *J. Org. Chem.* **1978**, *43*, 3684–3687.
4. Ogata, Y.; Adachi, K. *J. Org. Chem.* **1982**, *47*, 1182–1184.

5. Olah, G. A.; Zadok, E.; Edler, R.; Adamson, D. H.; Kasha, W.; Prakash, G. K. S. *J. Am. Chem. Soc.* **1989**, *111*, 9123–9124.

3.7 KETENES FROM OXIDATION OF ALKYNES

The chemical oxidation of alkynes proceeding through presumed oxirene intermediates leads to ketenes. Thus oxidation of [1-^{2}H] biphenylylacetylene (**1**, Ar = 4-PhC$_6$H$_4$) with *m*-chloroperbenzoic acid in methanol gave the methyl ester **2** with retention of the deuterium label, indicating the intermediacy of the oxirene **3** and ketene **4**.[1] When the oxidation was conducted enzymatically in aqueous medium, the deuterated acid corresponding to **2** was formed, implicating the intermediacy of **3** and **4** in this process as well. There are other reports implicating ketene intermediates in alkyne metabolism.[1–3]

$$\text{ArC}{\equiv}\text{CD} \xrightarrow{\text{Ar}^1\text{CO}_3\text{H}} \underset{\mathbf{3}}{\text{oxirene}} \longrightarrow \underset{\mathbf{4}}{\text{ArCD}{=}\text{C}{=}\text{O}} \xrightarrow{\text{MeOH}} \underset{\mathbf{2}}{\text{ArCDHCO}_2\text{Me}} \quad (1)$$

Phenylacetylene and diphenylacetylene reacted with peracids to give mixtures of products, some of which are attributed to the intervention of ketene intermediates.[4,5] The oxidation of some cyclic alkynes with *m*-chloroperbenzoic acid yields variable amounts of ring-contracted cycloalkanones, together with other products attributed to the intervention of ketocarbene intermediates (equation 2).[6,7] The cycloalkanones were shown to result from oxidation of the corresponding ketenes (equation 3).[6] Wolff rearrangements of the corresponding diazo ketones gave the same products but in different ratios, showing that the product-forming intermediates were related, but not identical.

(2) cyclooctyne + MCPBA → oxirene → cycloheptanone (22%) + bicyclic ketone (29%) + cycloheptenone (6%)

(3) cyclic ketene + MCPBA → cyclic ketone

The oxidation of alkynes with H$_3$PO$_5$ was proposed to lead to oxirene type intermediates that formed ketenes or were further oxidized to diketones.[8] These reactions were proposed not to involve ketocarbenes.[8] Oxidation of diphenylacetylene and phenylacetylene with dimethyldioxirane **5** and methyltrifluorodioxirane **6** also formed mostly ketene-derived products.[9] Reaction of 2-butyne and 4,4-dimethyl-2-pentyne with **5** led to mixtures of products, some of which were attributed to further oxidation of intermediate ketenes.[10]

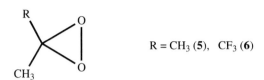

The photoreaction between nitrobenzene and diphenylacetylene was proposed to proceed by photoexcitation of nitrobenzene, which then transferred an oxygen atom to the alkynyl moiety leading to diphenylketene by migration of a phenyl group.[11] The autooxidations of some 1-phenylalkynes ArC≡CR are proposed to involve oxirenes leading to ketene intermediates.[12]

References

1. Ortiz de Montellano, P. R.; Kunze, K. L. *J. Am. Chem. Soc.* **1980**, *102*, 7373–7375.
2. Sullivan, H. R.; Roffey, P.; McMahon, R. E. *Drug Metabol. Dispos.* **1979**, *7*, 76–80; *Chem. Abstr.* **1979**, *91*, 68232m.
3. Wade, A.; Symons, A. M.; Martin, L.; Parke, D. V. *Biochem. J.* **1980**, *188*, 867–872.
4. Stille, J. K.; Whitehurst, D. D. *J. Am. Chem. Soc.* **1964**, *86*, 4871–4876.
5. McDonald, R. N.; Schwab, P. A. *J. Am. Chem. Soc.* **1964**, *86*, 4866–4871.
6. Concannon, P. W.; Ciabattoni, J. *J. Am. Chem. Soc.* **1973**, *95*, 3284–3289.
7. Ciabattoni, J.; Campbell, R. A.; Renner, C. A.; Concannon, P. W. *J. Am. Chem. Soc.* **1970**, *92*, 3826–3828.
8. Ogata, Y.; Sawaki, Y.; Ohno, T. *J. Am. Chem. Soc.* **1982**, *104*, 216–219.
9. Curci, R.; Fiorentino, M.; Fusco, C.; Mello, R.; Ballistreri, F. P.; Failla, S.; Tomaselli, G. A. *Tetrahedron Lett.* **1992**, *33*, 7929–7932.
10. Murray, R. W.; Singh, M. *J. Org. Chem.* **1993**, *58*, 5076–5080.
11. Scheinbaum. M. L. *J. Org. Chem.* **1964**, *29*, 2200–2203.
12. Dao, L. T. A.; Blau, K.; Pirtzkow, W.; Schmidt-Renner, W.; Voerckel, V.; Willecke, L. *J. Prakt. Chem.* **1984**, *326*, 73–80.

3.8 OTHER ROUTES TO KETENES

Reaction of β-peroxypropiolactones with triphenylphosphine gave ketene which was trapped with various alcohols.[1]

(1)

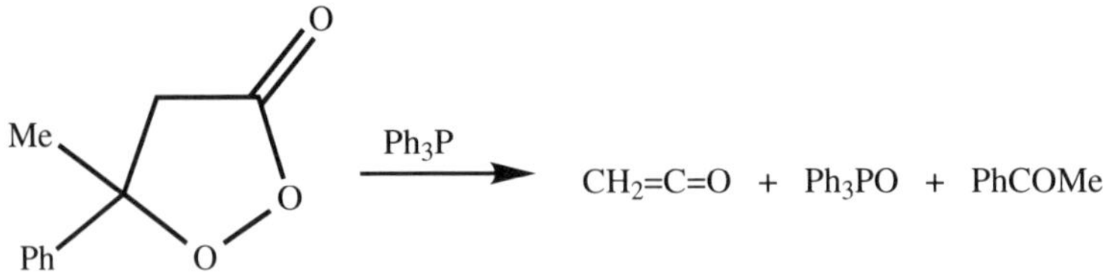

Ketene has been proposed as an intermediate in the H-ZSM-5–zeolite-catalyzed methanol to gasoline conversion.[2] The reaction is proposed to proceed with generation of CO, which reacts with an activated methanol to give an acylium ion which is reversibly deprotonated, as in equation 2. Higher acylium ions and ketenes are formed successively as long hydrocarbon chains are built up.

$$CH_3OH_2^+ \xrightarrow[-H_2O]{CO} CH_3C{\equiv}O^+ \rightleftharpoons CH_2{=}C{=}O \xrightarrow{CH_3OH_2^+} CH_3CH_2C{\equiv}O^+ \longrightarrow \text{etc.} \quad (2)$$

Ketenes are also proposed as intermediates in the formation of methanol from carbon monoxide and hydrogen promoted by rhodium catalysts,[3] and as discussed in Section 5.3 may form from the reaction of carbenes with carbon monoxide (equation 3).

$$R_2C{:} \xrightarrow{CO} R_2C{=}C{=}O \quad (3)$$

Ketene-forming reactions from cyclopropenones, as shown in equations 4–6, have been proposed,[4–6] and the IR absorption attributed to the ketene in equation 5 was observed.[5]

(4)[4]

(5)[5]

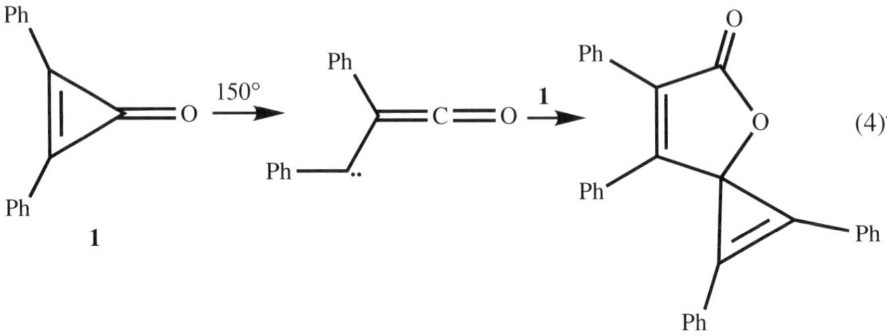

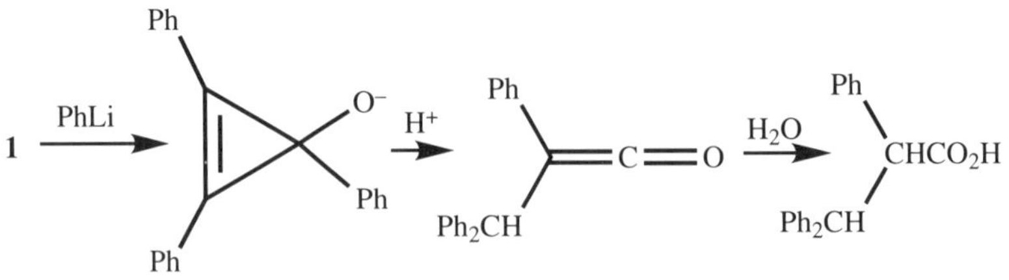

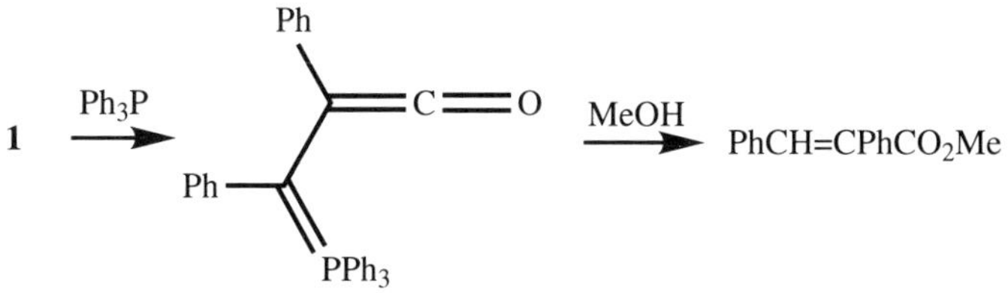

$$\mathbf{1} \xrightarrow{Ph_3P} \underset{\underset{PPh_3}{Ph}}{\overset{Ph}{\diagdown}}C=C=O \xrightarrow{MeOH} PhCH=CPhCO_2Me \qquad (6)^6$$

Photolysis of the sulfur ylide **2** in EtOH gave a 26% yield of ethyl phenylacetate, a result attributed to formation of phenylketene via a Wolff-type rearrangement (equation 7).[7]

$$\underset{\mathbf{2}}{PhCO\overset{-}{C}\overset{+}{H}SMe_2} \xrightarrow{h\nu} PhCOCH: \longrightarrow PhCH=C=O \xrightarrow{EtOH} PhCH_2CO_2Et$$

(7)

Reaction of arc-generated carbon atoms with *n*-butanal followed by addition of water led to the isolation of some pentanoic acid, which was proposed to arise from carbon atom insertion into the C=O bond (equation 8).[8]

$$n-PrCH=O + C \longrightarrow n-PrCH=C=O \xrightarrow{H_2O} n-BuCO_2H \qquad (8)$$

References

1. Adam, W.; Ramirez, R. J.; Tsai, S. *J. Am. Chem. Soc.* **1969**, *91*, 1254–1256.
2. Jackson, J. E.; Bertsch, F. M. *J. Am. Chem. Soc.* **1990**, *112*, 9085–9092.
3. Wang, H.; Liu, J.; Fu, J.; Wan, H.; Tsai, K. *Catal. Lett.* **1992**, *12*, 87–96.
4. Breslow, R.; Eicher, T.; Krebs, A.; Peterson, R. A.; Posner, J. *J. Am. Chem. Soc.* **1965**, *87*, 1320–1325.
5. Ciabattoni, J.; Kocienski, P. J.; Melloni, G. *Tetrahedron Lett.* **1969**, 1883–1887.
6. Hamada, A.; Takizawa, T. *Tetrahedron Lett.* **1972**, 1849–1850.
7. Trost, B. M. *J. Am. Chem. Soc.* **1966**, *88*, 1587–1588.
8. Dewar, M. J. S.; Nelson, D. J.; Shevlin, P. B.; Biesiada, K. A. *J. Am. Chem. Soc.* **1981**, *103*, 2802–2807.

CHAPTER 4

TYPES OF KETENES

4.1 CARBON-SUBSTITUTED KETENES

4.1.1 Alkylketenes

The family of alkylketenes begins with the parent $CH_2=C=O$, and includes aldoketenes such as *n*-butylketene (**1**), ketoketenes such as $Et_2C=C=O$, cycloalkylidene ketenes such as **2**, and cycloalkylketenes such as **3**. Cyclopropylketenes such as *c*-PrCH=C=O are considered separately in Section 4.1.5.

n–BuCH=C=O

1

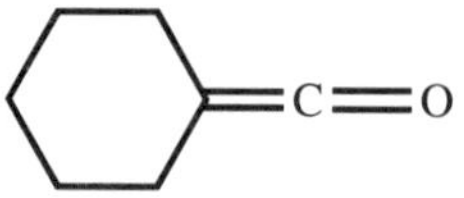

(cyclohexylidene)=C=O

2

(cyclohexyl)–CH=C=O

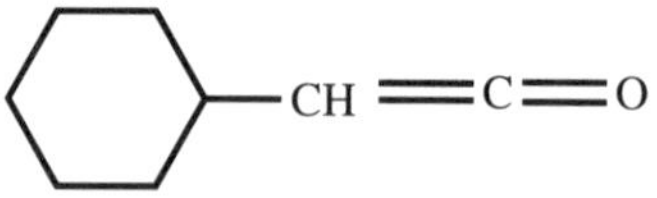

3

The preparation of dimethylketene from the zinc debromination route (equation 1) was reported in 1906 in Staudinger's second publication on ketenes.[1] As Staudinger pointed out,[1] this ketene had already been generated by dehydrochlorination of $Me_2CHCOCl$ by Et_3N followed by dimerization to the diketone by Wedekind and Weisswange as described in a dissertation in 1904, but these authors had not assigned the ketene structure.

$$Me_2CBrCOBr \xrightarrow{Zn} Me_2C=C=O \qquad (1)$$

Thermodynamically the stabilities of $CH_2=C=O$ and $CH_3CH=C=O$ compared to the corresponding alkenes are indicated by the isodesmic reaction of equation 2, and show that alkyl groups are less stabilizing of ketenes relative to alkenes, as compared to hydrogen.[2]

$$CH_3CH=C=O + CH_2=CH_2 \xrightarrow{\Delta E = -3.3 \text{ kcal/mol}} CH_2=C=O + CH_3CH=CH_2 \quad (2)$$

The aldoketenes, RCH=C=O, generally form unsymmetrical dimers **4** rather readily,[3,4] although some dimers prepared from acyl chlorides and Et$_3$N were originally misassigned as acylketenes.[3,4] Dialkyl ketenes R$_2$C=C=O with sterically undemanding alkyl groups are longer lived and form symmetrical dimers **5** (see Section 5.4.1.1).[5]

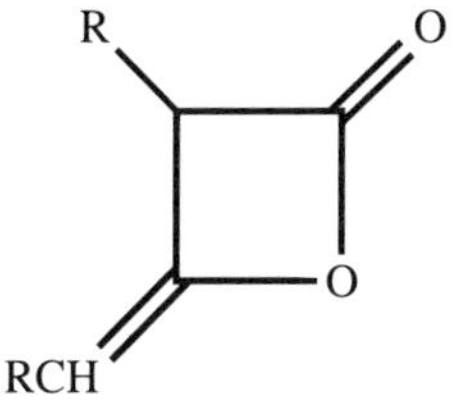

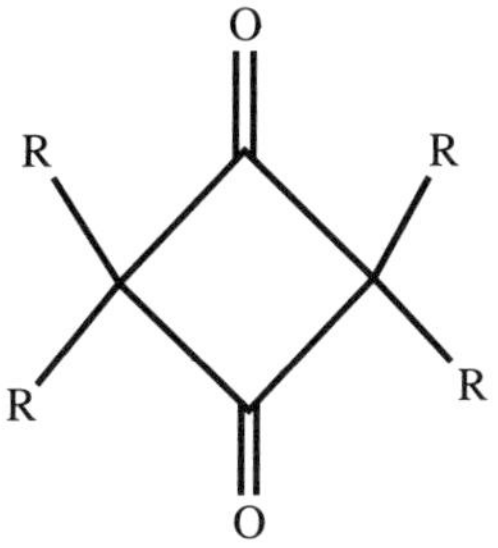

Ketenes substituted with bulky groups become much less reactive. The most notable example is di-*tert*-butylketene (**6**),[6–10] which shows no tendency to dimerize, and reacts only slowly with H$_2$O (see Section 5.5.1.2). However **6** undergoes cycloaddition with the imine **7**, and reaction rates for this process were measured at 100 and 120 °C (equation 3).[10]

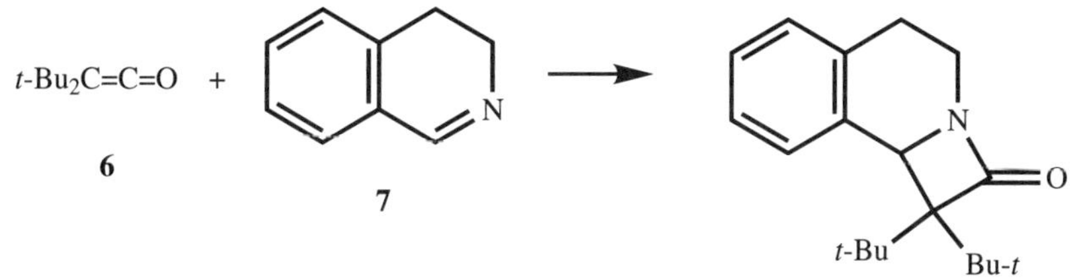

Other highly crowded ketenes which appear resistant to dimerization and are rather unreactive to nucleophiles are **8–13**. Ketenes substituted with bulky groups have been prepared by a variety of methods. Thus di-*tert*-butylketene (**6**) has been prepared by dehydrochlorination of the acid chloride,[6,7] dehydration of the acid with DCC,[8] and dehalogenation of *t*-Bu$_2$CClCOBr with Zn.[9] Wolff rearrangement of **14**, however, gave the ketene **6** in only 0–3% yield.[11]

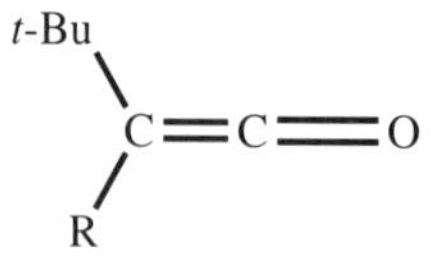

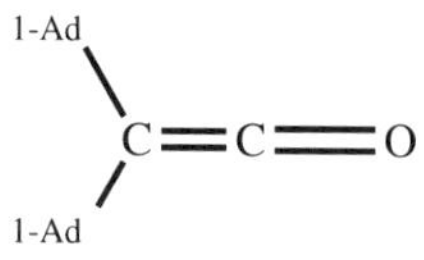

TYPES OF KETENES

R = i-Pr (**8**)[13]
R = 1-Ad (**9**)[6] (1-Ad = 1-adamantyl)
R = PhMe$_2$C (**10**)[15]

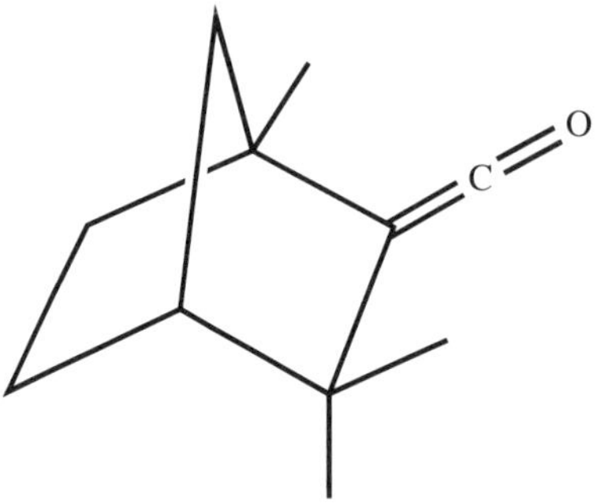

12[16] 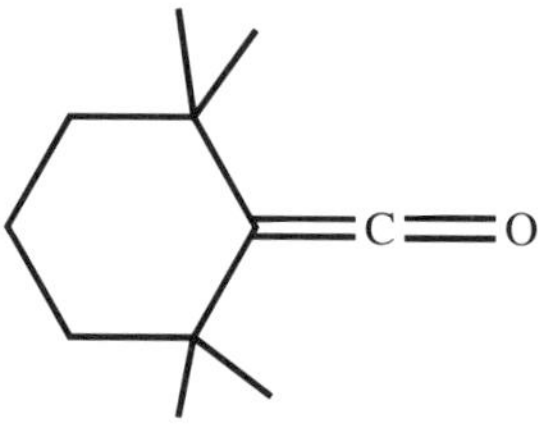   **13**[16] 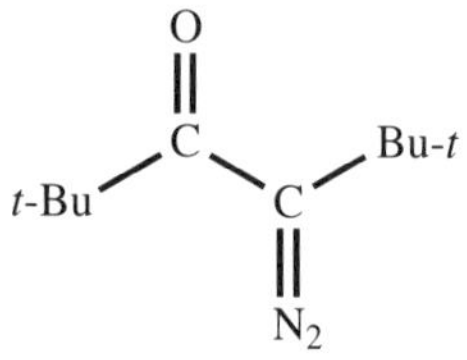   **14**

This latter result is consistent with an *s-E* geometry for the diazoketone precursor **14** for **6** which precludes concerted or nonconcerted *tert*-butyl migration to the backside of the diazo carbon.[12] However, in the cyclic analogue **15** such migration is possible and the ketene **16** is formed in 96% yield on photolysis (equation 4).[12]

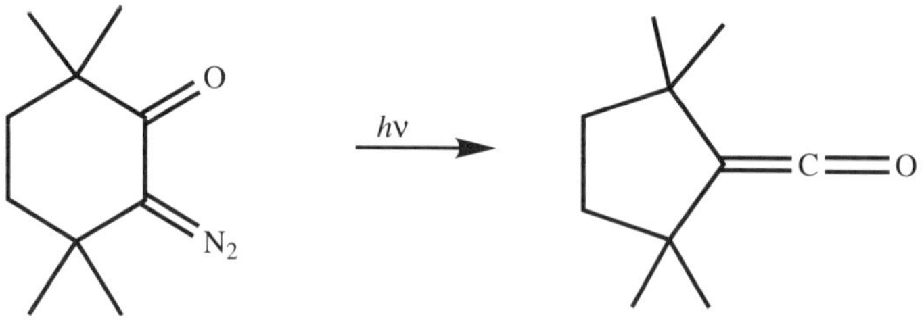

15 $\xrightarrow{h\nu}$ **16**   (4)

Flash vacuum pyrolysis of crowded silylated thionocarboxylic acid derivatives gave the ketenes as major products (equation 5).[14]

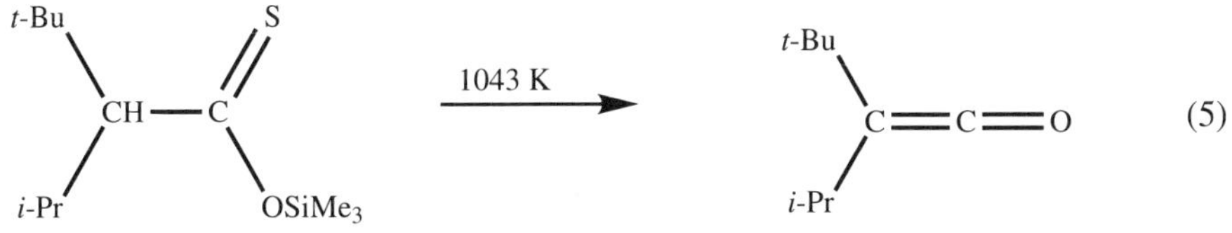

$$\underset{i\text{-Pr}}{\overset{t\text{-Bu}}{>}} CH - C \underset{OSiMe_3}{\overset{S}{\diagup\!\!\!\!\diagdown}} \quad \xrightarrow{1043 \text{ K}} \quad \underset{i\text{-Pr}}{\overset{t\text{-Bu}}{>}} C{=}C{=}O \qquad (5)$$

8

Some large ring cycloalkylideneketenes such as **17**[17] are also quite resistant to dimerization and hydrolyze slowly. A variety of such ketenes have been formed by ring contraction via Wolff rearrangement, and for the 8- to 11-membered rings the compounds were sufficiently stable for isolation by gas chromatography and measurement of IR bands between 2101 and 2105 cm^{-1}.[18] Sensitized photolysis greatly reduces ketene formation.[19] Although the geometries of these ketenes have not been elucidated, it seems likely that the alkyl side chains block the approach of other molecules to the ketenyl group.

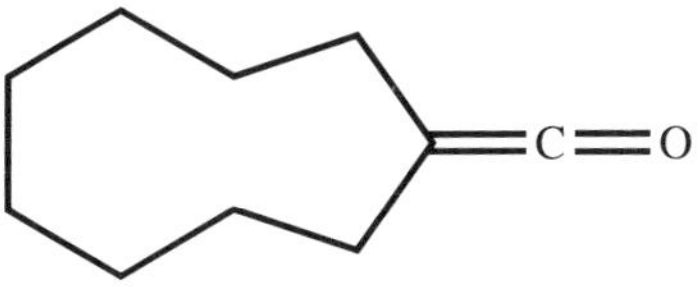

17

As discussed in Sections 3.1 and 3.2.1, dialkylketenes can also be prepared from thermolysis of the dimers, and from polymeric malonic anhydrides.

tert-Butylketene is not nearly so stable as ketenes with two bulky substituents, and readily undergoes dimerization and cycloaddition with cyclopentadiene.[20] This ketene has been prepared by dehydrochlorination of the acyl chloride,[20,20a] by the flash vacuum pyrolysis route related to equation 5,[14] and by dehalogenation (see Table 2, Section 2.1).[58]

Cyclohexylideneketene **18** is available by dehydrochlorination of the acyl chloride,[21] and adamantylideneketene **19** is available by the dehalogenation of the α-bromo acyl chloride[22] and by Wolff rearrangement.[23]

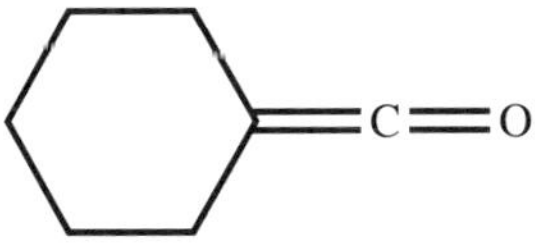

18 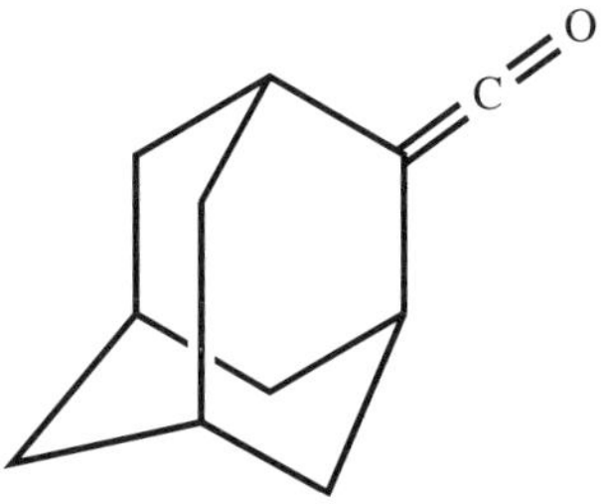   **19**

TYPES OF KETENES

The chemical and physical properties of alkylketenes as well as the structures and spectral properties are discussed in the appropriate sections. Not surprisingly there is little evidence for any interaction of the alkyl side chains with the ketenyl moiety unless these former contain reactive functional groups, such as OH, NH$_2$, alkenyl, or others. Such functionalized side chains may also be utilized in further synthetic transformations. Thus 3-bromo-n-propylketene (**20**) gave the cyclobutanone (**21**) by [2 + 2] cycloaddition, and further reaction with n-Bu$_3$SnH gave ring-expanded bicyclic derivatives **22** and other products (equation 6).[24]

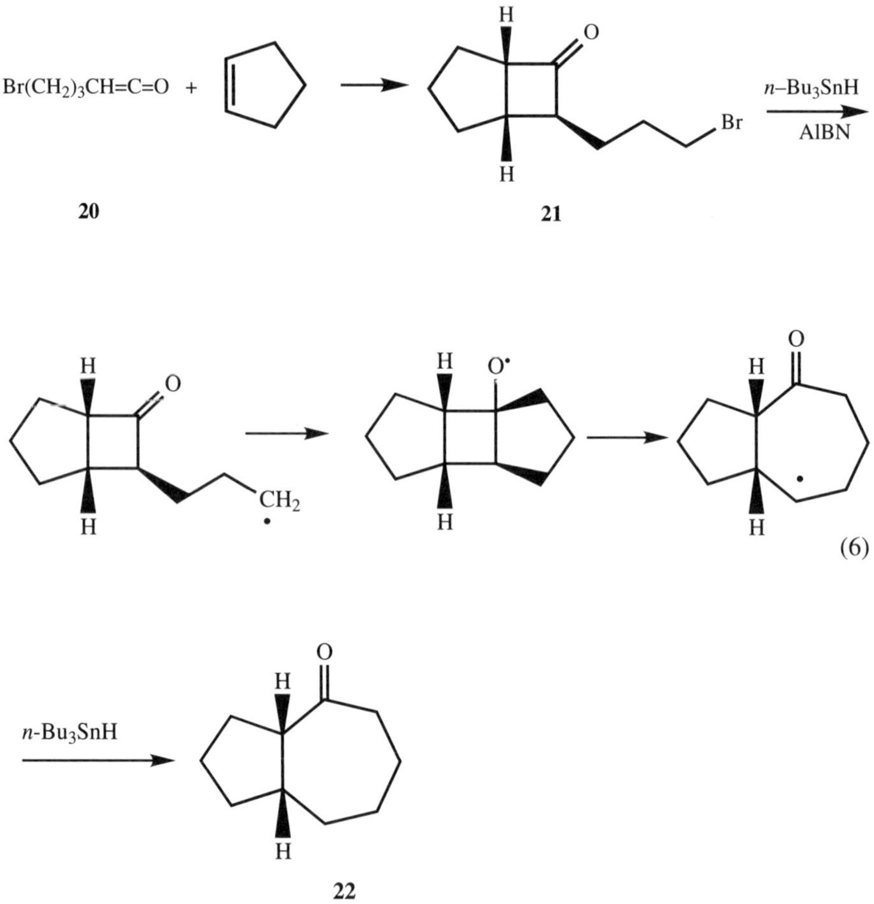

The steroidal ketene **23** was formed from the acyl chloride and Et$_3$N.[25] Reaction of **23** with MeOH gave the methyl ester, while ozonolysis gave **24** (equation 7).

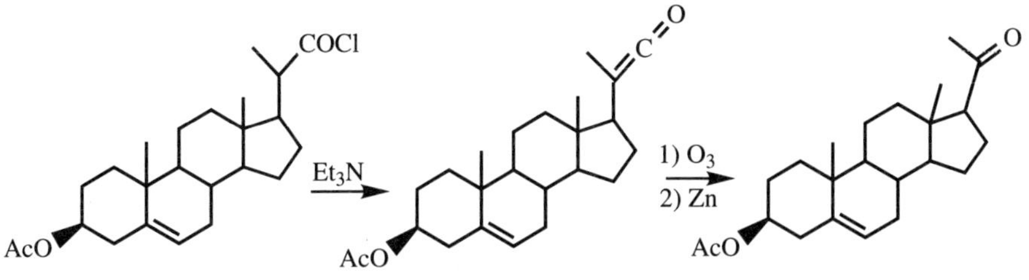

4.1.1 ALKYLKETENES

(7)

Photolysis of 2-diazo-3-hydroxy-1-cyclohexanones in H_2O–THF gave carboxylic acids, indicating formation of the hydroxy-substituted ketenes (equation 8).[26]

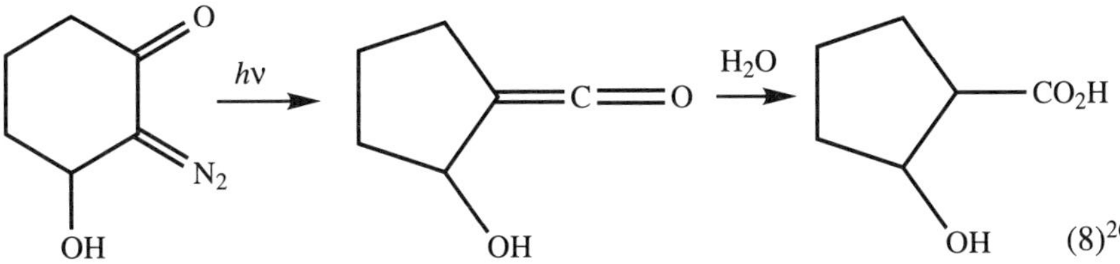

(8)[26]

The proposed[27] preparation of the ketene **25** by a fragmentation route (equation 9) has been discounted.[28]

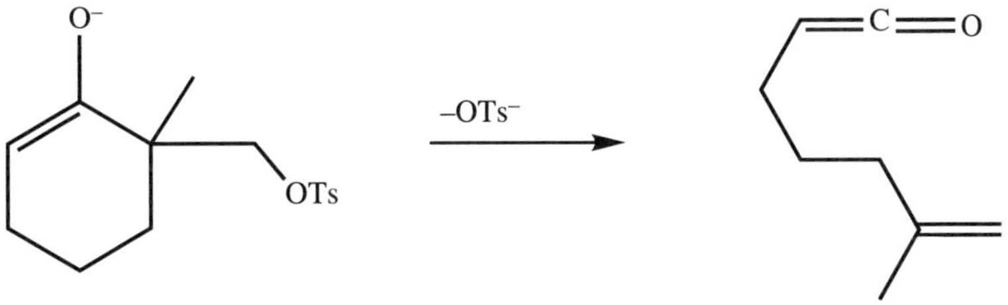

Reported alkylketenes, cycloalkylideneketenes, and cycloakylketenes are listed in Tables 1–3, respectively.

TABLE 1. Alkylketenes

Ketene	Formula	Preparation	Registry No.	Ref.
$CD_2=C=O$	C_2D_2O	Acetone pyrolysis	4789-21-3	29
	$(C_2D_2O)_2$		14523-82-1	30
$CD_2=C=^{18}O$	C_2D_2O	Anhydride pyrolysis	127867-16-7	31
$CHD=C=O$	C_2HDO	Anhydride pyrolysis	60032-17-9	31
$CHD=C=^{18}O$	C_2HDO	Anhydride pyrolysis	127867-15-6	31
$CHT=C=O$	C_2HOT	Anhydride pyrolysis	15727-74-9	32
$CH_2=C=O$	C_2H_2O	Dimer pyrolysis	463-51-4	33
	$(C_2H_2O)_2$		6842-10-0	34
	$(C_2H_2O)_2$		6144-29-2	33
$^{13}CH_2=C=O$	C_2H_2O	Acid pyrolysis	107573-88-6	35
$CH_2=^{13}C=O$	C_2H_2O	Acid pyrolysis	107573-87-5	35
$CH_2=C=^{18}O$	C_2H_2O	Acid pyrolysis	67071-63-0	35
$CH_2=^{11}C=O$	C_2H_2O	Acid pyrolysis	135909-23-8	36
$^{14}CH_2=C=O$	C_2H_2O	Anhydride pyrolysis	20216-86-8	37
$CH_2=^{14}C=O$	C_2H_2O		119149-21-2	37a
$CT_2=C=O$	C_2T_2O		20216-85-7	37
$CD_3CH=C=O$	C_3HD_3O	Anhydride pyrolysis	5982-12-7	38
$^+CH_2CH=C=O$	C_3H_3O	Theory	105028-86-2	39
$CH_2DCH=C=O$	C_3H_3DO	Anhydride pyrolysis	5917-71-5	38
$CH_3CD=C=O$	C_3H_3DO	Anhydride pyrolysis	5917-70-4	38
$CH_3CH=C=O$	C_3H_4O	Dehalogenation	6004-44-0	40
	$(C_3H_4O)_2$		86156-35-6	41
$CH_3CH=C=^{18}O$	C_3H_4O	Anhydride pyrolysis	5982-11-6	38
$^{13}CH_3CH=C=O$	C_3H_4O	Anhydride pyrolysis	5982-10-15	38
$CH_3{}^{13}CH=C=O$	C_3H_4O	Anhydride pyrolysis	5982-09-2	38
$CH_3CH=^{13}C=O$	C_3H_4O	Anhydride pyrolysis	5982-08-1	38

Compound	Formula	CAS	Method	Ref
O=CHCH$_2$CH=C=O	C$_4$H$_4$O$_2$	138720-22-6	Mass spectrom	42
(CD$_3$)$_2$C=C=O	C$_4$D$_6$O	50905-93-6	Malonic anhydride	43
C$_2$F$_5$CH=C=O	C$_4$HF$_5$O	82515-15-9	Wolff	44
CD$_3$CMe=C=O	C$_4$H$_3$D$_3$O	50905-92-5	Malonic anhydride	43
CH$_2$DCMe=C=O	C$_4$H$_5$DO	50905-94-7	Malonic anhydride	43
^{13}CH$_3$CMe=C=O	C$_4$H$_6$O	50905-95-8	Malonic anhydride	43
(CH$_3$)$_2$C=C=O	C$_4$H$_6$O	598-26-5	Dehalogenation	40
	(C$_4$H$_6$O)$_2$	7686-84-2	(Unsymmetrical)	45
EtCH=C=O	C$_4$H$_6$O	20334-52-5	Dehalogenation	46
EtCH=^{11}C=O	C$_4$H$_6$O		Acid pyrolysis	47
$^-$O$_2$CCOCH$_2$CH=C=O	C$_5$H$_3$O$_4^-$	138720-53-3	Mass spectrum	42
HO$_2$CCOCH$_2$CH=C=O	C$_5$H$_4$O$_4$	138720-21-5	Mass spectrum	42
ClOCCHMeCH=C=O	C$_5$H$_5$ClO$_2$	31537-46-9	Dehydrohalogenation	48
CH$_2$=CHCH$_2$CH=C=O	C$_5$H$_6$O	93039-42-0	Rearrangement	49
i-PrCH=C=O	C$_5$H$_8$O	69909-29-6	Dehydrochlorination	50
n-PrCH=C=O	C$_5$H$_8$O	10248-82-5	Dehydrochlorination	50
EtCMe=C=O	C$_5$H$_8$O	36854-53-2	Dehydrochlorination	50
CH$_2$=CHCH$_2$CH$_2$CH=C=O	C$_6$H$_8$O	70689-91-7	Index error	51
CH$_2$=CHCH$_2$CMe=C=O	C$_6$H$_8$O		Malonic anhydride	52,53
EtO$_2$CCH$_2$CH$_2$CH=C=O	C$_6$H$_8$O$_3$			54
i-PrCH$_2$CH=C=O	C$_6$H$_{10}$O	69802-75-1	Dehydrochlorination	50
n-C$_4$H$_9$CH=C=O	C$_6$H$_{10}$O	116373-98-9	Dehydrochlorination	50,55
	(C$_6$H$_{10}$O)$_2$	89723-54-6		
		108487-36-1		
n-PrCMe=C=O	C$_6$H$_{10}$O	29336-29-6	Dehydrochlorination	50,56
EtCHMeCH=C=O	C$_6$H$_{10}$O	89723-56-8	Dehydrochlorination	50

TABLE 1. (*Continued*)

Ketene	Formula	Preparation	Registry No.	Ref.
i-PrPrMe=C=O	$C_6H_{10}O$	Dehydrochlorination	59005-30-0	57
t-BuCH=C=O	$C_6H_{10}O$	Dehydrochlorination	59005-31-1	20,44
$Et_2C=C=O$	$C_6H_{10}O$	Dehydrochlorination	24264-08-2	50,58
$Me_3SiCH_2CH=C=O$	$C_6H_{12}OSi$	Dehydrochlorination	61063-48-7	59
$RCH_2CH_2CH=C=O$	$C_7H_8O_2$	Photolysis	66688-18-14	60
R = O=CHCH=CH				
CH_2=CHCHEtCH=C=O	$C_7H_{10}O$	Wolff	51550-09-5	61
EtO_2CCH_2CMe=C=O	$C_7H_{10}O_3$		89723-55-7	62
n-C_5H_{11}CH=C=O	$C_7H_{12}O$	Dehydrochlorination	65113-44-2	50
n-PrCEt=C=O	$C_7H_{12}O$	Wolff	118406-38-5	63
$(MeO)_3SiCH_2CH_2CH$=C=O	$C_7H_{14}O_4Si$		83740-58-3	64
$Me_3GeOCRR^1CH$=C=O	$C_8H_{10}Cl_3F_3GeO_2$	Alkyne pyrolysis		65
R = Cl_2F, $R^1 = CClF_2$				
$RC(CCl_2F)(CClF_2)CH$=C=O	$C_8H_{10}Cl_3F_3O_2Si$	Alkyne pyrolysis	83740-53-8	65
R = Me_3SiO				
$Me_3SiOC(CF_2Cl)_2CH$=C=O	$C_8H_{10}Cl_2F_4O_2Si$	Alkyne pyrolysis	79305-64-9	65
$Me_3GeOC(CF_3)_2CH$=C=O	$C_8H_{10}F_6Ge_2$	Alkyne pyrolysis	83740-54-9	65
$Me_3SiOC(CF_3)_2CH$=C=O	$C_8H_{10}F_6O_2Si$	Alkyne pyrolysis	83740-52-7	65
$(CH_2$=CHCH$_2)_2C$=C=O	$C_8H_{10}O$	Malonic anhydride		52
$C_8H_{10}O_2(E)$		Photolysis	66688-26-4	58
$R(CH_2)_3CH$=C=O				
R = O=CHCH=CH	Z			
CH_2=CMe$(CH_2)_3$CH=C=O	$C_8H_{12}O$	Photolysis	66688-25-3	58
	$C_8H_{14}O$	Incorrect structure	40942-42-3	27,28
n-HxCH=C=O			28988-82-1	67
$(C_8H_{14}O)_2$			113509-77-6	4

158

Structure	Formula	CAS	Method	Ref
n-Pr$_2$C=C=O	C$_8$H$_{14}$O	58844-38-5	Malonic anhydride	52,68
n-BuCEt=C=O	C$_8$H$_{14}$O	17139-73-0	Anhydride pyrolysis	69
n-BuCHMeCH=C=O	C$_8$H$_{14}$O	90370-65-3	Abstract error	70
t-BuSCH$_2$C(CH$_3$)=C=O	C$_8$H$_{14}$OS	82893-71-8	Dehydrochlorination	71
PhCH$_2$CH=C=O	C$_9$H$_8$O	87101-44-8	Dehydrochlorination	72
PhOCH$_2$CH=C=O	C$_9$H$_8$O$_2$	11306-20-8	Wolff	73
C$_9$H$_8$O$_2$$^{+\bullet}$		11306-21-9		73
⬠–CH=C=O	C$_9$H$_{10}$O		Wolff	74
RCH$_2$CMe=C=O R = Me$_2$C=CHCH$_2$	C$_9$H$_{14}$O	90611-14-6	Abstract error	75
RCHOHCH$_2$CH=C=O R = Me$_2$C=CHCH$_2$	C$_9$H$_{14}$O$_2$		Photolysis	76
t-BuC(Pr-i)=C=O	C$_9$H$_{16}$O	71106-63-3	Dehydrochlorination	13,77
PhCH$_2$CMe=C=O	C$_{10}$H$_{10}$O	42955-14-6		57a
PhSCH$_2$CMe=C=O	C$_{10}$H$_{10}$OS	82893-70-7	Dehydrochlorination	71
⬠–CMe$_2$CH=C=O	C$_{10}$H$_{14}$O		Wolff	61
RCH$_2$CH=C=O R = n-PrCH=CH(CHOH)$_2$	C$_{10}$H$_{16}$O$_3$		Photolysis	76
n-C$_8$H$_{17}$CH=C=O	C$_{10}$H$_{18}$O	112992-10-6	Dehydrochlorination	78
	(C$_{10}$H$_{18}$O)$_2$	112992-11-7		78
RCHMeCH=C=O R = i-Pr(CH$_2$)$_3$	C$_{10}$H$_{18}$O	66131-81-5	Photolysis	79
n-Bu$_2$C=C=O	C$_{10}$H$_{18}$O	36638-69-4	Malonyl peroxide	80
t-Bu$_2$C=C=O	C$_{10}$H$_{18}$O	19824-34-1	Dehydrochlorination	7
PhCH$_2$CEt=C=O	C$_{11}$H$_{12}$O	70633-90-8	Dehydrochlorination	55
PhCH(CMe=CF$_2$)CH=C=O	C$_{12}$H$_{12}$O	36920-31-7	Photolysis	81

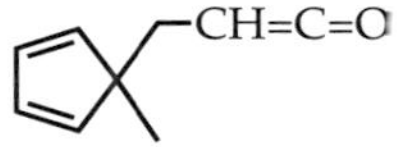

TABLE 1. (Continued)

Ketene	Formula	Preparation	Registry No.	Ref.
PhCH$_2$C(Pr-i)=C=O	C$_{12}$H$_{14}$O	Dehydrochlorination	70633-91-9	57
Ph(CH$_2$)$_4$CH=C=O	C$_{12}$H$_{14}$O	Photolysis	71260-91-8	82
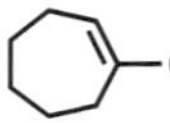 ⟨cycloheptylidene⟩—CMe$_2$CH=C=O	C$_{12}$H$_{18}$O	Wolff		61
RCMe=C=O	C$_{12}$H$_{18}$O$_2$	Thermolysis		83
R = Me$_2$C=MeCMeAc				
n-C$_{10}$H$_{21}$CH=C=O	C$_{12}$H$_{22}$O	Dehydrochlorination	74388-77-5	4
	(C$_{12}$H$_{22}$O)$_2$		74388-78-6	4
(t-BuCH$_2$)$_2$C=C=O	C$_{12}$H$_{22}$O	Dehydrochlorination	38182-98-8	84
RCHOHCH$_2$CH$_2$CH=C=O	C$_{13}$H$_{14}$O$_2$	Photolysis		76
R = PhCH=CH				
i-Pr$_3$SiOCHMeCH=C=O	C$_{13}$H$_{26}$O$_2$Si	Acyl chloride	122358-05-8	85
2-TolCHRCH=C=O	C$_{13}$H$_{14}$O	Photolysis	36920-33-9	81
R = CMe=CH$_2$				
Ph(CH$_2$)$_5$CH=C=O	C$_{13}$H$_{16}$O	Photolysis	71260-92-9	82
R(CH$_2$)$_3$CH=C=O	C$_{14}$H$_{16}$O	Photolysis	118744-75-5	86
R = PhCH(CH=CH$_2$)				
Ph(CH$_2$)$_6$CH=C=O	C$_{14}$H$_{18}$O	Photolysis	71260-93-0	82
(PhCH$_2$)$_2$C=C=O	C$_{16}$H$_{14}$O	Malonic anhydride		52
RC(Bu-t)=C=O	C$_{16}$H$_{22}$O	Dehydrochlorination	80959-02-0	15
R = PhCH$_2$CMe$_2$				
n-C$_{14}$H$_{29}$CH=C=O	C$_{16}$H$_{30}$O		4752-51-6	88
	(C$_{16}$H$_{30}$O)$_2$		7256-26-0	88
(n-C$_7$H$_{15}$)$_2$C=C=O	C$_{16}$H$_{30}$O	Dehydrochlorination		5
CH$_2$=CPhCHPhCH=C=O	C$_{17}$H$_{14}$O	Photolysis	38464-78-8	89
n-C$_{15}$H$_{31}$CH=C=O	C$_{17}$H$_{32}$O	Acyl chloride	121177-86-4	90

Structure	Formula	CAS Number	Notes	Ref
$R(CH_2)_3CH=C=O$	$(C_{17}H_{32}O)_2$	121177-87-5		90
	$C_{17}H_{25}NO_3$	54313-30-3	Abstract error	91
R = (bicyclic structure with Ac, Me, OHC, N)				
$R(CH_2)_6CH=C=O$	$C_{18}H_{32}O$	122215-06-9		92
R = n-$C_8H_{17}CH=CH$	$C_{18}H_{32}O$ (Z)	68556-24-1		92
	$(C_{18}H_{32}O)_2$	122215-07-0		92
	$(C_{18}H_{32}O)_2$	68556-25-2		92
n-$C_{16}H_{33}CH=C=O$	$C_{18}H_{34}O$	4752-50-5		93
	$(C_{18}H_{34}O)_2$	7049-56-1		94
$EtCR=C=O$ R = (pyrrole with MeO$_2$CN, N(CH$_2$)$_3$, Ph)	$C_{19}H_{22}N_2O_3$	73367-47-2	Photolysis	95
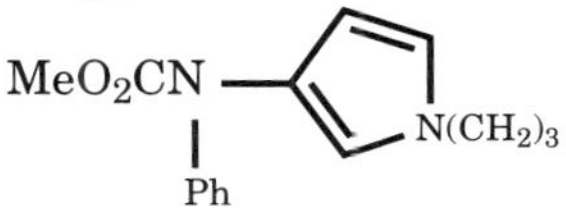	$(C_{19}H_{22}N_2O_3)_2$	73367-48-3		95
$EtCR=C=O$ R = $Ph_2C=CHCHOMeCHNH_2$	$C_{20}H_{21}NO_2$		Photolysis	76
n-$C_{18}H_{37}CH=C=O$	$C_{20}H_{38}O$	45263-60-3		96

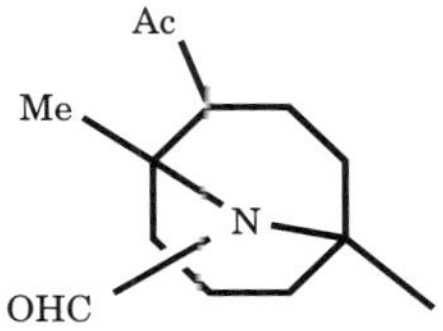

TABLE 1. (*Continued*)

Ketene	Formula	Preparation	Registry No.	Ref.
	$(C_{20}H_{38}O)_2$		14646-94-7	97
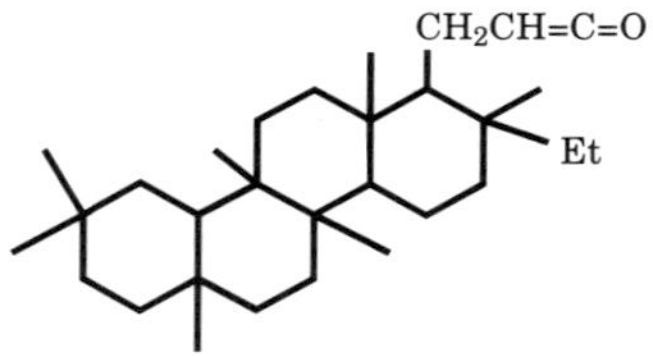 (steroid structure with CH₂CH=C=O and Et substituents)	$C_{30}H_{50}O$	Photolysis	53163-80-7	98
	$C_{28}H_{22}O$	Pyrolysis	96746-59-7	99
$Ph_2CHCPh_2CH=C=O$	$C_{38}H_{74}O$		53895-68-4	100
$(n\text{-}C_{18}H_{37})_2C=C=O$	$(C_{38}H_{74}O_2)_2$		53895-69-5	100
	$C_{46}H_{90}O$		94421-72-4	101
$(n\text{-}C_{22}H_{45})_2C=C=O$	$(C_{46}H_{90}O)_2$		94449-09-9	101

TABLE 2. Cycloalkylideneketenes

Ketene	Formula	Preparation	Registry No.	Ref.
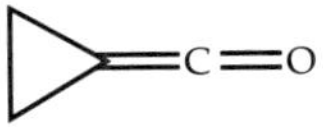	C_4H_4O	Dehalogenation	59078-44-3	40
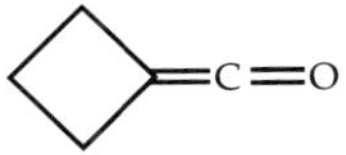	C_5H_6O	Dehalogenation	59078-45-4	40
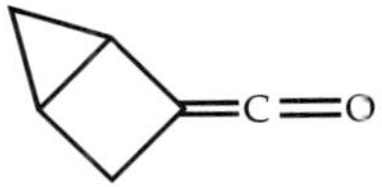	C_6H_6O	Wolff		102
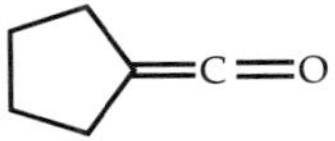	C_6H_8O	Dehydrochlorination	29713-50-6	23,103
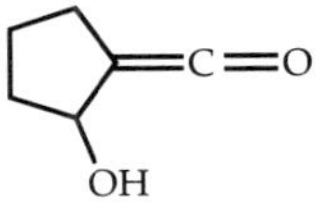	$C_8H_8O_2$	Wolff		26
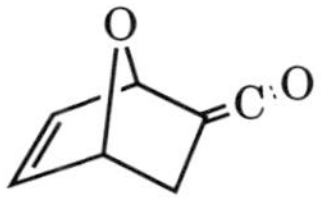	$C_7H_6O_2$	Wolff		104
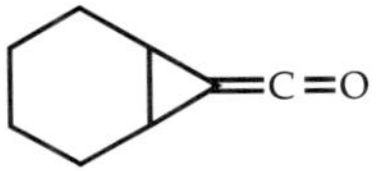	C_7H_8O	Dehydrochlorination	85763-24-2	105
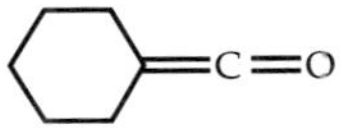	$C_7H_{10}O$	Dehydrochlorination	225589-13-5	21,23,106

TABLE 2. (Continued)

Ketene	Formula	Preparation	Registry No.	Ref.
Me₃Si—⟨△⟩—C=O	C₇H₁₂OSi	Zn dehalogenation		107
(norbornenyl ketene)	C₈H₈O	Dehydrochlorination	51317-50-1	108
(bicyclic ketene)	C₈H₁₀O	Wolff		109
(bicyclic ketene)	C₈H₁₂O	Wolff	45052-35-7	17,18
(cycloheptyl ketene)	C₈H₁₂O	Wolff		
(hydroxy cyclopentyl ketene)	C₈H₁₂O₂	Wolff		26

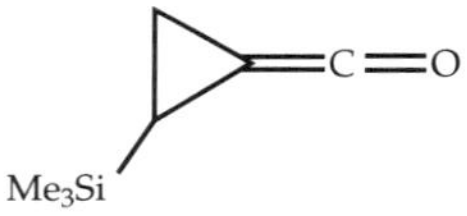

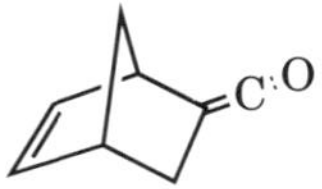

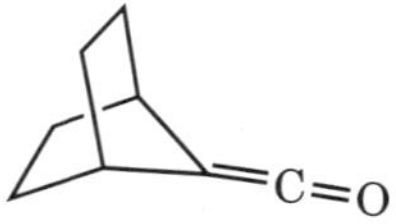

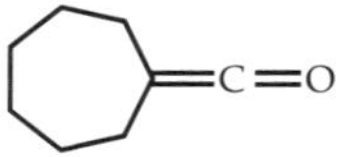

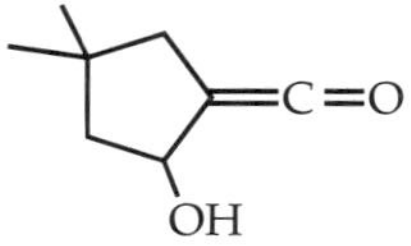

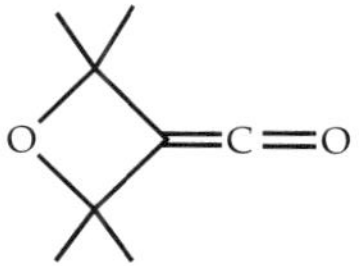 $C_8H_{12}O_2$	Wolff		110
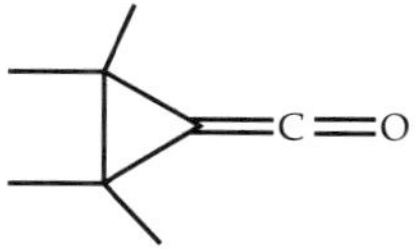 $C_8H_{12}O$	Dehalogenation	96929-29-2	107
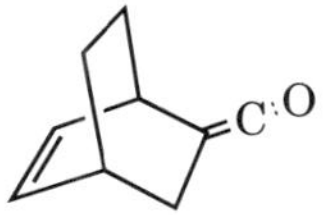 $C_9H_{10}O$	Pyrolysis		111
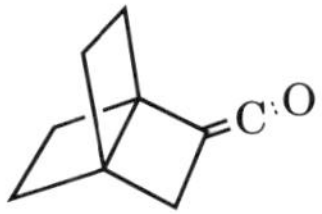 $C_9H_{10}O$	Wolff		112
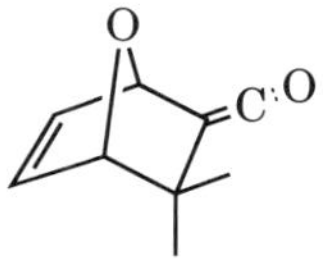 $C_9H_{10}O_2$	Wolff		105
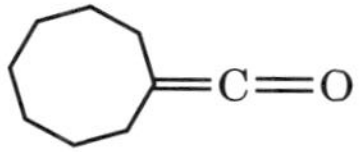 $C_9H_{14}O$	Wolff	42159-51-3	17,18

TABLE 2. (Continued)

Ketene	Formula	Preparation	Registry No.	Ref.
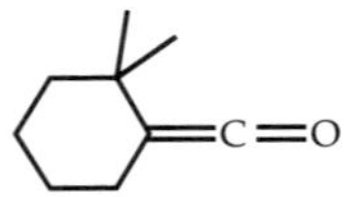	$C_9H_{14}O$	Dehalogenation	97234-76-9	113
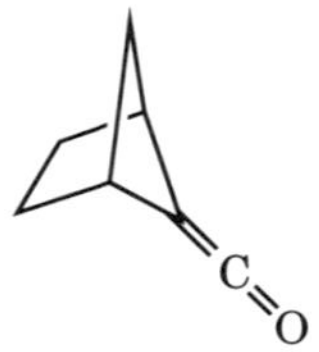	$C_{10}H_{12}O$	Wolff	132736-29-9	113a
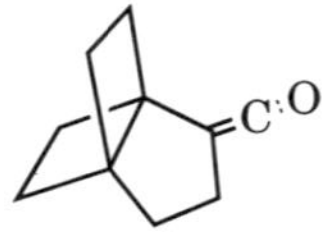	$C_{10}H_{12}O$	Wolff		112
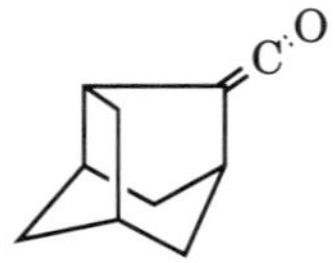	$C_{10}H_{12}O$	Wolff		23
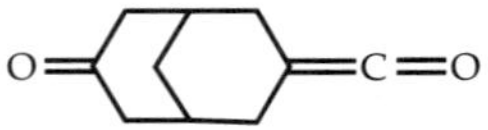	$C_{10}H_{12}O_2$	Pyrolysis		114

Structure	Formula	Method	CAS	Ref
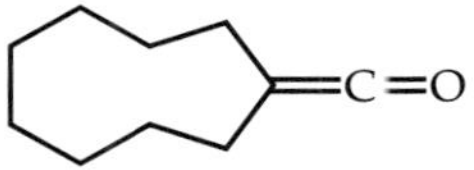 cyclodecanone	$C_{10}H_{16}O$	Wolff	40661-68-5	17,18
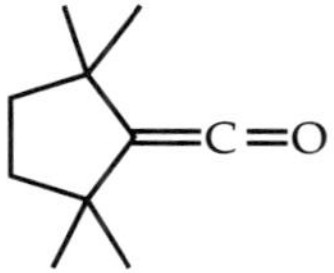 2,2,5,5-tetramethylcyclopentanone	$C_{10}H_{16}O$	Wolff	71269-06-2	12
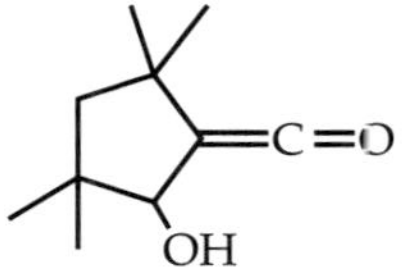 2-hydroxy-tetramethylcyclopentanone	$C_{10}H_{16}O_2$	Wolff		26
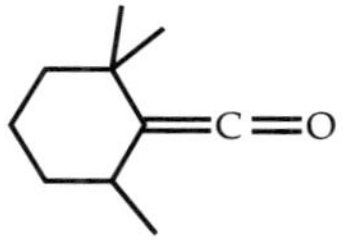 2,2,6,6-tetramethylcyclohexanone	$C_{10}H_{16}O$	Dehydrochlorination	78672-65-8	16
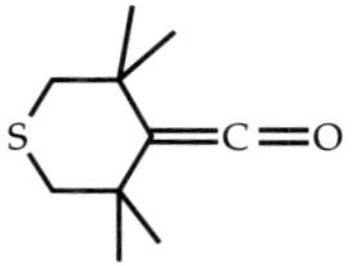 tetramethylthiane-one	$C_{10}H_{16}OS$	Wolff		115

TABLE 2. (Continued)

Ketene	Formula	Preparation	Registry No.	Ref.
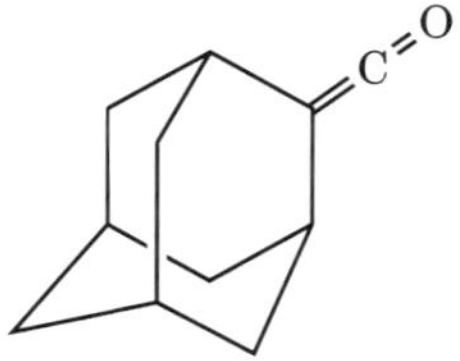 (adamantylidene ketene)	$C_{11}H_{14}O$	Dehydrochlorination	54781-13-4	22,23
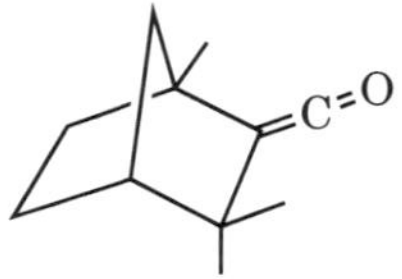 (bornylidene ketene)	$C_{11}H_{16}O$	Dehydrochlorination	78672-59-0	16
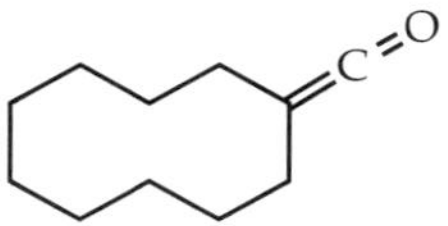 (cyclodecylidene ketene)	$C_{11}H_{18}O$	Wolff		18

Structure	Formula	Method	CAS	Ref
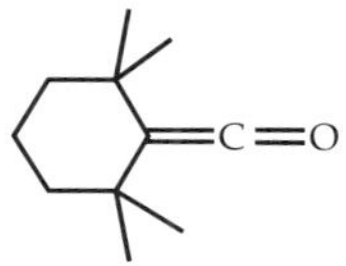 2,2,6,6-tetramethylcyclohexylidene ketene	$C_{11}H_{18}O$	Dehydrochlorination	73683-20-2	16
$(CH_2)_{10}C=C=O$	$C_{12}H_{20}O$	Wolff	42052-37-9	17–19
$(CH_2)_{11}C=C=O$	$C_{13}H_{22}O$	Dehydration	50834-66-7	115a
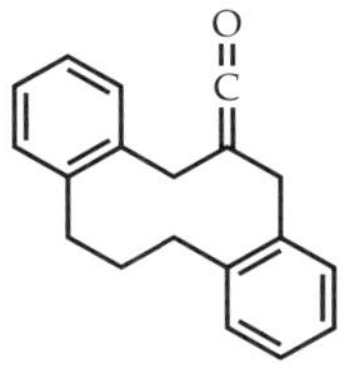 cyclophane ketone	$C_{19}H_{18}O$	Photolysis		116

TABLE 3. Cycloalkylketenes

Ketene	Formula	Preparation	Registry No.	Ref.
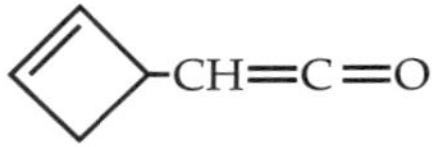 cyclobutenyl-CH=C=O	C_6H_6O	Pyrolysis		117
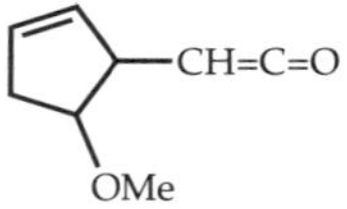 (methoxy-cyclopentenyl)-CH=C=O (OMe)	$C_8H_{10}O_2$	Wolff	90107-31-6	66
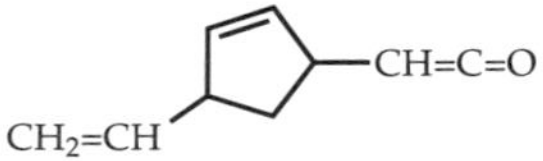 (vinylcyclopentenyl)-CH=C=O	$C_9H_{10}O$	Photolysis	35150-67-5	118
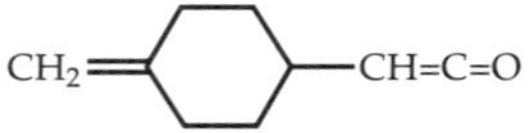 (methylenecyclohexyl)-CH=C=O	$C_9H_{12}O$	Photolysis	92144-45-1	119
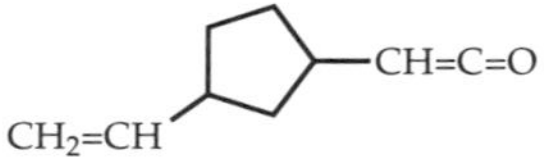 (vinylcyclopentyl)-CH=C=O	$C_9H_{12}O$	Photolysis	35150-68-6	118
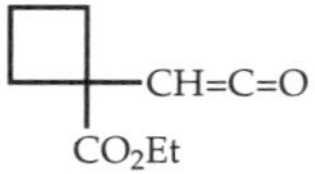 cyclobutyl(CO$_2$Et)-CH=C=O	$C_9H_{12}O_3$	Photolysis	106366-57-8	120
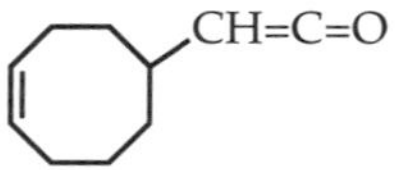 cyclooctenyl-CH=C=O	$C_{10}H_{14}O$	Dehydrochlorination		121
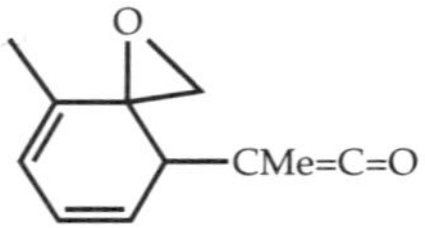 spiro epoxide-CMe=C=O	$C_{11}H_{12}O_2$	Dehydrochlorination	68317-57-7	122

170

Structure	Formula	Method	CAS	Ref
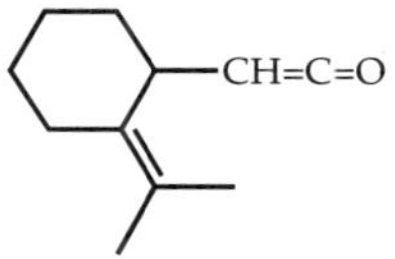 (cyclohexylidene-isopropyl)CH=C=O	$C_{11}H_{16}O$	Wolff	40323-68-0	66
1-AdCH=C=O	$C_{12}H_{16}O$	Dehydrochlorination		22
2-AdCH=C=O	$C_{12}H_{16}O$	Dehydrochlorination		22
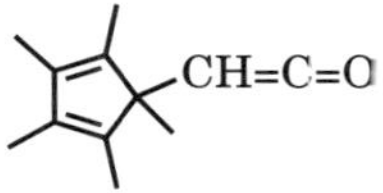 (tetramethylcyclopentadienyl)CH=C=O	$C_{12}H_{16}O$	Photolysis	81052-48-4	123
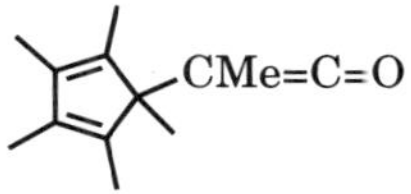 (tetramethylcyclopentadienyl)CMe=C=O	$C_{13}H_{18}O$	Photolysis	81052-45-1	123
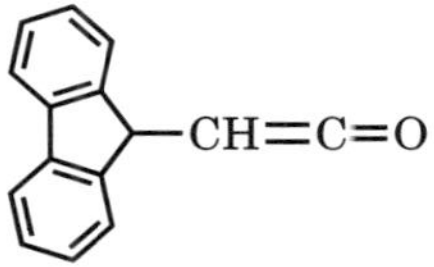 9-fluorenyl-CH=C=O	$C_{15}H_{10}O$	Dehydrochlorination[a]	24168-55-6	87
1-AdC(Bu-t)=C=O	$C_{16}H_{24}O$	Dehydrochlorination		8

TABLE 3. (Continued)

Ketene	Formula	Preparation	Registry No.	Ref.
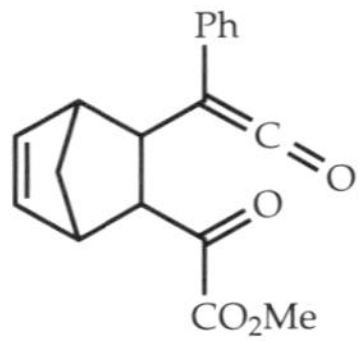 Ph–C(=O)–...–CO₂Me (norbornene)	$C_{18}H_{16}O_4$	Fragmentation	115410-84-9	124
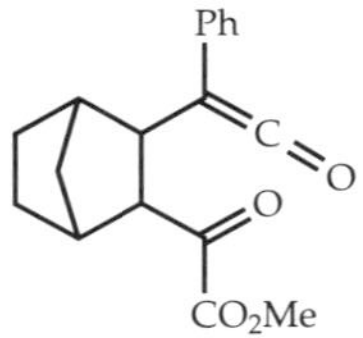 Ph–C(=O)–...–CO₂Me (norbornane)	$C_{18}H_{18}O_4$	Fragmentation	98171-28-9	124
$(1\text{-Ad})_2C{=}C{=}O$	$C_{22}H_{30}O$	Dehydrochlorination	88246-70-2	6, 8
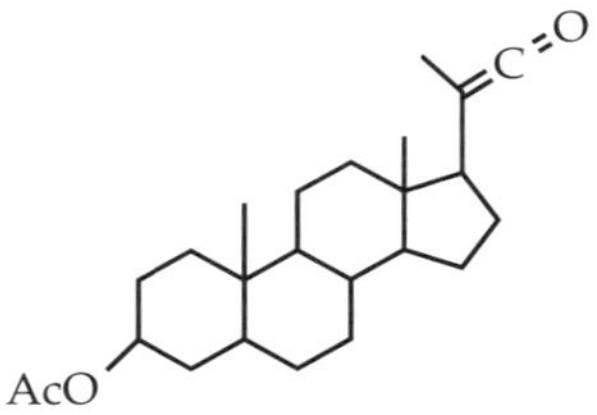 AcO–steroid–C=O	$C_{24}H_{36}O_3$	Dehydrochlorination		25
THPO–cyclopentane (R¹, CH=C=O, O₂CR)	$C_{27}H_{42}O_7$		65537-62-4	125
R = Me; R¹ = THPOCHPnCH=CH				
R = t-Bu	$C_{30}H_{48}O_7$		65537-66-8	125

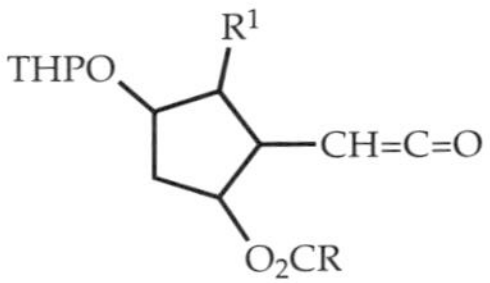

[a]The structure in the CA index appears to be incorrect.

References

1. Staudinger, H; Klever, H. W. *Chem. Ber.* **1906**, *39*, 968–971.
2. Gong, L.; McAllister, M. A.; Tidwell, T. T. *J. Am. Chem. Soc.* **1991**, *113*, 6021–6028.
3. Sauer, J. C. *J. Am. Chem. Soc.* **1947**, *69*, 2444–2448.
4. Sauer, J. C. *U.S. Patent* 2,369,919; *Chem. Abstr.* **1945**, *39*, 4086.
5. Sauer, J. C. *U.S. Patent* 2,268,169; *Chem. Abstr.* **1942**, *36*, 2737.
6. Olah, G. A.; Wu, A.; Farooq, O. *Synthesis*, **1989**, 566–567.
7. Newman, M. S.; Arkell, A.; Fukunaga, T. *J. Am. Chem. Soc.* **1960**, *82*, 2498–2501.
8. Olah, G. A.; Wu, A.; Farooq, O. *Synthesis*, **1989**, 568.
9. Hofmann, P.; Moya, L. A. P.; Kain, I. *Synthesis*, **1986**, 43–44.
10. Schaumann, E. *Chem. Ber.* **1976**, *109*, 906–921.
11. Newman, M. S.; Arkell, A. *J. Org. Chem.* **1959**, *24*, 385–387.
12. Kaplan, F.; Mitchell, M. L. *Tetrahedron Lett.* **1979**, 759–762.
13. Schaumann, E.; Ehlers, J. *Chem. Ber.* **1979**, *112*, 1000–1011.
14. Carlsen, L.; Egsgaard, H.; Schaumann, E.; Mrotzek, H.; Klein, W.-R. *J. Chem. Soc., Perkin Trans. 2*, **1980**, 1557–1562.
15. Lambrecht, J.; Zsolnai, L.; Huttner, G.; Jochims, C. *Chem. Ber.* **1982**, *115*, 172–184.
16. Kirmse, W.; Spaleck, W. *Angew. Chem., Int. Ed. Engl.* **1981**, *20*, 776–777.
17. Concannon, P. W.; Ciabattoni, J. *J. Am. Chem. Soc.* **1973**, *95*, 3284–3289.
18. Regitz, M.; Rüter, J. *Chem. Ber.* **1969**, *102*, 3877–3890.
19. Stojiljkovic, A.; Tasovac, R. *Tetrahedron* **1977**, *33*, 65–67.
20. Rellensmann, W.; Hafner, K. *Chem. Ber.* **1962**, *95*, 2579–2580.
20a. Brady, W. T.; Ting, P. L. *J. Org. Chem.* **1976**, *41*, 2336–2339.
21. Brady, W. T.; Ting, P. L. *J. Org. Chem.* **1974**, *39*, 763–765.
22. Strating, J.; Scharp, J.; Wynberg, H. *Recl. Trav. Chim. Pays-Bas*, **1970**, *89*, 23–31.
23. Majerski, Z.; Vinkovic, V. *Synthesis*, **1989**, 559–560.
24. Dowd, P.; Zhang, W. *J. Org. Chem.* **1992**, *57*, 7163–7171.
25. Rust, J. B. *U.S. Patent 2,623,055. Chem. Abstr.* **1953**, *47*, 10560g.
26. Nikolaev, V. A.; Zhdanova, O. V.; Korobitsyna, I. K. *Zh. Org. Khim.* **1982**, *18*, 559–572. *Engl. Transl.* **1982**, *18*, 488–501.
27. Bisceglia, R. H.; Cheer, C. J. *J. Chem. Soc., Chem. Commun.* **1973**, 165–166.
28. Wolff, S.; Agosta, W. C. *J. Chem. Soc., Chem. Commun.* **1973**, 771.
29. Hegelund, F.; Kauppinen, J.; Winther, F. *Mol. Phys.* **1973**, *61*, 261–273.
30. Durig, J. R.; Willis, J. N., Jr. *Spectrochim. Acta* **1966**, *22*, 1299–1313.
31. Hochstrasser, R.; Wirz, J. *Angew. Chem., Int. Ed. Engl.* **1990**, *29*, 411–413.
32. Lee, P. S.; Rowland, F. S. *J. Phys. Chem* **1980**, *84*, 3243–3249.
33. Andreades, S.; Carlson, H. D. *Organic Syntheses*, Baumgarten, H. E., Ed.; Wiley: New York, 1973; Coll. Vol. V, pp. 679–684.
34. Van Dunren, B. L.; Langseth, L.; Orris, L.; Teebor, H. G. U.; Nelson, N.; Kuschner, M. *J. Natl. Cancer Inst.* **1966**, *37*, 825–838; *Chem. Abstr.* **1966**, *66*, 17607p.
35. Brown, R. D.; Godfrey, P. D.; McNaughton, D.; Pierlot, A. P.; Taylor, W. H. *J. Mol. Spectrosc.* **1990**, *140*, 340–352.

36. Imahori, Y.; Fujii, A.; Nakabashi, H. *Jpn. Kokai Tokkyo Koho* JP 03005437; *Chem. Abstr.* **1991**, *115*, 135519q.
37. Russell, R. L.; Rowland, F. S. *J. Am. Chem. Soc.* **1968**, *90*, 1671–1673.
37a. Blyholder, G. D.; Emmett, P. H. *J. Phys. Chem.* **1960**, *64*, 470–472.
38. Bak, B.; Christiansen, J.; Nielsen, J. T. *Acta Chem. Scand.* **1965**, *19*, 2252–2253.
39. Sonveaux, E.; Andre, J. M.; Delhalle, J.; Fripiat, J. G. *Bull. Soc. Chim. Belg.* **1985**, *94*, 831–847.
40. Masters, A. P.; Sorensen, T. S. *Tetrahedron Lett.* **1989**, *30*, 5869–5872.
41. Woodward, R. B.; Small, G. *J. Am. Chem. Soc.* **1950**, *72*, 1297–1304.
42. Takhistov, V. V.; Muftakhov, M. V.; Krivoruchko, A. A.; Mazunov, V. A. *Izv. Akad. Nauk SSSR, Ser. Khim.* **1991**, 2049–2055; *Chem. Abstr.* **1992**, *116*, 58567v.
43. Nair, K. P. R.; Rudolph, H. D.; Dreizler, H. *J. Mol. Spectrosc.* **1973**, *48*, 571–591.
44. Maier, G.; Reisenauer, H. P.; Sayrac, T. *Chem. Ber.* **1982**, *115*, 2192–2201.
45. Martin, J. C.; Burpitt, R. D.; Hostettler, H. U. *J. Org. Chem.* **1967**, *32*, 210–231.
46. Masters, A. P.; Sorensen, T. S.; Zeigler, T. *J. Org. Chem.* **1986**, *51*, 3558–3559.
47. Imahori, Y.; Fujii, R.; Ido, J. *J. Labelled Compd. Radiopharm.* **1989**, *27*, 1025–1034.
48. Sterk, H.; Tritthart, P.; Ziegler, E. *Monatsh. Chem.* **1970**, *101*, 1056–1058.
49. Sponsler, M. B.; Dougherty, D. A. *J. Org. Chem.* **1984**, *49*, 4978–4984.
50. Maquestiau, A.; Flammang, R.; Pauwels, P. *Org. Mass Spectrom.* **1983**, *18*, 547–552.
51. Gajewski, J. J.; Conrad, N. D. *J. Am. Chem. Soc.* **1979**, *101*, 2747–2748.
52. Staudinger, H.; Schneider, H.; Schotz, P.; Strong, P. M. *Helv. Chim. Acta* **1923**, *6*, 291–303.
53. Staudinger, H. *Helv. Chim. Acta,* **1925**, *8*, 306–311.
54. Bojarska-Dahlig, H.; Naperty, S.; Bierdrzycki, M.; Drzewinski, W.; Ledwoch, S. *Pol. Patent* PL 94625; *Chem. Abstr.* **1979**, *90*, 152557h.
55. Allen, A. D.; Kresge, A. J.; Schepp, N. P.; Tidwell, T. T. *Can. J. Chem.* **1987**, *65*, 1719–1723.
56. Brady, W. T.; Roe, R., Jr. *J. Am. Chem. Soc.* **1970**, *92*, 4618–4621.
57. Dehmlow, E. V.; Pickardt, J.; Slopianka, M.; Fastabend, U.; Drechsler, K.; Soufi, J. *Liebigs Ann. Chem.* **1987**, 377–379.
57a. Dehmlow, E. V.; Slopianka, M.; Pickardt, J. *Liebigs Ann. Chem.* **1979**, 572–593.
58. Allen, A. D.; Tidwell, T. T. *J. Am. Chem. Soc.* **1987**, *109*, 2774–2780.
59. Brady, W. T.; Cheng, T. C. *J. Org. Chem.* **1977**, *42*, 732–734.
60. Ayral-Kaloustian, S.; Wolff, S.; Agosta, W. S. *J. Org. Chem.* **1978**, *43*, 3314–3319.
61. Smith, A. B., III; Toder, B. H.; Branca, S. J. *J. Am. Chem. Soc.* **1984**, *106*, 3995–4001.
62. Urbonas, A.; Valentukeviciene, G.; Vitenaite, G.; Navickaite, K.; Jasinskas, L. *Liet. TSR Aukst. Mokyklu Mosklo Darb., Chem. Chem. Technol.* **1972**, 153–157; *Chem. Abstr.* **1973**, *80*, 59909h.
63. Cormier, R. A. *Tetrahedron Lett.* **1980**, *21*, 2021–2024.
64. Takai, H.; Sakiyama, T.; Matsuzaki, K.; Nozaki, T.; Okumura, Y.; Imai, C. *Eur. Pat. Appl.* EP 288,286; *Chem. Abstr.* **1989**, *110*, 57853z.
65. Zaitseva, G. S.; Livantsova, L. I.; Orlova, N. A.; Baukov, Yu. I.; Lutsenko, I. F. *Zh. Obshch. Khim.* **1982**, *52*, 2076–2084; *Engl. Transl.* **1982**, *52*, 1847–1854.

66. Smith, A. B., III; Toder, B. H.; Richmond, R.; Branca, S. J. *J. Am. Chem. Soc.* **1984**, *106*, 4001–4009.
67. Wojtczak, J. *Poznan Tow. Przyj. Nauk, Pr. Kom. Mat.-Przyr., Pr. Chem.* **1973**, *13*, 13–20; *Chem. Abstr.* **1973**, *79*, 59948z.
68. Grzegorzewska, U.; Leplawy, M.; Redlinski, A. *Rocz. Chem.* **1975**, *49*, 1859–1863; *Chem. Abstr.* **1976**, *84*, 135059f.
69. Oku, J.; Padias, A. B.; Hall, H. K., Jr.; East, A. J. *Macromolecules* **1987**, *20*, 2314–2315.
70. Andrac, M. *Ann. Chim.* **1964**, *9*, 287–315; *Chem. Abstr.* **1965**, *62*, 3919a.
71. Ohshiro, Y.; Ishida, M.; Shibata, J.; Minami, T.; Agawa, T. *Chem. Lett.* **1982**, 587–590.
72. Singh, S.; Baird, M. C. *J. Organomet. Chem.* **1988**, *338*, 255–260.
73. Selva, A.; Ferrario, F.; Saba, A. *Org. Mass. Spectrom.* **1987**, *22*, 189–196.
74. Burger, U.; Zellweger, D. *Helv. Chim. Acta* **1986**, *69*, 676–682.
75. Crowley, K. J.; Schneider, R. A.; Meinwald, J. *J. Chem. Soc. (B)* **1966**, 571–572.
76. Rahman, S. S.; Wakefield, B. J.; Roberts, S. M.; Dowle, M. D. *J. Chem. Soc., Chem. Commun.* **1989**, 303–305.
77. Allen, A. D.; Baigrie, L. M.; Gong, L. *Can. J. Chem.* **1991**, *69*, 138–145.
78. Nightingale, D. V.; Turley, R. H., Jr. *J. Org. Chem.* **1961**, *26*, 2656–2658.
79. Baslas, R. K. *Indian Perfum.* **1976**, *20*, 131–136; *Chem. Abstr.* **1978**, *88*, 152789u.
80. Adam, W.; Diehl, J. W. *J. Chem. Soc., Chem. Commun.* **1972**, 797–798.
81. Wolff, S.; Agosta, W. C. *J. Chem. Soc., Chem. Commun.* **1972**, 226–227.
82. Doubleday, C., Jr. *Chem. Phys. Lett.* **1979**, *64*, 67–70.
83. Hart, H.; Shih, E. *J. Org. Chem.* **1976**, *21*, 3377–3381.
84. Crandall, J. K.; Sojka, S. A.; Komin, J. B. *J. Org. Chem.* **1974**, *39*, 2172–2175.
85. Lynch, J. E.; Riseman, S. M.; Laswell, W. L.; Volante, R. P.; Smith, G. B.; Shinkai, I.; Tschaen, D. M. *J. Org. Chem.* **1989**, *54*, 3792–3796.
86. Koppes, M. J. C. M.; Cerfontain, H. *Recl. Trav. Chem. Pays-Bas*, **1988**, 412–417.
87. Koenig, T. W.; Barklow, T. *Tetrahedron* **1969**, *25*, 2875–4886.
88. Takei, F.; Murai, K.; Akazome, G. *Kogyo Kagaku Zasshi* **1957**, *60*, 1271–1273; *Chem. Abstr.* **1959**, *53*, 13889d.
89. Zimmermann, H. E.; Little, R. D. *J. Chem. Soc., Chem. Commun.* **1972**, 698–700.
90. Higuchi, H.; Nakamura, S.; Takahashi, H.; Takahashi, F.; Jimichi, K. *Jpn. Kokai Tokkyo Koho* JP 63,264,545; *Chem. Abstr.* **1989**, *111*, 23378d.
91. Corey, E. J.; Balanson, R. D. *J. Am. Chem. Soc.* **1974**, *96*, 6516–6517.
92. Balan, V. *Ind. lemnului* **1958**, *7*, 467–469; *Chem. Abstr.* **1959**, *53*, 10763h.
93. Hertenstein, U.; Hunig, S. *Angew. Chem., Int. Ed. Engl.* **1975**, *14*, 179–180.
94. Ashikaga, T.; Tanaka, K.; Maeda, U. *Ger. Offen.* DE 1919843; *Chem. Abstr.* **1970**, *72*, 44747a.
95. Schultz, A. G.; Sha, C. *J. Org. Chem.* **1980**, *45*, 2040–2041.
96. Somerville, I. C.; Fisher, R. F. M. *U.S. Patent* 2,824,816; *Chem. Abstr.* **1958**, *52*, 12440h.
97. Reynolds, W. F., Jr. *U.S. Patent* US 3332834; *Chem. Abstr.* **1967**, *67*, 74682d.

98. Aoyagi, R.; Tsuyuki, T.; Takai, M.; Takahashi, T.; Kohen, F.; Stevenson, R. *Tetrahedron* **1973**, *29*, 4331–4340.
99. Adam, W.; Berkessel, A.; Peters, E. M.; Peters, K.; von Schnering, H. G. *J. Org. Chem.* **1985**, *50*, 2811–2815.
100. Kao Corp. *Jpn. Kokai Tokkyo Koho* JP 59,164,560; *Chem. Abstr.* **1985**, *102*, 70168j.
101. Kao Corp. *Jpn. Kokai Tokkyo Koho* JP 59,157,655; *Chem. Abstr.* **1985**, *102*, 53916j.
102. Brook, P. R.; Brophy, B. V. *J. Chem. Soc., Perkin Trans. 1* **1985**, 2509–2513.
103. Hill, C. M.; Senter, G. W. *J. Am. Chem. Soc.* **1949**, *71*, 364–365.
104. Brahms, J. C.; Dailey, W. P. *Tetrahedron Lett.* **1990**, *31*, 1381–1384.
105. Wentrup, C.; Gross, G.; Marquestiau, A.; Flammang, R. *Angew. Chem., Int. Ed. Engl.* **1983**, *22*, 542–543.
106. Hill, C. M.; Hill, M. E. *J. Am. Chem. Soc.* **1953**, *75*, 2765–2766.
107. Wulff, J.; Hoffmann, H. M. R. *Angew. Chem., Int. Ed. Engl.* **1985**, *24*, 605–606.
108. Lasne, M. C.; Ripoll, J. L.; Denis, J. M. *Tetrahedron* **1981**, *37*, 503–508.
109. Yates, P.; Crawford, R. J. *J. Am. Chem. Soc.* **1966**, *88*, 1562–1563.
110. Korobitsyna, I. K.; Rodina, L. L. *Zh. Org. Khim.* **1965**, *1*, 932–938; *Chem. Abstr.* **1965**, *63*, 6939a.
111. Brown, R. F. C.; Eastwood, F. W.; McMullen, G. L. *Austr. J. Chem.* **1977**, *30*, 179–193.
112. Eaton, P. E.; Temme, G. H., III *J. Am. Chem. Soc.* **1973**, *95*, 7508–7510.
113. Seikaly, H. R.; Tidwell, T. T. *J. Am. Chem. Soc.* **1985**, *107*, 5391–5396.
113a. Rau, H.; Bokel, M. *J. Photochem. Photobiol. A*, **1990**, *53*, 311 322.
114. Renzoni, G. E.; Yin, T.-K.; Borden, W. T. *J. Am. Chem. Soc.* **1986**, *108*, 7121–7122.
115. de Groot, A.; Boerma, J. A.; de Valk, J.; Wynberg, H. *J. Org. Chem.* **1968**, *33*, 4205–4209.
115a. Guiffre, L.; Righi, F.; Matera, G. *U.S. Patent* 3,914,313; *Chem. Abstr.* **1976**, *84*, 58777j.
116. Mataka, S.; Lee, S. T.; Tashiro, M. *J. Chem. Soc., Perkin Trans. 2*, **1990**, 2017–2021.
117. Ho, J.; Bond, F. T. *J. Am. Chem. Soc.* **1974**, *96*, 7355–7356.
118. Miller, R. D.; Abraitys, V. Y. *J. Am. Chem. Soc.* **1972**, *94*, 663–665.
119. Eaton, P. E.; Jobe, P. G.; Reingold, I. D. *J. Am. Chem. Soc.* **1984**, *106*, 6437–6439.
120. Miller, R. D.; Theis, W. *Tetrahedron Lett.* **1986**, *27*, 2447–2450.
121. Moon, S.; Kolesar, T. F. *J. Org. Chem.* **1974**, *39*, 995–998.
122. Nitta, M.; Nakatani, H. *Chem. Lett.* **1978**, 957–960.
123. Hart, H.; Chen, S.; Nitta, M. *Tetrahedron* **1981**, *37*, 3323–3328.
124. Christl, M.; Lanzendörfer, U.; Grötsch, M. M.; Ditterich, E.; Hegmann, J. *Chem. Ber.* **1990**, *123*, 2031–2037.
125. UpJohn Co. *Neth. Appl.* NL 7,608,823; *Chem. Abstr.* **1978**, *88*, 104766n.

4.1.2 Alkenylketenes

Alkenylketenes contain the 1,3-butadiene unit, and from the measured gas phase dipole moment[1] the parent vinylketene (**1**) adopts the *anti*-conformation shown.

Molecular orbital calculations[2,3] confirm that this is the most stable geometry, and is 1.7 kcal/mol more stable than the *syn* conformation. The parent is formed by pyrolysis of vinylacetic acid at 750 K or of *E*-crotonic acid at 820 K, and its photoelectron spectrum has been measured in the gas phase (equation 1).[4] Pyrolysis of crotonic anhydride and collection of the volatile product at −196 °C followed by addition of $CS_2/CDCl_3$ permits measurement of the NMR spectrum at −70 °C[5], which showed the ^{13}C shift of the CH_2 at δ 109, consistent with negative charge delocalization to C_δ as shown in **1a**.

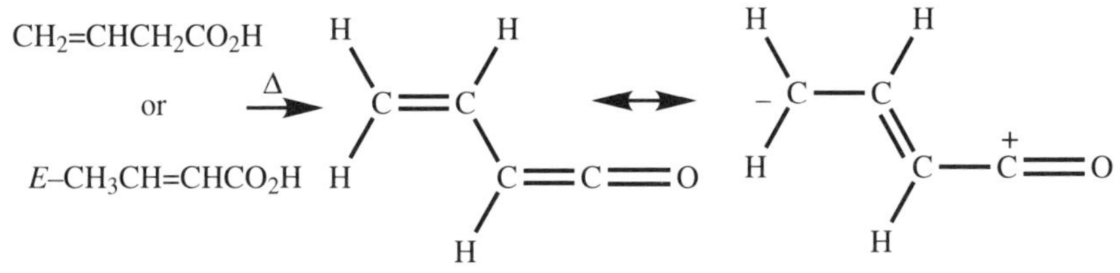

$$CH_2=CHCH_2CO_2H \text{ or } E\text{-}CH_3CH=CHCO_2H \xrightarrow{\Delta} \mathbf{1} \leftrightarrow \mathbf{1a} \quad (1)$$

The NMR chemical shifts of the methylene groups of vinyl ethers have been proposed[6] as a measure of the degree of π electron donation by the alkoxy group to the vinyl group. Using this criterion, from the chemical shifts shown the electron donor power of the ketenyl group to the vinyl group as in **1a** may be judged to be significant, but not as strong as for the methoxy group (Figure 1). Thus there is an upfield shift for both ^1H and ^{13}C of C_δ in **1** compared to $CH_2=CH_2$.

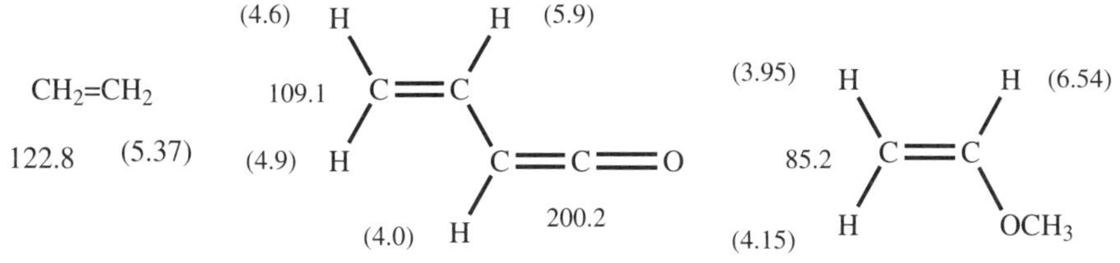

Figure 1. ^{13}C and ^1H (in parentheses) NMR chemical shifts (δ) of alkenes, ketenes, and vinyl ethers.[5,6]

It has been proposed[1] that the rather low dipole moment of vinylketene is also indicative of a significant contribution to the structure by the dipolar resonance structure **1a**. Calculated isodesmic energies are influenced by too many factors to quantify the energetic stabilization of vinylketene relative to methylketene, but these studies do not exclude stabilization by several kcal/mol.[3] Vinylketenes are quite reactive in a variety of processes such as hydration, and this has been ascribed to conjugative stabilization of the transition states by the vinyl group.[7] α-Trimethylsilylvinylketenes (**2**)[8] and (**3**)[9] are however quite long-lived, as is found for other α-trimethylsilylketenes (Section 4.5). The crowded vinylketene **4** is resistant to dimerization but is rather reactive towards H_2O.[10]

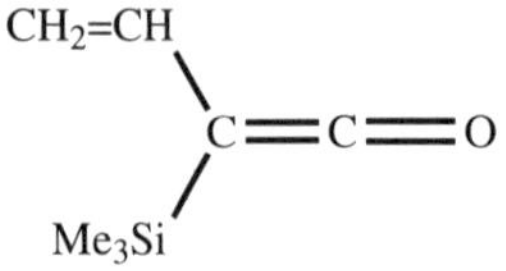

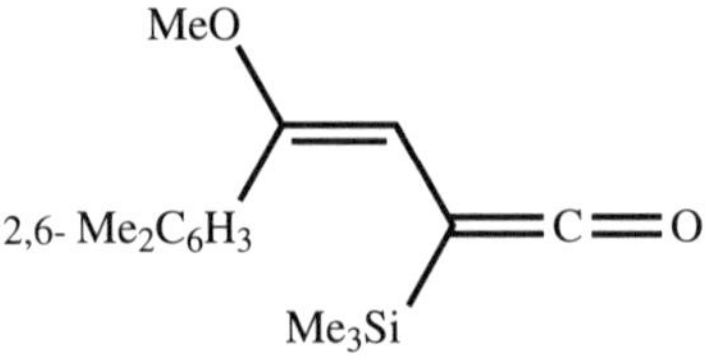

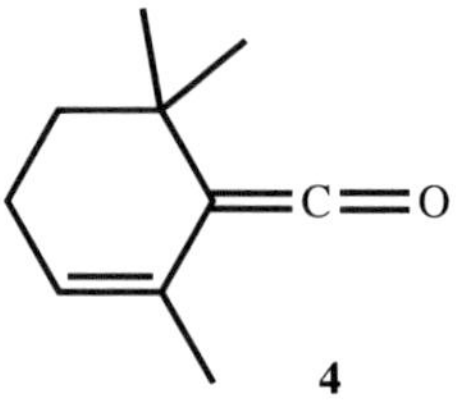

Vinylketene (**1**) generated from the reaction of crotonyl chloride with Et_3N is trapped by alcohols or active alkenes such as cyclopentadiene (equation 2).[11-19] In the absence of such trapping agents dimerization to 2-pyrones occurs (equation 3). The generation and trapping by many simple alkenes of a number of alkyl-substituted vinylketenes from the acyl chlorides and Et_3N have been examined.[11,12,14,15] Vinylketenes generated by this route are also captured by enamines.[16-18]

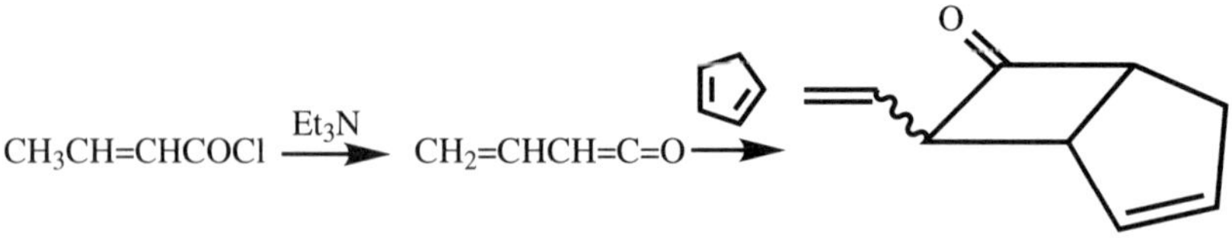

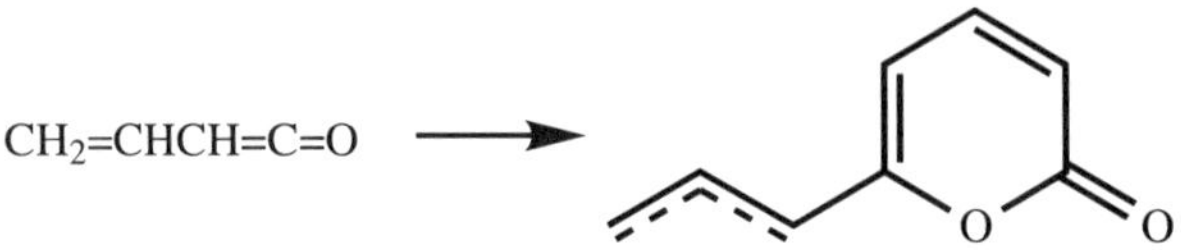

Vinylketenes are available from many other routes as discussed elsewhere, including Wolff rearrangements (Section 3.3), cyclobutenone and cyclohexadienone ring openings (Sections 3.4.1 and 3.4.3, respectively), metal carbene complexes (Section 3.5), ene reactions of dienylaldehydes (equation 4,[20] see Sections 3.4.6 and 5.4.4), and Meldrum's acid derivatives (equation 5).[21] The parent **1** was also obtained in high yield from gas phase thermal reaction of CH_2C=CHC≡$COEt$,[22] and its formation from flash vacuum pyrolysis of spiro[2.3]hexan-4-ones was suggested (equation 6).[23] Vinylketenes generated by flash vacuum pyrolysis of unsaturated acyl chlorides were identified by mass spectroscopy and low temperature IR.[24]

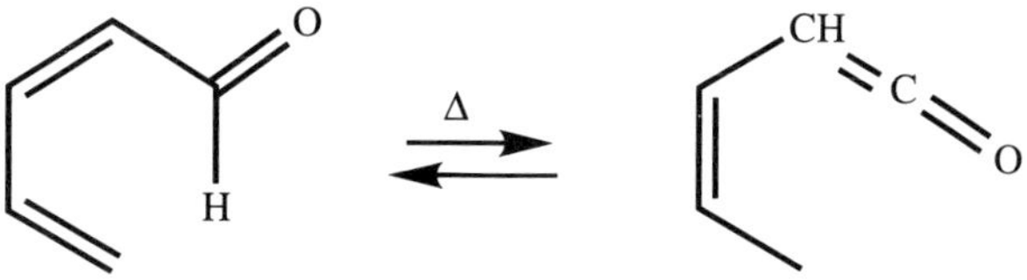

(4)

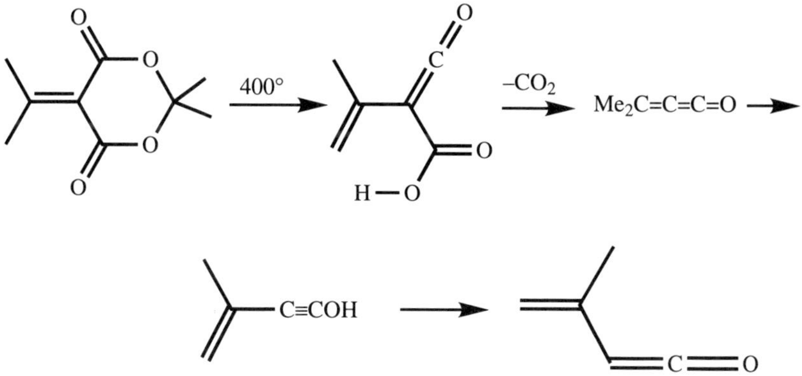

(5)

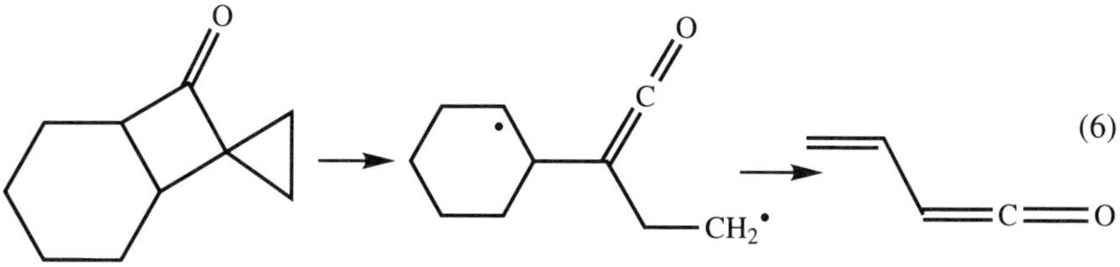

(6)

Photolysis of 5-acyl-3H-pyrazoles gives vinylketenes in low yields by Wolff-type rearrangements.[25,26]

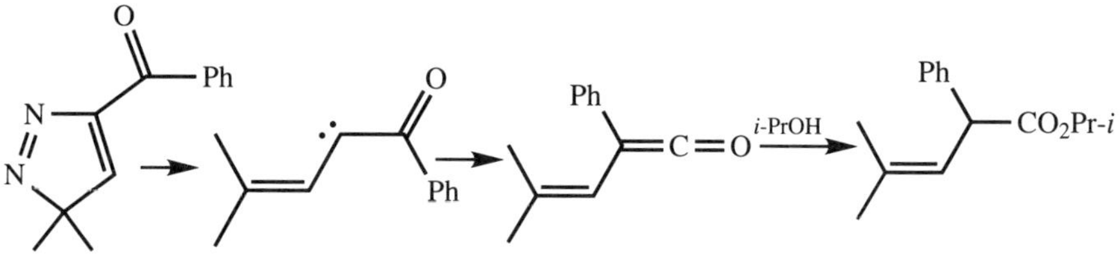

(7)

Distinctive reactions of alkenylketenes include [2 + 2] intermolecular cycloadditions (equation 2),[27,28] intramolecular [4 + 2] cycloadditions of β-arylvinylketenes and dienylketenes (Section 5.4.4), and electrophilic reactions at C_δ (Section 5.6.1).[10] Alkenylketenes are very reactive in hydration reactions.[7] Reaction of the 2-chloro-3-butenoic acids shown in equation 8 with acetic anhydride gave 20–30% of cyclobutenones, presumably via intramolecular cyclization of the alkenylketenes, but similar reactions of some other such acids did not give detectable yields of cyclobutenones.[29]

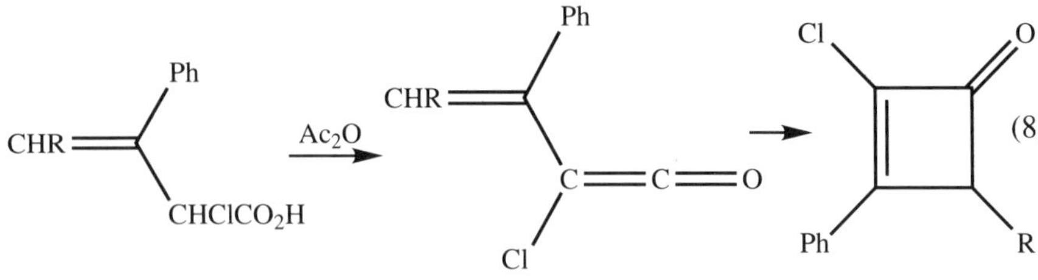

Formation of 1,3-pentadiene-1,5-dione (**5**) occurs on photolysis of α-pyrone labeled with ^{13}C at the position shown (equation 9), and at 8 K in an Ar matrix **5** is proposed to exist in the four conformers **5a–5d**, with IR absorptions at 2068, 2074, 2081, and 2089 cm^{-1}.[30] These ketenes have ^{13}C in the carbonyl carbon of the ketene, but on further photolysis hydrogen transfer occurs, leading to scrambling of the label to the formyl carbon (equation 10). The rearranged ketenes have IR absorptions at 2121, 2129, 2133, and 2136 cm^{-1}.[30] Further photolysis of **5** results in the disappearance of its IR spectrum.[31] This reaction is also considered in Section 3.4.6.

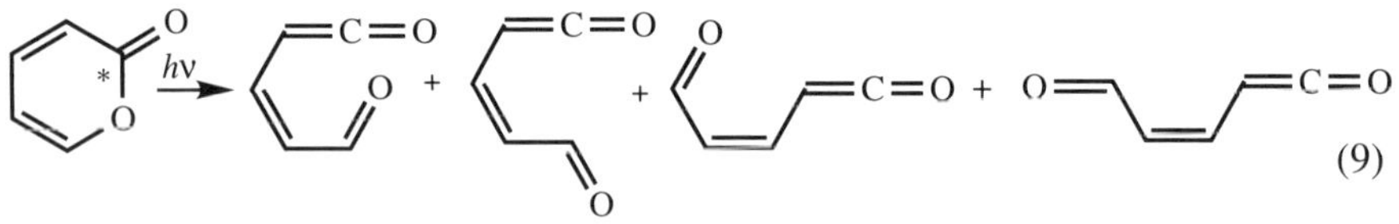

Photolysis at λ > 300 nm of o-benzoquinone in an Ar matrix at 10 K gives the bisketene **6** characterized by IR bands at 2105 and 2115 cm^{-1} [32] and further photolysis at 254 nm results in decarbonylation and formation of cyclopentadienone,[32] possibly by the carbene shown (equation 11).

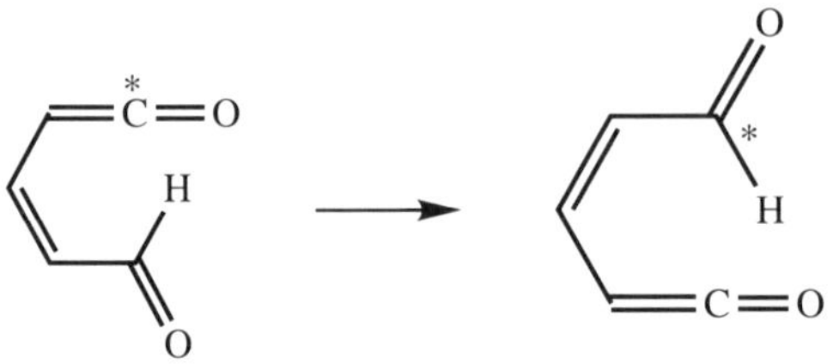

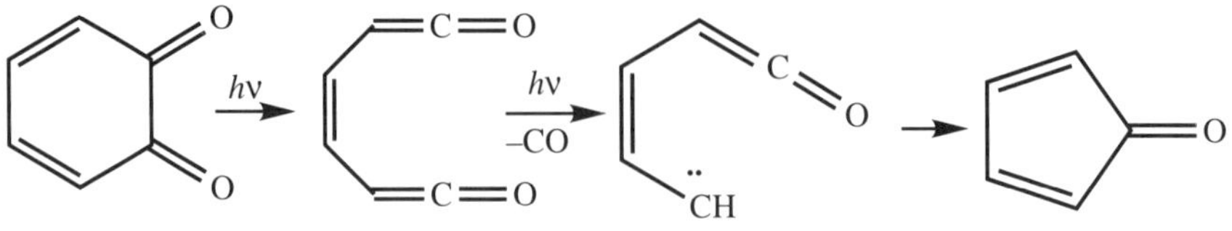

4.1.2 ALKENYLKETENES

Photolysis of the azide **7** in hexane gave the dienylketene **8** as evidenced by its UV spectrum [λ_{max} 315 nm (ε 15,500) 323, sh] which disappeared with a half-life at 22 °C of 4 min, presumably involving cyclization and dimerization in a slow process (equation 12).[33] For R = Me, cyclization of **9** was estimated to be 10^3 times faster than for **8**, and this was attributed to steric effects.[33] However as noted in Section 4.1.1 methyl groups are destabilizing to ketenes relative to hydrogen, and there may be an electronic factor present as well.

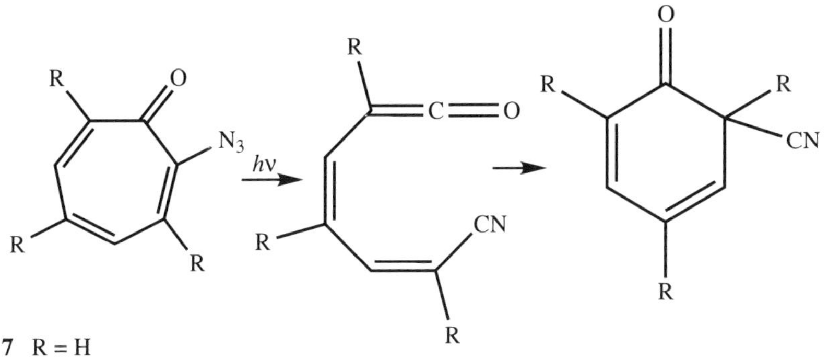

(12)

7 R = H
R = Me

8 R = H
9 R = Me

The parent dienylketene **10** was reported to be formed by flash vacuum pyrolysis of the acyl chloride, as evidenced by formation of phenol (42%) and **11** (25%) (equation 13).[34]

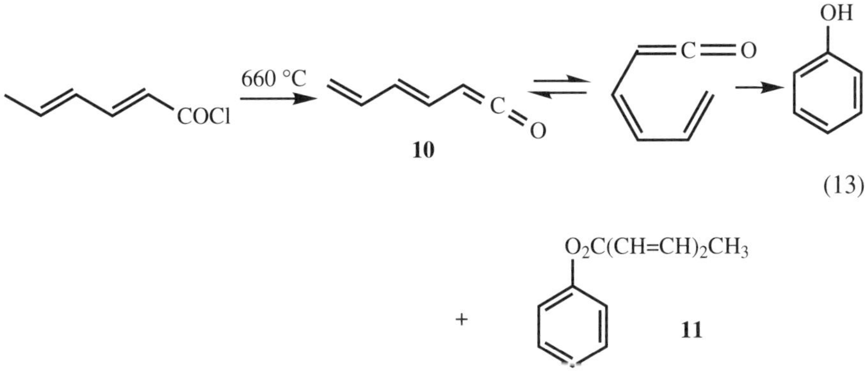

(13)

$O_2C(CH=CH)_2CH_3$

11

Photolysis of **12** in an Ar matrix containing 20% O_2 at 12 K led to formation of a blue species identified as **13** on the basis of an intense O—O vibration at 931 cm^{-1}, which on further irradiation led to a species tentatively identified as the ketene **14** (IR 2124, 1706 cm^{-1}) (equation 14).[35]

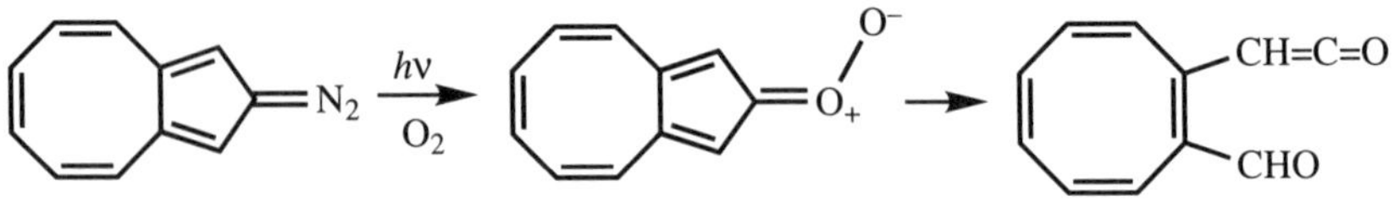

182 TYPES OF KETENES

$$\text{(14)}$$

Pyrolysis of 4-hydroxy and 4-alkoxy-4-alkynylcyclobutenones leads to hexa-1,3-diene-5-yne-1-ones which cyclize to quinones, as discussed in Section 3.4.1 (equation 15).[36,37]

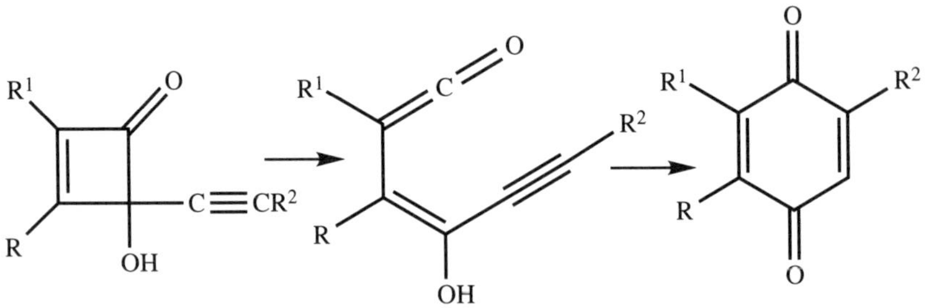

$$\text{(15)}$$

Pyrolysis of the cyclobutenone **15** was proposed to give the ketene **16**, which formed **17** by an ene reaction, and also formed **18** (equation 16).[38] It was shown that **17** did not give **18**, but the mode of formation of **18** was not discussed further.

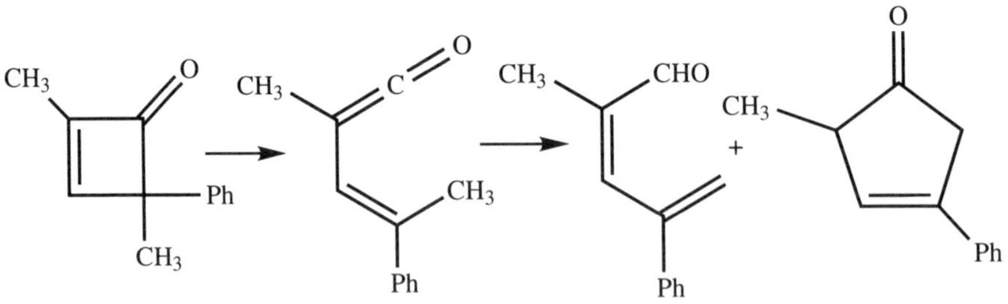

$$\text{(16)}$$

Reported alkenylketenes are listed in Table 1.

TABLE 1. Alkenylketenes

Ketene	Formula	Preparation	Registry No.	Ref.
$CCl_2=CClCCl=C=O$	C_4Cl_4O	Pyrolysis	78270-50-5	39
$CCl_2=CHCCl=C=O$	C_4HCl_3O	Pyrolysis	78270-49-2	39
$\cdot CH=CHCH=C=O$	$C_4H_3O^\cdot$	Mass spec	138720-36-2	40
$CH_2=\overset{+}{C}CH=C=O$	$C_4H_3O^+$	Flame	79530-60-2	40a
$-CH=CHCH=C=O$	$C_4H_3O^-$	Mass spec	138720-37-3	40
$CH_2=CHCH=C^{+\cdot}$	$C_4H_4O^{+\cdot}$	Mass spec	71793-85-6	41
$CH_2=CHCH=C=O$	C_4H_4O	Pyrolysis	50888-73-8	3,4,5
$O=CHCBr=CHCH=C=O$	$C_5H_3BrO_2$	Pyrolysis		42
$O=CHCD=CHCH=C=O$	$C_5H_3DO_2$	Pyrolysis		42
$NCCH=CHCH=C=O$	C_5H_3NO	Index entry	122935-07-3	43
$\cdot O_2CCH=CHCH=C=O$	$C_5H_3O_3$	Mass spec	138720-42-0	40
$^-O_2CCH=CHCH=C=O$	$C_5H_3O_3^-$	Mass spec	138720-39-5	40
$HO_2C\overset{\cdot}{C}=CHCH=C=O$	$C_5H_3O_3$	Mass spec	138720-38-4	40
$O=CHCH=CHCH=C=O$	$C_5H_4O_2$	Photolysis	39763-18-3	42,44
$HO_2CCH=CHCH=C=O$	$C_5H_4O_3$	Mass spec	138720-25-9	40
$CH_2=CH(HO_2C)C=C=O$	$C_5H_4O_3$	Photolysis	106007-69-6	21
$HN=CHCH=CHCH=C=O$	C_5H_5NO	Theory		45
$MeCH=CHCH=C=O$	C_5H_6O	Theory	52509-40-7	46
$MeCH=CHCH=C=O$ (E)	C_5H_6O	Pyrolysis	36566-95-7	20
$MeCH=CHCH=C=O$ (Z)	C_5H_6O	Pyrolysis	36566-94-6	20,47,48
$CH_2=CMeCH=C=O$	C_5H_6O	Dehydrochlorination	51637-55-9	49–51
$CH_2=CHCMe=C=O$	C_5H_6O	Dehydrochlorination	83897-55-6	11,15
$E\text{-}MeOCH=CHCH=C=O$	$C_5H_6O_2$	Wolff		52

TABLE 1. (Continued)

Ketene	Formula	Preparation	Registry No.	Ref.
$Cl_2C=CClCR=C=O$ $R = CO_2Me$	$C_6H_3Cl_3O_3$	Wolff	95142-03-3	53
	$(C_6H_3Cl_3O_3)_2$		95142-04-4	53
$O=C=CHCH=CHCH=C=O$	$C_6H_4O_2$	Photolysis	98850-20-5	32
$\cdot CH=CHCH=CHCH=C=O$	C_6H_5O	Index entry	91184-55-3	
$CH_2=CHCH=CHCH=C=O$	C_6H_6O	Pyrolysis	101055-58-7	34
$CH_2=CHCH=CHCH=C=O$	C_6H_6O	Mass spec	82944-25-0	54
$E\text{-}CH_2=CHCH=CHCH=C=O$	C_6H_6O	Theory	95333-12-3	55
$(CH_2=CH)_2C=C=O$	C_6H_6O	Theory	141272-01-7	56
$O=CHCMe=CHCH=C=O$	$C_6H_6O_2$	Pyrolysis		42
$CH_2=CMe(HO_2C)=C=O$	$C_6H_6O_3$	Pyrolysis	112946-26-6	21
$CH_2=CHCEt=C=O$	C_6H_8O	Dehydrochlorination	89237-34-3	11,12
$MeCH=CHCMe=C=O$	C_6H_8O	Rearrangement	64731-48-2	57
$CH_2=CMeCMe=C=O$	C_6H_8O	Rearrangement		50
$NC(CH=CH_2)_2CH=C=O$	C_7H_5NO	Photolysis	20599-09-1	33,58
1-cyclopentenylCH=C=O	C_7H_8O	Pyrolysis	96913-91-6	59,60
$RCH=CHCH=C=O$ $R = CH_2=CHCH_2$	C_7H_8O	Dehydrochlorination	104923-51-5	61,62
$Me_2C=CHCMe=C=O$	$C_7H_{10}O$	Rearrangement	55701-65-0	38
$R(HO_2C)C=C=O$ R = 1-cyclopentenyl	$C_8H_8O_3$	Pyrolysis	96913-89-2	59
$RC(CO_2Et)=C=O$ $R = O=CHCH=CH$	$C_8H_8O_4$	Pyrolysis		63

Compound	Formula	CAS	Method	Ref
CH_2=CRCH=C=O				
R = CH_2=CHCF$_2$CH$_2$	$C_8H_{10}O$	104923-41-3	Dehydrochlorination	61
RCH=CHCH=C=O				
R = Me_2C=CH	$C_8H_{12}O$	59358-84-8	Rearrangement	64
CH_2=C(Bu-n)CH=C=O	$C_8H_{12}O$		Cyclobutenone	50
RCH=CHCH=C=O	$C_9H_{10}O_3$	71256-67-2	Photolysis	65
R = AcOCMe=CH				
RCH=CHCH=C=O	$C_9H_{12}O$	104923-52-6	Dehydrochlorination	61
R = EtCH=CHCH$_2$				
RCH=CHCH=C=O	$C_9H_{12}O$	79801-44-8	Photolysis	66
R = Me_2C=CHCH$_2$				
CH_2=CRCH=C=O	$C_9H_{12}O$	104923-40-2	Dehydrochlorination	61
R = CH_2CH_2CMe=CH$_2$				
RCH=CMeCH=C=O	$C_9H_{12}O$	50401-19-9	Photolysis	67
R = Me_2C=CH				
CH_2=CHCR=C=O	$C_9H_{12}O$	104923-57-1	Dehydrochlorination	61
R = CH_2=CH(CH$_2$)$_3$				
AcCMe=CMeCMe=C=O	$C_9H_{12}O_2$	80673-77-4	Photolysis	68
CCl$_2$=CPhCH=C=O	$C_{10}H_6Cl_2O$	109384-88-5	Photolysis	69
CHCl=CPhCCl=C=O	$C_{10}H_6Cl_2O$		Pyrolysis	70
PhCH=CHCH=C=O (E)	$C_{10}H_8O$	138541-72-7	Wolff	71
NC(CMe=CH)$_2$CMe=C=O	$C_{10}H_{11}NO$	20599-10-4	Photolysis	33

Compound	Formula	CAS	Method	Ref
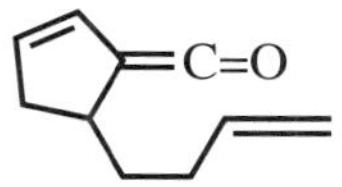 RCH=CHCMe=C=O	$C_{10}H_{12}O$		Dehydrochlorination	72
R = AcOCMe=CH	$C_{10}H_{12}O_3$	71256-68-3	Photolysis	65
RCH=CHCH=C=O				
R = AcOCMe=CMe	$C_{10}H_{12}O_3$	71256-69-4	Photolysis	65

TABLE 1. (*Continued*)

Ketene	Formula	Preparation	Registry No.	Ref.
RCMe=CMeCH=C=O R = Me$_2$C=CH	C$_{10}$H$_{14}$O	Rearrangement	21428-58-0	73
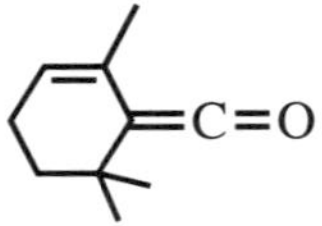 (2,6,6-trimethylcyclohex-2-enylidene ketene)	C$_{10}$H$_{14}$O	Dehydrochlorination	61899-98-7	10, 74
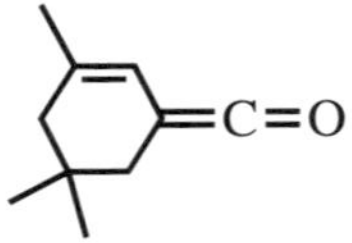 (3,5,5-trimethylcyclohex-2-enylidene ketene)	C$_{10}$H$_{14}$O	Index error	124165-71-5	75
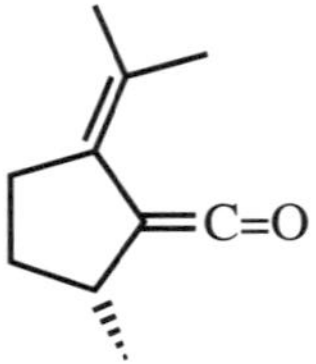 (isopropylidene methylcyclopentylidene ketene)	C$_{10}$H$_{14}$O	Dehydrochlorination		74
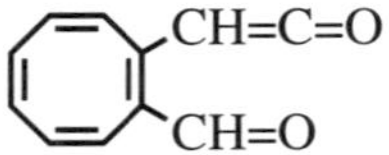 CH=C=O / CH=O	C$_{11}$H$_8$O$_2$	Oxidation	106689-38-7	35
RCMe=C=O (2: E,Z) R = AcO(CMe=CH)$_2$	C$_{11}$H$_{14}$O$_3$	Photolysis	21811-25-6	65 76

Structure	Formula	Reaction	CAS	Ref
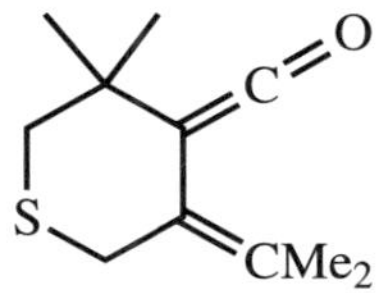 (thiopyran with C=O and CMe₂)	$C_{11}H_{16}OS$	Index error	55849-02-0	77
CH₂=CRCH=C=O, R = CONEt₃⁺Cl⁻	$C_{11}H_{18}NO_2 \cdot Cl$	Dehydrochlorination	49580-81-6	78
Me₂C=CHCPh=C=O	$C_{12}H_{12}O$	Wolff	55701-61-6	25
PhCMe=CHCMe=C=O (E)	$C_{12}H_{12}O$ (E)	Rearrangement	55701-58-1	38
(Z)		Rearrangement		38
RCMe=CMeCMe=C=O, R = AcOCMe=CH	$C_{12}H_{16}O_3$ (E, Z)	Photolysis	106689-39-8	65
RCMe=CHCMe=C=O	$C_{12}H_{17}N_3O$	Photolysis	86013-55-0	79
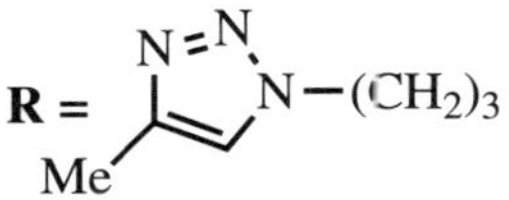 R = (pyrazole N=N N–(CH₂)₃ Me)				
RC(CN)=CHCR=C=O, R = t-Bu	$C_{13}H_{19}NO$	Pyrolysis	102343-60-2	80
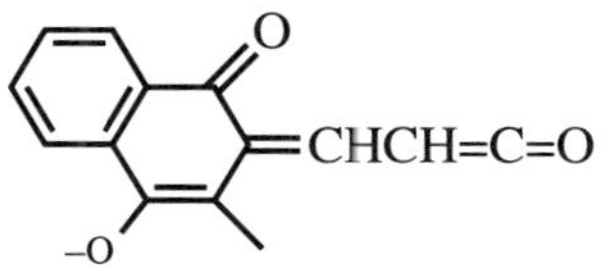 (naphthoquinone CHCH=C=O)	$C_{14}H_9O_3$	Photolysis	99477-45-9	81
RCH=CHCH=C=O, R = AcOCPh=CH (Z,Z)	$C_{14}H_{12}O_3$	Pyrolysis	58065-45-4	82
R = AcOCPh=CH (E,Z)		Pyrolysis	58065-44-4	82

TABLE 1. (*Continued*)

Ketene	Formula	Preparation	Registry No.	Ref.
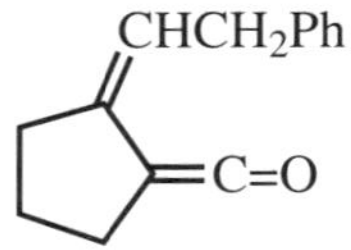 CHCH₂Ph, C=O (methylenecyclopentanone ketene)	$C_{14}H_{14}O$	Photolysis	38503-98-9	83
NCCR=C(OEt)CCl=C=O R = PhCH₂CH₂	$C_{15}H_{14}ClNO_2$	Pyrolysis		80
R₂NCH=CPhCCl=C=O R₂ = (CH₂)₅	$C_{15}H_{16}ClNO$	Pyrolysis		70
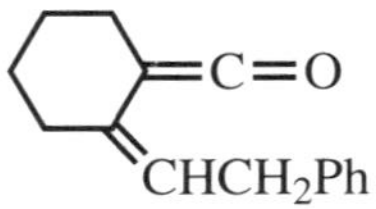 C=O, CHCH₂Ph (methylenecyclohexanone ketene)	$C_{15}H_{16}O$	Photolysis	22524-22-7	83
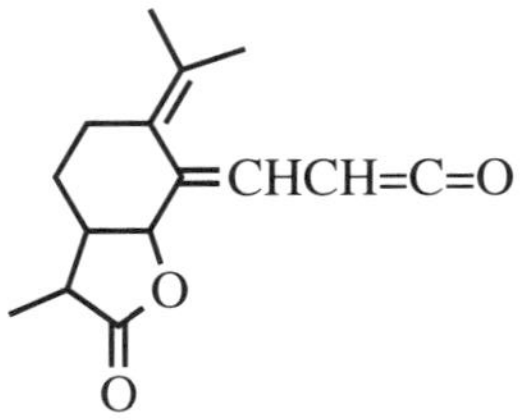 (bicyclic lactone with CHCH=C=O and isopropylidene)	$C_{15}H_{18}O_3$		120681-57-4	83a
CH₂=CRCH=C=O	$C_{15}H_{22}O$	Dehydrochlorination	103739-08-8	84

R =	Formula	Method	CAS	Ref
Ph$_2$C=CHCH=C=O	C$_{16}$H$_{12}$O	Pyrolysis		38
MeOCR=CHCR'=C=O	C$_{16}$H$_{22}$O$_2$Si	Metal carbene	88563-62-6	9
R = 2,6-Me$_2$C$_6$H$_3$; R' = Me$_3$Si				
NCCCl=PhCCPh=C=O	C$_{17}$H$_{10}$ClNO	Pyrolysis	97315-70-3	43
Ph$_2$C=CHCMe=C=O	C$_{17}$H$_{14}$O	Pyrolysis		38
Ph$_2$C=CHCEt=C=O	C$_{18}$H$_{16}$O	Pyrolysis		38
RCH=CHCH=C=O	C$_{19}$H$_{26}$O	Photolysis	87081-72-9	58
RCH=CHCPh=C=O	C$_{20}$H$_{16}$O$_3$	Pyrolysis	58296-80-3	82
R = PhC(OAc)=CH	2: E,Z; Z,Z	Pyrolysis	58296-79-0	82
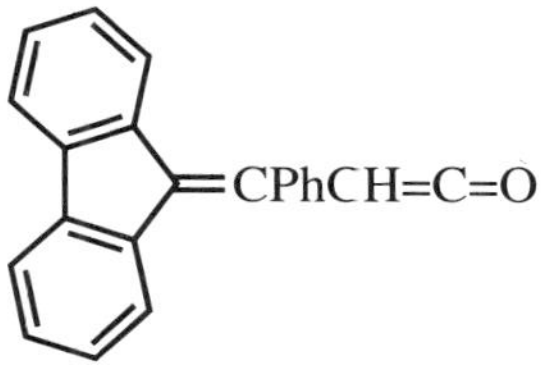 =CPhCH=C=O	C$_{22}$H$_{14}$O	Photolysis		85
Ph$_2$C=CPhCH=C=O	C$_{22}$H$_{16}$O	Photolysis		85
Ph(PhO)C=CRCH=C=O	C$_{22}$H$_{20}$N$_2$O$_2$	Photolysis	92670-90-1	86

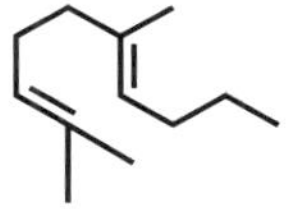

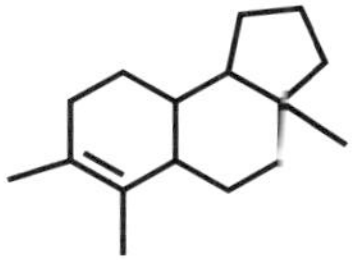

TABLE 1. (Continued)

Ketene	Formula	Preparation	Registry No.	Ref.
R= 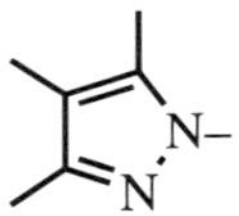 (3,4,5-trimethylpyrazol-1-yl)				
NCCPh=CPhCPh=C=O	$C_{23}H_{15}NO$	Pyrolysis	115476-62-5	87
RCPh=CHCPh=C=O	$C_{26}H_{20}O_3$	Photolysis	38389-57-0	88
R = PhC(OAc)=CH				
t-BuCR=CHC(Bu-t)=C=O	$C_{29}H_{40}O_3$	Photolysis	121359-33-9	89
R= 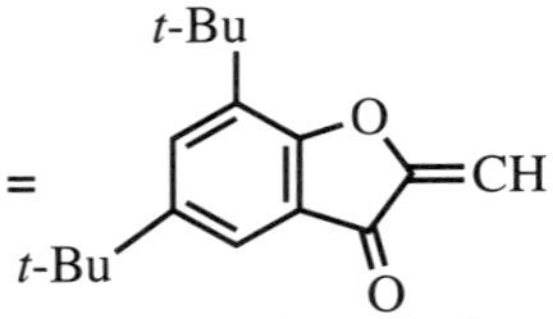 (5,7-di-t-butyl-3-oxo-benzofuran-2-ylidene CH)				
PhC(OPh)=CRCR=C=O	$C_{30}H_{20}O_4$	Photolysis	112148-07-9	90
R = PhCO				
PhCR=CPhCPh=C=O	$C_{37}H_{24}NO$	Pyrolysis		91
R= 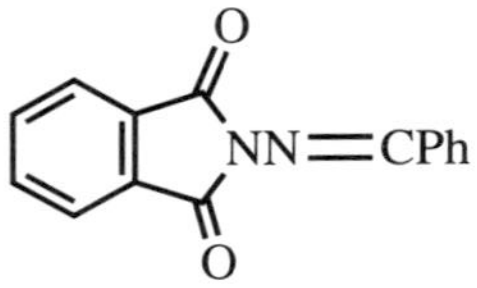 (phthalimido-N=CPh)				

References

1. Brown, R. D.; Godfrey, P. D.; Woodruff, M. *Austr. J. Chem.* **1979**, *32*, 2103–2109.
2. Nguyen, M. T.; Ha, T. K.; More O'Ferrall, R. A. *J. Org. Chem.* **1990**, *55*, 3251–3256.
3. Gong, L.; McAllister, M. A.; Tidwell, T. T. *J. Am. Chem. Soc.* **1991**, *113*, 6021–6028.
4. Mohmand, S.; Hirabayashi, T.; Bock, H. *Chem. Ber.* **1981**, *114*, 2609–2621.
5. Trahanovsky, W. S.; Surber, B. W.; Wilkes, M. C.; Preckel, M. M. *J. Am. Chem. Soc.* **1982**, *104*, 6779–6781.
6. Herberhold, M.; Wiedersatz, G. O.; Kreiter, C. G.; *Z. Naturforsch. B* **1976**, *31b*, 35–38.
7. Allen, A. D.; Andraos, J.; Kresge, A. J.; McAllister, M. A.; Tidwell, T. T. *J. Am. Chem. Soc.* **1992**, *114*, 1878–1879.
8. Danheiser, R. L.; Sard, H. *J. Org. Chem.* **1980**, *45*, 4810–4812.
9. Tang, P. C.; Wulff, W. D. *J. Am. Chem. Soc.* **1984**, *106*, 1132–1133.
10. Allen, A. D.; Stevenson, A.; Tidwell, T. T. *J. Org. Chem.* **1989**, *54*, 2843–2848.
11. Jackson, D. A.; Rey, M.; Dreiding, A. S. *Helv. Chim. Acta* **1983**, *66*, 2330–2341.
12. Jackson, D. A.; Rey, M.; Dreiding, A. S. *Tetrahedron Lett.* **1983**, *24*, 4817–4820.
13. Truce, W. E.; Bailey, P. S., Jr. *J. Org. Chem.* **1969**, *34*, 1341–1345.
14. Payne, G. B. *J. Org. Chem.* **1966**, *31*, 718–721.
15. Paquette, L. A.; Colapret, J. A.; Andrews, D. R. *J. Org. Chem.* **1985**, *50*, 201–205.
16. Hickmott, P. W.; Miles, G. J.; Sheppard, G.; Urbani, R.; Yoxall, C. T. *J. Chem. Soc., Perkin Trans. 1* **1973**, 1514–1519.
17. Hickmott, P. W.; Hargreaves, J. R. *Tetrahedron* **1967**, *23*, 3151–3159.
18. Gelin, R.; Gelin, S.; Dolmazon, R. *Bull. Soc. Chim. Fr.* **1973**, 1409–1416.
19. Lombardo, L. *Tetrahedron Lett.* **1985**, *26*, 381–384.
20. Schiess, P.; Radimerski, P. *Helv. Chim. Acta* **1974**, *57*, 2583–2597.
21. Wentrup, C.; Lorencak, P. *J. Am. Chem. Soc.* **1988**, *110*, 1880–1883.
22. Terlouw, J. K.; Burgers, P. C.; Holmes, J. L. *J. Am. Chem. Soc.* **1979**, *101*, 225–226.
23. Rousseau, G.; Bloch, R.; LePerchec, P.; Conia, J. M. *J. Chem. Soc., Chem. Commun.* **1973**, 795–796.
24. Marquestiau, A.; Pauwels, P.; Flammang, R.; Lorencak, P.; Wentrup, C. *Org. Mass Spectrom.* **1986**, *21*, 259–265.
25. Franck-Neumann, M.; Buchecker, C. *Tetrahedron Lett.* **1973**, 2875–2878.
26. Day, A. C.; McDonald, A. N.; Anderson, B. F.; Bartczak, T. J.; Hodder, O. J. R. *J. Chem. Soc., Chem. Commun.* **1973**, 247–248.
27. Holder, R. W.; Freiman, H. S.; Stefanchik, M. F. *J. Org. Chem.* **1976**, *41*, 3303–3307.
28. Barbaro, G.; Battaglia, A.; Giorgianni, P. *J. Org. Chem.* **1987**, *52*, 3290–3296.
29. Silversmith, E. F.; Kitahara, Y.; Roberts, J. D. *J. Am. Chem. Soc.* **1958**, *80*, 4088–4089.
30. Huang, B.-S.; Pong, R. G. S.; Laureni, J.; Krantz, A. *J. Am. Chem. Soc.* **1977**, *99*, 4154–4156.
31. Arnold, B. R.; Radziszewski, J. G.; Campion, A.; Perry, S. S.; Michl, J. *J. Am. Chem. Soc.* **1991**, *113*, 692–694.
32. Maier, G.; Franz, L. H.; Hartan, H. G.; Lanz, K.; Reisnauer, H. P. *Chem. Ber.* **1985**, *118*, 3196–3204.

33. Hobson, J. D.; Al Holly, M. M.; Malpass, J. R. *Chem. Commun.* **1968**, 764–765.
34. Dehmlow, E. V.; Slopianka, M. *Angew. Chem., Int. Ed. Engl.* **1979**, *18*, 170.
35. Murata, S.; Tomioka, H.; Kawase, T.; Oda, M. *J. Org. Chem.* **1990**, *55*, 4502–4504.
36. Moore, H. W.; Decker, O. H. W. *Chem. Rev.* **1986**, *86*, 821–830.
37. Liebeskind, L. S.; Foster, B. S. *J. Am. Chem. Soc.* **1990**, *112*, 8612–8613.
38. Mayr, H. *Angew. Chem., Int. Ed. Engl.* **1975**, *14*, 500–501.
39. Martin, P.; Greuter, H.; Bellus, D. *Helv. Chim. Acta* **1981**, *64*, 64–77.
40. Takhistov, V. V.; Muftakhov, M. V.; Krivoruchko, A. A.; Mazunov, V. A. *Izv. Akad. Nauk SSSR* **1991**, 2049–2055; *Chem. Abstr.* **1992**, *116*, 58567v.
40a. Olson, D. B.; Calcote, H. F. *Symp. (Int.) Combust., [Proc.]* 18th, 1980, 453–464; *Chem. Abstr.*, **1982**, *96*, 54782g.
41. Holmes, J. L.; Terlouw, J. K. *J. Am. Chem. Soc.* **1979**, *101*, 4973–4975.
42. Pirkle, W. H.; Seto, H.; Turner, M. V. *J. Am. Chem. Soc.* **1970**, *92*, 6984–6985.
43. Fishbein, P. L.; Moore, H. W. *J. Org. Chem.* **1985**, *50*, 3226–3228.
44. Chapman, O. L.; Hess, T. C. *J. Am. Chem. Soc.* **1984**, *106*, 1842–1843.
45. Kuzuya, M.; Ito, S.; Miyake, F.; Okuda, T. *Chem. Pharm. Bull.* **1982**, *30*, 1980–1985.
46. Fratev, F.; Tadzher, A. *J. Mol. Struct.* **1975**, *27*, 185–193.
47. Rey, M.; Dunkelblum, E.; Allain, R.; Dreiding, A. S. *Helv. Chim. Acta* **1970**, *53*, 2159–2175.
48. Funk, B. L.; Mossman, C. J.; Zeller, W. E. *Tetrahedron Lett.* **1984**, *25*, 1655–1658.
49. Berge, J. M.; Rey, M.; Dreiding, A. S. *Helv. Chim. Acta* **1982**, *65*, 2230–2241.
50. Danheiser, R. L.; Gee, S. K.; Sard, H. *J. Am. Chem. Soc.* **1982**, *104*, 7670–7672.
51. Palomo, C.; Aizpurua, J. M.; Lopez, M. C.; Aurrekoetxea, N.; Oiarbide, M. *Tetrahedron Lett.* **1990**, *31*, 6425–6428.
52. Danheiser, R. L.; Cha, D. D. *Tetrahedron Lett.* **1990**, *31*, 1527–1530.
53. Roedig, A.; Fahr, E.; Aman, H. *Chem. Ber.* **1964**, *97*, 77–79.
54. Turecek, F.; Drinkwater, D. E.; Maquestiau, A.; McLafferty, F. W. *Org. Mass. Spectrom.* **1989**, *24*, 669–672.
55. Kuzuya, M.; Miyake, F.; Okuda, T. *J. Chem. Soc., Perkin Trans. 2*, **1984**, 1471–1477.
56. McAllister, M. A.; Tidwell, T. T. *J. Am. Chem. Soc.* **1992**, *114*, 5362–5368.
57. Ficini, J.; Falou, S.; D'Angelo, J. *Tetrahedron Lett.* **1977**, 1931–1934.
58. Quinkert, G.; Englert, H.; Cech, F.; Stegk, A.; Haupt, E.; Leibfritz, D.; Rehm, D. *Chem. Ber.* **1979**, *112*, 310–348.
59. Tseng, J.; McKee, M. L.; Shevlin, P. B. *J. Am. Chem. Soc.* **1987**, *109*, 5474–5477.
60. Wentrup, C.; Gross, G.; Berstermann, H. M.; Lorencak, P. *J. Org. Chem.* **1985**, *50*, 2877–2881.
61. Kulkarni, Y. S.; Burbaum, B. W.; Snider, B. B. *Tetrahedron Lett.* **1985**, *26*, 5619–5622.
62. Snider, B. B.; Allentoff, A. J.; Kulkarni, Y. S. *J. Org. Chem.* **1988**, *53*, 5320–5328.
63. Engel, C. R.; de Krassny, A. F.; Belanger, A.; Dionne, G. *Can. J. Chem.* **1973**, *51*, 3263–3271.
64. Quinkert, G.; Bronstert, B.; Egert, D.; Michaelis, P.; Jürges, P.; Prescher, G.; Syldatk, A.; Perkampus, H. H. *Chem. Ber.* **1976**, *109*, 1332–1345.

65. Quinkert, G.; Kleiner, E.; Freitag, B. J.; Glenneberg, J.; Billhardt, U. M.; Cech, F.; Schmieder, K. R.; Schudok, C.; Steinmetzer, H. C.; Bats, J. W.; Zimmermann, G.; Dürner, G.; Rehm, D. *Helv. Chim. Acta* **1986**, *69*, 469–637.

66. Catalan, C. A. N.; Merep, D. J.; De Delfini, M. L. T.; Retamar, J. A. *Riv. Ital. EPPOS* **1981**, *63*, 209–215; *Chem. Abstr.* **1981**, *95*, 204163b.

67. Chapman, O. L.; Clardy, J. C.; McDowell, T. L.; Wright, H. E. *J. Am. Chem. Soc.* **1973**, *95*, 5086–5087.

68. Maier, G.; Reisenauer, H. P. *Chem. Ber.* **1981**, *114*, 3959–3964.

69. Hassner, A.; Naidorf, S. *Tetrahedron Lett.* **1986**, *27*, 6389–6392.

70. Jenny, E. F.; Druey, J. *J. Am. Chem. Soc.* **1960**, *82*, 3111–3117.

71. Allen, A. D.; Andraos, J.; Kresge, A. J.; McAllister, M. A.; Tidwell, T. T. *J. Am. Chem. Soc.* **1992**, *114*, 1878–1879.

72. Veenstra, S. J.; De Mesmaeker, A.; Ernst, B. *Tetrahedron Lett.* **1988**, *29*, 2303–2306.

73. Griffiths, J.; Hart, H. *J. Am. Chem. Soc.* **1968**, *90*, 5296–5298.

74. Wuest, J. D.; Madornik, A. M.; Gordon, D. C. *J. Org. Chem.* **1977**, *42*, 2111–2113.

75. Konishi, K.; Kuragano, T. *Nippon Noyaku Gakkaishi* **1989**, *14*, 351–357; *Chem. Abstr.* **1990**, *112*, 3530c.

76. Quinkert, G. *Photochem. Photobiol.* **1968**, *7*, 783–789.

77. Krebs, A.; Kimling, H. *Liebigs Ann. Chem.* **1974**, 2074–2084.

78. Sakai, M.; Kawarabayashi, N.; Sakakibara, Y.; Uchino, N.; Oka, S. *Nippon Kagaku Kaishi* **1973**, 1715–1718; *Chem. Abstr.* **1979**, *79*, 125781t.

79. Schultz, A. G.; Myong, S. O. *J. Org. Chem.* **1983**, *48*, 2432–2434.

80. Dorsey, D. A.; King, S. M.; Moore, H. W. *J. Org. Chem.* **1986**, *51*, 2814–2816.

81. Canfield, L. M.; Davy, L. A.; Tomer, K. B. *Photobiochem. Photobiophys.* **1985**, *10*, 23–33.

82. Dannenberg, W.; Perst, H. *Liebigs Ann. Chem.* **1975**, 1873–1894.

83. Schiess, P.; Suter, C. *Helv. Chim. Acta* **1971**, *54*, 2636–2645.

83a. Reisch, J.; Henkel, G.; Topaloglu, Y.; Simon, G. *Pharmazie* **1988**, *43*, 15–17; *Chem. Abstr.* **1989**, *110*, 231899f.

84. Corey, E. J.; Desai, M. C. *Tetrahedron Lett.* **1985**, *26*, 3535–3538.

85. Schrader, L.; Hartmann, W. *Tetrahedron Lett.* **1973**, 3995–3996.

86. Lohray, B. B.; Kumar, C. V.; Das, P. K.; George, M. V. *J. Org. Chem.* **1984**, *49*, 4647–4656.

87. Chow, K.; Moore, H. W. *Tetrahedron Lett.* **1987**, *28*, 5013–5016.

88. Lemmer, D.; Perst, H. *Tetrahedron Lett.* **1972**, 2735–2738.

89. Izuoka, A.; Miya, S.; Sugawara, T. *Tetrahedron Lett.* **1988**, *29*, 5673–5676.

90. Rubin, M.; Sander, W. *Tetrahedron Lett.* **1987**, *28*, 5137–5140.

91. Hoesch, L. *Chimia* **1975**, *29*, 531–532.

4.1.3 Alkynylketenes and Cyanoketenes

The alkynyl substituent is calculated to have the same stabilizing effect on a ketene as methyl,[1] but no persistent alkynylketenes are known. (Phenylethynyl)ketene (**1**) was proposed to be formed as an intermediate from the photochemical Wolff rearrangement of the diazo ketone **2** (equation 1).[2] Ketene **1** was not actually observed under the reaction conditions, but was proposed to undergo photochemical decarbonylation to the carbene which abstracted hydrogen from the solvent, leading to the observed products 1-phenylpropyne, phenylallene, and benzylacetylene.

$$PhC\equiv CCOCHN_2 \xrightarrow{h\nu} PhC\equiv CCH=C=O \xrightarrow[-CO]{h\nu} PhC\equiv C\ddot{C}H \longrightarrow PhC\equiv CCH_3 \quad (1)$$
$$\quad\quad\quad\quad\quad 2 \quad\quad\quad\quad\quad\quad\quad\quad 1$$

Further study of the flash photolysis of **2** in H_2O led to the observation of UV absorption attributed to **1**, which rapidly decayed by a path attributed to the hydration of the ketene.[3] This ketene is 15 times more reactive than $PhCH=C=O$ and 700 times more reactive than $n\text{-}BuCH=C=O$, and its high reactivity was attributed to conjugation in an enolate-like transition state for addition of H_2O (equation 2).[3]

$$PhC\equiv CCH=C=O \xrightarrow{H_2O} PhC\equiv CCH=C\begin{matrix}O^-\\ \diagup\\ \diagdown\\ \overset{+}{OH_2}\end{matrix} \quad (2)$$
$$\quad 1$$

Alkynylketenes have been generated by Moore et al.,[4-9] as reactive intermediates by a high temperature retro Diels-Alder reaction (equation 3)[5] and by the rather mild thermolysis of 2,5-dialkyl-3,6-diazidoquinones (equation 4).[6,7] This latter route provided a variety of α-cyanoalkynylketenes **4** whose reactivity, especially in cycloadditions, was examined.[6,7]

$$\text{[structure with } Me_3SiO, MeO, OMe, C\equiv CPh \text{ substituents]} \xrightarrow{220\ °C} \underset{Me_3SiO}{\overset{Ph}{\diagdown}}C\equiv C-C=C=O$$
$$\quad\quad\quad\quad\quad\quad\quad\quad\quad\quad\quad\quad\quad\quad\quad\quad\quad\quad\quad\quad\quad\quad 3$$

$$(3)[8]$$

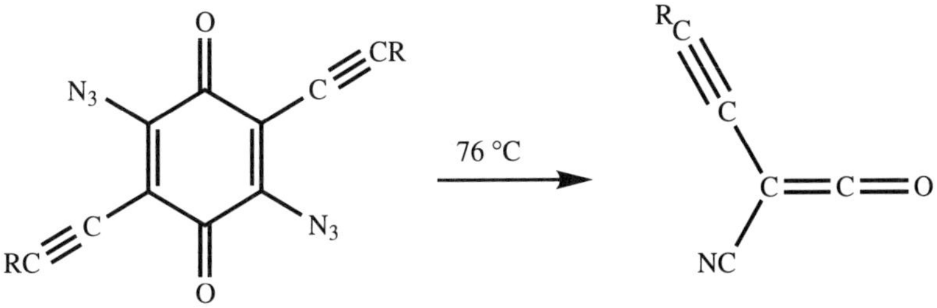

4.1.3 ALKYNYLKETENES AND CYANOKETENES

(4)[6]

Reaction of acyl chlorides **5–8** with Et$_3$N in the presence of cyclopentadiene was proposed to lead to the ketenes **9–12**, as evidenced by the formation of the bicyclic adducts (equations 5 and 6).[10] The ketenes themselves were not observed. All of the observed adducts had the *exo*-alkynyl groups as shown.

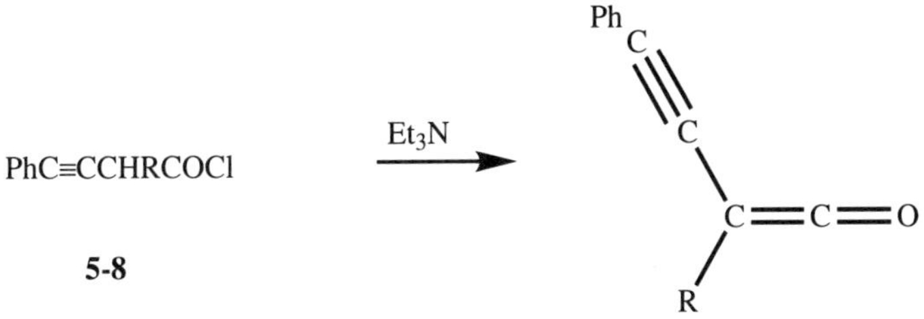

PhC≡CCHRCOCl (5)

5-8

R = Me (**9**), Et (**10**), *i*-Pr (**11**), *t*-Bu (**12**)

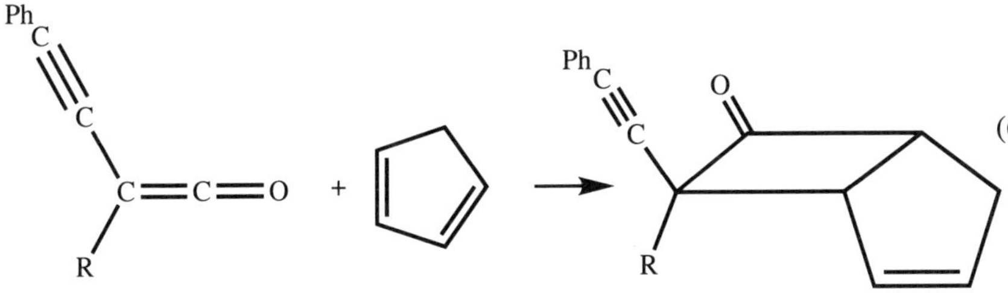

(6)

1-*tert*-Butoxy-1,3-butadiynes **13** on heating in benzene give alkynylketenes **14**, which were trapped intramolecularly by pendant hydroxy groups to give lactones (equation 7).[11] This general route is described in Section 3.4.5.

$$RC≡C-C≡COBu-t \xrightarrow[\Delta]{\text{benzene}} RC≡C-CH=C=O \quad (7)$$

13 **14**

Reaction of ketene **15** with alcohol was suggested to lead by an intramolecular rearrangement to the allene **16**, detected by an IR band at 1950 cm^{-1} in the crude reaction mixture, which added alcohol to give the ester **17** (equation 8).[9]

$$\underset{15}{\underset{NC}{\overset{n\text{-Bu}}{\Big|}}C{\equiv}C\Big\rangle C{=}C{=}O} \xrightarrow{ROH} \underset{NC}{\underset{\Big|}{\overset{n\text{-Bu}}{\Big|}}C{\equiv}C\Big\rangle \overset{H}{\underset{O}{\overset{\Big|}{C}}}{-}\overset{\overset{+}{O}{-}R}{\underset{\Big\|}{C}}} \longrightarrow \underset{n\text{-Bu}}{\overset{H}{\Big\rangle}}C{=}C{=}\underset{CN}{\overset{CO_2R}{\Big\langle}} \qquad (8)$$

$$\underset{16}{16} \xrightarrow{ROH} n\text{-BuCH}_2\text{C(OR)=C(CN)CO}_2\text{R}$$

$$17$$

Cyanoketenes have been widely studied but are rather reactive species.[12–15] The parent NCCH=C=O (**18**) has been generated in the gas phase by pyrolysis of the acid chloride and characterized spectroscopically (equation 9).[16,17] Kinetic analysis of the hydrolysis of 2-nitrophenyl cyanoacetate led to the conclusion that the ketene was formed as an intermediate,[18] and capture of NCCH=C=O from the 4-nitrophenyl ester when this reaction is conducted in EtOH confirms this hypothesis (equation 10).[19]

$$NCCH_2COCl \xrightarrow{920\ K} \underset{18}{NCCH{=}C{=}O} \qquad (9)^{16,17}$$

$$NCCH_2CO_2C_6H_4NO_2\text{-}4 \xrightarrow{Et_3N} \underset{18}{NCCH{=}C{=}O} \xrightarrow{EtOH} NCCH_2CO_2Et \qquad (10)^{18,19}$$

The reaction of cyanoacetyl chloride with benzalaniline led to the cycloaddition product **19**, although it was suggested that this reaction may have proceeded stepwise through the species **20** (isolated for reaction of PhCH=NCH$_3$) as shown and not through the free ketene (equation 11).[20] A published report[21,22] of the preparation of NCCH=C=O by the reaction of NCCH$_2$CO$_2$Et with P$_2$O$_5$ at 90 °C, and the claimed[21,22] isolation of this ketene as a neat liquid appears surprising. The isolation of material assigned the 1,3,5-trihydroxy-2,4,6-tricyanobenzene structure (**21**) from this reaction (equation 12)[22] is consistent with formation of NCCH=C=O as a free intermediate in the reaction, although stepwise process may also be envisaged. The formation of NCCH=C=O as a reactive intermediate from pyrolytic reactions of NCCH$_2$CO$_2$Me[23] and NCCH$_2$CO$_2$Et[24] has also been proposed.

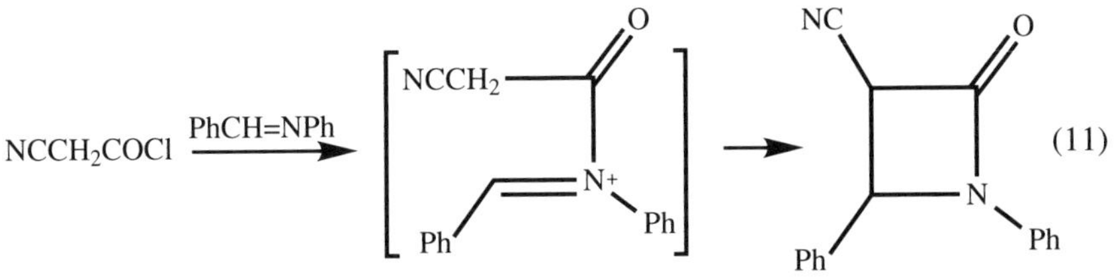

4.1.3 ALKYNYLKETENES AND CYANOKETENES

$$NCCH_2COCl \xrightarrow{PhCH=NPh} \left[\begin{array}{c}\text{NCCH}_2-\overset{O}{\underset{\underset{Ph}{\parallel}}{C}}\\ \overset{}{\underset{Ph}{\diagup}}\text{N}^+\diagdown Ph\end{array}\right] \longrightarrow \underset{\text{Ph}}{\overset{\text{NC}}{\beta\text{-lactam}}}\quad (11)$$

20, **19**, 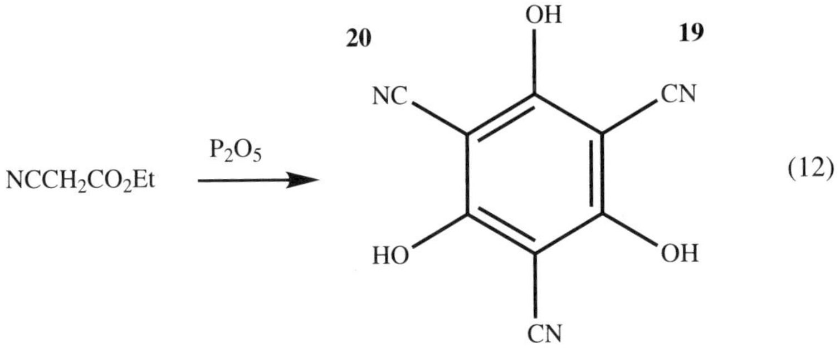 **21**

$$NCCH_2CO_2Et \xrightarrow{P_2O_5} \text{hexasubstituted benzene (21)} \quad (12)$$

The evident high reactivity of cyanoketene does not arise from ground state effects, as this ketene is calculated (Section 1.1.3) to have essentially the same ground state stability as alkylketenes, which have at least modest longevity. However the cyano group is small and provides little steric protection and also stabilizes many reaction transition states, causing the ephemeral nature of this species.

A general route to cyanoketenes via the pyrolysis of azido quinones has been thoroughly examined.[12–15,25–28] However only ketenes with bulky substituents such as *tert*-butylcyanoketene (**22**) prepared in this way (equation 13) have any appreciable kinetic stability, so that spectra may be recorded in solution at room temperature.[13–15] On attempted isolation of **22** polymerization occurs rapidly. The cleavage is proposed to proceed with initial formation of cyanocyclopentenediones **23**,[25] which may be isolated in some cases, and which then give the cyanoketenes (equation 13).

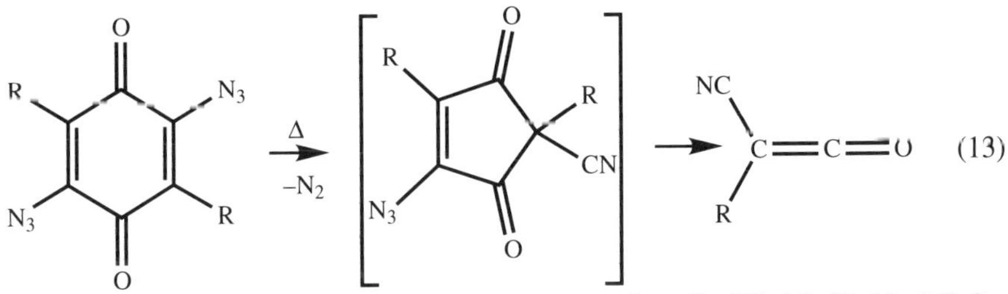

$$\text{azidoquinone} \xrightarrow[-N_2]{\Delta} [\textbf{23}] \longrightarrow \underset{R}{\overset{NC}{\diagdown}}C=C=O \quad (13)$$

R = *t*-Bu (**22**), *i*-Pr, Ph, Me, RC≡C

Dicyanoketene **24** has been obtained from dicyanodiazidoquinone (equation 14), and its reactions with various reagents have been examined.[26–28] Flash vacuum thermolysis of methyl dicyanoacetate at 500 °C also led to **24,** which was detected by mass spectrometry and by the IR band at 2175 cm^{-1} when deposited alone or in an Ar matrix at 12 K.[29] However at 60 K the absorption due to **24** disappeared. There is also evidence for formation of **24** from photolysis of **25** (equation 15).[30]

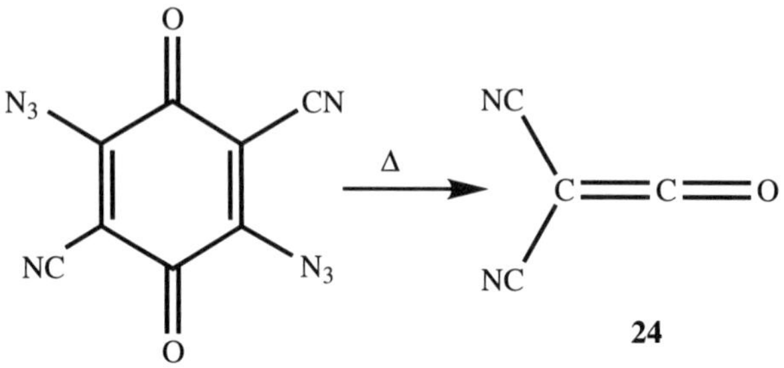

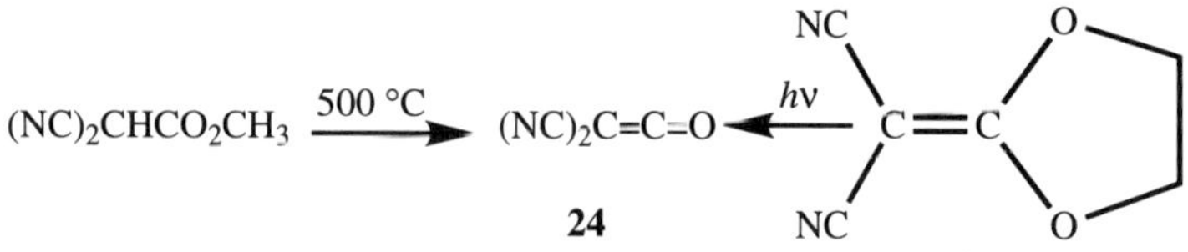

$$(NC)_2CHCO_2CH_3 \xrightarrow{500 \,°C} (NC)_2C=C=O \xleftarrow{h\nu} \text{[25]} \quad (15)$$

Chlorocyanoketene (**26**) could not be prepared by the azidoquinone route but was generated and trapped from the furanone **27** (equation 16).[31–33] Bromo- and iodocyanoketenes were formed similarly.[34]

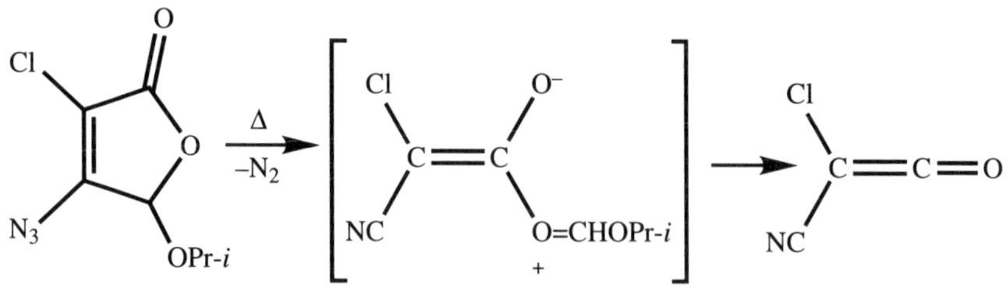

Reaction of the cyclobutenedione **28** with NaN$_3$ was proposed to involve formation of an azidocyclobutene which cleaved to phenylcyanoketene **29** (equation 17).[35,36]

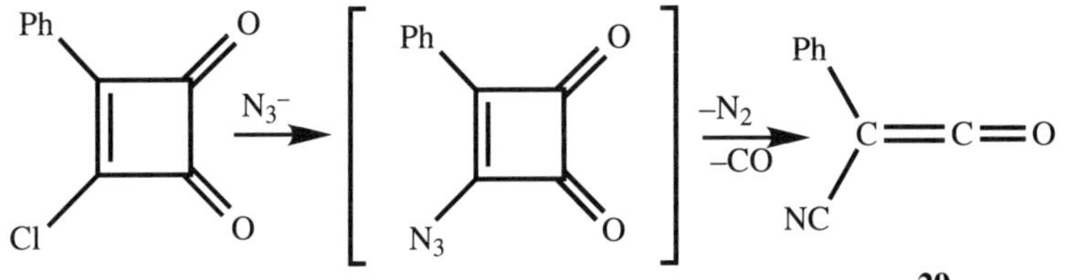

4.1.3 ALKYNYLKETENES AND CYANOKETENES

[Equation (17): Conversion of **28** via azide intermediate to cyanoketene **29**]

Capodative cyanoketenes, that is, cyanoketenes also bearing π-donor substituents including Me_2N, RO, and RS, have been proposed[37] as intermediates from substituted cyclobutenediones **30**, as shown in equation 18. Reaction of **30** with NaN_3 between 0 and 50 °C proceeds with evolution of N_2 and carbon monoxide and presumably involves the azidocyclobutenediones **31** which form the ketenes **32**, which are captured in situ by cycloaddition with imines to give cyano β-lactams **33**.[37]

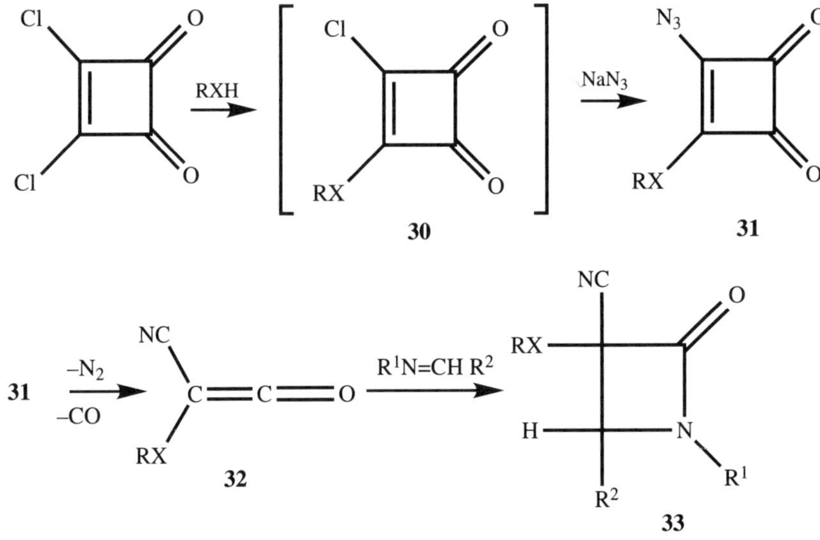

[Equation (18): Dichlorocyclobutenedione → **30** (RX substituted) → **31** (azido) → ketene **32** → β-lactam **33**]

RX = Me_2N, t-BuCH$_2$O, CH$_2$=C(Me)CH$_2$CH$_2$O, CH$_2$=CHCH$_2$O, EtS

The cycloaddition chemistry of cyanoketenes has been extensively examined, as discussed in Section 5.4.

TABLE 1. Alkynyl and Cyano Ketenes

Ketene	Formula	Preparation	Registry No.	Ref.
NCCBr=C=O	C_3BrNO	Pyrolysis	67767-48-0	34
NCCCl=C=O	C_3ClNO	Pyrolysis		31
NCCH=C=O	C_3HNO	Pyrolysis	4452-08-8	16
NCCI=C=O	C_3INO	Pyrolysis	67767-49-1	34
HC≡CCH=C=O	C_4H_2O	Theory	12881-22-7	1
NCCMe=C=O	C_4H_3NO	Pyrolysis	57681-10-4	38
$(NC)_2C=C=O$	C_4N_2O	Pyrolysis	4361-47-1	26
HC≡CC(CN)=C=O	C_5HNO	Pyrolysis	107202-29-9	7
NCC(SEt)=C=O	C_5H_5NOS	Pyrolysis		37
$NCC(NMe_2)=C=O$	$C_5H_6N_2O$	Pyrolysis		37
NCC(Pr-i)=C=O	C_6H_7NO	Pyrolysis	57681-11-5	14
$NCC(OCH_2CH_2CH=CH_2)=C=O$	$C_7H_7NO_2$	Pyrolysis		37
NCC(Bu-t)=C=O	C_7H_9NO	Pyrolysis	29342-22-1	14
NCCR=C=O R = CH_2=CMeC≡C R = OCH_2CH_2CMe=CH_2	C_8H_5NO	Pyrolysis		6
$NCC(CMe_2Et)=C=O$	$C_8H_9NO_2$	Pyrolysis		37
	$C_8H_{11}NO$	Pyrolysis	57681-09-1	14
$NCC(OCH_2Bu-t)=C=O$	$C_8H_{11}NO_2$	Pyrolysis		37
NCCPh=C=O	C_9H_5NO	Pyrolysis	24309-53-3	14

n-BuC≡CC(CN)=C=O	C_9H_9NO	Pyrolysis	94921-88-7	6,38
PhC≡CCH=C=O	$C_{10}H_6O$	Wolff	138541-71-6	3
PhC≡CC(OH)=C=O	$C_{10}H_6O_2$	Pyrolysis	119622-67-2	8
PhC≡CC(CN)=C=O	$C_{11}H_5NO$	Pyrolysis		6
PhC≡CCMe=C=O	$C_{11}H_8O$	Dehydrochlorination		10
THPOCH$_2$C≡CC(CN)=C=O	$C_{11}H_{11}NO_2$	Pyrolysis		6
o-TolC≡CC(CN)=C=O	$C_{12}H_7NO$	Pyrolysis		6
o-AnisC≡CC(CN)=C=O	$C_{12}H_7NO_2$	Pyrolysis		6
PhC≡CCEt=C=O	$C_{12}H_{10}O$	Dehydrochlorination		10
PhCH$_2$CH$_2$C≡CC(CN)=C=O	$C_{13}H_9NO$	Pyrolysis		6
PhC≡CC(Pr-i)=C=O	$C_{13}H_{12}O$	Dehydrochlorination		10
PhC≡CC(OSiMe$_3$)=C=O	$C_{13}H_{14}O_2Si$	Pyrolysis	119622-68-3	5
PhC≡CC(Bu-t)=C=O	$C_{14}H_{14}O$	Dehydrochlorination		10

References

1. Gong, L.; McAllister, M. A.; Tidwell, T. T. *J. Am. Chem. Soc.* **1991**, *113*, 6021–6028.
2. Selvarajan, R.; Boyer, J. H. *J. Org. Chem.* **1971**, *36*, 1679–1682.
3. Allen, A. D.; Andraos, J.; Kresge, A. J.; McAllister, M. A.; Tidwell, T. T. *J. Am. Chem. Soc.* **1992**, *114*, 1878–1879.
4. Moore, H. W.; Decker, O. H. W. *Chem. Rev.* **1986**, *86*, 821–830.
5. Pollart, D. J.; Moore, H. W. *J. Org. Chem.* **1989**, *54*, 5444–5448.
6. Nguyen, N. V.; Chow, K.; Moore, H. W. *J. Org. Chem.* **1987**, *52*, 1315–1319.
7. Chow, K.; Nguyen, N. V.; Moore, H. W. *J. Org. Chem.* **1990**, *55*, 3876–3880.
8. Fernandez, M.; Pollart, D. J.; Moore, H. W. *Tetrahedron Lett.* **1988**, *29*, 2765–2768.
9. Nguyen, N. V.; Moore, H. W. *J. Chem. Soc., Chem. Commun.* **1984**, 1066–1067.
10. Allen, A. D.; Gong, L.; Tidwell, T. T. *J. Am. Chem. Soc.* **1990**, *112*, 6396–6397.
11. Margriotis, P. A.; Vourloumis, D.; Scott, M. E.; Tarli, A. *Tetrahedron Lett.* **1993**, *34*, 2071–2074.
12. Moore, H. W.; Gheorghiu, M. D. *Chem. Soc. Rev.* **1981**, *10*, 289–328.
13. Moore, H. W.; Weyler, W., Jr. *J. Am. Chem. Soc.* **1970**, *92*, 4132–4133.
14. Weyler, W., Jr.; Duncan, W. G.; Moore, H. W. *J. Am. Chem. Soc.* **1975**, *97*, 6187–6192.
15. Weyler, W., Jr.; Duncan, W. G.; Liewen, M. B.; Moore, H. W. *Organic Syntheses*, Col. Vol. VI, Noland, W. E., Ed.; Wiley: New York, 1988; pp. 210–215.
16. Bock, H.; Hirabayashi, T.; Mohmand, S. *Chem. Ber.* **1981**, *114*, 2595–2608.
17. Hahn, M.; Bodenseh, H. In *Tenth Colloquium on High Resolution Molecular Spectroscopy;* Dijon, September 1987, paper M16; H. Bodenseh, private communication.
18. Holmquist, B.; Bruice, T. C. *J. Am. Chem. Soc.* **1969**, *91*, 3003–3009.
19. Allen, A. D.; Zhao, D.; Tidwell, T. T., unpublished results.
20. Böhme, H.; Ebel, S.; Hartke, K. *Chem. Ber.* **1965**, *98*, 1463–1464.
21. Zavlin, P. M.; Efremov, D. A. *Zh. Obshch. Khim.* **1988**, *58*, 2403–2404; *Engl. Transl.* **1988**, *58*, 2139.
22. Efremov, D. A.; Zavlin, P. M., Essentseva, N. S.; Tebby, J. C. *J. Chem. Soc., Perkin Trans. 1*, **1994**, 3163–3168.
23. den Tonkelaar, W. A. M.; Louw, R.; Kooyman, E. C. *Recl. Trav. Chim. Pays-Bas*, **1968**, *87*, 1281–1289.
24. Jaworski, T.; Kwiatkowski, S. *Rocz. Chem.* **1970**, *44*, 691–693; *Chem. Abstr.* **1970**, *73*, 130746r.
25. Moore, H. W. *Acc. Chem. Res.* **1979**, *12*, 125–132.
26. Neidlein, R.; Leidholdt, R. *Chem. Ber.* **1986**, *119*, 844–849.
27. Neidlein, R.; Bernhard, E. *Liebigs Ann. Chem.* **1979**, 959–964.
28. Neidlein, R.; Bernhard, E. *Angew. Chem., Int. Ed. Engl.* **1978**, *17*, 369–370.
29. Gano, J. E.; Jacobson, R. H.; Wettach, R. H. *Angew. Chem., Int. Ed. Engl.* **1983**, *22*, 165–166.
30. Footnote 2 in reference 29.
31. Moore, H. W.; Hernandez, L.; Sing, A. *J. Am. Chem. Soc.* **1976**, *98*, 3728–3730.
32. Fishbein, P. L.; Moore, H. W. *J. Org. Chem.* **1984**, *49*, 2190–2194.

33. Fishbein, P. L.; Moore, H. W. *Org. Synth.* **1990**, *68*, 205–211.
34. Kunert, D. M.; Chambers, R.; Mercer, F.; Hernandez, L. Jr.; Moore, H. W. *Tetrahedron Lett.* **1978**, 929–932.
35. De Selms, R. C. *Tetrahedron Lett.* **1969**, 1179–1180.
36. Schmidt, A. H.; Ried, W. *Tetrahedron Lett.* **1969**, 2431–2434.
37. Labille, M.; Janousek, Z.; Viehe, H. G. *Tetrahedron* **1991**, *47*, 8161–8166.
38. Moore, H. W.; Hughes, G.; Srinivasachar, K.; Fernandez, M.; Nguyen, N. V.; Schoon, D.; Tranne, A. *J. Org. Chem.* **1985**, *50*, 4231–4238.

4.1.4 Arylketenes

Diphenylketene (**1**) was the first ketene to be reported,[1] and is the ketene most cited in the literature, excluding ketene itself. Arylketenes are stabilized by conjugation of the aryl group with the alkenyl moiety of the ketene, and the bulk of the aryl group also provides steric protection against attack on the ketene. However the transition states for attack on the carbonyl carbon are usually stabilized by conjugation with the aryl group, and so these ketenes may be quite reactive. Thus the isodesmic stabilization energy ΔE of phenylketene (**2**) for reaction 1 is calculated at the 6-31G*//6-31G* level to be 0.7 kcal/mol.[2] This result indicates that the net stabilizing effect of the phenyl group on ketenes and alkenes is essentially the same in comparison to methyl, and that the major interaction of phenyl with the ketene is with the alkenyl double bond, and is unaffected by the carbonyl group.

$$\text{PhCH=C=O} + \text{CH}_3\text{CH=CH}_2 \xrightarrow{\Delta E = 0.7 \text{ kcal/mol}} \text{PhCH=CH}_2 + \text{CH}_3\text{CH=C=O} \quad (1)$$

2

Arylketenes are prepared quite readily by a variety of methods, including acyl chloride dehydrohalogenation,[3,4] α-halo acyl halide dehalogenation,[1] decarboxylation of malonic anhydrides,[3] and Wolff rearrangements,[5] but they are usually quite reactive, particularly toward nucleophiles. This is ascribed as being due to the enolate character of the transition states, which are stabilized by the conjugative effect of the aryl group (equation 2).[2]

$$\underset{R}{\overset{Ar}{>}}\!\!C\!=\!C\!=\!O \xrightarrow{Nu} \underset{R}{\overset{Ar}{>}}\!\!C\!=\!C\!\underset{Nu\ \delta+}{\overset{O\ \delta-}{<}} \quad (2)$$

Arylaldoketenes ArCH=C=O are particularly reactive, but have been generated as unobserved intermediates by reaction of the acyl chlorides with Et$_3$N and trapped with active alkenes,[6,7] and by Wolff rearrangements for kinetic studies of hydration.[2,8,9] In the latter studies the decrease in the UV absorption attributed to the ketene has been monitored (Section 5.5.1.2).[8] Reaction of PhCHClCOCl with Zn in ether gave PhCH=C=O, which could only be obtained in dilute solution because of its ease of decomposition and polymerization.[10]

Diarylketenes such as diphenylketene are slow to dimerize and have some stability, and in cases where the aryl groups are rather bulky, such as dimesitylketene, the ketenes are sterically protected from attack and are rather unreactive.[11–13] Arylalkylketenes ArCR=C=O are also rather resistant to dimerization, and may be kept for brief periods in the pure state.[3,14–18] These ketenes become less reactive toward dimerization, and in reaction with nucleophiles, as the size of the alkyl group increases. It has been suggested that with larger alkyl groups the aryl groups become more twisted out of conjugation with the C—C double bond.[17]

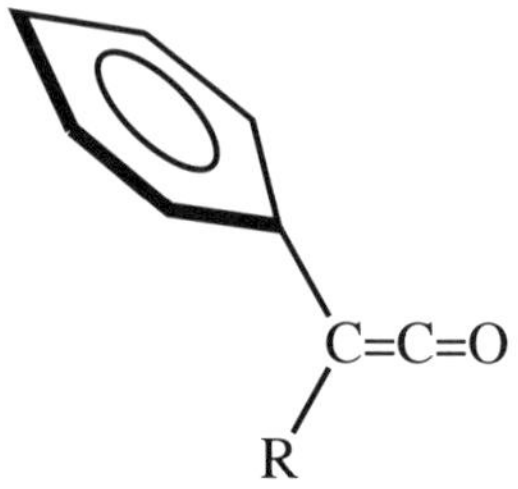

The perhalo aryl ketenes $(C_6F_5)_2C=C=O$[19] and $(C_6Cl_5)_2C=C=O$[20] have both been prepared and appear resistant to dimerization, but both of these ketenes react readily with water to form the corresponding carboxylic acids.

Capture of 1-naphthylketene with cyclopentadiene gave **3** which rearranged with acid to give **4** (equation 3).[21] This ketene can be prepared by Wolff rearrangement,[22] and the naphthylketenes (1-naphthyl)CPh=C=O, (2-naphthyl)$_2$C=C=O, and (1-naphthyl)$_2$C=C=O were prepared by zinc dehalogenation of the α-haloacyl halides.[23] The cycloaddition of these ketenes and of Ph$_2$C=C=O with EtOC≡CH proceeds at least partially by cyclization involving the aryl ring (equation 4),[23] as discussed further in Section 5.4.1.6.

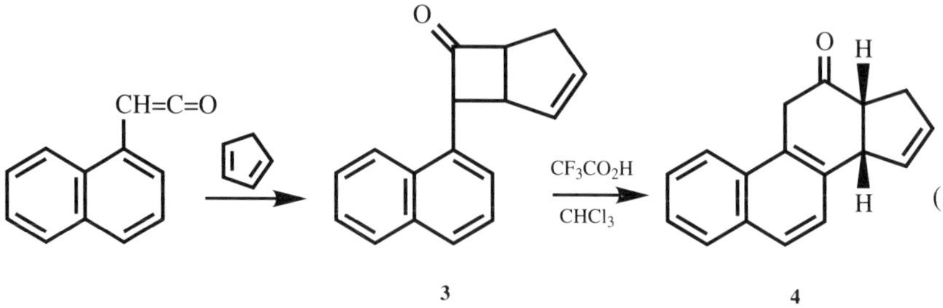

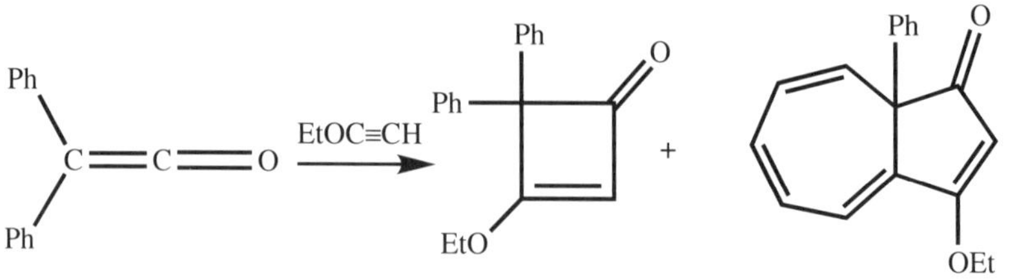

Intramolecular cycloadditions of α-arylketenes to alkenes have been carried out (equation 5).[24]

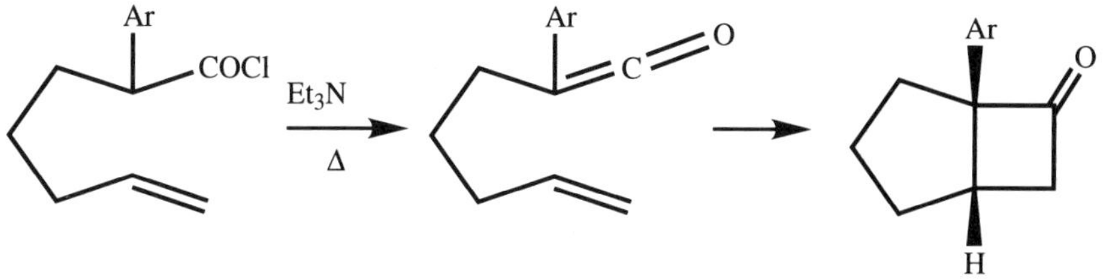

$$\text{(5)}$$

Ketenes such as dibenzofulvenone (**5**) have the possibility of stabilization by the aromatic character implied in the significant resonance contributing structure **6**. The fulvenones are discussed in more detail in Section 4.1.10. Photolysis of the diazoketone **7** gave evidence for the ketene **8** (equation 6).[25]

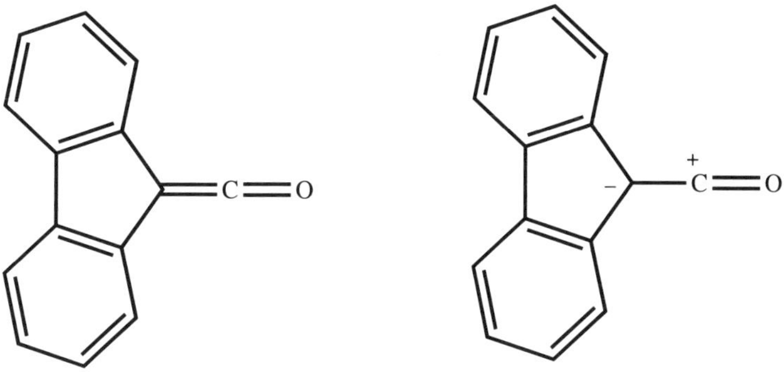

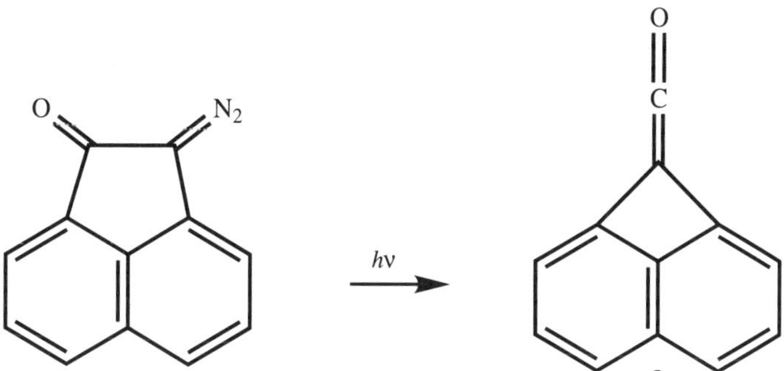

$$\text{(6)}$$

Only a few ketenes with pendant heterocyclic rings have been reported. 2-Furylketene was prepared by Wolff rearrangement,[25] 3-pyridylketene (**9**) was formed in a matrix both by Wolff rearrangement at 10 K and by addition of CO to the carbene **10** at 40 K (equation 7),[26] and 2-pyridylketene was formed by pyrolysis of ethyl 2-pyridylacetate.[27] An α-furylketene was also prepared by the dehydrochlorination route of equation 5,[24] and ArC(OMe)=C=O (Ar = 2- and 3-pyridyl) were formed by Wolff rearrangements in Ar matrices.[27a] Formation of ferrocenylketenes by Wolff rearrangement has been proposed.[27b]

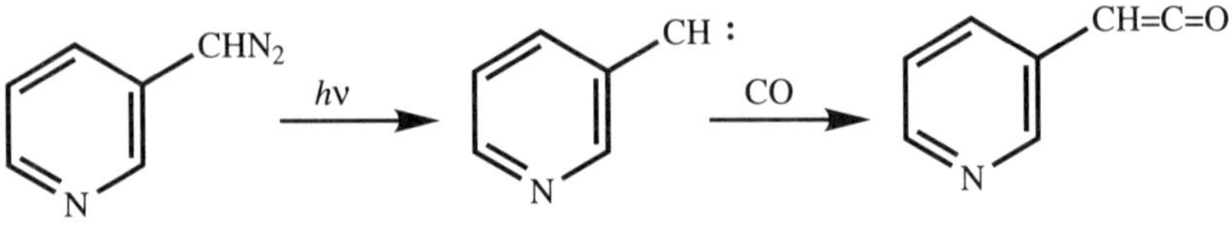

(7)

The addition of oxygen nucleophiles and proton donors to α-alkylarylketenes generates α-aryl carboxylic acid derivatives with a new chiral center, and as described in Section 5.9, the enantioselectivity of this process has been a topic of great interest, particularly because these compounds have pharmacological applications. A variety of chiral auxillaries have been used.[28–30] In the example of equation 8 cyclo(D-phenylalanyl-D-histidine) was the catalyst.[30]

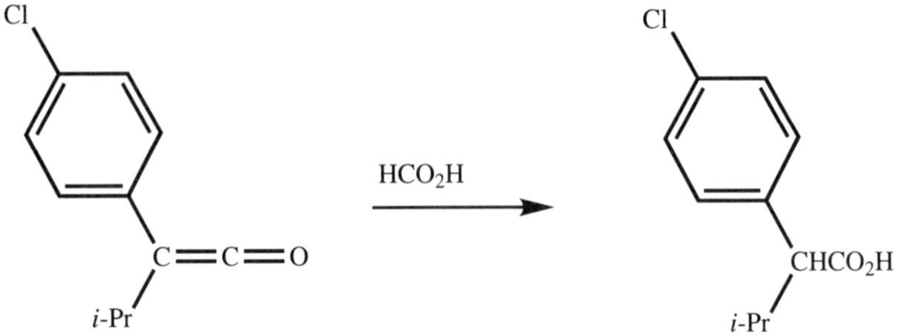

(8)

The photolysis of **11** in an Ar matrix at 10 K led to an IR absorption at 2120 cm^{-1} attributed to the ketene **12**, as well as the benzopyran-3-one **13**.[31] Flash vacuum pyrolysis of **11** gave **14**, which was proposed to form via **12** which rearranged to the ketene **15** and then to **14** (equations 9 and 10).[31]

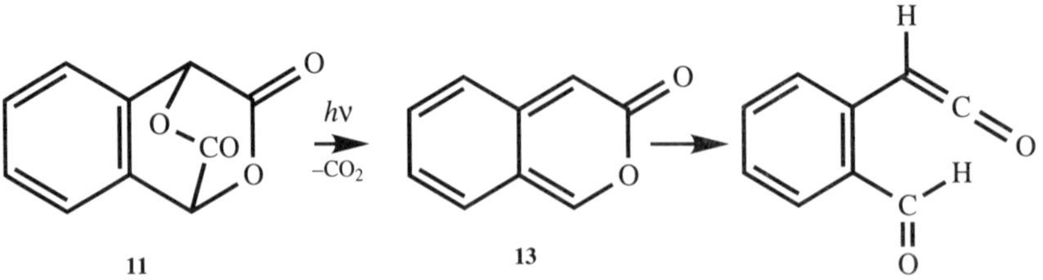

(9)

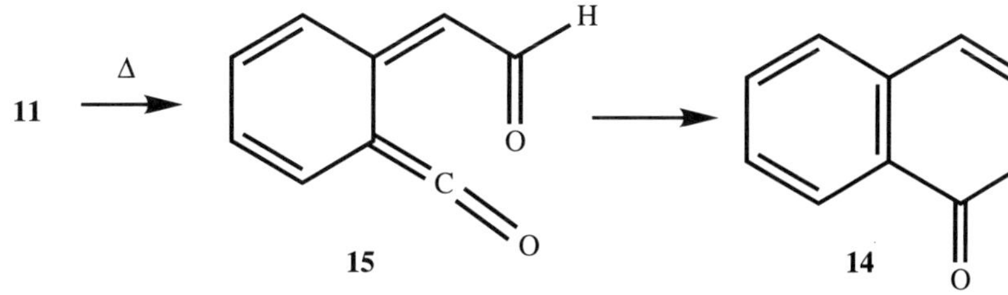

4.1.4 ARYLKETENES

The reaction of bicyclodiazo compounds such as **16** with formation of arylketene **17** (equation 11) is discussed in more detail in Section 5.4.5.[32–35] In some cases the ketenes were observed directly, and in others their presence was inferred from the formation of other products.

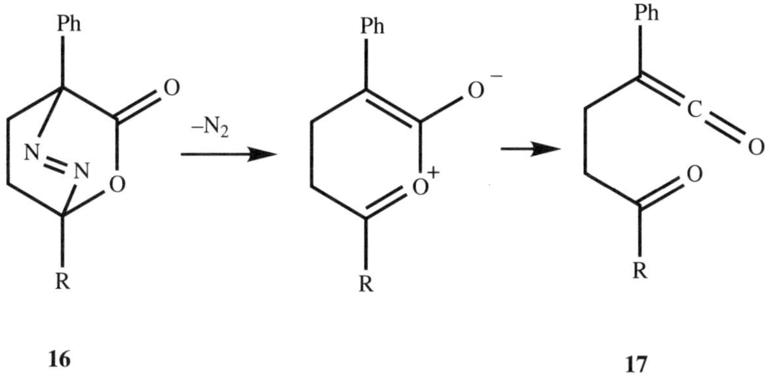

Three dimers of diphenylketene (**1**) have been reported.[36] The 1,3-cyclobutanedione dimer **18** is obtained from **1** and quinoline, while dimerization induced by sodium methoxide gives the β-lactone dimer **19**, while heating of **1** at 100 °C in benzoyl chloride gave **20**.[36]

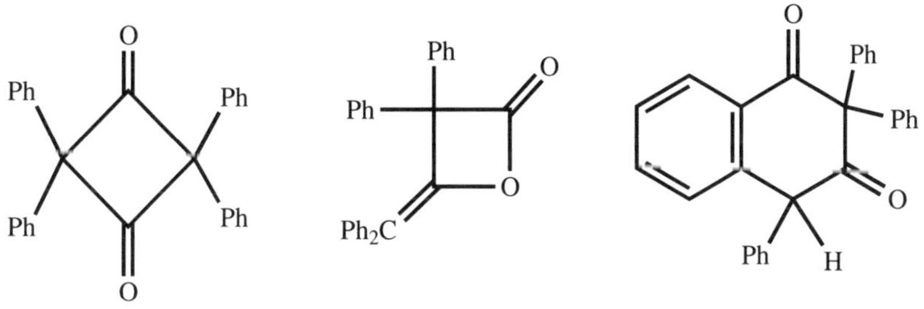

TABLE 1. Arylketenes

Ketene	Formula	Preparation	Registry No.	Ref.
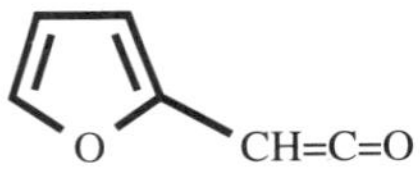 2-furyl-CH=C=O	$C_6H_4O_2$	Wolff	82515-16-0	25
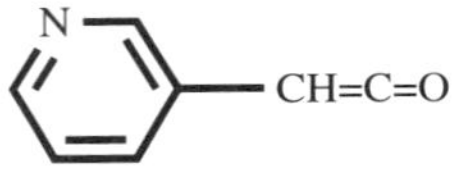 3-pyridyl-CH=C=O	C_7H_5NO	Wolff	70972-97-3	26
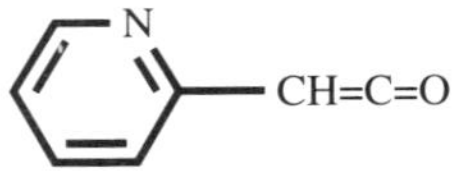 2-pyridyl-CH=C=O	C_7H_5NO	Ester pyrolysis		27
2,4-$Cl_2C_6H_3$CH=C=O	$C_8H_4Cl_2O$	Dehydrochlorination, Wolff	98556-40-2	30
4-ClC_6H_4CH=C=O	C_8H_5ClO	Wolff	58784-40-0	6,9
4-FC_6H_4CH=C=O	C_8H_5FO	Wolff, acyl halide	58784-38-6	9
4-$O_2NC_6H_4$CH=C=O	$C_8H_5NO_3$	Wolff	58784-38-6	7,9
PhCH=C=O	C_8H_6O	Dehydrochlorination	3496-32-0	9
PhC(COCl)=C=O	$C_9H_5ClO_2$	Wolff		37
4-NCC_6H_4CH=C=O	C_9H_5NO		58784-39-7	9
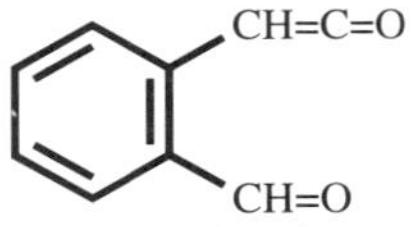 2-(CH=C=O)-C₆H₄-CH=O	$C_9H_6O_2$	Thermolysis	89002-82-4	31
4-TolCH=C=O	C_9H_8O	Wolff	58784-43-3	9
PhCMe=C=O	C_9H_8O	Malonic anhydride, dehydroclorination	3156-07-8	3,29
4-$MeOC_6H_4$CH=C=O	$C_9H_8O_2$	Dehydrochlorination, Wolff	58784-42-2	6,9
PhC(CH_2COCl)=C=O	$C_{10}H_7ClO_2$	Dehydrochlorination	17423-41-5	37

Structure	Formula	CAS	Method	Ref
PhCEt=C=O	$C_{10}H_{10}O$	20452-67-9	Malonic anhydride	3
4-TolC(CH$_3$)=C=O	$C_{10}H_{10}O$	84819-02-3	Not specified	38
4-CH$_3$OC$_6$H$_4$C(CH$_3$)=C=O	$C_{10}H_{10}O_2$	84819-03-4	Not specified	38
4-Me$_2$$\overset{+}{S}C_6H_4$C=C=O	$C_{10}H_{11}OS^+$		Elimination	38a
(1-naphthalenylidene ketene)	$C_{11}H_8O$	76966-08-0	Dehydrochlorination	23
(3,4-dihydronaphthalenylidene ketene)	$C_{11}H_{10}O$		Dehydrochlorination	39
4-ClC$_6$H$_4$C(Pr-i)=C=O	$C_{11}H_{11}ClO$	92742-11-5	Elimination	28
2,4,6-Me$_3$C$_6$H$_2$CHC=O	$C_{11}H_{12}O$		Dehydrochlorination	39a
PhC(Pr-n)=C=O	$C_{11}H_{12}O$	57768-76-0	Dehydrochlorination	40
PhC(Pr-i)=C=O	$C_{11}H_{12}O$	38082-08-5	Dehydrochlorination	17,18
(acenaphthylenylidene ketene)	$C_{12}H_6O$		Wolff	25
1-NaphthCH=C=O	$C_{12}H_8O$	67279-06-5	Wolff	21,22
C$_6$Cl$_5$C(Bu-t)C=O	$C_{12}H_9Cl_5O$		Dehalogenation	40a
2-RC$_6$H$_4$CH=C=O R = MeCHOHC≡C	$C_{12}H_{10}O$		Wolff	41
4-RC$_6$H$_4$C(Pr-i)=C=O R = CHF$_2$O	$C_{12}H_{12}F_2O$	102822-53-7		28

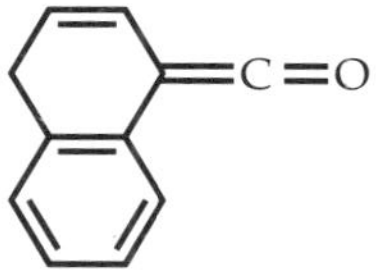

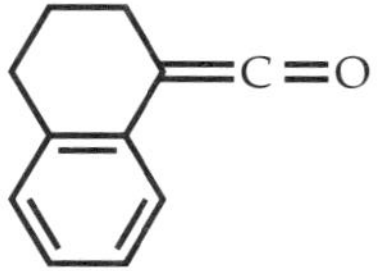

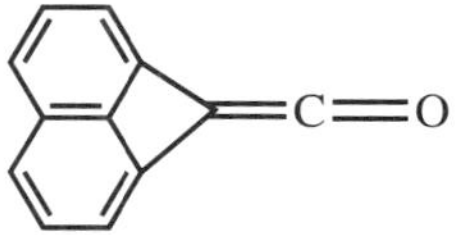

TABLE 1. (Continued)

Ketene	Formula	Preparation	Registry No.	Ref.
PhC(Bu-n)=C=O	$C_{12}H_{14}O$	Malonic anhydride	64498-08-4	42
PhC(Bu-t)=C=O	$C_{12}H_{14}O$	Zn debromination	57768-77-1	14
1-(5-CH_3Naphth)CH=C=O	$C_{13}H_{10}O$	Wolff		43
4-(2-BuC_6H_4)CMe=C=O	$C_{13}H_{16}O$	Dehydrochlorination		29
$(C_6Cl_5)_2$C=C=O	$C_{14}Cl_{10}O$	Dehydrochlorination	107846-85-5	20
$(C_6F_5)_2$C=C=O	$C_{14}F_{10}O$	Acid dehydration	36691-72-2	19
(4-$O_2NC_6H_4)_2$C=C=O	$C_{14}H_8N_2O_5$	Wolff	18627-17-3	44
4-BrC_6H_4CPh=C=O	$C_{14}H_9BrO$	Wolff	33351-00-7	45
4-ClC_6H_4CPh=C=O	$C_{14}H_9ClO$	Wolff	33350-99-1	45
4-(2-FC_6H_4)C_6H_4CH=C=O	$C_{14}H_9FO$	Alkyne oxidation	70740-83-9	46
4-$O_2NC_6H_4$CPh=C=O	$C_{14}H_9NO_3$	Wolff	20758-41-2	44
Ph_2C=C=O	$C_{14}H_{10}O$	Zn debromination	525-06-4	1,4,5
	$(C_{14}H_{10}O)_2$		103006-94-6	36
(2-RC_6H_4)CMe=C=O	$C_{14}H_{12}O$	Wolff		41
R = CH_2=CMeC≡C				
R = n-PrC≡C	$C_{14}H_{14}O$	Wolff		41
	$C_{14}H_{12}O_2$	Dehydrochlorination		29
1-(5-MeONaphth)CMe=C=O	$C_{15}H_{11}NO_4$	Wolff	20765-29-1	44
4-MeOC_6H_4C(C$_6H_4NO_2$-4)=C=O	$C_{15}H_{12}O$	Wolff	42955-17-9	18
PhC(CH$_2$Ph)=C=O	$C_{15}H_{12}O$	Wolff		45
4-MeC_6H_4CPh=C=O	$C_{15}H_{12}O_2$	Wolff	33350-98-0	45
4-MeOC_6H_4CPh=C=O	$C_{15}H_{12}O_2$	Wolff	33350-97-9	45
2-RC_6H_4CH=C=O				41
R = PhOCH_2				
2-RC_6H_4CH=C=O	$C_{16}H_{10}O$	Wolff		41
R = PhC≡C				
2-Tol_2C=C=O	$C_{16}H_{14}O$	Zn dehalogenation		45
4-Tol_2C=C=O	$C_{16}H_{14}O$	Wolff	40195-52-6	47,48

2-PhCH$_2$CH$_2$C$_6$H$_4$CH=C=O	C$_{16}$H$_{14}$O		Wolff	41
(4-MeOC$_6$H$_4$)$_2$C=C=O	C$_{16}$H$_{14}$O$_3$		Wolff	44
2-RC$_6$H$_4$CMe=C=O	C$_{17}$H$_{12}$O	18627-15-1	Wolff	41
R = PhC≡C				
4-EtO$_2$CC$_6$H$_4$CPh=C=O	C$_{17}$H$_{14}$O$_3$	33351-01-8	Wolff	45
3-Br-2,4,6-Me$_3$C$_6$HCFh=C=O	C$_{17}$H$_{15}$BrO		Dehydrochlorination	49
2,4,6-(MeO)$_3$C$_6$H$_2$C(C$_6$H$_4$NO$_2$-4)=C=O	C$_{17}$H$_{15}$NO$_6$	20758-42-3	Wolff	44
2,4,6-Me$_3$C$_6$H$_2$CPh=C=O	C$_{17}$H$_{16}$O	20458-66-6	Dehydrochlorination	12,50
2,4,6-(MeO)$_3$C$_6$H$_2$CP$_1$=C=O	C$_{17}$H$_{16}$O$_4$	58133-92-9	Index error	50
PhCR=C=O	C$_{17}$H$_{16}$O$_4$	115410-87-2	Fragmentation	33
R = ![structure with COCO$_2$Me]				
1-NaphthylClPhC=C=O	C$_{18}$H$_{12}$O	7277-01-2	Zn dehalogenation	23,39
PhCR=C=O	C$_{18}$H$_{14}$O	9802-75-6	Rearrangement	51
	C$_{18}$H$_{14}$O (1:S)	5230-75-1		51
R 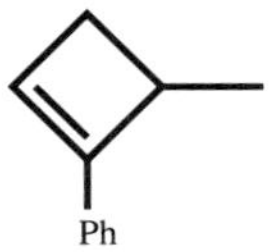 = ![cyclobutene structure with Ph]				
2,4,6-Me$_3$C$_6$H$_2$C(Tol-4)=C=O	C$_{18}$H$_{18}$O	101789-22-4	Dehydrochlorination	52
2,3,5,6-Me$_4$C$_6$H$_4$CPh=C=O	C$_{18}$H$_{18}$O		Dehydrochlorination	49
PhCR=C=O	C$_{18}$H$_{18}$O$_4$	98171-28-9	Fragmentation	32,34

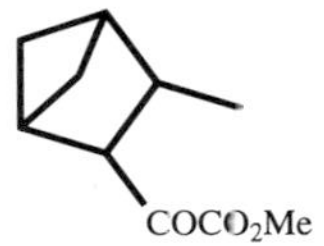

TABLE 1. (*Continued*)

Ketene	Formula	Preparation	Registry No.	Ref.
R 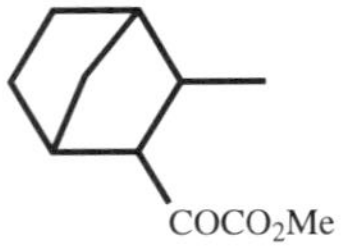 = [structure with COCO₂Me]				
MesCAr=C=O	$C_{20}H_{13}D_9O$	Dehydrochlorination	91110-63-3	11
Ar = 2,4,6-$(CD_3)_3C_6H_2$				
4-PhC_6H_4CPh=C=O	$C_{20}H_{14}O$	Dehalogenation		53
PhCRC=C=O	$C_{21}H_{16}O_2$	Fragmentation	78371-26-3	32
R  = [structure with COPh]				
R = [structure with COPh]				
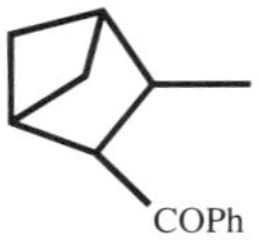 (1-Naphthyl)$_2$C=C=O	$C_{21}H_{18}O_2$	Fragmentation	115410-89-4	33
(2-Naphthyl)$_2$C=C=O	$C_{22}H_{14}O$	Zn dehalogenation	76966-03-5	23
2-RC_6H_4CH=C=O	$C_{22}H_{14}O$	Zn dehalogenation	76965-98-5	23
	$C_{22}H_{16}O$	Fragmentation	52706-56-6	54

R =

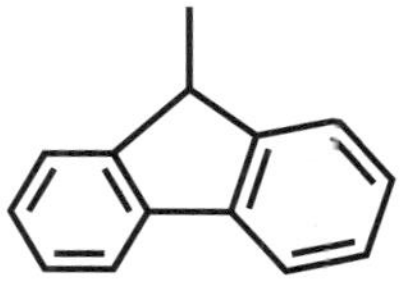

PhCR=C=O

R =

R— CH=C=O

R =

MesCAr=C=O
Ar = 2,6-Me$_2$-4-t-BuC$_6$H$_2$
C$_6$H$_4$(CH$_2$CH$_2$C$_6$H$_4$)$_2$C=C=O
(C$_6$Me$_5$)$_2$C=C=O
(4-PhC$_6$H$_4$)$_2$C=C=O
PhCR=C=O
R = Ph$_3$P=CPh

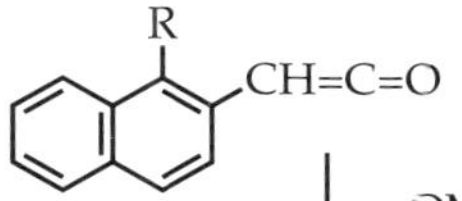

C$_{22}$H$_{20}$O$_2$	Thermolysis	98386-45-9	32
		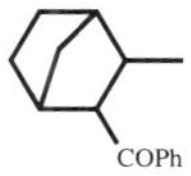 78371-28-5	35
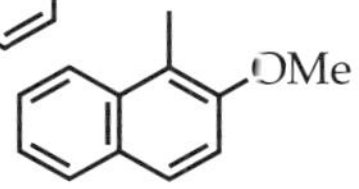 C$_{23}$H$_{16}$O$_2$	Wolff		55
C$_{23}$H$_{28}$O	Dehydrochlorination	91110-64-4	11
C$_{24}$H$_{20}$O	Wolff		56
C$_{24}$H$_{30}$O	Dehydrochlorination	112752-37-1	13
C$_{26}$H$_{18}$O	Dehalogenation		57
C$_{33}$H$_{25}$OP	Cleavage	37521-01-0	58

TABLE 1. (Continued)

Ketene	Formula	Preparation	Registry No.	Ref.
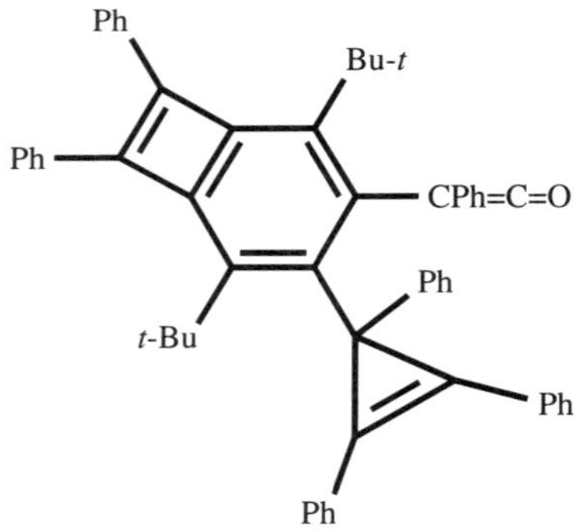 (structure)	$C_{57}H_{48}O$	Photolysis	62788-81-2	59
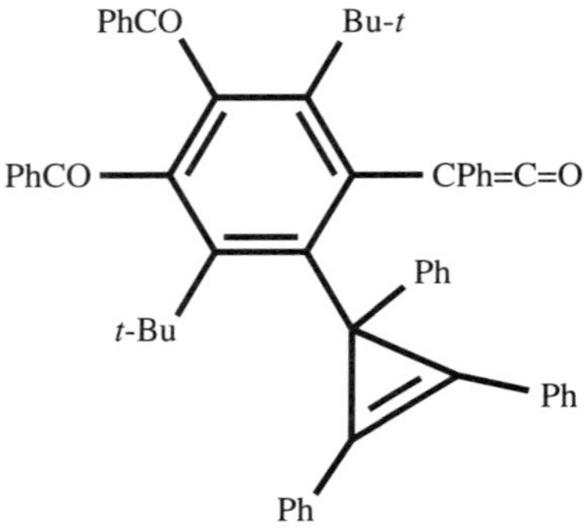 (structure)	$C_{57}H_{48}O_3$	Photolysis	62788-78-7	59

Structure	Formula	Method	CAS	Ref
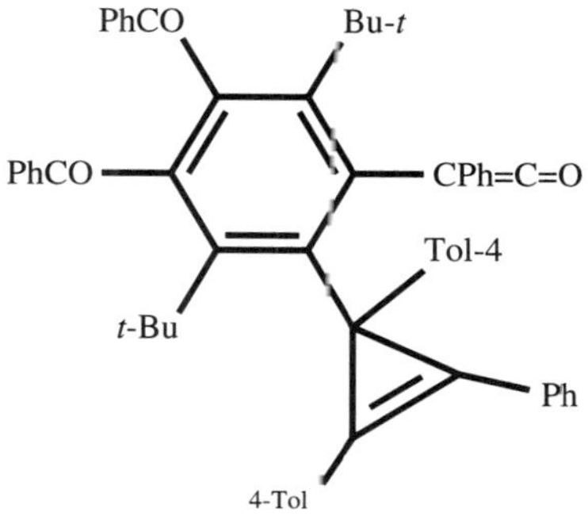 (PhCO, Bu-t, PhCO, CPh=C=O, Tol-4, t-Bu, Ph, 4-Tol substituted benzene with cyclopropene)	$C_{59}H_{52}O_3$	Photolysis	62828-42-6	59
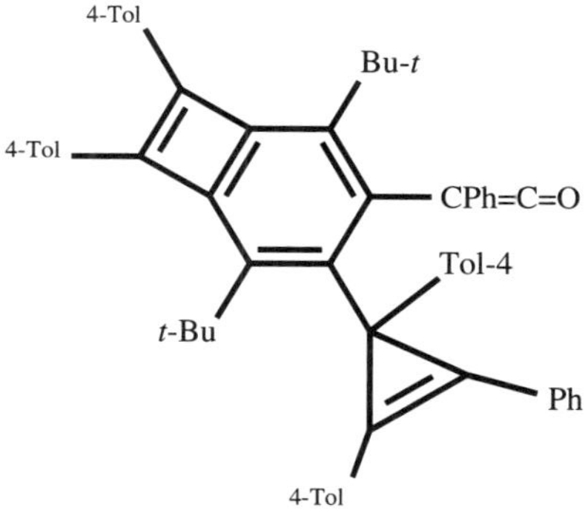 (Bu-t, CPh=C=O, Tol-4, t-Bu, Ph, 4-Tol, 4-Tol substituted benzocyclobutene with cyclopropene)	$C_{61}H_{52}O$	Photolysis	62788-82-3	59

References

1. Staudinger, H. *Chem. Ber.* **1905**, *38*, 1735–1739.
2. Allen, A. D.; Andraos, J.; Kresge, A. J.; McAllister, M. A.; Tidwell, T. T. *J. Am. Chem. Soc.* **1992**, *113*, 1878–1879.
3. Kresze, G.; Runge, W.; Ruch, E. *Liebigs Ann. Chem.* **1972**, *756*, 112–127.
4. Taylor, E. C.; McKillop, A.; Hawks, G. H. *Organic Syntheses,* Coll. Vol. VI; Noland, W. E.; Ed.; Wiley: New York, 1988; pp. 549–551.
5. Smith, L. I.; Hoehn, H. H. *Organic Syntheses,* Coll. Vol. III; Hornung, E. C.; Ed.; Wiley: New York, 1965; pp. 356–359.
6. Bellus, D. *J. Am. Chem. Soc.* **1978**, *100*, 8026–8028.
7. Law, K.-Y.; Bailey, F. C. *J. Chem. Soc., Chem. Commun.* **1991**, 1156–1158; *Can. J. Chem.* **1993**, *71*, 494–505.
8. Allen, A. D.; Kresge, A. J.; Schepp, N. P.; Tidwell, T. T. *Can. J. Chem.* **1987**, *65*, 1719–1723.
9. Bothe, E.; Meier, H.; Schulte-Frohlinde, D.; von Sonntag, C. *Angew. Chem., Int. Ed. Engl.* **1976**, *15*, 380.
10. Staudinger, H. *Chem. Ber.* **1911**, *44*, 533–543.
11. Biali, S. E.; Rappoport, Z. *J. Am. Chem. Soc.* **1984**, *106*, 5641–5653.
12. Nadler, E. B.; Zipory, E. S.; Rappoport, Z. *J. Org. Chem.* **1991**, *56*, 4241–4246.
13. Clarke, L. F.; Hegarty, A. F.; O'Neill, P. *J. Org. Chem.* **1992**, *57*, 362–366.
14. Allen, A. D.; Stevenson, A.; Tidwell, T. T. *J. Org. Chem.* **1989**, *54*, 2843–2848.
15. Baigrie, L. M.; Seiklay, H. R.; Tidwell, T. T. *J. Am. Chem. Soc.* **1985**, *107*, 5391–5396.
16. Gong, L.; Leung-Toung, R.; Tidwell, T. T. *J. Org. Chem.* **1990**, *55*, 3634–3639.
17. Allen, A. D.; Baigrie, L. M.; Gong, L.; Tidwell, T. T. *Can. J. Chem.* **1991**, *69*, 138–145.
18. Dehmlow, E. V.; Slopianka, M.; Pickardt, J. *Liebigs Ann. Chem.* **1979**, 572–593.
19. Lubenets, E. G.; Gerasimova, T. N.; Barkhash, V. A. *Zh. Org. Khim.* **1972**, *8*, 654. Engl. Trans. **1972**, *8*, 663.
20. O'Neill, P.; Hegarty, A. F. *J. Org. Chem.* **1992**, *57*, 4421–4426.
21. Dao, L. H.; Hopkinson, A. C.; Lee-Ruff, E. *Tetrahedron Lett.* **1978**, 1413–1414.
22. Lee, V.; Newman, M. S. *Organic Syntheses,* Coll. Vol. VI; Noland, W. E.; Ed.; Wiley: New York, 1988; pp. 613–615.
23. Wuest, J. D. *Tetrahedron* **1980**, *36*, 2291–2296.
24. Snider, B. B.; Niwa, M. *Tetrahedron Lett.* **1988**, *29*, 3175–3178.
25. Maier, G.; Reisenauer, H. P.; Sayrac, T. *Chem. Ber.* **1982**, *115*, 2192–2201.
26. Chapman, O. L.; Sheridan, R. S. *J. Am. Chem. Soc.* **1979**, *101*, 3690–3692.
27. Jaworski, T.; Kwiatkowski, S. *Rocz. Chem.* **1970**, *44*, 691–693; *Chem. Abstr.* **1970**, *73*, 130746r.
27a. Tomioka, H.; Ichikawa, N.; Komatsu, K. *J. Am. Chem. Soc.* **1993**, *115*, 8621–8626.
27b. Hisatome, M.; Watanabe, J.; Yamashita, R.; Yoshida, S.; Yamakawa, K. *Bull. Chem. Soc. Jpn.* **1994**, *67*, 490–494.
28. Stoutamire, D. W. *U.S. Patent* 4,570,017; *Chem. Abstr.* **1986**, *105*, 78661x.

29. Bellucci, G.; Berti, G.; Bianchini, R.; Vecchiani, S. *Gazz. Chim. Ital.* **1988**, *118*, 451–456.
30. Stoutamire, D. W.; Tieman, C. H. *U.S. Patent* 4,560,515; *Chem. Abstr.* **1986**, *105*, 42497j.
31. Bleasdale, D. A.; Jones, D. W.; Maier, G.; Reisenauer, H. P. *J. Chem. Soc., Chem. Commun.* **1983**, 1095–1096.
32. Christl, M.; Lanzendörfer, U.; Hegmann, J.; Peters, K.; Peters, E. M.; Von Schnering, H. G. *Chem. Ber.* **1985**, *118*, 2940–2973.
33. Reuchlein, H.; Kraft, A.; Christl, M.; Peters, E.; Peters, K.; von Schnering, H. G. *Chem. Ber.* **1991**, *124*, 1435–1444.
34. Christl, M.; Lanzendörfer, U.; Groetsch, M. M.; Ditterich, E.; Hegmann, J. *Chem. Ber.* **1990**, *123*, 2031–2037.
35. Christl, M.; Lanzendörfer, U.; Freund, S. *Angew. Chem., Int. Ed. Engl.* **1981**, *20*, 674–675.
35a. Feineis, E.; Schwarz, H.; Hegmann, J.; Christl, M.; Peters, E.-M.; Peters, K.; von Schnering, H. G. *Chem. Ber.* **1993**, *126*, 1743–1748.
36. Das, H.; Kooyman, E. C. *Recl. Trav. Chim. Pays-Bas* **1965**, *84*, 965–978.
37. Ziegler, E.; Sterk, H. *Monatsh. Chem.* **1967**, *98*, 1104–1107.
38. Jähme, J.; Rüchardt, C. *Tetrahedron Lett.* **1982**, *23*, 4011–4014.
38a. Casanova, J., Jr.; Werner, N. D.; Kiefer, H. R. *J. Am. Chem. Soc.* **1967**, *89*, 2411–2416.
39. Pracejus, H.; Wallura, G. *J. Prakt. Chem.* **1963**, *19*, 33–36.
39a. Fuson, R. C.; Armstrong, L. J.; Schenk, W. J., Jr. *J. Am. Chem. Soc.* **1944**, *66*, 964–967.
40. Huisgen, R.; Mayr, H. *Tetrahedron Lett.* **1975**, 2969–2972.
40a. Hegarty, A. F.; O'Neill, P. *Synthesis* **1993**, 606–610.
41. Padwa, A.; Chiacchio, U.; Fairfax, D. J.; Kassir, J. M.; Litrico, A.; Semones, M. A.; Xu, S. L. *J. Org. Chem.* **1993**, *58*, 6429–6437.
42. Turro, N. J.; Chow, M. *J. Am. Chem. Soc.* **1980**, *102*, 5058–5064.
43. Danheiser, R. L.; Casebier, D. S.; Loebach, J. L. *Tetrahedron Lett.* **1992**, *33*, 1149–1152.
44. Melzer, A.; Jenny, E. F. *Tetrahedron Lett.* **1968**, 4503–4506.
45. Teufel, H.; Jenny, E. F. *Tetrahedron Lett.* **1971**, 1769–1772.
46. Sullivan, H. R.; Roffey, P.; McMahon, R. E. *Drug. Metab. Dispos.* **1979**, *7*, 76–80; *Chem. Abstr.* **1979**, *91*, 68232m.
47. Taylor, G. *J. Chem. Soc., Perkin Trans. 1* **1975**, 1001–1009.
48. Gilman, H.; Adams, C. E. *Recl. Trav. Chim. Pays-Bas* **1929**, *48*, 464–465. *Chem. Abstr.* **1929**, *23*, 2950.
49. Fuson, R. C.; Foster, R. E.; Shenk, W. J., Jr.; Maynert, E. W. *J. Am. Chem. Soc.* **1945**, *67*, 1937–1939.
50. Hastings, J. S.; Heller, H. G.; Salisbury, K. *J. Chem. Soc., Perkin Trans. 1*, **1975**, 1995–1998.
51. Masamune, S.; Fukumoto, K. *Tetrahedron Lett.* **1965**, 4647–4654.
52. Fuson, R. C.; Maynert, E. W.; Tan, T.; Trumball, E. R.; Wassmundt, F. W. *J. Am. Chem. Soc.* **1957**, *79*, 1938–1941.

53. Burmistrov, S. I.; Shilov, E. A. *Zh. Obshch. Khim.* **1946**, *16*, 295–299. *Chem. Abstr.* **1947**, *41*, 474c.
54. Iwamura, H.; Yoshimura, K. *J. Am. Chem. Soc.* **1974**, *96*, 2652–2654.
55. Kropf, H.; Reichwaldt, R. *J. Chem. Res. (S)*, **1992**, 284–285.
56. Hatta, T.; Mataka, S.; Tanimoto, S.; Tashiro, M.; Tsuge, O. *Bull. Chem. Soc. Jpn.* **1992**, *65*, 1469–1471.
57. Shilov, E. A.; Burmistrov, S. I. *Chem. Ber.* **1935**, *68B*, 582–584.
58. Hamada, A.; Takizawa, T. *Tetrahedron Lett.* **1972**, 1849–1850.
59. Toda, F.; Tanaka, K. *J. Chem. Soc., Chem. Commun.* **1976**, 1010–1011.

4.1.5 Cyclopropylketenes

Cyclopropylketenes have been frequently generated and trapped, and recently some of these species have been isolated. Based on ab initio calculated isodesmic exchange reactions cyclopropylketenes are indicated to have the stability of normal alkylketenes.[1,2] However, the cyclopropyl group has steric bulk intermediate between ethyl and isopropyl groups, and this offers some steric protection, so that when an additional group is present such as phenyl, cyclopropyl, or *tert*-butyl the ketenes are rather persistent.[1]

The parent cyclopropylketene **1** was generated and trapped in 47% yield by heating the diazo ketone in MeOH with silver oxide,[3] while later workers used a temperature of 155–160 °C in benzyl alcohol and collidine,[4] with the ketene being trapped as the benzyl ester (equation 1).

$$c\text{-PrCOCHN}_2 \xrightarrow{\Delta} c\text{-PrCH=C=O} \xrightarrow{\text{PhCH}_2\text{OH}} c\text{-PrCH}_2\text{CO}_2\text{CH}_2\text{Ph} \quad (1)$$

$$\mathbf{1}$$

Ketene **1** was also generated by reaction of the acyl chloride with Et_3N and obtained as the dimer **1a** in 41% yield.[5] Pyrolyses of the dimer or of c-PrCH$_2$CO$_2$Et at 550 and 600 °C, respectively, were claimed to reform the ketene, which rearranged to cyclopentenone at this temperature (equation 2).[5]

4.1.5 CYCLOPROPYLKETENES

Cyclopropenylketene (**2**) was obtained by photolysis at $\lambda > 310$ nm of the diazoketone in an Ar matrix, and the ketene band at 2108 cm^{-1} was observed.[6] Photolysis of the ketene at 254 nm led to cyclobutadiene, which was also observed in the matrix (equation 3).[6] Formation of c-Pr$_2$C=C=O by CO addition to the carbene is noted in Section 3.4.6, equation 18.

$$\text{cyclopropenyl-CCHN}_2 \xrightarrow{h\nu} \text{cyclopropenyl-CH=C=O} \xrightarrow[-\text{CO}]{h\nu} \text{cyclobutadiene} \quad (3)$$

2

2,2-Diphenylcyclopropyl ketene (**3**) was formed by photolysis of the diazoketone in H$_2$O/CH$_3$CN and trapped as the acid.[7] Photolysis of 4,4-diphenylcyclopentene-3-one (**4**) also led to **3**.[7] Photolysis of 4,4-dimethylcyclopentene-3-one in pentane and IR examination of the solution showed a band at 2110 cm^{-1} assigned to the ketene **5**.[7] This identification was confirmed by reaction with MeOH and isolation of the methyl ester.[7,8] These and other cyclopentenone cleavages were proposed to proceed by initial α-cleavage and rebonding (equation 5).[8]

$$\text{Ph}_2\text{-cyclopentenone (4)} \text{ or Ph}_2\text{-cyclopropyl-CCHN}_2 \xrightarrow{h\nu} \text{Ph}_2\text{-cyclopropyl-CH=C=O} \quad (4)$$
3

$$\text{R,R}^1\text{-cyclopentenone} \xrightarrow{h\nu} \text{R}^1\text{RC:} \cdots \text{O=C} \longrightarrow \text{R,R}^1\text{-cyclopropyl-CH=C=O} \quad (5)$$
5

R = R^1 = Me; R = n-Pr, Et, EtO, R^1 = H

Photoreactions forming cyclopropylketenes has been observed with a variety of other polycyclic enones,[9–18] including the examples in equations 6 and 7. Ketene **7** was observed by the IR absorption at 2110 cm^{-1} when **6** was photolyzed as a thin film at -190 °C, and **8** was formed from **7** on warming.[17] Thermolysis of **8** at 100 °C reforms **6**.[17] A chlorine-substituted analogue of **6** has also been examined.[18]

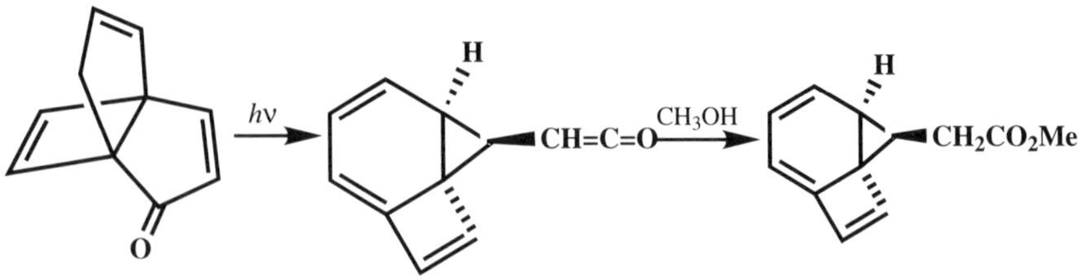

(6)[10]

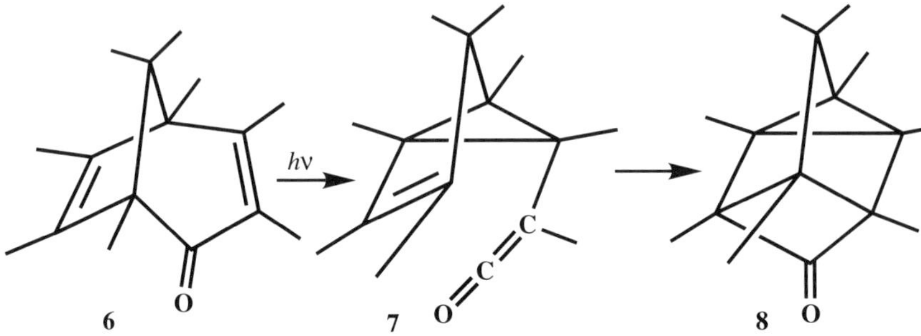

(7)

Cyclopropylcarboethoxyketene **9**[19,20] was formed by Wolff rearrangement with cyclopropyl migration, and was distilled at 3 mm pressure.[19] The ketene reacted with EtOH to give the diethyl ester (equation 8).[19,20]

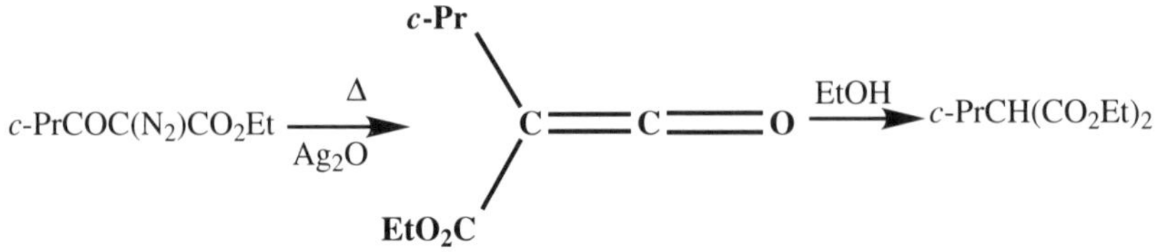

$$c\text{-PrCOC(N}_2)\text{CO}_2\text{Et} \xrightarrow[\text{Ag}_2\text{O}]{\Delta} \underset{\text{EtO}_2\text{C}}{\overset{c\text{-Pr}}{>}}\!\!\text{C}\!=\!\text{C}\!=\!\text{O} \xrightarrow{\text{EtOH}} c\text{-PrCH(CO}_2\text{Et)}_2 \quad (8)$$

9

Metal complexed cyclopropylketenes have been prepared.[21] Photolysis of cyclopropylcarbene complexes **10** gives the products expected from the corresponding cyclopropylketenes (equation 9).[22]

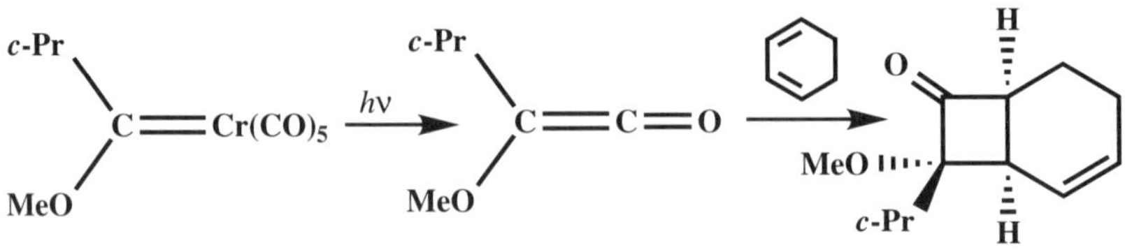

(9)

10   **11**

The ketenes **12–14** were generated from the acyl chlorides and were distilled as neat liquids (equation 10).[1,2] Studies of the hydrolytic reactivity and reactions with organolithium reagents of these ketenes are included in Section 5.5.

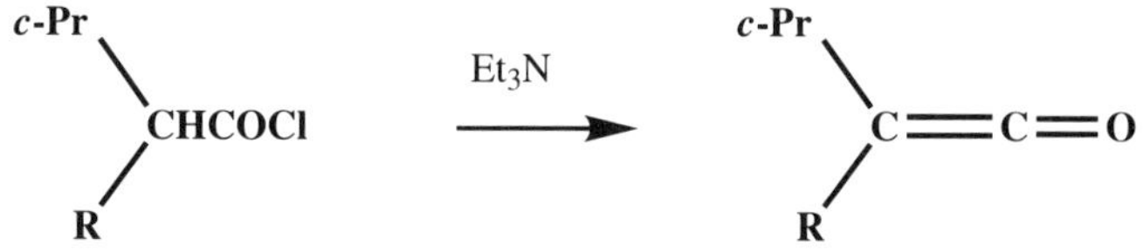

$$\underset{R}{\overset{c\text{-Pr}}{>}}\text{CHCOCl} \xrightarrow{\text{Et}_3\text{N}} \underset{R}{\overset{c\text{-Pr}}{>}}\text{C}=\text{C}=\text{O} \qquad (10)$$

R = *t*–Bu (**12**), Ph (**13**), *c*–Pr (**14**)

Ketene **15** was obtained by Wolff rearrangement and trapped by *N*-methylaniline as a mixture of epimers (equation 11).[23] The photochemical reaction in alcohols to give esters was also studied.[23]

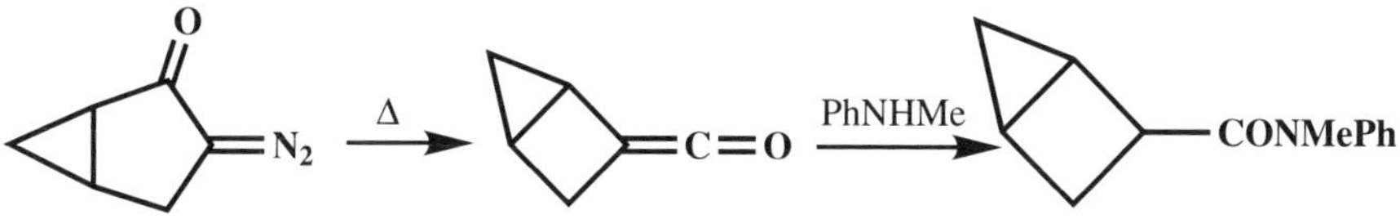

$$\text{[structure]}=\text{N}_2 \xrightarrow{\Delta} \text{[structure 15]}=\text{C}=\text{O} \xrightarrow{\text{PhNHMe}} \text{[structure]}-\text{CONMePh}$$

(11)

Photolysis of **16** produced ketene **17** by Wolff rearrangement, and this rearranged to the bicyclic dienone **18** (equation 12).[24]

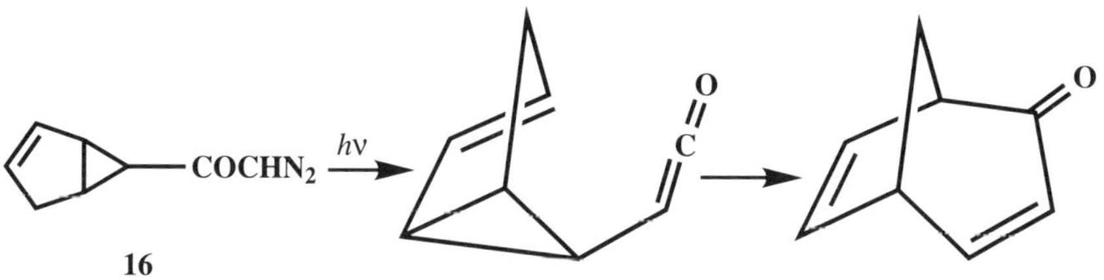

$$\underset{\mathbf{16}}{\text{[structure]}}-\text{COCHN}_2 \xrightarrow{h\nu} \underset{\mathbf{17}}{\text{[structure]}} \rightarrow \underset{\mathbf{18}}{\text{[structure]}}$$

(12)

Reaction of **19** with Et₃N was suggested to give the cycloheptatrienylketene **20**, which was proposed to be in equilibrium with the norcaradiene ketene **21**, as evidenced by the isolation of **22** and **23** in 44 and 40% yields, respectively (equations 13–15).[25] Possible routes from **21** to **22** and **23** are shown, but these could also arise directly from **20**. The presumed dimer of **20** was also isolated.[25]

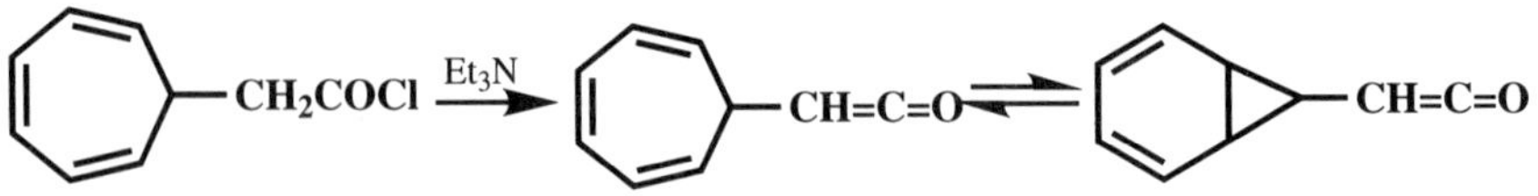

222 TYPES OF KETENES

(13)

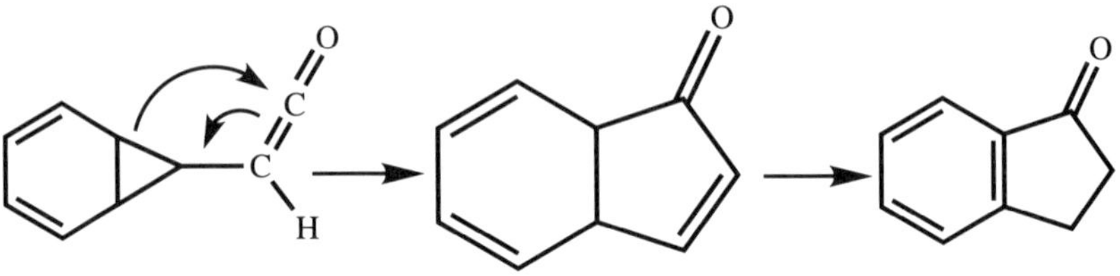

(14)

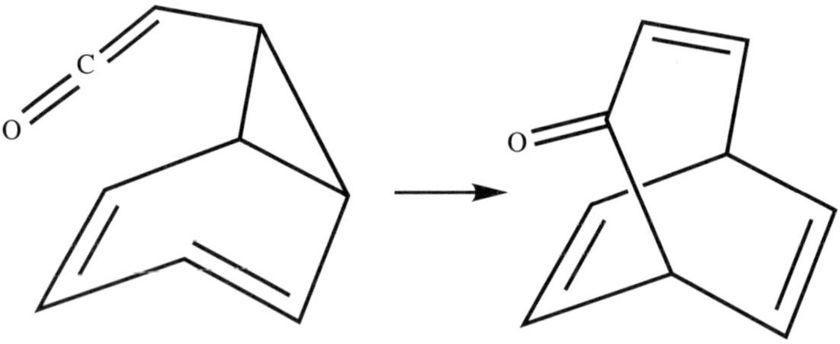

(15)

Oxiranylketenes are analogous to cyclopropylketenes and upon generation by Wolff rearrangement attack of EtOH on the ketene **24** proceeded with ring opening (equation 16).[26] When the ketene **25** was generated in the absence of nucleophiles, rearrangement to butenolide **26** occurred (equation 17).[27]

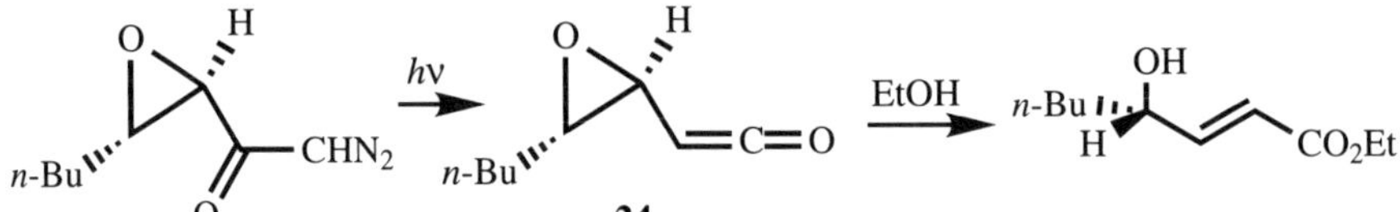

(16)[26]

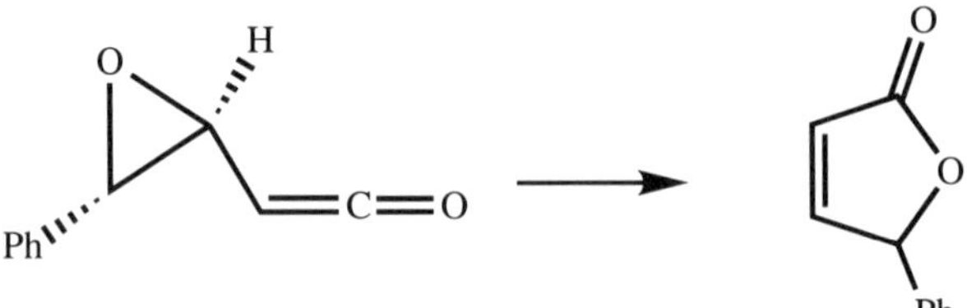

(17)[27]

TABLE 1. Cyclopropyl, Cyclopropenyl, and Oxiranyl Ketenes

Ketene	Formula	Preparation	Registry No.	Ref.
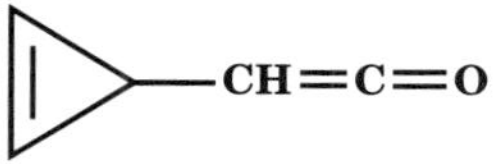 cyclopropyl–C≡C–CH=C=O	C_5H_4O	Wolff	94923-07-6	6
c-PrCH=C=O	C_5H_6O	Wolff	128871-21-6	3-5
	$(C_5H_6O)_2$	Dehydrochlorination		5
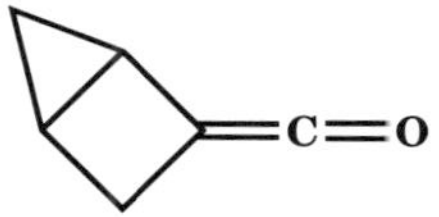 bicyclic =C=O	C_6H_6O	Wolff		23
c-PrC(OMe)=C=O	C_6H_8O	Metal complex		22
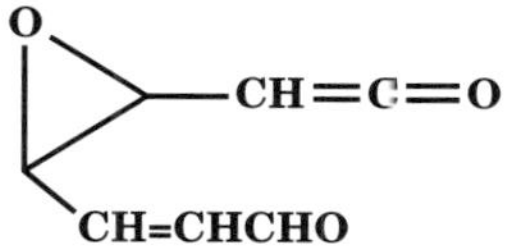 oxiranyl-CH=CHCHO (CH=C=O)	$C_7H_6O_3$	Photolysis	111248-62-5	28
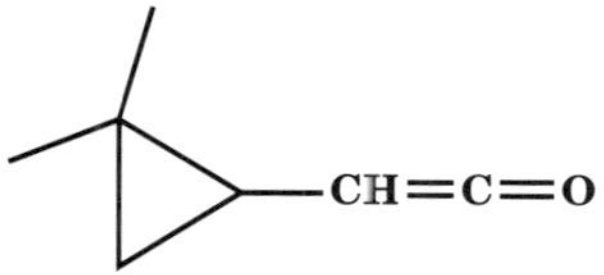 dimethylcyclopropyl-CH=C=O	$C_7H_{10}O$	Photolysis		7,8
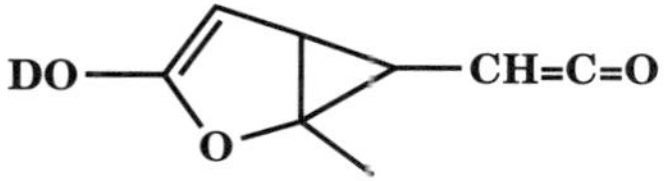 DO-furan-CH=C=O	$C_8H_7DO_3$	Photolysis	118620-65-8	29

TABLE 1. (Continued)

Ketene	Formula	Preparation	Registry No.	Ref.
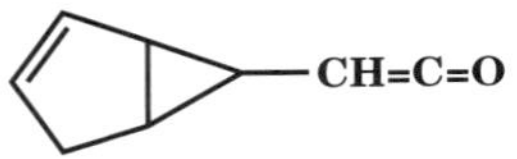 ⟨bicyclic⟩–CH=C=O	C_8H_8O	Wolff		24
HO-furanone with CH=C=O	$C_8H_8O_3$ $C_8H_{10}O$ $C_8H_{10}O_3$	Photolysis Dehydrochlorination Wolff	128871-24-9	29 1 19,20
$(c\text{-Pr})_2C=C=O$				
$c\text{-PrC}((CO_2Et))=C=O$				
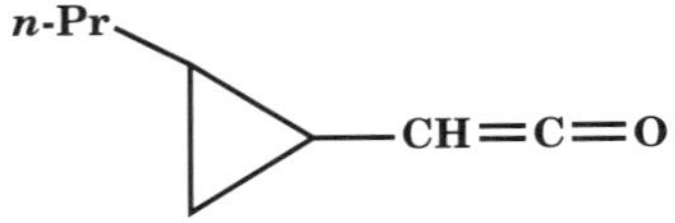 $n\text{-Pr}$–cyclopropyl–CH=C=O	$C_8H_{12}O$	Photolysis		8
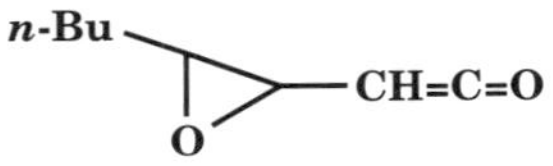 $n\text{-Bu}$–epoxide–CH=C=O	$C_8H_{12}O_2$	Wolff		26
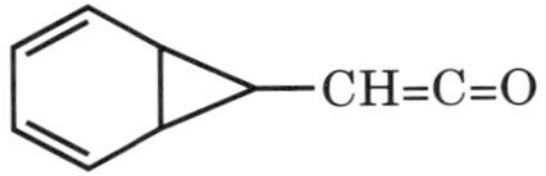 bicyclic–CH=C=O	C_9H_8O $C_9H_{14}O$	Thermolysis Dehydrochlorination	133658-76-1	28 1
$c\text{-PrC}(Bu\text{-}t)=C=O$				

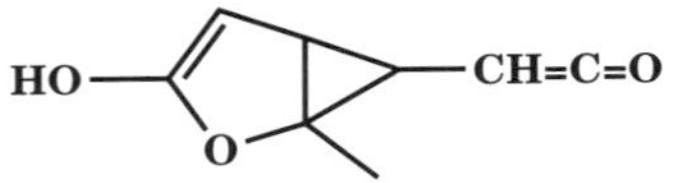

224

	$C_{10}H_8O_2$	Wolff	27
	$C_{11}H_8O$	Photolysis	10
	$C_{11}H_{10}O$	Dehydrochlorination	1
	$C_{13}H_{12}O_5$		24666-25-9
	$C_{17}H_{14}O$	Photolysis	7
	$C_{21}H_{36}O$	Photolysis	66809-04-9 30

References

1. Allen, A. D.; Baigrie, L. M.; Gong, L.; Tidwell, T. T. *Can. J. Chem.* **1991**, *69*, 138–145.
2. Allen, A. D.; Gong, L.; Tidwell, T. T. *J. Am. Chem. Soc.* **1990**, *112*, 6396–6397.
3. Turnbull, J. H.; Wallis, E. S. *J. Org. Chem.* **1956**, *21*, 663–666.
4. Basnak, I.; Farkas, J. *Coll. Czech. Chem. Commun.* **1976**, *41*, 311–316.
5. Berkowitz, W. F.; Ozorio, A. A. *J. Org. Chem.* **1975**, *40*, 527–528.
6. Maier, G.; Hoppe, M.; Lanz, K.; Reisenauer, H. P. *Tetrahedron Lett.* **1984**, *25*, 5645–5648.
7. Agosta, W. C.; Smith, A. B., III, Kende, A. S.; Eilerman, R. G.; Benham, J. *Tetrahedron Lett.* **1969**, 4517–4520.
8. Agosta, W. C.; Smith, A. B., III *J. Am. Chem. Soc.* **1971**, *93*, 5513–5520.
9. Hart, H.; Dean, D. L.; Buchanan, D. N. *J. Am. Chem. Soc.* **1973**, *95*, 6294–6301.
10. Ohkita, M.; Tsuji, T.; Suzuki, M.; Murakami, M.; Nishida, S. *J. Org. Chem.* **1990**, *55*, 1506–1513.
11. Uyehara, T.; Ogata, K.; Yamada, J.; Kato, T. *J. Chem. Soc., Chem. Commun.* **1983**, 17–18.
12. Chapman, O. L.; Lassila, J. D. *J. Am. Chem. Soc.* **1968**, *90*, 2449–2450.
13. Chapman, O. L.; Kane, M.; Lassila, J. D.; Loeschen, R. L.; Wright, H. E. *J. Am. Chem. Soc.* **1969**, *91*, 6856–6858.
14. Houk, K. N. *Chem. Rev.* **1976**, *76*, 1–74.
15. Kende, A. S.; Goldschmidt, Z.; Izzo, P. T. *J. Am. Chem. Soc.* **1969**, *91*, 6858–6860.
16. Ciabattoni, J.; Crowley, J. E.; Kende, A. S. *J. Am. Chem. Soc.* **1967**, *89*, 2778–2779.
17. Hart, H.; Love, G. M. *J. Am. Chem. Soc.* **1971**, *93*, 6266–6267.
18. Goldschmidt, Z.; Gutman, U.; Bakal, Y.; Worchel, A. *Tetrahedron Lett.* **1973**, 3759–3762.
19. Smith, L. I.; McKenzie, S., Jr. *J. Org. Chem.* **1950**, *15*, 74–80.
20. Keating, J. T.; Skell, P. S. *J. Am. Chem. Soc.* **1969**, *91*, 695–699.
21. Kreissl, F. R.; Sieber, W. J.; Alt, H. G. *Chem. Ber.*, **1984**, *117*, 2527–2530.
22. Söderberg, B. C.; Hegedus, L. S.; Sierra, M. A. *J. Am. Chem. Soc.* **1990**, *112*, 4364–4374.
23. Brook, P. R.; Brophy, B. V. *J. Chem. Soc., Perkin Trans. 2*, **1985**, 2509–2513.
24. Freeman, P. K.; Kuper, D. G. *Chem. Ind. (London)* **1965**, 424–425.
25. Goldstein, M. J.; Odell, B. G. *J. Am. Chem. Soc.* **1967**, *89*, 6356–6357.
26. Thijs, L.; Dommerholt, F. J.; Leemhuis, F. M. C.; Zwanenburg, B. *Tetrahedron Lett.* **1990**, *31*, 6589–6592.
27. van Haard, P. M. M.; Thijs, L.; Zwanenburg, B. *Tetrahedron Lett.* **1975**, 803–806.
28. Mori, A.; Takeshita, H. *Bull. Chem. Soc. Jpn.* **1987**, *60*, 3037–3038.
29. Mori, A.; Kasai, S.; Takeshita, H. *Bull. Chem. Soc. Jpn.* **1988**, *61*, 2259–2261.
30. Maier, G.; Pfriem, S.; Schäfer, U.; Malash, K.; Matusch, R. *Chem. Ber.* **1981**, *114*, 3965–3987.

4.1.6 Acylketenes

Ketene dimer (**1**) was originally proposed in 1908 to have the acetylketene (α-oxoketene) structure **2.**[1] However, the first authentic acylketene **3** was prepared in 1909 as an ethereal solution by the zinc dehalogenation of **4** (equation 1).[2]

1

2

4 **3** (1)

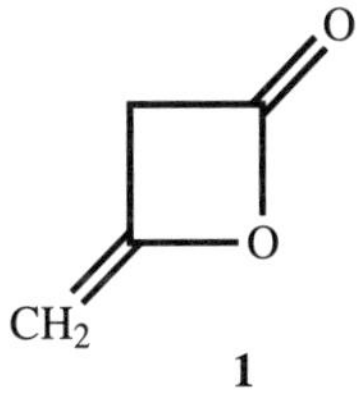

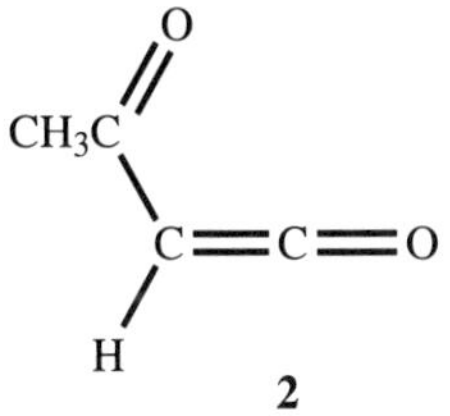

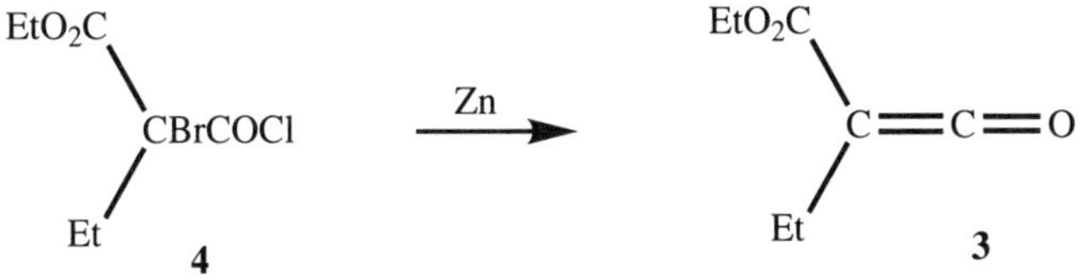

Decomposition of diazomalonaldehyde (**5**) with $PdCl_2$ gave only polymer,[3] but on photolysis in an Ar matrix at 10 K the parent formylketene **6** was observed by IR as *syn*/*anti* forms which are interconverted photochemically (equation 2).[4] Thermal generation of **6** was achieved by refluxing **5** in *n*-butyl vinyl ether, with trapping by [4 + 2] cycloaddition to give **7**, which was converted to **8** (equations 3 and 4).[5]

5 *anti*-**6** *syn*-**6** (2)

6 **7** **8** (3)

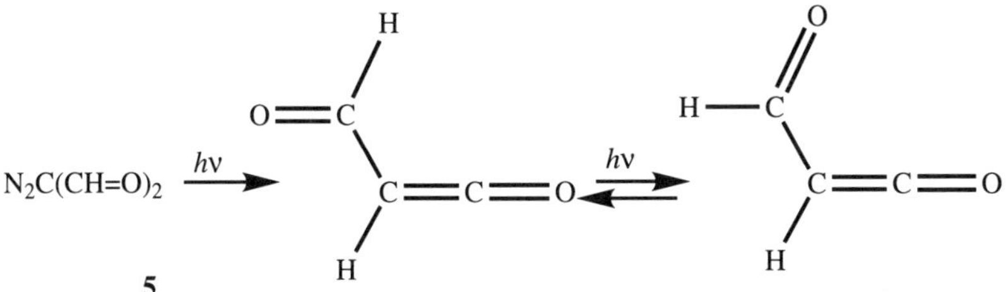

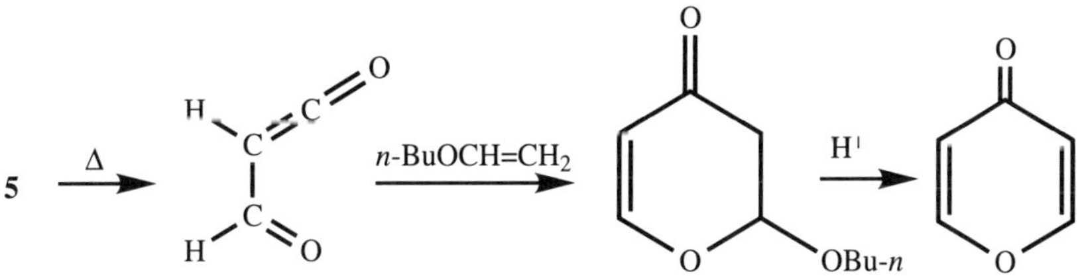

Acylketenes are known in a variety of structural types, including those with formyl (**9a**), keto (**9b**), ester (**9c**), acyl chloride (**9d**), and amido (**9e**) groupings.[5a] The aldoketenes in this series (R^1 = H) dimerize very rapidly but the ketoketenes are somewhat more stable, although they are only isolable when bulky substituents are present. The carboxy derivatives **9c–e** appear to be less reactive than **9a** and **b**.

9a: R = H
9b: R = alkyl
9c: R = RO
9d: R = Cl
9e: R = NR_2

On the basis of 6-31G*//6-31G* ab initio molecular orbital calculations, the *syn* conformer of formylketene **6** is the more stable by 1.0 kcal/mol.[6,7] As noted above, both conformers of **6** were observed in an Ar matrix, and two IR bands attributable to two conformations of **5** were also observed.[4] The structures, conformations, and IR spectra of a number of other acylketenes have been calculated by ab initio and semiempirical methods.[7a–c]

Acylketenes are thermodynamically stabilized relative to alkyl, vinyl, cyano, and other carbon substituted ketenes.[7] Thus 6-31G*//6-31G* calculations of the isodesmic reaction in equation 4 show that formylketene is stabilized relative to acrolein (**10**) by 3.6 kcal/mol, while $HO_2CCH=C=O$ is similarly stabilized by 4.8 kcal/mol.[7] This stabilization is attributed[7] to delocalization, as indicated by the zwitterionic resonance structure **6a**. The greater stability of the *syn* conformer probably arises from the favorable interaction of opposite charges in **6a**. However, the acyl group also stabilizes the transition states for the reactions of acylketenes and so many of these are rather reactive.

$$O=CHCH=C=O + CH_3CH=CH_2 \rightarrow CH_3CH=C=O + O=CHCH=CH_2 \quad (4)$$
$$\mathbf{10}$$

Several acylketenes with bulky substituents have been reported to be long-lived (persistent) as neat materials at room temperature. Thus the persistent ketene **11** was formed by the Wolff rearrangement (equation 5),[8] by pyrolysis of t-BuCOCH(t-Bu)CO$_2$Et above 400 °C,[9] and was also obtained[10] as a minor product in RuO$_4$ oxidation of an allenic ketone (equation 5). Still other routes to **11** were by reaction of t-BuCOCH(t-Bu)COCl with Et$_3$N,[11] and of t-BuCH(COCl)$_2$ with t-Bu$_2$(CuCN)Li$_2$.[12]

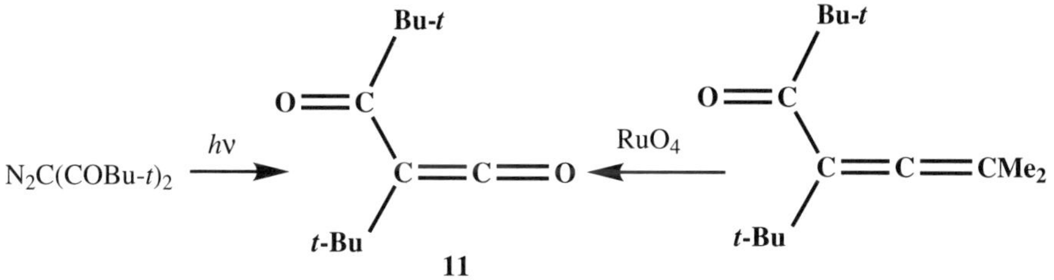

$$N_2C(COBu\text{-}t)_2 \xrightarrow{h\nu} \underset{\textbf{11}}{(t\text{-Bu})(t\text{-BuCO})C=C=O} \xleftarrow{RuO_4} (t\text{-Bu})(t\text{-BuCO})C=C=CMe_2 \quad (5)$$

The ketene **11** is proposed to exist preferentially in the *anti* conformation shown for steric reasons,[12] whereas **12** is suggested to prefer the *syn* conformation.[13]

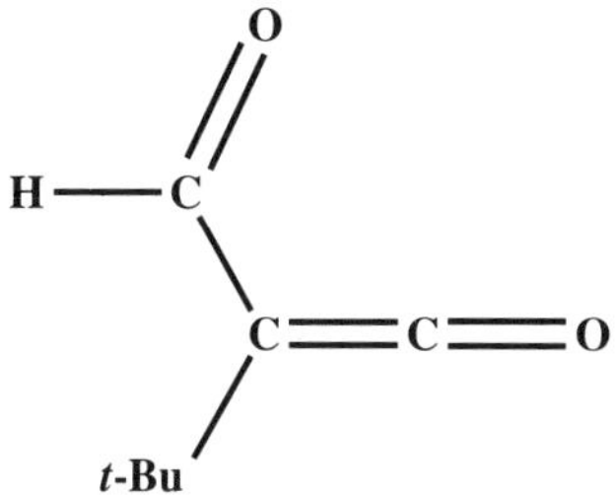

12: (HCO)(t-Bu)C=C=O

The ketenes **13** could be isolated as rather stable species from Wolff rearrangements, and reacted with alcohols or amines to give the esters or amides respectively (equation 6).[14]

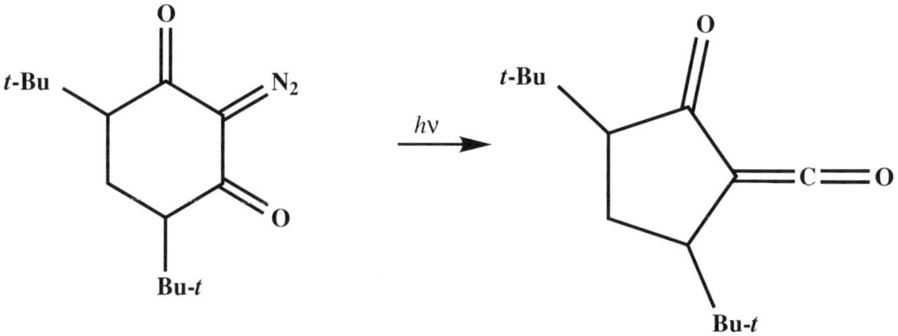

$$E/Z \text{ diazocyclohexanedione} \xrightarrow{h\nu} E/Z\text{-}\textbf{13} \quad (6)$$

The persistent carboethoxyketene **14** has been generated from the acid chloride by reaction with Et$_3$N (equation 7),[15] and the corresponding methyl ester was obtained similarly.[12] The related ketene RC(CO$_2$Et)=C=O (R = 1-adamantyl) was prepared by Wolff rearrangement.[15a]

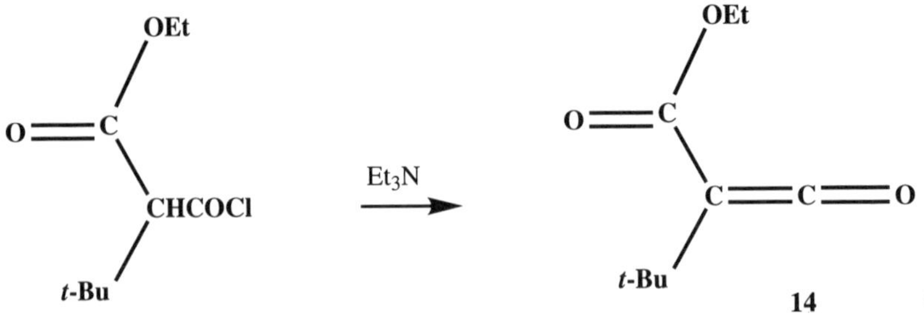

The diacylketene **15,** prepared by thermolysis at 350 °C of the furandione resulting from the reaction of dipivalylmethane with oxalyl chloride (equation 8), is stable in solution at −20 °C but reacts rapidly with MeOH to give the ester and undergoes self-reaction by a [4 + 2] cycloaddition to give another persistent ketene **16** (equation 9).[16] Reaction of **16** with arylamines led to **17** by an addition–decarboxylation route (equation 10),[17] while reaction with $PhCH_2NH_2$ gave $(t\text{-}BuCO)_2CHNHCH_2Ph$.[17] The X-ray structure of **16** was determined,[16] and is consistent with a contribution of resonance structure **6a.**

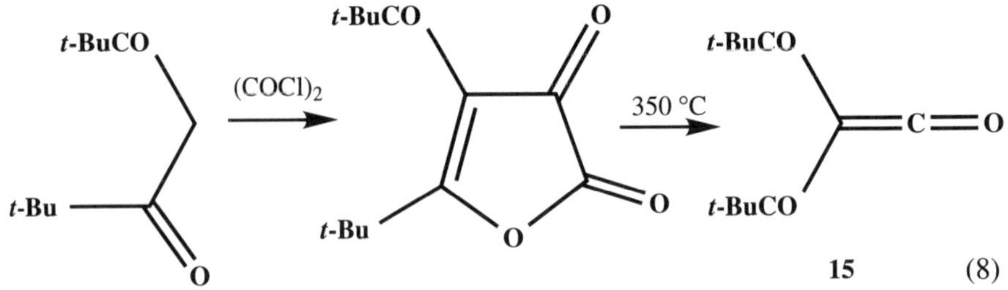

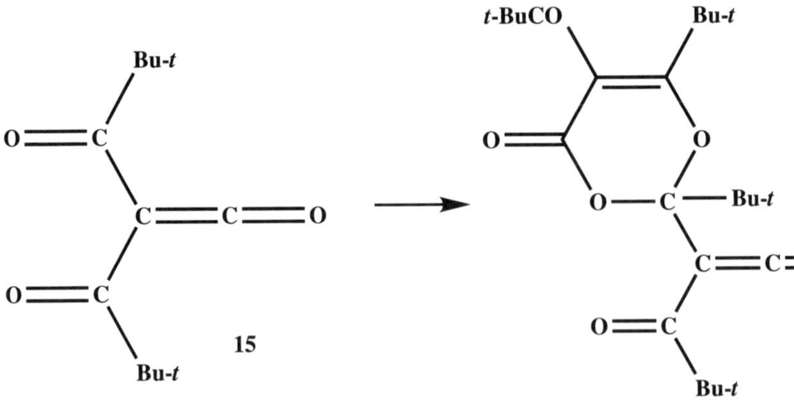

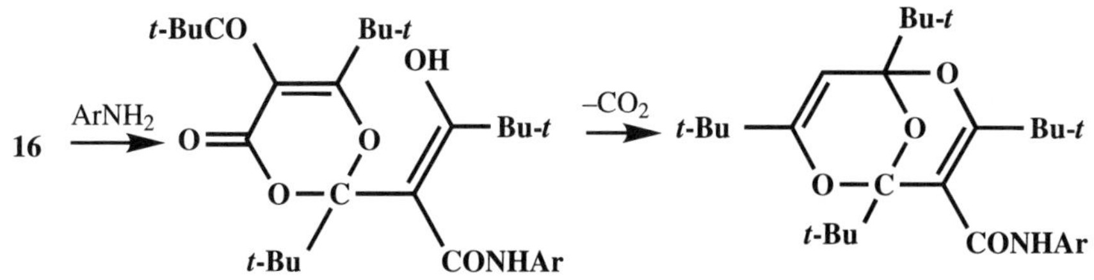

4.1.6 ACYLKETENES

$$16 \xrightarrow{ArNH_2} \text{[structure]} \xrightarrow{-CO_2} \text{[structure 17]} \quad (10)$$

In the presence of DMSO or $n\text{-}Bu_3PO$ **15** dimerizes across the C=O bond to give **18** (equation 11).[18] The former result is attributed to formation of the complex **19**. With many dienophiles **15** gives [4 + 2] cycloadditions involving the enone system,[18,19] but with carbodiimides **16** gives [2 + 2] cycloadducts,[18] which are unusual for acylketenes.

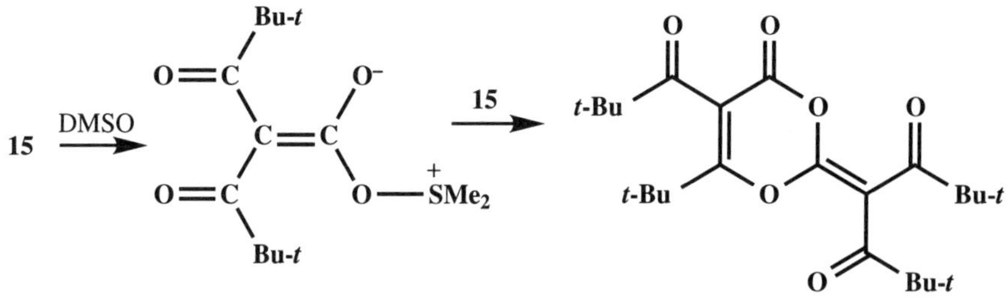

$$15 \xrightarrow{DMSO} \text{[structure 19]} \xrightarrow{15} \text{[structure 18]} \quad (11)$$

Acylketenes are formed by heating of β-keto esters **20**,[9,20–22] and flash vacuum pyrolysis, and Ar matrix isolation of the products leads to acylketenes **21**, as observed by IR spectroscopy (equation 12).[9,12] It was proposed that the ketene formation took place from the enol tautomers **20a** as the more enolized substrates formed ketenes more readily, and the unreacted starting material was enriched in the keto form.[9] The ketenes existed in both the *s-E* and *s-Z* forms, and on warming in the matrix reformed the keto esters. Ketenes **22–24** were also formed from the esters,[9] but **25**, available from the acyl chloride, did not form from the ester, which was not detectably enolized.[9]

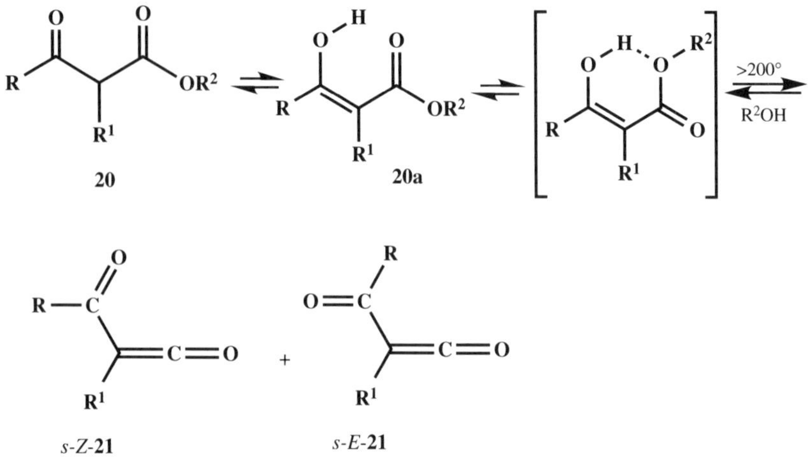

232 TYPES OF KETENES

(12)

$R^1 = H$; $R = Me$, t-Bu, 1-Ad; $R^2 = Me$, Et

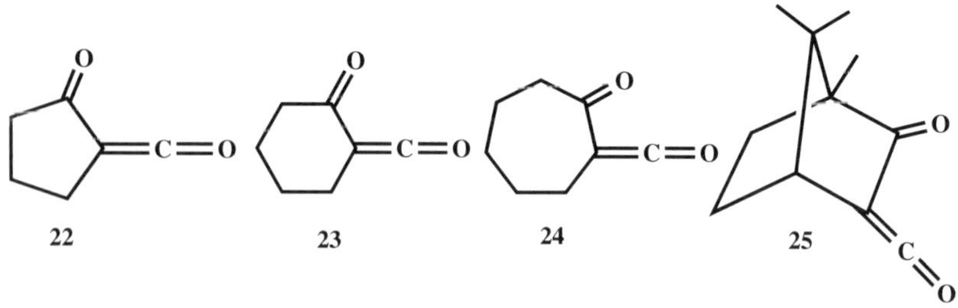

The decarbonylation of 2,4-diketoesters **26** at 170–190 °C is proposed to occur by enolization to the intermediates **27** followed by pericyclic loss of ROH and CO to give acylketenes **28** (equation 13), and then readdition of the alcohol to give **30** (equation 14).[23,24] When the esters **26** or **30** or the parent acids corresponding to **31** are heated in H_2O in a sealed reactor the ketones **32** are formed, and for the esters this is proposed to occur by formation of the ketenes **28** which are hydrated to the acids **31,** which then decarboxylate (equation 15). The stretching frequencies of the ketenes at 2120–2140 cm^{-1} are observed by GC–FTIR analysis of the reactions at 280 °C.[24]

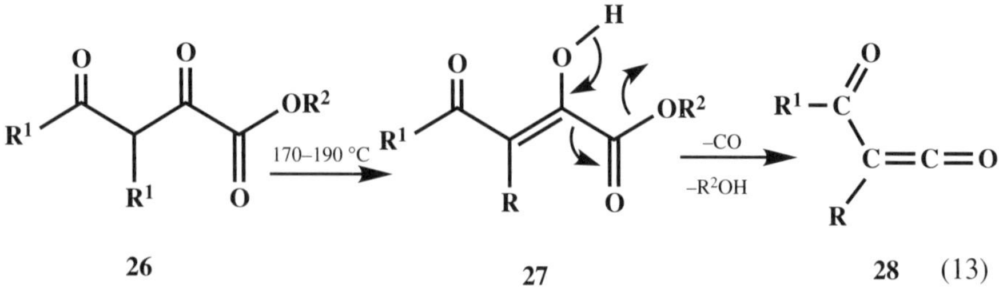

(13)

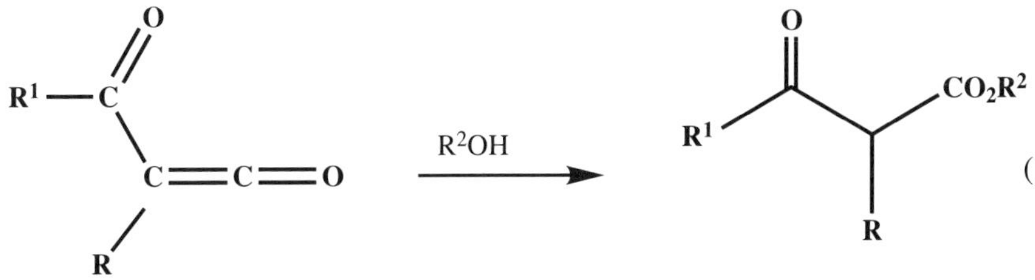

$$\underset{28}{\overset{R^1-\overset{O}{\underset{R}{C}}}{\underset{C=C=O}{\parallel}}} \xrightarrow{R^2OH} \underset{30}{\overset{O}{\underset{R}{\overset{R^1}{\bigvee}}\overset{CO_2R^2}{\bigvee}}} \quad (14)$$

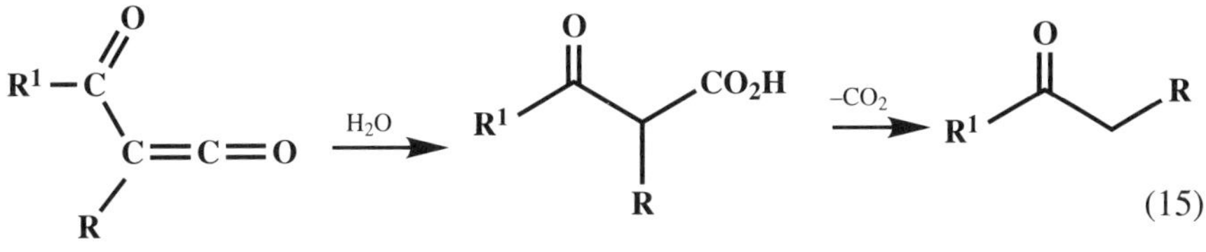

$$\underset{28}{\overset{R^1-\overset{O}{\underset{R}{C}}}{\underset{C=C=O}{\parallel}}} \xrightarrow{H_2O} \underset{31}{\overset{O}{\underset{R}{\overset{R^1}{\bigvee}}\overset{CO_2H}{\bigvee}}} \xrightarrow{-CO_2} \underset{32}{\overset{O}{R^1\bigvee R}} \quad (15)$$

Reaction of the 2-keto-3-methylsuccinate ester **33** at 105 °C was proposed to involve preferential formation of the ketene **34** by a process analogous to equation 12 (see Section 3.2.3.2), while at 280 °C decarbonylation and formation of ketene **35** was also observed (equation 16).[23,24]

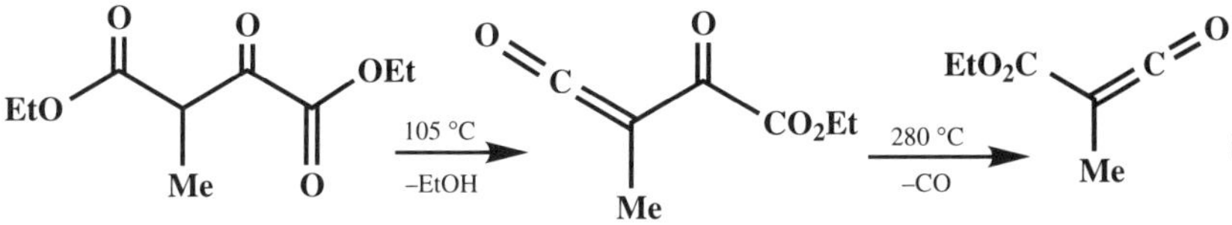

$$\underset{33}{\text{EtO}\overset{O}{\underset{Me}{\bigvee}}\overset{O}{\underset{O}{\bigvee}}\overset{OEt}{\bigvee}} \xrightarrow[-EtOH]{105\ °C} \underset{34}{\overset{O}{\underset{Me}{=C}}\overset{O}{\underset{}{\bigvee}}CO_2Et} \xrightarrow[-CO]{280\ °C} \underset{35}{\overset{EtO_2C}{\underset{Me}{\bigvee}}C=O} \quad (16)$$

Ketoketenes are also generated by thermolysis of 1,3-dioxin-4-ones (equation 17),[25–31] and the cyclization of hydroxyalkylketenes made in this way to lactones is a useful synthetic method (equation 18).[26,29–31] Cyclization to the 5-membered rings **36** via β-hydroxyacylketenes occurs in high yield (equation 19),[29] but the corresponding γ-hydroxy ketenes do not form 6-membered lactones, but give dimers and trimers instead.[29]

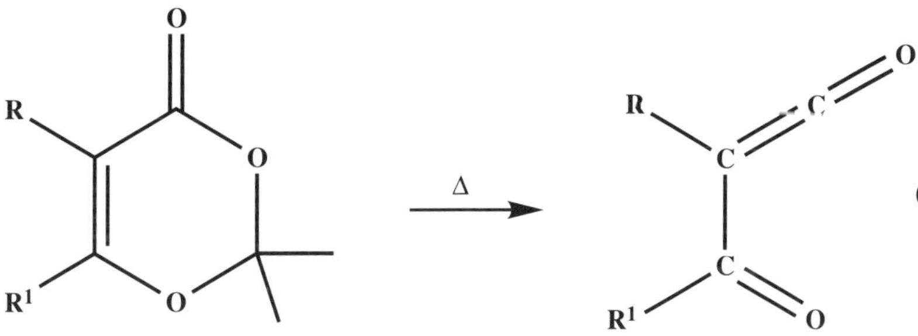

$$\underset{R^1}{\overset{R}{\bigvee}}\overset{O}{\underset{O}{\bigvee}}\overset{}{\underset{}{\bigvee}} \xrightarrow{\Delta} \underset{R^1}{\overset{R}{\bigvee}}\overset{C=O}{\underset{C=O}{\bigvee}} \quad (17)$$

(18)

R = H, Me

36 (19)

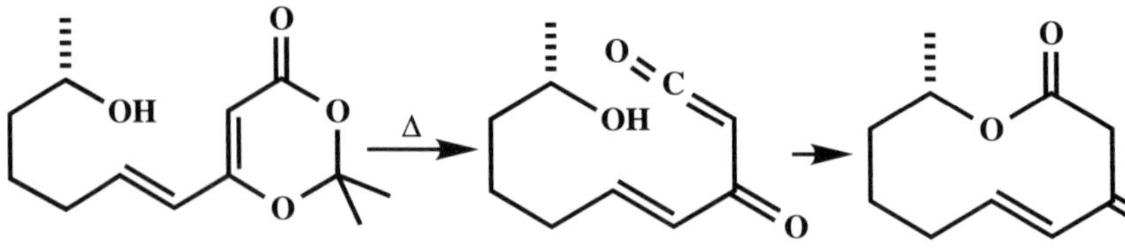

Arylfurandiones (**37**) form aroylketenes (**38**) on thermolysis in solution[32,33] and upon either flash vacuum pyrolysis or matrix photolysis (equation 20) the ketenes are identified by their low-temperature IR spectra.[34] These ketenes dimerize to pyrones, and give [4 + 2] cycloadditions with alkynes.

(20)

37 **38**

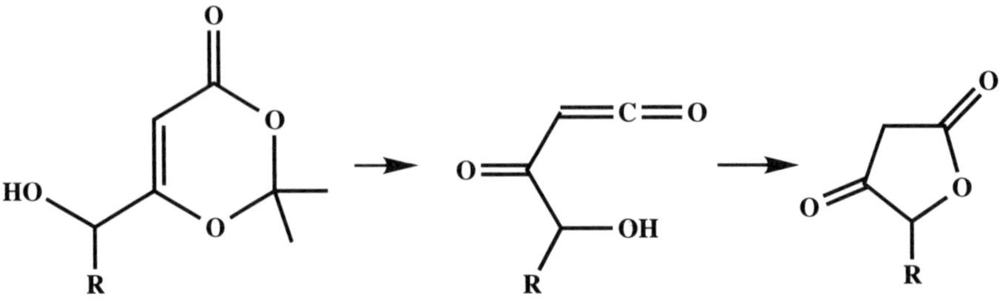

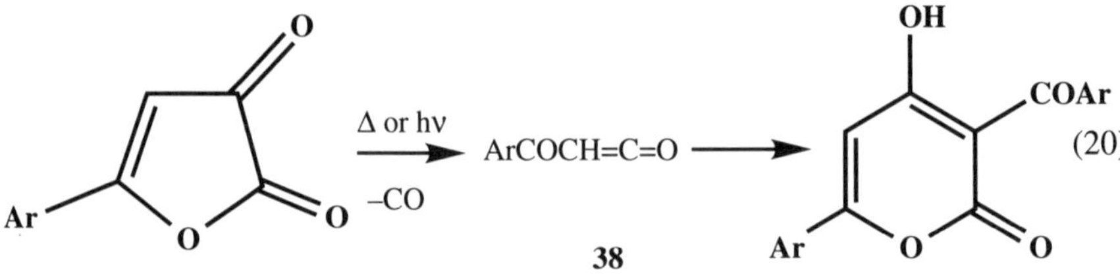

A strong IR band at 2147 cm^{-1} for PhCOCH=C=O (**39**) in an Ar matrix was ascribed to the *s*-*Z* conformation, whereas a shoulder at 2134 cm^{-1} was assigned to the *s*-*E* conformation.[34] The Ar matrix IR absorptions are typically 10 cm^{-1} higher than for neat solids.

s-*Z* **39** *s*-*E* **40**

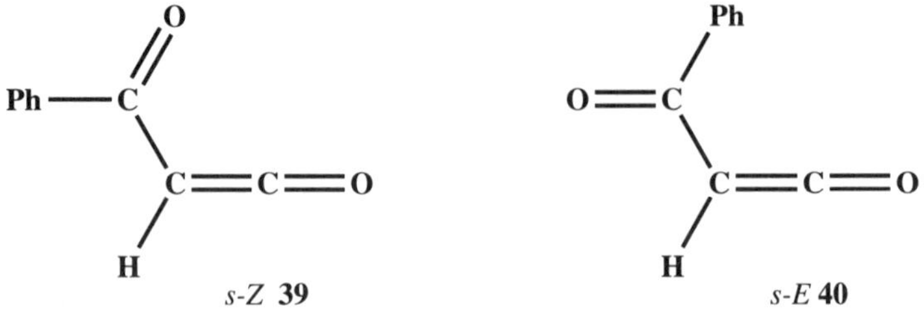

4.1.6 ACYLKETENES

Chlorocarbonylketenes have been made by thermolysis of malonyl dichlorides to give ketenes such as **41** as isolable intermediates (equation 21).[36–38] These were reported to react preferentially at the ketenyl moiety with nucleophiles.[36]

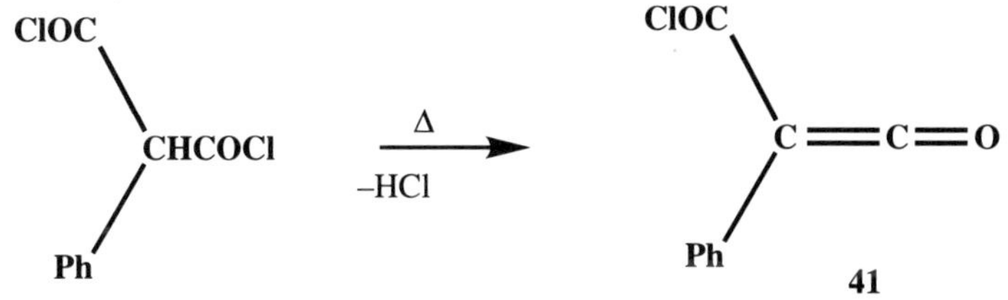

$$\text{(ClOC)(Ph)CHCOCl} \xrightarrow[-\text{HCl}]{\Delta} \text{(ClOC)(Ph)C=C=O} \quad \textbf{41} \tag{21}$$

Chloroacylketenes have been widely used intermediates in heterocyclic synthesis, as in the reactions with imidazoles (equation 22).[38] The intermediate N-acylated imidazole was isolated as a solid on mixing the reagents. Cyclization also occurs on reaction with enol ethers (equation 23)[39] and reactions with cyanamides have been studied.[40]

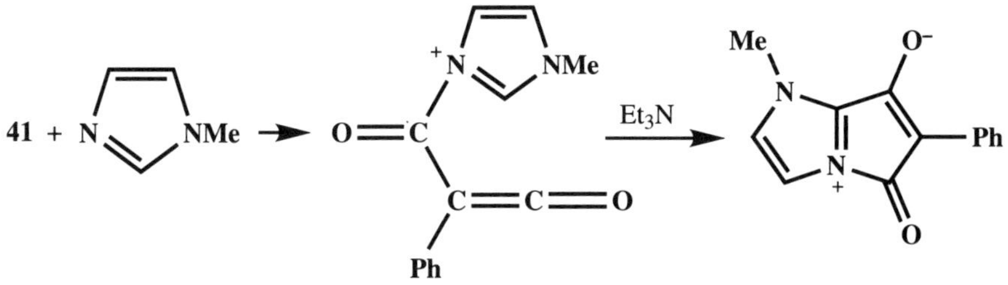

$$\textbf{41} + \text{imidazole-NMe} \longrightarrow \text{acyl intermediate} \xrightarrow{\text{Et}_3\text{N}} \text{bicyclic product} \tag{22}$$

$$\text{cyclohexenyl-OMe} \xrightarrow[\text{MeOH, Et}_3\text{N}]{\text{ClOCCH=C=O}} \text{intermediate} \longrightarrow \text{bicyclic adduct}, \quad R = \text{MeO} \tag{23}$$

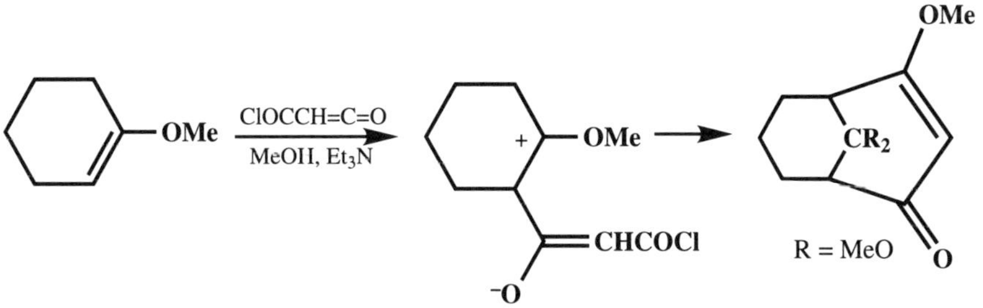

Some other reactions of acylketenes include [2 + 2] cycloadditions (equation 24),[41,42] [4 + 2] cycloadditions (equation 25),[27,28] and addition of organometallic reagents[43] to the ketene carbonyl. Further reaction of the [4 + 2] cycloadducts of acylketenes with vinyl ethers leads to 2,6-dideoxy sugars (equation 25).[28]

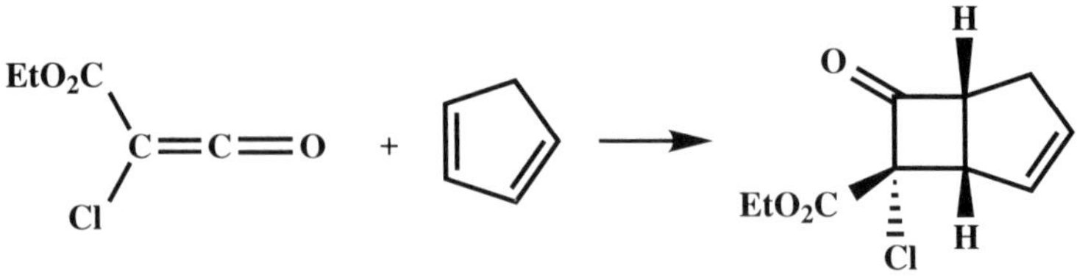

Nucleophilic addition of alcohols and H_2O to acylketenes is unique in that addition to the ciscoid conformation via a cyclic transition state leading to an enol occurs. Thus the addition of the phenolic hydroxy group of a novolac resin to a ketene to give the enol directly observed by IR (equation 26) was proposed to occur via a cyclic six-membered transition state.[44] The formation of enols was also shown directly in the generation of ketenes from β-acyl esters when the ketenes were isolated in an Ar matrix as shown in equation 12.[9] When the ketene reacted with the alcohol in the matrix, the ester enol **20a** was observed directly.

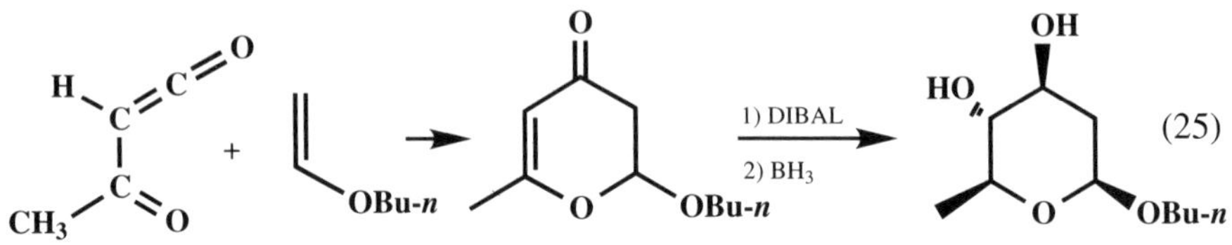

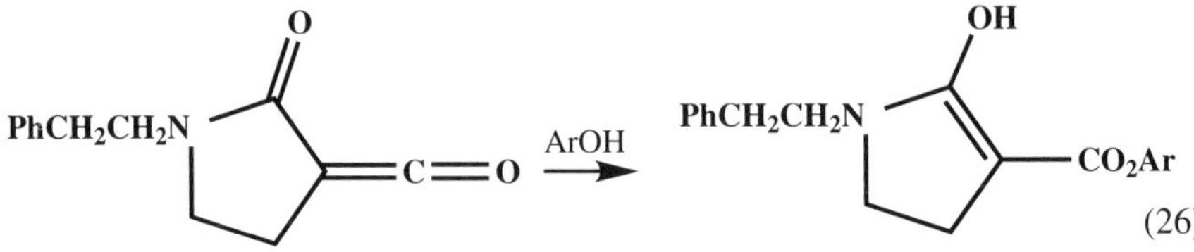

Further evidence for this pathway was found by Meier et al.[45,46] in that thermolysis of the diazodione **42** in EtOH led to an E/Z mixture of enols **43** but that the formation of the Z isomer was preferred kinetically.[45] This was attributed to 1,4-addition via a 6-membered cyclic transition state as shown (equation 27). The same mechanism was proposed to occur for hydration and for other acyl ketenes,[46] while addition through a noncyclic transition state occurred for ketenes with an *anti* disposition of the β-acyl and ketenyl groups (equation 28).

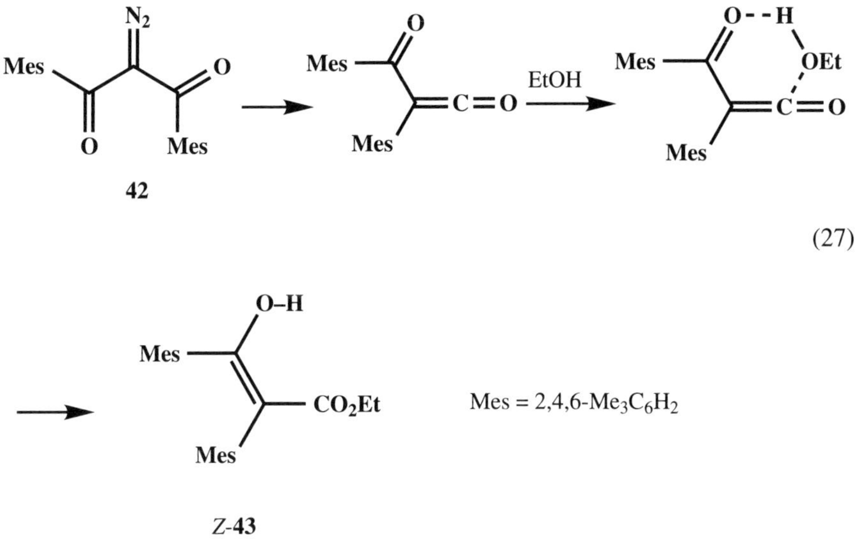

4.1.6 ACYLKETENES

(equations 27, 28, 29 shown as structural schemes)

Mes = 2,4,6-Me$_3$C$_6$H$_2$

(27)

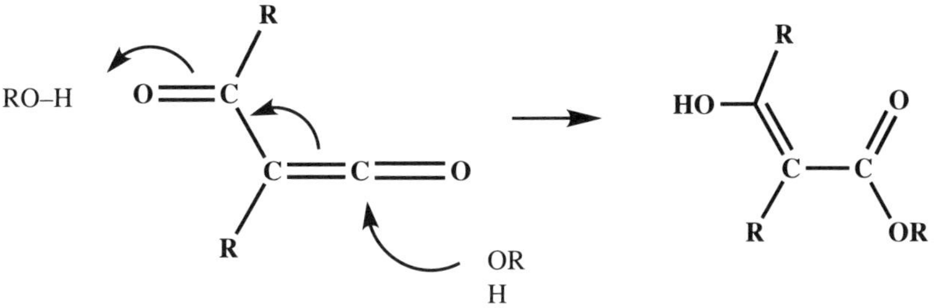

(28)

Additional support for this pathway is provided by the results of Witzeman,[21] who measured the kinetics of the thermal transesterification of acetoacetate esters (equation 12). The rate of this reaction was independent of the concentration of the alcohol R^{2}OH, but was faster for *tert*-butyl or *tert*-amyl esters. An acetylketene intermediate was proposed,[21] and a route similar to that of equation 12 involving an enol intermediate provides the best explanation of the results. In a related process the silyl enol ester **44** was proposed to react on flash vacuum pyrolysis to give the ketene **45**, which was trapped by alcohols and amines (equation 29).[47]

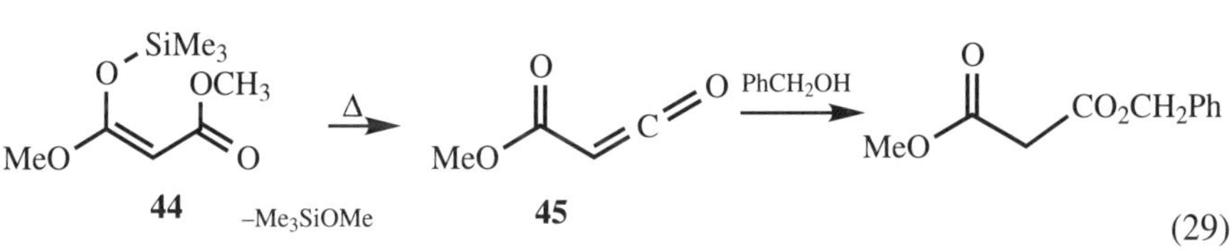

(29)

The rate of hydration of **14** is unique among ketenes for which this process has been studied, in that the rate in H$_2$O/CH$_3$CN is almost unchanged between 2 and 100% H$_2$O, and shows a shallow maximum in the rate between 50 and 100% H$_2$O.[48] This suggests that a nonpolar transition state is involved, consistent with the cyclic transition state involving attack of water to give the enol (Scheme 3).[48] This process has been examined by molecular orbital calculations.[48,48a]

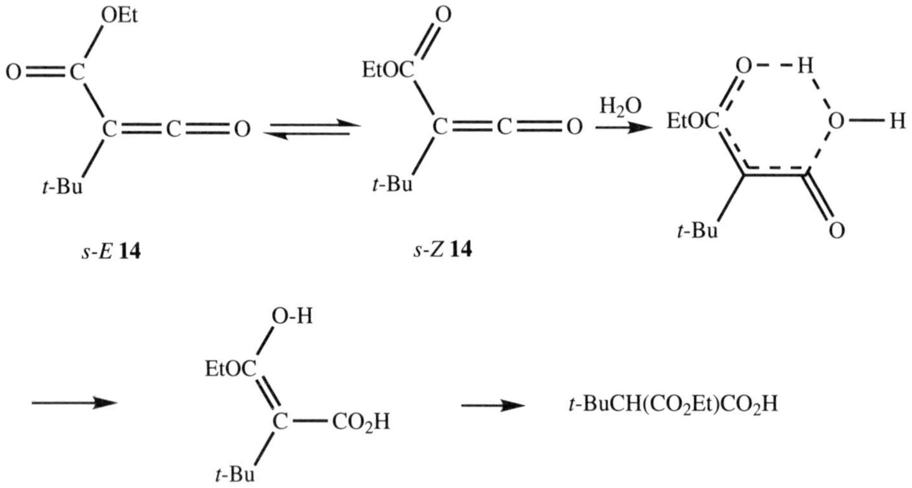

Scheme 3

The intramolecular cyclization of formylketene according to equation 30 has been studied using molecular orbital calculations, and found to be endothermic by 17 kcal/mol.[6] The gas phase reaction of ynolate ion **46**, generated from vinylene carbonate **47**, gives reversible addition of CS$_2$, COS, and CO$_2$, and isotope labeling studies suggest that these ions undergo reversible cyclization to form acylketenes, as in equation 31.[49]

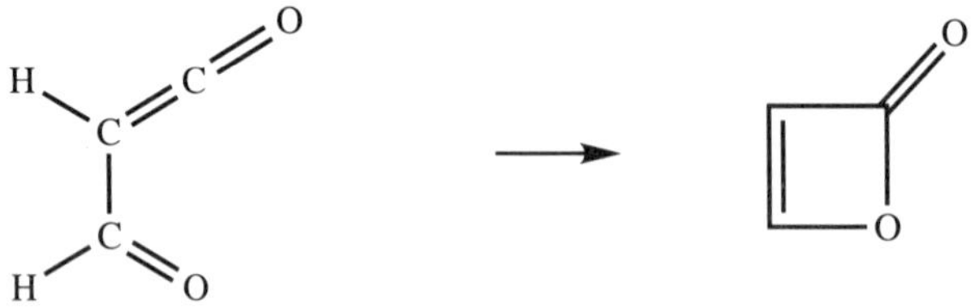

(30)

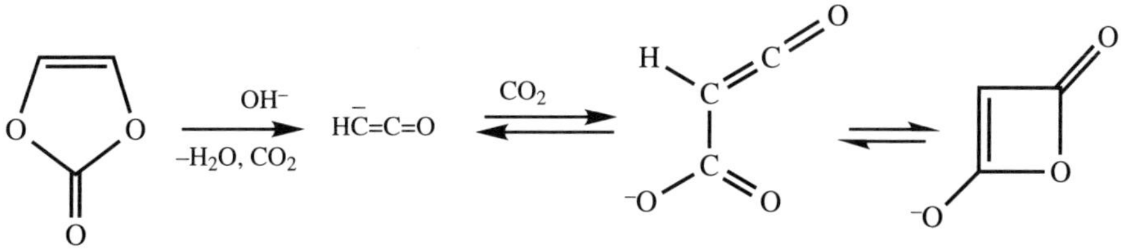

4.1.6 ACYLKETENES

(31)

The cyclization of equation 32 is suggested to occur in pyrolysis of **48** at 250 °C, which led to 18% of CO_2 and 24% of propyne. This was attributed to formation of ketene **49** which cyclized to **50** to give the observed products (equation 32).[50] The formation of ketenes by reaction of diketones with atomic carbon was also proposed (equations 33 and 34).[50]

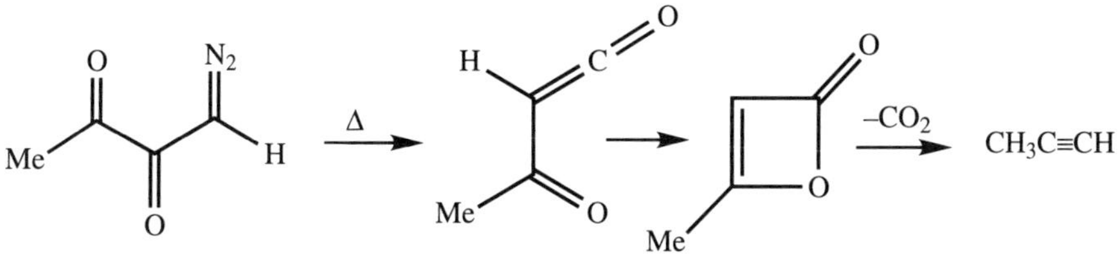

(32)

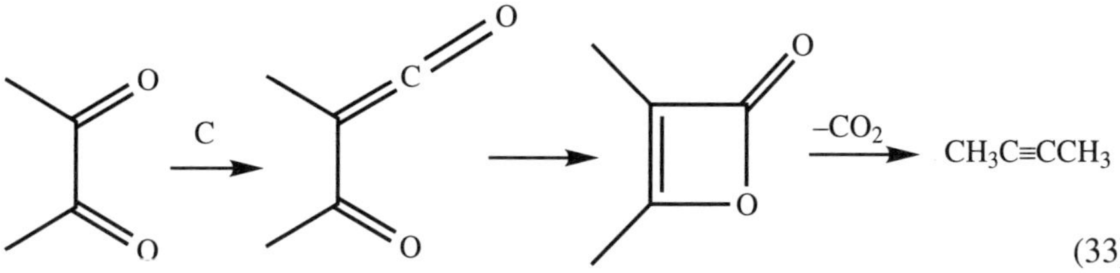

(33)

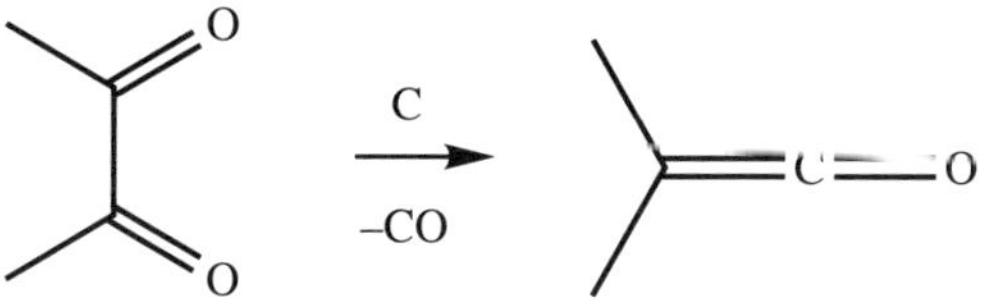

(34)

The evidence for vinylogous acylketenes such as **51**[51] and **52**[52] is considered in more detail in Section 4.1.10.4.

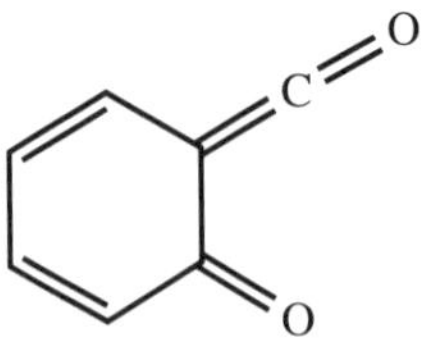

240 TYPES OF KETENES

51

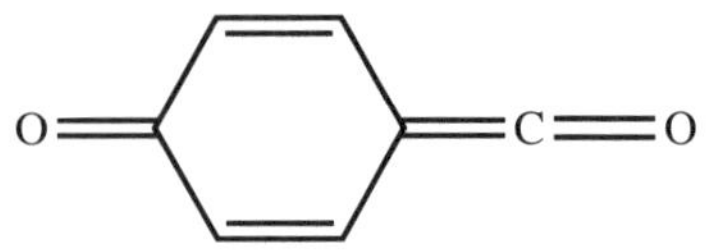

52

In 1916 Staudinger and Hirzel[53] prepared bis(carboethoxy)ketene (**53**) by thermal Wolff rearrangement (equation 35) as a distillable oil that underwent hydration by atmospheric moisture and dimerized slowly at room temperature.[54] This ketene reacted with silylalkynes in [4 + 2] cycloadditions (equation 35),[55] and in allene-forming reactions with Wittig reagents.[56] Reaction of **53** with 4-dimethylaminopyridine (DMAP) gave a solid adduct identified as **54** which gave **55** on heating to 110 °C (equation 36).[56]

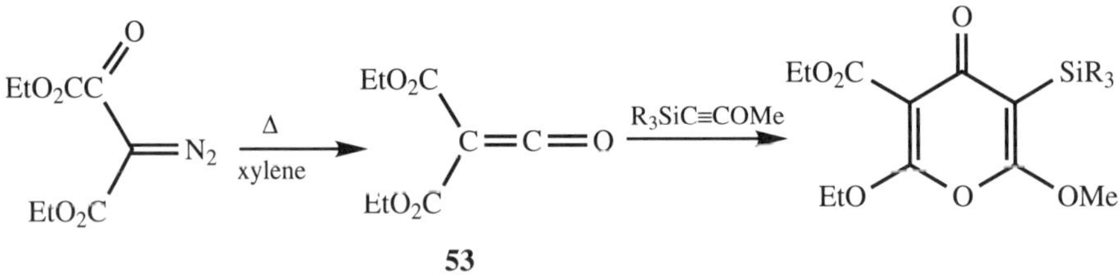

(35)

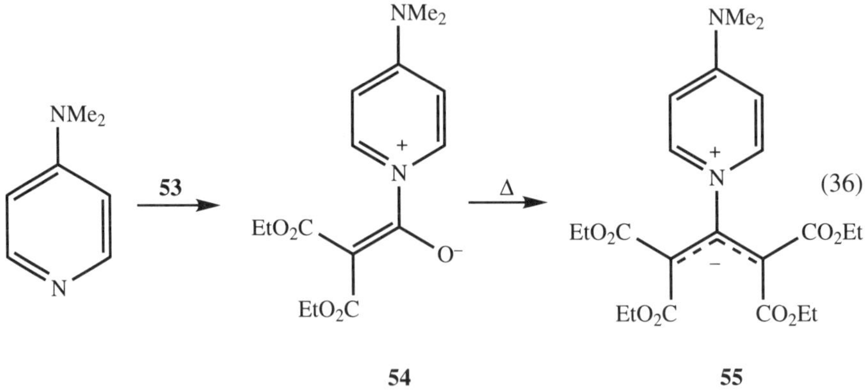

(36)

Dibenzoylketene (**56**) was prepared from the furandione route (equation 8) and trapped with imines,[57] and its IR band at 2140 cm^{-1} was observed in a matrix at −196 °C.[58] On warming **56** was suggested to dimerize by [4 + 2] cycloaddition, leading to the dimer **58** and the decarboxylation product **59** (equation 37).[58]

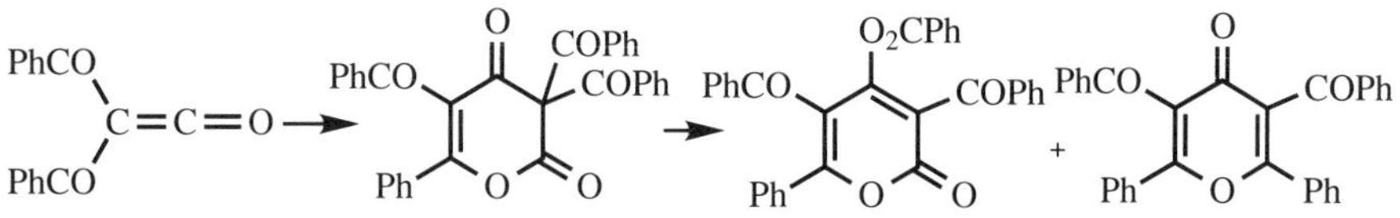

4.1.6 ACYLKETENES

(37)

Reaction of acyl Meldrum's acid derivatives **60** in refluxing benzene has been proposed to give ketenes **61** which were trapped with imines (equation 38).[59] The thermal decarboxylation of **61** to form ketenes **62** has also been proposed (equation 39).[39]

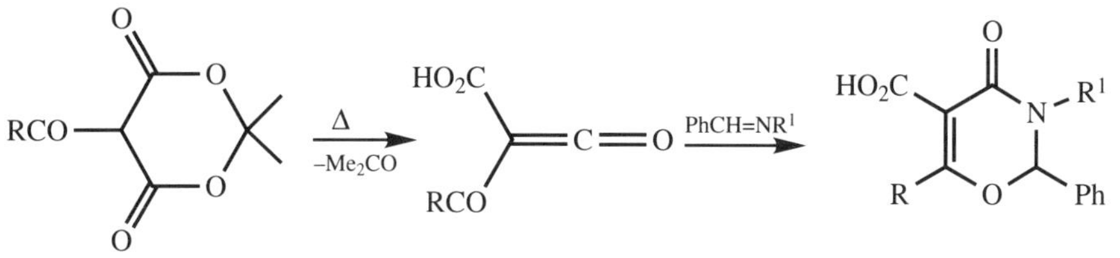

(38)

(39)

The formation by α-elimination and a Wolff-type rearrangement of acylketene **22** which then formed the dimer has been reported (equation 40).[60]

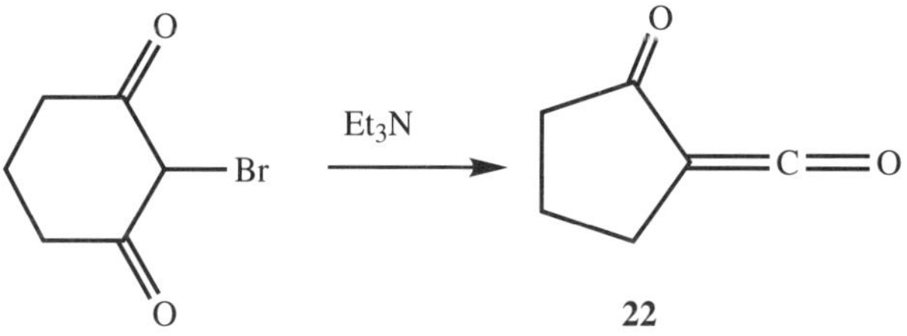

(40)

TABLE 1. Acylketenes

Ketene	Formula	Preparation	Registry No.	Ref.
O=CClCH=C=O	C_3HClO_2	Dehydrochlorination		39
O=CHCH=C=O	$C_3H_2O_2$	Wolff	50659-16-0	3–5
O=CHCH=C=O	$C_3H_2O_2$		36832-93-6	
O=CHCH=C=O	$C_3H_2O_2$		4484-43-9	
HS(O)CCH=C=O	$C_3H_2O_2S$	Index entry	6115-11-3	None
HO$_2$CCH=C=O	$C_3H_2O_3$	Theory	4452-03-3	7
H$_2$NCOCH=C=O	$C_3H_3NO_2$	Theory	4484-44-0	
CF$_3$COC(SO$_2$F)=C=O	$C_4F_4O_4S$	Sulfonation	104693-21-2	61
CF$_3$COCH=C=O	$C_4HF_3O_2$	Elimination	83341-89-3	62
N$_2$=CHCOCH=C=O	$C_4H_2N_2O_2$	Wolff	82515-19-3	4
HO$_2$CCOCH=C=O	$C_4H_2O_4$	Index entry	5809-64-3	None
(β-propiolactone ketene structure)				
(HO$_2$C)$_2$C=C=O	$C_4H_2O_5$	Wolff	4379-38-8	63
ClOCCMe=C=O	$C_4H_3ClO_2$	Index entry	17118-72-8	None
O=CHNHCOCH=C=O	$C_4H_3NO_3$	Dehydrochlorination	106104-56-7	64
CH$_3$COCH=C=O	$C_4H_4O_2$	Theory	691-45-2	65
CH$_3$O$_2$CCH=C=O	$C_4H_4O_3$	Dehalogenation		9
HOCH$_2$COCH=C=O	$C_4H_4O_3$	Pyrolysis	89238-09-5	12,47
MeO$_2$CC(CF$_3$)=C=O	$C_5H_3F_3O_3$	Dioxinone		29
HO$_2$CC(CH=CH$_2$)=C=O	$C_5H_4O_3$	Dehydration	106007-69-6	66
ClOCCEt=C=O	$C_5H_5ClO_2$	Pyrolysis	17423-43-7	67
MeC(S)CMe=C=O	C_5H_6OS	Dehydrochlorination	77771-49-4	64
		Wolff		68

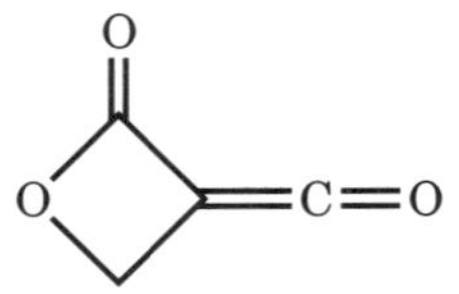

Structure	Formula	CAS	Method	Ref
EtCOCH=C=O	$C_5H_6O_2$	6704-80-9	Index entry	None
MeCOCMe=C=O	$C_5H_6O_2$		Pyrolysis	9,24
$HO_2CCEt=C=O$	$C_5H_6O_3$	29417-74-1	Index entry	29
HOCHMeCOCH=C=O	$C_5H_6O_3$		Dioxinone	29
$MeO_2CCMe=C=O$	$C_5H_6O_3$	36832-93-6	Pyrolysis	24
2-cyclopentanone-1-ylidene ketene	$C_6H_6O_2$	3491-01-8	Wolff	9
$CH_3COCH_2COCH=C=O$	$C_6H_6O_3$		Dioxinone	29
2,2-dimethyl-1,3-dioxolan-4-ylidene ketene	$C_6H_6O_4$	73454-18-9	Wolff	42
$MeO_2CCOCMe=C=O$	$C_6H_6O_4$	135147-56-7	Pyrolysis	24
$(MeO_2C)_2C=C=C$	$C_6H_6O_5$		Wolff	69
EtCOCMe=C=O	$C_6H_8O_2$	89533-71-1	Erroneous[a]	70
$EtO_2CCMe=C=O$	$C_6H_8O_3$	36277-50-6	Dehydrochlorination	71
$CH_3CHOHCOCH=C=O$	$C_6H_8O_3$		Dioxinone	29
2-selenoxocyclohexan-1-ylidene ketene (Se=C, C=O on cyclohexane)	C_7H_8OSe	122794-90-5	Metal complex	72a
RCOCH=C=O	$C_7H_8O_2$	38082-30-3	Pyrolysis	72
$R = CH_2=CHCH_2CH_2$				

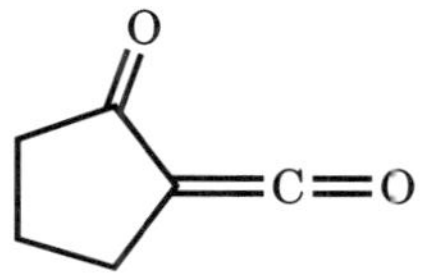

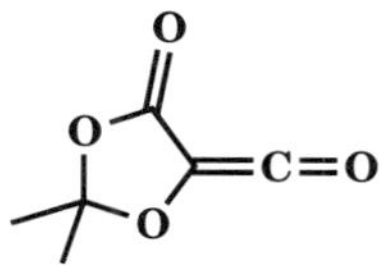

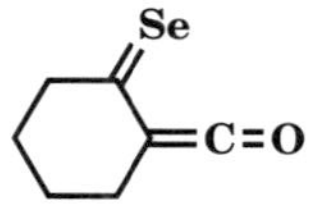

TABLE 1. (Continued)

Ketene	Formula	Preparation	Registry No.	Ref.
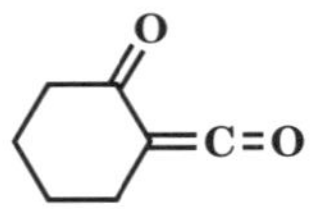 (cyclohexanone C=O)	$C_7H_8O_2$	Pyrolysis		9
$EtO_2CC(COMe)=C=O$	$C_7H_8O_4$	Pyrolysis	135147-55-6	24
$EtO_2CC(CO_2Me)=C=O$	$C_7H_8O_5$	Wolff	72091-45-3	56
$ClOCC(Bu-n)=C=O$	$C_7H_9ClO_2$	Dehydrochlorination	17423-42-6	64
$t\text{-}BuCOCH=C=O$	$C_7H_{10}O_2$	Thermolysis	135147-55-6	12,24
$EtO_2CCEt=C=O$	$C_7H_{10}O_3$	Dehalogenation	24489-48-3	2,43
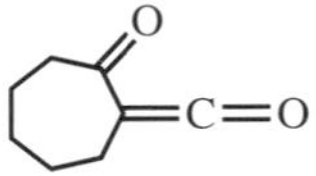 (cycloheptanone C=O)	$C_8H_{10}O_2$	Pyrolysis		9
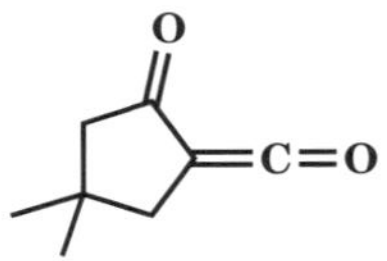 (4,4-dimethylcyclopentanone C=O)	$C_8H_{10}O_2$	Wolff	85526-49-4	8
$RCOCH=C=O$ $R = CH_2=CMeCH_2CH_2$	$C_8H_{10}O_2$	Pyrolysis	59534-96-2	72
$(EtO_2C)_2C=C=O$	$C_8H_{10}O_5$	Wolff	72091-41-9	53–56
$n\text{-}PrCOCEt=C=O$	$C_8H_{12}O_2$	Erroneous[a]		70
$MeO_2CC(Bu\text{-}t)=C=O$	$C_8H_{12}O_3$	Pyrolysis	922-11-2	12
$RCOCH=C=O$	$C_8H_{12}O_3$	Dioxinone		31

R = CH$_3$CHOH(C-I$_2$)$_3$			
Et$_2$NCOCMe=C=O	C$_8$H$_{13}$NO$_2$	Dehydrochlorination	36277-42-6	71
Me$_2$NCOC(GeMe$_3$)=C=O	C$_8$H$_{15}$GeNO$_2$	Addition	110698-78-7	73

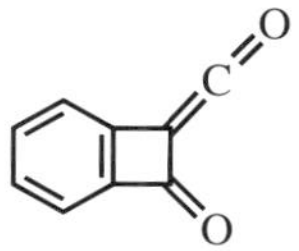

(structure: benzocyclobutenone with C=C=O)

	C$_9$H$_4$O$_2$	Wolff		74
4-FC$_6$H$_4$COCH=C=O	C$_9$H$_5$FO$_2$	Pyrolysis	84905-29-3	75
ClOCCPh=C=O	C$_9$H$_5$ClO$_2$	Dehydrochlorination	17118-70-6	64
4-ClC$_6$H$_4$COCH=C=O	C$_9$H$_5$ClO$_2$	Pyrolysis	84905-28-2	75,76
PhCOCH=C=O	C$_9$H$_6$O$_2$	Pyrolysis	4663-28-9	9,77
Ph^{13}COCH=C=O	C$_9$H$_6$O$_2$	Pyrolysis	92284-10-1	77
PhCOCH=^{13}C=O	C$_9$H$_6$O$_2$	Pyrolysis	92270-05-8	77
O=CHCPh=C=O	C$_9$H$_6$O$_2$	Index entry	27762-40-9	
HO$_2$CCPh=C=O	C$_9$H$_6$O$_3$	Index entry	4438-21-5	
PhNHCOCH=C=O	C$_9$H$_7$NO$_2$	Index entry	4452-15-7	
RCOCH=C=O	C$_9$H$_{12}$O$_2$	Pyrolysis	59534-98-4	72
R = MeCH=CMeCH$_2$CH$_2$				
RCOCH=C=O	C$_9$H$_{12}$O$_2$	Pyrolysis	59534-97-3	72
R = Me$_2$C=CHCH$_2$CH$_2$				
i-PrCOC(Pr-i)=C=O	C$_9$H$_{14}$O	Wolff	67398-50-9	8
EtO$_2$CC(Bu-t)=C=O	C$_9$H$_{14}$O$_3$	Dehydrochlorination	51552-63-7	16
ArO$_2$CCMe=C=C	C$_{10}$H$_5$Cl$_3$O$_3$	Dehydrochlorination	17118-68-2	64
Ar = 2,4,6-Cl$_3$C$_6$H$_2$				
PhC(S)C(CN)=C=O	C$_{10}$H$_5$NOS	Desulfurization	61628-70-4	78

245

TABLE 1. (Continued)

Ketene	Formula	Preparation	Registry No.	Ref.
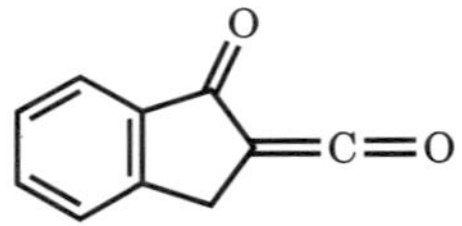 (indanone-C=O structure)	$C_{10}H_6O_2$	Index error	57558-69-7	79
ClOCC(CH$_2$Ph)=C=O	$C_{10}H_7ClO_2$	Dehydrochlorination	57421-93-9	37
4-MeC$_6$H$_4$COCH=C=O	$C_{10}H_8O_2$	Pyrolysis	84905-26-0	75,76
4-MeOC$_6$H$_4$COCH=C=O	$C_{10}H_8O_3$	Pyrolysis	84905-27-1	75,76
MeO$_2$CCPh=C=O	$C_{10}H_8O_3$	Wolff	32084-13-2	54
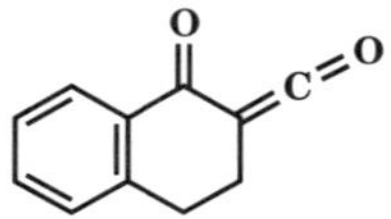 (tetralone-C=O structure)	$C_{11}H_8O_2$	Pyrolysis	57558-70-0	79
ClOCC(CH$_2$CH$_2$Ph)=C=O	$C_{11}H_9ClO_2$	Dehydrochlorination	57421-94-0	36
EtO$_2$CCPh=C=O	$C_{11}H_{10}O_3$	Dehydrochlorination	36277-53-9	71
PhCOCH$_2$COCH=C=O	$C_{11}H_8O_3$	Pyrolysis		29
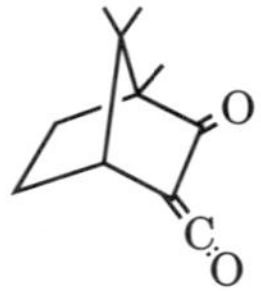 (bicyclic diketone-C=O structure)	$C_{11}H_{14}O_2$	Dehydrochlorination	57558-85-7	80
t-BuCOC(Bu-t)=C=O	$C_{11}H_{18}O_2$	Wolff	67832-67-1	8,9

Ph (cyclopentanone structure with C=O)	$C_{12}H_{10}O_2$	Wolff	117135-79-2	81
PhC(S)C(CO$_2$Et)=C=O	$C_{12}H_{10}O_3S$	Desulfurization	61628-69-1	78
CH$_3$COCR=C=O	$C_{12}H_{11}NO_2$	Pyrolysis	90368-10-8	82
R = CMe=NPh				
EtO$_2$CCR=C=O				
R = (MeO)$_3$PC(CO$_2$Et)	$C_{12}H_{19}PO_8$	Rearrangement		83
RO$_2$CCPh=C=O				
R = (CH$_2$)$_2$NCH$_2$CH$_2$	$C_{13}H_{13}NO_3$	Acylation	26129-24-8	84
R = Me$_2$NCH$_2$CH$_2$	$C_{13}H_{15}NO_3$	Acylation	26132-98-9	84
R = (CH$_2$)$_3$NCH$_2$CH$_2$	$C_{14}H_{15}NO_3$	Acylation	25994-40-5	84
R = Me$_2$NCH$_2$CHMe	$C_{14}H_{17}NO_3$	Acylation	25998-21-4	84
R = Me$_2$N(CH$_2$)$_3$	$C_{14}H_{17}NO_3$	Acylation	25994-28-9	84
R = (CH$_2$)$_4$NCH$_2$CH$_2$	$C_{15}H_{17}NO_3$	Acylation	26129-23-7	84
R = Et$_2$NCH$_2$CH$_2$	$C_{15}H_{19}NO_3$	Acylation	25994-01-8	84
R = O(CH$_2$CH$_2$)$_2$NCH$_2$CH$_2$	$C_{15}H_{17}NO_4$	Acylation	25994-31-4	84
R = (CH$_2$)$_4$N(CH$_2$)$_3$	$C_{16}H_{19}NO_3$	Acylation	25994-34-7	84
R = (CH$_2$)$_4$NCHMeCH$_2$	$C_{16}H_{19}NO_3$	Acylation	25994-39-2	84
R = (CH$_2$)$_4$NCH$_2$CHMe	$C_{16}H_{19}NO_3$	Acylation	25994-37-0	84
R = (CH$_2$)$_5$NCH$_2$CH$_2$	$C_{16}H_{19}NO_3$	Acylation	25994-30-3	84
R = O(CH$_2$CH$_2$)$_2$N(CH$_2$)$_3$	$C_{16}H_{19}NO_4$	Acylation	25994-36-9	84
R = O(CH$_2$CH$_2$)$_2$NCH$_2$CHMe	$C_{16}H_{19}NO_4$	Acylation	25998-22-5	84
R = O(CH$_2$CH$_2$)$_2$NCHMeCH$_2$	$C_{16}H_{19}NO_4$	Acylation	25998-19-0	84
R = (CH$_2$)$_5$NCHMeCH$_2$	$C_{17}H_{21}NO_3$	Acylation	25994-41-6	84
R = (CH$_2$)$_5$NCH$_2$CHMe	$C_{17}H_{21}NO_3$	Acylation	25994-38-1	84
R = (CH$_2$)$_5$N(CH$_2$)$_3$	$C_{17}H_{21}NO_3$	Acylation	25994-35-8	84
R = CH$_3$CH(CH$_2$)$_4$NCH$_2$CH$_2$	$C_{17}H_{21}NO_3$	Acylation	25994-33-6	84

TABLE 1. (Continued)

Ketene	Formula	Preparation	Registry No.	Ref.
R = $(CH_2CHMe)_2NCH_2CH_2$	$C_{17}H_{21}NO_3$	Acylation	25994-32-5	84
R = $n\text{-}Pr_2NCH_2CH_2$	$C_{17}H_{23}NO_3$	Acylation	25994-25-6	84
R = $i\text{-}Pr_2NCH_2CH_2$	$C_{17}H_{23}NO_3$	Acylation	25994-26-7	84
R = $CH_2(CH_2CHMe)_2NCH_2CH_2$	$C_{18}H_{23}NO_3$	Acylation	25994-42-7	84
R = $n\text{-}Pr_2NCH_2CHMe$	$C_{18}H_{25}NO_3$	Acylation	25994-27-0	84
R = $n\text{-}Pr_2NCHMeCH_2$	$C_{18}H_{25}NO_3$	Acylation	25994-42-7	84
R = $n\text{-}Bu_2NCH_2CH_2$	$C_{19}H_{27}NO_3$	Acylation	25994-27-8	84
$PhCOC(Bu\text{-}t)=C=O$	$C_{13}H_{14}O_2$	Ynolate		85
$t\text{-}BuCOCPh=C=O$	$C_{13}H_{14}O_2$	Ynolate		86
$Et_2NCOCPh=C=O$	$C_{13}H_{15}NO_2$	Dehydrochlorination	36277-45-9	71
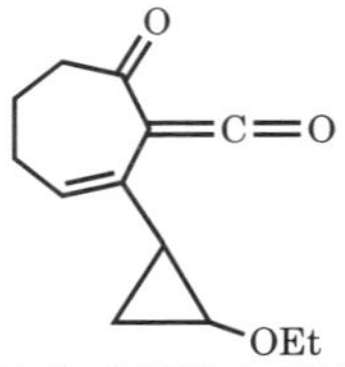 ![structure] cycloheptenone ketene with cyclopropyl-OEt	$C_{13}H_{16}O_3$	Thermolysis	98065-96-4	87
$EtO_2CC(CMe_2Ph)=C=O$	$C_{14}H_{16}O_3$	Dehydrochlorination	64287-40-7	15, 88
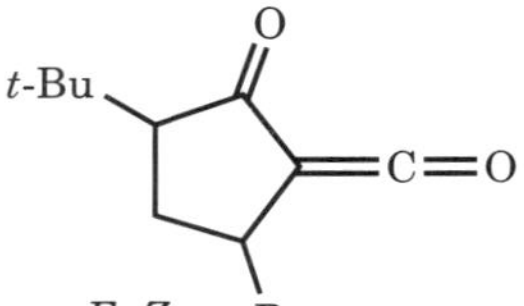 ![structure] $t\text{-Bu}$ ... $Bu\text{-}t$ cyclopentanone ketene, E, Z	$C_{14}H_{22}O_2$	Wolff		14

			Index entry	
n-C$_{11}$H$_{23}$COCH=C=O	C$_{14}$H$_{24}$O$_2$	5981-32-8		None
ArO$_2$CCPh=C=O				
Ar = 2,4,6-Cl$_3$C$_6$H$_2$	C$_{15}$H$_7$Cl$_3$O$_3$	17423-40-4	Dehydrochlorination	64
Ar = 2,4-Cl$_2$C$_6$H$_3$	C$_{15}$H$_8$Cl$_2$O$_3$	17118-69-3	Dehydrochlorination	64
ArCOCAr=C=O	C$_{15}$H$_8$Cl$_2$O$_2$	85526-48-3	Wolff	89
Ar = ClC$_6$H$_4$				
4-O$_2$NC$_6$H$_4$O$_2$CCPh=C=O	C$_{15}$H$_9$NO$_5$	69166-64-9	Dehydrochlorination	90
PhCOCPh=C=O	C$_{15}$H$_{10}$O$_2$	75508-81-5	Wolff	89
PhO$_2$CCPh=C=O	C$_{15}$H$_{10}$O$_3$	42309-43-3	Dehydrochlorination	91
R(CH$_2$)$_2$COCH=C=O	C$_{15}$H$_{19}$O$_3$Si		Dioxinone	30
R = MeCHO(TBDMS)				
RC(CO$_2$Et)=C=O	C$_{15}$H$_{20}$O$_4$		Wolff	15a
R = 1-adamantyl				
PhCOC(CSPh)=C=O	C$_{16}$H$_{10}$O$_2$S	85601-41-8	Pyrolysis	58
(PhCO)$_2$C=C=O	C$_{16}$H$_{10}$O$_3$	85601-40-7	Pyrolysis	58
PhCOC(CPh=NH)=C=O	C$_{16}$H$_{11}$NO$_2$	90368-12-0	Pyrolysis	82
PhNHCOCBn=C=O	C$_{16}$H$_{13}$NO$_2$	63766-82-5	Cleavage	92
RCH$_2$COCR=C=O	C$_{16}$H$_{28}$O$_2$	93158-00-0^a	Erroneous	93
R = n-C$_6$H$_{13}$				
PhCOC(CPh=NMe)=C=O	C$_{17}$H$_{13}$NO$_2$	90368-11-9	Pyrolysis	82
PhCOC(CMe=NPh)=C=O	C$_{17}$H$_{13}$NO$_2$	90368-09-5	Pyrolysis	82
ArCOCAr=C=O	C$_{17}$H$_{14}$O$_4$	85526-47-2	Wolff	89
Ar = 4-MeOC$_6$H$_4$				
[structure: 2-phenoxy-5-phenyl cyclopentanone with =C=O substituent]	C$_{18}$H$_{14}$O$_2$	117083-12-2	Wolff	81

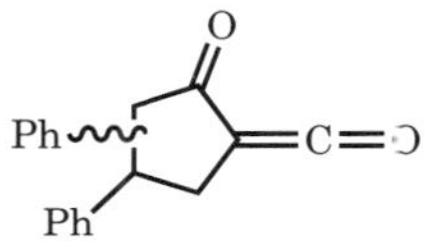

TABLE 1. (Continued)

Ketene	Formula	Preparation	Registry No.	Ref.
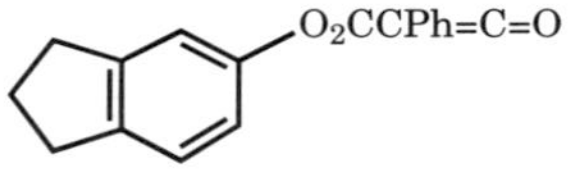 (indane-fused aryl ester) $O_2CCPh=C=O$	$C_{18}H_{14}O_3$	Dehydrochlorination	58137-69-2	94
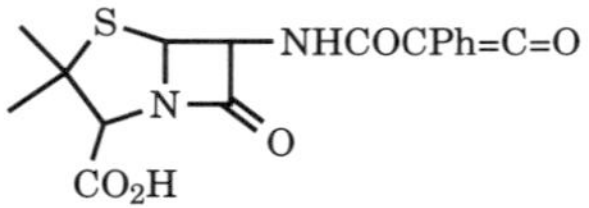 (β-lactam) NHCOCPh=C=O, CO_2H	$C_{17}H_{16}N_2O_5S \cdot 2Na$	Acylation	27261-69-4	95
ArCOCAr=C=O Ar = 2,4,6-Me$_3$C$_6$H$_2$	$C_{21}H_{22}O_2$	Wolff	95694-43-2	96
PhCOC(CPh=NPh)=C=O	$C_{22}H_{15}NO_2$	Pyrolysis	90368-08-4	82
RCH$_2$COCR=C=O R = n-C$_{16}$H$_{33}$	$C_{36}H_{68}O_2$	Erroneous[a]	66472-91-1	97

[a] The reported compound is actually the isomeric dimer of an alkylketene.

References

1. Chick, F.; Wilsmore, N. T. M. *J. Chem. Soc.* **1908**, *93*, 946–950.
2. Staudinger, H.; Bereza, S. *Chem. Ber.* **1909**, *42*, 4908–4918.
3. Arnold, Z.; Sauliova, J. *Coll. Czech. Chem. Commun.* **1973**, *38*, 2641–2647.
4. Maier, G.; Reisenauer, H. P.; Sayrac, T. *Chem. Ber.* **1982**, *115*, 2192–2201.
5. Wenkert, E.; Ananthanarayan, T. P.; Ferreira, V. F.; Hoffmann, M. G.; Kim, H. *J. Org. Chem.* **1990**, *55*, 4975–4976.
5a. Wentrup, C.; Heilmayer, W.; Kollenz, G. *Synthesis* **1994**, 1219–1248.
6. Nguyen, M. T.; Ha, T.; More O'Ferrall, R. A. *J. Org. Chem.* **1990**, *55*, 3251–3256.
7. Gong, L.; McAllister, M. A.; Tidwell, T. T. *J. Am. Chem. Soc.* **1991**, *113*, 6021–6028.
7a. Janoschek, R.; Fabian, W. M. F.; Kollenz, G.; Kappe, C. O. *J. Comput. Chem.* **1994**, *15*, 132–143.
7b. Birney, D. M. *J. Org. Chem.* **1994**, *59*, 2557–2564.
7c. Wong, M. W.; Wentrup, C. *J. Org. Chem.* **1994**, *59*, 5279–5285.
8. Nikolaev, V. A.; Frenkh, Yu.; Korobitsyna, I. K. *Zh. Org. Khim.* **1978**, *14*, 1433–1441, 1147–1160; *Engl. Transl.* **1978**, *14*, 1338–1346, 1069–1080.
9. Freiermuth, B.; Wentrup, C. *J. Org. Chem.* **1991**, *56*, 2286–2289.
10. Wolff, S.; Agosta, W. C. *Can. J. Chem.* **1984**, *62*, 2429–2434.
11. Allen, A. D., University of Toronto, unpublished results.
12. Leung-Tong, R.; Wentrup, C. *Tetrahedron* **1992**, *48*, 7641–7654.
13. Leung-Toung, R.; Wentrup, C. *J. Org. Chem.* **1992**, *57*, 4850–4858.
14. Nikolaev, V. A.; Korneev, S. M.; Terent'eva, I. V.; Korobitsyna, I. K. *Zh. Org. Khim.* **1991**, *27*, 2085–2100; *Engl. Transl.* **1991**, *27*, 1845–1858.
15. Newman, M. S.; Zuech, E. A. *J. Org. Chem.* **1962**, *27*, 1436–1438.
15a. Ohno, M.; Itoh, M.; Ohashi, T.; Eguchi, S. *Synthesis*, **1993**, 793–796.
16. Kappe, C. O.; Evans, R. A.; Kennard, C. H. L.; Wentrup, C. *J. Am. Chem. Soc.* **1991**, *113*, 4234–4237.
17. Kappe, C. O.; Färber, G.; Wentrup, C.; Kollenz, G. *Tetrahedron Lett.* **1992**, *33*, 4553–4556.
18. Kappe, C. O.; Färber, G.; Wentrup, C.; Kollenz, G. *J. Org. Chem.* **1992**, *57*, 7078–7083.
19. Kappe, C. O.; Wentrup, C.; Kollenz, G. *Monatsh. Chem.* **1993**, *124*, 1133–1141.
20. Campbell, D. S.; Lawrie, C. W. *Chem. Commun.* **1971**, 355–356.
21. Witzeman, J. S. *Tetrahedron Lett.* **1990**, *31*, 1401–1404.
22. Witzeman, J. S.; Nottingham, W. D. *J. Org. Chem.* **1991**, *56*, 1713–1718.
23. Emerson, D. W.; Titus, R. L.; González, R. M. *J. Org. Chem.* **1990**, *55*, 3572–3576.
24. Emerson, D. W.; Titus, R. L.; Gonzalez, R. M. *J. Org. Chem.* **1991**, *56*, 5301–5307.
25. Clemens, R. J.; Witzeman, J. S. *J. Am. Chem. Soc.* **1989**, *111*, 2186–2193.
26. Boeckman, R. K., Jr.; Pruitt, J. R. *J. Am. Chem. Soc.* **1989**, *111*, 8286–8288.
27. Coleman, R. S.; Grant, E. B. *Tetrahedron Lett.* **1990**, *31*, 3677–3680.
28. Coleman, R. S.; Fraser, J. R. *J. Org. Chem.* **1993**, *58*, 385–392.
29. Sato, M.; Sakaki, J.; Sugita, Y.; Yasuda, S.; Sakoda, H.; Kaneko, C. *Tetrahedron* **1991**, *47*, 5689–5708.
30. Sakaki, J.; Sugita, Y.; Sato, M.; Kaneko, C. *Tetrahedron* **1991**, *47*, 6197–6214.

31. Petasis, N. A.; Patane, M. A. *J. Chem. Soc., Chem. Commun.* **1990**, 836–837.
32. Nekrasov, D. D.; Andreichikov, Yu. S.; Rakitin, O. A. *Zh. Org. Khim.* **1992**, *28*, 1319–1320; *Chem. Abstr.* **1993**, *118*, 124672v.
33. Murai, S.; Hasegawa, K.; Sonoda, N. *Angew. Chem., Int. Ed. Engl.* **1975**, *14*, 636–637.
34. Andreichikov, Yu. S.; Kollenz, G.; Kappe, C. O.; Leung-Toung, R.; Wentrup, R. *Acta Chem. Scand.* **1992**, *46*, 683–685.
35. Kolesnikova, O. N.; Livantsova, L. I.; Shurov, S. N.; Zaitseva, G. S.; Andreichikov, Yu. S. *Zh. Obshch. Khim.* **1990**, *60*, 467–468; *Engl. Transl.* **1990**, *60*, 406–407.
36. Nakanishi, S.; Butler, K. *Org. Prep. Proc. Int.* **1975**, *7*, 155–158.
37. Potts, K. T.; Kanemasa, S.; Zvilichovsky, G. *J. Am. Chem. Soc.* **1980**, *102*, 3971–3972.
38. Potts, K. T.; Murphy, M.; Kuehnling, W. R. *J. Org. Chem.* **1988**, *53*, 2889–2898.
39. Schönwälder, K.-H.; Kollat, P.; Strezowski, J. J.; Effenberger, F. *Chem. Ber.* **1984**, *117*, 3270–3279.
40. Reid, W.; Nenninger, H. *Synthesis,* **1990**, 167–170.
41. Goldstein, S.; Vannes, P.; Houge, C.; Frisque-Hesbain, A. M.; Wiaux-Zamar, C.; Ghosez, L.; Germain, G.; Dedorca, J. P.; Van Meerssche, M.; Arrieta, J. M. *J. Am. Chem. Soc.* **1981**, *103*, 4616–4618.
42. Stevens, R. V.; Bisacchi, G. S.; Goldsmith, L. Strouse, C. E. *J. Org. Chem.* **1980**, *45*, 2708–2709.
43. Hurd, C. D.; Jones, R. N.; Blunck, F. H. *J. Am. Chem. Soc.* **1935**, *57*, 2033–2036.
44. Willson, C. G.; Miller, R. D.; McKean, D. R.; Pederson, L. A.; Regitz, M. *Proc. SPIE-Int. Opt. Eng.* **1987**, *777*, 2–10; *Chem. Abstr.* **1987**, *107*, 225797t.
45. Meier, H.; Wengenroth, H.; Lauer, W.; Vogt, W. *Chem. Ber.* **1988**, *121*, 1643–1646.
46. Meier, H.; Wengenroth, H.; Lauer, W.; Krause, V. *Tetrahedron Lett.* **1989**, *30*, 5253–5256.
47. Jullien, J.; Pechine, J. M.; Perez, F. *Tetrahedron Lett.* **1983**, *24*, 5525–5526.
48. Allen, A. D.; McAllister, M. A.; Tidwell, T. T. *Tetrahedron Lett.* **1993**, *34*, 1095–1098.
48a. Birney, D. M.; Wagenseller, P. E. *J. Am. Chem. Soc.* **1994**, *116*, 6262–6270.
49. Robinson, M. S.; Davico, G. E.; Bierbaum, V. M.; DePuy, C. H. *Int. J. Mass. Spec. Ion Proc.* **1994**, *137*, 107–119.
50. Biesiada, D. A.; Shevlin, P. B. *J. Org. Chem.* **1984**, *49*, 1151–1153.
51. Schulz, R.; Schweig, A. *Tetrahedron Lett.* **1979**, 59–62.
52. Cevasco, G.; Guanti, G.; Hopkins, A. R.; Thea, S.; Williams, A. *J. Org. Chem.* **1985**, *50*, 479–484.
53. Staudinger, H.; Hirzel, H. *Chem. Ber.* **1916**, *49*, 2522–2529.
54. Staudinger, H.; Hirzel, H. *Chem. Ber.* **1917**, *50*, 1024–1035.
55. Himbert, G.; Henn, L. *Liebigs Ann. Chem.* **1987**, 381–383, 771–776.
56. Gompper, R.; Wolf, U. *Liebigs Ann. Chem.* **1979**, 1388–1405.
57. Zeigler, E.; Kollenz, G.; Ott, W. *Synthesis* **1973**, 679–680.
58. Wentrup, C.; Winter, H. W.; Gross, G.; Netsch, K. P.; Kollenz, G.; Ott, W.; Biedermann, A. G. *Angew. Chem., Int. Ed. Engl.* **1984**, *23*, 800–802.
59. Yamamoto, Y.; Watanabe, Y.; Ohnishi, S. *Chem. Pharm. Bull.* **1987**, *35*, 1860–1870.
60. Selvarajan, R.; Narasimhan, K.; Swaminathan, S. *Tetrahedron Lett.* **1967**, 2089–2092.

61. Krespan, C. G.; Dixon, D. A. *J. Org. Chem.* **1986,** *51,* 4460–4466.
62. German, L. S.; Sterlin, S. R.; Cherstkov, V. F. *Izv. Akad. Nauk. SSSR, Ser. Khim.* **1982,** 1657–1659; *Engl. Transl.* **1982,** 1476–1477.
63. Chapman, O. L.; Miller, M. D.; Pitzenberger, S. M. *J. Am. Chem. Soc.* **1987,** *109,* 6867–6868.
64. Ziegler, E.; Sterk, H. *Monatsh. Chem.* **1967,** *98,* 1104–1107.
65. Pericas, M. A.; Serratosa, F.; Valenti, E.; Font-alba, M.; Jolans, X. *J. Chem. Soc., Perkin Trans. 2* **1986,** 961–967.
66. Krespan, C. G. *J. Fluorine Chem.* **1976,** *8,* 105–114.
67. Wentrup, C.; Lorencak, P. *J. Am. Chem. Soc.* **1988,** *110,* 1880–1883.
68. Torres, M.; Strausz, O. P. *Nouv. J. Chim.* **1980,** *4,* 703–705.
69. Gallucci, R. R.; Jones, M., Jr. *J. Org. Chem.* **1985,** *50,* 4404–4405.
70. F. Hoffmann-LaRoche & Co. A. G. *Brit. Patent* 948,752; 948,755; *Chem. Abstr.* **1964,** *60,* 13147b, 15738a.
71. Ficini, J.; Pouliquen, J. *Tetrahedron Lett.* **1972,** 1135–1138.
72. Leyendecker, F. *Tetrahedron* **1976,** *32,* 349–353.
72a. Pannell, K. H.; Mayr, A. J.; Carrasco-Flores, B. *J. Organomet. Chem.* **1988,** *354,* 97–101.
73. Ganis, P.; Paiaro, G.; Pandolfo, L.; Valle, G. *Organometallics,* **1988,** *7,* 210–214.
74. Murata, S.; Sugiyama, K.; Tomioka, H. *J. Org. Chem.* **1993,** *58,* 1976–1978.
75. Andreichikov, Yu. S.; Ionov, Yu. V. *Zh. Org. Khim.* **1982,** *18,* 2430–2435; *Engl. Trans.* **1981,** *18,* 2154–2158.
76. Andreichikov, Yu.; Nekrasov, D. D. *Zh. Org. Khim.* **1984,** *20,* 217–218; *Engl. Trans.* **1984,** *20,* 197–198.
77. Wentrup, C.; Netsch, K. P. *Angew. Chem., Int. Ed. Engl.* **1984,** *23,* 802.
78. Goerdeler, J.; Koehler, K. H. *Tetrahedron Lett.* **1976,** 2961–2962.
79. Minami, T.; Yamauchi, Y.; Ohshiro, Y.; Agawa, T.; Murai, S.; Sonoda, N. *J. Chem. Soc., Perkin Trans. 1,* **1977,** 904–908.
80. Staudinger, H.; Schotz, S. *Chem. Ber.* **1920,** *53B,* 1105–1124; *Chem. Abstr.* **1920,** *14,* 3424.
81. Ulbricht, M.; Thurner, J. U.; Siegmund, M.; Tomaschewski, G. *Z. Chem.* **1988,** *28,* 102–103; *Chem. Abstr.* **1988,** *109,* 190282h.
82. Briehl, H.; Lukosch, A.; Wentrup, C. *J. Org. Chem.* **1984,** *49,* 2772–2779.
83. Caesar, J. C.; Griffiths, D. V.; Griffiths, P. A.; Tebby, J. C. *J. Chem. Soc, Perkin Trans. 1* **1989,** 2425–2430.
84. Butler, K. *S. African Patent* 6900060; *Chem. Abstr.* **1970,** *72,* 111465m.
85. Zhdankin, V. V.; Stang, P. J. *Tetrahedron Lett.* **1993,** *34,* 1461–1462.
86. Hoppe, I.; Schöllkopf, U. *Liebigs Ann. Chim.* **1979,** 219–226.
87. Keyaniyan, S.; Apel, M.; Richmond, J. P.; DeMeijere, A. *Angew. Chem., Int. Ed. Engl.* **1985,** *24,* 763–764.
88. Musierowicz, S.; Wroblewski, A. E. *Tetrahedron* **1980,** *36,* 1375–1380.
89. Capuano, L.; Mörsdorf, P.; Scheidt, H. *Chem. Ber.* **1983,** *116,* 741–750.
90. Barth, W. E. *U.S. Patent* 4,115,385; *Chem. Abstr.* **1978,** *90,* 152170b.

91. Butler, K. *U.S. Patent* 3,790,620; *Chem. Abstr.* **1974**, *80*, 95499h.
92. Stadlbauer, W.; Schmut, O.; Kappe, T. *Monatsh. Chem.* **1977**, *108*, 367–379.
93. Sauer, J. C. *U.S. Patent* 2,369,919; *Chem. Abstr.* **1945**, *39*, 4086.
94. Carroll, R. D.; Reed, L. L. *Tetrahedron Lett.* **1975**, 3435–3438.
95. Love, D. A.; Neal, D. P. J.; Pidgeon, E. *Ger. Offen.* DE 1952021; *Chem. Abstr.* **1970**, *73*, 25454w.
96. Meier, H.; Wengenroth, H.; Lauer, W.; Vogt, W. *Chem. Ber.* **1988**, *121*, 1643–1646.
97. Griggs, W. H. *Res. Discl.* **1978**, *167*, 26–27; *Chem. Abstr.* **1978**, *88*, 180209y.

4.1.7 Imidoylketenes

Imidoylketenes (iminoketenes) (**1**) have attracted considerable interest because the valence isomeric azetinones (**2**) are potential intermediates in the synthesis of β-lactam antibiotics. Ab initio calculations indicate that cyclization of **1** to **2** is endothermic by 12.0 kcal/mol at the HF/6-31G*//HF/6-31G* level.[1] Similar results are found by MNDO for **1/2** and various substituted derivatives.[2]

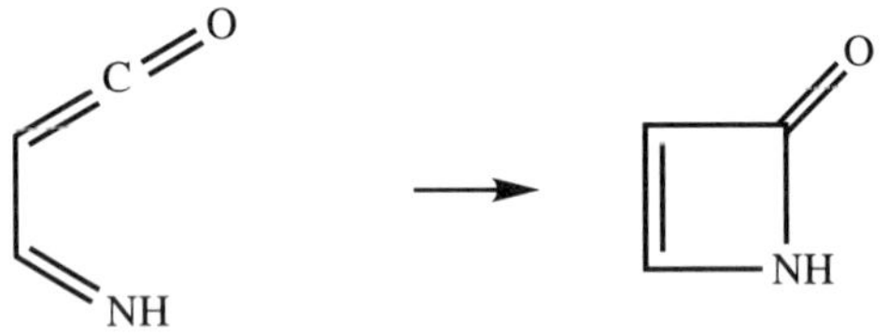

Photolysis of oxazinones **3** gave ketenes **4** which could be observed spectroscopically at temperatures below −160 °C.[3] Hydrogen transfer of **4** leading to isocyanates **5** has been suggested.[4] Pyrimidones give imino analogues of **4** on matrix photolysis.[4a]

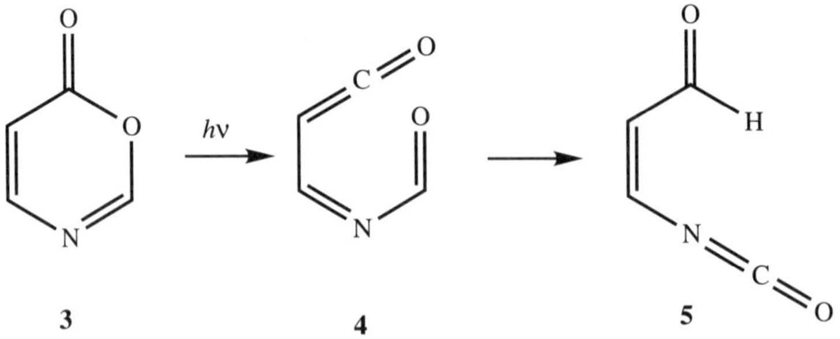

The formation of an iminoketene **6** leading to **7** was proposed to occur in the reaction of equation 3.[2] This type of reaction is also considered in Section 5.4.5 (equation 19).

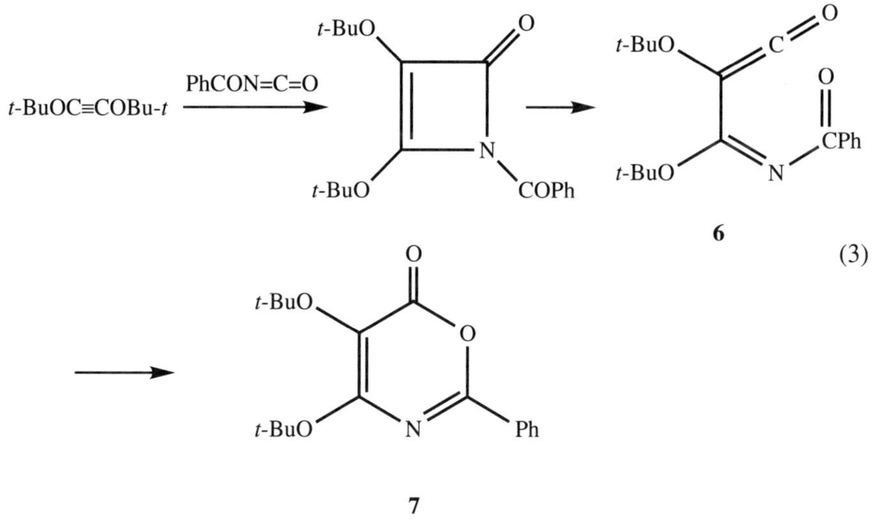

$$t\text{-BuOC}\equiv\text{COBu-}t \xrightarrow{\text{PhCON=C=O}} \text{[4-membered ring intermediate]} \longrightarrow \mathbf{6} \quad (3)$$

$$\longrightarrow \mathbf{7}$$

Imidoylketenes are generated in a variety of high-temperature pyrolytic methods (equations 4–7).[5–7] Reversible MeS migration is noted in **8** (equation 4).[5] The photolysis of 2-phenylisoxazol-5(2H)-one was proposed to lead to the ketene **9**, as evidenced by the capture by methanol as the ester **10** (equation 8).[8] The formation of **10** was proposed to occur by photolysis of the methanol addition product.

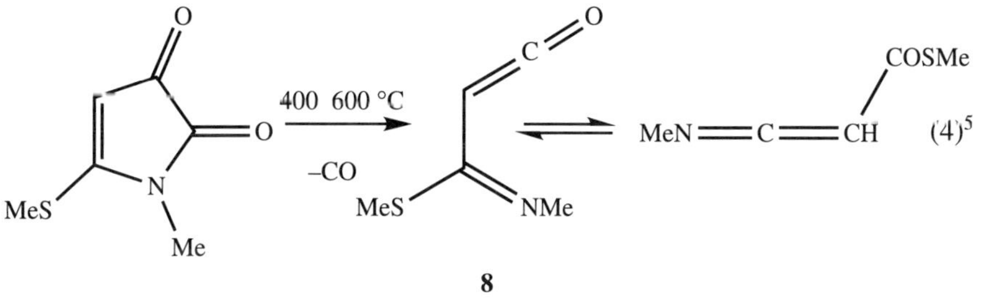

$$\xrightarrow{400-600\,°\text{C}}_{-\text{CO}} \mathbf{8} \rightleftharpoons \text{MeN}=\text{C}=\text{CH(COSMe)} \quad (4)^5$$

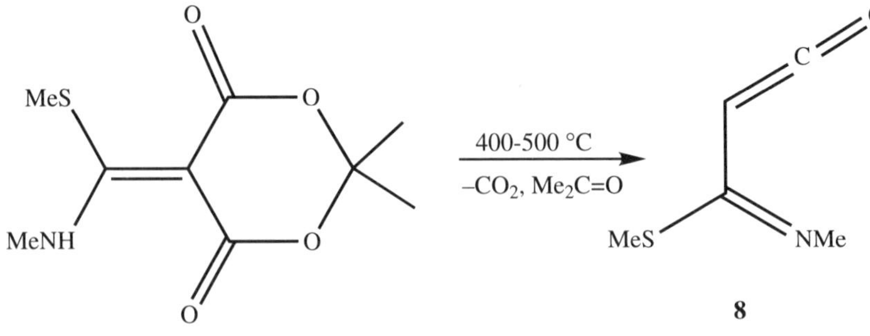

MeS
MeNH
O
O
O
O
400-500 °C
–CO2, Me2C=O
C
O
MeS
NMe

8

(5)[5]

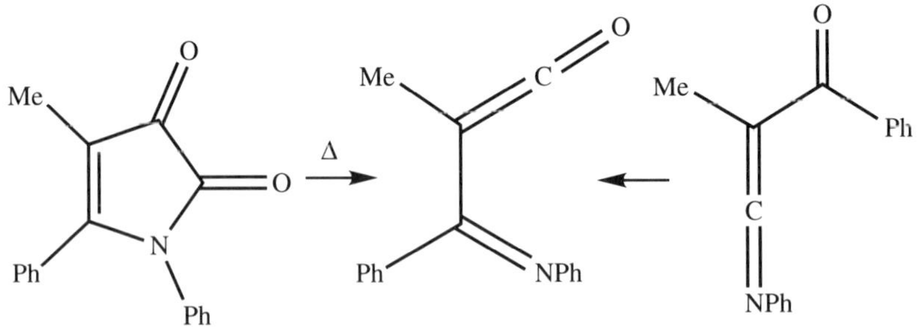

Me
Ph
O
O
N
Ph
Δ
Me
C
O
Ph
NPh
Me
O
Ph
C
NPh

(6)

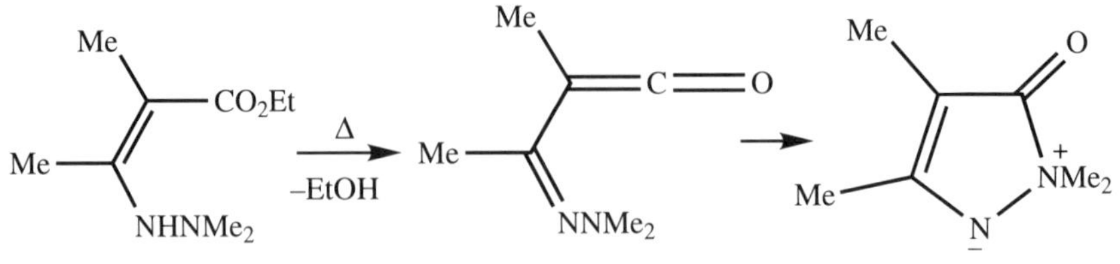

Me
CO2Et
Me
NHNMe2
Δ
–EtOH
Me
C
O
Me
NNMe2
Me
O
Me
NMe2
N

(7)[5]

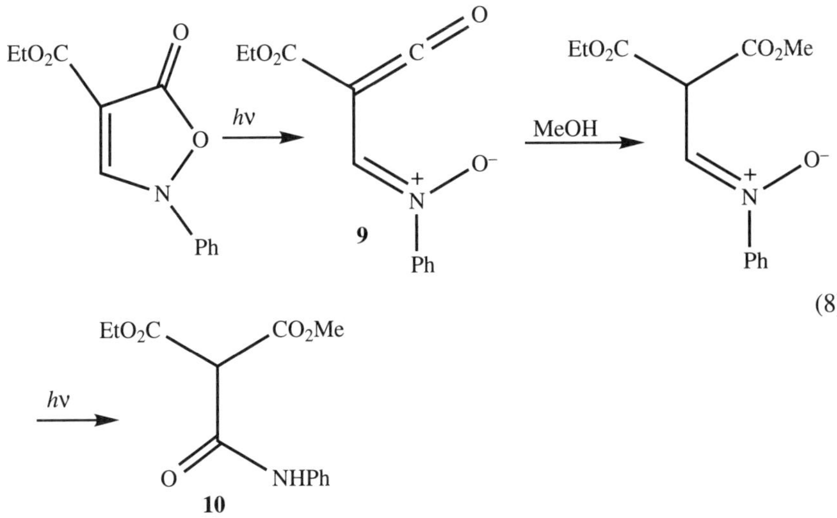

The reaction of diphenylcyclopropenone **11** with *N*-aminopyridinium iodide in MeOH was proposed to involve the ketene **12**, which was trapped as the ester (equation 9).[9] Other such reactions of **11** are known.[10]

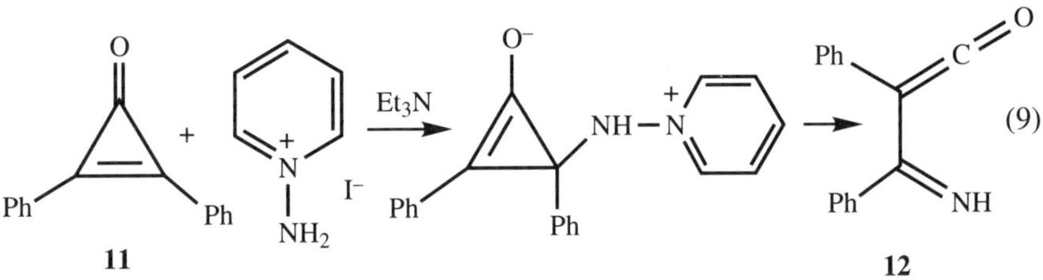

Several iminoketenes of the iminoquinone methide type **13** are mentioned in Section 4.1.10. 1,2,3-Triazoles have also been proposed to form imidoylketenes on pyrolysis.[10a] Reported imidoylketenes are listed in Table 1.

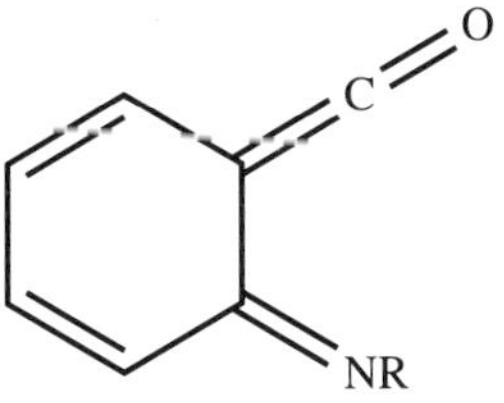

TABLE 1. Imidoylketenes

Ketene	Formula	Source	Registry No.	Ref.
HN=CHCH=C=O	C_3H_3NO	Theory	112532-08-8	1,12
HN=CHCH=C=O•H$^+$	$C_3H_3NO•H^+$	MS	141406-36-2	3,12
O=CHN=CHCH=C=O	$C_4H_3NO_2$	Theory	106104-55-6	2
		Pyrolysis	63801-28-5	3
CH_2=CHN=CHCH=C=O	$C_5H_5O_3$	Theory	82944-29-9	11
MeN=C(SMe)CH=C=O	C_5H_7NOS	Pyrolysis	141075-41-4	5
CF_3CON=CMeCH=C=O	$C_6H_4F_3NO_2$		63801-27-4	13
AcN=CMeCH=C=O	$C_6H_7NO_2$	Theory	74617-01-9	3
O=CHN=CMeCMe=C=O	$C_6H_7NO_2$	Theory	74617-00-8	3
AcN=CMeCMe=C=O	$C_7H_9NO_2$	Theory	74617-02-0	3
t-BuN=CHCH=C=O	$C_7H_{11}NO$	Pyrolysis		7
c-C_5H_9N=CHCD=C=O	$C_8H_{10}DNO$	Pyrolysis	90368-25-5	14
c-C_5H_9N=CHCH=C=O	$C_8H_{11}NO$	Pyrolysis	90368-14-2	14
t-BuCON=CMeCH=C=O	$C_9H_{13}NO_2$	Pyrolysis	63801-26-3	3
4-TolN=CHCH=C=O	$C_{10}H_9NO$	Pyrolysis	90368-05-1	14
PhN=C(SMe)CH=C=O	$C_{10}H_9NOS$	Pyrolysis	141075-43-6	5
PhNO=CHC(CO$_2$Et)=C=O	$C_{12}H_{11}NO_4$	Photolysis		8
PhN=CPhCBr=C=O	$C_{15}H_{10}BrNO$	Pyrolysis	141259-06-5	15
PhN=CPhCH=C=O	$C_{15}H_{11}NO$	Pyrolysis	141259-03-2	15
HN=CPhCPh=C=O	$C_{15}H_{11}NO$	Rearrangement		9
PhN=CPhC(CN)=C=O	$C_{16}H_{10}N_2O$	Pyrolysis	141259-07-6	15
PhN=CPhCMe=C=O	$C_{16}H_{13}NO$	Pyrolysis	141259-04-3	15
MeO$_2$CN=CPhCPh=C=O	$C_{17}H_{13}NO_3$	Rearrangement		10
PhN=CPhCEt=C=O	$C_{17}H_{15}NO$	Pyrolysis	141259-05-4	15
EtO$_2$CN=CPhCPh=C=O	$C_{18}H_{15}NO_3$	Rearrangement		10
PhN=CPhC(CO$_2$Et)=C=O	$C_{18}H_{15}NO_3$	Pyrolysis	141259-08-7	15
1-AdN=CPhCMe=C=O	$C_{20}H_{23}NO$	Pyrolysis	141258-98-2	16
PhCON=CPhCPh=C=O	$C_{22}H_{15}NO_2$	Rearrangement		10
Ph$_3$P=CPhCPh=C=O	$C_{33}H_{25}OP$	Addition		17

References

1. Nguyen, M. T.; Ha, T.; More O'Ferrall, R. A. *J. Org. Chem.* **1990,** *55,* 3251–3256.
2. Pericas, M. A.; Serratosa, F.; Valenti, E.; Font-Altaba, M.; Solans, X. *J. Chem. Soc., Perkin Trans. 2,* **1986,** 961–967.
3. Maier, G.; Schäfer, U. *Liebigs Ann. Chem.* **1980,** 798–813.
4. Huang, B. S.; Pong, R. G. S.; Laureni, J.; Krantz, A. *J. Am. Chem. Soc.* **1977,** *99,* 4154–4156.
4a. Lapinski, L.; Nowak, M. J., Les, A.; Adamowicz, L. *J. Am. Chem. Soc.* **1994,** *116,* 1461–1467.
5. Kappe, C. O.; Kollenz, G.; Leung-Toung, R.; Wentrup, C. *J. Chem. Soc., Chem. Commun.* **1992,** 487–488.
6. Coqueret, X.; Bourelle-Wargnier, F.; Chuche, J. *Tetrahedron* **1986,** *42,* 2263–2273.
7. Cheikh, A. B.; Chuche, J.; Manisse, N.; Pommelet, J. C.; Netsch, K.; Lorencak, P.; Wentrup, C. *J. Org. Chem.* **1991,** *56,* 970–975.
8. Ang, K. H.; Prager, R. H. *Tetrahedron* **1992,** *48,* 9073–9084.
9. Kascheres, A.; Marchi, D., Jr. *J. Org. Chem.* **1975,** *40,* 2985–2987.
10. Sasaki, T.; Kanematsu, K.; Kakehi, A. *J. Org. Chem.* **1971,** *36,* 2451–2453.
10a. Clarke, D.; Mares, R. W.; McNab, H. *J. Chem. Soc., Chem. Commun.* **1993,** 1026–1027.
11. Kuzuya, M.; Ito, S.; Miyake, F.; Okuda, T. *Chem. Pharm. Bull.* **1982,** *30,* 1980–1985.
12. Brown, J.; Flammang, R.; Govaert, Y.; Plisnier, M.; Wentrup, C.; van Haverbeke, Y. *Rapid Commun. Mass Spectrom.* **1992,** *6,* 249–253.
12a. Wentrup, C.; Briehl, H.; Lorencak, P.; Vogelbacher, U. J.; Winter, H.; Marquestiau, A.; Flammang, R. *J. Am. Chem. Soc.* **1988,** *110,* 1337–1343.
13. Maier, G.; Schaefer, U. *Tetrahedron Lett.* **1977,** 1053–1056.
14. Briehl, H.; Lukosch, A.; Wentrup, C. *J. Org. Chem.* **1984,** *49,* 2772–2779.
15. Kappe, C. O.; Kollenz, G.; Wentrup, C. *J. Chem. Soc., Chem. Commun.* **1992,** 485–486.
16. Kappe, C. O.; Kollenz, G.; Netsch, K. P.; Leung-Toung, R.; Wentrup, C. *J. Chem. Soc., Chem. Commun.* **1992,** 488–490.
17. Hamada, A.; Takizawa, T. *Tetrahedron Lett.* **1972,** 1849–1850.

4.1.8 Cumulene-Substituted Ketenes

These species include allenylketenes (**1**), keteniminylketenes (**2**), and diazomethylketenes (**3**). Bisketenes are considered in Section 4.9, and isocyanatoketenes are mentioned in Section 4.2.

$$CH_2=C=C\begin{smallmatrix}H\\ \\C=C=O\\H\end{smallmatrix} \qquad RN=C=C\begin{smallmatrix}H\\ \\C=C=O\\H\end{smallmatrix} \qquad ^-N=\overset{+}{N}=C\begin{smallmatrix}H\\ \\C=C=O\\H\end{smallmatrix}$$

      1   **2**   **3**

There are few reports of allenylketenes, but photolysis of the diazoketone **4** (equation 1) in an Ar matrix[1] gave ketene **1** and methylenebicyclobutanone (**5**). These were evidently directly observed at 8 K, but no further details have been published.

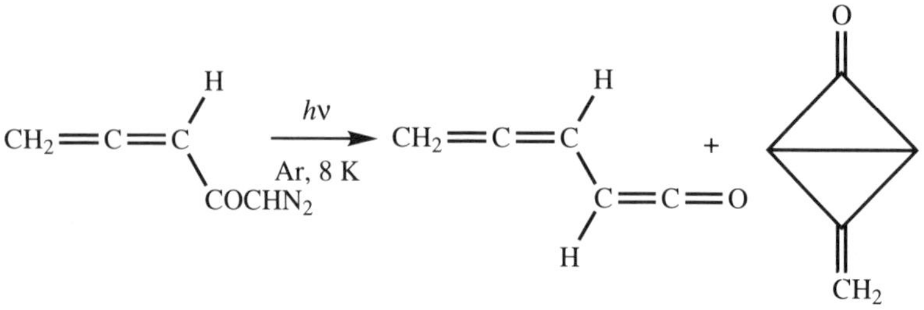

$$CH_2=C=C\begin{array}{c}H\\\\COCHN_2\end{array} \xrightarrow[\text{Ar, 8 K}]{h\nu} CH_2=C=C\begin{array}{c}H\\\\C=C=O\\H\end{array} + \mathbf{5} \quad (1)$$

4 **1** **5**

Pyrolysis of the furan **6** gave methylenecyclobutenone **7**,[2] and this transformation may have also involved the intermediacy of **1** (equation 2).

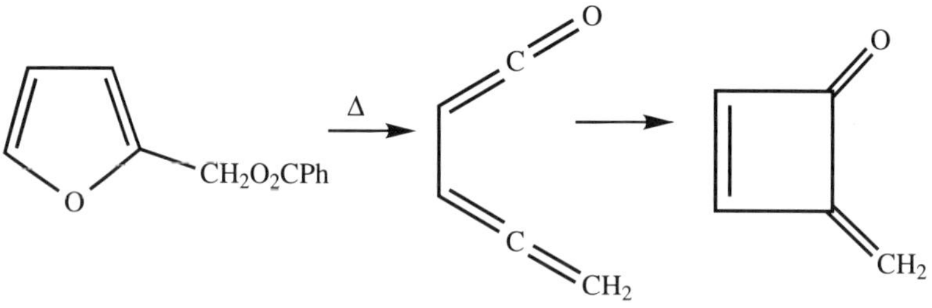

$$\mathbf{6} \xrightarrow{\Delta} \mathbf{1} \longrightarrow \mathbf{7} \quad (2)$$

Photolysis of the cyclopentadienone **8** leads to the allenylketene **9**, detected by its IR bands at 1890 and 2080 cm^{-1}, which on further photolysis forms **10** (equation 3).[3]

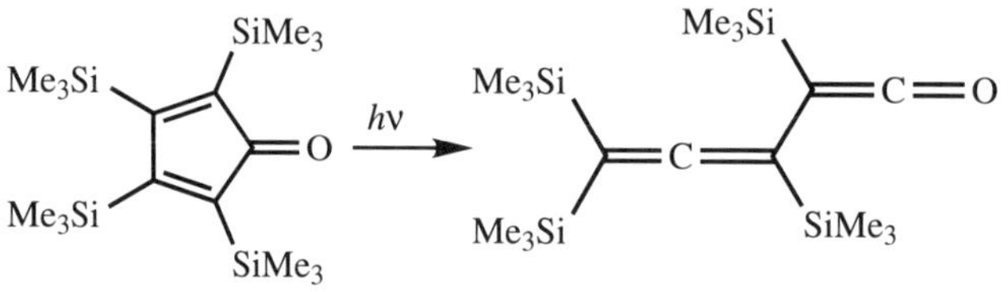

$$\mathbf{8} \xrightarrow{h\nu} \mathbf{9} \xrightarrow{h\nu} \mathbf{10} \quad (3)$$

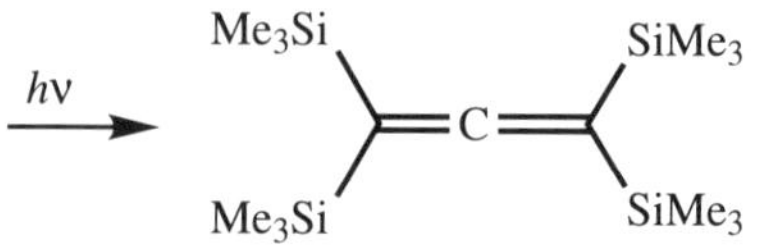

Diazomethylketenes have been observed at 8 K when formed by photolysis of 1,3-bisdiazoketones (equations 3 and 4).[4–7] Photolysis of **11** gave **12** (IR 2118, 2120 cm^{-1}), which gave **13** on further photolysis.[6] Photolysis of **14** gave **15,** identified by its IR band at 2106 cm^{-1} and UV λ_{max} at 507 nm.[7] Further irradiation of **15** gave photoproducts tentatively identified as **16** and **17** on the basis of the IR and UV spectra observed.[7]

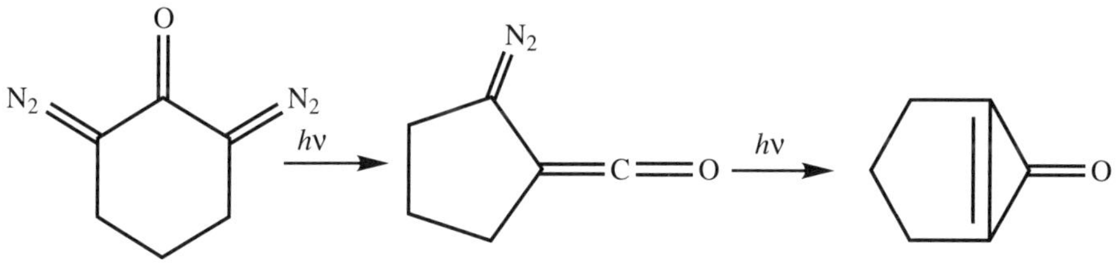

11 **12** **13**

(4)[6]

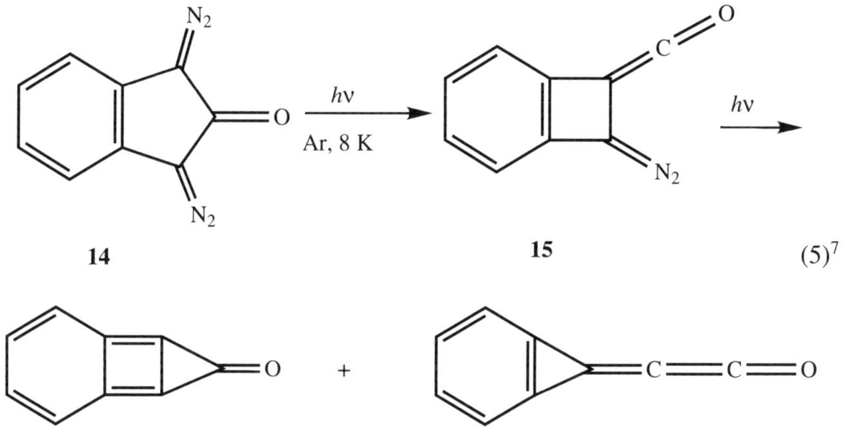

14 **15** (5)[7]

16 **17**

Photolysis of **18** in Ar at 15 K gave **19,**[8] presumably via a diazoketene.

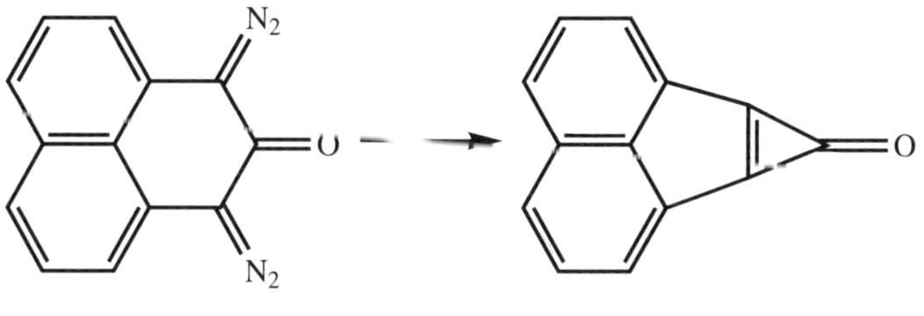

(6)

18 **19**

Reaction of the cyclobutenedione **20** with an arylisocyanide was proposed to result in formation of the keteniminylketene **21** which led to **22** (equation 7).[9,10]

(7)

References

1. Unpublished results of Chapman, O. L.; Chang, C.-C.; and Rosenquist, N., cited in Chapman, O. L. *Pure and Appl. Chem.* **1979**, *51*, 331–339.
2. Trahanovsky, W. S.; Park, M. *J. Am. Chem. Soc.* **1973**, *95*, 5412.
3. Maier, G.; Lage, H. W.; Reisenauer, H. P. *Angew. Chem., Int. Ed. Engl.* **1981**, *20*, 976–977.
4. Unpublished results of Chapman, O. L. and Gano, J. cited in Chapman, O. L. *Pure and Appl. Chem.* **1979**, *51*, 331–339.
5. Unpublished results of Chapman, O. L. and Roberts, D. C. cited in Chapman, O. L. *Pure and Appl. Chem.* **1979**, *51*, 331–339.
6. Chapman, O. L.; Gano, J.; West, P. R.; Regitz, M.; Maas, G. *J. Am. Chem. Soc.* **1981**, *103*, 7033–7036.
7. Murata, S.; Yamamoto, T.; Tomioka, H.; Lee, H.; Kim, H.; Yabe, A. *J. Chem. Soc., Chem. Commun.* **1990**, 1258–1260.
8. Trost, B. M.; Whitman, P. J. *J. Am. Chem. Soc.* **1974**, *96*, 7421–7429.
9. Obata, N.; Takizawa, T. *J. Chem. Soc. D* **1971**, 587–588.
10. Obata, N.; Takizawa, T. *Tetrahedron Lett.* **1969**, 3403–3406.

4.1.9 Ketenes with Charged, Radical, or Carbenic Sidechains

Ketenes with carbocation, radical, or carbanion centers at the γ-position are allylic intermediates and thus should have some conjugative stabilization, although there has not been a systematic theoretical assessment of the magnitude of this effect for the entire series. Because of its acylium ion character the carbocationic species should be the most favorable and indeed is the best characterized of this family. Other charged and radical ketenyl species are included in Section 4.10.

Reaction of a series of acrylyl fluorides **1** with SbF_5 in SO_2 gave the acylium ions **2**, which were directly observed by ^1H NMR and IR.[1] These have IR bands at 2210–2250 cm^{-1} for the C=O stretch,[1] which are reported[1] to be less than the values for alkyl and cycloalkyl acylium ions which range from 2205 to 2302 cm^{-1}.[2–4] These IR bands were interpreted[1] as supporting a contribution from the ketene structure **2a**. The ^1H NMR shifts of CH_3 at R^1 and R^2 were lower than for R^2, also supporting a contribution from **2a**.[1]

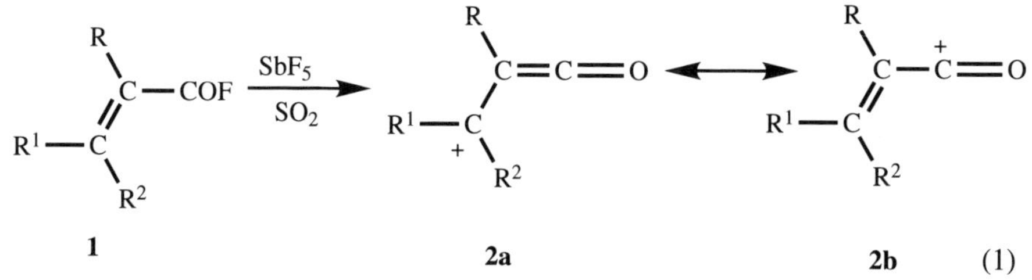

4.1.9 KETENES WITH CHARGED, RADICAL, OR CARBENIC SIDECHAINS

(1)

R = R^1 = R^2 = H
R = R^1 = Me, R^2 = H

R = Me, R^1 = R^2 = H
R = R^2 = H, R^1 = Me, Ph

R = H, R^1 = R^2 = Me

The acylium ion E-3 was generated from E-crotonic acid in H$_2$SO$_4$/SO$_3$, and its ^{13}C NMR spectrum showed very low field absorption for both C$_1$ and C$_3$, consistent with major positive charge development on these carbons.[5] Photolysis of E-3 produced a spectrum attributed to Z-3.[5] No isomerization of Z-3 to E-3 was observed over several hours at room temperature and so it was suggested that there was no ketene character 2a as opposed to acylium ion character 2b for these species.[5] However the chemical shifts shown for C-4 support some positive charge character, although the acylium ion contributor may be dominant.

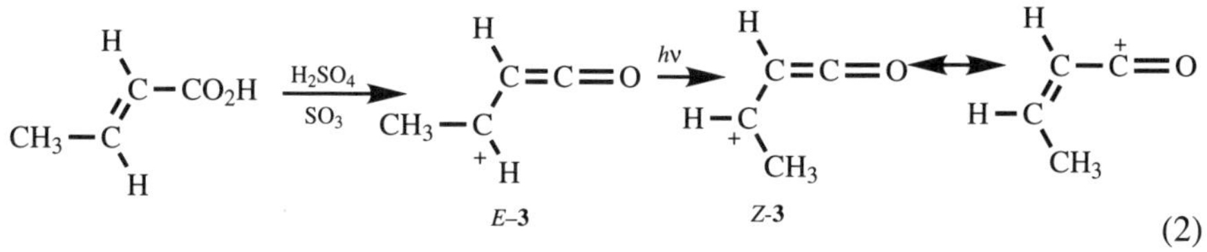

(2)

		E-3	^{13}C	^{1}H	Z-3	^{13}C	^{1}H
NMR		C^1	151.3			157.3	
		C^2	84.7	6.5		84.4	6.5
(δ)		C^3	203.8	9.0		202.9	9.0
		C^4	22.0	2.6		26.1	2.7

The reaction of either maleic or fumaric acids with H$_2$SO$_4$/SO$_3$ at room temperature led to formation of O-protonated maleic anhydride.[5] The isomerization of the fumaric to the maleic structure was proposed to involve an intermediate ketenyl structure **4**, but other mechanisms are possible, and the proposal[5] of oxygen participation in the isomerization appears unnecessary.

Z-**4** (3)

E-**4** (4)

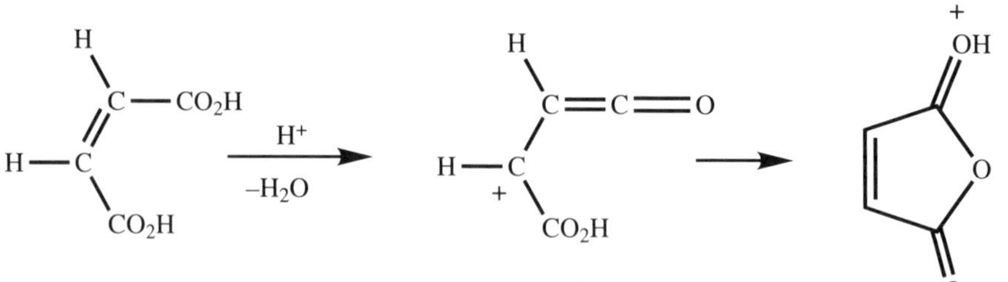

A theoretical study of the protonation of vinyl ketene **5** led to the prediction that addition to the δ-carbon leading to the γ-carbocationic system **3** is favored by 15.8 kcal/mol relative to protonation at the β-carbon to give **6.**[6] This was confirmed experimentally in the reaction of **7,** which gave the acid **8,** which is less stable than the nonconjugated product **9** which would result from β-protonation.[7]

$$CH_2{=}CHCH{=}C{=}O \xrightarrow{H^+} CH_3\overset{+}{C}H{-}CH{=}C{=}O \qquad CH_2{=}CHCH_2\overset{+}{C}{=}O$$

5 **3** **6**

7 **8** **9**

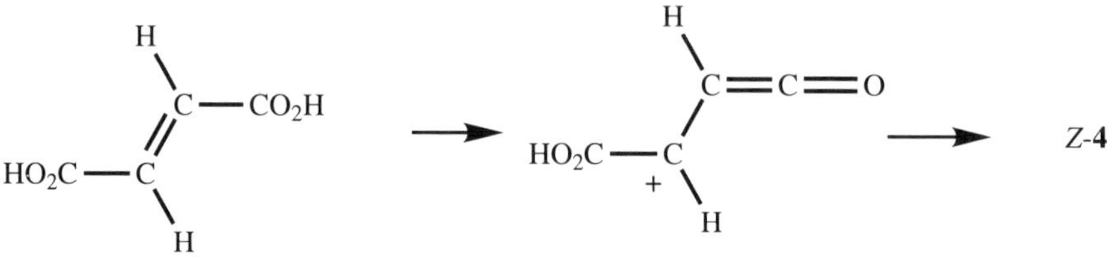

The reaction of methylmalonic acid or its dimethyl ester in $SbF_5{-}FSO_3H$ gave rise to an ^1H NMR signal at δ 3.05 and an IR band at 2170 cm^{-1}.[8] These signals were attributed[8] to formation of the ketenyl acylium **10** ion shown in equation 5, but further evidence for this structural assignment is desirable. Formation of a ketenylvinylacylium ion from an azulene-1,3-dicarboxylic acid or ester was observed by ^1H NMR (equation 6).[9]

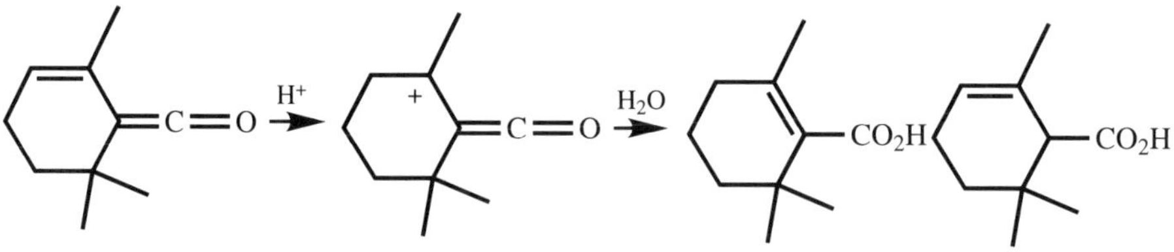

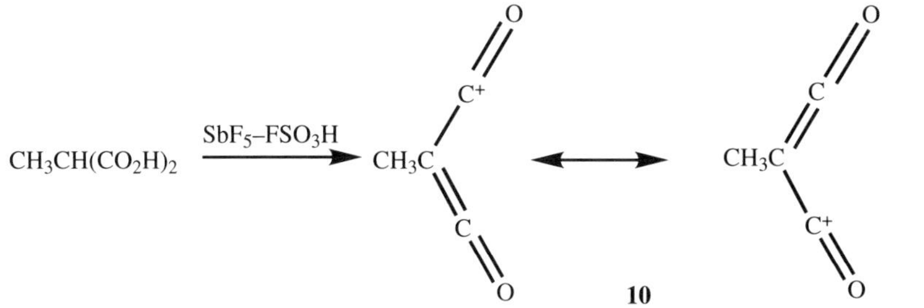

4.1.9 KETENES WITH CHARGED, RADICAL, OR CARBENIC SIDECHAINS

$$CH_3CH(CO_2H)_2 \xrightarrow{SbF_5-FSO_3H} \mathbf{10} \quad (5)$$

$$\xrightarrow{FSO_3H, SbF_5, SO_2} \quad (6)$$

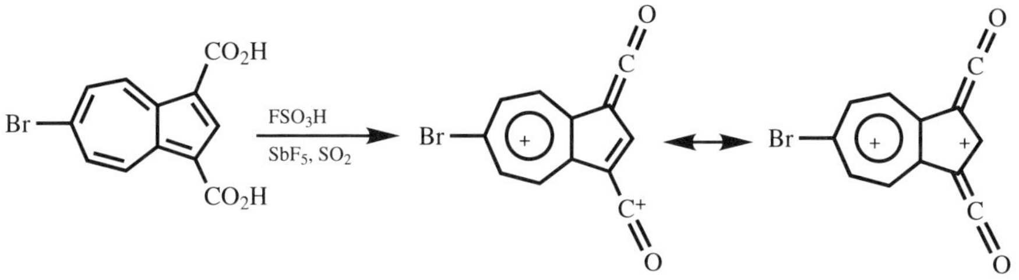

Vinylacyl radicals have been generated from a variety of precursors (equations 7–9).[10–14] Observations of the ESR spectra of the acyl radicals in equations 7 and 8 led to the assignment of these as σ radicals, with bent structures.[10,11]

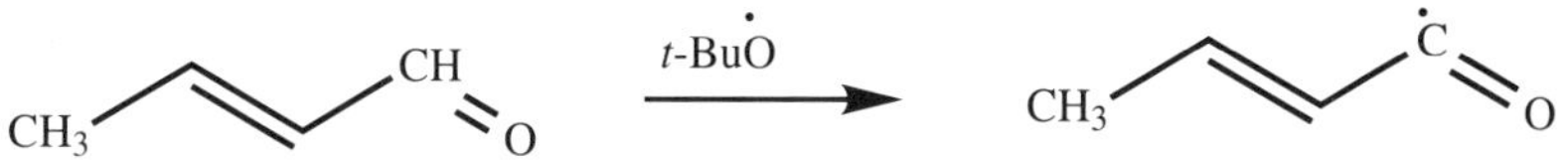

$$\xrightarrow{t\text{-BuO}\cdot} \quad (7)^{10}$$

$$MeOCH=CHCH=O \xrightarrow{t\text{-BuO}\cdot} MeOCH=CH\dot{C}=O \quad (8)^{11}$$

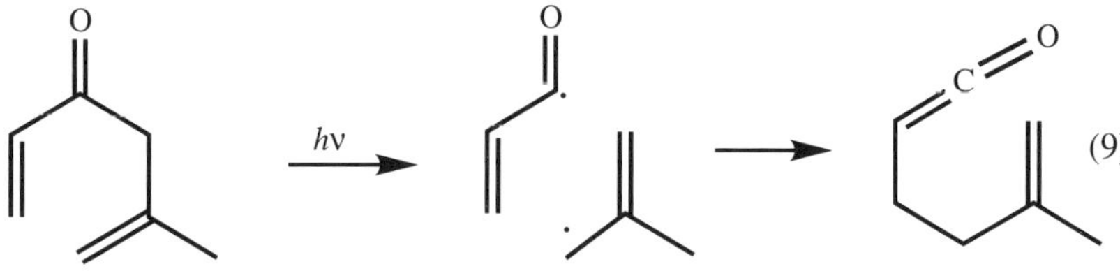

$$\xrightarrow{h\nu} \mathbf{11} \longrightarrow \quad (9)^{12-14}$$

In the case of equation 9 the radicals were not observed directly by ESR, although the cleavage/recombination process was clearly established by deuterium labeling studies (see Section 3.4.2).[12] The question of the precise electronic changes in the conversion of the initial σ radical to the product ketene has not been addressed.

In the photolysis of equation 10 the π-radical structure was claimed for **12**,[15] based on the expected conjugative ability of the phenyls.[15] Calculations on the unsubstituted acrylyl radical do not clearly predict whether the π-radical structure is favored, and better calculations are needed.[16,17] The formation of a small amount of methyl 3,3-diphenylacrylate **14** was ascribed to the hydrogen abstraction from **12**, forming the cumulenone **15**.[15] The IR spectrum of ketene **13** could be observed prior to addition of MeOH.

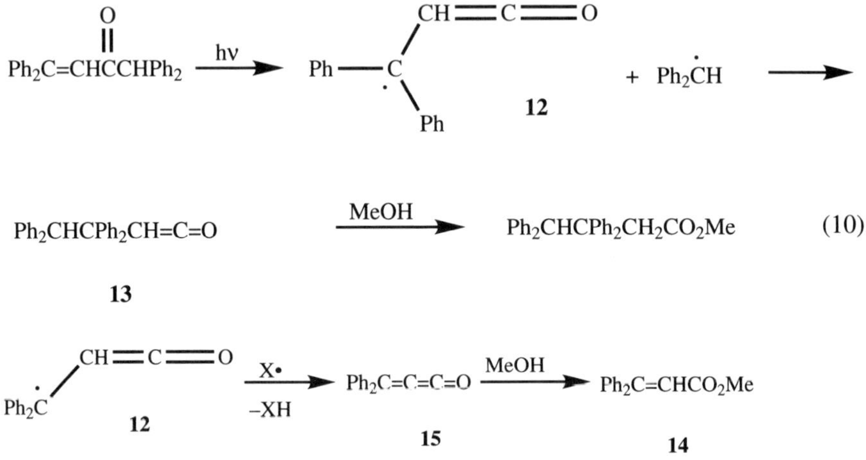

$$Ph_2C=CHCCHPh_2 \xrightarrow{h\nu} Ph\underset{Ph}{\overset{CH=C=O}{-C\cdot}} \quad \mathbf{12} \quad + \; Ph_2\dot{C}H \longrightarrow$$

$$Ph_2CHCPh_2CH=C=O \xrightarrow{MeOH} Ph_2CHCPh_2CH_2CO_2Me \quad (10)$$

13

$$Ph_2\dot{C}\overset{CH=C=O}{\diagup} \xrightarrow[-XH]{X\cdot} Ph_2C=C=C=O \xrightarrow{MeOH} Ph_2C=CHCO_2Me$$
12   **15**   **14**

The alkynylacyl radical **16a** has been suggested to have the bent structure **16b**, suggestive of the character of a cumulenone radical.[18]

$$HC\equiv C-\dot{C}=O \qquad \qquad \underset{H}{\overset{\cdot}{\diagup}}C=C=C=O$$
16a   **16b**

Formation of a ketene with a γ-carbanionic center has been proposed in the addition of PhLi to diphenylcyclopropenone **17**, which was suggested to give **18**, which opened to the ketenyl anion **19**.[19] Upon quenching of the solution with mesitoic acid a transient IR absorption attributed to the ketene **20** was observed, and upon hydrolysis α,β,β-triphenylpropionic acid **21** was isolated. It was pointed out that **20** could be formed directly from **18** and that **19** need not be an intermediate in these reactions.[18] A related example is shown in Section 3.4.6.

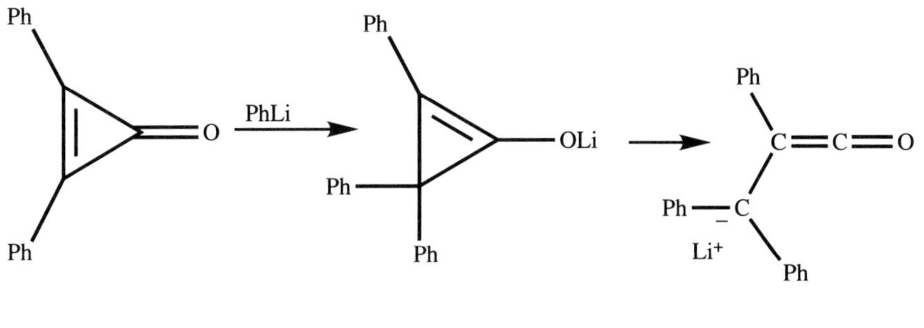

4.1.9 KETENES WITH CHARGED, RADICAL, OR CARBENIC SIDECHAINS

[Structures 17, 18, 19 with reaction scheme] (11)

[Structures 20, 21 with reaction scheme]

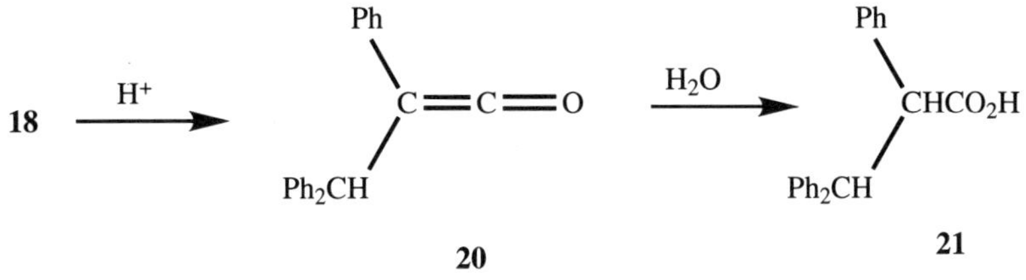

Other nucleophilic additions to **17**, especially by nitrogen ylides and imines, have also been proposed to occur by 1,4-addition and subsequent cleavage, but although these reactions can be envisaged to involve α-carbanionic ketenes, there is no good evidence for the intervention of such species.[20,21]

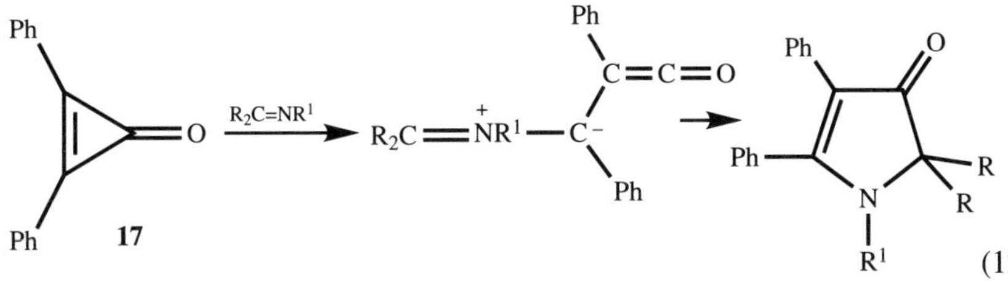

[Reaction scheme with structure 17] (12)

Reaction of the vinyllithium compound **22** with carbon monoxide proceeded by cyclization, and was suggested to possibly involve γ-lithioketene **23** (eq. 13).[22] Gas phase reaction of the cyclopropenyl anion **24** with CO_2 gave the conjugate base **25** of triafulvenone **26** (eq. 14).[23] Triafulvenone is calculated to be 40 kcal/mol more acidic than cyclopropene.

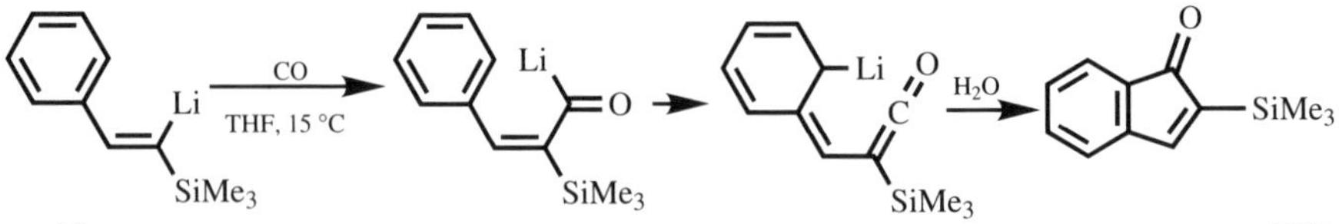

In several instances α-diazoketenes have been generated in matrices, and irradiated to give cyclopropenones, identified by their spectra in the matrix.[24,25] These may form via α-carbenylketenes, as in the conversion shown (equation 15).[24] The photolysis and pyrolysis of 1,2-bisketenes also may lead to α-carbenylketenes (Section 4.9).

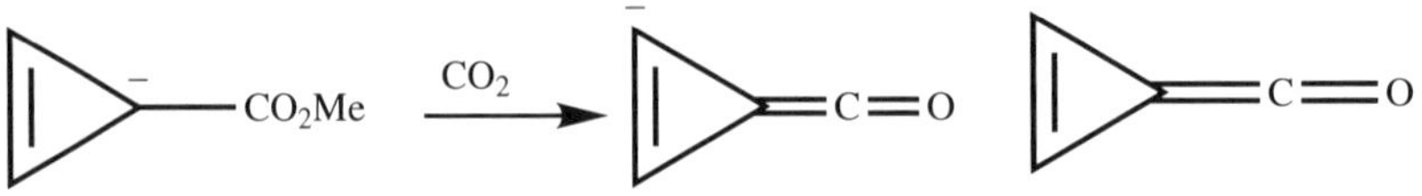

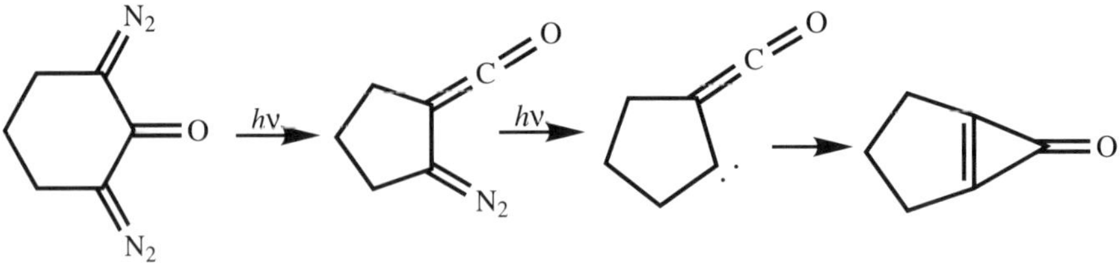

References

1. Olah, G. A.; Comisarow, M. B. *J. Am. Chem. Soc.* **1967**, *89*, 2694–2697.
2. Olah, G. A.; Comisarow, M. B. *J. Am. Chem. Soc.* **1966**, *88*, 4442–4447.
3. Olah, G. A.; Tolgyesi, W. S.; Kuhn, S. J.; Moffatt, M. E.; Bastein, I. J.; Baker, E. B. *J. Am. Chem. Soc.* **1963**, *85*, 1328–1334.
4. Olah, G. A.; Kuhn, S. J.; Tolgyesi, W. S.; Baker, E. B. *J. Am. Chem. Soc.* **1962**, *84*, 2733–2740.
5. Amat, A. M.; Asensio, G.; Castello, M. J.; Miranda, M. A.; Simon-Fuentes, A. *Tetrahedron* **1987**, *43*, 905–910.
6. Leung-Toung, R.; Peterson, M. R.; Tidwell, T. T.; Csizmadia, I. G. *J. Mol. Struct. (THEOCHEM)* **1989**, *183*, 319–330.
7. Allen, A. D.; Stevenson, P.; Tidwell, T. T. *J. Org. Chem.* **1989**, *54*, 2843–2848.
8. Conrow, K.; Morris, D. L. *J. Chem. Soc., Chem. Commun.* **1973**, 5–6.
9. McDonald, R. N.; Morris, D. L.; Petty, H. E.; Hoskins, T. L. *J. Chem. Soc., D* **1971**, 743–744.
10. Davies, A. G.; Sutcliffe, R. *J. Chem. Soc., Perkin Trans. 2,* **1980**, 819–824.
11. Davies, A. G.; Hawari, J. A.; Muggleton, B.; Tse, M. *J. Chem. Soc., Perkin Trans. 2,* **1981**, 1132–1137.

12. Matlin, A. R.; Wolff, S.; Agosta, W. C. *Tetrahedron Lett.* **1983,** *24,* 2961–2964.
13. Wolff, S.; Agosta, W. C. *J. Am. Chem. Soc.* **1983,** *105,* 1292–1299.
14. Dauben, W. G.; Cogen, J. M.; Ganzer, G. A.; Behar, V. *J. Am. Chem. Soc.* **1991,** *113,* 5817–5824.
15. Adam, W.; Berkessel, A.; Peters, E.; Peters, K.; von Schnering, H. G. *J. Org. Chem.* **1985,** *50,* 2811–2814.
16. Salem, L. *J. Am. Chem. Soc.* **1974,** *96,* 3486–3501.
17. Dauben, W. G.; Salem, L.; Turro, N. J. *Acc. Chem. Res.* **1975,** *8,* 41–54.
18. Cooksy, A. L.; Watson, J. K. G.; Gottlieb, C. A.; Thaddeus, P. *J. Mol. Struct.* **1992,** *153,* 610–626.
19. Ciabattoni, J.; Kocienski, P. J.; Melloni, G. *Tetrahedron Lett.* **1969,** 1883–1887.
20. Halton, B.; Banwell, M. G. In *The Chemistry of the Cyclopropyl Group;* Rappoport, Z., Ed.; Wiley: London, 1987.
21. Kascheres, A.; Marchi, D., Jr.; Rodriques, J. A. R. *J. Org. Chem.* **1978,** *43,* 2892–2896.
22. Ryu, I.; Hayama, Y.; Hirai, A.; Sonoda, N.; Orita, A.; Ohe, K.; Murai, S. *J. Am. Chem. Soc.* **1990,** *112,* 7061–7063.
23. Sachs, R. K.; Kass, S. R. *J. Am. Chem. Soc.* **1994,** *116,* 783–784.
24. Chapman, O. L.; Gano, J.; West, P. R.; Regitz, M.; Maas, G. *J. Am. Chem. Soc.* **1981,** *103,* 7033–7036.
25. Murata, S.; Yamamoto, T.; Tomioka, H.; Lee, H.; Kim, H.; Yabe, A. *J. Chem. Soc., Chem. Commun.* **1990,** 1258–1260.

4.1.10 Fulvenones

The fulvenones are the oxo-analogues of the well-studied fulvenes,[1–3] and include triafulvenone (**1**), pentafulvenone (**2**), and heptafulvenone (**3**). These are considered in separate sections, and some related ketene-analogues of quinone methides and related species are included in Section 4.1.10.4.

4.1.10.1 Triafulvenones So far **1** and its derivatives are known only theoretically.[4,5] Attempted preparation of the diphenyl analogue **4** by the route of equation 1 instead gave **5**.[6] Photolysis of **6** in the presence of 1-diethylaminopropyne gave a quantitative yield of N_2 and a 30% yield of **8,** which was proposed to result from the ketene **7** (equation 2).[6] Other pathways from **6** to **8** can be envisaged, and so the existence of **7** is not proven.

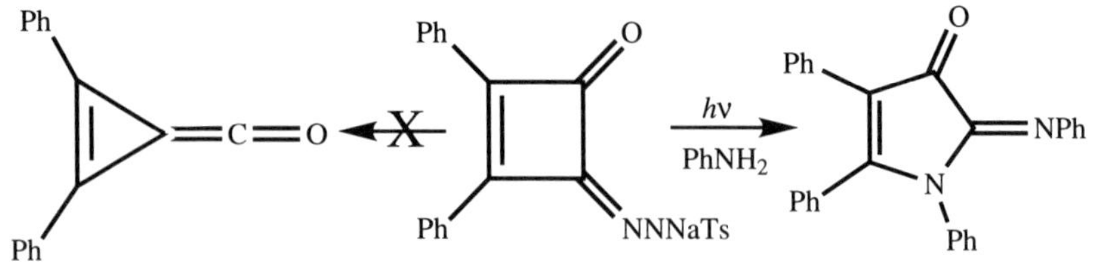

As discussed in Section 1.1.3 triafulvenone (**1**) is interpreted[5] as being destabilized relative to triafulvene **9** because of antiaromatic destabilization due to the negative charge density at C_β of ketenes. The resulting 4-π electron character in the ring is strongly destabilizing, and the same effect evidently inhibits formation of the diazo compound **10** and perhaps the carbene **11**. However, in the cyclic compounds **6** and **7** this antiaromatic destabilization is probably greatly attenuated.

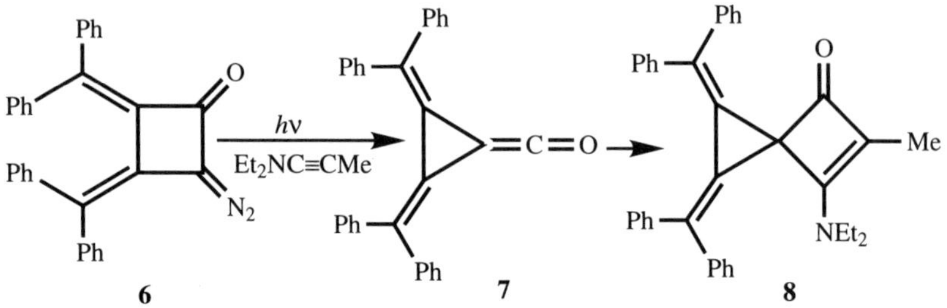

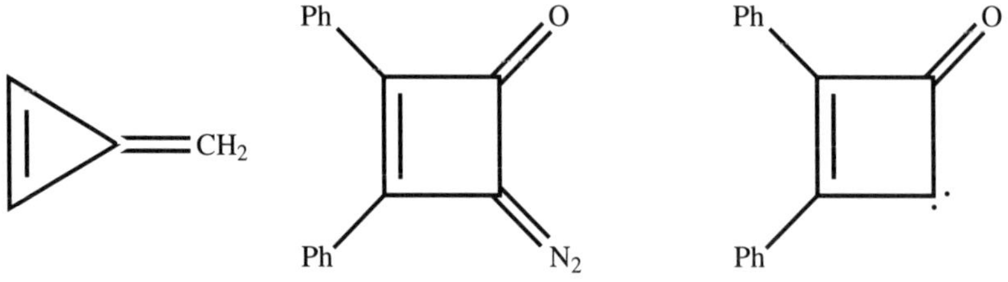

4.1.10.2 Pentafulvenones Pentafulvenone (**2**) is a conjugated species that is the ketene analogue of the well-studied fulvene (**12**).[1–3] As shown below the resonance structures **2a** and **2b** combine an aromatic cyclopentadienyl anion unit together with an acylium cation fragment, suggesting considerable aromatic stabilization of **2**.

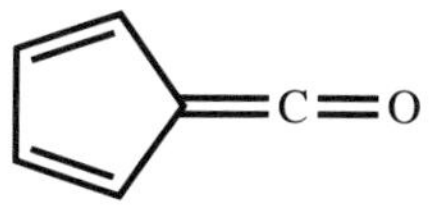

4.1.10 FULVENONES

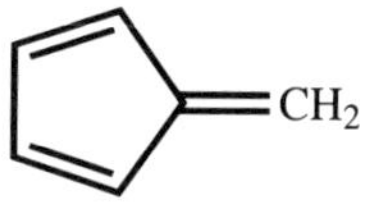

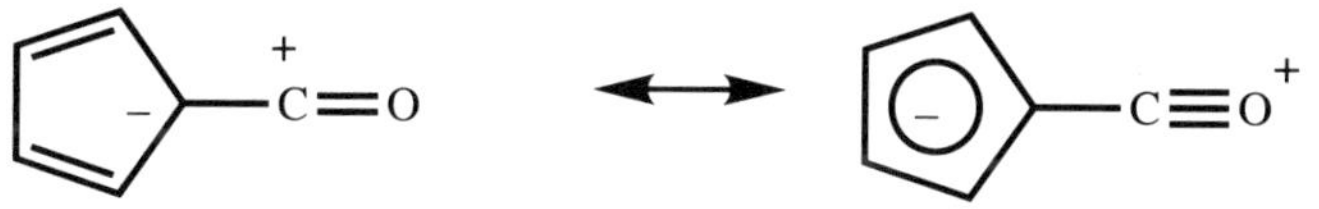

[Structures 2 and 12 shown: cyclopentadienyl=C=O (2) and cyclopentadienyl=CH₂ (12)]

[Resonance structures 2a and 2b shown]

Isodesmic reactions such as that of equation 3 using 6-31G*//6-31G* calculated energies for the species involved suggest there is significant aromatic stabilization of **2** relative to **12** owing to the contribution of the resonance structures **2a** and **2b**.[5]

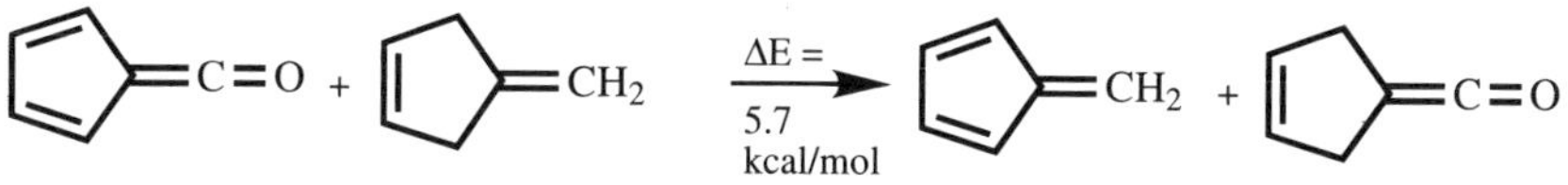

$$2 + 12 \xrightarrow{\Delta E = 5.7 \text{ kcal/mol}} 12 + 2 \quad (3)$$

The generation of **2** from the photolysis of *o*-quinone diazide (**13**) in aqueous solution and the hydration to cyclopentadienyl carboxylic acid (**15**) was established by Süs in 1944,[7] and Urwyler and Wirz[8] have demonstrated that this process occurs by initial formation of 6,6-dihydroxyfulvene (**14**) (equation 4).

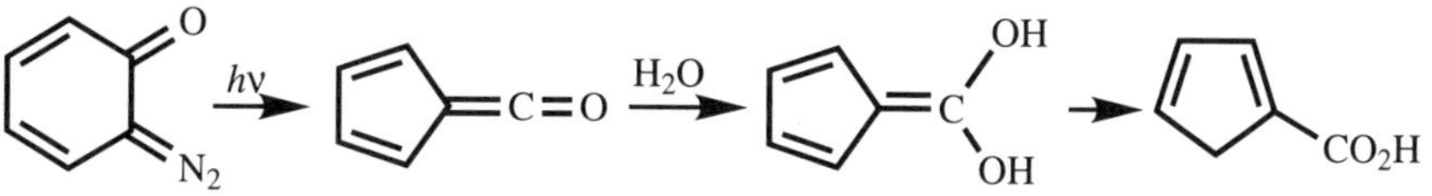

$$13 \xrightarrow{h\nu} 2 \xrightarrow{H_2O} 14 \rightarrow 15 \quad (4)$$

Many other substrates also form **2** on photolysis,[9-20] including *o*-chloro[13] and *o*-bromophenol,[8] and methyl salicylate.[9] Many of these reactions are interpreted in terms of initial formation of a ketocarbene **16**, and the adduct from combination of **2** and **16** has been isolated and identified as **17** (equation 5).[18] The formation of **2** from the combination of cyclopentadienylidene with carbon monoxide has also been reported, and in the presence of CO and O₂ there is competitive formation of **2** and the carbonyl oxide (equation 6).[15,16] The structure of **2** was confirmed by its IR and UV spectra, which are consistent with the ketene structure.[16] The assignment of the ketene IR absorption of **2** has been a matter of debate.[19a,19b]

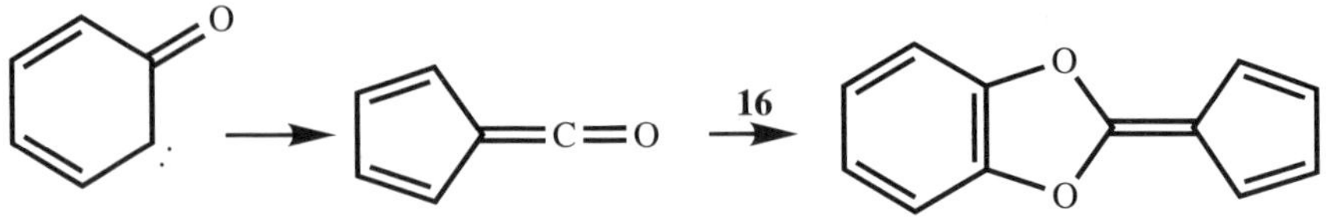

These processes are also involved in the photolithographic industry.[21–25] The technical literature on this subject is quite extensive and should be consulted for further details.

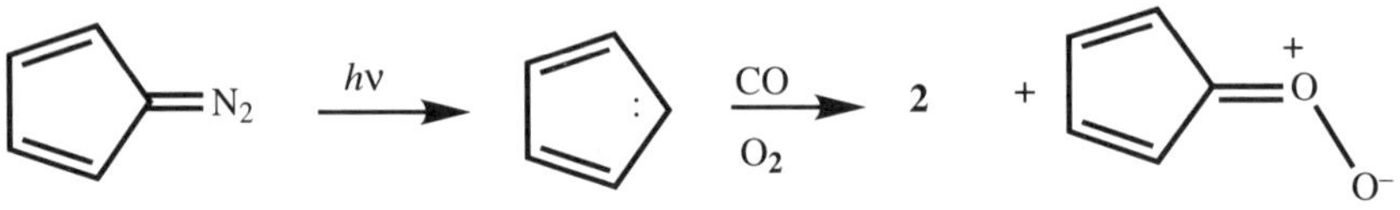

The rate constant for hydration of **2** in H_2O of ionic strength $\mu = 0.1$ at 25 °C is $(9.0 \pm 0.4) \times 10^5 \text{ s}^{-1}$, and k_{OH^-} is $(4.8 \pm 0.1) \times 10^7 \text{ M}^{-1} \text{ s}^{-1}$.[8] These rates and $k_{OH^-}/k_{H_2O} = 190 \text{ M}^{-1}$ are comparable to those of phenylketene, and the high reactivity in both cases suggest that even though both of these ground states are stabilized by conjugation, the stabilization of the transition states is even greater (Section 5.5.1). The transition state from **2** resembles the cyclopentadienyl anion (equation 7), and provides a strong driving force for hydration.

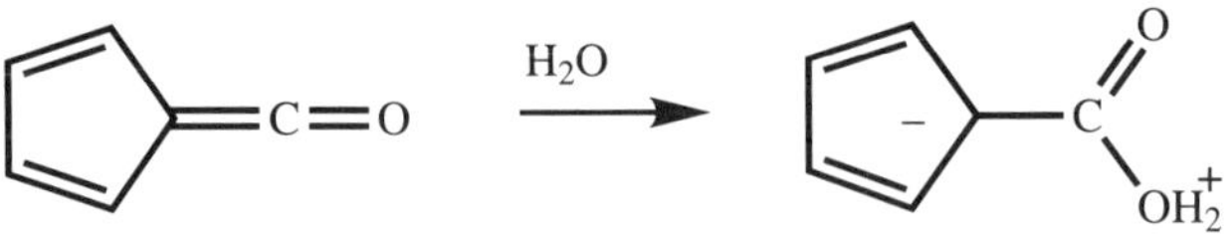

These structural assignments by Urwyler and Wirz[8] are a revision of the previous conclusions of Tanigaki and Ebbesen.[19] The latter authors proposed that the first intermediate observed in the photolysis of **13** is the oxirene **18**, which then forms the ketene **2** (equation 8).

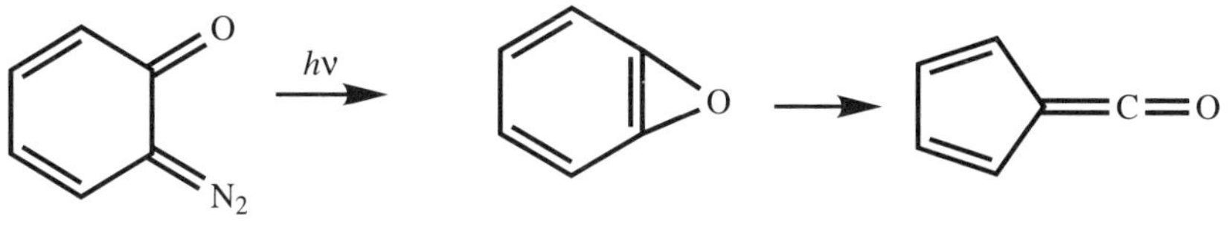

However the UV spectrum reported[19] for the first intermediate from **13** and the rate of its reaction in H$_2$O agree with those reported[8] for the ketene **2** and the assignment[16] of this UV to **2** appears secure. Thus it appears that Tanigaki and Ebbesen misassigned their spectra,[19] and that there is no evidence for the formation of **18** from **13** in H$_2$O.

The photolysis of 1,2-naphthoquinonediazide (**19**) produces the indenylketene **20** (equation 9).[26–34] The reaction of **20** with H$_2$O also proceeds with formation of an enediol,[8,31] although in another report[27] this has also been misidentified. Possible ketocarbene or oxirene intermediates have not been identified in this process, and it has been suggested that the conversion of the diazoketone to the ketene by Wolff rearrangement as in equation 9 is a concerted process on the nanosecond time scale,[31] as discussed further in Section 5.5.1.2.

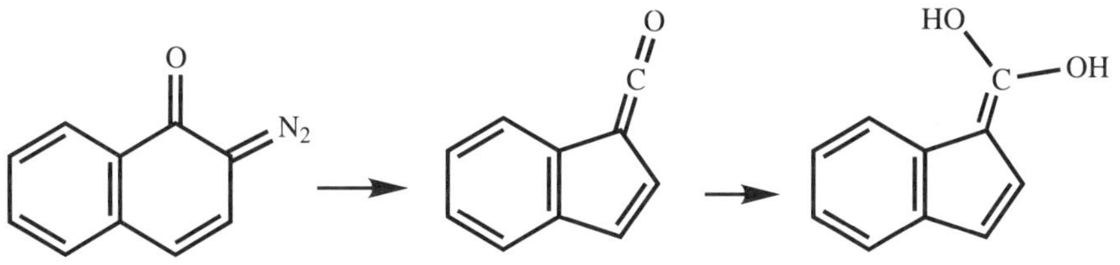

Derivatives of **19** substituted by sulfonate ester groups on photolysis at 77 K in matrices show the formation of UV bands at 313 nm attributed to ketenes analogous to **20,** and on warming in the presence of moisture the ketene band disappears and OH IR absorptions appear, but not carbonyl absorption.[32] On further warming the spectra of the carboxylic acids appear, and so the first intermediate is assigned the structure of a enediol analogous to **21**.[32] A large number of other substituted derivatives of **20** have also been studied, and are discussed in the photoresist literature.[21–25]

Photolysis of the azoindene **22** at 77 K in the presence of carbon monoxide also gave **20**, and **23** was formed analogously.[35] The interesting isomer **24** of **20** has been observed by mass spectrometry[71] and in matrix isolation,[71a] and has been trapped from Wolff rearrangement.[71b]

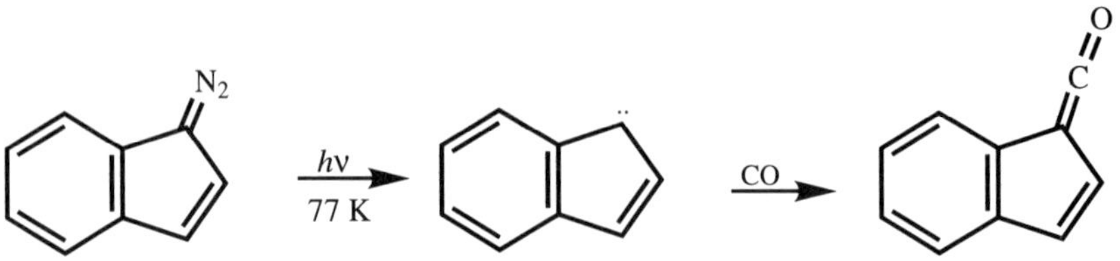

(10)

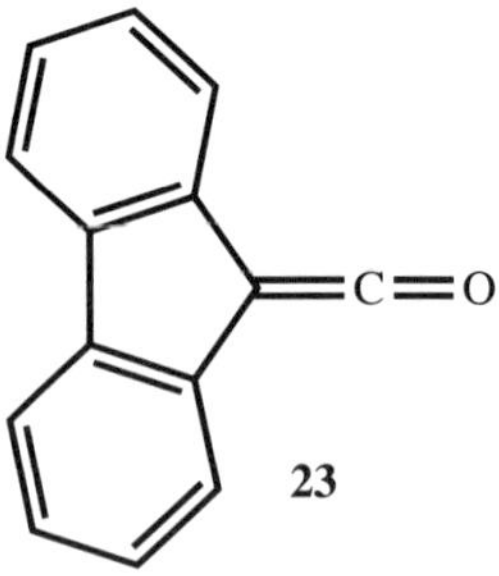

A variety of heterocyclic fulvenones have also been prepared, usually by pyrolytic methods, as listed in Table 1. Photolysis[36] of the anhydride **25** in an Ar matrix at 12 K gave the unique thiafulvene ylide **26,** with an IR band at 2157 cm^{-1}. Evidence for the proposed structure included agreement between the observed IR spectrum and that calculated at the SCF/6-31G* level.[36]

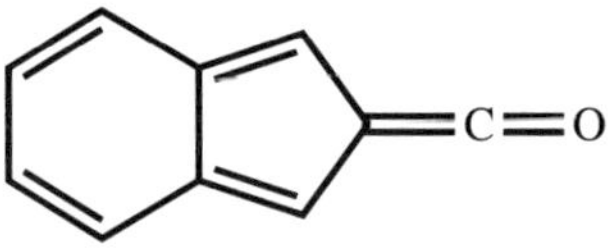

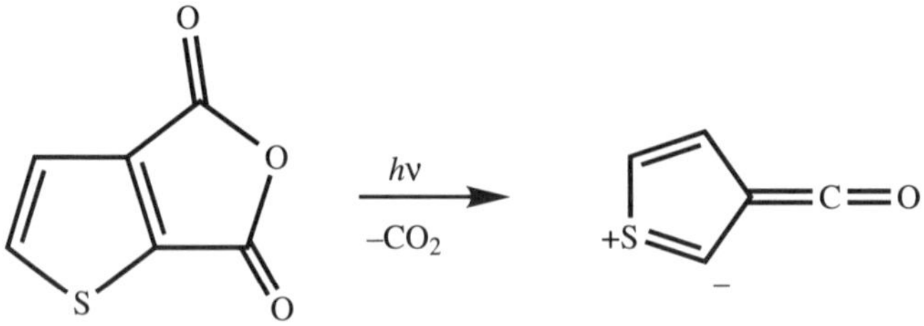

4.1.10.3 Heptafulvenones

Heptafulvenone **3** is the ketene analogue of heptafulvene **27**,[1-3,37,38] and it is known that **27** is stabilized by electron withdrawing groups on the exocyclic double bond, and so the possibility of stabilization of **3** by contributions of the resonance structure **3b** and **3c** merits consideration. Thus **3b** and **3c** combine a 6 π-electron tropylium cation with an acyl anion or carbene fragment, while **3d** contains the 8 π-electron cycloheptatrienyl anion combined with an acyl cation. Isodesmic energy calculations[5] indicate 5 kcal/mol antiaromatic destabilization for **3**, suggesting a net antiaromatic destabilization due to **3d**, which is the normal electron distribution for ketenes, but this effect is greatly attenuated relative to triafulvenone.

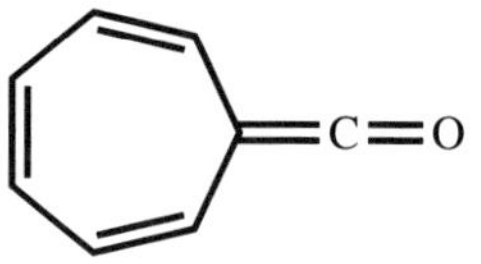

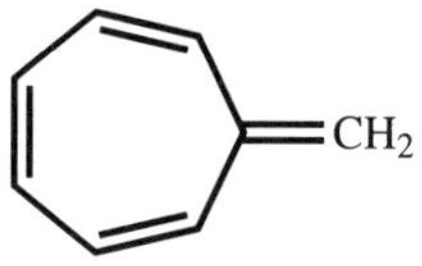

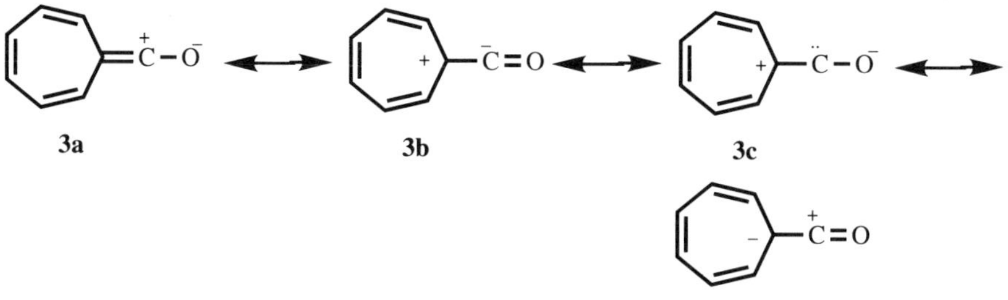

The generation of **3** was first reported by Asao, Morita, and Kitahara[3,37] from the reaction of the acyl chloride with Et$_3$N and trapping of **3** with cyclopentadiene to yield **28** (equation 11). Trapping of **3** with other alkenes and with ketones[39] has also been reported, as well as the formation of dimers.[38]

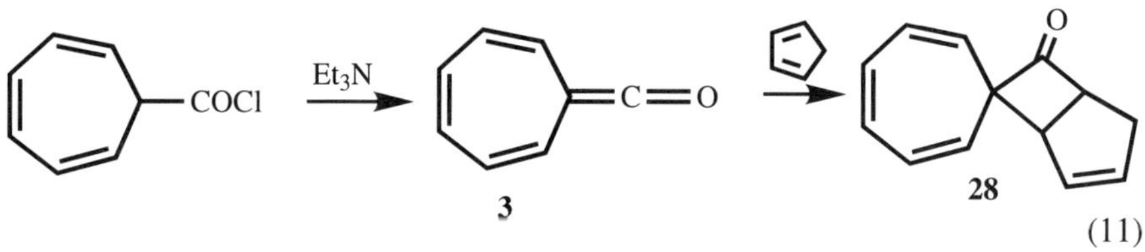

(11)

Photolysis of diazocycloheptatriene (**29**) in an argon matrix at 21 K containing carbon monoxide gave a material identified as the triplet carbene **30** on the basis of its IR, UV, and ESR spectra, along with a trace of material with an IR band at 2103 cm^{-1} assigned to **3**.[40] On warming to 35 K the IR band attributed to **30** disappeared, and that assigned to **3** increased concomitantly (equation 12). This

IR frequency is identical to that calculated for **3** (Section 2.3), but more definitive experimental observation of **3** is desirable.

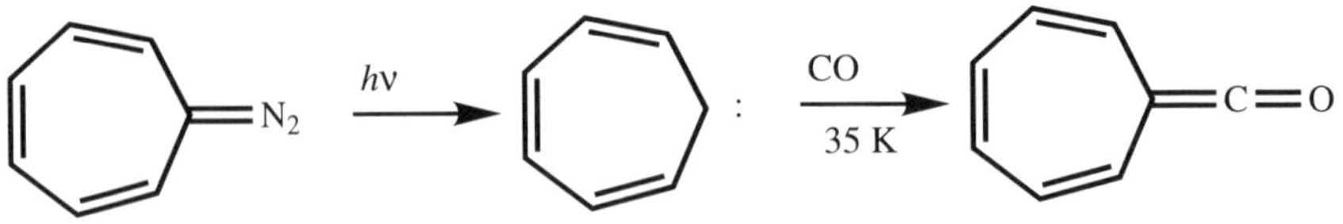

$$\text{29} \xrightarrow{h\nu} \text{30} \xrightarrow[35\text{ K}]{CO} \text{3} \qquad (12)$$

4.1.10.4 Oxoquinone Methides and Related Species

Quininoid ketenes are known of both the oxoquinone type **31**[41–43] and **33**, and the methanoquinone type **32** and **34**. The 1,2-isomers are valence tautomers of propenolactones **35** and benzocyclobutenones **36**.

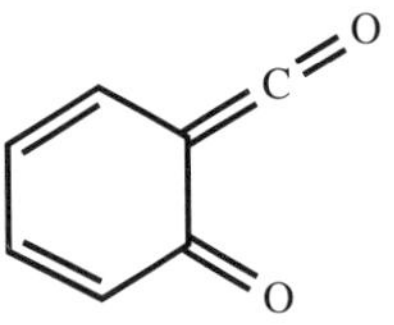

31 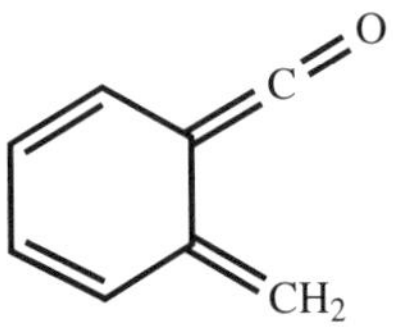   **32**

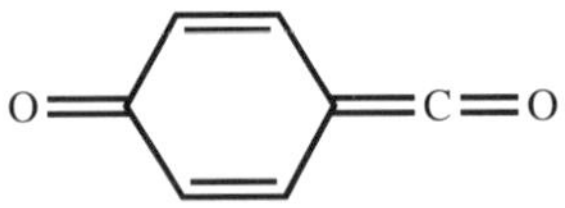

33 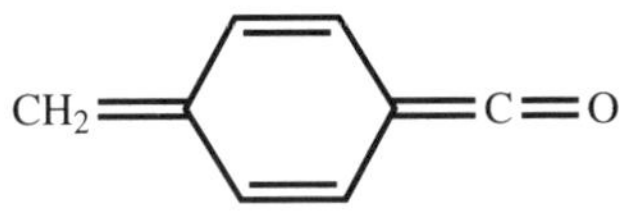   **34**

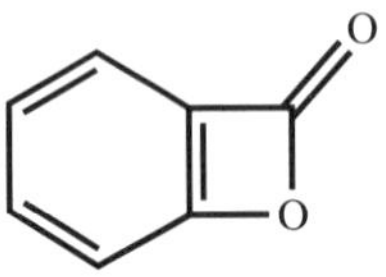

35 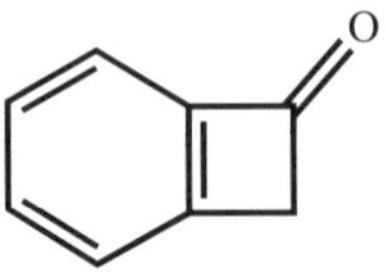   **36**

Thus photolysis of **36** in alcoholic solutions gives **32** as evidenced by the formation of o-toluate esters (equation 13),[44] and **32** is directly observable by IR at 2120 cm^{-1} when the photolysis is carried out at 20 K in an argon matrix.[45,46]

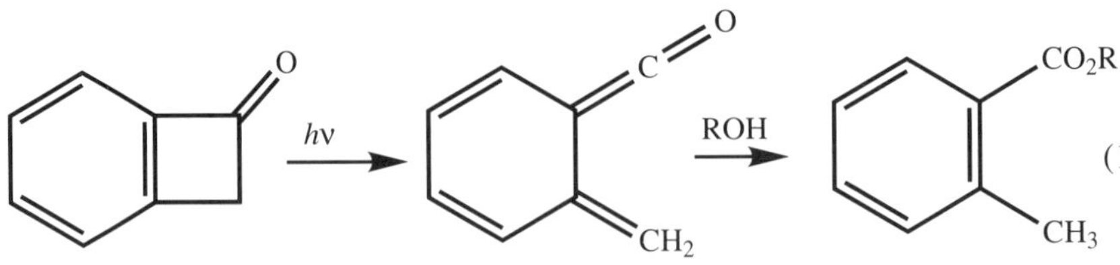

$$\text{36} \xrightarrow{h\nu} \text{32} \xrightarrow{ROH} \qquad (13)$$

Reaction of the o–trimethylsilyl substituted benzoyl chlorides with anhydrous CsF also gives **32** and substituted derivatives (equation 14).[47–49] These α-oxo-o-quinodimethanes can be trapped with a variety of dienophiles.[47–49] The reaction

4.1.10 FULVENONES

of equation 15 with Mo(CO)$_6$ and other group 6 metals was proposed to proceed through substituted derivatives of **32**.[50]

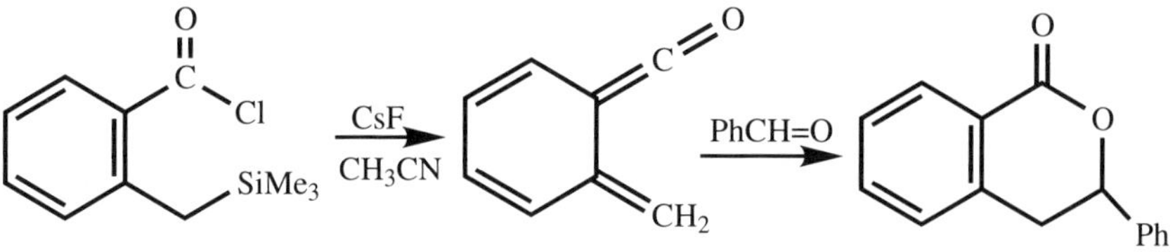

(14)

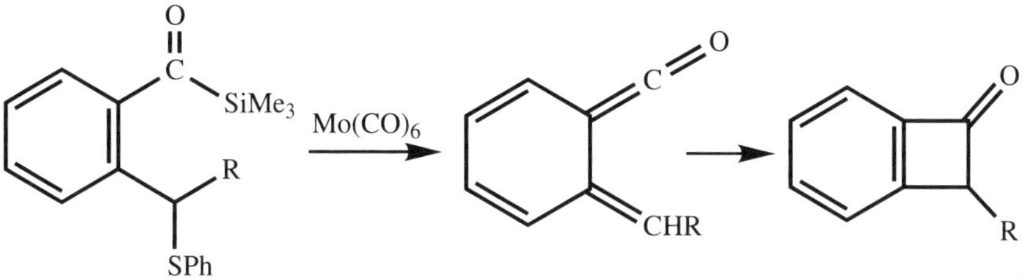

(15)

Photolysis of phthalaldehyde (**37**) produces the fulvenone **38**. As shown in equation 16 this process is proposed to occur by intramolecular hydrogen transfer producing a diradical intermediate **39** which reforms **37** or undergoes bond rotation to form **40** which gives **38**. The isomer of **38** with a *syn*-OH group was not observed, and is proposed to rapidly reform **37** by hydrogen atom transfer.[51,52] The structures **38** and **40** could differ both in electronic multiplicity and geometry.

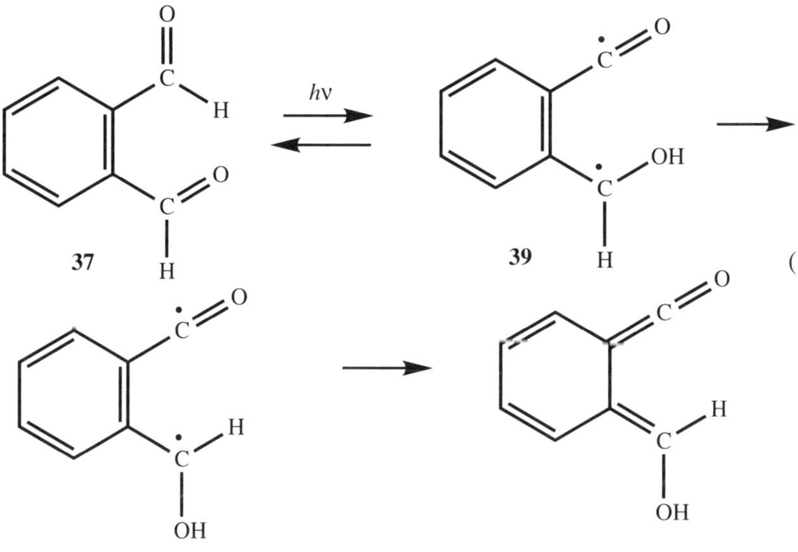

(16)

The photolysis of **41** gave the methoxy-substituted derivative **42** (equation 17).[52] The hydration of an analogue of **38** and **42** is discussed in Section 5.5.1.2.

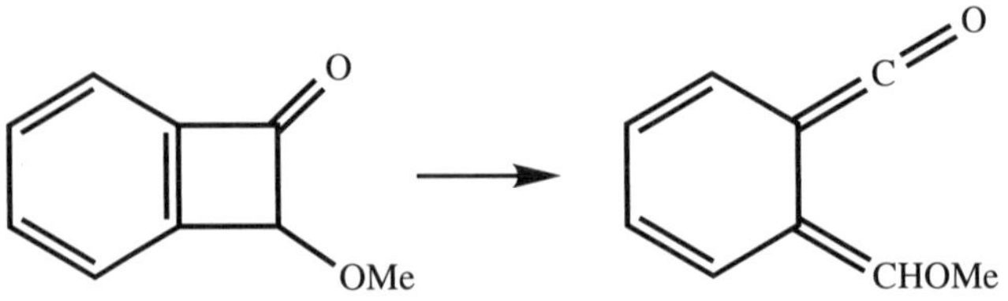

$$\text{41} \longrightarrow \text{42} \qquad (17)$$

Flash photolysis of *o*-nitrobenzaldehyde (**43**) in benzene gave a species with a strong UV λ_{max} near 450 nm identified as the fulvenone **44** (equation 18).[53] This species formed *o*-nitrosobenzoic acid either unimolecularly or in a reaction accelerated by H_2O. The X-ray irradiation of **43** in an argon matrix also led to **44**, as detected by its IR spectrum.[54]

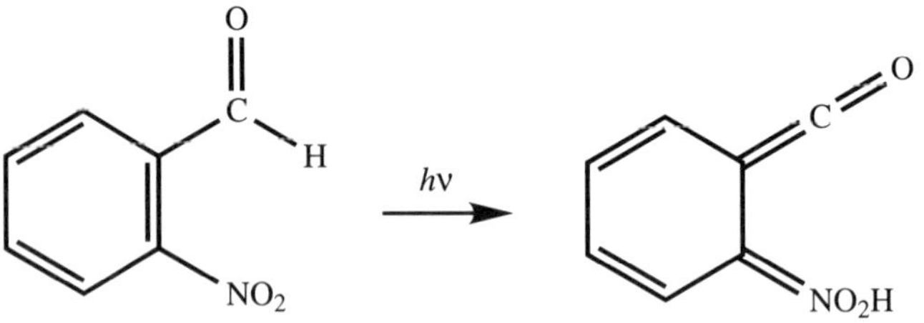

$$\text{43} \xrightarrow{h\nu} \text{44} \qquad (18)$$

Photolysis of *o*-nitrosobenzaldehyde proceeded with formation of quininoid derivative **45**, which evidently reacted further by OH migration to yield 3-carboxy-1-aza-1,2,4,6-cycloheptatetraene (**46**), as identified by its IR spectrum.[55] The methyl ether analogue of the oxime **45** has also been observed.[55a]

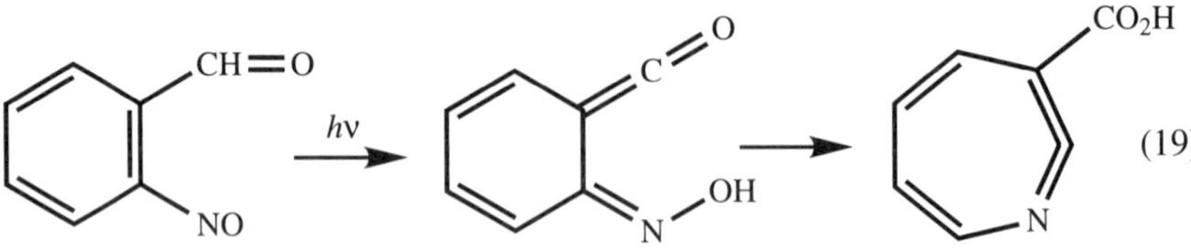

$$\xrightarrow{h\nu} \text{45} \longrightarrow \text{46} \qquad (19)$$

4.1.10 FULVENONES

Photolysis of the *o*-nitroaryldiazomethane **47** resulted in oxygen atom transfer to give the α-oximinoketene **48**, as detected by its IR band at 2094 cm^{-1} in Ar at 10 K.[55] The intermediacy of the charged species **49** has been proposed in the mass spectral decomposition of 4-aminoacetophenone (equation 21).[56]

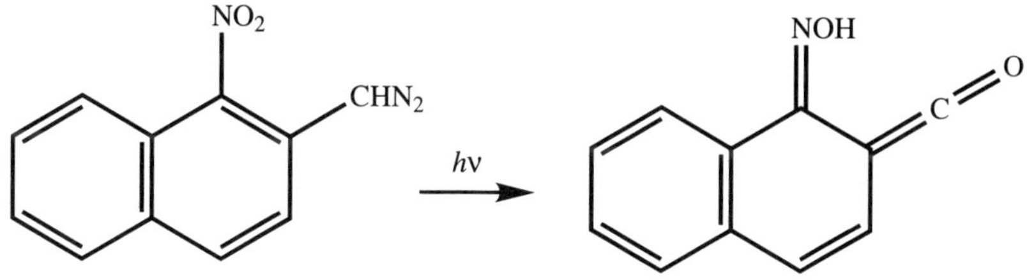

(20)

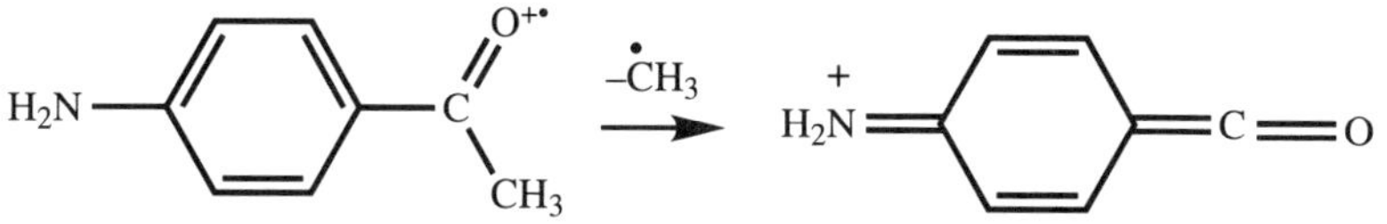

(21)

The reaction of **50** with Et$_3$N was proposed to lead to the unobserved iminoketene **51**, which cyclized to **52** (equation 22).[57,58] It was proposed that **51** was in thermal equilibrium with **52**, and was also formed by photolysis of **52**, as evidenced by the capture with alcohols or amines.[58] Reported fulvenones are listed in Table 1.

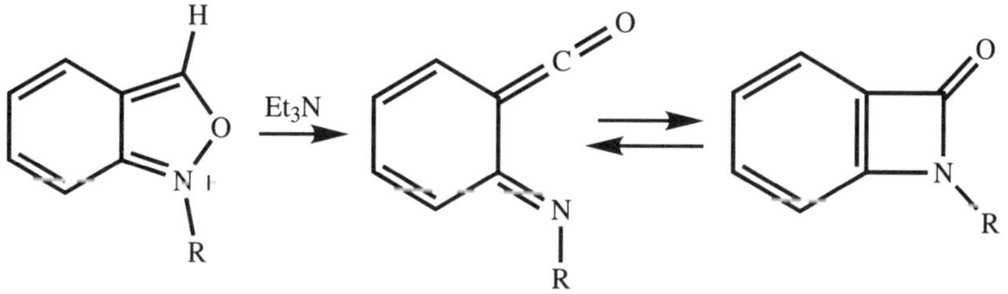

R = *t*-Bu, 1-Ad

(22)

TABLE 1. Fulvenones

Fulvenone	Formula	Source	Registry No.	Ref.
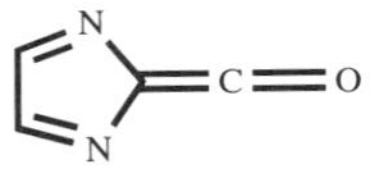 imidazol-2-ylidene ketene (N,N positions 1,3)	$C_4H_2N_2O$	Thermolysis	99560-57-3	59
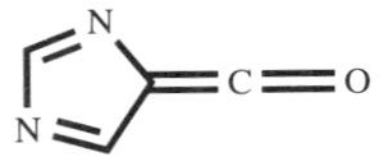 imidazol-4-ylidene ketene (N,N positions 1,3)	$C_4H_2N_2O$	Thermolysis	99560-56-2	59
 cyclopropenylidene ketene	C_4H_2O	Theory	109284-38-0	5
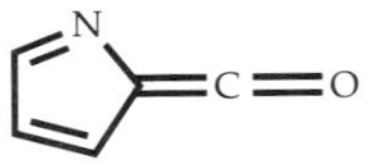 pyrrol-2-ylidene ketene	C_5H_3NO	Pyrolysis	82700-90-1	60
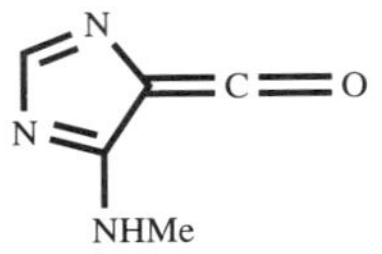 5-(NHMe)-imidazol-4-ylidene ketene	$C_5H_5N_3O$	Mass spectroscopy	70263-60-1	61
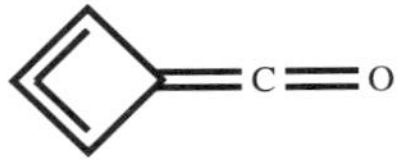 cyclobutadienylidene ketene (C_5O)	C_5O	Theory	113659-90-8	62

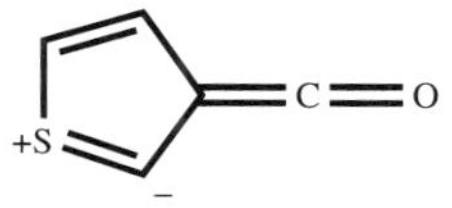 (structure)	C_5H_2OS	Theory	137040-79-0	36
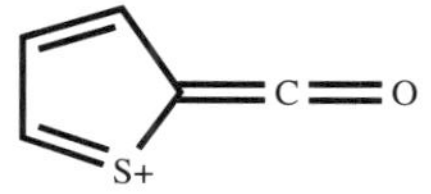 (structure)	C_5H_2OS	Theory		36
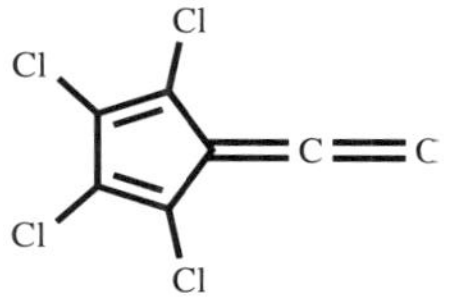 (structure)	C_6Cl_4O	Carbene	97745-68-1	35
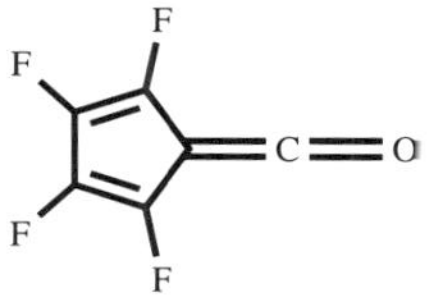 (structure)	C_6F_4O	Wolff	137348-77-7	62a
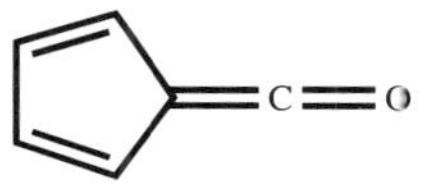 (structure)	C_6H_4O	Wolff	4727-22-4	7,8

TABLE 1. (Continued)

Fulvenone	Formula	Source	Registry No.	Ref.
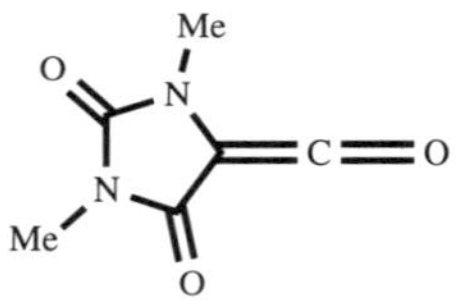	$C_6H_6N_2O_3$	Wolff	117135-80-5	63
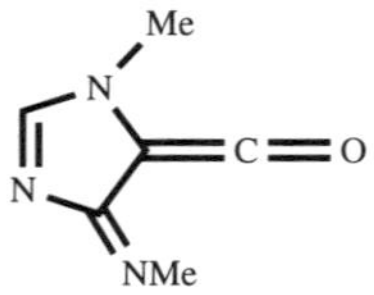	$C_6H_7N_3O$	Mass spectroscopy	70263-61-5	61
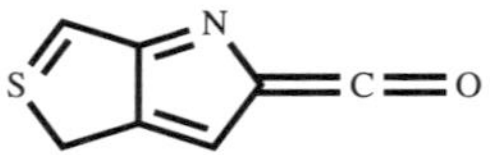	C_7H_3NOS	Pyrolysis	82700-93-4	60
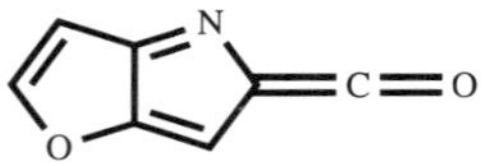	$C_7H_3NO_2$	Pyrolysis	82700-92-3	60
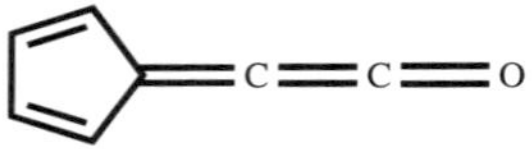	C_7H_4O			62

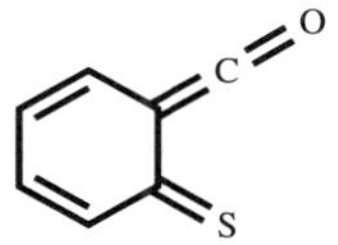

Structure	Formula	CAS	Method	Value
(6-thioxocyclohexa-2,4-dien-1-ylidene)methanone	C₇H₄OS	75126-75-9		None
6-oxocyclohexa-2,4-dien-1-ylidene methanone	C₇H₄O₂	21083-33-0	Photolysis	41–43
(6-iminocyclohexa-2,4-dien-1-ylidene)methanone	C₇H₅NO	81305-65-9	Theory	79
4-(methylenyl)cyclohexa-2,5-dien-1-one	C₇H₄O₂	94324-06-8		64–66
nitro-substituted	C₇H₅NO₃		Photolysis	54
benzoxete	C₈H₄O₂	57912-00-2	Photolysis	67

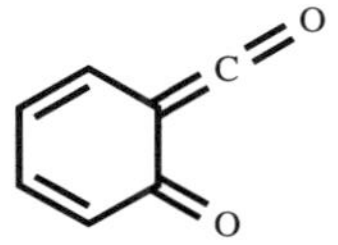

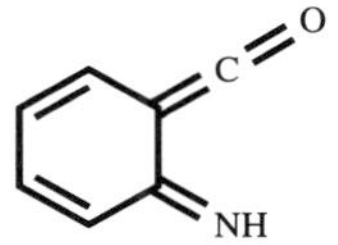

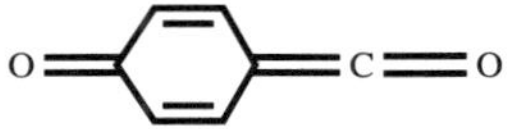

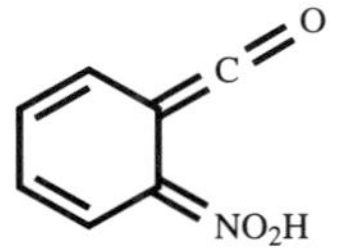

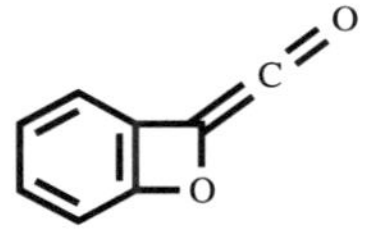

283

TABLE 1. (*Continued*)

Fulvenone	Formula	Source	Registry No.	Ref.
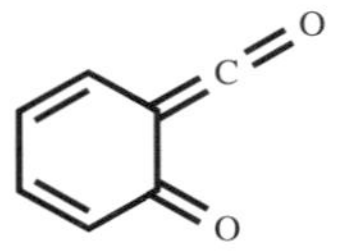 (cyclohexadienone with C=O and =O)	$C_8H_4O_2$	Photolysis	57912-01-3	67
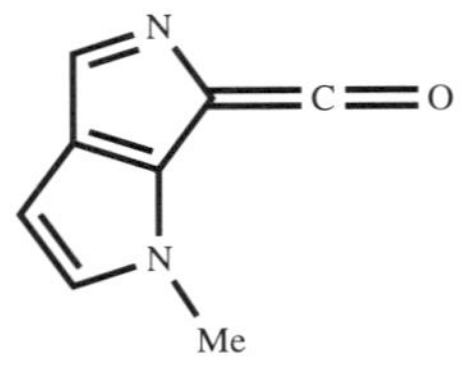 (pyrrolo-pyrrole with N, N-Me, C=O)	$C_8H_6N_2O$		82700-94-5	60
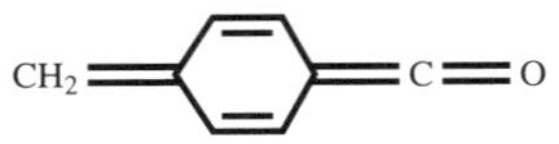 (quinoid with =CH$_2$ and C=O)	C_8H_6O	Theory	83148-10-1	68
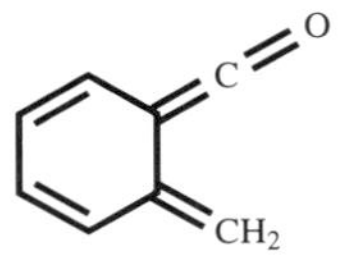 (cyclohexadiene with =CH$_2$ and C=O)	C_8H_6O	Photolysis	38382-55-7	61,68
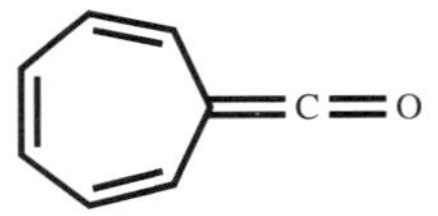 (cycloheptatriene with C=O)	C_8H_6O	Dehydrochlorination	36374-18-2	37

Structure	Formula	Method		CAS	Ref
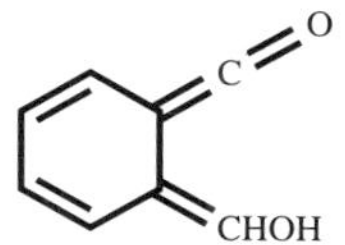 (6-oxocyclohexa-2,4-dienylidene)methanol, CHOH	$C_8H_6O_2$	Photolysis	E	118908-61-5	51
			Z	118908-60-4	51
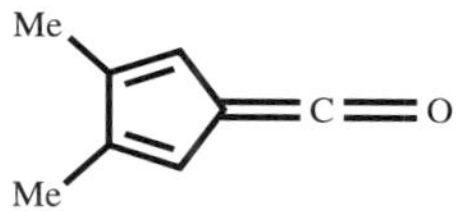 3,4-dimethylcyclopentadienylidene ketene	C_8H_8O	Wolff		137348-74-4	62a
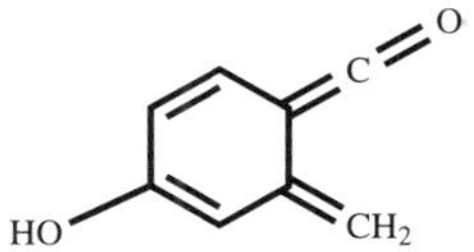 4-hydroxy-3-methylenecyclohexa-2,5-dienylidene ketene	$C_8H_6O_2$			98122-68-0	68
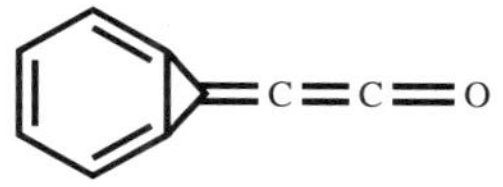 cycloheptatrienylidene ketene	C_9H_4O			119463-23-9	69
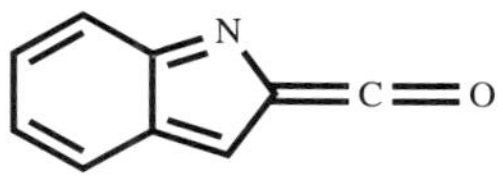 2H-indol-2-ylidene ketone	C_9H_5NO			82700-89-8	60

285

TABLE 1. (Continued)

Fulvenone	Formula	Source	Registry No.	Ref.
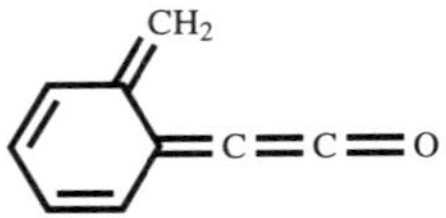 (o-quinoid with CH₂ and C=C=O)	C_9H_6O	Photolysis	57912-04-6	67
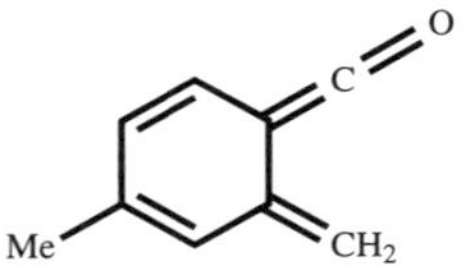 (Me-substituted quinoid with C=O and CH₂)	C_9H_8O		91416-35-2	70
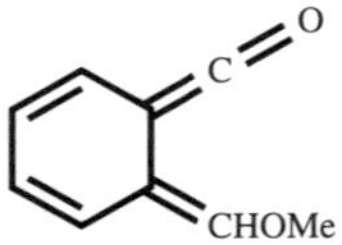 (quinoid with C=O and CHOMe)	$C_9H_8O_2$	Pyrolysis		53
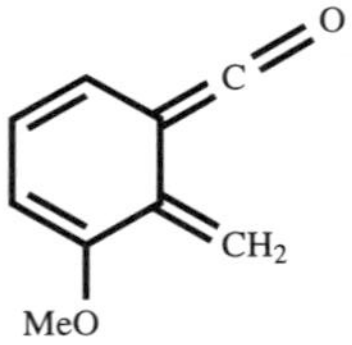 (MeO-substituted quinoid with C=O and CH₂)	$C_9H_8O_2$	Desilylation		47

Structure	Formula	Method	CAS	Ref
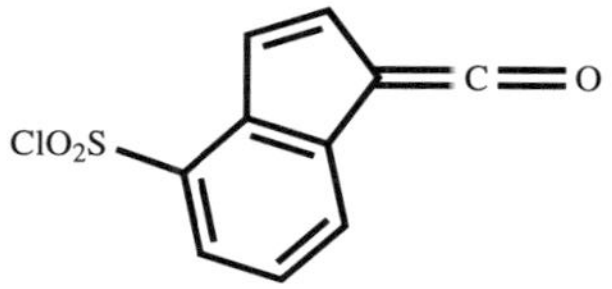 4-ClO₂S-indene-1-ketene	$C_{10}H_5ClO_3S$	Wolff	82656-19-7	46
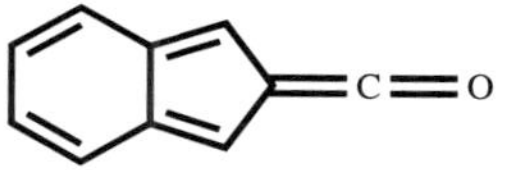 indene-2-ketene	$C_{10}H_6O$	Pyrolysis	49839-72-7	71, 71a
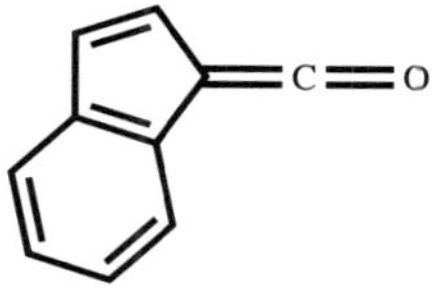 indene-1-ketene	$C_{10}H_6O$	Mass spectroscopy	49839-71-6	46, 71
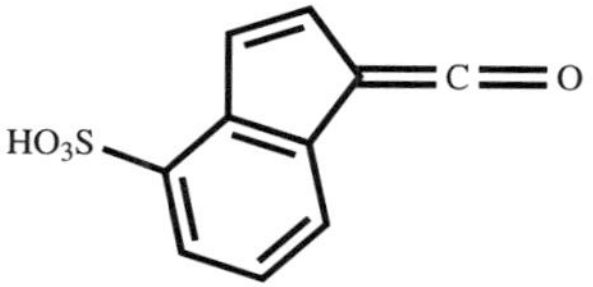 4-HO₃S-indene-1-ketene	$C_{10}H_6O_4S$	Wolff	38469-72-6	28
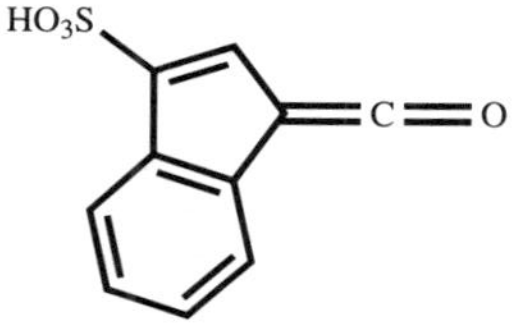 3-HO₃S-indene-1-ketene	$C_{10}H_6O_4S$	Wolff	38469-71-5	28

TABLE 1. (*Continued*)

Fulvenone	Formula	Source	Registry No.	Ref.
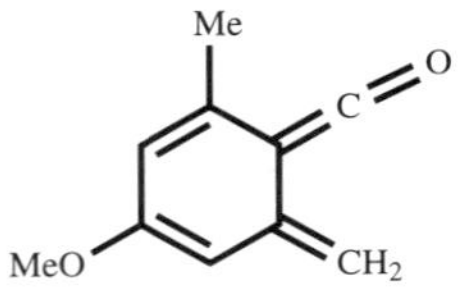 Me–, MeO–, =C=O, =CH$_2$ substituted cyclohexadiene	$C_{10}H_{10}O_2$	Pyrolysis	91416-38-5	70
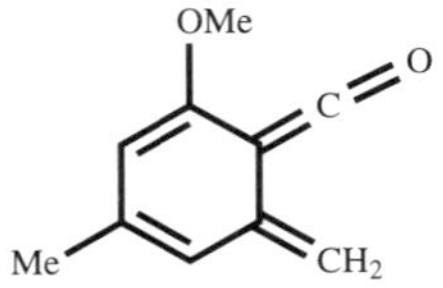 OMe–, Me–, =C=O, =CH$_2$ substituted cyclohexadiene	$C_{10}H_{10}O_2$	Pyrolysis	91416-39-6	70
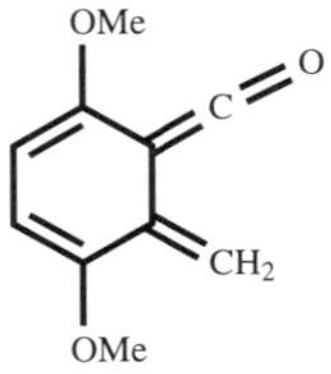 OMe–, OMe–, =C=O, =CH$_2$ substituted cyclohexadiene	$C_{10}H_{10}O_3$	Pyrolysis	116324-53-9	72

$C_{10}H_{10}O_3$	Desilylation		47
$C_{11}H_7NO_2$	Photolysis		52
$C_{11}H_{12}O$	Photolysis	91416-40-9	70
$C_{11}H_{13}NO$	Elimination		57,58
$C_{12}H_{14}O$		91416-41-0	70

TABLE 1. (Continued)

Fulvenone	Formula	Source	Registry No.	Ref.
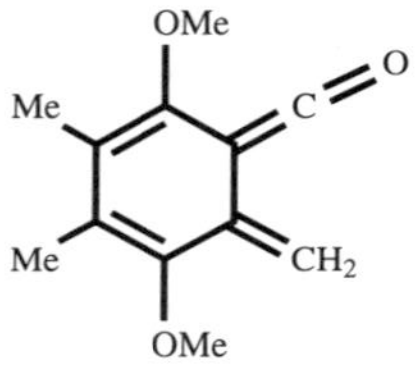 (structure: 2,3-dimethyl-5,6-dimethoxy with C=O and =CH₂)	$C_{12}H_{14}O_3$		91416-42-1	70
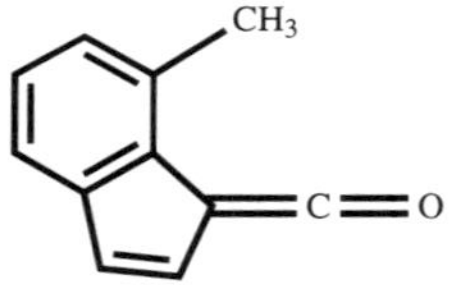 (7-methylindene-1-ylidene ketene)	$C_{12}H_8O$	Pyrolysis	115349-48-9	73
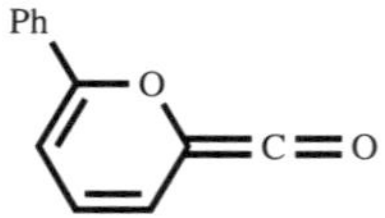 (6-phenyl-2H-pyran-2-ylidene ketene)	$C_{12}H_8O_2$		25552-11-8	74
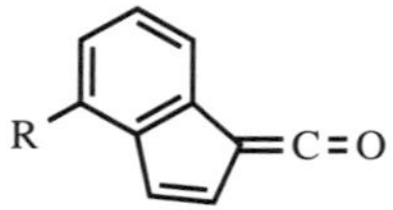 (4-R-indene-1-ylidene ketene)				
R = SO₃Et	$C_{12}H_{10}O_4S$	Wolff		75
R = SO₃Ph	$C_{16}H_{10}O_4S$	Wolff		75
R = SO₃C₆H₄Me-4	$C_{17}H_{12}O_4S$	Wolff		75
R = SO₃C₆H₄CO₂Pr-4	$C_{20}H_{16}O_4S$	Wolff		75

Structure	Formula	Method	CAS	Ref
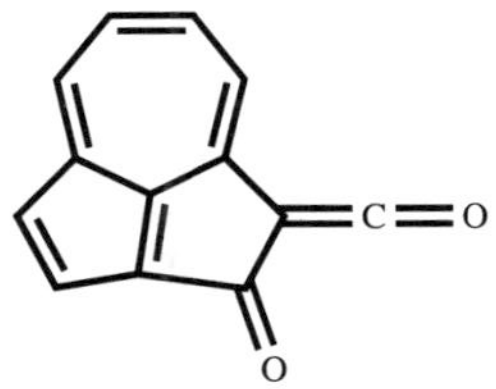 (cyclopenta-fused azulene ketene ketone)	$C_{13}H_6O_2$		98361-86-5	76
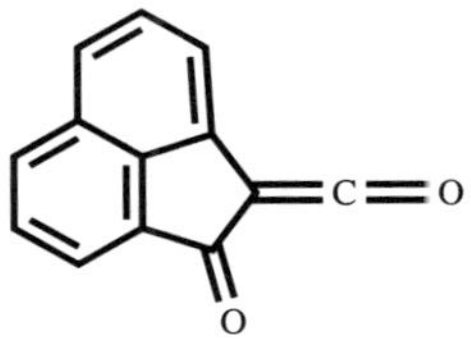 (acenaphthylene ketene ketone)	$C_{13}H_6O_2$		98361-83-2	76
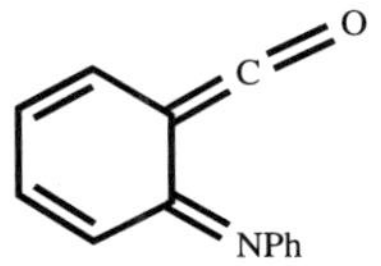 O=C=C=C(=NPh) cyclohexadienylidene	$C_{13}H_9NO$	Wolff	94605-10-4	77
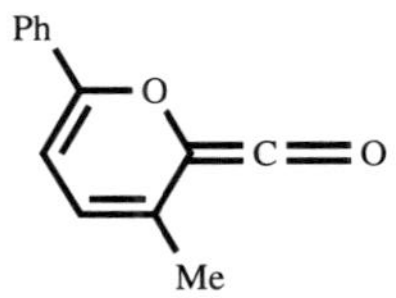 6-Ph-3-Me-2H-pyran-2-ylidene ketene	$C_{13}H_{10}O_2$		25552-13-0	74

TABLE 1. (Continued)

Fulvenone	Formula	Source	Registry No.	Ref.
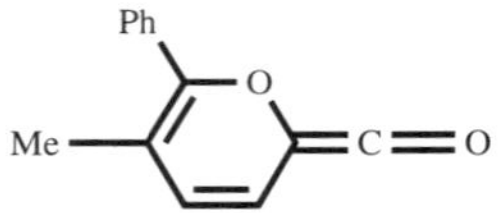 Ph–O–C(Me)=CH–C=O (2-Ph-3-Me pyranone fulvenone)	$C_{13}H_{10}O_2$		25552-12-9	74
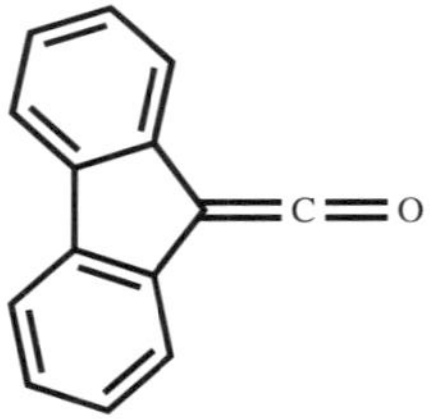 9-fluorenylidene ketene (C=O)	$C_{14}H_8O$	Zn dehalogenation	40012-77-9	35
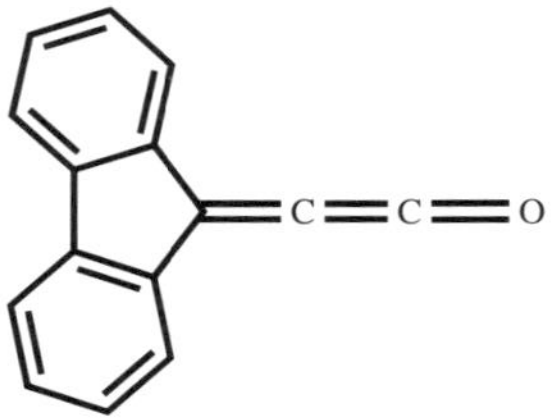 9-fluorenylidene (C=C=C=O)	$C_{15}H_8O$	Index error	57602-32-1	78

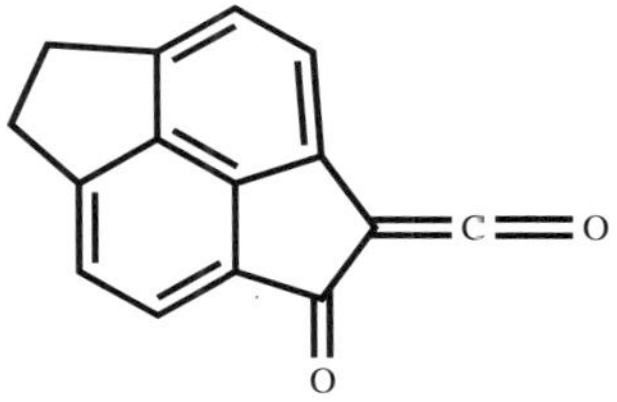 $C_{15}H_8O_2$	Photolysis	98361-84-3	76
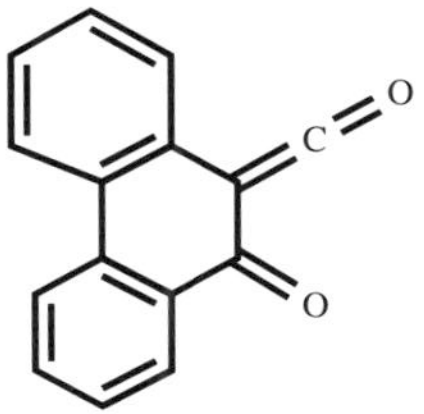 $C_{15}H_8O_2$	Wolff	85526-50-7	79
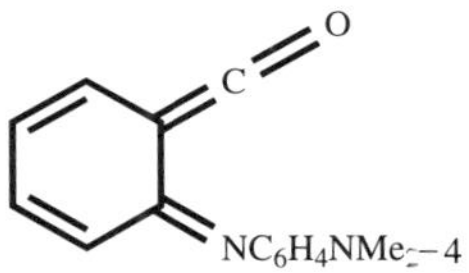 $C_{15}H_{14}N_2O$	Theory	82944-23-8	80

TABLE 1. (*Continued*)

Fulvenone	Formula	Source	Registry No.	Ref.
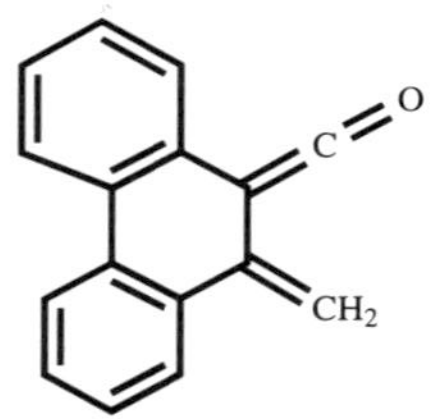 (phenanthrene-9-ylidene ketene with =CH₂)	$C_{16}H_{10}O$	Photolysis	87180-88-9	46
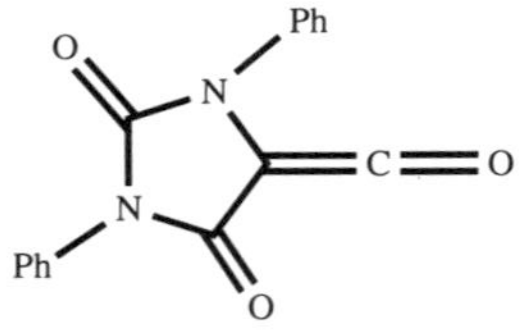 (N,N'-diphenyl hydantoin ketene)	$C_{16}H_{10}N_2O_3$	Wolff	117135-82-7	63
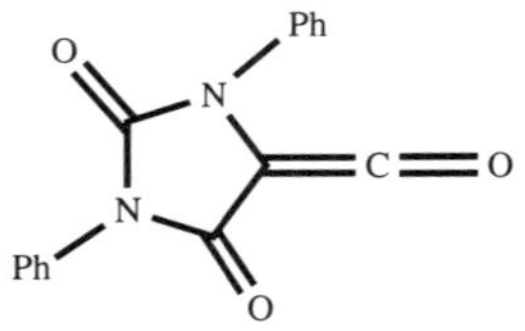 (N,N'-diphenyl hydantoin ketene isomer)	$C_{16}H_{10}N_2O_3$	Wolff	117135-81-6	63
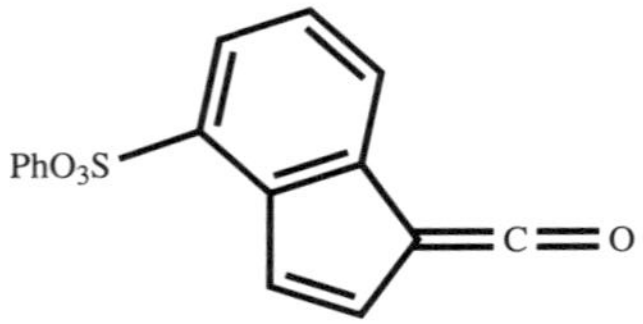 (4-PhO₃S-indenylidene ketene)	$C_{16}H_{10}O_4S$	Wolff	117714-42-8	81

Structure	Formula	Method	CAS	Ref
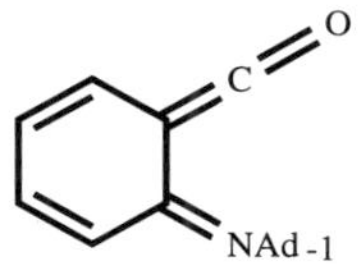 (ketene with NAd-1)	$C_{17}H_{19}NO$	Elimination		57,58
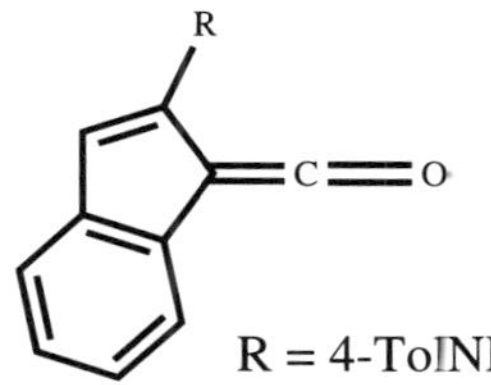 (indene ketene, R = 4-TolNHCO)	$C_{18}H_{13}NO_2$	Wolff	82813-15-8	82
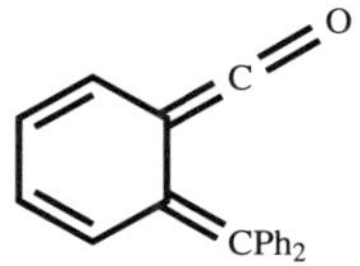 (ketene with CPh₂)	$C_{20}H_{14}O$	Photolysis	14596-71-5	83
(indene ketene, R = 4-t-BuC$_6$H$_4$O$_3$S)	$C_{20}H_{18}O_4S$	Wolff	81924-19-8	84

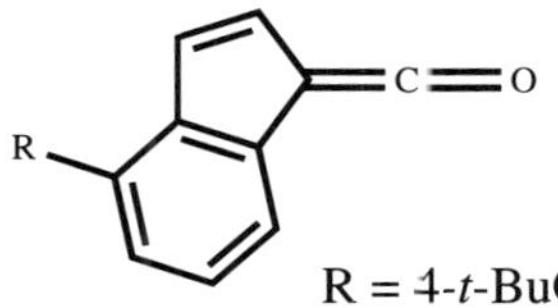

TABLE 1. (Continued)

Fulvenone	Formula	Source	Registry No.	Ref.
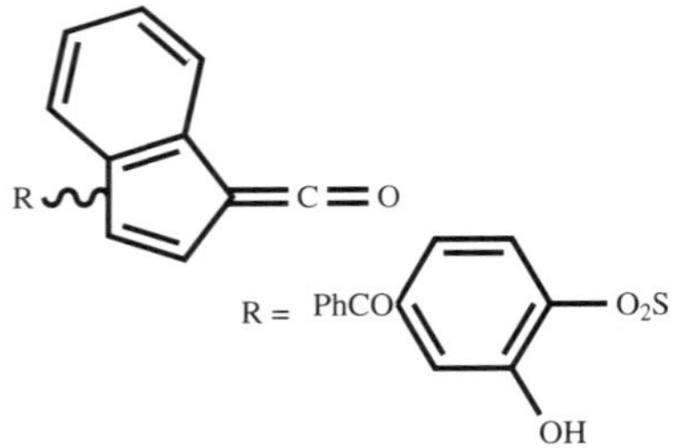 (indene-C=O, R substituent; R = PhCO, with OH and O$_2$S on phenyl)	C$_{23}$H$_{14}$O$_5$S	Wolff	122502-43-6	85
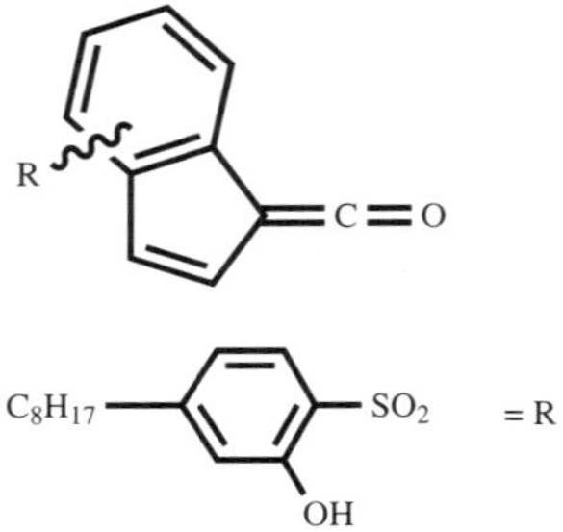 (indene-C=O, R substituent; R = SO$_2$-phenyl-OH with C$_8$H$_{17}$)	C$_{24}$H$_{26}$O$_4$S	Wolff	122502-45-8	85
		Wolff	122502-44-7	85
		Wolff	122502-43-6	85
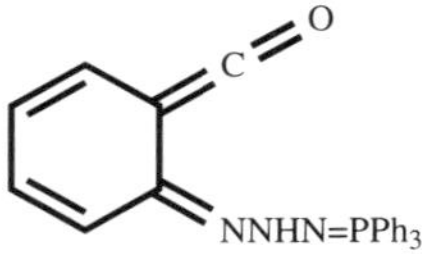 (cyclohexadienone C=O with NNHN=PPh$_3$)	C$_{25}$H$_{20}$N$_3$OP	Rearrangement		86

References

1. N. Neuenschwander, In *The Chemistry of Functional Groups. Supplement A. The Chemistry of Double-Bonded Functional Groups,* Volume 2, Part 2; Patai, S., Ed.; Wiley: New York, 1989.
2. Yates, P. *Adv. Alicyclic Chem.* **1968,** *2,* 59–184.
3. Bergmann, E. D. *Chem. Rev.* **1968,** *68,* 41–84.
4. Trinquier, G.; Malrieu, J. *J. Am. Chem. Soc.* **1987,** *109,* 5303–5315.
5. McAllister, M. A.; Tidwell, T. T. *J. Am. Chem. Soc.* **1992,** *114,* 5362–5368.
6. Nickels, H.; Dürr, H.; Toda, F. *Chem. Ber.* **1986,** *119,* 2249–2260.
7. Süs, O. *Liebigs Ann. Chem.* **1944,** *556,* 65–84; *Chem. Abstr.* **1946,** *40,* 5420.
8. Urwyler, B.; Wirz, J. *Angew. Chem., Int. Ed. Engl.* **1990,** *29,* 790–791.
9. Mamer, O. A.; Rutherford, K. G.; Seidewand, R. J. *Can. J. Chem.* **1974,** *52,* 1983–1987.
10. Bloch, R. *Tetrahedron Lett.* **1978,** 1071–1072.
11. Schulz, R.; Schweig, A. *Tetrahedron Lett.* **1979,** 59–62.
12. Schweig, A.; Zittlau, W. *Chem. Phys.* **1986,** *103,* 375–382.
13. Schulz, R.; Schweig, A. *Angew. Chem., Int. Ed. Engl.* **1984,** *23,* 509–511.
14. Yagihara, M.; Kitahara, Y.; Asao, T. *Chem. Lett.* **1974,** 1015–1019.
15. Chapman, O. L.; Hess, T. C. *J. Am. Chem. Soc.* **1984,** *106,* 1842–1843.
16. Baird, M. S.; Dunkin, I. R.; Hacker, N.; Poliakoff, M.; Turner, J. J. *J. Am. Chem. Soc.* **1981,** *103,* 5190–5195.
17. Chapman, O. L. *Pure Appl. Chem.* **1979,** *51,* 331–339.
18. Clinging, R.; Dean, F. M.; Mitchell, G. H. *Tetrahedron* **1974,** *30,* 4065–4067.
19. Tanigaki, K.; Ebbesen, T. W. *J. Phys. Chem.* **1989,** *93,* 4531–4536.
19a. Liu, R.; Zhou, X. *Chem. Phys. Lett.* **1993,** *212,* 702–704.
19b. Scott, A. P.; Radom, L. *Chem. Phys. Lett.* **1993,** *212,* 705–706.
20. Guyon, C.; Boule, P.; Lemaire, J. *Tetrahedron Lett.* **1982,** *23,* 1581–1584.
21. Thompson, L. F.; Willson, C. G.; Bowden, M. J. S. *Introduction to Microlithography;* American Chemical Society: Washington, D.C. 1984.
22. Reiser, A. *Photoreactive Polymers: The Science and Technology of Resists;* Wiley: New York, 1989.
23. Ershov, V. V.; Nikiforov, G. A.; de Jonge, C. R. H. I. *Quinone Azides. Studies in Organic Chemistry 7;* Elsevier: New York, 1981.
24. Steppan, H.; Buhr, G.; Vollmann, H. *Angew. Chem., Int. Ed. Engl.* **1982,** *21,* 455–469.
25. DeForest, W. S. *Photoresist Materials and Processes;* McGraw Hill: New York, 1975.
26. Shibata, T.; Koseki, K.; Yamaoka, T., Yoshizawa, M.; Uchiki, H.; Kobayashi, T. *J. Phys. Chem.* **1988,** *92,* 6269–6272.
27. Delaire, J. A.; Faure, J.; Hassine-Renou, F.; Soreau, M.; Mayeux, A. *New J. Chem.* **1987,** *11,* 15–19.
28. Nakamura, K.; Udagawa, S.; Honda, K. *Chem. Lett.* **1972,** 763–766.
29. Yates, P.; Robb, E. W. *J. Am. Chem. Soc.* **1957,** *79,* 5760–5768.
30. Horner, L.; Spietschka, E.; Gross, A. *Liebigs Ann. Chem.* **1951,** *573,* 17–30.
31. Barra, M.; Fisher, T. A.; Cernigliaro, G. J.; Sinta, R.; Scaiano, J. C. *J. Am. Chem. Soc.* **1992,** *114,* 2630–2634.

32. Bendig, J.; Sauer, E.; Polz, K.; Schopf, G. *Tetrahedron* **1992,** *48,* 9207–9216.
33. Hacker, N. P.; Kasai, P. H. *J. Am. Chem. Soc.* **1993,** *115,* 5410–5413.
34. Rosenfeld, A.; Mitzner, R.; Baumbach, B.; Bendig, J. *J. Photochem. Photobiol. A: Chem.* **1990,** *55,* 259–268.
35. Bell, G. A.; Dunkin, I. R. *J. Chem. Soc., Farad. Trans. 2,* **1985,** *81,* 725–734.
36. Teles, J. H.; Hess, B. A., Jr.; Schaad, L. J. *Chem. Ber.* **1992,** *125,* 423–431.
37. Asao, T.; Morita, N.; Kitahara, Y. *J. Am. Chem. Soc.* **1972,** *94,* 3655–3657.
38. Kitahara, Y. *Pure Appl. Chem.* **1975,** *44,* 833–859.
39. Asao, T.; Morita, N.; Ojima, J.; Fujiyoshi, M.; Wada, K.; Hamai, S. *Bull. Chem. Soc. Jpn.* **1986,** *59,* 1713–1721.
40. McMahon, R. J.; Chapman, O. L. *J. Am. Chem. Soc.* **1986,** *108,* 1713–1714.
41. Chapman, O. L.; McIntosh, C. L. *J. Am. Chem. Soc.* **1970,** *92,* 7001–7002.
42. Dvorak, V.; Kolc, J.; Michl, J. *Tetrahedron Lett.* **1972,** 3443–3446.
43. Chapman, O. L.; McIntosh, C. L.; Pacansky, J.; Calder, G. V.; Orr, G. *J. Am. Chem. Soc.* **1973,** *95,* 4061–4062.
44. Arnold, D. R.; Hedeya, E.; Merritt, V. Y.; Karnischky, L. A.; Kent, M. E. *Tetrahedron Lett.* **1972,** 3917–3920.
45. Krantz, A. *J. Am. Chem. Soc.* **1974,** *96,* 4992–4993.
46. Hacker, N. P.; Turro, N. J. *J. Photochem.* **1983,** *22,* 131–135.
47. Kessar, S. V.; Singh, P.; Vohra, R.; Kaur, N. P.; Venugopal, D. *J. Org. Chem.* **1992,** *57,* 6716–6720.
48. Kessar, S. V.; Singh, P.; Venugopal, D. *J. Chem. Soc., Chem. Commun.* **1985,** 1258–1259.
49. Aono, M.; Terao, Y.; Achiwa, K. *Chem. Lett.* **1985,** 339–340.
50. Kang, J.; Choi, Y. R.; Kim, B. J.; Jeong, J. U.; Lee, S.; Lee, J. H.; Pyun, C. *Tetrahedron Lett.* **1990,** *31,* 2713–2716.
51. Gebicki, J.; Kuberski, S.; Kaminski, R. *J. Chem. Soc., Perkin Trans. 2,* **1990,** 765–769.
52. Tomioka, H.; Kobayashi, N.; Murata, S.; Ohtawa, Y. *J. Am. Chem. Soc.* **1991,** *113,* 8771–8778.
53. George, M. V.; Scaiano, J. C. *J. Phys. Chem.* **1980,** *84,* 492–495.
54. Michalak, J.; Gebicki, J.; Bally, T. *J. Chem. Soc., Perkin Trans 2* **1993,** 1321–1325.
55. Tomioka, H.; Ichikawa, N.; Komatsu, K. *J. Am. Chem. Soc.* **1992,** *114,* 8045–8053.
55a. Tomioka, H.; Ichikawa, N.; Komatsu, K. *J. Am. Chem. Soc.* **1993,** *115,* 8621–8626.
56. Kotiaho, T.; Shay, B. J.; Cooks, R. G.; Eberlin, M. N. *J. Am. Chem. Soc.* **1993,** *115,* 1004–1014.
57. Olofson, R. A.; Vander Meer, R. K.; Hoskin, D. H.; Bernheim, M. Y.; Stournas, S.; Morrison, D. S. *J. Org. Chem.* **1984,** *49,* 3367–3372.
58. Olofson, R. A.; Vander Meer, R. K. *J. Org. Chem.* **1984,** *49,* 3377–3379.
59. Maquestiau, A.; Tommasetti, A.; Pedregal-Freire, C.; Elguero, J.; Flammang, R.; Wiersum, U. E.; Bender, H.; Wentrup, C. *J. Org. Chem.* **1986,** *51,* 306–309.
60. Gross, G.; Wentrup, C. *J. Chem. Soc., Chem. Commun.* **1982,** 360–361.
61. Goeber, B.; Kraft, R. *Pharmazie* **1978,** *33,* 717–720; *Chem. Abstr.* **1979,** *91,* 19230a.
62. Brown, R. D.; McNaughton, D.; Dyall, K. G. *Chem. Phys.* **1988,** *119,* 189–192.
62a. Schweig, A.; Baumgartl, H.; Schulz, R. *J. Mol. Struct.* **1991,** *247,* 135–171.

63. Ulbricht, M.; Thurner, J. U.; Siegmund, M.; Tomaschewski, G. *Z. Chem.* **1988,** *28,* 102–103; *Chem. Abstr.* **1988,** *109,* 190282h.
64. Cevasco, G.; Guanti, G.; Hopkins, A. R.; Thea, S.; Williams, A. *J. Org. Chem.* **1985,** *50,* 479–484.
65. Sander, W.; Bucher, G.; Reichel, F.; Cremer, D. *J. Am. Chem. Soc.* **1991,** *113,* 5311–5322.
66. Pardini, V. L.; Smith, C. Z.; Utley, J. H. P.; Vargas, R. R.; Viertler, H. *J. Org. Chem.* **1991,** *56,* 7305–7313.
67. Chapman, O. L.; Chang, C. C.; Kolc, J.; Rosenquist, N. R.; Tomioka, H. *J. Am. Chem. Soc.* **1975,** *97,* 6586–6588.
68. Kuzuya, M.; Miyake, F.; Okuda, T. *J. Chem. Soc., Perkin Trans. 2,* **1984,** 1471–1477.
69. Murata, S.; Yamamoto, T.; Tomioka, H. *J. Am. Chem. Soc.* **1993,** *115,* 4013–4023.
70. Schiess, P.; Eberle, M.; Huys-Francotte, M.; Wirz, J. *Tetrahedron Lett.* **1984,** *25,* 2201–2204.
71. Grützmacher, H. F.; Hübner, J. *Liebigs Ann. Chem.* **1973,** 793–798.
71a. Wentrup, C.; Gross, G. *Angew. Chem., Int. Ed. Engl.* **1983,** *22,* 543.
71b. Blocher, A.; Zeller, K.-P. *Chem. Ber.* **1994,** *127,* 551–555.
72. Krohn, K.; Rieger, H.; Broser, E.; Schiess, P.; Chen, S.; Strubin, T. *Liebigs Ann. Chem.* **1988,** 943–948.
73. Brown, R. F. C.; Coulston, K. J.; Dobney, B. J.; Eastwood, F. W.; Fallon, G. D. *Austr. J. Chem.* **1987,** *40,* 1687–1694.
74. Montiller, J. P.; Dreux, J. *Bull. Soc. Chim. Fr.* **1969,** 3638–3647.
75. Baumbach, B.; Bendig, J.; Nagel, T.; Dubsky, B. *J. Prakt. Chem.* **1991,** *333,* 625–635.
76. McMahon, R. J.; Chapman, O. L.; Hayes, R. A.; Hess, T. C.; Krimmer, H. P. *J. Am. Chem. Soc.* **1985,** *107,* 7597–7606.
77. Maslivets, A. N.; Andreichikov, Yu. S. *Zh. Org. Khim.* **1988,** *24,* 1564–1565; *Engl. Trans.* **1988,** *24,* 1410–1411.
78. Schweng, J.; Zbiral, E. *Tetrahedron* **1975,** *31,* 1823–1825.
79. Capuano, L.; Moersdorf, P.; Scheidt, H. *Chem. Ber.* **1983,** *116,* 741–750.
80. Kuzuya, M.; Ito, S.; Miyake, F.; Okuda, T. *Chem. Pharm. Bull.* **1982,** *30,* 1980–1985.
81. Pasch, H.; Much, H. *J. Inf. Rec. Mater.* **1988,** *16,* 313–320; *Chem. Abstr.* **1988,** *109,* 240473u.
82. Zaitseva, L. G.; Sarkisyan, A. Ts.; Erlikh, R. D. *Zh. Org. Khim.* **1982,** *18,* 1330–1331; *Engl. Transl.* **1982,** *18,* 1156–1157.
83. Rigaudy, J.; Paillous, N. *Tetrahedron Lett.* **1966,** 4825–4831.
84. Kabalin, G. A.; Askerov, D. B.; Dyumaev, K. M. *Zh. Org. Khim.* **1982,** *18,* 456–457; *Engl. Transl.* **1982,** *18,* 400–401.
85. Kol'tsov, Yu. I.; Yudina, V. I.; Solomonenko, G. V. *Zh. Prikl. Khim.* **1989,** *62,* 191–194; *Chem. Abst.* **1989,** *111,* 114816m.
86. Molina, P.; Arques, A.; Cartagena, I.; Obon, R. *Tetrahedron Lett.* **1991,** *32,* 2521–2524.

4.2 NITROGEN-SUBSTITUTED KETENES

Amino ketenes react with imines to form β-lactams (equation 1), and because of the utility of these products in the synthesis of penicillins the generation of nitrogen-substituted ketenes and their reactions with imines[1-9] and alkenes[10] have been extensively investigated.[1-10] These cycloaddition reactions are discussed in Section 5.4.1.7. Nitrogen, oxygen, and sulfur substituted ketenes have been reviewed.[10a]

$$R_2^1N\overset{2}{C}=C=O + R_2^2C=NR^3 \longrightarrow \begin{array}{c} R_2^1N-\underset{R^2}{\overset{R}{\mid}}-\underset{R^2}{\overset{O}{\mid}} \\ \mid \\ NR^3 \end{array} \quad (1)$$

Ketenes substituted with electronegative substituents such as NH_2 are calculated to be unstable compared to ketenes with more electropositive substituents such as alkyl groups.[11] Furthermore, the preferred geometry of aminoketene (**1**) is calculated to be that shown with the lone pair of electrons in the ketene plane.[11] This has been interpreted to indicate that n-π donation of the nitrogen lone pair to the C=C π system of the ketene is destabilizing.[11] It has, however, also been suggested that this conformation may be preferred because of a favorable interaction between the lone pair and the electron deficient p orbital on the carbon of the carbonyl group.[12,13] In an extreme form such an interaction would lead to the bridged structure **2**, but the computational evidence regarding **1** does not support this possibility.[11] Thus the N—C—C bond angle is wider than for many other ketenes, and there is no evidence of a bonding interaction between nitrogen and the carbonyl carbon.[11,36e]

1

2

Aminoketenes which react by cycloaddition with imines have usually been generated *in situ* by the reaction of activated *N*-acyl amino acids with bases, particularly tertiary amines. The acids have usually been activated as the acyl chlorides (equation 2),[5-8] but mixed carboxylate–tosylate anhydrides (equation 3) have also been used.[3,10]

$$R_2NCH_2COCl \xrightarrow{Et_3N} R_2NCH=C=O \quad (2)$$

$$R_2NCH_2CO_2H \xrightarrow[Et_3N]{TsCl} [R_2NCH_2\overset{O}{\overset{\|}{C}}OTs] \xrightarrow{Et_3N} R_2NCH=C=O \quad (3)$$

The benzoylated aminoketenes **3**[14-20] have been suggested to exist in equilibrium with the oxazolium-5-olate tautomers **4** (equation 4), which are typically the species detected spectroscopically.[14] Thermolysis of **4** (Ar = Ph) gives the allene **5**, which can be represented as arising through a reaction of **3** with **4** (equation 5). Similarly reaction of **4** (Ar = Ph) with imines, diisopropylcarbodiimide, and an enamine gave products of [2 + 2] cycloaddition to the C=C bond of **3**. It was argued that these reactions arose from **3**, which is in equilibrium with **4**, since different products were expected from reaction of **4**. Structures **4** are mesoionic anhydro-5-hydroxyoxazolium hydroxides, or "münchnones," and their chemistry has been extensively examined and reviewed.[15–18] Many of the other amino ketenes that have been used for imine cycloadditions are *N*-acylated, but as these species are in general not directly observed it is not known if they also exist as the cyclic tautomers.

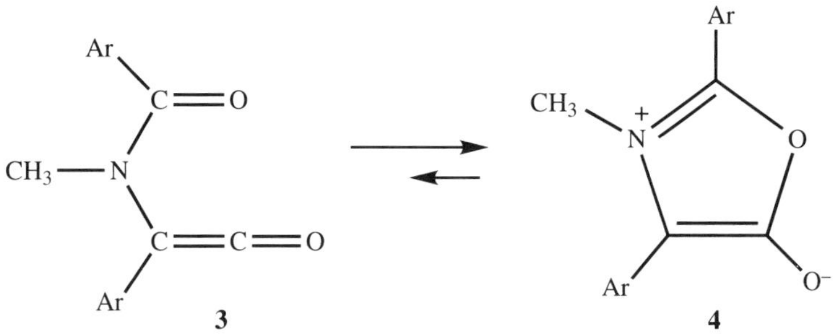

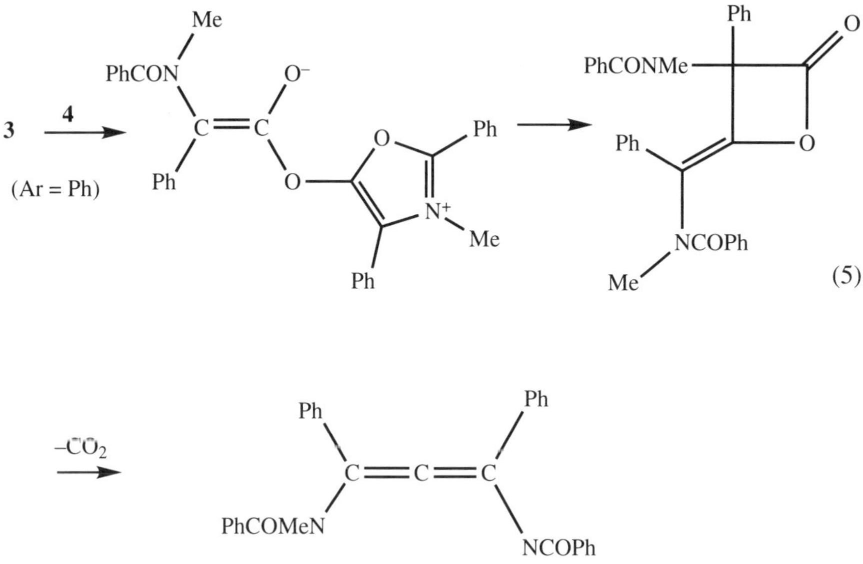

Phthalimido-*tert*-butylketene (**5**), obtained by dehydrochlorination of the acyl chloride with Me$_3$N, is a persistent ketene obtained as a crystalline solid,[21] mp 96–98 °C, with the carbonyl stretch at 2134 cm^{-1}. This result suggests that cyclization of acylated aminoketenes to form the cyclic structures **4** is dependent upon the particular structural features involved.

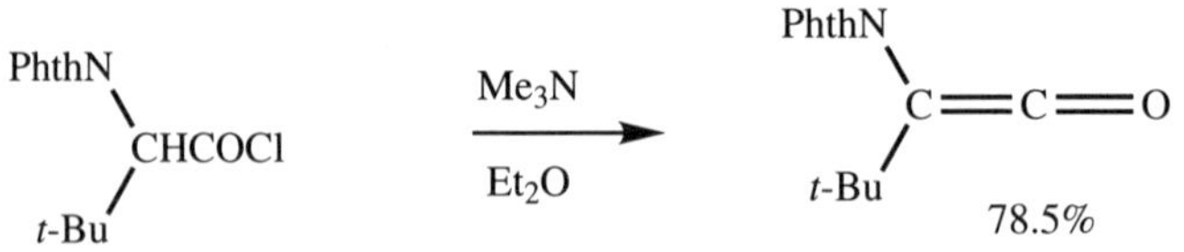

$$\underset{t\text{-Bu}}{\overset{\text{PhthN}}{\diagdown}}\!\!\!\text{CHCOCl} \quad \xrightarrow[\text{Et}_2\text{O}]{\text{Me}_3\text{N}} \quad \underset{t\text{-Bu}}{\overset{\text{PhthN}}{\diagdown}}\!\!\!\text{C}=\text{C}=\text{O} \quad \mathbf{5} \qquad 78.5\%$$

(6)

The chiral ketene **6**, generated in situ, reacts with imines to give β-lactams with high diastereoselectivity, as discussed further in Section 5.9.[7,8,22]

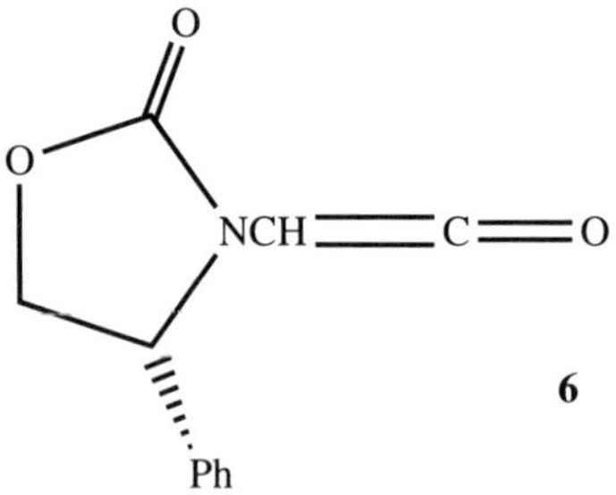

6

Generation of the ketene **7** for cycloaddition with imines has been reported (equation 7).[22] The potassium salt **8** is known as a Dane salt, and these have been utilized frequently in ketene preparation.

$$\text{MeO}_2\text{CCH}=\text{CHMeNHCH}_2\text{CO}_2\text{K} \quad \xrightarrow[\text{Et}_3\text{N}]{\text{PhOPOCl}_2} \quad \text{MeO}_2\text{CCH}=\text{CMeNHCH}=\text{C}=\text{O}$$

8   **7** (7)

Photochemical ring opening of **9** to an *N*-nitroso-*N*-(3-pyridyl)aminoketene **10** has also been proposed (equation 8).[23] Compounds such as **9** are known as "sydnones," and their chemistry has been extensively examined and reviewed.[17,24]

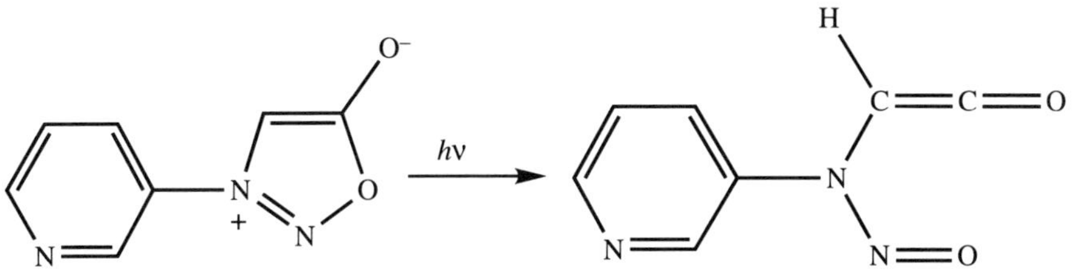

[Structure of compounds 9 and 10 shown, with equation (8)]

The group electronegativity of the *N*-formyl group calculated by the bond critical point model is 3.18, as compared to that of 3.12 for NH_2.[25] Thus on this basis an acylamino function is predicted to be at least as destabilizing as NH_2. However the π-donor destabilizing effect of NH_2 would be reduced by acylation, and intramolecular cyclization to **4** might also stabilize such ketenes. Theoretical analysis of this situation appears warranted.

A few alkylamino ketenes have been prepared by Wolff rearrangement (Section 3.3), as in the example shown where $Et_2NCH=C=O$ was captured by MeOH in yields up to 12% (equation 9).[26] The effect of the conformation of the diazoamide on the degree of rearrangement was discussed.[26]

$$N_2CHCONEt_2 \xrightarrow{h\nu} Et_2NCH=C=O \xrightarrow{MeOH} Et_2NCH_2CO_2Me \quad (9)$$

Azidoketene $N_3CH=C=O$ has been generated by reaction of the acyl chloride N_3CH_2COCl and Et_3N with in situ trapping with imines.[27–29] Calculations on azidoketene indicate the azido substituent is strongly destabilizing, as expected from the great electronegativity of this group.[11] This substituent is also a strong π donor,[30,31] and this effect is also not stabilizing for ketenes.[11]

$$N_3CH_2COCl \xrightarrow{Et_3N} [N_3CH=C=O] \quad (10)$$

Azoketenes have also been prepared from the acyl chlorides and trapped by cycloaddition reactions with imines and additions of alcohols and amines (equation 11).[32,33]

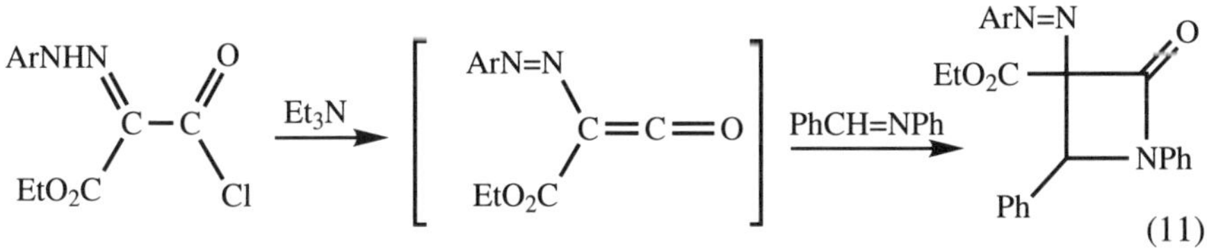

[Equation 11 showing ArNHN=C(CO2Et)–C(=O)Cl → ArN=N–C(CO2Et)=C=O → β-lactam product with PhCH=NPh]

Diazoketene **11** was distilled from the reaction of diazoacetyl chloride (**12**) and diazabicyclooctane (DABCO) in ether and identified by its reaction with EtOH followed by HBr (equation 12).[34]

$$N_2CHCOCl \xrightarrow{DABCO} N_2=C=C=O \xrightarrow{EtOH} N_2CHCO_2Et \xrightarrow{HBr} BrCH_2CO_2Et$$
$$\quad\quad\mathbf{12} \quad\quad\quad\quad \mathbf{11}$$

(12)

Pyrolysis of the 4-nitropyrroledione **13** was proposed to involve the nitroketene **14**, which gave the observed indole product **15**.[35] This is the only example of a nitroketene which has been reported. The $O_2NCH=C=O$ structure is cited in *Chemical Abstracts,* but this citation is derived from derivatives $(RS)_2C=CHNO_2$.[36]

[Structures of **13**, **14**, **15**]

(13)

Reaction of **16** with Et_3N could proceed through the iminoketene **17** which was trapped to give the observed **18** (eq. 14)[36a], but other pathways are also possible. Similarly reaction of **19** could involve the isocyanoketene **20**, leading to the observed **21** (eq. 15).[36b] Reaction of the isocyanate **22** was proposed to form the isocyanatoketene **23** which led to crosslinked and oligomeric products (eq. 16).[36c] Heating of hydroxyimino Meldrum's acid **24** in cyclohexanone in refluxing toluene for 2 hour evidently gives nitrosoketene (**25**), as evidenced by capture as the nitrone **27** in 60% yield (eq. 17).[36d] The reaction was proposed to involve initial [4 + 2] cycloaddition to **26**, which rearranged to **27**.[36d] The structures and energies of **25**,[36e] isocyanoketene (**28**),[36e] and isocyanovinylketene (**29**)[36f] have been calculated at the HF/6-31G*//HF/6-31G* level.

$$(MeS)_2C=NCH_2COCl \xrightarrow{Et_3N} (MeS)_2C=NCH=C=O \xrightarrow{Ph_2C=NPh} \mathbf{18}$$
$$\quad\quad\mathbf{16} \quad\quad\quad\quad\quad\quad \mathbf{17}$$

(14)

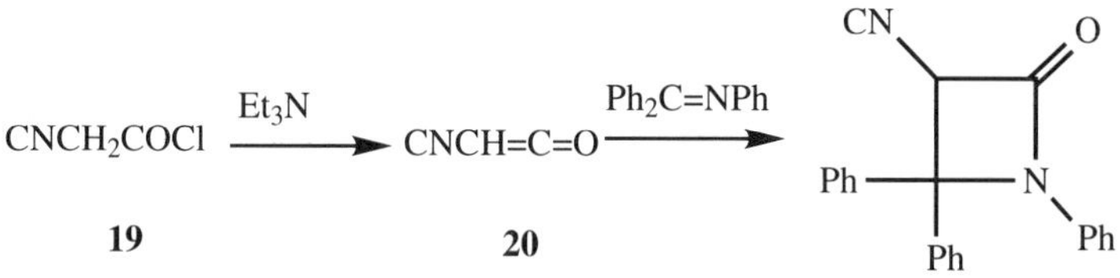

4.2 NITROGEN-SUBSTITUTED KETENES   **305**

$$\text{CNCH}_2\text{COCl} \xrightarrow{\text{Et}_3\text{N}} \text{CNCH=C=O} \xrightarrow{\text{Ph}_2\text{C=NPh}} \mathbf{21} \quad (15)$$

19   **20**

where **21** is a 4-membered β-lactam ring bearing CN, Ph, Ph substituents and N–Ph.

$$\text{O=C=NCH}_2\text{COCl} \xrightarrow{\text{Et}_3\text{N}} \text{O=C=NCH=C=O} \longrightarrow \text{polymer} \quad (16)$$

22   **23**

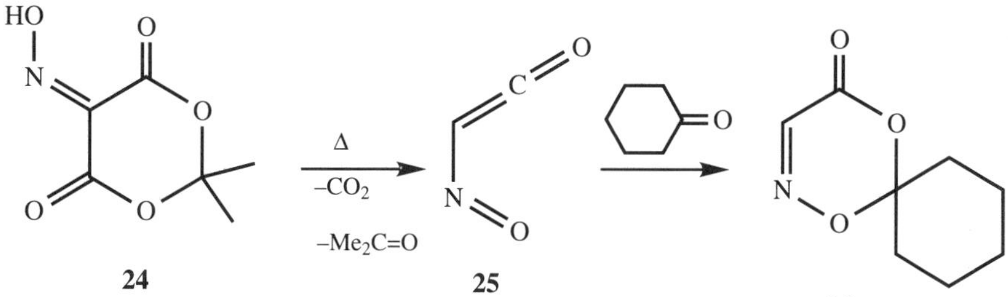

(17)

Structures **24** → **25** (O=C=CH–N=O) → **26** via cyclohexanone, with loss of CO_2 and $\text{Me}_2\text{C=O}$ (Δ).

26 → **27**

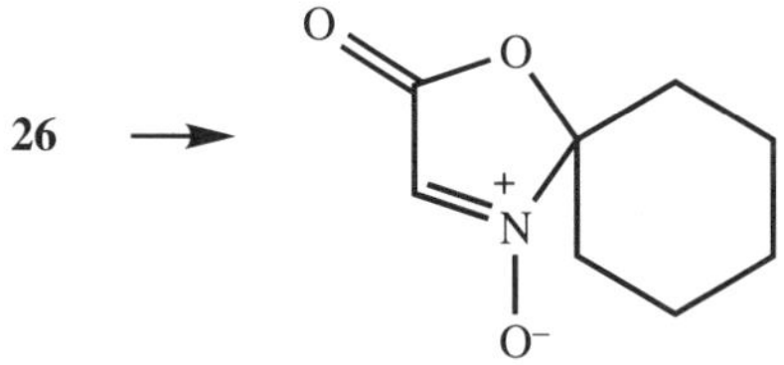

CNCH=C=O

28

CNCH=CHCH=C=O

29

Reported nitrogen-substituted ketenes are listed in Table 1.

TABLE 1. Nitrogen-Substituted Ketenes

Ketene	Formula	Preparation	Registry No.	Ref.
ONCH=C=O	C_2HNO_2	Theory	134736-48-4	11
O_2NCH=C=O	C_2HNO_3	Index entry	72751-68-9	36
N_3CH=C=O	C_2HN_3O	Dehydrochlorination	73786-06-8	28
H_2NCH=C=O	C_2H_3NO	Theory	126959-92-0	11
$(H_2N)_2C$=C=O	$C_2H_4N_2O$	Theory	111624-86-3	37
$ON\overset{+}{C}$=C=O$^{\bullet}$	$C_2NO_2^{\bullet}$	Mass spectrum	137846-94-7	38
$ON\overset{+}{C}$=C=O	$C_2NO_2^+$	Mass spectrum	38263-96-6	39
$(O_2N)_2C$=C=O	$C_2N_2O_5$	Theory	111647-29-1	37
N_2C=C=O	C_2N_2O	Dehydrochlorination	72228-30-9	34
$(N_2)_2C$=C=O^{2+}	C_2N_4O	Theory	111624-83-0	37
CNCH=C=O	C_3HNO	Theory	134736-47-3	11
O=C=NCH=C=O	C_3HNO_2		113366-01-1	40
CH_2=NCH=C=O	C_3H_3NO	Index entry	114214-75-4	
ONMeNCH=C=O(H$^+$)	$C_3H_4N_2O_2\cdot H$	Theory	81112-11-0	41
$Me_2NC(CN)$=C=O	$C_5H_6N_2O$	Thermolysis		42
Et_2NCH=C=O	$C_6H_{11}NO$	Wolff		26
R(ON)NCH=C=O	$C_7H_5N_3O_2$	Photolysis	102091-38-3	23
R = 3-pyridyl				
MeO_2CCH=CMeNHCH=C=O	$C_7H_9NO_3$	Elimination		22
Ph(ON)NCH=C=O	$C_8H_6N_2O_2$	Theory	99348-61-5	43
i-$Pr_2NC(OH)$=C=O$\cdot$Li	$C_8H_{15}NO_2\cdot Li$	Carbonylation	68986-64-1	44

Structure	Formula	Method	Ref.
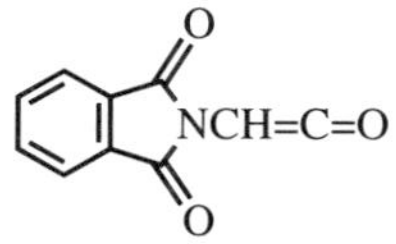 Phthalimide-NCH=C=O	$C_{10}H_5NO_3$	Dehydrochlorination	45
PhCONHCMe=C=C	$C_{10}H_9NO_2$	Index entry	None
PhCH$_2$CONHCH=C=O	$C_{10}H_9NO_2$	Dehydrochlorination	46
PhOCH$_2$CONHCH=C=O	$C_{10}H_9NO_3$	Dehydrochlorination	46
n-C$_7$H$_{15}$CONHCH=C=O	$C_{10}H_{17}NO_2$	Dehydrochlorination	46
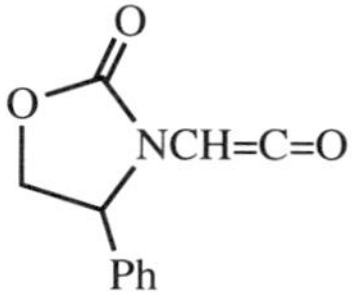 4-Ph-oxazolidinone-NCH=C=O	$C_{11}H_9NO_3$	Dehydrochlorination	5,7
			4
4-O$_2$NC$_6$H$_4$N=NC(CO$_2$Et)=C=O	$C_{11}H_9N_3O_5$	Dehydrochlorination	32
	$(C_{11}H_9N_3O_5)_3$		33
PhOCHMeCONHCH=C=O	$C_{11}H_{11}NO_3$	Dehydrochlorination	46
PhCHRCONHCH=C=O R = NHCOCHCl$_2$	$C_{12}H_{10}Cl_2N_2O_3$	Dehydrochlorination	46
PhOCHEtCONHCH=C=O	$C_{12}H_{13}NO_3$	Dehydrochlorination	46
Isoxazole-Ar/CCONHCH=C=O/CH$_2$OH			

307

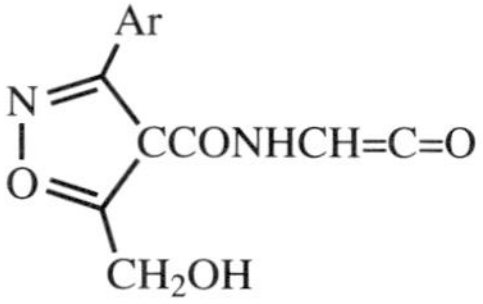

TABLE 1. (Continued)

Ketene	Formula	Preparation	Registry No.	Ref.
Ar = 2-F-6-ClC$_6$H$_3$	C$_{13}$H$_8$ClFN$_2$O$_4$	Mass spectroscopy	96315-63-8	47
= 2,6-Cl$_2$C$_6$H$_3$	C$_{13}$H$_8$Cl$_2$N$_2$O$_4$	Mass spectroscopy	96315-60-5	47
= 2-ClC$_6$H$_4$	C$_{13}$H$_9$ClN$_2$O$_4$	Mass spectroscopy	96315-56-9	47
= Ph	C$_{13}$H$_{10}$N$_2$O$_4$	Mass spectroscopy	96315-54-7	47
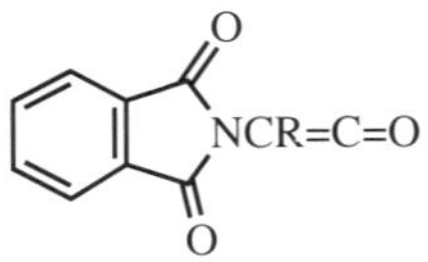 (phthalimido)NCR=C=O				
R = i-Pr	C$_{13}$H$_{11}$NO$_3$	Dehydrochlorination		21
R = t-Bu	C$_{14}$H$_{13}$NO$_3$	Dehydrochlorination	5511-74-0	21
R = Ph	C$_{16}$H$_9$NO$_3$	Dehydrochlorination		21
R = PhCH$_2$	C$_{17}$H$_{11}$NO$_3$	Dehydrochlorination		21
(PhCO)$_2$NCH=C=O	C$_{16}$H$_{11}$NO	Carbene complex		48
PhCONMeCPh=C=O	C$_{16}$H$_{13}$NO$_2$	Thermolysis	16108-16-0	20
Ar(BnO$_2$C)NC(Pr-i)=C=O	C$_{20}$H$_{17}$ClF$_3$NO$_3$		102822-54-8	49
Ar = 3-Cl-4-CF$_3$-C$_6$H$_3$				
RCOHNCR'=C=O	C$_{20}$H$_{20}$N$_2$O$_4$S	Speculative	83333-36-2	50
R = BnO$_2$CNHCH$_2$ R' = CH$_2$SBn				

References

1. Ikota, N.; Hanaki, A. *Heterocycles,* **1984,** *22,* 2227–2230.
2. Bose, A. K.; Kapur, J.; Sharma, S.; Manhas, M. S. *Tetrahedron Lett.* **1973,** 2319–2320.
3. Brady, W. T.; Gu, Y. Q. *J. Org. Chem.* **1989,** *54,* 2838–2842.
4. Ojima, I.; Chen, H.-J. C.; Qui, X. *Tetrahedron,* **1988,** *44,* 5307–5318.
5. Ojima, I.; Chen, H. C. *J. Chem. Soc., Chem. Commun.* **1987,** 625–626.
6. Evans, D. A.; Williams, J. M. *Tetrahedron Lett.* **1988,** *29,* 5065–5068.
7. Evans, D. A.; Sjogren, E. B. *Tetrahedron Lett.* **1985,** *26,* 3783–3786.
8. Evans, D. A.; Sjogren, E. B. *Tetrahedron Lett.* **1985,** *26,* 3787–3790.
9. Hegedus, L. S.; Imwinkelreid, R.; Alarid-Sargent, M.; Dvorak, D.; Satoh, Y. *J. Am. Chem. Soc.* **1990,** *112,* 1109–1117.
10. Brady, W. T.; Gu, Y. Q. *J. Org. Chem.* **1989,** *54,* 2834–2838.
10a. Reichen, W. *Chem. Rev.* **1978,** *78,* 569–588.
11. Gong, L.; McAllister, M.; Tidwell, T. T. *J. Am. Chem. Soc.* **1991,** *113,* 6021–6028.
12. Brady, W. T.; Dad, M. M. *J. Org. Chem.* **1991,** *56,* 6118–6122.
13. Footnote 36, ref. 12.
14. Bayer, H. O.; Huisgen, R.; Knorr, R.; Schaefer, F. C. *Chem. Ber.* **1970,** *103,* 2581–2597.
15. Boyd, G. V. In *Comprehensive Heterocyclic Chemistry,* Vol. 6; Potts, K. T., Ed.; Pergamon: New York, 1984; pp. 177–233.
16. Huisgen, R. In *Aromaticity,* Chemical Society Special Publication No. 21; The Chemical Society: London, 1967; pp. 51–73.
17. Ollis, W. D.; Ramsden, C. A. *Adv. Hetereocyclic Chem.* **1976,** *19,* 1–192.
18. Gingrich, H. L.; Baum, J. S. In *The Chemistry of Heterocyclic Compounds. Oxazoles,* Vol. 45; Turchi, I. J. Ed.; Wiley: New York, 1986; pp. 731–961.
19. Funke, E.; Huisgen, R. *Chem. Ber.* **1971,** *104,* 3222–3228.
20. Lukac, J.; Heimgartner, H. *Helv. Chim. Acta* **1979,** *62,* 1236–1245.
21. Winter, S.; Pracejus, H. *Chem. Ber.* **1966,** *99,* 151–159.
22. Palomo, C.; Cossio, F. P.; Cuevas, C.; Lecea, B.; Mielgo, A.; Roman, P.; Luque, A.; Martinez-Ripoll, M. *J. Am. Chem. Soc.* **1992,** *114,* 9360–9369.
23. Nespurek, S.; Böhm, S.; Kuthan, J. *THEOCHEM,* **1986,** *29,* 261–273.
24. Clapp, L. B. In *Comprehensive Heterocyclic Chemistry,* Vol. 6; Potts, K. T., Ed.; Pergamon: New York, 1984; pp. 365–391.
25. Boyd, R. J.; Boyd, S. L. *J. Am. Chem. Soc.* **1992,** *114,* 1652–1654.
26. Tomioka, H.; Kitagawa, H.; Izawa, Y. *J. Org. Chem.* **1979,** *44,* 3072–3075.
27. Bose, A. K.; Anjaneyulu, B.; Bhattacharya, S. K.; Manhas, M. S. *Tetrahedron* **1967,** *23,* 4769–4776.
28. Ojima, I,; Nakahashi, K.; Brandstadter, S. M.; Hatanaka, N. *J. Am. Chem. Soc.* **1987,** *109,* 1798–1805.
29. Sullivan, D. F.; Scopes, D. I. C.; Kluge, A. F.; Edwards, J. A. *J. Org. Chem.* **1976,** *41,* 1112–1117.
30. Hoz, S.; Wolk, J. L. *Tetrahedron Lett.* **1990,** *31,* 4085–4088.
31. Amyes, T. L.; Richard, J. P. *J. Am. Chem. Soc.* **1991,** *113,* 1867–1869.

32. Bodnar, V. N.; Lozindskii, M. O. *Ukr. Khim. Zh.* **1983**, *49*, 301–303; *Chem. Abstr.* **1983**, *98*, 197909f.
33. Lozinskii, M. O.; Bodnar, V. N.; Pel'kis, P. S. *Zh. Org. Khim.* **1980**, *16*, 228–229; *Chem. Abstr.* **1980**, *93*, 46125t.
34. Bestmann, H. J.; Soliman, F. M. *Angew. Chem., Int. Ed. Engl.* **1979**, *18*, 947–948.
35. Kappe, C. O.; Kollenz, G.; Wentrup, C. *J. Chem. Soc., Chem. Commun.* **1992**, 485–486.
36. Rajappa, S. *Indian Patent* IN 144277; *Chem. Abstr.* **1980**, *92*, 76546c.
36a. Hoppe, D.; Raude, E. *Liebigs Ann. Chem.* **1979**, 2076–2088.
36b. Hoppe, I.; Schöllkopf, U. *Chem. Ber.* **1976**, *109*, 482–487.
36c. Mormann, W.; Hoffmann, S.; Hoffmann, W. *Chem. Ber.* **1987**, *120*, 285–290.
36d. Katagiri, N.; Kurimoto, A.; Yamada, A.; Sato, H.; Katsuhara, T.; Takagi, K.; Kaneko, C. *J. Chem. Soc., Chem. Commun.* **1994**, 281–282.
36e. McAllister, M. A.; Tidwell, T. T. *J. Org. Chem.* **1994**, *59*, 4506–4515.
36f. Nguyen, M. T.; Hajnal, M. R.; Vanquickenborne, L. G. *J. Chem. Soc., Perkin Trans. 2*, **1994**, 169–170.
37. Gano, J. E.; Jacob, E. J. *Spectrochim. Acta* **1987**, *43A*, 1023–1025.
38. Suelzle, D.; O'Bannon, P. E.; Schwarz, H. *Chem. Ber.* **1992**, *125*, 279–283.
39. Jolly, W. L.; Gin, C. *Int. J. Mass. Spectrom. Ion Phys.* **1977**, *25*, 27–37.
40. Oda, R. *Kagaku* **1987**, *42*, 782–783; *Chem. Abstr.* **1988**, *108*, 130951v.
41. Eckert-Maksic, M.; Maksic, Z. B. *J. Chem. Soc., Perkin Trans. 2*, **1981**, 1462–1466.
42. Labille, M.; Janousek, Z.; Viehe, H. G. *Tetrahedron* **1991**, *47*, 8161–8166.
43. Eckert-Maksic, M.; Maksic, Z. B. *Croat. Chem. Acta* **1985**, *58*, 15–27; *Chem. Abstr.* **1986**, *104*, 5212w.
44. Rautenstrauch, V.; Joyeux, M. *Angew. Chem., Int. Ed. Engl.* **1979**, *18*, 83–85.
45. Bellus, D. *J. Org. Chem.* **1979**, *44*, 1208–1211.
46. Krebs, R. *Ger. (East) Patent* DD. 84384; *Chem. Abstr.* **1971**, *78*, 72130a.
47. Murai, Y.; Nakagawa, T.; Uno, T. *Chem. Pharm. Bull.* **1985**, *33*, 383–387.
48. Hegedus, L. S.; de Weck, G.; D'Andrea, S. *J. Am. Chem. Soc.* **1988**, *110*, 2122–2126.
49. Stoutamire, D. W.; Tieman, C. H. *U.S. Patent* US 4560515; *Chem. Abstr.* **1986**, *105*, 42497j.
50. Kovacs, J.; Hsieh, Y. *J. Org. Chem.* **1982**, *47*, 4996–5002.

4.3 OXYGEN-SUBSTITUTED KETENES

The electronegative OH substituent is calculated to cause an isodesmic destabilization of ketenes of -14.2 kcal/mol, almost as great as that of -17.2 kcal/mol caused by fluorine.[1] It was concluded that ketenes are destabilized by both electronegative groups and by π donors,[1] and OH is destabilizing by both mechanisms. Thus it is to be expected that such oxygen-substituted ketenes will be quite reactive if low barrier reaction pathways are available. Alkoxy and silyloxy groups have similar electronic properties to OH and are also expected to be strongly destabilizing to ketenes. The situation with acyloxy groups is not clear, as the π donor ability

4.3 OXYGEN-SUBSTITUTED KETENES

of oxygen is greatly reduced by acylation, but the calculated group electronegativities of OH, OCH$_3$, and O$_2$CCH$_3$ are 3.55, 3.53, and 3.57, respectively.[2] The close similarities of these values means that group electronegativities by themselves are not a helpful guide to the relative stabilities in this series.

Hydroxyketene (HOCH=C=O, **1**) is a valence tautomer of dihydroxyacetylene (HOC≡COH) and glyoxal (O=CHCH=O), and the interconversion of these isomers has been considered theoretically.[3] The X-ray structure of tantalum coordinated HOC≡COH has been determined,[4] and the detection of **1** by IR spectroscopy from the photolysis of CH$_2$=O in a CO matrix has been reported, and its formation by the process of equation 1 was proposed.[5] There is a long history of the study of the dianion C$_2$O$_2^{2-}$, as discussed in Section 4.8.1.[3]

$$CH_2=O \xrightarrow{h\nu} HO\ddot{C}H \xrightarrow{CO} HOCH=C=O \quad (1)$$

1

The formation of PhC(OH)=C=O (**2**) as a transient species by the photolysis of pyruvate esters has been proposed,[6,7] and photolysis of PhCOCO$_2$Me or PhCN$_2$CO$_2$H (equations 2 and 3) in H$_2$O leading to **2** has been reported, with characterization of **2** by UV spectroscopy and formation of mandelic acid.[8] The hydration kinetics of **2** were also measured by UV spectroscopy.[8] The mass spectrometric formation of ketenes RC(OH)=C=O derived from the fragmentation of deprotonated methyl α-hydroxy esters RCH(OH)CO$_2$Me has also been proposed.[9] There has been speculation that **2** could be formed by decarboxylation of a diketo acid, but there was no positive evidence for this pathway (equation 4).[10] The photolyses of amides of pyruvic acids also give hydroxyketenes by a process analogous to that of equation 2.[11] Reaction of **2** with imines leads to hydroxy β-lactams.[11]

$$(2)$$

$$(3)$$

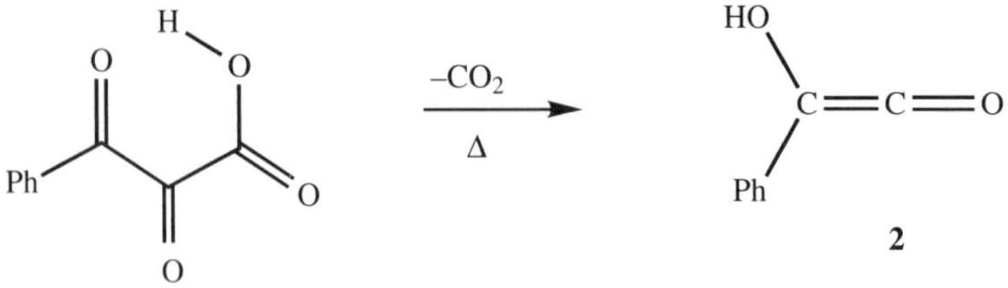

[Equation (4) shows a reaction: Ph-CO-CO-O-CHO (with H-O) undergoes $-CO_2$, Δ to give HO(Ph)C=C=O (**2**)]

Photolysis of **3** and **4** in $CHCl_3$ containing EtOH gave the ketenes **5** and **6**, as evidenced by isolation of the esters **7** and **8**.[12]

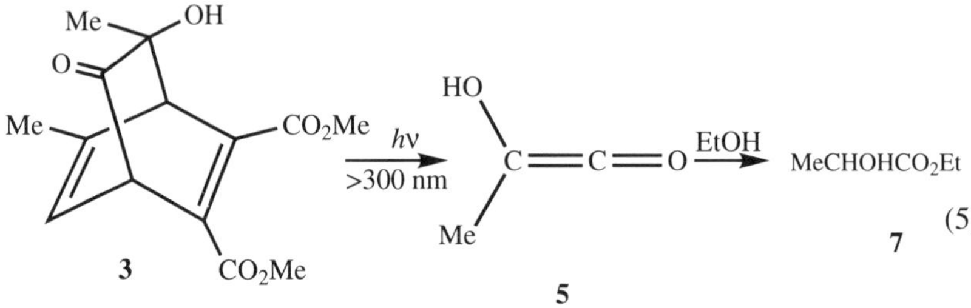

[Equation (5)[11]: Compound **3** (bicyclic diene with Me, OH, CO₂Me, CO₂Me substituents) $\xrightarrow{h\nu, >300\text{ nm}}$ HO(Me)C=C=O (**5**) $\xrightarrow{EtOH}$ $MeCHOHCO_2Et$ (**7**)]

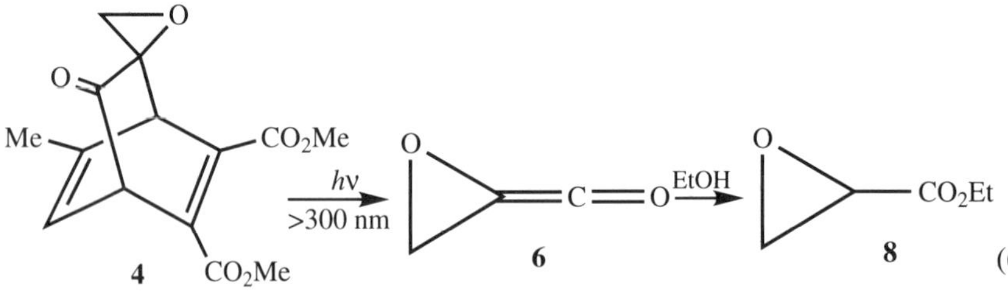

[Equation (6)[11]: Compound **4** (bicyclic with epoxide, Me, CO₂Me, CO₂Me) $\xrightarrow{h\nu, >300\text{ nm}}$ cyclopropyl-substituted ketene **6** $\xrightarrow{EtOH}$ cyclopropyl-CO_2Et (**8**)]

Reaction of lithium diisopropylamide with CO gives a 35% yield of **9** after hydrolysis, and this was attributed to the intermediacy of the acyl anion **10** and the ketenyl enolate **11**.[13]

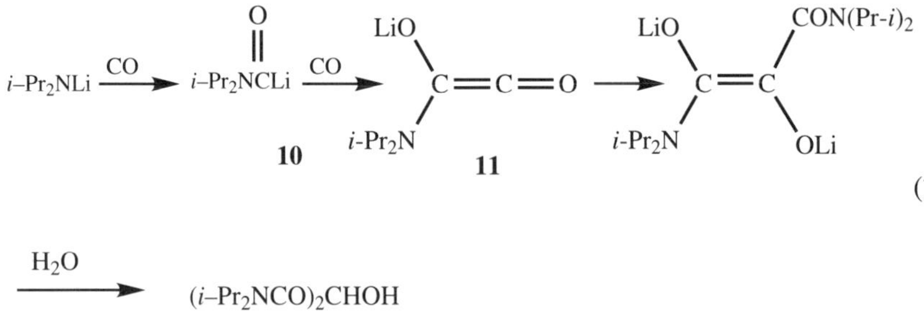

[Equation (7): $i\text{-}Pr_2NLi \xrightarrow{CO} i\text{-}Pr_2NCLi(=O)$ (**10**) $\xrightarrow{CO}$ ($i\text{-}Pr_2N$)(LiO)C=C=O (**11**) $\rightarrow$ ($i\text{-}Pr_2N$)(LiO)C=C(OLi)(CON(Pr-i)₂) $\xrightarrow{H_2O}$ ($i\text{-}Pr_2NCO)_2CHOH$ (**9**)]

Ketenes with alkoxy and aryloxy substituents have been generated by the reaction of acyl chlorides with Et_3N (equation 8),[14–20] the reaction of acids RCO_2H with Ac_2O and NaOAc via the generation of mixed anhydrides,[17] by

4.3 OXYGEN-SUBSTITUTED KETENES

photo-Wolff rearrangements (equation 9),[21,22] and by the photolysis of chromium-alkoxycarbenes (equation 10).[23] The latter process is actually thought to involve a ketene–chromium complex and not the free carbene. With suitable chiral substituents, R, R^1 the formation of **12** is highly diastereoselective.[23]

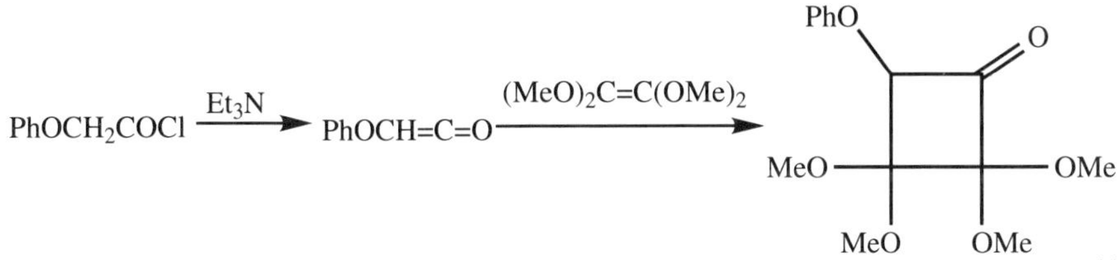

$$PhOCH_2COCl \xrightarrow{Et_3N} PhOCH=C=O \xrightarrow{(MeO)_2C=C(OMe)_2} \text{(cyclobutanone product)} \quad (8)^{16}$$

$$N_2CHCO_2Et \xrightarrow{h\nu} EtOCH=C=O \xrightarrow{E\text{-MeCH=CHMe}} \text{(cyclobutanone product)} \quad (9)^{21}$$

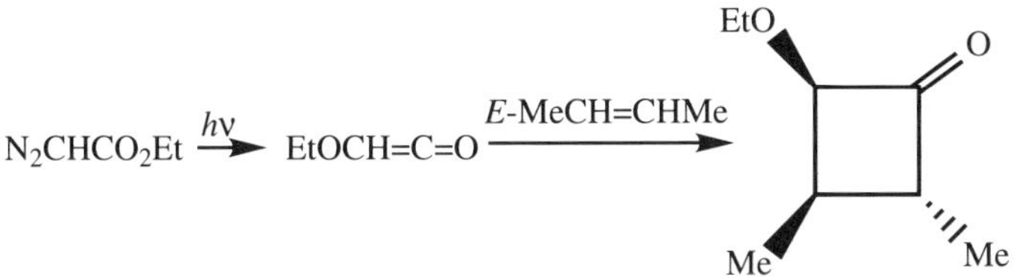

$$\underset{Me}{\overset{MeO}{>}}C=Cr(CO)_5 \xrightarrow[CO]{h\nu} \underset{Me}{\overset{MeO}{>}}C=C=O \cdot Cr(CO)_4 \xrightarrow{RR^1NCH=CH_2} \mathbf{12} \quad (10)^{23}$$

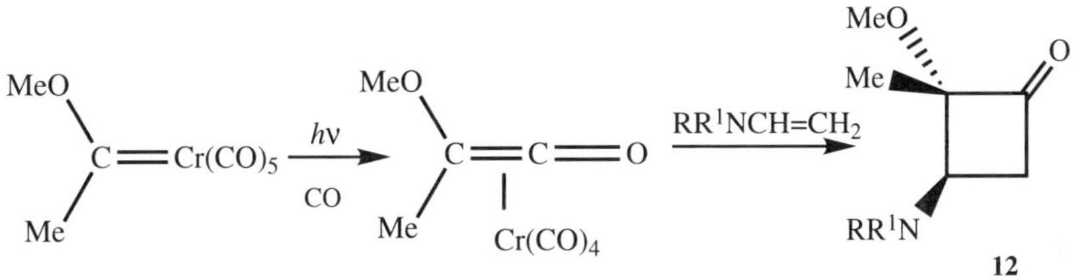

The thermolysis of di-*tert*-butoxyethyne (**13**) provides a route to *tert*-butoxyketene (**14**), which then cycloadds to **13** to give a high yield of a cyclobutenone (equation 11).[3] Disubstituted ketenes such as $CCl_2=C=O$ cycloadd to **13** in much lower yield, a result attributed to steric hindrance to the in-plane approach to $CCl_2=C=O$.[3] It has also been proposed that the structure assigned as KOC≡COK reacts with MeI to give the ketene **15**, which dimerizes to **16** (equation 12).[3]

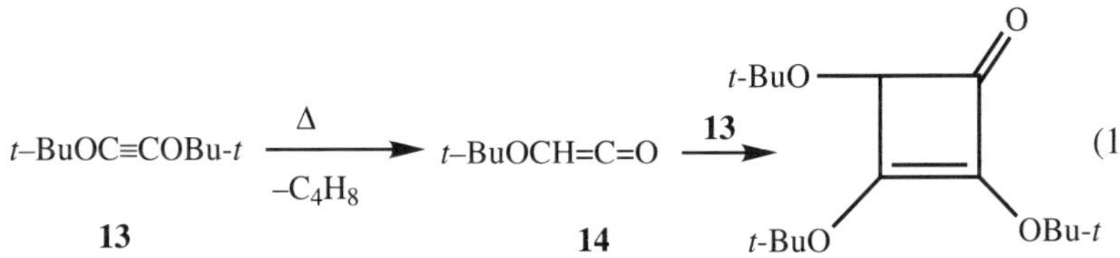

$$t\text{-BuOC}\equiv\text{COBu-}t \xrightarrow[-C_4H_8]{\Delta} t\text{-BuOCH=C=O} \xrightarrow{\mathbf{13}} \text{(cyclobutenone)} \quad (11)^3$$
13 **14**

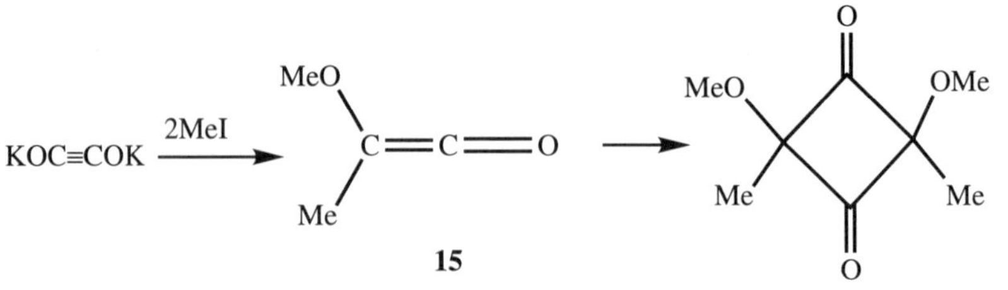

In the reactions shown in equation 13–15 the ketenes were not been directly observed but were trapped in situ. Previous reports of the isolation of free phenoxy ketenes evidently involved dimers or oligomers.[24–27] Often the oxygen–substituted ketenes are trapped with imines to give β-lactams (equation 13).[14,15] Others react in intramolecular [2 + 2] cycloadditions with carbonyl groups to give β-propiolactones (equation 14)[17] or with alkenyl groups to give cyclobutanones (equation 15).[18]

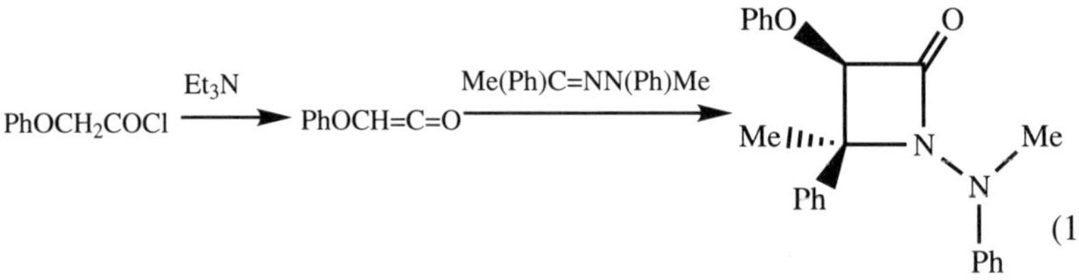

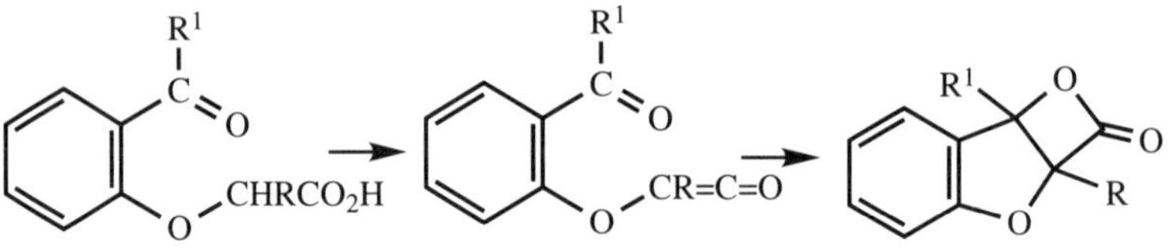

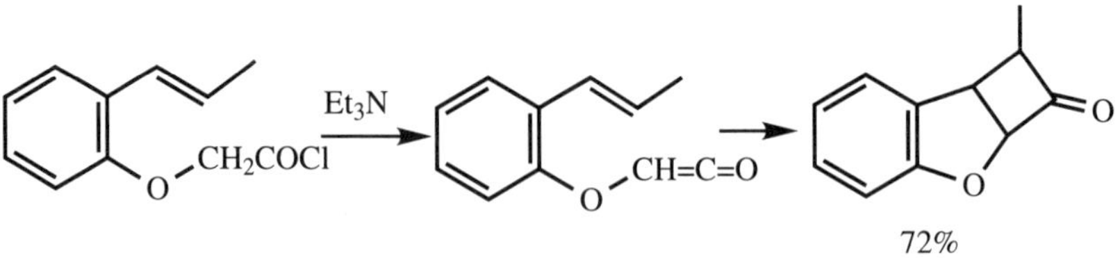

The alkoxyketene $Ph_3SiC(OEt)=C=O$, which has the powerfully stabilizing, and bulky, Ph_3Si substituent, was generated from metal carbene complexes and trapped in situ.[28]

Acetoxyketene (**17**) has been generated from the acyl chloride[16,29,30] and trapped with cyclopentadiene to give the *exo/endo* isomers of **18** in an isolated ratio of 1/22 (equation 16).[29] The ketene PhC(OAc)=C=O (**19**) generated similarly reacted with biacetyl to produce the β-propiolactone **20** (equation 17).[31] The preferred formation of the stereoisomer **20** was rationalized in terms of a $[_\pi 2_a + _\pi 2_s]$ transition state **21** in which there was a favorable secondary orbital interaction between the two carbonyl groups.[31] A stepwise process would involve an α-acylcarbocation **22** that is, however, stabilized by an α-oxygen.

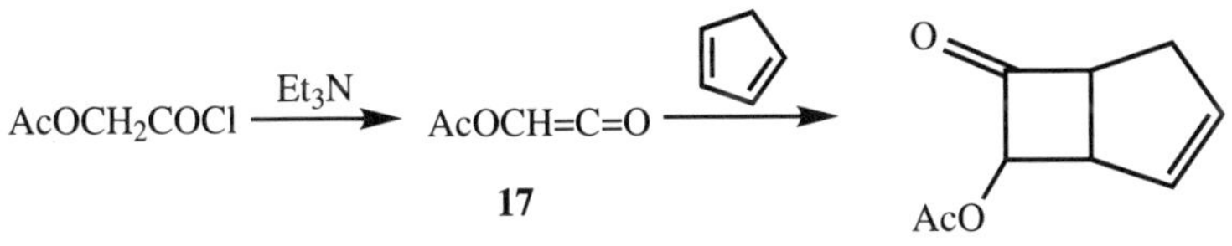

$$\text{AcOCH}_2\text{COCl} \xrightarrow{\text{Et}_3\text{N}} \text{AcOCH=C=O} \longrightarrow \mathbf{18} \quad (16)^{29}$$

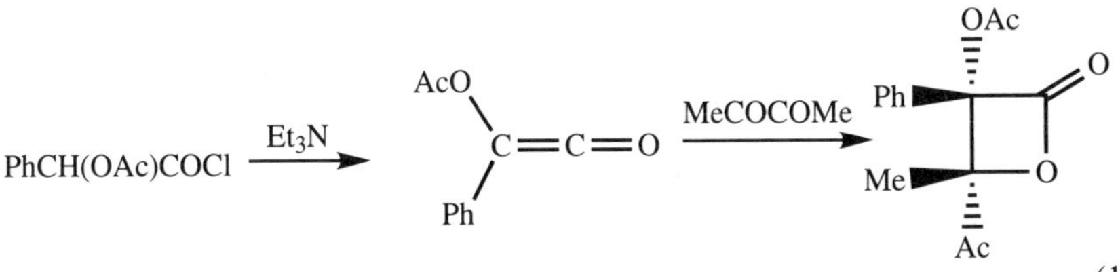

$$\text{PhCH(OAc)COCl} \xrightarrow{\text{Et}_3\text{N}} \mathbf{19} \xrightarrow{\text{MeCOCOMe}} \mathbf{20} \quad (17)^{31}$$

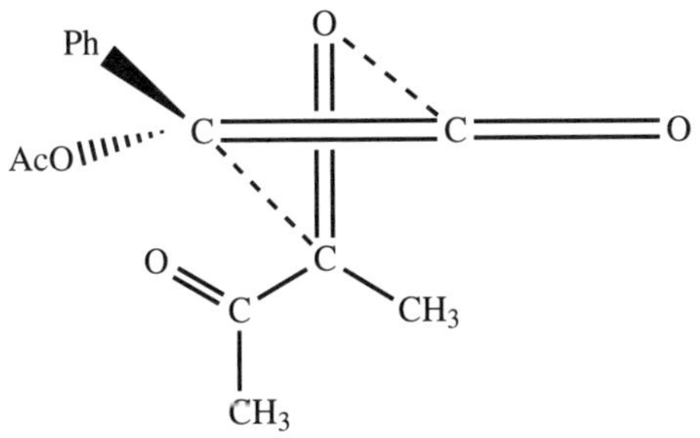

21

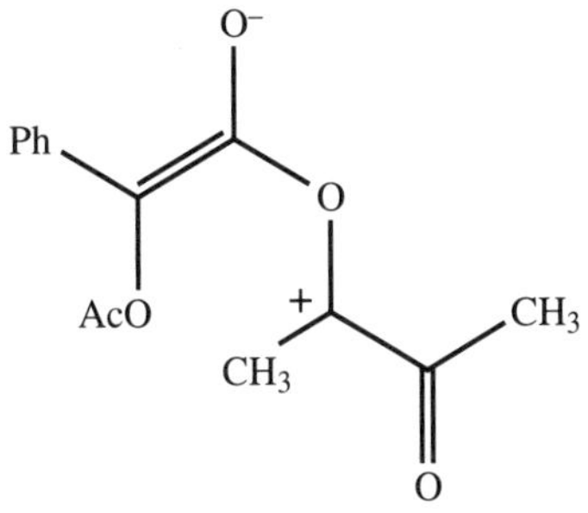

22

Gas phase pyrolysis of methyl diazomalonate was proposed to lead successively to ketenes **23** and **24** as intermediates in the formation of the final products **25–28** (equations 18–21).[32]

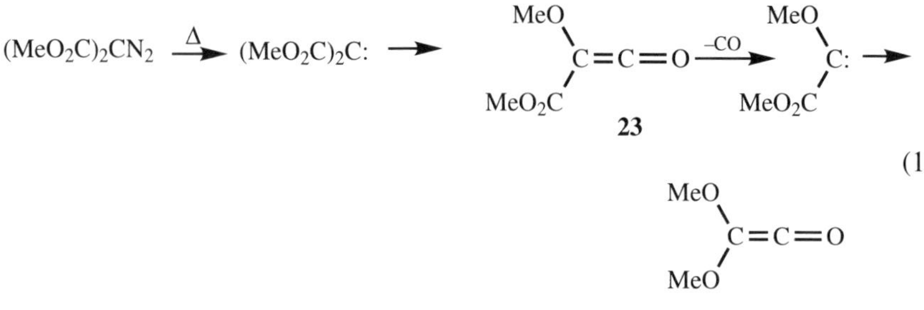

$$(MeO_2C)_2CN_2 \xrightarrow{\Delta} (MeO_2C)_2C: \longrightarrow \underset{\underset{MeO_2C}{\diagup}}{\overset{\overset{MeO}{\diagdown}}{C}}=C=O \xrightarrow{-CO} \underset{\underset{MeO_2C}{\diagup}}{\overset{\overset{MeO}{\diagdown}}{C}}: \longrightarrow \qquad (18)$$

$$\underset{\underset{MeO}{\diagup}}{\overset{\overset{MeO}{\diagdown}}{C}}=C=O$$

24

Structures showing:

$\underset{\underset{MeO_2C}{\diagup}}{\overset{\overset{MeO}{\diagdown}}{C:}} \longrightarrow$ [β-lactone with MeO, MeO substituents] $\xrightarrow{-CO_2}$ MeOCH=CH$_2$ $\qquad \underset{\underset{MeO_2C}{\diagup}}{\overset{\overset{MeO}{\diagdown}}{C:}} \longrightarrow MeO_2CCOCH_3$

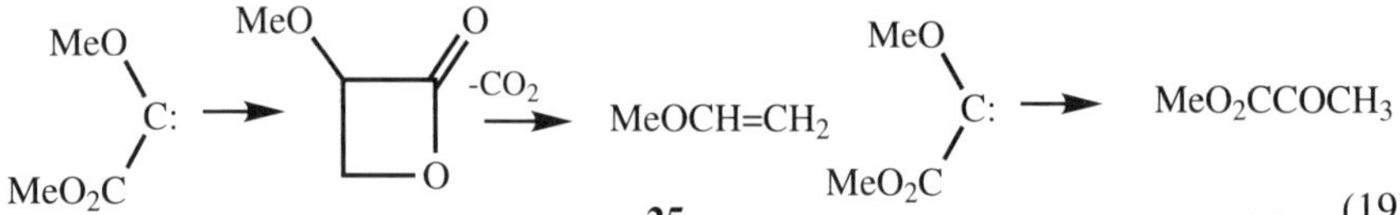

25 **26** (19)

$(MeO_2C)_2C: \longrightarrow$ [β-lactone with MeO$_2$C] $\xrightarrow{-CO_2}$ CH$_2$=CHCO$_2$Me

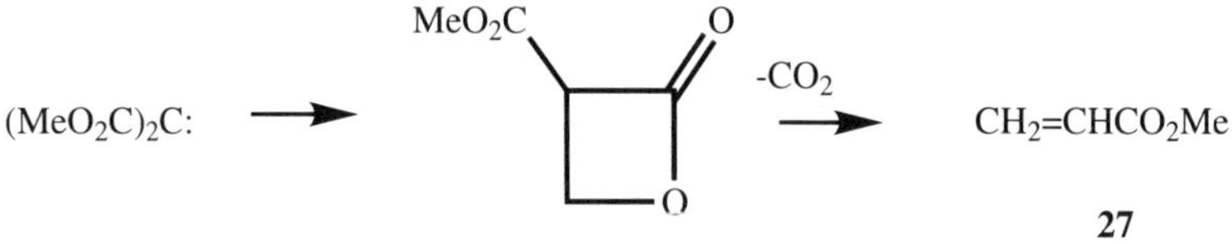

27 (20)

$$\mathbf{24} \xrightarrow{-CO} (MeO)_2C: \longrightarrow MeCO_2Me \qquad (21)$$

28

Pyrolysis of the dehydration product of **29** may have given **30**, as evidenced by capture by Ph$_2$C=NPh, giving the β-lactam (equation 22).[33]

$$(PhO)_2C(CO_2H)_2 \xrightarrow{\Delta} (PhO)_2C=C=O \qquad (22)$$

29 **30**

Pyrolysis of the Meldrum's acid derivative **31** at 460 °C gave the ketene **32** as evidenced by the observed formation of **33** (equation 23).[34]

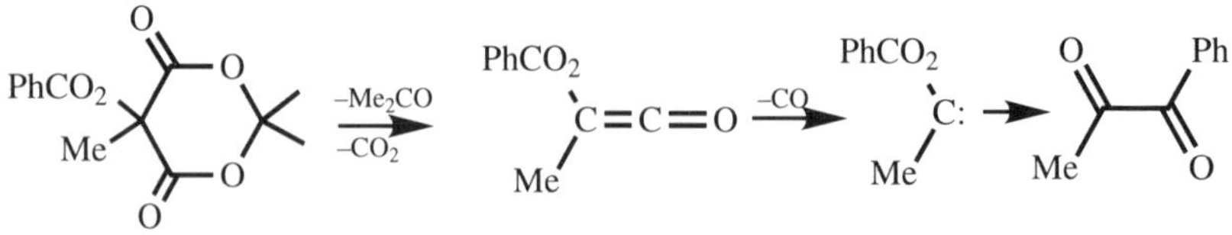

[Meldrum's acid derivative with PhCO$_2$ and Me] $\xrightarrow[-CO_2]{-Me_2CO}$ $\underset{\underset{Me}{\diagup}}{\overset{\overset{PhCO_2}{\diagdown}}{C}}=C=O \xrightarrow{-CO} \underset{\underset{Me}{\diagup}}{\overset{\overset{PhCO_2}{\diagdown}}{C:}} \longrightarrow$ [product with Me, Ph, O, O]

31 **32** **33** (23)

TABLE 1. Oxygen-Substituted Ketenes

Ketene	Formula	Preparation	Registry No.	Ref.
HOCH=C=O	$C_2H_2O_2$	Photolysis	74936-20-2	5
(-O)(-O)C=C=O				
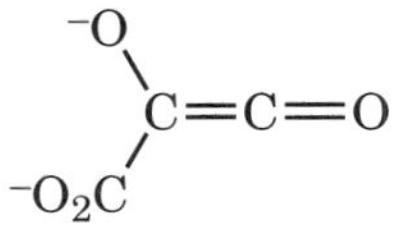 ^-O_2C-△-C=O	C_3O_4	Photolysis	126829-11-6	35
O-△-C=O				
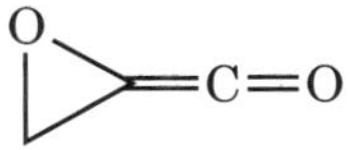 MeOCH=C=O	$C_3H_2O_2$	Photolysis	57091-51-7	12
HOC(CH=C=O)=C=O	$C_3H_4O_2$	Dehydrochlorination	54276-52-7	16
	$C_4H_2O_3$	Photolysis	118919-78-1	36
HOC(CH=NCHO)=C=O	$C_4H_3NO_3$	Theory	106104-57-8	37
HOC(CONHCHO)=C=O	$C_4H_3NO_4$	Theory	106104-58-9	37
AcOCH=C=O	$C_4H_4O_3$	Dehydrochlorination	55778-27-3	16,29
EtOCH=C=O	$C_4H_6O_2$	Dehydrochlorination	28288-35-9	16
MeOCMe=C=O	$C_4H_6O_2$	Dehydrochlorination	140666-37-1	38
(MeO)$_2$C=C=O	$C_4H_6O_3$	Wolff, theory	111624-85-2	32,39
CH$_2$=CHCH$_2$OCH=C=O	$C_5H_6O_2$	Dehydrochlorination	114081-53-7	40
AcOCMe=C=O	$C_5H_6O_3$	Pyrolysis		34
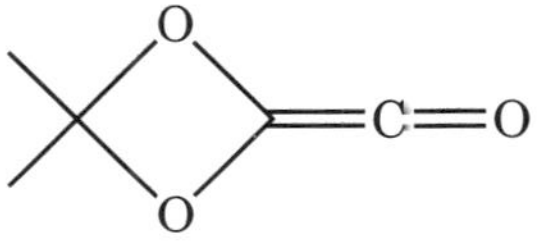 (O-□-O with C=O)	$C_5H_6O_3$	Wolff		41

TABLE 1. (Continued)

Ketene	Formula	Preparation	Registry No.	Ref.
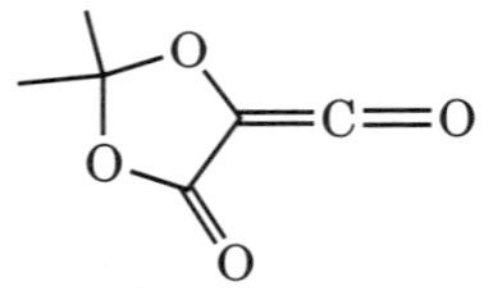 (dioxolane structure with C=O)	$C_6H_6O_4$	Wolff	73454-18-9	41,42
t-BuOCH=C=O	$C_6H_{10}O_2$	Alkoxyalkyne	66478-65-7	43
ROCH=C=O R = MeOCH$_2$CH$_2$OCH$_2$	$C_6H_{10}O_4$	Dehydrochlorination	114094-26-7	40
Me$_3$SiOCMe=C=O	$C_6H_{12}O_2Si$	Photolysis	114790-10-2	44
ROC(CN)=C=O R = CH$_2$=CMeCH$_2$	$C_7H_7NO_2$	Pyrolysis		45
2,4-Cl$_2$C$_6$H$_3$OCH=C=O	$C_8H_4Cl_2O_2$	Dehydrochlorination	98556-40-2[a]	25
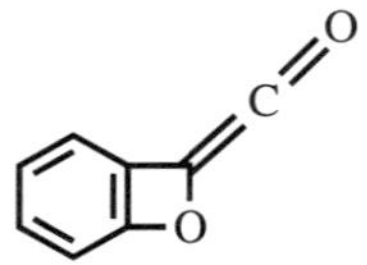 (benzofuranylidene ketene structure)	C_8H_4O	Photolysis	57912-00-2	46
2-ClC$_6$H$_4$OCH=C=O	$C_8H_5ClO_2$	Dehydrochlorination		47
PhOCH=C=O	$C_8H_6O_2$	Dehydrochlorination	107855-45-8	14,20
ROC(CN)=C=O R = CH$_2$=CMeCH$_2$CH$_2$	$C_8H_9NO_2$	Pyrolysis		45
t-BuCH$_2$OC(CN)=C=O	$C_8H_{11}NO_2$	Pyrolysis		45
LiOCN(Pr-i)$_2$=C=O	$C_8H_{14}NO_2 \cdot$Li		68986-64-1	13

Compound	Formula	CAS	Method	Ref
MeOCPh=C=O	$C_9H_8O_2$	98380-71-3	Wolff	48,49
PhCH$_2$OCH=C=O	$C_9H_8O_2$		Dehydrochlorination	30,40,50
t-BuMe$_2$SiOCMe=C=O	$C_9H_{18}O_2Si$	114790-11-3	Photolysis	44
HOC(C≡CPh)=C=O	$C_{10}H_6O_2$	119622-67-2	Pyrolysis	51
ArOCEt=C=O	$C_{10}H_8Cl_2O_2$	109570-90-3	Dehydrochlorination[a]	24
Ar = 2,4-Cl$_2$C$_6$H$_3$	$(C_{10}H_8Cl_2O_2)_2$	109570-91-4		24
AcOCPh=C=O	$C_{10}H_8O_3$	85539-76-0	Dehydrochlorination	31
PhCO$_2$CMe=C=O	$C_{10}H_8O_3$		Pyrolysis	34
PhOCEt=C=O	$C_{10}H_{10}O_2$		Malonic anhydride	33
4-t-BuC$_6$H$_4$OCH=C=O	$C_{12}H_{14}O_2$	100121-79-7[a]	Dehydrochlorination	27
	$(C_{12}H_{14}O_2)_2$	121177-44-4		27
Me$_3$SiOCR=C=O	$C_{13}H_{14}O_2Si$	119622-68-3	Pyrolysis	51,52
R = PhC≡C				
2-R-5-MeC$_6$H$_3$OCMe=C=O	$C_{13}H_{14}O_2$	123870-45-1	Dehydrochlorination	53
R = CH$_2$=CMe				
MenthylOCMe=C=O	$C_{13}H_{22}O_2$		Dehydrochlorination	54
(PhO)$_2$C=C=O	$C_{14}H_{10}O_3$		Malonic anhydride	33
4-t-BuC$_6$H$_4$OCEt=C=O	$C_{14}H_{18}O_2$	100971-94-6[a]	Dehydrochlorination	27
	$(C_{14}H_{18}O_2)_2$	119338-94-2		27
PhCO$_2$CPh=C=O	$C_{15}H_{10}O_3$		Pyrolysis	34
4-t-BuC$_6$H$_4$OC(R)=C=O	$C_{15}H_{20}O_2$	109648-69-3[a]	Dehydrochlorination	27
R = n-Pr	$(C_{15}H_{20}O_2)_2$	123702-50-1		27
4-t-BuC$_6$H$_4$OCR=C=O	$C_{16}H_{22}O_2$	101277-12-7[a]	Dehydrochlorination	27
R = n-Bu	$(C_{16}H_{22}O_2)_2$	120857-56-9		27
PhCO$_2$CR=C=O	$C_{17}H_{14}O_4$		Pyrolysis	34
R = 4-MeOC$_6$H$_4$CH$_2$				
EtOC(SiPh$_3$)=C=O	$C_{22}H_{20}O_2Si$	122300-87-2	Cr Carbene	28

[a] The structural assignment as a stable monomer is evidently incorrect.

References

1. Gong, L.; McAllister, M. A.; Tidwell, T. T. *J. Am. Chem. Soc.* **1991**, *113*, 6021–6028.
2. Boyd, R. J.; Boyd, S. L. *J. Am. Chem. Soc.* **1992**, *114*, 1652–1655.
3. Serratosa, F. *Acc. Chem. Res.* **1983**, *16*, 170–176.
4. Vrtis, R. N.; Rao, C. P.; Bott, S. G.; Lippard, S. J. *J. Am. Chem. Soc.* **1988**, *110*, 7564–7566.
5. Lee, E. K. C.; Sodeau, J. R.; Diem, M.; Shibuya, K. *Proc. Yamada Conf. Free Radicals, 3rd,* **1979**, 98–112; *Chem. Abstr.* **1980**, *93*, 140857e.
6. Huyser, E. S.; Neckers, D. C. *J. Org. Chem.* **1964**, *29*, 276–278.
7. Encinas, M. V.; Lissi, E. A.; Zanocco, A.; Stewart, L. C.; Scaiano, J. C. *Can. J. Chem.* **1984**, *62*, 386–391.
8. Chiang, Y.; Kresge, A. J.; Pruszynski, P.; Schepp, N. P.; Wirz, J. *Angew. Chem., Int. Ed. Engl.* **1990**, *29*, 792–794.
9. Eichinger, P. C. H.; Hayes, R. N.; Bowie, J. H. *J. Chem. Soc., Perkin Trans. 2,* **1990**, 1815–1820.
10. Dahn, H.; Rotzler, G. *J. Org. Chem.* **1991**, *56*, 3080–3082.
11. Aoyama, H.; Sakamoto, M.; Kuwabara, K.; Yoshida, K.; Omote, Y. *J. Am. Chem. Soc.* **1983**, *105*, 1958–1964.
12. Becker, H.; Ruge, B. *Angew. Chem., Int. Ed. Engl.* **1975**, *14*, 761–762.
13. Rautenstrauch, V.; Joyeux, M. *Angew. Chem., Int. Ed. Engl.* **1979**, *18*, 83–84.
14. Sharma, S. D.; Pandhi, S. B. *J. Org. Chem.* **1990**, *55*, 2196–2200.
15. Borer, B. C.; Balogh, D. W. *Tetrahedron Lett.* **1991**, *32*, 1039–1040.
16. Bellus, D. *J. Org. Chem.* **1979**, *44*, 1208–1211.
17. Brady, W. T.; Gu, Y.-Q. *J. Org. Chem.* **1988**, *53*, 1353–1356.
18. Brady, W. T.; Giang, Y. F. *J. Org. Chem.* **1985**, *50*, 5177–5179.
19. Snider, B. B.; Hui, R. A. H. F. *J. Org. Chem.* **1985**, *50*, 5167–5176.
20. Borrmann, D.; Wegler, R. *Chem. Ber.* **1966**, *99*, 1245–1251.
21. DoMinh, T.; Strausz, O. P. *J. Am. Chem. Soc.* **1970**, *92*, 1766–1768.
22. Chaimovich, H.; Vaughan, R. J.; Westheimer, F. H. *J. Am. Chem. Soc.* **1968**, *90*, 4088–4093.
23. Hegedus, L. S.; Bates, R. W.; Soderberg, B. F. *J. Am. Chem. Soc.* **1991**, *113*, 923–927.
24. Hill, C. M.; Schofield, H. I.; Spriggs, A. S.; Hill, M. E. *J. Am. Chem. Soc.* **1951**, *73*, 1660–1662.
25. Hill, C. M.; Senter, G. W.; Hill, M. E. *J. Am. Chem. Soc.* **1950**, *72*, 2286–2287.
26. Hill, C. M.; Hill, M. E.; Williams, A. O.; Shelton, E. M. *J. Am. Chem. Soc.* **1953**, *75*, 1084–1086.
27. Hill, C. M.; Woodberry, R.; Hill, M. E.; Williams, A. O. *J. Am. Chem. Soc.* **1959**, *81*, 3372–3374.
28. Kron, J.; Schubert, U. *J. Organomet. Chem.* **1989**, *373*, 203–219.
29. Russell, G. A.; Schmitt, K. D.; Mattox, J. *J. Am. Chem. Soc.* **1975**, *97*, 1882–1891.
30. Ojima, I.; Yamato, T.; Nakahashi, K. *Tetrahedron Lett.* **1985**, *26*, 2035–2038.
31. Dominguez, D.; Cava, M. P. *Tetrahedron Lett.* **1982**, *23*, 5513–5516.

32. Richardson, D. C.; Hendrick, M. E.; Jones, M., Jr. *J. Am. Chem. Soc.* **1971**, *93*, 3790–3791.
33. Staudinger, H.; Schneider, H. *Helv. Chim. Acta* **1923**, *6*, 304–315; *Chem. Abstr.* **1923**, *17*, 1953.
34. Brown, R. F. C.; Eastwood, F. W.; Lim, S. T.; McMullen, G. L. *Aust. J. Chem.* **1976**, *29*, 1705–1712.
35. Zecchina, A.; Coluccia, S.; Spoto, G.; Scarano, D.; Marchese, L. *J. Chem. Soc., Faraday Trans.* **1990**, *86*, 703–709.
36. Hochstrasser, R.; Wirz, J. *Angew. Chem., Int. Ed. Engl.* **1989**, *28*, 181–183.
37. Pericas, M. A.; Serratosa, F.; Valenti, E.; Font-Altaba, M.; Solans, X. *J. Chem. Soc., Perkin Trans. 2*, **1986**, 961–967.
38. Köbbing, S.; Mattay, J. *Tetrahedron Lett.* **1992**, *33*, 927–930.
39. Gano, J. E.; Jacob, E. J. *Spectrochim Acta.* **1987**, *43A*, 1023–1025.
40. Nagao, Y.; Kumagai, T.; Takao, S.; Abe, T.; Ochiai, M.; Taga, T.; Inoue, Y.; Fujita, E. *Nippon Kagaku Kaishi*, **1987**, 1447–1456; *Chem. Abstr.* **1988**, *108*, 186360u.
41. Kammula, S. L.; Tracer, H. L.; Shevlin, P. B.; Jones, M., Jr. *J. Org. Chem.* **1977**, *42*, 2931–2932.
42. Winnik, M. A.; Wang, F.; Nivaggioli, T.; Hruska, Z.; Fukumura, H.; Masuhara, H. *J. Am. Chem. Soc.* **1991**, *113*, 9702–9704.
43. Pericas, M. A.; Serratosa, F. *Tetrahedron Lett.* **1977**, 4437–4438.
44. Wright, B. B. *J. Am. Chem. Soc.* **1988**, *110*, 4456–4457.
45. Labille, M.; Janousek, Z.; Viehe, H. G. *Tetrahedron* **1991**, *47*, 8161–8166.
46. Chapman, O. L.; Chang, C.-C.; Kolc, J.; Rosenquist, N. R.; Tomioka, H. *J. Am. Chem. Soc.* **1975**, *97*, 6586–6588.
47. Metzger, C.; Wegler, R. *Chem. Ber.* **1968**, *101*, 1120–1130.
48. Tomioka, H.; Kobayashi, N.; Murata, S.; Ohtawa, Y. *J. Am. Chem. Soc.* **1991**, *113*, 8771–8778.
49. Tomioka, H.; Okuno, H.; Izawa, Y. *J. Chem. Soc., Perkin Trans. 2*, **1980**, 1636–1641.
50. Kobayashi, Y.; Takemoto, Y.; Kamijo, T.; Harada, H.; Ito, Y.; Terashima, S. *Tetrahedron* **1992**, *48*, 1853–1868.
51. Fernandez, M.; Pollart, D. J.; Moore, H. W. *Tetrahedron Lett.* **1988**, *29*, 2765–2768.
52. Pollart, D. J.; Moore, H. W. *J. Org. Chem.* **1989**, *54*, 5444–5448.
53. Kher, S. M.; Kulkarni, G. H.; Mitra, R. B. *Synth. Commun.* **1989**, *19*, 597–604.
54. Frater, G.; Müller, U.; Günther, W. *Helv. Chim. Acta* **1986**, *69*, 1858–1861.

4.4 HALOGEN-SUBSTITUTED KETENES

4.4.1 Fluoroketenes, Perfluoroalkylketenes, and Perfluoroarylketenes

Fluoroketene (**1**) is calculated to be the least stable of all monosubstituted ketenes by the isodesmic exchange reaction (Section 1.1), but nevertheless this species has been intensively studied and is well characterized. The ketene was generated for gas phase microwave studies by pyrolysis of the acid anhydride (equation 1),[1] and photolysis of CH_2FCOCl (**2**) in a matrix gave the ketene as observed by IR.[2] For reactions in solution **1** has been generated by reaction of the acyl chloride **2** with Et_3N with in situ capture by reactive alkenes (equation 2),[3-7] or by imines $R^1CH=NR$ (Section 5.4.1.7).[8]

$$(CH_2FCO)_2O \xrightarrow{\Delta} CFH=C=O \qquad (1)^1$$
$$\phantom{(CH_2FCO)_2O \xrightarrow{\Delta}} \mathbf{1}$$

$$CH_2FCOCl \xrightarrow{Et_3N} CFH=C=O \longrightarrow$$
$$ \mathbf{2} \phantom{\xrightarrow{Et_3N}} \mathbf{1}$$

E/Z

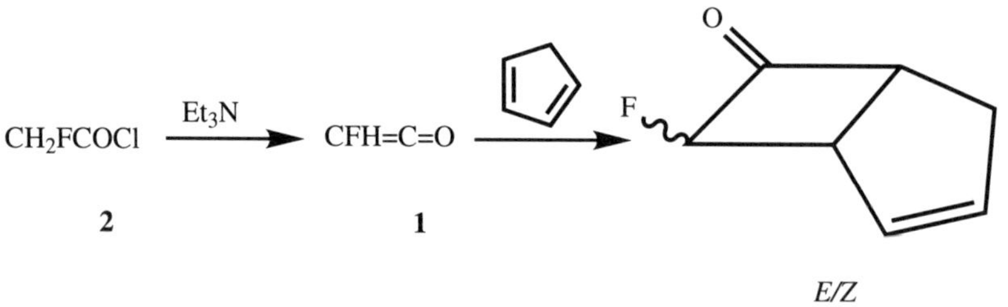

$$(2)^{3-7}$$

Dehydrohalogenation of acyl chlorides has been used to prepare a number of other fluoroketenes $CFR=C=O$, where R = Cl, Me, Et, Ph, and CF_3.[7-9] These were also not observed directly but trapped by imines $R^1CH=NR^8$ or by cyclopentadiene.[9] There have been several studies directed toward the preparation of $CF_2=C=O$, but this species has not been directly observed, and the dissociation into CF_2 and CO (equation 3) is calculated by MP3/6-31G*//6-31G* molecular orbital methods to be essentially thermoneutral.[10,11] The barrier to this dissociation has been calculated to be only 10 kcal/mol.[10]

$$CF_2=C=O \longrightarrow CF_2 + CO \qquad (3)$$

Efforts to prepare $CF_2=C=O$ (**3**) by dehydrochlorination of CHF_2COCl with in situ trapping with active alkenes were unsuccessful.[7] It was reported that this ketene was generated by zinc dehalogenation of $CBrF_2COCl$ in acetone at −10 to −5 °C with in situ trapping by acetone to form the lactone **4** in 50% yield (equation 4).[12] Earlier evidence for the experimental generation of $CF_2=C=O$

involved reaction with NH_3 of an ether solution distilled from the reaction of $CBrF_2COCl$ with Zn, resulting in isolation of a low yield of CHF_2CONH_2.[12,13] However this evidence was felt to be less conclusive for generation of free $CF_2=C=O$ because other intermediates might give the same product.[12]

$$CBrF_2COCl \xrightarrow{Zn} \underset{3}{CF_2=C=O} \xrightarrow{(CH_3)_2C=O} \underset{4}{\text{[β-lactone with F, F, CH}_3\text{, CH}_3\text{]}} \quad (4)$$

The formation of carbon monoxide was observed from $CBrF_2COCl$ and Zn even at -5 °C, and at 35 °C the yield of carbon monoxide increased, and $CF_2=CF_2$, presumably derived from CF_2, was also observed.[12] Photolysis of perfluorocyclobutanone in the gas phase was also proposed to give $CF_2=C=O$ as an unobserved intermediate which dissociated to CF_2 and CO.[12a] Thus the dissociation of $CF_2=C=O$ (equation 3) is evidently quite facile, and until this ketene is observed directly its existence cannot be proven. As is often the case in ketene chemistry, the isolation of cycloadducts that would be derived from ketenes does not prove that the free ketenes were actually formed and trapped. Reaction of the trapping agents with the ketene precursor or its reaction products is often conceivable and not easily ruled out.

More recently it has been reported that $CF_2=C=O$ (**3**) can be generated at 40 °C by the zinc dehalogenation route, with trapping by 1,2- or 1,1-bis(trimethylsilyloxy)alkenes (equations 5 and 6).[14,15] The reaction of equation 6 was interpreted in terms of a two-step process involving a zwitterionic intermediate.[15]

$$CClF_2COCl \xrightarrow[40 \text{ °C}]{Zn} \underset{3}{CF_2=C=O} \xrightarrow{\text{[cyclobutene with OSiMe}_3\text{, OSiMe}_3\text{]}} \text{[cyclobutanone with F, F, OSiMe}_3\text{, OSiMe}_3\text{]} \quad (5)$$

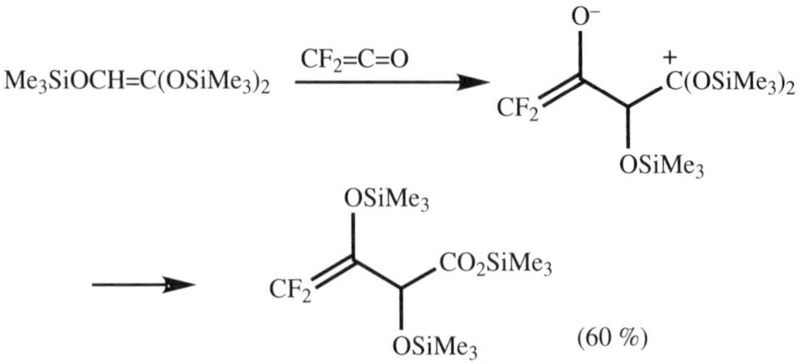

$$Me_3SiOCH=C(OSiMe_3)_2 \xrightarrow{CF_2=C=O} \text{[intermediate]} \longrightarrow \text{[product]} \quad (60\%) \quad (6)$$

These reactions were carried out[14] by addition of $CClF_2COCl$ to the Zn and alkene in EtOAc. It is possible that free $CF_2=C=O$ may be generated and trapped before dissociation under these conditions, but the formation of the observed products from other intermediates derived from the reagents cannot be excluded.

A unique generation of a fluoroketene **5** with added fluoride ion was proposed to account for the formation of **7** from **6**.[16]

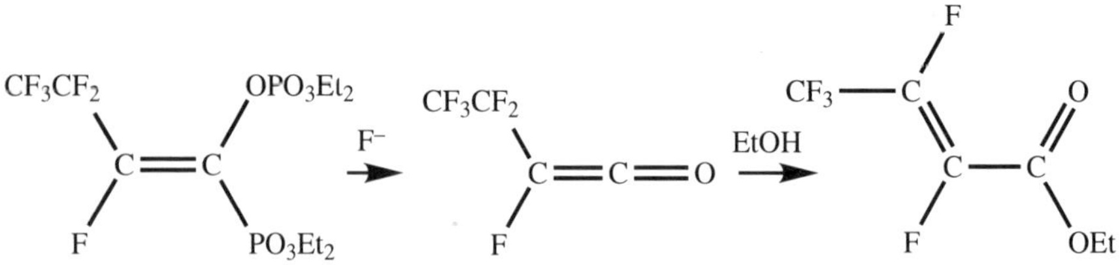

$$\mathbf{6} \xrightarrow{F^-} \mathbf{5} \xrightarrow{EtOH} \mathbf{7} \quad (7)$$

Fluoroacylketenes **8** were generated as unobserved reactive intermediates by thermolysis or photolysis of dioxinones and were trapped by alcohols as the esters (equation 8).[17]

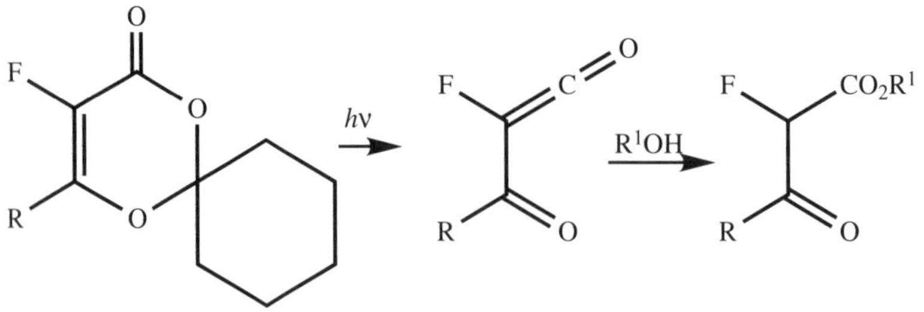

$$\text{dioxinone} \xrightarrow{h\nu} \mathbf{8} \xrightarrow{R^1OH} \text{ester} \quad (8)$$

8 $R^1 = H, Ph, Me, CF_3$

4.4 HALOGEN-SUBSTITUTED KETENES

Ketenes bearing perfluoroalkyl groups have interesting properties. Photolysis of CF_3COCHN_2 in EtOH gave the ester, and this reaction can be formulated as proceeding through ketene **9** (equation 9).[18,19] Confirmation of this was obtained by observing UV absorption attributed to **9** in H_2O from this reaction, and measuring the kinetics of hydration of **9**, which was quite reactive toward H_2O.[20]

$$CF_3COCHN_2 \xrightarrow{h\nu} CF_3CH=C=O \xrightarrow{EtOH} CF_3CH_2CO_2Et \quad (9)$$
$$\textbf{9}$$

Ketenes $CF_3C(CO_2R)=C=O$ (**10**) were obtained by elimination from malonate esters with P_2O_5 (equation 10).[21] These acylketenes were thermally stable but were reactive toward both electrophilic and nucleophilic reagents.[21] A malonyl fluoride gave the acyl ketene **11** (equation 11),[22] and SO_3 was used to induce elimination from a mixture of vinyl ethers to give **12** (equation 12).[23]

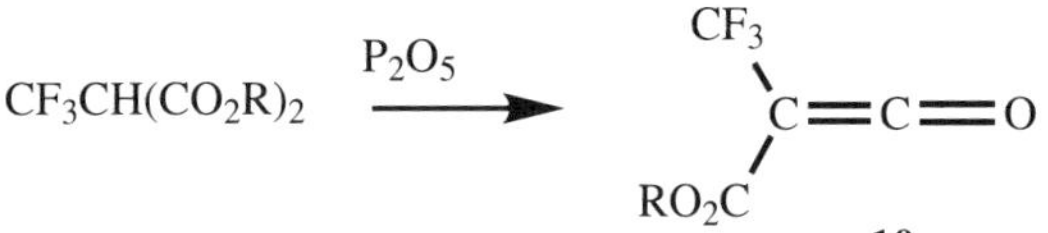

$$CF_3CH(CO_2R)_2 \xrightarrow{P_2O_5} \underset{RO_2C}{\overset{CF_3}{>}}C=C=O \quad (10)$$
$$\textbf{10}$$

$$CF_3CH(COF)_2 \xrightarrow[-(FSO_2)_2O]{SO_3} \underset{FOC}{\overset{CF_3}{>}}C=C=O \quad (11)$$
$$\textbf{11}$$

$$\underset{C_2F_5(MeO)CF}{\overset{CF_3}{>}}C=CFOMe + \underset{C_2F_5(MeO)C}{\overset{CF_3}{>}}CCF_2OMe \xrightarrow{SO_3} \underset{C_2F_5CO}{\overset{CF_3}{>}}C=C=O \quad (12)$$
$$\textbf{12}$$

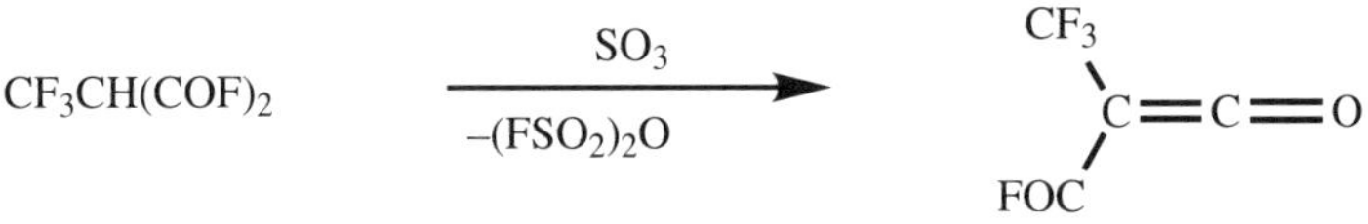

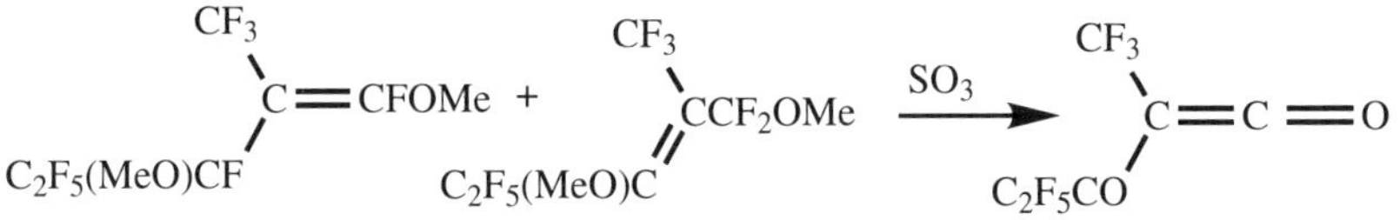

The extraordinarily stable ketene $(CF_3)_2C=C=O$ (**13**) was obtained by zinc dehalogenation of $(CF_3)_2CBrCOCl$,[24] from the acid and P_2O_5 (equation 13),[25] and from CO addition to the carbene $(CF_3)_2C$: at 12 K in an Ar matrix.[26] Some other stable perfluoroalkyl ketenes are **14**[27–29] and the aldoketene **15**, also prepared by dehydration of the acid.[30]

$(CF_3)_2CHCO_2H \xrightarrow{P_2O_5} (CF_3)_2C=C=O$ (13)[25]

13

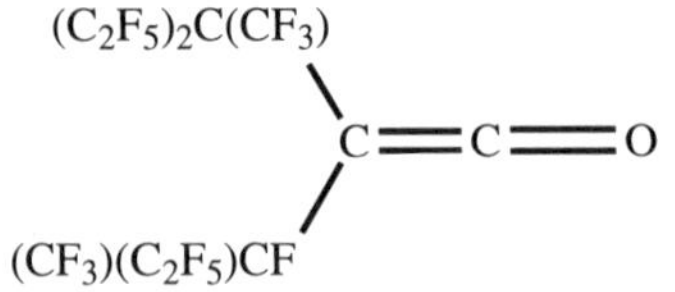

$(C_2F_5)_2C(CF_3)$ and $(CF_3)(C_2F_5)CF$ groups on $C=C=O$ → $(C_2F_5)_2C(CF_3)CH=C=O$

14   **15**

The ketene $PhC(CF_3)=C=O$ (**16**) has been obtained by zinc debromination of the dibromide,[31] and by reaction of the carbene with CO (equation 14).[32] Photolysis of **17** was proposed to give the perfluorobutadienylketene **18** (equation 15).[33]

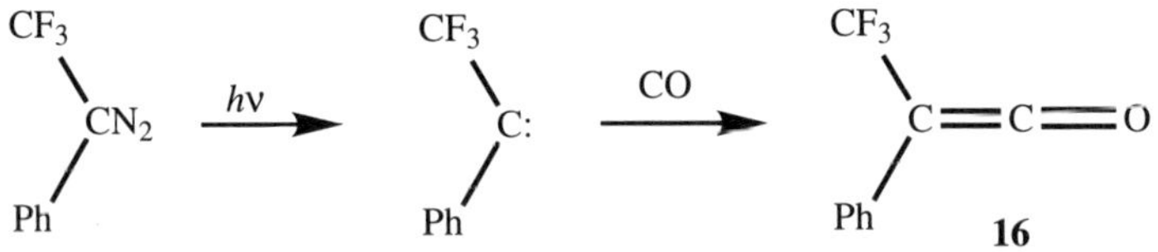

$$\begin{array}{c} CF_3 \\ Ph \end{array}\!\!\!>\!\!CN_2 \xrightarrow{h\nu} \begin{array}{c} CF_3 \\ Ph \end{array}\!\!\!>\!\!C: \xrightarrow{CO} \begin{array}{c} CF_3 \\ Ph \end{array}\!\!\!>\!\!C=C=O \quad (14)$$

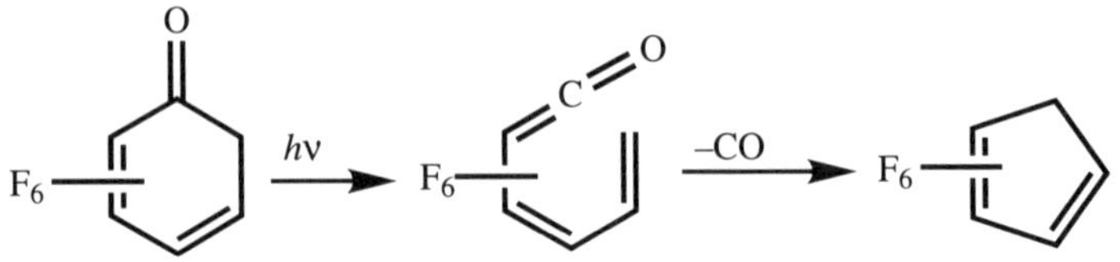

16

Equation (15): F_6-cyclohexadienone **17** $\xrightarrow{h\nu}$ F_6-ketene intermediate **18** $\xrightarrow{-CO}$ F_6-cyclopentadiene   (15)

The pyrolysis of alkynyl ethers **19** was reported to give formation of the aldoketene **20** and related derivatives as stable distillable liquids (equation 16).[34] The structures of **20** and the other ketenes were supported by the IR stretching frequencies at 2160–2170 cm^{-1}, and their stability was ascribed to a combination of the presence of electron-withdrawing fluorines and the organometallic residue.[34] However, the way in which these ketenes would actually be stabilized by these factors is not apparent and further study of these compounds is warranted.

$Me_3SiOC(CF_3)_2C\equiv COEt \xrightarrow{120-130\ °C} Me_3SiOC(CF_3)_2CH=C=O$ (16)

19   **20**

Reaction of the disulfide **21** with 1,4-diazabicyclo[2.2.2]octane (DABCO) in THF with 5 equivalents of cyclopentadiene gave the thioketene **22** which was not observed directly but was trapped as the cycloadduct **23**.[35] Ketene **22** was also generated and trapped from the cysteine *S*-conjugate **24** and the enzyme model *N*-dodecylpyridoxal.[35]

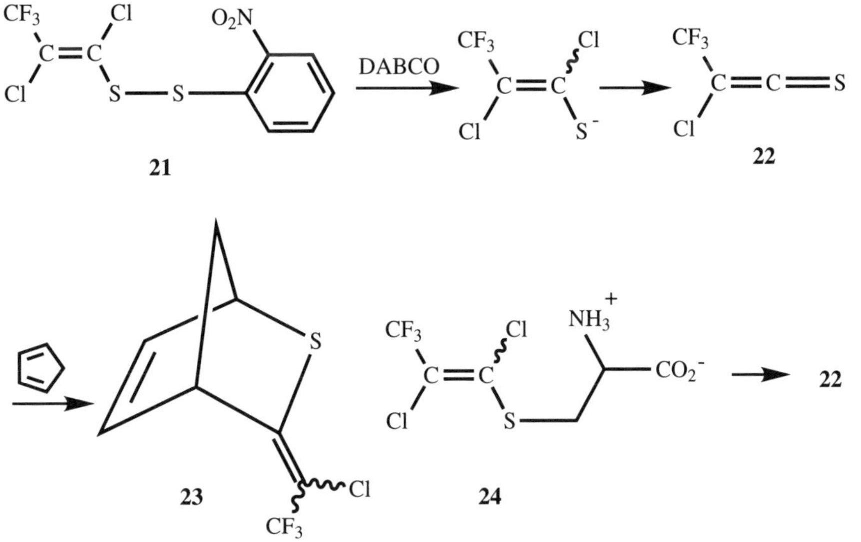

(17)

The ketene **13** reacts with electrophiles such as PhSCl only in polar solvents (equation 18).[36] The low reactivity of **13** was interpreted on the basis of MNDO calculations.[36]

$$(CF_3)_2C=C=O \xrightarrow[CH_3CN]{PhSCl} (CF_3)_2C(SPh)COCl \quad 93\% \qquad (18)$$

13

Ketene **13** reacts with many alkenes,[37,38] and with norbornene gives **25** (equation 19),[38] while reaction with norbornadiene gives **26** (equation 20).[38] This ketene gives cyclobutenones on reaction with alkynes (equation 21),[39] and pyrolysis of **27** was proposed[39] to give **28**, which rearranged to **29** and led to the observed products **30–32** (equation 22).[39]

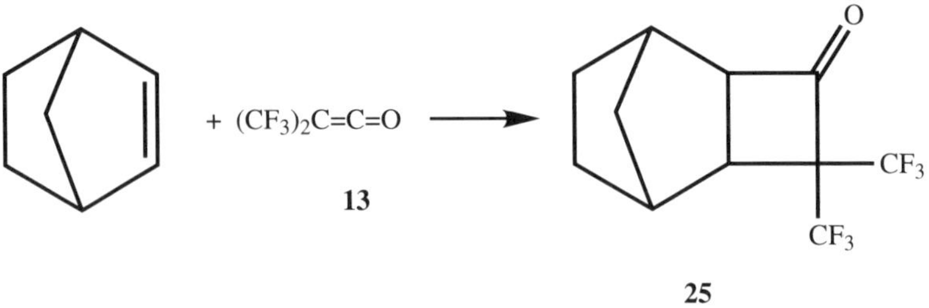

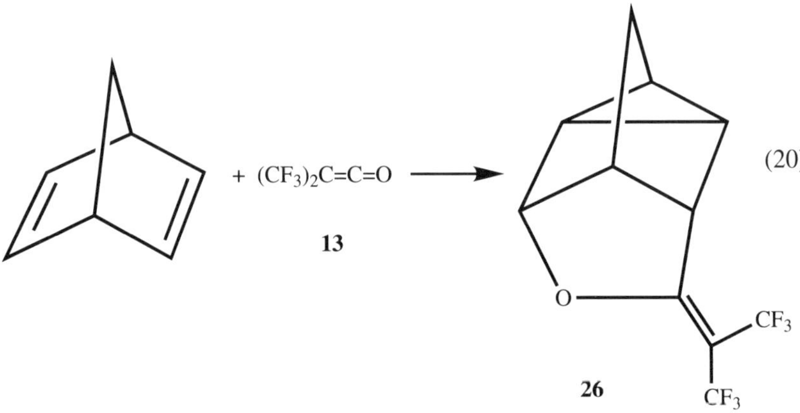

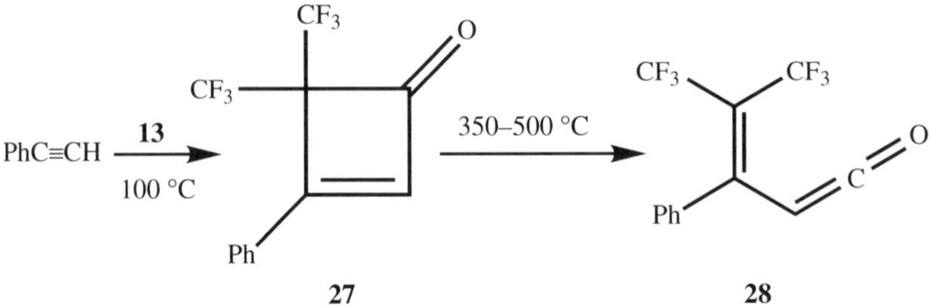

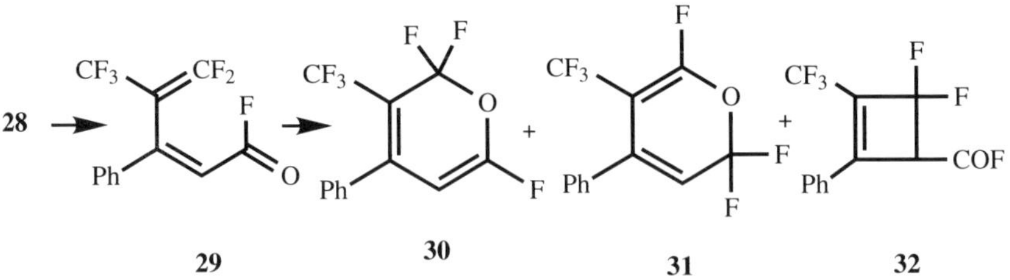

Reaction of **13** with vinyl ethers and with vinyl esters give oxetanes **33** (equation 23),[40] while 1,3-dimethylallene gave both a cyclobutanone and an oxetane (equation 24),[41] and tetramethylallene gave noncyclized products (equation 25).[41] Reaction with *N*-sulfinylamines gave 1,2-thiazetan-3-one 1-oxides (equation 26).[42]

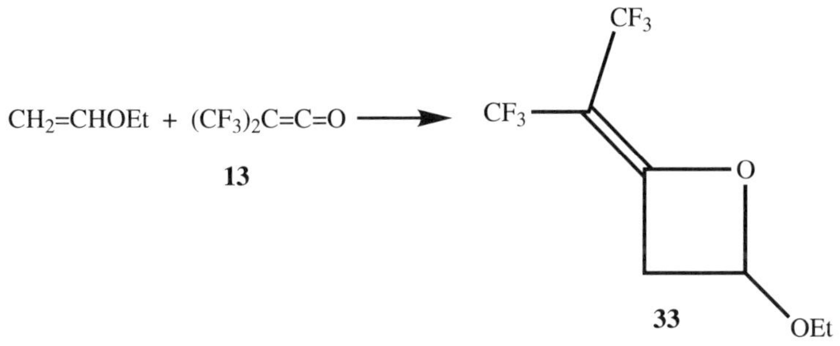

CH2=CHOEt + (CF3)2C=C=O
13
CF3
CF3
O
33
OEt

(23)[40]

CH3CH=C=CHCH3
13
CF3
CF3
O
CH3
CHCH3
+
CF3
CF3
O
CH3CH
CH3

(24)[41]

Me2C=C=CMe2
13
(CF3)2CHCO
+
O2CCH(CF3)2
(CF3)2C=C

(25)[41]

MeN=S=O
13
CF3
CF3
O
S
N
O
Me

(26)[42]

The addition of fluoride anion to **13** gave the stable crystalline salt **34** (equation 27), which in the ^{19}F NMR showed nonequivalent CF_3 groups even at 75 °C, implying a barrier to rotation of at least 19 kcal/mol in the enolate.[43]

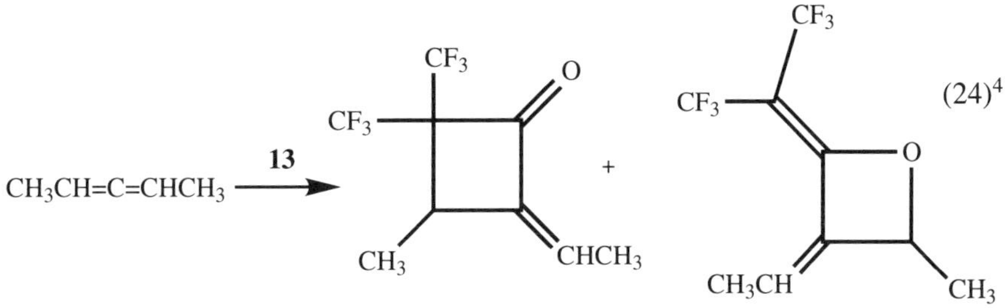

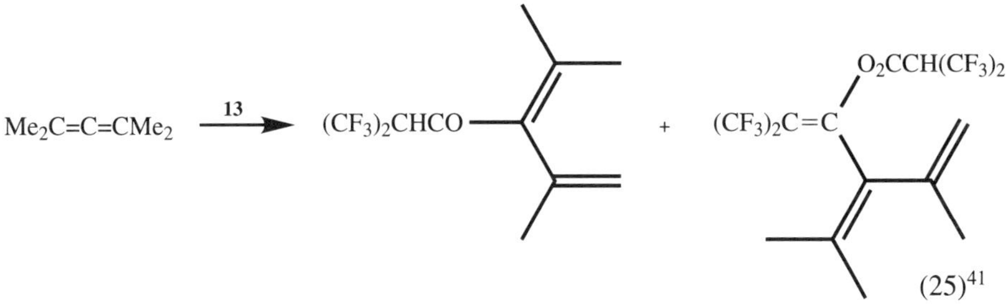

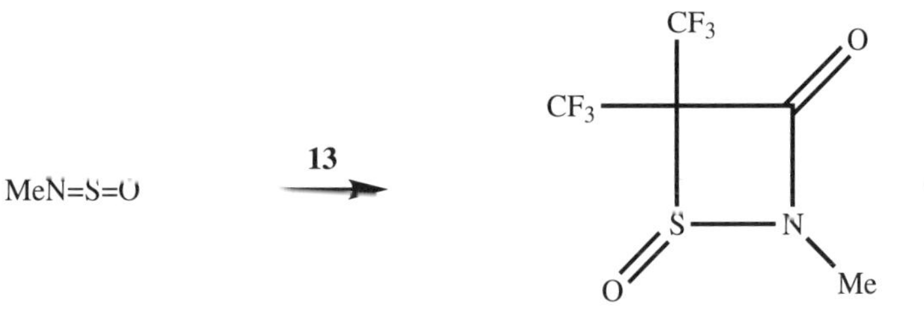

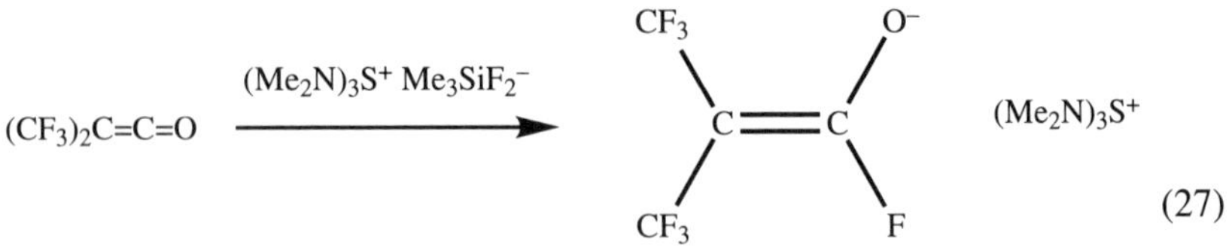

330 TYPES OF KETENES

$$(CF_3)_2C=C=O \xrightarrow{(Me_2N)_3S^+ Me_3SiF_2^-} \underset{\textbf{34}}{\overset{CF_3}{\underset{CF_3}{>}}C=C\overset{O^-}{\underset{F}{<}}} (Me_2N)_3S^+ \quad (27)$$

13

The delocalized cation **35** from ionization of **13** was directly observed by NMR (equation 28).[44] Reaction of this cation with $CF_2=CF_2$ gives the products in equation 29.[45,46]

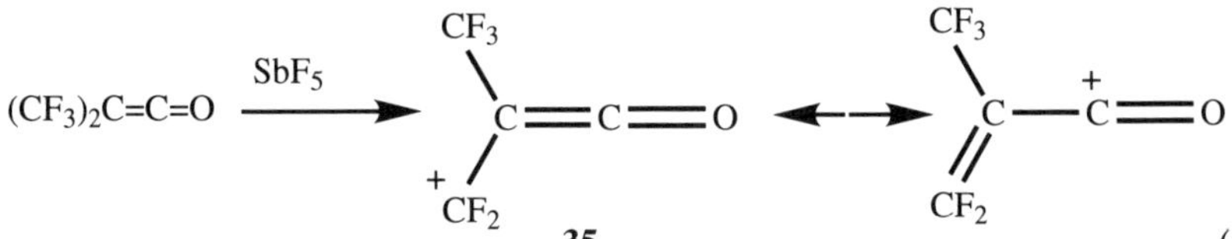

$$(CF_3)_2C=C=O \xrightarrow{SbF_5} \underset{\textbf{35}}{\overset{CF_3}{\underset{{}^+CF_2}{>}}C=C=O} \longleftrightarrow \overset{CF_3}{\underset{CF_2}{>}}\overset{+}{C}-C=O \quad (28)$$

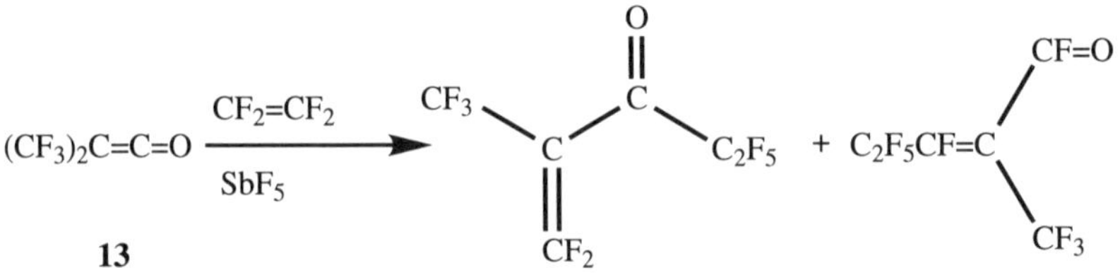

$$(CF_3)_2C=C=O \xrightarrow[SbF_5]{CF_2=CF_2} \underset{\textbf{}}{\overset{CF_3}{\underset{CF_2}{>}}C\overset{\overset{O}{\parallel}}{\underset{}{C}}C_2F_5} + C_2F_5CF=C\overset{CF=O}{\underset{CF_3}{<}} \quad (29)^{45}$$

13

Pyrolysis of difluoromaleic anhydride was proposed to form the ketenyl carbene **35**, which formed difluoropropadiene **36**, a nonlinear structure observable by IR on matrix isolation (equation 30).[47]

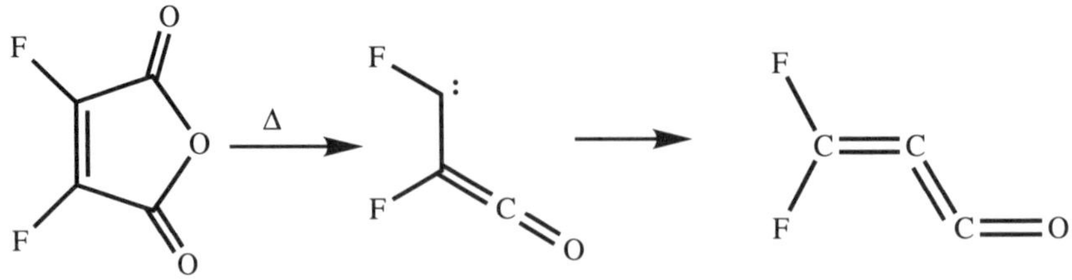

$$\text{(difluoromaleic anhydride)} \xrightarrow{\Delta} \underset{\textbf{35}}{\text{(ketenyl carbene)}} \longrightarrow \underset{\textbf{36}}{\text{(difluoropropadiene)}} \quad (30)$$

TABLE 1. Fluoroketenes

Ketenes	Formula	Preparation	Registry No.	Ref.
CFCl=C=O	C_2ClFO	Dehydrohalogenation		9
CF_2=C=O	C_2F_2O	Dehalogenation	683-54-5	4
CHF=C=O	C_2HFO	Pyrolysis	37580-39-5	1
CHF=^{13}C=O	C_2HFO	Pyrolysis	122615-64-9	1
^{13}CHF=C=O	C_2HFO	Pyrolysis	122615-63-8	1
CHF=C=^{18}O	C_2HFO	Pyrolysis	122327-94-0	1
FC(CF)=C=O	C_3F_2O	Pyrolysis	123675-00-3	47
$CF_3C(SO_2F)$=C=O	$C_3F_4O_3S$	Sulfonation	60956-76-5	22
CF_3CF=C=O	C_3F_4O	Dehydrochlorination	22758-59-4	9
O=CHCF=C=O	C_3HFO_2	Dioxinone		17
CF_3CH=C=O	C_3HF_3O	Wolff	134736-46-2	18–20
CH_3CF=C=O	C_3H_3FO	Dehydrochlorination		7
$CF_3C(COBr)$=C=O	$C_4BrF_3O_2$	Elimination	62935-56-2	21
$CF_3C(COCl)$=C=O	$C_4ClF_3O_2$	Elimination	62935-55-1	21
CF_3COCF=C=O	$C_4F_4O_2$	Dioxinone		17
$CF_3C(COF)$=C=O	$C_4F_4O_2$	Elimination	60001-86-7	22
$CF_3C(CF_2^+)$=C=O SbF_5^-	$C_4F_5O^+SbF_6^-$	Ionization	106915-54-2	44
	$C_4F_5O^+$	Ionization	106915-53-1	44
$(CF_3)_2C$=C=O	C_4F_6O	Dehydration	684-22-0	24,25
	$(C_4F_6O)_2$		18324-84-0	47–50
	C_4F_6O	Elimination		16
CF_3CF_2CF=C=O	$C_4HF_3O_2$			60
$CF_3C(CHO)$=C=O	$C_4HF_3O_3$	Elimination	63167-30-6	50
$CF_3C(CO_2H)$=C=O	C_4HF_5O	Wolff	82515-15-9	51
C_2F_5CH=C=O	$C_4H_3FO_2$	Dioxinone		17
CH_3COCF=C=O	C_4H_5FO	Dehydrochlorination		9
C_2H_5CF=C=O				

TABLE 1. (*Continued*)

Ketenes	Formula	Preparation	Registry No.	Ref.
$CF_3C(COMe)=C=O$	$C_5H_3F_3O_2$	Dioxinone	60956-77-6	60
$CF_3C(CO_2Me)=C=O$	$C_5H_3F_3O_3$	Elimination	58105-77-4	21,51
$CF_2=C(CF_3)CCF_3=C=O$	C_6F_8O	Elimination	58105-77-4	52,53
$CF_3C(COC_2F_5)=C=O$	$C_6F_8O_2$	Elimination	53352-88-8	23,54,55
$(CF_3)_2CHCCF_3=C=O$	C_6HF_9O	Elimination	58105-78-5	52
$CF_3C(CO_2CH_2CH_2NO_2)=C=O$	$C_6H_2F_5NO_5$	Elimination	63296-95-7	56
$CF_3C(CO_2Et)=C=O$	$C_6H_5F_3O_2$	Elimination	63009-25-6	21
$CF_3C(CON=C(CF_3)_2)=C=O$	$C_7F_9NO_2$	Elimination	62935-57-3	21
$CF_3C(CO_2N=C(CF_3)_2)=C=O$	$C_7F_9NO_3$	Elimination	63176-26-1	21
$CF_3C(CO_2CH(CF_3)_2)=C=O$	$C_7HF_9O_3$	Elimination	62935-43-7	21
$CF_3C(CO_2CH_2CF_2CHF_2)=C=O$	$C_7H_3F_3O_3$	Elimination	62935-44-8	21
$CF_3C(CO_2Pr-i)=C=O$	$C_7H_7F_3O_3$	Elimination	62935-34-6	21
$CF_3C(CO_2Pr-n)=C=O$	$C_7H_7F_3O_3$	Elimination	62935-33-5	21
$PrCF_2C(CF_3)=C=O$	$C_7H_7F_5O$	Dehydration	66567-02-0	57
$n-C_6F_{13}CF=C=O$	$C_8F_{14}O$	Elimination		16
$CF_3(C_2F_5)_2CCH=C=O$	$C_8HF_{13}O$	Elimination	24293-81-0	30,58
$CF_3C(CO_2(CH_2)_3CF_3)=C=O$	$C_8H_6F_6O_3$	Elimination	63296-96-8	56
$CF_3C(CO_2Bu-i)=C=O$	$C_8H_9F_3O_3$	Elimination	62935-36-8	21
$CF_3C(CO_2Bu-n)=C=O$	$C_8H_9F_3O_3$	Elimination	62935-35-7	21

Compound	Formula	CAS	Method	Ref
BuCF$_2$C(CF$_3$)=C=O	C$_8$H$_9$F$_5$O	66567-01-9	Elimination	57
CF$_2$C(CO$_2$P(OEt)$_2$)=C=O	C$_8$H$_{10}$F$_3$O$_5$P	62935-58-4	Elimination	21
Me$_3$SiOC(CF$_3$)$_2$CH=C=O	C$_8$H$_{10}$F$_6$O$_2$Si	83740-52-7	Alkynyl ether	34
FC(COPh)=C=O	C$_9$H$_5$FO$_2$		Dioxinone	17
CF$_3$CPh=C=O	C$_9$H$_5$F$_3$O	40916-23-2	Dehalogenation	31
n-C$_8$F$_{17}$CF=C=O	C$_{10}$F$_{18}$O		Elimination	16
C$_2$F$_5$CF(CF$_3$)C(C$_4$F$_9$-t)=C=O	C$_{10}$F$_{18}$O	26730-22-3		None
CF$_3$CR=C=O	C$_{10}$HF$_{15}$O$_4$	71309-13-2	Elimination	59
R = (CF$_3$)$_2$C(OH)C(CF$_3$)$_2$O$_2$C				
R = CF$_2$H(CF$_2$)$_4$CH$_2$O$_2$C	C$_{10}$H$_3$F$_{13}$O$_3$	63296-97-9	Elimination	56
CF$_3$C(COPh)=C=O	C$_{10}$H$_5$F$_3$O$_2$		Dioxinone	60
(CF$_3$)$_2$C=CPhCH=C=O	C$_{12}$H$_6$F$_6$O		Pyrolysis	39
CF$_3$(C$_2$F$_5$)$_2$CCF(CF$_3$)C(C$_2$F$_5$)=C=O	C$_{12}$F$_{22}$O	27930-64-9		27
(C$_6$F$_5$)$_2$C=C=O	C$_{14}$F$_{10}$O	36691-72-2	Dehydration	61
n-C$_{12}$F$_{25}$CH=C=O	C$_{14}$HF$_{25}$O	50698-47-0		62
(C$_{14}$HF$_{25}$O)$_2$		50698-48-1		62
(CF$_3$)$_2$C=CPhCH=CPhCH=C=O	C$_{20}$H$_{12}$F$_6$O		Pyrolysis	39

References

1. Brown, R. D.; Godfrey, P. D.; Wiedenmann, K. H. *J. Mol Spectrosc.* **1989**, *136*, 241–249.
2. Davidovics, G.; Monnier, M.; Allouche, A. *Chem. Phys.* **1991**, *150*, 395–403.
3. Brady, W. T.; Hoff, E. F., Jr. *J. Am. Chem. Soc.* **1968**, *90*, 6256.
4. Brady, W. T.; Hoff, E. F., Jr.; Roe, R., Jr.; Parry, F. H., Jr. *J. Am. Chem. Soc.* **1969**, *91*, 5679–5680.
5. Rey, M.; Roberts, S. M.; Dreiding, A. S.; Roussel, A.; Vanlierde, H.; Toppet, S.; Ghosez, L. *Helv. Chim. Acta* **1982**, *65*, 703–720.
6. Jacobson, B. M.; Bartlett, P. D. *J. Org. Chem.* **1973**, *38*, 1030–1041.
7. Dolbier, W. R., Jr.; Lee, S. K.; Phanstiel, O., IV *Tetrahedron* **1991**, *47*, 2065–2072.
8. Welch, J. T.; Arakai, K.; Kawecki, R.; Wichtowski, J. A. *J. Org. Chem.* **1993**, *58*, 2454–2462.
9. Cheburkov, Yu. A.; Platoshkin, A. M.; Knunyants, I. L. *Dokl. Akad. Nauk SSSR*, **1967**, *173*, 1117–1120; *Engl. Transl.* **1967**, *173*, 369–372.
10. Berson, J. A.; Birney, D. M.; Dailey, W. P., III; Liebman, J. F. In *Molecular Structure and Energetics* Vol. 6, Chap. 9, VCH Publishers, New York, 1988, pp. 391–441.
11. Dailey, W. P. ACS Ninth Winter Fluorine Conference; St. Petersburg, FL; January 29–February 3, 1989, Abstract 3.
12. England, D. C.; Krespan, C. G. *J. Org. Chem.* **1968**, *33*, 816–819.
12a. Lewis, R. S.; Lee, E. K. C. *J. Phys. Chem.* **1975**, *79*, 187–191.
13. Yarovenko, N. N.; Motornyi, S. P.; Kirenskaya, L. I. *Zh. Obshch. Khim.* **1957**, *27*, 2796–2799; *Engl. Transl.* **1957**, *27*, 2832–2834.
14. Habibi, M. H.; Saidi, K.; Sams, L. C. *J. Fluorine Chem.* **1987**, *37*, 177–181.
15. Saidi, K.; Habibi, M. H. *J. Fluorine Chem.* **1991**, *51*, 217–222.
16. Ishihara, T.; Yamasaki, Y.; Ando, T. *Tetrahedron Lett.* **1986**, *27*, 2879–2880.
17. Iwaoka, T.; Murohashi, T.; Sato, M.; Kaneko, C. *Synthesis* **1992**, 977–981.
18. Brown, F.; Musgrave, W. K. R. *J. Chem. Soc.* **1953**, 2087–2089.
19. Park, J. D.; Larsen, E. R.; Holler, H. V.; Lacher, J. R. *J. Org. Chem.* **1958**, *23*, 1166–1169.
20. Allen, A. D.; Andraos, J.; Kresge, A. J.; McAllister, M. A.; Tidwell, T. T. *J. Am. Chem. Soc.* **1992**, *114*, 1878–1879.
21. Kryukova, L. Yu.; Kryukov, L. N.; Truskanova, T. D.; Isaev, V. L.; Sterlin, R. N.; Knunyants, I. L. *Dokl. Akad. Nauk SSSR* **1977**, *232*, 1311–1313; *Engl. Transl.* **1977**, *232*, 90–92.
22. Krespan, C. G. *J. Fluorine Chem.* **1976**, *8*, 105–114.
23. England, D. C. *J. Org. Chem.* **1981**, *46*, 147–153.
24. Cherburkov, Y. A.; Bargamova, M. D. *Izv. Akad. Nauk SSSR* **1967**, 833–840; *Engl. Transl.* **1967**, 801–806.
25. England, D. C.; Krespan, C. G. *J. Am. Chem. Soc.* **1966**, *88*, 5582–5587.
26. Mal'tsev, A. K.; Zuev, P. S.; Nefedov, O. M. *Izv. Akad. Nauk SSSR*, **1985**, 957–958; *Engl. Transl.* **1985**, 876.
27. Coe, P. L.; Owen, I. R.; Sellers, A. *J. Chem. Soc., Perkin Trans. 1*, **1989**, 1097–1103.

28. Coe, P. L.; Sellars, A.; Tatlow, J. C.; Fielding, H. C.; Whittaker, G. *J. Fluorine Chem.* **1986**, *32*, 135–150.
29. Coe, P. L.; Sellars, A.; Tatlow, J. C.; Fielding, H. C.; Whittaker, G. *J. Fluorine Chem.* **1986**, *32*, 151–161.
30. Coe, P. L.; Ray, N. C. *J. Fluorine Chem.* **1989**, *45*, 90.
31. Anders, E.; Ruch, E.; Ugi, I. *Angew. Chem., Int. Ed. Engl.* **1973**, *12*, 25–29.
32. Sander, W. W. *J. Org. Chem.* **1988**, *53*, 121–126.
33. Soelch, R. R.; Mauer, G. W.; Lemal, D. M. *J. Org. Chem.* **1985**, *50*, 5845–5852.
34. Zaitseva, G. S.; Livantsova, L. I.; Orlova, N. A.; Baukov, Yu. I.; Lutsenko, I. F. *Zh. Obshch. Khim.* **1982**, *52*, 2076–2084; *Engl. Transl.* **1982**, *52*, 1847–1855.
35. Dekant, W.; Urban, G.; Görsmann, C.; Anders, M. W. *J. Am. Chem. Soc.* **1991**, *113*, 5120–5122.
36. Shkurak, S. N.; Ezhov, V. V.; Kolomiets, A. F.; Fokin, A. V. *Izv. Akad. Nauk SSSR, Ser. Khim.* **1984**, 1371–1378; *Engl. Transl.* **1984**, 1261–1267.
37. Cheburkov, Yu. A.; Mukhamadaliev, N.; Knunyants, I. L. *Tetrahedron* **1968**, *24*, 1341–1356.
38. England, D. C.; Krespan, C. G. *J. Org. Chem.* **1970**, *35*, 3300–3307.
39. England, D. C.; Krespan, C. G. *J. Org. Chem.* **1970**, *35*, 3308–3312.
40. England, D. C.; Krespan, C. G. *J. Org. Chem.* **1970**, *35*, 3312–3322.
41. England, D. C.; Krespan, C. G. *J. Org. Chem.* **1970**, *35*, 3322–3327.
42. Jäger, U.; Schwab, M.; Sundermeyer, W. *Chem. Ber.* **1986**, *119*, 1127–1132.
43. Farnham, W. B.; Middleton, W. J.; Fultz, W. C.; Smart, B. E. *J. Am. Chem. Soc.* **1986**, *108*, 3125–3127.
44. Snegirev, V. F.; Galakhov, M. V.; Petrov, V. A.; Makarov, K. N.; Bakhmutov, V. I. *Izv. Akad. Nauk SSSR,* **1986**, 1318–1325; *Engl. Transl.* **1986**, 1194–1200.
45. Chepik, S.; Belen'kii, G.; Cherstkov, V.; Sterlin, S. R.; German, L. S. *J. Fluorine Chem.* **1991**, *54*, 304.
46. Chepik, S. D.; Belen'kii, G. G.; Cherstkov, V. F.; Sterlin, S. R.; German, L. S. *Izv. Akad. Nauk SSSR, Ser. Khim.* **1991**, 513–516; *Engl. Transl.* **1991**, 446–448.
47. Brahms, J. C.; Dailey, W. P. *J. Am. Chem. Soc.* **1989**, *111*, 8940–8941.
48. Cheburkov, Yu. A.; Bargamova, M. D.; Knunyants, I. L. *Izv. Akad. Nauk SSSR, Ser. Khim.* **1967**, 2124–2125; *Engl. Transl.* **1967**, 2052–2053.
49. Sasaki, S.; Matsumoto, M. *Jpn. Kokai Tokkyo Koho JP 03* 11,027; *Chem. Abstr.* **1991**, *114*, 246859z.
50. Knunyants, I. L.; Sterlin, R. N.; Isaev, V. L.; Mal'kevich, L. Yu.; Kryukov, L. N.; Truskanova, T. D. U.S.S.R. SU 539865; *Chem. Abstr.* **1977**, *87*, 22429v.
51. Maier, G.; Reisenauer, H. P.; Sayrac, T. *Chem. Ber.* **1982**, *115*, 2192–2201.
52. Kaz'mina, N. B.; Krasnikova, G. S.; Lur'e, E. P.; Mysov, E. I.; Knunyants, I. L. *Izv. Akad. Nauk SSSR, Ser. Khim.* **1975**, 2525–2529; *Engl. Transl.* **1975**, 2410–2415.
53. Bryce, M. R.; Chambers, R. D.; Lindley, A. A.; Fielding, H. C. *J. Chem. Soc., Perkin Trans. 1* **1983**, 2451–2454.
54. England, D. C. *J. Org. Chem.* **1984**, *49*, 4007–4008.
55. England, D. C. *J. Org. Chem.* **1981**, *46*, 153–157.

56. Kryukova, L. Yu.; Kryukov, L. N.; Isaev, V. L.; Sterlin, R. N.; Knunyants, I. L. *Zh. Vses. Khim. O-va* **1977,** *22,* 231–233; *Chem. Abstr.* **1977,** *87,* 38799b.
57. Isaev, V. L.; Truskanova, T. D.; Sterlin, R. N.; Knunyants, I. L. *Zh. Vses. Khim. O-va* **1978,** *23,* 113; *Chem. Abstr.* **1978,** *88,* 190039g.
58. Deem, W. R. Brit. Patent GB 1,395,751; *Chem. Abstr.* **1975,** *83,* 96462q.
59. Kryukova, L. Yu.; Ermolov, A. F.; Kryukov, L. N.; Sterlin, R. N.; Knunyants, I. L. *Zh. Vses, Khim. O-va* **1979,** *24,* 297–298; *Chem. Abstr.* **1979,** *91,* 123374m.
60. Iwaoka, T.; Sato, M.; Kaneko, C. *J. Chem. Soc., Chem. Commun.* **1991,** 1241–1242.
61. Lubenets, E. G.; Gerasimova, T. N.; Barkhash, V. A. *Zh. Org. Khim.* **1972,** *8,* 654; *Engl. Transl.* **1972,** *8,* 663.
62. Turbak, A. F.; Rose, H. J. U.S. Patent US 3,753,740; *Chem. Abstr.* **1973,** *79,* 135518b.

4.4.2 Chlorine, Bromine, and Iodine-Substituted Ketenes

Chloroketenes are ready prepared, highly reactive, and extensively utilized in synthesis because they provide a facile route to cyclobutanones, which are themselves versatile chemical intermediates. Reviews devoted specifically to the topic of haloketenes have appeared.[1,2]

The first reports of the preparation and trapping of chloroketenes appeared in 1965–1966, and involved both the dehydrohalogenation of $CHCl_2COCl$, and the dehalogenation of CCl_3COBr (equations 1 and 2).[3–5] Dichloroketene (**1**) reacts readily by [2 + 2] cycloaddition with cyclopentadiene, cyclohexadiene, cyclohexene, dihydropyran, indene, norbornene, and norbornadiene.[6]

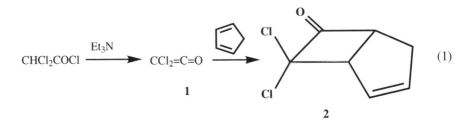

Chloroketene (**3**) is usually generated by the dehydrochlorination route analogous to equation 1,[1] but **3** and methylchloroketene were also generated by the procedure of equation 3, and the [2 + 2] cycloadduct of **3** and cyclohexene was obtained (equation 4).[7] These ketenes could be collected at −80 °C and their ^{13}C NMR spectra measured.[7] When mixed silylphosphines $Me_3SiPPhR$ were used in the procedure of equation 3, the intermediate chloroketenes were not observed but were trapped with HCl.[8]

$$CHCl_2COCl \xrightarrow[-80\,°C]{Me_3SiPPh_2} CHCl_2COPPh_2 \xrightarrow[-Ph_2PCl]{80\,°C} CHCl=C=O \quad (3)$$

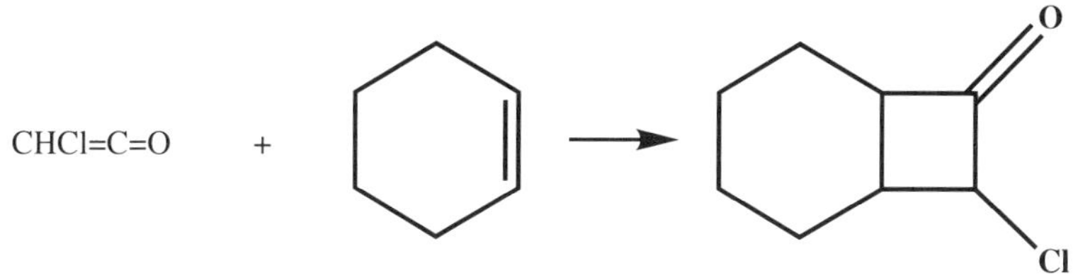

$$\mathbf{3}$$

CHCl=C=O + [cyclohexene] → [bicyclic cyclobutanone with Cl] (4)

The ketenes CHCl=C=O and CCl_2=C=O have also been studied in the gas phase and their molecular structures determined.[9–11] Photolysis of $ClCH_2COCl$ in a matrix led to CHCl=C=O, observed by IR.[12] Pyrolysis of dichlorovinylene carbonate apparently gave CCl_2=C=O, captured in an Ar matrix at 8 K, as evidenced by the IR absorption at 2155 cm^{-1}.[13] The formation of CCl_2=C=O was proposed to occur by decarboxylation and Wolff rearrangement.[13]

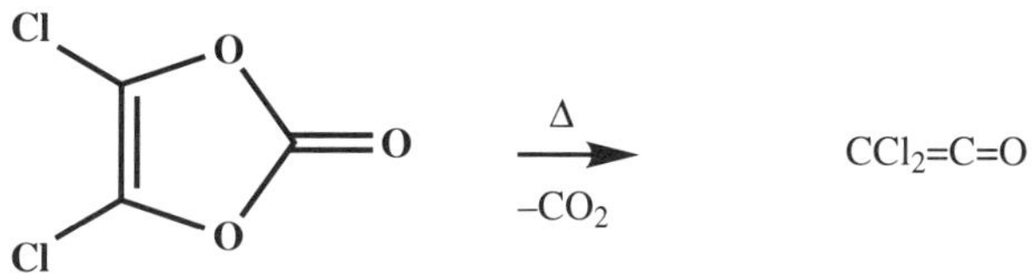

[dichlorovinylene carbonate] $\xrightarrow[-CO_2]{\Delta}$ CCl_2=C=O

Calculations of isodesmic reactions have now clarified that the electronegative substituent causes chloroketenes to be highly destabilized in the thermodynamic sense, compared to ordinary alkenes.[14] However, the ketenes can still be readily formed (equations 1 and 2) and their low stability renders them highly reactive, particularly in cycloadditions, which is their most common application. Still, even with CCl_2=C=O, nucleophilic alkenes are required for reaction, and electrophilic alkenes, such as acrylonitrile, are unreactive.[2] The facile removal of the chlorine from chloroketene cycloadducts by Zn/HOAc or n-Bu$_3$SnH greatly enhances the synthetic utility of these products (equation 5).[6,15,16]

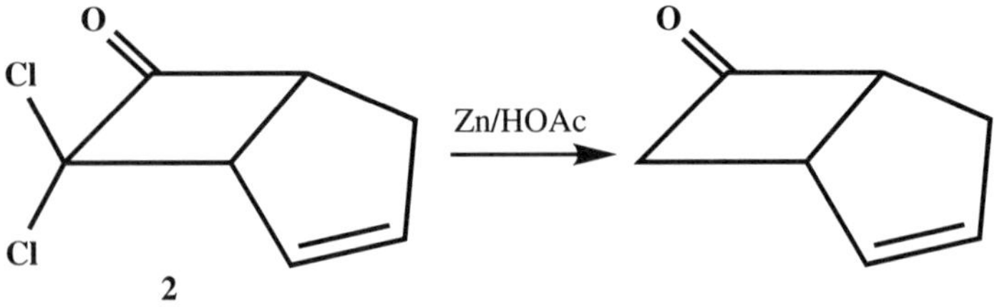

The hydrolysis of the adduct **2** gives a simple synthesis of tropolone (equation 6).[3,17] Cycloaddition of 6,6-dimethylfulvenes with $CCl_2=C=O$ followed by hydrolysis gives isopropenyltropolones (equation 6a).[17a]

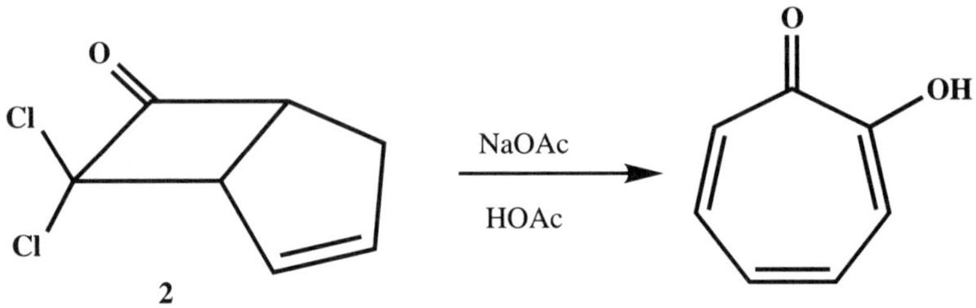

The reaction of $CHCl_2COCl$ with Et_3N in refluxing ether produces **4** and **5**.[18] The mechanism suggested for the formation of these products is shown (Scheme 1), and might involve the dimer **6** of $CCl_2=C=O$. Dimers of $CCl_2=C=O$ have not been observed, although mixed dimers with other ketenes have been obtained.[2] The X-ray structure of **5** was reported.[19]

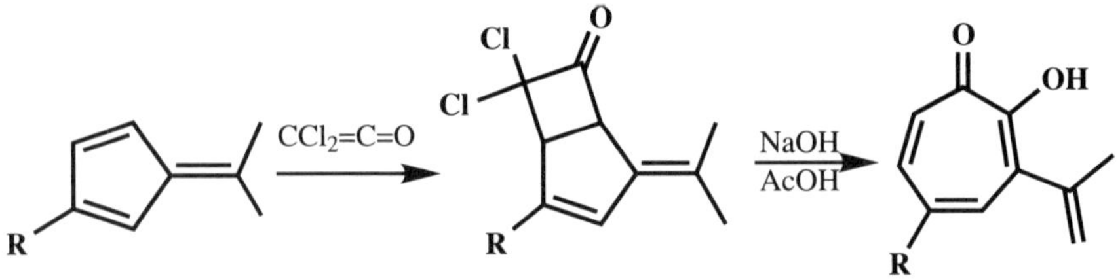

The reaction of $CH_3CHClCOCl$ with Et_3N evidently gives $CH_3CCl=C=O$ as a reactive intermediate and also gives the enol ester $CH_3CHClCO_2CCl=CClCH_3$, presumably by a process analogous to the formation of **8** (Section 3.2.2).[18] The presence of $CH_3CCl=C=O$ in this reaction could not be detected even by stopped-flow NMR.[18]

Dichloroketene reacts with unactivated alkynes to give dichlorocyclobutenones,[20–23] and procedures have been developed to convert these to dechlorinated cyclobutenones, which are very useful synthetic intermediates. Reduction

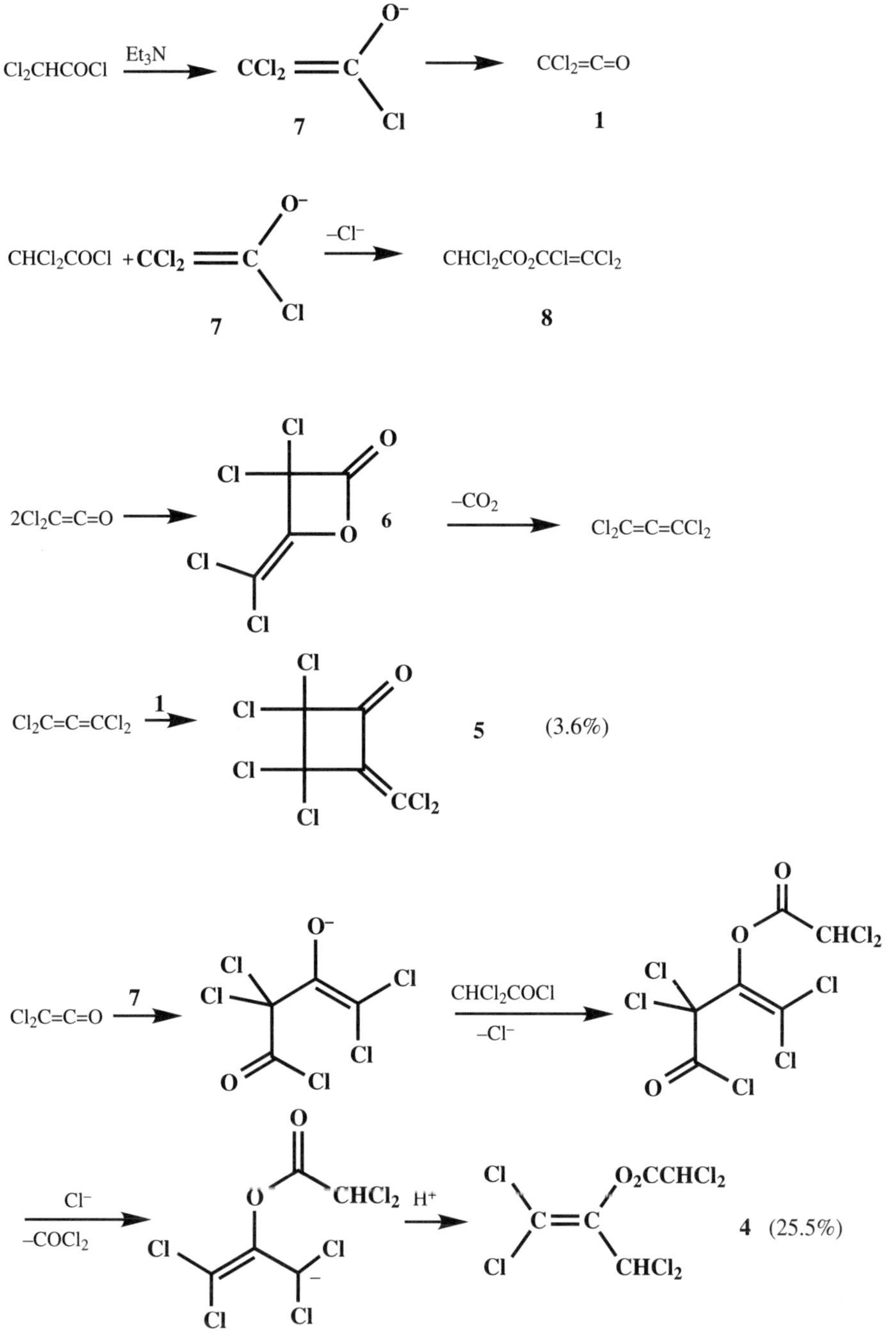

4.4 HALOGEN-SUBSTITUTED KETENES

$$Cl_2CHCOCl \xrightarrow{Et_3N} CCl_2=C(O^-)(Cl) \longrightarrow CCl_2=C=O$$
$$\phantom{Cl_2CHCOCl \xrightarrow{Et_3N}} \quad\quad\quad 7 \quad\quad\quad\quad\quad\quad 1$$

$$CHCl_2COCl + CCl_2=C(O^-)(Cl) \xrightarrow{-Cl^-} CHCl_2CO_2CCl=CCl_2$$
$$ \quad\quad 7 \quad\quad\quad\quad\quad\quad\quad\quad 8$$

$$2Cl_2C=C=O \longrightarrow \text{[β-lactone 6]} \xrightarrow{-CO_2} Cl_2C=C=CCl_2$$

$$Cl_2C=C=CCl_2 \xrightarrow{1} \text{[cyclobutanone 5]} \quad (3.6\%)$$

$$Cl_2C=C=O \xrightarrow{7} \text{[intermediate]} \xrightarrow[-Cl^-]{CHCl_2COCl} \text{[intermediate]}$$

$$\xrightarrow[-COCl_2]{Cl^-} \text{[intermediate]} \xrightarrow{H^+} \text{[product 4]} \quad (25.5\%)$$

Scheme 1

may be achieved as in equation 7,[20] or by using zinc dust in EtOH with AcOH/TMEDA.[22] Ultrasound greatly assists in the generation of dichloroketene from CCl_3COCl and zinc for cycloaddition,[24] and thus the time-consuming preparation of the Zn–Cu couple can be avoided. Simple heating is also useful for activation of the zinc for use in this reaction.[25]

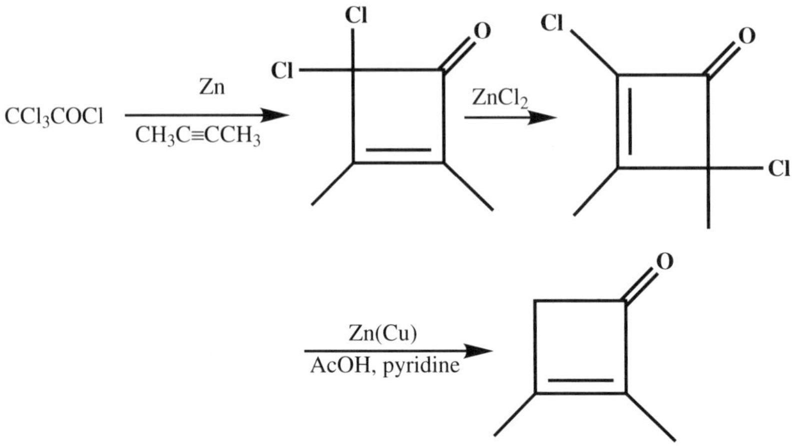

$$CCl_3COCl \xrightarrow[CH_3C\equiv CCH_3]{Zn} \text{[cyclobutenone intermediate]} \xrightarrow{ZnCl_2} \text{[product]} \quad \xrightarrow[AcOH, pyridine]{Zn(Cu)} \text{[product]} \qquad (7)$$

In the reaction of $CCl_2=C=O$ with E,E– and E,Z–2,4-hexadiene the latter diene was found to be at least 170 times more reactive, and gave as the only detectable product the cyclobutanone **9** from addition to the *cis*-double bond (equation 8).[26] The E,E-diene gave **10** as the only detectable product (equation 9).

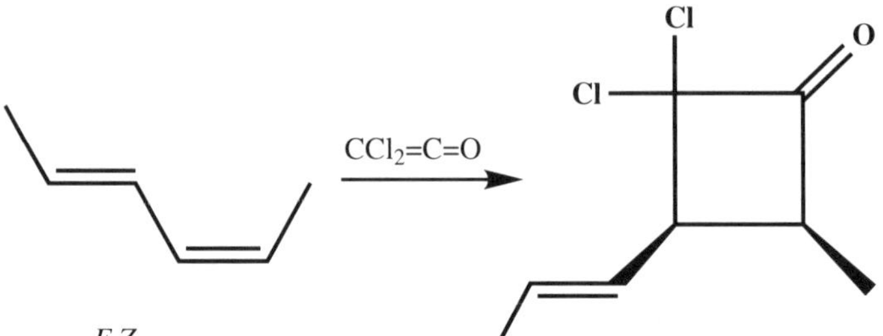

$$E,Z \xrightarrow{CCl_2=C=O} \mathbf{9} \qquad (8)$$

$$E,E \xrightarrow{CCl_2=C=O} \mathbf{10} \qquad (9)$$

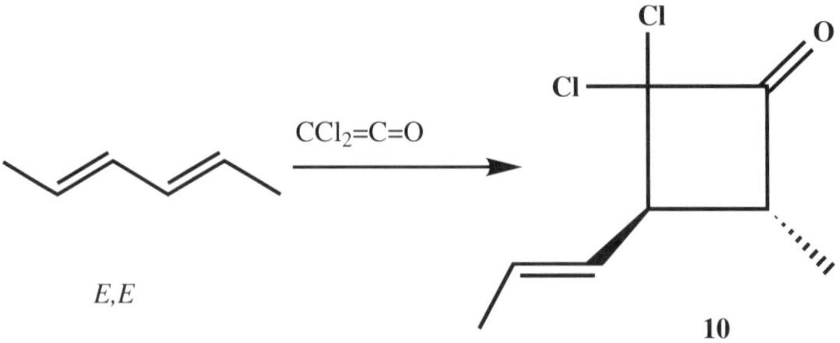

On treatment of **9** with Et₃N there was transient formation of **10,** and the rapid formation of the two stereoisomers of **11** by chlorine rearrangement so that the same stereoisomeric mixture arises from either **9** or **10**.[26] It was suggested that these products might arise from the same zwitterionic intermediate **12** (equation 10). Similar isomerizations were observed in the reaction of the products of $CCl_2=C=O$ cycloaddition to the isomeric 2-butenes.[26]

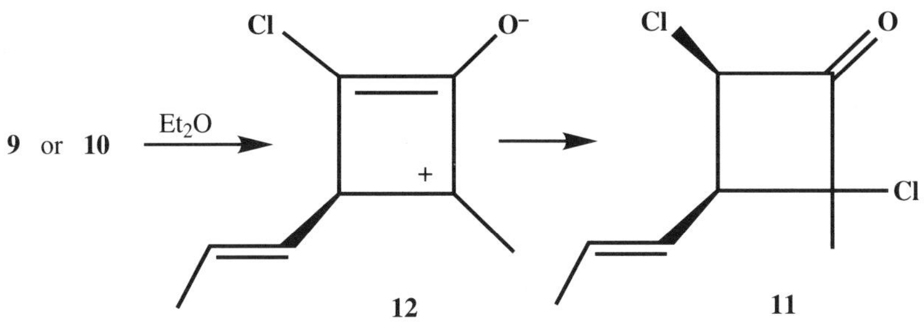

$$\textbf{9 or 10} \xrightarrow{Et_2O} \textbf{12} \longrightarrow \textbf{11} \tag{10}$$

The chloroketenes **13** and **14** are highly reactive in cycloaddition reactions and have been used in the synthesis of pyrethroids.[27]

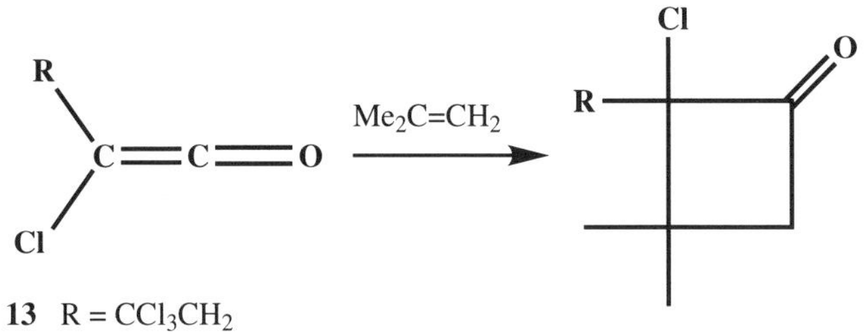

$$\tag{11}$$

13 R = CCl₃CH₂
14 R = CCl₂=CCl

Chloroketene **15** gave the cycloadduct **16**, which was converted to the iodide and ring-expanded by a free-radical path to the spiro product **18** (equation 12).[28] A related process is shown in equation 6, Section 4.1.1.

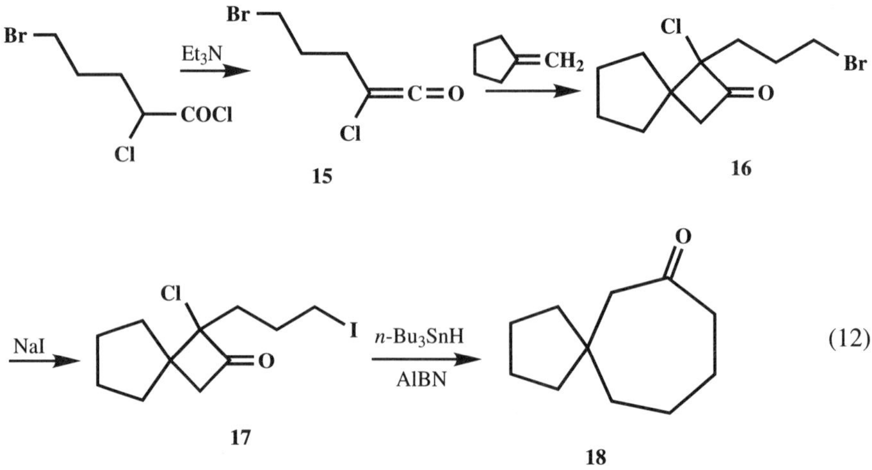

[Equation 12 scheme]

Reaction of 1 equivalent of dichloroketene with 1,5-dienes gave monoadducts which on reduction with tri-n-butyltin hydride gave radical formation, cyclization, and reduction of the remaining chlorine (equation 13).[29,30]

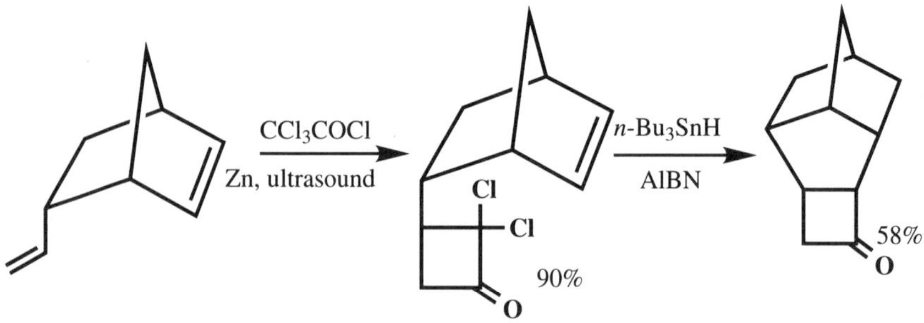

[Equation 13 scheme]

The chloroketene **19** was useful in the synthesis of α-methylenecyclobutanones (equation 14).[31]

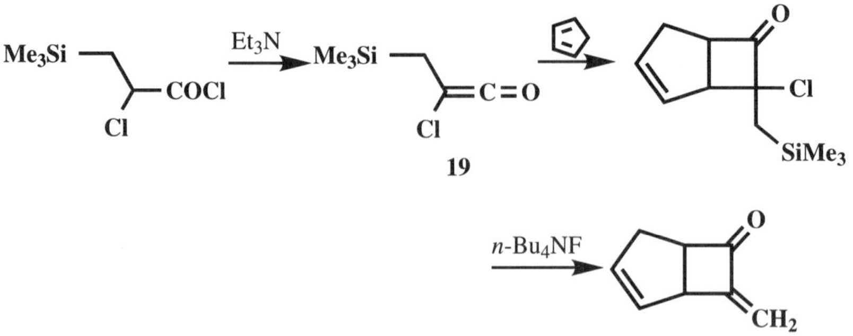

[Equation 14 scheme]

A large number of arylbromoketenes **20** have been generated which gave 1,3-cycloaddition with 3-oxidopyridinium betaines (equation 15).[32]

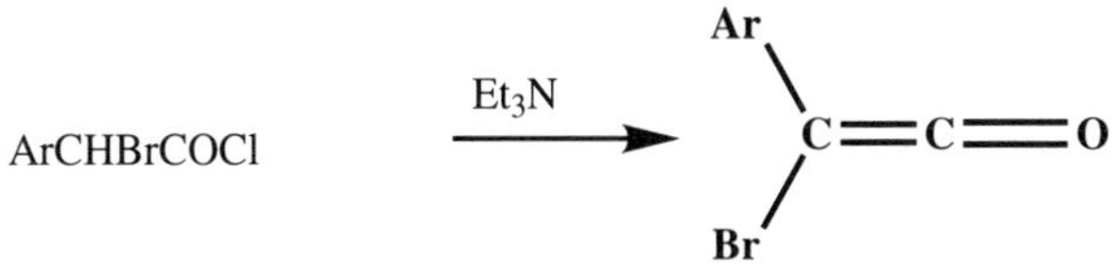

$$\text{ArCHBrCOCl} \xrightarrow{\text{Et}_3\text{N}} \underset{\text{Br}}{\overset{\text{Ar}}{\text{C}}}=\text{C}=\text{O}$$

20

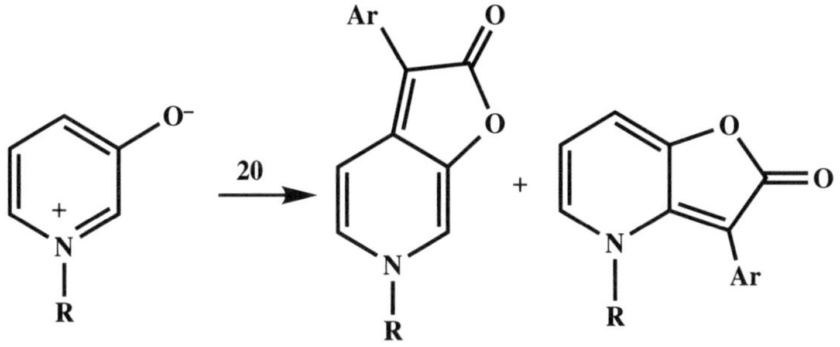

(15)

The intramolecular cycloadditions of haloketenes with remote alkenyl groups has also been utilized in synthesis (equation 16).[33,34]

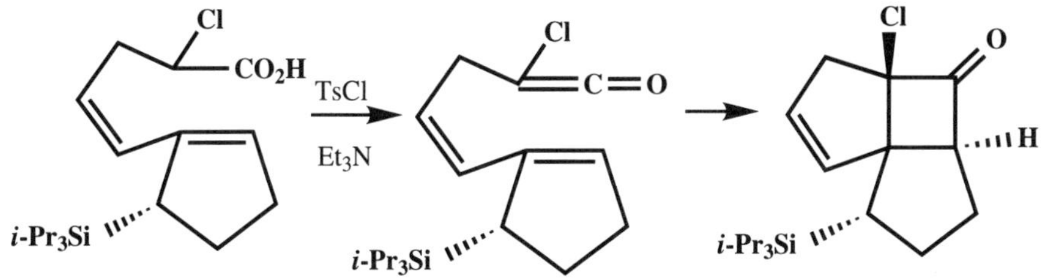

(16)

Reaction of α-halo-α-alkylacetyl halides **21** with Et$_3$N gave α-haloketenes **22** with react with **R**-pantolactone to give esters with diastereoselectivities of 75 to >95% (equation 17).[35]

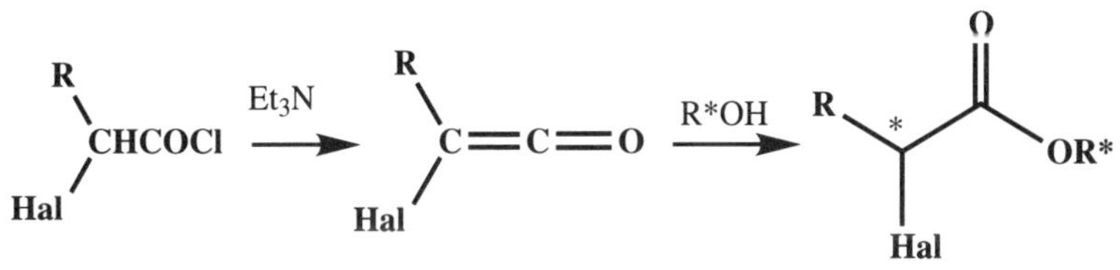

(17)

21

22 Hal = Br, I

TABLE 1. Chlorine, Bromine, and Iodine-Substituted Ketenes

Ketene	Formula	Preparation	Registry No.	Ref.
CClBr=C=O	C$_2$BrClO	Dehydrochlorination	42915-26-4	36
CBr$_2$=C=O	C$_2$Br$_2$O	Dehalogenation	10547-07-6	37,38
CCl$_2$=C=O	C$_2$Cl$_2$O	Dehalogenation	4591-28-0	39
CI$_2$=C=O	C$_2$I$_2$O	Theory	111647-28-0	40
CHBr=C=O	C$_2$HBrO	Dehalogenation	78957-22-9	41
CH^{79}Br=C=O	C$_2$HBrO	Dehalogenation	122607-77-6	41
CH^{81}Br=C=O	C$_2$HBrO	Dehalogenation	122607-76-5	41
CHCl=C=O	C$_2$HClO	Dehydrochlorination	29804-89-5	7,25,42
CD^{37}Cl=C=O	C$_2$ClDO	Dehydrochlorination	88193-14-0	9
CD^{35}Cl=C=O	C$_2$ClDO	Dehydrochlorination	88193-13-9	9
CH^{37}Cl=C=O	C$_2$HClO	Dehydrochlorination	88193-12-8	9
CH^{35}Cl=C=O	C$_2$HClO	Dehydrochlorination	88193-11-7	9
CBr(CN)=C=O	C$_3$BrNO	Pyrolysis	67767-48-0	43,44
CCl(CN)=C=O	C$_3$ClNO	Pyrolysis	60010-89-1	45
CI(CN)=C=O	C$_3$INO	Pyrolysis	67767-49-1	43
CH$_3$CBr=C=O	C$_3$H$_3$BrO	Dehydrochlorination	29264-45-7	46
CH$_3$CCl=C=O	C$_3$H$_3$ClO	Dehydrochlorination	13363-86-5	27
CCl(CCl=CCl$_2$)=C=O	C$_4$Cl$_4$O	Dehydrochlorination	78270-50-5	27
CCl(CH=CCl$_2$)=C=O	C$_4$HCl$_3$O	Dehydrochlorination	78270-49-2	27
CBr(CH$_2$CCl$_3$)=C=O	C$_4$H$_2$BrCl$_3$O	Dehydrochlorination	78270-48-1	27
CBr(CH$_2$CBr$_3$)=C=O	C$_4$H$_2$Br$_4$O	Dehydrochlorination	76706-76-8	27
CCl(CH$_2$CBr$_3$)=C=O	C$_4$H$_2$Br$_3$ClO	Dehydrochlorination	76694-38-7	27
CCl(CH$_2$CClF$_2$)=C=O	C$_4$H$_2$Cl$_2$F$_2$O	Dehydrochlorination	76694-39-8	27
CCl(CH$_2$CCl$_3$)=C=O	C$_4$H$_2$Cl$_4$O	Dehydrochlorination	71855-74-8	27
CBr(CH=CH$_2$)=C=O	C$_4$H$_3$BrO	Dehydrochlorination	85219-16-5	47
CCl(CO$_2$Me)=C=O	C$_4$H$_3$ClO$_3$	Dehydrochlorination	79060-34-7	48

Compound	Formula	Reaction	CAS#	Ref.
CCl(CH$_2$CHClF)=C=C	C$_4$H$_3$Cl$_2$FO	Dehydrochlorination	78270-47-0	27
CCl(CH$_2$CHCl$_2$)=C=O	C$_4$H$_3$Cl$_3$O	Dehydrochlorination	78270-46-9	27
EtCBr=C=O	C$_4$H$_5$BrO	Dehydrochlorination	29264-46-8	35,46,49
EtCCl=C=O	C$_4$H$_5$ClO	Dehydrochlorination	29264-44-6	46
CBr(CMe=CH$_2$)=C=O	C$_5$H$_5$BrO	Dehydrochlorination	85219-17-6	47
CCl[(CH$_2$)$_3$Br]=C=O	C$_5$H$_6$ClBrO	Dehydrochlorination		28
CBr(Pr-i)=C=O	C$_5$H$_7$BrO	Dehydrochlorination	29264-47-9	27,35
CCl(Pr-i)=C=O	C$_5$H$_7$ClO	Dehydrochlorination	29336-28-5	50
CI(Pr-i)=C=O	C$_5$H$_7$IO	Dehydrochlorination		35
CBr(SiMe$_3$)=C=O	C$_5$H$_9$BrOSi	Dehydrochlorination	60366-59-8	51
CBr(Bu-t)=C=O	C$_6$H$_9$BrO	Dehydrochlorination	29264-48-0	35,50,52
CCl[(CH$_2$)$_4$Br]=C=O	C$_6$H$_8$ClBrO	Dehydrochlorination		28
CCl(Bu-i)=C=O	C$_6$H$_9$ClO	Dehydrochlorination	78270-51-6	27
CCl(Bu-t)=C=O	C$_6$H$_9$ClO	Dehydrochlorination	52920-17-9	52
CCl(Bu-n)=C=O	C$_6$H$_9$ClO	Dehydrochlorination	60010-89-1	53
			58216-35-6	53
CCl(PO$_3$Et$_2$)=C=O	C$_6$H$_{10}$ClO$_4$P	Dehydrochlorination	34255-80-6	54
CCl(CH$_2$SiMe$_3$)=C=O	C$_6$H$_{11}$ClOSi	Dehydrochlorination	89121-60-8	31
CCl(CO$_2$Bu-t)=C=O	C$_7$H$_9$ClO$_3$	Dehydrochlorination	79060-33-6	48
CI(Pn-c)=C=O	C$_7$H$_9$IO	Dehydrochlorination		35
CBr(C$_6$H$_4$NO$_2$-4)=C=O	C$_8$H$_4$BrNO$_3$	Dehydrochlorination	69896-67-9	32
CBr(C$_6$H$_4$Br-4)=C=O	C$_8$H$_4$Br$_2$O	Dehydrochlorination	69875-42-9	32
CBrPh=C=O	C$_8$H$_5$BrO	Dehydrochlorination	69875-41-8	32
CClPh=C=O	C$_8$H$_5$ClO	Dehydrochlorination	29804-92-0	55
CBr(C$_6$H$_4$OMe-4)=C=O	C$_9$H$_7$BrO$_2$	Dehydrochlorination	69875-43-0	32
ClCH$_2$Ph=C=O	C$_9$H$_7$IO	Dehydrochlorination		35
CI(CH$_2$Hx-c)=C=O	C$_9$H$_{13}$IO	Dehydrochlorination		35

TABLE 1. (Continued)

Ketene	Formula	Preparation	Registry No.	Ref.
CCl[C(OEt)=CRCN]=C=O				
R = 2-thienyl	$C_{11}H_8ClNO_2S$	Pyrolysis	115476-57-8	56
R = Ph	$C_{13}H_{10}ClNO_2$	Pyrolysis	115476-54-5	56
R = 2-(1-tosylpyridyl)	$C_{18}H_{15}ClN_2O_4S$	Pyrolysis	115476-59-0	56
Cl(CHPh$_2$)=C=O	$C_{15}H_{11}IO$	Dehydrochlorination		35
CCl(CH$_2$CH=CHR)=C=O	$C_{20}H_{31}ClO_2Si$	Elimination		33
R = *i*-Pr$_3$SiO—				

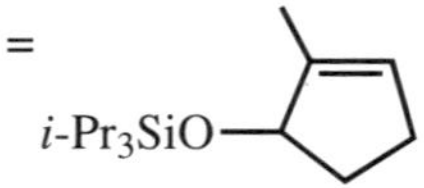

References

1. Brady, W. T. *Synthesis* **1971**, 415–422.
2. Brady, W. T. *Tetrahedron* **1981**, *37*, 2949–2966.
3. Stevens, H. C.; Reich, D. A.; Brandt, D. R.; Fountain, K. R.; Gaughan, E. J. *J. Am. Chem. Soc.* **1965**, *87*, 5257–5269.
4. Ghosez, L.; Montaigne, R.; Mollet, P. *Tetrahedron Lett.* **1966**, 135–139.
5. Brady, W. T.; Liddell, H. G.; Vaughn, W. L. *J. Org. Chem.* **1966**, *31*, 626–628.
6. Ghosez, L.; Montaigne, R.; Vanlierde, H.; Dumay, F. *Angew. Chem., Int. Ed. Engl.* **1968**, *7*, 643.
7. Lindner, E.; Steinward, M.; Hoehne, S. *Angew. Chem., Int. Ed. Engl.* **1982**, 355–356; *Angew. Chem. Suppl.* **1982**, 893–906; *Chem. Ber.* **1982**, *115*, 2181–2191.
8. Lindner, E.; Merkle, R. D.; Mayer, H. A. *Chem. Ber.* **1986**, *119*, 645–658.
9. Gerry, M. C. L.; Lewis-Bevan, W.; Westwood, N. P. C. *J. Chem. Phys.* **1983**, *79*, 4655–4663.
10. Rozsondai, B.; Tremmel, J.; Hargittai, I.; Khabashesku, V. N.; Kagramanov, N. D.; Nefedov, O. M. *J. Am. Chem. Soc.* **1989**, *111*, 2845–2849.
11. Bock, H.; Hirabayashi, T.; Mohmand, S. *Chem. Ber.* **1981**, *114*, 2595–2608.
12. Davidovics, G.; Monnier, M.; Allouche, A. *Chem. Phys.* **1991**, *150*, 395–403.
13. Torres, M.; Ribo, J.; Clement, A.; Strausz, O. P. *Nouv. J. Chem.* **1981**, 351–352.
14. Gong, L.; McAllister, M. A.; Tidwell, T. T. *J. Am. Chem. Soc.* **1991**, *113*, 6021–6028.
15. Ghosez, L.; Montaigne, R.; Roussel, A.; Vanlierde, H.; Mollet, P. *Tetrahedron* **1971**, *27*, 615–633.
16. Rey, M.; Huber, U. A.; Dreiding, A. S. *Tetrahedron Lett.* **1968**, 3583–3588.
17. Minns, R. A. *Organic Synthesis,* Coll. Vol. VI; Noland, W. E., Ed.; Wiley: New York, 1988, pp. 1037–1040.
17a. Imafuku, K., Arai, K. *Synthesis* **1989**, 501–505.
18. Cocivera, M.; Effio, A. *J. Org. Chem.* **1980**, *45*, 415–420.
19. Hopf, H.; Stamm, R.; Jones, P. G. *Chem. Ber.* **1991**, *124*, 1291–1294.
20. Ammann, A. A.; Rey, M.; Dreiding, A. S. *Helv. Chim. Acta* **1987**, *70*, 321–328.
21. Danheiser, R. L.; Savariar, S. *Tetrahedron Lett.* **1987**, *29*, 3299–3302.
22. Danheiser, R. L.; Savariar, S.; Cha, D. D. *Org. Synth.* **1989**, *68*, 32–40.
23. Depres, J.; Greene, A. E. *Org. Synth.* **1989**, *68*, 41–48.
24. Mehta, G.; Rao, H. S. P. *Synth. Commun.* **1985**, *15*, 991–1000.
25. Stenstrøm, Y. *Synth. Commun.* **1992**, *22*, 2801–2810.
26. Bartlett, P. D. *Pure Appl. Chem.* **1971**, *27*, 597–609.
27. Martin, P.; Greuter, H.; Bellus, D. *Helv. Chim. Acta* **1981**, *64*, 64–77.
28. Dowd, P.; Zhang, W. *J. Org. Chem.* **1992**, *57*, 7163–7171.
29. Dowd, P.; Zhang, W. *J. Am. Chem. Soc.* **1992**, *114*, 10084–10085.
30. Zhang, W.; Dowd, P. *Tetrahedron Lett.* **1992**, *33*, 3285–3288.
31. Paquette, L.; Valpey, R. S.; Annis, G. D. *J. Org. Chem.* **1984**, *49*, 1317–1319.
32. Katritzky, A. R.; Cutler, A. T.; Dennis, N.; Sabongi, G. J.; Rahimi-Rastgoo, S.; Fischer, G. W.; Fletcher, I. J. *J. Chem. Soc., Perkin Trans. 2,* **1980**, 1176–1189.

33. Corey, E. J.; Rao, K. S. *Tetrahedron Lett.* **1991**, *32*, 4623–4626.
34. Snider, B. B.; Kulkarni, Y. S. *J. Org. Chem.* **1987**, *52*, 307–310.
35. Durst, T.; Koh, K. *Tetrahedron Lett.* **1992**, *33*, 6799–6802.
36. Brady, W. T.; Stockton, J. D.; Patel, A. D. *J. Org. Chem.* **1974**, *39*, 236–238.
37. Birkofer, L.;. Lueckenhaus, W. *Liebigs Ann. Chem.* **1984**, 1193–1204.
38. Saidi, K.; Habibi, M. H. *J. Fluorine Chem.* **1991**, *51*, 217–222.
39. Depres, J. P.; Navarro, B.; Greene, A. E. *Tetrahedron* **1989**, *45*, 2989–2998.
40. Gano, J. E.; Jacob, E. J. *Spectrochim. Acta* **1987**, *43A*, 1023–1025.
41. Westwood, N. P. C.; Lewis-Bevan, W.; Gerry, M. C. L. *J. Mol. Spectrosc.* **1989**, *136*, 93–104.
42. Marino, J. P.; Perez, A. D. *J. Am. Chem. Soc.* **1984**, *106*, 7643–7644.
43. Kunert, D. M.; Chambers, R.; Mercer, F.; Hernandez, L., Jr.; Moore, H. W. *Tetrahedron Lett.* **1978**, 929–932.
44. Moore, H. W.; Mercer, F.; Kunert, D.; Albaugh, P. *J. Am. Chem. Soc.* **1979**, *101*, 5435–5436.
45. Dehmlow, E. V.; Birkhahn, M. *Liebigs Ann. Chem.* **1987**, 701–704.
46. Brady, W. T.; Roe, R., Jr. *J. Am. Chem. Soc.* **1970**, *92*, 4618–4621.
47. Berge, J. M.; Rey, M.; Dreiding, A. S. *Helv. Chim. Acta* **1982**, *65*, 2230–2241.
48. Goldstein, S.; Vannes, P.; Houge, C.; Frisque-Hesbain, A. M.; Wiaux-Zamar, C.; Ghosez, L.; Germain, G.; Declercq, J. P.; Van Meerssche, M.; Arrieta, J. M. *J. Am. Chem. Soc.* **1981**, *103*, 4616–4618.
49. Staudinger, H.; Schneider, H.; Schotz, P.; Strong, P. M. *Helv. Chim. Acta* **1923**, *6*, 291–303; *Chem. Abstr.* **1923**, *17*, 1953.
50. De Selms, R. C.; Delay, F. *J. Org. Chem.* **1972**, *37*, 2908–2910.
51. Brady, W. T.; Owens, R. A. *Tetrahedron Lett.* **1976**, 1553–1556.
52. Brady, W. T.; Ting, P. L. *J. Org. Chem.* **1976**, *41*, 2336–2339.
53. Bellus, D.; *Helv. Chim. Acta* **1975**, *58*, 2509–2511.
54. Motoyshiya, J.; Hirata, K. *Chem. Lett.* **1988**, 211–214.
55. Minami, T.; Agawa, T. *J. Org. Chem.* **1974**, *39*, 1210–1215.
56. Chow, K.; Moore, H. W. *Tetrahedron Lett.* **1987**, *28*, 5013–5016.

4.5 SILYL-, GERMYL-, AND STANNYLKETENES

(Trimethylsilyl)ketene (**1**) is a remarkably stable ketene that was first prepared in 1965[1,2] by the pyrolysis of (trimethylsilyl)ethoxyacetylene (equation 1), and has also been obtained by the dehydrohalogenation of (trimethylsilyl)acetyl chloride (equation 2),[3] and by dehydration of trimethylsilylacetic acid by dicyclohexylcarbodiimide (DCC) (equation 3).[4] The first of these methods has been the most used for the preparation of this ketene, which, along with other silylketenes, appears to be unique among monosubstituted ketenes because of its resistance to dimerization,[1-8] which permits the compound to be stored for long periods of time. The related $Et_3SiCH=C=O$ was prepared by pyrolysis of triethylsilylacetic anhydride (equation 4).[9]

$$Me_3SiC\equiv COEt \xrightarrow{120°} Me_3SiCH=C=O \quad (90\%) \qquad (1)[1]$$
$$\mathbf{1}$$

$$Me_3SiCH_2COCl \xrightarrow{\Delta} \mathbf{1} \qquad (2)[3]$$

$$Me_3SiCH_2CO_2H \xrightarrow{DCC,\ Et_3N} \mathbf{1} \quad (63\%) \qquad (3)[4]$$

$$(Et_3SiCH_2CO)_2O \xrightarrow{\Delta} Et_3SiCH=C=O \ \mathbf{1a} \ (80\%) \qquad (4)[9]$$

Bis(trimethylsilyl)ketene (**2**) was prepared as shown in equation 5,[10,11] as well as by the procedures of equations 6–9.[12-15] Other examples of the preparation of silylketenes are shown in equations 10–32.[16-39] Thermal stability is a characteristic shared by many of the silylketenes, but not by the β-silylketene **3**,[16] the α-bromo ketene **4**,[17] the α-alkoxyketene **48**,[39] or the silyloxyketenes **35–37**.[35,36]

$$BrMgC\equiv COBu\text{-}n \xrightarrow{Me_3SiBr} Me_3SiC\equiv COBu\text{-}n \xrightarrow[MgBr_2]{Me_3SiBr} (Me_3Si)_2C=C=O$$
$$\mathbf{2}$$
$$(35\%)$$
$$(5)[10,11]$$

$$Me_3SiCH=C=O \xrightarrow[-100\ °C]{n\text{-}BuLi} Me_3SiC\equiv COLi \xrightarrow{Me_3SiCl} \mathbf{2} \quad (80\%) \qquad (6)[12]$$

$$(Me_3Si)_2CHCO_2Bu\text{-}t \xrightarrow[-78\ °C]{LDA} (Me_3Si)_2CLiCO_2Bu\text{-}t \xrightarrow{25\ °C} \mathbf{2} \quad (60\%) \qquad (7)[13]$$

$$CH_2=C=O \xrightarrow{Me_3SiOTf} \mathbf{1} \xrightarrow{Me_3SiOTf} \mathbf{2} \quad (48\%) \qquad (8)[14]$$

$$Me_3SiSiMe_2C(N_2)COMe \xrightarrow{h\nu} \mathbf{2} \qquad (9)[15]$$

$$\text{Me}_3\text{SiCH}_2\text{CH}_2\text{COCl} \xrightarrow{\text{Et}_3\text{N}} \underset{\mathbf{3}}{\text{Me}_3\text{SiCH}_2\text{CH=C=O}} \qquad (10)^{16}$$

$$\text{Me}_3\text{SiCH=C=O} \xrightarrow{\text{Br}_2} \text{Me}_3\text{SiCHBrCOBr} \xrightarrow{\text{Et}_3\text{N}} \underset{\mathbf{4}}{\text{Me}_3\text{SiCBr=C=O}} \qquad (11)^{17}$$

$$\text{Me}_3\text{SiCH}_2\text{Ph} \xrightarrow[\text{2) CO}_2,\ 3)\ \text{H}^+]{\text{1) }n\text{-BuLi}} \text{Me}_3\text{SiCHPhCO}_2\text{H} \xrightarrow[\text{2) Br}_2\ 3)\ \text{Zn}]{\text{1) (COCl)}_2} \underset{\mathbf{5}}{\overset{\text{Me}_3\text{Si}}{\underset{\text{Ph}}{>}}\!\!\text{C=C=O}} \quad (41\%)$$

$$(12)^{18}$$

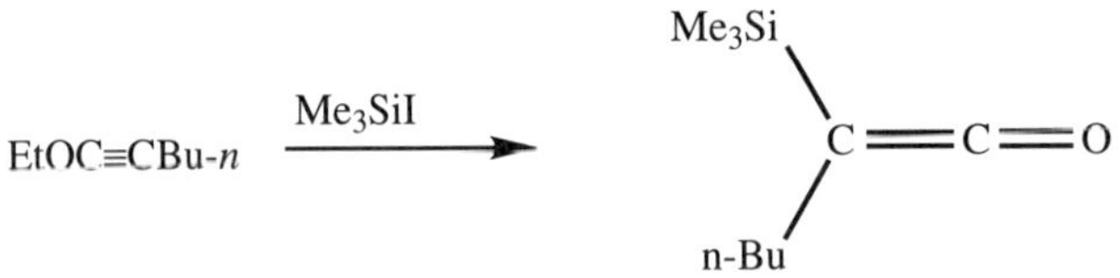

$$\text{EtOC≡CBu-}n \xrightarrow{\text{Me}_3\text{SiI}} \underset{\mathbf{6}}{\overset{\text{Me}_3\text{Si}}{\underset{n\text{-Bu}}{>}}\!\!\text{C=C=O}} \quad (57\%) \qquad (13)^{19}$$

$$\text{MeC≡CSiMe}_3 \xrightarrow[\text{2) MeLi 3) CO}_2]{\text{1) DIBAL}} Z\text{-MeCH=C(SiMe}_3)\text{CO}_2\text{H} \quad (68\%) \qquad (14)^{20}$$

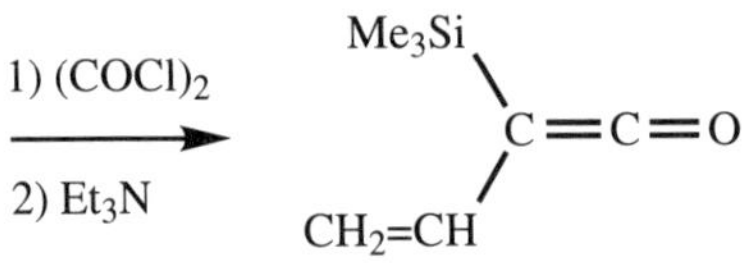

$$\xrightarrow[\text{2) Et}_3\text{N}]{\text{1) (COCl)}_2} \underset{\mathbf{7}}{\overset{\text{Me}_3\text{Si}}{\underset{\text{CH}_2=\text{CH}}{>}}\!\!\text{C=C=O}} \quad (39\text{–}50\%)$$

$$\text{Me}_3\text{SiCH}_2\text{CO}_2\text{H} \xrightarrow[\text{3) H}_3\text{O}^+]{\text{1) LDA 2) EtI}} \text{Me}_3\text{SiCHEtCO}_2\text{H} \xrightarrow[\text{2) Et}_3\text{N}]{\text{1) (COCl)}_2} \underset{\mathbf{8}}{\overset{\text{Me}_3\text{Si}}{\underset{\text{Et}}{>}}\!\!\text{C=C=O}} \qquad (15)^{21}$$

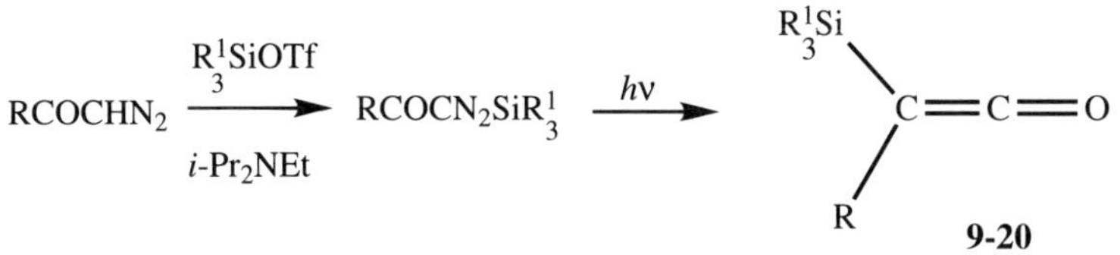

4.5 SILYL-, GERMYL-, AND STANNYLKETENES

$$RCOCHN_2 \xrightarrow[i\text{-}Pr_2NEt]{R^1_3SiOTf} RCOCN_2SiR^1_3 \xrightarrow{h\nu} \underset{R}{\overset{R^1_3Si}{>}}C=C=O \quad \textbf{9-20} \qquad (16)^{22,23}$$

Ketene	R	R^1_3Si	Ketene	R	R^1_3Si
9	Ph	Et$_3$Si (87%)	15	t-Bu	t-BuPh$_2$Si (26%)
10	2-thienyl	i-Pr$_3$Si (94%)	16	t-Bu	i-Pr$_3$Si (38%)
11	Me	Et$_3$Si (54%)	17	Ph$_2$CH	i-Pr$_3$Si (71%)
12	Me	i-Pr$_3$Si (49)	18	1-Ad	t-BuMe$_2$Si (72%)
13	t-Bu	t-BuMe$_2$Si (50%)	19	1-Ad	t-BuPh$_2$Si (71%)
14	t-Bu	t-Bu$_2$MeSi (39%)	20	1-Ad	i-Pr$_3$Si (62%)

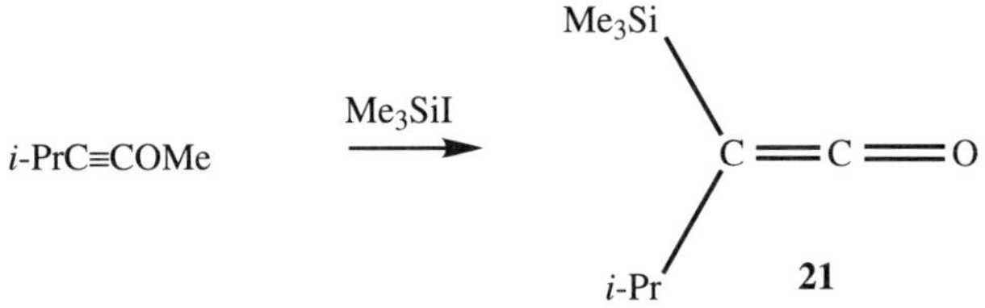

$$i\text{-}PrC\equiv COMe \xrightarrow{Me_3SiI} \underset{i\text{-}Pr}{\overset{Me_3Si}{>}}C=C=O \quad \textbf{21} \quad (90\%) \qquad (17)^{24}$$

By a procedure analogous to that of equation 17 there were prepared **2** (81%), Me$_3$SiC(GeMe$_3$)=C=O (**22**) (95%), and Me$_3$SiC(SnEt$_3$)=C=O (**23**) (83%).[24]

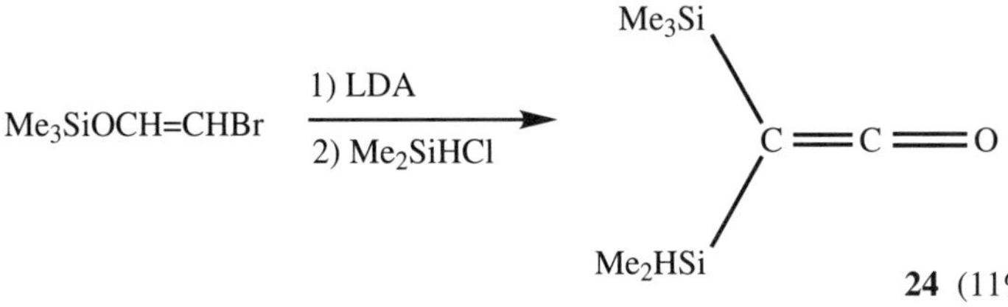

$$Me_3SiOCH=CHBr \xrightarrow[2)\ Me_2SiHCl]{1)\ LDA} \underset{Me_2HSi}{\overset{Me_3Si}{>}}C=C=O \quad \textbf{24}\ (11\%) \qquad (18)^{25,26}$$

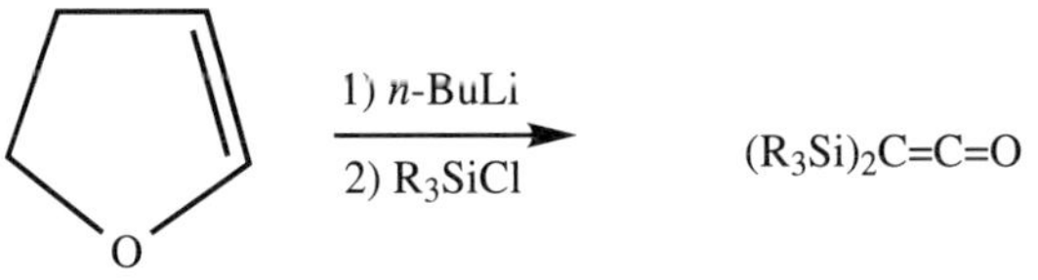

$$\text{(2,3-dihydrofuran)} \xrightarrow[2)\ R_3SiCl]{1)\ n\text{-}BuLi} (R_3Si)_2C=C=O \qquad (19)^{27}$$

R$_3$Si = Me$_3$Si, 21% (**2**), Me$_2$HSi, 28%(**25**),

t-BuMe$_2$Si, 40%(**26**)

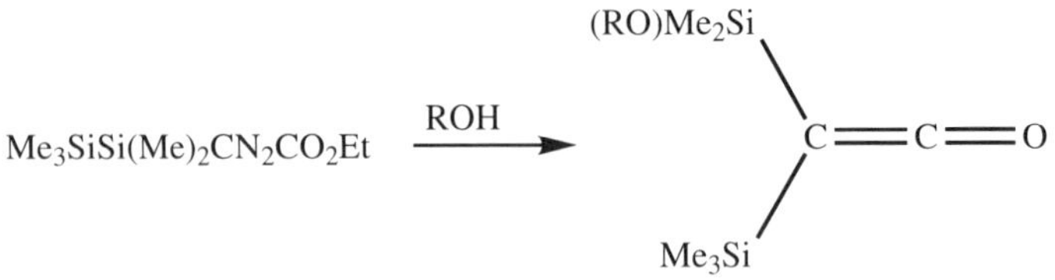

Me$_3$SiSi(Me)$_2$CN$_2$CO$_2$Et $\xrightarrow{\text{ROH}}$ (RO)Me$_2$Si\(C\)=C=O / Me$_3$Si (20)[28,29]

R = Me (**27**), Et (**28**), t-Bu (**29**)

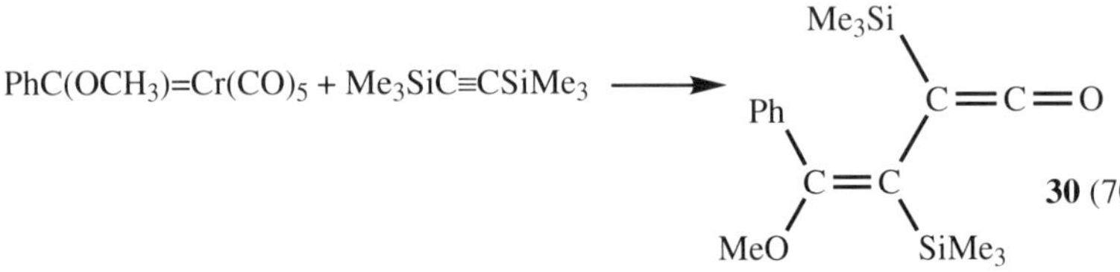

PhC(OCH$_3$)=Cr(CO)$_5$ + Me$_3$SiC≡CSiMe$_3$ ⟶ Ph–C(=C(OMe)(SiMe$_3$))–C(SiMe$_3$)=C=O (21)[30–32]

30 (70%)

31 [complex with Cr(CO)$_3$, 52%]

R$_3$SiC≡COBu-t $\xrightarrow{\Delta}$ R$_3$SiCH=C=O (80–90%) (22)[33]

R$_3$Si = Me$_3$Si (**1**), t-BuMe$_2$Si (**32**), t-BuPh$_2$Si (**33**)

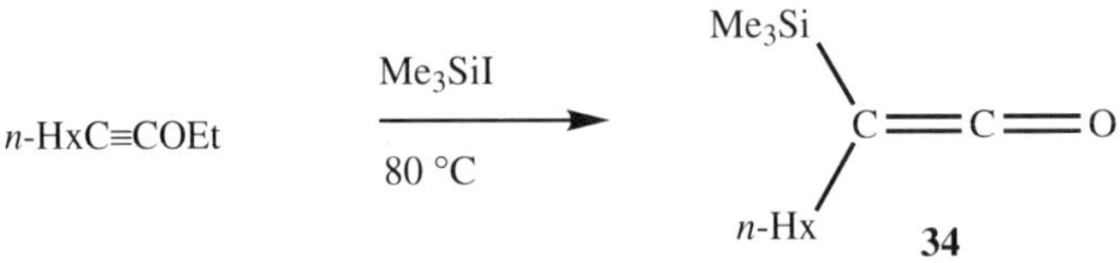

n-HxC≡COEt $\xrightarrow[80\ °C]{\text{Me}_3\text{SiI}}$ Me$_3$Si\(C\)=C=O / n-Hx   (50%)   (23)[34]

34

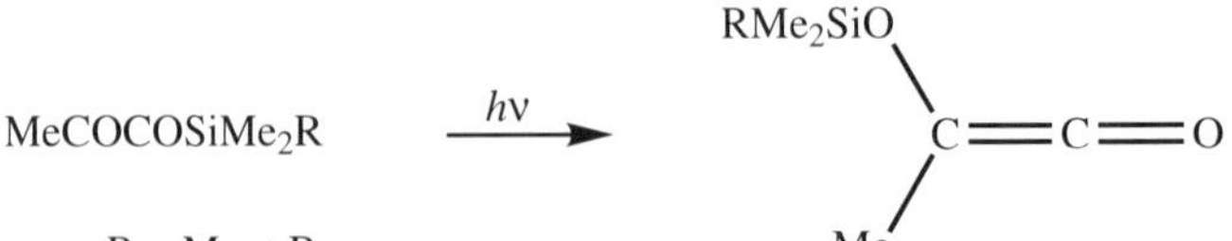

MeCOCOSiMe$_2$R $\xrightarrow{hv}$ RMe$_2$SiO\(C\)=C=O / Me   (24)[35]

R = Me, t-Bu

35, 36

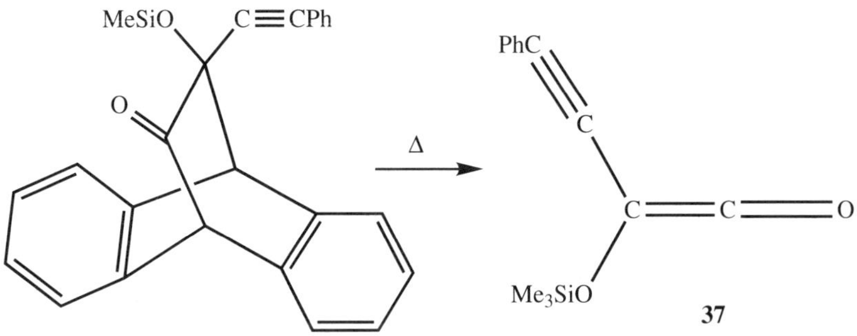

4.5 SILYL-, GERMYL-, AND STANNYLKETENES

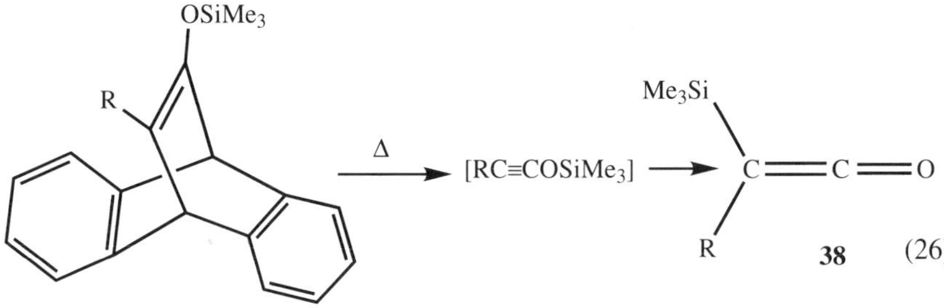

(equation 25)[36] — product **37**: PhC≡C–C(OSiMe$_3$)=C=O

(equation 26)[37] — [RC≡COSiMe$_3$] → **38**: Me$_3$Si(R)C=C=O

R = H (**1**), Ph (**5**), Me (**39**)

Photolysis of diazo(pentamethyldisilanyl)methyl ketones in acetone gave the (pentamethyldisilanyl)ketenes **40,** along with products from the silenes **41,** by Wolff rearrangement with migration of the alkyl groups R (equation 27).[38] These ketenes could be isolated in an impure state and identified by their IR bands at 2075–2080 cm^{-1}, and were trapped by MeOH as the methyl esters.

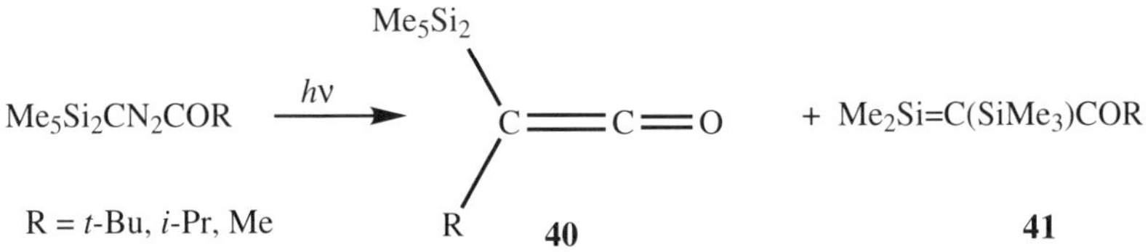

$$Me_5Si_2CN_2COR \xrightarrow{h\nu} Me_5Si_2(R)C=C=O \;+\; Me_2Si=C(SiMe_3)COR$$

R = *t*-Bu, *i*-Pr, Me   **40**   **41**   (27)[38]

Photolysis of the diazoester **42** in MeOH led to the esters **43–46** which were proposed to result from the intermediates shown (equations 28–31).[39] Pyrolysis of **42** at 360 °C led to trapping of the ketene **49** in 5–11% yield (equation 32).[39] The catalytic decomposition of other silyl diazoesters analogous to **42** giving ketenes analogous to **49** has been reported.[40]

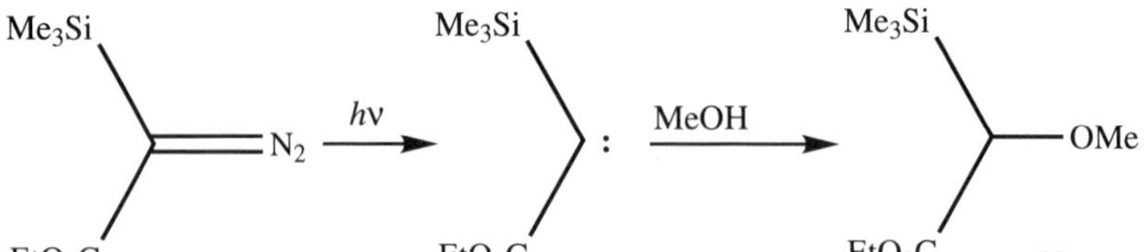

354 TYPES OF KETENES

$$\underset{42}{\underset{EtO_2C}{Me_3Si}}\!=\!N_2 \xrightarrow{h\nu} \underset{47}{\underset{EtO_2C}{Me_3Si}}: \xrightarrow{MeOH} \underset{43}{\underset{EtO_2C}{Me_3Si}}\!\!\!\!\text{—OMe} \quad (28)^{39}$$

$$47 \longrightarrow \underset{OEt^-}{Me_3Si\overset{+}{C}\!\!=\!\!C\!\!=\!\!O} \xrightarrow{MeOH} \underset{44}{Me_3SiCH(OMe)CO_2Me} \quad (29)$$

$$47 \longrightarrow \underset{48}{Me_3SiC(OEt)\!\!=\!\!C\!\!=\!\!O} \xrightarrow{MeOH} \underset{45}{Me_3SiCH(OEt)CO_2Me} \quad (30)$$

$$47 \longrightarrow Me_2Si\!\!=\!\!CMeCO_2Et \xrightarrow{MeOH} \underset{46}{Me_2Si(OMe)CHMeCO_2Et} \quad (31)$$

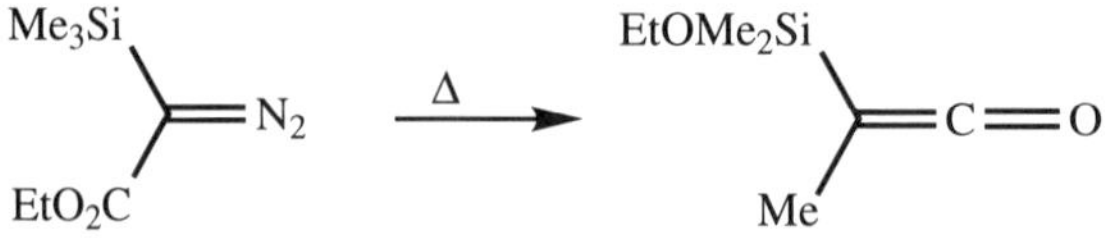

$$\underset{42}{\underset{EtO_2C}{Me_3Si}}\!\!\!\!\!\!\!\!\!\!\!\!\!=\!N_2 \xrightarrow{\Delta} \underset{49}{\underset{Me}{EtOMe_2Si}}\!\!\!\!\!\!\!\!\!\!\!\!\!=\!C\!\!=\!\!O \quad (32)^{39}$$

It was suggested by Brady and Cheng[16] that the extraordinary resistance of trimethylsilylketene (**1**) towards dimerization was due to ground state stabilization through hyperconjugative electron donation from the C–Si bond, as shown in **50**. This explanation appears plausible since the C–Si bond and the carbonyl pi bond in **1** are in the same plane, and the power of the C–Si bond as a hyperconjugative electron donor has been thoroughly documented.[41-46]

$$\underset{1}{\underset{H}{Me_3Si}}\!\!\!\!\!\!\!\!\!\!\!\!\!C\!\!=\!\!C\!\!=\!\!O \longleftrightarrow \underset{50}{\underset{H}{Me_3Si^+}}\!\!\!\!\!\!\!\!\!\!\!\!\!C\!\!\equiv\!\!C\!\!-\!\!O^-$$

It was proposed however by Runge,[47] on the basis of CNDO/S calculations, that the trimethylsilyl group acted as an electron acceptor, with back donation from the ketene π to the *d* orbitals on silicon. The resonance structures **51** and **52** were proposed, leading to a negative charge on silicon and partial Si—C double bond character.

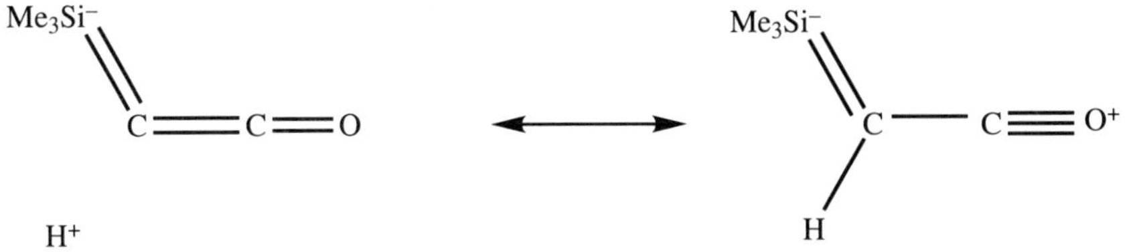

Me₃Si⁻\
    \\\
     C=C=O   ⟷   Me₃Si⁻\
                                   \\\
                                    C—C≡O⁺\
H⁺   H\
51   **52**

However, ab initio molecular orbital calculations at the 6-31G*//6-31G* level show that the C=C and C=O lengths in Me₃SiCH=C=O and CH₃CH=C=O are almost the same (Figure 1).[48] This suggests that any major contribution by resonance structures such as **52** does not lead to shorter C—O and longer C—C bond lengths. The C—Si bond length in SiH₃CH=C=O (1.860 Å) is shorter than that of SiH₃CH=CH₂ (1.874 Å), suggesting that the contribution of structure **50** does not lead to the lengthening of this bond.

Examination of the bond lengths of some ketenes proposed to involve significant pi donation to the substituent, namely H₂BCH=C=O and O=CHCH=C=O, shows that lengthening of the C—C bond and shortening of the C—O bond does occur in these compounds (Figure 1). Also *d*–π conjugation between silicon and carbon is not thought to be significant in silicon chemistry, arguing against the importance of **51** and **52**.[42–46] Thus there is no evidence substantiating any significant role for the resonance structures **51** and **52**, and these appear to be unimportant.

Some analogy with structure **50** involving σ–π donation from the C—Si bond is found in the case of the Li and Na substituents, which are expected to be very

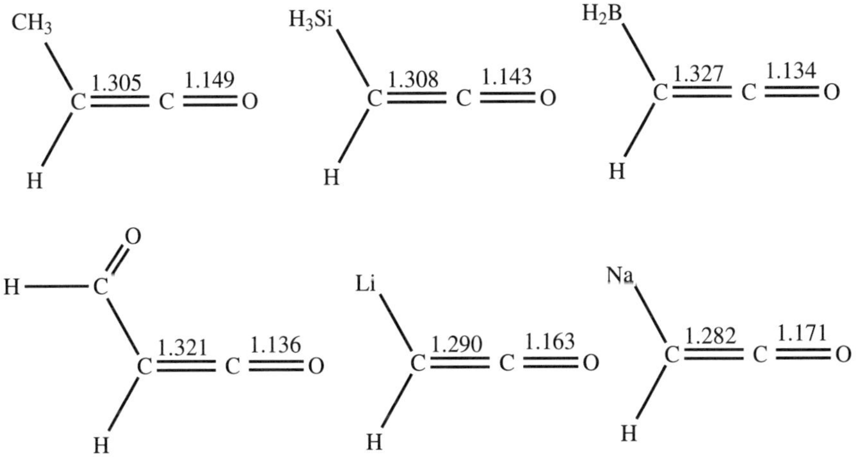

Figure 1. Calculated (6-31G*//6-31G*) bond lengths in ketenes.[48]

strong $\sigma-\pi$ donors from the C—M bonds, and which show a shortening of the C=C bond and lengthening of the C—O bond.[48] The ^{13}C NMR chemical shift of C_α of ketenes is distinctly shifted upfield by Me$_3$Si substituents at C_β (Table 1, Section 2.1), and this is also consistent with electron donation to C_α by the silyl substituent. Thus there is some evidence for SiH$_3$ acting as a $\sigma-\pi$ donor in ketenes, and this phenomenon is of such pervasive importance in other aspects of silicon chemistry that it is likely to play a role in ketene chemistry.

Studies of the hydration of Me$_3$SiCH=C=O showed that this ketene was less reactive than n-BuCH=C=O and t-BuCH=C=O by factors of 100–400.[49] However the ratios k_{OH^-}/k_{H_2O} and k_{H^+}/k_{H_2O} for Me$_3$SiCH=C=O were 2.4×10^4 and 1.7×10^5, as compared to values of 40–4700 for n-BuCH=C=O and t-BuCH=C=O. Thus the reactivity of Me$_3$SiCH=C=O is enhanced in the H$^+$ and OH$^-$ reactions relative to the slow H$_2$O reaction.[49]

These results may mean that the lower reactivity of Me$_3$SiCH=C=O in the H$_2$O reaction reflects the ground state stability of this ketene and a relatively nonpolar transition state. However, the OH$^-$ and H$^+$ reactions have the more polar transition states shown in equations 33 and 34, and the ability of silicon to stabilize adjacent negative charge and β-positive charge, enhances the rates of the respective processes.

$$\underset{\underset{H}{\overset{Me_3Si}{\diagdown}}{\overset{\diagup}{C}}}{}=C=O \quad \xrightarrow{OH^-} \quad \left[\underset{\underset{H}{\overset{Me_3Si}{\diagdown}}{\overset{\diagup}{C}}}{} \cdots\cdots \underset{OH}{\overset{O^-}{C}} \right] \quad (33)$$

$$\mathbf{1}$$

$$\mathbf{1} \xrightarrow{H^+} \underset{Me_3Si}{\overset{H^{\delta+}}{\diagdown}}\!\!\!\!\underset{}{\overset{}{CH}}=\!\!=\overset{\delta+}{C}=O \quad \longrightarrow \quad Me_3SiCH_2\overset{+}{C}=O \quad (34)$$

The kinetics of reaction of Me$_3$SiCH=C=O and of t-BuMe$_2$SiCH=C=O with a variety of oxygen nucleophiles have been measured,[50] as given in Table 1. The addition of alcohols to Me$_3$SiCH=C=O in organic media is reported to be catalyzed by lipase.[50a]

TABLE 1. Reactivity of Silylketenes with Oxygen Nucleophiles at 25 °C[50]

Solvent	k_{obs} (s^{-1})	
	Me$_3$SiCH=C=O	t-BuMe$_2$SiCH=C=O
H$_2$O	0.254	0.115
H$_2$O/CH$_3$CN (50%)	2.58×10^{-3}	9.20×10^{-4}
MeOH	6.00×10^{-3}	5.83×10^{-3}
EtOH (80%)		1.72×10^{-3}
(95%)	9.86×10^{-4}	5.18×10^{-4}

Preparative additions of a number of reagents to Me$_3$SiCH=C=O have been carried out and yielded the products indicated in parentheses: CH$_3$OH (Me$_3$SiCH$_2$CO$_2$Me, 81%), EtOH (Me$_3$SiCH$_2$CO$_2$Et, 77.5%), n-BuOH (Me$_3$SiCH$_2$CO$_2$Bu-n, 85%), Br$_2$ (Me$_3$SiCHBrCOBr, 85%), Et$_3$SiOH (Me$_3$SiCH$_2$- CO$_2$SiEt$_3$, 77%), PhNH$_2$ (Me$_3$SiCH$_2$CONHPh), and H$_2$O (Me$_3$SiCH$_2$CO$_2$H).[1] In the last two examples the yields were not given.[1] Desilylated products are also formed in these reactions.[49,50] Alkoxystannanes, including n-Bu$_3$SnOEt, add to silylketenes RMe$_2$SiCH=C=O (R = Me, t-Bu) to form esters n-Bu$_3$Sn-CH(SiMe$_2$R)CO$_2$Et.[51]

Amines add readily to silylketenes to give amides (equation 35)[5,33] and catalysis of the reaction of *tert*-butanol and other hindered alcohols with Me$_3$SiCH=C=O resulted in rapid formation of the esters (equation 36).[5] The ZnCl$_2$-catalyzed reaction of alcohols and ZnI$_2$-catalyzed reaction of phenols have been used in the preparation of α-silylacetates, which are sometimes difficult to obtain by other methods (equation 37).[52] This ketene gave Wittig reactions with a stabilized phosphorous ylide (equation 38)[5] and with a sulfur ylide gave **53** (equation 39).[5]

$$\text{Me}_3\text{SiCH=C=O} \xrightarrow{i\text{-Pr}_2\text{NH}} \text{Me}_3\text{SiCH}_2\text{CON(Pr-}i)_2 \quad (\sim 100\%) \quad (35)^5$$

$$\text{Me}_3\text{SiCH=C=O} \xrightarrow[\text{BF}_3\cdot\text{OEt}_2]{t\text{-BuOH}} \text{Me}_3\text{SiCH}_2\text{CO}_2\text{Bu-}t \quad (93\%) \quad (36)^5$$

Me$_3$SiCH=C=O + [Ph(Me)C=C(OH)] $\xrightarrow{\text{ZnCl}_2}$ [Ph(Me)C=C(O$_2$CCH$_2$SiMe$_3$)] (89%) (37)[52]

Me$_3$SiCH=C=O + Ph$_3$P=CHCO$_2$Et $\longrightarrow$ Me$_3$SiCH=C=CHCO$_2$Et (85%) (38)[5]

$$Me_3SiCH=C=O + Me_2S=CHCO_2Et \longrightarrow Me_3SiCH_2COC(CO_2Et)=SMe_2$$

$$\mathbf{53} \tag{39}^5$$

Reaction of **1** with salts of salicylaldehyde and substituted derivatives leads to coumarins by the route of equation 40.[53]

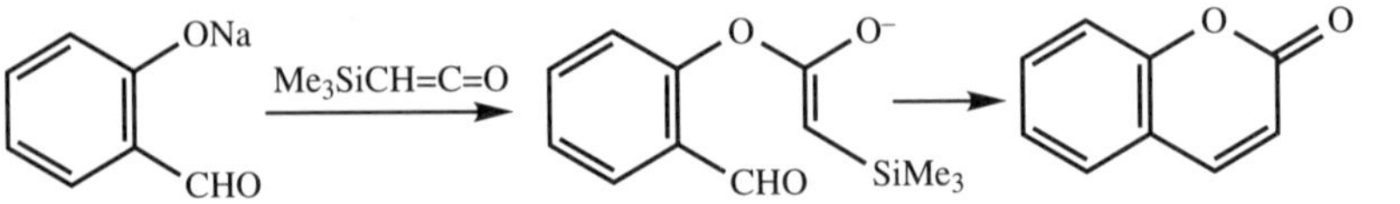

(40)

Reaction of $(Me_3Si)_2C=C=O$ with n-BuLi gives $(Me_3Si)_2CHCOBu-n$,[12,54] and the addition reactions of $Me_3SiCEt=C=O$ with n-BuLi, $CH_2=CHLi$, and $HC\equiv CLi$ have been observed (equation 41).[21] However, the addition of organolithium reagents to silylketenes $R_3SiCH=C=O$ with α-hydrogens results in proton abstraction.[12,13] The use of the less basic organocerium reagents allows addition to $R_3SiCH=C=O$, and the resulting enolates can be either quenched with H_2O (equation 42) or reacted with alkyl halides to give ketones (equation 43).[55]

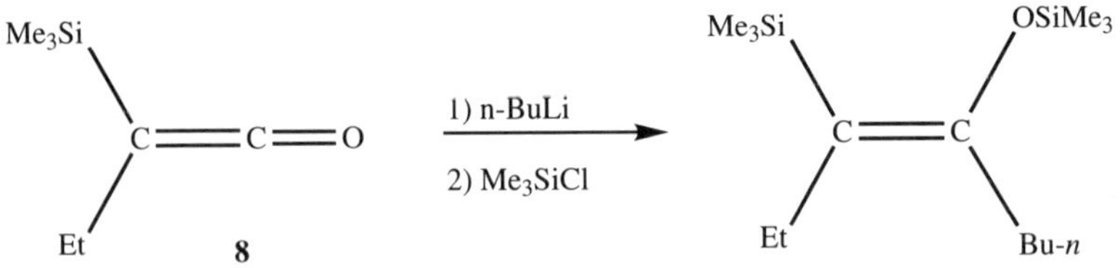

(41)

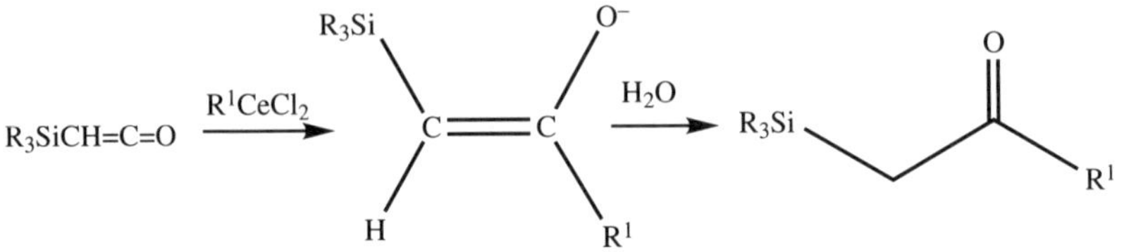

(42)

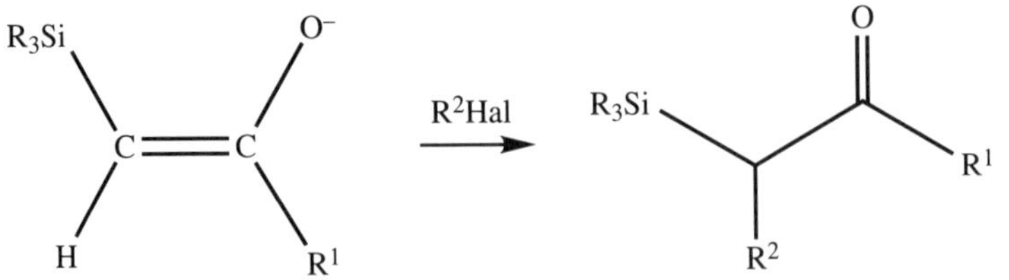

(43)

The cycloaddition reaction of $Me_3SiCH=C=O$ with many alkenes is not successful,[6,7] but reaction does occur with aldehydes to give oxetanones (equation 44)[6] and with tetraalkoxyalkenes to give cyclobutanones (equation 45).[7] Reaction with methoxydi(trimethylsilyloxy)ethylene did not lead to ring closure but gave the ester shown (equation 46).[8]

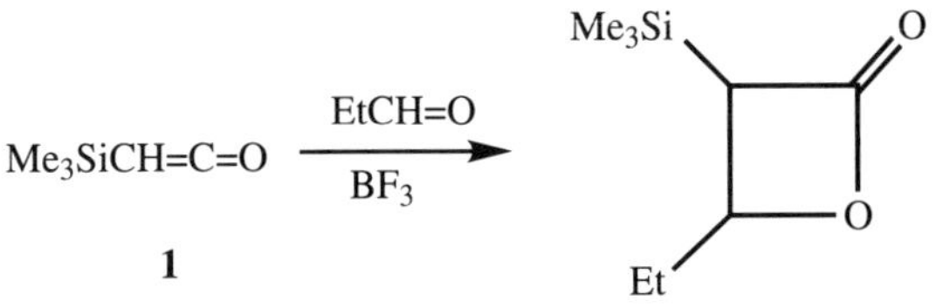

4.5 SILYL-, GERMYL-, AND STANNYLKETENES

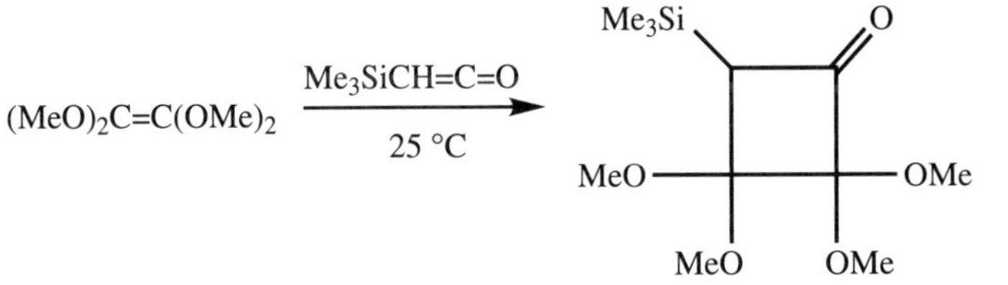

$$Me_3SiCH=C=O \xrightarrow[BF_3]{EtCH=O} \underset{Et}{\overset{Me_3Si}{\beta\text{-lactone}}} \quad (44)^6$$

1

$$(MeO)_2C=C(OMe)_2 \xrightarrow[25\,°C]{Me_3SiCH=C=O} \text{cyclobutanone with } Me_3Si, MeO, MeO, OMe, OMe \quad (70\%) \quad (45)^7$$

$$MeOCH=C(OSiMe_3)_2 \xrightarrow{\mathbf{1}} Me_3SiCH=C\overset{OSiMe_3}{\underset{CH(OMe)CO_2SiMe_3}{\big|}} \quad (46)^8$$

An interesting reaction of $Me_3SiCH=C=O$ with silenes has been observed, and a possible pathway for the reaction was suggested as shown in equation 47.[56]

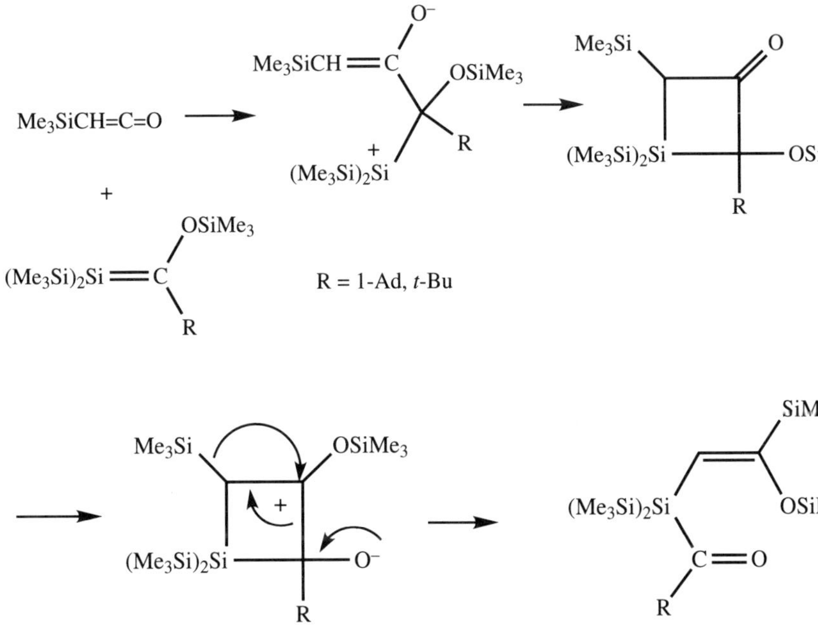

$$Me_3SiCH=C=O + (Me_3Si)_2Si=C\overset{OSiMe_3}{\underset{R}{\big|}} \longrightarrow Me_3SiCH=C\overset{O^-}{\underset{+ (Me_3Si)_2Si}{\big|}}\overset{OSiMe_3}{\underset{R}{\big|}} \longrightarrow \text{cyclobutanone intermediate}$$

R = 1-Ad, t-Bu (47)

$$\longrightarrow \text{rearrangement} \longrightarrow (Me_3Si)_2Si(R)-C(=O)-C(OSiMe_3)=C(SiMe_3)$$

Silylketenes react with diazomethanes to give cyclopropanones and cyclobutanones (equation 48).[18,57]

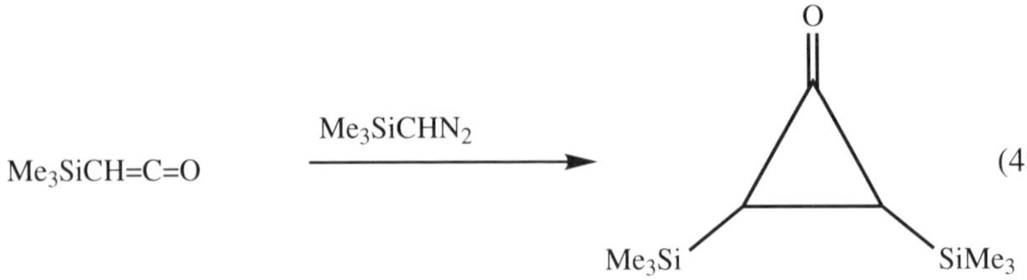

$$Me_3SiCH=C=O \xrightarrow{Me_3SiCHN_2} \text{(cyclopropanone with Me}_3\text{Si and SiMe}_3\text{)} \quad (48)^{57}$$

The ketene $Ph_3SiC(OEt)=C=O$ is more reactive and gives cycloaddition reactions when generated in situ with reactive alkenes.[58] Silylketenes $RMePhSiCH=C=O$ react with highly activated aminoalkynes.[59]

Photolysis of the cyclopentadienone **54** leads to the allenylketene **55**, detected by its IR bands at 1890 and 2080 cm^{-1}, which on further photolysis forms **56** (equation 49).[60] The cycloaddition reactivity[61] of other unsaturated silylketenes[62] prepared from metal complexes has also been examined.

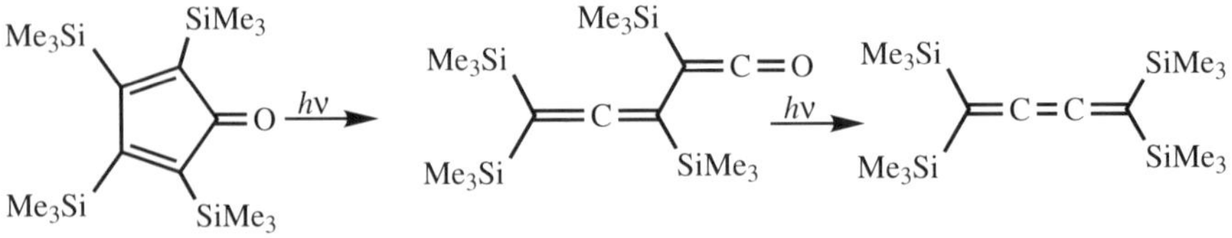

$$\mathbf{54} \xrightarrow{h\nu} \mathbf{55} \xrightarrow{h\nu} \mathbf{56} \quad (49)$$

As listed in Table 2, quite a few ketenes substituted with germanium and tin substituents have also been prepared, and some of this work has been reviewed.[63] Many of these are prepared as shown in equations 50 and 51 by reactions of alkynyl ethers.[10] The formation of the disubstituted derivative **59** is assumed to proceed through an intermediate alkyne **58** which reacts further as shown.[10] Ketenes bearing two different organometal substituents were also prepared by these procedures, but these dimetalated ketenes are highly reactive and rather unstable.[10]

$$Et_3GeBr \xrightarrow[40\%]{LiC\equiv COEt} Et_3GeC\equiv COEt \xrightarrow[93\%]{120-130\ °C} Et_3GeCH=C=O$$

57 (50)

$$Me_3GeBr \xrightarrow[\Delta]{BrMgC\equiv COEt} [Me_3GeC\equiv COEt] \xrightarrow{Me_3GeBr} (Me_3Ge)_2C=C=O \quad (51)$$
$$\phantom{Me_3GeBr \xrightarrow[\Delta]{BrMgC\equiv COEt}} \mathbf{58} \mathbf{59}$$

The reaction of ynolates derived from alkynyl tosylates gives germanium- and tin-substituted ketenes such as **60** (equation 52).[64]

$$t\text{-BuC}\equiv\text{COTs} \xrightarrow{MeLi} t\text{-BuC}\equiv\text{COLi} \xrightarrow{Et_3GeCl} \begin{array}{c} Et_3Ge \\ \diagdown \\ C=C=O \\ \diagup \\ t\text{-Bu} \quad \mathbf{60} \end{array} \quad (52)$$

Reaction of the alkyne **61** with trialkylsilyl or trialkylgermyl halides gave the ketenes **62** and **63** as stable products identified by their characteristic spectral properties (equations 53 and 54).[64a] The ketene **63** was also obtained by the route of equation 55.[64a]

$$t\text{-BuCOC}\equiv\text{COEt} \xrightarrow{Me_3SiI} \begin{array}{c} t\text{-Bu} \quad\quad OEt \\ \diagdown \diagup \\ C=C=C \\ \diagup \diagdown \\ Me_3SiO \quad\quad I \end{array} \longrightarrow \begin{array}{c} t\text{-BuCO} \\ \diagdown \\ C=C=O \\ \diagup \\ Me_3Si \quad \mathbf{62} \end{array} \quad (53)$$
$\mathbf{61}$

$$\mathbf{61} \xrightarrow{Me_3GeI} \begin{array}{c} t\text{-Bu} \quad\quad OEt \\ \diagdown \diagup \\ C=C=C \\ \diagup \diagdown \\ Me_3GeO \quad\quad I \end{array} \longrightarrow \begin{array}{c} t\text{-BuCO} \\ \diagdown \\ C=C=O \\ \diagup \\ Me_3Ge \quad \mathbf{63} \end{array} \quad (54)$$

$$Me_3GeC\equiv COEt \xrightarrow{t\text{-BuCOBr}} \begin{array}{c} t\text{-BuCO} \quad\quad Br \\ \diagdown \diagup \\ C=C \\ \diagup \diagdown \\ Me_3Ge \quad\quad OEt \end{array} \xrightarrow{120\,°C} \mathbf{63} \quad (55)$$

TABLE 2. Silicon-, Germanium-, and Tin-Substituted Ketenes

Ketene	Formula	Preparation	Registry No.	Ref.
SiCl$_3$CH=C=O	C$_2$HCl$_3$OSi	Alkynyl ether	19060-99-2	2
SiH$_3$CH=C=O	C$_2$H$_4$OSi	Theory	6544-46-3	48
Me$_2$SiHCH=C=O	C$_4$H$_8$OSi	Pyrolysis	99016-53-2	26
Me$_3$SiCLi=C=O	C$_5$H$_9$OSiLi	Deprotonation	65213-29-8	12
Me$_2$(ClCH$_2$)SiCH=C=O	C$_5$H$_9$ClOSi		67354-28-3	65
Me$_3$GeC(GeCl$_3$)=C=O	C$_5$H$_9$Cl$_3$Ge$_2$O	Alkynyl ether	32278-92-5	10
Me$_3$SiC(SnCl$_3$)=C=O	C$_5$H$_9$Cl$_3$OSiSn	Alkynyl ether	75669-17-9	66
Me$_3$SiCBr=C=O	C$_5$H$_9$BrOSi	Dehydrochlorination	60366-59-8	17
(CD$_3$)$_3$SiCH=C=O	C$_5$HD$_9$OSi	Alkynyl ether	19331-05-6	2
Me$_3$GeCH=C=O	C$_5$H$_{10}$GeO	Alkynyl ether	32278-84-5	10
Me$_3$SiCH=C=O	C$_5$H$_{10}$OSi	Alkynyl ether	4071-85-6	1
Me$_3$SiCMe=C=O	C$_6$H$_{12}$OSi	Rearrangement	104992-49-6	37
Et$_2$SiHCH=C=O	C$_6$H$_{12}$OSi	Alkynyl ether	19060-97-0	2
MeOMe$_2$SiCMe=C=O	C$_6$H$_{12}$O$_2$Si	Wolff	56510-33-9	39,40
(Me$_2$SiH)$_2$C=C=O	C$_6$H$_{14}$OSi$_2$	Pyrolysis	98991-81-2	26
Me$_3$SiC(CH=CH$_2$)=C=O	C$_7$H$_{12}$OSi	Dehydrochlorination	75232-81-4	20
MeEt$_2$SiCH=C=O	C$_7$H$_{14}$OSi		67354-27-2	65,67
Me$_3$SiCEt=C=O	C$_7$H$_{14}$OSi	Dehydrochlorination	97234-78-1	21
EtOMe$_2$SiCMe=C=O	C$_7$H$_{14}$O$_2$Si	Wolff	72227-92-0	39
Me$_3$SiC(SiHMe$_2$)=C=O	C$_7$H$_{16}$OSi$_2$	Pyrolysis	98991-82-3	26
GeCl$_3$CPh=C=O	C$_8$H$_5$Cl$_3$GeO	Elimination	27008-65-7	68
Me$_3$GeC(CONMe$_2$)=C=O	C$_8$H$_{15}$GeNO$_2$		110698-78-7	69
Me$_3$SnC(CONMe$_2$)=C=O	C$_8$H$_{15}$NO$_2$Sn		108202-43-3	69
Et$_3$GeCH=C=O	C$_8$H$_{16}$GeO	Alkynyl ether	21803-13-4	68,70
Et$_3$SiCH=C=O	C$_8$H$_{16}$OSi	Dehydrochlorination	19060-98-1	3
t-BuMe$_2$SiCH=C=O	C$_8$H$_{16}$OSi	Alkynyl ether	104992-44-1	33

Me$_3$SiC(Pr-i)=C=O	C$_8$H$_{16}$OSi	Alkynyl ether	101911-92-8	24
Me$_3$SiC(GeMe$_3$)=C=O	C$_8$H$_{18}$GeOSi	Ester elimination	38860-05-8	71
Me$_3$GeC(SnMe$_3$)=C=O	C$_8$H$_{18}$GeOSn	Alkynyl ether	32278-90-3	10
(Me$_3$Ge)$_2$C=C=O	C$_8$H$_{18}$Ge$_2$O	Alkynyl ether	32329-69-4	24,71
Me$_3$SiC(SnMe$_3$)=C=O	C$_8$H$_{18}$OSiSn	Alkynyl ether	32278-93-6	10,24
(Me$_3$Si)$_2$C=C=O	C$_8$H$_{18}$OSi$_2$	Ester elimination	19061-00-8	15
Me$_5$Si$_2$CMe=C=O	C$_8$H$_{18}$OSi$_2$	Wolff		38
Me$_3$Si(MeO)MeSiCMe=C=O	C$_8$H$_{18}$O$_2$Si$_2$	Wolff	79251-26-6	40
Me$_3$Si(SiMe$_2$OMe)=C=O	C$_8$H$_{18}$O$_2$Si$_2$	Wolff	32278-87-8	28
(Me$_3$Sn)$_2$C=C=O	C$_8$H$_{18}$O$_2$Sn$_2$	Alkynyl ether	108836-29-9	10
Me$_3$SiCR=C=O	C$_9$H$_{16}$O$_2$Si	Metal complex		30–32
R = CH=C(OMe)Me				
Et$_3$GeCMe=C=O	C$_9$H$_{18}$GeO	Wolff	62299-61-0	72
Et$_3$SiCMe=C=O	C$_9$H$_{18}$OSi	Wolff		23
Me$_3$SiC(Bu-n)=C=O	C$_9$H$_{18}$OSi	Alkynyl ether	71985-41-6	19
MeOEt$_2$SiCEt=C=O	C$_9$H$_{18}$O$_2$Si	Wolff		23
MeO(t-Bu)MeSiCMe=C=O	C$_9$H$_{18}$O$_2$Si	Wolff		40
Me$_3$SiC(SiMe$_2$OEt)=C=O	C$_9$H$_{20}$O$_2$Si$_2$	Wolff	79251-25-5	28
PhMe$_2$SiCH=C=O	C$_{10}$H$_{12}$OSi	Alkynyl ether	42414-77-7	59
Et$_3$GeCEt=C=O	C$_{10}$H$_{20}$GeO	Wolff	62299-59-6	73
Me$_5$Si$_2$C(Pr-i)=C=O	C$_{10}$H$_{22}$OSi$_2$	Wolff		38
Me$_3$SiCPh=C=O	C$_{11}$H$_{14}$OSi	Dehalogenation	64545-16-0	18
MeOMe$_2$SiCPh=C=O	C$_{11}$H$_{14}$O$_2$Si	Wolff	72227-93-1	39
Et$_3$GeC(Pr-i)=C=O	C$_{11}$H$_{22}$GeO	Alkynyl ether	101911-03-9	24
Me$_3$SiC(Hx-n)=C=C	C$_{11}$H$_{22}$OSi	Alkynyl ether	125564-91-2	34
Me$_3$SiC(GeEt$_3$)=C=O	C$_{11}$H$_{24}$GeOSi	Alkynyl ether	38860-06-9	54
Me$_3$GeC(SnEt$_3$)=C=O	C$_{11}$H$_{24}$GeOSn	Alkynyl ether	32278-91-4	10
Me$_3$SiC(SnEt$_3$)=C=O	C$_{11}$H$_{24}$OSiSn	Alkynyl ether	32278-94-7	10
Me$_5$Si$_2$C(Bu-t)=C=O	C$_{11}$H$_{24}$OSi$_2$	Wolff		38
t-BuOMe$_2$SiC(SiMe$_3$)=C=O	C$_{11}$H$_{24}$O$_2$Si$_2$	Wolff	79257-68-4	28

TABLE 2. (Continued)

Ketene	Formula	Preparation	Registry No.	Ref.
$Et_3GeC(Bu\text{-}t)=C=O$	$C_{12}H_{24}GeO$	Ynolate		64
$Et_3GeC(Bu\text{-}s)=C=O$	$C_{12}H_{24}GeO$	Ynolate		64
$i\text{-}Pr_3SiC(Me)=C=O$	$C_{12}H_{24}OSi$	Wolff	126364-75-8	23
$Me_3SiCR=C=O$	$C_{12}H_{24}O_2Si_2$	Wolff	108836-30-2	32
R = $Me(MeO)C=CSiMe_3$				
R = $Ph(MeO)C=CH$	$C_{14}H_{18}O_2Si$	Metal complex	74011-94-2	62
$Me_3SiC(CH_2CH_2Ph)=C=O$	$C_{13}H_{18}OSi$	Ynolate	104875-67-4	74
$Et_3GeCPh=C=O$	$C_{14}H_{20}GeO$	Wolff	62299-60-9	73
$Et_3SiCPh=C=O$	$C_{14}H_{20}OSi$	Wolff	96845-82-8	22
$Et_3GeC(SiEt_3)=C=O$	$C_{14}H_{30}GeOSi$	Wolff	59466-91-0	73
$Et_3GeC(SnEt_3)=C=O$	$C_{14}H_{30}GeOSn$	Wolff	59466-92-1	73
$(Et_3Ge)_2C=C=O$	$C_{14}H_{30}Ge_2O$	Wolff	21803-14-5	70,73
$(t\text{-}BuMe_2Si)_2C=C=O$	$C_{14}H_{30}OSi_2$	Alkynyl ether	111268-97-4	27
$(Et_3Si)_2C=C=O$	$C_{14}H_{30}OSi_2$	Wolff	59466-90-9	73
$Me_3SiC(CH=COMePh)=C=O$	$C_{14}H_{18}O_2Si$	Metal complex	74011-94-2	62
$Et_3SiC(GeEt_3)=C=O$	$C_{14}H_{30}GeOSi$	Wolff	59466-91-0	73
$(Et_3Sn)_2C=C=O$	$C_{14}H_{30}OSn_2$	Alkynyl ether	32278-88-9	10
$(Et_3Si)_2C=C=O$	$C_{14}H_{30}OSi_2$	Wolff	59466-90-9	73
$MePh_2SiCH=C=O$	$C_{15}H_{14}OSi$	Alkynyl ether	92177-88-3	59
$i\text{-}Pr_3SiCR=C=O$	$C_{15}H_{24}OSSi$	Wolff	96845-83-9	23
R = 2-thienyl				
$Me_3SiCR=C=O$	$C_{16}H_{22}O_2Si$	Metal complex	88563-62-6	75
R = $2,6\text{-}Me_2C_6H_3C(OMe)=CH$				
$Me_3SiCR=C=O$	$C_{16}H_{30}O_3Si_2$	Wolff	83547-75-5	29

R = SiMe$_2$O-7-EtO-7-norbornyl				
MeO(Ph)(t-Bu)SiCPh=C=O	C$_{17}$H$_{22}$O$_2$Si		Wolff	40
Me$_3$SiCR=C=O	C$_{17}$H$_{26}$O$_2$Si$_2$	72207-50-2	Metal complex	31
R = Z-Ph(MeO)C=C(SiMe$_3$)				
n-Bu$_3$SnC(CONMe$_2$)=C=O	C$_{17}$H$_{33}$NO$_2$Sn	108202-45-5		69
Me$_3$SiCR=C=O	C$_{17}$H$_{36}$OSi$_4$	79593-36-5	Metal complex	60
R = (Me$_3$Si)$_2$C=C=C(SiMe$_3$)				
(PhMe$_2$Si)$_2$C=C=O	C$_{18}$H$_{22}$OSi$_2$	37170-37-9	Alkynyl ether	76
Me$_3$SiCR=C=O	C$_{18}$H$_{25}$F$_3$O$_2$Si$_2$	76833-11-9	Metal complex	61
R = Z-CF$_3$C$_6$H$_4$C(OMe)=C(SiMe$_3$)				
Me$_3$SiCR=C=O	C$_{18}$H$_{28}$O$_2$Si$_2$	74011-97-5	Metal complex	30
R = 4-TolC(OMe)=C(SiMe$_3$)				
Me$_3$SiCR=C=O	C$_{18}$H$_{28}$O$_3$Si$_2$	74011-95-3	Metal complex	62
R = Z-4-MeOC$_6$H$_4$C(OMe)=C(SiMe$_3$)				
t-BuPh$_2$SiCH=C=O	C$_{18}$H$_{20}$OSi	74011-96-4	Metal complex	62
t-BuMe$_2$SiC(Ad-1)=C=O	C$_{18}$H$_{30}$OSi	124537-27-5	Alkynyl ether	33
n-Bu$_3$SnC(Bu-t)=C=O	C$_{18}$H$_{36}$OSn	126364-88-3	Wolff	23
Ph(1-naph)MeSiCH=C=O	C$_{19}$H$_{16}$OSi		Ynolate	63
Ph$_3$SiCH=C=O	C$_{20}$H$_{16}$OSi	51042-09-2	Alkynyl ether	77
(n-Pr$_3$Ge)$_2$C=C=O	C$_{20}$H$_{42}$Ge$_2$O	42414-76-6	Dehydrochlorination	78
(n-Pr$_3$Sn)$_2$C=C=O	C$_{20}$H$_{42}$OSn$_2$	38860-04-7	Alkynyl ether	54
i-Pr$_3$SiC(Ad-1)=C=C	C$_{21}$H$_{36}$OSi	32278-89-0	Wolff	10
Ph$_3$SiC(SEt)=C=O	C$_{22}$H$_{20}$OSSi	126364-91-8	Wolff	23
Ph$_3$SiC(OEt)=C=O	C$_{22}$H$_{20}$O$_2$Si	98014-97-2	Metal complex	58
t-BuPh$_2$SiC(Bu-t)=C=O	C$_{22}$H$_{28}$OSi	122300-87-2	Metal complex	58
(Me$_3$Si)$_3$SiMe$_2$SiC(Ad-1)=C=O	C$_{23}$H$_{48}$OSi$_5$	126364-82-7	Wolff	23
i-Pr$_3$SiC(CHPh$_2$)=C=O	C$_{24}$H$_{32}$OSi	123725-27-9	Wolff	15
t-BuPh$_2$SiC(Ad-1)=C=O	C$_{28}$H$_{34}$OSi	126364-86-1	Wolff	23
		126364-90-7	Wolff	23
1-Ad = 1-adamanyl (C$_{10}$H$_{15}$)				

References

1. Shchukovskaya, L. L.; Pal'chik, R. I.; Lazarev, A. N. *Dokl. Akad. Nauk SSSR,* **1965**, *164,* 357–360; *Engl. Transl.* **1965**, *164,* 884–890; *Chem. Abstr.* **1965**, *63,* 18138g.
2. Shchukovskaya, L. L.; Kol'tsov, A. I.; Lazarev, A. N.; Pal'chik, R. I. *Dokl. Akad. Nauk SSSR* **1968**, *179,* 892–895; *Engl. Transl.* **1968**, *179,* 318–320.
3. Lutsenko, I. F.; Baukov, Yu. I.; Kostyuk, A. S.; Savelyeva, N. I.; Krysina, V. K. *J. Organomet. Chem.* **1969**, *17,* 241–262.
4. Olah, G. A.; Wu, A.; Farooq, O. *Synthesis,* **1989**, 568.
5. Ruden, R. A. *J. Org. Chem.* **1974**, *39,* 3607–3608.
6. Brady, W. T.; Saidi, K. *J. Org. Chem.* **1979**, *44,* 733–737.
7. Brady, W. T.; Saidi, K. *J. Org. Chem.* **1980**, *45,* 727–729.
8. Brady, W. T.; Saidi, K. *J. Org. Chem.* **1990**, *55,* 4215–4216.
9. Kostyuk, A. S.; Dudukina, O. V.; Burlachenko, G. S.; Baukov, Yu. I.; Lutsenko, I. F. *Zh. Obshch. Khim.* **1969**, *39,* 467; *Engl. Transl.* **1969**, *39,* 441.
10. Ponomarev, S. V.; Erman, M. B.; Lebedev, S. A.; Pechurina, S. Ya.; Lutsenko, I. F. *Zh. Obshch. Khim.* **1971**, *41,* 127–133; *Engl. Transl.* **1971**, *41,* 122–127.
11. Pal'chik, R. I.; Shchukovskaya, L. L.; Kol'stov, A. I. *Zh. Obshch. Khim.* **1969**, *39,* 1792–1796; *Engl. Transl.* **1969**, *39,* 1756.
12. Woodbury, R. P.; Long, N. R.; Rathke, M. W. *J. Org. Chem.* **1978**, *43,* 376.
13. Sullivan, D. F.; Woodbury, R. P.; Rathke, M. W. *J. Org. Chem.* **1977**, *42,* 2038–2039.
14. Uhlig, W.; Tzschach, A. *Z. Chem.* **1988**, *28,* 409–410; *Chem. Abstr.* **1989**, *111,* 194849n.
15. Schneider, K.; Daucher, B.; Fronda, A.; Maas, G. *Chem. Ber.* **1990**, *123,* 589–594.
16. Brady, W. T.; Cheng, T. C. *J. Org. Chem.* **1977**, *42,* 732–734.
17. Brady, W. T.; Owens, R. A. *Tetrahedron Lett.* **1976**, 1553–1556.
18. Brady, W. T.; Cheng, T. C. *J. Organomet. Chem.* **1977**, *137,* 287–292.
19. Sakurai, H.; Shirahata, A.; Sasaki, K.; Hosomi, A. *Synthesis* **1979**, 740–741.
20. Danheiser, R. L.; Sard, H. *J. Org. Chem.* **1980**, *45,* 4810–4812.
21. Baigrie, L. M.; Seikaly, H. R.; Tidwell, T. T. *J. Am. Chem. Soc.* **1985**, *107,* 5391–5396.
22. Maas, G.; Brückmann, R. *J. Org. Chem.* **1985**, *50,* 2801–2802.
23. Brückmann, R.; Schneider, K.; Maas, G. *Tetrahedron* **1989**, *45,* 5517–5530.
24. Efimova, I. V.; Kazanokova, M. A.; Lutsenko, I. F. *Zh. Obshch. Khim.* **1985**, *55,* 1647–1649; *Engl. Transl.* **1985**, *55,* 1465–1466.
25. Barton, T. J.; Paul, G. C. *J. Am. Chem. Soc.* **1987**, *109,* 5292–5293.
26. Barton, T. J.; Groh, B. L. *J. Am. Chem. Soc.* **1985**, *107,* 7221–7222.
27. Groh, B. L.; Magrum, G. R.; Barton, T. J. *J. Am. Chem. Soc.* **1987**, *109,* 7568–7569.
28. Ando, W.; Sekiguchi, A.; Sato, T. *J. Am. Chem. Soc.* **1981**, *103,* 5573–5574.
29. Ando, W.; Sekiguchi, A.; Sato, T. *J. Am. Chem. Soc.* **1982**, *104,* 6830–6831.
30. Dötz, K. H.; Mühlemeier, J.; Trenkle, B. *J. Organomet. Chem.* **1985**, *289,* 257–262.
31. Dötz, K. H. *Angew. Chem., Int. Ed. Engl.* **1979**, *18,* 954–955.
32. Xu, Y.-C.; Wulff, W. D. *J. Org. Chem.* **1987**, *52,* 3263–3275.
33. Valentí, E.; Pericas, M. A.; Serratosa, F. *J. Org. Chem.* **1990**, *55,* 395–397.

34. Pons, J.-M.; Kocienski, P. *Tetrahedron Lett.* **1989**, *30*, 1833–1836.
34a. Pons, J.-M.; Pommier, A.; Lerpiniere, J.; Kocienski, P. *J. Chem. Soc., Perkin Trans. 1,* **1992**, 1549–1551.
35. Wright, B. B. *J. Am. Chem. Soc.* **1988**, *110*, 4456–4457.
36. Pollart, D. J.; Moore, H. W. *J. Org. Chem.* **1989**, *54*, 5444–5448.
37. Jabry, Z.; Lasne, M.; Ripoll, J. *J. Chem. Res. (S),* **1986**, 188–189.
38. Maas, G.; Alt, M.; Schneider, K.; Fronda, A. *Chem. Ber.* **1991**, *124*, 1295–1300.
39. Ando, W.; Sekiguchi, A.; Hagiwara, T.; Migita, T.; Chowdhry, V.; Westheimer, F. H.; Kammula, S. L.; Green, M.; Jones, M., Jr. *J. Am. Chem. Soc.* **1979**, *101*, 6393–6398.
40. Maas, G.; Gimmy, M.; Alt, M. *Organometallics,* **1992**, *11*, 3813–3820.
41. Traylor, T. G.; Hanstein, W.; Berwin, H. J.; Clinton, N. A.; Brown, R. S. *J. Am. Chem. Soc.* **1971**, *93*, 5715–5725.
42. *The Chemistry of Organic Silicon Compounds;* Patai, S.; Rappoport, Z., Eds., Wiley: New York, 1989.
43. Weber, W. P. *Silicon Reagents for Organic Synthesis;* Springer-Verlag: Berlin, 1983.
44. Colvin, E. W. *Silicon in Organic Synthesis,* Butterworths: London, 1981.
45. Fleming, I. In *Comprehensive Organic Chemistry,* Vol. 3, Jones, D. N., ed, Pergamon, New York 1979; pp. 541–686.
46. Armitage, D. A. In *Comprehensive Organomettalic Chemistry,* Vol. 2, Wilkinson, G., ed, Pergamon, New York, 1982; pp. 1–203.
47. Runge, W. *Prog. Phys. Org. Chem.* **1981**, *13*, 315–484.
48. Gong, L.; McAllister, M. A.; Tidwell, T. T. *J. Am. Chem. Soc.* **1991**, *113*, 6021–6028.
49. Allen, A. D.; Tidwell, T. T. *Tetrahedron Lett.* **1991**, *32*, 847–850.
50. Zhao, D.; Allen, A. D.; Tidwell, T. T. *J. Am. Chem. Soc.* **1993**, *115*, 10097–10103.
50a. Yamamoto, Y.; Ozasa, N.; Sawada, S. *Chem. Express* **1993**, *8*, 305–308. *Chem. Abstr.* **1993**, *119*, 139310v.
51. Akai, S.; Tsuzuki, Y.; Matsuda, S.; Kitagaki, S.; Kita, Y. *J. Chem. Soc., Perkin Trans. 1,* **1992**, 2813–2820.
52. Kita, Y.; Sekihachi, J.; Hayashi, Y.; Da, Y.; Yamamoto, M.; Akai, S. *J. Org. Chem.* **1990**, *55*, 1108–1112.
53. Taylor, R. T.; Cassell, R. A. *Synthesis* **1982**, 672–673.
54. Lebedev, S. A.; Ponomarev, S. V.; Lutsenko, I. F. *Zh. Obshch. Chem.* **1972**, *42*, 647–651; *Engl. Transl.* **1972**, *42*, 643–647.
55. Kita, Y.; Matsuda, S.; Kitagaki, S.; Tsuzuki, Y.; Akai, S. *Synlett* **1991**, 401–402.
56. Brook, A. G.; Baumegger, A. *J. Organomet. Chem.* **1993**, *446*, C9–C11.
57. Fedorenko, E. N.; Zaitseva, G. S.; Baukov, Yu. I.; Lutsenko, I. F. *Zh. Obshch. Khim.* **1986**, *56*, 2431–2432, *Engl. Transl.* **1986**, *56*, 2150–2151.
58. Kron, J.; Schubert, U. *J. Organomet. Chem.* **1989**, *373*, 203–219.
59. Himbert, G.; Henn, L. *Liebigs Ann. Chem.* **1984**, 1358–1366.
60. Maier, G.; Lage, H. W.; Reisenauer, H. P. *Angew. Chem., Int. Ed. Engl.* **1981**, *20*, 976–977.
61. Dötz, K. H.; Trenkle, B.; Schubert, U. *Angew. Chem., Int. Ed. Engl.* **1981**, *20*, 287.
62. Dötz, K. H.; Fügen-Köster, B. *Chem. Ber.* **1980**, *113*, 1449–1457.
63. Moreau, J.-L. In *The Chemistry of Ketenes, Allenes and Related Compounds;* Patai, S., Ed.; Wiley: New York, 1980; pp. 363–413.

64. Stang, P. J.; Roberts, K. A. *J. Am. Chem. Soc.* **1986**, *108*, 7125–7127.
64a. Lukashev, N. V.; Fil'chikov, A. A.; Kazankova, M. A.; Beletskaya, I. *Heteroatom. Chem.* **1993**, *4*, 403–407.
65. Zaitseva, G. S.; Vasil'eva, L. I.; Vinkurova, N. G.; Safronova, O. A.; Baukov, Yu. I. *Zh. Obshch. Khim.* **1978**, *48*, 1363–1368; *Engl. Transl.* **1978**, *48*, 1249–1253.
66. Kazankova, M. A.; Ilyushin, V. A.; Ladeishchikova, E. V.; Lutsenko, I. F. *Zh. Obshch. Khim.* **1980**, *50*, 692–693; *Chem. Abstr.* **1980**, *93*, 239546q.
67. Zaitseva, G. S.; Lutsenko, I. F.; Kisin, A. V.; Baukov, Yu. I.; Lorberth, J. *J. Organomet. Chem.* **1988**, *345*, 253–262.
68. Baukov, Yu. I.; Burlachencko, G. S.; Kostyuk, A. S.; Lutsenko, I. F. *Zh Obshch. Khim.* **1970**, *40*, 707; *Engl. Transl.* **1970**, *40*, 681.
69. Ganis, P.; Paiaro, G.; Pandolfo, L.; Valle, G. *Organometallics*, **1988**, *7*, 210–214.
70. Mazerolles, P.; Laporterie, A.; Lesbre, M. *C. R. Acad. Sci. Paris, Ser. C* **1969**, *268*, 361–364.
71. Inoue, S.; Sato, Y.; Suzuki, T. *Organometallics*, **1988**, *7*, 739–743.
72. Gostevskii, B. A.; Kruglaya, O. A.; Albanov, A. I.; Vyazankin, N. S. *J. Organomet. Chem.* **1980**, *187*, 157–166.
73. Kruglaya, O. A.; Fedot'eva, I. B.; Fedot'ev, B. V.; Kalikhman, I. D.; Brodskaya, E. I.; Vyazankin, N. S. *J. Organomet. Chem.* **1977**, *142*, 155–164.
74. Kowalski, C. J.; Lal, G. S.; Haque, M. S. *J. Am. Chem. Soc.* **1986**, *108*, 7127–7128.
75. Tang, P. C.; Wulff, W. D. *J. Am. Chem. Soc.* **1984**, *106*, 1132–1133.
76. Shchukovskaya, L. L.; Pal'chik, R. I.; Petrushina, T. A. *Zh. Obshch. Khim.* **1972**, *42*, 240; *Engl. Transl.* **1972**, *42*, 234.
77. Vodolazskaya, V. M.; Baukov, Yu. I. *Zh. Obshch. Khim.* **1973**, *43*, 2088–2089; *Engl. Transl.* **1973**, *43*, 2076.
78. Baukov, Yu. I.; Burlachenko, G. S.; Kostyuk, A. S.; Lutsenko, I. F. *Dokl. Vses. Konf. Khim. Atsetilena, 4th*, **1972**, *2*, 130–133; *Chem. Abstr.* **1973**, *79*, 78904y.

4.6 PHOSPHOROUS- AND ARSENIC-SUBSTITUTED KETENES

Phosphorous is a rather electropositive element, and phosphinylketene ($H_2PCH=C=O$) is predicted to be quite stabilized.[1] In accord with this expectation a number of phosphorous-substituted ketenes have been prepared which show significant stability.

The kinetics of the thermal decomposition of $Ph_2P(O)C(N_2)COPh$, which could lead to the ketene **1**, were examined by Regitz and co-workers (equation 1).[2,3]

$$Ph_2P(O)C(N_2)COPh \xrightarrow{h\nu} Ph_2P(O)\ddot{C}COPh \longrightarrow \underset{Ph}{\overset{Ph_2(O)P}{>}}C=C=O$$

1   (1)

Phosphoryl migration occurred on photolysis of the diazo compound **2** to give the ketene **3**, which was observed by its IR band at 2160 cm^{-1}.[4] On photolysis or heating of **2** in MeOH the ester **4** was obtained in high yield from the ketene (equation 2).

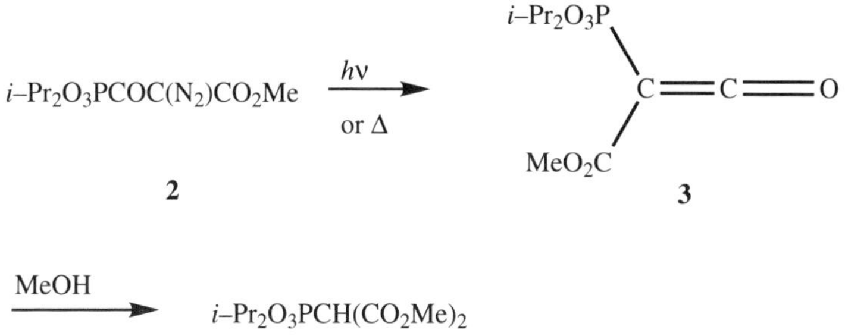

$$i\text{-}Pr_2O_3PCOC(N_2)CO_2Me \xrightarrow[\text{or } \Delta]{h\nu} \underset{\underset{3}{MeO_2C}}{\overset{i\text{-}Pr_2O_3P}{\diagdown}}C=C=O \quad (2)$$

$$\underset{2}{} \xrightarrow{MeOH} \underset{4}{i\text{-}Pr_2O_3PCH(CO_2Me)_2}$$

Irradiation of **5** in MeOH gave the products **8** and **9**, derived from capture of the carbene **6** and the ketene **7**, respectively, in a 1:1 ratio.[5] The sodium salt **10** gave almost exclusively products derived from the ketene, and it was proposed that the presence of the anionic group suppressed reaction from the triplet state of the carbene.[5] Photolysis of **5** in an Ar matrix at 10 K also led to **7**, as observed by IR, and **7** was also formed by photolysis of $PhCN_2PO_3Me_2$ at 10 K to give the carbene followed by reaction with CO at 35 K.[6] The ketenes $R_2P(Bu\text{-}n)=C=O$ (R = EtO, Ph) were formed by Wolff rearrangement, and trapped with EtOH.[6a]

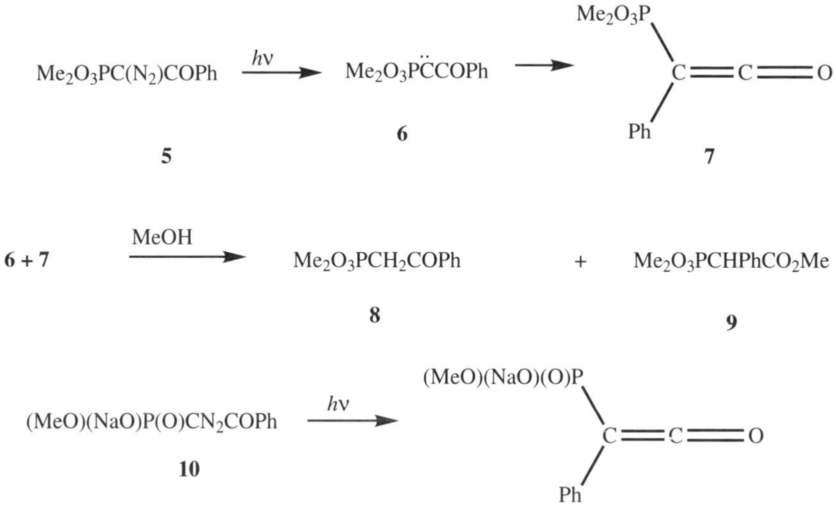

$$\underset{5}{Me_2O_3PC(N_2)COPh} \xrightarrow{h\nu} \underset{6}{Me_2O_3P\ddot{C}COPh} \longrightarrow \underset{\underset{7}{Ph}}{\overset{Me_2O_3P}{\diagdown}}C=C=O$$

$$6 + 7 \xrightarrow{MeOH} \underset{8}{Me_2O_3PCH_2COPh} + \underset{9}{Me_2O_3PCHPhCO_2Me}$$

$$\underset{10}{(MeO)(NaO)P(O)CN_2COPh} \xrightarrow{h\nu} \underset{\underset{}{Ph}}{\overset{(MeO)(NaO)(O)P}{\diagdown}}C=C=O$$

Phosphinylketenes with silicon or germanium substituents are formed together with alkenes from alkoxyalkynes (equation 3).[7] A rather unstable diphosphinoketene is also obtained by this route (equation 4).[7]

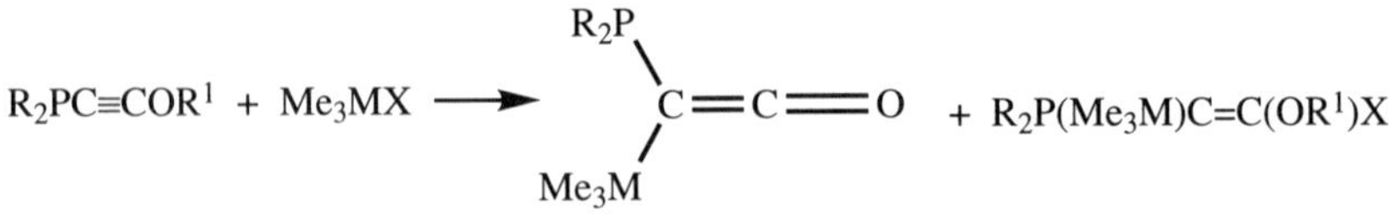

$$R_2PC \equiv COR^1 + Me_3MX \longrightarrow \underset{Me_3M}{\overset{R_2P}{>}}C=C=O + R_2P(Me_3M)C=C(OR^1)X \quad (3)$$

R = i-Pr, t-Bu, C_6F_5; R^1 = Me, Et; M = Si, Ge

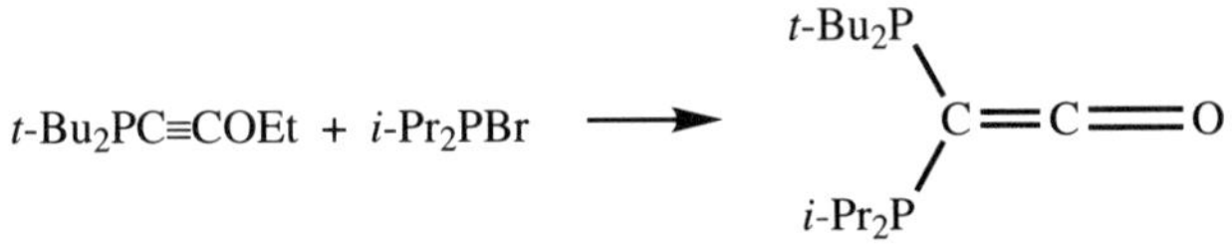

$$t\text{-}Bu_2PC \equiv COEt + i\text{-}Pr_2PBr \longrightarrow \underset{i\text{-}Pr_2P}{\overset{t\text{-}Bu_2P}{>}}C=C=O \quad (4)$$

Chloro and bromo phosphorous ylides are reported to react with CO_2 to give phosphoryl-substituted ketenes which can be observed directly or trapped with MeOH (equation 5).[8] In the case of the fluoro analogue **11** the fluorooxaphosphetane **12** was observed by IR and NMR at room temperature, and the ketene **13** was formed on heating (equation 6).[9] Carbomethoxyphosphorylketenes were obtained by a process shown (equation 7).[10] The monosubstituted ketene **14** was obtained similarly (equation 8).[11]

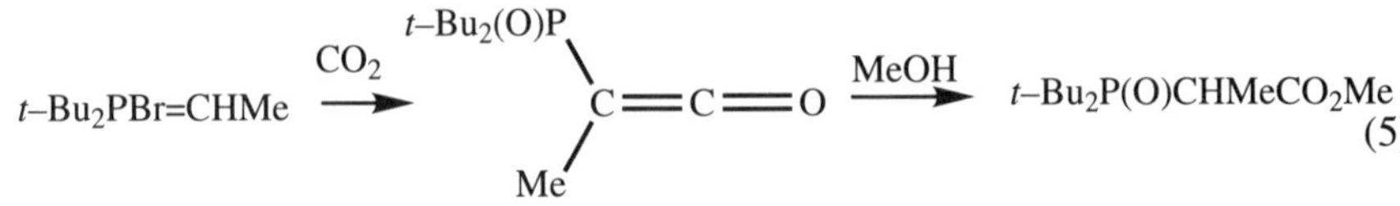

$$t\text{-}Bu_2PBr=CHMe \xrightarrow{CO_2} \underset{Me}{\overset{t\text{-}Bu_2(O)P}{>}}C=C=O \xrightarrow{MeOH} t\text{-}Bu_2P(O)CHMeCO_2Me \quad (5)$$

$$t\text{-}Bu_2PF=CHPr\text{-}n \underset{-HF}{\overset{CO_2}{\longrightarrow}} \underset{\underset{F}{|}}{\overset{O-}{t\text{-}Bu_2P}}\!\!\!\overset{\overset{O}{\|}}{\underset{}{\rule{2em}{0.4pt}}}\!\!\!\text{Pr-}n \xrightarrow{\Delta} \underset{n\text{-}Pr}{\overset{t\text{-}Bu_2(O)P}{>}}C=C=O \quad (6)$$

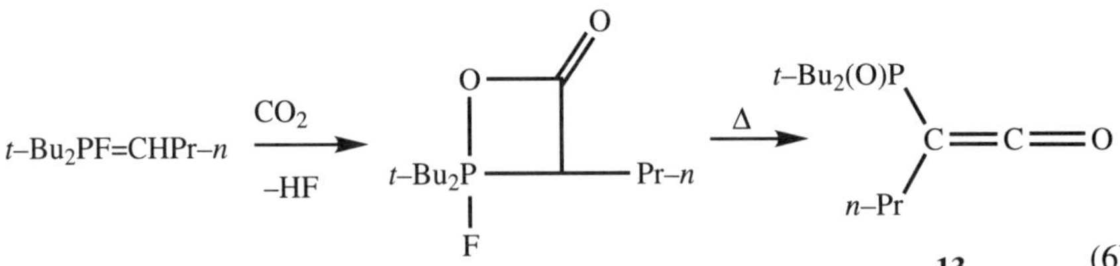

11   **12**   **13**

$$R_2PCH(CO_2Me)_2 \xrightarrow{CBr_4} R_2PBr=C(CO_2Me)_2 \underset{-MeBr}{\xrightarrow{20°}} \underset{MeO_2C}{\overset{R_2(O)P}{>}}C=C=O \quad (7)$$

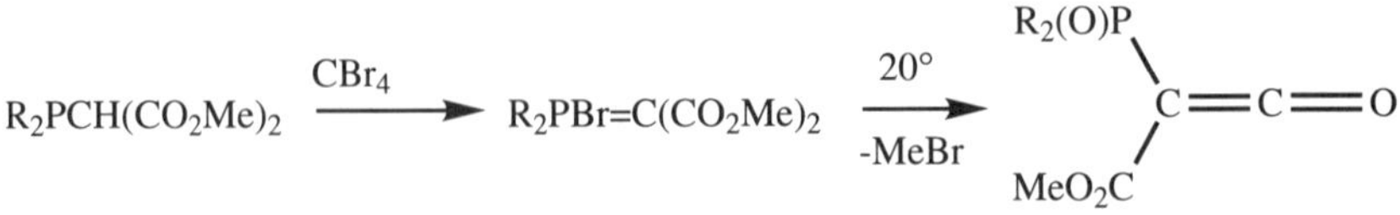

4.6 PHOSPHOROUS- AND ARSENIC-SUBSTITUTED KETENES

$$(EtO)_2PBr=CHCO_2Et \xrightarrow[-EtBr]{>0°} (EtO)_2P(O)CH=C=O \quad (8)$$
$$\mathbf{14}$$

Heating of acid chlorides also produced phosphorylketenes (equation 9).[12]

$$R_2O_3PCHR^1COCl \xrightarrow[-HCl]{\Delta} \begin{array}{c} R_2O_3P \\ \diagdown \\ C=C=O \\ \diagup \\ R^1 \end{array} \quad (9)$$

R^1 = Ph, 4–CH$_3$C$_6$H$_4$, CH$_3$; R = Et, i–Pr

Reaction of the alkynyl ether **15** with MeCOBr gave an ethyl α-bromovinyl ether which underwent α-elimination of EtBr at 80 °C to give the ketene **16** (equation 10).[13] This ketene was sufficiently stable to be distilled in vacuo and the structure was confirmed by the ketene band in the IR at 2100 cm^{-1}, and the value of the ^{13}C chemical shift of C$_\beta$ of δ 83.05 was attributed to conjugation between the C=O and C=C=O groups.[13] Data is given in Table 1, Section 2.1, for comparison of this chemical shift to those for other acyl-substituted ketenes.

$$t\text{-Bu}_2P(S)C\equiv COEt \xrightarrow{MeCOBr} \begin{array}{c} t\text{-Bu}_2(S)P \\ \diagdown \\ C=C(OEt)Br \\ \diagup \\ MeCO \end{array} \xrightarrow{80\ °C} \begin{array}{c} t\text{-Bu}_2(S)P \\ \diagdown \\ C=C=O \\ \diagup \\ MeCO \end{array} \quad (10)$$

15   **16**

Reaction of **17** with MeCOBr was proposed to give ketene **18** and CH$_2$=C=O, which combined to give **19** (equation 11).[13] However, no direct evidence for the formation of **18** was obtained.

$$t\text{-Bu}_2PC\equiv COEt \xrightarrow{MeCOBr} \begin{array}{c} O \\ \parallel \\ MeC \\ \diagdown \\ C=C=O \\ \diagup \\ t\text{-Bu}_2P \end{array} \xrightarrow{CH_2=C=O} \begin{array}{c} \text{MeCO} \\ \diagdown \\ \text{C}-\text{C} \\ t\text{-Bu}_2P \diagup \diagdown O \\ \diagup \\ CH_2 \end{array} \quad (11)$$

17   **18**   **19**

In addition to reactions with H_2O, alcohols, and amines, phosphoryl-substituted ketenes, generated by dehydrochlorination, give 1,2-addition of NOCl (equation 12)[14] and [2 + 2] cycloadditions with dienes or imines (equation 13).[15]

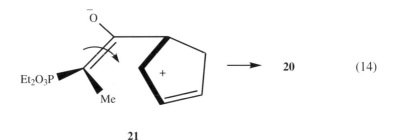

It is interesting that the cycloaddition products, such as **20,** were assigned the configuration with the $(EtO)_2PO$ in the *exo* position,[15] which was analogous to the result for the reaction of $(MeO_2C)CCl=C=O$ with cyclopentadiene. This formation of what is apparently the more stable product (Section 5.4.1.2) may be explained by the formation of phosphorous-stabilized zwitterions such as **21,** which form products under thermodynamic control by the rotation shown in equation 14 rather than by the opposite rotation which is sterically easier but leads to the presumably less stable product.

The kinetics of the addition of substituted phenols to the ketene **22** in CCl_4 have been measured (equation 15),[16] and the data are given in Table 1. The reactions are first order in the phenol, and for the parent phenol reacted with **22** more rapidly than with $Ph_2C=C=O$ by a factor of 2000.[16] As noted in Section 5.5.1.2, $Ph_2C=C=O$ is somewhat less reactive toward nucleophiles than is the highly reactive $PhCH=C=O$, a result attributed to stabilization of the enolate-like transition state from $PhCH=C=O$, whereas the corresponding transition state from $Ph_2C=C=O$ is both cross-conjugated and crowded. In the case of the Et_2O_3P

4.6 PHOSPHOROUS- AND ARSENIC-SUBSTITUTED KETENES

substituent the crowding may be less due to the longer C—P bond, and since phosphorous stabilizes enolates by polarization effects that are not conformationally dependent, the transition state stabilization may be enhanced.

$$\underset{\substack{\text{Ph} \\ \mathbf{22}}}{\overset{\text{Et}_2\text{O}_3\text{P}}{\diagdown}}\text{C}=\text{C}=\text{O} \quad \xrightarrow{\text{4-XC}_6\text{H}_4\text{OH}} \quad \underset{\substack{\text{Ph}}}{\overset{\text{Et}_2\text{O}_3\text{P}}{\diagdown}}\text{CHCO}_2\text{C}_6\text{H}_4\text{X-4} \quad (15)$$

TABLE 1. Relative Rate Constants for the Reaction of Equation 15

X	MeO	F	Cl	CH$_3$	H	SCF$_3$
k_{rel}	23.6	10.1	4.4	8.0	4.2	1.0

The rate constants listed in Table 1 did not give a linear correlation with the σ_p values of the substituents, but there is some tendency for greater reactivity with π-donor and electronegative substituents.[16] Two mechanisms were considered, namely attack of the phenol at the carbonyl carbon, to give an enol, which can be represented as forming from a zwitterionic intermediate **23**, and a [2 + 2] cycloaddition involving **24**.

23 — Et$_2$O$_3$P, Ph, C=C, O$^-$, $^+$OAr, H

24 — Et$_2$O$_3$P, Ph, C=C=O, H-----OAr

The latter possibility was considered because of a claim for an analogous process for the reaction of Me$_2$C=C=O and PhOH.[17] However, as detailed in Section 5.5.1 on nucleophilic additions to ketenes, this possibility is not in good accord with the available data, whereas a zwitterionic transition state resembling **23** fits the data for the reaction with phenols, and is consistent with the accepted mechanism for nucleophilic additions to ketenes. The remarkably high reactivity of **22** indicates significant stabilization of the enolate-like transition state by the Et_2O_3P group, and is strong evidence for the zwitterionic transition state **23**.

The isolable phosphaketene ylide **25** is readily prepared as shown in equation 16[18,19] and also by other routes.[20] This compound is a crystalline solid with a $P-C_\beta-C_\alpha$ angle of 145.5°.[18] Another route to phosphaketene ylides is shown in equation 17.[21,22] Synthetic applications of **25** include HCl-promoted dimerization,[23] many additions to the C=O bond,[24] and a particularly interesting type involves Wittig cyclization of the products of nucleophilic addition (equation 18).[25–27] These ketenes also react with α,β-unsaturated carbonyl compounds by [4 + 2] cycloadditions (equation 19).[28]

$$Ph_3PCHCO_2Me \xrightarrow{NaN[Si(Pr\text{-}i)_3]_2} Ph_3P=C=C=O \longleftrightarrow Ph_3\overset{+}{P}-\overset{-}{C}=C=O \quad (16)$$

$$\textbf{25}$$

$$i\text{-}Pr_2PC\equiv COMe \xrightarrow{\Delta} i\text{-}Pr_2Me\overset{+}{P}-\overset{-}{C}=C=O \quad (17)^{21}$$

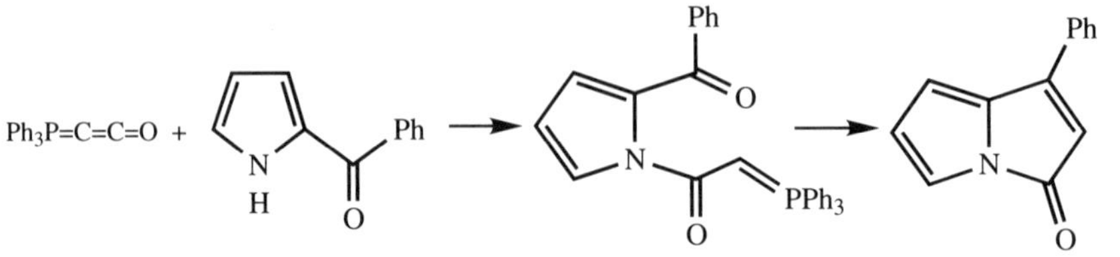

$$Ph_3P=C=C=O + \underset{H}{\underset{|}{\overset{}{\underset{N}{\bigcirc}}}}\!\!\!\!\!\!-\!C(O)Ph \longrightarrow \text{(pyrrole intermediate with PPh}_3\text{)} \longrightarrow \text{(bicyclic product)} \quad (18)^{26}$$

$$Ph_3\overset{+}{P}-\overset{-}{C}=C=O + CH_3COCH=CH_2 \longrightarrow \text{(dihydropyranone with =PPh}_3\text{)} \quad (19)$$

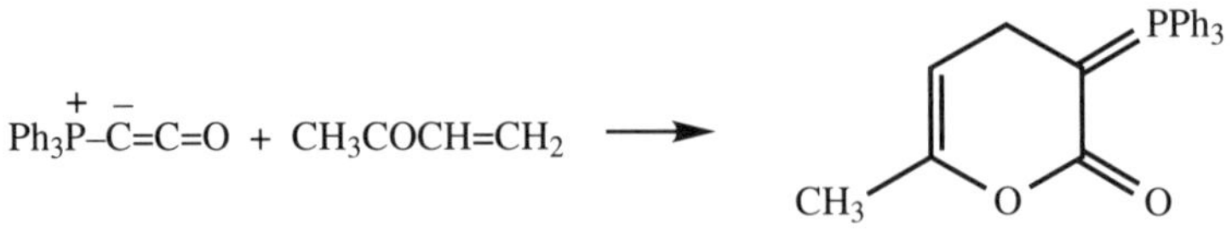

Pyrolysis of the ester **26** also led to the formation of **25**, perhaps through the intermediate **27** (equation 20).[29]

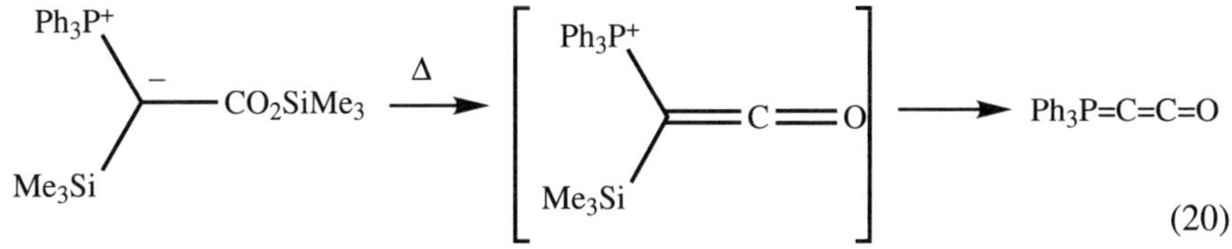

$$\underset{26}{\underset{Me_3Si}{\overset{Ph_3P^+}{>}}C{-}CO_2SiMe_3} \xrightarrow{\Delta} \left[\underset{27}{\underset{Me_3Si}{\overset{Ph_3P^+}{>}}C{=}C{=}O}\right] \longrightarrow \underset{25}{Ph_3P{=}C{=}C{=}O} \quad (20)$$

Reaction of the ynol ether **28** with Ph_2AsCl gave the bis(arsenic) substituted ketene **29** (equation 21).[30] Similar reactions gave $Ph_2As(Ph_2MeSi)C{=}C{=}O$ and $Ph_2As(Ph_3Ge)C{=}C{=}O$.[30] The "surprisingly stable" ketene **30** was isolated after liberation from a tungsten complex.[31,32] This stability is completely consistent with the electropositive character of arsenic.

$$\underset{28}{Ph_2AsC{\equiv}COMe} + Ph_2AsCl \longrightarrow \underset{29}{(Ph_2As)_2C{=}C{=}O} \quad (21)$$

$$\underset{4\text{-}CH_3C_6H_4}{\overset{Me_3As}{>}}C{=}C{=}O \quad \mathbf{30}$$

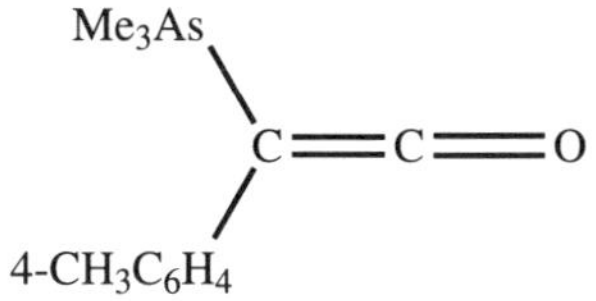

Triphenylarsoranylideneketene (**31**) was obtained as a crystalline solid (equation 22).[33] Several reactions of **31** were observed, including the reaction with MeI (equation 23).[33]

$$Ph_3As{=}CHCO_2Me \xrightarrow{NaN[Si(Pr\text{-}i)_3]_2} \underset{31}{Ph_3\overset{+}{As}{-}\overset{-}{C}{=}C{=}O} \quad (22)$$

$$\mathbf{31} \xrightarrow{MeI} \underset{Me}{\overset{Ph_3As^+}{>}}C{=}C{=}O \xrightarrow{\mathbf{31}} Ph_3\overset{+}{As}\underset{O}{\overset{Me \quad O}{\underset{\Box}{\Box}}}AsPh_3 \quad (23)$$

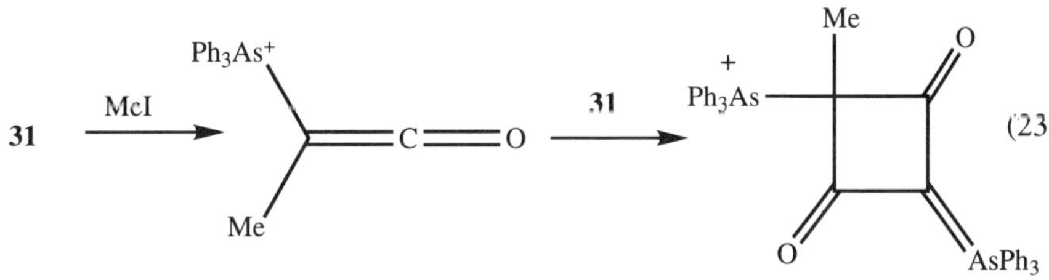

Reported phosphorous and arsenic substituted ketenes are listed in Table 1.

TABLE 2. Phosphorous- and Arsenic-Substituted Ketenes

Ketene	Formula	Preparation	Registry No.	Ref.
$H_3P=C=C=O$	C_2H_3OP	Index entry	98647-31-5	34
	C_2H_3OP	Theory	73703-06-7	35
$H_2PCH=C=O$		Theory	134736-44-0	1
$(HO)_2(S)PCH=C=O$	$C_2H_3O_3PS$	Index entry	34421-22-2	None
$H_2O_3PCH=C=O$	$C_2H_3O_4P$	Index entry	34421-21-1	None
$Cl_2(O)PC(Pr-i)=C=O$	$C_5H_7Cl_2O_2P$	Alkyne pyrolysis	88011-20-5	36
$Me(MeO)_2PC=C=O$	$C_5H_9O_3P$		140682-66-2	22
$Et_2O_3PCCl=C=O$	$C_6H_{10}ClO_4P$	Dehydrochlorination	34255-80-6	15
$Et_2O_3PCH=C=O$	$C_6H_{11}O_4P$	Pyrolysis	67683-20-9	11
$Et_2O_3PCMe=C=O$	$C_7H_{13}O_4P$	Dehydrochlorination	70706-09-1	14
$Et(EtO)(O)PCEt=C=O$	$C_8H_{15}O_3P$	Alkyne pyrolysis	88011-23-8	36
$Et_2O_3PC(CO_2Et)=C=O$	$C_9H_{15}O_6P$	Dehydrochlorination	67683-17-4	12
i-$Pr_2MePC=C=O$	$C_9H_{17}OP$	Alkyne pyrolysis	95111-99-2	21
			95111-88-9	
$Et(EtO)(O)PC(Pr-i)=C=O$	$C_9H_{17}O_3P$	Alkyne pyrolysis	88011-22-7	36
$Me(i\text{-}PrO)_2PC=C=O$	$C_9H_{17}O_3P$	Alkyne pyrolysis	140682-67-3	22
$(EtO)_2(S)PC(Pr-i)=C=O$	$C_9H_{17}O_3PS$	Alkyne pyrolysis	88011-24-9	36
$Et_2O_3PC(Pr-n)=C=O$	$C_9H_{17}O_4P$	Anhydride pyrolysis	84589-95-7	16
$Et_2O_3PC(Pr-i)=C=O$	$C_9H_{17}O_4P$	Alkyne pyrolysis	88011-21-6	36
$Me_2PCPh=C=O$	$C_{10}H_{11}PO_4$	Wolff		6
i-$Pr_2(S)PC(CO_2Me)=C=O$	$C_{10}H_{17}O_3PS$	Elimination	75568-44-4	10
i-$Pr_2(O)PC(CO_2Me)=C=O$	$C_{10}H_{17}O_4P$	Elimination	75568-38-6	10
i-$Pr_2O_3PC(CO_2Me)=C=O$	$C_{10}H_{17}O_6P$	Wolff	72946-12-4	4
i-$Pr_2EtPC=C=O$	$C_{10}H_{19}OP$	Alkyne pyrolysis	95112-02-0	21
			95111-91-4	
t-$Bu_2(O)PCH=C=O$	$C_{10}H_{19}O_2P$	Ylide addition	76711-29-0	8

Me$_2$AsC(C$_6$H$_4$Me-4)=C=O	C$_{11}$H$_{13}$AsO	Metal complex	87954-19-6	31
i-Pr(t-Bu)(S)PC(CO$_2$Me)=C=O	C$_{11}$H$_{19}$O$_3$PS	Elimination	75568-45-5	10
i-Pr(t-Bu)(O)PC(CO$_2$Me)=C=O	C$_{11}$H$_{19}$O$_4$P	Elimination	75568-39-7	10
i-Pr$_2$O$_3$PC(CO$_2$Et)=C=O	C$_{11}$H$_{19}$O$_6$P	Elimination	67683-18-5	11
t-Bu$_2$MePC=C=O	C$_{11}$H$_{21}$OP	Alkyne rearrangement	95112-01-9	21
			95111-90-3	
t-Bu$_2$(O)PCMe=C=O	C$_{11}$H$_{21}$O$_2$P	Ylide addition	76711-30-3	8
i-Pr$_2$(S)PC(SiMe$_3$)=C=O	C$_{11}$H$_{23}$OPSSi	Alkyne addition	127044-77-3	37
i-Pr$_2$(S)PC(GeMe$_3$)=C=O	C$_{11}$H$_{23}$GeOPS	Alkyne addition	127044-79-5	37
Et$_2$O$_3$PCPh=C=O	C$_{12}$H$_{15}$O$_4$P	Dehydrochlorination	67683-19-6	12
t-Bu$_2$(S)PC(COCH$_3$)=C=O	C$_{12}$H$_{21}$O$_2$PS	Elimination	137789-41-4	13
t-Bu$_2$EtPC=C=O	C$_{12}$H$_{23}$OP	Alkyne rearrangement	95112-03-1	21
			95111-92-5	
Et$_2$O$_3$PC(Tol-4)=C=O	C$_{13}$H$_{17}$O$_4$P	Dehydrochlorination	70706-08-0	12
t-Bu$_2$(O)PC(Pr-n)=C=O	C$_{13}$H$_{25}$O$_2$P	Ylide addition	76711-31-4	9
t-Bu$_2$(S)PC(GeMe$_3$)=C=O	C$_{13}$H$_{27}$GeOPS	Alkyne addition	127044-82-0	37
t-Bu$_2$(S)PC(SiMe$_3$)=C=O	C$_{13}$H$_{27}$OPSSi	Alkyne addition	127044-81-9	37
Ph$_2$(O)PCH=C=O	C$_{14}$H$_{11}$O$_2$P	Dehydrochlorination	67683-21-0	11
i-Pr$_2$O$_3$PCPh=C=O	C$_{14}$H$_{19}$O$_4$P	Dehydrochlorination	70706-07-9	12
t-Bu$_2$(n-Bu)PC=C=O	C$_{14}$H$_{27}$OP	Alkyne rearrangement	95112-04-2	21
			95112-93-6	
i-Pr$_2$(S)PC(GeEt$_3$)=C=O	C$_{14}$H$_{29}$GeOPS	Alkyne addition	127044-80-8	37
i-Pr$_2$(S)PC(SiEt$_3$)=C=O	C$_{14}$H$_{29}$OPSSi	Alkyne addition	127044-78-4	37
RP(O$_2$-Et)CPh=C=O				
R = 1-piperidinyl	C$_{15}$H$_{20}$NO$_3$P		140848-55-1	38
R = 2-Mepiperidinyl	C$_{16}$H$_{22}$NO$_3$P		140848-56-2	38
c-Hx$_2$MePC=C=O	C$_{15}$H$_{25}$OP	Alkyne rearrangement	95112-00-8	21
			95111-89-0	
t-Bu$_2$(S)PC(GeEt$_3$)=C=O	C$_{16}$H$_{33}$GeOPS	Alkyne addition	127070-71-7	37
t-Bu$_2$(S)PC(SiEt$_3$)=C=O	C$_{16}$H$_{33}$OPSSi	Alkyne addition	127070-70-6	37

TABLE 2. (*Continued*)

Ketene	Formula	Preparation	Registry No.	Ref.
Ph$_3$AsC=C=O	C$_{20}$H$_{15}$AsO	Elimination	80717-17-5	33
Ph$_3$PC=C=O	C$_{20}$H$_{15}$OP	Elimination	73818-55-0	19
Ph$_2$(O)PCPh=C=O	C$_{20}$H$_{15}$O$_2$P	Wolff	111337-44-1	2,39
Ph$_3\overset{+}{\text{P}}$CH=C=O Cl$^-$	C$_{20}$H$_{16}$OPCl	Protonation	62126-70-9	23
Ph$_2$AsC(SiPhMe$_2$)=C=O	C$_{22}$H$_{21}$AsOSi	Alkyne addition	92136-09-9	30
(Ph$_2$As)$_2$C=C=O	C$_{26}$H$_{20}$As$_2$O	Alkyne addition	92136-07-7	30
Ph$_2$AsC(GePh$_3$)=C=O	C$_{32}$H$_{25}$AsGeO	Alkyne addition	92136-08-8	30

References

1. Gong, L.; McAllister, M. A.; Tidwell, T. T. *J. Am. Chem. Soc.* **1991,** *113,* 6021–6028.
2. Regitz, M.; Bartz, W. *Chem. Ber.* **1970,** *103,* 1477–1485.
3. Regitz, M.; Anschütz, W.; Bartz, W.; Liedhegener, A. *Tetrahedron Lett.* **1968,** 3171–3174.
4. Polozov, A. M.; Pavlov, V. A.; Polezhaeva, N. A.; Liorber, B. G.; Tarzivolova, T. A.; Arbuzov, B. A. *Zh. Obshch. Khim.* **1986,** *56,* 1217–1220; *Engl. Transl.* **1986,** *56,* 1072–1074.
5. Tomioka, H.; Hirai, K. *J. Chem. Soc., Chem. Commun.* **1990,** 1611–1613.
6. Tomioka, H.; Komatsu, K.; Shimizu, M. *J. Org. Chem.* **1992,** *57,* 6216–6222.
6a. Corbel, B.; Hernot, D.; Haelters, J.-P.; Sturtz, G. *Tetrahedron Lett.* **1987,** *28,* 6605–6608.
7. Lukashev, N. V.; Artyushin, O. I.; Lazhko, E. I.; Luzikova, E. V.; Kazankova, M. A. *Zh. Obshch. Khim.* **1990,** *60,* 1539–1549; *Engl. Transl.* **1990,** *60,* 1374–1382.
8. Kolodiazhnyi, O. I. *Tetrahedron Lett.* **1980,** 3983–3986.
9. Kolodyazhnyi, O. I. *Zh. Obshch. Khim.* **1987,** *57,* 821–827; *Engl. Transl.* **1987,** *57,* 724–730.
10. Kolodyazhnyi, O. I. *Zh. Obshch. Khim.* **1980,** *50,* 1485–1498; *Engl. Transl.* **1980,** *50,* 1198–1209.
11. Kolodyazhnyi, O. I.; Kukhar, V. P. *Zh. Org. Khim.* **1978,** *14,* 1340; *Engl. Transl.* **1978,** *14,* 1244–1245.
12. Kolodyazhnyi, O. I. *Zh. Obshch. Khim.* **1979,** *49,* 716–717; *Engl. Transl.* **1979,** *49,* 621–622.
13. Lukashev, N. V.; Fil'chikov, A. A.; Zhichkin, P. E.; Kazankova, M. A.; Beletskaya, I. P. *Zh. Obshch. Khim.* **1991,** *61,* 1014–1016; *Engl. Transl.* **1991,** *61,* 920–921.
14. Kashemirov, B. A.; Skoblikova, L. I.; Khokhlov, P. S. *Zh. Obshch. Khim.* **1988,** *58,* 1672–1673; *Engl. Transl.* **1988,** *58,* 1492–1493.
15. Motoyoshiya, J.; Hirata, K. *Chem. Lett.* **1988,** 211–214.
16. Vdovenko, S. I.; Yakovlev, V. I.; Kolodyazhnyi, O. I.; Kukhar, V. P. *Zh. Obshch. Khim.* **1982,** *52,* 2223–2227; *Engl. Transl.* **1982,** *52,* 1978–1981.
17. Lillford, P. J.; Satchell, D. P. N. *J. Chem. Soc. B.* **1968,** 889–897.
18. Bestmann, H. J. *Angew. Chem., Int. Ed. Engl.* **1977,** *16,* 349–364.
19. Bestmann, H. J.; Sandmeier, D. *Chem. Ber.* **1980,** *113,* 274–277.
20. Mathews, C. N.; Birum, G. M. *Acc. Chem. Res.* **1969,** *2,* 373–379.
21. Lukashev, N. V.; Artyushin, O. I.; Kazankova, M. A.; Lutsenko, I. F. *Zh. Obshch. Khim.* **1984,** *54,* 2391–2393; *Engl. Transl.* **1984,** *54,* 2137–2138.
22. Lukashev, N. V.; Fil'chikov, A. A.; Kozlov, A. I.; Luzikov, Yu. N.; Kazankova, M. A. *Zh. Obshch. Khim.* **1991,** *61,* 1739–1743; *Engl. Transl.* **1991,** *61,* 1600–1603.
23. Bestmann, H. J.; Schmid, G.; Sandmeier, D.; Kisielowski, L. *Angew. Chem., Int. Ed. Engl.* **1977,** *16,* 268–269.
24. Bestmann, H. J.; Schmid, G.; Sandmeier, D. *Chem. Ber.* **1980,** *113,* 912–918.
25. Bestmann, H. J.; Schobert, R. *Angew. Chem., Int. Ed. Engl.* **1983,** *22,* 780–782.
26. Nickisch, K.; Klose, W.; Nordhoff, E.; Bohlmann, F. *Chem. Ber.* **1980,** *113,* 3086–3088.

27. Klose, W.; Nickisch, K.; Bohlmann, F. *Chem. Ber.* **1980**, *113*, 2694–2698.
28. Bestmann, H. J.; Schmid, G. *Tetrahedron Lett.* **1984**, *25*, 1441–1444.
29. Bestmann, H. J.; Dostalek, R.; Zimmermann, R. *Chem. Ber.* **1992**, *125*, 2081–2084.
30. Himbert, G.; Henn, L. *Tetrahedron Lett.* **1984**, *25*, 1357–1358.
31. Kreissl, F. R.; Wolfgruber, M.; Sieber, W. *Angew. Chem., Int. Ed. Engl.* **1983**, *22*, 1001–1002.
32. Wolfgruber, M.; Sieber, W.; Kreissl, F. R. *Chem. Ber.* **1984**, *117*, 427–433.
33. Bestmann, H. J.; Bansal, R. K. *Tetrahedron Lett.* **1981**, *22*, 3839–3842.
34. Artyushin, O. I.; Lukashev, N. V.; Kazankova, M. A.; Lutsenko, I. F. *U.S.S.R. SU* 1,154,286; *Chem. Abstr.* **1985**, *103*, 160697f.
35. Albright, T. A.; Hofmann, P.; Rossi, A. R. *Z. Naturforsch. B: Anorg. Chem., Org. Chem.* **1980**, *35B*, 343–351.
36. Kazankova, M. A.; Lutsenko, I. F. *Vestn. Mosk. Univ., Ser. 2: Khim.*, **1983**, *24*, 315–331; *Chem. Abstr.* **1984**, *100*, 6718v.
37. Fil'chikov, A. A.; Kozlov, A. I.; Lukashev, N. V.; Kazankova, M. A. *U.S.S.R. SU* 1,532,562; *Chem. Abstr.* **1990**, *112*, 217276k.
38. Afarinkia, K.; Cadogan, J. I. G.; Rees, C. W. *J. Chem. Soc., Chem. Commun.* **1992**, 285–287.
39. Willson, C. G.; Miller, R. D.; McKean, D. R.; Pederson, L. A.; Regitz, M. *Proc. SPIE-Int. Soc. Opt. Eng.* **1987**, *771*, 2–10; *Chem. Abstr.* **1987**, *107*, 225797t.

4.7 SULFUR-SUBSTITUTED KETENES

Sulfur-substituted ketenes are predicted from isodesmic energy calculations[1] to have reasonable stability and a number of these species are known. As noted below, ketenes $(RS)_2C=C=O$ are indeed quite remarkably stable, and this entire class deserves further study. Photolysis of (ethylthio)diazoacetate is suggested to involve formation of (ethylthio)ketene **1,** as indicated by the capture of esters from reactions conducted in alcohols (equation 1).[2]

$$N_2CHCO(SEt) \xrightarrow{h\nu} EtSCH=C=O \xrightarrow{ROH} EtSCH_2CO_2R \quad (1)$$
$$\mathbf{1}$$

Photolysis of the diazodione **2** gave exclusive sulfur migration in the intermediate carbene, and **3** was directly observed by IR and NMR (equation 2), and was captured by hydration or cycloaddition with $PhCH=NPh$.[3] The ketene $CH_3SCPh=C=O$ was prepared similarly from $PhCN_2COSCH_3$, and its hydration kinetics measured.[4] Wolff rearrangement with sulfur migration also occurred in the reaction of equation 3.[5]

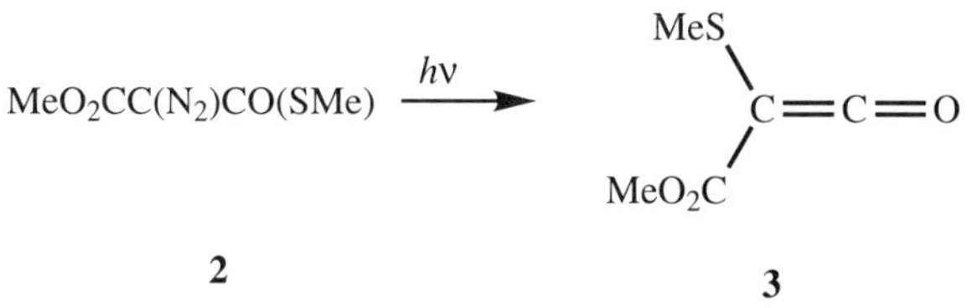

4.7 SULFUR-SUBSTITUTED KETENES

$$MeO_2CC(N_2)CO(SMe) \xrightarrow{h\nu} \underset{MeO_2C}{\overset{MeS}{>}}C=C=O \qquad (2)^3$$

2 **3**

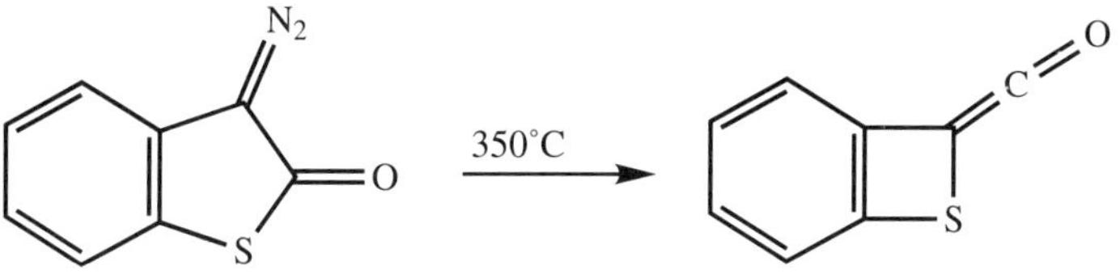

(equation 3)[5] — benzothiophene-derived diazo ketone heated at 350°C to give the corresponding sulfur-containing ketene.

Thio-substituted ketenes have also been prepared by dehydrochlorination of acyl halides or dehydration of carboxylic acids, as in the examples of equations 4–6.[6–14] Ketene **4** was also trapped with imines.[6] The cyclobutanone in equation 5 reacted further.[9] The silylketene **5** was isolable, in common with other silylketenes.[13] The ketenes from equation 6 were also isolable.[12]

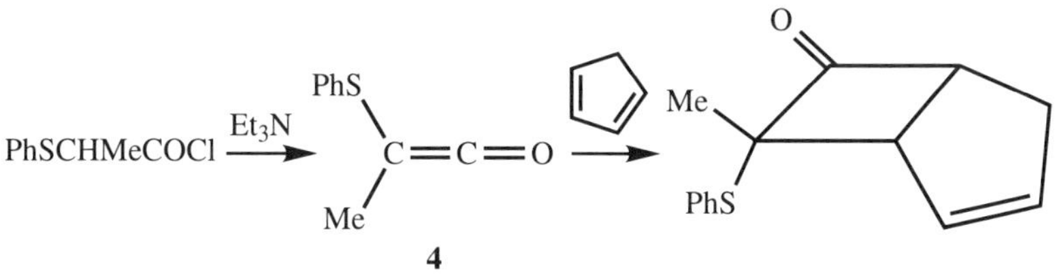

$$PhSCHMeCOCl \xrightarrow{Et_3N} \underset{Me}{\overset{PhS}{>}}C=C=O \xrightarrow{\text{cyclopentadiene}} \text{(bicyclic adduct with Me, PhS)} \qquad (4)^{6-10}$$

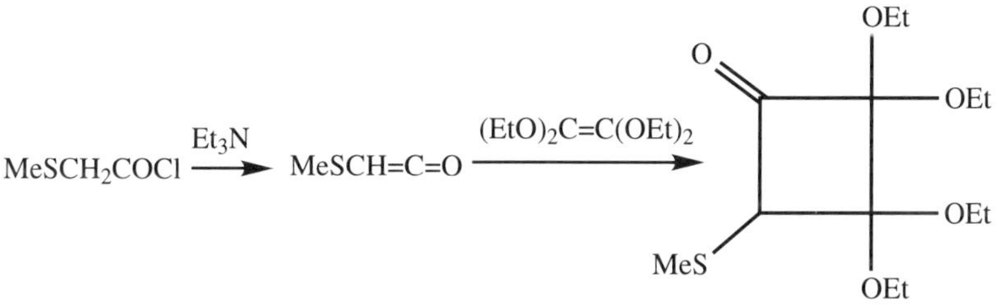

4

$$MeSCH_2COCl \xrightarrow{Et_3N} MeSCH=C=O \xrightarrow{(EtO)_2C=C(OEt)_2} \text{(cyclobutanone with MeS, OEt groups)} \qquad (5)^{10}$$

$$SF_5CHRCO_2H \xrightarrow{P_2O_5} SF_5CR=C=O \qquad (6)^{12}$$

R = H, Me, Cl, Br

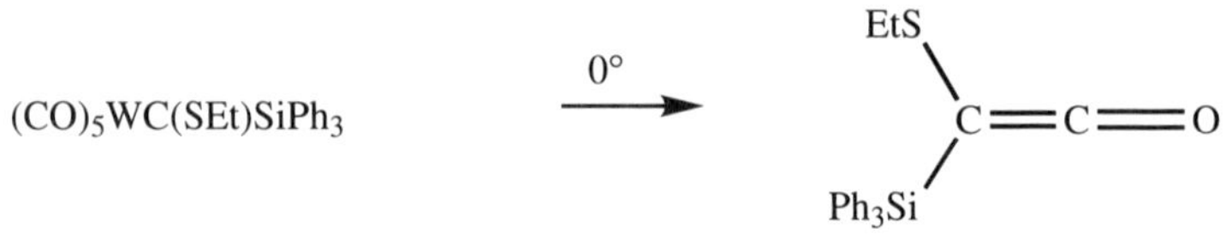

$$(CO)_5WC(SEt)SiPh_3 \xrightarrow{0°} \underset{Ph_3Si}{\overset{EtS}{>}}C=C=O \quad \mathbf{5} \qquad (7)^{13}$$

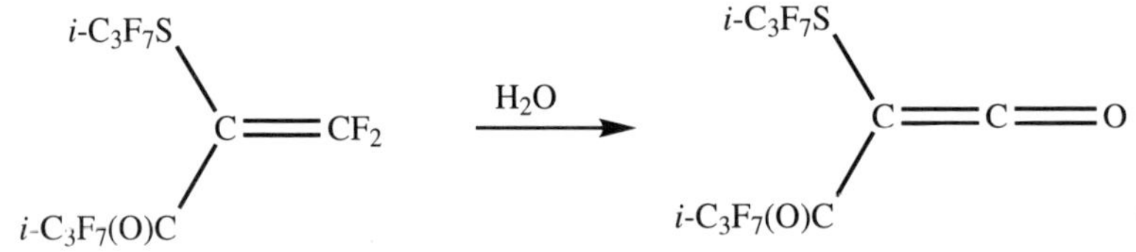

$$\underset{i\text{-}C_3F_7(O)C}{\overset{i\text{-}C_3F_7S}{>}}C=CF_2 \xrightarrow{H_2O} \underset{i\text{-}C_3F_7(O)C}{\overset{i\text{-}C_3F_7S}{>}}C=C=O \qquad (8)^{14}$$

Sulfonylketenes have been prepared as unisolated intermediates from a Wolff rearrangement (equation 9)[15,16] and elimination from a malonate ester (equation 10).[17] Fluorinated sulfonylketenes were isolable (equations 11, 12).[18,19]

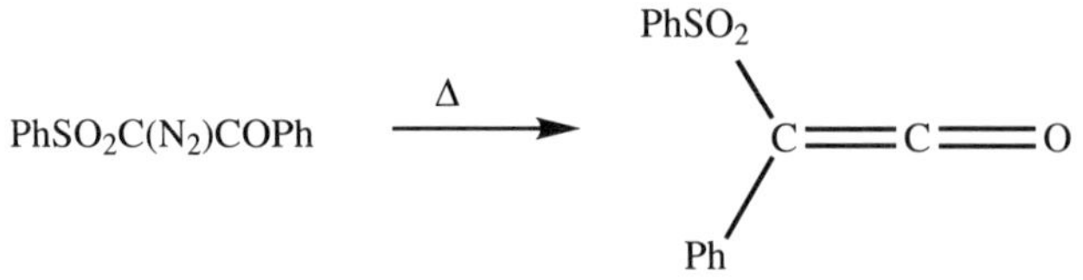

$$PhSO_2C(N_2)COPh \xrightarrow{\Delta} \underset{Ph}{\overset{PhSO_2}{>}}C=C=O \qquad (9)^{15}$$

$$R_fSO_2CH(CO_2Et)_2 \xrightarrow{P_2O_5} \underset{EtO_2C}{\overset{R_fSO_2}{>}}C=C=O \qquad (10)^{17}$$

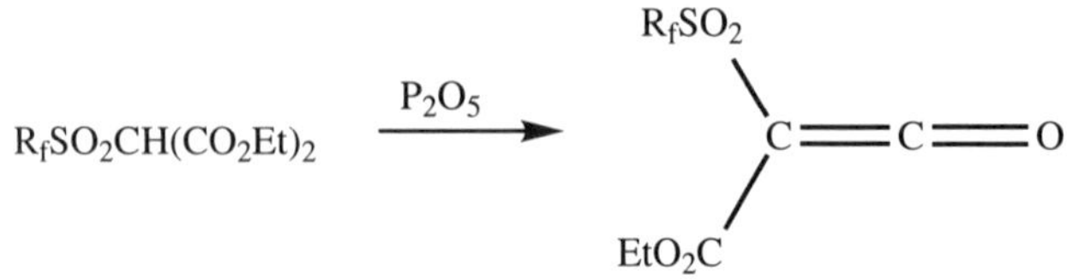

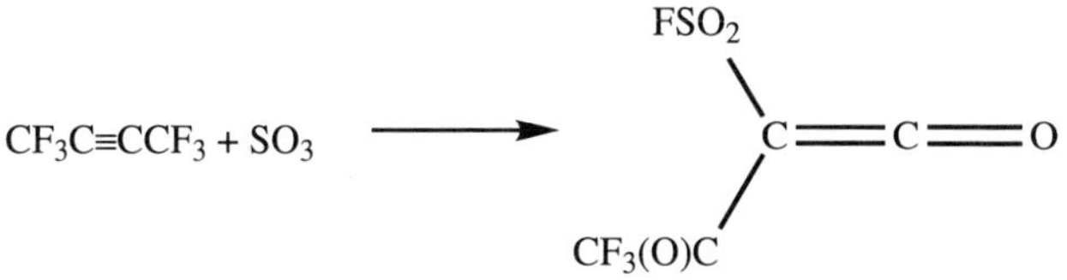

4.7 SULFUR-SUBSTITUTED KETENES

$$CF_3C\equiv CCF_3 + SO_3 \longrightarrow \underset{CF_3(O)C}{\overset{FSO_2}{\diagdown}}C=C=O \qquad (11)^{18}$$

$$i\text{-}C_3F_7C\equiv CF + SO_3 \longrightarrow \underset{i\text{-}C_3F_7}{\overset{FSO_2}{\diagdown}}C=C=O \qquad (12)^{19}$$

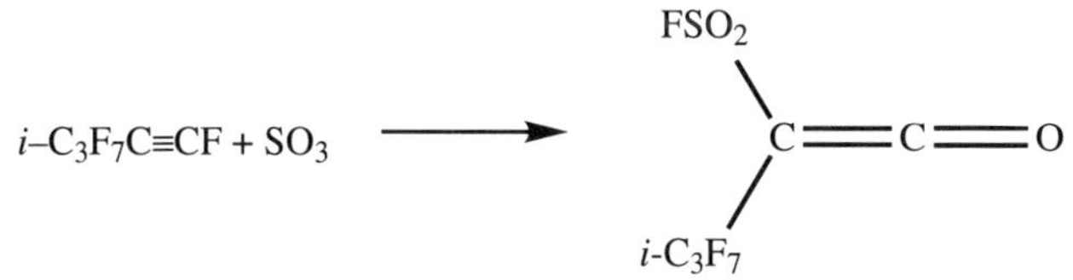

A number of disulfur-substituted ketenes have been prepared and are sometimes surprisingly stable (equations 13–16).[20–26] Thus the UV spectra of the ketenes **6** were measured in MeOH! This high stability suggests that sulfur may be even more stabilizing than expected from the electronegativity correlations. The ketene formed in equation 15 showed strong UV absorption near 220 nm (ε 16,600) and 325 nm (ε 730), and on photolysis was proposed to undergo decarbonylation to the carbene.[24a] The conformation of **6a** has the lone pairs on sulfur near the ketene plane,[25a] as predicted by calculation.[1]

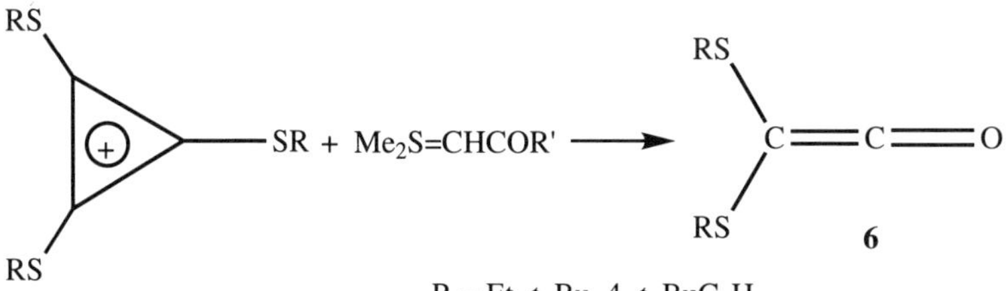

$$\text{(cyclopropenyl cation with RS groups)} + Me_2S=CHCOR' \longrightarrow \underset{RS}{\overset{RS}{\diagdown}}C=C=O \qquad (13)^{20}$$

$$\text{R = Et, }t\text{-Bu, }4\text{-}t\text{-BuC}_6H_4$$

$$\text{(1,3-dithiolane-2-yl)}-COCl \xrightarrow{Et_3N} \text{(dithiolane)}C=C=O \longrightarrow \text{(spiro bicyclic ketone product)} \qquad (14)^{9,21,22}$$

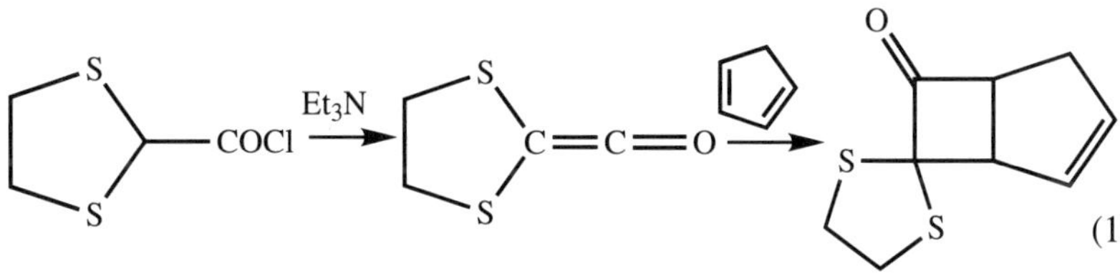

$$(CF_3S)_2CHCO_2H \xrightarrow{P_2O_5} (CF_3S)_2C=C=O \quad (15)^{24}$$

$$(FOS_2)_2CHCOF \xrightarrow{KF} (FO_2S)_2C=C=O \quad (16)^{26}$$

The selenium-substituted ketene **7** was generated but trapped in situ because of its instability (equation 17).[27] The ketene $(CF_3Se)_2C=C=O$ was formed in a reaction analogous to that of equation 15, and also by dehydrochlorination of the acyl chloride and the reaction of equation 18 (see also Section 4.8.3).[25] This ketene was thermally stable and did not dimerize.

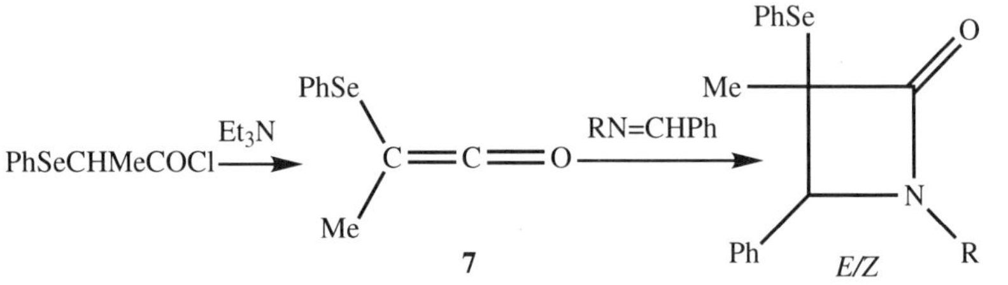

$$\text{PhSeCHMeCOCl} \xrightarrow{Et_3N} \underset{\underset{Me}{|}}{\overset{\overset{PhSe}{|}}{C}}=C=O \xrightarrow{RN=CHPh} \text{[azetidinone product, E/Z]} \quad (17)^{27}$$

$$Ag_2C=C=O \xrightarrow{CF_3SeCl} (CF_3Se)_2C=C=O \quad (18)^{25}$$

The formation of "thioisomünchnone" **8** via a possible mechanism involving the thioketene **9** has been reported (equation 19).[27a]

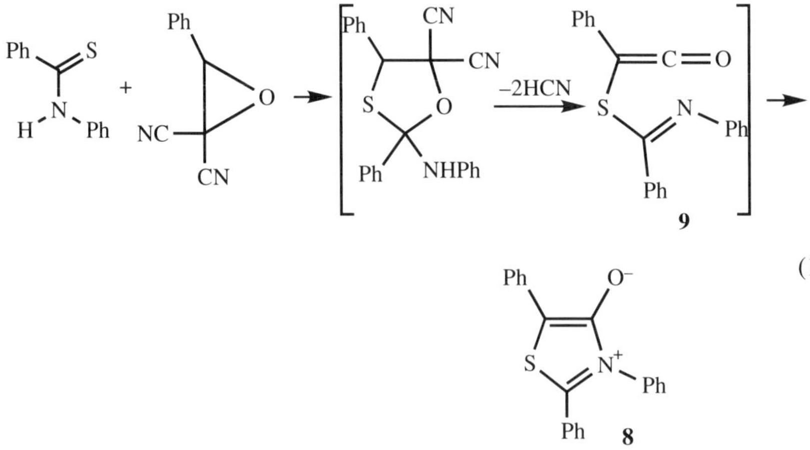

$$\text{(equation 19 scheme)} \quad (19)$$

TABLE 1. Sulfur- and Selenium-Substituted Ketenes

Ketene	Formula	Preparation	Registry No.	Ref.
$F_5SCBr=C=O$	C_2BrF_5OS	Dehydration	128632-49-5	12
$F_5SCCl=C=O$	C_2ClF_5OS	Dehydration	128632-48-4	12
$(FO_2S)_2C=C=O$	$C_2F_2O_5S_2$	Dehydrofluorination	74765-84-7	26
$F_5SC(SO_2F)=C=O$	$C_2F_6O_3S$	Dehydrofluorination	117527-66-9	28
$F_5SCH=C=O$	C_2HF_5OS	Dehydration	109907-07-5	11
$(S^-)CH=C=O$	C_2HOS	Gas phase		23
$HSCH=C=O$	C_2H_2OS	Theory	134736-45-1	1
$FO_2SC(CF_3)=C=O$	$C_3F_4O_3S$		60956-76-5	
$F_5SCMe=C=O$	$C_3H_3F_5OS$	Dehydration	128632-47-3	12
$MeSCH=C=O$	C_3H_4OS	Dehydrochlorination	58216-36-7	10
$FO_2SC(COCF_3)=C=O$	$C_4F_4O_4S$	Sulfonation	104693-21-2	18
$(CF_3S)_2C=C=O$	$C_4F_6OS_2$	Dehydration	125420-67-9	24
$(CF_3Se)_2C=C=O$	$C_4F_6OSe_2$	Dehydration	137946-92-9	25
$(CF_3O_2S)_2C=C=O$	$C_4F_6O_5S_2$	Dehydration	137946-93-1	25
$(CH_2S)_2C=C=O$	$C_4H_4OS_2$	Dehydrochlorination	56380-64-4	9
$EtSCH=C=O$	C_4H_6OS	Wolff		2
$(MeO_2S)_2C=C=O$	$C_4H_6O_5S_2$	Theory	111647-30-4	29
$FO_2SC(C_3F_{7}-i)=C=C$	$C_5F_8O_3S$	Sulfonation	113273-16-8	19
$EtSC(CN)=C=O$	C_5H_5NOS	Fragmentation		30
$CH_3(CH_2)_2S_2C=C=O$	$C_5H_6OS_2$	Dehydrochlorination	54235-70-0	9,21,22
$MeSC(CO_2Me)=C=O$	$C_5H_6O_3S$	Wolff	72867-19-7	3
$FO_2SC(C_4F_{9}-i)=C=O$	$C_6F_{10}O_3S$	Sulfonation	113273-17-9	19
$CF_3SO_2C(CO_2Et)=C=O$	$C_6H_5F_3O_5S$	Elimination	75988-05-5	17
$(EtS)_2C=C=O$	$C_6H_{10}OS_2$	Addition/rearrangement	3122-33-6	20

TABLE 1. (Continued)

Ketene	Formula	Preparation	Registry No.	Ref.
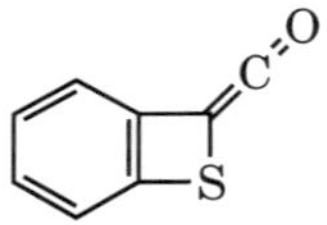 (benzothiophene-ketene structure)	C_8H_4OS	Wolff	58150-61-1	5
ONSCPh=C=O	$C_8H_5NO_2S$	Photolysis	39091-90-2	31,32
•SCPh=C=O	C_8H_5OS	Photolysis	113424-58-1	31,32
PhSCH=C=O	C_8H_6OS	Dehydrochlorination	127055-27-0	9
n-$C_4F_9O_2$SCR=C=O, R = CO_2Et	$C_9H_5F_9O_5S$	Dehydration	75988-06-6	17
i-C_3F_7SCR=C=O, R = COC_3F_7-i	$C_9F_{14}O_2S$	Hydrolysis	75790-42-0	14
PhSCMe=C=O	C_9H_8OS	Dehydrochlorination	66977-59-1	6
CH_3SCPh=C=O	C_9H_8OS	Wolff		4
PhSeCMe=C=O	C_9H_8OSe	Dehydrochlorination	76893-63-5	27
(t-BuS)$_2$C=C=O	$C_{10}H_{18}OS_2$	Addition/rearrangement	85296-01-1	20
2-NaphthylSCH=C=O	$C_{12}H_8OS$	Dehydrochlorination		33
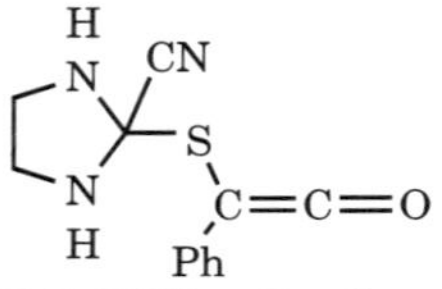 (pyrrolidine-CN structure with Ph, C=O)	$C_{12}H_{11}N_3OS$	Addition/rearrangement	76808-29-2	34
PhO$_2$SCPh=C=O	$C_{14}H_{10}O_3S$	Wolff	85526-51-8	15
EtSC(SiPh$_3$)=C=O	$C_{22}H_{20}OSSi$	Metal complex	98014-97-2	13
(4-t-BuC$_6$H$_4$S)$_2$C=C=O	$C_{22}H_{26}OS_2$	Addition/rearrangement	85296-02-2	20

References

1. Gong, L.; McAllister, M. A.; Tidwell, T. T. *J. Am. Chem. Soc.* **1991**, *113*, 6021–6028.
2. Orphanides, G. G., Ph.D. Thesis, Ohio State University, 1972, quoted by Ando, W. in *Photochemistry of Diazonium and Diazo Groups*, in *The Chemistry of Diazonium and Diazo Groups;* Patai, S., ed.; Wiley: New York, 1978; p. 473.
3. Georgian, V.; Boyer, S. K.; Edwards, B. *J. Org. Chem.* **1980**, *45*, 1686–1688.
4. Jones, J., Jr.; Kresge, A. J. *J. Org. Chem.* **1992**, *57*, 6467–6469.
5. Schulz, R.; Schweig, A. *Tetrahedron Lett.* **1980**, *21*, 343–346.
6. Ishida, M.; Minami, T.; Agawa, T. *J. Org. Chem.* **1979**, *44*, 2067–2073.
7. Minami, T.; Ishida, M.; Agawa, T. *J. Chem. Soc., Chem. Commun.* **1978**, 12–13.
8. Palomo, C.; Cossio, F. P.; Ordiozola, J. M.; Oiarbide, M.; Ontoria, J. M. *Tetrahedron Lett.* **1989**, *30*, 4577–4580; **1990**, *31*, 2218; *J. Org. Chem.* **1991**, *56*, 4418–4428.
9. Michel, P.; O'Donnell, M.; Biname, R.; Hesbain-Frisque, A. M.; Ghosez, L.; Declercq, J. P.; Germain, G.; Arte, E.; Van Meerssche, M. *Tetrahedron Lett.* **1980**, *21*, 2577–2580.
10. Bellus, D. *J. Org. Chem.* **1979**, *44*, 1208–1211.
11. Krügerke, T.; Seppelt, K. *Chem. Ber.* **1988**, *121*, 1977–1981.
12. Bittner, J.; Seppelt, K. *Chem. Ber.* **1990**, *123*, 2187–2190.
13. Schubert, U.; Kron, J.; Hörnig, H. *J. Organomet. Chem.* **1988**, *355*, 243–256.
14. England, D. C. *J. Org. Chem.* **1981**, *46*, 153–157.
15. Capuano, L.; Mörsdorf, P.; Scheidt, H. *Chem. Ber.* **1983**, *116*, 741–750.
16. Regitz, M.; Bartz, W. *Chem. Ber.* **1970**, *103*, 1477–1485.
17. Ogoiko, P. I.; Nazaretyan, V. P.; Il'chenko, A. Ya.; Yagupol'skii, L. M. *Zh. Org. Khim.* **1980**, *16*, 1397–1401; *Engl. Transl.* **1980**, *16*, 1200–1203.
18. Krespan, C. G.; Dixon, D. A. *J. Org. Chem.* **1986**, *51*, 4460–4466.
19. Galakhov, M. V.; Cherstkov, V. F.; Sterlin, A. R.; German, L. S. *Izv. Akad. Nauk SSSR* **1987**, 958; *Engl. Transl.* **1987**, 886.
20. Inoue, S.; Hori, T. *Bull. Chem. Soc. Jpn.* **1983**, *56*, 171–174.
21. Cossement, E.; Biname, R.; Ghosez, L. *Tetrahedron Lett.* **1974**, 997–1000.
22. Bellus, D. *Helv. Chim. Acta* **1975**, *58*, 2509–2511.
23. Kass, S. R.; Guo, H.; Dahlke, G. D. *J. Am. Soc. Mass Spectrom.* **1990**, *1*, 366–371.
24. Haas, A.; Lieb, M.; Praas, H.-W. *J. Fluorine Chem.* **1989**, *44*, 329–337.
24a. Dorra, M.; Haas, A. *J. Fluorine Chem.* **1994**, *66*, 91–94.
25. Haas, A.; Praas, H.-W. *Chem. Ber.* **1992**, *125*, 571–579.
25a. Oberhammer, H., et al, *J. Mol. Struct.*, **1995**, in press.
26. Eleev, A. F.; Sokol'skii, G. A.; Knunyants, I. L. *Izv. Akad. Nauk SSSR* **1979**, 892–896; *Engl. Transl.* **1979**, 641–644.
27. Agawa, T.; Ishida, M.; Ohshiro, Y. *Synthesis*, **1980**, 933–935.
27a. Baudy, M.; Robert, A.; Foucaud, A. *J. Org. Chem.* **1978**, *43*, 3732–3742.
28. Winter, R.; Gard, G. L. *J. Fluorine Chem.* **1991**, *52*, 73–98.
29. Gano, J. E.; Jacob, E. *Spectrochim. Acta* **1987**, *43A*, 1023–1025.

30. Labille, M.; Janousek, Z.; Viehe, H. G. *Tetrahedron* **1991**, *47*, 8161–8166.
31. Harrit, N.; Holm, A.; Dunkin, I. R.; Poliakoff, M.; Turner, J. J. *J. Chem. Soc., Perkin Trans. 2* **1987**, 1227–1238.
32. Tono, M.; Aoto, H.; Matsudaira, Y.; Sugiyama, T.; Kajitani, M.; Akiyama, T.; Sugimori, A. *Tetrahedron Lett.* **1991**, *32*, 4023–4026.
33. Borrmann, D.; Wegler, R. *Chem. Ber.* **1966**, *99*, 1245–1251.
34. Baudy, M.; Robert, A. *Tetrahedron Lett.* **1980**, *21*, 2517–2520.

4.8 METAL-SUBSTITUTED KETENES

4.8.1 Lithium Ketenes (Lithium Ynolates) and Ynols

Ynols **1** are valence isomers of ketenes (**2**) and ynolates **3** are conjugate bases of either species. Ynols are known in the gas phase, for example, **1** was generated by decarbonylation and neutralization of the radical cation of $HC\equiv CCO_2H$ using tandem mass spectrometry.[1] Irradiation of **4** in an Ar matrix at 12 K also led to **1** by the reaction sequence shown (equation 1).[2] The various species involved were identified by their characteristic IR bands.[2]

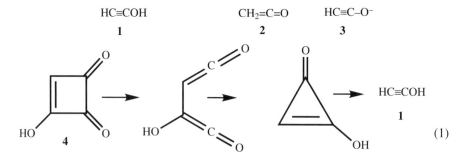

Flash photolysis of arylhydroxycyclopropenones **5** (Ar = Ph and 2,4,6-trimethylphenyl) in aqueous solution led to transients which could be observed by their UV absorption and identified as the ynols **6** (equation 2).[3,4] The reactions of **6** with H_2O were proposed to occur by ionization to the ynolates and protonation to the ketenes, which were observed by UV (equation 2).[3,4]

$$\underset{\underset{5}{Ar}}{\triangle}\underset{OH}{\overset{O}{\parallel}} \xrightarrow[-CO]{h\nu} \underset{6}{ArC\equiv COH} \longrightarrow ArC\equiv CO^- \xrightarrow{H_2O} ArCH=C=O \quad (2)$$

Molecular orbital calculations suggest that ynols are remarkably acidic,[5] and ynolate ions such as **3** are readily observed in the gas phase. Thus reaction of OH^-

with acyl chlorides in negative ion mass spectrometry is proposed to give both the ketene **8** and the ynolate **9** (equations 3 and 4).[6] Fragmentation of enolates in collisional activation mass spectra also produces ynolates (equations 5 and 6),[7,8] as have reactions with O^- (equation 7).[9]

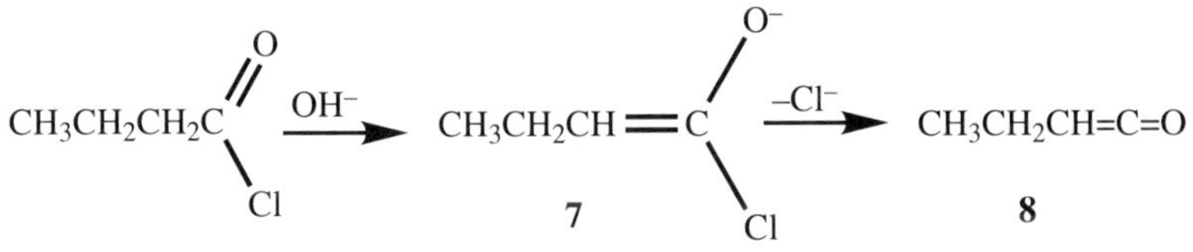

$$CH_3CH_2CH_2C(=O)Cl \xrightarrow{OH^-} CH_3CH_2CH=C(O^-)(Cl) \xrightarrow{-Cl^-} CH_3CH_2CH=C=O \quad (3)$$
$$\phantom{CH_3CH_2CH_2C(=O)Cl \xrightarrow{OH^-}} \mathbf{7} \phantom{\xrightarrow{-Cl^-}} \mathbf{8}$$

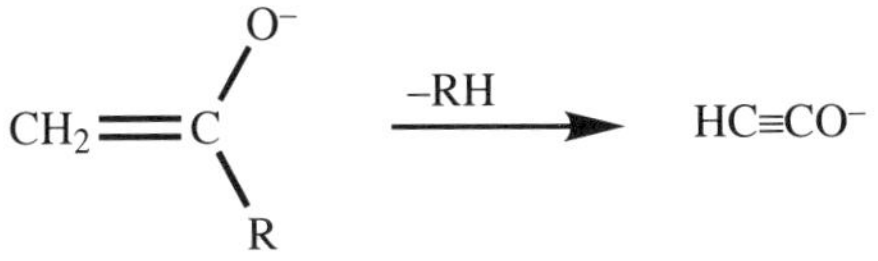

$$\mathbf{7} \xrightarrow{-HCl} CH_3CH_2C\equiv CO^- \quad (4)$$
$$\phantom{\mathbf{7} \xrightarrow{-HCl}} \mathbf{9}$$

$$CH_2=C(O^-)(R) \xrightarrow{-RH} HC\equiv CO^- \quad (5)^7$$

$$CH_3COCOCH_2^- \longrightarrow HC\equiv CO^- \quad (6)^8$$

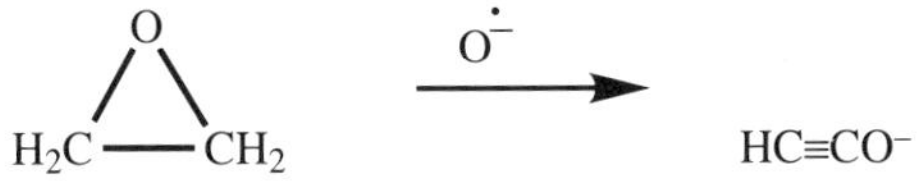

$$H_2C\text{—}CH_2\,(\text{epoxide}) \xrightarrow{O^{\cdot-}} HC\equiv CO^- \quad (7)^9$$

The gas phase acidity of ketene in a flowing afterglow device according to equation 8 was determined using NH_2^-, MeO^-, and other bases.[10] Slow formation of the adduct shown on reaction of the ynolates with ketene was also observed (equation 8).[10] The gas phase reactions of ^{13}C-labeled $HCCO^-$ and of CH_3CCO^- generated from vinylene carbonate (equation 9) with CS_2, COS, and CO_2 have also been studied.[11]

$$CH_2=C=O \xrightarrow{B} \overline{HC}=C=O \xrightarrow{CH_2=C=O} [CH_2CO\ HCCO]^- \quad (8)$$

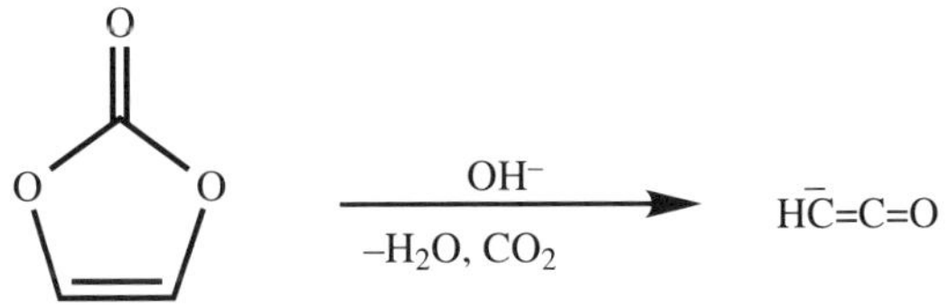

$$\text{(vinylene carbonate)} \xrightarrow[-H_2O,\ CO_2]{OH^-} \overline{HC}=C=O \quad (9)$$

A series of aryl ynolates were generated in the gas phase by the negative ion Wolff rearrangement.[12] Loss of nitrogen from the deprotonated diazoketones occurred with rearrangement to give the ynolates (equation 10).[12] The structural assignment for Ar = Ph was confirmed by comparison to the collisional activation and charge reversal mass spectra of the authentic ion prepared according to equation 11. For Ar = 4-MeOC$_6$H$_4$, further cleavage according to equation 12 is proposed.[12]

$$\text{Ar}\bar{\text{C}}\text{OCN}_2 \longrightarrow \text{Ar}\ddot{\text{C}}\text{OC}^- \longrightarrow \text{ArC}\equiv\text{C-O}^- \quad (10)$$

$$\text{Ph}\bar{\text{C}}\text{HCO}_2\text{Et} \longrightarrow [(\text{PhCH}=\text{C}=\text{O})\text{OEt}^-] \longrightarrow \text{PhC}\equiv\text{C-O}^- + \text{EtOH} \quad (11)$$

$$4\text{-MeO-C}_6\text{H}_4\text{-C}\equiv\text{C-O}^- \xrightarrow{-\text{Me}^\bullet} [\text{O}=\text{C}_6\text{H}_4=\text{C}=\text{C}=\text{O}]^{\cdot -} \quad (12)$$

Calculations at the MP2/6-31G*//6-31G* level indicate that, in the absence of solvent, the C-lithiated structure CHLi=C=O is more stable than the O-lithiated isomer HC≡COLi by 4 kcal/mol.[13] The favored structure in solution has not been determined. For Li$_2$C$_2$O$_2$ the bis(ynolate) structure LiOC≡COLi and the bicyclic structure **11** are calculated to be of equal stability.[14] The acetylene diolate structure KOC≡COK has been proposed for "potassium carbonyl"[15] and an introduction to the long history of this substance is given by Serratosa.[16] Other recent studies of C$_2$O$_2^{2-}$ species have appeared.[17,18]

$$\begin{array}{c} \text{Li}-\text{C}\equiv\text{O} \\ | \quad\quad | \\ \text{O}\equiv\text{C}-\text{Li} \end{array} \quad \mathbf{11}$$

Some other aspects of the gas phase chemistry of ynolate ions are included in Section 4.10.

Solution phase generation of ynolate ions such as **12** includes the rearrangement of dianions generated by deprotonation of α-haloenolates (equation 13).[19] These ynolates are alkylated on carbon by ketones and can give β-lactones on protonation (equation 14).[19] Reaction of esters with CH$_2$Br$_2$ and strong bases also led to ynolates by a related route.[20]

$$\text{PhCOCHBr}_2 \xrightarrow{\text{LiN(SiMe}_3)_2} \underset{\text{PhC=CBr}_2}{\overset{\text{OLi}}{|}} \xrightarrow{t\text{-BuLi}} \underset{\text{PhC=CLiBr}}{\overset{\text{OLi}}{|}} \longrightarrow \underset{\mathbf{12}}{\text{PhC}\equiv\text{COLi}} \quad (13)$$

4.8 METAL-SUBSTITUTED KETENES

$$PhC{\equiv}COLi \quad \xrightarrow{\text{cyclohexanone}} \quad \text{(intermediate with Ph, C=O, OLi)} \quad \longrightarrow \quad \text{(β-lactone spiro product)} \quad (14)$$

12

An optimized route for ester homologation on a moderate scale (25 mmol) involves reaction of LiCHBr$_2$ (generated from CH$_2$Br$_2$ and lithium tetramethylpiperidide) to the ester to generate an intermediate **13** which was most efficiently converted to the ynolate **12** through the intermediates shown by the specified reagents (equation 15).[21,22] Reaction of **12** with HCl in ethanol led through the ketene to the homologated ester (equation 16).[21]

$$PhCO_2Et \xrightarrow{LiCHBr_2} \underset{\underset{OEt}{|}}{\overset{\overset{OLi}{|}}{PhCCHBr_2}} \xrightarrow{\substack{1)\ LiN(SiMe_3)_2 \\ 2)\ s\text{-BuLi} \\ 3)\ n\text{-BuLi}}} PhC{\equiv}COLi \quad (15)$$

13   **12**

$$PhC{\equiv}COLi \xrightarrow{H^+} PhCH{=}C{=}O \xrightarrow{EtOH} PhCH_2CO_2Et \quad (16)$$

12   78%

Ketenes were also proposed to be formed in similar processes by the reaction of the ynolate **14** with n-BuBr (resulting from metal–halogen exchange) to give ketene **15**, which reacted with LiOEt (from the starting ester) to give the ester **16** after hydrolysis (equation 17).[21] Also, when **12** was reacted at −78 °C with EtOH containing HCl, the initially formed PhCH=C=O was proposed to react with **12** to give the further ketene intermediates **17** and **18** (equation 18), which were suggested as possible precursors to the products **19** and **20** (equations 19 and 20).[21] The reaction of ketenes with enolates is reported in Section 5.5.2.4.

$$PhCH_2CH_2C{\equiv}COLi \xrightarrow{n\text{-BuBr}} \underset{n\text{-Bu}}{\overset{PhCH_2CH_2}{>}}C{=}C{=}O \xrightarrow{\substack{1)\ LiOEt \\ 2)\ H_2O}} \underset{n\text{-Bu}}{\overset{PhCH_2CH_2}{>}}CHCO_2Et \quad (17)$$

14   **15**   **16**

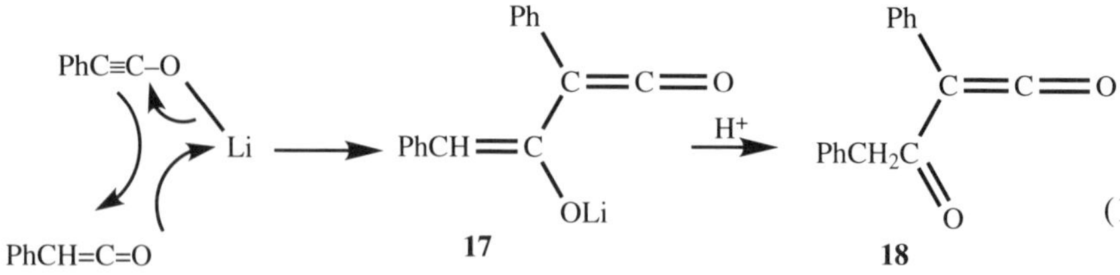

$$PhC\equiv C-O^- \quad PhCH=C=O \xrightarrow{Li} PhCH=C\begin{matrix}Ph\\ \| \\ OLi\end{matrix}C=C=O \xrightarrow{H^+} PhCH_2C\begin{matrix}Ph\\ \| \\ O\end{matrix}C=C=O \quad (18)$$

$$\mathbf{17} \qquad \mathbf{18}$$

$$\mathbf{17} \longrightarrow PhCH=C=C\begin{matrix}Ph\\ \\ CO_2Li\end{matrix} \xrightarrow{EtOH} PhCH=C=C\begin{matrix}Ph\\ \\ CO_2Et\end{matrix} \quad (19)$$

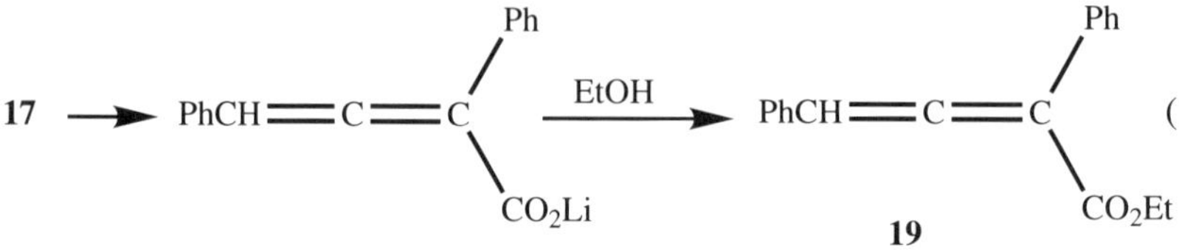

$$\mathbf{19}$$

$$\mathbf{18} \xrightarrow{EtOH} PhCH_2COCHPhCO_2Et \quad (20)$$

$$\mathbf{20}$$

Reaction of these ynolates with LiH was suggested to proceed by addition to form **21**, which on reaction with acetic anhydride gave **22** (equation 21).[20,23]

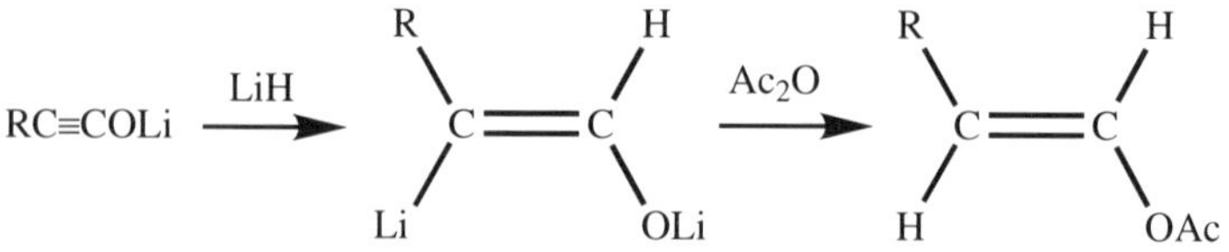

$$RC\equiv COLi \xrightarrow{LiH} \begin{matrix}R\\ \\ Li\end{matrix}C=C\begin{matrix}H\\ \\ OLi\end{matrix} \xrightarrow{Ac_2O} \begin{matrix}R\\ \\ H\end{matrix}C=C\begin{matrix}H\\ \\ OAc\end{matrix} \quad (21)$$

$$\mathbf{21} \qquad \mathbf{22}$$

Deprotonation of the ketene **23** gave the ynolate **24**, as evidenced by capture with Me_3SiCl to give **25** in 80–90% yield (equation 22).[24]

$$Me_3SiCH=C=O \xrightarrow{n\text{-BuLi}} Me_3SiC\equiv COLi \xrightarrow{Me_3SiCl} (Me_3Si)_2C=C=O \quad (22)$$

$$\mathbf{23} \qquad \mathbf{24} \qquad \mathbf{25}$$

Another route to ynolates is the reaction of 3,4-diphenylisoxazole with n-BuLi (equation 23),[25] as evidenced by the capture with carbonyl compounds (equation 14), and Me_3SiCl (equation 24). Reaction of ynolates with acyl chlorides gives acylketenes, which can be isolated (equation 25),[26] or obtained as their dimers or reaction products with the ynolate (equation 26).[25] The ynolate **12** reacted with imines by a [2 + 2] cycloaddition pathway giving β-lactams.[25a]

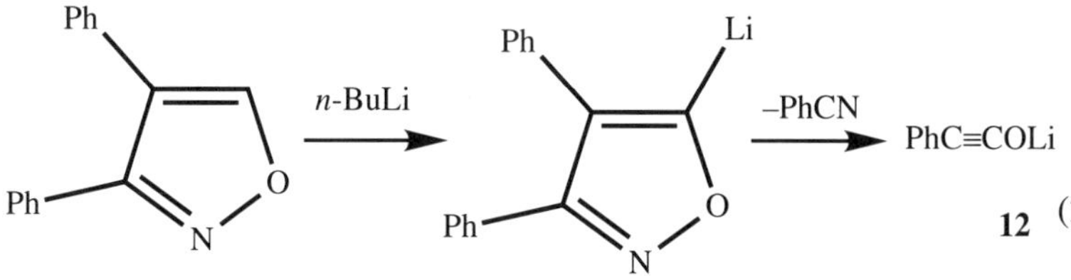

4.8 METAL-SUBSTITUTED KETENES

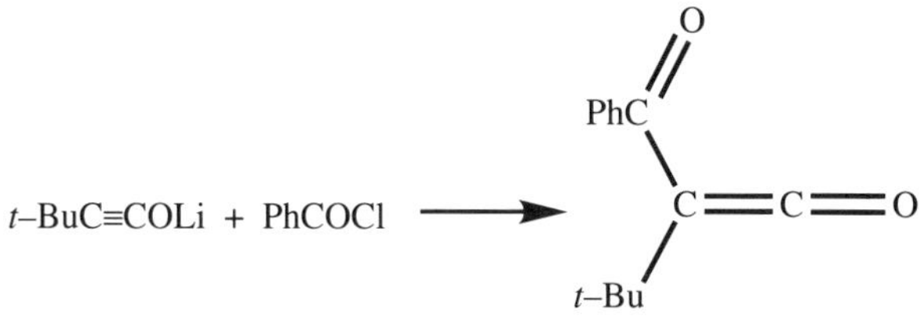

[Scheme showing 3,4-diphenyl-5-lithio-isoxazole formation from 3,4-diphenylisoxazole with n-BuLi, then loss of PhCN to give PhC≡COLi **12**] (23)[25]

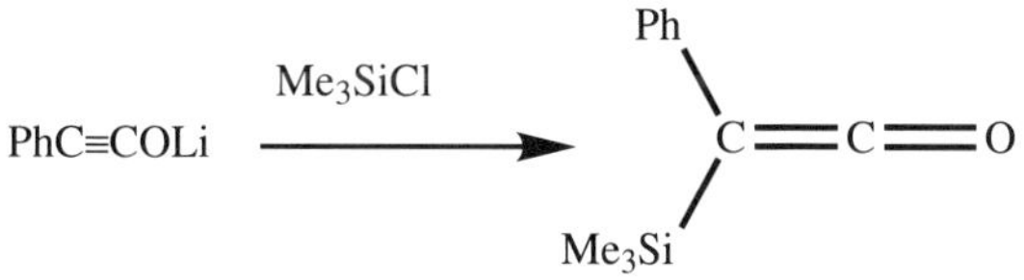

[Scheme: PhC≡COLi + Me₃SiCl → Ph(Me₃Si)C=C=O] (24)[25]

[Scheme: t-BuC≡COLi + PhCOCl → PhC(O)(t-Bu)C=C=O (53%)] (25)[26]

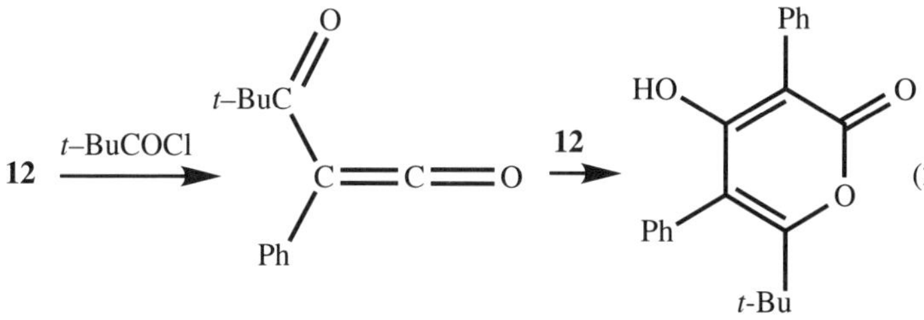

[Scheme: **12** + t-BuCOCl → t-BuC(O)(Ph)C=C=O → 4-hydroxy-3,6-diphenyl-6-t-butyl-2H-pyran-2-one (28%)] (26)[25]

Reaction of alkynyl tosylates with MeLi gave the ynolates, which undergo silylation on oxygen with t-BuMe₂SiCl (equation 27).[27] Attack on carbon of these ynolates occurs with Et₃GeCl or n-Bu₃SnCl to give new ketenes.[27]

$$\text{RC≡COTs} \xrightarrow{2\text{MeLi}} \text{RC≡COLi} \xrightarrow{t\text{-BuMe}_2\text{SiCl}} \text{RC≡COSiMe}_2\text{Bu-}t \quad (27)$$

Cleavage of triisopropylsilyl and tert-butyldimethylsilyl ynol ethers also gives ynolates, and these were also resilylated on oxygen (equation 28).[28]

$$\text{RC≡COSi(Pr-}i)_3 \xrightarrow{\text{MeLi}} \text{RC≡COLi} \xrightarrow{t\text{-BuMe}_2\text{SiCl}} \text{RC≡COSiMe}_2\text{Bu-}t \quad (28)$$

The neutral, basic, and enzymatic hydrolysis of alkynyl esters have been studied,[29–31] and these reactions may potentially involve ketene formation. The reaction of diethyl 1-hexynyl phosphate (**26**) with phosphotriesterase was found to inactivate the enzyme, and this was attributed to hydrolysis of **26** to ketene **27**, which then acylated and deactivated the enzyme (equation 29).[29]

$$n\text{-BuC}\equiv\text{COPO}_3\text{Et}_2 \xrightarrow[\text{(ENZ)X}]{\text{H}_2\text{O}} n\text{-BuCH=C=O} \xrightarrow{\text{(ENZ)X}} n\text{-PnCOX(ENZ)} \quad (29)$$

26   **27**

Ynolates complexed with other metals are also known. Thus tungsten complexes containing the PhC≡CO⁻ ligand have been prepared,[32] as well as a vanadium complex of HC≡CO⁻.[33]

Anionic ketenyl species including ynolates are included in Table 1, Section 4.10.

References

1. von Baar, B.; Weiske, T.; Terlouw, J. K.; Schwarz, H. *Angew. Chem., Int. Ed. Engl.* **1986**, *25*, 282–284.
2. Hochstrasser, R.; Wirz, J. *Angew. Chem., Int. Ed. Engl.* **1989**, *28*, 181–183.
3. Chiang, Y.; Kresge, A. J.; Hochstrasser, R.; Wirz, J. *J. Am. Chem. Soc.* **1989**, *111*, 2355–2357.
4. Kresge, A. J. *Acc. Chem. Res.* **1990**, *23*, 43–48.
5. Smith, B. J.; Radom, L.; Kresge, A. J. *J. Am. Chem. Soc.* **1989**, *111*, 8297–8999.
6. Lloyd, J. R.; Agosta, W. C.; Field, F. H. *J. Org. Chem.* **1980**, *45*, 1614–1619.
7. Bowie, J. H.; Stringer, M. B.; Currie, G. J. *J. Chem. Soc., Perkin Trans. 2*, **1986**, 1821–1825.
8. Chowdhury, S.; Harrison, A. G. *Org. Mass Spectrom.* **1989**, *24*, 123–127.
9. Lee, J.; Grabowski, J. J. *Chem. Rev.* **1992**, *92*, 1611–1647.
10. Oakes, J. M.; Jones, M. E.; Bierbaum, V. M.; Ellison, G. B. *J. Phys. Chem.* **1983**, *87*, 4810–4815.
11. Robinson, M. S.; Depuy, C. H. *Abstracts of Papers, 205th National Meeting of the American Chemical Society;* Denver, CO; March–April, 1993; ORGN 105.
12. Lebedev, A. T.; Hayes, R. N.; Bowie, J. H. *J. Chem. Soc., Perkin Trans. 2*, **1991**, 1127–1129.
13. Gong, L.; McAllister, M. A.; Tidwell, T. T. *J. Am. Chem. Soc.* **1991**, *113*, 6021–6028.
14. Cioslowski, J. *J. Am. Chem. Soc.* **1990**, *112*, 6536–6538.
15. Büchner, W. *Helv. Chim. Acta* **1963**, *46*, 2111–2120.
16. Serratosa, F. *Acc. Chem. Res.* **1983**, *16*, 170–176.
17. Zecchina, A.; Coluccia, S.; Spoto, G.; Scarano, D.; Marchese, L. *J. Chem. Soc., Farad. Trans.* **1990**, *86*, 703–709.
18. Ayed, O.; Manceron, L.; Silvi, B. *J. Phys. Chem.* **1988**, *92*, 37–45.
19. Kowalski, C. J.; Fields, K. W. *J. Am. Chem. Soc.* **1982**, *104*, 321–323.
20. Kowalski, C. J.; Haque, M. S.; Fields, K. W. *J. Am. Chem. Soc.* **1985**, *107*, 1429–1430.
21. Kowalski, C. J.; Reddy, R. E. *J. Org. Chem.* **1992**, *57*, 7194–7208.

22. Reddy, R. E.; Kowalski, C. J. *Org. Synth.* **1992**, *71*, 146–157.
23. Kowalski, C. J.; Lal, G. S. *J. Am. Chem. Soc.* **1986**, *108*, 5356–5357.
24. Woodbury, R. P.; Long, N. R.; Rathke, M. W. *J. Org. Chem.* **1978**, *43*, 376.
25. Hoppe, I.; Schöllkopf, U. *Liebigs Ann. Chem.* **1979**, 219–226.
25a. Barratt, A. G. M.; Quayle, P. J. *Chem. Soc., Perkin Trans. 1*, **1982**, 2193–2196.
26. Zhdankin, V. K.; Stang, P. J. *Tetrahedron Lett.* **1993**, *34*, 1461–1462.
27. Stang, P. J.; Roberts, K. A. *J. Am. Chem. Soc.* **1986**, *108*, 7125–7127.
28. Kowalski, C. J.; Lal, G. S.; Haque, M. S. *J. Am. Chem. Soc.* **1986**, *108*, 7127–7128.
29. Blankenship, J. N.; Abu-Soud, H.; Francisco, W. A.; Raushel, F. M.; Fischer, D. R.; Stang, P. J. *J. Am. Chem. Soc.* **1991**, *113*, 8530–8561.
30. Allen, A. D.; Kitamura, T.; Roberts, K. A.; Stang, P. J.; Tidwell, T. T. *J. Am. Chem. Soc.* **1988**, *110*, 622–624.
31. Stang, P. J. *Acc. Chem. Res.* **1991**, *24*, 304–310.
32. Mayr, A.; McDermott, G. A.; Dorries, A. M.; Holder, A. K.; Fultz, W. C.; Rheingold, A. L. *J. Am. Chem. Soc.* **1986**, *108*, 310–311.
33. Jubb, J.; Gambarotta, S. *J. Am. Chem. Soc.* **1993**, *115*, 10410–10411.

4.8.2 Boron-Substituted Ketenes

The BH_2 group can act as a π-acceptor from the ketenyl moiety, as illustrated by the resonance structure **1a**. The importance of this effect is also indicated by the stabilization of **1** by 17 kcal/mol according to the isodesmic exchange reaction with an alkene.[1]

Boron-substituted ketenes are not abundant, and their stability suggests this is at least partly due to a lack of effort directed toward their preparation, and a lack of well-documented routes to their synthesis. Also, there may well be decomposition pathways with low barriers for these materials.

One route to boron-substituted ketenes is via a substitution reaction for tin on a silyl- or germyl-substituted ketene (equation 1).[2,3] A second route involves reaction of boron halides with alkynyl ethers, and is postulated to proceed via an intermediate addition product (equations 2 and 3).[4] A third route consists of the preparation of diaminoborylalkynyl ethers and their conversion to ketenes (equation 4).[5] There was IR evidence supporting the formation of the alkyne **5** and not a vinyl ether during the reaction of equation 4, but this was not proven.[5] Ketene **3** is also converted to **4** on reaction with 4 equivalents of Me_2NH.[4]

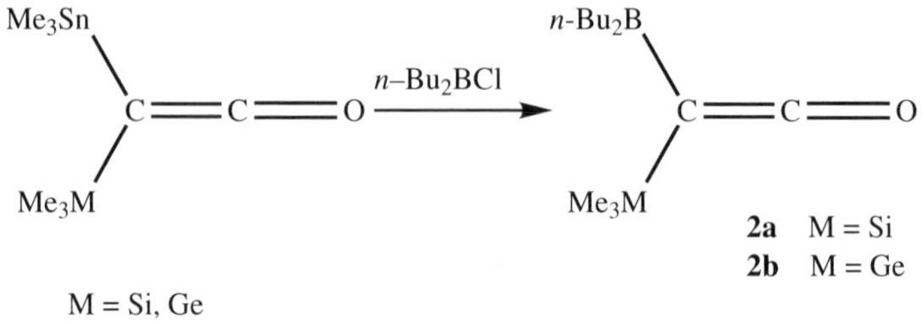

$$Me_3Sn\diagdown C=C=O \xrightarrow{n-Bu_2BCl} n\text{-}Bu_2B\diagdown C=C=O \quad (1)^{2,3}$$
$$Me_3M\diagup \qquad\qquad Me_3M\diagup$$

$$M = Si, Ge \qquad\qquad \textbf{2a}\ M = Si$$
$$\textbf{2b}\ M = Ge$$

$$Me_3SiC{\equiv}COEt + BX_3 \xrightarrow{-50°C} [Me_3SiC(BX_2){=}C(OEt)X] \longrightarrow \begin{matrix} X_2B\diagdown \\ C=C=O \\ Me_3Si\diagup \end{matrix} \quad (2)^4$$

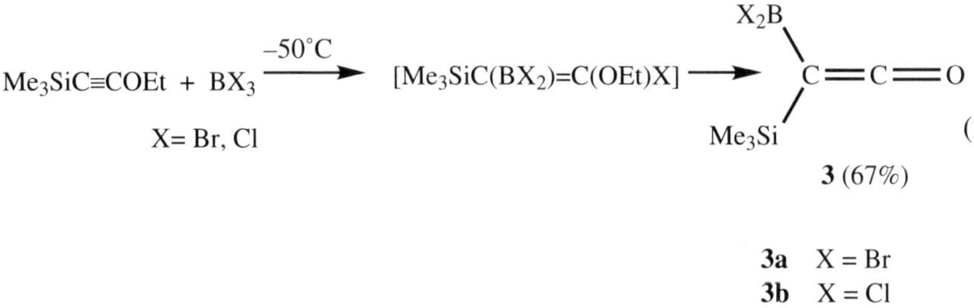

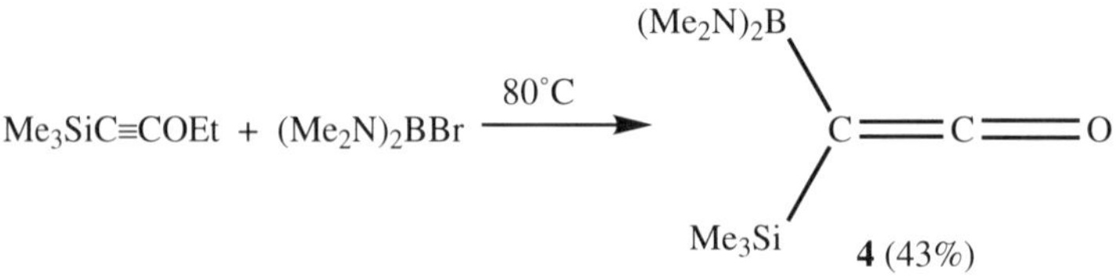

$$X = Br, Cl \qquad\qquad\qquad\qquad \textbf{3 (67\%)}$$

$$\textbf{3a}\ X = Br$$
$$\textbf{3b}\ X = Cl$$

$$Me_3SiC{\equiv}COEt + (Me_2N)_2BBr \xrightarrow{80°C} \begin{matrix} (Me_2N)_2B\diagdown \\ C=C=O \\ Me_3Si\diagup \end{matrix} \quad (3)^5$$

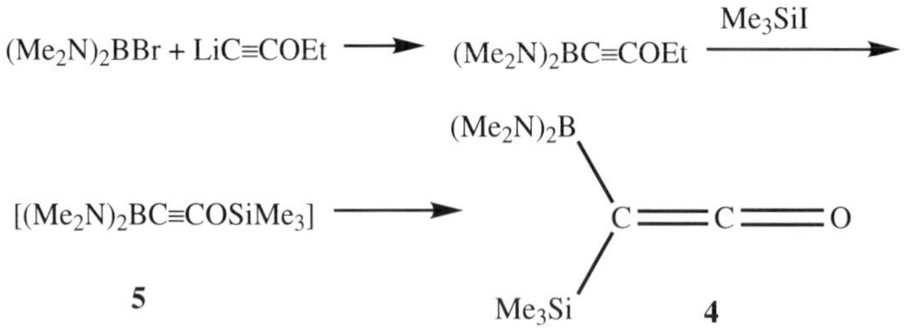

$$\textbf{4 (43\%)}$$

$$(Me_2N)_2BBr + LiC{\equiv}COEt \longrightarrow (Me_2N)_2BC{\equiv}COEt \xrightarrow{Me_3SiI}$$

$$[(Me_2N)_2BC{\equiv}COSiMe_3] \longrightarrow \begin{matrix} (Me_2N)_2B\diagdown \\ C=C=O \\ Me_3Si\diagup \end{matrix} \quad (4)^5$$

$$\textbf{5} \qquad\qquad\qquad \textbf{4}$$

These boron-substituted ketenes react readily with MeOH with cleavage of the B—C bond (equation 5).[2] Ketene **4** reacts similarly with MeOH but does not react with Et_2NH.[4] With Me_2NH **4** reacts as in equation 6.[6] Ketene **2** is unstable at room temperature, but **4** is long lived in the absence of air.[4]

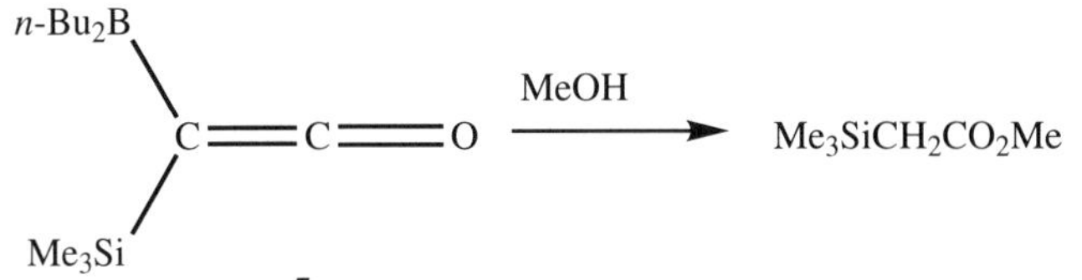

$$\underset{\underset{Me_3Si}{\big/}}{\overset{\overset{n\text{-}Bu_2B}{\big\backslash}}{C}}{=}C{=}O \quad \xrightarrow{MeOH} \quad Me_3SiCH_2CO_2Me \qquad (5)^2$$

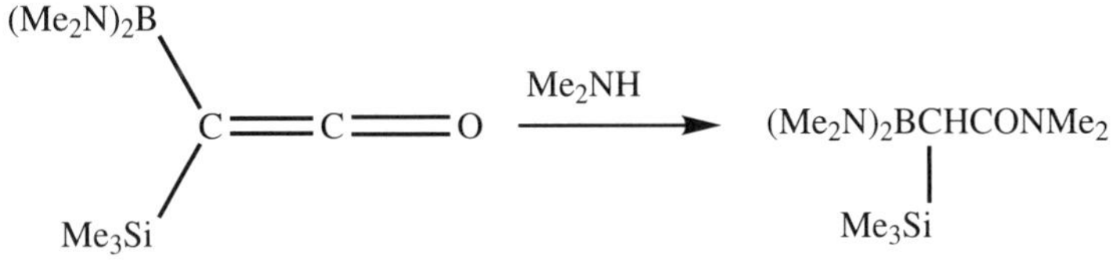

5

$$\underset{\underset{Me_3Si}{\big/}}{\overset{\overset{(Me_2N)_2B}{\big\backslash}}{C}}{=}C{=}O \quad \xrightarrow{Me_2NH} \quad \underset{\underset{Me_3Si}{|}}{(Me_2N)_2BCHCONMe_2} \qquad (6)$$

4

$$\xrightarrow{\Delta} (Me_2N)_2BCH{=}C(OSiMe_3)NMe_2 + Me_3SiCH{=}C(NMe_2)OB(NMe_2)_2$$

E/Z

The ketenes formed by all these sequences also have silyl or germyl substituents which are known to be stabilizing to ketenes in their own right. Thus the possible existence of ketenes bearing only boron substituents as the only heteroatom group cannot be evaluated from these results.

Wolff rearrangement of 3-*o*-carboranyl diazomethyl ketone (**6**) with Ag$_2$O in aqueous dioxane at 70 °C evidently gave the boron-substituted ketene **7**, which led to the carboxylic acid **8**, isolated in 33% yield (equation 7).[7]

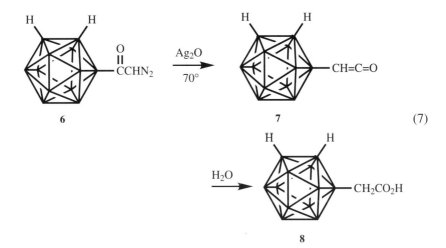

(7)

TABLE 1. Boron-Substituted Ketenes

Structure	Formula	Preparation	Registry No.	Ref.
$H_2BCH=C=O$	C_2H_3BO	Theory	134736-41-7	1
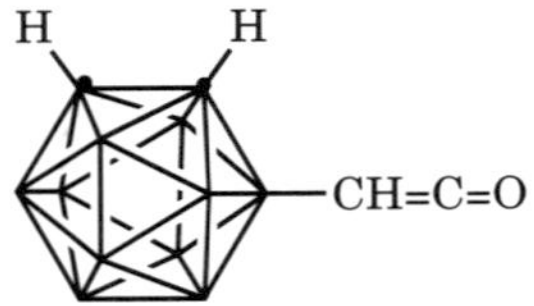 (carborane)-CH=C=O	$C_4H_{12}B_{10}C$	Wolff		7
$Br_2BC(SiMe_3)=C=O$	$C_5H_9BBr_2OSi$	Addition/elimination	94363-48-1	4
$Cl_2BC(SiMe_3)=C=O$	$C_5H_9BCl_2OSi$	Addition/elimination	94317-96-1	4
$Br_2BC(SiEt_3)=C=O$	$C_8H_{15}BBr_2OSi$	Addition/elimination	94317-97-2	4
$(Me_2N)_2BC(SiMe_3)=C=O$	$C_9H_{21}BN_2OSi$	Addition/elimination	94317-98-3	4
$(Me_2N)_2BC(SiEt_3)=C=O$	$C_{12}H_{27}BN_2OSi$	Substitution	105551-54-0	6
$n\text{-}Bu_2BC(GeMe_3)=C=O$	$C_{13}H_{27}BGeO$	Substitution	37590-96-8	2
$n\text{-}Bu_2BC(SiMe_3)=C=O$	$C_{13}H_{27}BOSi$	Substitution	37499-08-4	2
$(Et_2N)_2BC(SiMe_3)=C=O$	$C_{13}H_{29}BN_2OSi$	Addition/elimination	94317-99-4	4

References

1. Gong, L.; McAllister, M. A.; Tidwell, T. T. *J. Am. Chem. Soc.* **1991**, *113*, 6021–6028.
2. Ponomarev, S. V.; Erman, M. B.; Gervits, L. L. *Zh. Obshch. Khim.* **1972**, *42*, 469–470; *Engl. Transl.* **1972**, *42*, 463.
3. Ponomarev, S. V. *Angew. Chem., Int. Ed. Engl.* **1973**, *12*, 675.
4. Ponomarev, S. V.; Nikolaeva, S. N.; Molchanova, G. N.; Kostyuk, A. S.; Grishin, Yu. K. *Zh. Obshch. Khim.* **1984**, *54*, 1817–1821; *Engl. Transl.* **1984**, *54*, 1620–1623.
5. Ponomarev, S. V.; Gromova, E. M.; Nikolaeva, S. N.; Zolotareva, A. S. *Zh. Obshch. Khim.* **1989**, *59*, 2277–2282; *Engl. Transl.* **1989**, *59*, 2040–2044.
6. Ponomarev, S. V.; Nikolaeva, S. N.; Kostyuk, A. S.; Grishin, Yu. K. *Zh. Obshch. Khim.* **1985**, *55*, 2801–2802; *Engl. Transl.* **1985**, *55*, 2493–2494.
7. Zakharkin, C. I.; Kalinin, V. N.; Gedymin, V. V. *Tetrahedron* **1971**, *27*, 1317–1322.

4.8.3 Other Metal-Substituted Ketenes and Metal Ketenides

Ketenes with electropositive substituents are predicted to be stabilized, and this has been demonstrated experimentally in the case of silicon-, germanium-, and tin-substituted ketenes, as discussed in Section 4.5. Lithium substitution is considered in Section 4.8.1. Boron substitution is predicted to be even more stabilizing, and while several members of this family are known and discussed in Section 4.8.2, the effect of the boron substituents cannot be differentiated from the other substituents present.

Ketenes with metal substituents that have been studied theoretically[1] include RCH=C=O with R = Li, Na, BeH, MgH, and AlH$_2$. In the future some of these may be amenable to experimental study if the reactivity of the carbon–metal bonds can be suppressed. Gas phase experiments may be useful in this regard.

Metal ketenides of the type M$_2$C=C=O where M = Ag, Cu, Au, and Hg have been studied experimentally.[2–8] Thus silver salts react with ketene to give the ketenides, or the ketene can be generated in situ by reaction of acetic anhydride, silver acetate, and triethylamine (equation 1).[6,7]

$$(CH_3CO)_2O \xrightarrow{Et_3N} CH_2=C=O \xrightarrow[Et_3N]{Ag^+} Ag_2C=C=O \quad (1)$$

The yellow monomeric Ag$_2$C=C=O is proposed to exist briefly in this reaction, although there is a strong tendency to form a polymeric precipitate which cannot be redissolved.[6,7] This reagent can be used in the synthesis of other ketenes (equation 2) in a process that has been little exploited. [9]

$$Ag_2C=C=O \xrightarrow{CF_3SeCl} (CF_3Se)_2C=C=O \quad (2)$$

Copper ketenide is prepared as in equation 3,[8] mercury derivatives as in equation 4,[3] and gold compounds as in equation 5, using (1,4-oxathian)gold(I), 2,6-dimethylpyridine, and Et_3N.[4,5]

$$CH_2=C=O \xrightarrow[CH_3CN, 25°C]{CuO_2CCF_3} Cu_2C=C=O \quad (3)$$

$$(CH_3CO)_2O \xrightarrow{Hg(OAc)_2} (AcOHg)_2C=C-O \quad (4)$$

$$CH_2=C=O + AuCl(C_4H_8OS) \longrightarrow Au_2C=C=O \quad (5)$$

The various metal ketenides display typical ketenyl IR bands between 2015[4] and 2070 cm^{-1}.[3]

Reaction of $[Fe_3(CO)_{11}]^{2-}$ with acetyl chloride produces a ketenylidene **1** (equation 6).[10] The three iron atoms are equivalently bonded to the ketenylidene moiety in this structure. Analogous cobalt and osmium compounds were also prepared.[10] Reported metal ketenides are listed in Table 1.

$$Fe_3(CO)_{11}{}^{2-} \xrightarrow{CH_3COCl} [(OC)_3Fe]_3C=C=O \quad (6)$$

1

TABLE 1. Metal Ketenides

Ketene	Formula	Preparation	Registry No.	Ref.
$Ag_2C=C=O$	C_2Ag_2O	Metallation	27378-01-4	7,8
$Au_2C=C=O$	C_2Au_2O	Metallation	54086-41-8	4
$Li_2C=C=O$	C_2Li_2O	Theory	69974-20-5	2
$Cu_2C=C=O$	C_2Cu_2O	Metallation	50869-69-7	8
$LiCH=C=O$	C_2HLiO	Theory	134736-39-3	1
$NaCH=C=O$	C_2HNaO	Theory	134736-49-5	1
$HBeCH=C=O$	C_2H_2BeO	Theory	134736-40-6	1
$HMgCH=C=O$	C_2H_2MgO	Theory	134736-42-8	1
$H_2AlCH=C=O$	C_2H_3AlO	Theory	134736-43-9	1
$AcOHgC(HgClO_4)=C=O$	$C_4H_3ClHg_2O_7$	Metallation	73399-68-5	3
$AcOHg_2C(Hg_2ClO_4)=C=O$	$C_4H_3ClHg_4O_7$	Metallation	73399-71-0	3
$AcOHgC(HgONO_2)=C=O$	$C_4H_3Hg_2NO_6$	Metallation	73399-67-4	3
$AcOHg_2C(Hg_2ONO_2)=C=O$	$C_4H_3Hg_4NO_6$	Metallation	73399-70-1	3
$(AcOHg)_2C=C=O$	$C_6H_6Hg_2O_5$	Metallation	73399-66-3	3
$(AcOHg_2)_2C=C=O$	$C_6H_6Hg_4O_5$	Metallation	73399-69-6	3
 ![pyridine]NAuAuC=C=O	$C_7H_5Au_2NO$	Metallation	52150-42-2	5
 ![pyridine]NAgAgC=C=O	$C_7H_5Ag_2NO$	Metallation	52150-43-3	5
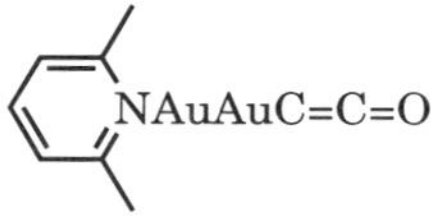 ![dimethylpyridine]NAuAuC=C=O	$C_9H_9Au_2NO$	Metallation	54699-75-1	4

References

1. Gong, L.; McAllister, M. A.; Tidwell, T. T. *J. Am. Chem. Soc.* **1991**, *113*, 6021–6028.
2. Hoppe, I.; Schöllkopf, U. *Liebigs Ann. Chem.* **1979**, 219–226.
3. Blues, E. T.; Bryce-Smith, D.; Karimpour, H. *J. Chem. Soc., Chem. Commun.* **1979**, 1043–1044.
4. Blues, E. T.; Bryce-Smith, D.; Lawston, I. W.; Wall, G. D. *J. Chem. Soc., Chem. Commun.* **1974**, 513–514.
5. Bryce-Smith, D.; Blues, E. T. *Ger. Offen.* DE 2,336,396; *Chem. Abstr.* **1974**, *80*, 108672m.
6. Bryce-Smith, D. *Chem. Ind. (London)* **1975**, 155–158.
7. Blues, E. T.; Bryce-Smith, D.; Shaoul, R.; Hirsch, H.; Simons, M. J. *J. Chem. Soc., Perkin Trans. 2*, **1993**, 1631–1642.
8. Blues, E. T.; Bryce-Smith, D.; Kettlewell, B.; Roy, M. *J. Chem. Soc., Chem. Commun.* **1973**, 921.
9. Haas, A.; Praas, H.-W. *Chem. Ber.* **1992**, *125*, 571–579.
10. Jensen, M. P.; Shriver, D. F. *J. Mol. Catal.* **1992**, *74*, 73–84.

4.8.4 Metal-Complexed Ketenes

Ketene complexes with metals exist in a variety of forms, with the metal attached by either σ or π bonding.[1,2] Such complexes can be formed by direct addition (equation 1),[3] carbonylation of metal ketenes (equation 2),[4] and other routes.[1,2] Sometimes the bonding involves coordination to other ketene substituents, as in the alkenylketene complex **1** (vide infra). This chemistry has been extensively reviewed[1,2] and only a few recent examples are mentioned here.

$$(C_5H_5)_2V + Ph_2C=C=O \longrightarrow (C_5H_5)_2V(Ph_2C=C=O) \quad (1)^3$$

$$C_5H_5(CO)_2Mn=CPh_2 \xrightarrow[650\ atm]{CO} C_5H_5(CO)_2Mn(Ph_2C=C=O) \quad (2)^4$$

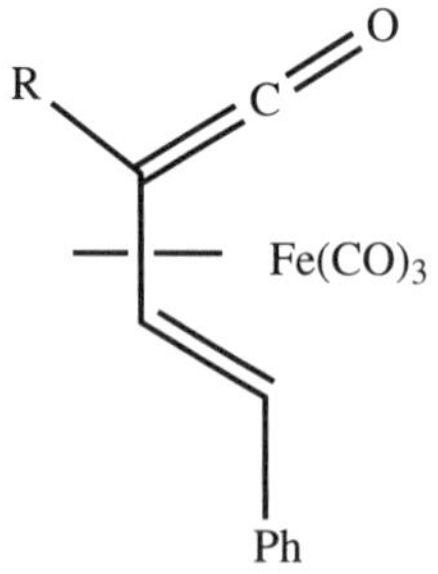

1

Many references to the preparation of ketenes from alkenylcarbene metal complexes are given in Section 3.5. For example, reaction of the chromium aminocarbene complex **2** with diphenylacetylene gave the enaminoketene complex **3** (equation 3),[5] while **4** gave **5** on heating (equation 4).[6]

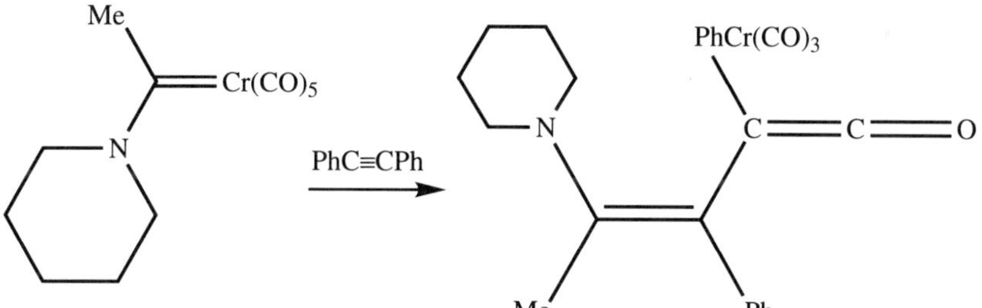

$$(3)^5$$

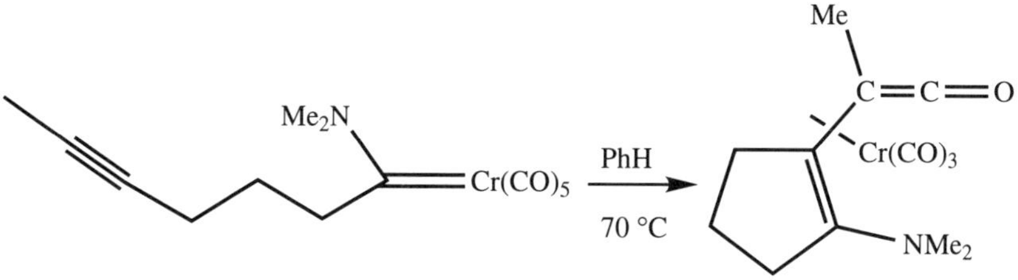

$$(4)^6$$

Vinylketenetricarbonyliron (0) complexes **1** reacted with nucleophiles including MeLi, LiCMe$_2$CO$_2$Et, PhCH$_2$NH$_2$, NaOMe, and t-BuSNa to form the addition products **6** (equation 5).[7] With isonitriles ketenimine complexes **7** were formed (equation 6),[8,9] after initial formation of isonitrile iron(0) complexes. Reactions of **1** with alkynes were also observed.[10,11,11a,11b]

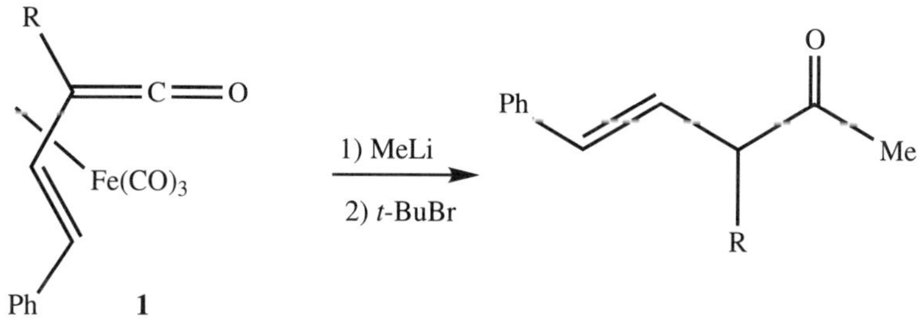

$$(5)^7$$

(6)[8,9]

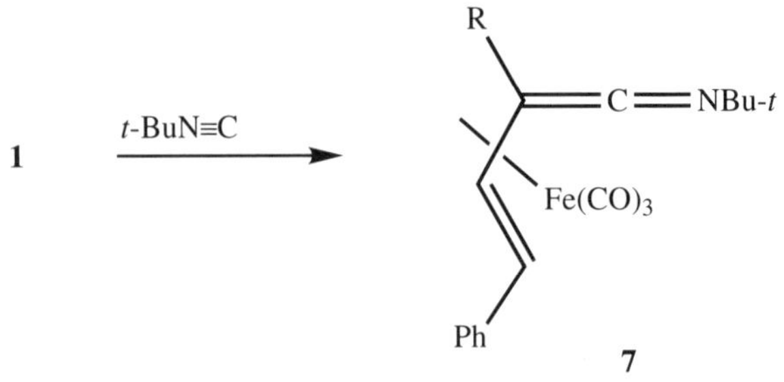

The cobalt complex **8** does not react with MeOH but with NaOMe gives **9** (equation 7).[12] The indenyl complex **10** gives phenols by cycloaddition to alkynes (equation 8),[13] and the complex **11** reacted with imines to give **12** (equation 9).[14] Chromium-complexed ketenes also give cyclizations on aromatic rings.[15]

(7)[12]

(8)[13]

In = indenyl

(9)[14]

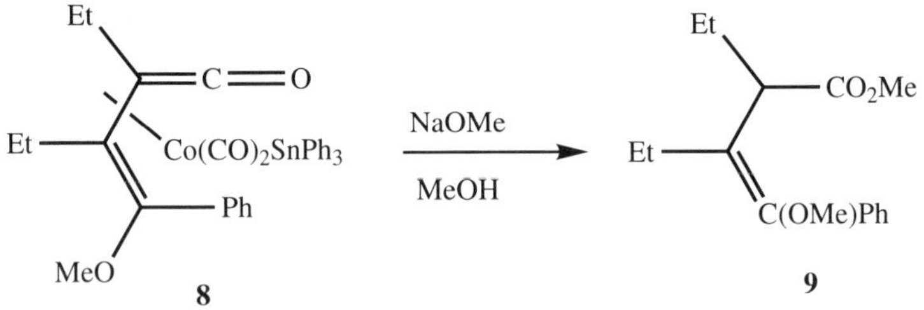

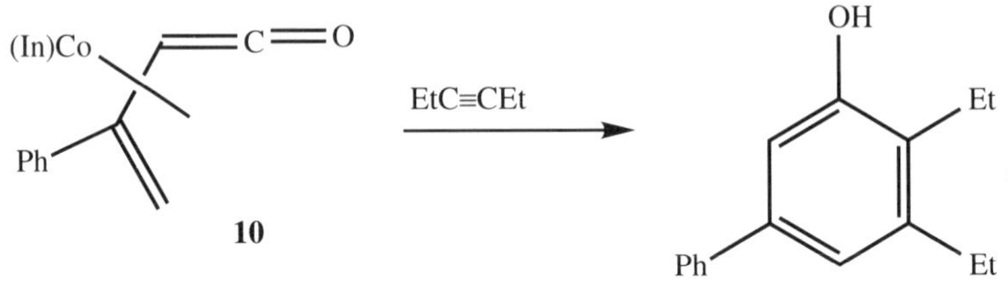

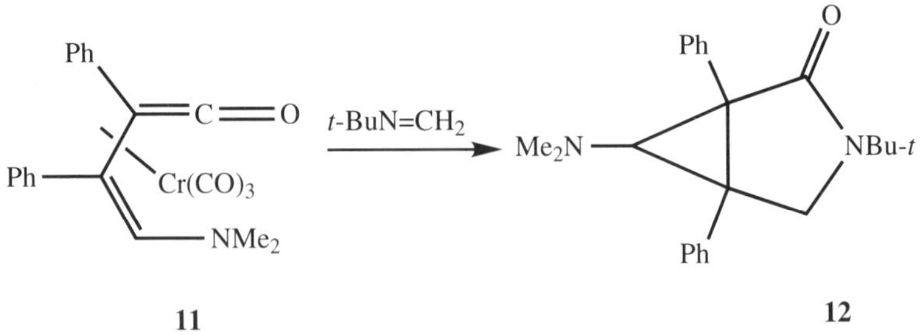

These examples are merely illustrative of the rich variety of chemistry of metal-complexed ketenes. For more complete discussions and further examples, reviews[1,2] and the original literature[3-15] should be consulted. Because of the large number and complexity of the various metal–ketene complexes cited in *Chemical Abstracts* these have not been compiled in this section.

References

1. Geoffroy, G. L.; Bassner, S. L. *Adv. Organomet. Chem.* **1988**, *28*, 1–83.
2. Jensen, M. P.; Shriver, D. F. *J. Mol. Catal.* **1992**, *74*, 73–84.
3. Gambarotta, S.; Pasquali, M.; Floriani, C.; Chiesi-Villa, A.; Guastini, C. *Inorg. Chem.* **1981**, *20*, 1173–1178.
4. Kelley, C.; Lugan, N.; Terry, M. R.; Geoffroy, G. L.; Haggerty, B. S.; Rheingold, A. L. *J. Am. Chem. Soc.* **1992**, *114*, 6735–6749.
5. Denise, B.; Goumont, R.; Parlier, A.; Rudler, H.; Daran, J. C.; Vaissermann, J. *J. Chem. Soc., Chem. Commun.* **1990**, 1238–1240.
6. Anderson, B. A.; Wulff, W. D.; Rheingold, A. L. *J. Am. Chem. Soc.* **1990**, *112*, 8615–8617.
7. Hill, L.; Richards, C. J.; Thomas, S. E. *J. Chem. Soc., Chem. Commun.* **1990**, 1085–1086.
8. Richards, C. J.; Thomas, S. E. *J. Chem. Soc., Chem. Commun.* **1990**, 307–309.
9. Hill, L.; Richards, C. J.; Saberi, S. P.; Thomas, S. E. *Pure Appl. Chem.* **1992**, *65*, 371–377.
10. Morris, K. G.; Saberi, S. P.; Slawin, A. M. Z.; Thomas, S. E.; Williams, S. E. *J. Chem. Soc., Chem. Commun.* **1992**, 1788–1791.
11. Morris, K. G.; Saberi, S. P.; Thomas, S. E. *J. Chem. Soc., Chem. Commun.* **1993**, 209–211.
11a. Morris, K. G.; Saberi, S. P.; Salter, M. M.; Thomas, S. E.; Ward, M. F.; Slawin, A. M. Z.; Williams, D. J. *Tetrahedron* **1993**, *49*, 5617–5634.
11b. Saberi, S. P.; Salter, M. M.; Slawin, A. M. Z.; Thomas, S. E.; Williams, D. J. *J. Chem. Soc., Perkin Trans. 1* **1994**, 167–171.
12. Wulff, W. D.; Gilbertson, S. R.; Springer, J. P. *J. Am. Chem. Soc.* **1986**, *108*, 520–522.
13. Huffman, M. A.; Liebeskind, L. S. *J. Am. Chem. Soc.* **1990**, *112*, 8617–8618.
14. Hegedus, L. S.; Miller, D. B., Jr. *J. Org. Chem.* **1989**, *54*, 1241–1244.
15. Merlic, C. A.; Xu, D.; Gladstone, B. G. *J. Org. Chem.* **1993**, *58*, 538–545.

4.9 BISKETENES

Bisketenes and polyketenes are molecules with two or more ketene groups, which may either be bonded directly or separated by other atoms. Cumulated bisketenes such as carbon suboxide ($O=C=C=C=O$) are considered in Section 4.11 on cumulenones.

TYPES OF KETENES

Published[1] ab initio calculations at the 6-31G*//6-31G* level for the planar forms of the parent bisketene, **1**, reveal this to be 1.6 kcal/mol more stable in the *transoid* form **1a** as compared to the *ciscoid* form **1b**, which is 8.5 kcal/mol less stable than cyclobutenedione **2**.[1] This bisketene is not conjugatively stabilized, as can be seen from the isodesmic energy changes of the reactions in equation 2, based on ab initio calculated energies of the species involved.[1] Further calculations[2] reveal that the most stable structure of this ketene is the twisted form **1c**, which is only 2.9 kcal/mol less stable than **2**. The preference for the twisted conformation **1c** confirms the destabilizing interaction between the two ketenyl groups.

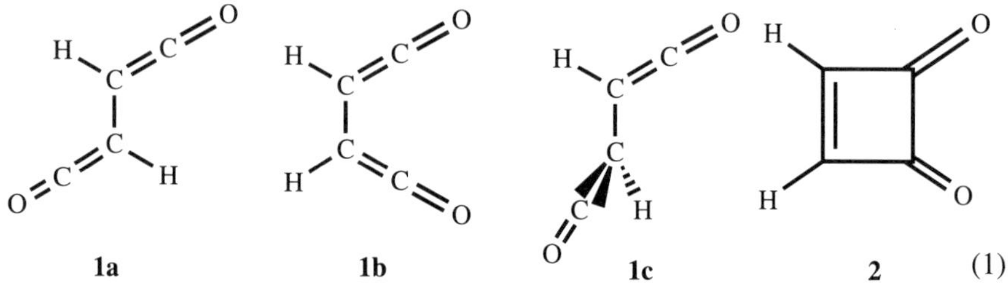

(1)

$$\Delta E = -11.9 \text{ kcal/mol}$$

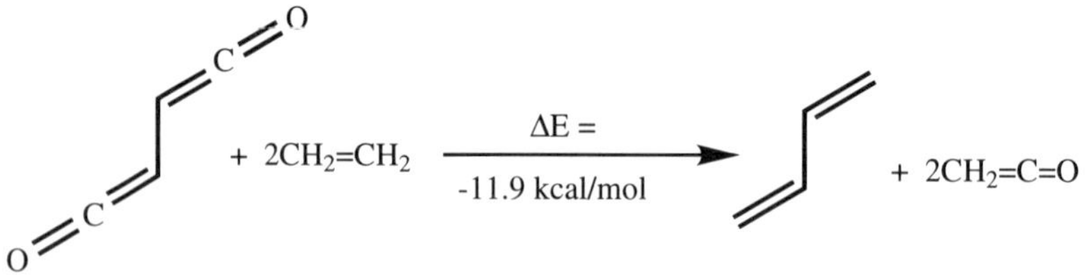

(2)

The preparation of **1** has been accomplished by independent routes. Pyrolysis of 4-cyclopentene-1,2,3-trione **3** led to **1b**, as evidenced by the formation of **2** in 9% yield (equation 3).[3] Photolysis of **4** in an argon matrix at 10 K gave a strong IR band at 2125 cm^{-1} attributed to **1b**.[4] The formation of **1** was proposed in the gas phase decomposition of the anion radical of 4-methoxy-2-furan carboxylic acid (equation 4).[5] The question of whether different conformers of **1** were present in these products was not addressed.[3–5]

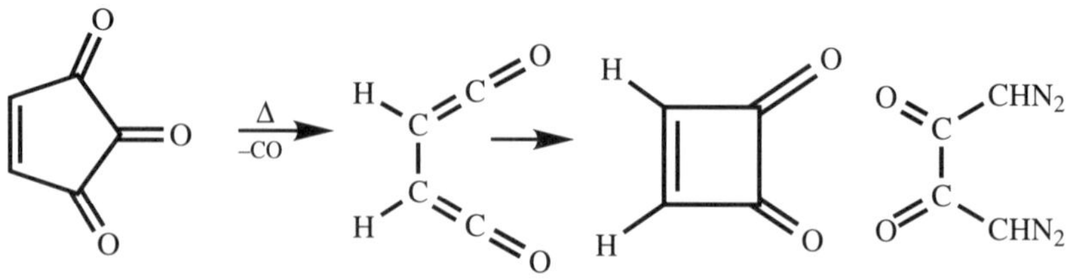

(3)[2]

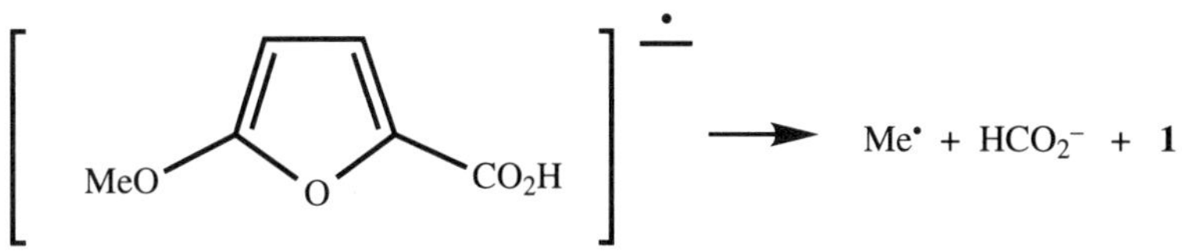

4.9 BISKETENES

$$[\text{MeO-furan-CO}_2\text{H}]^{\bullet-} \longrightarrow \text{Me}^{\bullet} + \text{HCO}_2^- + \mathbf{1} \qquad (4)$$

Formation of the α-hydroxy derivative **5** of **1** from hydroxycyclobutenedione (equation 5) has also been reported from photolysis in an Ar matrix.[6] The bisketene displayed IR bands at 2112 and 2135 cm^{-1}, and on further photolysis formed successively hydroxycyclopropenone, ethynol, and ketene (Section 4.8.1).

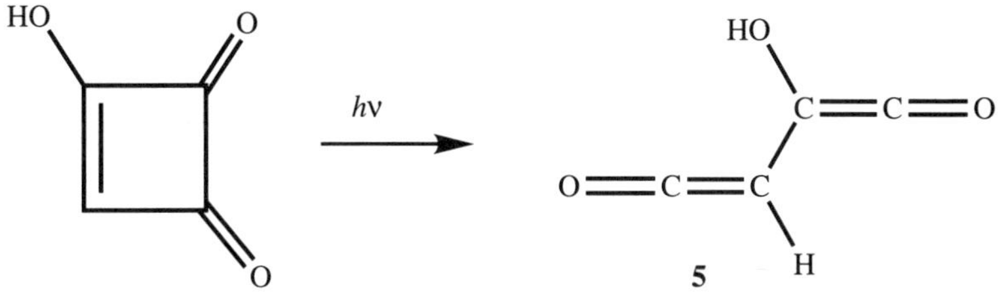

$$\text{hydroxycyclobutenedione} \xrightarrow{h\nu} \mathbf{5} \qquad (5)^6$$

The metal complexed bisketene **6** was obtained as a crystalline solid by a photochemical decarbonylation (equation 6), and its structure was established by X-ray crystallography.[7] In this structure the four carbons of the bisketene define a plane with the cobalt below the plane. The C—C—O angles are 137.8° and 138.5°, with the oxygens above the carbon plane and the carbonyl carbons substantially rehybridized from sp towards sp^2. Thus the bisketene character of the complex is not strong. Reaction of **6** with alkynes gives quinone complexes.[7]

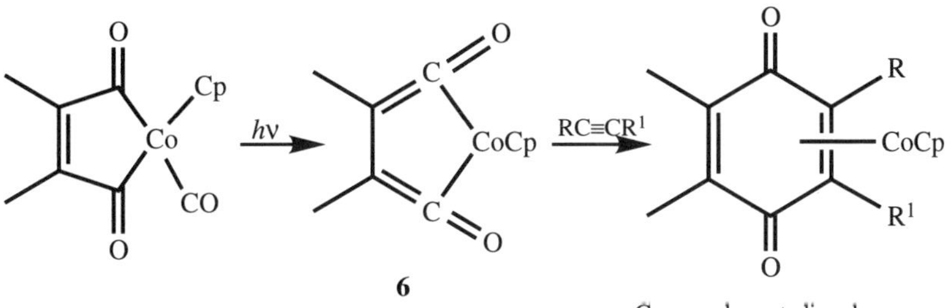

$$\xrightarrow{h\nu} \mathbf{6} \xrightarrow{RC\equiv CR^1} \text{quinone complex} \qquad (6)$$

Cp = cyclopentadienyl

Refluxing of 3-phenyl-1,2-cyclobutenedione (**7**) in CH$_3$OH led to the diester **9** among other products, and this may have occurred by ring opening to the bisketene **8** (equation 7).[8] Alternatively it was suggested that all of the products could arise by initial addition of CH$_3$OH to **7** and ring opening to a vinylketene (equations 8 and 9; see also Section 3.4.1).

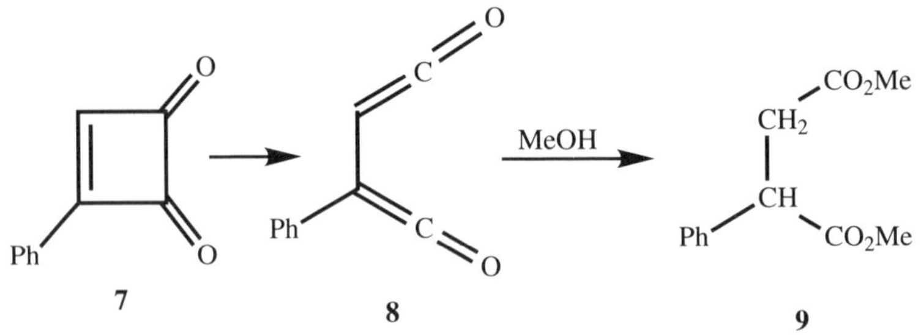

Photolysis of 3,4-diphenyl-1,2-cyclobutenedione (**11a**) in THF at −78 °C gave the bisketene **12a** as observed by its IR band at 2103 cm^{-1}, which persisted for several days at −78 °C (equation 10).[9-11] On standing in EtOH at room temperature **11a** was converted to the diester **13a**,[12] and this may have been through the bisketene formed thermally, or may have involved attack of EtOH on **11a**.

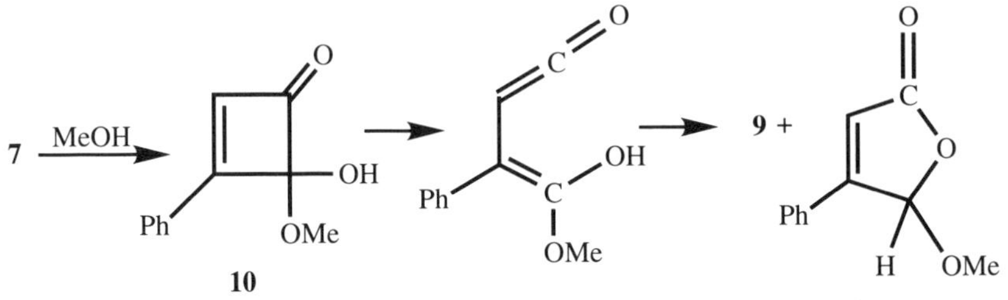

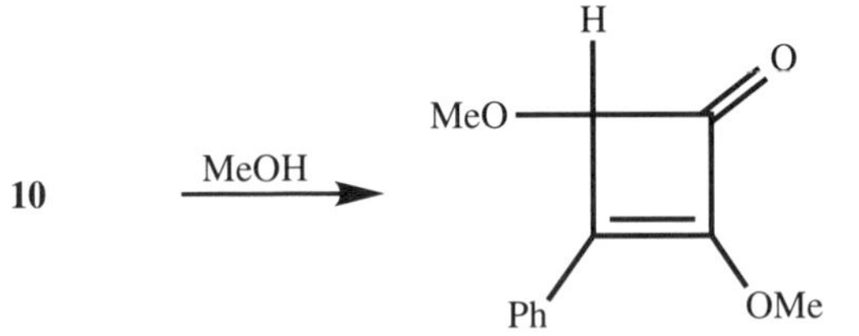

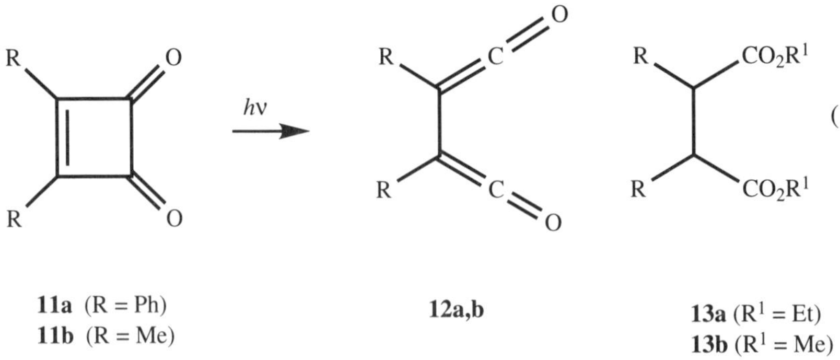

11a (R = Ph)
11b (R = Me)

12a,b

13a (R^1 = Et)
13b (R^1 = Me)

Irradiation of **11a** as a film at 77 K gave IR bands at 2100 and 2112 cm^{-1} attributed to **12a**.[11] Photolysis of **11b** at 77 K in MeOH gave IR bands at 2096 and 2117 cm^{-1} attributed to **12b,** and at -80 °C the decay of these bands and the concurrent formation of **13b** (R^1 = Me) was observed.[11]

The reaction of **12a** from **11a** with oxygen forms diphenylmaleic anhydride **14** (equation 11).[10] It was also found that photolysis of **11a** in the presence of O_2 led to **15** and **16**,[13] and that these latter two products were also formed from photolysis of **14**, which was suggested to form from **11a**. No mechanism was proposed for these reactions, which may involve reaction of either **11a** or **12a** with O_2. A mechanistic proposal for formation of substituted maleic anhydrides from oxidation of a bisketene is given in equation 19 (vide infra).

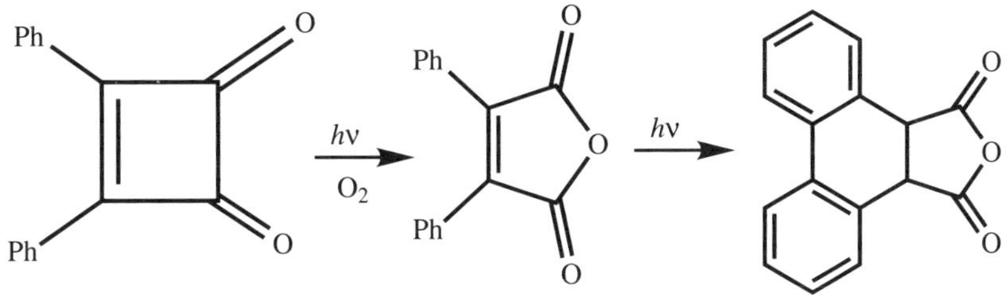

(11)

Photolysis of 3,4-diacetoxycyclobutene-1,2-dione **17** was proposed to give the bisketene **18** as an unobserved intermediate which formed the bisanhydride **19** (equation 12).[14] Photolysis of the diethoxy analogue **17a** in ether gave diethoxycyclopropenone by the proposed route shown (equation 12a).[14a] In ethanol the bisketene gave the diester.[14a]

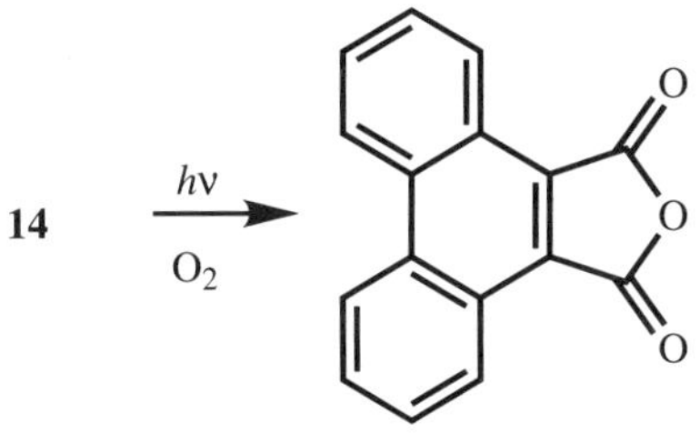

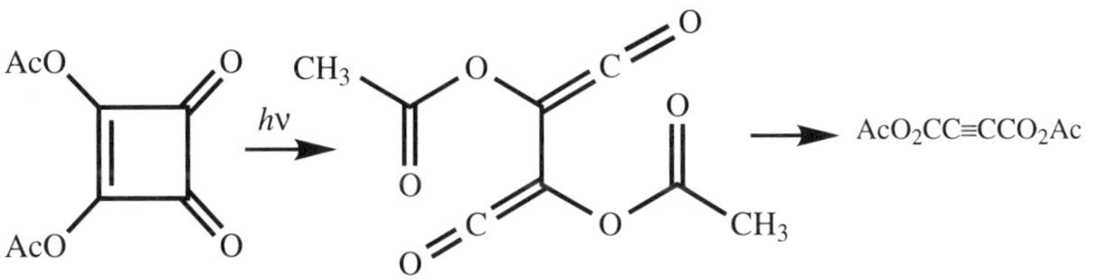

(12)[14]

$AcO_2CC{\equiv}CCO_2Ac$

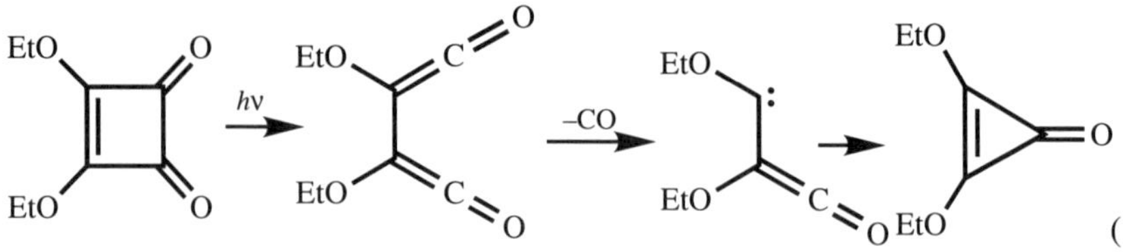

$$\text{(12a)}$$

Photolyses of cyclobutene-1,2-diones **7, 11a, 11b,** and **20** in the presence of cyclopentadiene were proposed to give bisketenes such as **21** which led to the adducts **22**.[15] This could occur by the route shown (equations 13 and 14). Alternatively the intermediate **24** could occur in the interconversion of **21** to **22**, and the formation of **22** from **23** could be by a concerted path. No evidence for the formation of oxacarbene intermediates such as **25** was found.[15] Other reactions of ketenylcyclobutanones such as **23** have been studied.[16]

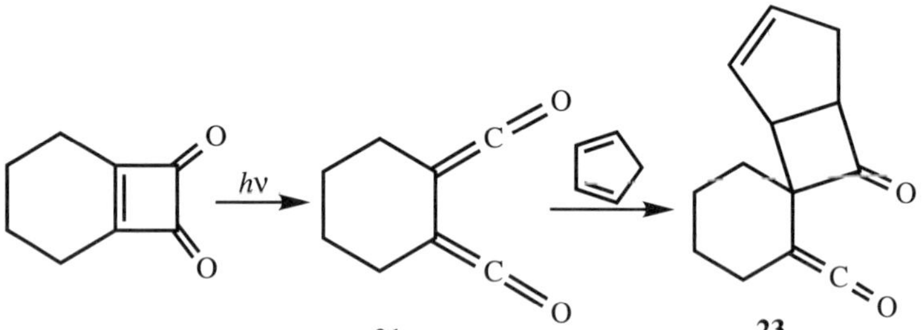

$$\text{(13)}$$

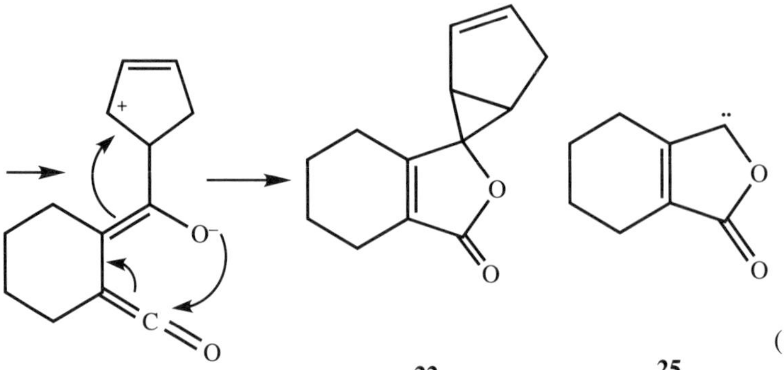

$$\text{(14)}$$

Thermolysis of the cyclobutenedione **26** at 720 K occurred with bisdecarbonylation, possibly through the bisketene **27**, and gave the alkyne **28** (equation 15).[17] Similar results were obtained for the bis(CH$_3$Se) and dichloro-substituted cyclobutenediones.[17]

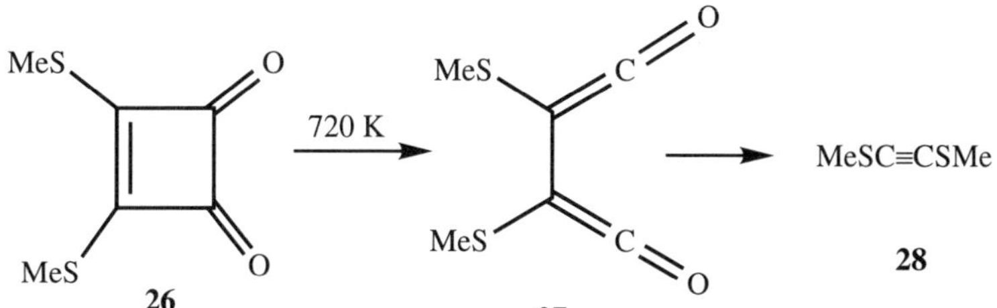

[Scheme for equation (15): compound 26 (MeS-substituted cyclobutenedione) → 720 K → 27 (bisketene) → 28 (MeSC≡CSMe)]

Flash pyrolysis at 650 °C of alkynyl-substituted cyclobutenediones (**29**)[18] leads to bisdecarbonylations and formation of new alkynes, and these reactions likely proceed through bisketene intermediates which undergo thermal decarbonylation (equation 16).[18] Irradiation at >338 nm of **30** in a matrix at 30 K gives quantitative generation of the hexaketene **31**, identified by its IR band at 2115 cm^{-1}.[19] Further irradiation of **31** at shorter wavelengths results in the disappearance of **31** and the formation of carbon monoxide, and perhaps the cyclic polyalkyne C_{18} (equation 17).[19] Laser desorption Fourier transform mass spectrometry of **30** and analogues with four and five cyclobutenedione rings produced cyclocarbon ions C_{18}^-, C_{24}^-, and C_{30}^-, respectively, and the corresponding positive ions were also observed.[20]

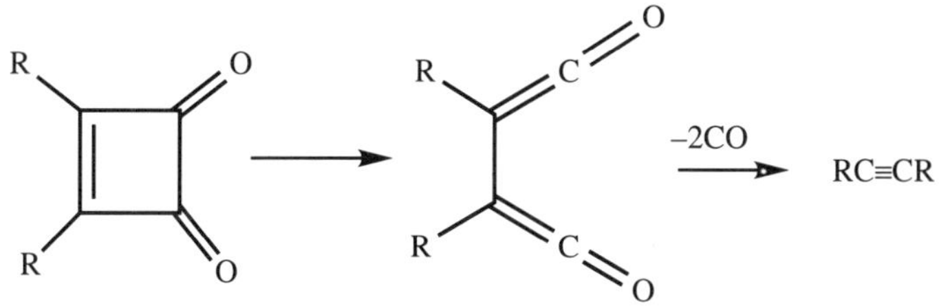

[Scheme for equation (16): compound 29 → bisketene intermediate → −2CO → RC≡CR]

29 R = PhC≡C, PhC≡CC≡C

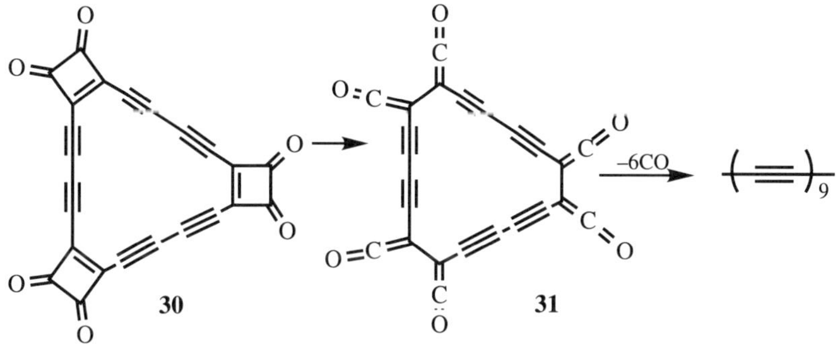

[Scheme for equation (17): compound 30 (cyclic trimer of cyclobutenedione linked by diynes) → 31 (hexaketene) → −6CO → (C≡C)$_9$ cyclic polyalkyne]

The calculated isodesmic ketene stabilizing energy of SiH_3 of 7.6 kcal/mol[1] and the 6.9 kcal/mol greater stability of **2** compared to **1a**[1] leads to the prediction[21] that suitable electropositive substituents could make the bisketene structure more stable than the cyclobutenedione. In confirmation of this prediction it has been found that 1,2-bis(trimethylsilyl)-3,4-cyclobutenedione **32** on heating or photolysis forms a quantitative yield of the bisketene **33** that is stable indefinitely in the absence of oxygen or water (equation 18).[21]

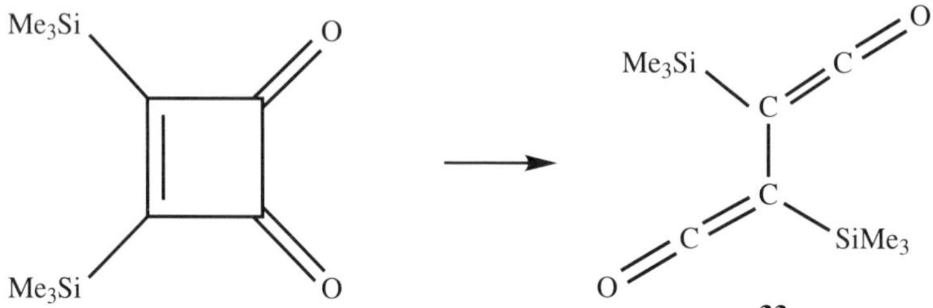

$$ \mathbf{32} \longrightarrow \mathbf{33} \tag{18} $$

The comparative reactivity of thermal ring opening of a series of cyclobutenes is shown in Table 1.[21] Thus the reactivity of **32** fits within the range of substituent effects on these compounds.

TABLE 1. Substituent Effects on E_{act} (kcal/mol) for Conversion of Cyclobutenes to Butadienes

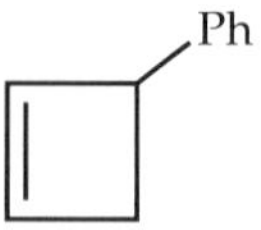

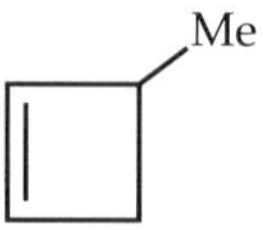

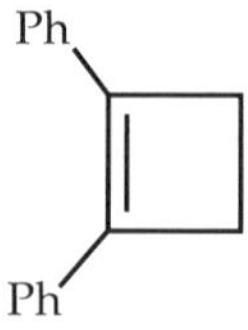

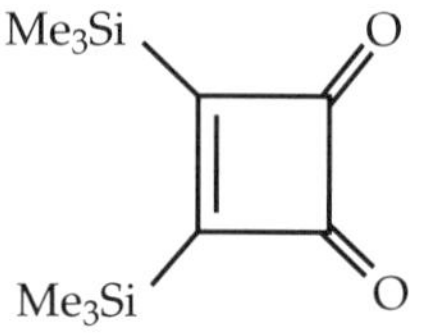

cyclobutene	Ph-cyclobutene	Me-cyclobutene	Ph,Ph-cyclobutene	Me₃Si,Me₃Si-cyclobutene	Me₃Si,Me₃Si-dione
32.9	26.0	31.6	32.0		29.7

The parent bisketene is predicted to have the nonplanar structure **1c**, and **33**, which has additional steric destablization of planar geometries, is calculated to have a nonplanar structure as well.[2] The reaction of **33** with O_2 gave the anhydride **34**, a process which can be explained by the route of equation 19.[21] The analogous formation of diphenylmaleic anhydride from the reaction of **12a** with oxygen has been reported (vide supra).[10]

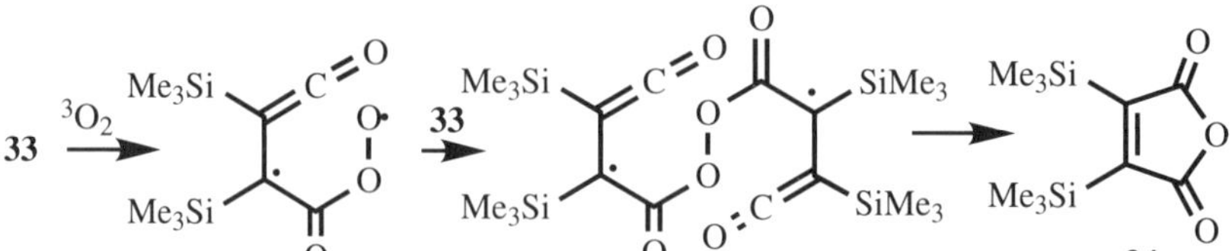

$$ \mathbf{33} \xrightarrow{^3O_2} \longrightarrow \mathbf{33} \longrightarrow \longrightarrow \mathbf{34} \tag{19} $$

4.9 BISKETENES

Water reacts with **33** to yield **35,** and this can be explained as shown in equation 20.[22] Both the *E* and *Z* isomers of **35** were isolated and the stereochemistry of both were proven from the X-ray crystal structures.[22]

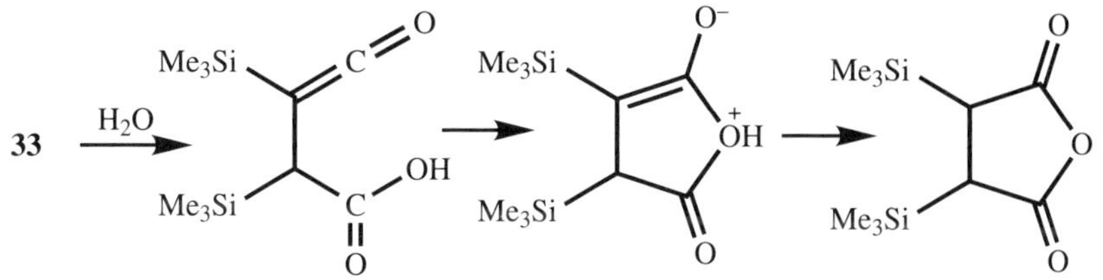

$$33 \xrightarrow{H_2O} \cdots \longrightarrow \cdots \longrightarrow 35 \quad (20)$$

The reaction of **33** with EtOH proceeded by an initial fast formation of the monoketene **36,** which could be isolated as a pure compound (equation 21). In a second slower step **37** was formed, resulting from desilylation and further addition (equation 21).[22]

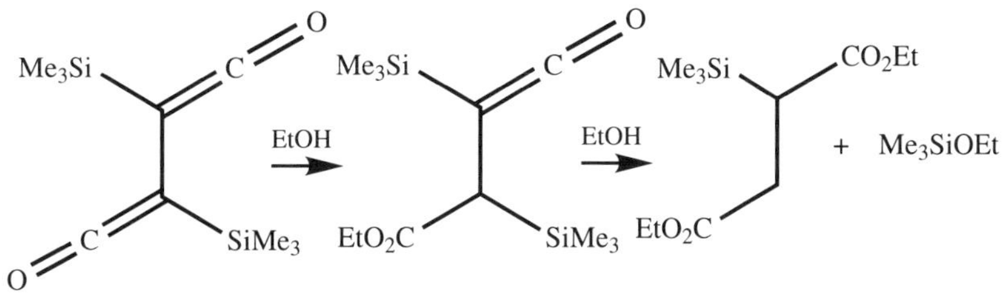

$$33 \xrightarrow{EtOH} 36 \xrightarrow{EtOH} 37 + Me_3SiOEt \quad (21)$$

The unsymmetrically substituted cyclobutenedione **38** gave bisketene **39** on photolysis, and the kinetics of the reversion of **39** to the more stable **38** were measured (equation 22).[23] The reaction of **39** with MeOH was much faster on the phenylketenyl moiety, yielding the monoketene **40.**

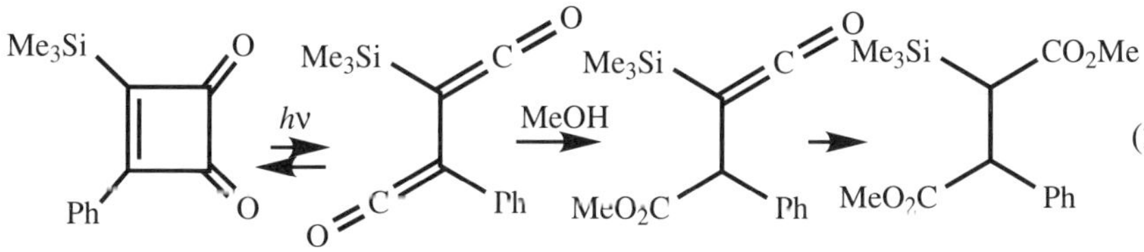

$$38 \xrightleftharpoons{h\nu} 39 \xrightarrow{MeOH} 40 \longrightarrow 41 \quad (22)$$

The presence of **39** in thermal equilibrium with **38** could be detected by both IR and ^{1}H NMR, and the percentage of **39** as measured by integration of the NMR signals varied from 0.7% at 100 °C to 2.2% at 161 °C.[23] An approximate calculation based on isodesmic stabilization energies (Section 1.1.3, Table 4) gives an estimate that Me$_3$Si and Ph substitution will stabilize monoketenes relative to alkenes by 5.0 kcal/mol (equation 23), and combined with the calculated greater stability of **2** relative to transoid **1b** of 6.9 kcal/mol it is predicted that **38** is 1.9 kcal/mol more stable than **39**.

$$\text{Me}_3\text{SiCH=C=O} + \text{PhCH=C=O} + 2\text{CH}_2\text{=CH}_2 \xrightarrow{\Delta E = 5.0 \text{ kcal/mol}}$$

$$\text{Me}_3\text{SiCH=CH}_2 + \text{PhCH=CH}_2 + 2\text{CH}_2\text{=C=O} \quad (23)$$

Benzocyclobutenedione (**42**) photolysis to form the bisketene **43** has been studied,[24–27] and **43** is reported to have a half life >100 μs at 25°, with UV λ_{max} 380 nm.[27] The reaction of **43** with dienophiles leads to [4 + 2] cycloadditions (equation 24).[24,27a] Pyrolysis of **44** also gave **43**, presumably by way of **45** (equation 25).[28]

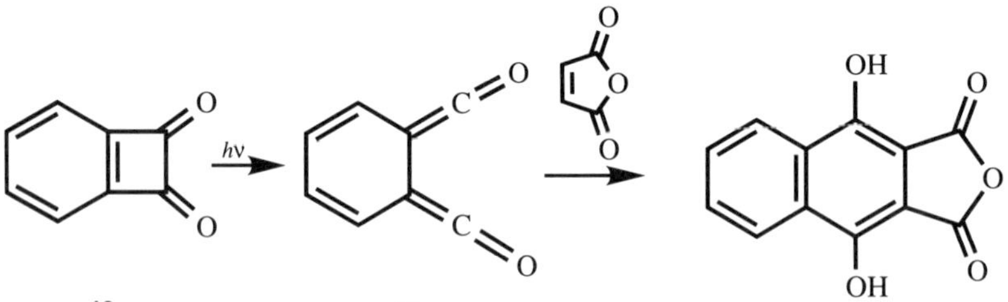

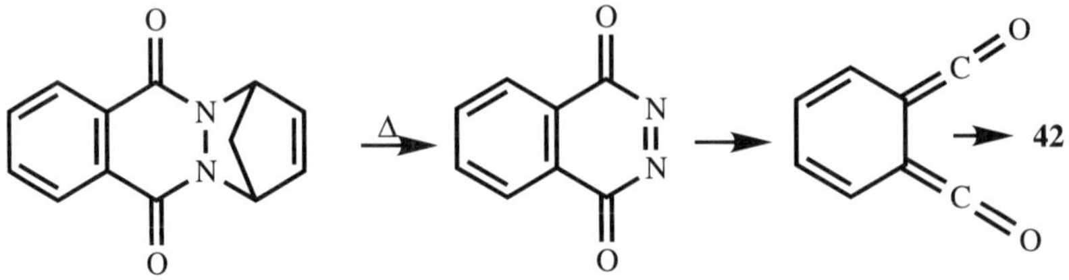

A reinvestigation of the photolysis of **42** in an Ar matrix at 11 K showed the formation of **43**, which displayed strong IR bands at 2138 and 2077 cm^{-1} in a 1:1 ratio ascribed to the coupled vibrations of the two ketenyl groups in **43**.[29] Further photolysis of **43** gives reformation of **42** and decarbonylation to benzyne, while the reaction of **43** with MeOH produced **46, 47,** and **48** in a ratio of 2.30/2.71/1.00.[29] The formation of **46** and **47** was ascribed to the paths of equations 26 and 27 respectively, while the diester **48** was proposed to result from an air oxidation. No evidence for the formation of **49** as an intermediate was found.[29] Other studies[15] have also argued against the frequently proposed (vide infra) formation of oxacarbenes such as **49**.

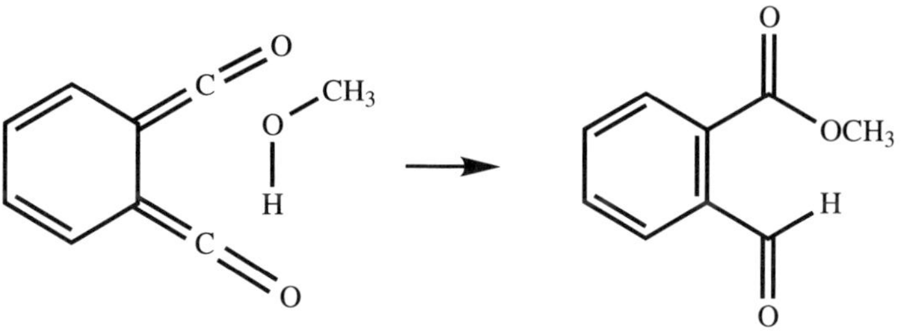

(26)

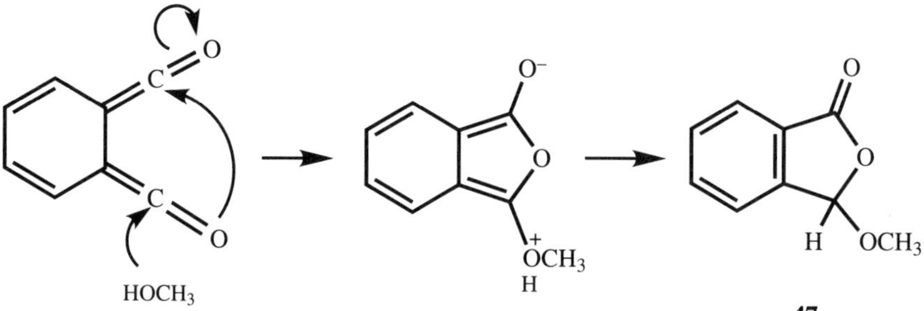

(27)

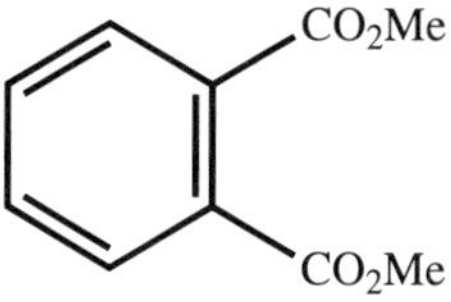

The reaction of acid anhydrides with phosphites has also been reported to lead to bisketene intermediates (equations 28–31), although the bisketenes have not actually been observed in these reactions.[30–32] As noted above, the formation of oxacarbenes from bisketenes has been questioned, and alternative routes to the observed products are conceivable.

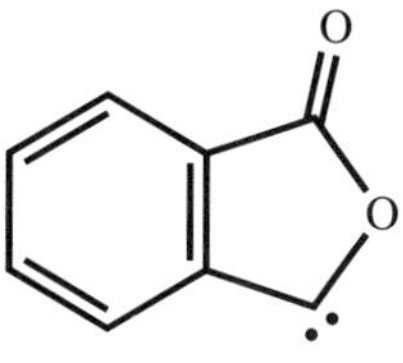

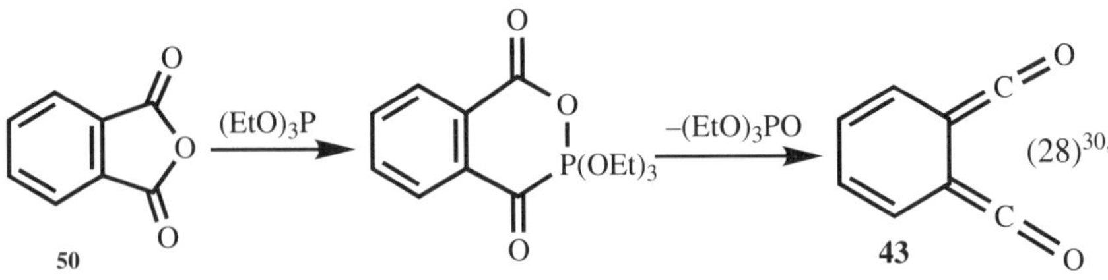

Dioxodibenzo-4-quinodimethane (**53**) was prepared by dehydrochlorination of the bisacyl chloride as a crystalline orange-red solid that was stable at room temperature (equation 32).[33] The bisketene had an IR band at 2083 cm^{-1}, and reacted rapidly with oxygen, water, air, and aniline.[33] The possible generation of **53a** and various analogues on electrochemical reduction of terephthaloyl halides has also been reported (equation 32a).[33a] These species were not directly observed but formed polymers.[33a]

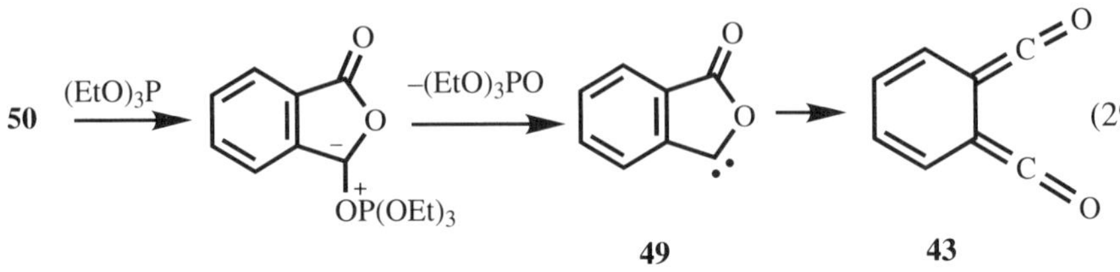

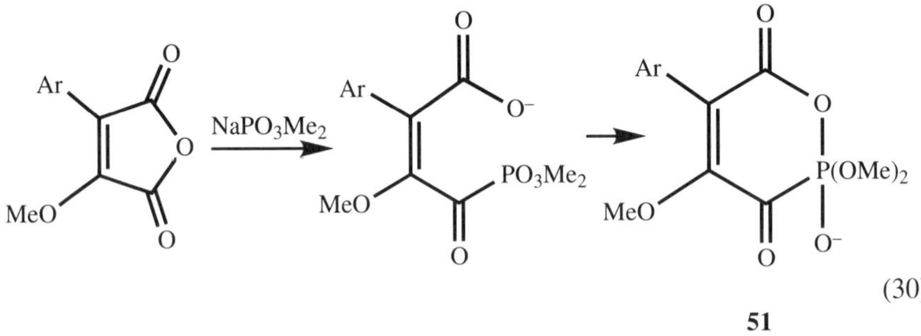

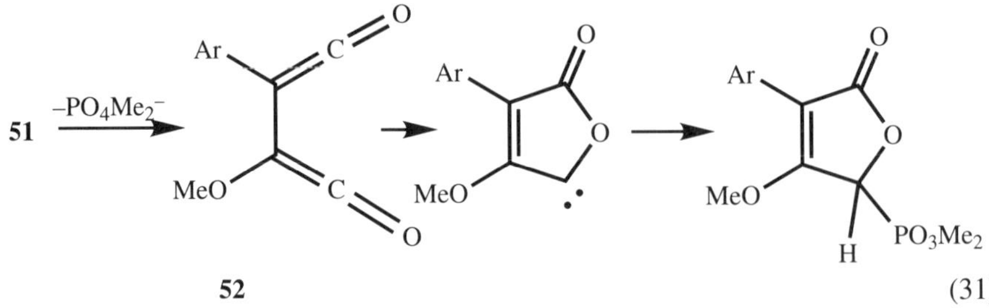

53

(32)[33]

53a

(32a)[33a]

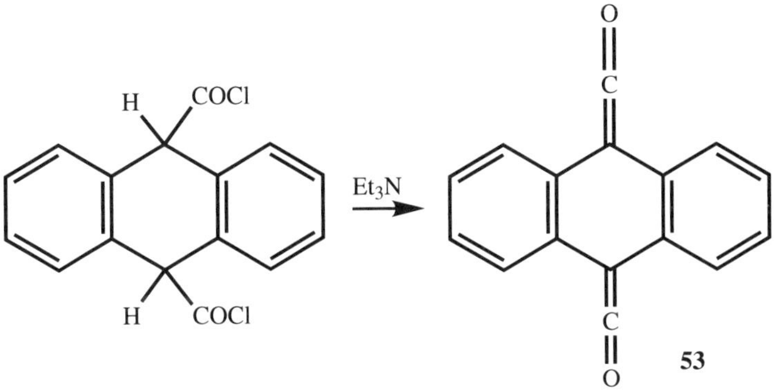

The photolysis of 4-*tert*-butyl-1,2-benzoquinone (**54**) in an Ar matrix at 8 K gave rise to IR bands at 2125 and 2156 cm^{-1}, assigned to the *syn* and *anti* rotamers **55** and **56** respectively (equation 33).[34] Photolysis of 1,2-benzoquinone at 10 K was proposed to give the unsubstituted parent of **55.**[34,35]

(33)[34]

54 **55** **56**

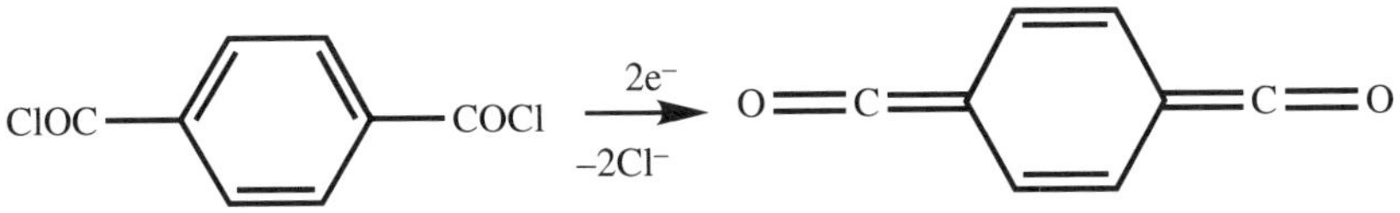

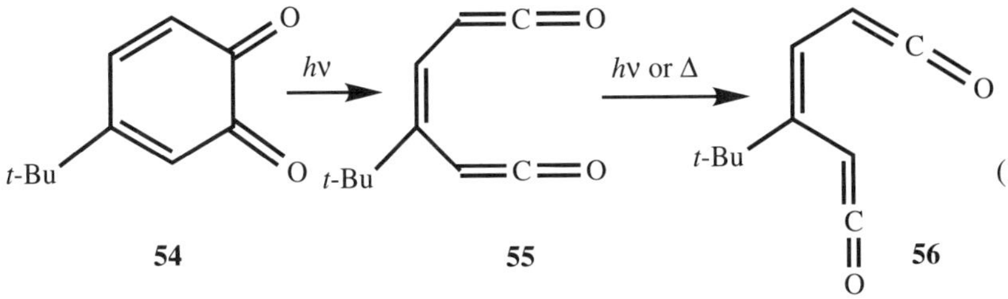

Dehydrochlorination of the acid chloride gave the bisketene **57** as a stable orange granular solid in 41% yield.[36] Other bisketenes prepared similarly include **58**[37] and **59.**[37]

57

58

59

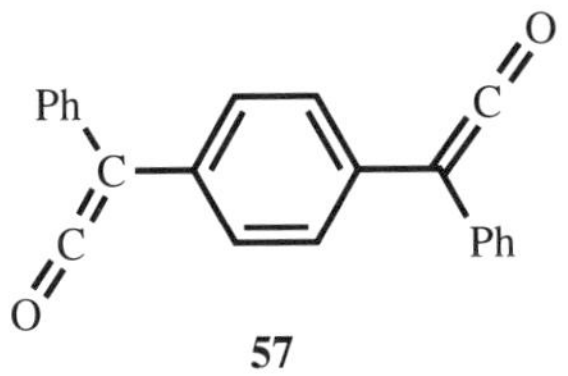

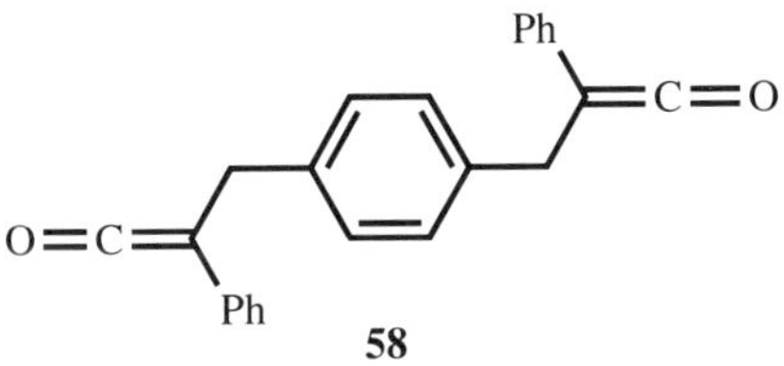

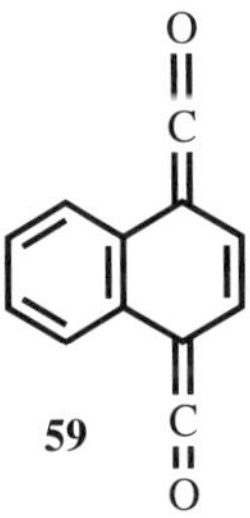

Photolysis of **60** in MeOH gave the diester **62**, perhaps by way of the bisketene **61** (equation 34).[38]

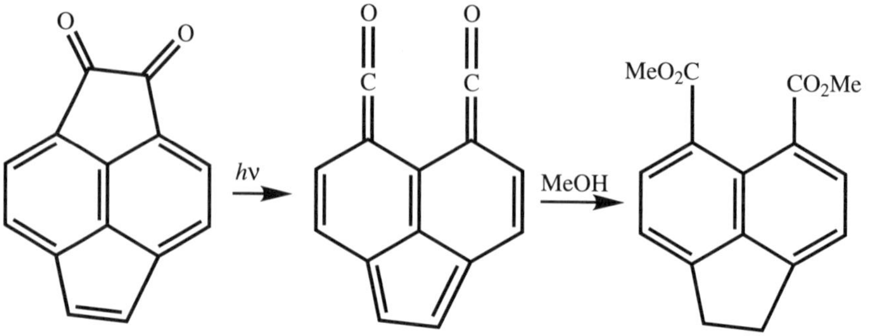

$$\mathbf{60} \xrightarrow{h\nu} \mathbf{61} \xrightarrow{MeOH} \mathbf{62} \qquad (34)^{38}$$

The reactions of a number of aliphatic bis(acyl chlorides) with Et_3N have been examined, and these may have involved formation of the bisketenes **63–67**, although these were usually not isolated.[39–49] These reactions are also proposed to proceed through the intermediacy of acylketenes (equation 34a).[39a] However **66** was obtained by sublimation as a yellow crystalline solid with an IR absorption at 2105 cm^{-1}, which spontaneously polymerized on slight warming.[45] In a later preparation **66** was obtained in benzene solution and observed by IR.[46] The reaction of pyridine adducts of **66** with polymers such as poly(ethylene oxide) containing OH and other nucleophilic groups gave rise to higher molecular weight polymers,[47] and with diamines polyamides are formed.[48]

$O=C=CH(CH_2)_2CH=C=O$ $O=C=CH(CH_2)_4CH=C=O$ $O=C=CH(CH_2)_6CH=C=O$

63[39,40]   **64**[41]   **65**[39, 42–44]

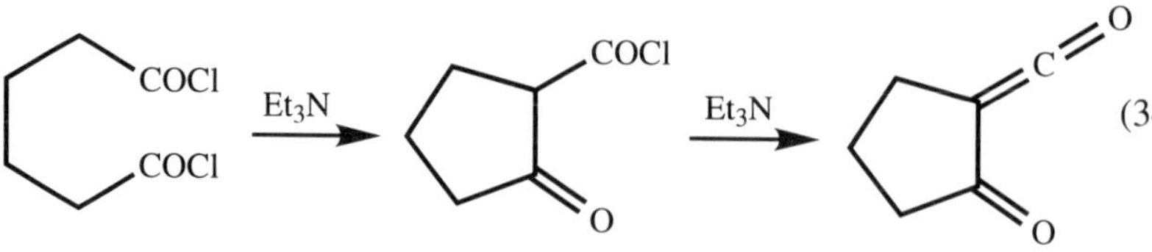

(34a)

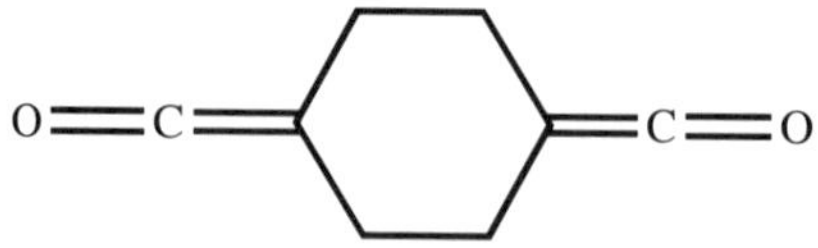

66[45–49] 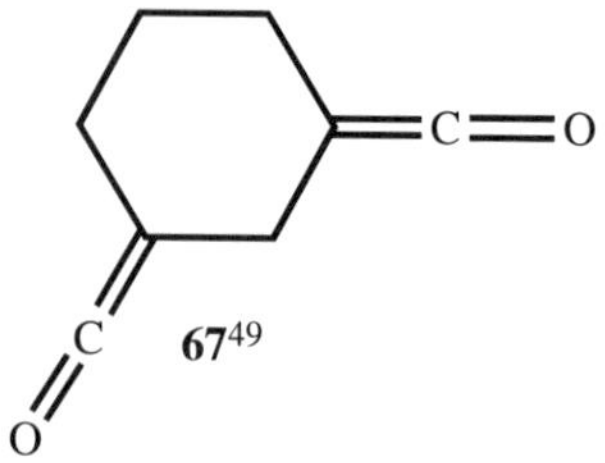   **67**[49]

The reaction of linear aliphatic dibasic acid chlorides with Et$_3$N has been suggested to form bisketenes which cyclize intramolecularly (equation 35).[42] Treatment of the cyclized materials with EtOK gives cyclic ketones and diketones (equations 36 and 37).[42] Bisketenes have not been directly observed in these reactions and so their formation is not established, but reactions involving bisketenes may be formulated as shown using the example of suberyl chloride (equation 35). Lactone **68** was isolated from this reaction and its structure established.[41] The pyrolysis of isopropenyl esters at 170 °C leads to ketenes (Section 3.2.3.2) and the reaction of diisopropenyl sebacate in the presence of diols and diamines led to polyesters and polyamides, respectively.[50] Nominally this reaction could proceed through the bisketene **65**, although a stepwise process is probably involved. The reaction of **65**, generated from the diacyl chloride, with diamines to form polyamides has also been studied.[43,44]

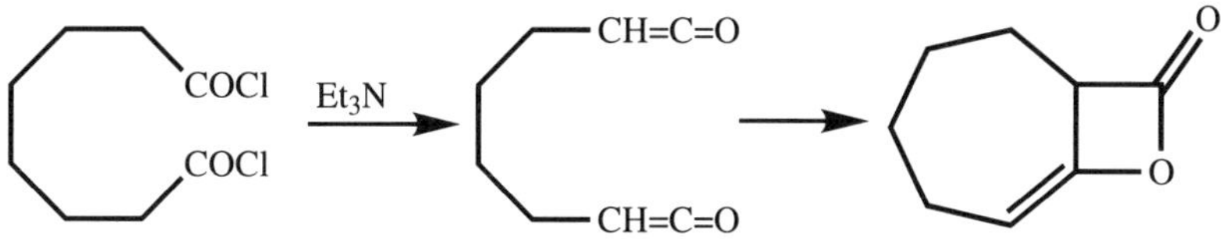

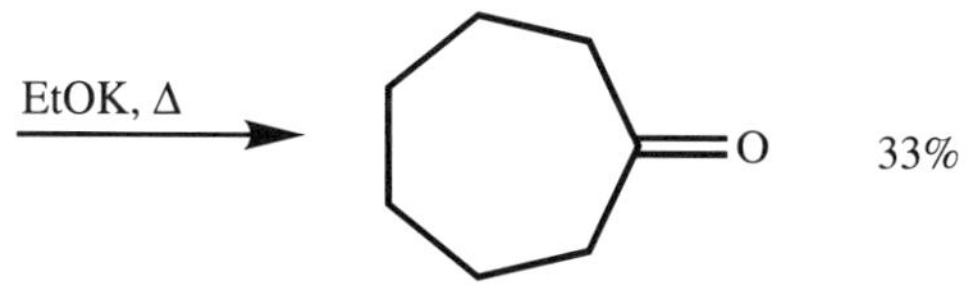

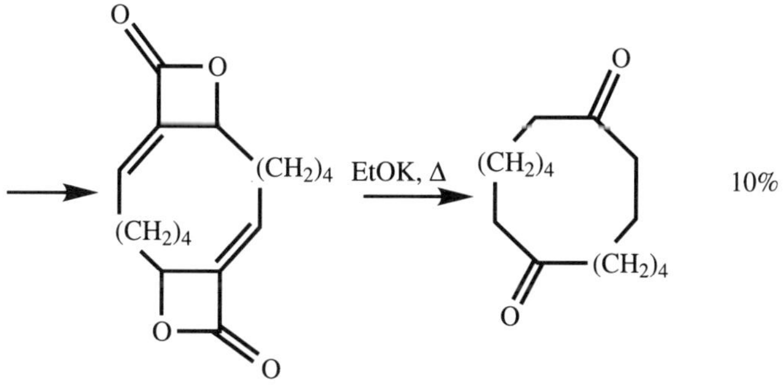

The photolysis of the bis(diazo ketone) **69** in methanol gave dimethyl suberate in a process that formally was shown as involving the bisketene **64** (equation 38),[51] although a stepwise process is probably involved. In the presence of 1,4-butanediol and 1,6-diaminohexane, photolysis of **69** gave a polyester and polyamide, respectively. Photolysis of **70** in MeOH gave tetramethyl ethanetetracarboxylate, and this was shown as proceeding formally through the bisketene **71** (equation 39).[51]

$$(CH_2CH_2COCHN_2)_2 \xrightarrow{h\nu} (CH_2CH_2CH=C=O)_2 \xrightarrow{MeOH} (CH_2CH_2CH_2CO_2Me)_2$$

$$\text{69} \qquad\qquad\qquad \text{64}$$

(38)

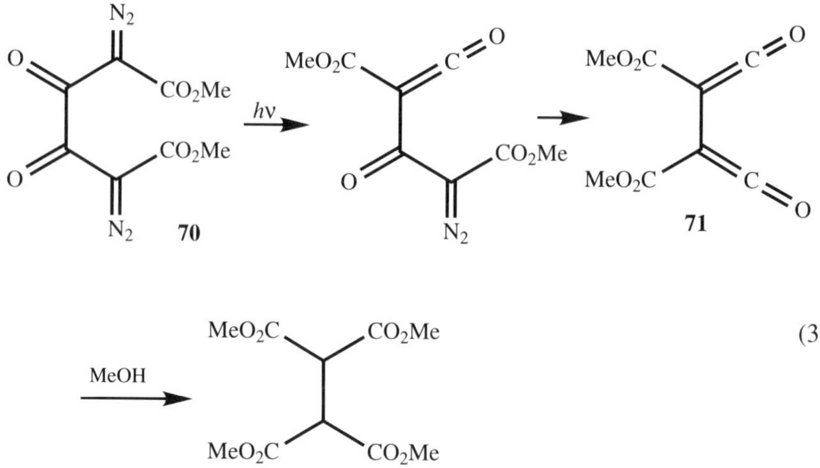

(39)

Reaction of norbornadiene with an excess of the oxadiazinone **72** at 80 °C proceeded with loss of N_2 and formation of the bisketene **73**, which was long lived (equation 40).[52] Thermolysis of the bisdioxinone **74** in the presence of alcohols gave diesters and may formally be represented as proceeding through the bisketene **75** (equation 41).[53,54]

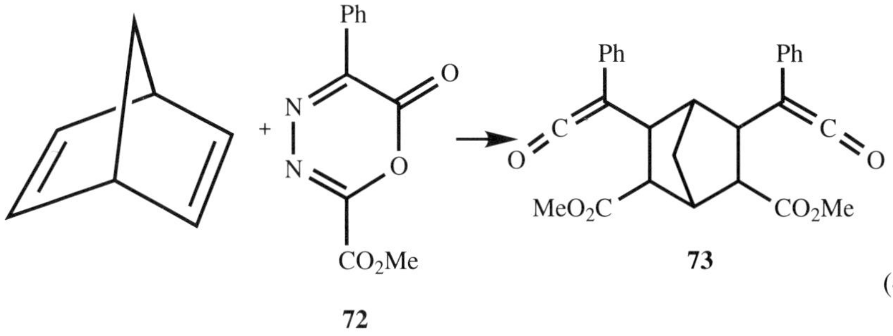

4.9 BISKETENES

[Structural diagrams for equation (40)[52] showing norbornadiene + compound **72** (with Ph, N=N, CO₂Me, O substituents) reacting to give **73** (bisketene with Ph and CO₂Me groups on norbornane scaffold).]

(40)[52]

[Structural diagrams for equation (41) showing compound **74** (bis-dioxinone) undergoing thermolysis (Δ) to a bisketene intermediate, then reaction with ROH to give **75** (diester diketone).]

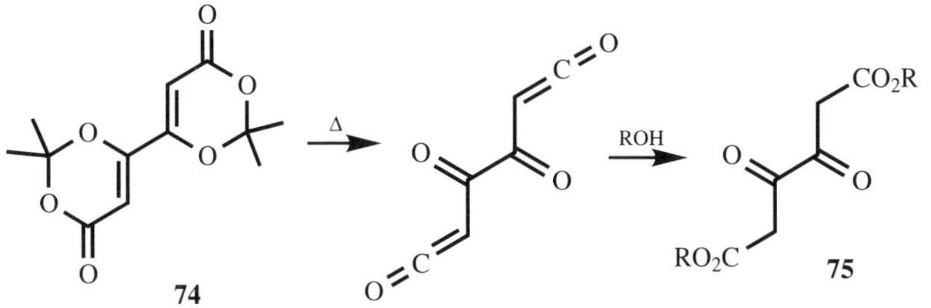

(41)

Photolysis of **76** was proposed to lead to the diradical bisketene **77,** which was suggested to react via **78** to give the observed product **79** (equation 42).[55,56] The formation of **79** from **76** involves an oxidation step, and could also proceed by addition of adventitious H_2O followed by oxidation. Photoreaction of **76** with benzylamine proceeds to the product **80** and this also involves an oxidation step (equation 43).[55,56]

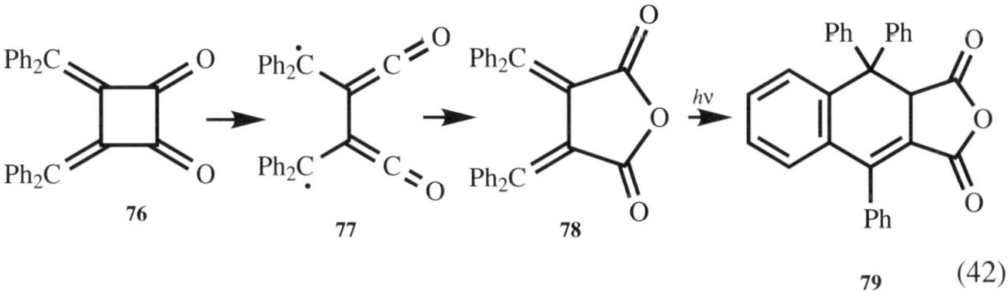

[Structural diagrams showing **76** (cyclobutanedione with Ph₂C= groups) → **77** (diradical bisketene) → **78** (anhydride) → (hv) → **79** (naphthalene-fused anhydride with Ph, Ph, Ph substituents).]

(42)

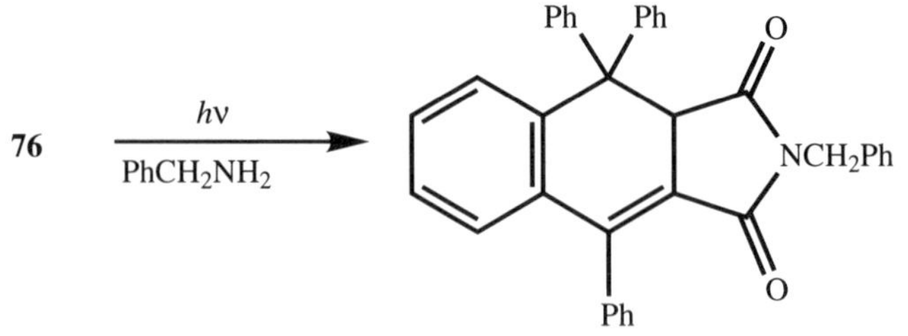

[Structure of compound **80** shown with photolysis of **76** in PhCH₂NH₂] (43)

Reaction of 6-bromoazulene-1,3-dicarboxylic acid **81** or its dimethyl ester in Magic Acid gave an ^{1}H NMR spectrum attributed to a dication with a bisketene resonance structure **82** (equation 44).[57]

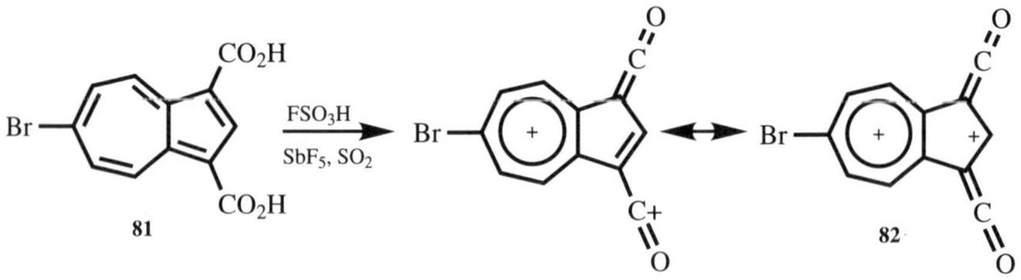

[Scheme showing **81** with FSO₃H, SbF₅, SO₂ giving dication **82**] (44)

Reaction of **63** formed in situ with chloral has also been suggested (equation 45).[58]

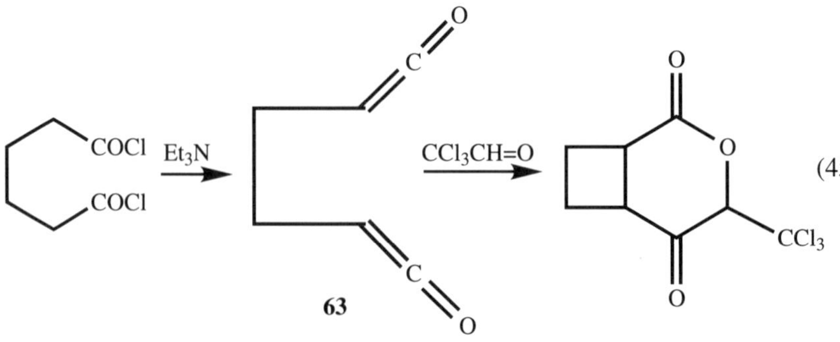

[Scheme showing cyclohexane-1,2-dicarbonyl dichloride with Et₃N forming bisketene **63**, then reacting with CCl₃CH=O to give bicyclic product with CCl₃ group] (45)

TABLE 2. Bisketenes

Ketene	Formula	Registry No.	Preparation	Ref.
(CH=C=O)$_2$	C$_4$H$_2$O$_2$	2829-38-1	Photolysis	1,3
(O=C=CHC(OH))=C=O	C$_4$H$_2$O$_3$	118919-78-1	Photolysis	6
(O=CCH=C=O)$_2$	C$_6$H$_2$O$_4$	4403-22-9	Thermolysis	53,54
O=C=CHCH=CHCH=C=O	C$_6$H$_4$O$_2$	98850-20-5	Photolysis	34,35
(CH$_3$C=C=O)$_2$	C$_6$H$_6$O$_2$	109609-84-9	Photolysis	15
(CH$_2$CH=C=O)$_2$	C$_6$H$_6$O$_2$	4403-21-8	Dehydrochlorination	39,40
	C$_8$H$_4$O$_2$	30839-70-4	Photolysis	24–27
	C$_8$H$_4$O$_2$	13221-13-1	Dehalogenation	33a,59
	C$_8$H$_4$O$_2^{2-}$	64554-28-5	Theory	59
	C$_8$H$_8$O$_2$	86261-27-0	Photolysis	15
	C$_8$H$_8$O$_2$		Dehydrochlorination	49

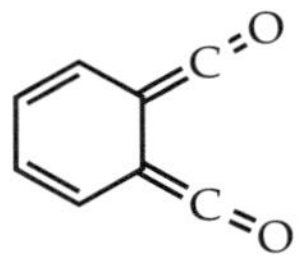

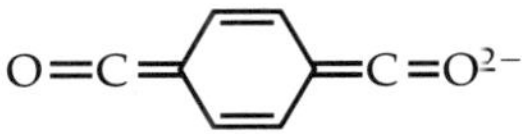

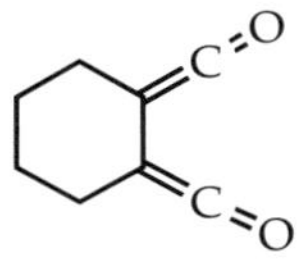

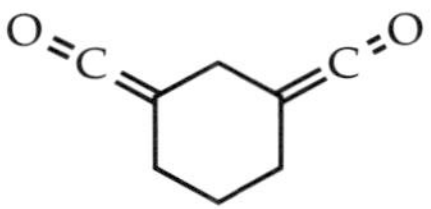

TABLE 2. (*Continued*)

Ketene	Formula	Preparation	Registry No.	Ref.
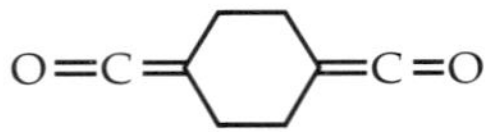 O=C—⟨cyclohexane⟩—C=O	$C_8H_8O_2$	Dehydrochlorination	66471-64-5	45
O=C=CHCH=C(Bu-*t*)CH=C=O	$C_{10}H_{12}O_2$	Photolysis		34
[(CH$_2$)$_3$CH=C=O]$_2$	$C_{10}H_{14}O_2$	Dehydrochlorination	45037-70-5	39,42–44
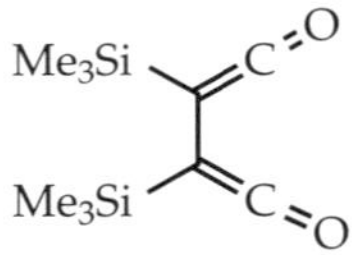 Me$_3$Si-C(=O)-C(=O)-SiMe$_3$ (with Me$_3$Si substituents)	$C_{10}H_{18}O_2Si_2$	Thermolysis	145178-53-6	21,22
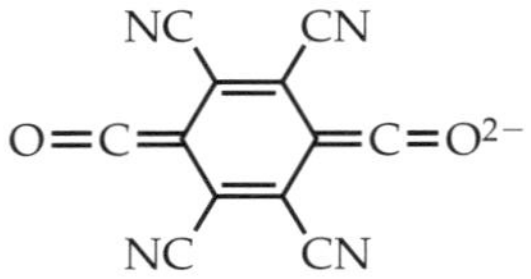 O=C=C(CN)-C(CN)=C(CN)-C(CN)=C=O (tetracyano) with C=O^{2-}	$C_{12}N_4O_2^{2-}$	Theory	64554-30-9	59
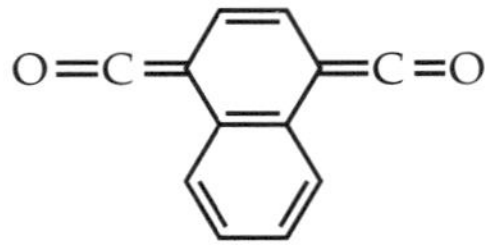 O=C=⟨naphthalene⟩=C=O	$C_{12}H_6O_2$	Dehydrochlorination	94504-42-4	37

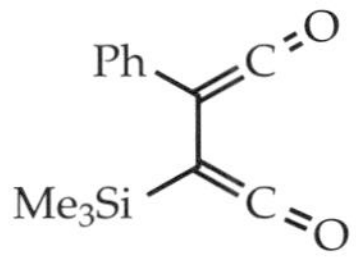

Structure	Formula	CAS / Method		Ref
Ph-C(=O)-C(=O)-SiMe₃	$C_{13}H_{14}O_2Si$	154473-77-5	Photolysis	23
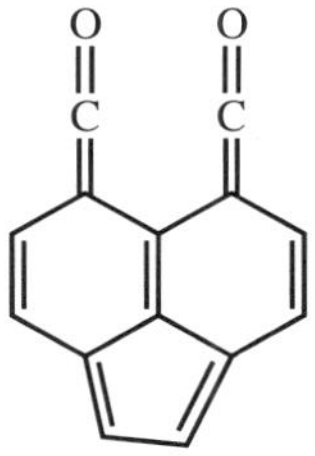 acenaphthylene-5,6-dione	$C_{14}H_6O_2$		Photolysis	38
 R-C₆H₄-R where R = CH₂CMe=C=O; [(CH₂)₄COCH=C=O]₂; (CPh=C=O)₂	$C_{14}H_{14}O_2$ $C_{14}H_{18}O_4$ $C_{16}H_{10}O_2$	46744-13-2 92005-69-1 34072-94-1	Dehydrochlorination Index error Photolysis	60 39,43[a] 9,12
t-BuMe₂Si-C(=C=O)-C(=C=O)-SiMe₂-t-Bu; (CH₂CPh=C=O)₂	$C_{16}H_{30}O_2Si_2$ $C_{18}H_{14}O_2$	152530-62-6 86047-17-8	Thermolysis Dehydrochlorination	22

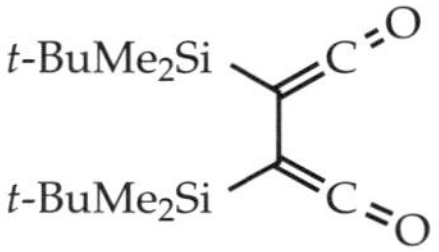

TABLE 2. (Continued)

Ketene	Formula	Preparation	Registry No.	Ref.
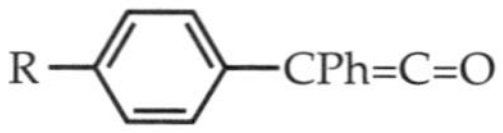 R—CPh=C=O, R = C(Hx-c)=C=O	$C_{22}H_{20}O_2$	Dehydrochlorination	101295-08-3	37
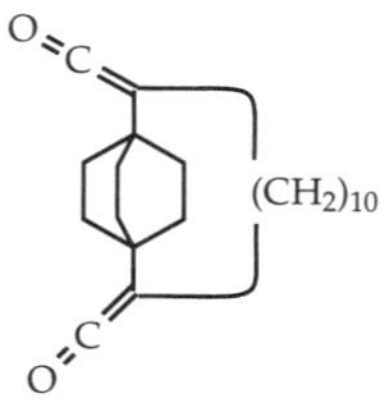 O=C=(CH₂)₁₀=C=O	$C_{22}H_{32}O_2$	Wolff		61
 R—C₆H₄—R, R = CH₂CPh=C=O	$C_{24}H_{18}O_2$	Dehydrochlorination	88856-07-9	37
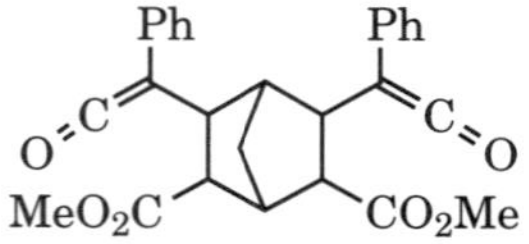 Ph, Ph, O=C, C=O, MeO₂C, CO₂Me	$C_{27}H_{24}O_6$	Thermolysis	127280-53-9	52

[a]The structure should be [(CH₂)₃CH=C=O]₂ (sebacylketene)

References

 1. Gong, L.; McAllister, M. A.; Tidwell, T. T. *J. Am. Chem. Soc.* **1991,** *113,* 6021–6028.
 2. McAllister, M. A., Tidwell, T. T. *J. Am. Chem. Soc.* **1994,** *116,* 7233–7238.
 3. Kasai, M.; Oda, M.; Kitahara, Y. *Chem. Lett.* **1978,** 217–218.
 4. Maier, G.; Reisenauer, H. P.; Sayrac, T. *Chem. Ber.* **1982,** *115,* 2192–2201.
 5. Takhistov, V. V.; Muftakhov, M. V.; Krivoruchko, A. A.; Mazunov, V. A. *Izv. Akad. Nauk S.S.S.R., Ser. Khim.* **1991,** 2049–2054; *Engl. Trans.* **1991,** 1812–1817.
 6. Hochstrasser, R.; Wirz, J. *Angew. Chem., Int. Ed. Engl.* **1989,** *28,* 181–183.
 7. Jewell, C. F.; Jr.; Liebeskind, L. S.; Williamson, M. *J. Am. Chem. Soc.* **1985,** *107,* 6715–6716.
 8. Mallory, F. B.; Roberts, J. D. *J. Am. Chem. Soc.* **1961,** *83,* 393–397.
 9. Obata, N.; Takizawa, T. *J. Chem. Soc. D, Chem. Commun.* **1971,** 587–588.
10. Obata, N.; Takizawa, T. *Bull. Chem. Soc. Jpn.* **1977,** *50,* 2017–2020.
11. Chapman, O. L.; McIntosh, C. L.; Barber, L. L. *J. Chem. Soc., Chem. Commun.* **1971,** 1162–1163.
12. Blomquist, A. T.; LaLancette, E. A. *J. Am. Chem. Soc.* **1961,** *83,* 1387–1391.
13. Bird, C. W. *Chem. Commun.* **1968,** 1537.
14. Maier, G.; Jung, W. A. *Tetrahedron Lett.* **1980,** *21,* 3875–3878.
14a. Dehmlow, E. V. *Tetrahedron Lett.* **1972,** 1271–1274.
15. Miller, R. D.; Kirchmeyer, S. *J. Org. Chem.* **1993,** *58,* 90–94.
16. Miller, R. D.; Theis, W.; Heilig, G.; Kirchmeyer, S. *J. Org. Chem.* **1991,** *56,* 1453–1463.
17. Bock, H.; Ried, W.; Stein, U. *Chem. Ber.* **1981,** *114,* 673–683.
18. Rubin, Y.; Lin, S. S.; Knobler, C. B.; Anthony, J.; Boldi, A. M.; Diederich, F. *J. Am. Chem. Soc.* **1991,** 113, 6943–6949.
19. Diederich, F.; Rubin, Y. *Angew. Chem., Int. Ed. Engl.* **1992,** *31,* 1101–1123.
20. McElvany, S. W.; Ross, M. M.; Goroff, N. S.; Diederich, F. *Science,* **1993,** *259,* 1594–1596.
21. Zhao, D.; Tidwell, T. T. *J. Am. Chem. Soc.* **1992,** *114,* 10980–10981.
22. Zhao, D.; Allen, A. D.; Tidwell, T. T. *J. Am. Chem. Soc.* **1993,** *115,* 10097–10103.
23. Allen, A. D.; Lai, W.-Y.; Ma, J.; Tidwell, T. T. *J. Am. Chem. Soc.,* **1994,** *116,* 2625–2626. *Heteroatom Chem.* **1994,** *5,* 235–244.
24. Staab, H. A.; Ipaktschi, J. *Chem. Ber.* **1968,** *101,* 1457–1472.
25. Chapman, O. L.; Mattes, C.; McIntosh, C. L.; Pacansky, J.; Calder, G. V.; Orr, G. *J. Am. Chem. Soc.* **1973,** *95,* 6134–6135.
26. Spangler, R. J.; Henscheid, L. G.; Buck, K. T. *J. Org. Chem.* **1977,** *42,* 1693–1697.
27. Boate, D. R.; Johnson, L. J.; Kwong, P. C.; Lee-Ruff, E.; Scaiano, J. C. *J. Am. Chem. Soc.* **1990,** *112,* 8858–8863.
27a. Lowe, J. A.; Jung, M. E. *J. Org. Chem.* **1977,** *42,* 2371–2373.
28. Forster, D. L.; Gilchrist, T. L.; Rees, C. W.; Stanton, E. *Chem. Commun.* **1971,** 695–696.
29. Mosandl, T.; Wentrup, C. *J. Org. Chem.* **1993,** *58,* 747–749.
30. Ramirez, F.; Yamanaka, H.; Basedow, O. H. *J. Am. Chem. Soc.* **1961,** *83,* 173–178.

31. Ramirez, F.; Ricci, J. S., Jr.; Tsuboi, H.; Marecek, J. F.; Yamanaka, H. *J. Org. Chem.* **1976**, *41*, 3909–3914.
32. Pattenden, G.; Turvill, M. W.; Chorlton, A. P. *J. Chem. Soc., Perkin Trans. 1*, **1991**, 2357–2361.
33. Blomquist, A. T.; Meinwald, Y. C. *J. Am. Chem. Soc.* **1957**, *79*, 2021–2022.
33a. Utley, J. H. P.; Gao, Y.; Lines, R. *J. Chem. Soc., Chem. Commun.* **1993**, 1540–1542.
34. Tomioka, H.; Fukao, H.; Izawa, Y. *Bull. Chem. Soc. Jpn.* **1978**, *51*, 540–543.
35. Maier, G.; Franz, L. H.; Hartan, H. G.; Lanz, K.; Reisnauer, H. P. *Chem. Ber.* **1985**, *118*, 3196–3204.
36. LeGoff, L., Ph.D. Dissertation, Cornell University, 1960, as cited in ref. 49.
37. Blomquist, A. T. *U.S. Patent* 3,002,024; *Chem. Abstr.* **1962**, *56*, 8657b.
38. Beringer, F. M.; Winter, R. E. K.; Castellano, J. A. *Tetrahedron Lett.* **1968**, 6183–6186.
39. Sauer, J. C. *J. Am. Chem. Soc.* **1947**, *69*, 2444–2448.
39a. Jäger, G. *Chem. Ber.* **1972**, *105*, 137–149.
40. Baldwin, J. E. *J. Org. Chem.* **1963**, *28*, 3112–3114.
41. Baldwin, J. E. *J. Org. Chem.* **1964**, *29*, 1882–1883.
42. Blomquist, A. T.; Spencer, R. D. *J. Am. Chem. Soc.* **1947**, *69*, 472–473; *ibid.* **1948**, *70*, 30–33.
43. Garner, D. P.; Fasulo, P. D. *J. Polym. Sci., Polym. Chem. Ed.* **1985**, *23*, 2177–2195.
44. Garner, D. P. *J. Polym. Sci., Polym. Chem. Ed.* **1982**, *20*, 2979–2988.
45. Hatchard, W. R.; Schneider, A. K. *J. Am. Chem. Soc.* **1957**, *79*, 6261–6263.
46. Franke, V. W. K. R.; Ahne, H. *Angew. Makromol. Chem.* **1972**, *21*, 195–205.
47. Chemische Werk Hüls A-G. British Patent 1,025,693, 1966; *Chem. Abstr.* **1966**, *65*, 5557d.
48. Yoda, N.; Hagihara, Y.; Ikeda, K.; Kuihara, M. Japan Patent 70 01,631; *Chem. Abstr.* **1970**, *72*, 133651w.
49. Unpublished results of A. T. Blomquist and Y. C. Meinwald quoted by R. A. Vierling, Ph.D. Thesis, Cornell University, 1962; *Chem. Abstr.* **1963**, *58*, 7838a.
50. Rothman, E. S. *J. Am. Oil Chem. Soc.* **1968**, *45*, 189–193.
51. Horner, L.; Spietschka, E. *Chem. Ber.* **1952**, *85*, 225–229.
52. Christl, M.; Lanzendörfer, U.; Grötsch, N. M.; Ditterich, E.; Hegman, J. *Chem. Ber.* **1990**, *123*, 2031–2037.
53. Stachel, H. D. *Arch. Pharm.* **1962**, *295*, 735–744.
54. Stachel, H. D. *Angew. Chem.* **1957**, *69*, 507.
55. Toda, F.; Nakaoka, H.; Yuwane, K.; Todo, E. *Bull. Chem. Soc. Jpn.* **1973**, *46*, 1737–1740.
56. Toda, F.; Garratt, P. *Chem. Rev.* **1992**, *92*, 1688–1707.
57. McDonald, R. N.; Morris, D. L.; Petty, H. E.; Hoskins, T. L. *Chem. Commun.* **1971**, 743–744.
58. Borrman, D.; Wegler, R. *Chem. Ber.* **1966**, *99*, 1245–1251.
59. Krivoshei, I. V.; Luong, P.; Savenkova, L. N. *Deposited Doc.*, **1975**, VINITI 2559–2575; *Chem. Abstr.* **1977**, *87*, 167302h.
60. Chemische Werke Hüls A.-G. French Patent 1,341,799; *Chem. Abstr.* **1964**, *60*, 6793g.
61. Eaton, P. E.; Leipzig, B. D. *J. Am Chem. Soc.* **1983**, *105*, 1656–1658.

4.10 KETENYL RADICALS, ANIONS, AND CATIONS

Ketenyl radicals, anions, and cations are **1–3**, respectively. A number of aspects of the chemistry of **2** are considered in Section 4.8.1. Electron transfer reactions of ketenes are included in Section 5.1. Ketenes with charged, radical, or carbenic side chains are considered in Section 4.1.9, but many of these structures are listed in the Table 1 at the end of this section.

$$H\overset{\bullet}{C}=C=O \qquad H\overset{-}{C}=C=O \qquad H\overset{+}{C}=C=O$$

$$\mathbf{1} \qquad\qquad \mathbf{2} \qquad\qquad \mathbf{3}$$

Laser photolysis of ketene at 193 nm produced ketenyl radical (**1**), whose high-resolution IR spectrum was measured and analyzed.[1] Preliminary examinations of the kinetics of the reaction of **1** were also carried out.[1] This species is of major interest as it is believed to be an intermediate in the formation of soot. Formation of $Ph\overset{\bullet}{C}=C=O$ was suggested as a possible cause of an ESR signal observed on photolysis of phenylbenzoylcarbene in a matrix at 77 K (equation 2).[2]

$$CH_2=C=O \xrightarrow{h\nu} H\overset{\bullet}{C}=C=O \longleftrightarrow HC\equiv C-O^{\bullet} \qquad (1)$$

$$\mathbf{1}$$

$$PhCOCN_2Ph \xrightarrow{h\nu} PhCO\overset{\bullet\bullet}{C}Ph \xrightarrow[-Ph\bullet]{h\nu} Ph\overset{\bullet}{C}=C=O \qquad (2)$$

As discussed in Section 4.8.1, the ketenyl anion **2**, which also has the character of the ynolate ion, may be generated in the gas phase by proton abstraction from ketene by various bases (equation 3).[3]

$$CH_2=C=O \xrightarrow{base} H\overset{-}{C}=C=O \longleftrightarrow HC\equiv CO^- \qquad (3)$$

$$\mathbf{2}$$

The photoelectron spectrum of **2** gave an electron affinity of the radical **1** EA(HCCO) of 2.350 ± 0.020 eV, and the use of this value and the $DH^{\circ}_{acid}(CH_2CO)$ of 365 ± 2 kcal/mol derived from the proton abstraction from ketene by bases (equation 3) gave the $DH^{\circ}_{298}(H-CHCO)$ of 105.9 ± 2.1 kcal/mol.[3,4]

From the appearance energies of cations in an energy-resolved electron beam the $DH^{\circ}_{298}(H-CHCO)$ of 92 kcal/mol was determined[5,6] with the ΔH_f of **3** ($CHCO^+$) of 262 kcal/mol, and a hydride affinity for ketene $D(CHCO^+ + H^-)$ of 308 kcal/mol.[6]

The activation energy for formation of cation **3** from the radical cation **4** has been estimated[7] as 68 kcal/mol from the ΔH_f (**4**) of 247 kcal/mol and the value given above of ΔH_f° (**3**).

$$HC \equiv COH^{+\bullet} \longrightarrow H\overset{\bullet}{C}=C=\overset{+}{O} + H\bullet \qquad (4)$$

$$\qquad\qquad 4 \qquad\qquad\qquad 3$$

The difference in the two values of DH_{298}° (H—CH=C=O) of 105.9[3] and 92[5] kcal/mol may arise from the different reactions studied in the two cases. Thus the photoelectron spectrum derived electron affinity of $H\overset{\bullet}{C}=C=O$ (**1**) is derived from the photoelectron spectrum of the anion, which is expected to be linear,[3] while calculations predict that the radical $H\overset{\bullet}{C}=C=O$ is bent.[8,9] On the other hand the heat of formation derived for **1** from the appearance energy relied on a heat of formation of the precursor calculated by group additivity,[5] and this appears to be less reliable.

It was proposed that photolysis of the diazoester **5** led to carbene **6**, which in methanol partially ionized to the ketenyl cation **7**, as evidenced by the formation of the product **8** (equation 5).[10]

$$Me_3SiCN_2CO_2Et \xrightarrow[-N_2]{h\nu} Me_3Si\overset{..}{\overset{\bullet}{C}}CO_2Et \xrightarrow{-EtO^-} Me_3Si\overset{+}{C}=C=O$$

$$\qquad 5 \qquad\qquad\qquad 6 \qquad\qquad\qquad 7$$

$$\xrightarrow{MeOH} Me_3SiCH(OMe)CO_2Me \qquad (5)$$

$$\qquad\qquad\qquad 8$$

Cationic carbon species C_n^+ (n= 2–6) react in the gas phase with one or two carbon monoxide molecules to form cationic species that may be formulated as ketenyl radical cations, as shown in equation 6.[11]

$$\overset{+}{\bullet}C=C: \xrightarrow{CO} \overset{+}{\bullet}C=C=C=O \xrightarrow{CO} O=\overset{+}{C}-\overset{\bullet}{C}=C=C=O \qquad (6)$$

Other species formed in these reactions include $C_3O_2^{+\bullet}$, $C_4O^{+\bullet}$, $C_5O^{+\bullet}$, $C_5O_2^{+\bullet}$, $C_6O^{+\bullet}$, $C_6O_2^{+\bullet}$, $C_7O^{+\bullet}$, and $C_7O_2^{+\bullet}$.[11] The reactions of C_2H^+, C_3H^+, C_4H^+, and C_5H^+ with one molecule of CO were also observed, as in equation 7, and gave species which can be formulated as C_3HO^+, C_4HO^+, C_5HO^+, and C_6HO^+.

$$HC \equiv C^+ \xrightarrow{CO} HC \equiv C-\overset{+}{C}=O \longleftrightarrow H\overset{+}{C}=C=C=O \qquad (7)$$

A number of cation radicals of ketenes have been identified by mass spectroscopy and some of these are reported in Section 2.5. Some of these species that have been given *Chemical Abstracts* Registry Numbers are listed in Table 1 and in Section 4.11 along with a variety of other charged species of diverse types.

TABLE 1. Cationic, Radical, and Carbanionic Ketenyl Species

Ketene	Formula	Preparation	Registry No.	Ref.
$CCl_2=C=O^+$	$C_2Cl_2O^+$	Gas phase	96607-18-0	13
$CD_2=C=O^+$	$C_2D_2O^+$	Photoelectron	64947-45-1	14
$CHF=C=O.H^+$	$C_2HFO.H^+$	Theory	114396-10-0	15
$\cdot CH=C=O$	C_2HO	Gas phase	51095-15-9	6
$\cdot CD=C=O$	C_2DO	Gas phase	102937-69-9	16
$CH=C=O^-$	C_2HO^-	Theory	41084-91-7	17
$^+CH=C=O$	C_2HO^+	Gas phase	92056-63-8	6,12
$^+CH=C=O.2H^+$	$C_2HO.2H^{3+}$	Theory	123641-75-8	18
$CH_2=C=O.2H^+$	$C_2H_2O.2H^{2+}$	Theory	97459-62-6	19
$CH_2=C=O^+\cdot$	$C_2H_2O^+$	Theory	64999-16-2	20
$CH_2=C=O.H^+$	$C_2H_2O.H^+$	Mass spec	62581-51-5	21
$CH_2=C=O.H^+$	$C_2H_2O.H^{2+}$	Mass spec	69784-90-3	21
$CH_2=C=O^{\cdot-}$	$C_2H_2O^-$	Theory	56391-79-8	22,23
$HOCH=C=O.2K$	$C_2H_2O_2.2K$	Matrix	3053-67-6	23a
$^+C=C=O$	C_2O^+	Gas phase	83483-97-0	11
$^-C=C=O$	C_2O^-	Gas phase	87191-88-6	3,13
$^{2-}C=C=O$	C_2O^{2-}	Theory	41084-90-6	17
$HN=CHCH=C=O^+$	$C_3H_3NO^+$	Mass spec	81451-31-2	24
$^+CH_2CH=C=O$	$C_3H_3O^+$	Theory	105028-86-2	25
$CH_3\overset{+}{C}=C=O$	$C_3H_3O^+$	Theory	101705-92-4	26
$CH_3CH=C=O^+$	$C_3H_4O^+$	γ-irradiation	84005-03-8	27
$CH_3CH=C=O.H^+$	$C_3H_4O.H^+$	Gas phase	77414-28-9	28
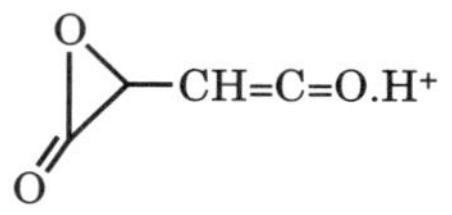 ![cyclic structure] O-CH=C=O.H^+	$C_4H_2O_3.H^+$	Photolysis	111830-27-4	29

TABLE 1. (Continued)

Ketene	Formula	Preparation	Registry No.	Ref.
$CH_2=\overset{+}{C}CH=C=O$	$C_4H_3O^+$	Flame	79530-60-2	30
$O=\overset{+}{C}CMe=C=O$	$C_4H_3O_2^+$	Dehydration	37803-05-7	31
$CH_3\overset{+}{C}CH=C=O$	$C_4H_4O^+$	Mass spec	98352-53-5	32
$CH_3CCH=C=O^\bullet$	$C_4H_4O^+$	Mass spec	71793-85-6	33
$CH_2=CHCH=C=O^+$	$C_4H_4O^+$	Mass spec	71793-85-6	33
$CH_3COCH=C=O.H^+$	$C_4H_4O_2.H^+$	Mass spec	77414-29-0	28
$EtCH=C=O^+$	$C_4H_6O^+$	γ-irradiation	87551-95-9	27
$Me_2C=C=O^+$	$C_4H_6O^+$	γ-irradiation	84005-06-1	27
$EtCH=C=O.H^+$	$C_4H_6O.H^+$	Theory	116232-23-6	34
$Me_2C=C=O.H^+$	$C_4H_6O_2^+$	Theory	116232-22-5	34
⌬=C=O	$C_7H_5O^-$	Mass spec	116932-46-8	35
$Ph\dot{C}=C=O$	C_8H_5O	Photolysis		2
$3,4\text{-}(MeO)_2C_6H_3\overset{-}{C}=C=O$ Li^+	$C_{10}H_9O_3^-Li^+$	Fragmentation	83933-43-1	36
R = H	$C_{11}H_7O^+$	Dehydration		37
R = CO_2H	$C_{12}H_6BrO_3.H^+$	Dehydration	34742-28-4	38
R = CO_2Me	$C_{13}H_8BrO_3.H^+$	Dehydration	34742-29-5	38

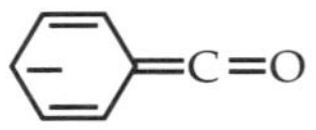

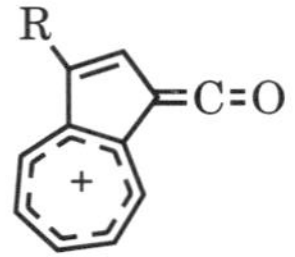

References

1. Unfried, K. G.; Glass, G. P.; Curl, R. F. *Chem. Phys. Lett.* **1991**, *177*, 33–38.
2. Murai, H.; Safarik, I.; Torres, M.; Strausz, O. P. *J. Am. Chem. Soc.* **1988**, *110*, 1025–1032.
3. Oakes, J. M.; Jones, M. E.; Bierbaum, V. M.; Ellison, G. B. *J. Phys. Chem.* **1983**, *87*, 4810–4815.
4. Berkowitz, J.; Ellison, G. B.; Gutman, D. *J. Phys. Chem.* **1994**, *98*, 2744–2765.
5. Holmes, J. L.; Lossing, F. P. *Int. J. Mass Spectrom. Ion Proc.* **1984**, *58*, 113–121.
6. Lossing, F. P.; Holmes, J. L. *J. Am. Chem. Soc.* **1984**, *106*, 6917–6920.
7. Aue, D. H.; Bowers, M. T. In *Gas Phase Ion Chemistry*, Vol. 2; Bowers, M. T., Ed.; Academic: New York, 1979; Chapter 9.
8. Hu, C.-H.; Schaeffer, H. F., III; Hou, Z.; Bayles, K. D. *J. Am. Chem. Soc.* **1993**, *115*, 6904–6907.
9. Harding, L. B. *J. Phys. Chem.* **1981**, *85*, 10–11.
10. Ando, W.; Sekiguchi, A.; Hagiwara, T.; Migita, T.; Chowdry, V.; Westheimer, F. H.; Kammula, S. L.; Green, M.; Jones, M., Jr. *J. Am. Chem. Soc.* **1979**, *101*, 6393–6398.
11. Bohme, D. K.; Wlodek, S.; Williams, L.; Forte, L.; Fox, A. *J. Chem. Phys.* **1987**, *87*, 6934–6938.
12. van Baar, B.; Weiske, T.; Terlouw, J. K.; Schwarz, H. *Angew. Chem., Int. Ed. Engl.* **1986**, *25*, 282–284.
13. Jacox, M. E. *J. Phys. Chem. Ref. Data*, **1984**, *13*, 945–1068; *Chem. Abstr.* **1984**, *103*, 45024x.
14. Hall, D.; Maier, J. P.; Rosmus, P. *Chem. Phys.* **1977**, *24*, 373–378.
15. Lien, M. H.; Hopkinson, A. C. *J. Am. Chem. Soc.* **1988**, *110*, 3788–3792.
16. Inoue, G.; Suzuki, M. *J. Chem. Phys.* **1986**, *84*, 3709–3716.
17. Hopkinson, A. C. *J. Chem. Soc., Perkin Trans. 2*, **1973**, 795–797.
18. Wong, M. W.; Radom, L. *J. Am. Chem. Soc.* **1989**, *111*, 6976–6983.
19. Koch, W.; Frenking, G.; Schwarz, H.; Maquin, F.; Stahl, D. *Int. J. Mass Spectrom. Ion Proc.* **1985**, *63*, 59–82.
20. Heinrich, N.; Koch, W.; Morrow, J. C.; Schwarz, H. *J. Am. Chem. Soc.* **1988**, *110*, 6332–6336.
21. Drewello, T.; Schwarz, H. *Int J. Mass Spectrom. Ion Proc.* **1989**, *87*, 135–140.
22. Kartasheva, L. I.; Pikaev, A. K. *Khim. Vys. Energ.* **1975**, *9*, 242–246; *Chem. Abstr.* **1975**, *83*, 88640f.
23. van Baar, B. L. M.; Heinrich, N.; Koch, W.; Postma, R.; Terlouw, J. K.; Schwarz, H. *Angew. Chem., Int. Ed. Engl.* **1987**, *26*, 140–142.
23a. Ayed, O.; Manceron, L.; Silvi, B. *J. Phys. Chem.* **1988**, *92*, 37–45.
24. Bouchoux, G.; Hoppilliard, Y. *Org. Mass Spectrom.* **1981**, *16*, 459–464.
25. Sonveaux, E.; Andre, J. M.; Delhalle, J.; Fripiat, J. G. *Bull. Soc. Chim. Belg.* **1985**, *94*, 831–847.
26. Bouchoux, G.; Hoppilliard, Y.; Flament, J.-P. *Org. Mass Spectrom.* **1985**, *20*, 560–564.
27. Shimokoshi, K.; Fujisawa, J.; Nakamura, K.; Sato, S.; Shida, T. *Chem. Phys. Lett.* **1983**, *99*, 483–486.

28. Reents, W. D.; Jr.; Murray, K. J.; Freisen, B. S. *Org. Mass Spectrom.* **1980**, *15*, 509–515.
29. Amat, A. M.; Asensio, G.; Castello, M. J.; Miranda, M. A.; Simon-Fuentes, A. *Tetrahedron* **1987**, *43*, 905–910.
30. Olson, B. D.; Calcote, H. F. *Symp. Combust. 1980*, **1981**, *18*, 453–464; *Chem. Abstr.* **1981**, *96*, 54782g.
31. Conrow, K.; Morris, D. L. *J. Chem. Soc., Chem. Commun.* **1973**, 5–6.
32. Bouchoux, G.; Dagaut, J.; Fillaud, J.; Burgers, P. C.; Terlouw, J. K. *Nouv. J. Chim.* **1985**, *9*, 25–31.
33. Holmes, J. L.; Terlouw, J. K. *J. Am. Chem. Soc.* **1979**, *101*, 4973–4975.
34. Bouchoux, G.; Hoppilliard, Y. *J. Phys. Chem.* **1988**, *92*, 5869–5874.
35. Raftery, M. J.; Bowie, J. H.; Sheldon, J. C. *J. Chem. Soc., Perkin Trans. 2*, **1988**, 563–569.
36. Barrett, A. G. M.; Quayle, P. *J. Chem. Soc., Perkin Trans. 1*, **1982**, 2193–2196.
37. Longridge, J. L.; Long, F. A. *J. Am. Chem. Soc.* **1968**, *90*, 3088–3092.
38. McDonald, R. N.; Morris, D. L.; Petty, H. E.; Hoskins, T. L. *J. Chem. Soc., Chem. Commun.* **1971**, 743–744.

4.11 CUMULENONES

The cumulenones cover a wide variety of structural types, including the higher oxides of carbon such as carbon suboxide (**1**), the methyleneketenes, such as propadienone (**2**), and carbenic species such as **3**. Some of these species are not known experimentally but have only been studied theoretically. These are arbitrarily grouped together, including some ionic and radical species (see also Section 4.10), and those such as **1** that could be classed as bisketenes.

$$O=C=C=C=O \qquad CH_2=C=C=O \qquad :C=C=O$$
$$\quad\quad \mathbf{1} \quad\quad\quad\quad\quad\quad \mathbf{2} \quad\quad\quad\quad\quad \mathbf{3}$$

Carbon suboxide (**1**) was prepared in 1906 by Diels and Wolf[1] by the dehydration of malonic acid (equation 1) and in 1908 by Staudinger and Bereza by a bis(dehalogenation) route (equation 2).[2] This compound has received continued attention which has been periodically reviewed[3–6] and continues[7–10] unabated.

$$CH_2(CO_2H)_2 \xrightarrow{P_2O_5} O=C=C=C=O \qquad (1)$$

$$CBr_2(COBr)_2 \xrightarrow{Zn} O=C=C=C=O \qquad (2)$$

The carbon dioxides with even numbers of carbons are less stable than those with odd numbers, and all attempts to prepare C_2O_2, the dimer of carbon monoxide have so far failed,[11,12,71] although evidence for C_4O_2 in an Ar matrix,[13] and the gas phase,[14] has been obtained. The odd number carbon dioxides C_5O_2[15] and C_7O_2[16] have been detected in Ar matrices.

The methyleneketenes, of which **2** is the parent, are usually rather unstable, and they have been reviewed.[17,18] Pyrolysis of **4** at 680 °C and isolation in an Ar matrix gave **2** (IR 2125 cm^{-1}) by the route of equation 3.[19]

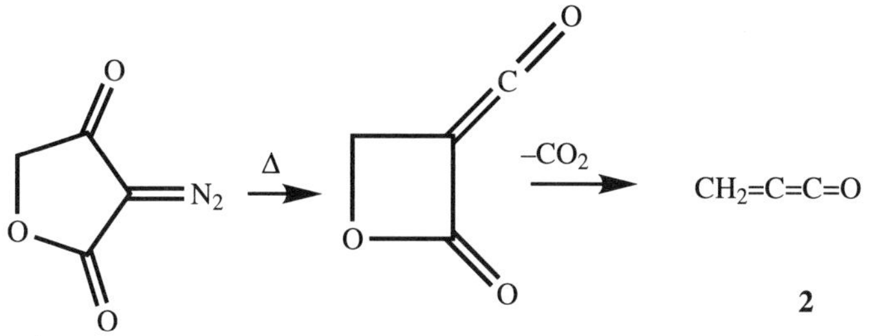

$$\mathbf{4} \xrightarrow{\Delta} [\text{intermediate}] \xrightarrow{-CO_2} CH_2=C=C=O \quad \mathbf{2} \tag{3}$$

Among the analogues of **2** is **7**, obtained by photolysis of the vinyl halide **5**, which was proposed to lead to the α-acyl vinyl cation **6** which lost a proton to give the cumulenone **7**, which was captured as the ester as 4% of the product mixture (equation 4).[20]

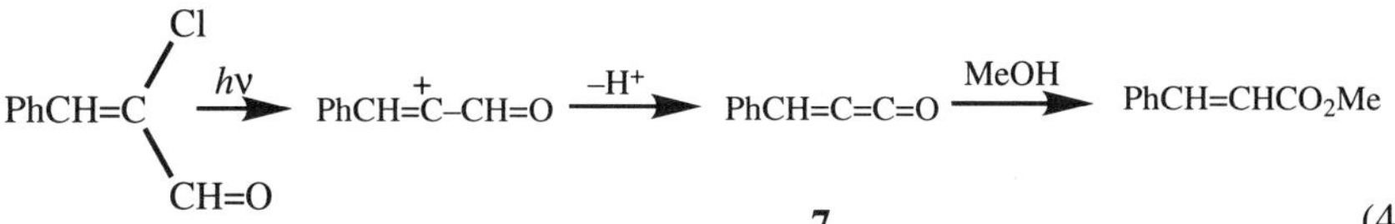

$$\underset{\mathbf{5}}{PhCH=C(Cl)(CH=O)} \xrightarrow{h\nu} \underset{\mathbf{6}}{PhCH=\overset{+}{C}-CH=O} \xrightarrow{-H^+} \underset{\mathbf{7}}{PhCH=C=C=O} \xrightarrow{MeOH} PhCH=CHCO_2Me \tag{4}$$

Pyrolysis or photolysis of phthalic anhydride or benzocyclobutadiene gave benzyne and the cumulated fulvenone **8**.[21] The latter has an IR band at 2085 cm^{-1} which was previously thought to result from benzyne. The structure of **8** has been determined by ab initio MO calculations.[22,23]

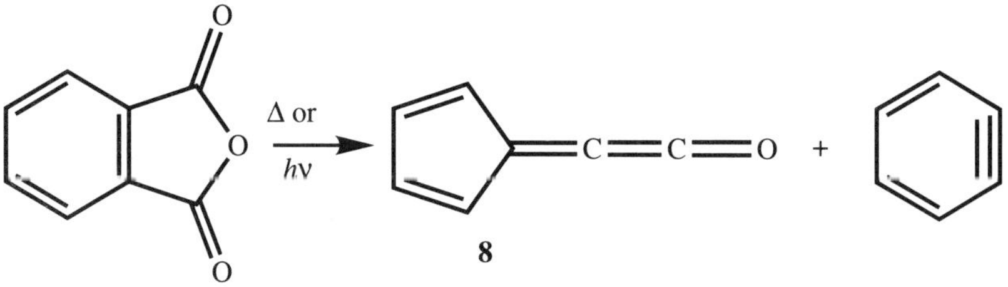

$$\text{phthalic anhydride} \xrightarrow{\Delta \text{ or } h\nu} \underset{\mathbf{8}}{[\text{fulvenylidene}]=C=C=O} + \text{benzyne} \tag{5}$$

The methyleneketene **9** was the first member of this class obtained as a stable compound in solution at room temperature, and for which a ^{13}C NMR spectrum was measured.[24] The ketene **10**, prepared similarly, was obtained in solution at -50 °C.[25,26]

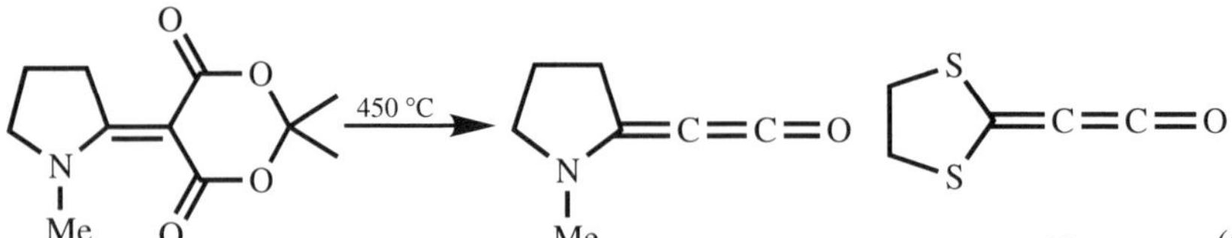

TYPES OF KETENES

[Equation (6) showing pyrolysis at 450 °C giving structure **9** and structure **10**]

Iminopropadienones **11** have been generated by flash vacuum pyrolysis of each of **12**, **13**, and **14**.[27]

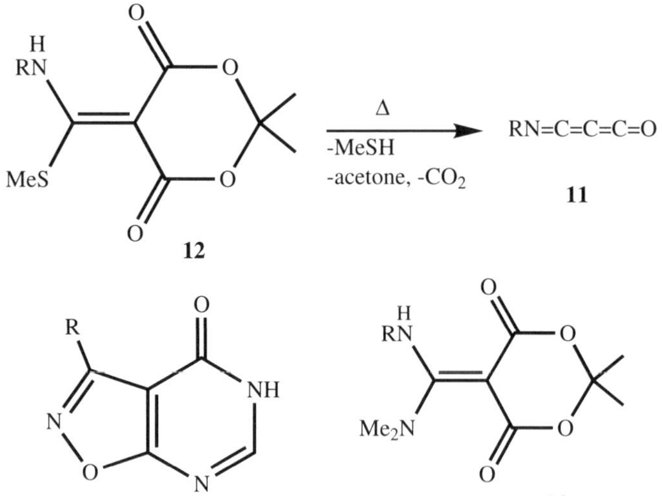

$$\text{12} \xrightarrow[\text{-acetone, -CO}_2]{\Delta, \text{-MeSH}} \text{RN=C=C=C=O} \quad \textbf{11}$$

(7)

[Structures **12**, **13**, and **14** shown]

Propadienones have also been obtained by low-temperature dehalogenation (equation 8),[28] and have been proposed as undetected intermediates from processes resembling E1cB type eliminations (equation 9).[29] Reported cumulenones are listed in Table 1.

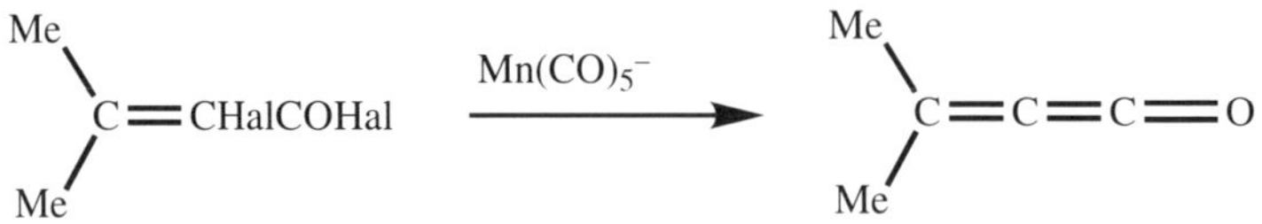

$$\text{Me}_2\text{C=CHalCOHal} \xrightarrow{\text{Mn(CO)}_5^-} \text{Me}_2\text{C=C=C=O} \quad (8)$$

$$\text{PhC≡CCO}_2\text{Me} \xrightarrow{i\text{-Pr}_2\text{NLi}} \underset{\text{Ph}}{\overset{i\text{-Pr}_2\text{N}}{\text{C=C=C}}}\text{(OMe)O}^- \rightarrow \underset{\text{Ph}}{\overset{i\text{-Pr}_2\text{N}}{\text{C=C=C=O}}} \rightarrow \underset{\text{Ph}}{\overset{i\text{-Pr}_2\text{N}}{\text{C=CHN(Pr-}i)_2}}$$

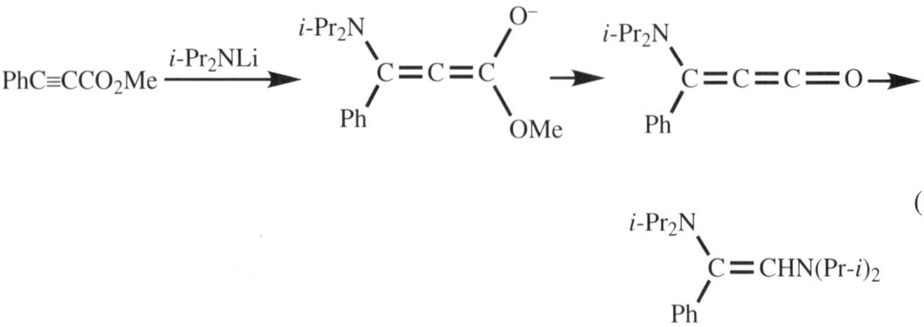

(9)

TABLE 1. Cumulenones

Ketene	Formula	Preparation	Registry No.	Ref.
C=C=O	C₂O	Theory	119754-08-4	30
C=C=¹⁸O	C₂O	Carbonylation	11127-14-3	31
¹⁸C=C=¹⁸O	C₂O	Carbonylation	11127-15-4	31
C=C=¹⁷O	C₂O	ESR	111915-86-7	32
S=C=C=O⁺	C₂OS⁺	MS	140895-21-2	33
O=C=¹³C=O⁺•	C₂O₂⁺•	Matrix	91777-09-2	34
¹⁷O=C=C=¹⁷O⁺	C₂O₂⁺•	Matrix	91777-12-7	34
¹⁷O=C=¹³C=O⁺	C₂O₂⁺•	Matrix	91777-13-8	34
O=¹³C=¹³C=O⁺	C₂O₂⁺•	Matrix	91777-10-5	34
O=C=C=¹⁷O⁺	C₂O₂⁺•	Matrix	91777-11-6	34
¹⁸O=C=C=¹⁸O⁺•	C₂O₂⁺•	Matrix	136232-98-9	35
¹⁸O=C=C=O⁺•	C₂O₂⁺•	Matrix	136232-99-0	35
¹⁸O=C=¹³C=O⁺•	C₂O₂⁺•	Matrix	136134-84-4	35
O=C=C=O²⁺	C₂O₂²⁺	Matrix, Theory	25535-27-7	71,72
O=¹³C=C=O⁻•	C₂O₂⁻•	Matrix	136232-97-8	35
O=¹³C=¹³C=O⁻•	C₂O₂⁻•	Matrix	136232-96-7	35
¹⁸O=C=C=O⁻•	C₂O₂⁻•	Matrix	136134-83-3	35
¹⁸O=C=C=¹⁸O⁻•	C₂O₂⁻•	Matrix	136134-82-2	35
O=C=C=O⁻•	C₂O₂⁻•	Theory	77783-15-4	36
¹⁸O=C=¹³C=O⁻•	C₂O₂⁻•	Matrix	136233-59-5	35
O=C=C=O.Na	C₂O₂⁻•Na⁺	Theory	111112-68-6	37
O=C=C=O.K	C₂O₂⁻•K⁺	Theory	111112-67-5	37
O=C=C=O²⁻	C₂O₂²⁻	Addition	138656-60-7	38
O₂C=C=O²⁻	C₂O₃²⁻	Addition	138452-69-4	38
CD₂=C=C=O	C₃D₂O	Pyrolysis	64918-71-4	39

TABLE 1. (*Continued*)

Ketene	Formula	Preparation	Registry No.	Ref.
$\overset{+}{F}C=C=O$	C_3FO^+	Theory	38264-01-6	40
$F_2C=C=O$	C_3F_2O	Pyrolysis	119820-21-2	41–43
$\overset{+}{C}H=C=O$	C_3HO^+	Theory		44
$\cdot CH=C=O$	$C_3HO^\cdot$	Theory		45
$CDH=C=O$	C_3HDO	Pyrolysis	64918-70-3	46
$CH_2=C=O$	C_3H_2O	Pyrolysis	61244-93-7	47
$CH_2=C=O$	C_3H_2O	Pyrolysis	109362-94-9	19
$CH_2=C=^{18}O$	C_3H_2O	Pyrolysis	79048-71-8	46
$^{13}CH_2=C=O$	C_3H_2O	Pyrolysis	79048-72-9	46
$CH_2=C=^{13}C=O$	C_3H_2O	Pyrolysis	96445-82-8	46
$CH_2=C=C=O^+$	$C_3H_2O^+$	Mass spec	87612-93-9	48
$CH_2=C=C=O(H^+)$	$C_3H_2O(H^+)$	Theory	101671-84-5	49
$HOCH=C=C=O$	$C_3H_2O_2$	Photolysis	82515-18-2	50
$H_2NCH=C=C=O$	C_3H_3NO	Thermolysis	112548-75-1	51
$H_2NCH=C=C=O^+$	$C_3H_3NO^+$	Mass spec	141305-77-3	51a
$C=C=C=^{18}O$	C_3O	Carbonylation	11127-21-2	31
$^{13}C=^{13}C=C=O$	C_3O	Carbonylation	11127-22-3	31
$C=C=^{13}C=O$	C_3O	Carbonylation	11127-20-1	31
$^{13}C=C=^{13}C=O$	C_3O	Carbonylation	11127-23-4	31
$C=^{13}C=C=O$	C_3O	Carbonylation	11127-19-8	31
$^{13}C=^{13}C=^{13}C=O$	C_3O	Carbonylation	11127-25-6	31
$^{13}C=C=C=O$	C_3O	Carbonylation	11127-18-17	31

Formula	Structure	Method	CAS	Ref
C_3O	$^{13}C=^{13}C=C=^{18}O$	Carbonylation	11127-26-7	31
C_3O	$C=C=C=O$	Carbonylation	11127-17-6	31
C_3O	$C=^{13}C=^{13}C=O$	Carbonylation	11127-24-5	31
C_3O^+	$^+C=C=C=O$	Carbonylation	116388-14-8	52
C_3O^-	$C=C=C=O^-$	Irradiation	95115-69-9, 70-1, 71-2	10
		Irradiation	109292-50-4	10
C_3OS	$S=C=C=C=O$	Gas phase	2219-62-7	53
C_3O_2	$O=C=C=C=O$	Dehalogenation	504-64-3	2,53
$C_3O_2^-$	$O=C=C=C=O^-$	Irradiation	109362-94-9	10
$C_3O_2^{2+}$	$O=C=C=C=O^{2+}$	Photolysis	51741-97-0	54
C_4D_2O	$CD_2=C=C=C=O$	Pyrolysis	71546-36-6	55
C_4HDO	$CHD=C=C=C=O$	Pyrolysis	71546-35-5	56
C_4H_2O	$CH_2=C=C=C=O$	Pyrolysis		56
C_4H_3NO	$MeN=C=C=C=O$	Pyrolysis		27
$C_4H_3NO.H^+$	$MeN=C=C=C=O.H^-$	Mass spec	141433-67-2	51a
C_4H_4O	$MeCH=C=C=O$	Pyrolysis	78957-08-1	57
$C_4H_4O^+$	$MeCH=C=C=O^+$	Mass spec	71793-84-5	58
$C_4H_4O_2$	$MeOCH=C=C=O$	Pyrolysis		26
C_4O^+	$^+C=C=C=C=O$	Carbonylation	116388-15-9	52
C_4O^+	$C=C=C=C=^{17}O$	Carbonylation	112022-65-8	59
C_4O	$^{13}C=^{13}C=^{13}C=C=O$	Carbonylation	112022-64-7	59
C_4O	$C=C=C=^{13}C=O$	Carbonylation	112022-63-6	59
C_4O_2	$O=C=C=^{13}C=O$	Pyrolysis	51799-35-0	13,14,60
C_5HDOS	$S=CHCD=C=C=C=O$	Pyrolysis	137040-75-6	61
C_5HDOS	$S=CDCH=C=C=C=O$	Pyrolysis	137040-76-7	61
C_5D_2OS	$S=CDCD=C=C=C=O$	Pyrolysis	137040-77-8	61
C_5H_2O	$CH_2=C=C=C=C=O$	Pyrolysis	87829-10-5	62
C_5H_2OS	$S=CHCH=C=C=^{18}C=O$	Pyrolysis	137040-78-9	61

TABLE 1. (*Continued*)

Ketene	Formula	Preparation	Registry No.	Ref.
S=CHCH=C=C=O	C_5H_2OS	Pyrolysis	137040-74-5	61
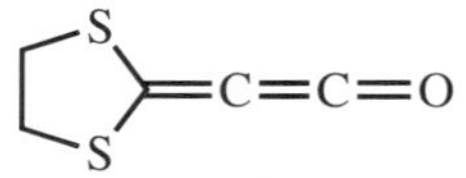 (dithiolane)=C=C=O	$C_5H_4OS_2$	Pyrolysis		25
$Me_2C=C=C=O$	C_5H_6O	Pyrolysis	63364-70-5	47
$Me(MeO)C=C=C=O$	$C_5H_6O_2$	Pyrolysis		26
$C=C=C=C=O$	C_5O	Theory	123340-79-4	30
$^+\!\cdot C=C=C=C=O$	$C_5O^{+\cdot}$	Gas phase	116410-64-1	52
$O=C=C=C=C=O$	C_5O_2	Theory	51799-36-1	15,60
$CH_2=C=C=C=C=O$	C_6H_2O	Theory	63766-92-7	63
$Me_2C=C=C=C=O$	C_6H_6O	Pyrolysis	135102-88-4	64
$^+\!\cdot C=C=C=C=C=O$	$C_6O^{+\cdot}$	Gas phase	116410-65-2	52
$C=C=C=C=C=C=O$	C_6O	Theory	112022-62-5	30,59
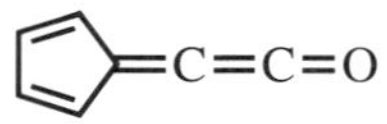 (cyclopentadienylidene)=C=O	C_7H_4O	Pyrolysis		21–23
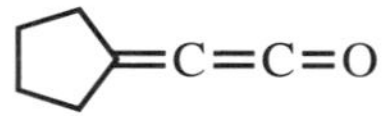 (cyclopentylidene)=C=O	C_7H_8O	Pyrolysis	96913-90-5	57,65

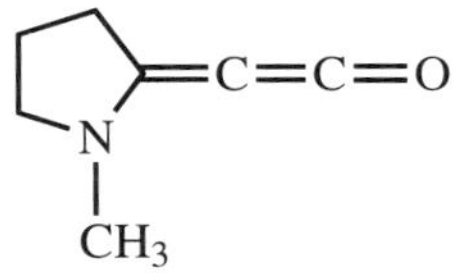

Structure	Formula	Method	Ref
pyrrolidine-N-C=C=O with CH₃	C₇H₉NO	Thermolysis	24
O=C(C=C)₃=O	C₇O₂	Pyrolysis	16,60
c-C₅H₉NDCH=C=C=O	C₈H₁₀DNO	Pyrolysis	66
c-C₅H₉NHCH=C=C=O	C₈H₁₁NO	Pyrolysis	66
PhN=C=C=C=O	C₉H₅NO	Pyrolysis	27
Me₂NC(Bu-t)=C=C=O	C₉H₁₅NO		67
4-TolNHCH=C=C=O	C₁₀H₉NO	Pyrolysis	66
PhNHC(SMe)=C=C=O	C₁₀H₉NOS	Pyrolysis	68
Ph₂C=C=C=O	C₁₂H₁₀O	Photolysis	69
i-Pr₂NCPh=C=C=O	C₁₅H₁₉NO	Elimination	29
Ph₂C=C=C=C=O	C₁₇H₁₀O	Synthetic goal	70

		63615-05-4	
		90368-24-4	
		90368-13-1	
		72393-90-0	
		90368-04-0	
		141075-42-4	
		115349-60-5	

References

1. Diels, O.; Wolf, B. *Chem. Ber.* **1906**, *39*, 689–697.
2. Staudinger, H.; Bereza, S. *Chem. Ber.* **1908**, *41*, 4461–4465.
3. Reyerson, L. H.; Kobe, K. *Chem. Rev.* **1930**, *7*, 479–492.
4. Ulrich, H. *Cycloaddition Reactions of Heterocumulenes;* Academic Press: New York, 1967; pp. 110–121.
5. Kappe, T.; Ziegler, E. *Angew. Chem., Int. Ed. Engl.* **1974**, *13*, 491–504.
6. Kappe, T. *Methoden der Organischen Chemie,* Vol. E15; Theime Verlag: Stuttgart, 1993.
7. Pandolfo, L.; Facchin, G.; Bertani, R.; Ganis, P.; Valle, G. *Angew. Chem., Int. Ed. Engl.* **1994**, *33*, 576–578.
8. Potts, K. T.; Murphy, P. M.; Kuehnling, W. R. *J. Org. Chem.* **1988**, *53*, 2889–2898.
9. Oakes, J. M.; Ellison, G. B. *Tetrahedron,* **1986**, *42*, 6263–6267.
10. Polanc, S.; Labille, M. C.; Janousek, Z.; Merenyi, R.; Vermander, M.; Viehe, H. G.; Tinant, B.; Piret-Meunier, J.; Declercq, J. P. *New J. Chem.* **1991**, *15*, 79–83.
11. Birney, D. M.; Berson, J. A. *J. Am. Chem. Soc.* **1985**, *107*, 4553–4554.
12. Berson, J. A.; Birney, D. M.; Dailey, W. P., III; Liebman, J. F. In *Molecular Structure and Energetics* Vol. 6, Chap. 9, VCH Publishers, New York, 1988.
13. Maier, G.; Reisenauer, H. P.; Balli, H.; Brandt, W.; Janoschek, R. *Angew. Chem., Int. Ed. Engl.* **1990**, *29*, 905–908.
14. Sülzle, D.; Schwarz, H. *Angew. Chem., Int. Ed. Engl.* **1990**, *29*, 908–909.
15. Maier, G.; Reisenauer, H. P.; Schäfer, U.; Balli, H. *Angew. Chem., Int. Ed. Engl.* **1988**, *27*, 566–568.
16. Meier, G.; Reisenauer, H. P.; Ulrich, A. *Tetrahedron Lett.* **1991**, *32*, 4469–4472.
17. Brown, R. F. C.; Eastwood, F. W. In *The Chemistry of Ketenes, Allenes, and Related Compounds;* Patai, S., Ed.; Wiley: New York, 1980; pp. 757–778.
18. Brown, R. F. C.; Eastwood, F. W. *Synlett.* **1993**, 9–19.
19. Chapman, O. L.; Miller, M. D.; Pitzenberger, S. M. *J. Am. Chem. Soc.* **1987**, *109*, 6868–6869.
20. Krijnen, E. S.; Lodder, G. *Tetrahedron Lett.* **1993**, *34*, 729–732.
21. Simon, J. G. G.; Münzel, N.; Schweig, A. *Chem. Phys. Lett.* **1990**, *170*, 187.
22. Scheiner, A. C.; Schaefer, H. F., III *J. Am. Chem. Soc.* **1992**, *114*, 4758–4762.
23. Scott, A. P.; Radom, L. *Chem. Phys. Lett.* **1992**, *200*, 15–20.
24. Lorencak, P.; Pommelet, J. C.; Chuche, J.; Wentrup, C. *J. Chem. Soc., Chem. Commun.* **1986**, 396–370.
25. Wentrup, C.; Kambouris, P.; Evans, R. A.; Owen, D.; Macfarlane, G.; Chuche, J.; Pommelet, J. C.; Ben Cheikh, A.; Plisnier, M.; Flammang, R. *J. Am. Chem. Soc.* **1991**, *113*, 3130–3135.
26. Ben Cheik, A.; Pommelet, J. C.; Chuche, J. *J. Chem. Soc., Chem. Commun.* **1990**, 615–617.
27. Mosandl, T.; Kappe, C. O.; Flammang, R.; Wentrup, C. *J. Chem. Soc., Chem. Commun.* **1992**, 1571–1573.
28. Masters, A. P.; Sorensen, T. S.; Tran, P. M. *Can. J. Chem.* **1987**, *65*, 1499–1502.

29. Shen, C. C.; Ainsworth, C. *Tetrahedron Lett.* **1979,** 89–92.
30. Ewing, D. W. *J. Am. Chem. Soc.* **1989,** *111,* 8809–8811.
31. DeKock, R. L.; Weltner, W., Jr. *J. Am. Chem. Soc.* **1971,** *93,* 7106–7107.
32. Van Zee, R. J.; Ferrante, R. F.; Weltner, W., Jr. *Chem. Phys. Lett.* **1987,** *139,* 426–430.
33. Sülzle, D.; Schwarz, H. *NATO ASI Ser., Ser. C,* **1991,** *347,* 237–248; *Chem. Abstr.* **1992,** *116,* 203753j.
34. Knight, L. B., Jr., Steadman, J.; Miller, P. K.; Bowman, D. E.; Davidson, E. R.; Feller, D. *J. Chem. Phys.* **1984,** *80,* 4593–4604.
35. Thompson, W. E.; Jacox, M. E. *J. Chem. Phys.* **1991,** *95,* 735–745.
36. Duke, B. J. *Theochem* **1990,** *64,* 279–285.
37. Ayed, O.; Manceron, L.; Silvi, B. *J. Phys. Chem.* **1988,** *92,* 37–45.
38. Zecchina, A.; Coluccia, S.; Spoto, G.; Scarano, D.; Marchese, L. *J. Chem. Soc. Faraday Trans.* **1990,** *86,* 703–709.
39. Blackman, G. L.; Brown, R. D.; Brown, R. F. C.; Eastwood, F. W.; McMullen, G. L. *J. Mol. Spectrosc.* **1977,** *68,* 488–491.
40. Jolly, W. L.; Gin, C. *Int. J. Mass Spectrom. Ion Phys.* **1977,** *25,* 27–37.
41. McNaughton, D.; Elmes, P. *Spectrochim. Acta, Part A* **1992,** *48A,* 605–611.
42. Tam, H. S.; Harmony, M. D.; Brahms, J. C.; Dailey, W. P. *J. Mol. Struct.* **1990,** *223,* 217–230.
43. Brahms, J. C.; Dailey, W. P. *J. Am. Chem. Soc.* **1989,** *11,* 3071–3073.
44. Goddard, J. D. *Chem. Phys. Lett.* **1984,** *109,* 170–174.
45. Tomasic, Z. A.; Scuseria, G. E. *J. Phys. Chem.* **1991,** *95,* 6905–6908.
46. Brown, R. D.; Champion, R.; Elmes, P. S.; Godfrey, P. D. *J. Am. Chem. Soc.* **1985,** *107,* 4109–4112.
47. Chuburu, F.; Lacombe, S.; Pfister-Guillouzo, G.; Ben Cheik, A.; Chuche, J.; Pommelet, J. C. *J. Am. Chem. Soc.* **1991,** *113,* 1954–1960.
48. Bouchoux, G.; Hoppilliard, Y.; Flament, J. P.; Terlouw, J. K.; van der Valk, F. *J. Phys. Chem.* **1986,** *90,* 1582–1585.
49. Hopkinson, A. C.; Lien, M. H. *J. Am. Chem. Soc.* **1986,** *108,* 2843–2849.
50. Maier, G.; Reisenauer, H. P.; Sayrac, T. *Chem. Ber.* **1982,** *115,* 2192–2201.
51. Wentrup, C.; Briehl, H.; Lorencak, P.; Vogelbacher, U. J.; Winter, H. W.; Maquestiau, A.; Flammang, R. *J. Am. Chem. Soc.* **1988,** *110,* 1337–1343.
51a. Brown, J.; Flammang, R.; Govaert, Y.; Plisnier, M.; Wentrup, C.; Van Haverbeke, Y. *Rapid Commun. Mass. Spectrom.* **1992,** *6,* 249–253.
52. Bohme, D. K.; Wlodek, S.; Williams, L.; Forte, L.; Fox, A. *J. Chem. Phys.* **1987,** *87,* 6934–6938.
53. Sülzle, D.; Schwarz, H. *J. Am. Chem. Soc.* **1991,** *113,* 48–51.
54. Karlsson, L.; Werme, L. O.; Bergmark, T.; Siegbahn, K. *J. Electron Spectrosc. Relat. Phenomena* **1974,** *3,* 181–189; *Chem. Abstr.* **1974,** *80,* 114397s.
55. Brown, R. D.; Godfrey, P. D.; Ball, M. J.; Godfrey, S.; McNaughton, D.; Rodler, M.; Kleiboemer, B.; Champion, R. *J. Am. Chem. Soc.* **1986,** *108,* 6534–6538.
56. Brown, R. F. C.; Coulston, K. J.; Eastwood, F. W.; Pullin, A. D. E.; Staffa, A. C. *Austr. J. Chem.* **1990,** *43,* 561–577.
57. Tseng, J.; McKee, M. L.; Shevlin, P. B. *J. Am. Chem. Soc.* **1987,** *109,* 5474–5477.

58. Holmes, J. L.; Terlouw, J. K. *J. Am. Chem. Soc.* **1979,** *101,* 4973–4975.
59. Van Zee, R. J.; Smith, G. R.; Weltner, W., Jr. *J. Am. Chem. Soc.* **1988,** *110,* 609–610.
60. Brown, L. D.; Lipscomb, W. N. *J. Am. Chem. Soc.* **1977,** *99,* 3968–3979.
61. Teles, J. H.; Hess, B. A., Jr.; Schaad, L. J. *Chem. Ber.* **1992,** *125,* 423–431.
62. Brown, R. F. C.; Coulston, K. J.; Eastwood, F. W.; Irvine, M. J.; Pulin, A. D. E. *Aust. J. Chem.* **1988,** *41,* 225–233.
63. Farnell, L.; Radom, L. *J. Am. Chem. Soc.* **1984,** *106,* 25–28.
64. Brown, R. F. C.; Coulston, K. J.; Eastwood, F. W.; Irvine, M. J. *Aust. J. Chem.* **1991,** *44,* 87–101.
65. Wentrup, C.; Gross, G.; Bestermann, H. M.; Lorencak, P. *J. Org. Chem.* **1985,** *50,* 2877–2881.
66. Briehl, H.; Lukosch, A.; Wentrup, C. *J. Org. Chem.* **1984,** *49,* 2772–2779.
67. Galloy, J.; Declercq, J. P.; Germain, G.; van Meerssche, M. *Bull. Soc. Chim. Belg.* **1979,** *88,* 343–344.
68. Kappe, C. O.; Kollenz, G.; Leung-Toung, R.; Wentrup, C. *J. Chem. Soc., Chem. Commun.* **1992,** 487–488.
69. Adam, W.; Berkessel, A.; Peters, E-M.; Peters, K.; von Schnering, H. G. *J. Org. Chem.* **1985,** *50,* 2811–2814.
70. Browne, N. R.; Brown, R. F. C.; Eastwood, F. W.; Fallon, G. D. *Aust. J. Chem.* **1987,** *40,* 1675–1686.
71. Adelheim, M. *Angew. Chem., Int. Ed. Engl.* **1969,** *8,* 516–517.
72. Prakash, G. K. S.; Bausch, J. W.; Olah, G. A. *J. Am. Chem. Soc.* **1991,** *113,* 3203–3205.
73. Sülze, D.; Weiske, T.; Schwarz, H. *Int. J. Mass. Spectr. Ion Proc.* **1993,** *125,* 75–79.

CHAPTER 5

REACTIONS OF KETENES

5.1 OXIDATION AND REDUCTION OF KETENES (ELECTRON TRANSFER)

Ketenes are susceptible to both oxidation and reduction, involving removal and addition of electrons, respectively. The parent radical cation $CH_2=C=O^{+\bullet}$ generated by one-electron removal from ketene has been examined by photoelectron spectroscopy,[1] and its rearrangement and cycloaddition reaction with ethylene have been studied.[2–4]

Further ionization of the cation radical $CH_2=C=O^{+\bullet}$, generated by 70-eV electron impact of ketene or by loss of H_2O from the molecular ion of acetic acid, led to the dication **2** by charge stripping in a mass spectrometer (equation 1).[5] This ion can be generated in the gas phase, although it is very unstable and is expected to be highly oxidizing toward other molecules. Theoretical calculations predict a planar structure, and at the MP2/6-31G*//4-31G level a 27.2 kcal/mol greater stability than the oxirene structure **3**.[5]

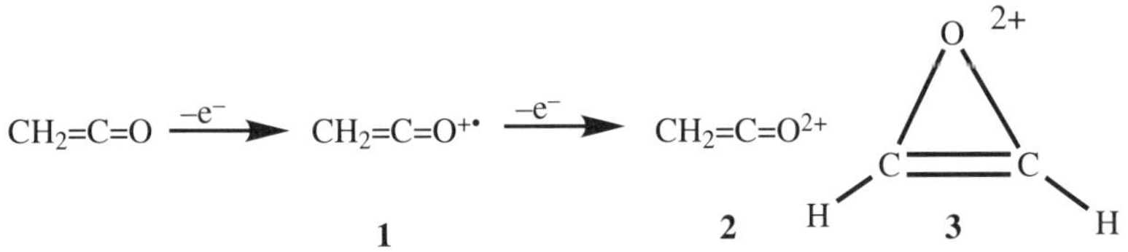

$$CH_2=C=O \xrightarrow{-e^-} CH_2=C=O^{+\bullet} \xrightarrow{-e^-} CH_2=C=O^{2+} \qquad (1)$$

$$\mathbf{1} \qquad\qquad \mathbf{2} \qquad \mathbf{3}$$

The electrochemical oxidation of diphenylketene in CH_3CN leads to benzophenone,[6] and could occur by attack by adventitious H_2O (equation 2) or by a related process including a Ritter reaction step with the solvent.[7]

$$Ph_2C=C=O \xrightarrow{-e^-} Ph_2C=C=O^{+\bullet} \xrightarrow[-H^+]{H_2O} Ph_2\overset{\bullet}{C}-\underset{OH}{\overset{}{C}}=O \xrightarrow{-CO} Ph_2\overset{\bullet}{C}OH \quad (2)$$

$$\phantom{Ph_2C=C=O \xrightarrow{-e^-} }\phantom{Ph_2C=C=O^{+\bullet}}\mathbf{4}$$

Oxidation potentials E_p (versus SCE) in CH_3CN have been obtained by cyclic voltametry for 4-MeOC$_6$H$_4$CMe=C=O (**5**, E_p = 0.91 V),[8,9] 4-MeC$_6$H$_4$CMe=C=O (**6**, E_p = 1.11 V),[8,9] and t-Bu$_2$C=C=O (**7**, E_p = 1.44 V).[10] The ease of the oxidation is clearly dependent on the electron donor ability of the ketene substituents. As described in Section 5.4.3, the radical cations of **5** and **6** gave [4 + 2] cycloadditions with pentamethylcyclopentadiene using tris(4-tolyl)aminium hexafluoroantimonate (4-Tol$_3$N$^{+\bullet}$SbF$_6^-$) as an electron transfer catalyst (equation 3).[8,9] The cation radical of Me$_2$C=C=O was generated by γ-irradiation in a solid solution at 77 K, and its ESR spectrum was observed.[11]

$$\underset{Me}{\overset{Ar}{>}}C=C=O \;+\; \text{(pentamethylcyclopentadiene)} \longrightarrow \text{(cycloadduct } \mathbf{8}\text{)} \quad (3)$$

5 (Ar = 4-MeOC$_6$H$_4$)

6 (Ar = 4-MeC$_6$H$_4$)

In contrast to the frequent observations of ketene oxidations, their reduction is much rarer. Attempted electrochemical reduction of the ketenes **7** and **9–12** was unsuccessful, as no reduction was observed up to -3 V, and the electrolyte began to decompose at higher potentials.[12] However, a reduction potential for Ph$_2$C=C=O at -2.06 V was observed, but an ESR spectrum of the radical ion could not be observed.[12]

t-Bu$_2$C=C=O t-BuC(Pr–i)=C=O

7   **9**

R = H; R^1,R^2 = Me (**10**)

R,R^1 = t–Bu; R^2 = H (**11**)

R,R^2 = H; R^2 = t–Bu (**12**)

5.1 OXIDATION AND REDUCTION OF KETENES (ELECTRON TRANSFER)

Evidence for the ketene radical anion $CH_2=C=O^{-\bullet}$ was obtained by reaction of vinylene carbonate radical anion **13**, generated using low-energy electrons in the gas phase.[13,14] This was interpreted as giving the radical anion **14**, which rearranges to **15**, which undergoes loss of two electrons to the ketocarbene radical cation **16**.[13,14] Formation of **15** as an intermediate from radiolysis of ethylene glycol has also been suggested.[15]

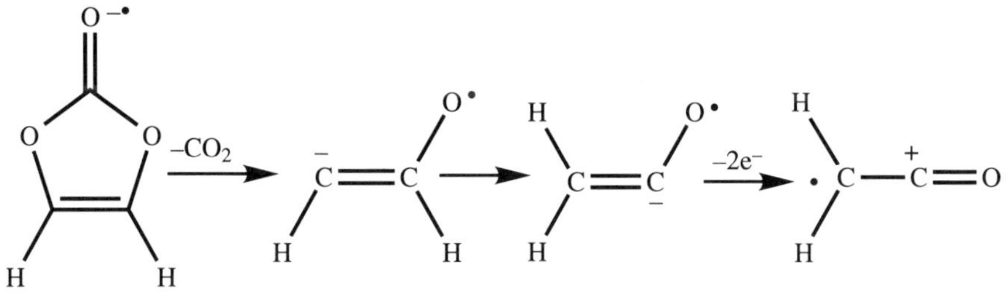

13 → (−CO_2) **14** → **15** → (−2e⁻) **16**

The radical anions of the thioketenes **17–19** were generated electrochemically and their ESR spectra measured.[12]

t-Bu_2C=C=S t–BuC(Pr–i)=C=S

17   **18**   **19**

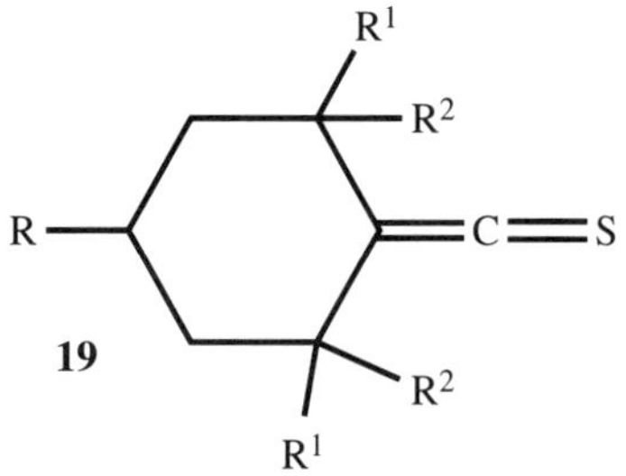

References

1. Hall, D.; Maier, J. P.; Rosmus, P. *Chem. Phys.* **1977**, *24*, 373–378.
2. Heinrich, N.; Koch, W.; Morrow, J. C.; Schwarz, H. *J. Am. Chem. Soc.* **1988**, *110*, 6332–6336.
3. Hop, C. E. C. A.; Holmes, J. L.; Terlouw, J. K. *J. Am. Chen. Soc.* **1989**, *111*, 441–445.
4. Dass, C Gross, M. L. *J. Am. Chem. Soc.* **1984**, *106*, 5775–5780.
5. Koch, W.; Maquin, F.; Stahl, D.; Schwarz, H. *J Chem Soc., Chem. Commun,* **1984**, 1679–1680.
6. Becker, J. Y.; Zinger, B. *J. Am. Chem. Soc.* **1982**, *104*, 2327–2329.
7. These mechanisms were suggested by Professor J. P. Dinnocenzo, University of Rochester.
8. Schmittel, M.; von Seggern, H. *Angew. Chem., Int. Ed. Engl.* **1991**, *30*, 999–1001.
9. Schmittel, M.; von Seggern, H. *J. Am. Chem. Soc.* **1993**, *115*, 2165–2177.
10. D. Wayner, NRC Canada, unpublished results.

11. Fujisawa, J.; Sato, S.; Shimokoshi, K.; Shida, T. *Bull. Chem. Soc. Jpn.* **1985**, *58*, 1267–1272.
12. Klages, C.; Köhler, S.; Schaumann, E.; Schmüser, W.; Voss, J. *J. Phys. Chem.* **1979**, *83*, 738–741.
13. van Baar, B. L. M.; Heinrich, N.; Koch, W.; Postma, R.; Terlouw, J. K.; Schwarz, H. *Angew. Chem., Int. Ed. Engl.* **1987**, *26*, 140–142.
14. Compton, R. N.; Reinhardt, P. W.; Schweinler, H. C. *Int. J. Mass. Spectrom. Ion Phys.* **1983**, *49*, 113–122.
15. Kartasheva, L. I.; Pikaev, A. K. *Khim. Vys. Energ.* **1975**, *9*, 242–246; *Chem. Abstr.* **1975**, *83*, 88640f.

5.2 PHOTOCHEMICAL REACTIONS

The photochemistry of ketenes is a classical research problem that has been pursued for more than 50 years, and the early work has been reviewed.[1–5] Much of the interest in this problem is due to the simple unimolecular dissociation that occurs to form methylene and carbon monoxide in this process (equation 1), and the desire to account for the energy change in the processs and for the singlet–triplet energy difference in the carbene. The photodissociation of ketene has a quantum yield near unity at low pressures, but this is only 0.1 at 3000 torr.[4,5] No light emission is detected, and so nonradiative decay processes are implicated.[4,5] The possibility of the interconversion of ketene, oxirene, and formylcarbene, and the relative stabilities of these isomeric species, has also attracted wide interest (equation 2).[6,7] Some information on the UV spectra of ketenes that is relevant to ketene photochemistry is contained in Section 2.2.

$$CH_2=C=O \xrightarrow{h\nu} CH_2 + CO \quad (1)$$

$$CH_2=C=O \rightleftharpoons \underset{H}{\overset{O}{\triangle}}\underset{H}{} \rightleftharpoons \underset{H}{\overset{O}{C-C}}\underset{H}{} \quad (2)$$

Photolysis of $^{14}CH_2=C=O$ leads to both ^{12}CO and ^{14}CO, attributed to formation of the oxirene **1** (equation 3), with the percentage of scrambling being greater at lower pressures.[6] A slow increase in the yield of ^{14}CO with time suggested that some of the oxirene was forming rearranged ketene (equation 4).

$$^{14}CH_2=C=O \xrightarrow{h\nu} \underset{H^{14}C \equiv\!=\!= CH}{\overset{O}{\triangle}} \longrightarrow {}^{14}CO \quad (3)$$

1

$$\mathbf{1} \longrightarrow CH_2={}^{14}C=O \qquad (4)$$

Rearrangement of highly vibrationally excited ketene labeled with ^{13}C occurs as shown in equation 5, followed by dissociation (equation 6).[7–9] The kinetics of these exchanges have been analyzed. The rotational-state distributions of the CO formed[10] and the singlet/triplet branching ratio[11,12] in ketene photodissociation have also been studied. The rate constants for dissociation of highly vibrationally excited ketene (equation 6) increase in a stepwise manner with increasing energy, consistent with the premise that the rate of a unimolecular reaction is controlled by the quantized transition state thresholds.[8,9]

$$^{13}CH_2=C=O \rightleftharpoons CH_2={}^{13}C=O \qquad (5)$$

$$CH_2=C=O \longrightarrow CH_2 + CO \qquad (6)$$

The addition of CH_2 produced by ketene photolysis to ketene to yield ethylene (equation 7) is a process that occurs on NaCl[13] and in the gas phase.[14] Hydrogen abstraction by triplet methylene has also been considered (equation 8),[15] as well as reaction of atomic hydrogen with ketene.[14]

$$CH_2 + CH_2=C=O \longrightarrow CH_2=CH_2 + CO \qquad (7)$$

$$^3CH_2 + CH_2=C=O \longrightarrow \dot{C}H_3 + H\dot{C}=C=O \qquad (8)$$

Photolysis of ketene in an Ar matrix with a 308-nm XeCl laser at 12 K led to isomerization to ethynol (**2**).[16] This irradiation would excite the n → π* absorption of ketene at 310 nm, and the reaction was proposed to occur by decarbonylation to CH_2 and CO and insertion of CO into the C—H bond of CH_2 (equation 9). Photolysis of $CH_2=C=O(NO)$ pairs isolated in an Ar matrix gave iminoxy radicals $CH_2=N-O^\bullet$ by a photochemical carbenoid reaction.[16a]

$$CH_2=C=O \xrightarrow[308\ nm]{h\nu} CH_2 + CO \longrightarrow HC\equiv COH \qquad (9)$$
$$\phantom{CH_2=C=O \xrightarrow[308\ nm]{h\nu} CH_2 + CO \longrightarrow}\mathbf{2}$$

Photolyses of higher ketenes include studies of $MeCH=C=O$,[17] $Me_2C=C=O$,[18–20] and $Ph_2C=C=O$.[21] These reactions all give the corresponding carbenes, and for methylketene and dimethylketene in the gas phase $CH_2=CH_2$ and $CH_3CH=CH_2$ are the major products, respectively.[17,18] Photolysis of $Me_2C=C=O$ in cyclohexane solution also gave $Me_2C=\dot{C}Me_2$, which was proposed to form via cyclopropanone **3** (equation 10) which was observed by IR.[19] The quantum yield of photodissociation of $CH_3CH=C=O$ is less than that for $CH_2=C=O$.[17] For $Me_2C=C=O$ the quantum yield for dissociation in the gas phase was unity at 254 nm, but for photolysis at 366 nm the quantum yield was pressure dependent, indicating the formation of a long-lived excited state.[18] Photolysis of $t\text{-}Bu_2C=C=O$ may also involve the carbene.[21a,b]

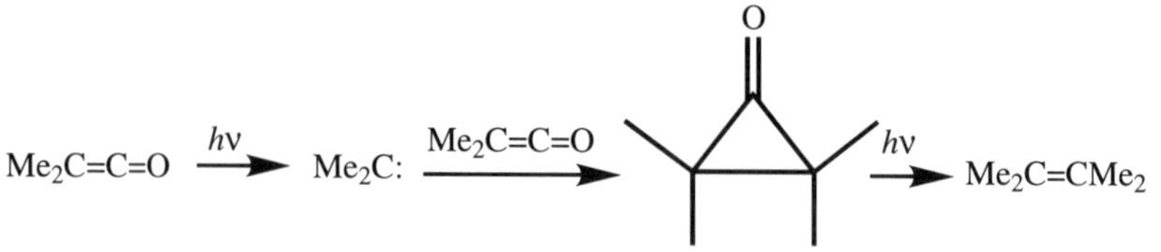

$$\text{Me}_2\text{C=C=O} \xrightarrow{h\nu} \text{Me}_2\text{C:} \xrightarrow{\text{Me}_2\text{C=C=O}} \mathbf{3} \xrightarrow{h\nu} \text{Me}_2\text{C=CMe}_2 \quad (10)$$

Dimethylketene was formed in an argon matrix at 10 K by photolysis at 254 nm or greater than 570 nm of dimethylpyrazolen-3,5-dione (equation 11), but the ketene was practically photostable in the matrix on further photolysis at these wavelengths.[20] However photolysis at 222 nm gave allene via dimethylcarbene (equation 11).

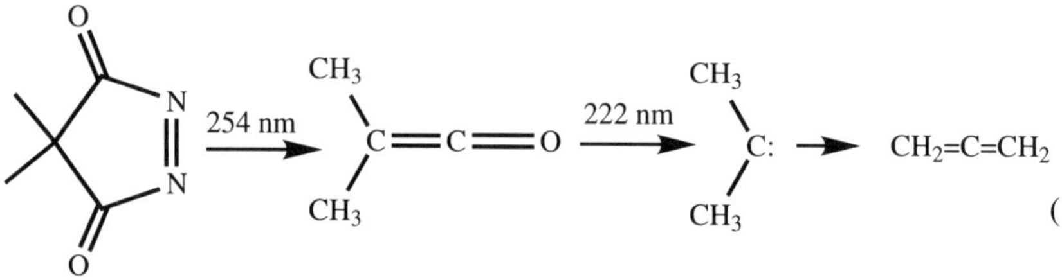

$$\text{(dimethylpyrazolendione)} \xrightarrow{254\text{ nm}} (\text{CH}_3)_2\text{C=C=O} \xrightarrow{222\text{ nm}} (\text{CH}_3)_2\text{C:} \longrightarrow \text{CH}_2\text{=C=CH}_2 \quad (11)$$

Photolysis of $\text{Ph}_2\text{C=C=O}$ in several solvents gave products expected from diphenylcarbene intermediates, but photolysis in THF gave the diphenylacetyl derivative **4**, which was proposed to form from the radical pair resulting from hydrogen atom abstraction by the photoexcited ketene from the solvent to give the radical pair **5**, which collapsed to **4** (equation 11).[21]

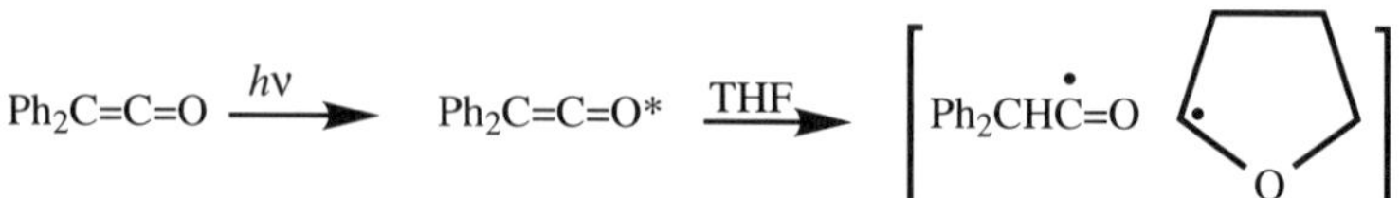

$$\text{Ph}_2\text{C=C=O} \xrightarrow{h\nu} \text{Ph}_2\text{C=C=O*} \xrightarrow{\text{THF}} [\text{Ph}_2\text{CHC=O} \cdot \text{(THF radical)}] \quad (12)$$
$$\mathbf{5}$$

$$\mathbf{5} \longrightarrow \text{Ph}_2\text{CHC(O)-(tetrahydrofuryl)}$$
$$\mathbf{4}$$

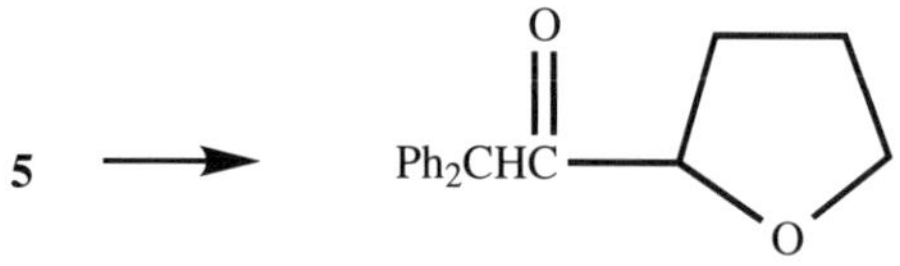

Photolysis of dimesitylketene gives the carbene which dimerizes to the tetraarylethylene (equation 13).[22]

$$\text{Mes}_2\text{C=C=O} \xrightarrow{h\nu} \text{Mes}_2\text{C:} \longrightarrow \text{Mes}_2\text{C=CMes}_2 \quad (13)^{22}$$

$$\text{Mes} = 2,4,6\text{-Me}_3\text{C}_6\text{H}_2$$

Photolysis of **6** in pentane occurs by decarbonylation to the carbene **7**, which forms **8** by C—H insertion.[23] However, when the photolysis is conducted in CH_3OH at -60 °C, the rearranged aldehyde **9** is formed along with **8** in a 1:1 ratio. Proton incorporation in **9** from solvent was established by an experiment using CH_3OD, and the formation of **9** was proposed to occur by protonation at C_α, which bears a partial negative charge in the excited state of **6** to give the cation **10**, which rearranges and deprotonates to form **9**.

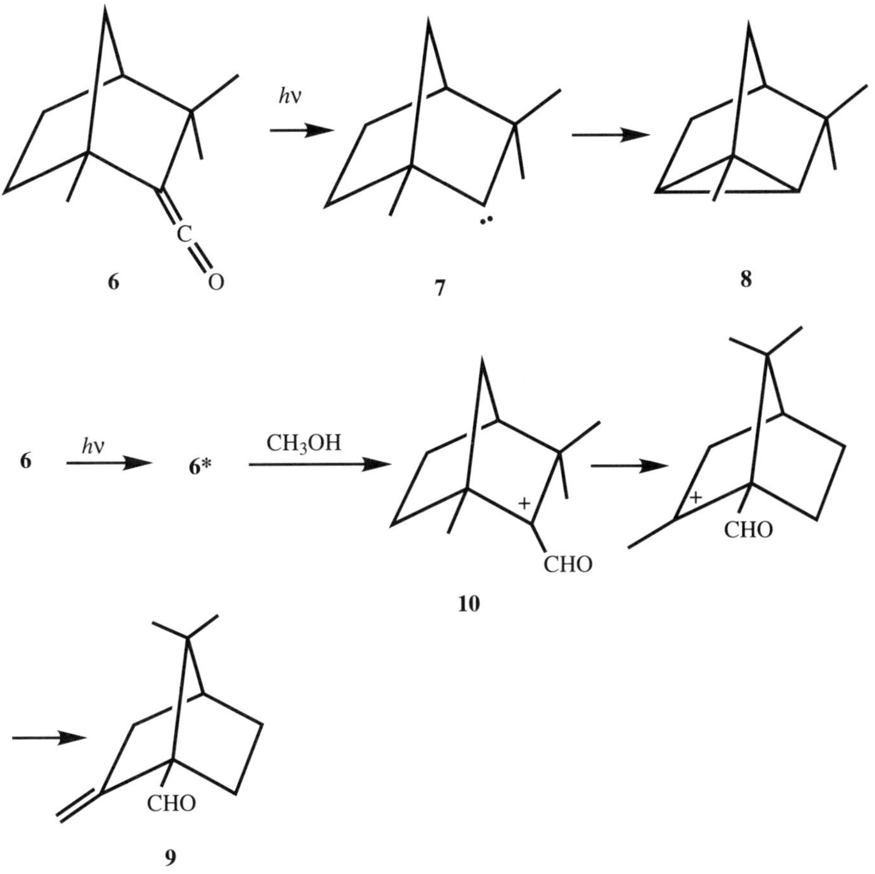

Photolysis of **11** in pentane gave the carbene **12** which formed **13** and **14** in a 30:1 ratio.[23] Photolysis in CH_3OH gave the alkenes **15** and **16** and a trace of the aldehyde **17** in addition to **13** and **14**. It was suggested that **11** also reacted in CH_3OH by protonation and rearrangement to the aldehydes **17** and **18**, which were further photolyzed to **15** and **16**.[23] Photolysis of **11** in CH_3OD led to **15** and **16** with deuterium labeling at the tertiary carbon.

11 12 13 14

15 16

17 18

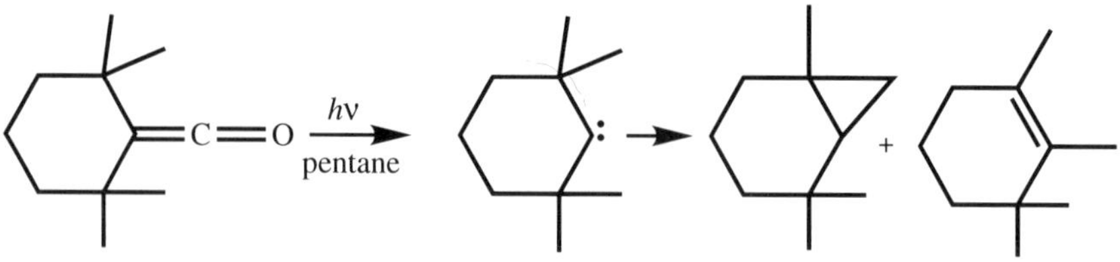

Photolysis of **19** in pentane and CH_3OH or CH_3OD led only to the alkene **22,** without deuterium incorporation, presumably by the carbene **21.** It was suggested that the excited ketene **19** gave the carbene faster than it was protonated by CH_3OH.[23]

19 21 22

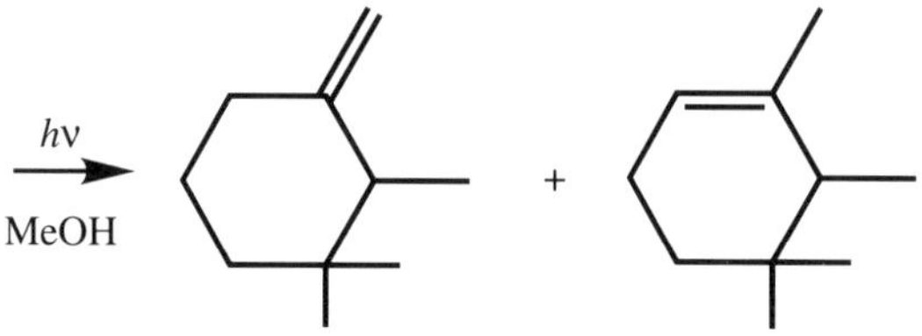

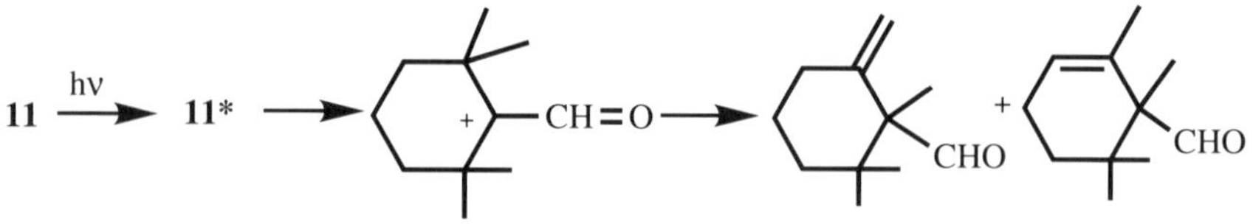

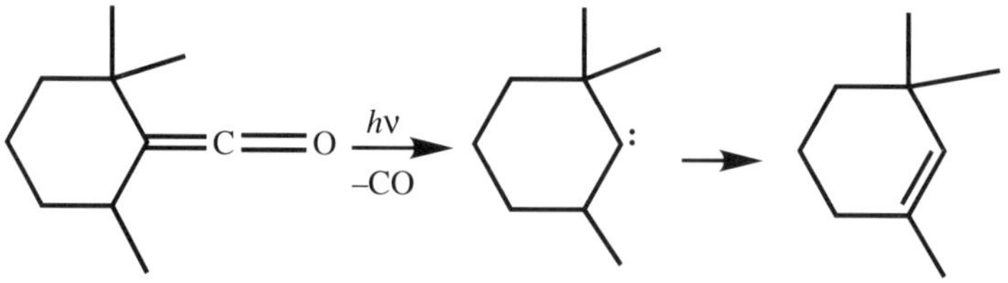

Photolysis of acylketenes can lead to formation of new ketenes by decarbonylation and Wolff rearrangement, as in equation 14.[24]

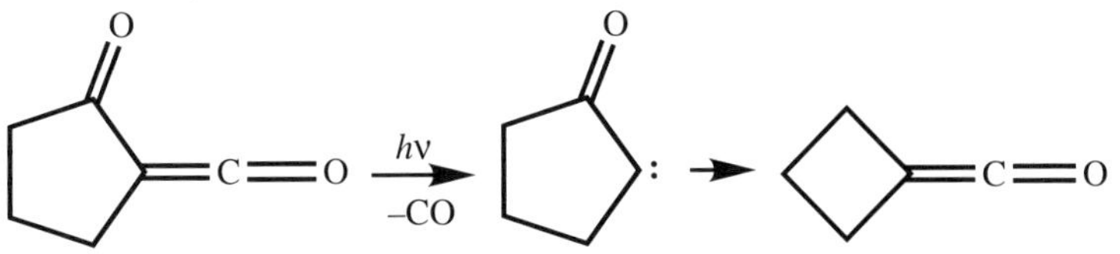

Photolysis of the bis(diazo)ketone **23** in an Ar matrix at 10 K gave the bisketene **24** and then cyclopropenone, presumably through the intermediacy of the ketenyl-carbene **25** (equation 15).[25] Photolysis of formylketene (OCHCH=C=O) under these conditions gave ketene.[25]

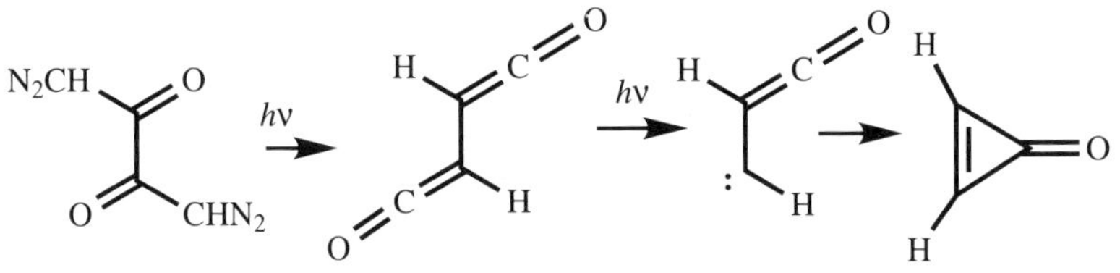

There are many photochemical routes to ketenes, and in some cases (cf. equation 5, Section 3.4.2) further photochemical reactions of the ketenes formed have been observed.

The photolysis of carbon suboxide (Section 4.11) in MeOH leads to methyl malonate (equation 16), as well as products derived from dissociation of the C_3O_2.[26]

$$C_3O_2 + MeOH \xrightarrow{h\nu} CH_2(CO_2Me) \qquad (16)$$

References

1. Kirmse, W. *Carbene Chemistry,* 2nd ed.; Academic: New York, 1971; Chapter 2.
2. Johnson, R. P. In *Organic Photochemistry;* Padwa, A., Ed.; Dekker: New York, 1985; Vol. 7, Chapter 2.
3. Zeller, K.-P.; Gugel, H. In *Methoden der Organischen Chemie,* 4th ed., Vol. E19b, *Carbene (Carbenoide);* Regitz, M., Ed.; Thieme: Stuttgart, 1989; pp. 325–336.
4. DeMore, W. B.; Benson, S. W. *Adv. Photochem.* **1964,** *2,* 219–261.
5. Lee, E. K. C.; Lewis, R. C. *Adv. Photochem.* **1980,** *12,* 1–96.
6. Russell, R. L.; Rowland, F. S. *J. Am. Chem. Soc.* **1970,** *92,* 7508–7510.
7. Lovejoy, E. R.; Kim, S. K.; Alvarez, R. A.; Moore, C. B. *J. Chem. Phys.* **1991,** *95,* 4081–4093.
8. Lovejoy, E. R.; Kim, S. K.; Moore, C. B. *Science* **1992,** *256,* 1541–1544.
8a. Lovejoy, E. R.; Moore, C. B. *J. Chem. Phys.* **1993,** *98,* 7846–7854.
9. Marcus, R. A. *Science,* **1992,** *256,* 1523–1524.
10. Chen, I.; Moore, C. B. *J. Phys. Chem.* **1990,** *94,* 263–269; 269–274.

11. Kim, S. K.; Choi, Y. S.; Pibel, C. D.; Zheng, Q.; Moore, C. B. *J. Chem. Phys.* **1991,** *94,* 1954–1960.
12. Green, W. A., Jr.; Mahoney, A. J.; Zheng, Q.; Moore, C. B. *J. Chem. Phys.* **1991,** *94,* 1961–1969.
13. Berg, O.; Ewing, G. E. *J. Phys. Chem.* **1991,** *95,* 2908–2916.
14. Becerra, R.; Canosa-Mas, C. E.; Frey, H. M.; Walsh, R. *J. Chem. Soc., Faraday Trans. 2,* **1987,** *83,* 435–448.
15. Banyard, S. A.; Canosa-Mas, C. E.; Ellis, M. D.; Frey, H. M.; Walsh, R. *J. Chem. Soc., Chem. Commun.* **1980,** 1156–1157.
16. Hochstrasser, R.; Wirz, J. *Angew. Chem., Int. Ed. Engl.* **1990,** *29,* 411–413.
16a. McCluskey, M.; Frei, H. *J. Phys. Chem.* **1993,** *97,* 5204–5207.
17. Chong, D. P.; Kistiakowsky, G. B. *J. Phys. Chem.* **1964,** *68,* 1793–1797.
18. Holroyd, R. A.; Blacet, F. E. *J. Am. Chem. Soc.* **1957,** *79,* 4830–4834.
19. Haller, I.; Srinivasan, R. *J. Am. Chem. Soc.* **1965,** *87,* 1144–1145.
20. Maier, G.; Heider, M.; Sierakowski, C. *Tetrahedron Lett.* **1991,** *32,* 1961–1962.
21. Nozaki, H.; Nakano, M.; Kondo, K. *Tetrahedron* **1966,** *22,* 477–481.
21a. Malatesta, V.; Forrest, D.; Ingold, K. U. *J. Phys. Chem.* **1978,** *82,* 2370–2373.
21b. Marfisi, C.; Verlaque, P.; Davidovics, G.; Pourcin, J.; Pizzala, L.; Aycard, J.-P.; Bodot, H. *J. Org. Chem.* **1983,** *48,* 533–537.
22. Zimmerman, H. E.; Paskovich, D. H. *J. Am. Chem. Soc.* **1964,** *86,* 2149–2160.
23. Kirmse, W.; Spaleck. W. *Angew. Chem., Int. Ed. Engl.* **1981,** *20,* 776–777.
24. Leung-Toung, R.; Wentrup, C. *J. Org. Chem.* **1992,** *57,* 4850–4858.
25. Maier, G.; Reisenauer, H. P.; Sayrac, T. *Chem. Ber.* **1982,** *115,* 2192–2201.
26. Hagelloch, G.; Feess, E. *Chem. Ber.* **1951,** *84,* 730–733.

5.3 THERMOLYSIS REACTIONS

With rare exceptions ketenes are thermodynamically stable relative to formation of a carbene and carbon monoxide. The notable exception is $CF_2=C=O$, which spontaneously forms CF_2 at room temperature (Section 4.4.1). However, at higher temperatures carbene formation from many ketenes becomes significant.

Very early Staudinger and Endle found that pyrolysis of $Ph_2C=C=O$ at 600–700 °C gives diphenylcarbene, which forms fluorene (equation 1).[1] Phenylketene formed by Wolff rearrangement decarbonylates at 800 °C and gives the fulvenallene **1** (equation 2).[2] Pyrolysis of $Me_2C=C=O$ gave propylene and tetramethylethylene, which were attributed to initial formation of Me_2C.[1]

$Ph_2C=C=O \xrightarrow{\Delta} Ph_2C: \longrightarrow$ [fluorene] (1)

$$\text{PhCOCHN}_2 \xrightarrow{\Delta} \text{PhCH=C=O} \xrightarrow{-CO} \text{Ph}\ddot{\text{C}}\text{H} \rightarrow \underset{\mathbf{1}}{\text{[cyclopentadienylidene]}}\!\!=\!\!\text{C=CH}_2 \qquad (2)$$

The thermal decarbonylation of ketenes to give carbenes is a reversible process, as shown by the formation of ketenes from carbenes generated from diazoalkanes in the presence of carbon monoxide (equation 3).[3–5] Other examples of ketene formation from CO addition to carbenes are known (cf. Sections 4.1.4 and 4.1.10).

$$\text{CH}_3\text{CHN}_2 \xrightarrow{h\nu} \text{CH}_3\ddot{\text{C}}\text{H} \xrightarrow{\text{CO}} \underset{\mathbf{2}}{\text{CH}_3\text{CH=C=O}} \qquad (3)$$

Thermolysis of methylketene (**2**) at 360–540 °C gave a complex mixture of products by two processes that are each kinetically three-halves order in **2** and were interpreted as each involving formation of the radical $\dot{\text{C}}\text{H}_2\text{CH=C=O}$ (**3**) in an initial surface-catalyzed reaction.[6] Major products included CO, CO_2, 2,3-pentadiene, and 2-butene. Some of the steps proposed are shown in equations 4–8.[6]

$$\underset{\mathbf{2}}{\text{CH}_3\text{CH=C=O}} \xrightarrow{\text{surface}} \underset{\mathbf{3}}{\dot{\text{C}}\text{H}_2\text{CH=C=O}} \qquad (4)$$

$$\underset{\mathbf{3}}{\dot{\text{C}}\text{H}_2\text{CH=C=O}} + \underset{\mathbf{2}}{\text{CH}_3\text{CH=C=O}} \longrightarrow \underset{\mathbf{4}}{\text{[β-lactone with CH}_3\text{ and }\dot{\text{C}}\text{H}_2\text{CH]}} \qquad (5)$$

$$\mathbf{4} \xrightarrow{-CO_2} \dot{\text{C}}\text{H}_2\text{CH=C=CHCH}_3 \xrightarrow{\mathbf{2}} \text{CH}_3\text{CH=C=CHCH}_3 \qquad (6)$$

$$\mathbf{3} \xrightarrow{-CO} \dot{\text{CH}}_2\text{-CH} \underset{-\mathbf{3}}{\xrightarrow{\mathbf{2}}} \underset{\mathbf{5}}{\text{CH}_3\text{CH=CH}\dot{\text{C}}\text{H}_2} \qquad (7)$$

$$\mathbf{5} + \text{CH}_3\text{CH=C=O} \longrightarrow \dot{\text{C}}\text{H}_2\text{CH=C=O} + \text{CH}_3\text{CH=CHCH}_3 \qquad (8)$$

Pyrolysis of the ketene **6** is proposed to occur by retro-Wolff rearrangement to give **7** by the series of steps shown (equation 9).[7]

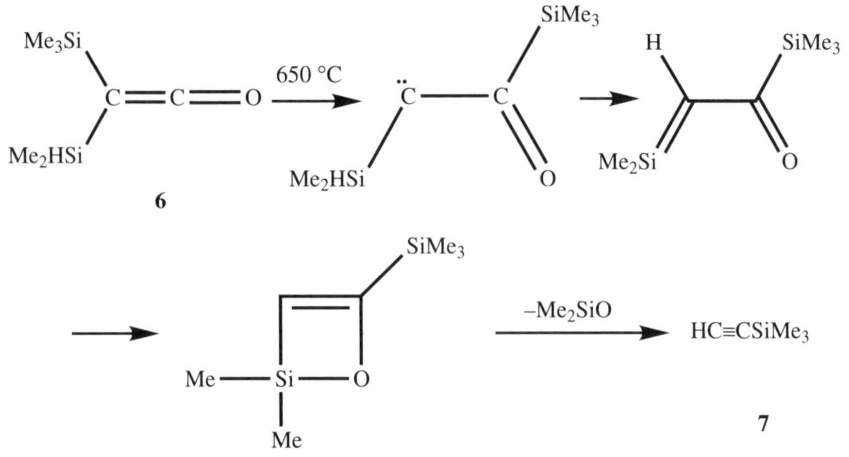

$$\text{Me}_3\text{Si}\text{C}=\text{C}=\text{O} \text{ (Me}_2\text{HSi)} \xrightarrow{650\,°\text{C}} \cdots \tag{9}$$

Pyrolysis of methyl diazomalonate **8** at 280–640 °C gave the products shown in amounts that varied with temperature (Scheme 1).[8] They were proposed to occur by processes including successive formation of the ketenes **9** and **10**.

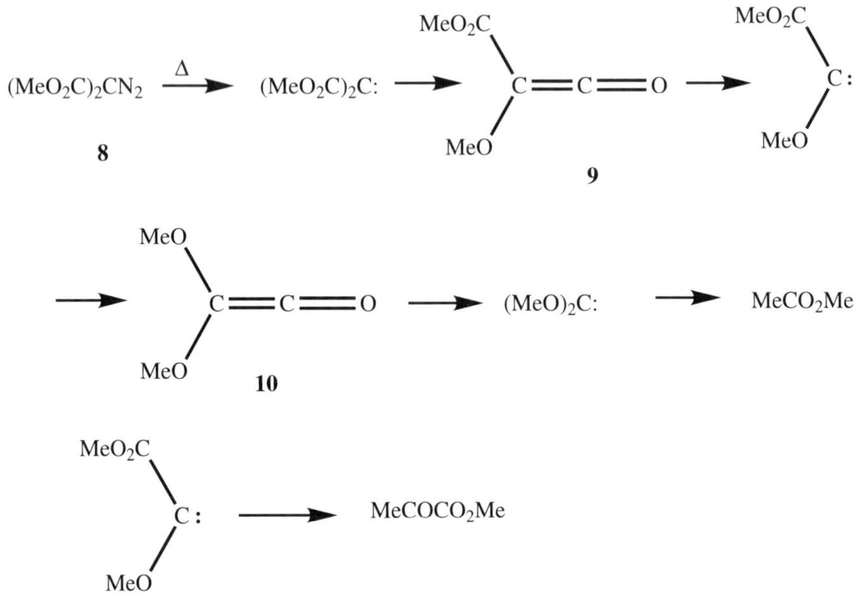

Scheme 1

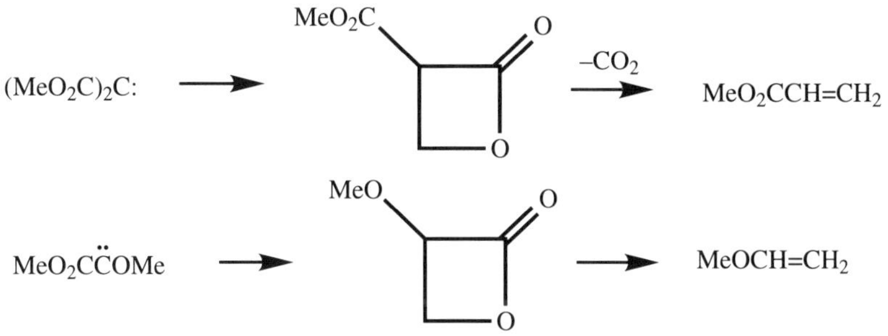

5.3 THERMOLYSIS REACTIONS

Scheme 1. *(Continued)*

Pyrolysis of **11** at 460 °C gave the diketone **12** with high yield, and this was proposed to involve decarbonylation of the ketene **13** (equation 10).[9]

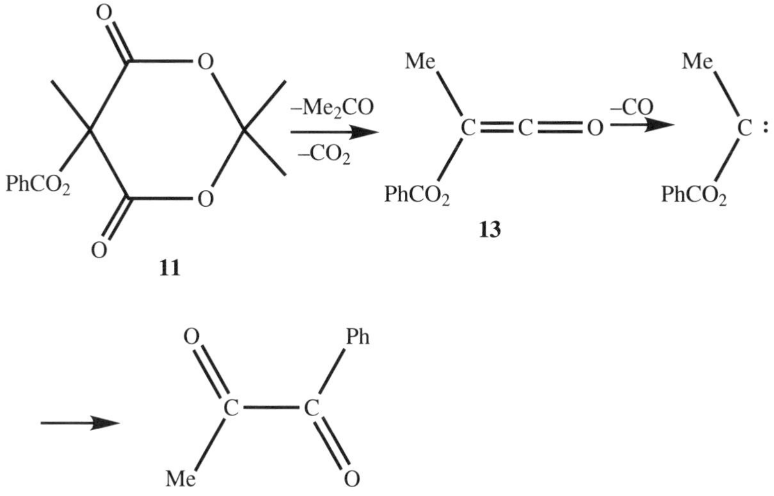

(10)

Pyrolysis of methyl α-diazophenylacetate gave acetophenone as the most abundant product, and this was proposed to occur through formation of the ketene and decarbonylation (equation 11).[10]

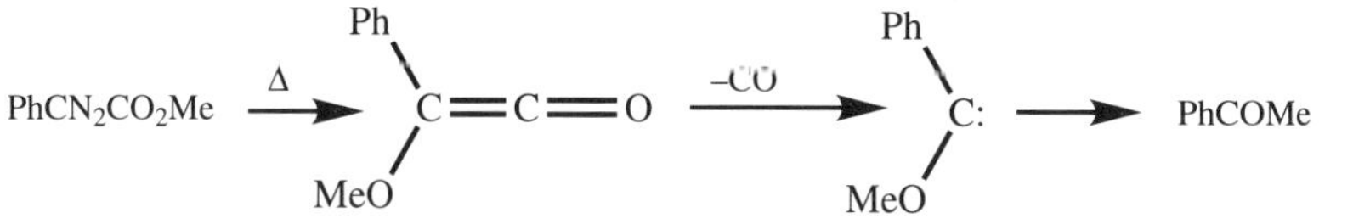

$$PhCN_2CO_2Me \xrightarrow{\Delta} \underset{MeO}{\overset{Ph}{>}}C=C=O \xrightarrow{-CO} \underset{MeO}{\overset{Ph}{>}}C: \longrightarrow PhCOMe \quad (11)$$

Thermolysis of β-keto esters such as **14** at 560 °C is proposed to give acylketene **15**, which undergoes intramolecular [2 + 2] cycloaddition of the ketene and alkenyl groups (equation 12) and also decarbonylation to an acylcarbene that leads to the cyclopropane **16** (equation 13).[11]

14 **15** 80% (12)

16 (20%) (13)

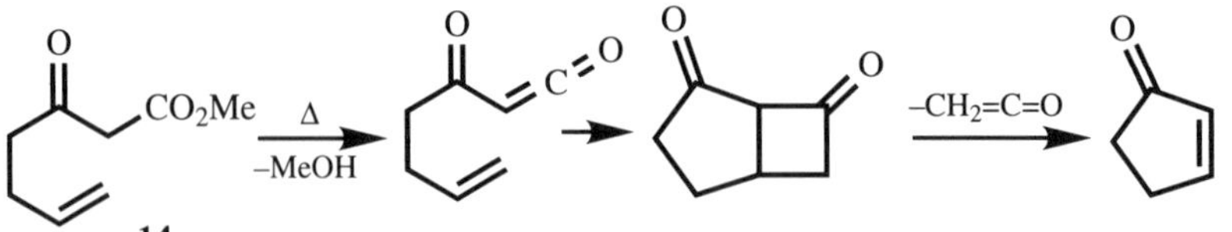

Pyrolysis of the diazoketones **17** is proposed to occur by Wolff rearrangement to the ketenes **18,** which are not observed directly, but undergo retro Diels–Alder cleavage to propadienones which are directly observed in a matrix at 22 K or trapped in cycloadditions (equation 14).[12]

17 R = H, Me **18** (14)

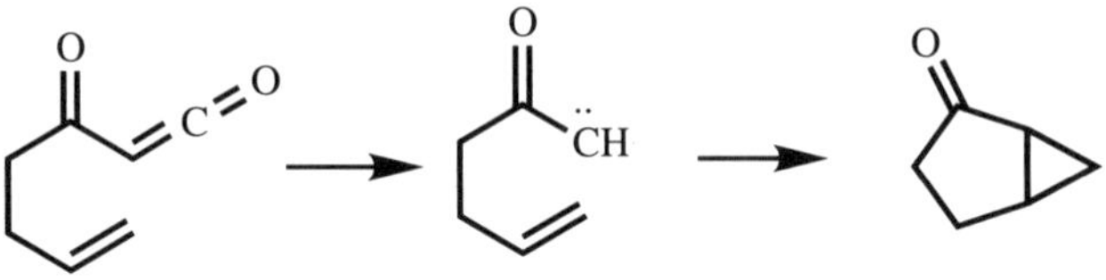

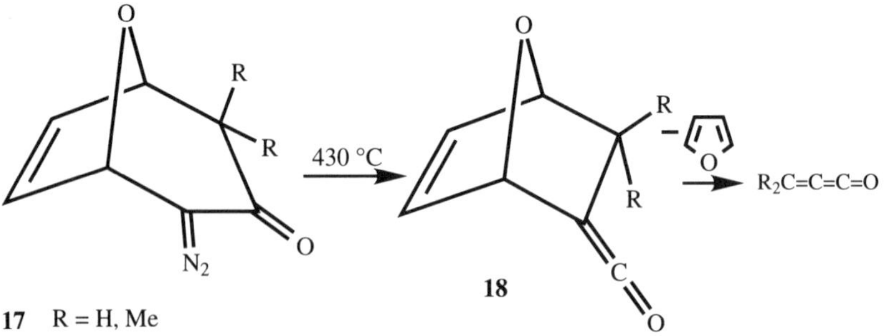

References

1. Staudinger, H.; Endle, R. *Chem. Ber.* **1913,** *46,* 1437–1442.
2. Wentrup, C. In *Methoden der Organischen Chemie,* 4th ed., Vol. E19b, *Carbene (Carbenoide);* Thieme: Stuttgart, 1989; p. 968.
3. Seburg, R. A.; McMahon, R. J. *J. Am. Chem. Soc.* **1992,** *114,* 7183–7189.
4. Ammann, J. R.; Subramanian, R.; Sheridan, R. S. *J. Am. Chem. Soc.* **1992,** *114,* 7592–7594.
5. Tomioka, H.; Komatsu, K.; Shimizu, M. *J. Org. Chem.* **1992,** *57,* 6216–6222.
6. Blake, P. G.; Hole, K. J. *J. Phys. Chem.* **1966,** *70,* 1464–1469.
7. Barton, T. J.; Groh, B. L. *J. Am. Chem. Soc.* **1985,** *107,* 7221–7222.
8. Richardson, D. C.; Hendrick, M. E.; Jones, M., Jr. *J. Am. Chem. Soc.* **1971,** *93,* 3790–3791.
9. Brown, R. F. C.; Eastwood, F. W.; Lim, S. T.; McMullen, G. L. *Aust. J. Chem.* **1976,** *29,* 1705–1712.

10. Tomioka, H.; Ohtawa, Y.; Murata, S. *Nippon Kagaku Kaishi,* **1989,** 1431–1439; *Chem. Abstr.* **1989,** *112,* 118034y.
11. Leyendecker, F. *Tetrahedron* **1976,** *32,* 349–353.
12. Brahms, J. C.; Dailey, W. P. *Tetrahedron Lett.* **1990,** *31,* 1381–1384.

5.4 CYCLOADDITION REACTIONS OF KETENES

The cycloaddition chemistry of ketenes is a large and diverse area that has been part of the field since the earliest period, with the isolation of the unsymmetrical ketene dimer **1** by Chick and Wilsmore[1,2] and the symmetrical dimer **2** of $Me_2C=C=O$ by Staudinger and Klever (equations 1 and 2).[3] Staudinger also found very early that ketene formed addition compounds with alkenes that were later shown to result from [2 + 2] cycloadditions.[4] Since that time cycloaddition has remained as the most distinctive, useful, and intellectually challenging aspect of ketene chemistry.

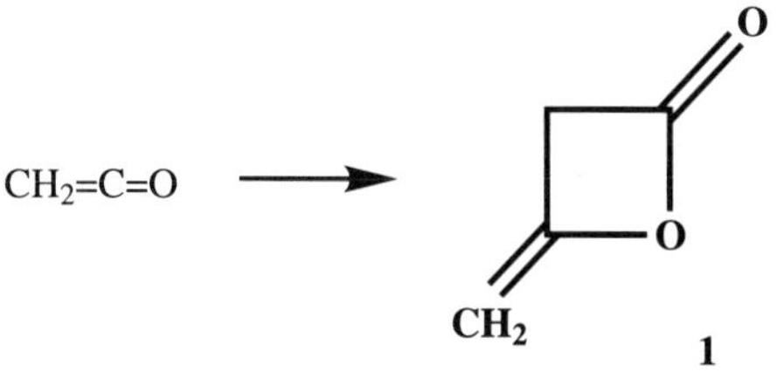

$$CH_2=C=O \longrightarrow \mathbf{1} \qquad (1)$$

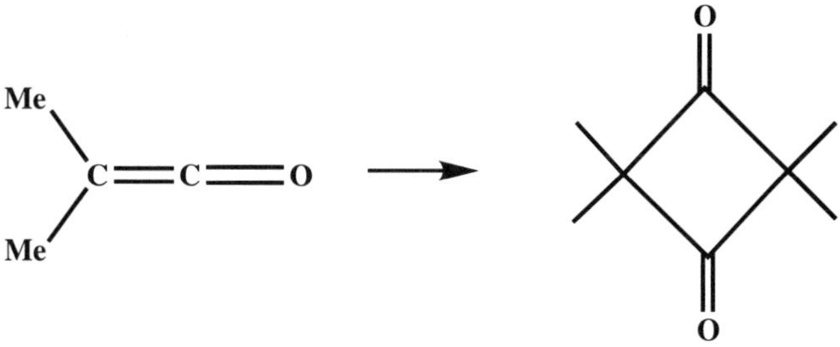

$$Me_2C=C=O \longrightarrow \mathbf{2} \qquad (2)$$

Ketenes are unique in their propensity in giving facile [2 + 2] cycloaddition reactions, even with other pathways are available. This feature of ketene chemistry is of major synthetic importance, in that the reaction with alkenes gives cyclobutanones, usually with a high degree of stereoselectivity, in a process of considerable synthetic utility. The reaction with imines gives the valuable β-lactams (Section 5.4.1.7). Ketenes also usually give [2 + 2] cycloadditions with dienes, and the [4 + 2] cycloadditions to give cyclohexenones are so rare that "ketene equivalents" are used to mimic the behavior of hypothetical ketenes that would react by this pathway.[5–7] A number of reviews have dealt with cycloaddition reactions of ketenes,[8–15] including a recent comprehensive treatment of preparative reactions.[14b]

The Woodward and Hoffmann analysis of ketene cycloadditions[16] revolutionized study in this field. It was suggested[16] that ketenes reacted with alkenes by the perpendicular arrangement of the two moieties as shown by **3**, with both bonds to the alkene being formed from the same side *(suprafacial)*, while the bonds to the two carbons of the ketene were formed from opposite sides *(antarafacial)*. The actual role of this process has been the subject of continuous debate, and is still unsettled, as discussed below. However there is strong evidence that many ketene cycloadditions are nonconcerted, that is, they proceed by a two-step process with formation of a zwitterionic intermediate. This latter process is also consistent with the stereospecific formation of the thermodynamically less stable products, without loss of the stereochemistry of the reactants. Thus, although the concerted pathway is consistent with the observed data in many cases, a stepwise process usually provides at least an equally convincing explanation of the result.

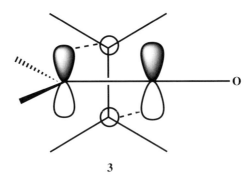

3

Ketene cycloaddition is reversible, and the reverse reaction is an important preparative route in the pyrolysis of unsymmetrical and symmetrical ketene dimers to ketenes (Section 3.1), and in the cleavage of cyclobutanones and cyclobutenones to ketenes (Section 3.4.1).

Other important cycloaddition reactions of ketenes besides [2 + 2] cycloaddition to carbon–carbon double bonds are the [2 + 2] cycloaddition to imines (Section 5.4.1.7) and to carbonyl groups (Section 5.4.5), [3 + 2] and [4 + 2] cycloadditions (Sections 5.4.2 and 5.4.3), and [2 + 1] cycloadditions (Section 5.10).

5.4.1 Intermolecular [2 + 2] Cycloaddition

5.4.1.1 Dimerization of Ketenes. Dimerization of ketenes is one of their most characteristic reactions, and many ketene monomers spontaneously dimerize by [2 + 2] cycloaddition. Ketene itself rapidly forms its dimer diketene (**1**), which was discovered in 1907[1,2] during the first preparation of ketene. The structure of this dimer was in doubt for almost 50 years before it was settled that it had the methylene-β-propiolactone structure **1**, instead of one of a variety of others that had been proposed (**2–5**). The final structure proof was based on chemical studies,[17] electron diffraction,[18] X-ray crystallography,[19,20] and the microwave spectrum[21] and there is an excellent review of the long controversy regarding this structure.[8]

The dimer **1** is commercially available and serves as a laboratory precursor for monomeric ketene,[22] and is a valuable synthetic reagent in its own right.[8,23]

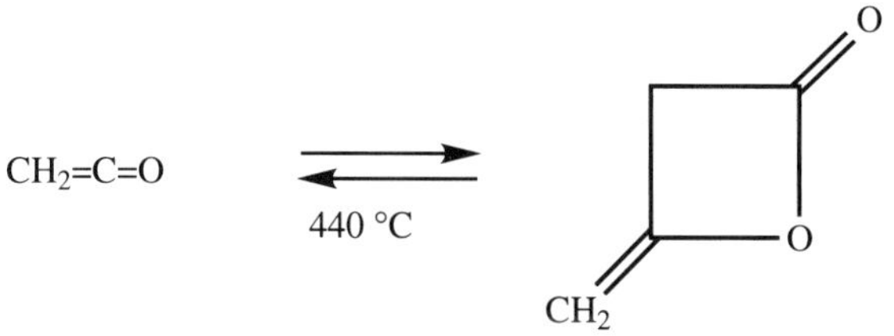

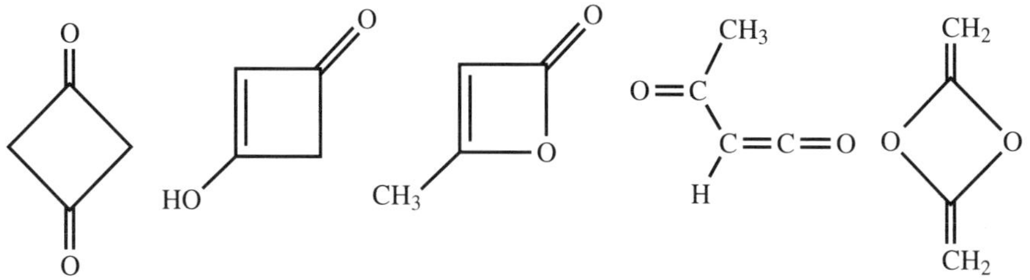

In Staudinger's second publication on ketenes[3] the preparation of $Me_2C=C=O$ and its dimer, now known to have structure **7**, were reported. Many monosubstituted ketenes ("aldoketenes") tend to form both lactone dimers of type **1** as well as substituted derivatives of enolized cyclobutanediones analogous to **3**,[24,25] whereas disubstituted ketenes ("ketoketenes") usually form cyclobutanedione dimers analogous to **2**.

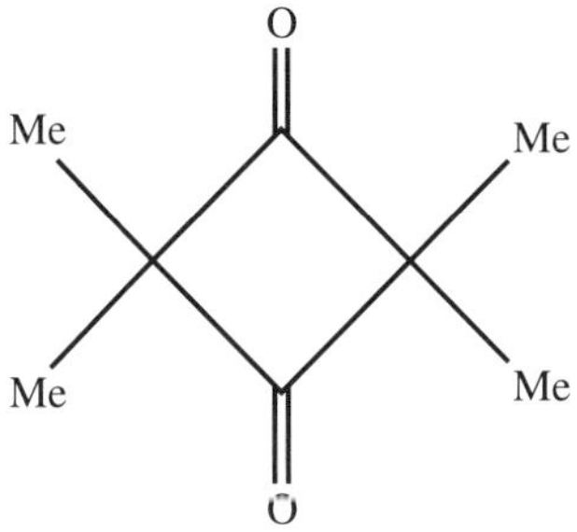

A detailed theoretical study has been carried out which compared the structures **1**, **2**, and **6**.[26] The calculations indicate that the cyclobutanedione structure **2** is 1 kcal/mol more stable than the lactone structure **1**, which is 32 kcal/mol more stable than the unknown isomer **6**. The agreement of the calculated geometries with experimental values was characterized as "only fair."[26]

Further calculations of the structure of **1** at higher levels of theory[27] led to the conclusion that the theoretical bond lengths, which agree well with experimental microwave results,[21] were more accurate than early electron diffraction[18] and X-ray[19,20] values. The comparative values (Å) are shown below, with electron diffraction data[18] in parentheses, X-ray results[19,20] in brackets, and the calculated values.[27]

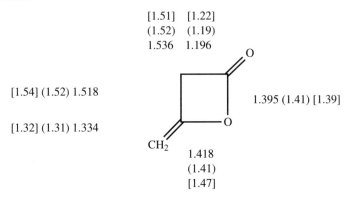

Dimers of monoalkylketenes may be prepared from the acyl chloride and Et$_3$N.[28–30] It was suggested that triethylamine hydrochloride favored formation of lactone dimers of type **8**, whereas in uncatalyzed reactions cyclobutanediones **2** or the enolized forms **9** are formed (equations 1 and 2).[24]

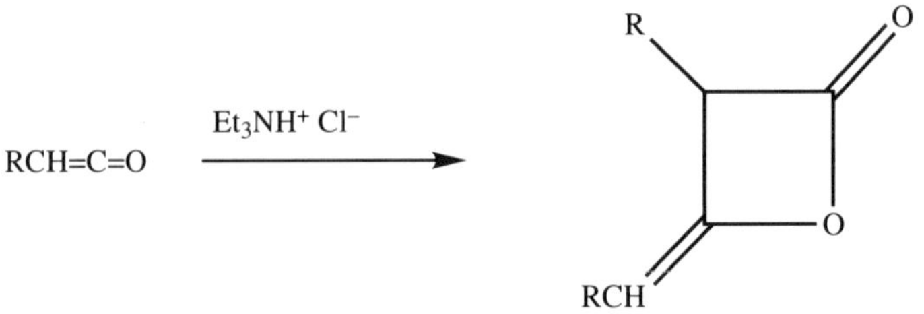

$$\text{RCH=C=O} \quad \xrightarrow{\text{Et}_3\text{NH}^+ \text{Cl}^-} \quad \mathbf{8} \tag{1}$$

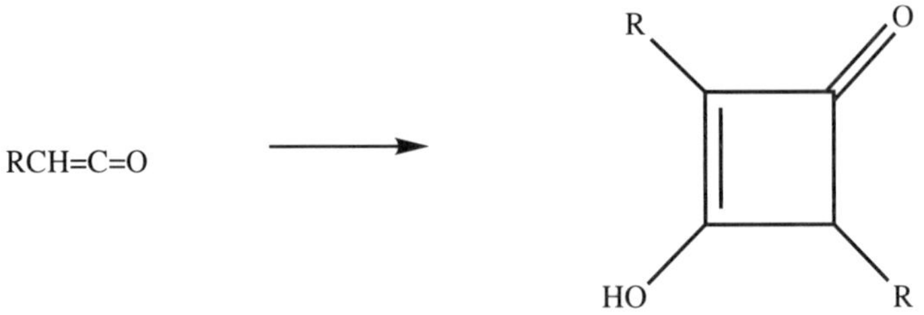

$$\text{RCH=C=O} \quad \longrightarrow \quad \mathbf{9} \tag{2}$$

In 1950 Woodward and Small[31] established the lactone structure **8** (R = CH$_3$) for the liquid dimer of methylketene, and the enolized 1,3-cyclobutanedione structure **9** for the crystalline, acidic dimer. In a prescient mechanistic analysis these authors pointed out that zwitterionic intermediates such as **10** could be involved in these ketene dimerizations, and that attack would occur on the C=C bond of ketenes perpendicular to the ketene plane, while attack on the carbonyl would be in the plane.

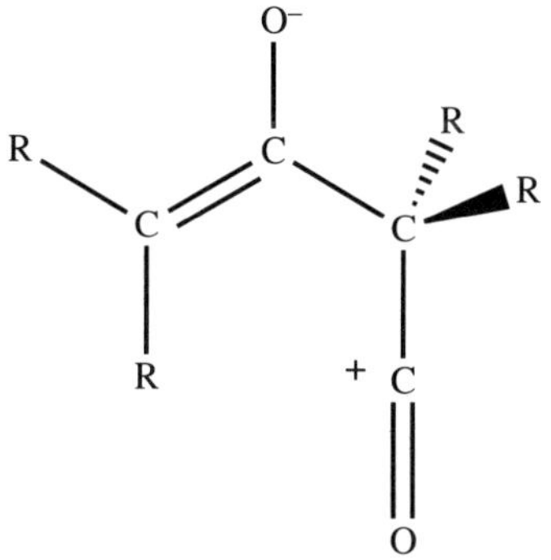

It has been suggested[24] that, with the exception of $CH_2=C=O$ itself, the primary uncatalyzed dimerization pathway for both aldo- and ketoketenes is formation of cyclobutanedione derivatives of type **2** or **3**. This could occur by a concerted $[_\pi 2_a + _\pi 2_s]$ cycloaddition (vide infra).[16] In the presence of acidic catalysts such as triethylammonium chloride or $ZnCl_2$, lactone dimers of type **1** or γ-pyrone type trimers (**11**) may be formed. Furthermore, it was demonstrated that in the presence of base the lactone dimers revert to cyclobutanedione structures **2** and **3**.[24]

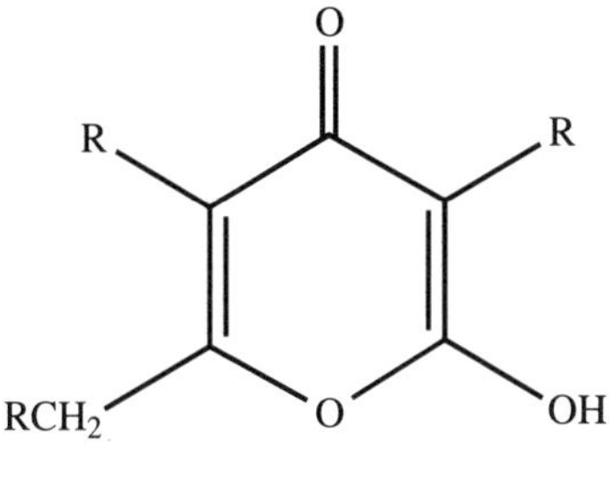

A rationalization of the effect of acidic catalysts in promoting the formation of lactone structures is shown in equation 3. The catalyst enhances the electrophilicity of the carbonyl carbon and assists nucleophilic attack by the carbonyl oxygen of a second ketene molecule. Formation of the zwitterionic intermediate **12** appears likely, as a concerted cycloaddition would not appear to be promoted by the acid catalyst.

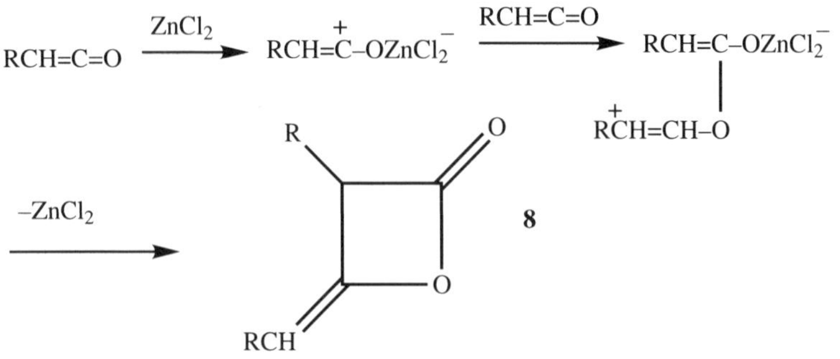

$$RCH=C=O \xrightarrow{ZnCl_2} RCH=\overset{+}{C}-OZnCl_2^- \xrightarrow{RCH=C=O} RCH=C-OZnCl_2^-$$

$$\downarrow \overset{+}{RCH}=CH-O \quad \mathbf{12} \quad (3)$$

$$\xrightarrow{-ZnCl_2} \mathbf{8}$$

Base-catalyzed isomerization of the lactone structure **8** has been suggested[24] to occur by enolization and generation of a ketene intermediate **13** which cyclizes to form **9** (equations 4 and 5). Trapping of **13** by another ketene monomer could lead to the pyrone **11**.

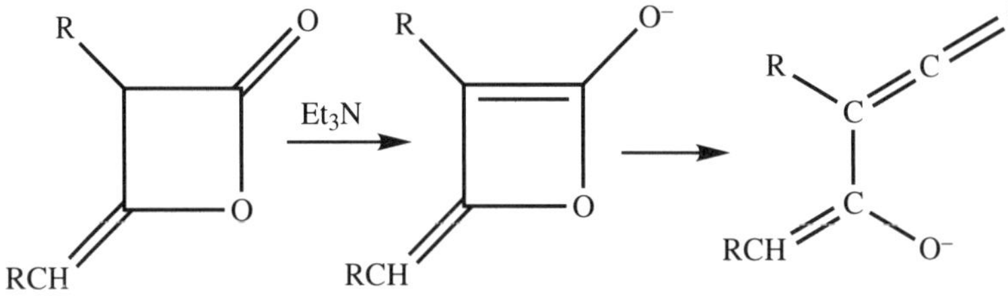

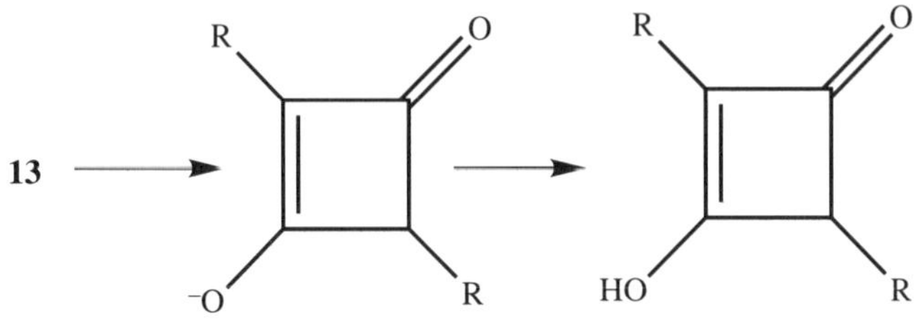

$$\mathbf{8} \xrightarrow{Et_3N} \longrightarrow \mathbf{13} \quad (4)$$

$$\mathbf{13} \longrightarrow \longrightarrow \mathbf{9} \quad (5)$$

Lactone dimers of type **8** with long-chain alkyl groups were originally incorrectly formulated as acylketenes.[28] On hydrolysis these dimers yield β-keto acids which decarboxylate to give practical syntheses of long-chain ketones.[29,30]

$$\mathbf{8} \xrightarrow[\Delta]{OH^-} [RCH(CO_2H)COCH_2R] \xrightarrow{-CO_2} (RCH_2)_2C=O$$

Reaction of the α-bromo acyl chloride **14** with Zn–CuCl in CH_3CN with ultrasonication gave the ketene **15**, which formed isomeric dimers **16** and **17** together with the trimer **18** (equation 6).[32] Reaction of the free ketene gives only the dimer **17**,[33] and treatment of **17** with Lewis acids did not lead to isomerization to **16**.[32] Evidently complexation with ZnCl$_2$ favored the formation of **16**.

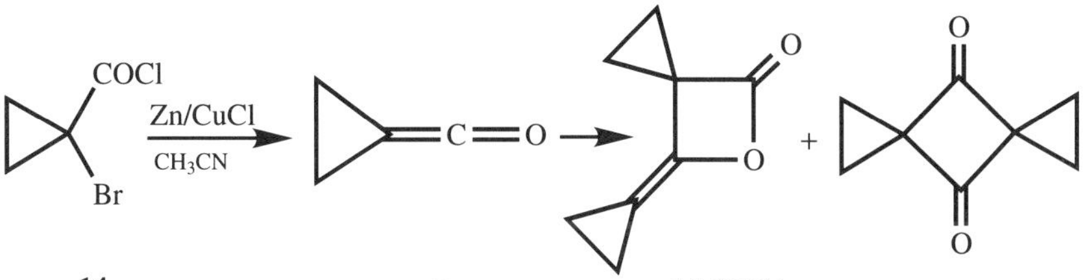

14   **15**   **16** (27%)   **17** (37%)

(6)

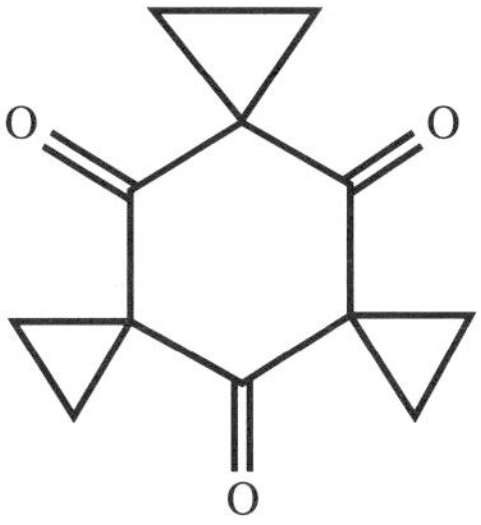

18 (5%)

Treatment of dialkylketene dimers with NaOMe leads to ketene trimers with 1,3,5-cyclohexanetrione structures (equation 7).[34]

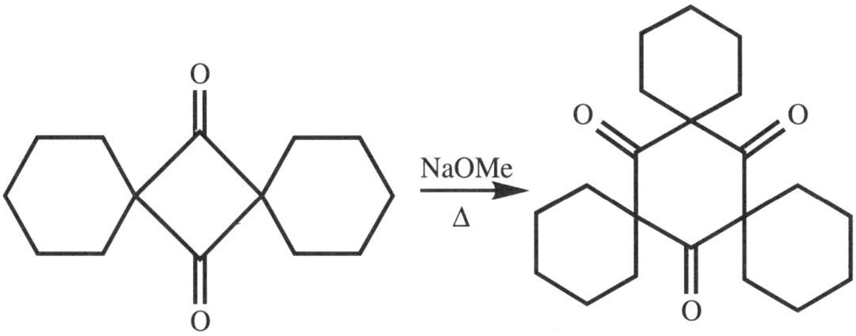

(7)

Treatment of Me$_2$C=C=O, or its dimer **7**, with a catalytic amount of NaOMe in refluxing toluene also leads to a trimer,[34] and dimethyleneketene (**15**) and trimethyleneketene give similar dimers and trimers (equation 6).[34,35]

Dimers and mixed dimers of some very reactive ketenes are obtained by using the metal anions Mn(CO)$_5^-$ and Cr(CO)$_4$NO$^-$ for the dehydrohalogenation of α-bromoacyl chlorides.[36] It was also noted that there was often a kinetic preference for formation of mixed dimers when two ketenes were cogenerated (equation 8),[36] and this is suggestive of an unsymmetrical transition state.

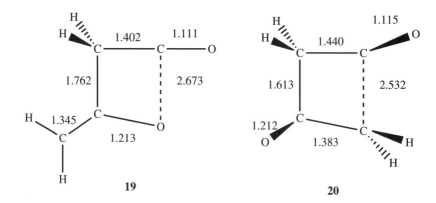

(8)

Mixed ketene dimers have also been formed by generation of haloketenes in the presence of $Me_2C=C=O$, by the mixing of solutions of two different ketenes, and by the cogeneration of two different ketenes.[37,38] When two unsymmetrical ketenes reacted, equal amounts of the stereoisomeric mixed dimers were formed.[37]

The interconversion of ketene and ketene dimer has attracted considerable attention. An experimental study of the thermal cleavage of ^2H and ^{18}O labeled diketene showed that scrambling of the labels did not occur and led to the conclusion that the reaction occurred in one step or that any intermediate did not reform the dimer with interconversion of the labeled groups.[39]

Initial semiempirical[40] and ab initio[41] theoretical studies of the dimerization of ketene have been followed by studies using much higher levels of theory.[42,43] Results for the formation of the almost isoenergetic dimers **1** and **2** indicate that these are formed by transition states **19** and **20** respectively, with the bond distances shown, and barriers of 32 kcal/mol for the formation of **1**, the observed product, and 36 kcal/mol for formation of **2**.[42] Neither of these transition structures is planar, but instead have ring dihedral angles ranging from 23.1 to 55.7°.[42]

The transition states are highly unsymmetrical, and were interpreted as indicating nonsynchronous but concerted reactions,[42] with "no ring formation in the transition state."[43a] For the transition state **19**, this was proposed to result from interaction of the HOMO of one ketene (the C=C π MO) and the LUMO of the other (the C=O π* MO).[42] Also, **19** has negative charge development on oxygen (**21**), which is more favorable than in **22**, where both the positive and negative charge development is on carbon.[42] It was suggested that methyl substituents would tend to favor concerted dimerization, whereas electron-withdrawing groups would favor stepwise processes.[42]

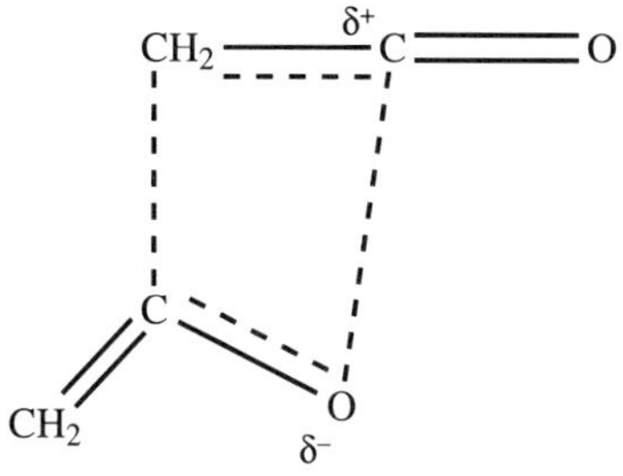

21 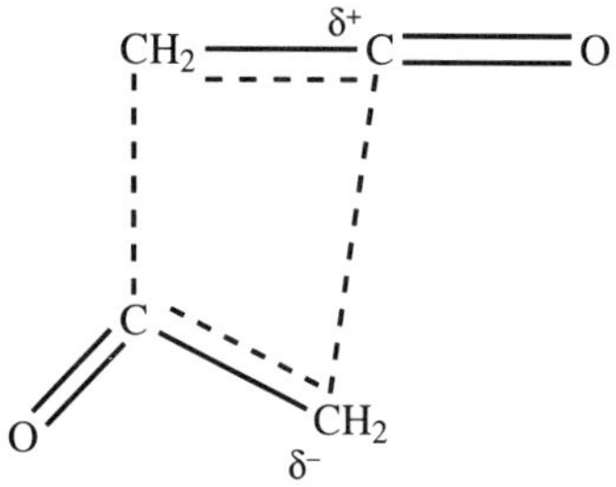   **22**

The structure **19** is proposed to resemble neither of the [$\pi 2_a + \pi 2_s$] transition states **23** and **24**, as suggested by Woodward and Hoffmann,[16] but instead more closely resembles the geometry **25**.[42]

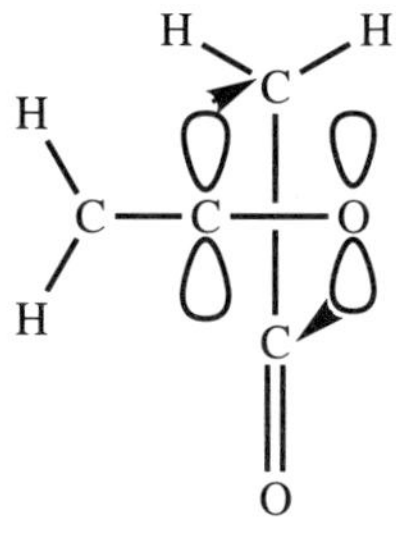

23 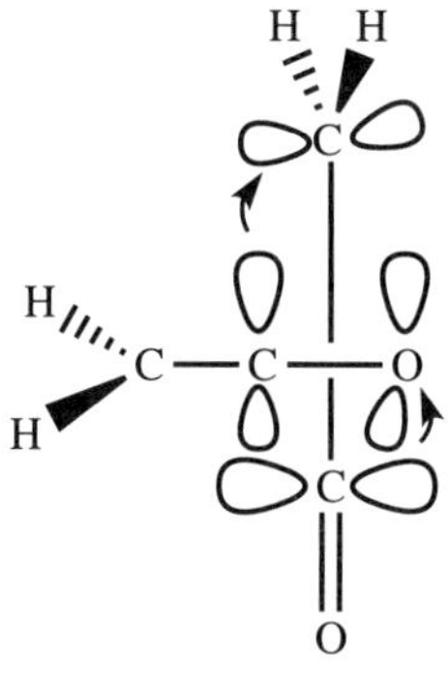   **24** 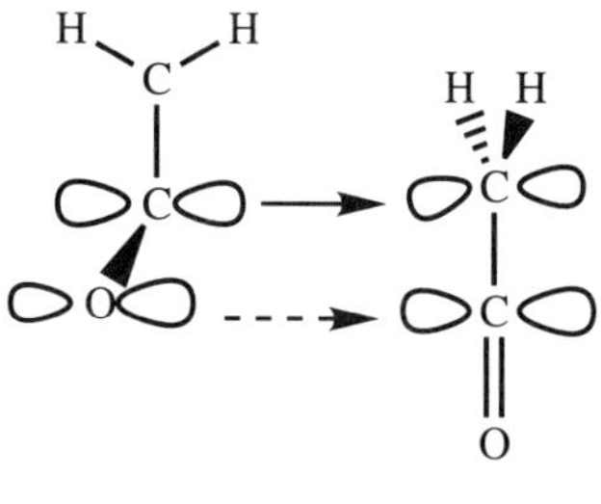   **25**

An independent high-level theoretical study of ketene dimerization[43] gave good agreement with this transition state geometry,[42] but led the authors to describe the reaction as resembling that proposed by Woodward and Hoffmann[16] for a [$\pi 2_s + \pi 2_a$] process.[43] The two groups were in agreement that the reaction is highly asynchronous.[42,43] The deuterium content on the two methylene carbons of **1** expected from the secondary isotope effect on a [2 + 2] cycloaddition was calculated,[43] and good agreement was obtained with experimental values.[44]

To summarize the theoretical conclusions,[42,43] there is agreement that in the absence of solvent the dimerization of unsubstituted $CH_2=C=O$ proceeds by a one-step process, but with highly unequal formation of the two new bonds. Inclusion of solvent effects in these calculations for a realistic comparison to the experimental situation remains a challenge for the future.

The stereochemistry of the products of dimerization of ketenes **26** according to Scheme 1 have been examined, as summarized in Table 1.[45,46] These results (Table 1) were discussed in terms of the $[_\pi 2_s + {}_\pi 2_a]$ mechanism with transition states **27** and **28** leading to the Z dimer **31,** and **29** and **30** leading to the E dimer **32.** It was proposed that as group R^1 became progressively larger, **28** and **30** were increasingly favored. However the trends in the results were not consistent, and particularly with neopentyl(methyl)ketene the *trans* isomer was favored, and it was concluded that these results are not definitive for establishing a $[_\pi 2_s + {}_\pi 2_a]$ mechanism.[46]

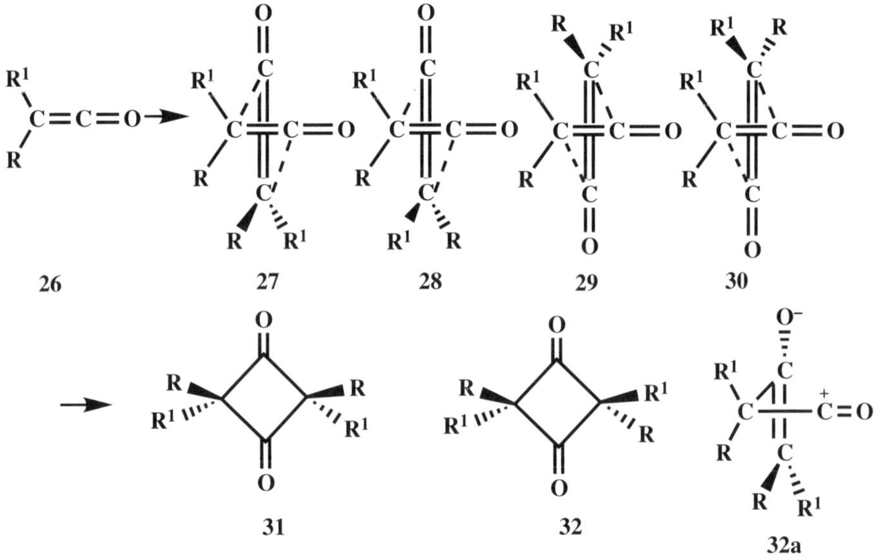

Scheme 1

TABLE 1. Dimerization of Ketenes to 31 and 32

R	Ph	Ph	Ph	Ph	PhCH$_2$	PhCH$_2$	PhCH$_2$	i-Pr	t-BuCH$_2$
R^1	Me	Et	i-Pr	PhCH$_2$	Me	Et	i-Pr	Me	Me
31:32	20:80	62:38	100:0	73:27	60:40	58:42	64:36	76:24	23:77

The steric situation in the four transition states **27–30** is rather complex owing to the different spatial requirements of the substituents in different conformations plus the greater stability of the ketene when the phenyl substituent can conjugate with the alkene grouping. With increasing size of R^1 the conformation of the Ph group will change, altering both the conjugative ability and the effective size of this group.

A stepwise process from, for example, **27** would give the zwitterion **32a**, and if the interaction between the two groups R is significant, it would appear that rotation of these groups away from one another, leading to **32**, would be favored. However, this would increase the interaction between the groups R and R^1, and the balancing of these two effects may be why the results do not follow a clear trend.

The dimerization of ketene **33** was found[47–50] to give the *syn* and *anti* dimers **34** and **35** in a 1:5 ratio (equation 9).[50] This preference for the less crowded *anti* isomer is unusual, and would not be predicted by the less hindered transition state **36** for a concerted $[_\pi 2_s + _\pi 2_a]$ process. Instead, this behavior is that expected for a stepwise pathway, in which the zwitterionic intermediate **37** undergoes preferential formation of the less hindered *anti* isomer **35** (equation 10). The orthogonal reaction complex **36** can also give the zwitterion **37**.

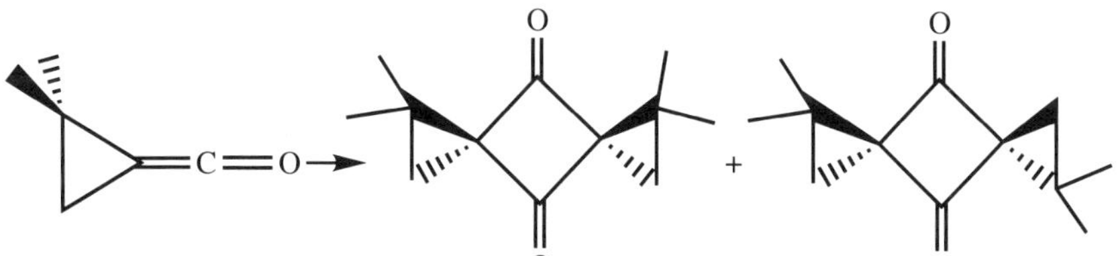

(9)

33

34 (*syn*)

35 (*anti*)

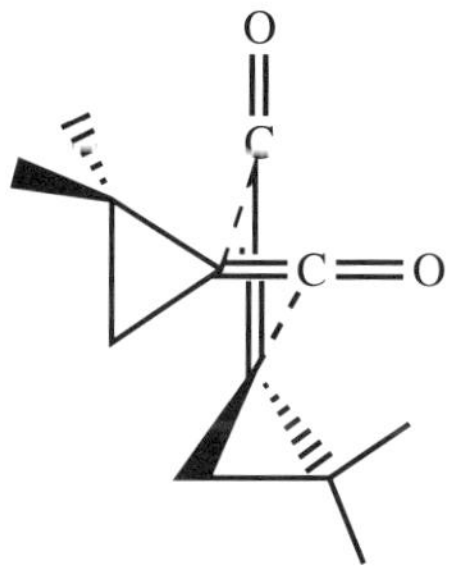

36

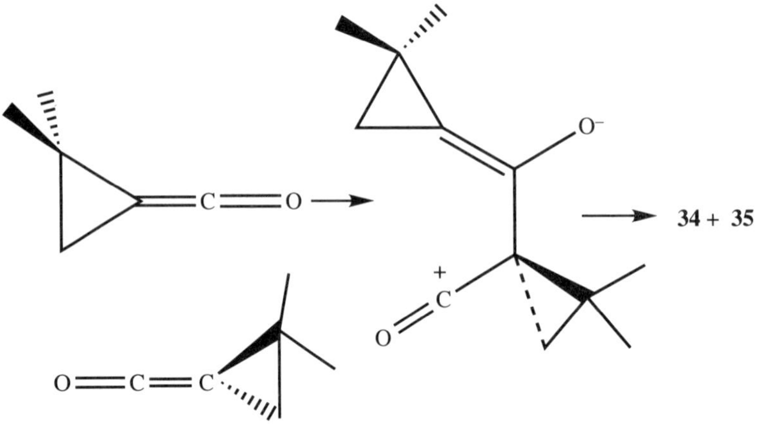

The reaction of **38** gave the *anti* dimer **39** (equation 11), whose structure was established by X-ray crystallography.[35,47] Just as in the case of **33**, this is not the product predicted by a concerted [$_\pi 2_a + _\pi 2_s$] cycloaddition, and a two-step process via a zwitterionic intermediate analogous to equation 10 is indicated. The reason that **33** and **38** give more of the *anti* dimer compared to **26** (Table 1) is not known, but may result from stabilization of the acylium ion by the cyclopropane ring, leading to a longer lifetime of **37**. The compact size of the cyclopropyl group may also reduce the steric interactions in the path to the *anti* product.

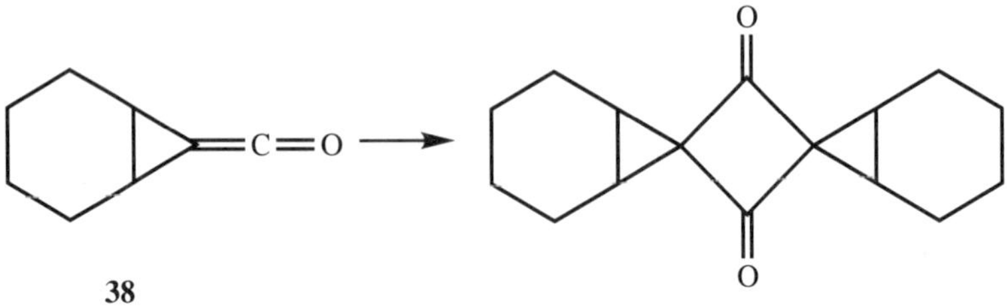

Moore and Wilbur[51,52] studied the reaction of *t*-BuC(CN)=C=O with CH_2=C=O, MeCH=C=O, Me_2C=C=O, and MeEtC=C=O. In the first two examples β-propiolactone dimers were formed, whereas the latter two formed cyclobutanediones (Scheme 2). These ketenes were proposed to react by the head-to-tail approach shown, giving initially the zwitterionic intermediate **40**. When R = H this isomerized to **41**, and **40** and **41** formed **42** and **43**, respectively. The same intermediates were also generated from thermolysis of the corresponding 4-azido-5-*tert*-butylcyclopentene-1,3-diones **44**. Thus these results were taken as showing unequivocally that for these highly polar ketenes, stepwise zwitterionic mechanisms were involved in the dimerizations.[51,52] The preference for formation of β-propiolactone-type dimers **43** of aldoketenes (R = H) would appear to arise from the lower barrier of conversion of **40** to **41,** or to a lower barrier to formation of **41** directly from the ketenes. The structures were represented[51,52] with the ring-forming atoms in an essentially coplanar arrangement, but it would appear that steric interactions would cause some skewing from this geometry to approach that shown in **32a** (Scheme 1), with little decrease in bonding efficiency.

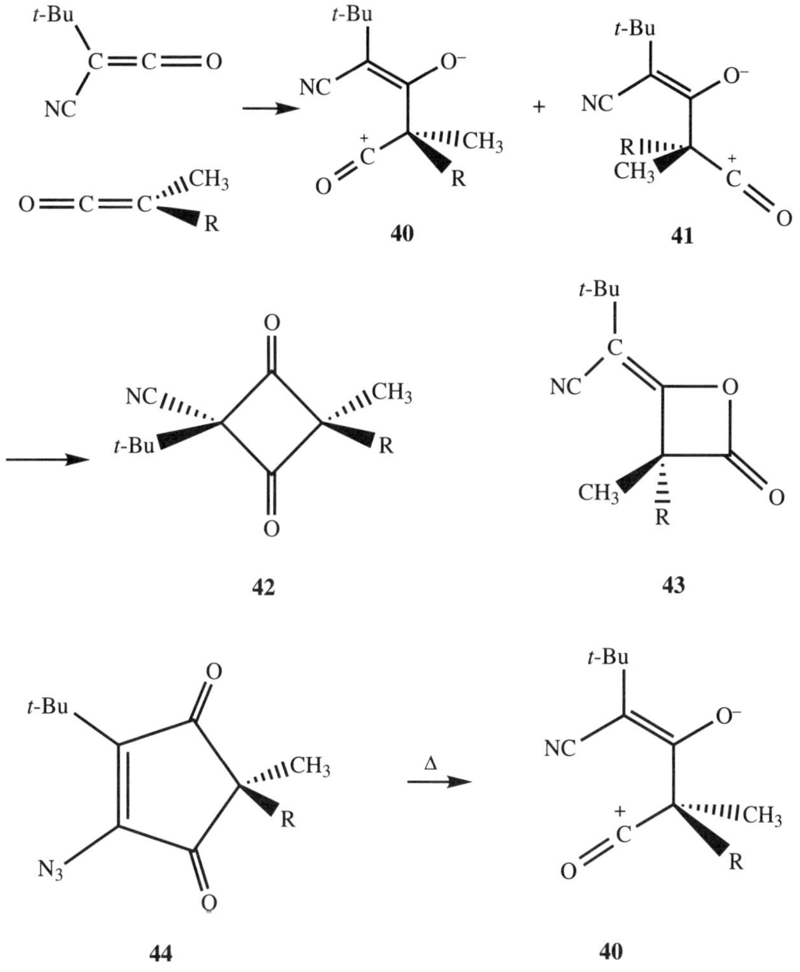

Scheme 2

In the reaction of MeEtC=C=O there was very little selectivity in the formation of the dimer, with a 46:54 ratio of the $E:Z$ isomers being observed. This lack of selectivity led Moore and Wilbur[51] to question if the products with high selectivity observed by Dehmlow et al. (Table 1)[45] were good evidence for a concerted pathway. As noted above, in later work Dehmlow et al. did not rule out stepwise processes.[46]

Bis(trifluoromethyl)ketene **45** does not dimerize thermally but is quite reactive towards other ketenes, reacting with Me$_2$C=C=O to form both cyclobutanedione and β-propiolactone-type dimers **46** and **47**, and with CH$_2$=C=O and MeCH=C=O to form only β-propiolactone-type dimers.[53] The reactivity was explained in terms of steric hindrance in a zwitterionic intermediate **48** which allowed competitive formation of **48a** leading to **47**, as shown in Scheme 3.

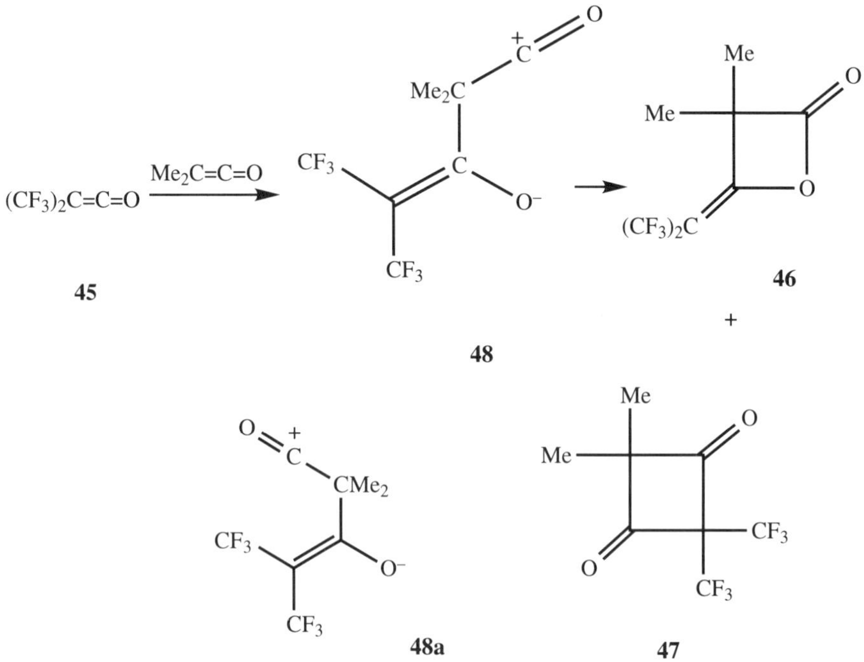

Scheme 3

Huisgen and Otto[54] measured the solvent dependence of the dimerization rate of Me$_2$C=C=O by ^{1}H NMR and found good second-order kinetics, with a variation in rate at 35 °C from 2.31 × 10^{-5} M^{-1} s^{-1} in CCl$_4$ to 67.4 × 10^{-5} M^{-1} s^{-1} in CH$_3$CN. Relative rates were 1 (CCl$_4$), 2.00 (C$_6$H$_6$), 2.81 (C$_6$H$_5$Cl), 10.3 (CDCl$_3$), 11.3 (CH$_2$Cl$_2$), 15.1 (PhCN), and 29.2 (CH$_3$CN). This variation in rate gave a moderately good correlation with the solvent polarity parameter E_T. The reaction in PhCN gave $\Delta H^{\ddagger} = 10.8$ kcal/mol and $\Delta S^{\ddagger} = -42$ eu, suggesting an ordered transition state. These results were interpreted as showing less of an effect of solvent polarity on the rate than was expected for a zwitterionic transition state **49**, and so a one-step reaction proceeding through a dipolar transition state **50** was favored. However, both transition states **49** and **50** are expected qualitatively to be stabilized in polar solvents and, furthermore, the reactant ketenes are quite polar. Hence the magnitude and even the direction of the solvent effect on the rates by either transition state are difficult to predict. As pointed out[54] the dipole moment of the product dione is zero and a product-like transition state would be expected to show an inverse dependence of rate on solvent polarity. Both processes are also expected to have large negative entropies of activation, and so quantitative criteria for differentiating the two are not available, and these data appear inconclusive for deciding which is correct.

In summary, the evidence appears compelling that at least some mixed ketene dimerizations involving ketenes with substituents of very different polarities proceed through two-step processes involving zwitterionic intermediates. For dimerizations of ketenes with similar polarities the dimerizations occur either through one-step processes with a high degree of polar character, or through two-step processes with zwitterionic intermediates. The one-step concerted process is not proven in any example.

5.4.1.2 Cycloadditions with Alkenes and Dienes.
Cycloaddition of ketenes with alkenes to form cyclobutanones is another characteristic ketene reaction. This reaction is reversible, and this process for forming ketenes is discussed in Section 3.4.1. In this section some of the features of the reaction are given, followed by a fuller discussion of some of the mechanistic applications in the next section.

Diphenylketene gives cyclobutanones on reaction with ethylene, propene, and 1-hexene at temperatures of 85–115 °C.[55,56] Under these conditions E- and Z-2-butene react stereospecifically with retention of the alkene stereochemistry, within the limits of ^{1}H NMR detection (4–6%).[55,56] Cycloaddition of these alkenes with $Me_2C=C=O$ is carried out at 105 °C,[57] while preparative cycloaddition reactions of n-BuCEt=C=O with cyclohexene were carried out at 180 °C.[58] In the reaction of the 2-butenes with $Me_2C=C=O$ the Z-isomer reacts about twice as fast and is stereospecific, whereas the less reactive E-isomer gives significant loss of the alkene stereochemistry (equations 1 and 2). This loss of stereochemistry is strong evidence for a stepwise process.[57]

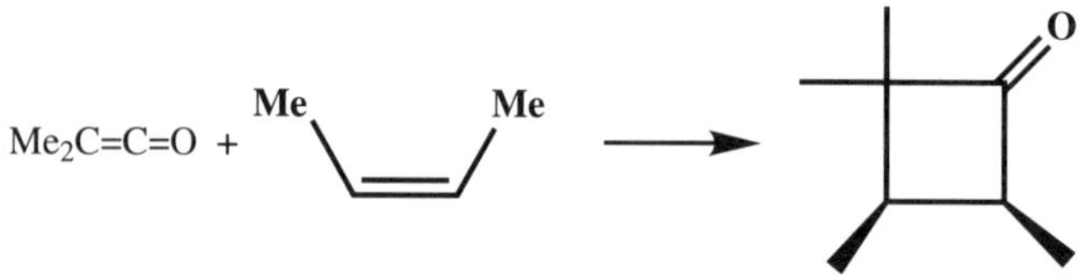

REACTIONS OF KETENES

$Me_2C=C=O$ + [Z-2-butene] → [cyclobutanone **1**] (1)

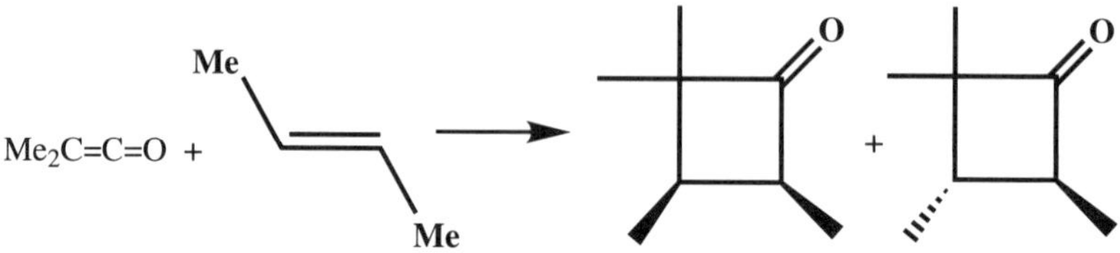

$Me_2C=C=O$ + [E-2-butene] → [**1**] + [**2**] (2)

$2/1 = 2/1$

The mechanisms of ketene cycloadditions are discussed in more detail in Section 5.4.1.3, but an explanation of the lower reactivity and loss of stereochemistry of the E-alkenes is provided by Scheme 1. The greater reactivity of the Z-alkene indicates a near orthogonal approach of the alkene to the ketene LUMO, and the loss of stereochemical integrity of the E-alkene occurs through an intermediate zwitterion **3a** that is converted to **3b** to reduce the steric interaction between the ketene moiety and the *syn* methyl of the alkene. Ring closure of **3a** and **3b** leads to **2** and **1**, respectively.

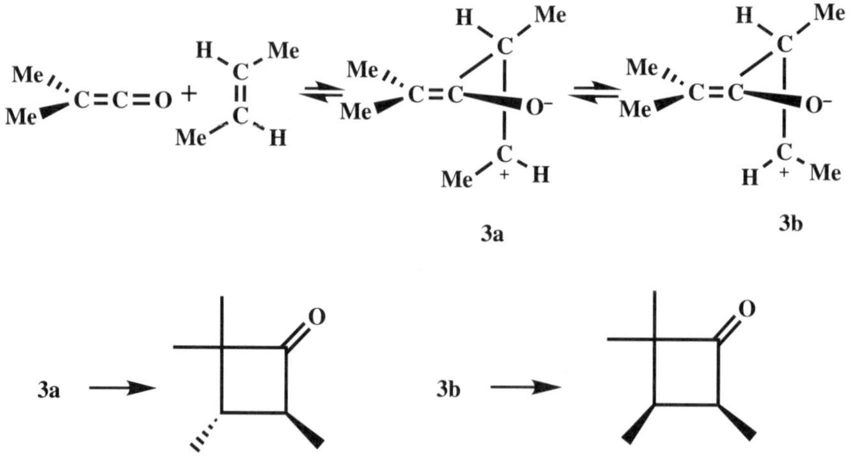

3a → **2**   **3b** → **1**

Scheme 1

The more crowded intermediate **3a** undergoes partial isomerization to **3b** and hence leads to two stereoisomeric products, whereas the more reactive Z-alkene gives the more stable intermediate **3b** directly and only one product is observed.

The cycloaddition of *t*-BuC(CN)=C=O with 2-methyl-2-butene proceeded regio- and stereoselectively, as shown by X-ray crystallography.[59] This reaction was interpreted[59] as occurring by a concerted pathway, but the results can also be explained by a pathway analogous to Scheme 1 in which the initial bond rotation in the intermediate **4** that determines the stereochemistry of the final product involves rotation of the cyano group away from the *syn* methyl.

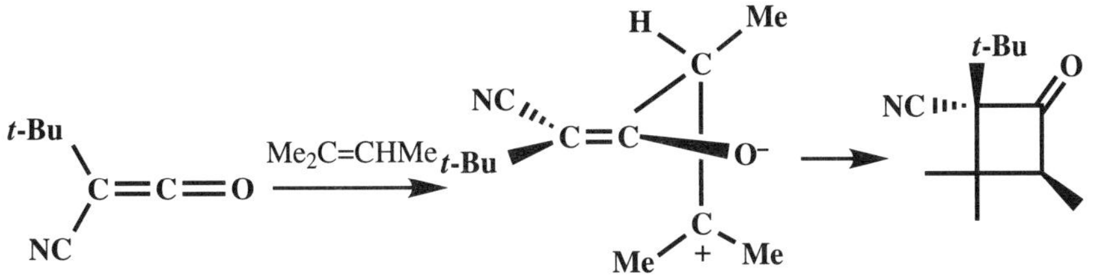

$$\text{(3)}$$

Reactive unsaturated groupings are usually required to trap the more reactive ketenes, which otherwise are consumed by self-reaction or other pathways. Dienes have been particularly useful in capturing many of the most reactive ketenes,[60–79] such as FCH=C=O, as shown in equation 4.[63,66] Electron-rich ketenophiles such as vinyl ethers[80] and imines are discussed in later sections.

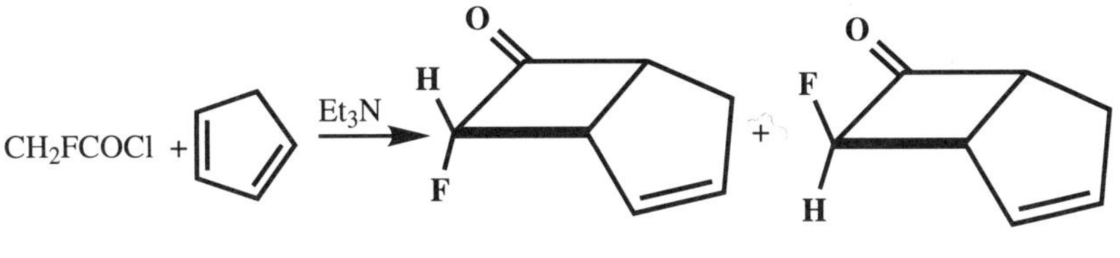

$$\text{(4)}$$

The preference of the larger substituent for the *endo* position in this cycloaddition is a hallmark of ketene chemistry, and has been taken as diagnostic evidence of the [$\pi 2_s + \pi 2_a$] mode of addition as shown in equation 5. The transition state **5** shows that the substituent in the least-hindered position in the transition state will occupy the *endo* position in the final product, which is seemingly more hindered. Bartlett[64] has used the phrase "masochistic steric effect" for this situation where the steric hindrance of a group is reversed between the transition state and the product.

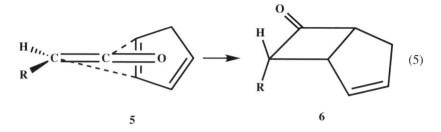

(5)

This simple picture is not as clear cut as it seems, as the *endo* isomer of **6** is not necessarily the least stable isomer, and indeed in many cases is more stable than the *exo* isomer. This was demonstrated by equilibration of the two isomers, in which the equilibrium product ratios shown in Figure 1 were found.[60–62]

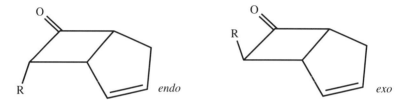

Figure 1. Product ratios at equilibrium for 7-substituted bicyclo[3.2.0]hept-2-en-6-ones.[60–62]

R	endo	exo
F	88.7	11.3
Cl	86.7	13.3
CH_3	76.3	23.7
Et	64.4	35.6
i-Pr	56.9	43.1
t-Bu	9.6	90.4
Ph	67	33

The explanation for this unanticipated result[60] is that the cyclobutanone ring in **6** is not planar but puckered, and exists in the conformation depicted in **7**. This was established by an analysis of the ^{1}H NMR spectra, in particular the various H—H coupling constants. The preference for **7** was explained by unfavorable steric interactions between the carbonyl group and H_1 and H_5 in the alternative conformation **8**, and possibly a favorable electronic interaction between H_5 and the carbonyl group in **7** as well. The *endo* position in **7** is more crowded, but this is offset for all but the *tert*-butyl substituent by a preference of H_7 for the *exo* position so as to be orthogonal to the carbonyl group.[60] A stabilizing hyperconjugative interaction of the C—H_7 bond and the carbonyl may be envisaged in this geometry.

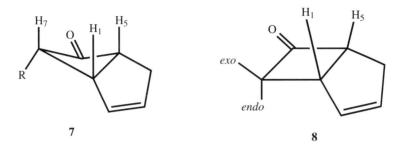

There is a strong solvent effect on the *exo/endo* product ratio of the reaction of ketenes $CH_3CX{=}C{=}O$ (X = Cl, Br) with cyclopentadiene, as shown in Table 1.[68] This result suggests that the formation of the less stable product in these reactions is not good evidence of a concerted mechanism. Further, this significant effect of solvent polarity may be taken as evidence that a polar zwitterionic intermediate is formed in the reaction, and that the partitioning of this intermediate between *exo* and *endo* products is affected by solvent polarity. The more polar solvents favor the *exo* isomer, and this effect may derive from a longer lifetime of a polar intermediate, favoring a more stable product.

TABLE 1. Cycloaddition Product Stereochemistry of $CH_3CX{=}C{=}O$ with Cyclopentadiene[68]

X	Solvent	*exo/endo*	X	Solvent	*exo/endo*
Cl	Hexane	1/4.3	Br	Hexane	1/0.71
Cl	Et_3N	1/2.2	Br	Et_3N	1/0.28
Cl	$CHCl_3$	1/1.6	Br	CH_3CN	1/0.14
Cl	CH_3CN	1/0.59			

The *exo/endo* stereochemical preference for cycloaddition is greatest for cyclopentadiene compared to some other alkenes (Table 2).[70] It has also been shown that for $MeCCl{=}C{=}O$ and $MeCBr{=}C{=}O$ cyclopentadiene and cyclopentene gave similar selectivities.[69]

TABLE 2. *Exo/Endo* Product Ratios for $PhCMe{=}C{=}O$ Reaction with Different Alkenes[70]

Alkene	*exo/endo* (Me)
Cyclopentadiene	>20
$EtOCH{=}CH_2$	2.3
Dihydropyran	1.7
Cyclohexene	2
Cyclooctene	1

The fact that for many alkenes the *exo/endo* product ratio for ketene cycloadditions is insensitive to alkene structure, and rather close to 1, is evidence that these cycloaddition transition states are not so highly ordered. It is also notable that EtOCH=CH$_2$ and cyclohexene, expected to involve extremes in polarity, give similar results.

Rate data for the reaction of Ph$_2$C=C=O with some alkenes and dienes are given in Table 3.[74–76] In the reaction with isoprene and chloroprene the ratio of addition to the substituted and the unsubstituted double bonds were 1.5/1 and 1/4, respectively.[75] These reactions were interpreted in terms of "near-concerted" processes,[75] but it may be seen that there is a very large accelerating effect for cyclopentadiene, which is well suited for conjugative stabilization in a polar transition state. The regiochemistry of the reaction of isoprene also shows a significant polar effect, directing the reaction to the more hindered double bond.

TABLE 3. Rates of Reaction of Ph$_2$C=C=O with Alkenes and Dienes (M^{-1} s^{-1} × 10^6)[74–76]

	k_2 (40 °C, PhCN)	k_2 (30 °C, THF)
Cyclopentene	1.25	
CH$_2$=CHCH=CH$_2$		1.17
Cyclopentadiene	3.73 × 10^4	280
Cyclohexadiene	53.0	
E-CH$_3$CH=CHCH=CH$_2$	17.5	
Z-CH$_3$CH=CHCH=CH$_2$	9.3	
CH$_2$=CMeCH=CH$_2$	4.9	
Dihydropyran	23.5^a	16.5

^{a}In THF, also 34.2 (toluene), 15.6 (n-PrCN), and 9.4 (DMF).[76]

In the reaction of t-BuC(CN)=C=O with Me$_2$C=CHCH=CMe$_2$ the adducts **9** and **10** were formed, and on heating in benzene containing 7% EtOH the proportion of **10** in the mixture increased from 67 to 85%.[77] The formation of two products, and the apparent isomerization of **9** and **10**, were interpreted as showing that a zwitterionic intermediate was involved in the reaction which was not trapped by EtOH. In the reaction of **11** the initial product showed a 5/1 preference for **12**, but on prolonged heating **13** was the only product observed.

(6)

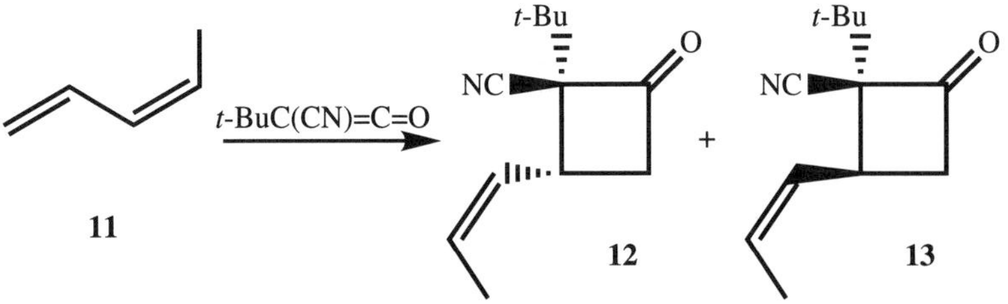

5.4 CYCLOADDITION REACTIONS OF KETENES

$$\text{(7)}$$

Even very stable ketenes such as diphenylketene react with unactivated alkyl-substituted alkenes such as ethylene[55,56] and cyclopentene (equation 8).[74] Thus there is a very wide span of reactivity of the partners over which reaction may be observed. Several factors need to be considered. Very unstable ketenes such as FCH=C=O may have several reaction channels available, so highly reactive ketenophiles are necessary to trap the ketenes as cycloadducts. Disubstituted ketenes such as $Ph_2C=C=O$ are relatively resistant to dimerization, but may have stabilized transition states for cycloaddition with unhindered alkenes, so that cycloaddition is still feasible even with deactivated alkenes (equation 8).

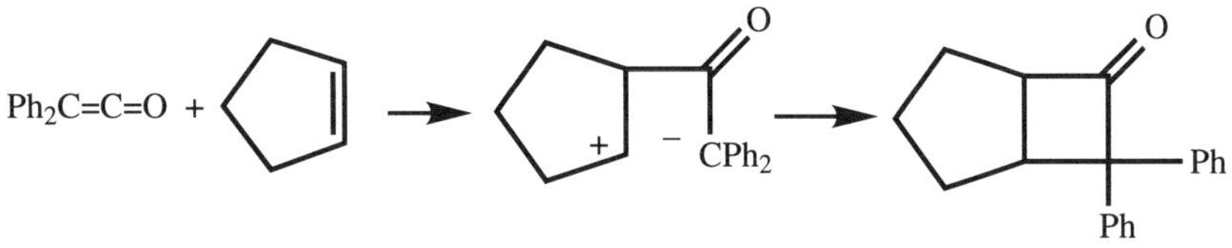

$$\text{(8)}$$

Interest in the mechanism of ketene cycloadditions dates at least to 1950, when Woodward and Small[31] made the seminal observation that nucleophilic reagents would approach ketenes in the ketene plane to add to the carbonyl carbon, whereas electrophilic attack would occur perpendicular to the ketene at the disubstituted carbon of the alkene grouping. Much confusion resulted later from mechanistic interpretations that ignored these stereoelectronic principles.

Already in 1970 the polar character of the ketene cycloaddition process was noted in the reaction of dichloroketene with the vinylcyclopropane **14**, in which the product formed with the regiochemistry and stereochemistry shown (equation 9).[81,82] This was attributed to the polar transition state **15**, in which positive charge buildup as shown was favored by the adjacent cyclopropane ring and possibly by the *syn*-OMe group as well.[81] Cyclopropane ring opening did not occur.

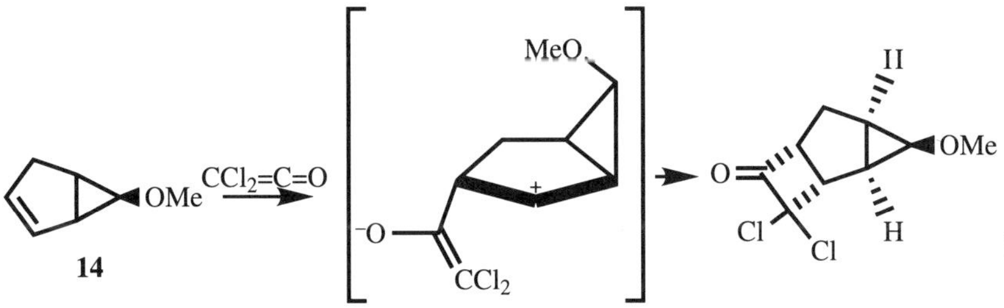

$$\text{(9)}$$

The normal inductive effect of the methoxy group gives the opposite regiochemistry to that in equation 9, as the cycloaddition of the 3-methoxycyclohexene **16** with $CCl_2=C=O$ gave the product regiochemistry shown in **17** with a preference of 95/5 (equation 10).[83] The electron-withdrawing effect of the CH_3O group inhibits positive charge development on the adjacent carbon.

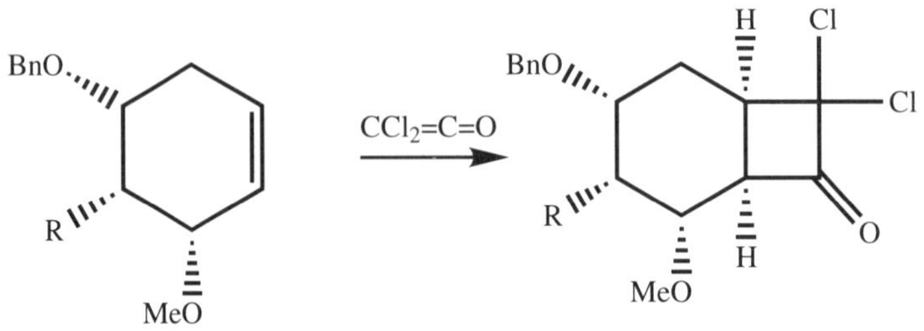

16 R = t-BuCO$_2$CH$_2$CHMe

(10)

Cycloadditions of ketenes to 3,3-dimethylcyclopentene show an interesting variation with ketene structure.[84] Thus, as shown in equations 11 and 12, there is a reversal in the regiochemistry of the preferred product for t-BuC(CN)=C=O as compared to $CCl_2=C=O$ or $Ph_2C=C=O$. These results were interpreted in terms of nonparallel approach of the reactants and concerted [2 + 2] processes, but can also be explained by stepwise processes. Thus formation of zwitterionic transition states (or intermediates) in equation 11 would be favored by both steric and polar factors. In equation 12 the most stable product is formed, and in a stepwise process this would involve initial formation of a bond at the most crowded carbon. However, this arrangement minimizes the interaction of the geminal dimethyl groups with the *tert*-butyl, and is evidently the most favorable situation.

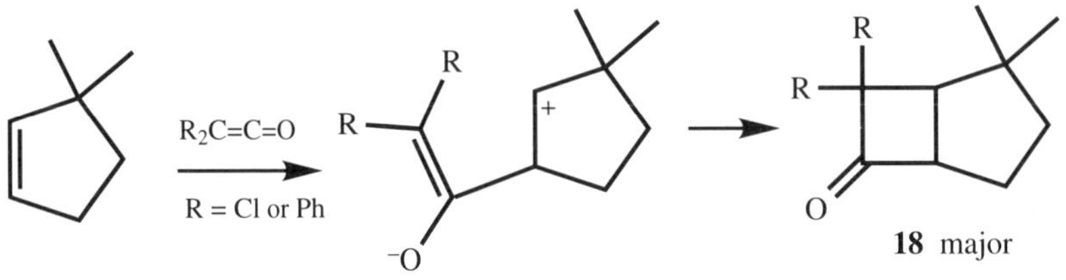

(11)

18 major

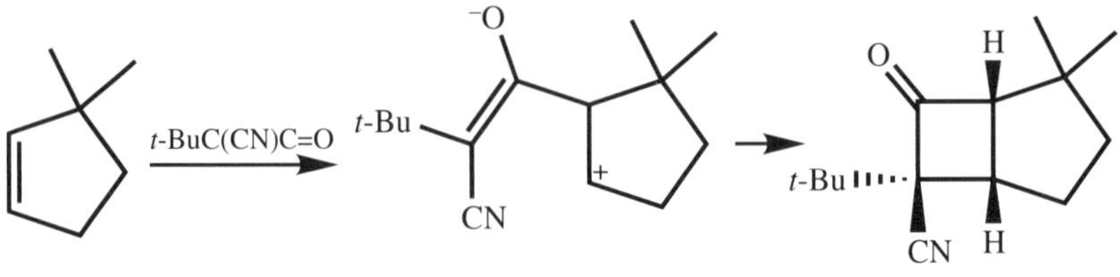

(12)

exclusive

The reaction of equation 11 was conducted with a 2 : 1 molar ratio of the neat alkene and $Ph_2C=C=O$ at 100 °C.[84] When **18** (R = Ph) was heated in hexachlorobutadiene, dissociation into the ketene and alkene was observed, with a half-life of about 15–20 min. The regioisomer of **18**, which constituted about 15% of the product mixture and was suggested to be more stable,[84] was not observed to react under these conditions.

The reaction of $CCl_2=C=O$ with 3,3-dimethylcyclohexene gave a 2.6 : 1 preference for the regioisomer shown (equation 13), a result attributed to stereoelectronic factors, namely, axial bonding by the carbonyl carbon.[84] However, this result may also be explained by a process involving zwitterionic intermediates, as discussed for equation 12.

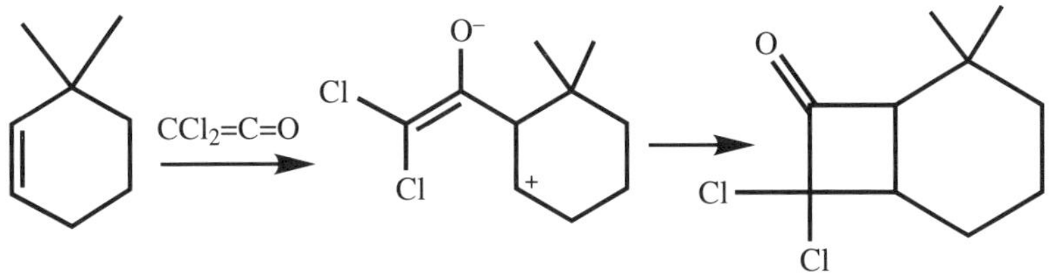

$$(13)$$

Cycloaddition of methylenecyclopropane with dichloroketene is not regiospecific, and gives a 5 : 1 mixture of the spirocyclohexanones **19** and **20** in overall 49% yield.[85]

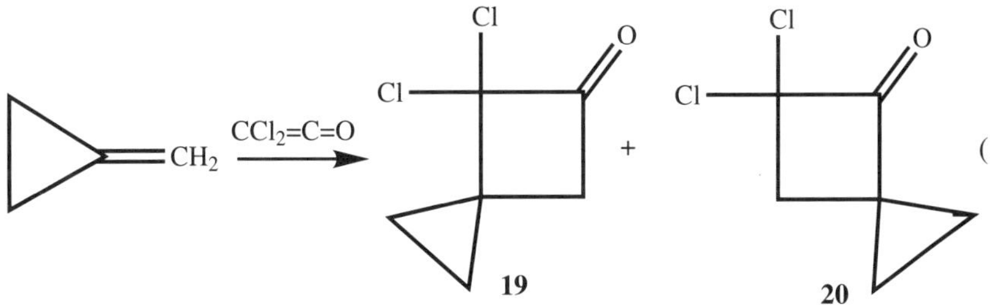

$$(14)$$

The [2 + 2] cycloaddition of $CCl_2=C=O$ to substituted methylenecyclopropanes **21** occurs with formation of **23**. The regiochemistry of this reaction suggests some cyclopropylmethylcarbinyl cation character (**22**) to the transition state for the reaction (equation 15).[86,87]

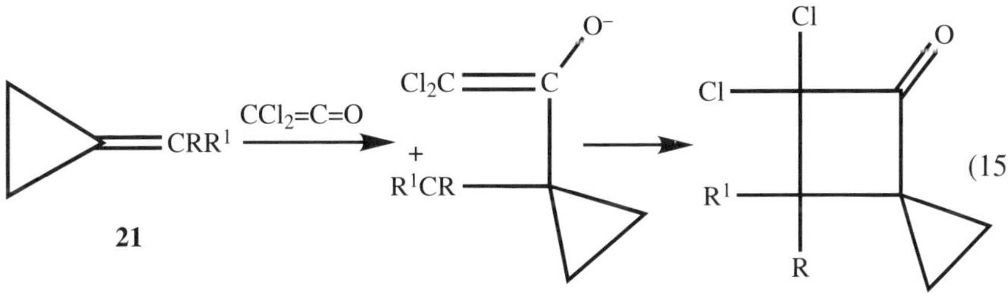

$$(15)$$

482 REACTIONS OF KETENES

The reaction of cyclopropylalkenes with dichloroketene does not give cyclopropane ring opening (equation 16).[88] Thus while the regiochemistry of the reaction is consistent with positive charge development adjacent to the cyclopropyl in the transition state,[88] this does not lead to ring opening, as is sometimes observed in cyclopropylcarbinyl cations. The reaction of equation 17 also did not give cyclopropyl ring opening.[84]

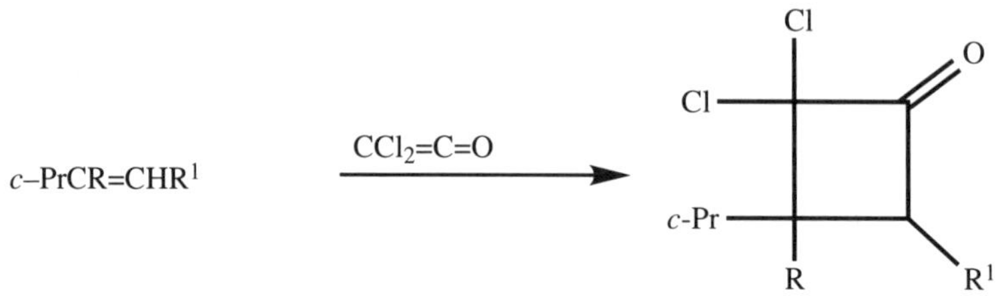

$$c\text{-PrCR=CHR}^1 \xrightarrow{\text{CCl}_2=\text{C=O}} \text{[cyclobutanone with Cl, Cl, c-Pr, R, R}^1\text{]} \quad (16)$$

$$\text{[cyclopentadiene-spirocyclopropane]} \xrightarrow{\text{CCl}_2=\text{C=O}} \text{[bicyclic product with Cl, Cl]} \quad (17)$$

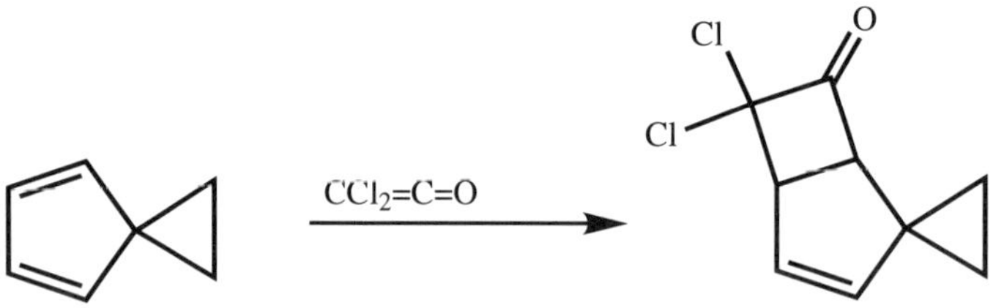

Norbornene reacts with diphenylketene in refluxing benzene to produce the 1:1 adduct **24**, but reaction of Ph$_2$C=C=O with excess norbornene leads to a 5:1 ratio of **24** and the 2:1 adduct **25**, but **24** is not converted to **25** on heating at 75 °C in cyclohexane.[89] Heating of **25** at 200 °C leads to the isomer **26**, which is also formed along with **24** from reaction of norbornene and Ph$_2$C=C=O at 70 °C in CH$_3$CN.[89] The pathway for formation of **25** was not established, but one possibility is initial formation of the zwitterion **27**,[89] which cyclizes both to **24** and in a competitive but reversible path to **28**, which then, depending upon the exact conditions, reacts with a second mole of Ph$_2$C=C=O to give **25** or **26** (equations 18–20).

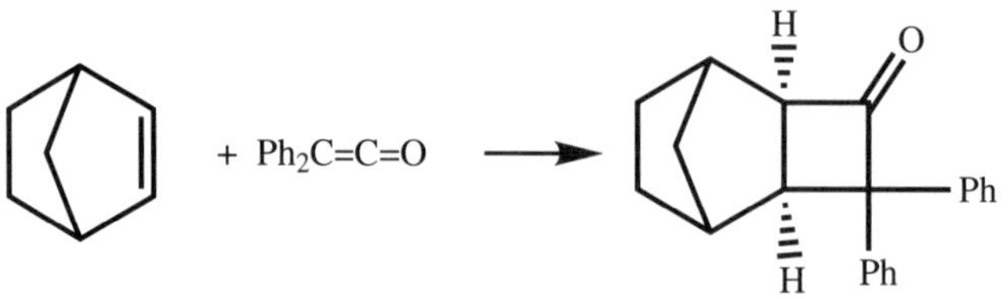

$$\text{[norbornene]} + \text{Ph}_2\text{C=C=O} \longrightarrow \text{[adduct with H, H, Ph, Ph]} \quad (18)$$

24

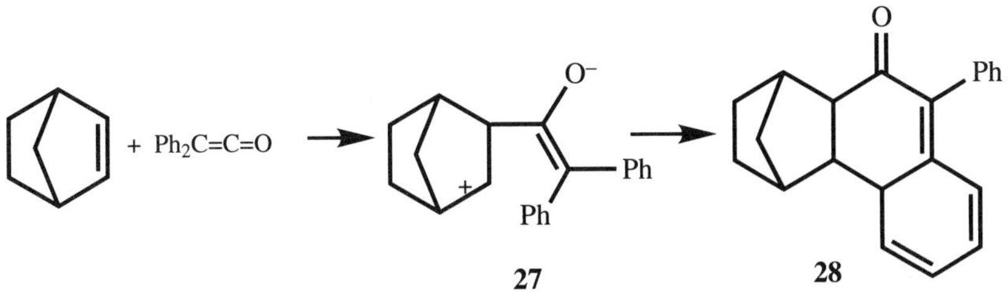

5.4 CYCLOADDITION REACTIONS OF KETENES

(19)

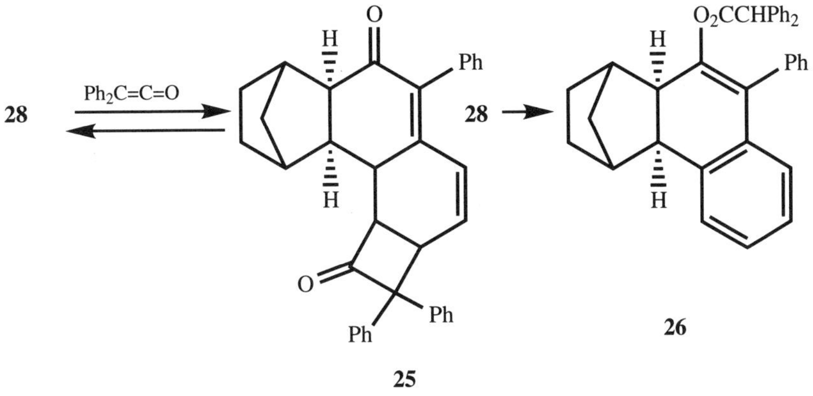

(20)

The [2 + 2] cycloaddition of t-BuC(CN)=C=O with norbornene and other bicyclo[2.2.1]heptenes[90–94] involves formation of adducts such as **29** with the *tert*-butyl group *syn* to the CH$_2$ bridge. Just as in the other cases discussed, this result is consistent with either a concerted addition or a stepwise process in which the initial bond rotation which determines the product stereochemistry is to move the cyano away from the congested norbornyl fragment (equation 21).

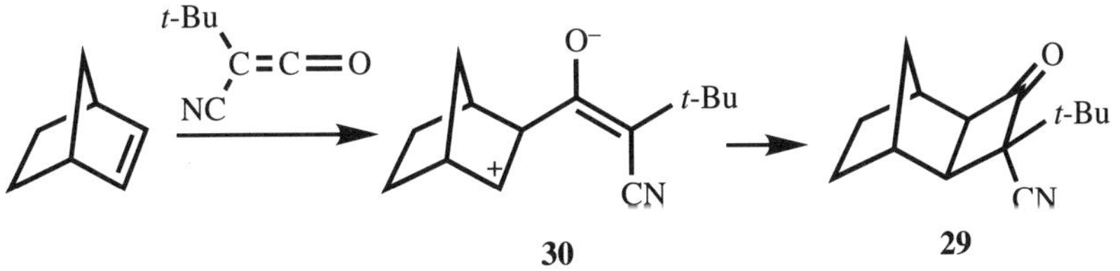

(21)

No rearrangement is detected in this process, and this was taken as evidence of a concerted process. Migration of the 1,6 bond in **30** followed by the most direct ring closure would lead to **31** (equation 22). For a stepwise process the close proximity of the groups in **30** would favor formation of the cyclobutanone adduct. The relative stability of **29** and **31** is not known.

484   REACTIONS OF KETENES

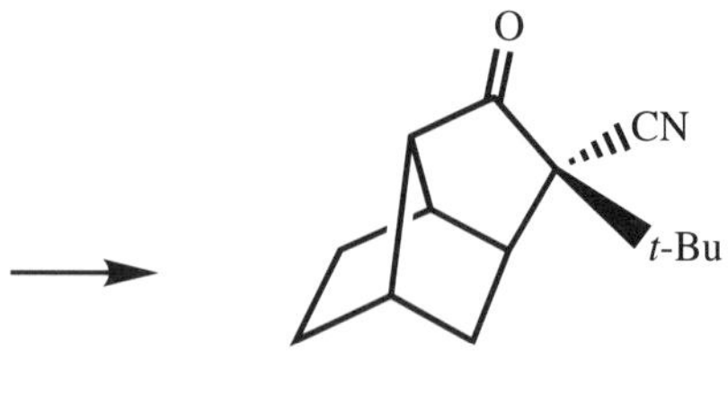

$$30 \longrightarrow 31 \qquad (22)$$

The reaction of t-BuC(CN)=C=O with norbornadiene leads to both the adducts **32** and **33**,[93] and these products arise from competing *exo* and *endo* attack of the ketene. The ratio of the two products showed little dependence on the solvent polarity, indicating that the two transition states had similar polarities. The ketene $(CF_3)_2C$=C=O gave only a product analogous to **33** on reaction with norbornadiene,[95] whereas CCl_2=C=O and i-PrCCl=C=O gave the product analogous to **32**,[96,97] as did MeCBr=C=O, but the latter ketene showed a solvent dependent stereochemistry of the substituents in the product.[97]

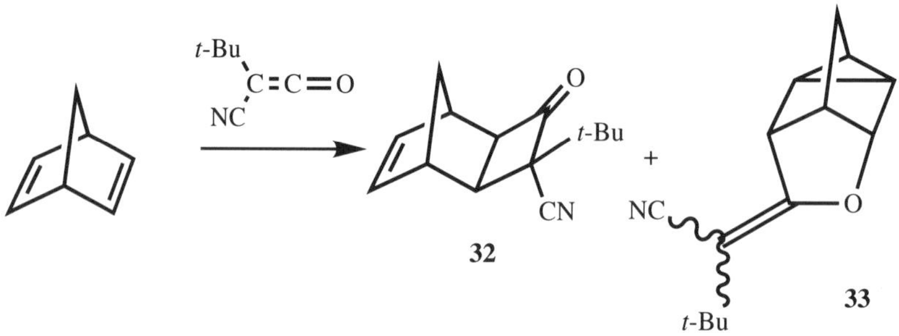

$$\qquad (23)$$

The cycloadditions of t-BuC(CN)=C=O and $(CF_3)_2C$=C=O with cyclopropenes lead to a variety of products, including those involving cleavage of the three-membered rings and ene reactions, showing that stepwise processes are involved (equations 24–27).[98,99] The mechanisms of these processes are not established, but initial formation of **34** could lead to **35** and **36** as shown.

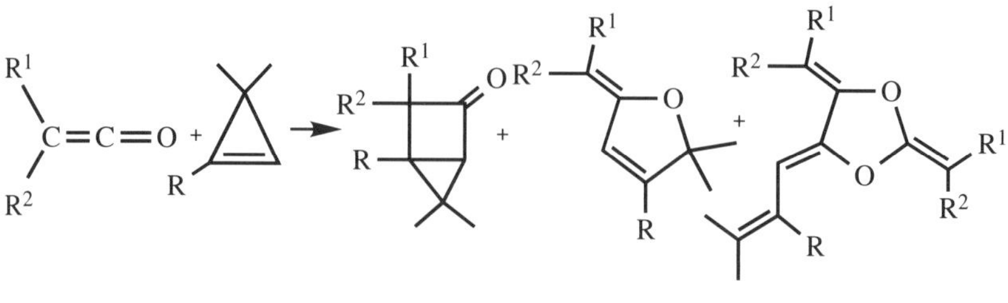

$$\qquad (24)$$

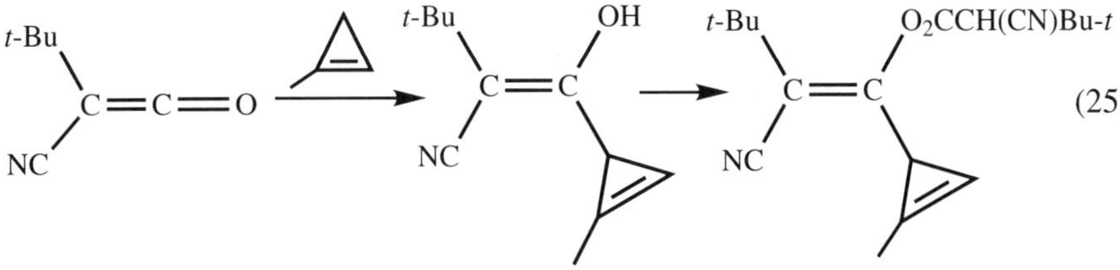

5.4 CYCLOADDITION REACTIONS OF KETENES

$$\underset{NC}{\overset{t\text{-}Bu}{>}}C=C=O \xrightarrow{\triangle} \underset{NC}{\overset{t\text{-}Bu}{>}}C=C\overset{OH}{\underset{\triangle}{<}} \longrightarrow \underset{NC}{\overset{t\text{-}Bu}{>}}C=C\overset{O_2CCH(CN)Bu\text{-}t}{\underset{\triangle}{<}} \quad (25)$$

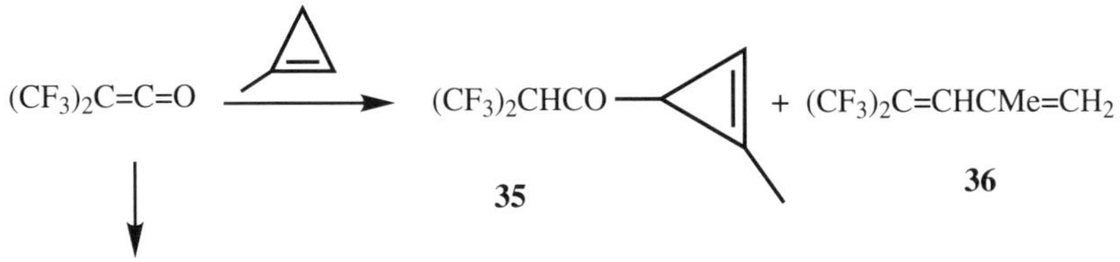

$$(CF_3)_2C=C=O \xrightarrow{\triangle} (CF_3)_2CHCO-\triangle + (CF_3)_2C=CHCMe=CH_2$$
$$\qquad\qquad\qquad\qquad\qquad \mathbf{35} \qquad\qquad\qquad \mathbf{36}$$
$$(26)$$

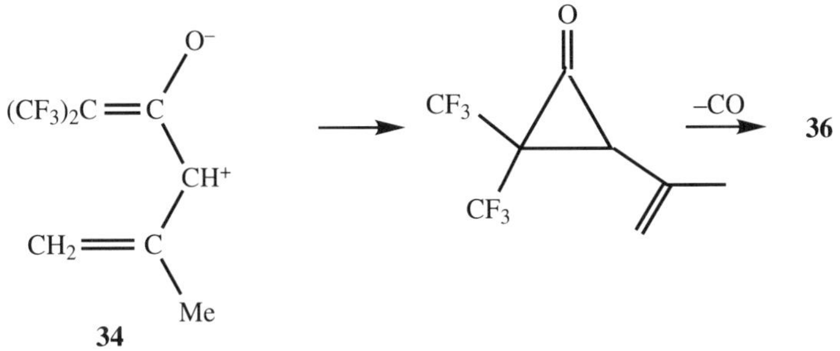

$$(CF_3)_2C=C\overset{O^-}{\underset{CH^+}{<}} \longrightarrow \underset{CF_3}{\overset{CF_3}{>}}\triangle \xrightarrow{-CO} \mathbf{36} \quad (27)$$
$$CH_2=C\overset{}{\underset{Me}{<}}$$
$$\mathbf{34}$$

Dichloroketene cycloaddition to 4-*tert*-butyl-1-methylidenecyclohexane **37**[100,101] occurs with a 4:1 preference on the axial face in a contrasteric fashion (equation 28), and to 2-methylene-5-phenyladamantane (**38**) with a 56:44 preference for formation of the *E*-product (equation 29).[102] Both of these results were explained by a greater susceptibility for nucleophilic attack by one face of the alkene leading to the observed facial diastereoselectivity.

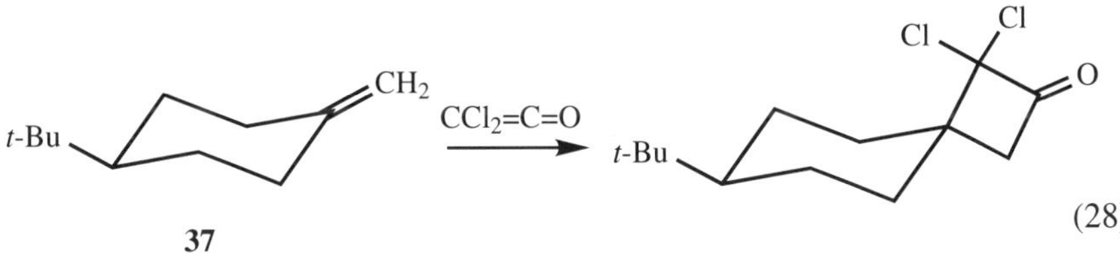

$$t\text{-}Bu-\text{cyclohexane}=CH_2 \xrightarrow{CCl_2=C=O} t\text{-}Bu-\text{cyclohexane-spiro-cyclobutanone}(Cl,Cl) \quad (28)$$
$$\qquad\qquad \mathbf{37}$$

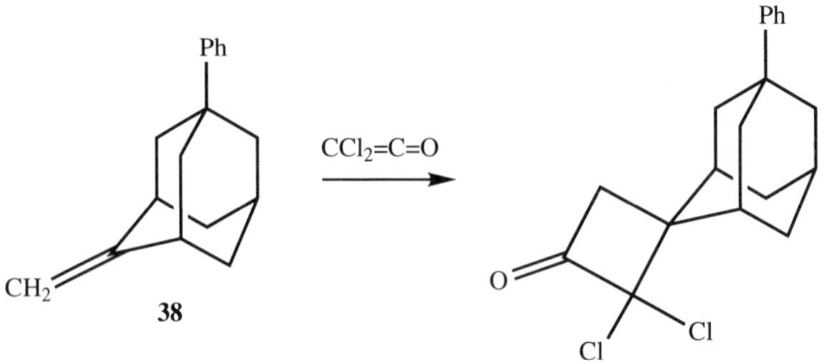

Cycloaddition of dichloroketene to the *exo*-methylenecyclohexanes **37** and **39** occurs preferentially from the axial face with 80 and 86% selectivity, respectively, whereas for **40** axial attack decreases to 21%.[103] Nevertheless, this is the greatest amount of axial attack on **39** seen for various reagents, and illustrates the low steric demands possible for ketenes during orthogonal approach to alkenes.[103]

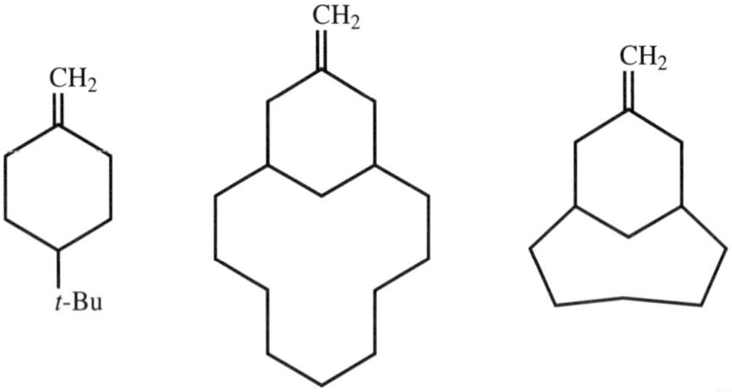

5.4.1.3 Mechanism of Ketene [2 + 2] Cycloadditions with Alkenes.

The facile [2 + 2] cycloaddition reaction of ketenes is a vital part of one of the classic mechanistic problems of organic chemistry, namely the readiness of many alkenes to undergo thermal concerted [4 + 2] cycloaddition with dienes, whereas thermal [2 + 2] cycloadditions, if they occur at all, are clearly stepwise. The appearance in the mid-1960s of the Woodward–Hoffmann rules[16] had an electrifying effect on the chemical community, and provided the explanation that $[_\pi 4_s + _\pi 2_s]$ cycloadditions were allowed by orbital symmetry, whereas $[_\pi 2_s + _\pi 2_s]$ processes were forbidden.

Ketene cycloadditions played a key role in the acceptance of these proposals, as the $[_\pi 2_a + _\pi 2_s]$ cycloaddition is an allowed process, and the existing data on ketene cycloadditions appeared to fit this proposal exactly, and it was stated with authority regarding this mechanism that "the evidence is now conclusive that it is concerted."[16]

The evident readiness of ketenes to react in this way, whereas ordinary alkenes do not, was explained by the proposal that the ketene transition state **1** was much more suitably disposed to the $[_\pi 2_s + _\pi 2_a]$ process than that for alkenes (**2**) because of the lesser steric demands of the ketene, and also because of the favorable bonding interactions between both carbons of the alkene and the electron-deficient *p* orbital of the carbonyl carbon.

1 **2**

However, even in 1970 a counter proposal was made that instead of "a (more or less) synchronous cycloaddition," as proposed by Woodward and Hoffmann, the cycloaddition of ketenes and alkenes occurred by a two-step reaction involving a dipolar intermediate.[104,105] This intermediate was proposed to be involved in an interaction between the cationic carbon and the enolate system of the dipolar intermediate, as shown in **3** (equation 1).[104,105]

As pointed out by Gompper and as discussed below, many experimental criteria have been applied to the differentiation of a concerted and a stepwise reaction pathway for ketene cycloaddition.[105] However, in most cases these tests are ambiguous, as unsymmetrical concerted processes mimic the characteristics of stepwise processes, and a two-step process in which one step or the other is rate limiting can produce a wide spectrum of observed behavior. However, the isolation or detection by spectroscopic, kinetic, or interception methods of an intermediate that lies on the cycloaddition pathway is unequivocal evidence for a stepwise process. Such intermediates have indeed been detected in many cycloaddition reactions of ketenes, as discussed below. Perhaps the best evidence for the occurrence of concerted processes in some examples is the failure to detect an intermediate by the best theoretical methods. However, even in these cases new theoretical studies may locate such intermediates, and in any event the theoretical studies so far have not dealt with reactions in solution. Thus the proof that any ketene cycloadditions proceed by concerted pathways remains an elusive goal.

As already noted,[57,81] and as discussed below, other evidence also appeared in 1970 that some ketene [2 + 2] cycloadditions are not concerted processes. Because of the widely different polarities of the reactants it is clear that ketene cycloadditions, even if concerted, are not "synchronous," that is, formation of both new bonds is not equally advanced in the transition state. Ketenes are highly polar species and are expected to have major dipolar character in their reaction transition states.

The dimerization of ketenes involves even less steric interaction than alkene–alkene or alkene–ketene cycloaddition and is therefore a sterically favorable situation for the concerted $[_\pi 2_a + {}_\pi 2_s]$ cycloaddition. However, as noted in Section 5.4.1.1, ketene itself dimerizes unsymmetrically, and proceeds by a highly asynchronous process.

The reasons for the tendency of ketenes to undergo [2 + 2] as opposed to [4 + 2] cycloadditions were put forward by Woodward and Hoffmann[16] in terms of a concerted pathway, but their arguments appear equally valid if the [2 + 2] cycloaddition is a nonconcerted reaction. First, ketenes are powerful electrophiles, specifically the $\pi^*_{C=O}$ orbital is quite low in energy and very susceptible to nucleophilic attack. This carbon is also only disubstituted, which facilitates its reactivity even more by the absence of steric interactions. Thus nucleophilic attack is greatly facilitated and takes place rapidly, usually resulting in [2 + 2] cycloaddition. However, the reactivity of the C=C linkage of the ketene in a concerted $[_\pi 4_s + {}_\pi 2_s]$ process is not facilitated by the carbonyl group, and so in this process the ketene resembles a normal alkene and is not very reactive. Only in unusual cases in which the intermediates are long lived does [4 + 2] cycloaddition result, but probably not by concerted processes (Section 5.4.3).

A variety of evidence has accumulated about the mechanism of ketene cycloaddition with alkenes, but as described below, in many cases the results appear compatible with dipolar process that may be either concerted or stepwise. The tremendous authority of Woodward and Hoffmann seems to have predisposed many authors to interpret their results in terms of concerted processes. Among the types of evidence available are: (a) theoretical analysis; (b) stereochemistry, (c) isotope effects; (d) substituent effects, (e) solvent effects, (f) pressure effects, and (g) activation parameters. A general mechanistic discussion is given, followed by further discussion in other sections of particular types of cycloadditions.

The propensity of ketenes to undergo [2 + 2] cycloadditions coupled with the analysis[16] that the cycloaddition of $CH_2=C=O$ with $CH_2=CH_2$ is allowed as an $[_\pi 2_s + {}_\pi 2_a]$ process involving the ketene as the antarafacial unit was initially very persuasive. However, at an early date it was suggested that this process was best described as a $[_\pi 2_s + (_\pi 2_s + _\pi 2_s)]$ pathway using the C=C and C=O π bonds of the ketene and the π bond of the alkene.[106,140] In essence, in this mechanism the major initial bond formation involves attack by the more nucleophilic of the alkene carbons on the carbonyl π-orbital at C_α of the ketene. To a lesser extent the other carbon of the alkene carries out electrophilic attack of C_β of the ketene. Thus addition to the ketene involves nucleophilic attack at C_α, and electrophilic attack on the orthogonal orbital at C_β.

There has been a great deal of theoretical study of the transition state of ketene cycloadditions,[107-119] and these studies are helping to define the possible variations in the transition states and the factors that influence them.

A central question remains as to whether these reactions are concerted or involve discrete intermediates. As yet theoretical methods cannot decide this question for solution phase reactions, as the interactions of the polar ketene molecules with solvent are too complex to be reliably described by existing methods. However, for gas phase reactions these theoretical studies are more reliable and deserve serious consideration.

One of the arguments for nonconcerted mechanisms is that studies of intramolecular [2 + 2] cycloadditions of ketenes with alkenes have revealed numerous examples which proceed quite readily by net $[_\pi 2_s + _\pi 2_s]$ stereochemistry (Section 5.4.4). It is argued that these processes cannot attain the geometries required for synchronous formation of both bonds and therefore must be stepwise. If stepwise processes are accepted as being important for these ketenes, they must therefore be considered as a possible route in intermolecular ketene cycloadditions unless proven otherwise. As noted in Section 5.4.1.2 and elsewhere, there are also numerous examples of assuredly nonconcerted reactions that involve rearrangement or that do not result in ring closure.

The regiochemistry of ketene [2 + 2] cycloaddition with alkenes is almost invariably that expected from a species with dipolar character, as shown by the examples of equations 2-4.[120] Perturbation theory predicts this regiochemistry, in that bonding occurs between the carbonyl carbon of the ketene, the atom with the highest coefficient in the LUMO, and the carbon of the alkene with the highest coefficient of the HOMO. Bonding between these atoms also produces the most stabilized zwitterionic intermediate **4,** which leads to the products shown in equations 2-4.

$$CCl_2=C=O + \text{[cyclopentadiene]} \longrightarrow \text{[bicyclic product with O, Cl, Cl]} \tag{2}$$

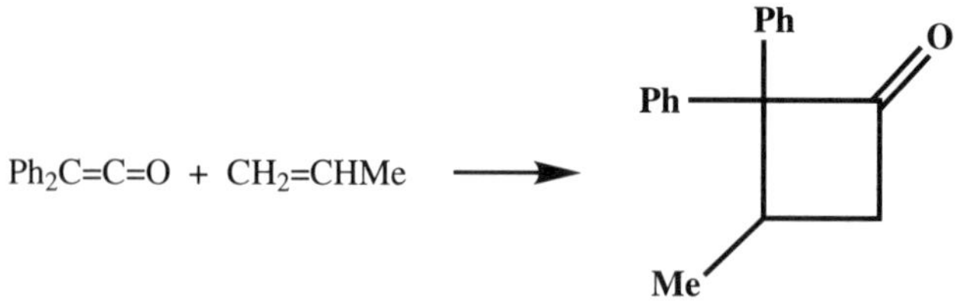

$Ph_2C=C=O + CH_2=CHMe \longrightarrow$ [cyclobutanone with Ph, Ph, Me substituents] (3)

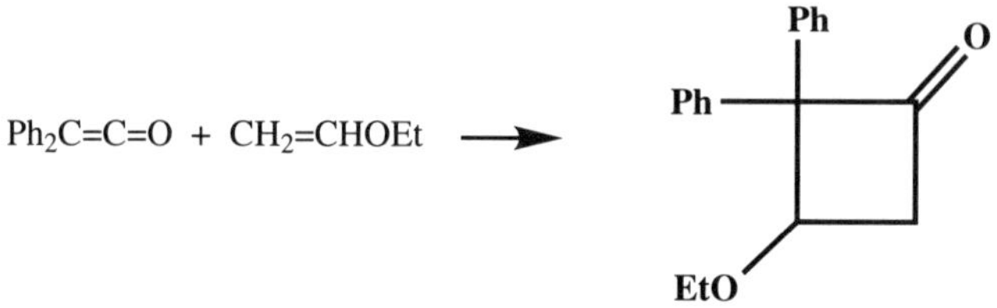

$Ph_2C=C=O + CH_2=CHOEt \longrightarrow$ [cyclobutanone with Ph, Ph, EtO substituents] (4)

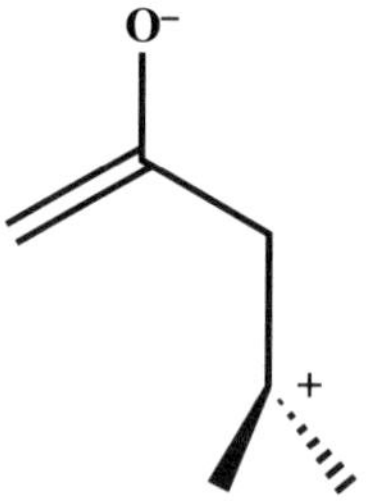

4

The stereochemistry of addition is such that the largest substituents on the ketene and alkene which are on adjacent carbons in the cycloadduct are almost always in the *syn* position on the cyclobutanone.[120–127] However, as discussed in Section 5.4.1.2, this product is also very often the thermodynamically most stable isomer. Z-Alkenes are also significantly more reactive than the E-isomers toward $Ph_2C=C=O$ (Tables 1–3)[121,122,126,127] and $Me_2C=C=O$ (Table 4).[128] The results are explained by an encounter complex **5** in which the more nucleophilic carbon C_1 attacks the carbonyl carbon in the ketene plane, and the more bulky C_β terminus of the ketene is directed away from the alkene substituents. For convenience this structure is usually represented with an orthogonal alignment of the ketene and alkene, but theoretical studies indicate that carbonyl carbon will tilt more closely toward the more nucleophilic carbon C_2, while C_β of the ketene is skewed toward C_1 as in **6** and is also higher above the ketene plane than C_α. A $[\pi 2_s + \pi 2_a]$ or $[\pi 2_s + (\pi 2_s + \pi 2_s)]$ pathway from **5** leads to a *cis* disposition of all three R groups in the product, as is usually observed. However, as noted below, a zwitterionic intermediate predicts the same product stereochemistry.

5.4 CYCLOADDITION REACTIONS OF KETENES

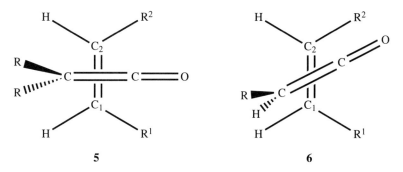

5 **6**

The reaction of t-BuC(CN)=C=O with Z- and E-cyclooctene forming **7a** and **7b,** respectively, provides an impressive example of the stereoselectivity of ketene cycloadditions.[123] In each case the reaction proceeds with retention of the cyclooctene configuration, and leads to the stereoisomer with the t-butyl in the seemingly less stable position *cis* to the alkyl group (equations 5 and 6). These results appear consistent with a concerted [$_\pi 2_s + _\pi 2_a$] cycloaddition, whereas free zwitterionic intermediates would seem to easily convert to a more stable isomer if formed. Dichloroketene also adds stereospecifically to Z- and E-cyclooctenes.[124]

Usually in [2 + 2] cycloadditions of alkenes with ketenes the stereochemistry of the starting alkene is preserved, as in equations 5 and 6, indicating that there is either a concerted process or formation of an intermediate which does not undergo free bond rotation. However, with some E-alkenes there is partial loss of stereochemistry, indicating that in these cases a two-step process is involved[57,126] as already shown in Scheme 1, Section 5.4.1.2.

The greater reactivity of Z relative to E alkenes is strongly supportive of an approximately orthogonal approach of the ketene to the alkene, as shown in **5** or **6**. This process could continue by the formation of two bonds leading directly to the cyclobutanone, or formation of only one bond leading to a zwitterionic intermediate **8**. Conversion of **8** to the product would occur by bond rotation by the least-hindered path in which the hydrogens pass each other, giving the same product stereochemistry as the one-step process (equation 7).

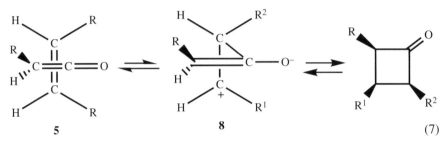

$$\text{5} \qquad \text{8} \qquad (7)$$

In the case of E-cyclooctene (equation 6) the formation of the Z-ring fused product would require significant molecular motions of the alkyl groups in a highly congested environment. This situation is sufficiently complex that advanced molecular mechanics calculations would be helpful to define the feasible reaction pathways.

The reactivity of $Ph_2C=C=O$ with alkenes in PhCN decreases by a factor of 10^7 as the alkene structure becomes less electron rich (Table 1).[127] Comparable kinetic data has been found for reactions of enol ethers with $Ph_2C=C=O$ in CCl_4 and $CDCl_3$ (Table 2).[128] The span of reactivities is quite large, and shows convincingly that there are very large polar effects present in these reactions. The very high reactivity of the enamines in particular strongly argues that zwitterionic intermediates are involved in these processes, as does the formation of noncyclized products in the reactions of some very nucleophilic alkenes. This evidence is discussed in Section 5.4.1.7.

TABLE 1. Rates of Cycloaddition of Alkenes with $Ph_2C=C=O$ (k_2, M^{-1} s^{-1}) in PhCH at 40.3 °C[127]

Alkene	$k_2 \times 10^4$	Alkene	$k_2 \times 10^4$
1-Pyrrolidinoisobutene	2.4×10^5	$PhC(OMe)=CH_2$	12
2,3-Dihydrofuran	1.04×10^3	E-n-$PrOCH=CHCH_3$	0.6
Cyclopentadiene	372	$EtOCH=CMe_2$	0.29
2-Ethoxypropene	227	$PhCH=CH_2$	0.23
1-Morpholinoisobutene	169	$PhCMe=CH_2$	0.038
Z-n-$PrOCH=CHCH_3$	110	Cyclopentene	0.0122
$EtOCH=CH_2$	45	$CH_2=CHCN$	No reaction
Z-$EtOCH=CHMe$	109[a]	Z-$EtOCH=CHPr$-i	117[a]
E-$EtOCH=CHMe$	1.29[a]	E-$EtOCH=CHPr$-i	0.742[a]
Z-$EtOCH=CHEt$	128[a]	Z-$EtOCH=CHBu$-t	3.6[a]
E-$EtOCH=CHEt$	1.20[a]	E-$EtOCH=CHBu$-t	0.054[a]

[a]Reference 122.

TABLE 2. Rates of Cycloaddition of Enol Ethers with $Ph_2C=C=O$ at 23 °C ($s^{-1} \times 10^6$)[128]

	$k(Z)$ (CCl_4)	$k(Z)$ ($CDCl_3$)	$k(E)$ (CCl_4)	$k(E)$ ($CDCl_3$)
MeOCH=CHMe	48	880	0.303	5.6
MeOCH=CHEt	73	732	0.483	3.8
EtOCH=CHMe	258	2050	2.2	13.5
EtOCH=CHEt	173	2530	1.48	8.58

The Z/E rate ratios in Table 2 vary from 115 to 160 in CCl_4 and 150 to 295 on $CDCl_3$, confirming the greater reactivity of *cis* alkenes found in other examples. The ratio $k(CDCl_3)/k(CCl_4)$ varies from 5.7 to 18.5, showing a significant polar solvent effect on these reactions.

The reaction of the Z- and E-1-ethoxypropenes with PhCR=C=O proceeded quantitatively to the cyclobutanones.[121] On the assumption that the products resulted exclusively from stereospecific $[_\pi 2_a + _\pi 2_s]$ reactions through the complexes **9** shown, the rate and product data were quantitatively separated into the individual rate constants shown in Table 3.[121]

TABLE 3. Partial Rate Constants (k_2, s^{-1} M^{-1} $\times 10^6$) for Reaction of Z- and E-2-Ethoxylpropene with PhCR=C=O in PhCN at 40 °C[121]

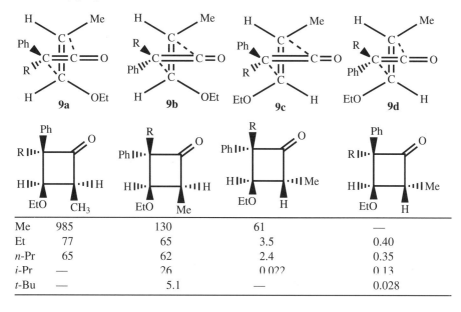

	9a	9b	9c	9d
Me	985	130	61	—
Et	77	65	3.5	0.40
n-Pr	65	62	2.4	0.35
i-Pr	—	26	0.022	0.13
t-Bu	—	5.1	—	0.028

The influence of R on the rates of the processes proceeding through the reaction complexes **9b** and **9d** was attributed both to the varying electronic properties of R and also to a remote steric effect in which the more bulky groups R affected the conformation of the phenyl and increased steric repulsions in the transition states.[121] Because the electronic properties of these R groups vary much less than their steric bulk, the latter factor would appear to be most important.

It has been argued that the rate effects[121,122,127] are explicable in terms of transition states resembling the orthogonal structures **9,** or somewhat twisted structures corresponding more closely to the [$_\pi 2_s$ + ($_\pi 2_s$ + $_\pi 2_s$)] formulation.[121] However, the large kinetic effects do not support the synchronous formation of the two bonds shown in **9;** indeed the data emphasize the dipolar nature of the transition state.[121] As already noted, the formation in some instances of zwitterionic intermediates is established, as evidenced by stereochemical randomization, racemization, or the failure of ring closure to occur. Furthermore, the stereoselective product formation observed is interpretable in terms of zwitterionic intermediates. It may be concluded that none of the data demands the intervention of a concerted pathway, and none of it refutes a zwitterionic path.

The relative reactivities of $Me_2C=C=O$ with alkenes determined by competitive techniques (Table 4)[129] are generally consistent with the rate constants reported in Tables 1–3, although the latter sets cover a wider range of reactivities. The relative reactivities also show large rate effects $k(EtOCH=CH_2)/k(PhCH=CH_2) = 89$ that argue for highly polar transition states, at least for the vinyl ether.

TABLE 4. Relative Reactivities at 100 °C of $Me_2C=C=O$ with Alkenes[129]

Alkene	k_{rel}	Alkene	k_{rel}
Z-2-butene	1.0	$PhCH=CH_2$	0.1
Z-2-pentene	1.0	$4\text{-}MeOC_6H_4CH=CH_2$	2.6
1-Butene	0.44	$4\text{-}MeC_6H_4CH=CH_2$	1.8
Isobutene	0.124	$4\text{-}ClC_6H_4CH=CH_2$	0.36
E-2-butene	8×10^{-4}	Cyclobutene	0.009
E-2-pentene	10^{-3}	Cyclopentene	0.190
1,3-Butadiene	2.64	Cyclohexene	0.068
2-Me-2-butene	0.016	Cycloheptene	0.06
$Me_2C=CMe_2$	No reaction	Cyclooctene	0.92
$EtOCH=CH_2$	8.9	Methylenecyclopropane	1.81
$n\text{-}BuOCH=CH_2$	9.0	$EtOC{\equiv}CH$	23.7

Cycloaddition rates of $Ph_2C=C=O$ with 1,1-diarylethylenes and substituted styrenes (Table 5) gave linear Hammett plots with the σ constants of the aryl substituents with ρ values of -0.78 and -0.73, respectively.[130] These reactions were all first order in both $Ph_2C=C=O$ and the alkene, but whereas the styrenes gave cyclobutanones the diphenylethylenes gave adducts containing two molecules of $Ph_2C=C=O$ and one of the alkene.[130] It was proposed that the rate-limiting step for the latter reaction was a [2 + 2] cycloaddition to form a cyclobutanone followed by reaction with a second ketene. However, the structure of the final product and the details of the reaction have evidently not been settled. The [2 + 2] cycloaddition of $Me_2C=C=O$ with substituted styrenes (Table 4) gave a ρ value of -1.4,[129] indicative of modest charge development.

TABLE 5. Rates of Reaction of $Ph_2C=C=O$ with $(4-XC_6H_4)_2C=CH_2$ and $4-XC_6H_4CH=CH_2$ at 120 °C in $PhBr$[130]

	$k(s^{-1} \times 10^4)$	
X	$Ar_2C=CH_2$	$ArCH=CH_2$
CF_3	0.0167	2.85
Cl	0.0863	7.82
H	0.113	7.00[a]
CH_3	0.212	9.43

[a] 6.66 (1,2-$Cl_2C_6H_4$), 6.50 ([$ClCH_2CH_2$]$_2$O), and 6.03 (DMF).

The very large rate differences observed with the structural variations shown in Tables 1–4 are indicative of a high degree of positive charge buildup on the alkene. The substituent effects of the styrenes noted are more modest, and overall these trends are less than those observed in alkene protonations in water.[131] Thus the $EtOCH=CH_2/PhCH=CH_2$ rate ratio for the latter reaction is 10^5, and the ρ value for styrene protonation is -2.9.[131]

The prediction of rates of ketene cycloadditions have been considered as part of a unified scheme in which known rates or product yields are fitted to predict usable reaction temperatures.[132] However, this does not provide a criterion of mechanism.

The kinetics of cycloreversion of the ketene–styrene adducts *E*-**10** and *Z*-**10** have been examined.[126] The reactions were carried out in the presence of CD_3OD and monitored by the appearance of the ester product **11**. It was found that *E*-**10**, which was more stable than *Z*-**10**, also formed **11** less rapidly. A 4-MeO substituent in *E*-**10** accelerated the reaction by a factor of 30. It was argued[126] that since the solvent effects of these reactions were small, the processes were concerted. However, the reactions evidently follow the same reaction path as the corresponding cycloadditions, which is accepted to be a stepwise process.[77]

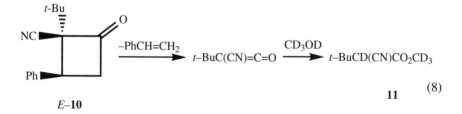

(8)

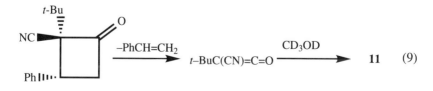

(9)

The negative entropies of activation of [2 + 2] ketene cycloadditions[76,127] are consistent with bimolecular processes proceeding by either concerted pathways or through zwitterionic intermediates.

Solvent effects on ketene cycloaddition reveal a rate ratio $k(CH_3CN)/k(cyclohexane)$ of 160 for the reaction of $Ph_2C=C=O$ with n-BuOCH=CH$_2$.[127] The comparable ratio for the same two solvents for the reaction of TCNE with n-BuOCH=CH$_2$ is about 2600.[133–135] The latter reaction is established to be stepwise, and there is a qualitative resemblance between the magnitudes of the solvent effects of the two processes.

The effect of solvent on the ground states of the reactants must also be considered, and, since $Ph_2C=C=O$ has a higher dipole moment than TCNE, greater ground state stabilization, and hence a lower $k(CH_3CN)/k(hexane)$ might be expected for the ketene for a comparable mechanism, as observed. It has been shown that solvent effects on the ground states have major effects in Diels–Alder reactions,[136–138] and it has been suggested that the polarity of the reactants in ketene cycloadditions is not very different from that of possible dipolar intermediates, which "makes the experimental distinction between a one-step or a multistep reaction difficult."[139]

Isotope effects on ketene cycloadditions have been intensively studied, but do not lead to unequivocal interpretations of the mechanism. For the reaction of $Ph_2C=C=O$ with styrene, intermolecular competition studies indicated that k_H/k_D was 0.91 for the β-carbon of styrene, consistent with a hybridization change from sp^2 to sp^3, but for k_α k_H/k_D was 1.23.[140] For the reaction of $Me_2C=C=O$ with PhCD=CH$_2$ k_H/k_D at k_α was found to be 0.80,[141] in disagreement with the other results.[140,142,143]

A further study of isotope effects involved all four possible $Ph_2C=C=O$ and PhCH=CH$_2$ structures monolabeled with ^{14}C in the alkene carbons, and also E and Z PhCH=CHD.[144] The k_H/k_D isotope effects were the same at 0.884 ± 0.006, in agreement with the results of Baldwin and Kapecki.[140] The $k(^{12}C)/k(^{14}C)$ isotope effects ranged from 1.0055 to 1.08, as shown in 12. These isotope effects were interpreted as consistent with significant negative charge buildup on oxygen, greater bond breaking than formation at the carbonyl carbon, and essentially equal bond making and breaking at the other carbons.[144] No steric compression contribution to isotope effects was noticeable.[144]

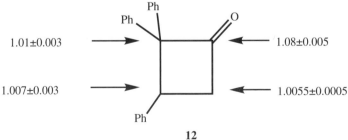

12

The reaction of $Ph_2C=C=O$ with the selectively deuterated 5,5-dimethylcyclopentadienes **13** and **14** showed the ratio of product formation of the nondeuterated and deuterated double bonds k_H/k_D to be 0.84 ± 0.02 for **13** and 1.0 ± 0.01 for **14**.[145] The large inverse isotope effect for **13** was interpreted as indicating that substantial bond formation occurred at C_1 in the transition state, with a change in hybridization from sp^2 to sp^3, whereas the absence of a substantial isotope effect for **14** was interpreted as showing there was little bond formation or change in hybridization at C_2.[145] Thus these results favored a stepwise mechanism, and the transition state was depicted with the "usual" orthogonal geometry, **15,** with the diphenyl carbon of the ketene raised further above the diene plane than the carbonyl carbon to minimize steric interactions.[145]

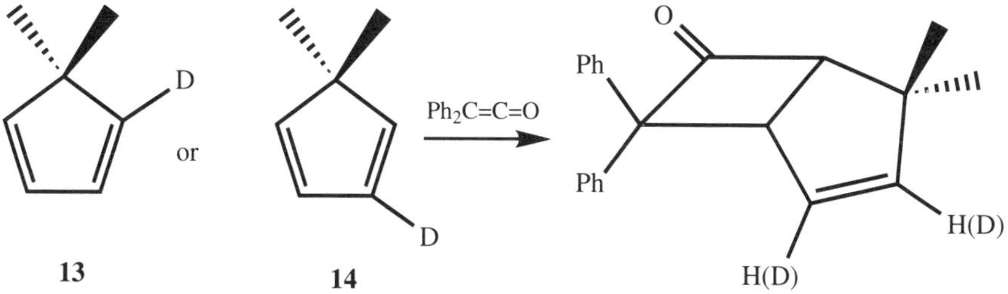

It is possible that this variation in isotope effect with alkene structures means there are significant variations in the transition state structures as the ketenophile is changed, and that isotope effects will be specific to particular substrates, with a gradation in mechanisms.

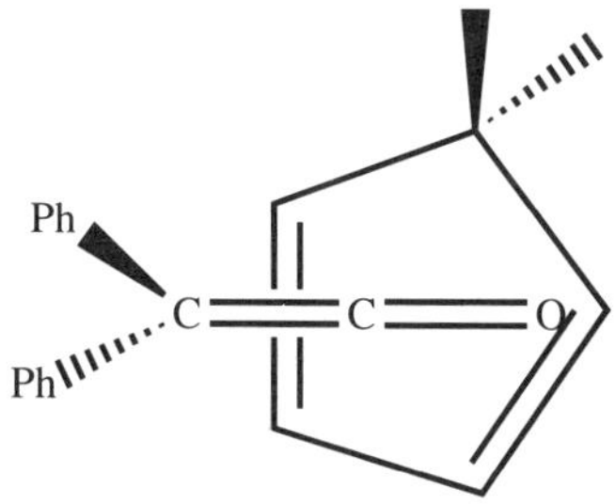

Wang and Houk[110] calculated the geometry and energy of the transition state for the [2 + 2] cycloaddition of ketene and ethylene at the MP2/6-31G* level and found a highly asymmetric geometry **16** (Scheme 1). The forming bond length to the carbonyl carbon is 1.78 Å while that of the second forming bond is 2.43 Å.

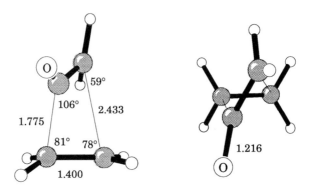

Scheme 1. Calculated MP2/6-31G* Transition State for [2 + 2] Cycloaddition of Ketene and Ethylene (reprinted from reference 111 with permission of the publishers).

This reaction was also studied by Bernardi et al.[109] at the 4-31G level. The authors reported one transition state resembling that shown in Scheme 1 formed by attack in the ketene plane. These authors also located a second transition state similar in energy to the first but with attack of the ethylene perpendicular to the ketene plane.

Both of these studies emphasized the highly asynchronous nature of the reaction.[109,110] One group reported that a diradical intermediate was formed,[109] while Wang and Houk suggested that there was considerable charge separation.[110] The carbonyl carbon of the ketene interacts strongly with both ethylene carbons.[110] This structure is also far removed from the geometry of the [4 + 2] cycloaddition product, and so this latter product is not formed[110] unless there is formation of a long-lived intermediate which can undergo extensive bond rotation. Mulliken population analysis shows an increase in negative charge on oxygen ($\Delta q = -0.16$) and a concomitant increase in positive charge on the weakly bonding carbon from ethylene ($\Delta q = +0.22$), showing the zwitterionic character of the structure.[110]

Two transition states for the [2 + 2] cycloaddition of $CH_3CH=C=O$ and $CH_2=CH_2$ were calculated at the RHF/3-21G level, and **17** for $CH_2=CH_2$ addition *anti* to the CH_3 group was 7.2 kcal/mol lower in energy.[110]

Four transition structures for the [2 + 2] cycloaddition of $CH_2=C=O$ with $CH_3CH=CH_2$ were found, and the most stable, by 4.5 cal/mol, is **18**.[110] In this structure steric interactions with the CH_3 group are minimized, and there is partial positive charge on the CH_3-substituted carbon and possibly a favorable electrostatic interaction between the CH_3 and the carbonyl oxygen.[110] On the assumption of additive CH_3 effects the activation energy for the cycloaddition of *trans*-2-butene was calculated from these results to exceed that for *cis*-2-butene by 2.4 kcal/mol, in qualitative agreement with experimental reactivity trends (Table 2).

5.4 CYCLOADDITION REACTIONS OF KETENES

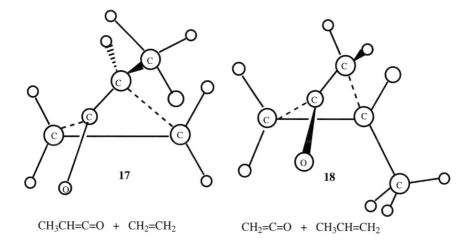

$CH_3CH=C=O + CH_2=CH_2$ $CH_2=C=O + CH_3CH=CH_2$

A theoretical analysis of substituent effects on ketene cycloadditions using the semiempirical AM1 method has been reported, and the calculated geometries and energies are given in Table 6.[113] This study concluded that the $[_\pi 2_s + (_\pi 2_s + _\pi 2_s)]$ mechanism was favored.[113] This process (Scheme 2) may be described for monosubstituted alkenes and ketenes as a one-step process in which the reactants approach in an orientation with the larger ketene substituent pointed away from the alkene. As shown in **19** the major bond-forming interaction is between the more nucleophilic carbon of the alkene and the carbonyl carbon of the ketene. In the transition state the alkene and ketene are twisted by about 50°, and bond formation is much more advanced between C_1 and C_4 (1.64–1.78 Å) compared to C_2–C_3 (2.47–2.81 Å). The zwitterionic character of the transition state is stabilized by electron donors on the alkene and electron-withdrawing groups on the ketene, with a range in AM1 calculated values of $\Delta H^\ddagger$ (kcal/mol) from 39.8 for $CH_2=C=O$ and $CH_2=CH_2$, and 25.7 for $CH_3CH=C=O$ and $MeOCH=CH_2$ (Table 6).[113]

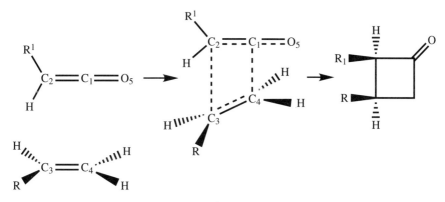

Scheme 2. Cycloaddition of $R^1CH=C=O$ with $RCH=CH_2$.

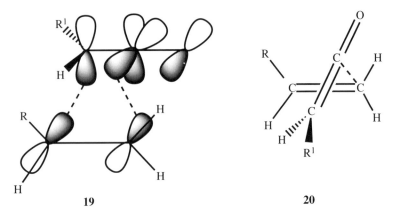

Scheme 2. *(Continued)*

TABLE 6. Calculated AM1 Transition States for Ketene Cycloadditions (Scheme 2)[113]

R_1	R	r_{12}	r_{23}	r_{34}	r_{41}	r_{15}	$\Delta H^\ddagger$ (kcal/mol)
H	H	1.377	2.474	1.404	1.714	1.229	39.8
H[a]	H[a]	1.394	2.433	1.400	1.776	1.216	26.7[a]
Cl	H	1.384	2.476	1.397	1.752	1.222	33.7
CH_3	H	1.384	2.542	1.403	1.698	1.231	37.5
CH_3[b]	H[b]	1.370	2.428	1.399	1.810	1.212	46.9[b]
OMe	H	1.398	2.535	1.389	1.776	1.225	28.0
$CH=CH_2$	H	1.389	2.587	1.404	1.683	1.231	36.2
Ph	H	1.390	2.589	1.405	1.682	1.232	36.1
Cl	CH_3	1.381	2.638	1.413	1.672	1.233	31.4
CH_3	CH_3	1.385	2.647	1.420	1.642	1.240	35.2
CH_3	MeO	1.372	2.747	1.425	1.639	1.245	25.7

[a]MP2/6-31G* (ref. 110).
[b]RHF 3-21G (ref. 110).

It should be noted that while this scheme was claimed to predict the trends in reactivity and the stereochemistry observed in ketene cycloadditions, a corresponding process with a zwitterionic intermediate does the same. As shown previously, this intermediate (**8**) possesses a geometry that would collapse to product by rotation of the hydrogens toward one another leading to the observed *syn* orientation of R and R^1. This picture has been supported in recent experimental studies of ketene cycloadditions with electron-rich alkenes (Section 5.4.1.4). The failure of these theoretical studies to detect a discrete reaction intermediate does not preclude the existence of such a species when the calculations are carried out at a higher level or in experiments in solution.

Theoretical study of the ketene–imine cycloaddition supports a two-step process involving a zwitterionic intermediate for this reaction, as discussed in Section 5.4.1.7.[114]

The course of the [2 + 2] cycloadditions of ketene with ethylene, methylenimine, and formaldehyde, were compared by MP2/6-31G* calculations of the transition states, and by intrinsic reaction coordinate examination of the approach to the transition states.[115] All three reactions are characterized by an approach of the reactants that is skewed slightly from the orthogonal, and give transition states with skewed geometries with activation barriers of 26.9, 30.8, and 24.9 kcal/mol, for ethylene, methylenimine, and formaldehyde, respectively. Quantitative frontier molecular orbital (FMO) analysis showed the reactions could be regarded as involving two one-center orbital interactions, as illustrated in Figure 1, so that the reactions were not concerned with [2 + 2] or [4 + 2] orbital symmetries. For the ethylene reaction bond formation of the ketene LUMO at C_α was far advanced, whereas for formaldehyde bond formation of the ketene HOMO at C_β was greatly enhanced. It was pointed out that a dominance of one or the other one-center interaction would lead to zwitterionic intermediates.[115]

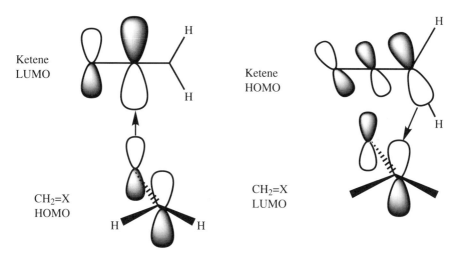

Figure 1. Orbital Interactions in Cycloadditions of Ketene with $CH_2=CH_2$, $CH_2=NH$, and $CH_2=O$.[115]

An alternative stepwise mechanism has been proposed by Halevi[146] and was illustrated for the reaction of t-BuC(CN)=C=O with Z-ROCH=CHMe (R = Me$_3$Si).[147,148] Halevi proposed that the reactants approached "on a path in which the four interacting C atoms are initially coplanar, but are induced by orbital symmetry conservation to bond along the diagonal." Initial bond formation occurred with bond rotation and generation of a zwitterionic intermediate **22**.

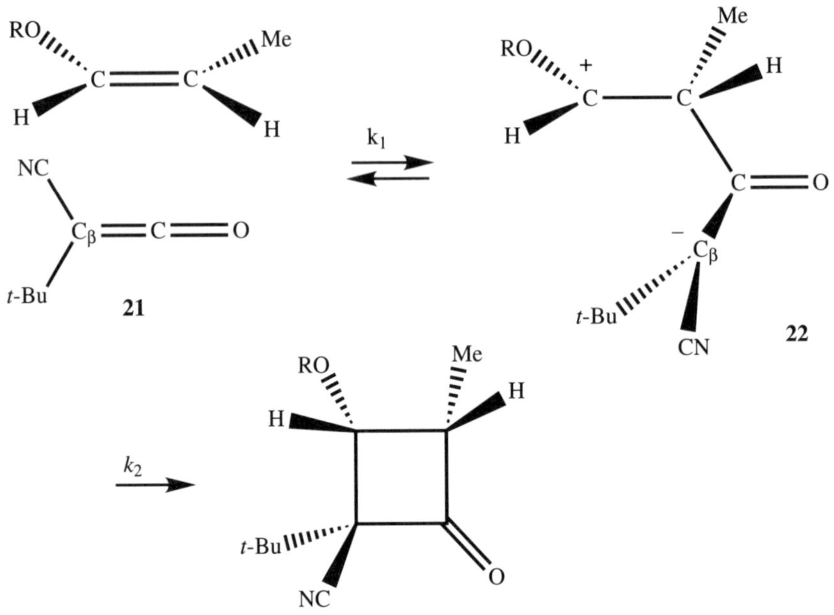

21 ⇌ **22** (via k_1)

22 → **23** (via k_2)

This direction of approach is contradicted by the intrinsic reaction coordinate analysis by Yamabe et al., discussed above.[115] Moreover, the arrangement in **21** suffers from a steric interaction of the cyano group which is directed toward the groups on the alkene. Since the negative charge in the developing zwitterion **22** is concentrated on oxygen, there is a large degree of charge separation in this arrangement. The ketene-ethylene complex has been directly observed in the gas phase and its structure determined by microwave spectroscopy,[148a] and was found to be "remarkably similar" to that calculated by Yamabe et al. for the early stages of the reaction.[115]

The transition states for the reaction of t-BuC(CN)=C=O with a series of alkenes RCH=CH$_2$ have also been calculated by the AM1 method, and show the nonplanar geometry represented by **20**.[116] Further study[116a] of the reaction of this ketene with PhCH=CH$_2$ using a CAS–MCSCF method predicts a stepwise process leading to a biradical intermediate, and formation of the less stable Z-product, as observed. Calculation of the effect of benzene solvent predicts that this will enhance the preference for the Z-product.[116a]

5.4.1.4 Cycloaddition of Ketenes with Nucleophilic Alkenes.

The cycloaddition of ketenes with strongly nucleophilic alkenes such as vinyl ethers and related species such as imines is a situation favoring formation of zwitterionic intermediates and there is evidence for this process in many different examples. This reaction involves attack of the nucleophilic species on the least-hindered side of the ketene, leading to a transition state with a high degree of enolate character. Attack occurs at C_α of the ketene, which has a high coefficient in the LUMO. A freely rotating zwitterionic intermediate is usually not formed, as the stereochemistry of the alkene is often preserved, and the addition is also usually stereospecific at C_β of the ketene to give in many cases the thermodynamically least stable cyclobutanone derivative. Such retention of stereochemistry and formation of the less stable product are however accommodated both by a concerted [$\pi 2_a + \pi 2_s$] process and by formation of an intermediate zwitterion where the zwitterionic groups are in proximity and rotation occurs in the least-hindered sense to form the less stable product.

Thus the [2 + 2] cycloaddition of $Me_2C=C=O$ with E-MeOCH=CHMe gave the cyclobutanone **1** with retention of the alkene stereochemistry (equation 1).[129] Similar results were found for a variety of enol ethers ROCH=CHR^1 with $Ph_2C=C=O$.[121,122,127] Kinetic data for some of these reactions are given in Tables 1–3, Section 5.4.1.3.

$Me_2C=C=O$ + MeO\/Me C=C /H \H → (structure **1**) (1)

However, the [2 + 2] cycloaddition of t-BuC(CN)=C=O with CH_2=CHOEt and CH_2=CHOAc did not give 100% stereoselectivity, but gave E/Z ratios in the product cyclobutanones **2** of 3/1 for EtO and CN and 3.5/1 for AcO and CN (equation 2).[147] With ethyl isopropenyl ether and isopropenyl acetate and benzoate there was clear evidence that zwitterionic intermediates were formed, as hydride or acyl transfer resulting in acyclic products **3** and **4** occurred, and not ring closure (equations 3 and 4). These latter results are strong evidence that the enol ether reaction of equation 1 also involved a zwitterionic intermediate.

t-Bu\/NC C=C=O $\xrightarrow{ROCH=CH_2}$ (zwitterion) → (structure **2**) (2)

$E/Z = 3/1$ (R = Et)

3.5/1 (R = Ac)

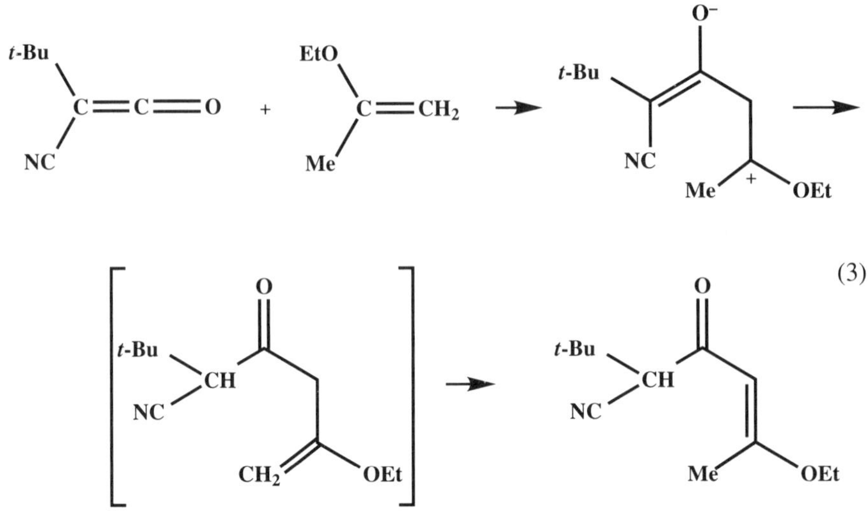

(3)

(4)

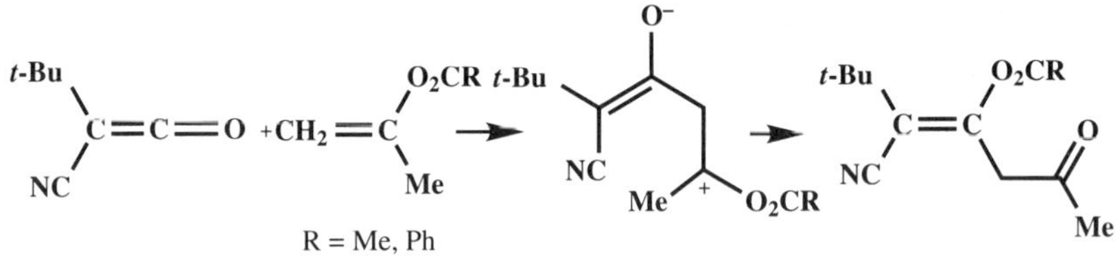

The [2 + 2] cycloadditions of silyl enol ethers $RMe_2SiOCH=CHMe$ make an informative comparison to the alkyl vinyl ethers. With R = Me (**5**) the Z-alkene was more reactive than the E-isomer, and each reaction was highly stereoselective (equations 5 and 6).[148] On heating the product **6** was converted to **8**, but **7** was stable.

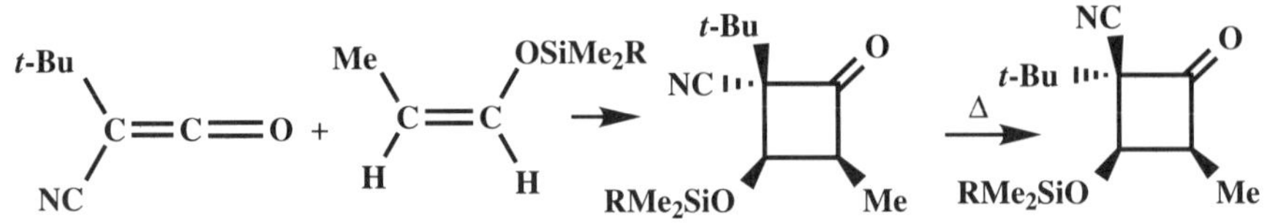

(5)

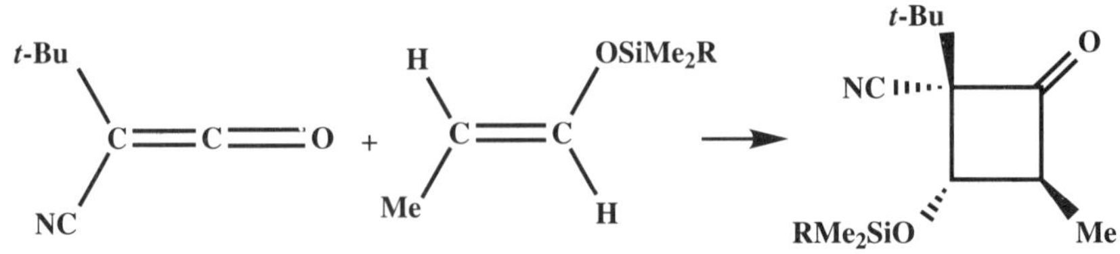

(6)

5.4 CYCLOADDITION REACTIONS OF KETENES

By contrast the E isomer of t-BuMe$_2$SiOCH=CHMe (**9**) was more reactive than the Z isomer, although the products **6** and **7** were still formed with high stereoselectivity. It was proposed[148] that the reason for the low reactivity of Z-**9** was that a zwitterionic intermediate **10** was being formed, and that ring closure of **10** was rate determining, implying that the formation of **10** was reversible (equation 7).[148] Presumably the bulk of the t-Bu group made ring closure to **6** slow.

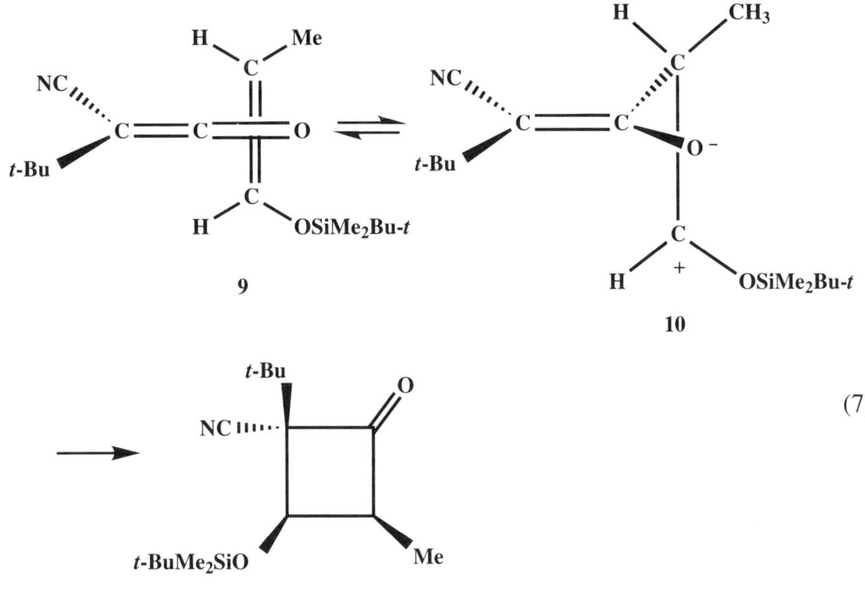

(7)

Confirmatory evidence that the reactions proceeded by zwitterionic intermediates was obtained from thermolysis of the azide **11** which gave the product **7**. It was proposed that the reaction of **11** gave **12**, which underwent bond rotation to **12a** followed by closure to **7** (equation 8).[148] Reaction of the Z-isomer of **11** in refluxing benzene gave **6**, **7**, and **8** in relative yields of 40, 50, and 5%, respectively. The same products were formed from the reaction of equation 5 under the same conditions.

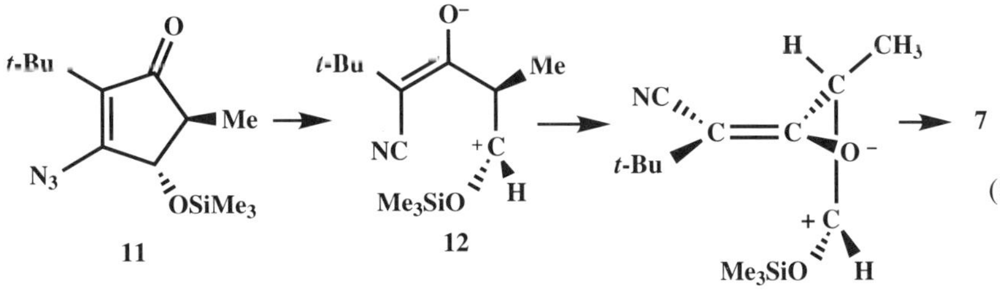

(8)

Other reactions of silyl enol ethers with alkenes sometimes gave mixtures of [2 + 2] cycloaddition products and acyclic products (equation 9).[149-151] In some of these cases the [2 + 2] cycloaddition products were stable under the reaction conditions and so stepwise processes are highly likely.

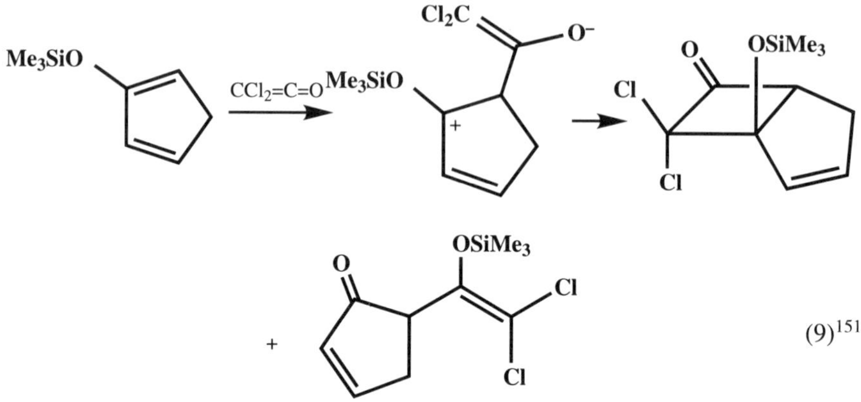

$$(9)^{151}$$

The reaction of chloroketenes with 1,1-dialkoxyalkenes gave [2 + 2] cycloaddition in some cases (equation 10)[152] and acyclic products in others (equation 11).[152] Diphenylketene reacted similarly with $CH_2=C(OMe)_2$.[153] 1,1-Disilyloxyalkenes gave acyclic products with ketenes (equation 12) in reactions interpreted as proceeding through zwitterionic intermediates.[152]

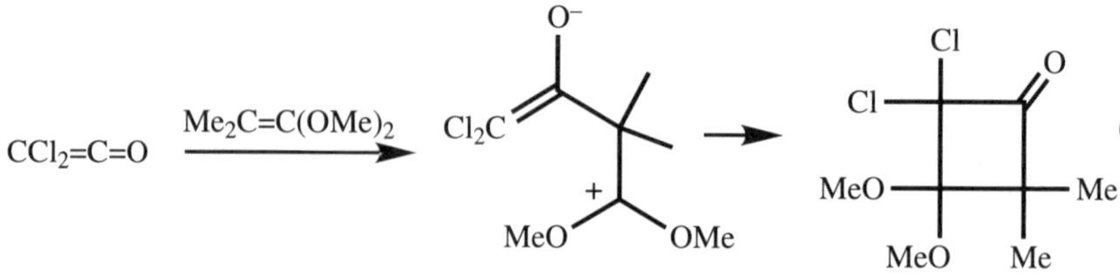

$$(10)$$

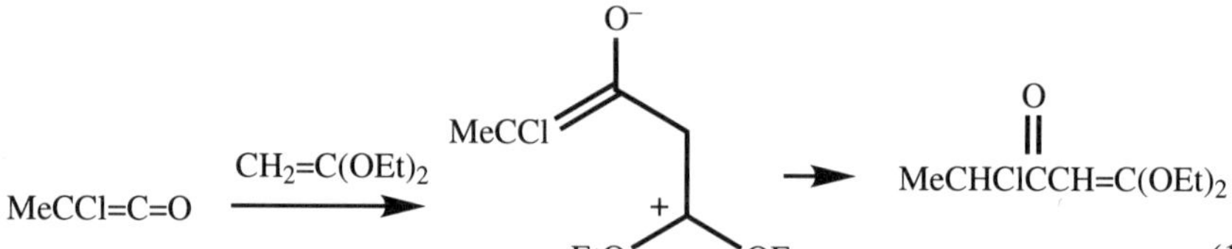

$$(11)$$

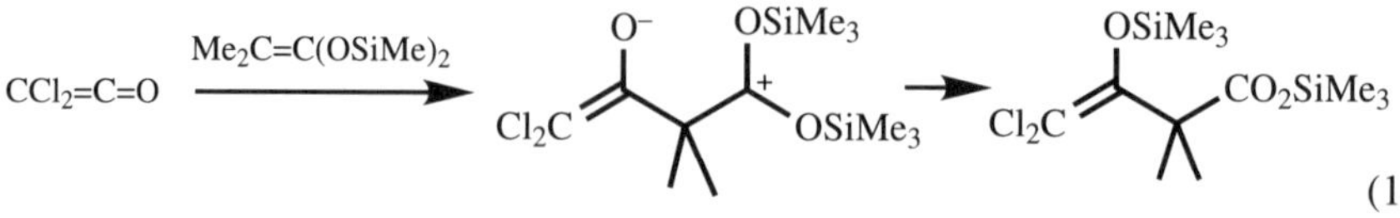

$$(12)$$

The reaction of phenylchloroketene with 4,5-dihydrofuran gave a 75% yield of the cycloaddition product **13**, along with 15% of the acyclic product **14** (equation 13).[154] Formation of the latter product was attributed to a zwitterionic intermediate,[154] which can lead to **13** as well.

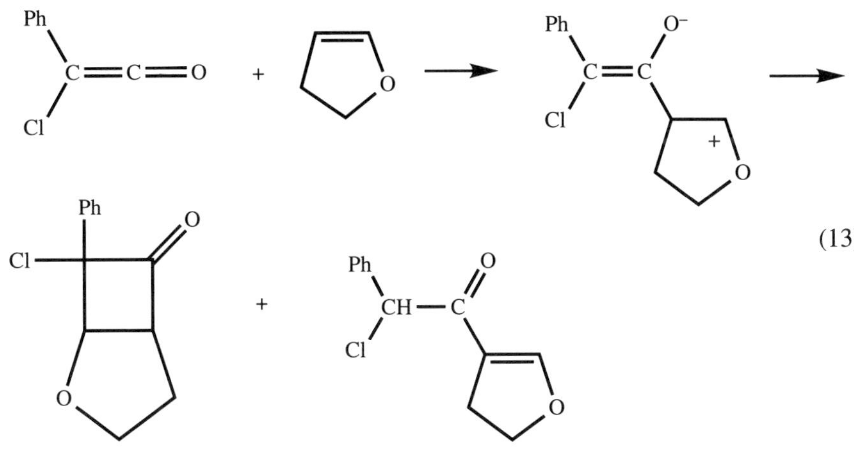

(13)

Further evidence for the formation of a zwitterionic species comes from other studies of the reaction of vinyl silyl ethers with ketenes in which rearrangement occurred that could be conveniently accounted for by the intermediacy of a zwitterion.[155] Thus $Ph_2C=C=O$ and $CH_2=CHOTBDMS$ at room temperature gave an 86% yield of the cycloadduct **16** (equation 14), while $Ph_2C=C=O$ and $CH_2=C(OTBDMS)(C_6H_4Cl$-$4)$ give a 99% yield of **17**, whose formation can be understood to involve silyl migration in the intermediate **18** (equation 15).[155] The cyclobutanone product was formed in 56% yield from $Et_2C=C=O$ and $CH_2=CHOTBDMS$, but the reaction of $Ph_2C=C=O$ was much faster, as expected for a transition state resembling **18**.[155]

$Ph_2C=C=O + CH_2=CHOTBDMS \longrightarrow$ (14)

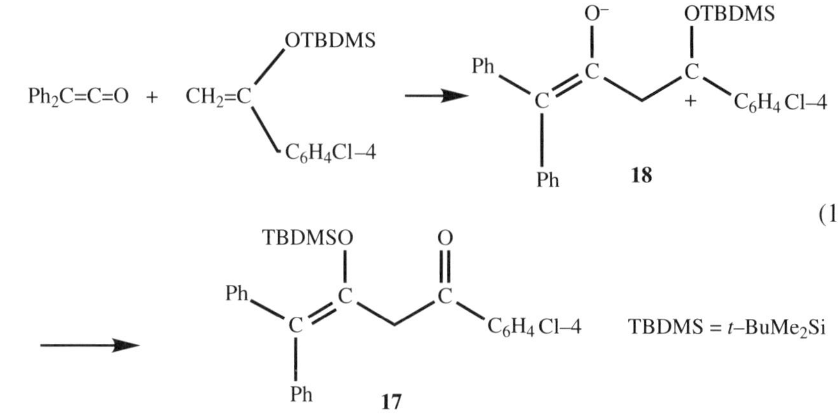

In the reaction of $Ph_2C=C=O$ and $CH_2C(OTBDMS)Bu$-t both silyl rearrangement to **19** and formation of **20** were observed, and the latter product would also result from the zwitterion **21** shown (equation 16). The formation of **20** was found to be reversible to the reactants and on prolonged heating was converted to **19**.[155]

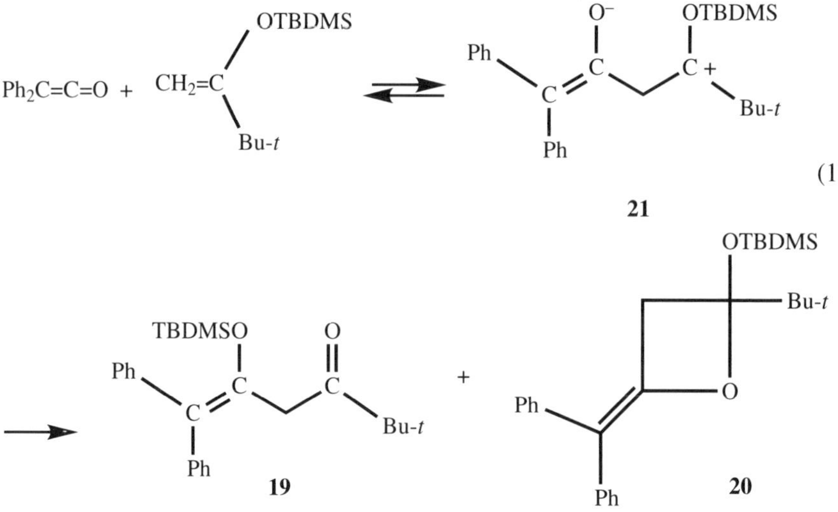

Cycloaddition of ketene with tetramethoxyethylene also gave an oxetane **22**, along with cyclobutanone **23**, and diphenylketene gave only oxetane, and these results can be explained by a stepwise process (equation 17).[156] The use of $ZnCl_2$ catalyst enhances the rate of formation of cyclobutanones from the cycloaddition of ketenes and vinyl ethers, including tetramethoxyethylene, and disfavors the formation of oxetanes.[157] Reaction of tetraalkoxyalkenes with a variety of other ketenes gives cyclobutanone products in high yields, with only small amounts of ring-opened products.[158]

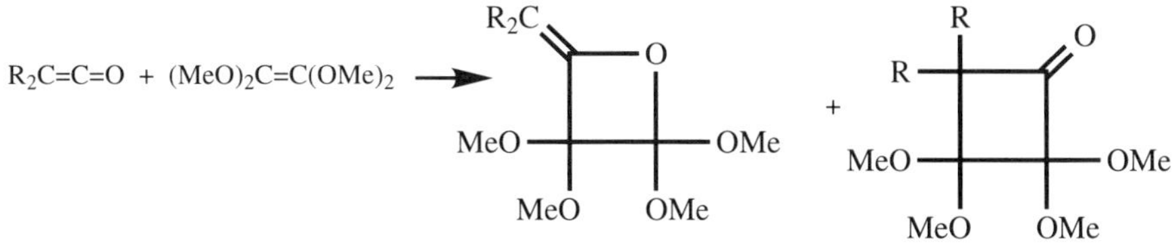

5.4 CYCLOADDITION REACTIONS OF KETENES

$R_2C=C=O + (MeO)_2C=C(OMe)_2 \longrightarrow$ **22** + **23**

(17)

It was argued[155] that previous proposals[55,159,160] for the concerted nature of the cycloaddition of ketenes with vinyl ethers based on reaction stereospecificity,[159] solvent effects,[55] activation parameters,[129] pressure effects,[161] and the absence of products like **19** and **20** were unconvincing, and that stepwise processes were viable routes in such reactions.

The reaction of $Me_2C=C=O$ (**24**) with N-isobutenyl pyrrolidine (**25**) gave the [2 + 2] cycloadduct **26** and the product **27** resulting from reaction of two molecules of ketene.[162] It was found that with a constant ratio [**24**]/[**25**] (**24** in excess) the ratio of **26** and **27** varied with dilution, but the results could not be fitted to a simple expression involving a competition between ring closure of a zwitterionic intermediate **28** to form **26** and reaction of **28** and **24** to form **27**. However, the results could be dissected into two concurrent processes, one involving a competition of intermediate **28** for ring closure to **26** via **29** and reaction with **24** to give **27,** and a separate process that was unaffected by the concentration of **24**. This latter process was ascribed to a concerted [2 + 2] cycloaddition (Scheme 1).[162]

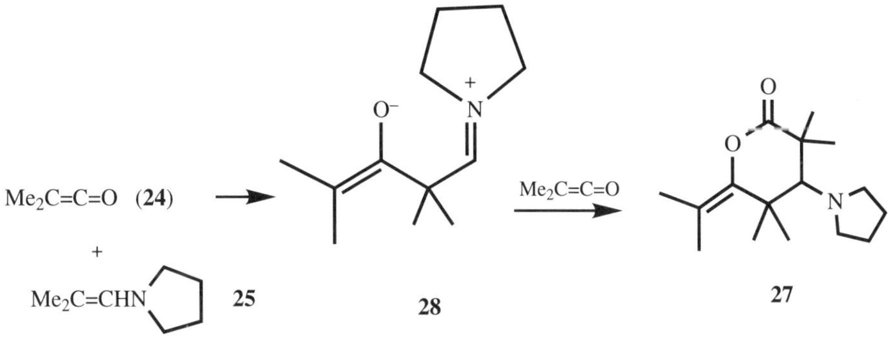

Scheme 1

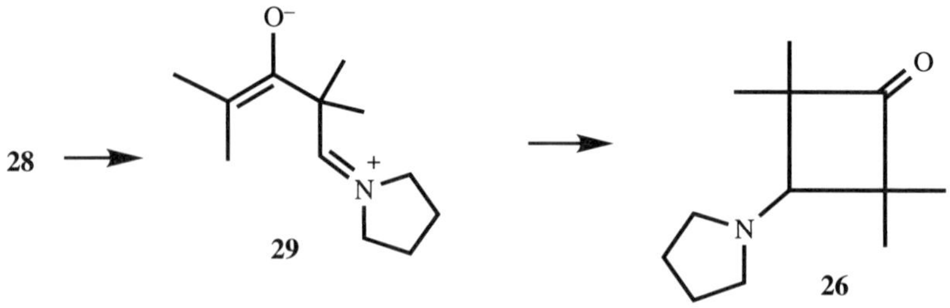

Scheme 1. *(Continued)*

Thus it was envisaged that one process involved a one-step formation of **26** from **24** and **25,** while a separate process involved formation of **28,** which reacted competitively to form **26** and **27.**[162] The formation of mixed products from ketene reactions with enamines has also been found in other examples.[163,164]

The explanation of Scheme 1 appeared plausible on the basis of the claim[162] that the concerted [2 + 2] cycloaddition was the exclusive pathway for ketene reactions with vinyl ethers, but because much evidence[155] has accumulated that stepwise processes are common, further consideration is required. The data appear equally consistent with a mechanism in which there is initial formation of perpendicular encounter complex **30** leading to zwitterion **31,** which undergoes competitive closure to **26** and formation of equilibrating zwitterions **28** and **29,** which form **27** and **26,** respectively (Scheme 2).

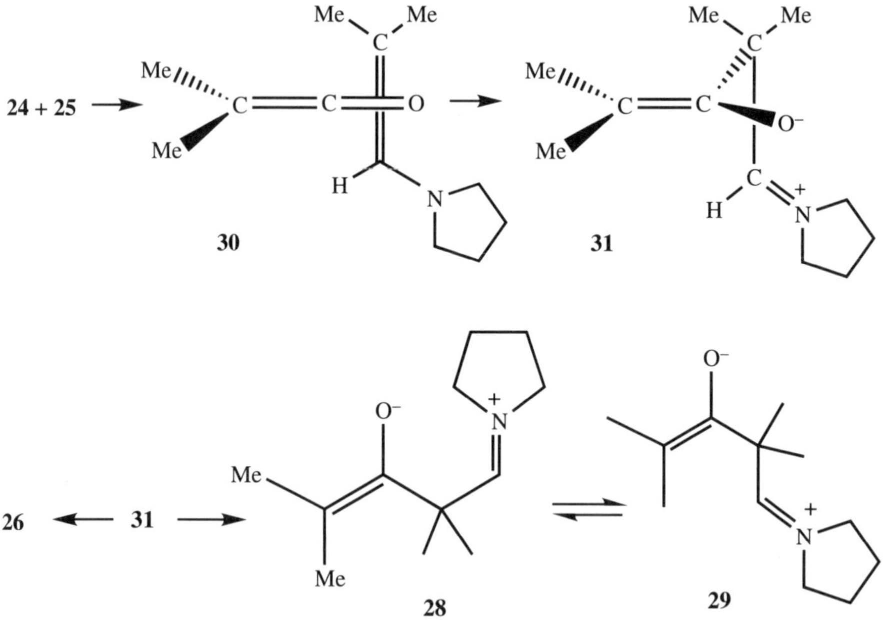

Scheme 2

Other studies of ketene reaction with enamines have been reported, and the formation of various noncyclobutanone products in these reactions show there is a role for zwitterionic intermediates in these reactions.[165,166] Cycloaddition of the ketenes $CCl_2=C=O$, $CHCl=C=O$, $CH_3CBr=C=O$, $CHBr=C=O$, $EtCH=C=O$, $EtOCH=C=O$, $MeOCH=C=O$, $ClCH_2CH_2CH=C=O$, and $PhCH=C=O$ (all generated in situ from the acyl chloride and Et_3N) with eneamides (**32**, R = alkyl) or enecarbamates (**33**, R = OR) gave cyclobutanones which could be converted to useful synthetic intermediates such as the Geissman–Waiss lactone **34** (equation 18).[167]

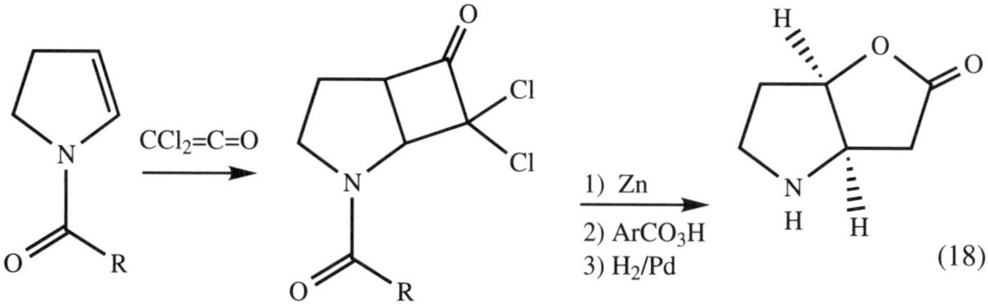

32, 33   **34**

5.4.1.5 Cycloaddition of Ketenes with Allenes.

The [2 + 2] cycloaddition of ketenes and allenes follows some fascinating pathways, and the results favor the occurrence of stepwise processes proceeding through zwitterionic intermediates. Results with dimethyl and tetramethylallene are shown in equations 1–3.[90,168,169]

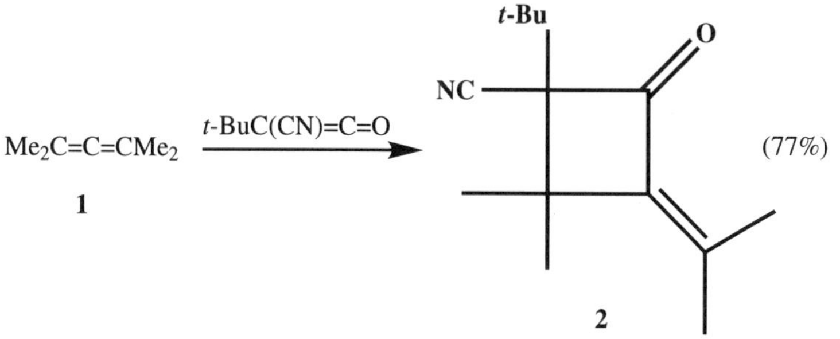

$Me_2C=C=CMe_2$ $\xrightarrow{t\text{-BuC(CN)}=C=O}$ (77%)   (1)

1 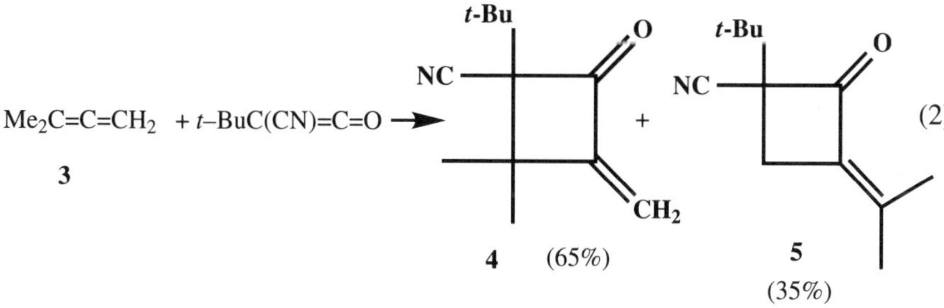   **2**

$Me_2C=C=CH_2 + t\text{-BuC(CN)}=C=O \longrightarrow$ +   (2)

3   **4** (65%)   **5** (35%)

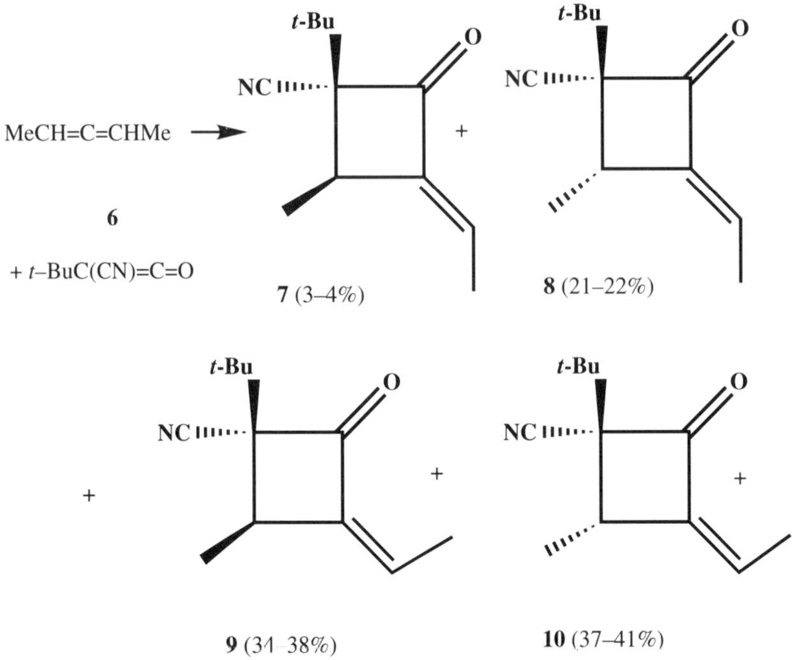

MeCH=C=CHMe ⟶

6

+ t-BuC(CN)=C=O

7 (3–4%)   **8** (21–22%)   (3)

9 (34–38%)   **10** (37–41%)

The fact that optically active **6** gave optically active **7** and **8** but racemic products **9** and **10** shows that an intermediate capable of losing chirality must be formed.[168] These reactions are proposed[90,168,169] to proceed via initial encounter complexes **11** and **12** (Scheme 1). Steric interactions between the ketene substituents and the allene are modest so there is little preference between the two. As shown for **11** (equation 4), bond formation results in a zwitterionic intermediate **13** which by bond rotations can form the achiral **14**. Product formation from **13** can give optically active products, while those from **14** are racemic. Analogous zwitterions form from **12**. The relative yields of the observed products have been rationalized on the basis of the anticipated steric interactions,[90,168,169] but in energetic terms these differences are small.

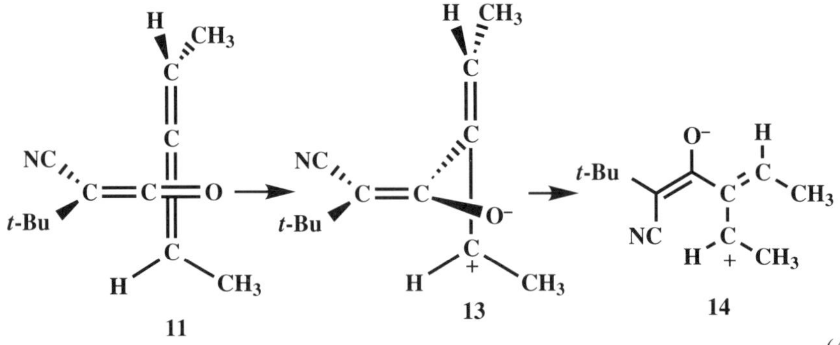

11   **13**   **14**

(4)

Scheme 1

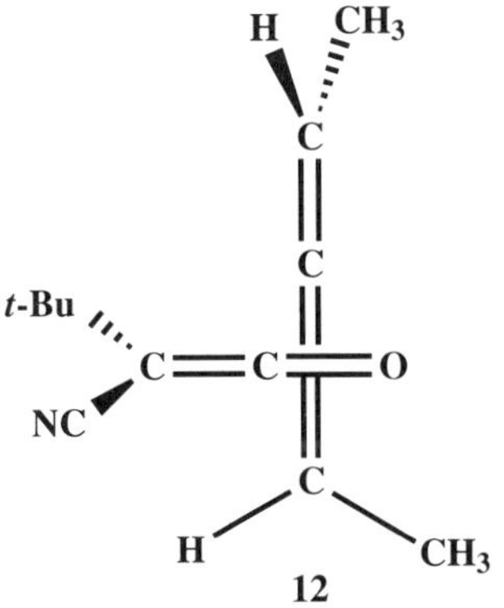

Scheme 1. *(Continued)*

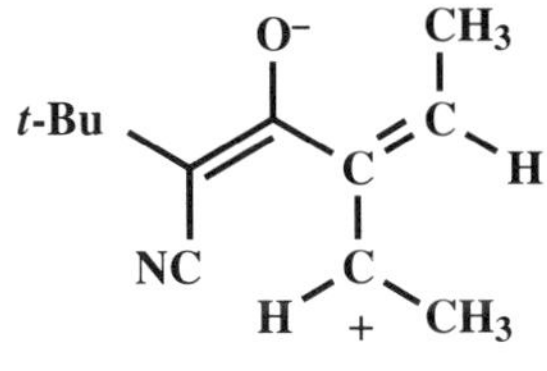

The cycloaddition of *t*-BuC(CN)=C=O with optically active 1,2-cyclononadiene **16** gave a 2/3 mixture of the *E/Z* stereoisomeric adducts **17** (equation 5).[123] Both of the products were optically active, although the optical purities were not established. The low *E/Z* stereoselectivity is consistent with formation of a zwitterionic intermediate **18**, and the results were interpreted[90] similarly to the reaction of 2,3-pentadiene (Scheme 1),[168] as shown in equation 4. Optically active products evidently arise from a chiral form of **18**.

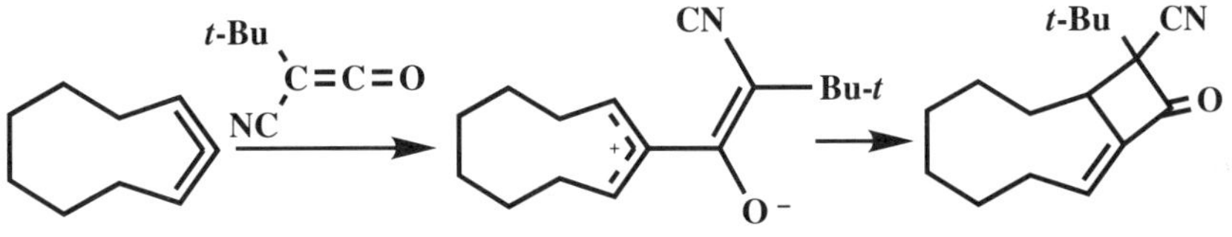

(5)

The reaction of optically active 1,3-diphenylallene with *t*-BuC(CN)=C=O gave only cyclobutanones with the *E* configuration of the phenylvinyl groups (*anti* to the carbonyl) analogous to **7** and **8**, and both were optically active.[170,171] These results were interpreted[90] in terms of the formation of two encounter complexes resembling **11** and **12** and analogous formation of chiral and achiral zwitterions. However the greater steric interactions with the phenyl groups prevented the formation of the Z-stereoisomers analogous to **9** and **10**.

The reactions of $Me_2C=C=O$[1/2,1/3] with a large number of allenes, and of other ketenes and allenes[174] have been studied. The allenes included symmetrically and unsymmetrically substituted and acyclic examples, and the reactions often formed all of the possible stereoisomeric products. The reaction pathways appear to be consistent with the stepwise processes outlined in Scheme 1. An interpretation of the reaction rates and product distributions for various ketene/allene pairs involving formation of uncharged species by bonding between the *sp* carbons of the ketene and the allene has been proposed.[175]

5.4.1.6 Cycloaddition of Ketenes with Alkynes.

Ketenes substituted with electronegative groups react with nonactivated alkynes to form cyclobutenones.[176,177] Thus dichloroketene reacts with 2-butyne, 3-hexyne, or 1-hexyne to form adducts **1** which are isomerized in situ by $ZnCl_2$ formed in the generation of the $CCl_2{=}C{=}O$ to an equilibrium mixture with **2** (equation 1), and both **1** and **2** can be reduced with zinc and acetic acid to **3**.[176,178] Hydrolysis of **1** gives cyclobutenediones. *tert*-Butylcyanoketene also reacts with a variety of alkyl and arylalkynes to give cyclobutenones,[179] and dialkylketenes give [2 + 2] cycloaddition to alkynyl ethers (equation 2).[180]

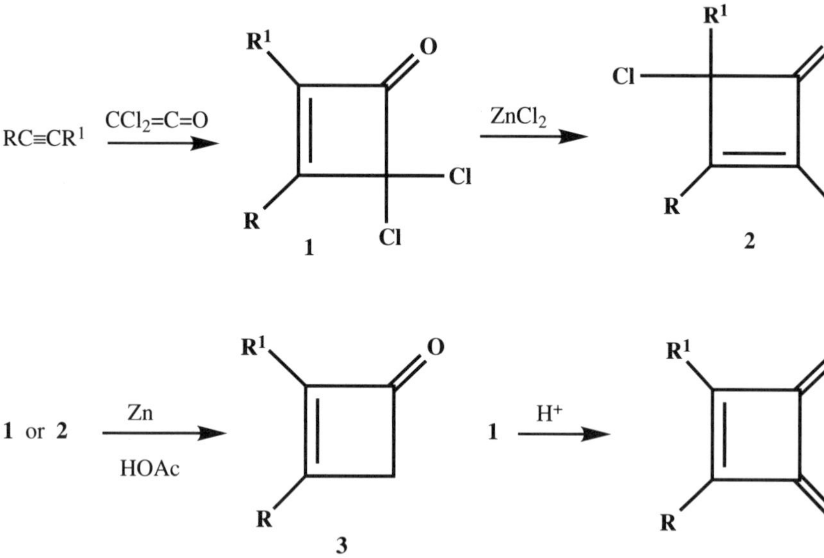

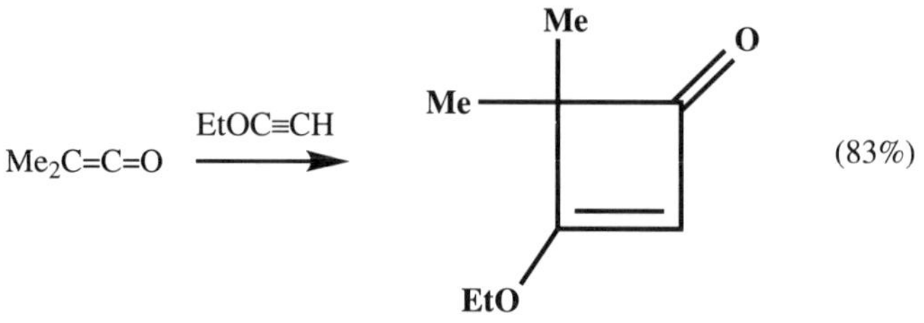

The reaction of trimethylsilylacetylene with dichloroketene gives mainly the product resulting from electrophilic attack at the nonsilicon-substituted carbon of the alkyne, contrary to the usual direction of electrophilic substitution on this alkyne (equation 3).[181] However, this carbon does have the highest coefficient of the HOMO of $SiH_3C{\equiv}CH$.[181] With $PhC{\equiv}CSiMe_3$ electrophilic attack occurs at the silylated carbon, in accord with the expected stability of the silylated intermediate (equation 4).[181]

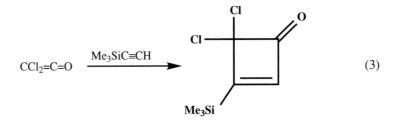

(3)

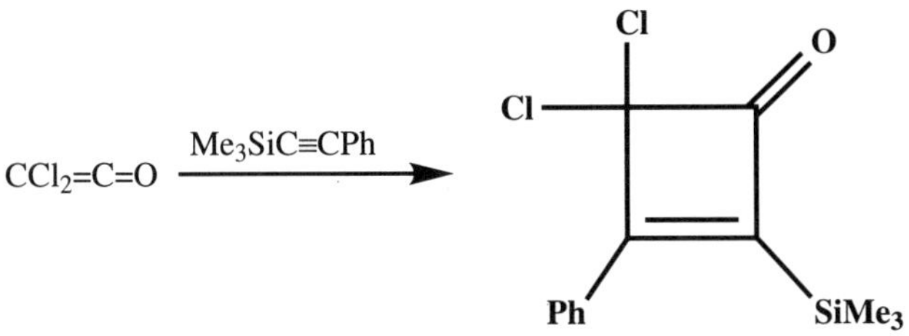

(4)

Reactive alkynes such as 1-triisopropylsilyloxyheptyne react with ketene to give 3-silyloxycyclobutenones **4** (equation 5).[182] 1-Ethoxyheptyne was less reactive toward ketene than the silyloxyheptyne. Reaction of **4** with MeLi followed by acid gave cyclobutenones **5** (equation 5).[182]

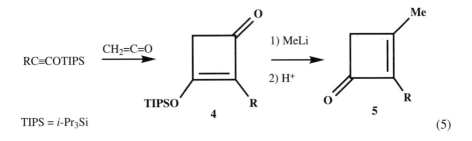

TIPS = i-Pr$_3$Si

(5)

An MNDO theoretical study of the [2 + 2] cycloaddition of hydroxyketene with dihydroxyethyne indicates that the reaction proceeds by formation of a zwitterionic intermediate **6** (equation 6).[112] The encounter complex **7** has the OH group of the ketene oriented towards the alkynyl group. The transition state for the closure of **6** to **8** is calculated to be only 1.06 kcal/mol higher in energy than **6,** although there is a large geometrical difference. The structure of the complex of acetylene and ketene in the gas phase has been examined by microwave spectroscopy and found to be planar, with a 25° tilting of the axes of the molecules.[182a]

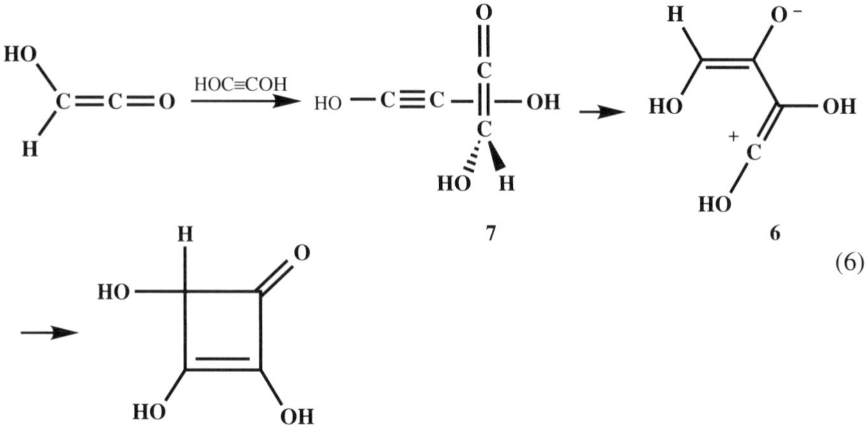

Monosubstituted diphenylketenes and ethoxyacetylene react in a process formulated as involving a concerted [4 + 2] cycloaddition to form the intermediate **9**, which gave azulene derivatives **10** as observed products as well as cyclobutenones from [2 + 2] cycloadditions.[183] There was a preference for attack on the ring bearing an electron-donating substituent (MeO, Me), but not an electron-withdrawing group (Cl, Br, CO_2Et). An alternative stepwise process involving initial formation of a zwitterion prior to formation of **9** is shown, and readily explains the preference for attack on an electron-rich aryl group (equations 7 and 8). The intermediate norcaradienes **9** are isolable in the reaction of 1-naphthylphenylketene, di-(2-naphthyl)ketene, and di-(1-naphthyl)ketene with EtOC≡CH.[184] The vinylketenes **11** and **12** gave only [2 + 2] cycloaddition products with HC≡COEt.[184]

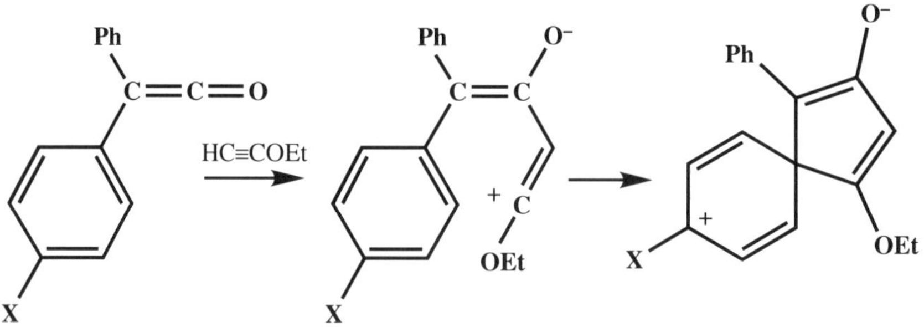

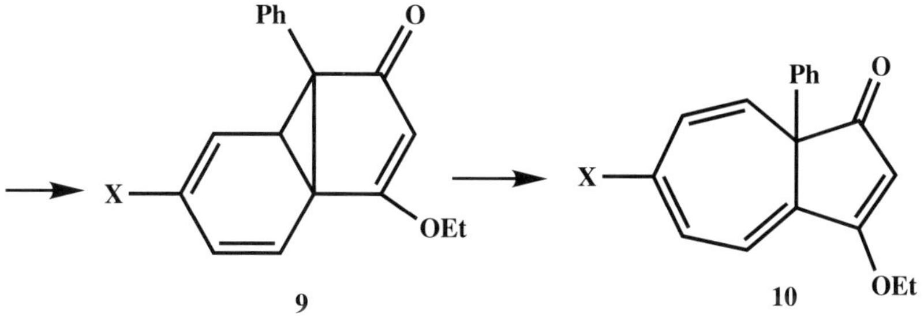

(structures 9, 10, 11, 12 shown)

11: CH₂=CHCH=C=O → $CH_2=CHCH=C=O$

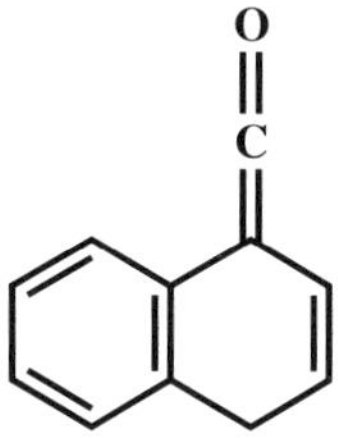

Carboethoxy and carboxamido ketenes react with alkynylamines to give cyclobutenones by [2 + 2] cycloaddition, pyrones by [4 + 2] cycloaddition, and allenes, depending upon the particular substrate (equations 9–11).[185,186] Vinylketenes also give allenes and other products on cycloaddition with alkynylamines, together with cyclobutenones.[187–189]

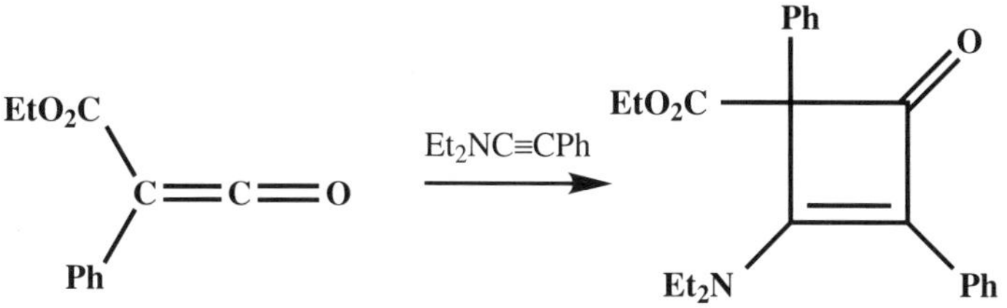

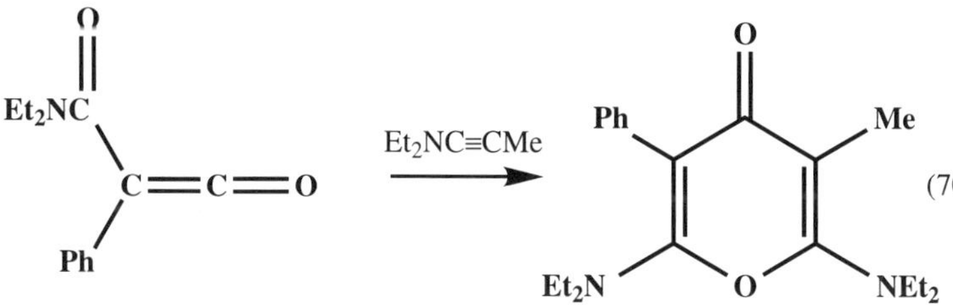

(11)

5.4.1.7 Cycloaddition of Ketenes with Imines.

The cycloaddition of ketenes with imines produces β-lactams, which are of wide interest because of the presence of this ring system in β-lactam antibiotics such as penicillin. Several reviews of this reaction are available,[11,190-195] and interest in this process has continued.[196-225] This reaction is often carried out with ketenes generated in situ, and when the ketene is formed from an acyl halide and Et_3N there is sometimes a question as to whether a free ketene has actually been formed or whether the reaction involves the acyl chloride or an acyl chloride/triethylamine complex.[200,201] Reactions between cyanoketenes generated by thermal routes and imines are proposed to be free of this ambiguity,[198,199] and it is usually believed that reactions conducted by adding an acyl chloride to a mixture of an amine and imine also involve free ketenes.[200]

Study of the ketene–imine cycloaddition at the RHF/6-31G*//RHF/6-31G* and RMP2/6-31G*//RMP2/6-31G* levels indicate an *anti* and near-planar geometry **1** for the first transition state, leading to the planar zwitterionic intermediate **2**.[114] Some salient bond angles are shown (Scheme 1). This reaction differs from ketene–alkene cycloadditions (Section 5.4.1.3) in that the lone electron pair on the imino nitrogen serves as the nucleophile, and the intermediate is a planar conjugated π-system. Product formation occurs by the transition state **3**, which is nonplanar and resembles the transition state for the cyclobutene–butadiene interconversion. At the RMP2/6-31G* level the energies of **1, 2, 3**, and **4** are 3.7, 3.5, 21.3, and −41.2 kcal/mol above the energy of the reactants, indicating that there is only a relatively small barrier to formation of the intermediate which is near the transition state in energy, and that the slow step of the reaction is the conrotatory electrocyclic ring closure.[114] Calculations at the 3-21G level did not locate the transition state **1**, and favored **3** as the first transition state.[114,117] The effect of substituents on the product stereochemistry has also been examined theoretically.[114a] (See also Section 5.9).

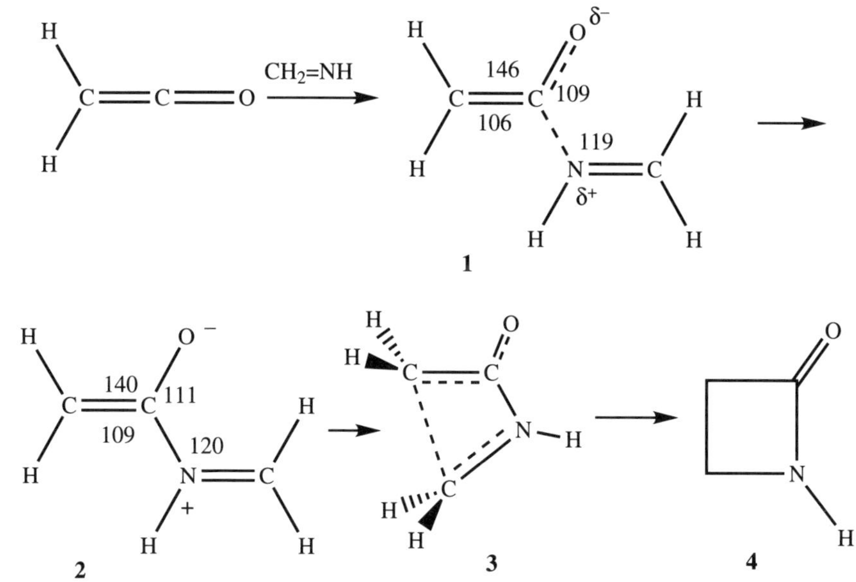

Scheme 1

Calculations using AM1 of the reaction of MeOCH=C=O with the chiral imine **5** forming the initial zwitterion **6** were used to predict the diastereoselectivity in the formation of the β-lactam **7** (equation 1).[211] These calculations were extended to several substituted ketenes and a number of substituted imines, and correctly predicted the observed stereoselectivity of the reaction.[212]

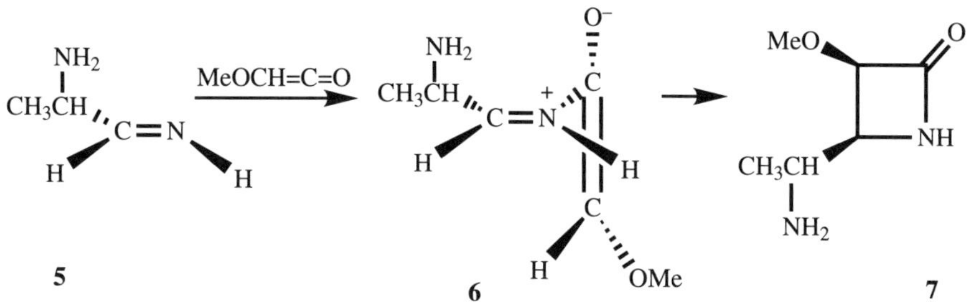

(1)

These calculations are consistent with the experimental observation of the zwitterionic intermediate by IR spectroscopy,[196] including kinetic analysis of the intermediates observed by FTIR.[197] In the former experiment irradiation of a film containing the diazo ketone **8**, imidazole (**9**), and a phenolic resin **10** at 10 K gave an IR spectrum which showed the presence of ketene **11** along with **9** and **10**, but at 140 K the ketene absorption at 2110 cm^{-1} was replaced by a band at 1635 cm^{-1}

attributed to the zwitterion **12** (Scheme 2). On further warming the absorption of **12** was replaced by that of the ester **13**.[196]

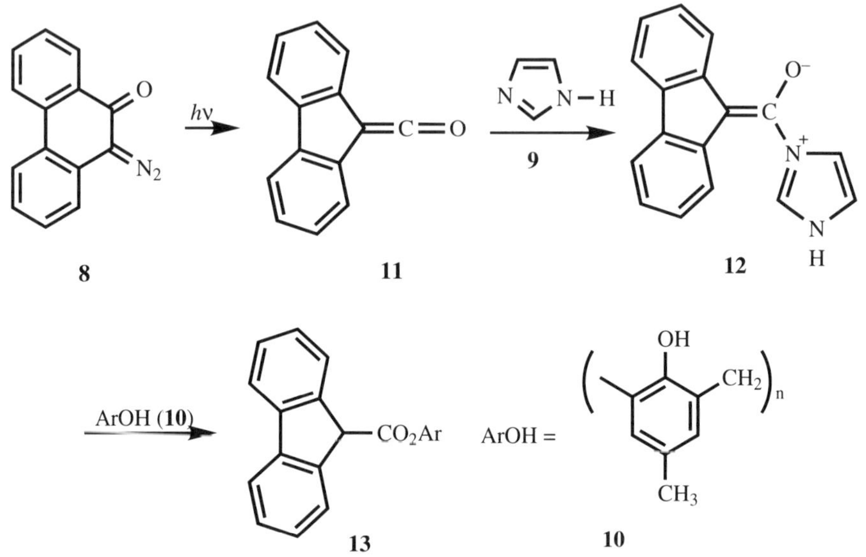

Scheme 2

The stereochemistry of the cycloaddition of ketenes with imines is variable and depends upon the substituents on the ketene and on the imine. Thus reaction of the ketenes $R^1C(CN)=C=O$ ($R^1 = t$-Bu, Cl, n-BuC≡C) with PhCH=NR gives exclusively or predominantly the stereoisomeric product **14** (equation 2) for R = Ar, n-Bu, or c-Hx, but when $R^1 = t$-Bu **15** is the predominant or exclusive product.[199] These results are explained[198] by the mechanism of Scheme 3 in which the imine and the ketene approach in a common plane so that the lone pair of electrons on the imino nitrogen can attack at the LUMO of the carbonyl oxygen. Initial transoid dienyl intermediates **16** may be favored, which form products after bond rotation to cisoid zwitterions **17** and **18**. The intermediate **17** is favored on steric grounds in all cases except when R is t-Bu, in which case both **17** and **18** are formed. In **18** the *syn* steric interaction between t-Bu and Ph is relieved, but at the expense of creating an interaction between Ph and CN. Product formation occurs by conrotatory motion of **17** and **18** to give **14** and **15**, respectively.

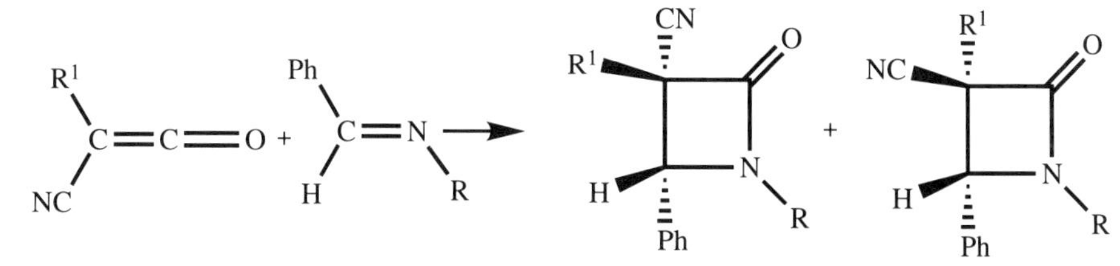

5.4 CYCLOADDITION REACTIONS OF KETENES

(2)

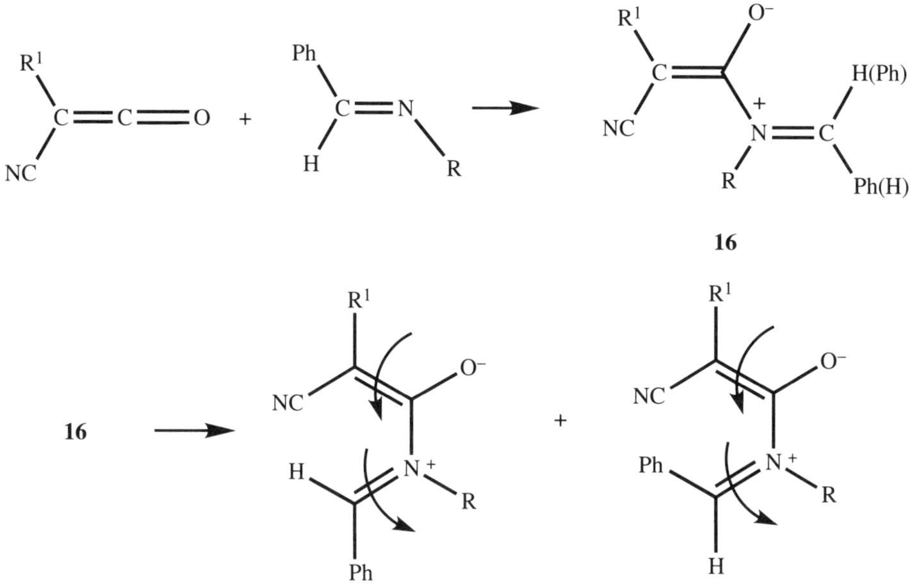

Scheme 3

This mechanism presupposes that the imine and the ketene are in near planarity in the transition state so that the lone pair of electrons of the sp^2 imino group is involved. Moreover interconversion of the *anti* imine to give the *syn* form leading to **18** is required. These assumptions appear reasonable and Scheme 3 explains the results and agrees with recent theoretical analyses of the reaction (Scheme 1).[114,211,212]

As discussed in Section 5.9, the reaction of ketenes with imines bearing chiral substituents can give β-lactams with virtually complete control of disastereoselectivity.[211]

When the imine (Scheme 3) is changed to PhCH=CHCH=NR or Ph$_2$C=CHCH=NR, the β-lactam products are **19** and **20**, with a preference for **19** except when R is t-Bu, when **20** is favored (equation 3).[198] These results are analogous to those in equation 2. The intervention of zwitterionic intermediate **21** is consistent with the formation of the cyclized products **22**, which are also formed (equation 4).[198,198a]

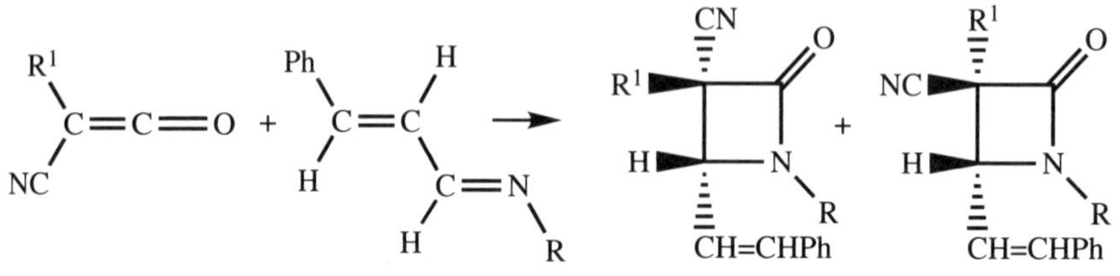

(3)

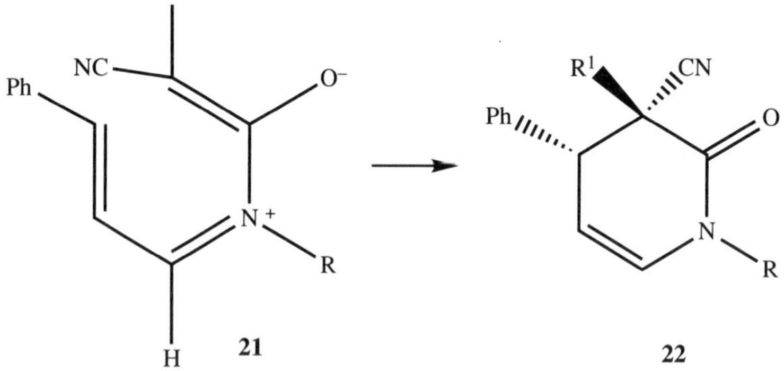

(4)

The reaction of the phenylthio-substituted ketene **23** with imine **24** gave [2 + 2] cycloaddition with preferential formation of the *cis* product **25**, whereas imine **26** gave **27** by [4 + 2] cycloaddition (equations 5–7).[206] More polar solvents gave less stereoselectivity in the formation of **25**. These results are consistent with formation of the zwitterions **28** and **29**, with preferred approach to **23** from the side of the PhS group.

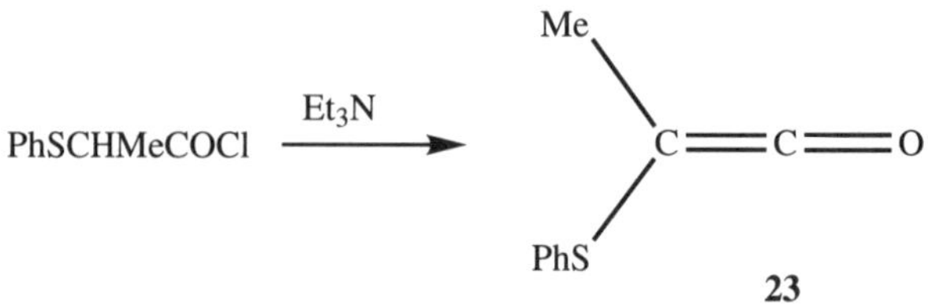

(5)

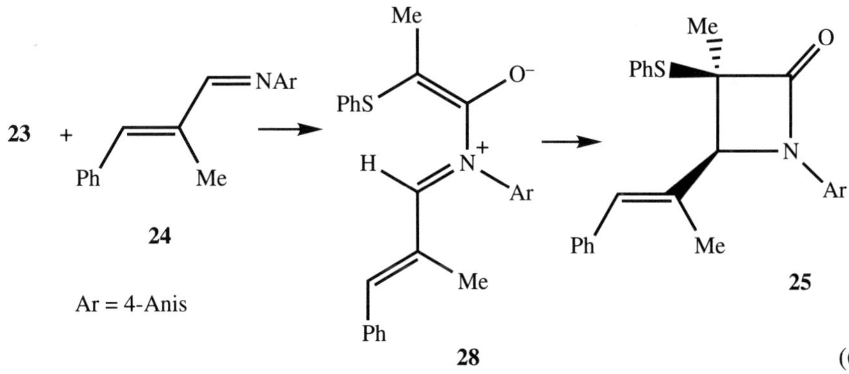

5.4 CYCLOADDITION REACTIONS OF KETENES

(6)

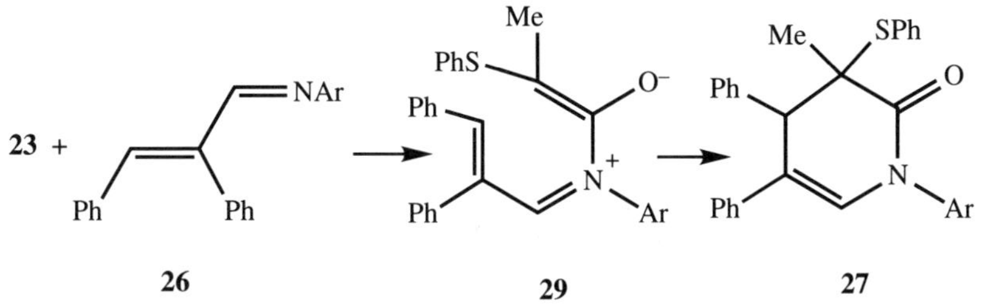

(7)

The [4 + 2] cycloaddition of $Ph_2C{=}C{=}O$ with vinylimine **30** (equation 8) and the [2 + 2] cycloaddition of $CCl_2{=}C{=}O$ with the diphenylvinylimine **31** (equation 9) can similarly be explained by a steric preference for the zwitterionic intermediates shown.[208]

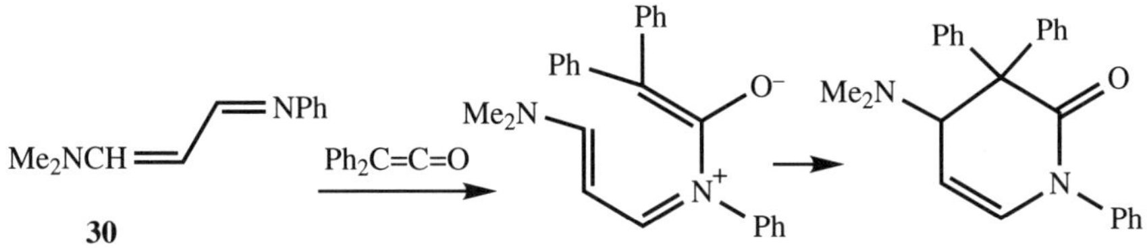

(8)

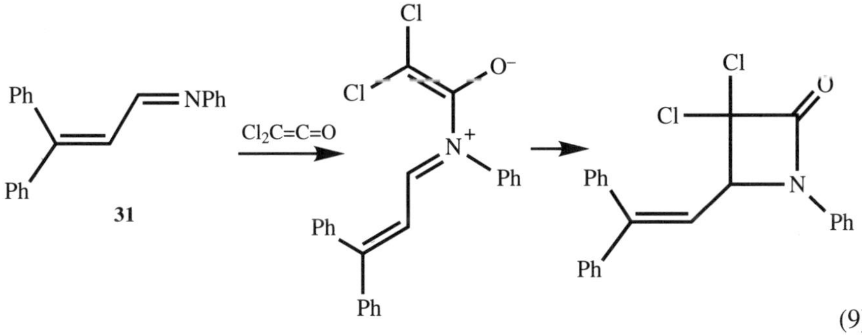

(9)

524 REACTIONS OF KETENES

The stereochemical results of β-lactam formation are consistent with those of Brady and Gu,[207] who reacted ketenes $R^1R^2NCH=C=O$ with imines $ArCH=NR$ and observed the *cis* product **32** as the only diastereoisomeric product in all cases except when Ar was 2– or 4–$CH_3OC_6H_4$. In these cases some of the *trans* isomer of **32** was also formed (equation 10). This was attributed to the reaction proceeding through the zwitterionic intermediate **33**, which ordinarily gave **32** by conrotatory ring closure. However when Ar was 2– or 4–$MeOC_6H_4$ the conrotatory control of the product formation was diminished owing to decreased positive charge on nitrogen and the more stable product was formed.

(10)

In a study[209,210] of the reactions of fluoroketenes $PhCF=C=O$ and $CHF=C=O$ with imines $R^1CH=NR$ (R^1 = Ar, Et, Me) it was found that the products almost exclusively had the *cis* stereochemistry, and high stereoselectivity was achieved using an imine derived from D-glyceraldehyde acetonide.[210] These results are consistent with the explanations given above, such that zwitterions **35** are formed from $CHF=C=O$ and **36** from $PhCF=C=O$, and form a *cis* product in either case by conrotation. The high selectivity for attack *anti* to the sterically undemanding fluorine was considered remarkable, and possible stereoelectronic control of the reaction was discussed.[210] Reactions of $PhCF=C=O$ proceeding through the intermediate **37** were considered, in which attack on the ketene was *anti* to the fluorine because of "simple dipolar effects or secondary orbital interactions" followed by conrotation.[209] However, it appears likely that these reactions proceeded with initial formation of transoid zwitterions **38** and **39**, which led to **35** and **36** on bond rotation.[210]

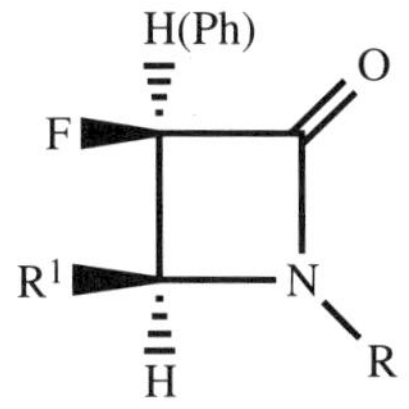

34, 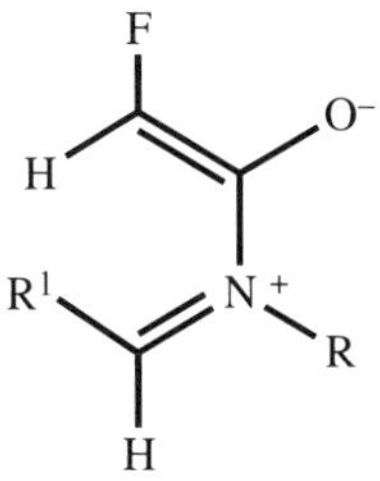 **35**, 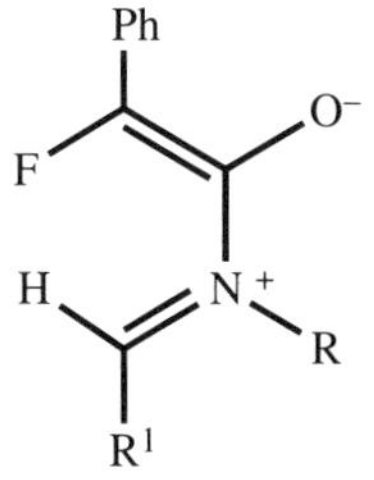 **36**, 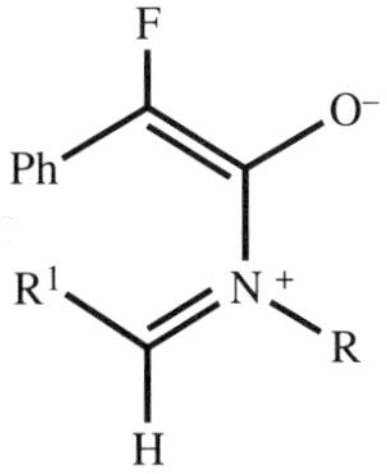 **37**

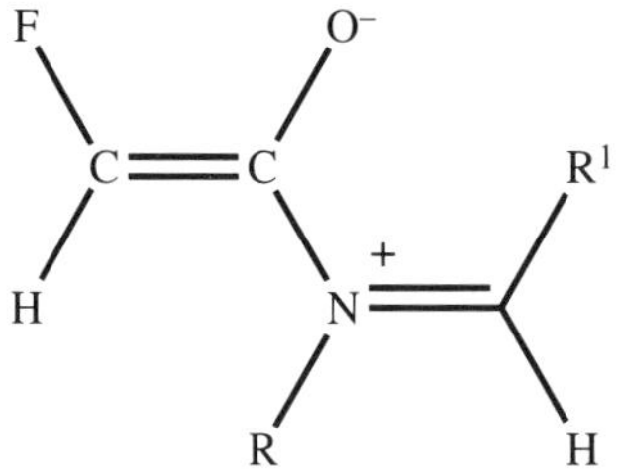

38, 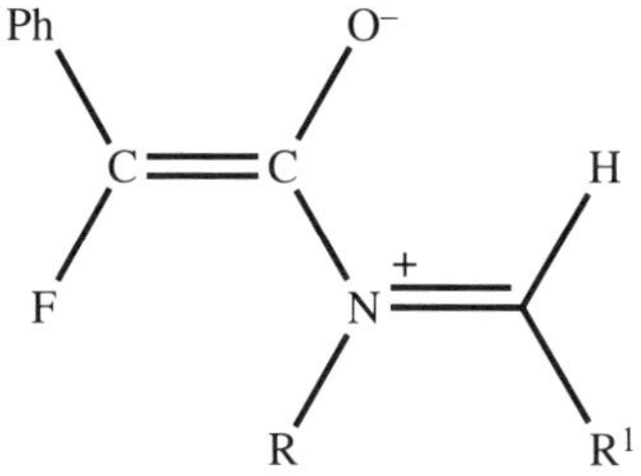 **39**

A comparison was made of the stereoselectivity of the reaction of imines RCH=NPh (R = MeO, Ph, PhCH=CH, Me) plus **40** and **41** with ketenes generated by photolysis of the oxazolidine– and oxazolidinone–chromonium carbene complexes **42** and **43** and from reaction of the oxazolidinone acid chloride **44**.[213] Some of the reactions proceeded to give high chemical yields and high diastereoselectivities, but differences in the results from different precursors led to the conclusion that chromium-complexed ketenes were involved in the reactions of **42** and **43**.

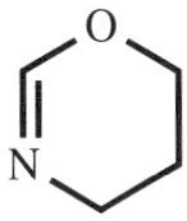

40,  **41**, 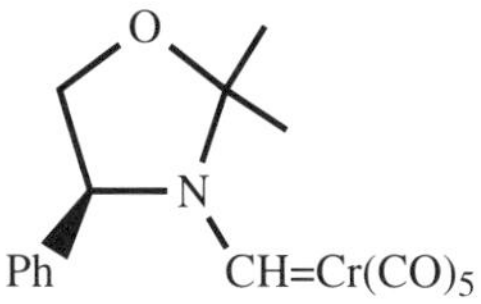 **42**, 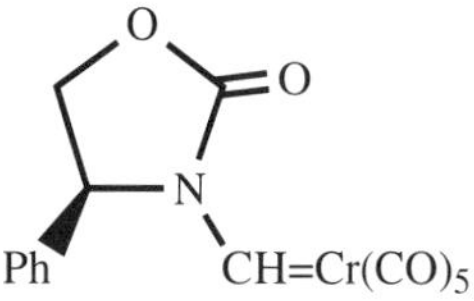 **43**

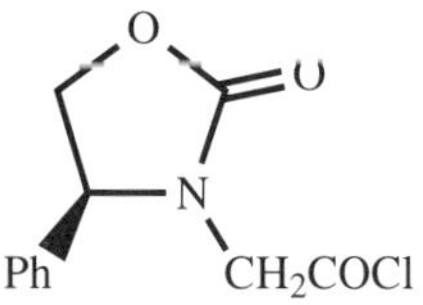

44

Photolysis of chromium carbene complexes in the presence of iminodithiocarbonates proceeds with formation of the β-lactams **45** through the ketene equivalents **46** (equation 11).[214] Products **45** (R^1 = PhO, R = H) were also prepared from acyl chlorides.[215]

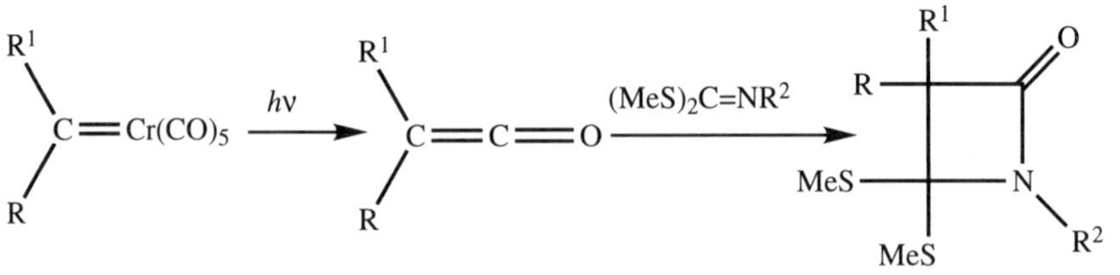

$$\underset{R}{\overset{R^1}{>}}C=Cr(CO)_5 \xrightarrow{h\nu} \underset{R}{\overset{R^1}{>}}C=C=O \xrightarrow{(MeS)_2C=NR^2} \underset{\mathbf{45}}{\text{β-lactam with } R^1, R, MeS, MeS, N-R^2, C=O}$$

46   R = MeO, R^1 = Me, Ph, c–Pr
R = PhCH$_2$O, R^1 = Me
R = i-PrO, R^1 = Me
(PhCH$_2$)$_2$N, R^1 = H

(11)

The reaction of galactose-derived imines with a variety of acyl chlorides and Et$_3$N gave a mixture of two cis β-lactams, typically in a 60:40 ratio (equation 12).[216] These absolute configurations of the lactams were elucidated by X-ray crystallography of the separated isomers. Asymmetric cycloadditions of ketenes with glucose-derived imines have also been studied.[217]

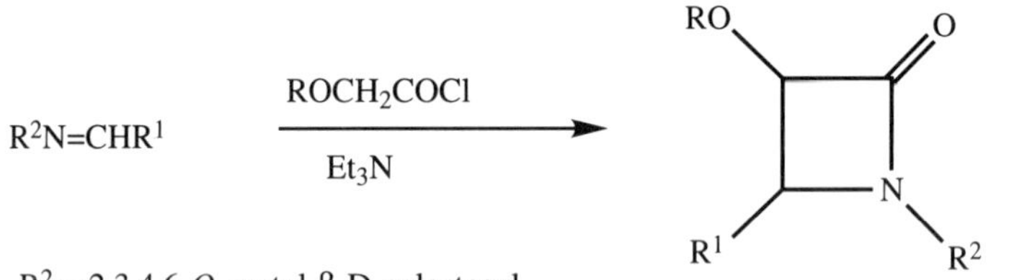

$$R^2N=CHR^1 \xrightarrow[\text{Et}_3\text{N}]{\text{ROCH}_2\text{COCl}} \text{β-lactam}$$

(12)

R^2 = 2,3,4,6-O-acetyl-β-D-galactosyl

Cycloaddition of the β-silylalkylketene **47** gave a 9/91 preference for the Z stereoisomers **48** (equation 13).[218] The silicon could be replaced in subsequent steps.

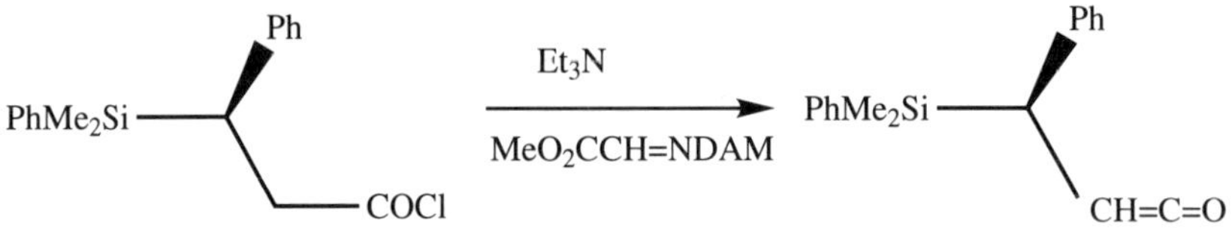

$$\text{PhMe}_2\text{Si}-\overset{\text{Ph}}{\underset{\text{COCl}}{\text{C}}} \xrightarrow[\text{MeO}_2\text{CCH=NDAM}]{\text{Et}_3\text{N}} \text{PhMe}_2\text{Si}-\overset{\text{Ph}}{\underset{\text{CH=C=O}}{\text{C}}}$$

DAM = (4-MeOC$_6$H$_4$)$_2$CH   **47**

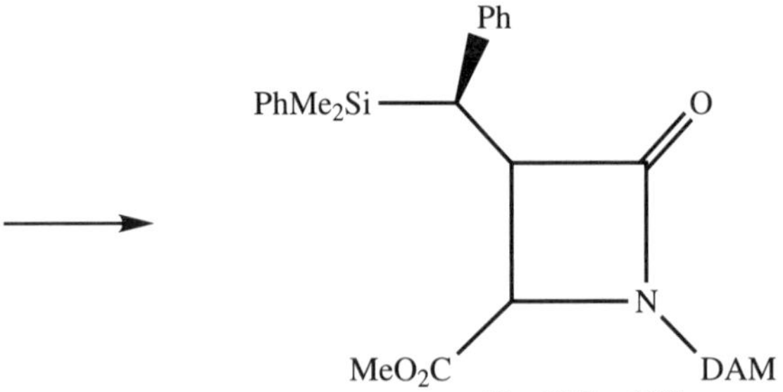

(13)

48 E/Z = 9/91

Cycloaddition of ketene with perfluoropropylisocyanate gave the azetidinedione.[219]

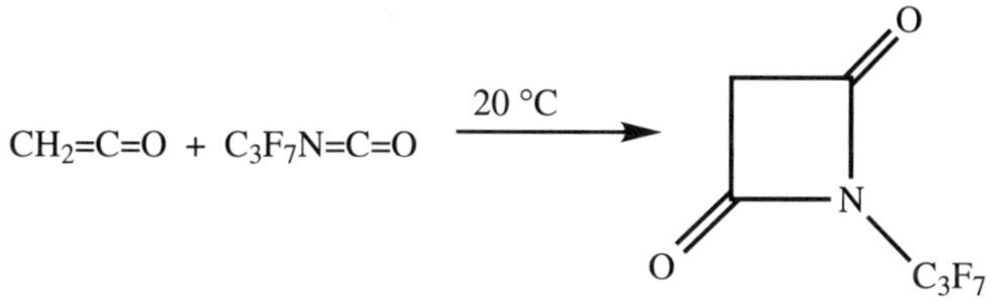

20 °C
CH_2=C=O + C_3F_7N=C=O
O
N
O
C_3F_7

(14)

The reaction of 1,4-diaza-1,3-butadienes (**49**) with ketenes generated from acyl chlorides gave 4-imino-2-azetidinones (**50**) and then 4,4′-bis(2-azetidinones) (**51**) by either one-pot or two-stage reactions (equation 15).[220,224]

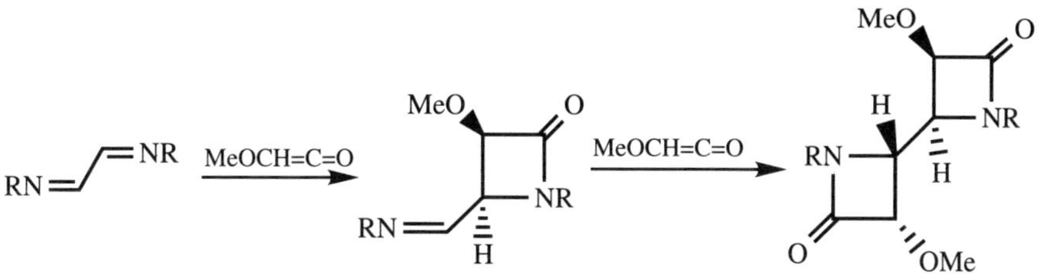

RN
NR
MeOCH=C=O
MeO
O
NR
RN
H
MeOCH=C=O
MeO
O
H
NR
RN
H
O
OMe

49 **50** **51**

R = 4-MeOC$_6$H$_4$ (15)

5.4.1.8 Cycloadditions of Ketenes with Other Substrates.

The cycloaddition of ketenes with azo compounds was first studied by Staudinger. The reaction of diphenylketene with *Z*-azobenzenes is rapid at room temperature in CCl_4, giving for Ar = Ph the product diazetidin-3-one (**1**) in 76% yield (equation 1).[226] The reaction of *E*-azobenzenes under these conditions was too slow to detect, but photoisomerization to the *Z*-isomer in situ led to an efficient reaction.

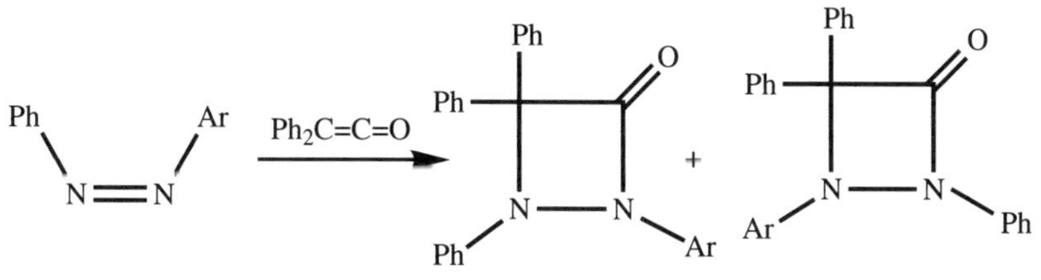

Ph
Ar
N
N
Ph_2C=C=O
Ph
Ph
O
N
N
Ph
Ar
+
Ph
Ph
O
N
N
Ar
Ph

(1)

1 **2**

The reactions of azo compounds PhN=NAr with different aryl substituents led to low regioselectivity for formation of the two isomers **1** and **2,** and the rates also showed only a small dependence on the aryl substituent and solvent (Table 1).[226,227]

TABLE 1. Reaction Rates and Positional Selectivity of $Ph_2C=C=O$ with Z-4-$XC_6H_4N=NPh$ in Benzene at 25 °C[226]

X	k (M^{-1} s^{-1} × 10^2)	1/2[a]
4-MeO	20	36/64
4-Me	7.8	46/54
H	5.4[c]	
4-Cl	2.5	41/59
4-CN[b]	21	36/64
4-NO$_2$	—	39/61

[a] 80 °C.
[b] 22 °C.
[c] 2.1 (cyclohexane), 6.9 (CH_2Cl_2), and 20 (CH_3CN).

The reaction of either isomer of PhN=NPr-i gave a mixture of products (equation 2). Both formed **3**, interpreted as the product of an ene reaction, while only the more reactive Z-isomer formed the diazetidinone **4**. The reactions of Z-PhN=NNMe$_2$ and PhN=NCO$_2$Et also gave diazetidinone products.

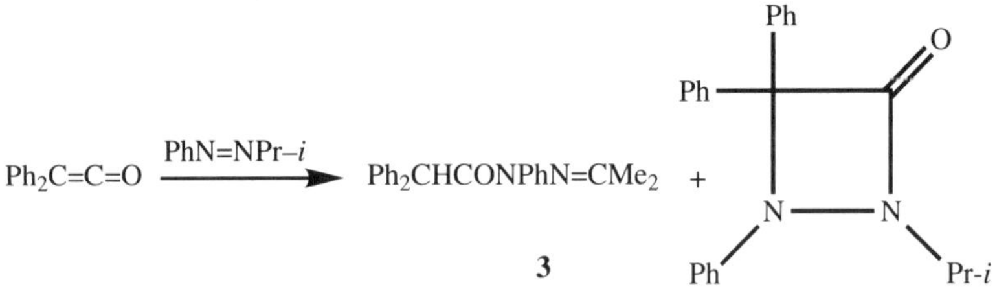

$$Ph_2C=C=O \xrightarrow{PhN=NPr-i} Ph_2CHCONPhN=CMe_2 + \mathbf{4} \qquad (2)$$

3

The formation of the diazetidinones was interpreted as occurring through concerted [2 + 2] cycloadditions based primarily on the small rate effects observed and low positional selectivity (Table 1). However, the formation of the product **3** and the strong positional selectivity observed with the azo compounds PhN=NR (R ≠ aryl) are more in accord with highly unsymmetrical transition states.

Cycloaddition of ketenes to nitroso compounds[228] and N-sulfinyl-4-toluene sulfonamide have also been observed (equations 3 and 4).[228,229] There are many reactions known with isocyanates, and these have been reviewed.[11,194]

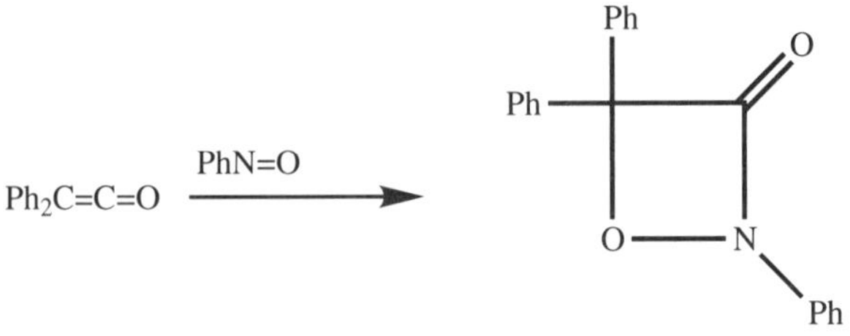

$$Ph_2C=C=O \xrightarrow{PhN=O} \qquad (3)$$

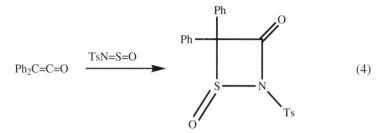

(4)

References

1. Wilsmore, N. T. M. *J. Chem. Soc.* **1907**, *91*, 1938–1941.
2. Chick, F.; Wilsmore, N. T. M. *J. Chem. Soc.* **1908**, *93*, 946–950.
3. Staudinger, H.; Klever, H. W. *Chem. Ber.* **1906**, *39*, 968–971.
4. Staudinger, H. *Chem. Ber.* **1907**, *40*, 1145–1148.
5. Ranganathan, S.; Ranganathan, D.; Mehrotra, A. K. *Synthesis*, **1977**, 289–296.
6. Williams, R. V.; Lin, X. *J. Chem. Soc., Chem. Commun.* **1989**, 1872–1873.
7. Wenkert, E.; Vial, C.; Näf, F. *Chimia* **1992**, *46*, 95–97.
8. Clemens, R. J. *Chem. Rev.* **1986**, *86*, 241–318.
9. Roberts, J. D.; Sharts, C. M. *Org. React.* **1962**, *12*, 1–56.
10. Ghosez, L.; O'Donnell, M. J. In *Pericyclic Reactions;* Marchand, A. P.; Lehr, R. E., Eds.; Academic: New York, 1977; pp. 79–140.
11. Ulrich, H. *Cycloaddition Reactions of Heterocumulenes,* Academic: New York, 1967; pp. 38–109.
12. Luknitskii, F. I.; Vovsi, B. A. *Russian Chem. Revs.* **1969**, *38*, 487–494.
13. Brady, W. T. *Tetrahedron,* **1981**, *37*, 2949–2966.

14a. Hanford, W. E.; Sauer, J. C. *Org. React.* **1946**, *3*, 108–140.

14b. Hyatt, J. A.; Raynolds, P. W. *Org. React.* **1994**, *45*, 159–646.

15. Ghosez, L.; Marchand-Brynaert, J. In *Comprehensive Organic Synthesis,* Vol. 5; Trost, B. M., Ed.; Pergamon: New York, 1991; pp. 85–122.
16. Woodward, R. B.; Hoffmann, R. *Angew. Chem., Int. Ed. Engl.* **1969**, *8*, 781–853. *The Conservation of Orbital Symmetry;* Verlag Chemie: Academic: New York, 1970.
17. Johnson, J. R.; Shiner, V. J., Jr. *J. Am. Chem. Soc.* **1953**, *75*, 1350–1355.
18. Bregman, J.; Bauer, S. H. *J. Am. Chem. Soc.* **1955**, *77*, 1955–1965.
19. Katz, L.; Lipscomb W. N. *J. Org. Chem.* **1952**, *17*, 515–517.
20. Kay, M. I.; Katz, L. *Acta Crystallogr.* **1958**, *11*, 897–898.
21. Mönnig, F.; Dreizler, H.; Rudolph, H. D. *Z. Naturforsch., A.* **1967**, *22*, 1471–1473.
22. Andreades, S.; Carlson, H. D. *Organic Synthesis,* Coll. Vol. V; Baumgarten, H. E., Ed.; Wiley: New York, 1973; pp. 679–684.
23. Kato, T. *Acc. Chem. Res.* **1974**, *7*, 265–271.
24. Farnum, D. G.; Johnson, J. R.; Hess, R. E.; Marshall, T. B.; Webster, B. *J. Am. Chem. Soc.* **1965**, *87*, 5191–5197.

25. Baldwin, J. E.; Roberts, J. D. *J. Am. Chem. Soc.* **1963**, *85*, 2444–2445.
26. Seidl, E. T.; Schaefer, H. F., III. *J. Am. Chem. Soc.* **1990**, *112*, 1493–1499.
27. Seidl, E. T.; Schaefer, H. F., III. *J. Phys. Chem.* **1992**, *96*, 657–661.
28. Sauer, J. C. *J. Am. Chem. Soc.* **1947**, *69*, 2444–2448.
29. Sauer, J. C. *Organic Syntheses,* Coll. Vol. IV; Rabjohn, N., Ed.; Wiley: New York, 1963; pp. 560–563.
30. Dehmlow, E. V.; Fastabend, U. *Synth. Commun.* **1993**, *23*, 79–82.
31. Woodward, R. B.; Small, G., Jr. *J. Am. Chem. Soc.* **1950**, *72*, 1297–1304.
32. Wulferding, A.; Wartchow, R.; Hoffmann, H. M. R. *Synlett.* **1992**, 476–479.
33. Baxter, G. J.; Brown, R. F. C.; Eastwood, F. W.; Harrington, K. J. *Tetrahedron Lett.* **1975**, 4283–4284.
34. Erickson, J. L. E.; Collins, F. E., Jr.; Owen, B. L. *J. Org. Chem.* **1966**, *31*, 480–484.
35. Hoffmann, H. M. R.; Eggert, U.; Walenta, A.; Weineck, E.; Schomburg, D.; Wartchow, R.; Allen, F. H. *J. Org. Chem.* **1989**, *54*, 6096–6100.
36. Masters, A. P.; Sorensen, T. S. *Tetrahedron Lett.* **1989**, *30*, 5869–5872.
37. Brady, W. T.; Ting, P. L. *J. Org. Chem.* **1976**, *41*, 2336–2339.
38. Brady, W. T.; Ting, P. L. *J. Org. Chem.* **1975**, *40*, 3417–3420.
39. Chickos, J. S.; Sherwood, D. E., Jr.; Jug, K. *J. Org. Chem.* **1978**, *43*, 1146–1150.
40. Jug, K.; Dwivedi, C. P. D.; Chickos, J. S. *Theor. Chim. Acta,* **1978**, *49*, 249–257.
41. Fu, X.; Decai, F.; Yanbo, D. *J. Mol. Struct. (Theochem.)* **1988**, *167*, 349–358.
42. Seidl, E. T.; Schaeffer, H. F., III *J. Am. Chem. Soc.* **1991**, *113*, 5195–5200.
43. Schaad, L. J.; Gutman, I.; Hess, B. A., Jr.; Hu, J. *J. Am. Chem. Soc.* **1991**, *113*, 5200–5203.
43a. Salzner, U.; Bachrach, S. M. *J. Am. Chem. Soc.* **1994**, *116*, 6850–6855.
44. Pascal, R. A., Jr.; Baum, M. W.; Wagner, C. K.; Rodgers, L. R.; Huang, D. *J. Am. Chem. Soc.* **1986**, *108*, 6477–6482.
45. Dehmlow, E. V.; Slopianka, M.; Pickardt, J. *Liebigs Ann. Chem.* **1979**, 572–593.
46. Dehmlow, E. V.; Pickardt, J.; Slopianka, M.; Fastabend, U.; Dreschsler, K.; Soufi, J. *Liebigs Ann. Chem.* **1987**, 377–379.
47. Hoffmann, H. M. R.; Wulff, J. M.; Kütz, A.; Wartchow, R. *Angew. Chem., Int. Ed. Engl.* **1982**, *21*, 83.
48. Wulff, J. M.; Hoffmann, H. M. R. *Angew. Chem., Int. Ed. Engl.* **1985**, *24*, 605–606.
49. Hoffmann, H. M. R.; Walenta, A.; Eggert, U.; Schomburg, D. *Angew. Chem., Int. Ed. Engl.* **1985**, *24*, 607–608.
50. Hoffmann, H. M. R.; Geschwinder, P. M.; Hollwege, H.-P.; Walenta, A. *Helv. Chim. Acta* **1988**, *71*, 1930–1936.
51. Moore, H. W.; Wilbur, D. S. *J. Org. Chem.* **1980**, *45*, 4483–4491.
52. Moore, H. W.; Wilbur, D. S. *J. Am. Chem. Soc.* **1978**, *100*, 6523–6525.
53. England, D. C.; Krespan, C. G. *J. Org. Chem.* **1970**, *35*, 3322–3327.
54. Huisgen, R.; Otto, P. *J. Am. Chem. Soc.* **1968**, *90*, 5342–5343.
55. Binsch, G.; Feiler, L. A.; Huisgen, R. *Tetrahedron Lett.* **1968**, 4497–4501.
56. Huigen, R.; Feiler, L. A. *Chem. Ber.* **1969**, *102*, 3391–3404.

57. Frey, H. M.; Isaacs, N. J. *J. Chem. Soc. B,* **1970,** 830–832.
58. Martin, J. C.; Gott, P. G.; Goodlett, V. W.; Hasek, R. H. *J. Org. Chem.* **1965,** *30,* 4175–4180.
59. Brook, P. R.; Eldeeb, A. F.; Hunt, K.; McDonald, W. S. *J. Chem. Soc., Chem. Commun.* **1978,** 10–11.
60. Rey, M.; Roberts, S. M.; Dreiding, A. S.; Roussel, A.; Vanlierde, H.; Toppett, S.; Ghosez, L. *Helv. Chim. Acta* **1982,** *65,* 703–720.
61. Brook, P. R.; Duke, A. J.; Harrison, J. M.; Hunt, K. *J. Chem. Soc., Perkin Trans. 1* **1974,** 927–932.
62. Brook, P. R.; Duke, A. J.; Griffiths, J. G.; Roberts, S. M.; Rey, M.; Dreiding, A. S. *Helv. Chim. Acta* **1977,** *60,* 1528–1544.
63. Jacobson, B. M.; Bartlett, P. D. *J. Org. Chem.* **1973,** *38,* 1030–1041.
64. Bartlett, P. D. *Pure Appl. Chem.* **1971,** *27,* 597–609.
65. Rey, M.; Roberts, S.; Diffenbacher, A.; Dreiding, A. S. *Helv. Chim. Acta* **1970,** *53,* 417–430.
66. Brady, W. T.; Hoff, E. F., Jr. *J. Am. Chem. Soc.* **1968,** *90,* 6256.
67. Brady, W. T.; Roe, R., Jr. *J. Am. Chem. Soc.* **1970,** *92,* 4618–4621.
68. Brady, W. T.; Roe, R. Jr.; Hoff, E. F.; Parry, F. H., III, *J. Am. Chem. Soc.* **1970,** *92,* 146–148.
69. Brady, W. T.; Roe, R., Jr. *J. Am. Chem. Soc.* **1971,** *93,* 1662–1664.
70. Brady, W. T.; Parry, F. H., III; Stockton, J. D. *J. Org. Chem.* **1971,** *36,* 1486–1489.
71. Brady, W. T.; Hoff, E. F.; Roe, R., Jr.; Parry, F. H., Jr. *J. Am. Chem. Soc.* **1969,** *91,* 5679–5680.
72. Brady, W. T.; Holifield, B. M. *Tetrahedron* **1967,** *23,* 4251–4255.
73. Brady, W. T.; Parry, F. H., III; Roe, R., Jr.; Hoff, E. F., Jr. *Tetrahedron Lett.* **1970,** 819–822.
74. Huisgen, R.; Otto, P. *Chem. Ber.* **1969,** *102,* 3475–3485.
75. Brady, W. T.; O'Neal, H. R. *J. Org. Chem.* **1967,** *32,* 2704–2707.
76. Brady, W. T.; O'Neal, H. R. *J. Org. Chem.* **1967,** *32,* 612–614.
77. Al-Husaini, A. H.; Khan, I.; Ali, S. A. *Tetrahedron* **1991,** *47,* 3845–3852.
78. Holder, R. W.; Freiman, H. S.; Stefanchik, M. T. *J. Org. Chem.* **1976,** *41,* 3303–3307.
79. Marvel, C. S.; Kohan, M. I. *J. Org. Chem.* **1951,** *16,* 741–745.
80. Martin, J. C.; Goodlett, V. W.; Burpitt, R. D. *J. Org. Chem.* **1965,** *30,* 4309–4311.
81. Corey, E. J.; Arnold, Z.; Hutton, J. *Tetrahedron Lett.* **1970,** 307–310.
82. Kelly, R. C.; VanRheenen, V.; Schletter, I.; Pillai, M. D. *J. Am. Chem. Soc.* **1973,** *95,* 2746–2747.
83. Wakamutsu, T.; Miyachi, N.; Ozaki, F.; Shibasaki, M.; Ban, Y. *Tetrahedron Lett.* **1988,** *29,* 3829–3832.
84. Hassner, A.; Cory, R. M.; Sartoris, N. *J. Am. Chem. Soc.* **1976,** *98,* 7698–7704.
85. Bessmertnykh, A. G.; Volkovich, S. V.; Donskaya, N. A.; Kisina, M. Yu.; Lukovskii, B. A. *Zh. Org. Khim.* **1991,** *27,* 1206–1209; *Engl. Transl.* **1991,** *27,* 1049–1051.
86. Donskaya, N. A.; Lukovskii, B. A. *Mendeleev Commun.* **1991,** 127–128.

87. Donskaya, N. A.; Bessmertnykh, A. G.; Lukovskii, B. A.; Kisina, M. Yu., Ryabova, M. A. *Zh. Org. Khim.* **1991,** *27,* 2528–2533; *Engl. Transl.* **1991,** *27,* 2249–2253.
88. Bessmertnykh, A. G.; Bubnov, Yu. N.; Voevodsdkaya, T. I.; Donskaya, N. A.; Zykov, A. Yu. *Zh. Org. Khim.* **1990,** *26,* 2348–2355; *Engl. Transl.* **1990,** *26,* 2027–2033.
89. Feiler, L. A.; Huisgen, R.; Koppitz, P. *J. Am. Chem. Soc.* **1974,** *96,* 2270–2271.
90. Moore, H. W.; Gheorghiu, M. D. *Chem. Soc. Revs.* **1981,** *10,* 289–328.
91. Gheorghiu, M. D.; Draghici, C.; Parvulescu, L. *Tetrahedron,* **1977,** *33,* 3295–3299.
92. Gheorghiu, M. D.; Filip, P.; Draghici, C.; Parvulescu, L. *J. Chem. Soc., Chem. Commun.* **1975,** 635–636.
93. Brook, P. R.; Hunt, K. *J. Chem. Soc., Chem. Commun.* **1974,** 989–990.
94. Gheorghiu, M. D.; Parvulescu, L.; Turdibekov, K. M.; Struchkov, Yu. T. *Rev. Roum. Chem.* **1990,** *30,* 427–435.
95. England, D. C.; Krespan, C. G. *J. Org. Chem.* **1970,** *35,* 3300–3307.
96. De Selms, R. C.; Delay, F. *J. Am. Chem. Soc.* **1973,** *95,* 274–276.
97. De Selms, R. C.; Delay, F. *J. Org. Chem.* **1972,** *37,* 2908–2910.
98. Aue, D. H.; Shellhamer, D. F.; Helwig, G. S. *J. Chem. Soc., Chem. Commun.* **1975,** 603–604.
99. Aue, D. H.; Helwig, G. S. *J. Chem. Soc., Chem. Commun.* **1975,** 604–605; **1974,** 925–927.
100. Dunkelbaum, E. *Tetrahedron* **1976,** *32,* 975–978.
101. Picard, P.; Moulines, J.; Lecoustre, M. *Bull. Chem. Soc. Fr.* **1984,** *II,* 65–70.
102. Li, H.; Silver, J. E.; Watson, W. H.; Kashyap, R. P.; le Noble, W. J. *J. Org. Chem.* **1991,** *56,* 5932–5939.
103. Paquette, L. A.; Underiner, T. L.; Gallucci, J. C. *J. Org. Chem.* **1992,** *57,* 86–96.
104. Wagner, H. U.; Gompper, R. *Tetrahedron Lett.* **1970,** 2819–2822.
105. Gompper, R. *Angew Chem., Int. Ed. Engl.* **1969,** *8,* 312–327.
106. Zimmerman, H. E. *Acc. Chem. Res.* **1971,** *4,* 272–280.
107. Burke, L. A. *J. Org. Chem.* **1985,** *50,* 3149–3155.
108. Bernardi, F.; Bottoni, A.; Olivucci, M.; Robb, M. A.; Schlegel, H. B.; Tonachini, G. *J. Am. Chem. Soc.* **1988,** *110,* 5993–5995.
109. Bernardi, F.; Bottoni, A.; Robb, M. A.; Venturini, A. *J. Am. Chem. Soc.* **1990,** *112,* 2106–2114.
110. Wang, X.; Houk, K. N. *J. Am. Chem. Soc.* **1990,** *112,* 1754–1756.
111. Houk, K. N.; Li, Y.; Evanseck, J. D. *Angew. Chem., Int. Ed. Engl.* **1992,** *31,* 682–708.
112. Pericas, M. A.; Serratosa, F.; Valenti, E. *J. Chem. Soc., Perkin Trans. 2,* **1987,** 151–158.
113. Valenti, E.; Pericas, M. A.; Moyano, A. *J. Org. Chem.* **1990,** *55,* 3582–3593.
114. Sordo, J. A.; Gonzalez, J.; Sordo, T. L. *J. Am. Chem. Soc.* **1992,** *114,* 6249–6251.
114a. Lopez, R.; Sordo, T. L.; Sordo, J. A.; Gonzales, J. *J. Org. Chem.* **1993,** *58,* 7036–7037.
115. Yamabe, S.; Minato, T.; Osamura, Y. *J. Chem. Soc., Chem. Commun.* **1993,** 450–452.
116. Rzepa, H. S.; Wylie, W. A. *Int. J. Quantum Chem.* **1992,** *44,* 469–476.

116a. Reguero, M.; Pappalardo, R. R.; Robb, M. A.; Rzepa, H. S. *J. Chem. Soc., Perkin 2,* **1993,** 1499–1502.
117. Fang, D.; Fu, X. *Int. J. Quantum Chem.* **1992,** *43,* 669–676.
118. Rzepa, H. S.; Wylie, W. A. *J. Chem. Soc., Perkin Trans. 2,* **1991,** 939–946.
119. Nguyen, M. T.; Ha, T.; More O'Ferrall, R. A. *J. Org. Chem.* **1990,** *55,* 3251–3256.
120. Holder, R. W. *J. Chem. Ed.* **1976,** *53,* 81–85.
121. Huisgen, R.; Mayr, H. *Tetrahedron Lett.* **1975,** 2969–2972.
122. Huisgen, R.; Mayr, H. *Tetrahedron Lett.* **1975,** 2965–2968.
123. Weyler, W., Jr.; Byrd, L. R.; Caserio, M. C.; Moore, H. W. *J. Am. Chem. Soc.* **1972,** *94,* 1027–1029.
124. Montaigne, R.; Ghosez, L. *Angew. Chem., Int. Ed. Engl.* **1968,** *7,* 221.
125. Huisgen, R.; Feiler, L. A.; Otto, P. *Chem. Ber.* **1969,** *102,* 3405–3427.
126. Al-Husaini, A. H.; Muqtar, M.; Ali, S. A. *Tetrahedron* **1991,** *47,* 7719–7726.
127. Huisgen, R.; Feiler, L. A.; Otto, P. *Chem. Ber.* **1969,** *102,* 3444–3459.
128. Effenberger, F.; Prossel, G.; Fischer, P. *Chem. Ber.* **1971,** *104,* 2002–2012.
129. Isaacs, N. S.; Stanbury, P. *J. Chem. Soc., Perkin 2,* **1973,** 166–169.
130. Baldwin, J. E.; Kapecki, J. A. *J. Am. Chem. Soc.* **1970,** *92,* 4868–4873.
131. Oyama, K.; Tidwell, T. T. *J. Am. Chem. Soc.* **1976,** *98,* 947–951.
132. Burnier, J. S.; Jorgensen, W. L. *J. Org. Chem.* **1984,** *49,* 3001–3020.
133. Steiner, G.; Huisgen, R. *J. Am. Chem. Soc.* **1973,** *95,* 5056–5068.
134. Huisgen, R. *Acc. Chem. Res.* **1977,** *10,* 117–124.
135. Huisgen, R. *Pure Appl. Chem.* **1980,** *52,* 2283–2302.
136. Baldwin, J. E. In *Comprehensive Organic Synthesis;* Trost, B. M., Ed.; Pergamon: New York, 1991; Vol. 5, Chapter 2.1.
137. Haberfeld, P.; Ray, A. K. *J. Org. Chem.* **1972,** *37,* 3093–3096.
138. Solomonov, B. N.; Antipin, I. S.; Konovalov, A. I. *Zh. Org. Khim.* **1977,** *13,* 2491–2495; *Engl. Transl.* **1977,** *13,* 2317–2319.
139. Sustmann, R.; Ansmann, A.; Vahrenholt, F. *J. Am. Chem. Soc.* **1972,** *94,* 8099–8105.
140. Baldwin, J. E.; Kapecki, J. A. *J. Am. Chem. Soc.* **1970,** *92,* 4874–4879.
141. Isaacs, N. S.; Hatcher, G. B. *J. Chem. Soc., Chem. Commun.* **1974,** 593–594.
142. Katz, T. J.; Dessau, R. *J. Am. Chem. Soc.* **1963,** *85,* 2172–2173.
143. Snyder, E. I. *J. Org. Chem.* **1970,** *35,* 4287–4288.
144. Collins, C. J.; Benjamin, B. M.; Kabalka, G. W. *J. Am. Chem. Soc.* **1978,** *100,* 2570–2571.
145. Holder, R. W.; Graf, N. A.; Duesler, E.; Moss, I. C. *J. Am. Chem. Soc.* **1983,** *105,* 2929–2931.
146. Halevi, E. A. *Orbital Symmetry and Reaction Mechanism;* Springer-Verlag: Berlin, 1992; pp. 149–159.
147. Becker, D.; Brodsky, N. C. *J. Chem. Soc., Chem. Commun.* **1978,** 237–238.
148. Al-Husaini, A. H.; Moore, H. W. *J. Org. Chem.* **1985,** *50,* 2595–2597.
148a. Lovas, F. J.; Suenram, K. D.; Gillies, C. W.; Gillies, J. Z., Fowler, P. W.; Kisiel, Z. *J. Am. Chem. Soc.* **1994,** *116,* 5285–5294.

149. Brady, W. T.; Lloyd, R. M. *J. Org. Chem.* **1979,** *44,* 2560–2564.
150. Brady, W. T.; Lloyd, R. M. *J. Org. Chem.* **1980,** *45,* 2025–2028.
151. Brady, W. T.; Lloyd, R. M. *J. Org. Chem.* **1981,** *46,* 1322–1326.
152. Brady, W. T.; Watts, R. D. *J. Org. Chem.* **1981,** *46,* 4047–4050.
153. Scarpati, R.; Sica, D.; Santacroce, C. *Tetrahedron* **1964,** *20,* 2735–2744.
154. Naidorf-Meir, S.; Hassner, A. *J. Org. Chem.* **1992,** *57,* 5102–5105.
155. Raynolds, P. W.; DeLoach, J. A. *J. Am. Chem. Soc.* **1984,** *106,* 4566–4570.
156. Hoffmann, R. W.; Bressel, U.; Gehlhaus, J.; Häuser, H. *Chem. Ber.* **1971,** *104,* 873–885.
157. Aben, R. W.; Scheeren, H. W. *J. Chem. Soc., Perkin Trans. 1,* **1979,** 3132–3138.
158. Bellus, D. *J. Am. Chem. Soc.* **1978,** *100,* 8026–8028.
159. Huisgen, R.; Feiler, L. A.; Binsch, G. *Chem. Ber.* **1969,** *102,* 3460–3474.
160. Huisgen, R.; Feiler, L.; Binsch, G. *Angew. Chem., Int. Ed. Engl.* **1964,** *3,* 753–754.
161. Swieton, G.; von Jouanne, J.; Kelm, H.; Huisgen, R. *J. Chem. Soc., Perkin Trans. 2,* **1983,** 37–43.
162. Huisgen, R.; Otto, P. *J. Am. Chem. Soc.* **1969,** *91,* 5922–5923.
163. Feiler, L. A.; Huisgen, R. *Chem. Ber.* **1969,** *102,* 3428–3443.
164. Hoch, H.; Hünig, S. *Chem. Ber.* **1972,** *105,* 2660–2685.
165. Martin, J. C.; Gott, P. G.; Hostettler, H. U. *J. Org. Chem.* **1967,** *32,* 1654–1655.
166. Hasek, R. H.; Gott, P. G.; Martin, J. C. *J. Org. Chem.* **1964,** *29,* 2513–2516.
167. de Faria, A. R.; Matos, C. R. R.; Correia, C. R. D. *Tetrahedron Lett.* **1993,** *34,* 27–30.
168. Duncan, W. G.; Weyler, W., Jr.; Moore, H. W. *Tetrahedron Lett.* **1973,** 4391–4394.
169. Bampfield, H. A.; Brook, P. R. *J. Chem. Soc., Chem. Commun.* **1974,** 171–172.
170. Bampfield, H. A.; Brook, P. R.; McDonald, W. S. *J. Chem. Soc., Chem. Commun.,* **1975,** 132–133.
171. Bampfield, H. A.; Brook, P. R.; Hunt, K. *J. Chem. Soc., Chem. Commun.,* **1976,** 146–147.
172. Bertrand, M.; Gras, J.-L.; Gore, J. *Tetrahedron* **1975,** *31,* 857–862.
173. Bertrand, M.; Maurin, R.; Gras, J. L.; Gil, G. *Tetrahedron* **1975,** *31,* 849–855.
174. Brady, W. T.; Stockton, J. D.; Patel, A. D. *J. Org. Chem.* **1974,** *39,* 236–238.
175. Gras, J.-L.; Bertrand, M. *Nouv. J. Chim.* **1981,** *5,* 521–530.
176. Ammann, A. A.; Rey, M.; Dreiding, A. S. *Helv. Chim. Acta* **1987,** *70,* 321–328.
177. Knoche, H. *Liebigs Ann. Chem.* **1969,** *722,* 232–233.
178. Danheiser, R. L.; Savariar, S. *Tetrahedron Lett.* **1987,** *28,* 3299–3302.
179. Gheorghiu, M. D.; Draghici, C.; Stanescu, L.; Avram, M. *Tetrahedron Lett.* **1973,** 9–12.
180. Haseck, R. H.; Gott, P. G.; Martin, J. C. *J. Org. Chem.* **1964,** *29,* 2510–2513.
181. Danheiser, R. L.; Sard, H. *Tetrahedron Lett.* **1983,** *24,* 23–26.
182. Kowalski, C. J.; Lal, G. S. *J. Am. Chem. Soc.* **1988,** *110,* 3693–3695.

182a. Gillies, C. W.; Gillies, J. Z.; Lovas, F. J.; Suenram, R. D. *J. Am. Chem. Soc.* **1993**, *15,* 9253–9262.
183. Teufel, H.; Jenny, E. F. *Tetrahedron Lett.* **1971,** 1769–1772.
184. Wuest, J. D. *Tetrahedron* **1980,** *36,* 2291–2296.
185. Delaunois, M.; Ghosez, L. *Angew. Chem., Int. Ed. Engl.* **1969,** *8,* 72–73.
186. Ficini, J.; Pouliquen, J. *Tetrahedron Lett.* **1972,** 1135–1138.
187. Dötz, K. H.; Trenkle, B.; Schubert, U. *Angew. Chem., Int. Ed. Engl.* **1981,** *20,* 287.
188. Dötz, K. H.; Mühlemeier, J.; Trenkle, B. *J. Organomet. Chem.* **1985,** *289,* 257–262.
189. Ficini, J.; Falou, S.; d'Angelo, J. *Tetrahedron Lett.* **1977,** 1931–1934.
190. Sheehan, J. C.; Corey, E. J. *Org. React.* **1957,** *9,* 388–408.
191. Holden, K. G. In *Chemistry and Biology of β-Lactam Antibiotics,* Vol. 2; Morin, R. B.; Gorman, M., Eds.; Academic: New York, 1982; Chapter 2.
192. Cooper, R. D. G.; Daugherty, B. W.; Boyd, D. B. *Pure Appl. Chem.* **1987,** *59,* 485–492.
193. Thomas, R. C. In *Recent Progress in the Chemical Synthesis of Antibiotics;* Lukacs, G.; Ohno, M. Eds.; Springer-Verlag: Berlin, 1990; p. 533.
194. Muller, L. L.; Hamer, J. *1,2-Cycloaddition Reactions. The Formation of Three- and Four-Membered Heterocycles;* Interscience: New York, 1967.
195. van der Steen, F. H.; van Koten, G. *Tetrahedron* **1991,** *47,* 7503–7524.
196. Pacansky, J.; Chang, J. S.; Brown, D. W.; Schwarz, W. *J. Org. Chem.* **1982,** *47,* 2233–2234.
197. Lynch, J. E.; Riseman, S. M.; Laswell, W. L.; Tschaen, D. M.; Volante, R. P.; Smith, G. B.; Shinkai, I. *J. Org. Chem.* **1989,** *54,* 3792–3796.
198. Moore, H. W.; Hughes, G.; Srivivasachar, K.; Fernandez, M.; Nguyen, N. V.; Schoon, D.; Tranne, A. *J. Org. Chem.* **1985,** *50,* 4231–4238.
198a. Arrastia, I.; Arrieta, A.; Ugalde, J. M.; Cossio, F. P.; Lecea, B. *Tetrahedron Lett.* **1994,** *35,* 7825–7828.
199. Moore, H. W.; Hernandez, L.; Jr.; Chambers, R. *J. Am. Chem. Soc.* **1978,** *100,* 2245–2247.
200. Duran, F.; Ghosez, L. *Tetrahedron Lett.* **1970,** 245–248.
201. Bose, A. K.; Manhas, M. S.; Chib, J. S.; Chawla, H. P. S.; Dagal, B. *J. Org. Chem.* **1974,** *39,* 2877–2884.
202. Wagle, D. R.; Garai, C.; Chiang, J.; Monteleone, M. G.; Kurys, B. E.; Strohmeyer, T. W.; Hegde, V. R.; Manhas, M. S.; Bose, A. J. *J. Org. Chem.* **1988,** *53,* 4227–4236.
203. Kunert, D. M.; Chambers, R.; Mercer, F.; Hernandez, L., Jr.; Moore, H. W. *Tetrahedron Lett.* **1978,** 929–932.
204. Ojima, I.; Chen, H. C.; Qui, X. *Tetrahedron* **1988,** *44,* 5307–5318.
205. Evans, D. A.; Williams, J. M. *Tetrahedron Lett.* **1988,** *29,* 5065–5068.
206. Palomo, C.; Cossio, F. P.; Odiozola, J. M.; Oiarbide, M.; Ontoria, J. M. *J. Org. Chem.* **1991,** *56,* 4418–4428.
207. Brady, W. T.; Gu, Y. Q. *J. Org. Chem.* **1989,** *54,* 2838–2842.
208. Brady, W. T.; Shieh, C. H. *J. Org. Chem.* **1983,** *43,* 2499–2502.

209. Araki, K.; Wichtowski, J. A.; Welch, J. T. *Tetrahedron Lett.* **1991,** *40,* 5461–5464.

210. Welch, J. T.; Araki, K.; Kaweki, R.; Wichtowski, J. A. *J. Org. Chem.* **1993,** *58,* 2454–2462.

211. Palomo, C.; Cossio, F. P.; Cuevas, C.; Lecea, B.; Mielgo, A.; Roman, P.; Luque, A.; Martinez-Ripoll, M. *J. Am. Chem. Soc.* **1992,** *114,* 9360–9369.

212. Cossio, F. P.; Ugalde, J. M.; Lopez, X.; Lecea, B.; Palomo, C. *J. Am. Chem. Soc.* **1993,** *115,* 995–1004.

213. Hegedus, L. S.; Montgomery, J.; Narukawa, Y.; Snustad, D. C. *J. Am. Chem. Soc.* **1991,** *113,* 5784–5791.

214. Alcaide, B.; Dominquez, G.; Plumet, J.; Sierra, M. A. *J. Org. Chem.* **1992,** *57,* 447–451.

215. Bari, S. S.; Trehan, I. R.; Sharma, A. K.; Manhas, M. S. *Synthesis,* **1992,** 439–442.

216. Georg, G. I.; Akgün, E.; Mashava, P. M.; Milstead, M.; Ping, H.; Wu, Z.; Vander Velde, D.; Takusagawa, F. *Tetrahedron Lett.* **1992,** *33,* 2113–2114.

217. Barton, D. H. R.; Gateau-Olesker, A.; Anaya-Mateos, J.; Cleophax, J.; Gero, S. D.; Chiaroni, A.; Riche, C. *J. Chem. Soc., Perkin Trans. 1* **1990,** 3211–3212.

218. Palomo, C.; Aizpurua, J. M.; Urchegui, R.; Iturburu, M. *J. Org. Chem.* **1992,** *57,* 1571–1579.

219. Deltsova, D. P.; Krasuskaya, M. P.; Knunyants, I. L. *Izv. Akad. Nauk SSSR, Ser. Khim.* **1967,** 2567–2569; *Engl. Transl.* **1967,** 2448–2450.

220. Alcaide, B.; Martin-Cantalejo, Y.; Perez-Castells, J.; Rodriques-Lopez, J.; Sierra, M. A.; Monge, A.; Perez-Garcia, V. *J. Org. Chem.* **1992,** *57,* 5921–5931.

221. Palomo, C.; Cossio, F. P.; Cuevas, C. *Tetrahedron Lett.* **1991,** *32,* 3109–3110.

222. Palomo, C.; Aizpurua, J. M.; Ontoria, J. M.; Iturburu, M. *Tetrahedron Lett.* **1992,** *33,* 4823–4826.

223. Grochowski, E.; Pupek, K. *Tetrahedron* **1991,** *47,* 6759–6768.

224. Ojima, I.; Zhao, M.; Yamato, T.; Nakahashi, K.; Yamashita, M.; Abe, R. *J. Org. Chem.* **1991,** *56,* 5263–5277.

225. Hubschwerlen, C.; Schmid, G. *Helv. Chim. Acta,* **1983,** *66,* 2206–2209.

226. Kerber, R. C.; Ryan, T. J.; Hsu, S. D. *J. Org. Chem.* **1974,** *39,* 1215–1221.

227. Cann, M. C. Ph.D. Thesis, State University of New York, Stony Brook, 1973; *Chem. Abstr.* **1974,** *81,* 3055e.

228. Kresze, G.; Trede, A. *Tetrahedron* **1963,** *19,* 133–136.

229. Kresze, G.; Maschke, A.; Albrecht, R.; Bederke, K.; Patzchke, H. P.; Smalla, H.; Trede, A. *Angew. Chem., Int. Ed. Engl.* **1962,** *1,* 89–98.

5.4.2 [3 + 2] Cycloaddition Reactions of Ketenes

There was an early review of 1,3-dipolar additions,[1] and many examples are known in which ketenes react as 1,3-dipolarophiles to give 5-membered ring products. These include the reaction of diphenyldiazomethane with diphenylketene (equation 1),[2] which could occur through a one-step process or a two-step reaction involving initial nucleophilic attack at the carbonyl carbon forming zwitterion **1** which cyclizes to **2** as shown. Cyclization of **1** to form a C—C bond would be hindered by the crowding in the resulting product.

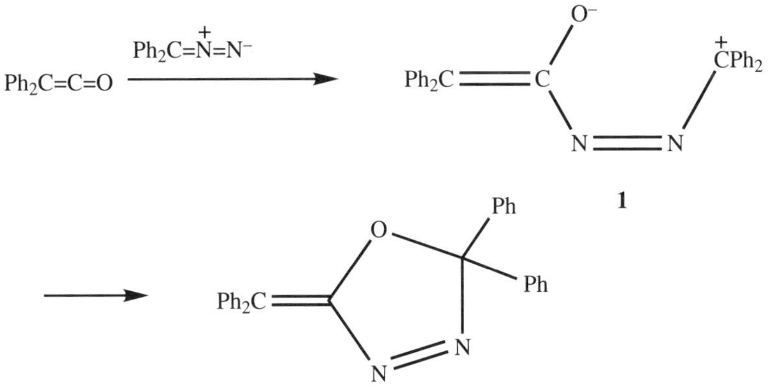

$$Ph_2C=C=O \xrightarrow{Ph_2C=\overset{+}{N}=N^-} Ph_2C=C\begin{pmatrix}O^-\\N=N\end{pmatrix}\overset{+}{C}Ph_2 \quad (1)$$

1

$$\longrightarrow \text{ [structure } \mathbf{2}\text{]}$$

2

Reaction of α-diazoacetophenone with silver benzoate–triethylamine gives the cycloadduct **3**, ascribed to [3 + 2] cycloaddition of phenylketene with the diazoketone (equation 2).[3]

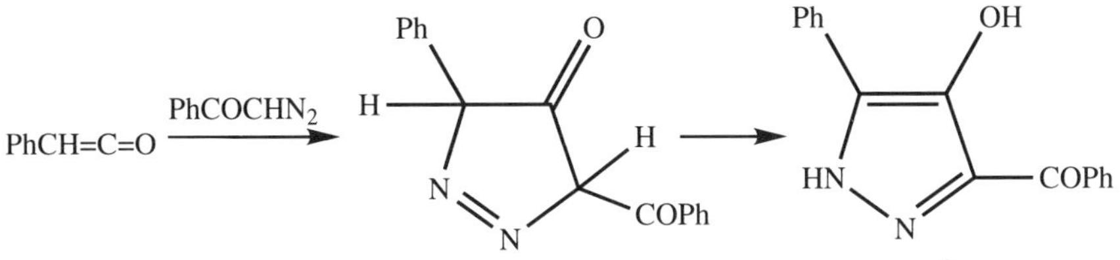

$$PhCH=C=O \xrightarrow{PhCOCHN_2} \text{[intermediate]} \longrightarrow \text{[structure } \mathbf{3}\text{]}$$

3

(2)

The product of reaction of *tert*-butylcyanide *N*-oxide with diphenylketene was originally formulated as **4**,[4] but the evidence was reinterpreted and the structure assigned as **5** (equation 3).[5] Reaction of trimethylbenzonitrile *N*-oxide with several ketenes also gave structures analogous to **5**, and the reaction with $Me_2C=C=O$ was interpreted as occurring via **6** as in equations 4 and 5.[6] Formation of a second product **7** was interpreted as occurring as shown in equation 6.[5]

538 REACTIONS OF KETENES

5.4 CYCLOADDITION REACTIONS OF KETENES

N-Substituted nitrones **8** react with a diverse variety of ketenes **9** to give either oxazolidinones **10** or isoxazolidinones **11** depending upon the electronic and steric properties of the substituents on the nitrone and the ketene.[7,8] The results were interpreted in terms of nucleophilic attack of the nitrone oxygen on the carbonyl carbon, yielding a zwitterion **12** which could close to **11** or rearrange to the zwitterion **13** which leads to **10**.[7]

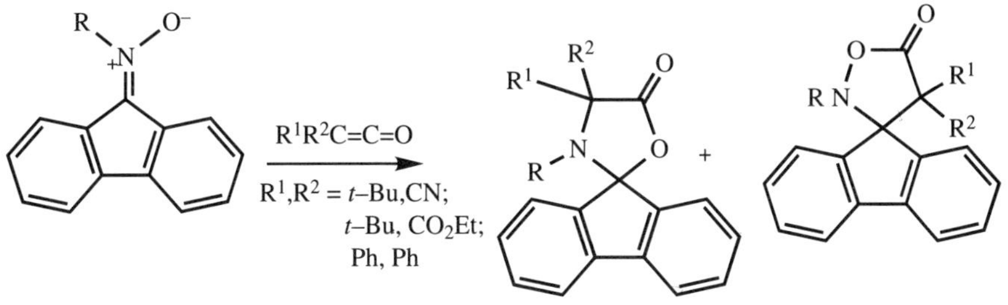

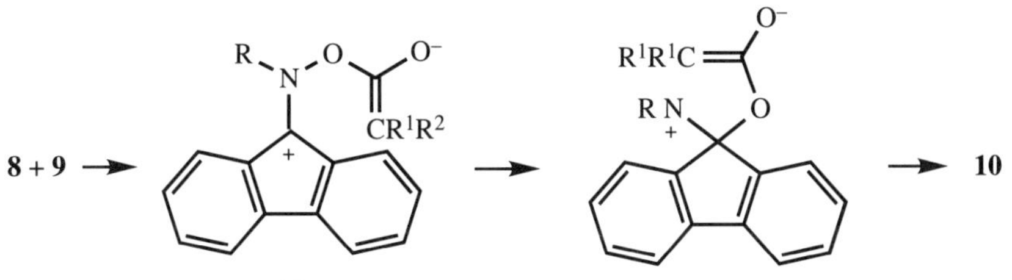

Reaction of pyridine *N*-oxides with dichloroketene gave products that appear to result from initial formation of the zwitterion **14** which cyclizes to **15** and **16** and then undergoes decarboxylation or reaction with additional $Cl_2C=C=O$.[9]

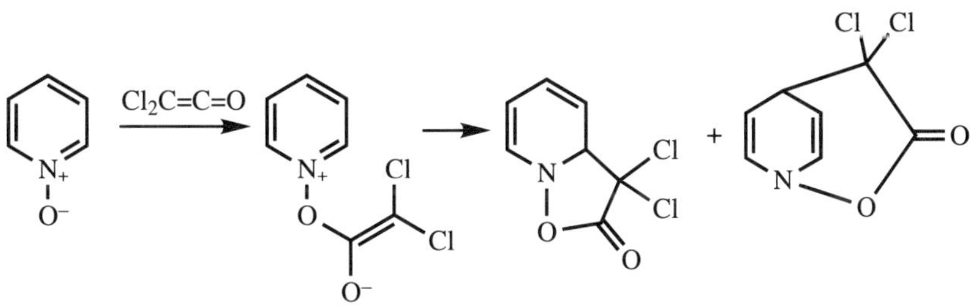

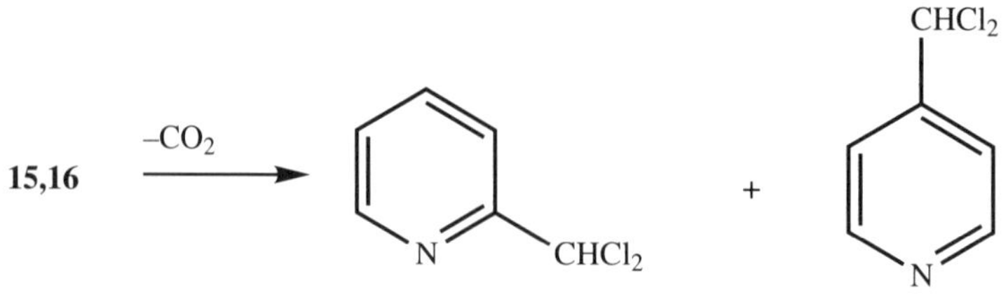

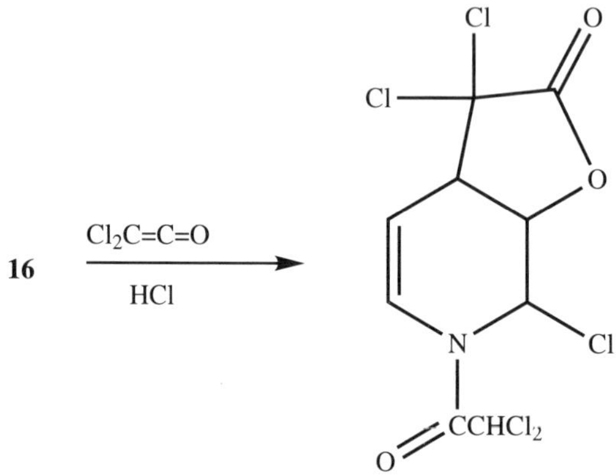

Reaction of the thiocarbonyl ylide precursors **17** with diphenylketene gave cycloadducts **18** with retention of stereochemistry (equation 7).[9] This result was interpreted as consistent with a $[\pi 4_s + \pi 2_s]$ concerted process or a stepwise process in which the stereochemistry of the zwitterionic intermediate **19** was retained.[10]

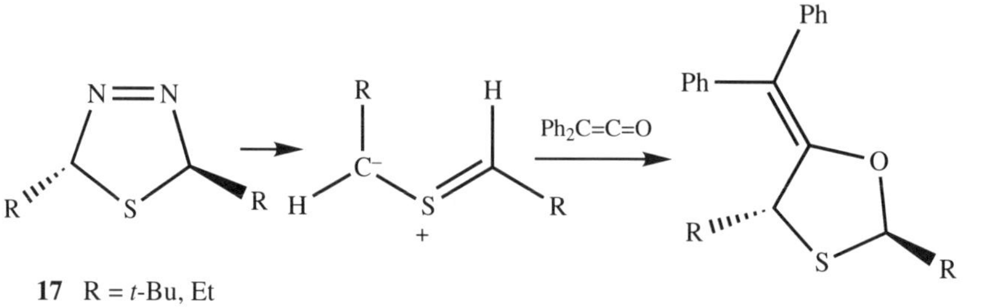

17 R = *t*-Bu, Et

(7)

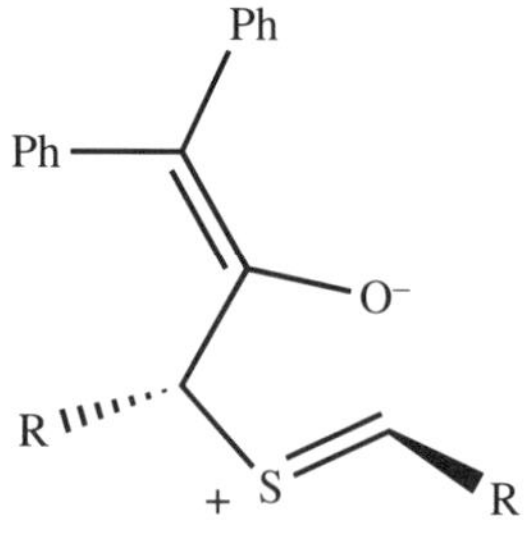

5.4 CYCLOADDITION REACTIONS OF KETENES

When ketenes are generated from α-diazocarbonyl compounds the formal products of 1,3-cycloaddition of acylcarbenes to ketenes are sometimes observed, as in the thermolysis of dimethyl diazomalonate to give **20** and the lactone **21** in a 5:1 ratio.[11–13] This reaction was proposed[11–13] to involve formation of a carbene **22** which gave Wolff rearrangement to the ketene **23**, which then combined with **22** by two regiochemical directions to give **20** and **21** (Scheme 1).

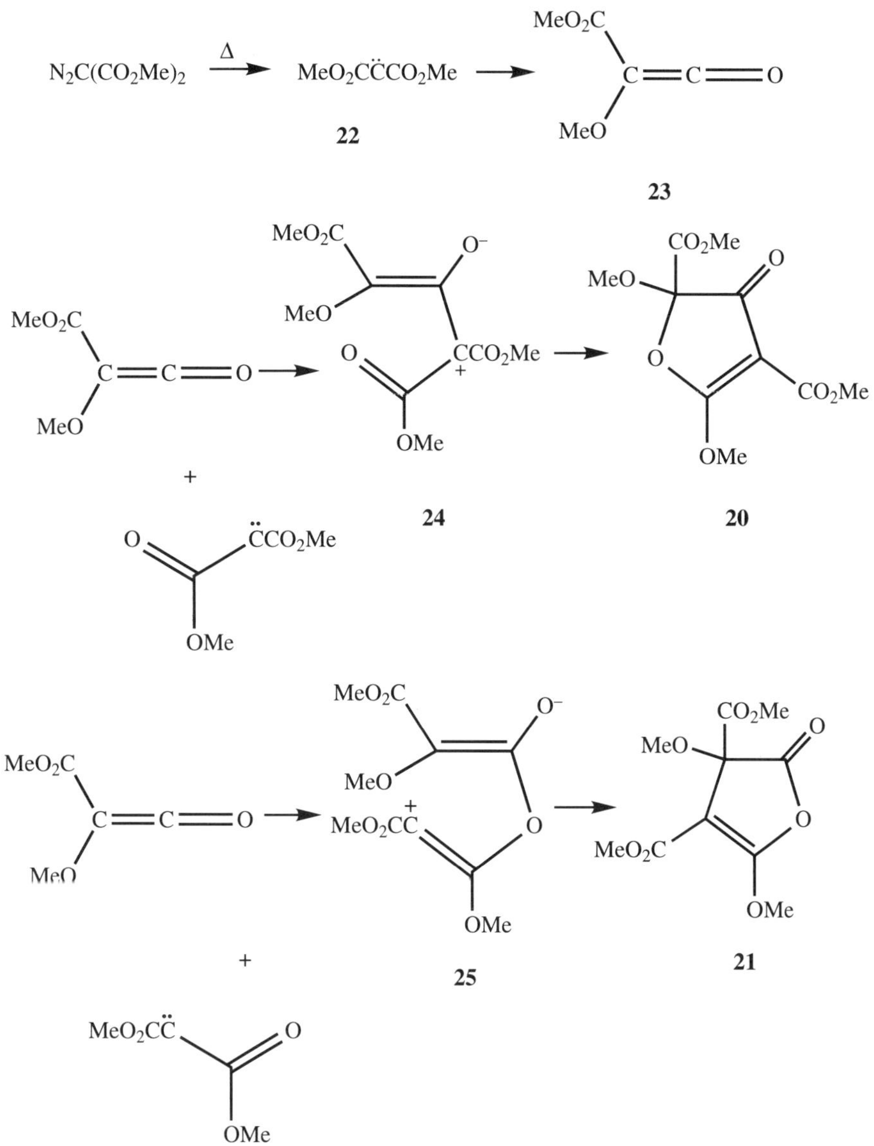

Scheme 1

As shown, this reaction may occur by nucleophilic attack on the carbonyl carbon to give zwitterions **24** and **25**, but both of these zwitterions possess positive charge adjacent to a carbonyl group, and this is not a highly stabilized situation.[14] Further it was stated[11] that the less abundant product **21** "would be the expected one based on other ketene cycloadditions," but no explanation for the predominance of **20** was available.

An alternative explanation of such cycloadditions is that they involve attack of the diazocarbonyl compound on the ketene,[15,16] as in the case of the photolysis of neat diazopinacolone (**26**) forming *tert*-butylketene **27**. The ketene concentration could reach a rather high level, and so reaction as shown in equations 8 and 9 could occur. The formation of **21** could occur by a similar path, while **20** would arise from the intermediate **30**. Reaction of $Ph_2C=C=O$ with aryl and alkyl diazo ketones was proposed to proceed similarly, and gave butenolides analogous to **29**.[16–19]

$$t\text{-BuCOCHN}_2 \xrightarrow{h\nu} t\text{-BuCH=C=O} \tag{8}$$

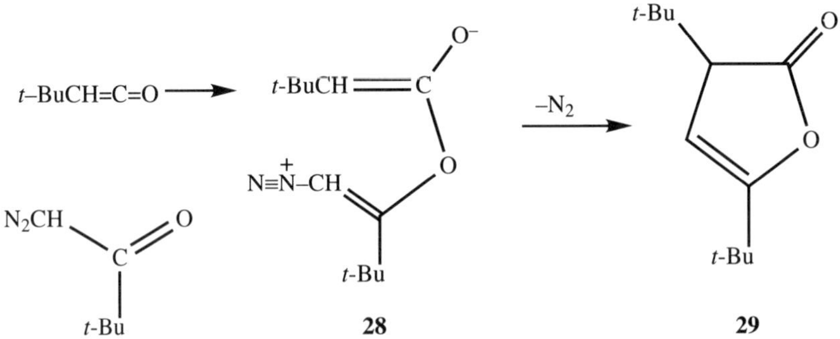

$$\textbf{26} \qquad\qquad \textbf{27}$$

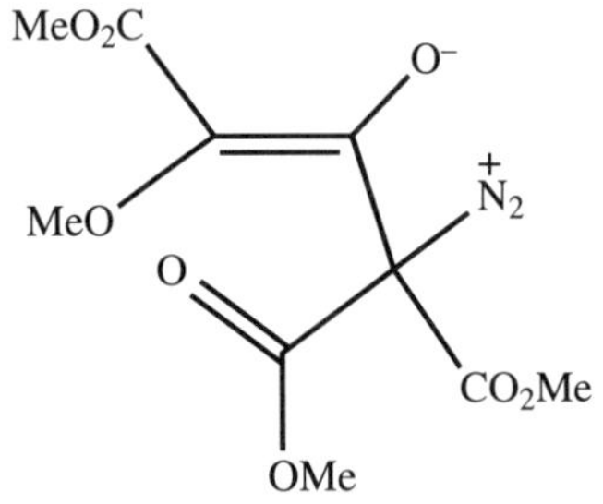

Some examples of similar reactions are shown in equations 10–12.[13,18,20] In these cases the ketene was not prepared in situ, but the possibility of the ketene reacting with either the diazoester or the carbene appears to exist in these reactions as well. In the case of equation 11 the ethyl diazoacetate is stable in the absence of the ketene. Ketenes $R_3SiC(OMe)=C=O$ generated by metal-catalyzed Wolff rearrangements give similar products (equations 12).[20]

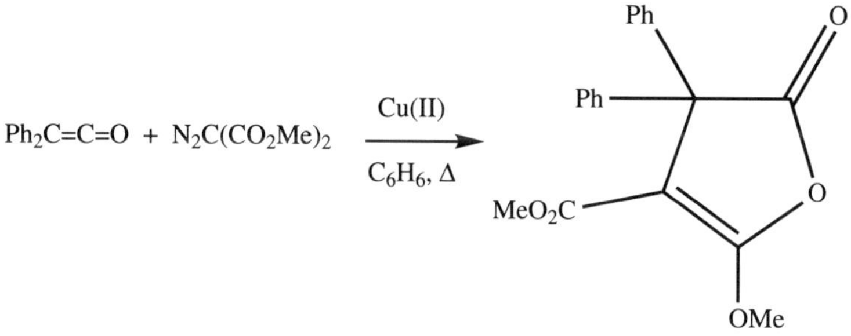

$Ph_2C=C=O + N_2C(CO_2Me)_2 \xrightarrow{Cu(II)}_{C_6H_6, \Delta}$ [product] (10)[13]

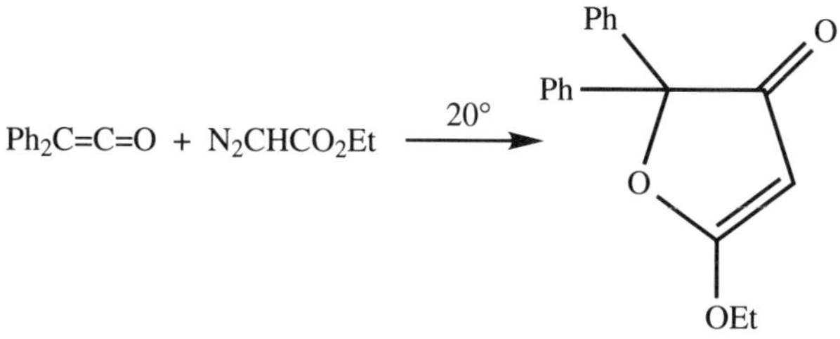

$Ph_2C=C=O + N_2CHCO_2Et \xrightarrow{20°}$ [product] (11)[18]

$Et_3SiC(OMe)=C=O \xrightarrow{Et_3SiCN_2CO_2Me}_{CuOTf}$ [product] → [product] (12)[20]

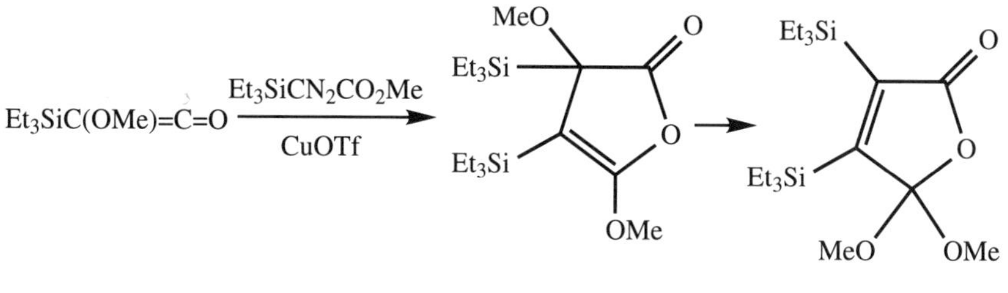

The thermolysis of aryldiazooxides leads to ketocarbenes which can undergo Wolff rearrangement to form ketenes. Reaction products formally derived from addition of either the generated ketenes or added ketenes may be formed (equations 13 and 14)[21,22] (see also Section 4.1.10.2), but there is again some ambiguity as to whether reaction of the ketenes with the diazooxides may have led to the observed products.

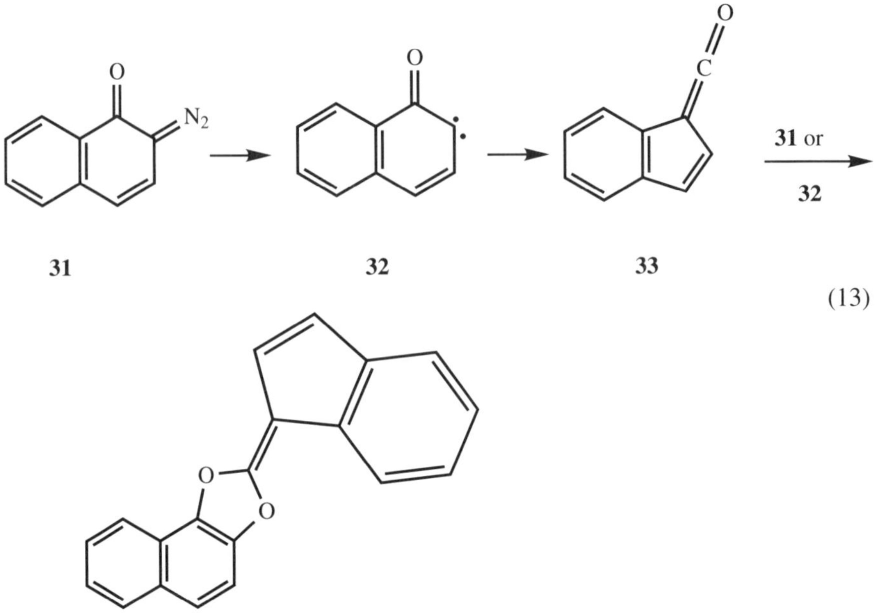

5.4.3 [4 + 2] Cycloadditions of Ketenes

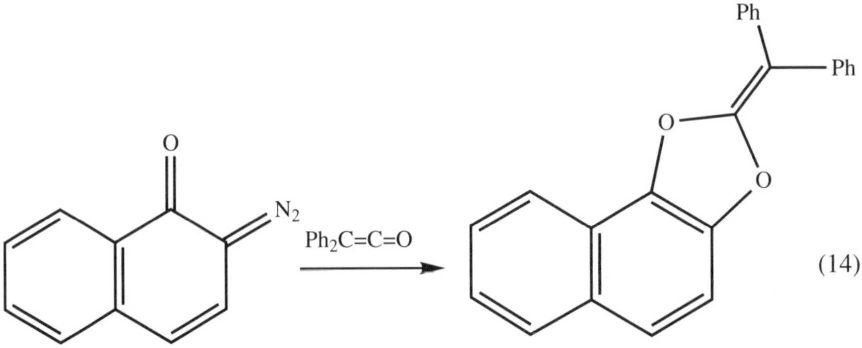

Reactions of the [4 + 2] type with ketenes are not normal, but there are examples in which either the ketene C=C or the C=O bonds take the role of the dienophile. Also reactions of acyl and vinylketenes have been observed in which the unsaturated ketene unit takes the part of the diene unit. Some intramolecular examples are shown on pp. 560–562.

Examples of [4 + 2] cycloadditions involving both the alkene and carbonyl units of $Ph_2C=C=O$ occur with the highly crowded diene **1**.[23] Both of these reactions can be understood as involving initial formation of a zwitterionic species that closes to give **2** and **3** (equation 1). Formation of cyclobutanone intermediates is evidently sterically prohibited.

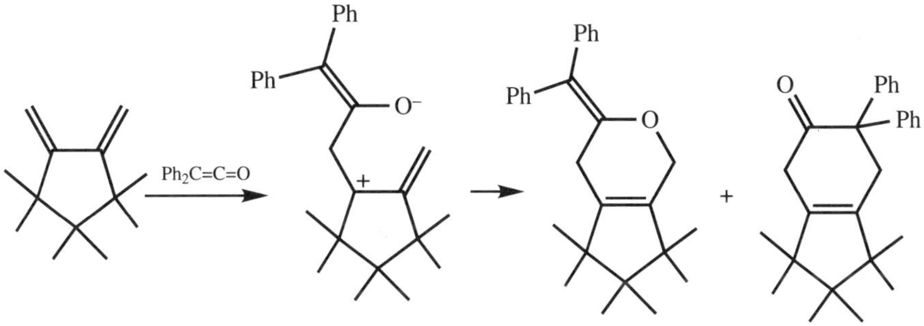

(1)

Diphenylketene reacts with the dienes **4** and **5** in THF to give mixtures of the [2 + 2] cycloadduct **6** and the [4 + 2] cycloadduct **7** in 56:44 and 10:90 ratios, respectively.[24,25] Reaction of **4** in hexane gave a product ratio of 62:38, indicating only a small effect of solvent polarity on the partitioning of the two products. Heating of **7** caused partial isomerization to **6**, establishing that **7** was a kinetic product.[25] Only [2 + 2] cycloadducts were formed from methylphenylketene and **4**, and from diphenylketene and 1,3-cyclohexadiene.[24,26]

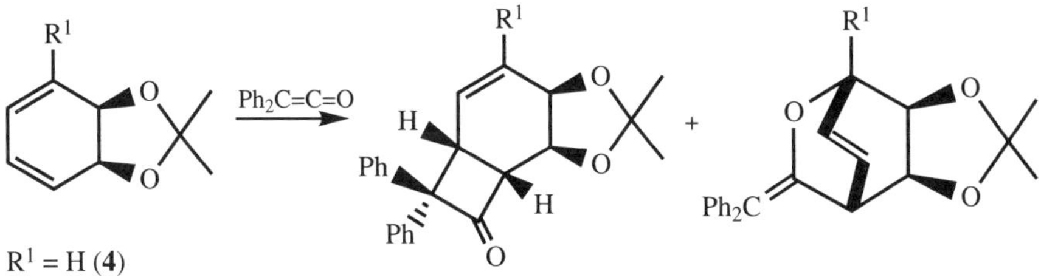

(2)

The greater propensity of **4** to form [4 + 2] cycloadducts compared to 1,3-cyclohexadiene could arise from a combination of conformational, steric, and electronic effects that have not yet been explained. The effect of fluorine could be to favor a zwitterionic intermediate **8** that preferentially closes to the [4 + 2] cycloadduct **7**. The details of these reactions and the factors that influence [4 + 2] versus [2 + 2] cycloadditions are by no means established, but the study of the systematic variation of diene substituents on the reaction course is a promising approach to the problem.

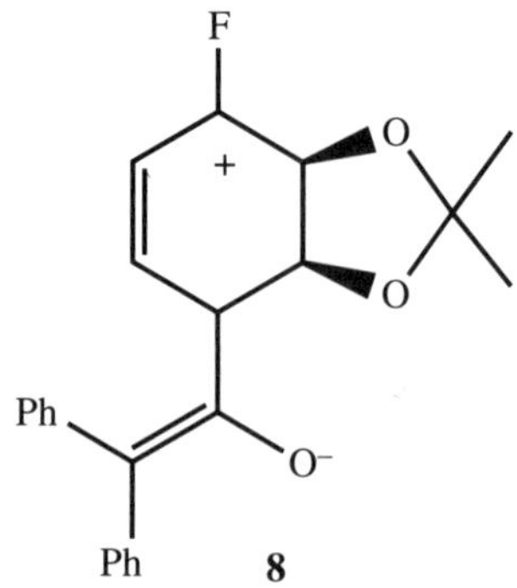

The only product obtained from reaction of diphenylketene with cycloheptatriene in refluxing benzene for 6 days was **9**, in 11% yield.[27] The formation of **9** was ascribed to the intervention of the zwitterion **10** (equation 3).[27]

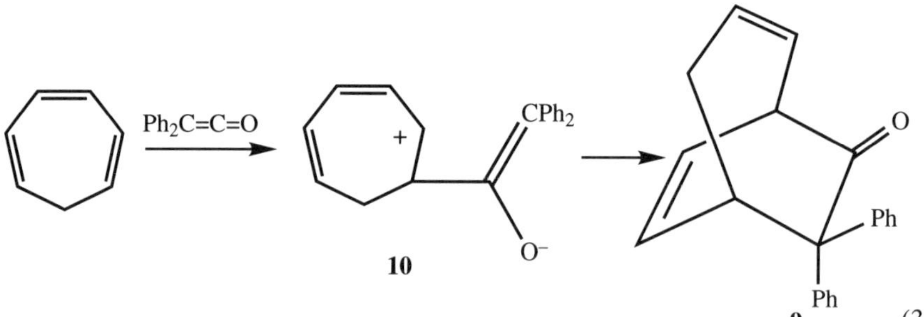

Aryl and alkyl ketenes generated in situ by thermolytic Wolff rearrangement react with o-quinones by [4 + 2] cycloaddition.[28]

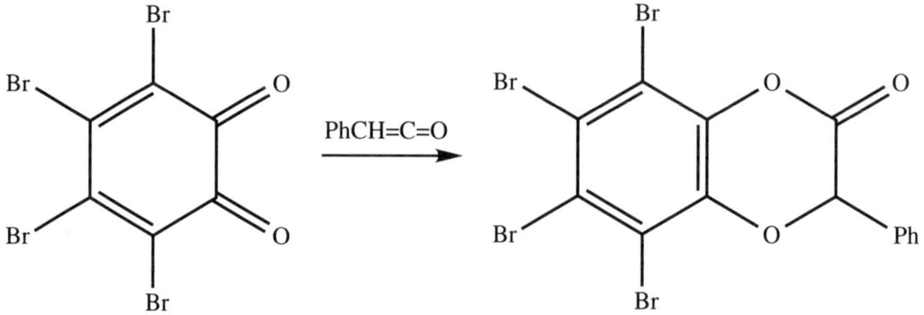

Vinyl imines of the type $RN=CHCR^1=CR^2R^3$ sometimes react with the C=C bond of ketenes by [4 + 2] cycloaddition, as discussed in Section 5.4.1.7 (equation 4).

1,3-Diazabutadienes react with ketenes to form [2 + 2] and also [4 + 2] cycloadducts as shown in equation 4.[29-31] These reactions may be formulated as involving initial nucleophilic attack by the diazadiene such as **11** to give a highly stabilized zwitterionic intermediate such as **12** which closes to **13**, which undergoes facile loss of dimethylamine to yield pyrimidones.[31] Studies of the reaction of **14** to give 3,4-dihydro-(1H)-quinoxalinones led to the conclusion that these reactions were examples of cycloadditions with moderately polar transition states.[31a]

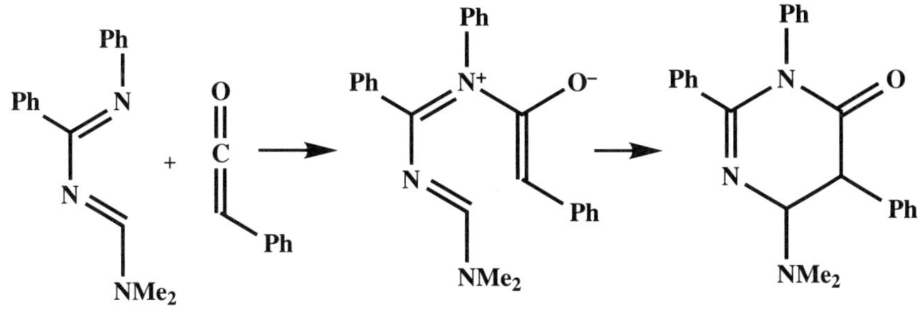

(4)

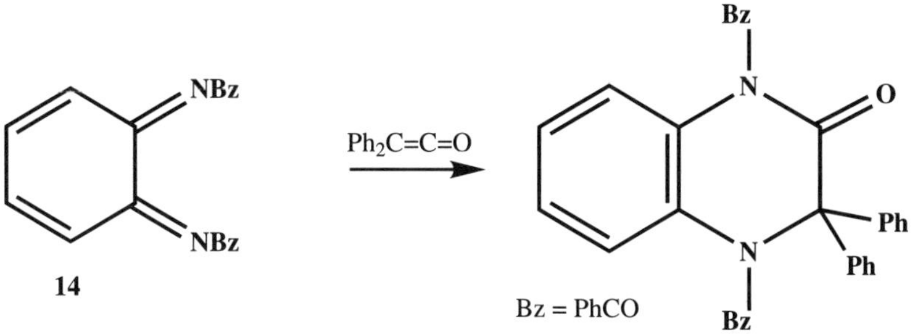

Bz = PhCO

The [4 + 2] cycloadditions of diphenylketene to electron-rich dienones **15** and **16** are proposed to be two-step processes.[32] Diphenylketene adds to chalcone to give the lactone **17**[33] and the ketene **18** gives **19** on heating.[34] The [4 + 2] and [2 + 2] cycloadditions of ketene with CH_2=CHCH=NH have been studied by semiempirical calculations.[34a]

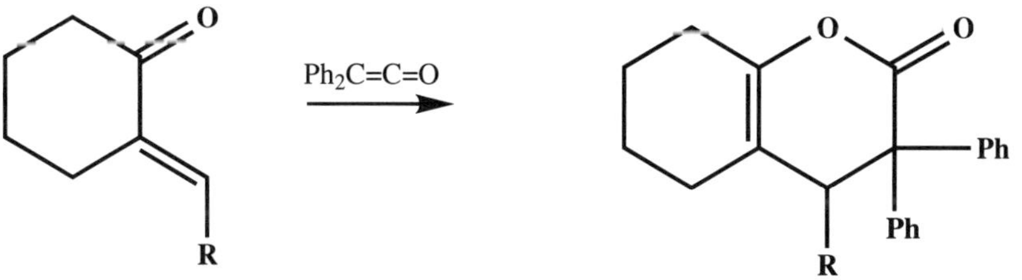

15 (R = N(CH$_2$CH$_2$)$_2$O, OEt)

(5)

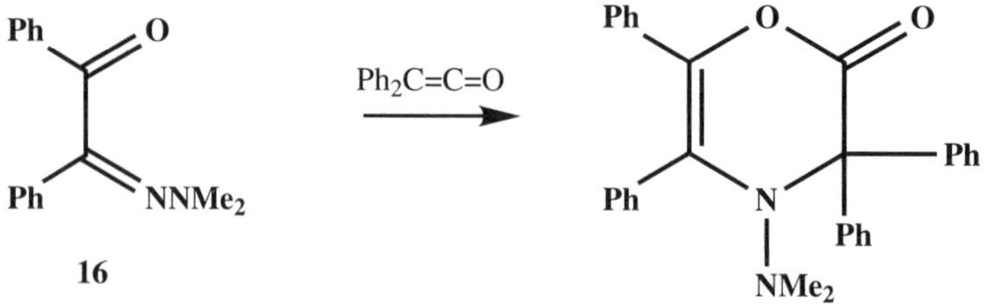

548 REACTIONS OF KETENES

$$\text{(6)}$$

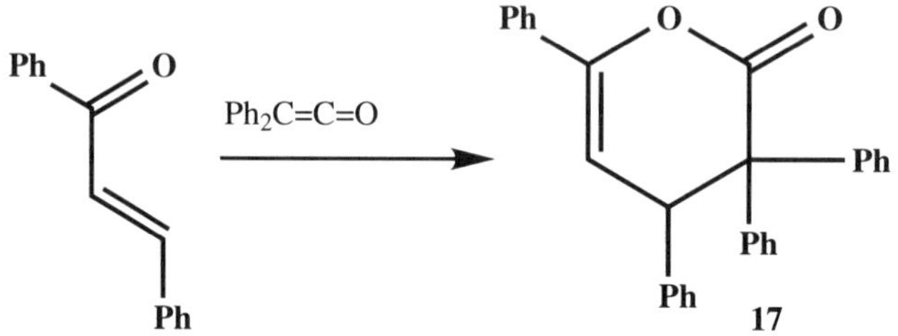

$$\text{(7)}$$

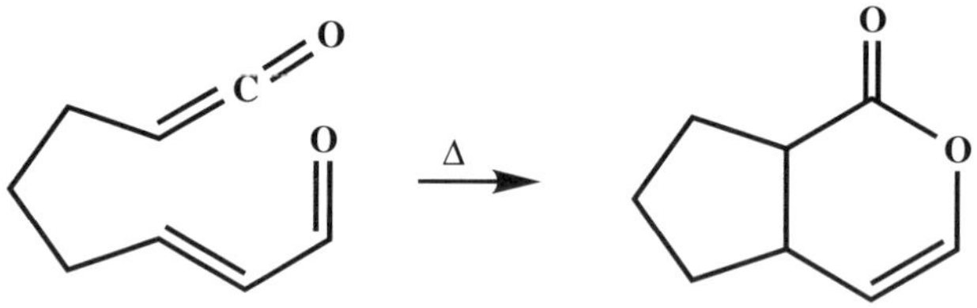

$$\text{(8)}$$

Ketenes react by [4 + 2] pathways with β-methoxy α,β-unsaturated ketones (equation 9).[35] Silyloxydienes give mixtures of [4 + 2] and [2 + 2] cycloaddition products in reactions explained by steric factors in closure of the zwitterionic intermediates (equation 10).[36,37] Silylketenes react by [4 + 2] cycloaddition to the ketenyl carbonyl with CH_2=CMeCH=$C(OSiMe_3)_2$ and with o-quinodimethanes.[37a]

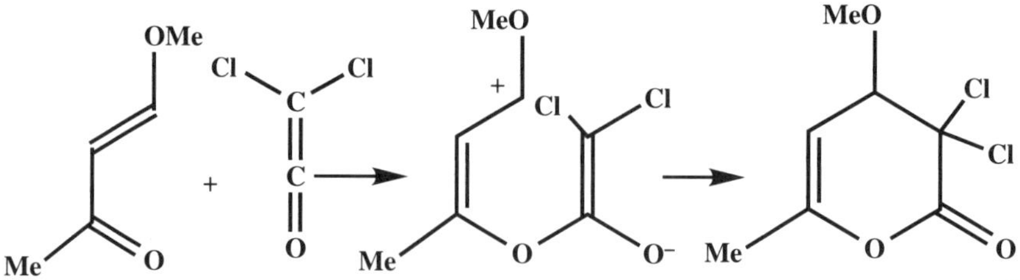

$$\text{(9)}$$

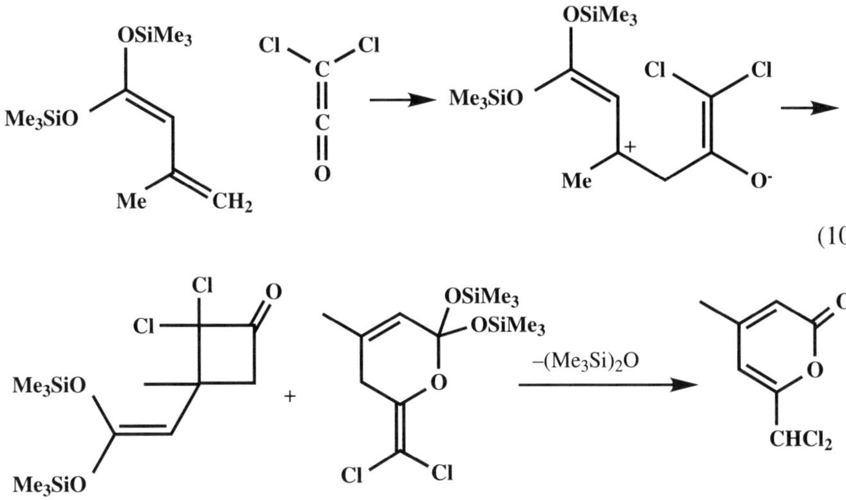

(10)

It was found that the electron-rich ketenes **20** and **21** gave [2 + 2] cycloadditions with **22** in benzene, but in the presence of the oxidant $(4\text{-Tol})_3\text{N}^+\cdot\text{SbF}_6^-$ [tris(4-toly)aminium hexafluoroantimonate] [4 + 2] cycloaddition occurred, forming **22a,** via the ketene radical cations.[38,39] The products **22b** were formed by protonation of the ketene, followed by [2 + 2] cycloaddition and rearrangement.[38]

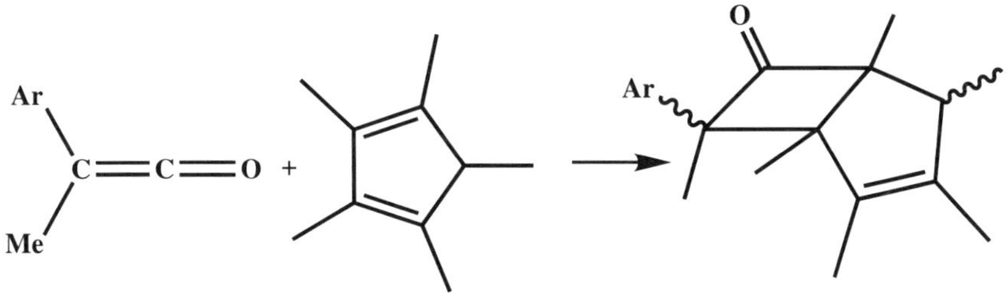

(11)

20 Ar = 4-MeOC$_6$H$_4$

21 Ar = 4-MeC$_6$H$_4$

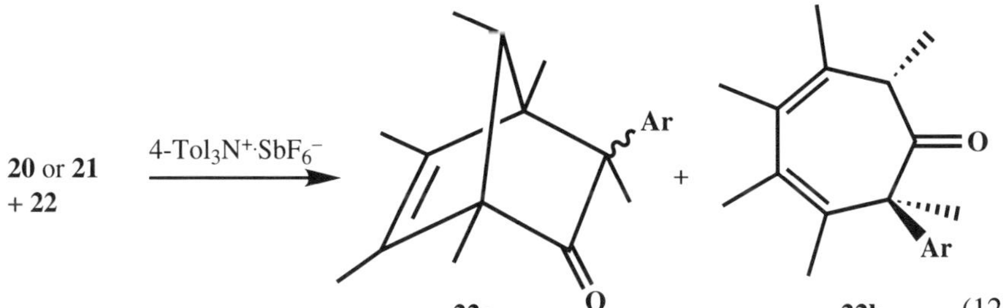

(12)

550 REACTIONS OF KETENES

Vinylketene (**23**) dimerizes by a [4 + 2] pathway in which the ketene plays the role of both the diene and dienophile (equation 13).[40] Vinylmethylketene (**24**) reacts with reactive dienophiles such as eneamines,[41] TCNE, and 4-phenyl-4H-1,2,4-triazole-3,5-dione (PTAD)[42] by a [4 + 2] pathway (equations 14–16), while with diethyl azodicarboxylate both [4 + 2] and [2 + 2] products were observed in a 2.3 : 1 ratio (equation 17).[42] Reaction of the Fe(CO)$_3$ complex of Me$_2$C=CHCMe=C=O with TCNE resulted in cleavage of the Fe(CO)$_3$ and formation of a 1 : 1 adduct that may have a structure similar to **25**.[43] Formylketene (O=CHCH=C=O) also acts as a diene in [4 + 2] cycloadditions (Section 4.1.6). Some related reactions are considered in Section 4.1.10 on fulvenones.

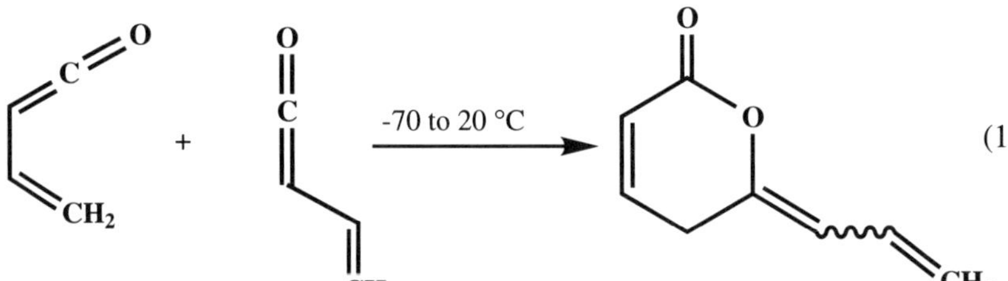

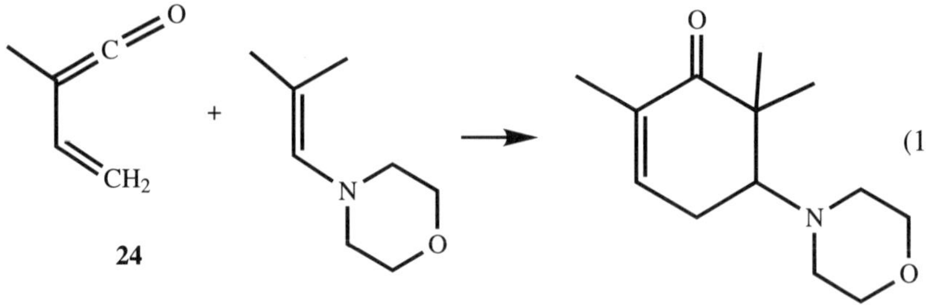

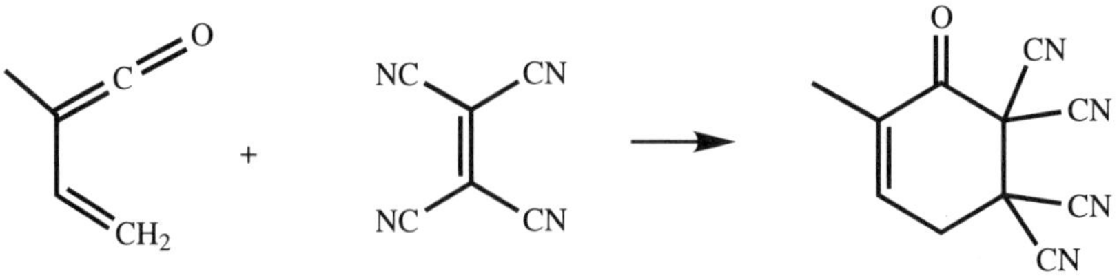

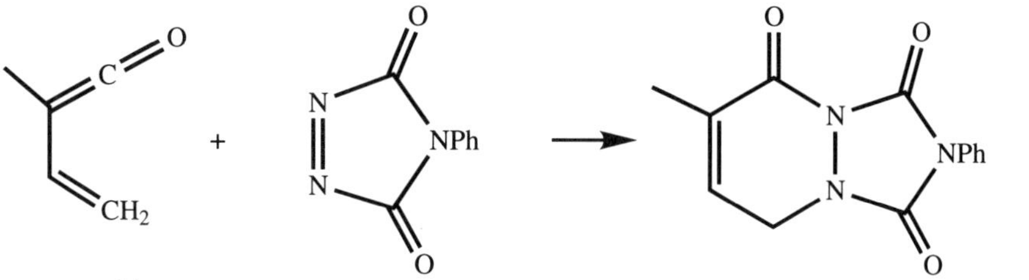

5.4 CYCLOADDITION REACTIONS OF KETENES

[Structure of diene-ketene **24** + triazolinedione-NPh → bicyclic adduct] (16)

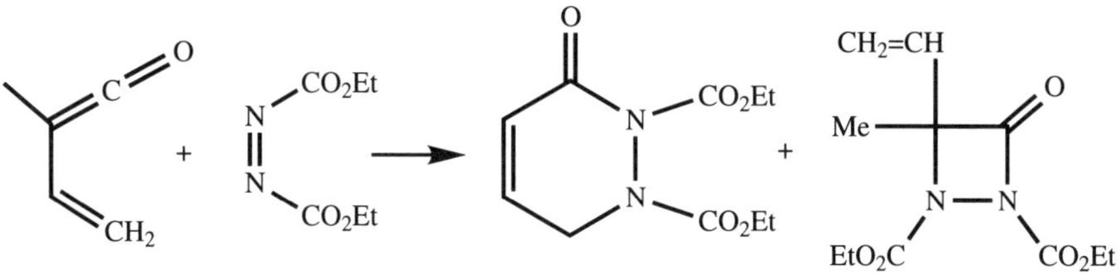

[Structure of **24** + azodicarboxylate (N=N with CO₂Et groups) → pyridazine product + four-membered ring with CH=CH₂, Me, EtO₂C, CO₂Et substituents] (17)

The [4 + 2] cycloaddition of an acylketene to benzonitrile has been reported (equation 18).[44] Asymmetric synthesis of an HIV protease inhibitor via an α-oxoketene/ketene [4 + 2] cycloaddition has been reported.[44a]

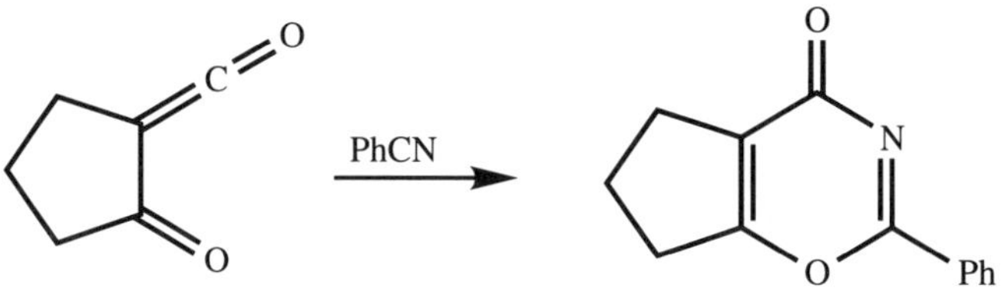

[Cyclopentanone-fused ketene + PhCN → bicyclic oxazine product] (18)

tert-Butylphosphaethyne (**26**) reacts with $Ph_2C=C=O$ to yield the adduct **27**, presumably derived from **28**.[45]

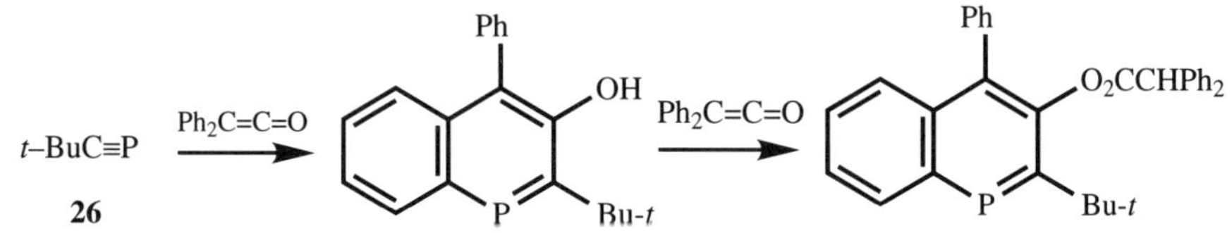

$t-BuC\equiv P$ **26** $\xrightarrow{Ph_2C=C=O}$ [phosphinine with Ph, OH, Bu-*t* substituents] **28** $\xrightarrow{Ph_2C=C=O}$ [phosphinine with Ph, O₂CCHPh₂, Bu-*t* substituents] **27**

The reaction of $Me_2C=C=O$ with a diazooxide constitutes a [5 + 2] cycloaddition (equation 19),[46] and a [6 + 2] pathway occurs for dichloroketene and bullvalene, in competition with a [2 + 2] reaction (equation 20).[47] A [6 + 2] cycloaddition of $Ph_2C=C=O$ occurs with a metal-complexed triene (equation 21).[48] The $Fe(CO)_3$ complexed triene (Section 5.5.2.4) and a $Cr(CO)_3$ complexed ketene[48a] behave similarly.

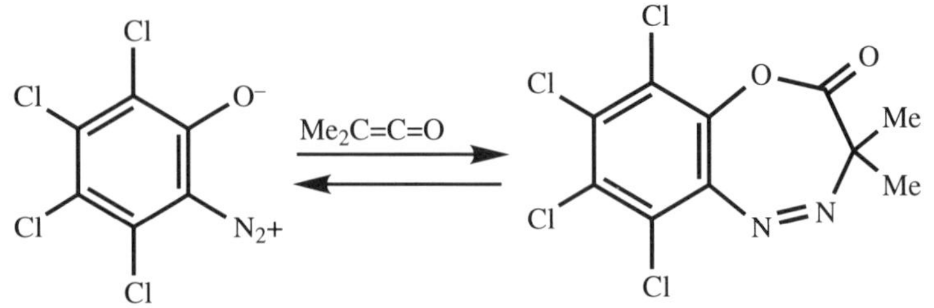

$$\text{(19)}$$

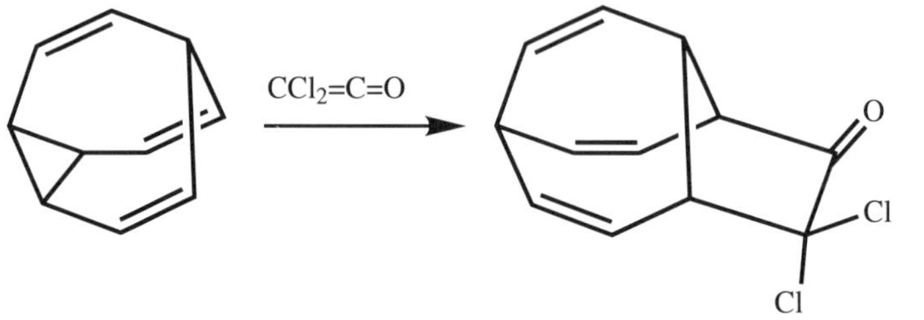

$$\text{(20)}$$

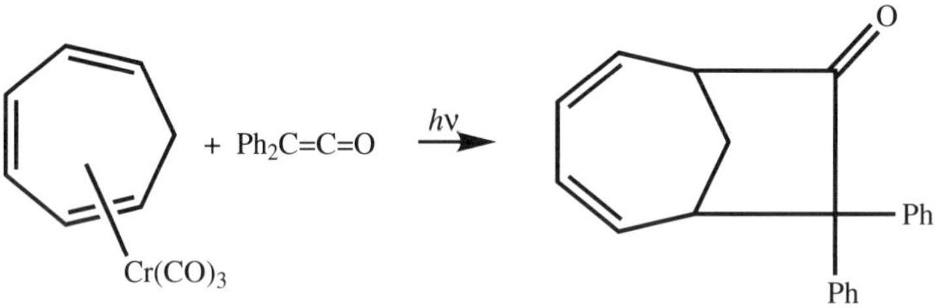

$$\text{(21)}$$

The reaction of tropothione (**28**) with $Ph_2C=C=O$ has been shown to occur by an [8 + 2] pathway (equation 21).[49] Reaction of *N*-arylcycloheptatrien-1-imines **29** with ketenes **30** (R = H, Cl, Ph) led to the formation of the [8 + 2] cycloadducts **31** (equation 23).[50]

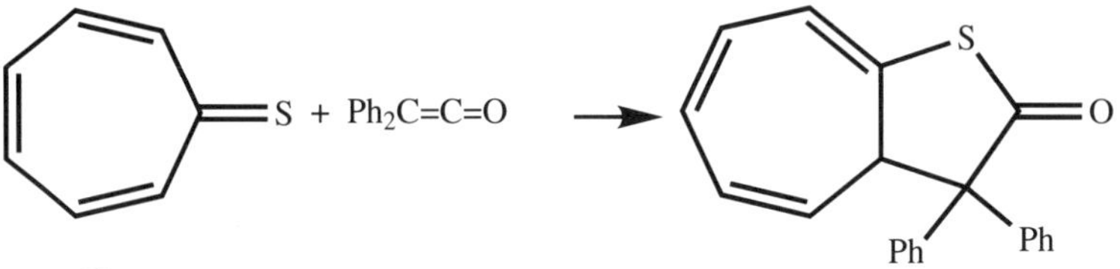

$$\text{(22)}$$

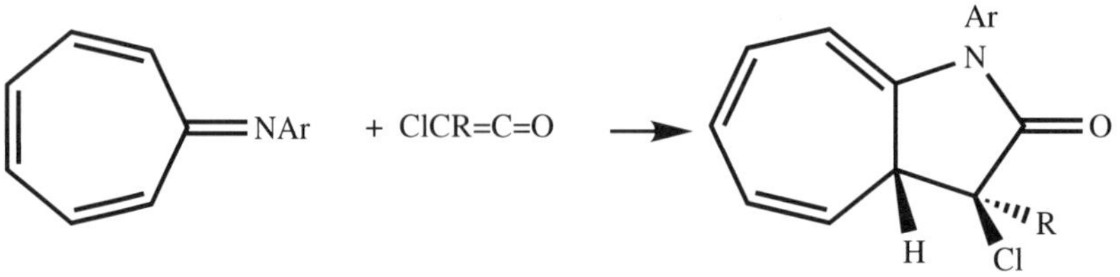

$$\text{(23)}$$

References

1. Huisgen, R. *Angew. Chem., Int. Ed. Engl.* **1963,** *2,* 565–598.
2. Kirmse, W. *Chem. Ber.* **1960,** *93,* 2357–2360.
3. Takebayashi, M.; Ibata, T. *Bull. Chem. Soc. Jpn.* **1968,** *41,* 1700–1707.
4. Scarpati, R.; Sorrentino, P. *Gazz. Chim. Ital.* **1959,** *89,* 1525–1533.
5. Evans, A. R.; Taylor, G. A. *J. Chem. Soc., Perkin Trans. 1* **1983,** 979–983.
6. Evans, A. R.; Taylor, G. A. *J. Chem. Soc., Perkin Trans. 1* **1987,** 567–589.
7. Abou-Gharbia, M. A.; Jouille, M. M. *J. Org. Chem.* **1979,** *44,* 2961–2966.
8. Gettins, A. F.; Stokes, D. P.; Taylor, G. A.; Judge, C. B. *J. Chem. Soc., Perkin Trans. 1* **1977,** 1849–1855.
9. Katagiri, N.; Niwa, R.; Furuya, Y.; Kato, T. *Chem. Pharm. Bull.* **1983,** *31,* 1833–1841.
10. Kellogg, R. M. *J. Org. Chem.* **1973,** *38,* 844–846.
11. Pomerantz, M.; Levanon, M. *Tetrahedron Lett.* **1991,** *32,* 995–998.
12. Maas, G.; Regitz, M. *Chem. Ber.* **1976,** *109,* 2039–2063.
13. Eichhorn, K.; Hoge, R.; Maas, G.; Regitz, M. *Chem. Ber.* **1977,** *110,* 3272–3278.
14. Creary, X. *Acc. Chem. Res.* **1985,** *18,* 3–8.
15. Wiberg, K. B.; Hutton, T. W. *J. Am. Chem. Soc.* **1954,** *76,* 5367–5371.
16. Ried, W.; Mengler, H. *Liebigs. Ann. Chem.* **1962,** *651,* 54–65; *Ibid.* **1964,** *678,* 113–126.
17. Ried, W.; Mengler, H. *Angew. Chem.* **1961,** *73,* 218–219.
18. Kende, A. S. *Chem. Ind. (London)* **1956,** 1053–1054.
19. Huisgen, R.; König, H.; Binsch, G.; Sturm, H. J. *Angew. Chem.* **1961,** *73,* 368–371.
20. Maas, G.; Gimmy, M.; Alt, M. *Organometallics* **1992,** *11,* 3813–3820.
21. Yates, P.; Robb, E. W. *J. Am. Chem. Soc.* **1957,** *79,* 5760–5768.
22. Huisgen, R.; Binsch, G.; König, H. *Chem. Ber.* **1964,** *97,* 2868–2883.
23. Mayr, H.; Heigl, U. W. *J. Chem. Soc., Chem. Commun.* **1987,** 1804–1805.
24. Downing, W.; Latouche, R.; Pittol, C. A.; Pryce, R. J.; Roberts, S. M.; Ryback, G.; Williams, J. O. *J. Chem. Soc., Perkin Trans. 1,* **1990,** 2613–2615.
25. Maruya, R.; Pittol, C. A.; Pryce, R. J.; Roberts, S. M.; Thomas, R. J.; Williams, J. O. *J. Chem. Soc., Perkin Trans. 1,* **1992,** 1617–1621.
26. Davies, H. G.; Rahman, S. S.; Roberts, S. M.; Wakefield, B. J.; Winders, J. A. *J. Chem. Soc., Perkin Trans. 1,* **1987,** 85–89.
27. Falshaw, C. P.; Lakoues, A.; Taylor, G. A. *J. Chem. Res. S* **1985,** 106.
28. Ried, W.; Radt, W. *Liebigs. Ann. Chem.* **1965,** *688,* 170–173; *Ibid.* **1964,** *676,* 110–114.
29. Luthardt, P.; Würthwein, E.-U. *Tetrahedron Lett.* **1988,** *29,* 921–924.
30. Ibnusaud, I.; Malar, E. J. P.; Sundaram, N. *Tetrahedron Lett.,* **1990,** *31,* 7357–7358.
31. Mazumdar, S. N.; Mahajan, M. P. *Tetrahedron* **1991,** *47,* 1473–1484.
31a. Friedrichsen, W.; Oeser, H.-G. *Chem. Ber.* **1975,** *108,* 31–47.
32. Gompper, R. *Angew. Chem., Int. Ed. Engl.* **1969,** *8,* 312–327.
33. Staudinger, H.; Endle, R. *Liebigs Ann. Chem.* **1913,** *401,* 263–292.
34. Ayral-Kaloustian, S.; Wolff, S.; Agosta, W. C. *J. Org. Chem.* **1978,** *43,* 3314–3319.
34a. Fabian, W. M. F.; Kollenz, G. *J. Phys. Org. Chem.* **1994,** *7,* 1–8.

35. Brady, W. T.; Agho, M. O. *J. Org. Chem.* **1983**, *48*, 5337–5341.
36. Brady, W. T.; Agho, M. O. *Synthesis* **1982**, 500–502.
37. Brady, W. T.; Agho, M. O. *J. Heterocyclic Chem.* **1983**, *20*, 501–506.
37a. Ito, T.; Aoyama, T.; Shioiri, T. *Tetrahedron Lett.* **1993**, *34*, 6583–6586.
38. von Seggern, H.; Schmittel, M. *Chem. Ber.* **1994**, *127*, 1269–1274.
39. Schmittel, M.; von Seggern, H. *J. Am. Chem. Soc.* **1993**, *115*, 2165–2177.
40. Trahanovsky, W. S.; Surber, B. W.; Wilkes, M. C.; Preckel, M. M. *J. Am. Chem. Soc.* **1982**, *104*, 6779–6781.
41. Berge, J. M.; Rey, M.; Dreiding, A. S. *Helv. Chim. Acta* **1982**, *65*, 2230–2241.
42. Barbaro, G.; Battaglia, A.; Giorgianni, P. *J. Org. Chem.* **1987**, *52*, 3289–3296.
43. Newton, M. G.; Pantaleo, N. S.; King, R. B.; Chu, C. *J. Chem. Soc., Chem. Commun.* **1979**, 10–12.
44. Stetter, H.; Kiehs, K. *Chem. Ber.* **1965**, *98*, 2099–2102.
44a. Gammill, R. B.; Judge, T. M.; Phillips, G.; Zhang, Q.; Sowell, C. G. Cheney, B. V.; Mizsak, S. A.; Dolak, L. A.; Seest, E. P. *J. Am. Chem. Soc.* **1994**, *116*, 12113–12114.
45. Märkl, G.; Kallmünzer, A.; Nöth, H.; Pohlmann, K. *Tetrahedron Lett.* **1992**, *33*, 1597–1600.
46. Ried, W.; Kraemer, R. *Liebigs Ann. Chem.* **1965**, *681*, 52–54.
47. Erden, I. *Tetrahedron Lett.* **1985**, *26*, 5635–5638.
48. Rigby, J. H.; Ahmed, G.; Ferguson, M. D. *Tetrahedron Lett.* **1993**, *34*, 5397–5400.
48a. Rigby, J. H.; Pigge, F. C.; Ferguson, M. D. *Tetrahedron Lett.* **1994**, *35*, 8131–8132.
50. Machiguchi, T.; Yamabe, S. *Tetrahedron Lett.* **1990**, *31*, 4169–4172.
51. Ito, K.; Saito, K.; Takahashi, K. *Bull. Chem. Soc. Jpn.* **1992**, *65*, 812–816.

5.4.4 Intramolecular Cycloadditions of Ketenes

The intramolecular cycloaddition of ketenes is a valuable synthetic procedure that has been gaining increasing attention. This topic has been recently reviewed[1–3] and greater detail is available from these sources. Various examples are also mentioned elsewhere in the text (e.g., equation 8, Section 5.4.3).

Perhaps the simplest intramolecular cyclization of ketenes is the formation of cyclobutenone and related derivatives from vinyl, imidoyl, or acylketenes, as shown in equations 1–3. These reactions were studied theoretically[4] with geometries calculated at the HF/6-31G** level with MP4SDQ/6-31G** energies and were found to be endothermic by 0, 10, and 17 kcal/mol, respectively, with barriers of 21, 25, and 31 kcal/mol respectively.

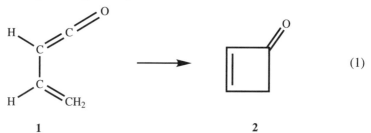

(1)

1   2

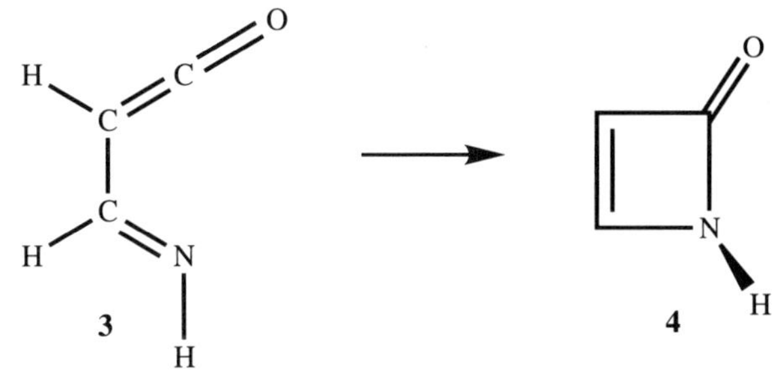

5.4 CYCLOADDITION REACTIONS OF KETENES

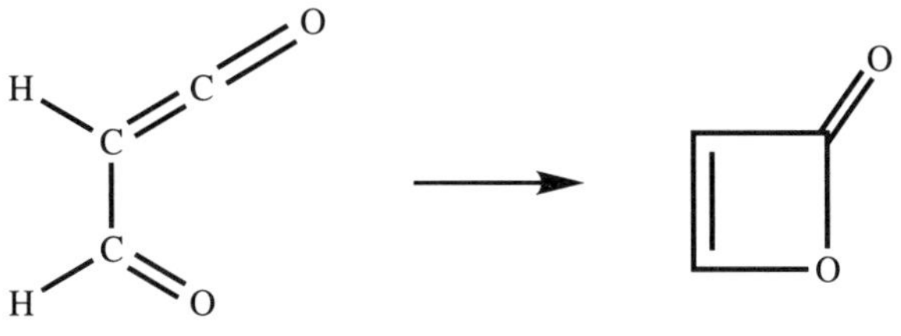

(2)

(3)

These calculations predict that cyclobutenones should be similar in stability to vinylketenes but will have a high barrier to interconversion. Indeed many derivatives of **1** and **2** are known, and the isomerization of substituted cyclobutenones to vinylketenes is a valuable synthetic pathway where the vinylketenes are usually trapped by further inter- or intramolecular reactions. Imidoylketenes **3** undergo cyclization from the geometry shown where the nitrogen lone pair is positioned to attack the carbonyl carbon. The azetone **4** adopts a pyramidal geometry at nitrogen to relieve destabilization due to 4π-electron *anti*-aromatic destabilization.[4] There is a high barrier for cyclization of formylketene to the oxetone **6**, which is less stable and has a low barrier to conversion back to the formylketene. As described in Section 4.1.6 acylketenes are rather stable, and oxetones **6** are evidently unknown.

Flash vacuum pyrolysis at 400–600 °C of **7** led to **8** and **9** that were trapped in an Ar matrix at 18 K and identified by their IR bands. At higher pyrolysis temperatures the ketene **8** predominated, presumably formed by ring opening of **9**.[5]

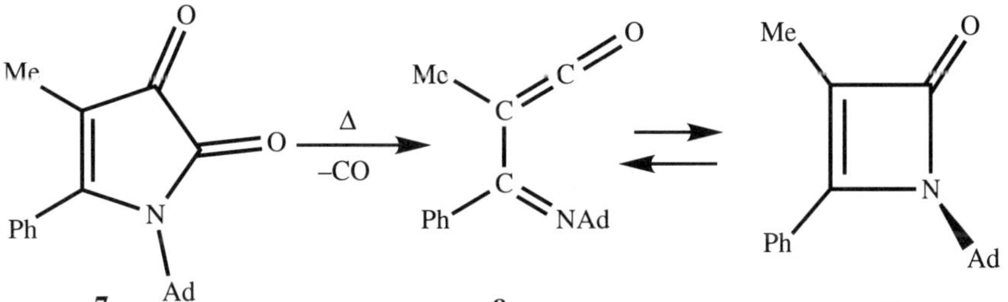

Ad = 1-adamantyl

(4)

Intramolecular cycloaddition of ketenes to alkenes has been observed with ketenes generated by Wolff rearrangement of diazoketones,[6–10] by Norrish Type I reactions of ketones,[11,12] from metal carbene complexes,[12a] and from acyl chlorides or anhydrides and base.[13–24] Some examples are shown in equations 5–8. The diradical **10** forms the ketene **11** with low efficiency, but preferentially forms **12**, which reforms **10** on photolysis.[12]

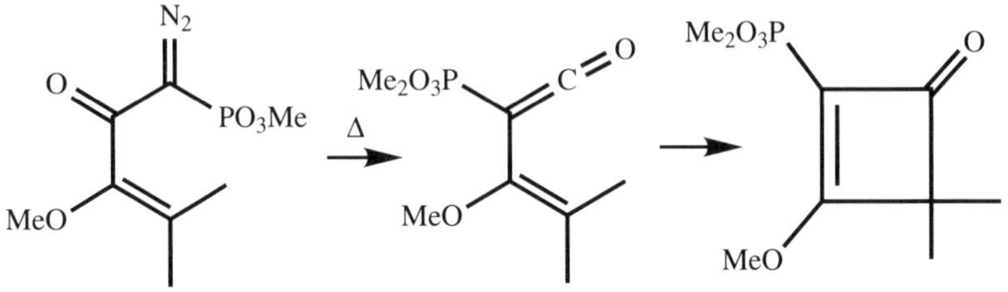

75% (4a)[10a]

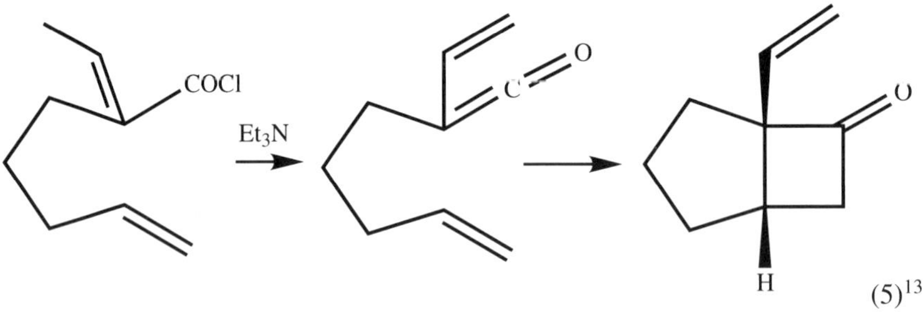

(5)[13]

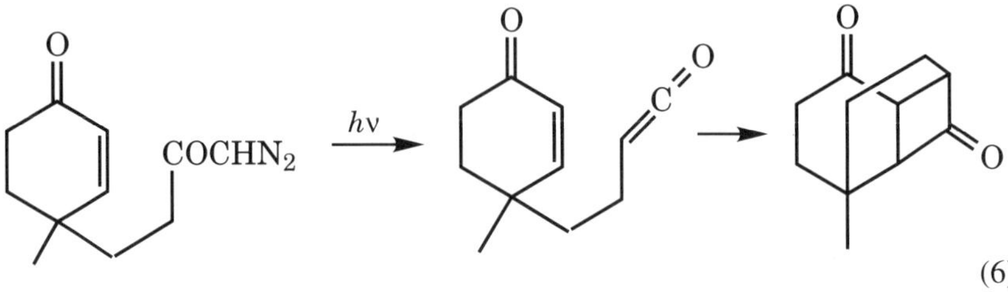

(6)[6]

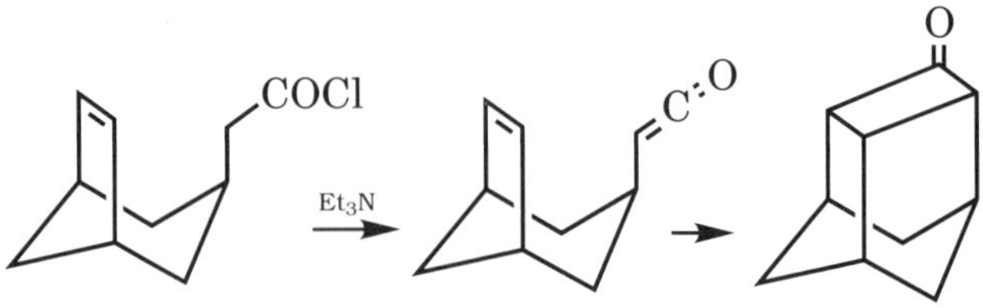

(7)[14]

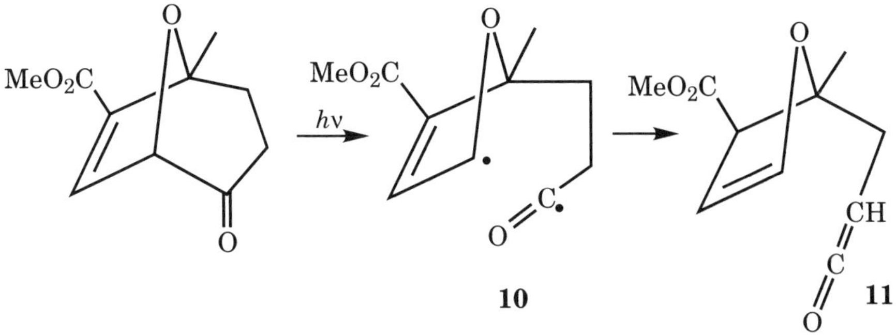

5.4 CYCLOADDITION REACTIONS OF KETENES

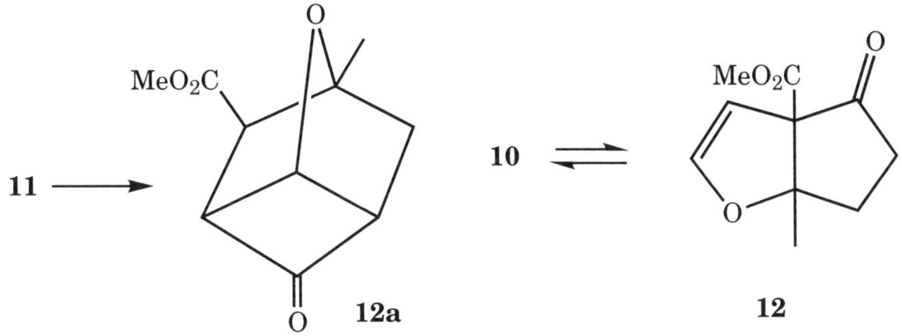

(8)[12]

There are several interesting structural influences on these reactions. Thus disubstituted alkenes are much more reactive than monosubstituted alkenes in cyclization, an effect that presumably arises from the greater nucleophilicity of the disubstituted alkenes (equation 9).[16] Also vinyl, alkoxy, and chloroketenes are much more effective than alkylketenes in intramolecular cycloadditions.[1,13] Cyclizations such as the conversion of **13** into **14,** which if concerted would be forbidden $[\pi 2_a + \pi 2_a]$ *syn* additions, proceed smoothly (equation 10), even though this would appear impossible to achieve by a $[\pi 2_s + \pi 2_a]$ concerted mechanism.[17] All of these facts are accommodated by a two-step process proceeding through a zwitterionic intermediate (equation 11).

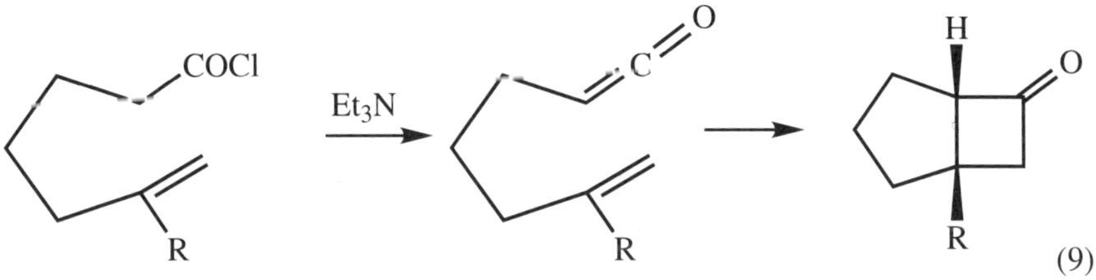

(9)

3% (R = H)
80% (R = Me)

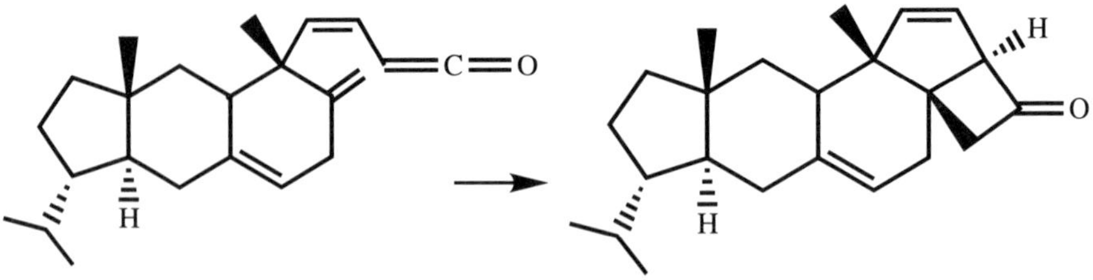

558 REACTIONS OF KETENES

[Structures **13** → **14**] (10)

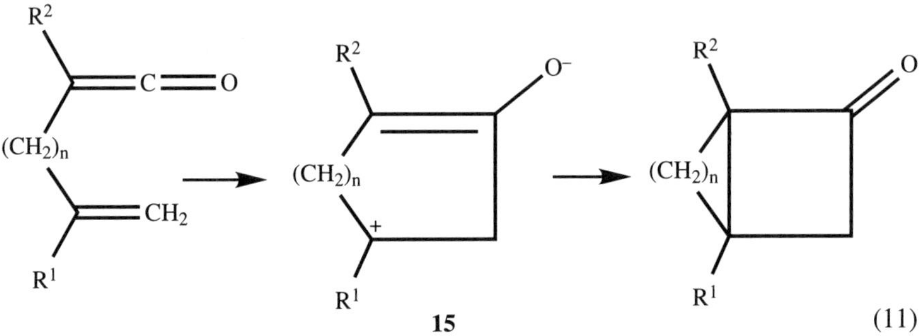

[Structures showing ketene cyclization via zwitterionic intermediate **15**] (11)

As shown in equation 11 the zwitterionic intermediate **15** would be stabilized by alkyl groups R^1 and vinyl groups R^2. The cyclization of **16** to **17** as part of a synthesis of β-*trans*-bergamotene was posed[18] as a test case of this hypothesis, and the cyclization of **16** occurs with *cis* bond formation, whereas an *anti* fusion would be prohibitively strained. Other formations of bicyclic rings by ketene cyclization have been demonstrated,[19–27] including the formation of **19** (equation 13).[24]

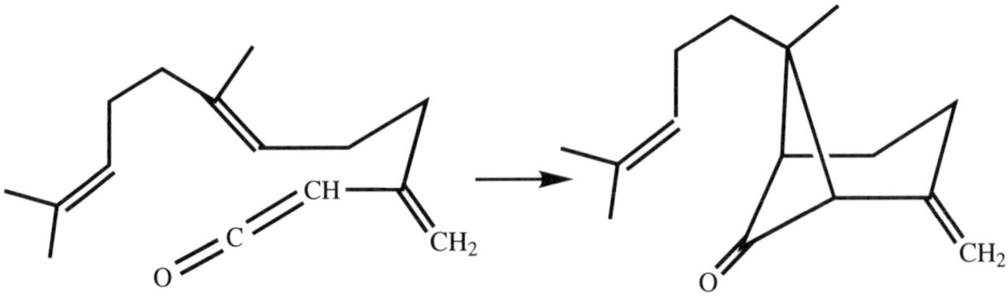

[Structures **16** → **17**] (12)

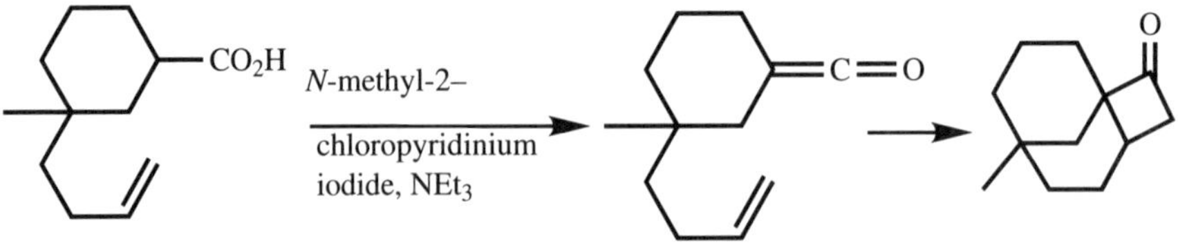

[Structures **18** → **19** via N-methyl-2-chloropyridinium iodide, NEt$_3$] (13)

5.4 CYCLOADDITION REACTIONS OF KETENES

Chloroketenes are reactive in intramolecular cyclizations with alkenes (equation 14).[25,26] A highly strained and labile ring system was generated from a diazoketone (equation 15).[27]

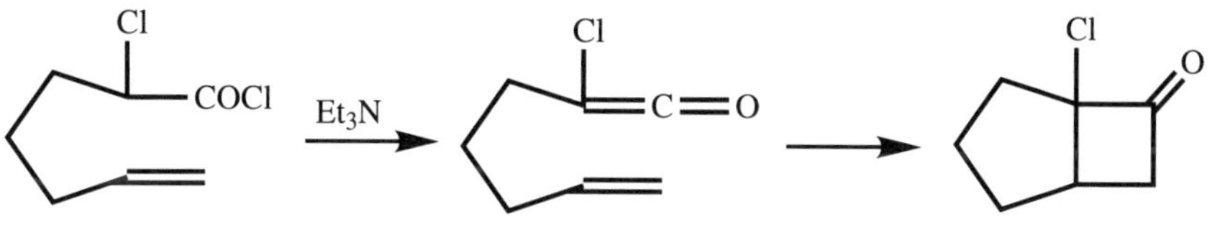

(68%)

(14)

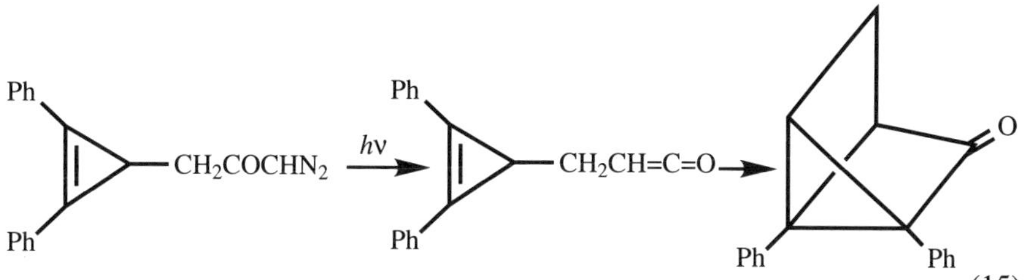

(15)

The generation of vinylketenes from alkynyl cyclobutenones and their cyclization to quinones is shown in equation 16.[28] Photolysis of **20** is proposed to give the isomeric ketenes **21a,** which form the lactones **22** (equation 17).[28] The thermolyses of 4-alkenyl and 4-phenylcyclobutenones such as **23** give ring opening to dienylketenes (**24**) which undergo intramolecular cyclization and tautomerization to phenols (equation 18).[29,30] Other examples of the cyclobutene ring opening are shown in Section 3.4.1. Equation 19 shows the preparation of dienylketenes from 3,5-hexadienylacyl chlorides.[30] This cyclization was also achieved by reacting hexadienyl carboxylic acids with $NaOAc/Ac_2O$.[31] The photochemical ring opening of cyclohexadienones to form dienylketenes is discussed in Section 3.4.3.

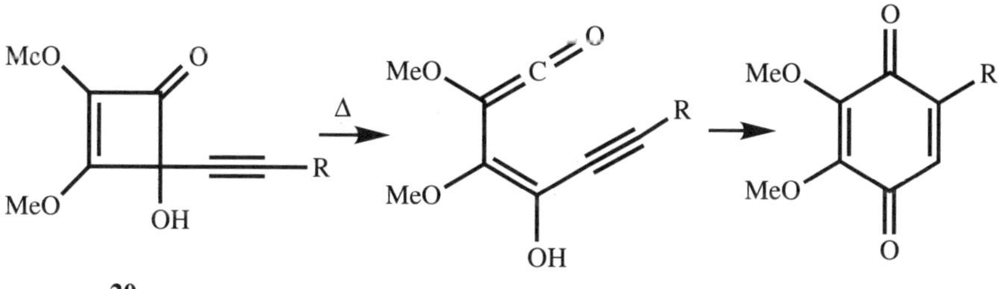

(16)[28]

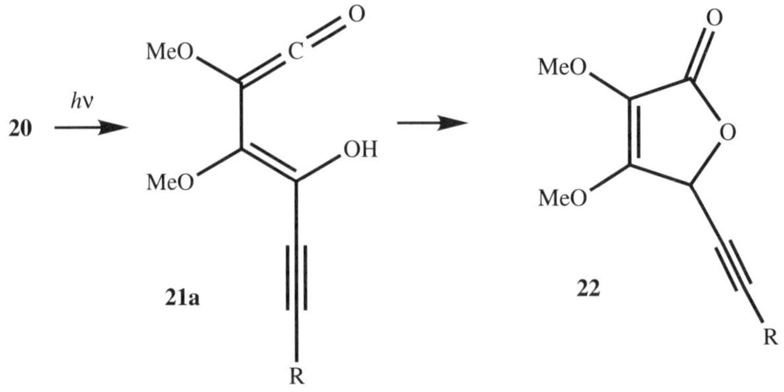

In the scheme of equation 20 a ketene is generated by Wolff rearrangement and reacts by [2 + 2] cycloaddition with an alkyne to give a cyclobutenone which on thermolysis or irradiation gives a dienylketene which undergoes cyclization.[32,33] The phenols from these reactions are often converted to quinones.

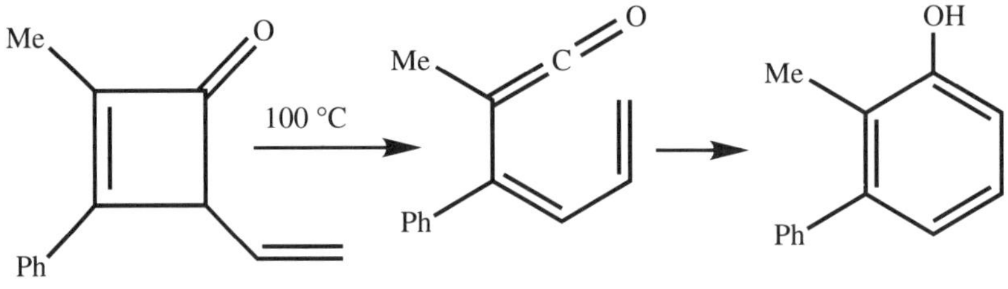

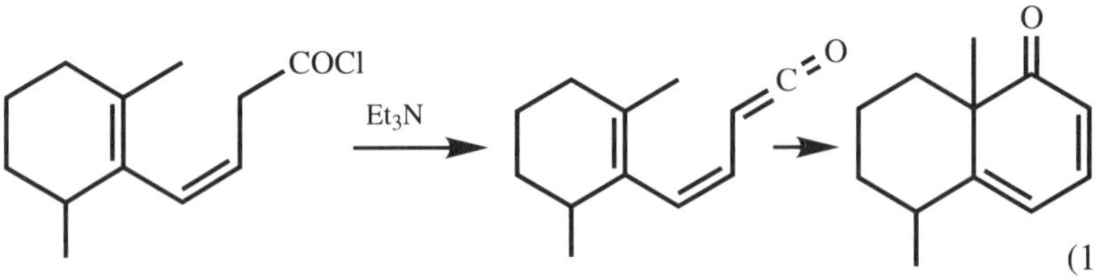

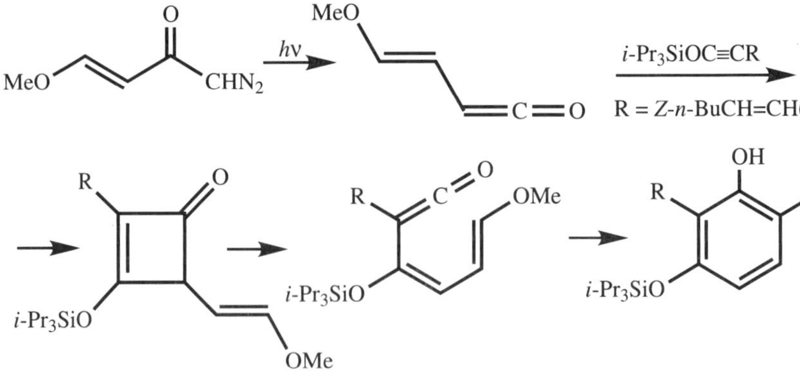

Cyclization of alkenylketenes with pendant aryl groups is closely related to the reaction with alkenes and provides a route to quinones, as in equation 21.[34] Pyrroles as pendant aryl groups react similarly.[35]

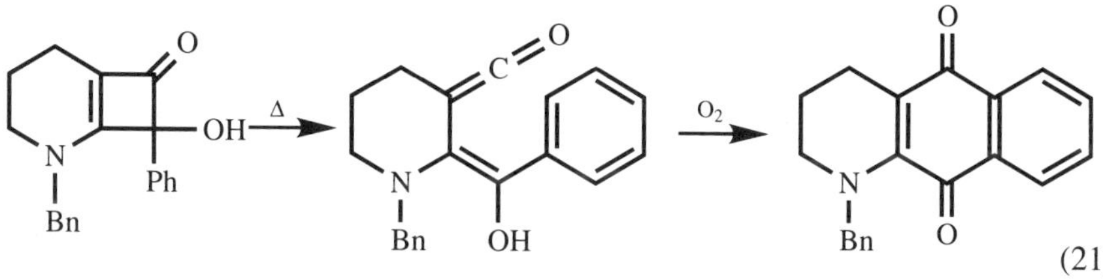

(21)

Cyclization of an alkenylketene with a pendant naphthyl group was a key step in the synthesis of diterpenoid quinones **25** in the Dan Shen family (equation 22).[36] A related phenalenone was prepared using a similar step.[36a]

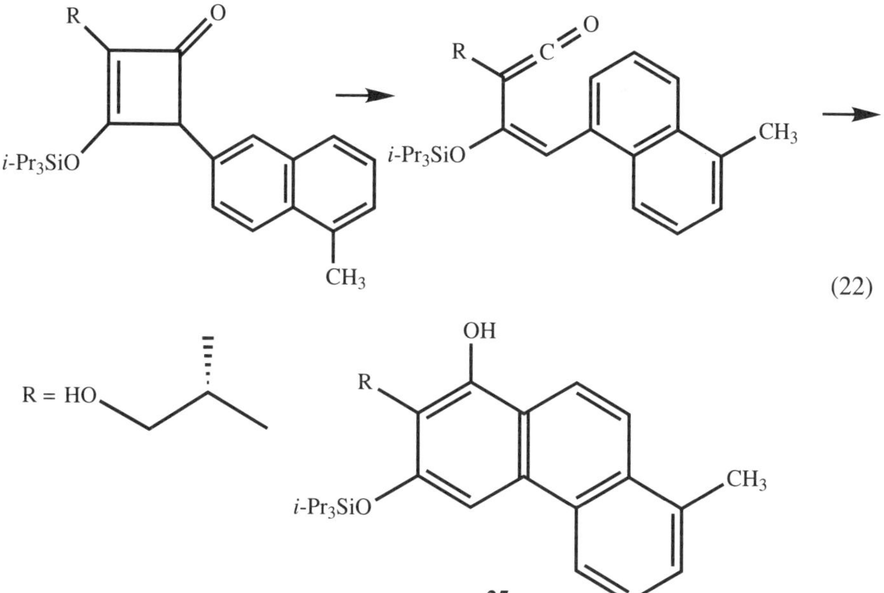

(22)

Photolysis of *o*-alkynylaryl diazo ketones **26** proceed by Wolff rearrangement to ketenes **27** which cyclize on the alkyne to give diradical intermediates which abstract hydrogens to give naphthols (equation 23) or react by a further intramolecular cyclization.[37] Evidence for the formation of the ketenes **27** was provided by trapping with MeOH. Diradical intermediates are also proposed to be formed from reaction of equation 16,[28] and the diradicals from cyclization of eneyneketenes formed by Wolff rearrangement analogous to equation 23,[37a] or by cyclobutenone thermolysis,[37b] were trapped by hydrogen atom donors, and were shown to cleave DNA.

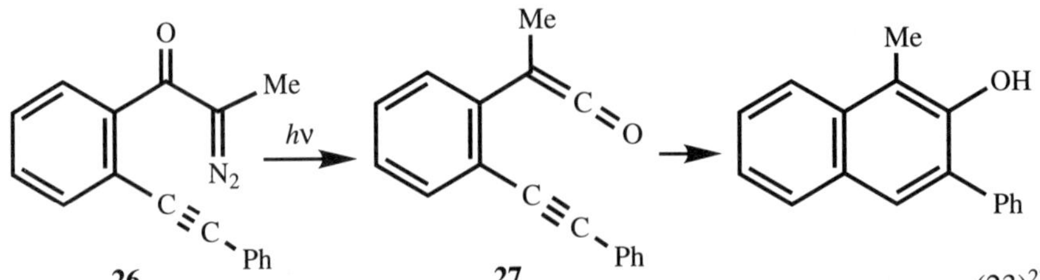

562 REACTIONS OF KETENES

Vinylketenes can undergo intramolecular ene reactions (equation 24),[38] (see also Sections 3.4.6 and 4.1.2). The reaction of equation 25 could involve a concerted process, or a zwitterionic intermediate **28**.[39]

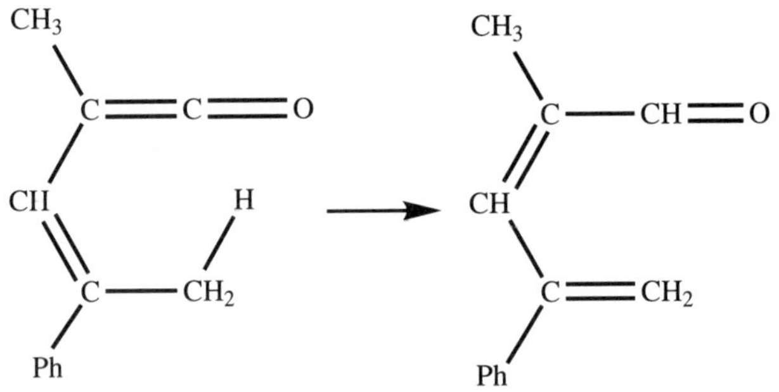

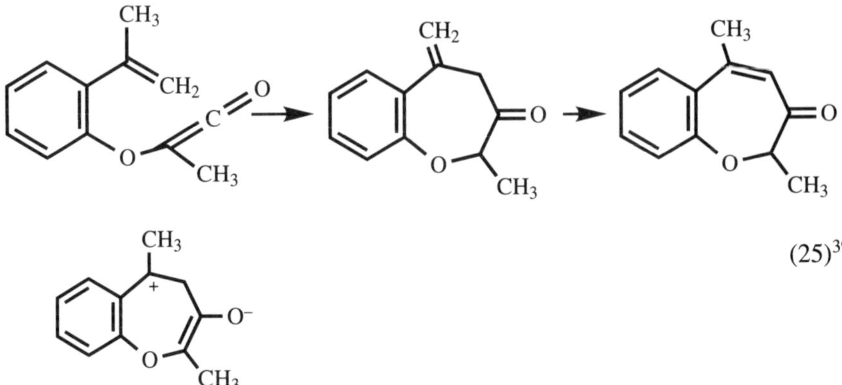

Pyrolysis of **29** was proposed to proceed via the methyleneketene **30** to the ketene **31** which cyclized to the observed product **32** (equation 26).[40]

References

1. Snider, B. B. *Chem. Rev.* **1988**, *88*, 793–811.
2. Snider, B. B. *Chemtracts-Organic Chem.* **1991**, *4*, 403–419.
3. Snider, B. B. In *Advances in Strain in Organic Chemistry*, Vol. 2; JAI Press: London, 1992; pp. 95–142.
4. Nguyen, M. T.; Ha, T.; More O'Ferrall, R. A. *J. Org. Chem.* **1990**, *55*, 3251–3256.
5. Kappe, C. O.; Kollenz, G.; Netsch, K.-P.; Leung-Toung, R.; Wentrup, C. *J. Chem. Soc., Chem. Commun.* **1992**, 488–490.
6. Becker, D.; Nagler, M.; Birnbaum, D. *J. Am. Chem. Soc.* **1972**, *94*, 4771–4773.
7. Becker, D.; Harel, Z.; Birnbaum, D. *J. Chem. Soc., Chem. Commun.* **1975**, 377–378.
8. Becker, D.; Birnbaum, D. *J. Org. Chem.* **1980**, *45*, 570–578.
9. Yates, P.; Fallis, A. G. *Tetrahedron Lett.* **1968**, 2493–2496.
10. Ireland, R. E.; Aristoff, P. A. *J. Org. Chem.* **1979**, *44*, 4323–4331.
10a. Andriamiadanarivo, R.; Pujol, B.; Chantegrel, B.; Deshayes, C.; Doutheau, A. *Tetrahedron Lett.* **1993**, *34*, 7923–7924.
11. Hart, H.; Love, G. M. *J. Am. Chem. Soc.* **1971**, *93*, 6266–6267.
12. Padwa, A.; Zhi, L.; Fryxell, G. E. *J. Org. Chem.* **1991**, *56*, 1077–1083.
12a. Kim, O. K.; Wulff, W. D.; Jiang, W.; Ball, R. G. *J. Org. Chem.* **1993**, *58*, 5571–5573.
13. Lee, S. Y.; Kulkarni, Y. S.; Burbaum, B. W.; Johnston, M. I.; Snider, B. B. *J. Org. Chem.* **1988**, *53*, 1848–1855.
14. Murray, R. K., Jr.; Goff, D. L.; Ford, T. M. *J. Org. Chem.* **1977**, *42*, 3870–3874.
15. Baeckström, P.; Li, L.; Polec, I.; Unelius, C. R.; Wimalasiri, W. R. *J. Org. Chem.* **1991**, *56*, 3358–3362.
16. Marko, I.; Ronsmans, B.; Hesbain-Frisque, A.; Dumas, S.; Ghosez, L.; Ernst, B.; Greuter, H. *J. Am. Chem. Soc.* **1985**, *107*, 2192–2194.
17. Corey, E. J.; Desai, M. C.; Engler, T. A. *J. Am. Chem. Soc.* **1985**, *107*, 4339–4341.
18. Corey, E. J.; Desai, M. C. *Tetrahedron Lett.* **1985**, *26*, 3535–3538.
19. Kulkarni, Y. S.; Niwa, M.; Ron, E.; Snider, B. B. *J. Org. Chem.* **1987**, *52*, 1568–1576.

20. Snider, B. B.; Allentoff, A. J.; Walner, M. B. *Tetrahedron* **1990**, *46*, 8031–8042.
21. Snider, B. B.; Hui, R. A. H. F.; Kulkarni, Y. S. *J. Am. Chem. Soc.* **1985**, *107*, 2194–2196.
22. Snider, B. B.; Beal, R. B. *J. Org. Chem.* **1988**, *53*, 4508–4515.
23. Snider, B. B.; Allentoff, A. J. *J. Org. Chem.* **1991**, *56*, 321–328.
24. Funk, R. L.; Novak, P. M.; Abelman, M. M. *Tetrahedron Lett.* **1988**, *29*, 1493–1496.
25. Corey, E. J.; Rao, K. S. *Tetrahedron Lett.* **1991**, *32*, 4623–4626.
26. Snider, B. B.; Kulkarni, Y. S. *J. Org. Chem.* **1987**, *52*, 307–310.
27. Masamune, S.; Fukumoto, K. *Tetrahedron Lett.* **1965**, 4647–4654.
28. Foland, L. D.; Karlsson, J. O.; Perri, S. T.; Schwabe, R.; Xu, S. L.; Patil, S.; Moore, H. W. *J. Am. Chem. Soc.* **1984**, *111*, 975–989.
29. Krysan, D. J.; Gurski, A.; Liebeskind, L. S. *J. Am. Chem. Soc.* **1992**, *114*, 1412–1418.
30. Barron, C. A.; Khan, N.; Sutherland, J. K. *J. Chem. Soc., Chem. Commun.* **1987**, 1728–1730.
31. Khodabocus, A.; Shing, T. K. M.; Sutherland, J. K.; Williams, J. G. *J. Chem. Soc., Chem. Commun.* **1989**, 783–784.
32. Danheiser, R. L.; Brisbois, R. G.; Kowalczyk, J. J.; Miller, R. F. *J. Am. Chem. Soc.* **1990**, *112*, 3093–3100.
33. Danheiser, R. L.; Cha, D. D. *Tetrahedron Lett.* **1990**, *31*, 1527–1530.
34. Liebeskind, L. S.; Zhang, J. *J. Org. Chem.* **1991**, *56*, 6379–6385.
35. Yerxa, B. R.; Moore, H. W. *Tetrahedron Lett.* **1992**, *33*, 7811–7814.
36. Danheiser, R. L.; Casebier, D. S.; Loebach, J. L. *Tetrahedron Lett.* **1992**, *33*, 1149–1152.
36a. Danheiser, R. L.; Helgason, A. L. *J. Am. Chem. Soc.* **1994**, *116*, 9471–9479.
37. Padwa, A.; Austin, D. J.; Chiacchio, U.; Kassir, J. M.; Rescifina, A.; Xu, S. L. *Tetrahedron Lett.* **1991**, *32*, 5923–5926.
37a. Nakatani, K.; Isoe, S.; Maekawa, S.; Saito, I. *Tetrahedron Lett.* **1994**, *35*, 605–608.
37b. Sullivan, R. W.; Coghlan, V. M.; Munk, S. A.; Reed, M. A.; Moore, H. W. *J. Org. Chem.* **1994**, *59*, 2276–2278.
38. Mayr, H. *Angew. Chem., Int. Ed. Engl.* **1975**, *14*, 500–501.
39. Kher, S.; Kulkarni, G. H.; Mitra, R. B. *Synth. Commun.* **1989**, *19*, 597–604.
40. Brown, R. F. C.; McMullen, G. L. *Aust. J. Chem.* **1974**, *27*, 2385–2391.

5.4.5 Intermolecular and Intramolecular Cycloaddition of Ketenes with Carbonyl Groups

The C=C bonds of ketenes also undergo [2 + 2] cycloadditions with carbonyl groups to form β-oxetanones (β-lactones, equation 1). This pathway is prominent in aldoketene dimerization (Section 5.4.1.1). Many examples of this reaction are reported in reviews,[1–5] and the process has the character of nucleophilic attack on C_α and electrophilic attack on C_β of the ketene.[6] Carbonyl groups with electron-withdrawing substituents are prone to this reaction, as in the case of quinones (equation 1) and dicyano ketone (equation 2).[1] Lewis acid catalysts promote this reaction.[1–5] Thiocarbonyl compounds also react with ketenes.[7]

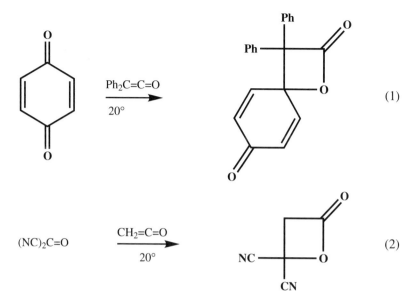

Strongly polarized carbonyl compounds and thiocarbonyl compounds react with ketenes to give β-lactones, which often decarboxylate to form alkenes with loss of CO_2.[1-17] The reactions can proceed by [2 + 2] cycloadditions of the ketene with the carbonyl compound involving either dominant nucleophilic attack at the LUMO on C_α of the ketene or electrophilic attack on the ketene HOMO at C_β followed by cyclization and decarboxylation (equations 3 and 4). A theoretical analysis and comparison of this reaction to alkene and methylenimine cycloadditions is given in Section 5.4.1.3.[18,19,19a] Some examples are shown in equations 5–12. Ketenes also react with tetramesityldisilene ($Mes_2Si=SiMes_2$) by [2 + 2] cycloaddition to the carbonyl group.[17a]

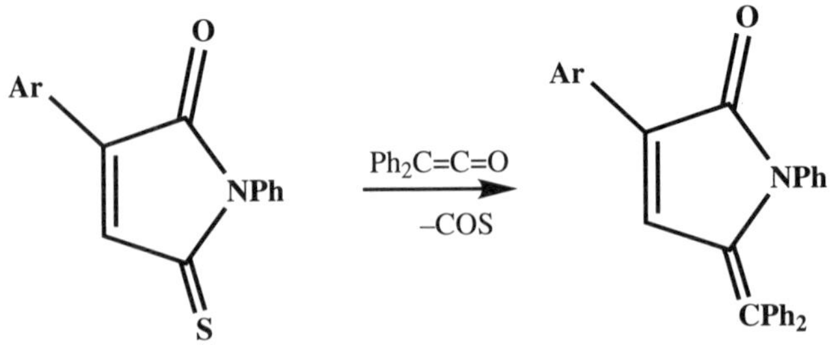

$$\text{Ar} \underset{S}{\overset{O}{\underset{\|}{\bigcirc}}} \text{NPh} \xrightarrow[-\text{COS}]{\text{Ph}_2\text{C=C=O}} \text{Ar} \underset{\text{CPh}_2}{\overset{O}{\underset{\|}{\bigcirc}}} \text{NPh} \qquad (5)^9$$

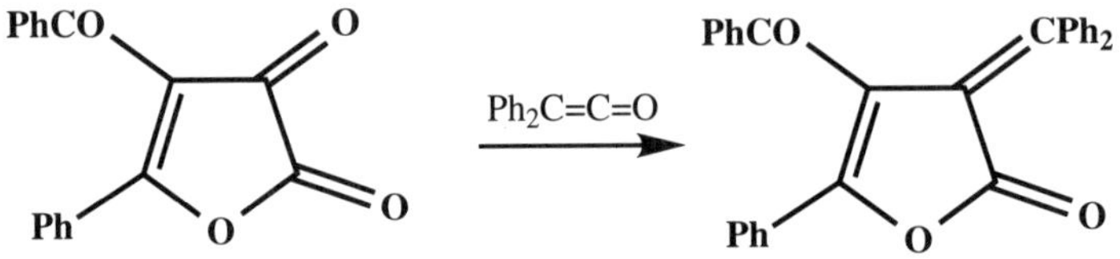

$$\text{PhCO} \cdots \text{O} \xrightarrow{\text{Ph}_2\text{C=C=O}} \text{PhCO} \cdots \text{CPh}_2 \qquad (6)^{10}$$

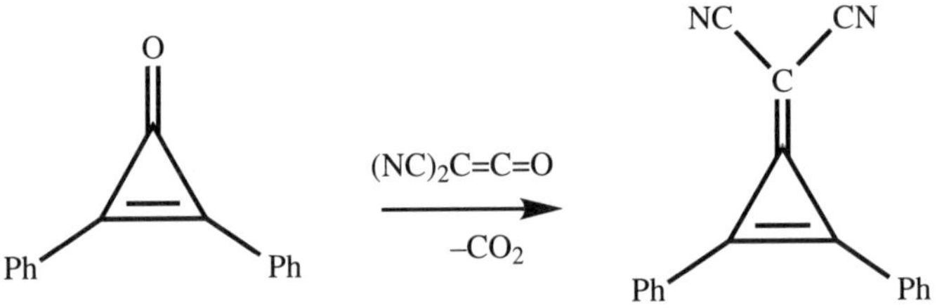

$$\text{Ph} \triangle \text{Ph} \xrightarrow[-\text{CO}_2]{(\text{NC})_2\text{C=C=O}} \text{Ph} \triangle \text{Ph} \qquad (7)^{11}$$

$$\text{(macrocycle with C=O)} \xrightarrow[-\text{CO}_2]{\text{Cl}_2\text{C=C=O}} \text{(macrocycle with =CCl}_2\text{)} \qquad (8)^{13}$$

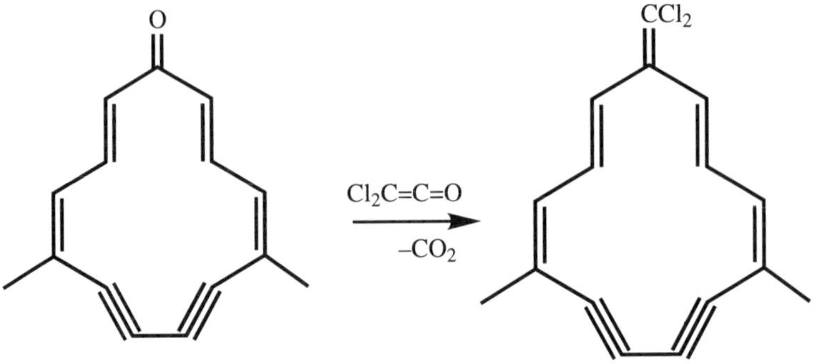

$$\underset{\text{Cl}}{\overset{\text{NC}}{>}}\text{C=C=O} \xrightarrow{\text{ArCH=O}} \underset{\text{Cl}}{\overset{\text{NC}}{>}}\text{C=C}\underset{\text{H}}{\overset{\text{Ar}}{<}} \qquad (9)^{14}$$

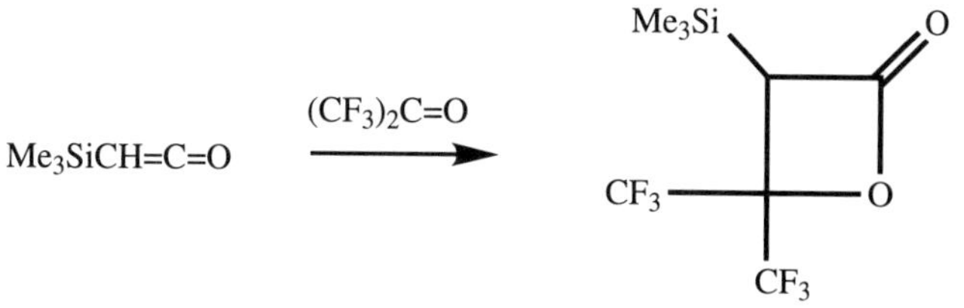

$Me_3SiCH=C=O$ $\xrightarrow{(CF_3)_2C=O}$ [β-lactone with Me₃Si, CF₃, CF₃ substituents] (10)[15]

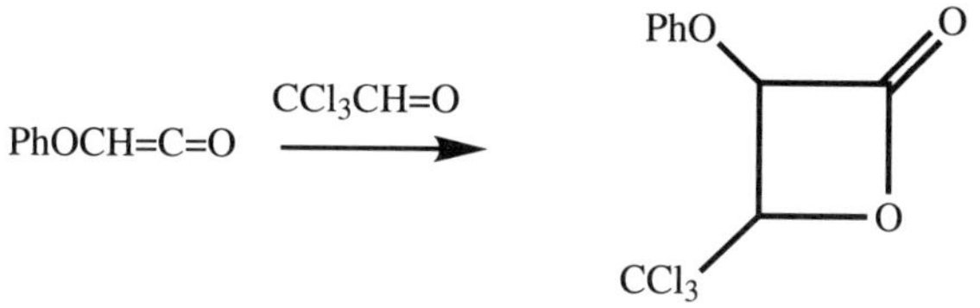

$PhOCH=C=O$ $\xrightarrow{CCl_3CH=O}$ [β-lactone with PhO, CCl₃ substituents] (11)[16]

$Ph_2C=C=O$ + Ar–CH=CH–C(Ph)=O (with O) $\xrightarrow{-CO_2}$ Ar–CH=CH–C(Ph)=CPh₂ (12)[17]

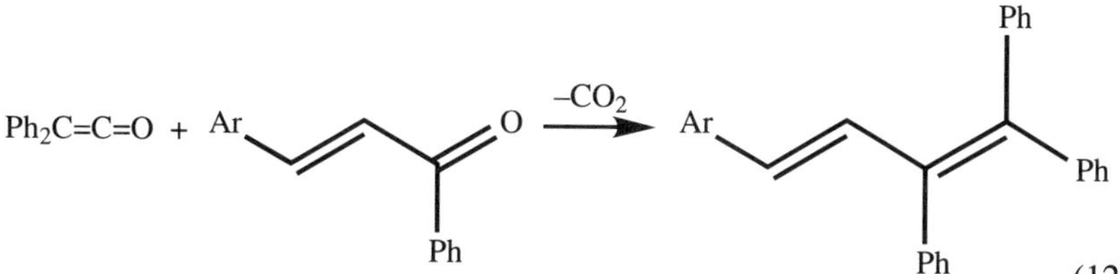

α-Ketenylcyclobutanones (**2**), formed by thermal or photochemical Wolff rearrangements from the corresponding diazoketones, undergo a cycloaddition process with formation of 5-spirocyclopropyl-$\Delta^{\alpha,\beta}$-butenolides **3** (equation 13).[20] When the reactions were carried out photochemically at λ > 330 nm in alcohols, capture of the ketenes occurred to yield cyclobutanone-substituted acetic esters (equation 14).[20] When the diazoketones were irradiated in methylcyclohexane between −40 and −50 °C the formation of the ketene band at 2120 cm^{-1} in the IR could be observed directly. On warming the kinetics of the rearrangement to the butenolides could be directly measured.[20] The transition state for the interconversion of **2** to **3** was studied by MNDO and found to involve a concerted process with almost complete formation of the new C—O bond from the cyclobutanone oxygen to the ketene carbon in the transition state.[20]

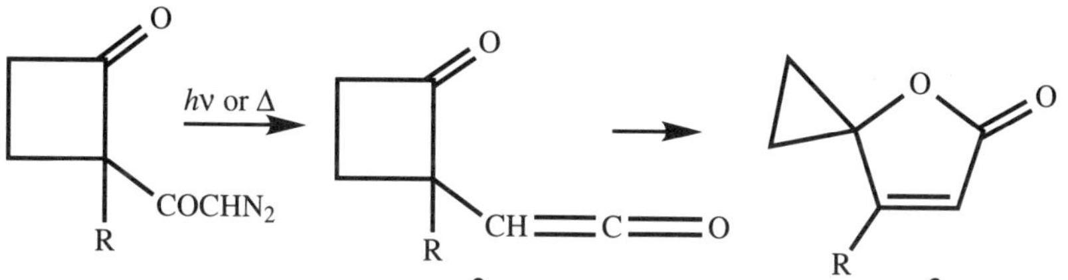

[cyclobutanone with COCHN₂ and R] $\xrightarrow{h\nu \text{ or } \Delta}$ [cyclobutanone with CH=C=O and R] (**2**) → [spirocyclopropyl butenolide with R] (**3**) (13)

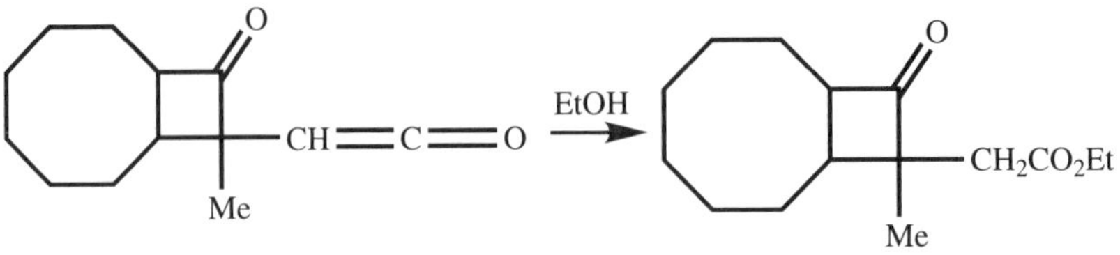

568 REACTIONS OF KETENES

$$\text{(14)}$$

The photolysis of tetrabenzoylethylene has been proposed to lead to ketene **4**, which also undergoes cyclization to a butenolide (equation 15).[21,22] Butenolide formation from cyclization of metal-complexed ketenes generated from alkenylcarbene metal complexes has also been proposed (equation 16).[23]

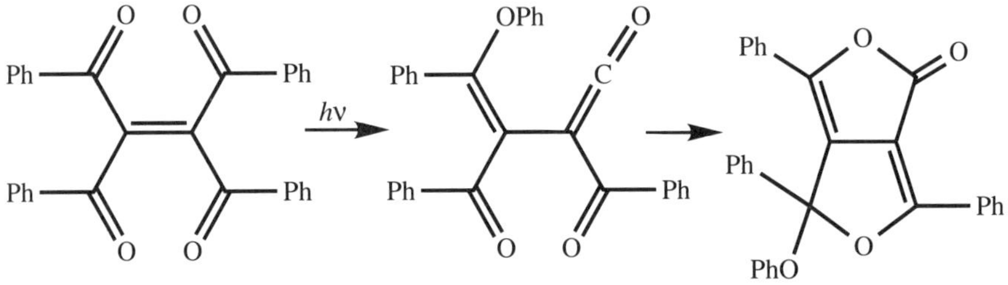

$$\text{(15)}$$

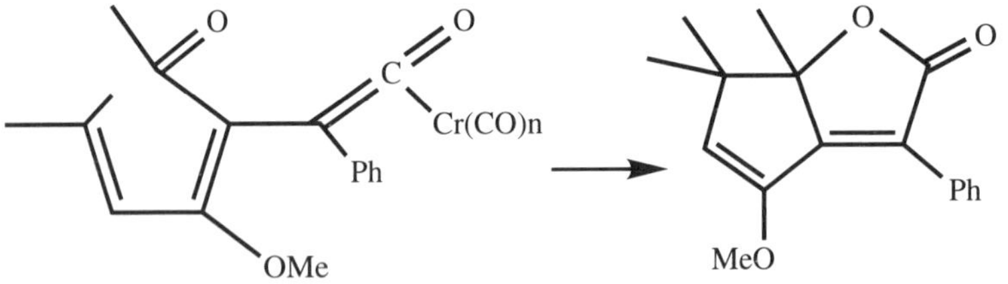

$$\text{(16)}$$

Reversible formation of an *N*-acylaminoketene occurs on heating 3-methyl-2,4-diphenyloxazolium-5-olate (**5**) (equation 3).[24] Ketene **6** can be trapped by a variety of reagents in [2 + 2] cycloadditions. This process is also considered in Section 4.2.

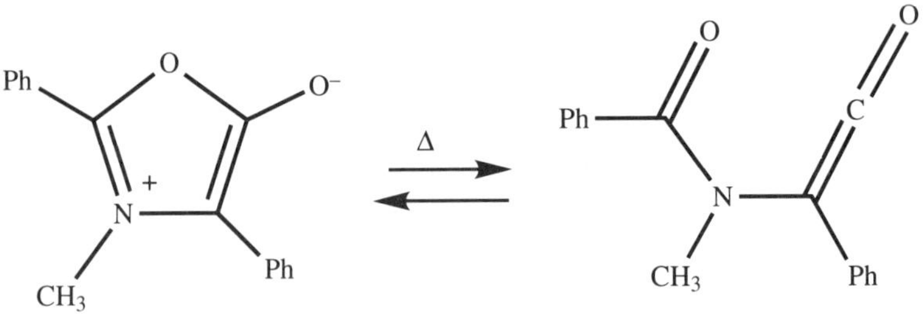

$$\text{(17)}$$

The ketene **7** undergoes thermal cyclization to cyclobutane-fused lactone **8** (equation 18).[25] Photochemical cyclization gives the opposite stereochemistry of the lactone fusion.[25] A bicyclo[2.1.1]hexyl analogue of **7** gave somewhat similar behavior.[26] Intramolecular [4 + 2] cycloadditions of imidoyl ketenes are also known (equation 19).[27] Other examples are given in Section 4.1.7.

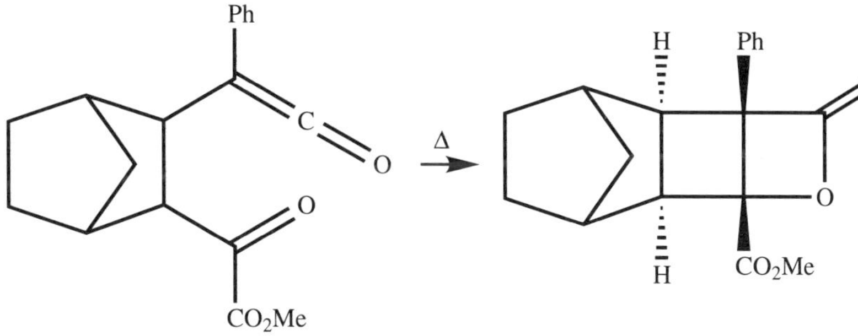

(18)

7   **8**

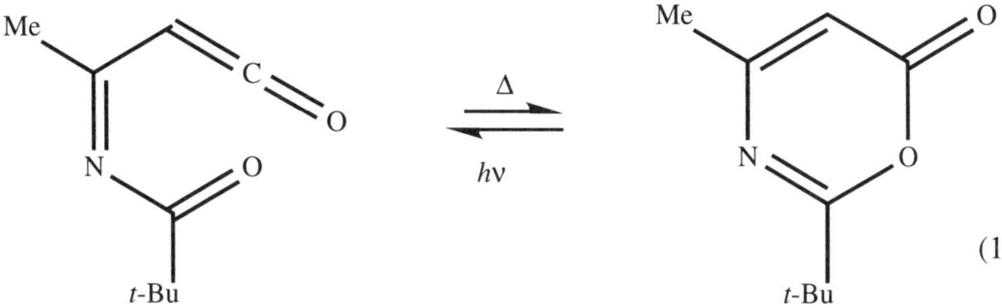

(19)

Intramolecular [2 + 2] cycloaddition of ketenes to carbonyl groups followed by loss of CO_2 was observed as in equation 20.[28,29]

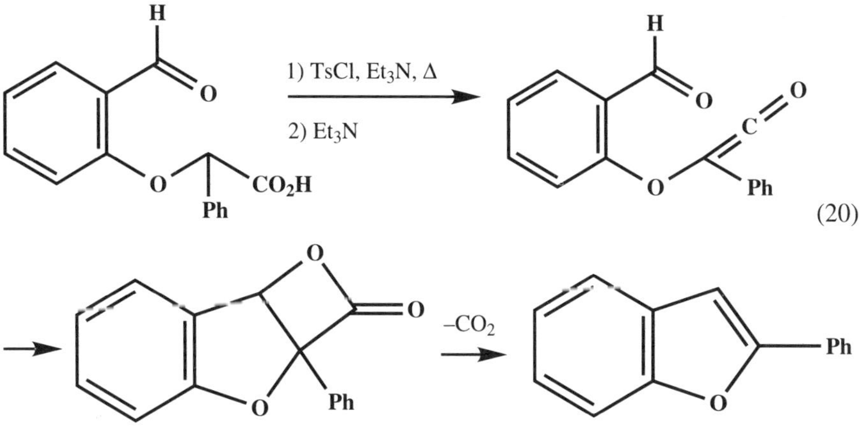

(20)

The ketene **9** generated by Wolff rearrangement undergoes efficient cycloaddition to the carbonyl oxygen (equation 21).[30] Further examples of the cyclization of 1,3-pentadiene-1,5-diones to α-pyrones are discussed in Section 4.1.2.

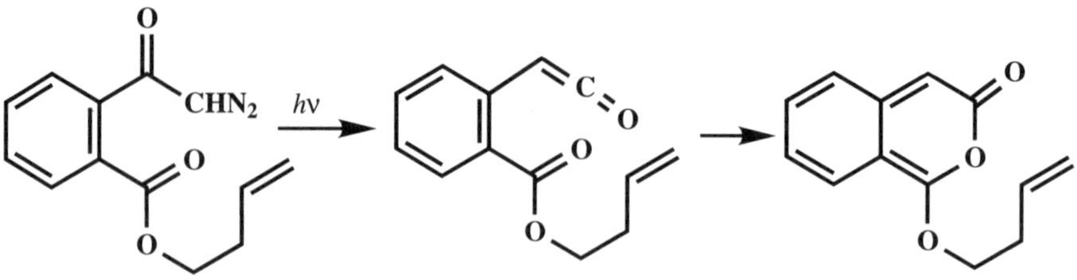

References

1. Ghosez, L.; Marchand-Brynaert, J. In *Comprehensive Organic Synthesis,* Vol. 5; Trost, B. M., Ed.; Pergamon: New York, 1991; Chapter 2.2.
2. Muller, L. L.; Hamer, J. *1,2-Cycloaddition Reactions. The Formation of Three- and Four-Membered Heterocycles;* Interscience: New York, 1967; Chapter 3.
3. Ghosez, L.; O'Donnell, M. J. In *Pericyclic Reactions;* Marchand, A. P.; Lehr, R. E., Eds.; Academic: New York, 1977; pp. 79–140.
4. Ulrich, H. *Cycloaddition Reactions of Heterocumulenes,* Academic: New York, 1967; pp. 38–109.
5. Pommier, A.; Pons, J.-M. *Synthesis* **1993**, 441–459.
6. Krabbenhoft, H. O. *J. Org. Chem.* **1978**, *43*, 1305–1311.
7. Kohn, H.; Charumilind, P.; Gopichand, Y. *J. Org. Chem.* **1978**, *43*, 4961–4965.
8. Grzegorzewska, U.; Leplawy, M.; Redlinski, A. *Rocz. Chem.* **1975**, *49*, 1859–1863; *Chem. Abstr.* **1976**, *84*, 135059f.
9. Augustin, A.; Köhler, M. *Z. Chem.* **1983**, *23*, 402–403.
10. Terpetschnig, E.; Penn, G.; Kollenz, G.; Peters, K.; Peters, E.; von Schnering, G. *Tetrahedron* **1991**, *47*, 3045–3058.
11. Neidlein, R.; Bernhard, E. *Angew. Chem., Int. Ed. Engl.* **1978**, *17*, 369–370.
12. Kuroda, S.; Ojima, J.; Kitatani, K.; Kirita, M.; Nakada, T. *J. Chem. Soc., Perkin Trans. 1,* **1983**, 2987–2995.
13. Ojima, J.; Itagawa, K.; Hamai, S.; Nakada, T.; Kuroda, S. *J. Chem. Soc., Perkin Trans. 1* **1983**, 2997–3004.
14. Moore, H. W.; Mercer, F.; Kunert, D.; Albaugh, P. *J. Am. Chem. Soc.* **1979**, *101*, 5435–5436.
15. Zaitseva, G. S.; Livantsova, L. I.; Bekker, R. A.; Baukov, Yu. I.; Lutsenko, I. F. *Zh. Obshch. Khim.* **1983**, *53*, 2068–2077; *Engl. Transl.* **1983**, *53*, 1867–1874.
16. Borrmann, D.; Wegler, R. *Chem. Ber.* **1966**, *99*, 1245–1251.
17. Regitz, M.; Eckes, H. *Tetrahedron* **1981**, *37*, 1039–1044.

17a. Fanta, A. D.; Belzner, J.; Powell, D. R.; West, R. *Organometallics* **1993**, *12*, 2177–2181.
18. Yamabe, S.; Minato, T.; Osamura, Y. *J. Chem. Soc., Chem. Commun.* **1993**, 450–452.
19. Fang, D. C.; Fu, X. Y. *Chin. Chem. Lett.* **1992**, *3*, 367–368; *Chem. Abstr.* **1993**, *118*, 59108z.
19a. Lecea, B.; Arrieta, A.; Roa, G.; Ugalde, J. M.; Cossío, F. P. *J. Am. Chem. Soc.* **1994**, *116*, 9613–9619.
20. Miller, R O.; Theis, W.; Helig, G.; Kirchmeyer, S. *J. Org. Chem.* **1991**, *56*, 1453–1463.
21. Rubin, M. B.; Sander, W. W. *Tetrahedron Lett.* **1987**, *28*, 5137–5140.
22. Cannon, J. R.; Patrick, V. A.; Raston, C. L.; White, A. H. *Aust. J. Chem.* **1978**, *31*, 1265–1283.
23. Brandvold, T. A.; Wulff, W. D.; Rheingold, A. L. *J. Am. Chem. Soc.* **1990**, *112*, 1645–1647.
24. Funke, E.; Huisgen, R. *Chem. Ber.* **1971**, *104*, 3222–3228.
25. Hegmann, J.; Christl, M.; Peters, K.; Peters, E.; von Schnering, H. G. *Tetrahedron Lett.* **1987**, *28*, 6429–6432.
26. Reuchlein, H.; Kraft, A.; Christl, M.; Peters, E.-M.; Peters, K.; von Schnering, H. G. *Chem. Ber.* **1991**, *124*, 1435–1444.
27. Maier, G.; Schafer, U. *Tetrahedron Lett.* **1977**, 1053–1056.
28. Brady, W. T.; Marchand, A. P.; Giang, Y. F.; Wu, A. *Synthesis* **1987**, 395–396.
29. Brady, W. T.; Gu, Y. *J. Heterocyclic Chem.* **1988**, *25*, 969–971.
30. Fairfax, D. J.; Austin, D. J.; Xu, S. L.; Padwa, A. *J. Chem. Soc., Perkin Trans. 1*, **1992**, 2837–2844.

5.5 NUCLEOPHILIC ADDITION TO KETENES

5.5.1 Mechanisms

The susceptibility to nucleophilic attack is one of the most characteristic properties of ketenes, and derives from the high positive charge present on the *sp*-hybridized C_α. The mechanisms of such additions have been discussed in detail in a previous review,[1] and this should be consulted for a more detailed discussion of earlier work. Only the most important and recent aspects are considered here.

The most common types of nucleophilic additions are those involving oxygen, nitrogen, and sulfur nucleophiles such as water, alcohols, amines, and thiols,[1] and the addition of carbon nucleophiles, especially organolithium species including lithium enolates.[2–7] Ketenes formed by Wolff rearrangements (Section 3.3) are usually trapped by H_2O, alcohols, or amines to give acids, esters, or amides, respectively. As discussed together below there is a strong mechanistic similarity between these two seemingly disparate classes of reactions. As considered in Section 5.4, cycloaddition reactions of ketenes are also commonly initiated by nucleophilic attack that bears a clear resemblance to the other nucleophilic additions.

5.5.1.1 Theoretical Studies. An introduction to theoretical studies of ketene reactions, including nucleophilic additions, is given in Section 1.1.2. The central feature which governs nucleophilic additions to ketenes is the large coefficient of the LUMO at C_α in the ketene plane which directs nucleophilic attack to occur at this carbon in the ketene plane. The presence of hydrogen or lithium cations can also provide electrophilic assistance by coordination to the ketene oxygen and thereby promote the addition.

Ab initio molecular orbital studies of the addition of water monomer and dimer[8–14] and of organolithium species such as methyllithium[15,16] and lithium enolates[3] give a good picture of how these additions occur. Shown in Figures 1 and 2 are the calculated energy profiles for these processes in the gas phase. In the case of H_2O dimer initial addition to the C=O bond giving an enediol or direct addition to the C=C bond giving acetic acid may be considered (Figure 1). Calculations by different groups are in reasonable agreement considering the different levels of theory used, but the highest level and most reliable calculations are those carried out by Skancke.[12] The first step in the addition to the C=O bond is the exothermic formation of a complex **1** involving coordination of a water oxygen to C_α of the ketene, and a hydrogen bonding interaction of a water proton to the carbonyl oxygen. The initial complex **2** for addition to the C=C bond involves a nonplanar arrangement of the two water molecules and the ketene skeleton, which at this level of theory is 1.8 kcal/mol above the complex for C=O addition. There is only a minor geometric change in the ketene structure in this step.

In the second step of the reaction there are major geometrical changes and significant barriers are surmounted to formation of the respective transition states, both of which are nonplanar. At the 6-31G*//6-31G* level the transition state for addition to the C=O bond lies 4.8 kcal/mol below that for addition to the C=C bond.[12] Then there is exothermic formation of the product enediol from addition to the C=O bond or acetic acid from addition to the C=C bond.

Energy calculations at the MP4 (SDTQ)/6-31G*//RHF/6-31G* with zero point energy corrections gave barriers of 19.9 and 21.8 kcal/mol for addition of H_2O dimer and 39.7 and 41.0 kcal/mol for addition of H_2O monomer to the C=O and C=C bonds, respectively.[12] The significant energy difference from the calculations at the 6-31G*//6-31G* level indicate that the levels of theory utilized so far are not sufficient to reliably describe this reaction. Geometry optimizations at even higher levels of theory are necessary before the preferred path for these gas phase reactions can be confidently assigned. Furthermore, it must be recognized that these calculations must be used with great care for interpreting the details of the reaction coordinate in aqueous solution, where many more H_2O molecules are involved. The latter situation is a much more complex problem that cannot now be reliably answered by molecular orbital calculations.

5.5 NUCLEOPHILIC ADDITION TO KETENES

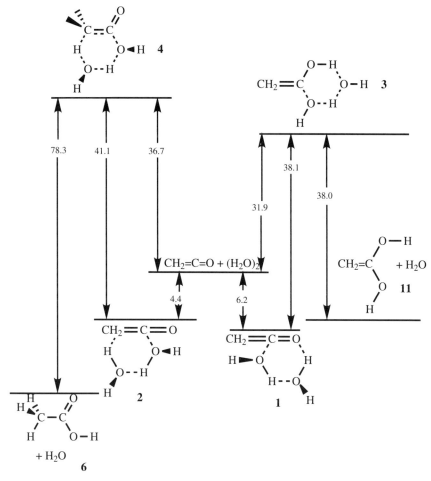

Figure 1. Calculated (6-31G*//6-31G*) energy profile (kcal/mol) for the addition of $(H_2O)_2$ to $CH_2=C=O$.[12]

The energetics of the calculated ketene hydration may be compared to the experimental $\Delta H^\ddagger$ and $\Delta S^\ddagger$ of the same reaction in H_2O at 25° of 10.3 kcal/mol and −16 eu, respectively.[17] Thus the calculated[12] ΔE of 19.9 kcal/mol for conversion of the ground state ketene and water dimer to the transition state (Figure 1) exceeds the experimental $\Delta H^\ddagger$ by 9.6 kcal/mol. If the assumption is made that the calculated energy difference is a reasonable approximation to the true gas phase barrier to hydration, then the additional solvation in water gives a differential transition state stabilization of about 10 kcal/mol. Thus the water dimer model for the hydration reaction (Figure 1)[9–14] is not to be taken as an accurate picture of ketene hydration in aqueous solution, as many other water molecules are demonstrably

574 REACTIONS OF KETENES

involved. However the picture of inplane nucleophilic attack at C_α with concomitant electrophilic assistance by hydrogen bonding on oxygen appears to be reliable.

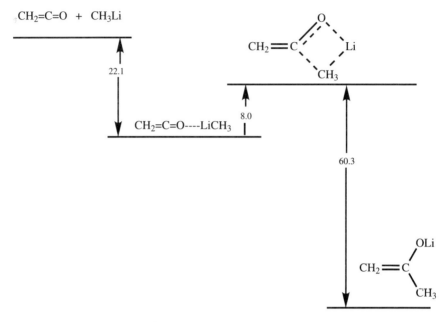

Figure 2. Calculated (3-21G//3-21G) energy profile (kcal/mol) for addition of CH_3Li to $CH_2{=}C{=}O$.[15]

Optimized structures have been calculated using the semiempirical AM1 method for the most stable conformers of the zwitterions derived from reaction of ketene with trimethylamine (**7**) and with pyridine (**8**), and for the complex **9** (Scheme I).[18] The calculated structures had similar dimensions to that determined for **9** by X ray, and it was concluded that the calculations not only reproduce the experimentally observed geometries and interactions, but can also explain the general enhanced reactivity of ketenes toward nucleophiles in the presence of tertiary amines.

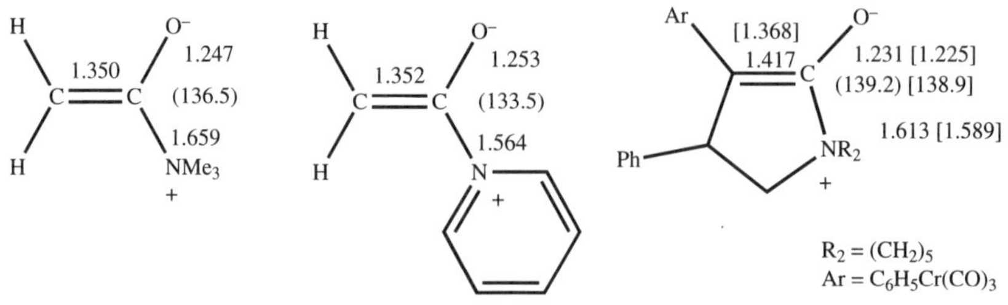

Scheme I. Bond distances (Å) and bond angles (deg, parentheses) for ketene complexes calculated by AM1.[18] Experimental (X-ray) values in brackets.

5.5 NUCLEOPHILIC ADDITION TO KETENES

Calculations at the 3-21G//3-21G level for addition for methyllithium to ketene show an initial complexation of the lithium to the carbonyl oxygen that is exothermic by 22.1 kcal/mol (Figure 2).[15,16] Then there is an 8.0 kcal/mol barrier to a planar transition state involving nucleophilic attack by carbon at C_α while maintaining lithium coordination to oxygen, followed by highly exothermic formation of the lithium enolate of acetone.

Calculations at the 3-21G//3-21G level for the addition of methyllithium to methylketene[15] give a transition state difference of 1.3 kcal/mol for *anti* addition leading to formation of the Z-product **10**, even though this is less stable than the isomer *E*-**10** by 3.1 kcal/mol (equation 1). This calculation of a kinetic preference for attack of the organolithium from the least-hindered side is in qualitative agreement with many experimental observations; for example, the 1.7/1 preference for formation of Z-**11** from reaction of methyllithium with methylethylketene (equation 2).[7]

$$CH_3Li + CH_3CH=C=O \longrightarrow \underset{Z\text{-}10}{\overset{CH_3}{\underset{H}{>}}C=C\overset{OLi}{\underset{CH_3}{<}}} + \underset{E\text{-}10}{\overset{CH_3}{\underset{H}{>}}C=C\overset{CH_3}{\underset{OLi}{<}}} \quad (1)$$

$$CH_3Li + CH_3C(C_2H_5)=C=O \longrightarrow \underset{Z\text{-}11}{\overset{C_2H_5}{\underset{CH_3}{>}}C=C\overset{OLi}{\underset{CH_3}{<}}} + \underset{E\text{-}11}{\overset{C_2H_5}{\underset{CH_3}{>}}C=C\overset{CH_3}{\underset{OLi}{<}}} \quad (2)$$

The addition of the lithium enolate of acetaldehyde to ketene was also studied theoretically at the 3-21G//3-21G level.[3] It was found that the energy profile first showed exothermic formation of a complex **12** with lithium coordinated to oxygen, but then there were separate pathways for *O*-acylation of the enolate by ketene via a 4-membered cyclic transition state **13** or *C*-acylation via a 6-membered cyclic transition state **14** to form **15** and **16**, respectively (Figure 3).[3] The barrier to the former process was lower but the product from the latter process was more favorable. This theoretical result agrees with experiment in that it was found that enolates undergo *O*-acylation by ketenes under kinetic control, whereas the equilibrium products are those of *C*-acylation.[2]

Thus *ab initio* calculations predict in-plane nucleophilic attack at C_α for ketenes, with assistance by electrophilic coordination at oxygen also occurring in the plane. As discussed below mechanistic studies are consistent with this interpretation. However the calculations also suggest that addition to the C=C bond, while apparently less favorable for $CH_2=C=O$ itself, is not prohibitively destabilized.

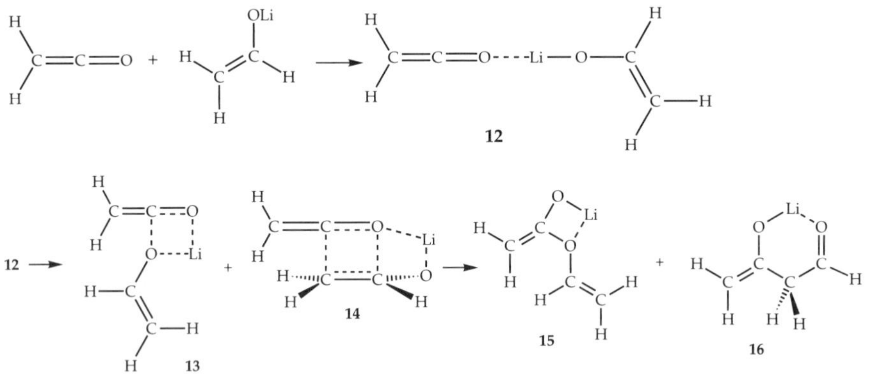

Figure 3. O- and C-Acylation of $CH_2=CHOLi$ y $CH_2=C=O$

5.5.1.2 Kinetics of Hydration: Neutral and Base Reactions.

Attempts to measure the kinetics of the neutral gas phase reaction of ketene with water were unsuccessful, as the reaction rate depended upon the surface area, and was concluded to occur on the surface of the reactor.[19] In the presence of acetic acid a homogeneous reaction that was first order in ketene, H_2O, and acetic acid was observed. Two mechanisms proposed as consistent with the kinetics are equation 3, a one-step process involving conversion of the three reactants to two molecules of acetic acid, and a two-step process with acetic acid and ketene first giving acetic anhydride (equation 4) which reacts further with H_2O to give acetic acid (equation 5). No definitive choice between these two mechanisms could be made.[19]

$$CH_2=C=O + H_2O + CH_3CO_2H \longrightarrow 2CH_3CO_2H \qquad (3)$$

$$CH_2=C=O + CH_3CO_2H \longrightarrow (CH_3CO)_2O \qquad (4)$$

$$(CH_3CO)_2 + H_2O \longrightarrow 2CH_3CO_2H \qquad (5)$$

The first studies of the kinetics of hydration of ketenes in pure water involved arylketenes 4-$RC_6H_4CH=C=O$ (**3**) generated by Wolff rearrangement.[20] It was found that the 4-NO_2 derivative was 13 times more reactive than the 4-CH_3 derivative, thus indicating a rather polar transition state. Together with studies of ketene itself[17] these were interpreted as involving water addition to the carbonyl group giving an enediol (ketene hydrate) which then was converted to the carboxylic acid. Enediols have been directly observed as reasonably long-lived intermediates in the hydrations of $Ar_2C=C=O$ (Ar = 2,4,6-$Me_3C_6H_2$ and 2,3,4,5,6-Me_5C_6).[13] The hydration reactivities of a number of other ketenes in highly aqueous solvents have now been measured and are presented in Table 1.[17,20-36] In some cases rates measured in mixed aqueous solvents have been extrapolated to give estimates of hydration rates in pure water. Empirical correlations of log k_{obsd} with [H_2O] have been found to be linear for many ketenes above 10.1 M (20% H_2O), and these correlations give estimated rates in pure H_2O that agree with available measured rates within a factor of 3.[22,23] These data show a variation in reactivity of 10^9 in the neutral reaction, and the importance of both acid- and base-catalyzed processes in certain cases. Major roles for conjugative, electronic, and steric factors in the hydration of ketenes are demonstrated and the results (Table 1) lead to the overall mechanism for neutral hydrolysis as depicted in Scheme II,[1,21-23] with a polar transition state **17**, and possible zwitterionic and enediol intermediates **18** and **19** leading to the acid product.

TABLE 1. Hydration Rates of Ketenes at 25 °C

Ketene	Solvent	k_{H^+} (M^{-1}s^{-1})	k_{H_2O} (s^{-1})	k_{OH^-} (M^{-1}s^{-1})	$\frac{k_{H_2O}}{k_{D_2O}}$	$\frac{k_{H^+}}{k_{H_2O}}$ (M^{-1})	$\frac{k_{OH^-}}{k_{H_2O}}$ (M^{-1})	Ref.
CH$_2$=C=O	H$_2$O	1.01×10^4	44, 36.5		1.59	276		17,30
n-BuCH=C=O	H$_2$O	3.98×10^3	99.4	3.29×10^4		40	330	24
4-RC$_6$H$_4$CH=C=O								
R = NO$_2$	H$_2$O		49.5×10^3					20
CN	H$_2$O		25.6×10^3					20
Cl	H$_2$O		9.6×10^3					20
F	H$_2$O		5.9×10^3					20
H	H$_2$O		4.77×10^3	1.22×10^6	1.44		260	20,24
OCH$_3$	H$_2$O		4.5×10^3					20
CH$_3$	H$_2$O		3.8×10^3					20
Me$_3$SiCH=C=O	H$_2$O		0.254			1.7×10^5	2.4×10^4	27
CF$_3$CH=C=O	H$_2$O		6×10^2					29
c-PrCH=C=O	H$_2$O		304	2.48×10^5			816	29
PhC≡CCH=C=O	H$_2$O	1.61×10^5	7.16×10^4			2.2		29
E-PhCH=CHCH=C=O	H$_2$O	2.09×10^4	5.76×10^3	2.31×10^6		3.6	401	29
CH$_3$SCPh=C=O	H$_2$O		1.33×10^2	3.09×10^2			232	32
HOCPh=C=O	H$_2$O		6.7×10^5		1.3			35
PhCOCPh=C=O	H$_2$O		6.29×10^3					29
t-BuCH=C=O	H$_2$O	4.77×10^3	14.6, (37.9)^a		2.01	327		22,30
Et$_2$C=C=O	H$_2$O		8.26^a					22

Compound	Solvent					Ref
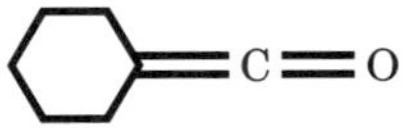 cyclohexylidene ketene	H_2O	2.78×10^3				30
$t\text{-}Bu_2C{=}C{=}O$	H_2O	434				22
$Ph_2C{=}C{=}O$	H_2O				220	24
$MeCPh{=}C{=}O$	H_2O		6.11×10^4	2.04	121	30
$i\text{-}PrCPh{=}C{=}O$	H_2O			1.74	2.8×10^6	23
$t\text{-}BuCPh{=}C{=}O$	H_2O			1.55		23
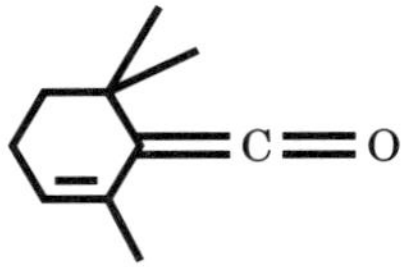 2,6,6-trimethyl-cyclohexenylidene ketene	H_2O	7.03	2.97×10^{-4}	2.32	2.4×10^4	23
	H_2O		0.124	1.48	350	25
	H_2O		2.35×10^{-4}	1.37		25
$t\text{-}BuC(CO_2Et){=}C{=}O$	20% CH_3CN		0.0780			22
$t\text{-}BuC(COBu\text{-}t){=}C{=}O$	20% CH_3CN	135			1.7×10^3	13
$t\text{-}BuCH{=}C{=}O$	20% CH_3CN		4.17			13
$Ph_2C{=}C{=}O$	50% CH_3CN		0.1625			13
$Mes_2C{=}C{=}O$	50% CH_3CN		4.26×10^{-2}			13
$(Me_6C_5)_2C{=}C{=}O$	50% CH_3CN		0.357	1.59	145^b	22
$Et_2C{=}C{=}O$	50% CH_3CN		0.406		790^b	4
$c\text{-}Pr_2C{=}C{=}O$	50% CH_3CN		0.0424^a	58.4	1400	23
$i\text{-}PrCPh{=}C{=}O$	50% CH_3CN	0.120	0.00997	46.7	4700	23
$t\text{-}BuCPh{=}C{=}O$	50% CH_3CN	0.488		1.67	2.8	23
$t\text{-}Bu_2C{=}C{=}O$	50% CH_3CN	3.6			49	22
$Me_3SiCH{=}C{=}O$	50% CH_3CN	542	3.23×10^{-3}	70	1.7×10^5	27
					2.1×10^4	

TABLE 1. (Continued)

Ketene	Solvent	k_{H^+} (M^{-1}s^{-1})	k_{H_2O} (s^{-1})	k_{OH^-} (M^{-1}s^{-1})	$\frac{k_{H_2O}}{k_{D_2O}}$	$\frac{k_{H^+}}{k_{H_2O}}$ (M^{-1})	$\frac{k_{OH^-}}{k_{H_2O}}$ (M^{-1})	Ref.
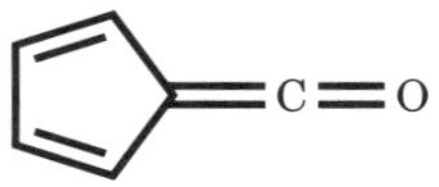 cyclopentadienylidene ketene	H$_2$O		9.0×10^5	4.8×10^7			53	33
indenylidene ketene	H$_2$O		1.61×10^7 1.1×10^{6c}	3.8×10^{7c}	1.32		36	34a 34b

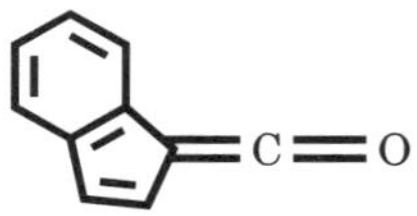

^{a}Extrapolated.
^{b}9.0 °C.
^{c}5 °C.

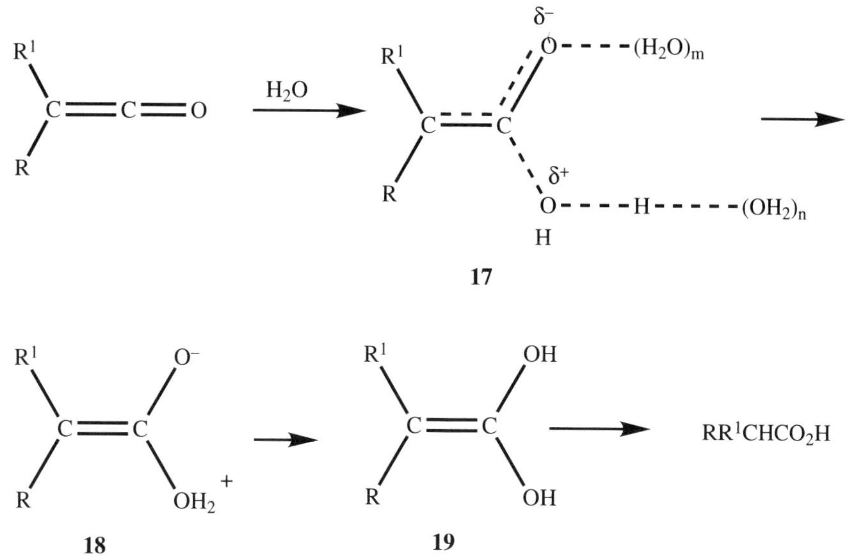

Scheme II. Hydration mechanism for ketenes in neutral solution.

Attack occurs in the ketene plane on the LUMO at C_α, and for the ketenes such as $CH_2=C=O$, n-BuCH=C=O, and t-BuCH=C=O in which one side is unhindered the relative reactivities are 1.0, 2.2, and 0.4, respectively.[21,30] These results show only minor effects on the reactivity for an alkyl group located on the opposite side of the ketene from the attacking nucleophile. The modest difference between the n-Bu and t-Bu groups may arise from some hindrance to solvation of the polar transition state by the bulkier group. However in t-Bu$_2$C=C=O in which there is a large *tert*-butyl group on each side of the ketene the relative reactivity is reduced to 4×10^{-6} relative to ketene.[21,22] The conjugating substituents Ph, E-PhCH=CH, and PhC≡C produce large accelerations of the rate, with reactivities relative to H of 10^2, 10^3, and 10^4, respectively.[29] These major effects due to conjugating substituents show the importance of enolate character in the transition state. There is evidence that these substituents also stabilize the ketene ground states, but the transition state effects are more important. The nonconjugating but electron-withdrawing CF_3 also produces acceleration by a factor of 10,[29] but this may arise partly from ground state effects.

As noted above, the fact that ketene is experimentally much more reactive in neutral hydration than is suggested by the MO calculations shows that many water molecules are involved in the hydration of ketenes, and these are depicted in the transition state **17**. The solvent isotope effects k_{H_2O}/k_{D_2O} for neutral hydration of 1.3–1.8 are in the range of those of reactions which have sometimes been interpreted as involving breaking of O—H bonds in the transition state,[22–24] and this may be represented as in **17** by water molecules acting as general acids and bases to assist the reaction. An enediol (ketene hydrate) has been detected as the initial product in several cases, and this is also shown in Scheme II.[13,30,33–35,57]

The decreased reactivities of the ketenes PhCH=C=O (R = Ph, i-Pr, t-Bu) relative to PhCR=C=O provide further evidence for the importance of steric effects due to bulky groups which inhibit the approach of water in the ketene plane. The groups may also twist the phenyl partly out of conjugation.[22,23]

The reactivities of many ketenes in the hydroxide ion induced hydration have also been measured. While in some cases these rates could not be determined because of difficulties in making the measurements, the observed values of k_{OH^-}/k_{H_2O} show an interesting constancy with values of 100–330 M^{-1} in the cases observed in pure H_2O. This reaction evidently closely resembles the neutral hydration and involves hydroxide attack in the plane leading a transition state with enolate character, which is rapidly converted to a carboxylate anion (equation 6). The magnitude of the k_{OH^-}/k_{H_2O} ratio is rather lower than for many other carbonyl group hydrations, but this has not yet been explained.

$$R^1R'C=C=O \xrightarrow{OH^-} \left[\begin{array}{c} R^1 \\ \diagdown \\ R \end{array} C=C \begin{array}{c} O^- \\ \diagup \\ OH \end{array} \right] \longrightarrow RR^1CHCO_2^- \quad (6)$$

In summary the evidence for the mechanism of in-plane nucleophilic addition to ketenes with nucleophilic attack on C_α with electrophilic participation at oxygen includes the following:

1. Theoretical calculations for addition of H_2O dimer, LiH, LiCH$_3$ or LiOCH=CH$_2$ favoring this pathway.
2. Measurement of the electronic effect of substituents at C_β favoring a polar transition state with negative charge development at this carbon.
3. Acceleration of the rate of hydration with increasing H_2O content in mixed solvents indicating a polar transition state.
4. Demonstration of the unique steric effect of substituents at C_β, in that one bulky group at this position has little effect on the reactivity, whereas a second bulky group causes a drastic decrease.
5. The rather low value and relative constancy of k_{OH^-}/k_{H_2O} ratios, indicating that both processes proceed by a similar mechanism.
6. The strong acceleration of the reaction by conjugating substituents at C_β, indicating negative charge development at this position.
7. Direct observation of enediol intermediates from ketene hydration in certain cases.

These mechanistic principles were originally described quite cogently in an early review by Lacey.[37] Unfortunately this was ignored in some later work,[38–45] leading to some confusion regarding the mechanism.

Other studies of the addition of water and alcohols to ketenes have also been reported.[13,38–45] Based on these data an alternative mechanism for the nucleophilic addition of water, alcohols, and amines to ketenes has been proposed by Satchell and co-workers,[38–45] and involves concerted additions to the C—C bonds by dimers or trimers of the nucleophiles perpendicular to the ketene plane, as depicted in structure **20**. The experimental conditions for these studies have usually involved low concentrations of nucleophiles in solvents such as benzene, in which the state of aggregation of H_2O has been a further reaction variable.

(7)

20

The evidence presented for this proposal has been discussed in detail elsewhere and a number of criticisms have been presented.[1] These include the fact that the proposed direction of attack is not on the LUMO of the ketene which is in the ketene plane, but rather involves the HOMO. Secondly the dependence of the rate of nucleophilic addition upon the concentration of water has been ascribed to the molecularity of water dimers and trimers with no consideration of the solvent polarity. However, as discussed above, the transition states for ketenes are highly polar. The most detailed theoretical studies of nucleophilic addition uniformly favor that addition occurs to the carbonyl group in the plane of the ketene. The mechanism of equation 7 also does not explain the steric and polar electronic substituent effects on the rate of reaction, as described above. Although calculations suggest that for the gas phase reaction of ketene with water dimer that the addition to the C=C double bond has a barrier only 1.9 kcal/mol higher than for addition to the carbonyl there is no positive evidence for such cyclic transition states in solution and no basis to suppose they play a significant role in nucleophilic additions to ketenes.

The reactivity in H_2O of pentafulvenone (**21**), generated by photochemical Wolff rearrangement could also be measured.[33] The formation of the ketene hydrate **22** was observed, and this clearly involves addition to the C=O bond (equation 8). The reaction of **21** with OH^- and CN^- was also observed, with rate constants $k_{OH^-} = 4.8 \times 10^7$ M^{-1} s^{-1} and $k_{KCN} = 1.6 \times 10^8$ M^{-1} s^{-1}.

(8)

21   **22**

The benzo analogue **23** was generated in CH_3CN and the reaction with added H_2O and MeOH up to 1.8 M was observed (equation 9).[34] These rates were fit better to second-order behavior in [ROH], and were interpreted as involving addition to the C=O bond by a solvent dimer through a similar transition state to that calculated for the addition of H_2O dimer to ketene. The values of k_2 were 5.7×10^5 and 2.5×10^6 M^{-2} s^{-1} for H_2O and CH_3OH, respectively. The reaction of **23** with pyridine was also observed,[34] and rate constants for the hydration of **23** and some sulfonated analogues have been measured.[34a-c] Relative rates obtained by competition experiments for the addition of nucleophiles to the 4-t-BuC$_6$H$_4$O$_3$S substituted derivative of **23** were 2.90 (MeOH), 2.08 (EtOH), 2.07 (n-PrOH), 2.07 (n-BuOH), 1.07 (i-PrOH), 1.00 (H$_2$O), and 0.66 (2-butanol).[36]

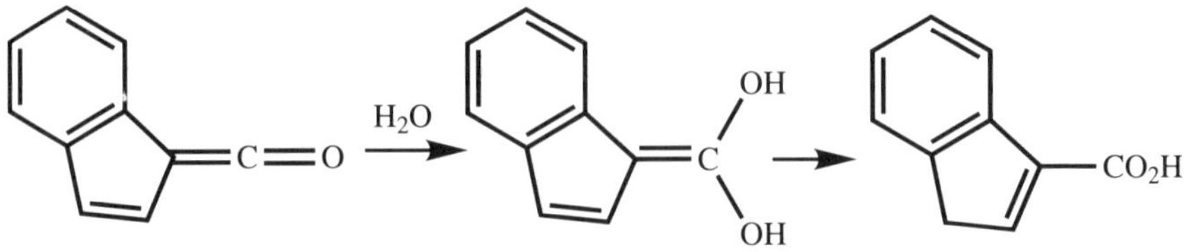

$$\text{23} \xrightarrow{\text{H}_2\text{O}} \text{[indene=C(OH)$_2$]} \longrightarrow \text{[indene-CO}_2\text{H]} \tag{9}$$

Thus, the pentafulvenones **21** and **23** are among the most reactive ketenes known in the hydration reaction, even their ground states are calculated to be stabilized by aromaticity effects (Section 4.1.10). The very fast reactions of **21** and **23** with nucleophiles may be attributed to a high degree of polar aromatic character in the transition states, as shown in **24,** which is even more stabilizing than the ground state effects.

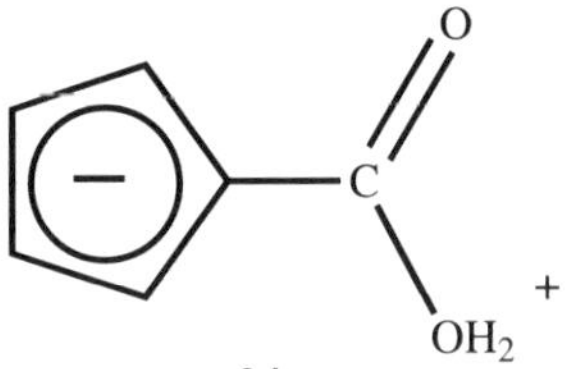

24

The stereochemistry of the addition of H$_2$O or alcohols to ketenes to form carboxylic acids or esters has been observed in a number of cases involving cycloalkylidiene ketenes generated by Wolff rearrangement,[46-50] as in the examples of equations 10 and 11. In the case of equation 10 the formation of the less stable product is consistent with the process already discussed of formation of an enolate or enol intermediate which is protonated from the less hindered side.[46] In the case of an acyl derivative, formation of the more stable isomer was attributed to equilibration of the initial product.[50]

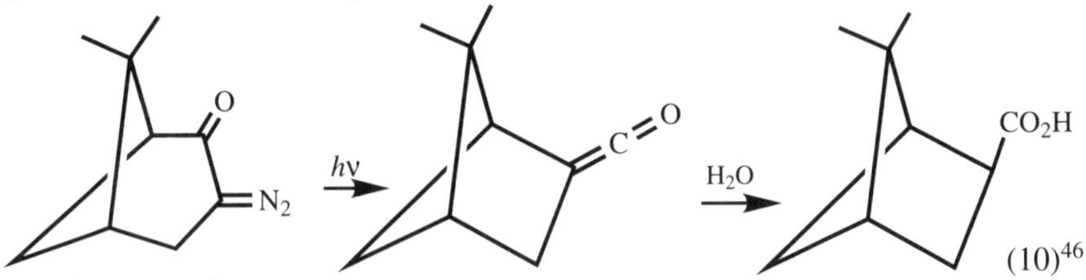

$$\tag{10}^{46}$$

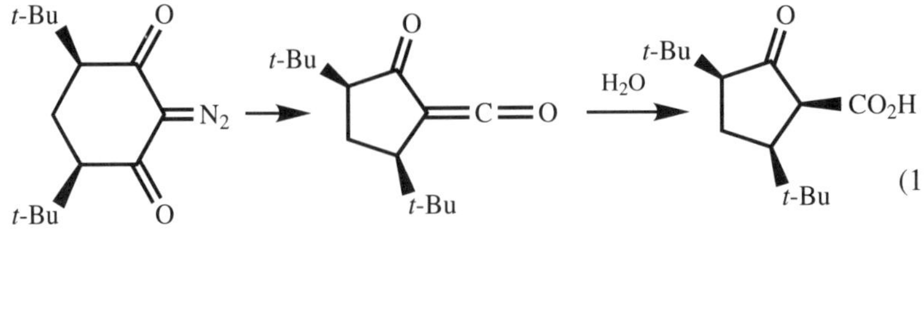

$$\tag{11}^{50}$$

Recently reversibility has been demonstrated in the hydration of arylketenes by carrying out the reaction in ^{18}O-labeled H_2O and observing ^{18}O incorporation in the recovered ketene (for Ar = 2,4,6-triisopropylphenyl) and the incorporation of two ^{18}O atoms in some of the product diarylacetic acid (equation 12).[50a] These reactions were interpreted as proceeding through initial formation of an enediol **25** which could revert to ^{18}O labeled ketene, which on rehydration gives doubly labeled enediol and then doubly labeled acid.[50a] As shown in Figure 1, Section 5.5.1.1, a calculated energy of the enediol $CH_2=C(OH)_2$ is only 6.1 kcal/mol below that of $CH_2=C=O$ plus water so that it is reasonable that appropriate substituents could tip the balance in favor of reformation of the ketene from the enediol.

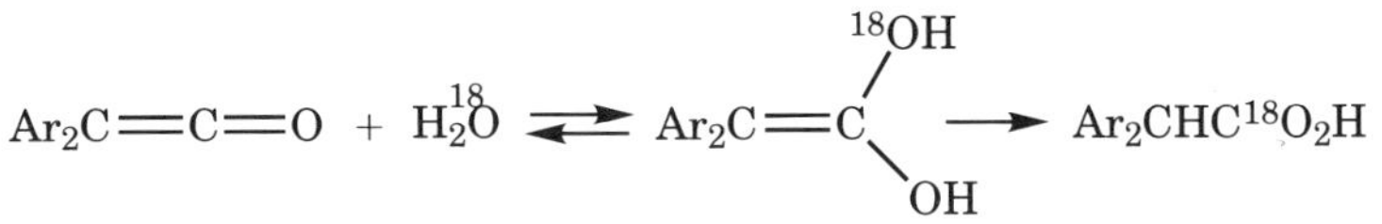

$$Ar_2C=C=O + H_2^{18}O \rightleftarrows Ar_2C=C\begin{smallmatrix}^{18}OH\\OH\end{smallmatrix} \longrightarrow Ar_2CHC^{18}O_2H$$

(12)

5.5.1.3 Acid-Catalyzed Hydration.
The acid-catalyzed reaction shows quite different behavior compared to the neutral and base hydrolyses, with small effects due to steric crowding and inhibition of the reaction by conjugating groups. This reaction is indicated to occur by rate-determining attack of a proton at C_β perpendicular to the ketene plane leading to an acylium ion (Scheme III). This pathway is supported by theoretical studies[51] and by gas phase results (Sections 2.5 and 5.6.1) demonstrating the formation of such acylium ions. Retarding effects are observed on the reactivity by conjugating groups such as phenyl which parallel the effects observed for protonation of ordinary alkenes.[23,24]

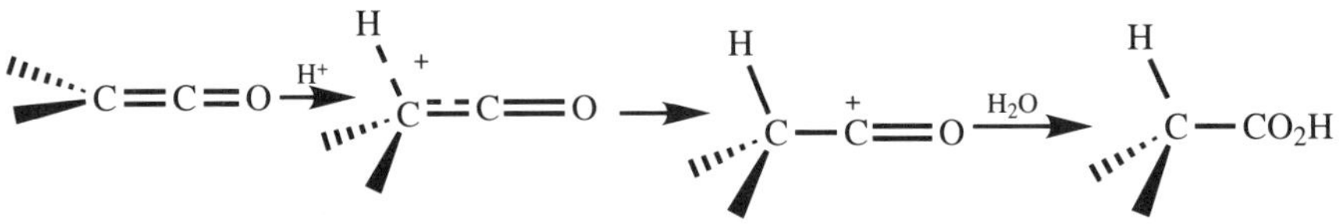

Scheme III. Acid-catalyzed hydration of ketenes.

Some rate constants for acid-catalyzed hydration of different ketenes are given in Table 1, Section 5.5.1.2. For many aryl ketenes whose reactivity in neutral H_2O were measured, acid catalysis could not be detected. This phenomenon apparently arises from experimental limitations in cases in which the neutral hydrations of the ketenes are quite rapid so that high acid concentrations are required to detect the acid-catalyzed hydration. However these high acid concentrations are incompatible with the experimental techniques, such as conductivity used to measure the hydration rates, or when the diazo ketones used as the ketene precursors are not stable at the high acidities.

For several ketenes for which acid-catalyzed hydration was detected, the measurements were made in mixed aqueous solvents either for solubility reasons or because the ketenes were too reactive in pure water for kinetics measurements. Thus these rate constants cannot be compared directly to those of other ketenes measured in pure H_2O.

The reactivities of different ketenes can be compared by the ratio k_{H^+}/k_{H_2O} (M^{-1}). Thus for n-BuCH=C=O, which can be taken as a reference structure, this ratio has a value of 40 (Table 1, Section 5.5.1.2). For the ketene Et_2C=C=O this ratio is somewhat larger (145) and for t-Bu_2C=C=O the ratio increases to 2.8×10^6. If the ratios $k(n$-BuCH=C=O$)/k(t$-Bu_2C=C=O) are compared directly the ratio for k_{H^+} is 9.2, while for k_{H_2O} the ratio is 6.3×10^5. Thus for t-Bu_2C=C=O there is a modestly reduced value of k_{H^+}, while for k_{H_2O} the reduction is much greater. These trends are consistent with protonation at C_β perpendicular to the ketene plane, as steric effects are modest for this reaction, but a major steric effect occurs for k_{H_2O}, where nucleophilic attack occurs in the ketene plane.

Acid catalysis was observed[23] for the hydration of the aryl ketenes PhC(Bu-t)=C=O and PhC(Pr-i)=C=O substituted with bulky alkyl groups, but not[24] for PhCH=C=O and Ph_2C=C=O. The failure to detect acid catalysis for the latter two substrates was attributed to retardation of the protonation at C_β bearing the aryl group(s).[24,30] Examples of such retardation in other substrates were given. For PhC(Bu-t)=C=O and PhC(Pr-i)=C=O protonation at the β-carbon was also retarded by the aryl group, but such retardation of protonation by the β-alkyl groups according to Scheme III is small. It was argued that these β-alkyl groups mainly retard the attack of H_2O in the plane as in Scheme II, and because of the increased k_{H^+}/k_{H_2O} ratio the acid-catalyzed process becomes detectable.

Protonation of the vinyl ketene **25** in aqueous acid was found to occur at C_δ of the vinyl group (equation 12).[23] This process gives the resonance stabilized allyl cation **26,** but interestingly the resulting conjugated acid **27** is not the most stable isomer. Molecular orbital calculations of the site of protonation of both ketene[9] and vinylketene[51] supported the paths of Scheme III and equation 12 as the most favorable ones.

(12)

The ketene $Me_3SiCH=C=O$ is much less reactive towards water than is n-$BuCH=C=O$ because of its greater ground state stability, but the ratio k_{H^+}/k_{H_2O} is greater by a factor of 4×10^3. This was ascribed[27] to the β-silicon effect stabilizing the acylium ion resulting from protonation at C_β (equation 13).

$$Me_3SiCH=C=O \xrightarrow{H^+} Me_3SiCH_2\overset{+}{C}=O \qquad (13)$$

In summary the available evidence is strongly supportive that acid hydration of nonconjugated ketenes occurs by protonation at C_β, and there is general agreement with this mechanism.[43]

5.5.1.4 Alcoholysis and Aminolysis.
There have been many kinetic studies of the reactions of ketenes with alcohols.[39–42,52–56] The kinetics of the reaction of 3-pentanol with a variety of ketenes in n-hexane at 25 °C gave the reactivity shown in Table 1.[52,52a]

TABLE 1. Reactivity of Ketenes with 3-Pentanol in Hexane, 25 °C

Ketene	$k_2 \times 10^4$ (M^{-1} s^{-1})	Ketene	$k_2 \times 10^4$ (M^{-1} s^{-1})
$Ph_2C=C=O$	74 (hexane)	$Me_2C=C=O$	36
	3.3 (ether)	$Et_2C=C=O$	17.6
	1.7 (CH_3CN)		
		n-BuCEt=C=O	4.3

The results provide evidence for accelerating effects due to aryl groups that is attributable to delocalization in an enolate-like transition state, and for steric retardation of the reaction with increasing crowding. The decreasing rates in more polar solvents show that polar solvation of the reactants is more important than in the transition state. The authors suggested that the reaction proceeded through a cyclic 4-membered transition state involving addition to the C=C bond,[52] but in view of the results for hydration, addition to the C=O bond appears more likely.

The rates of addition of MeOH to the quinomethane ketene **28** show both steric and electronic effects on the reactivity.[56] These results were interpreted in terms of in-plane attack on C_α of the ketene, which was sterically hindered by substituents at C_1.[56] The pseudo first-order rate constants k_1 for reaction of **28** with MeOH (equation 14) are compared in Table 2 to the bimolecular rate constants for reaction with maleic anhydride (equation 15). It is notable that there is a large decelerating effect of substituents at C_1 on the reaction with CH_3OH, but a much smaller effect on the rate of the Diels–Alder reaction with maleic anhydride. It was concluded that the two reactions follow a different course, with MeOH reacting by in-plane attack on C_α, which is subject to steric interference by the substituents at C_1, whereas the attack on the ketene in the cycloaddition process occurs above the plane and is not subject to steric retardation.

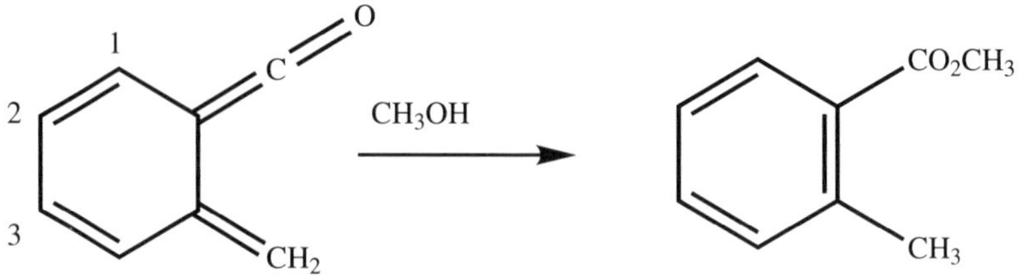

The stereoisomeric ketenes **29** and **30** have been detected on photolysis of the corresponding aldehyde (equation 16).[57] The latter ketene was estimated to react 10 times as fast as **29** with water, and was suggested to lead to the ketene hydrate **31**.[57]

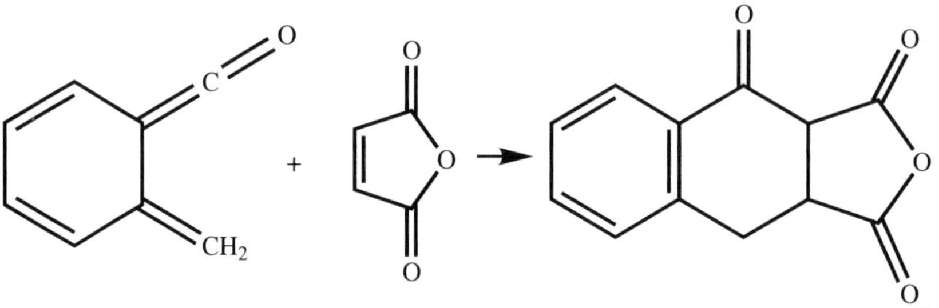

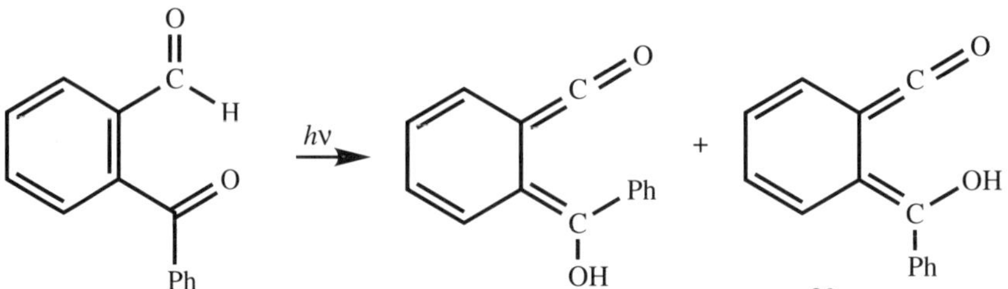

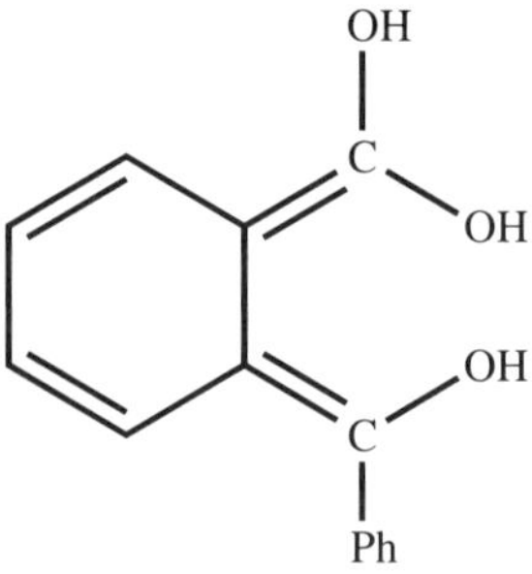

TABLE 2. Effects of Substituents in 28 on Reactivity with MeOH (Equation 14) and Maleic Anhydride (Equation 15)[56]

C_1	C_2	C_3	C_4	k_{MeOH} (s^{-1})	k_2 (Maleic anhydride) (×10^{-3})
H	H	H	H	172	14
H	H	CH$_3$	H	118	12
H	H	CH$_3$O	H	137	30
CH$_3$	H	CH$_3$	H	≤0.1	17
CH$_3$	H	CH$_3$O	H	≤1	26
CH$_3$O	H	CH$_3$	H	≤0.1	6
H	CH$_3$	CH$_3$	CH$_3$	67	52
CH$_3$	CH$_3$	CH$_3$	CH$_3$	≤10	46
CH$_3$O	CH$_3$	CH$_3$	CH$_3$O	≤0.1	5

Catalysis of the addition of EtOH to diphenylketene catalyzed by macrocyclic "concave" pyridines such as **32** has been examined.[58] These extremely hindered bases were interpreted as enhancing the reactions by general-base catalysis as shown in equation 17. The addition of alcohol to Me$_3$SiCH=C=O, including enzyme catalyzed alcoholysis, is considered in Section 4.5.

$$Ph_2C=C=O \xrightarrow[\text{EtOH}]{B} Ph_2C\cdots\overset{\overset{\delta-}{O}}{\underset{\underset{Et}{O\text{---}H\text{---}B}}{C}}{}^{\delta+} \longrightarrow Ph_2CHCO_2Et \quad (17)$$

[Structure **32**: pyridine with R substituent at 4-position, with macrocyclic bridge NCH$_2$(CH$_2$OCH$_2$)$_2$CH$_2$N and CO(CH$_2$)$_7$OC linker]

The effects of 30 different bases, mostly amines, on the kinetics of the reaction of Ph$_2$C=C=O in aqueous solution have been studied.[31] The rate accelerations due to the added bases were attributed to direct nucleophilic attack of the bases, as illustrated in equation 18 for reaction with ammonia, in competition with uncatalyzed reaction of Ph$_2$C=C=O with H$_2$O to give Ph$_2$CHCO$_2$H.[31] The reactions in the presence of NH$_3$ and morpholine gave quantitative rate/product correlations for the competitive formation of amides and acid.[31]

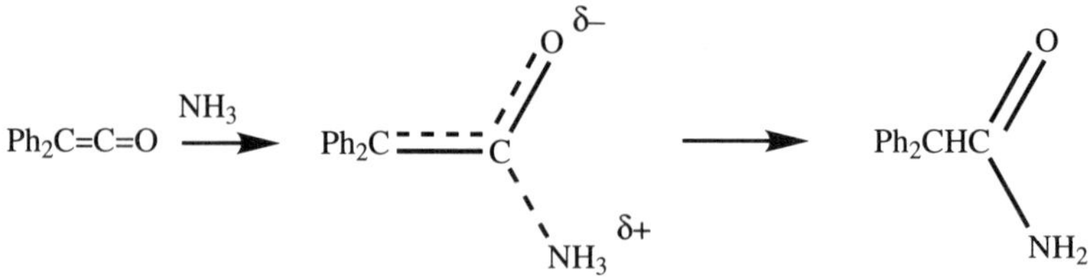

5.5.2 Nucleophilic Additions to Ketenes: Preparative Aspects

5.5.2.1 Hydride Addition.
Ketenes are reduced by $LiAlH_4$ to enolates which have been trapped as the enol esters by CH_3COCl (equation 1)[59] or as silyl enol ethers by Me_3SiCl (equation 2).[60] In the latter investigation there was a 10:1 preference for formation of the evidently more crowded Z isomer due to attack of the hydride in the plane from the less-hindered side. Reduction of dimesitylketene followed by hydrolysis gave the stable enol (equation 3).[61] Reduction of ketene **25** (Section 5.5.1.3) with $LiAlH_4$ followed by silylation of the resulting enolate gave the isomeric dienyl ethers in a 2:1 ratio.[62] A multistep sequence shown in equations 4 and 5 involved conjugate addition of n-BuLi to an aryl naphthalene carboxylate to give an enolate which formed ketene **1** by phenolate expulsion (Section 3.2.3.1).[63,64] The ketene was reduced in situ by $LiBEt_3H$ to a new enolate which was alkylated with MeI to give an aldehyde which was reduced with $NaBH_4$.[64]

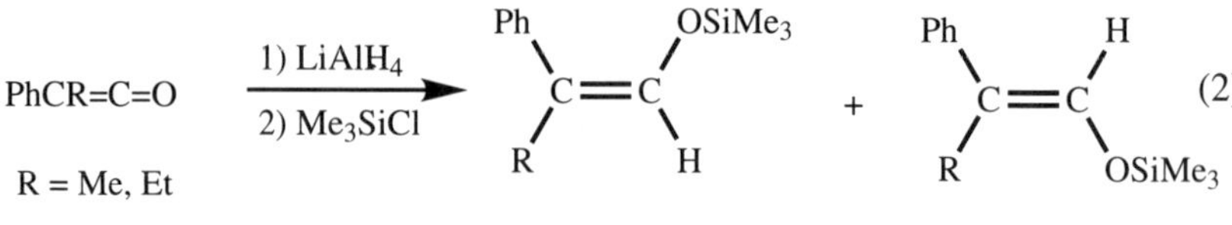

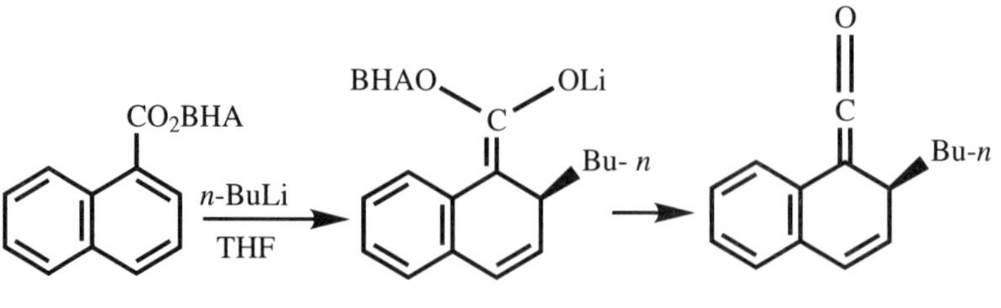

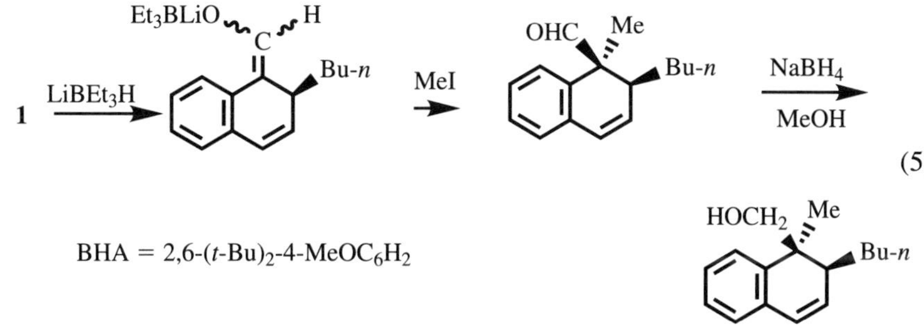

5.5 NUCLEOPHILIC ADDITION TO KETENES

(5)

BHA = 2,6-(t-Bu)$_2$-4-MeOC$_6$H$_2$

The reduction of ketenes by silanes,[65] germanes,[66] and stannanes[67] gives silyl, germyl, and stannyl enol ethers respectively (equations 6–8). The gas phase reaction of BH$_3$ with ketene is noted in Section 5.6.3.

$$PhCR=C=O + R^1_3SiH \xrightarrow[140°C]{H_2PtCl_6} PhCR=CHOSiR^1_3 \quad (6)^{65}$$

R = Et, Ph, n-Pr, n-Bu

$$Ph_2C=C=O + R^1_3GeH \longrightarrow Ph_2C=CHOGeR^1_3 \quad (7)^{66}$$

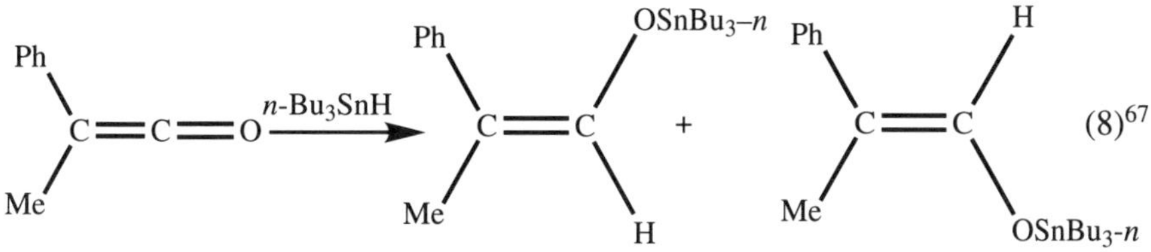

(8)[67]

There was a modest preference for the Z isomer in equation 8 under both kinetic and thermodynamic conditions, and the results were interpreted in terms of a concerted addition of metal hydride to the carbonyl group with little steric preference.[67]

Crowded ketenes frequently undergo reduction rather than addition on reaction with bulky Grignard or organolithium reagents with β-hydrogens.[4,6,68,69] Examples are shown in equations 9 and 10 including the formation of stable enols when very bulky aryl groups are present.

$$t\text{-BuCR=C=O} \xrightarrow[2)\ H_2O]{1)\ t\text{-BuLi}} t\text{-BuCHRCBu-}t + t\text{-BuRCHCH=O} \quad (9)^{4,6}$$
$$\hspace{6cm} \overset{O}{\|}$$

R = i-Pr, t-Bu

$$ArAr^1C=C=O \xrightarrow[2)\ H_2O]{1)\ RMgBr} ArAr^1C=CHOH \quad (10)^{68,69}$$

Ar = Ph, Mes; Ar^1 = Mes; 3-BrMes; 2,3,5,6-Me$_4$C$_6$H; R = t-Bu, i-Pr

Thermal intramolecular hydride attack occurs by rearrangement in a vinyl ketene (equation 11).[70] Other examples of this reaction are known (Section 3.4.6).

$$\text{(11)}$$

The reversal of hydride addition to ketene has been proposed in the reduction of nonenolizable aldehydes by the lithium enolate of acetaldehyde in THF at 25 °C (equation 12).[71] After quenching of the reaction with H_2O, substantial amounts of acetic acid were found in the aqueous phase, and this was taken as evidence of ketene formation. However ketene is known to react with $CH_2=CHOLi$,[2] and so further verification of this mechanism is required.

$$PhCH=O + CH_2=CHOLi \longrightarrow PhCH_2OLi + CH_2=C=O \quad (12)$$

5.5.2.2 Oxygen Nucleophiles.

The mechanism of the hydration of ketenes to carboxylic acids is discussed in Section 5.5.1.2. The preparation of esters, by the reaction of ketenes with alcohols (equation 1), has been the subject of kinetic studies[52-58] and presumably follows a pathway similar to that of hydration (Section 5.5.1.4). These reactions are of major preparative use in the Wolff rearrangement (Section 3.3). In some cases esterification of alcohols by acyl halides and base proceeds by the E1cB mechanism (Section 3.2.3.1). Of particular interest is the generation of optically active esters by the reaction of unsymmetrical ketenes with optically active alcohols or achiral alcohols in the presence of chiral catalysts, as discussed in Section 5.9. The reaction of vinylketene with alcohols is useful in the preparation of esters (equation 2).[72]

$$RR^1C=C=O \xrightarrow{ROH} RR^1CHCO_2R \quad (1)$$

$$CH_3CH=CHCOCl \xrightarrow{Et_3N} CH_2=CHCH=C=O \xrightarrow{ROH} CH_2=CHCH_2CO_2R \quad (2)^{72}$$

Hydroxyketenes that undergo preparatively useful cyclization to lactones have been generated by thermolysis of dioxinones (equation 3),[73] by photolysis of cyclobutanones (equation 4),[74-76] by thermolysis of alkynyl ethers (Section 3.4.5),[77] and by photolysis of carbene–chromium complexes (equation 5).[78] Thermolysis of cyclobutenones with hydroxyalkyl sidechains also gave intermediate (hydroxyalkyl) ketenes which cyclized to lactones (equation 6).[79] The formation of dimeric and trimeric lactones in these reactions has also been noted.[79a]

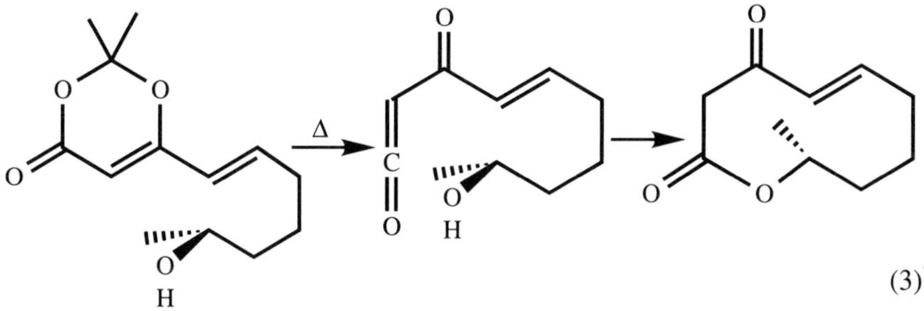

5.5 NUCLEOPHILIC ADDITION TO KETENES

(3)[73]

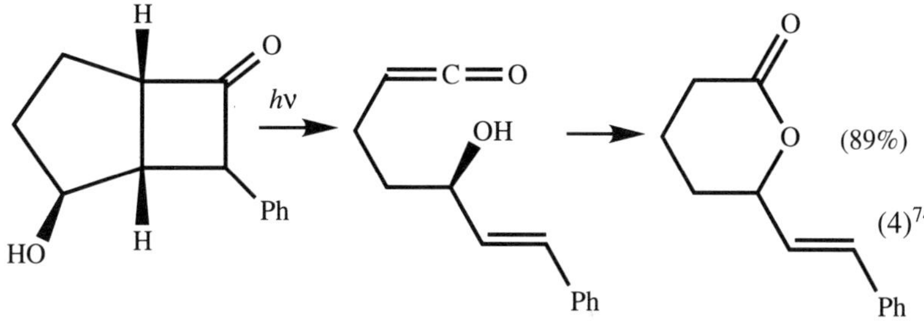

(4)[74–76] (89%)

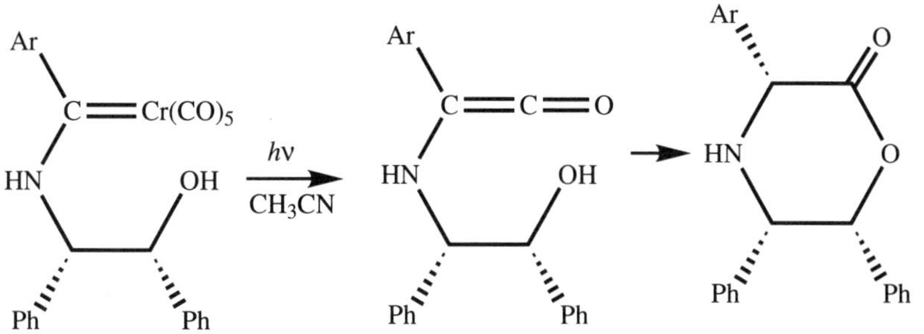

(5)

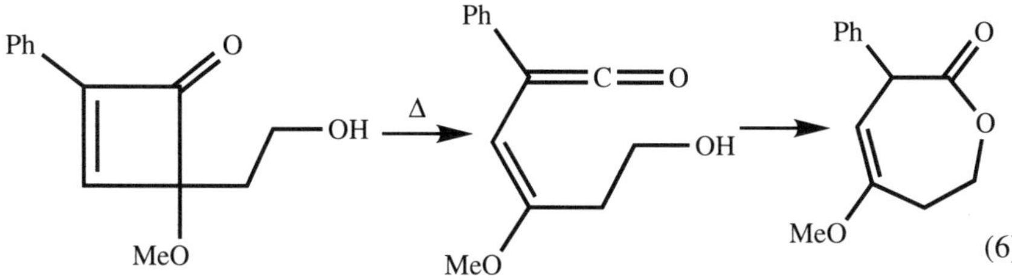

(6)

Enolates react with ketenes with a kinetic preference for nucleophilic attack by the enolate oxygen, but the thermodynamic products involve *C*-acylation of the enolate (equations 7 and 8).[2,3,80] The further reaction of **2** with ketene can lead to polyacetals (see Section 5.8).

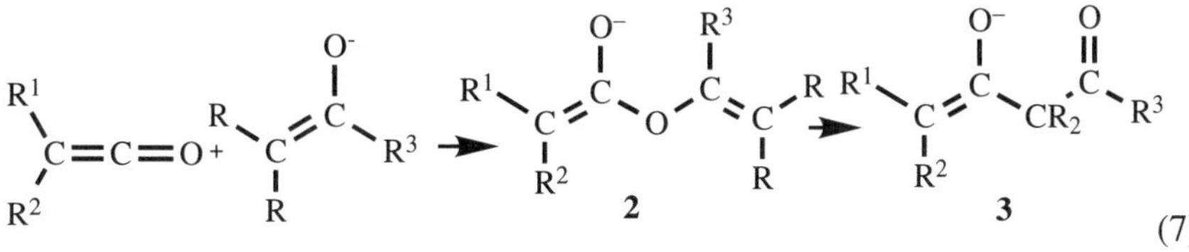

(7)

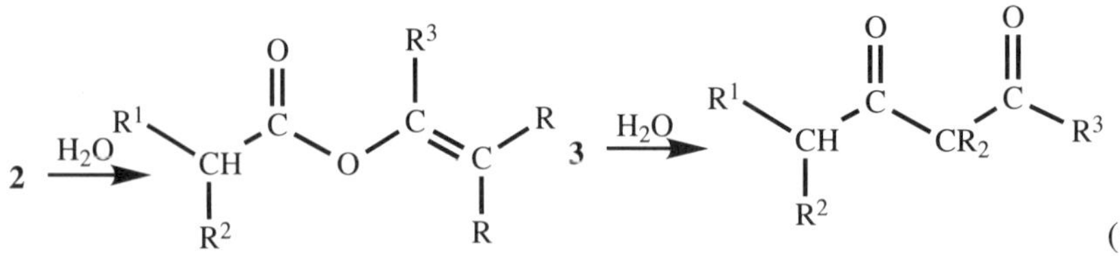

The addition of LiOMe to the highly hindered ketene **4** provides a route to the corresponding enolate (equation 9).[81,82]

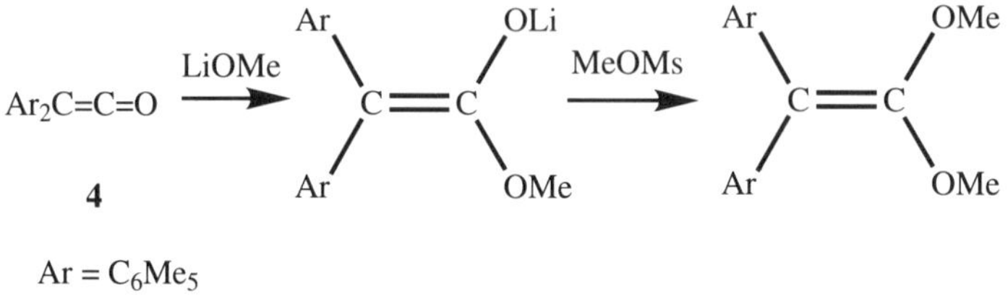

Ar = C_6Me_5

The reaction of formylketene (**5**), prepared from formyl Meldrum's acid (**6**), with alcohols gives a general synthesis of formylacetic esters **7** (equation 10).[83] Alcohol addition to **6** to give **7** without the intervention of **5** was not however excluded.[83]

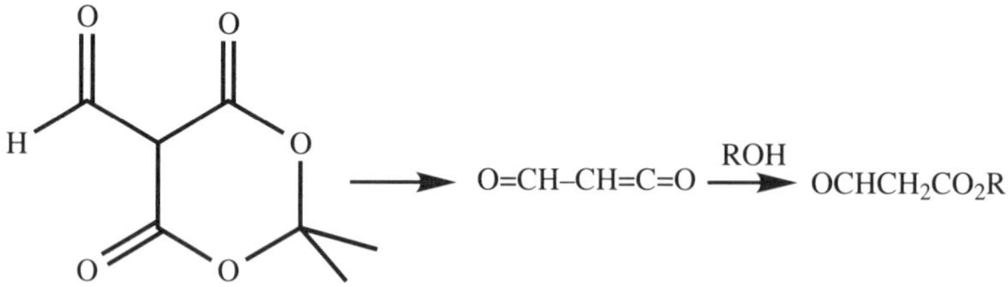

Addition of methanol to ketene **9**, generated photochemically from **8**, led to the aldehyde ester **10**, proposed to form from conjugate addition of the ester enolate to the unsaturated aldehyde (equation 11).[84]

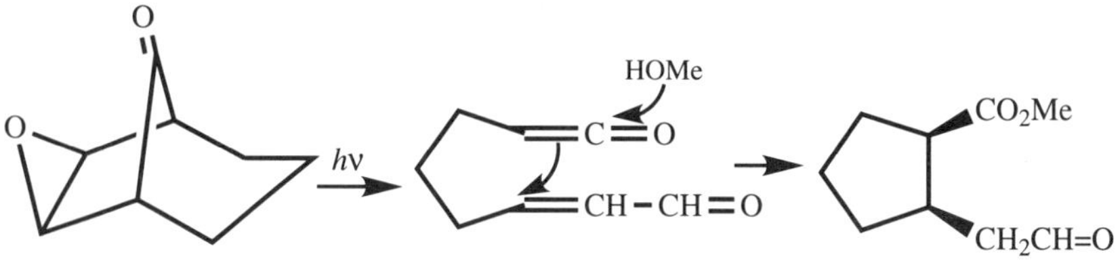

Reaction of the diazine **11** with styrene gave the acylketene **12**, which was proposed to react with methanol by the Dieckmann-like reaction with electrophilic attack by the ester carbonyl carbon on C_β of the ketene as shown to give the enol **13** in 19% yield after chromatography on silica gel (equation 12).[85]

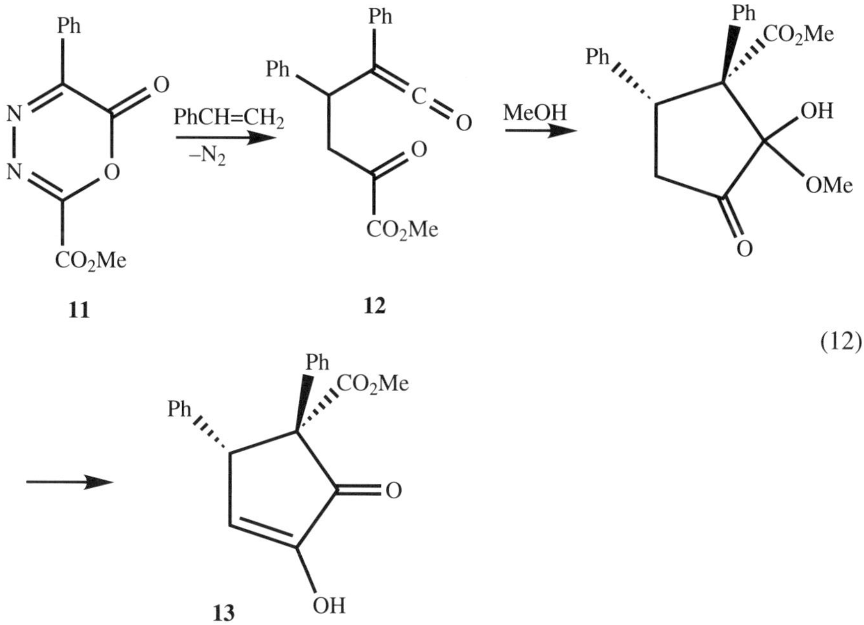

5.5 NUCLEOPHILIC ADDITION TO KETENES

(12)

The reaction of allyl ethers and allyl thioethers with haloketenes proceeds by initial nucleophilic attack followed by Claisen rearrangement to give unsaturated esters.[86] The reaction proceeded efficiently with $CCl_2=C=O$ and $CCl_3CCl=C=O$ while other electrophilic ketenes were not as successful. A variety of other substrates were examined (equations 13–15).[86,87]

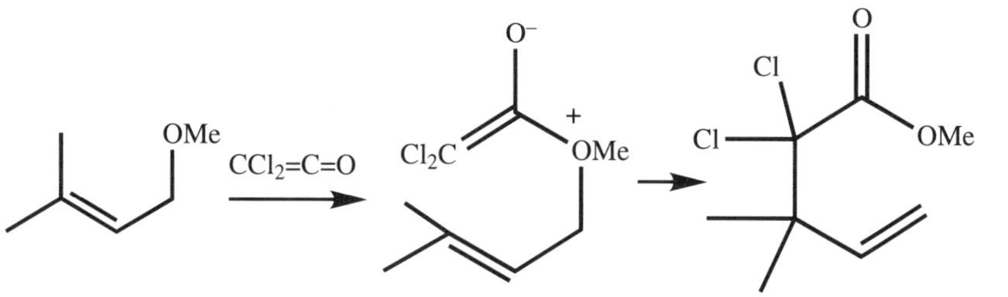

(48%)   (13)[86]

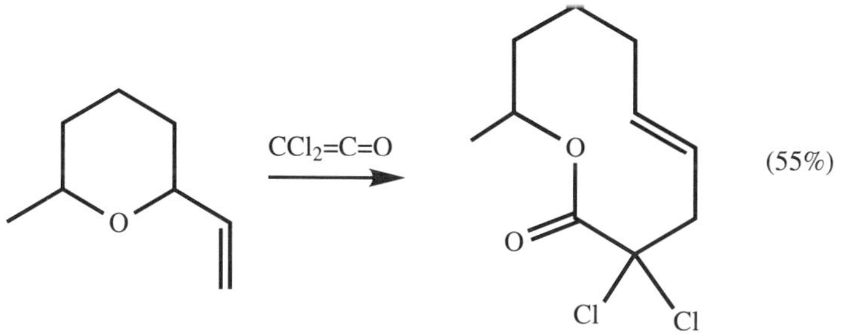

(55%)

(14)[86]

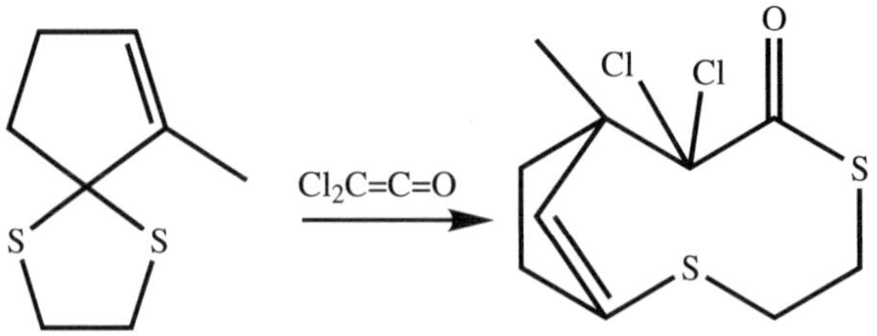

Sulfoxide oxygen serves as the nucleophile in the reaction of equation 16.[88–92] This reaction has been termed an additive Pummerer rearrangement.[90]

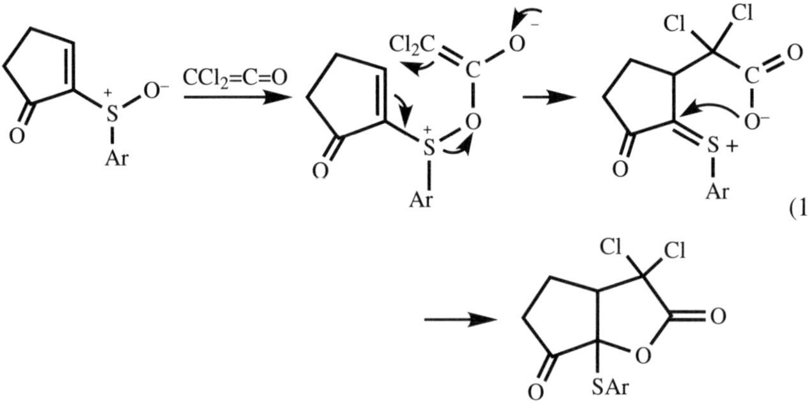

Vinyloxirane addition to $CCl_2=C=O$ can lead to oxepin-2-ones by nucleophilic attack by oxygen (equation 17).[93] Vinylethylenimines react similarly.[93] Oxiranes also react with ketenes intramolecularly (equation 18).[94] Ketene **14** was identified by its IR band at 2120 cm^{-1} when formed by ketone photolysis at 77 K, and gave no incorporation of deuterium in the formation of **15** in CH_3OD, and so this process was envisaged to occur with concerted intramolecular proton transfer.[94]

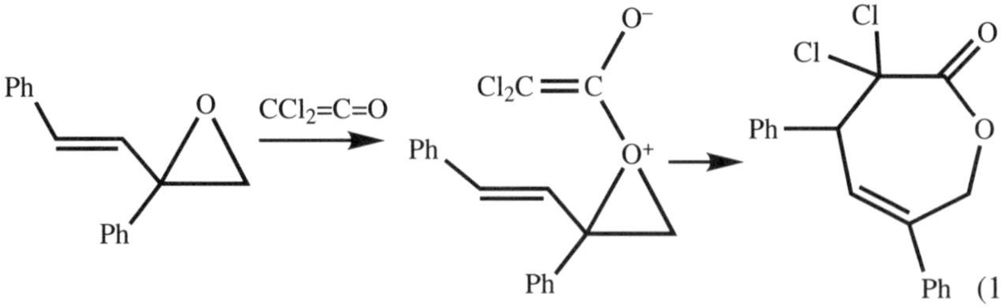

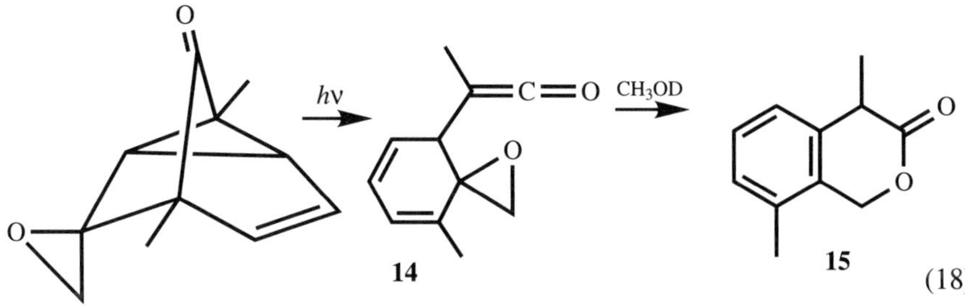

5.5 NUCLEOPHILIC ADDITION TO KETENES

(18)

In another intramolecular ketene–oxirane reaction the product **17** was favored in benzene, but in the presence of 0.2 M MeOH the ketene **16** gave only 1–2% of **17**, along with the MeOH capture product of **16** and other products (equation 19).[75]

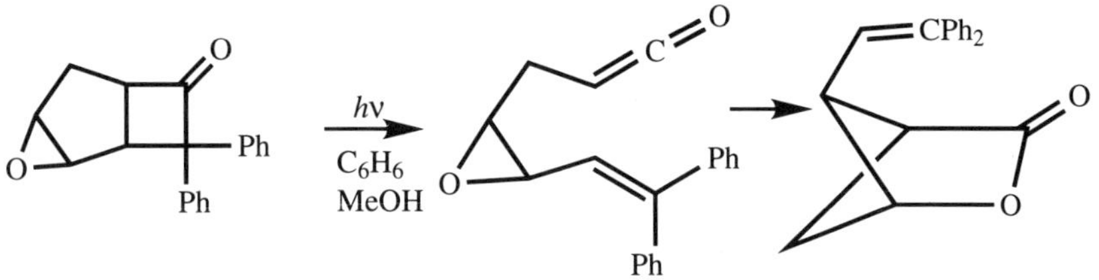

(19)

The reaction of phosphorous acids with ketenes to give mixed anhydrides is thought to involve initial protonation of the ketene (equation 20).[95] Mixed anhydrides of sulfonic and carboxylic acids are prepared similarly.[96]

$$(EtO)_2P(S)OH \xrightarrow{CH_2=C=O} (EtO)_2P(S)OAc \qquad (20)$$

The reaction of phenols with ketene is promoted by adsorption of the phenol onto silica gel or alumina.[97]

5.5.2.3 Nitrogen Nucleophiles.

The reaction of ketenes with amines to give amides is a familiar process (equation 1), and these reactions occur by in-plane nucleophilic attack on C_α of the ketene, as discussed in Section 5.5.1.4 (equation 18).[31] Mechanistic[98,99] and theoretical[100] studies which have been interpreted in terms of addition of amines across the carbon–carbon double bond of ketenes do not appear plausible, as discussed elsewhere,[1] and in Section 5.5.1. Chiral induction in the reaction of unsymmetrical ketenes with chiral amines is discussed in Section 5.9.

$$RR^1C=C=O \xrightarrow{R^2NH_2} RR^1CHCONHR^2 \qquad (1)$$

Examples of the many reported reactions of ketenes with nitrogen nucleophiles are shown in equations 2–10. The reaction of equation 2 proceeded diastereoselectively with amines with chiral sidechains.[101]

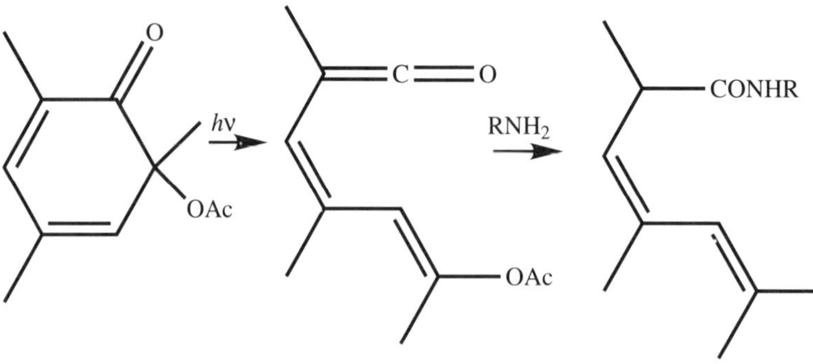

598 REACTIONS OF KETENES

$$\text{[2,4-dimethyl-6-acetoxy cyclohexadienone]} \xrightarrow{h\nu} \text{[dienyl ketene with OAc]} \xrightarrow{RNH_2} \text{[dienyl amide with OAc]} \quad (2)^{101}$$

$$CH_2=C=O + RCONH_2 \longrightarrow RC(O)NHAc \quad (3)^{102}$$

$$Me_2C=C=O + PhCH=NEt \longrightarrow \text{[1,3-oxazinone]} \quad (4)^{103}$$

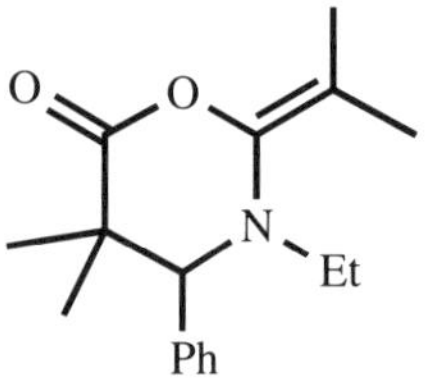

$$Ph_2C=C=O + Ph_2C=NSiMe_3 \longrightarrow Ph_2C=C(OSiMe_3)N=CPh_2 \quad (5)^{104}$$

$$CH_2=C=O + Et_3SiNMe_2 \longrightarrow CH_2=C(OSiEt_3)(NMe_2) \longrightarrow Et_3SiCH_2CONMe_2 \quad (6)^{105}$$

$$Ph_2C=C=O + (Et_3Ge)_2NH \longrightarrow [Ph_2C=C(OGeEt_3)]_2NH \quad (7)^{106}$$

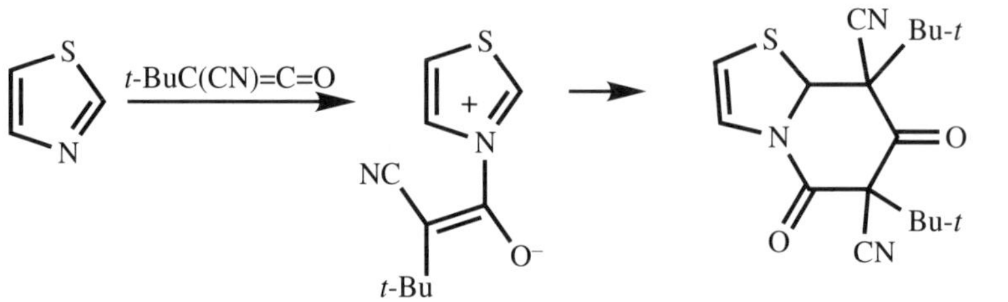

$$\text{[thiazole]} \xrightarrow{t\text{-BuC(CN)=C=O}} \text{[zwitterion intermediate]} \longrightarrow \text{[bicyclic product]} \quad (8)^{107}$$

$$Ph_2C=C=O \xrightarrow{Me_3SiN_3} [Ph_2C=C(OSiMe_3)(N_3)] \longrightarrow \text{[tetraphenyl succinimide]} \quad (9)^{108}$$

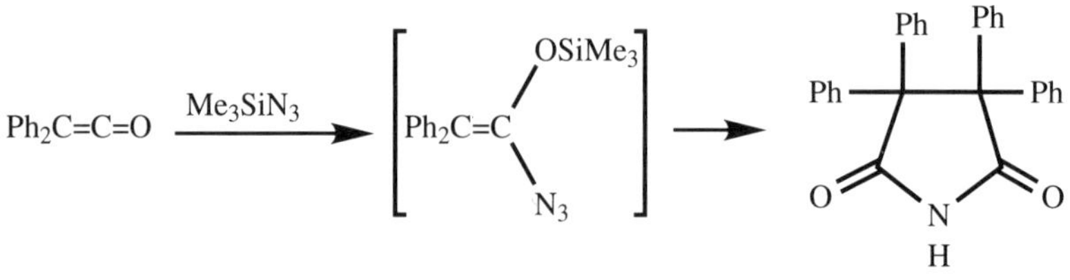

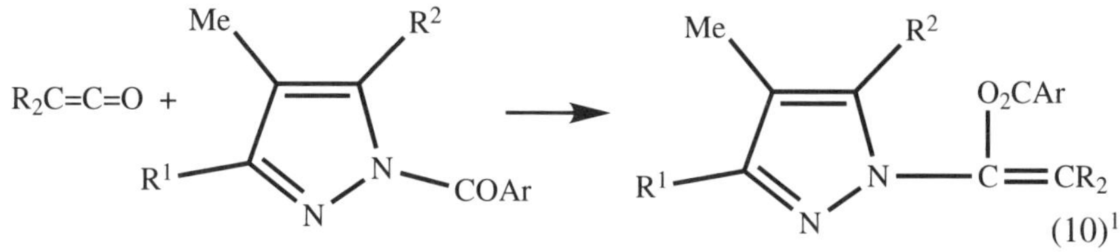

$R_2C=C=O$ + [pyrazole with Me, R^2, R^1, N-COAr] → [product with O_2CAr, $C=CR_2$] (10)[109]

Photolytic coupling of chromium carbene complexes in the presence of amino acids as shown in equation 11 proceeds with high diastereoselection in the formation of the new chiral center, and evidently involves a chromium-complexed ketene.[110]

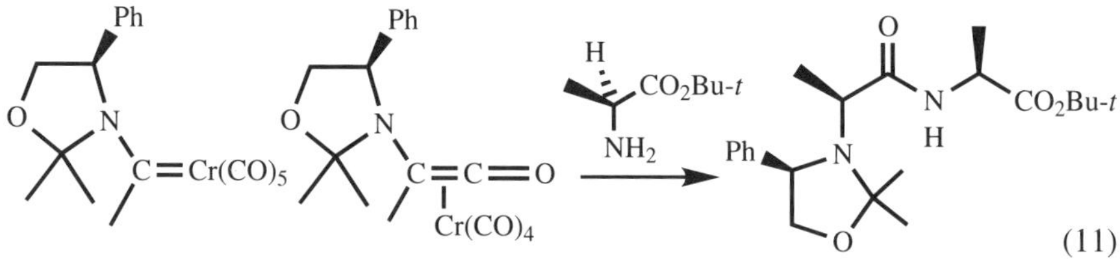

[Scheme for equation 11 showing oxazolidine-Cr(CO)$_5$ carbene → ketene-Cr(CO)$_4$ complex, reacting with H-C(CO$_2$Bu-t)(NH$_2$)(Ph) to give amide product with CO$_2$Bu-t] (11)

Nucleophilic attack by nitrogen on dichloroketene gave a zwitterionic species observed by NMR which then rearranged to an azacyclodecane (equation 12).[111] Such products are useful in the synthesis of indolizidines and quinolizidines.[111] A similar rearrangement was proposed for the zwitterionic intermediate in the reaction of equation 13.[112,113]

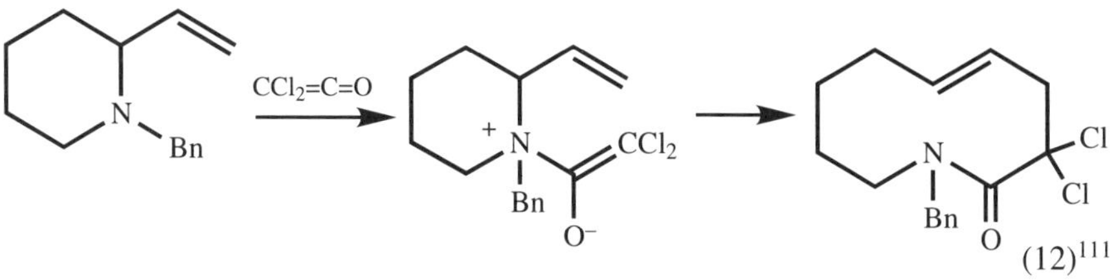

[Scheme for equation 12: N-Bn-2-vinylpiperidine + $CCl_2=C=O$ → zwitterion → azacyclodecane with N-Bn, C=O, CHCl$_2$] (12)[111]

[Scheme for equation 13: $Ph_2C=C=O$ + N-methyl azanorbornene → zwitterionic intermediate → bicyclic product with Ph, Ph, N-Me, C=O] (13)[112]

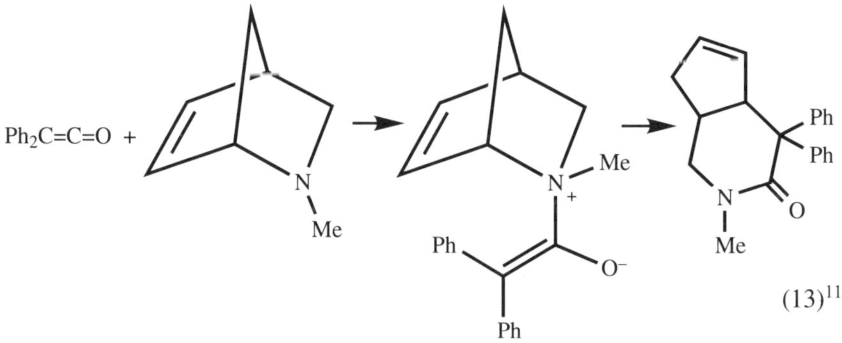

Reaction of 4-dimethylaminopyridine with di(carboethyoxy)ketene gave an isolable ylide, which reacted with a variety of substrates (equation 14).[114]

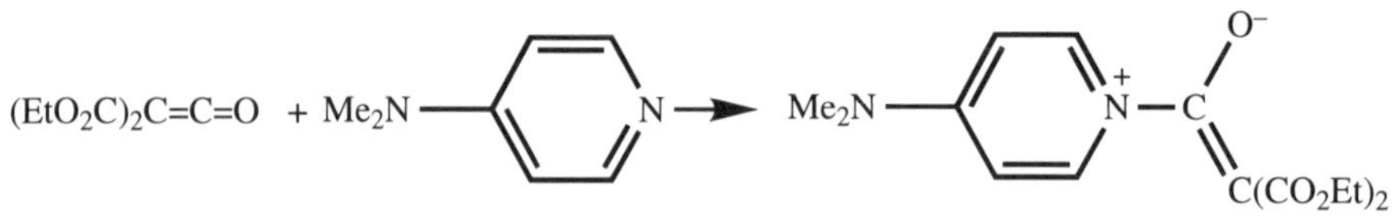

$$(EtO_2C)_2C=C=O + Me_2N-\underset{}{\bigcirc}-N \longrightarrow Me_2N-\underset{}{\bigcirc}-\overset{+}{N}-\underset{\underset{C(CO_2Et)_2}{\|}}{C}-O^- \quad (14)^{114}$$

The reaction of the ketene **18** with pyridine also gave a zwitterion, observed by UV spectroscopy (equation 15).[34]

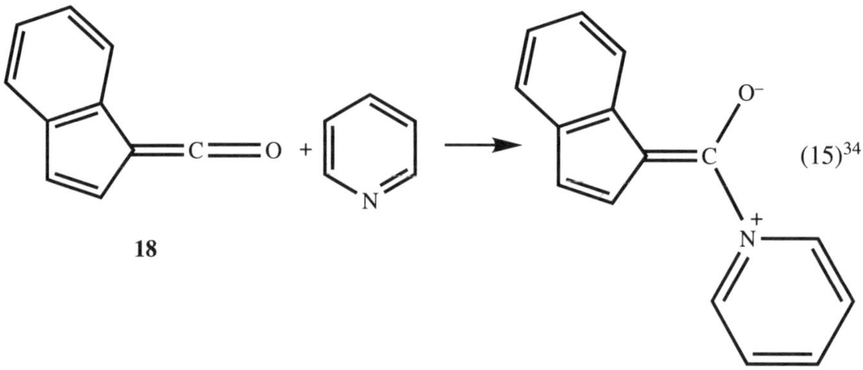

[Structure of indene=C=O + pyridine → zwitterion] (15)[34]

18

The reaction of sulfilimines with excess ketenes provides oxazolinones (equation 16).[115]

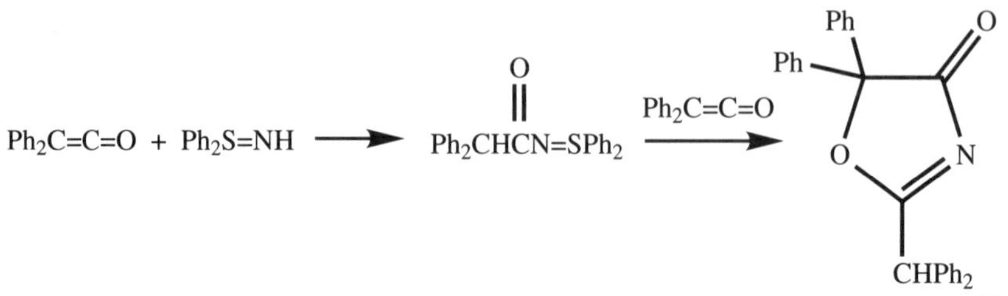

$$Ph_2C=C=O + Ph_2S=NH \longrightarrow Ph_2CHCN=SPh_2 \xrightarrow{Ph_2C=C=O} \text{[oxazolinone with Ph, Ph, CHPh}_2\text{]} \quad (16)$$

The reaction of **19** with $CCl_2=C=O$ (equation 17) can proceed through nitrogen attack to give **20** and **21**.[116] The reaction of equation 18 also involves nucleophilic attack by nitrogen.[117]

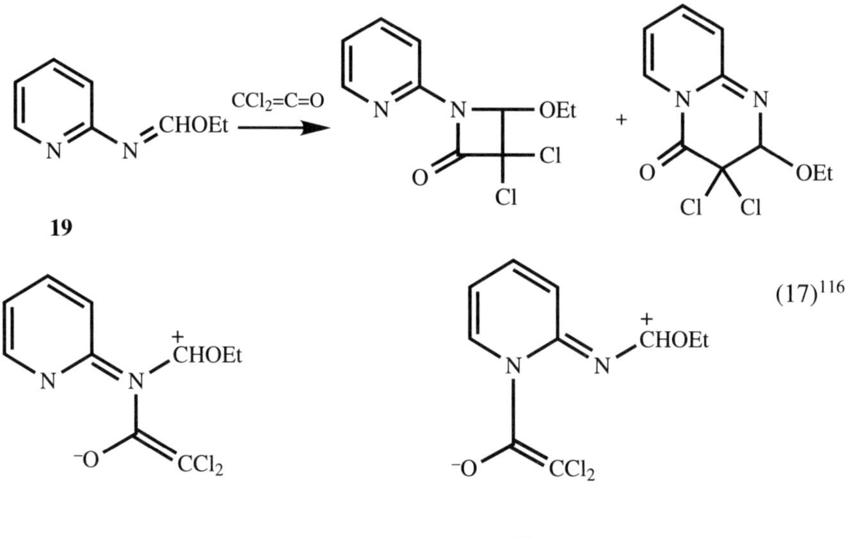

19 20 21

(17)[116]

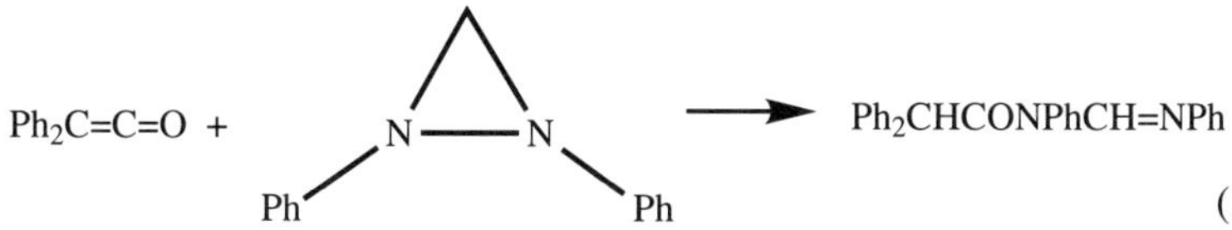

(18)[117]

The reaction of the azirine **22** with diphenylketene is proposed to proceed through a zwitterion to give the observed products **23–26** (equation 19).[118]

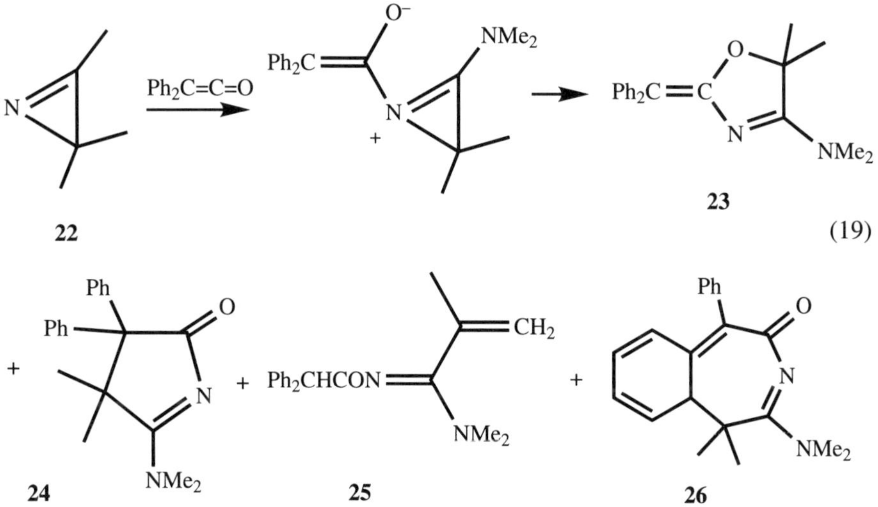

22 23 24 25 26

(19)

602 REACTIONS OF KETENES

Photochemical generation of ketenes containing amino groups has proven to be a convenient method for the formation of lactams (equations 20 and 21).[74,119] Ketenes generated by Wolff rearrangement and thermal methods reacted similarly.[119a-c] Cyclization of amidoketenes was also proposed (equations 22 and 23).[120,121]

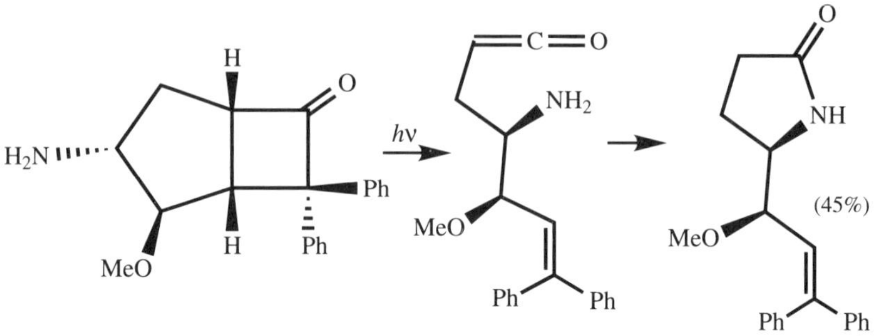

(20)[74]

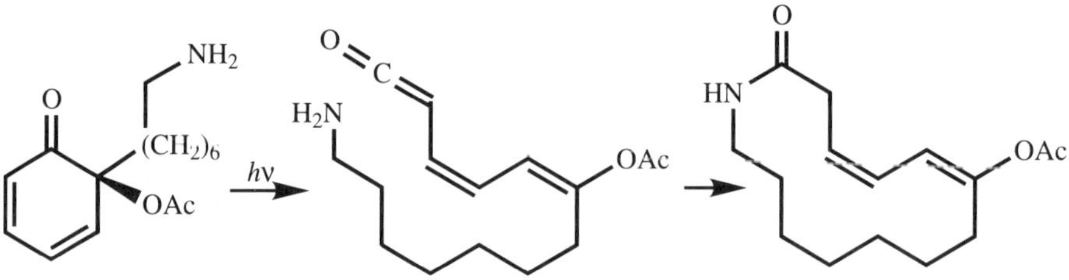

(21)[119]

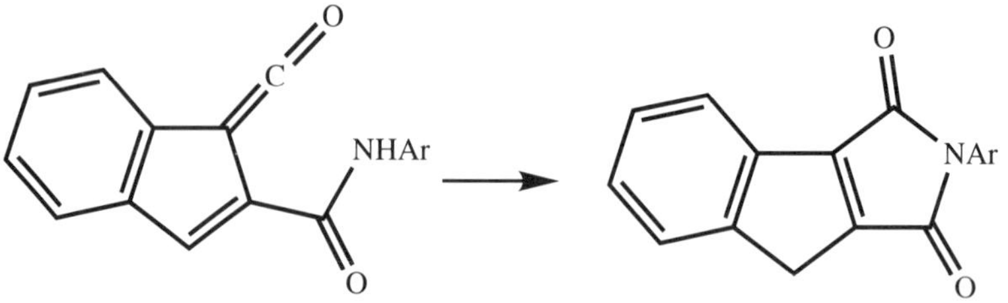

Ar = 4-Tol

(22)[120]

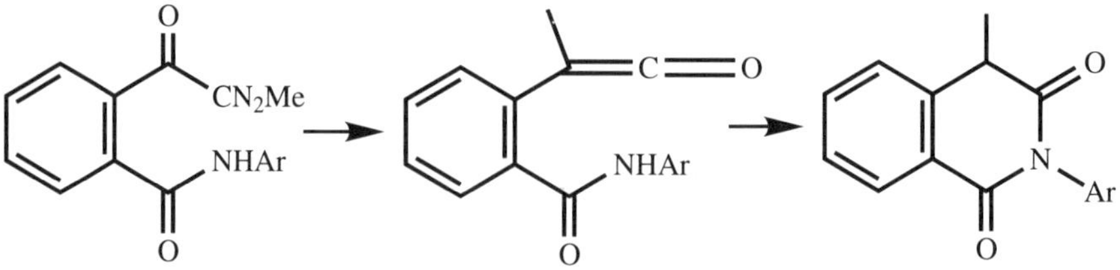

Ar = 4-O$_2$NC$_6$H$_4$

(23)[121]

5.5 NUCLEOPHILIC ADDITION TO KETENES

Both the nitrogen and carbon of β-acylenamines serve as nucleophiles in the addition to diphenylketene (equation 24).[122] N-Acylation was observed in the reaction of equation 25.[123]

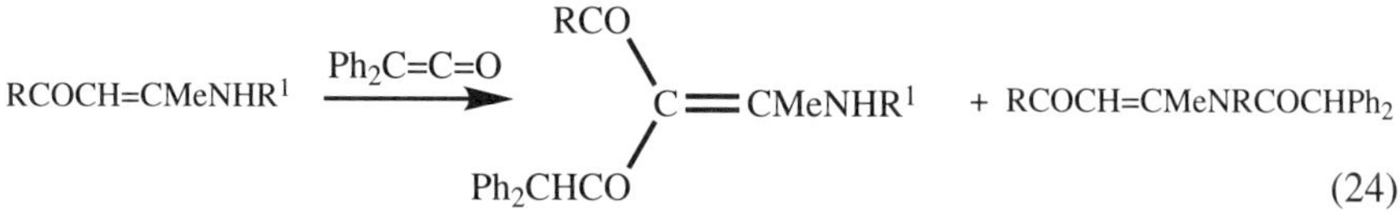

$$\text{RCOCH=CMeNHR}^1 \xrightarrow{\text{Ph}_2\text{C=C=O}} \underset{\text{Ph}_2\text{CHCO}}{\overset{\text{RCO}}{\text{C}}}=\text{CMeNHR}^1 + \text{RCOCH=CMeNRCOCHPh}_2 \quad (24)$$

$$\underset{\text{R}^1\text{R}^2\text{NC}\equiv\text{CC}=\text{NR}^3}{\overset{\text{NHR}}{|}} \xrightarrow{\text{Ph}_2\text{C=C=O}} \underset{\text{R}^1\text{R}^2\text{NC}\equiv\text{CC}=\text{NR}^3}{\overset{\text{Ph}_2\text{CHCONR}}{|}} \quad (25)$$

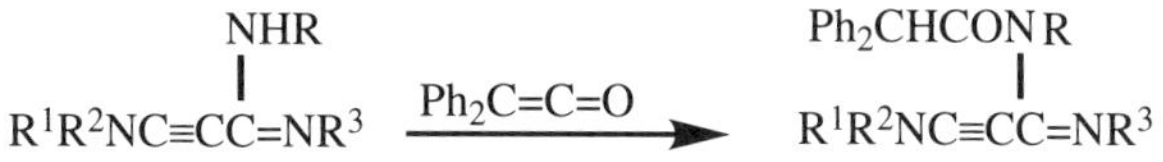

The reaction of MeLi/HMPA with dimesitylketene or mesitylphenylketene gave the products of the addition of Me_2N^-, Me_2NPO^-, and H^- to the ketenes, along with other products, as shown in equation 26.[124] Initial cleavage of the HMPA by MeLi to give the active nucleophiles was proposed.[124]

$$\text{Mes}_2\text{C=C=O} \xrightarrow[\text{MeLi}]{(\text{Me}_2\text{N})_3\text{PO}} \text{Mes}_2\text{CHCONMe}_2 + \text{Mes}_2\text{C=C(OH)PO(NMe}_2)_2 + \text{Mes}_2\text{C=CHOH} \quad (26)$$

The dimerization of $t\text{-BuC(CN)=C=O}$ (**27**) does not occur thermally, but is catalyzed by Et_3N, with reaction in the presence of 0.1 equivalent of amine proposed to occur by the path of equations 27 and 28.[125]

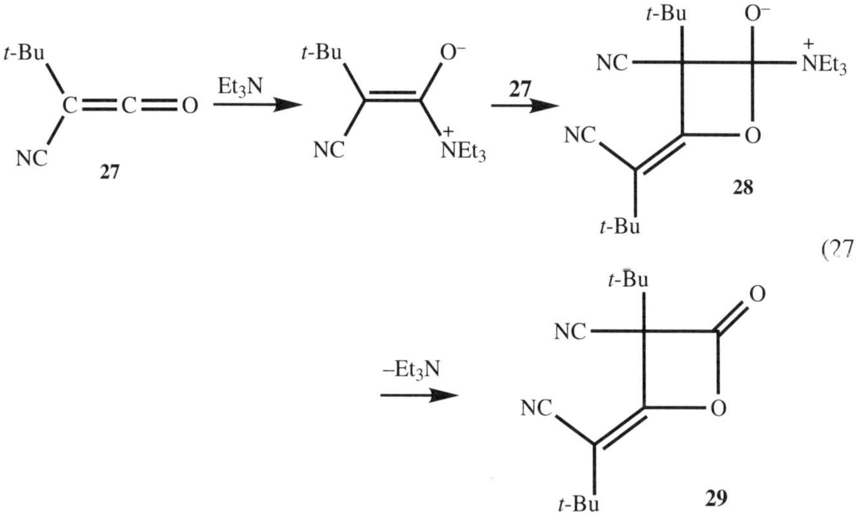

(27)

$$28 \longrightarrow \underset{\substack{\text{CN} \quad t\text{-Bu}}}{\overset{\substack{\text{O} \\ \| + \\ \text{OCNEt}_3 \\ t\text{-Bu} \qquad \text{CN}}}{\diagdown\!\!\diagup}} \xrightarrow[-\text{Et}_3\text{N}]{-\text{CO}_2} \underset{\substack{\text{NC} \qquad \text{Bu-}t \\ \mathbf{30}}}{\overset{t\text{-Bu} \qquad \text{CN}}{\text{C}=\text{C}=\text{C}}} \qquad (28)$$

5.5.2.4 Carbon Nucleophiles.

One of the first reactions observed for ketenes was the addition of a Grignard reagent,[126] and this was shown to give an enolate that could be trapped by acylation (equation 1).[127]

$$\text{Ph}_2\text{C}=\text{C}=\text{O} \xrightarrow{\text{PhMgBr}} \underset{\text{Ph}}{\text{Ph}_2\text{C}=\text{C}} \!\!\diagup\!\! ^{\text{OMgBr}} \xrightarrow{\text{PhCOCl}} \underset{\text{Ph}}{\text{Ph}_2\text{C}=\text{C}} \!\!\diagup\!\! ^{\text{O}_2\text{CPh}} \qquad (1)$$

Surveys of the reaction of ketenes with Grignard reagents and organolithiums have been presented,[1,128] and the reactions with organolithium reagents are summarized in Table 1.[129-131]

Reaction of $t\text{-Bu}_2\text{C}=\text{C}=\text{O}$ with t-BuLi provides the enolate **31**, which can be trapped by Me$_3$SiCl or MeI to give the respective enol ethers or hydrolyzed to the ketone **32**.[6] Attempts to enolize the ketone **32** to regenerate the enolate were unsuccessful and so nucleophilic attack on the ketene is the only available route to this enolate (equation 2).

$$\underset{t\text{-Bu}}{\overset{t\text{-Bu}}{\diagdown}}\text{C}=\text{C}=\text{O} \xrightarrow{t\text{-BuLi}} \underset{t\text{-Bu}}{\overset{t\text{-Bu}}{\diagdown}}\text{C}=\text{C}\!\!\diagup\!\!\overset{\text{O}^-}{\underset{\text{Bu-}t}{}} \xrightarrow{\text{H}_2\text{O}} \underset{t\text{-Bu}}{\overset{t\text{-Bu}}{\diagdown}}\text{CH--C}\!\!\diagup\!\!\overset{\text{O}}{\underset{\text{Bu-}t}{\|}} \qquad (2)$$

$$\qquad\qquad\qquad\qquad\qquad\qquad \mathbf{31} \qquad\qquad\qquad \mathbf{32}$$

When unsymmetrical ketenes react with organolithium reagents and the product enolates are trapped by silylation or acetylation the resulting enol ethers or esters show there was a high stereoselectivity for attack of the organolithium reagent *anti* to the larger substituent in the ketene. In the case of PhCR=C=O (R = t-Bu or

TABLE 1. Ketene Reactions with Organolithium Reagents

Ketene	RLi	Ref.
EtMeC=C=O	MeLi, PhCH$_2$Li	7
i-PrMeC=C=O	MeLi, n-BuLi, PhCH$_2$Li	7
t-BuMeC=C=O	MeLi, n-BuLi, PhCH$_2$Li	7
Me$_2$C=C=O	MeLi, PhC≡CLi	7
Et$_2$C=C=O	n-BuLi	7
t-Bu$_2$C=C=O	PhLi	129
t-Bu$_2$C=C=O	t-BuLi	6
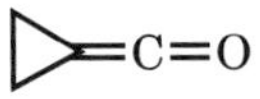 (cyclopropylidene)C=O	n-BuLi, PhLi, PhCH$_2$Li	7
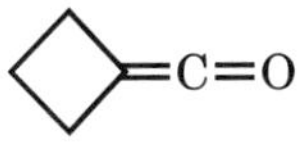 (cyclobutylidene)C=O	PhCH$_2$Li	7
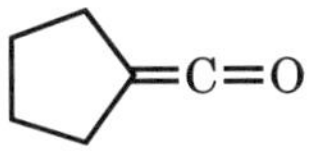 (cyclopentylidene)C=O	PhCH$_2$Li	7
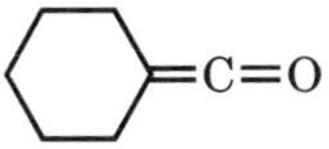 (cyclohexylidene)C=O	PhCH$_2$Li	7
(Me$_3$M)$_2$C=C=O, M = Si,Ge,Sn	MeLi, n-BuLi, PhLi	130,131
Mes$_2$C=C=O	i-PrLi, t-BuLi	69
PhCMe=C=O	n-BuLi	5
PhCEt=C=O	n-BuLi	5
PhC(Pr-i)=C=O	n-BuLi, t-BuLi	4
PhC(Bu-t)=C=O	n-BuLi, t-BuLi	4
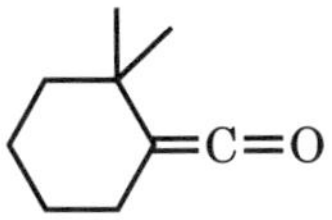 (2,2-dimethylcyclohexylidene)C=O	PhLi	5
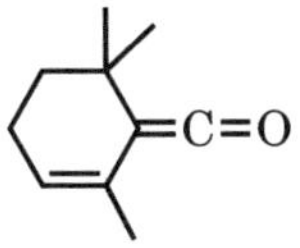 (6,6-dimethylcyclohex-2-enylidene)C=O	PhLi	62,135

i-Pr) the favored product stereochemistry depends on the identity of the organolithium reagents.[4] The steric demand of the phenyl group is variable depending upon whether this group is in a planar or twisted conformation, and the product stereochemistry evidently results from a complex interplay of steric and electronic factors (Table 2).

Reaction of silylketenes with organocerium reagents followed by hydration or alkylation with alkylhalides gave a variety of α-silyl ketones.[132]

TABLE 2. Stereoselectivity in Addition of Organometallic Reagents to Ketenes $R^1R^2C=C=O$

R^1	R^2	RLi	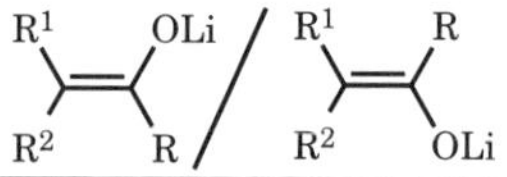 $\underset{R^2}{\overset{R^1}{>}}\!\!=\!\!\underset{R}{\overset{OLi}{<}}\ /\ \underset{R^2}{\overset{R^1}{>}}\!\!=\!\!\underset{OLi}{\overset{R}{<}}$	Ref.
t-Bu	H	t-BuLi	>(95/5)	5
Et	Me	MeLi	1.7/1	7
i-Pr	Me	MeLi	7.0/1	7
t-Bu	Me	MeLi	>(99/1)	7
t-Bu	i-Pr	t-BuLi	>(95/5)	4
Ph	Me	n-BuLi	>(95/5)	5
Ph	Et	n-BuLi	>(95/5)	5
Ph	Et	Me$_3$SiLi	>(95/5)	2
Ph	i-Pr	t-BuLi	1/4	4
Ph	t-Bu	t-BuLi	>(5/95)	4
Ph	c-Pr	n-BuLi	79/21	4
Ph	c-Pr	t-BuLi	9/91	4
Ph	c-C$_5$H$_9$	n-BuLi	88/12	4
Ph	c-C$_6$H$_{11}$	n-BuLi	56/44	4
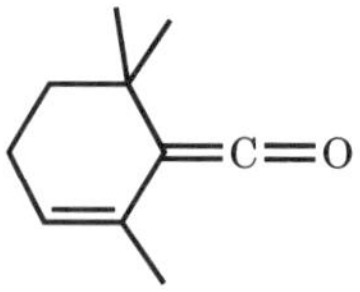 (cyclohexenyl ketene structure)		CH$_2$=CHCH$_2$MgCl	9/1	62
Et	Me$_3$Si	n-BuLi	>(5/95)	5
		CH$_2$=CHLi	>(5/95)	5
		HC≡CLi	>(5/95)	5

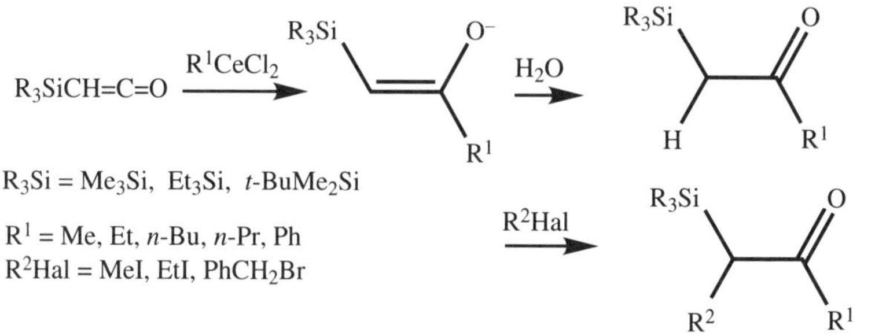

$$R_3SiCH=C=O \xrightarrow{R^1CeCl_2} \underset{R^1}{\overset{R_3Si}{>}}\!\!=\!\!\overset{O^-}{<} \xrightarrow{H_2O} \underset{H}{\overset{R_3Si}{>}}\!\!\overset{O}{\underset{R^1}{\diagdown}} \quad (3)$$

$$\xrightarrow{R^2Hal} \underset{R^2}{\overset{R_3Si}{>}}\!\!\overset{O}{\underset{R^1}{\diagdown}}$$

$R_3Si = Me_3Si,\ Et_3Si,\ t\text{-}BuMe_2Si$

$R^1 = Me,\ Et,\ n\text{-}Bu,\ n\text{-}Pr,\ Ph$

$R^2Hal = MeI,\ EtI,\ PhCH_2Br$

Alkylarylketenes undergo SmI_2-mediated allylation or benzylation with $CH_2=CHCH_2I$ or $PhCH_2Br$ to give enantiomerically enriched ketones **33** after protonation with chiral proton sources (equation 4).[133] Substituted allyl halides give regioselective allylation.[133a]

5.5 NUCLEOPHILIC ADDITION TO KETENES

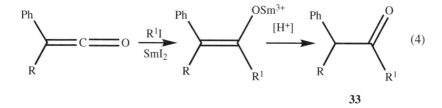

(4)

33

The addition of Me₃SiCN to ketenes was proposed to involve a [$_\pi 2 + {}_\pi 2 + {}_\sigma 2$] cycloaddition (equation 5).[134] The kinetics of the addition of CN⁻ to pentafulvenone have also been studied (Section 5.5.1.2)[33] but the addition reaction of CN⁻ to ketenes has been little utilized in synthesis. Some other reactions of ketenes with carbon nucleophiles are shown in equations 5–11 and in Table 1.[135–142]

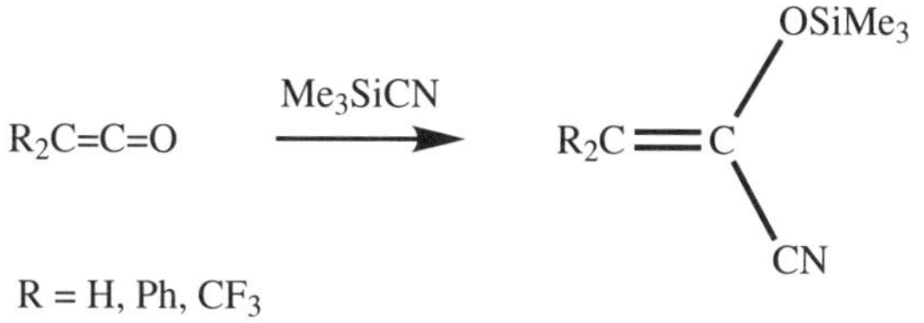

$$R_2C=C=O \quad \xrightarrow{Me_3SiCN} \quad R_2C=C{\overset{OSiMe_3}{\underset{CN}{\diagup}}} \qquad (5)^{134}$$

R = H, Ph, CF₃

$$CH_2=C=O \quad \xrightarrow{Et_2Zn} \quad CH_3COEt \qquad (6)^{136}$$

$$Ph_2C=C=O \quad \xrightarrow{Ph_2CHK} \quad Ph_2CHCOCHPh_2 \qquad (7)^{137}$$

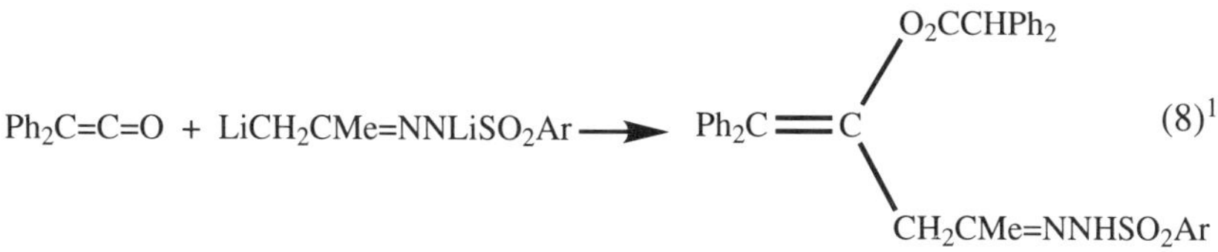

$$Ph_2C=C=O + LiCH_2CMe=NNLiSO_2Ar \longrightarrow Ph_2C=C{\overset{O_2CCHPh_2}{\underset{CH_2CMe=NNHSO_2Ar}{\diagup}}} \qquad (8)^{138}$$

$$Ph_2C=C=O + t\text{-}BuN\equiv C \longrightarrow Ph_2C=C{\overset{O^-}{\underset{+\ C=NBu\text{-}t}{\diagup}}} \qquad (9)^{139}$$

Ph₂C=C=O ⟶ [dioxole ring with Ph₂C, O, =CPh₂, =NBu-t, O]

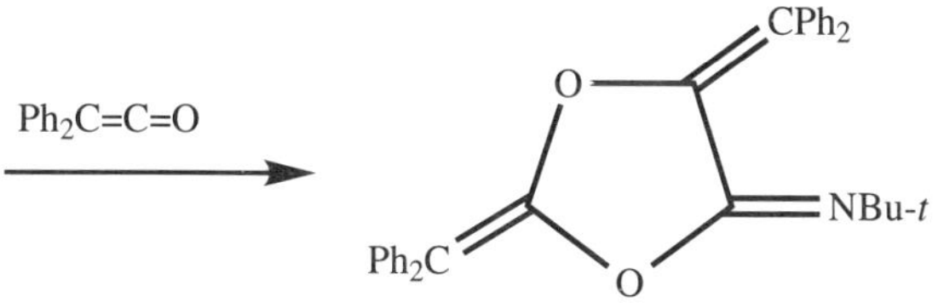

$$(CF_3)_2C=C=O + \bar{C}F(CF_3)SO_2F \xrightarrow{H^+} (CF_3)_2CHCOCF(CF_3)SO_2F \quad (10)^{140}$$

$$Ph_2C=C=O + R^1R^2NC\equiv CMR_3 \longrightarrow Ph_2C=C\begin{smallmatrix}OMR_3\\ \\C\equiv CNR^1R^2\end{smallmatrix} \quad (11)^{141,142}$$

$$M = Si, Ge$$

The reaction of equation 12 is a catalyzed Friedel–Crafts reaction,[143] while that in equation 13 involves an initial cycloaddition followed by an uncatalyzed Friedel–Crafts process.[144]

$$Ph_2C=C=O + ArH \xrightarrow{AlCl_3} Ph_2CHCOAr \quad (12)^{143}$$

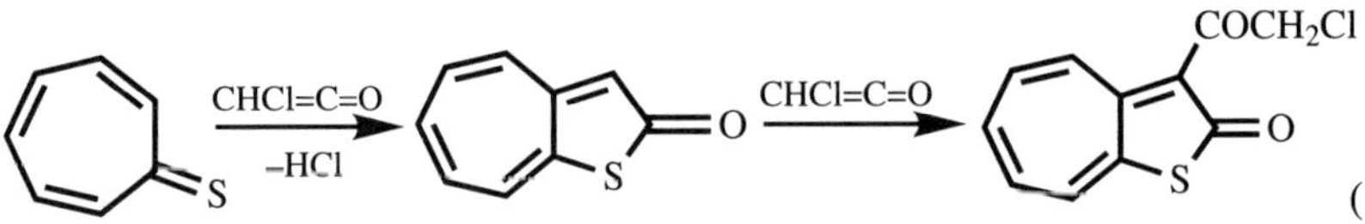

(13)[144]

The addition of ketene acetals to ketenes involves net nucleophilic addition to the C=O bond of ketenes (equation 14).[145] Silyl enol ethers give both cycloaddition and acyclic products analogous to **34**.[146] Further examples of these reactions are given in Section 5.4.1.4.

$$Me_2C=C=O + CH_2=C\begin{smallmatrix}OMe\\ \\OSiEt_3\end{smallmatrix} \longrightarrow Me_2C=C\begin{smallmatrix}OSiEt_3\\ \\CH_2CO_2Me\end{smallmatrix} \quad (14)^{145}$$

34

Enamines also serve as carbon nucleophiles toward ketenes[147,148] (Section 5.4.1.4). The reaction of the acyl chloride **35** with Et$_3$N was proposed to give the ketene **36**, which in the presence of the morpholino enamine **37** led to the zwitterion **38** which gave products **39** and **40** as shown in equations 15 and 16. Intramolecular reactions of ketenes with alkenes in related systems are discussed in Section 3.2.2, equations 30–34.

$$CH_3CH=CMeCOCl \xrightarrow{Et_3N} CH_2=CH-CMe=C=O \quad (15)$$

35   **36**

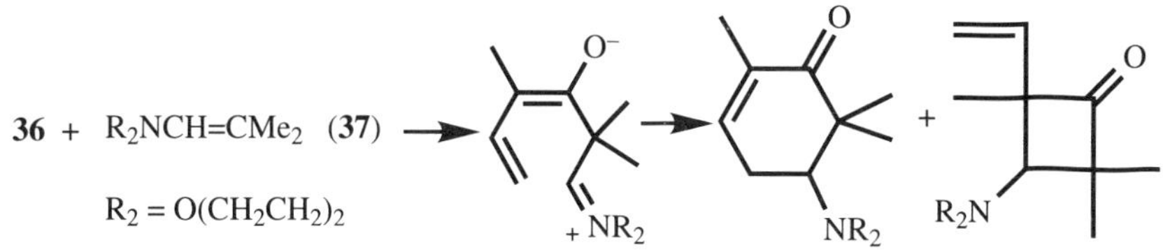

5.5 NUCLEOPHILIC ADDITION TO KETENES

36 + $R_2NCH=CMe_2$ (37) →

[Structures 38, 39, 40 shown]

$R_2 = O(CH_2CH_2)_2$ (16)

Enolates also serve as carbon nucleophiles in attacks on ketenes, although as discussed in Section 5.5.2.2, the oxygen of the enolate is kinetically the most important nucleophile. Another example is given in Section 3.2.3.1, equation 8. The reaction of ynolates with ketenes is mentioned in Section 4.8.1. Acetic anhydride and Et_3N react with $Ph_2C=C=O$ with cyclization as shown in equation 17.[149]

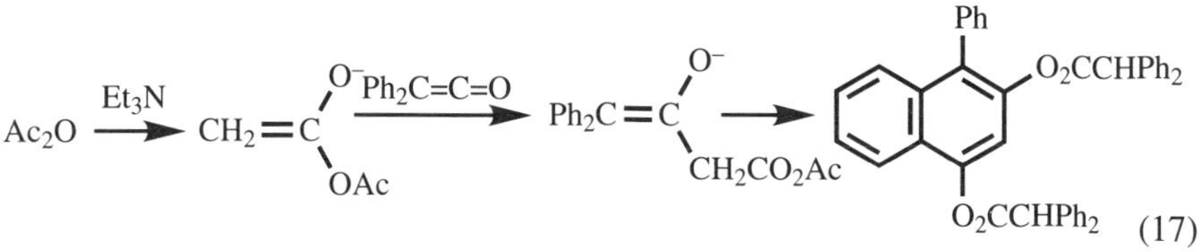

$Ac_2O \xrightarrow{Et_3N} CH_2=C(O^-)(OAc) \xrightarrow{Ph_2C=C=O} Ph_2C=C(O^-)(CH_2CO_2Ac) \longrightarrow$ [naphthalene product with Ph, O_2CCHPh_2, O_2CCHPh_2] (17)

The reaction of ketenes with Pd complexes can result in C—C bond formation to C_α or C_β of ketenes (equations 18 and 19).[150,151] Decarbonylation of the acyl chloride and arylpalladium addition is proposed for equation 19.[151]

$Ph_2C=C=O + CH_2=CHCH_2PdOMe \longrightarrow Ph_2C(CH_2CH=CH_2)CO_2Me$ (18)[150]

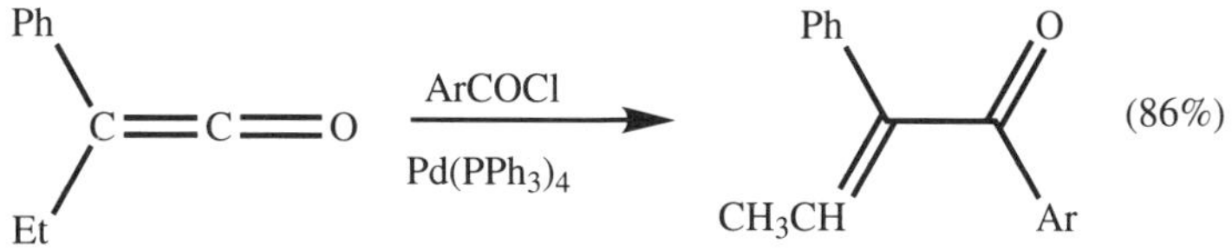

$\underset{Et}{\overset{Ph}{>}}C=C=O \xrightarrow[Pd(PPh_3)_4]{ArCOCl} \underset{CH_3CH}{\overset{Ph}{>}}C=C(O)Ar$ (86%) (19)[151]

Ketenes react with some allyl and 2-alkynylmetal complexes in reactions interpreted as involving zwitterionic intermediates (equations 20 and 21)[152,153], but in the examples of equations 22 and 23 the formation of the *endo* adduct was interpreted as indicating a concerted process.[154]

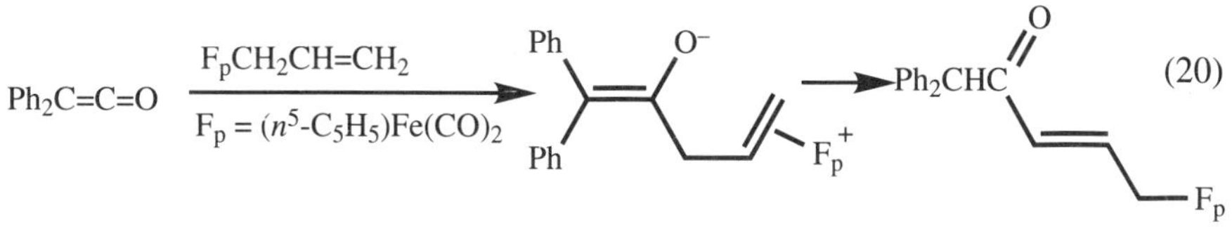

$Ph_2C=C=O \xrightarrow[F_p = (\eta^5-C_5H_5)Fe(CO)_2]{F_pCH_2CH=CH_2}$ [intermediate with Ph, Ph, O^-, F_p^+] $\longrightarrow Ph_2CHC(O)$—[vinyl-$F_p$] (20)

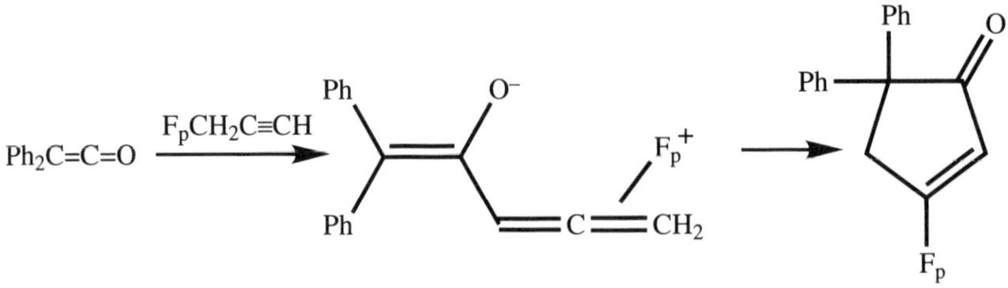

$Ph_2C=C=O \xrightarrow{F_pCH_2C\equiv CH}$ [intermediate with Ph, Ph, O⁻, and C=C=CH₂-F_p⁺] → cyclopentenone product (21)

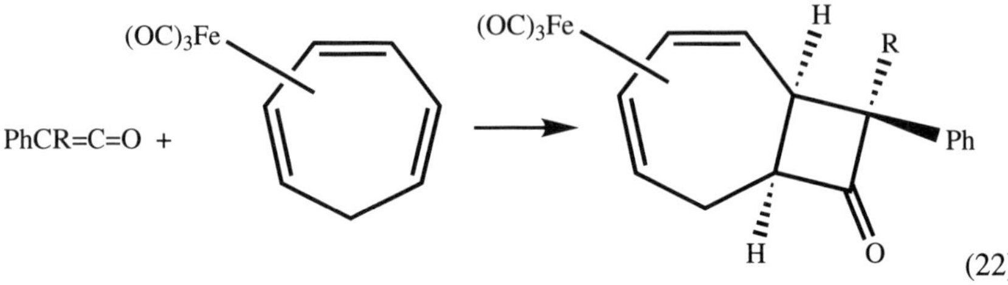

$PhCR=C=O +$ (OC)₃Fe-cyclooctatetraene → bicyclic cycloadduct (22)

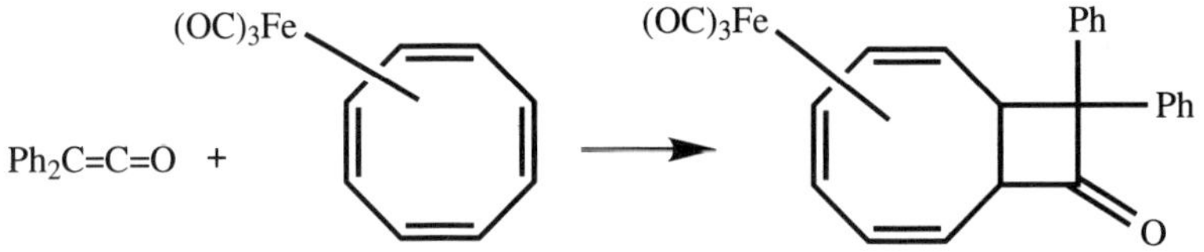

$Ph_2C=C=O +$ (OC)₃Fe-cyclooctatetraene → bicyclic cycloadduct (23)

The reaction of the iron acetylides **41** and **42** with $Ph_2C=C=O$ gave cycloadducts such as **43** (equation 24), whereas **44** gave the acylvinylidene **45** (equation 25).[155]

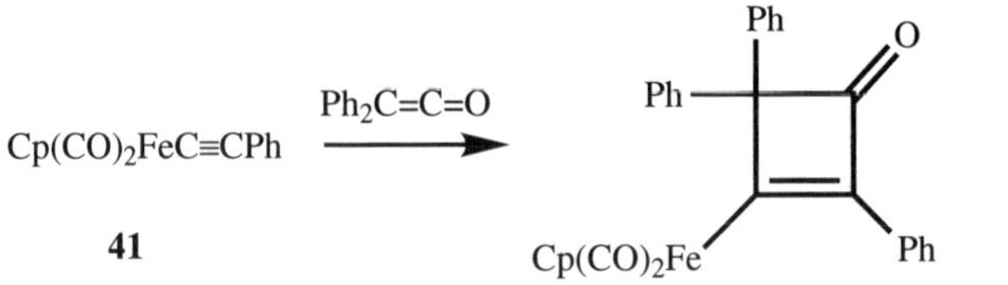

$Cp(CO)_2FeC\equiv CPh \xrightarrow{Ph_2C=C=O}$ cyclobutenone **43** (24)

41

$(MeO)_3P(OC)(Cp)FeC\equiv CPh$

42

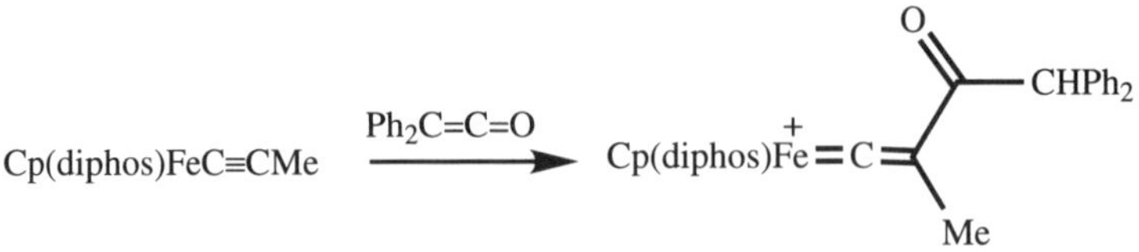

$Cp(diphos)FeC\equiv CMe \xrightarrow{Ph_2C=C=O} Cp(diphos)\overset{+}{Fe}=C=C(Me)C(O)CHPh_2$ (25)

44   **45**

5.5 NUCLEOPHILIC ADDITION TO KETENES

The vinylidenetitanocene **46** reacted with t-Bu$_2$C=C=O to give the crystalline metalloxetane **47** (equation 26).[156]

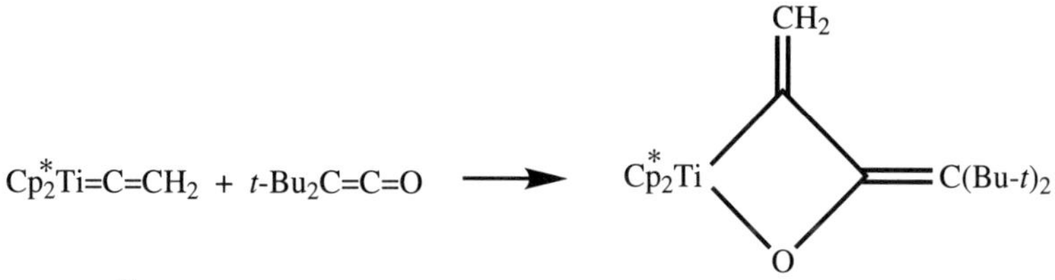

$$Cp_2^*Ti=C=CH_2 + t\text{-}Bu_2C=C=O \longrightarrow \text{metalloxetane } \mathbf{47}$$

46 $Cp^* = C_5Me_5$ **47**   (6)

Enol ethers also act as carbon nucleophiles on ketenes (equation 27),[157] and the degenerate rearrangement in equation 28 involves attack by carbon.[158]

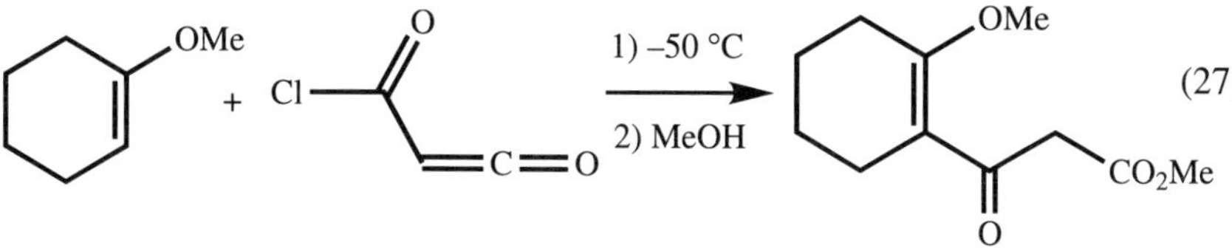

(27)

(28)

5.5.2.5 Other Nucleophiles.

The addition of Me$_3$SiLi and Me$_3$SiMe$_2$SiLi to ketenes such as dimesitylketene has been observed (equation 1).[2,159] Both of these reagents can be generated during the reaction of hexamethyldisilane with MeLi, and mixtures of the products observed. Reactions of some other nucleophiles are shown in equations 2–14.[66,136,160–169] In the reaction of equation 9, selectivity for attack on the least crowded side of the ketene was observed.[165]

$$Me_3SiSiMe_3 \xrightarrow{MeLi} Me_3SiLi + Me_3SiMe_2SiLi \xrightarrow{Mes_2C=C=O}$$

$$Mes_2C=C(OH)SiMe_3 + Mes_2C=C(OH)SiMe_2SiMe_3 \quad (1)$$

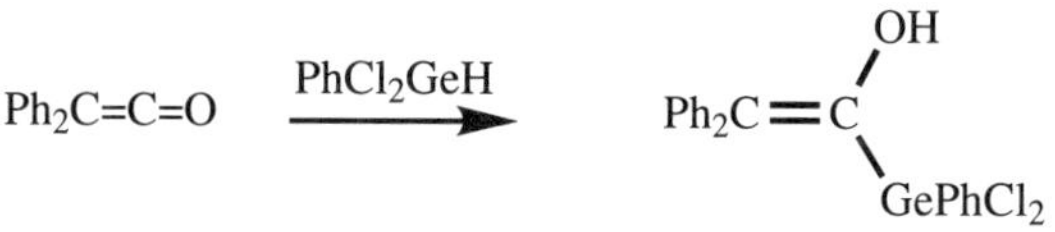

$$Ph_2C=C=O \xrightarrow{PhCl_2GeH} Ph_2C=C(OH)(GePhCl_2) \quad (2)^{66}$$

$$(CF_3)_2C=C=O \xrightarrow{PhCH_2SH} (CF_3)_2CHCOSCH_2Ph \quad (3)^{160}$$

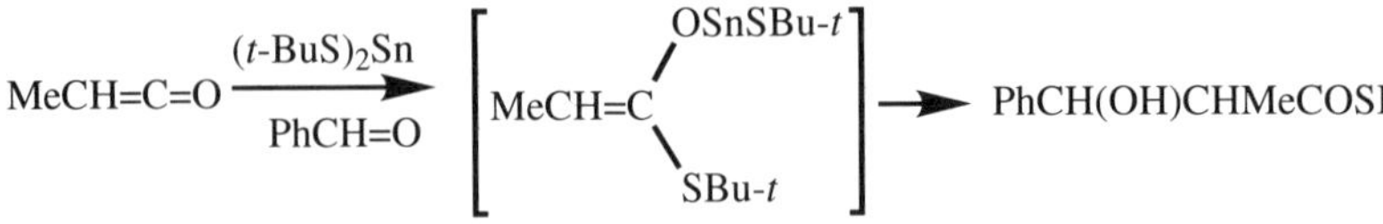

REACTIONS OF KETENES

$$\text{MeCH=C=O} \xrightarrow[\text{PhCH=O}]{(t\text{-BuS})_2\text{Sn}} \left[\text{MeCH=C} \begin{array}{c} \text{OSnSBu-}t \\ \text{SBu-}t \end{array} \right] \longrightarrow \text{PhCH(OH)CHMeCOSBu-}t \quad (4)^{161}$$

$$\text{CH}_2\text{=C=O} \xrightarrow{\text{EtSH}} \text{CH}_3\text{COSEt} \quad (5)^{136}$$

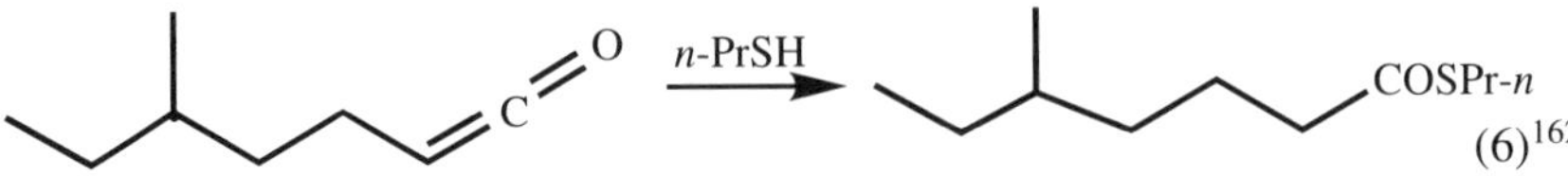

(6)[162] — reaction of branched alkyl ketene with n-PrSH to give branched alkyl COSPr-n

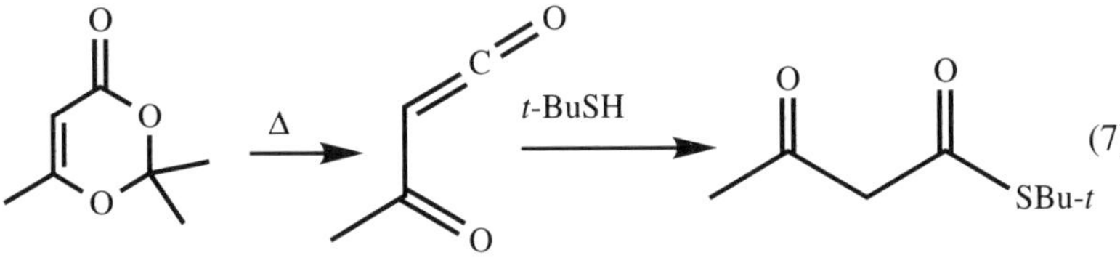

(7)[163] — dioxinone → acylketene + t-BuSH → β-ketothioester SBu-t

$$\text{CH}_2\text{=C=O} \xrightarrow{\text{CH}_3\text{CS}_2\text{H}} \underset{\substack{\text{O}\quad\text{S}\\ \|\quad\|}}{\text{CH}_3\text{CSCCH}_3} \quad (8)^{164}$$

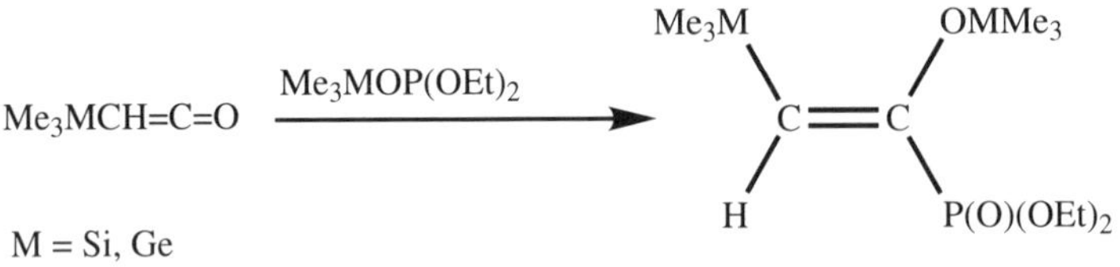

$$\text{Me}_3\text{MCH=C=O} \xrightarrow{\text{Me}_3\text{MOP(OEt)}_2} \underset{\text{H}}{\overset{\text{Me}_3\text{M}}{\text{C}}}=\underset{\text{P(O)(OEt)}_2}{\overset{\text{OMMe}_3}{\text{C}}} \quad (9)^{165}$$

M = Si, Ge

$$\text{Ph}_2\text{C=C=O} \xrightarrow{\text{RR}^1\text{PSiMe}_3} \text{Ph}_2\text{C=C(OSiMe}_3)\text{PRR}^1 \quad (10)^{166}$$

$$\text{Ph}_2\text{C=C=O} \xrightarrow{t\text{-BuPH}_2} \text{Ph}_2\text{CHCOPHBu-}t \quad (11)^{166}$$

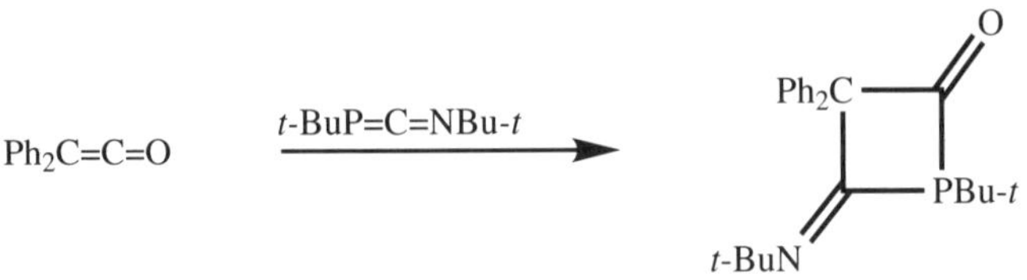

(12)[167] — Ph₂C=C=O + t-BuP=C=NBu-t → four-membered ring with Ph₂C, C=O, PBu-t, C=NBu-t

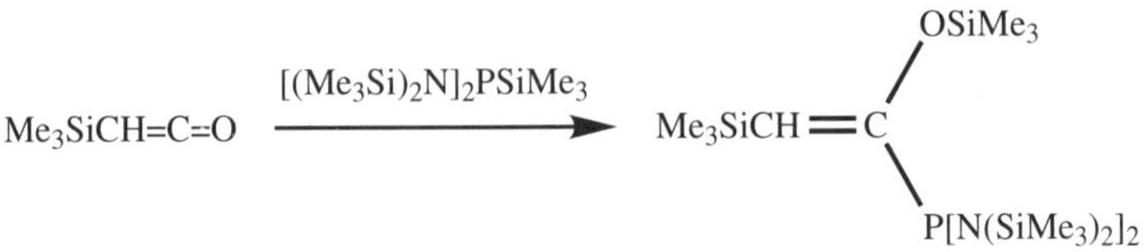

$$\text{Me}_3\text{SiCH=C=O} \xrightarrow{[(\text{Me}_3\text{Si})_2\text{N}]_2\text{PSiMe}_3} \text{Me}_3\text{SiCH}=\underset{\text{P[N(SiMe}_3)_2]_2}{\overset{\text{OSiMe}_3}{\text{C}}} \quad (13)^{168}$$

$$\text{Me}_3\text{SiCH=C=O} \xrightarrow{(\text{Me}_2\text{N})_2\text{C=PSiMe}_3} \text{Me}_3\text{SiCH=C}\begin{matrix}\text{OSiMe}_3 \\ \\ \text{P=C(NMe}_2)_2\end{matrix} \quad (14)^{169}$$

$$\underset{\text{Me}}{\overset{\text{Ph}}{>}}\text{C=C=O} \xrightarrow{\text{Me}_2\text{POSiMe}_3} \underset{\text{Me}}{\overset{\text{Ph}}{>}}\text{C=C}\underset{\text{OSiMe}_3}{\overset{\text{P(O)Me}_2}{<}} \quad (15)^{170}$$

The addition of niobium complexes to the C=O bond of ketenes to give stereoisomeric complexes has been observed (equation 16).[171]

$$\text{PhCR=C=O} + \text{Cp}'_2\text{NbCl} \longrightarrow \underset{R}{\overset{Ph}{>}}\text{C=C}\underset{\text{NbCp}'_2\text{Cl}}{\overset{O}{<|}} + \underset{Ph}{\overset{R}{>}}\text{C=C}\underset{\text{NbCp}'_2\text{Cl}}{\overset{O}{<|}} \quad (16)^{171}$$

$$\text{Cp}' = \text{C}_5\text{Me}_4\text{SiMe}_3$$

References

1. Seikaly, H. R.; Tidwell, T. T. *Tetrahedron* **1986**, *42*, 2587–2613.
2. Gong, L.; Leung-Toung, R.; Tidwell, T. T. *J. Org. Chem.* **1990**, *55*, 3534–3639.
3. Leung-Toung, R.; Tidwell, T. T. *J. Am. Chem. Soc.* **1990**, *112*, 1042–1048.
4. Allen, A. D.; Baigrie, L. M.; Gong, L.; Tidwell, T. T. *Can. J. Chem.* **1991**, *69*, 138–145.
5. Baigrie, L. M.; Seikaly, H. R.; Tidwell, T. T. *J. Am. Chem. Soc.* **1985**, *107*, 5391–5396.
6. Baigrie, L. M.; Lenoir, D.; Seikaly, H. R.; Tidwell, T. T. *J. Org. Chem.* **1985**, *50*, 2105–2109.
7. Häner, R.; Laube, T.; Seebach, D. *J. Am. Chem. Soc.* **1985**, *107*, 5396–5403.
8. Nguyen, M. T.; Hegarty, A. F. *J. Am. Chem. Soc.* **1984**, *106*, 1552–1557.
9. Andraos, J.; Kresge, A. J.; Peterson, M. R.; Csizmadia, I. G. *J. Mol. Struct. (Theochem)* **1991**, *232*, 155–177.
10. Andraos, J.; Kresge, A. J. *J. Mol. Struct. (Theochem)* **1991**, *233*, 165–184.
11. McAllister, M. A.; Kresge, A. J.; Csizmadia, I. G. *J. Mol. Struct. (Theochem)* **1992**, *258*, 399–400.
12. Skancke, P. N. *J. Phys. Chem.* **1992**, *96*, 8065–8069.
13. Allen, B. M.; Hegarty, A. F.; O'Neill, P.; Nguyen, M. T. *J. Chem. Soc., Perkin Trans. 2* **1992**, 927–934.
14. Nguyen, M. T.; Ruelle, P. *Chem. Phys. Lett.* **1987**, *138*, 486–488.
15. Leung-Toung, R. Ph.D. Thesis, University of Toronto, 1990.

16. Schleyer, P. v. R. Unpublished results privately communicated.
17. Bothe, E.; Dessouki, A. M.; Schulte-Frohlinde, D. *J. Phys. Chem.* **1980**, *84*, 3270–3272.
18. Chelain, E.; Goumont, R.; Hamon, L.; Parlier, A.; Rudler, M.; Rudler, H.; Daran, J.; Vaissermann, J. *J. Am. Chem. Soc.* **1992**, *114*, 8088–8098.
19. Blake, P. G.; Davies, H. H. *J. Chem. Soc., Perkin Trans. 2* **1972**, 321–323.
20. Bothe, E.; Meier, H.; Schulte-Frohlinde, D.; von Sonntag, C. *Angew. Chem., Int. Ed. Engl.* **1976**, *15*, 380.
21. Tidwell, T. T. *Acc. Chem. Res.* **1990**, *23*, 273–279.
22. Allen, A. D.; Tidwell, T. T. *J. Am. Chem. Soc.* **1987**, *109*, 2774–2780.
23. Allen, A. D.; Stevenson, A.; Tidwell, T. T. *J. Org. Chem.* **1989**, *54*, 2843–2848.
24. Allen, A. D.; Kresge, A. J.; Schepp, N. P.; Tidwell, T. T. *Can. J. Chem.* **1987**, *65*, 1719–1723.
25. Allen, A. D.; McAllister, M. A.; Tidwell, T. T. *Tetrahedron Lett.* **1993**, *34*, 1095–1098.
26. Kabir, S. H.; Seikaly, H. R.; Tidwell, T. T. *J. Am. Chem. Soc.* **1979**, *101*, 1059–1060.
27. Allen, A. D.; Tidwell, T. T. *Tetrahedron Lett.* **1991**, *32*, 847–850.
28. Allen, A. D.; Gong, L.; Tidwell, T. T. *J. Am. Chem. Soc.* **1990**, *112*, 6396–6397.
29. Allen, A. D.; Andraos, J.; Kresge, A. J.; McAllister, M. A.; Tidwell, T. T. *J. Am. Chem. Soc.* **1992**, *114*, 1878–1879.
30. Andraos, J.; Kresge, A. J. *J. Photochem. Photobiol. A: Chem.* **1991**, *57*, 165–173.
31. Andraos, J.; Kresge, A. J. *J. Am. Chem. Soc.* **1992**, *114*, 5643–5646.
32. Jones, J., Jr.; Kresge, A. J. *J. Org. Chem.* **1992**, *57*, 6467–6469.
33. Urwyler, B.; Wirz, J. *Angew. Chem., Int. Ed. Engl.* **1990**, *29*, 790–791.
34. Barra, M.; Fisher, T. A.; Cernigliaro, G. J.; Sinta, R.; Scaiano, J. C. *J. Am. Chem. Soc.* **1992**, *114*, 2630–2634.
34a. Andraos, J.; Chiang, Y.; Huang, C.-G.; Kresge, A. J.; Scaiano, J. C. *J. Am. Chem. Soc.* **1993**, *115*, 10605–10610.
34b. Almstead, J. K.; Urwyler, B.; Wirz, J. *J. Am. Chem. Soc.* **1994**, *116*, 954–960.
34c. Andraos, J.; Kresge, A. J.; Popic, V. V. *J. Am. Chem. Soc.* **1994**, *116*, 961–967.
35. Chiang, Y.; Kresge, A. J.; Pruszynski, P.; Schepp, N. P.; Wirz, J. *Angew. Chem., Int. Ed. Engl.* **1990**, *29*, 792–794.
36. Kabalina, G. A.; Askerov, D. B.; Dyumaev, K. M. *Zh. Org. Khim.* **1982**, *18*, 456–457; *Engl. Transl.* **1982**, *15*, 400–401.
37. Lacey, R. N. *The Chemistry of Alkenes;* Patai, S., Ed.; Wiley: New York, 1964; pp. 1161–1227.
38. Satchell, D. P. N; Satchell, R. S. *Chem. Soc. Rev.* **1975**, *4*, 231–250.
39. Donohoe, G.; Satchell, D. P. N; Satchell, R. S. *J. Chem. Soc., Perkin Trans. 2* **1990**, 1671–1674.
40. Poon, N. L.; Satchell, D. P. N. *J. Chem. Soc., Perkin Trans. 2* **1984**, 1083–1087.
41. Satchell, D. P. N., Satchell, M. J. *Z. Naturforsch. B* **1991**, *46*, 391–392.
42. Poon, N. L.; Satchell, D. P. N. *J. Chem. Soc., Perkin Trans. 2* **1985**, 1551–1554.
43. Poon, N. L.; Satchell, D. P. N. *J. Chem. Soc., Perkin Trans. 2* **1983**, 1381–1383.

44. Poon, N. L.; Satchell, D. P. N. *J. Chem. Soc., Perkin Trans. 2* **1986,** 1485–1490.
45. Lillford, P. J.; Satchell, D. P. N. *J. Chem. Soc. B,* **1968,** 889–897.
46. Meinwald, J.; Gassman, P. G. *J. Am. Chem. Soc.* **1960,** *82,* 2857–2863.
47. Meinwald, J.; Lewis, A.; Gassman, P. G. *J. Am. Chem. Soc.* **1962,** *84,* 977–983.
48. Brook, P. R.; Brophy, B. V. *J. Chem. Soc., Perkin Trans. 1* **1985,** 2509–2513.
49. Fessner, W.; Sedelmeier, G.; Spurr, P. R.; Rihs, G.; Prinzbach, H. *J. Am. Chem. Soc.* **1987,** *109,* 4626–4642.
50. Nikolaev, V. A.; Korneev, S. M.; Terent'eva, I. V.; Korobitsyna, I. K. *Zh. Org. Khim.* **1991,** *27,* 2085–2100. *Engl. Transl.* **1991,** *27,* 1845–1858.
50a. Frey, J.; Rappoport, Z. *J. Am. Chem. Soc.* **1995,** *117,* 1161–1162.
51. Leung-Toung, R.; Peterson, M. R.; Tidwell, T. T.; Csizmadia, I. G. *J. Mol. Struct. (Theochem)* **1989,** *183,* 319–330.
52. Brady, W. T.; Vaughn, W. L.; Hoff, E. F. *J. Org. Chem.* **1969,** *34,* 843–845.
52a. Reactivity data reported in reference (52) for haloketenes is probably unreliable, as it is doubtful if these ketenes were actually isolated in solution: Prof. W. T. Brady, private communication.
53. Tille, A.; Pracejus, H. *Chem. Ber.* **1967,** *100,* 196–210.
54. Jähme, J.; Rüchardt, C. *Tetrahedron Lett.* **1982,** *23,* 4011–4014.
55. Jähme, J.; Rüchardt, C. *Angew. Chem., Int. Ed. Engl.* **1981,** *20,* 885–887.
56. Schiess, P.; Eberle, M.; Huys-Francotte, M.; Wirz, J. *Tetrahedron Lett.* **1984,** *25,* 2201–2204.
57. Netto-Ferreira, J. C.; Scaiano, J. C. *Can. J. Chem.* **1993,** *71,* 1209–1215.
58. Lüning, U.; Baumstark, R.; Schyja, W. *Leibigs Ann. Chem.* **1991,** 999–1002.
59. Micovic, V. M.; Rogic, M. M.; Mihailovic, M. L. *Tetrahedron* **1957,** *1,* 340–342.
60. Lam, C. Y., University of Toronto, unpublished results.
61. Biali, S. E.; Rappoport, Z. *J. Am. Chem. Soc.* **1984,** *106,* 5641–5653.
62. Naef, F.; Decorzant, R. *Tetrahedron* **1986,** *42,* 3245–3250.
63. Tomioka, K.; Shindo, M.; Koga, K. *Tetrahedron Lett.* **1990,** *31,* 1739–1740.
64. Tomioka, K.; Shindo, M.; Koga, K. *J. Org. Chem.* **1990,** *55,* 2276–2277.
65. Frainnet, E.; Causse, J. *Bull. Soc. Chim. Fr.* **1968,** 3034.
66. Satge, J.; Riviere, P. *J. Organomet. Chem.* **1969,** *16,* 71–82.
67. Hneihen, A. S.; Bruno, J. W.; Huffman, J. C. *J. Organomet. Chem.* **1990,** *382,* 361–373.
68. Fuson, R. C.; Foster, R. E.; Shenk, W. J., Jr.; Maynert, E. W. *J. Am. Chem. Soc.* **1945,** *67,* 1937–1939.
69. Nugiel, D. A.; Rappoport, Z. *J. Am. Chem. Soc.* **1985,** *107,* 3669–3676.
70. Wuest, J. D.; Madonik, A. M.; Gordon, D. C. *J. Org. Chem.* **1977,** *42,* 2111–2113.
71. Di Nunno, L.; Scilimati, A. *Tetrahedron* **1988,** *44,* 3639–3644.
72. Lombardo, L. *Tetrahedron Lett.* **1985,** *26,* 381–384.
73. Boeckman, R. K., Jr.; Pruitt, J. R. *J. Am. Chem. Soc.* **1989,** *111,* 8286–8288.
74. Rahman, S. S.; Wakefield, B. J.; Roberts, S. M.; Dowle, M. D. *J. Chem. Soc., Chem. Commun.* **1989,** 303–304.

75. Davies, H. G.; Rahman, S. S.; Roberts, S. M.; Wakefield, B. J.; Winders, J. A. *J. Chem. Soc., Perkin Trans. 1* **1987**, 85–89.
76. Butt, H. G.; Davies, H. G.; Dawson, M. J.; Lawrence, G. C.; Leaver, J.; Roberts, S. M.; Turner, M. K.; Wakefield, B. J.; Wall, W. F.; Winders, J. A. *J. Chem. Soc., Perkin Trans. 1* **1987**, 903–907.
77. Liang, L.; Ramadeshan, M.; MaGee, D. I. *Tetrahedron* **1993**, *49*, 2159–2168.
78. Vernier, J.-M.; Hegedus, L. S.; Miller, D. B. *J. Org. Chem.* **1992**, *57*, 6914–6920.
79. Naidorf-Meir, S.; Hassner, A. *J. Org. Chem.* **1992**, *57*, 5102–5105.
79a. Chen, C.; Quinn, E. K.; Olmstead, M. M.; Kurth, M. *J. Org. Chem.* **1993**, *58*, 5011–5014.
80. Yoshida, K.; Yamishita, Y. *Tetrahedron Lett.* **1966**, 693–696.
81. O'Neill, P.; Hegarty, A. F. *J. Org. Chem.* **1987**, *52*, 2113–2114.
82. O'Neill, P.; Hegarty, A. F. *J. Chem. Soc., Chem. Commun.* **1987**, 744–745.
83. Sato, M.; Yoneda, N.; Katagiri, N.; Watanabe, H.; Kaneko, C. *Snythesis* **1986**, 672–674.
84. Ayral-Kaloustian, S.; Wolff, S.; Agosta, W. C. *J. Org. Chem.* **1978**, *43*, 3314–3319.
85. Christl, M.; Lanzendörfer, U.; Grötsch, M. M.; Hegmann, J.; Ditterich, E.; Hüttner, G.; Peters, E.; Peters, K.; von Schnering, H. G. *Chem. Ber.* **1993**, *126*, 797–802.
86. Malherbe, R.; Rist, G.; Bellus, D. *J. Org. Chem.* **1983**, *48*, 860–869.
87. Rosini, G.; Spineti, G. G.; Foresti, E.; Pradella, G. *J. Org. Chem.* **1981**, *46*, 2228–2230.
88. Marino, J. P.; Neisser, M. *J. Am. Chem. Soc.* **1981**, *103*, 7687–7689.
89. Marino, J. P.; Perez, A. D. *J. Am. Chem. Soc.* **1984**, *106*, 7643–7644.
90. Posner, G. H.; Asirvatham, E.; Ali, S. F. *J. Chem. Soc., Chem. Commun.* **1985**, 542–543.
91. Marino, J. P.; Kim, M. W. *Tetrahedron Lett.* **1987**, *28*, 4925–4928.
91a. Arjona, O.; Fernandez, P.; de la Pradilla, R. B.; Morente, M.; Plumet, J. *J. Org. Chem.* **1993**, *58*, 3172–3175.
92. Kosugi, H.; Tagami, K.; Takahashi, A.; Kanna, H.; Uda, H. *J. Chem. Soc., Perkin Trans. 1* **1989**, 935–943.
93. Ishida, M.; Muramaru, H.; Kato, S. *Synthesis,* **1989**, 562–564.
94. Nitta, M.; Nakatani, H. *Chem. Lett.* **1978**, 957–960.
95. Mikolajczyk, M.; Omelanczuk, J.; Michalski, J. *Bull. Acad. Pol. Sci., Ser. Sci. Chim.* **1969**, *17*, 155–156; *Chem. Abstr.* **1969**, *71*, 60633j.
96. Tempesti, G.; Giuffre, L.; Sioli, G.; Fornaroli, M.; Airoldi, G. *J. Chem. Soc., Perkin Trans. 1* **1974**, 771–773.
97. Chihara, T.; Teratani, S.; Ogawa, H. *J. Chem. Soc., Chem. Commun.* **1981**, 1120.
98. Briody, J. M.; Satchell, D. P. N. *Tetrahedron* **1966**, *22*, 2649–2653.
99. Lillford, P. J.; Satchell, D. P. N. *J. Chem. Soc. B* **1967**, 360–365; **1968**, 54–57; **1970**, 1016–1019.
100. Lee, I.; Song, C. H.; Uhm, T. S. *J. Phys. Org. Chem.* **1988**, *1*, 83–90.
101. Schultz, A. G.; Kulkarni, Y. S. *J. Org. Chem.* **1984**, *49*, 5202–5206.
102. Dunbar, R. E.; White, G. C. *J. Org. Chem.* **1958**, *23*, 915–916.

103. Martin, J. C.; Hoyle, V. A., Jr.; Brannock, K. C. *Tetrahedron Lett.* **1965**, 3589–3594.
104. Birkofer, L.; Schramm, J. *Liebigs Ann. Chem.* **1977**, 760–766.
105. Lutsenko, I. F.; Baukov, Yu. I.; Kostyuk, A. S.; Salvelyeva, N. I.; Krysina, V. K. *J. Organomet. Chem.* **1969**, *17*, 241–262.
106. Satge, J.; Rivere-Baudet, M. *Bull. Soc. Chim. Fr.* **1968**, 4093–4096.
107. Medici, A.; Fantin, G.; Fogagnoio, M.; Pedrini, P.; Dondoni, A.; Andreetti, G. D. *J. Org. Chem.* **1984**, *49*, 590–596.
108. Tsuge, O.; Urano, S.; Oe, K. *J. Org. Chem.* **1980**, *45*, 5130–5136.
109. Mitkidou, S.; Papadopoulos, S.; Stephanidou-Stephanatou, J.; Terzis, A.; Mentzafos, D. *J. Chem. Soc., Perkin Trans. 1* **1990**, 1025–1031.
110. Miller, J. R.; Pulley, S. R.; Hegedus, L. S.; DeLombaert, S. *J. Am. Chem. Soc.* **1992**, *114*, 5602–5607.
111. Edstrom, E D. *J. Am. Chem. Soc.* **1991**, *113*, 6690–6692.
112. Maurya, R.; Pittol, C. A.; Pryce, R. J.; Roberts, S. M.; Thomas, R. J.; Williams, J. O. *J. Chem. Soc., Perkin Trans. 1,* **1992**, 1617–1621.
113. Roberts, S. M.; Smith, C.; Thomas, R. J. *J. Chem. Soc., Perkin Trans. 1* **1990**, 1493–1495.
114. Gompper, R.; Wolf, U. *Liebigs. Ann. Chem.* **1979**, 1388–1405.
115. Abou-Gharbia, M.; Kecha, D. M.; Zacharias, D. E.; Swern, D. *J. Org. Chem.* **1985**, *50*, 2224–2228.
116. Katagiri, N.; Niwa, R.; Kato, T. *Chem. Pharm. Bull.* **1983**, *31*, 2899–2904.
117. Carboni, B.; Toupet, L.; Carrie, R. *Tetrahedron* **1987**, *43*, 2293–2302.
118. Schaumann, E.; Grabley, S.; Henriet, M.; Ghosez, L.; Touillaux, R.; Declercq, J. P.; Germain, G.; Van Meerssche, M. *J. Org. Chem.* **1980**, *45*, 2951–2955.
119. Quinkert, G.; Nestler, H. P.; Schumacher, B.; del Grosso, M.; Dürner, G.; Bats, J. W. *Tetrahedron Lett.* **1992**, *33*, 1977–1980.
119a. Dieterich, P.; Young, D. W. *Tetrahedron Lett.* **1993**, *34*, 5455–5458.
119b. Boeckman, R. K., Jr.; Perni, R. B. *J. Org. Chem.* **1986**, *51*, 5486–5489.
119c. MaGee, D. I.; Ramaseshan, M. *Synlett* **1994**, 743–744.
120. Zaitseva, L. G.; Sarkisyan, A. Ts.; Erlikh, R. D. *Zh. Org. Khim.* **1982**, *18*, 1330–1331; *Engl. Transl.* **1982**, *18*, 1156–1157.
121. Benati, L.; Montevecchi, P. C.; Spagnolo, P.; Foresti, E. *J. Chem. Soc., Perkin Trans. 1,* **1992**, 2845–2850.
122. Eberlin, M. N.; Takahata, Y.; Kascheres, C. *J. Org. Chem.* **1990**, *55*, 5150–5155.
123. Himbert, G.; Schwickerath, W. *Liebigs Ann. Chem.* **1984**, 85–97.
124. Nadler, F. B.; Zipory, E. S.; Rappoport, Z. *J. Org. Chem.* **1991**, *56*, 4241–4246.
125. Moore, H. W.; Duncan, W. G. *J. Org. Chem.* **1973**, *38*, 156–158.
126. Staudinger, H. *Chem. Ber.* **1907**, *40*, 1145–1148.
127. Gilman, H.; Heckert, L. C. *J. Am. Chem. Soc.* **1920**, *42*, 1010–1014.
128. Combret, J.-C. *Ann. Chim. (Paris)* **1969**, 481–496.
129. Schaumann, E.; Walter, W. *Chem. Ber.* **1974**, *107*, 3562–3573.
130. Woodbury, R. P.; Long, N. P.; Rathke, M. W. *J. Org. Chem.* **1977**, *43*, 376.

131. Lebedev, S. A.; Ponomarev, S. V.; Lutsenko, I. F. *Zh. Obshch. Khim.* **1972**, *42*, 647–651; *Engl. Transl.* **1972**, *42*, 643–647.
132. Kita, Y.; Matsuda, S.; Kitagaki, S.; Tsuzuki, Y.; Akai, S. *Synlett* **1991**, 401–402.
133. Takeuchi, S.; Miyoshi, N.; Ohgo, Y. *Chem. Lett.* **1992**, 551–554.
133a. Miyoshi, N.; Takeuchi, S.; Ohgo, Y. *Bull. Chem. Soc. Jpn.* **1994**, *67*, 445–451.
134. Hertenstein, U.; Hünig, S.; Reichelt, H.; Schaller, R. *Chem. Ber.* **1982**, *115*, 261–287; *Ibid.* **1986**, *119*, 699–721.
135. Fehr, C.; Galindo, J. *J. Org. Chem.* **1988**, *53*, 1828–1830.
136. Hurd, C. D.; Williams, J. W. *J. Am. Chem. Soc.* **1936**, *58*, 962–968.
137. Dean, D. O.; Dickinson, W. B.; Quayle, O. R.; Lester, C. T. *J. Am. Chem. Soc.* **1950**, *72*, 1740–1741.
138. Adlington, R. M.; Barrett, A. G. M. *J. Chem. Soc., Chem. Commun.* **1979**, 1122–1123.
139. El Gomati, T.; Firl, J.; Ugi, I. *Chem. Ber.* **1977**, *110*, 1603–1605.
140. Ermolov, A. F.; Eleev, A. F. *Zh. Org. Khim.* **1983**, *19*, 1340–1341; *Engl. Transl.* **1983**, *19*, 1198–1199.
141. Himbert, G. *Liebigs Ann. Chem.* **1979**, 1828–1846.
142. Henn, L.; Himbert, G. *Chem. Ber.* **1981**, *114*, 1015–1026.
143. Fountain, K.R.; Heinze, P.; Maddex, D.; Gerhart, G.; John, P. *Can. J. Chem.* **1980**, *58*, 1939–1946.
144. Machiguchi, T.; Yamabe, S. *Chem. Lett.* **1990**, 1511–1512.
145. Burlachenko, G. S.; Baukov, Yu. I.; Lutsenko, I. F. *Zh. Obshch. Khim.* **1965**, *35*, 933–934; **1968**, *38*, 2815–2816. *Engl. Transl.* **1965**, *35*, 939; **1968**, *38*, 2716.
146. Raynolds, P. W.; DeLoach, J. A. *J. Am. Chem. Soc.* **1984**, *106*, 4564–4570.
147. Hickmott, P. W.; Miles, G. J.; Sheppard, G.; Urbani, R.; Yoxall, C. T. *J. Chem. Soc., Perkin Trans. 1* **1973**, 1514–1519.
148. Hickmott, P. W.; Hargreaves, J. R. *Tetrahedron* **1967**, *23*, 3151–3159.
149. Feiler, L. A.; Huisgen, R.; Koppitz, P. *J. Chem. Soc., Chem. Commun.* **1974**, 405–406.
150. Mitsudo, T.; Kadokura, M.; Watanabe, Y. *J. Chem. Soc., Chem. Commun.* **1986**, 1539–1541.
151. Mitsudo, T.; Kadokura, M.; Watanabe, Y. *J. Org. Chem.* **1987**, *52*, 3186–3192.
152. Bucheister, A.; Klemarczyk, P.; Rosenblum, M. *Organometallics*, **1982**, *1*, 1679–1684.
153. Chen, L. S.; Lichtenberg, D. W.; Robinson, P. W.; Yamamoto, Y.; Wojcicki, A. *Inorg. Chim. Acta* **1977**, *25*, 165–172.
154. Goldschmidt, Z.; Antebi, S.; Cohen, D.; Goldberg, I. *J. Organomet. Chem.* **1984**, *273*, 347–359.
155. Barrett, A. G. M.; Carpenter, N. E.; Mortier, J.; Sabat, M. *Organometallics* **1990**, *9*, 151–156.
156. Beckhaus, R.; Strauss, I.; Wagner, T.; Kiprof, P. *Angew. Chem., Int. Ed. Engl.* **1993**, *32*, 264–266.
157. Schönwalder, K.; Kollat, P.; Strezowski, J. J.; Effenberger, F. *Chem. Ber.* **1984**, *117*, 3280–3296.

158. Wentrup, C.; Netsch, K. *Angew. Chem., Int. Ed. Engl.* **1984**, *23*, 802.
159. Nadler, E. B.; Rappoport, Z. *Tetrahedron Lett.* **1990**, *31*, 555–558.
160. Bekker, R. A.; Rozov, L. A.; Popkova, V. Ya.; Knunyants, I. L. *Izv. Akad. Nauk SSSR, Ser. Khim.* **1982**, 2408–2410; *Engl. Transl.* **1982**, 2123–2125.
161. Mukaiyama, T.; Yamasaki, N.; Stevens, R. W.; Murakami, M. *Chem. Lett.* **1986**, 213–216.
162. Vasi, I. G.; Nanavati, N. T. *J. Indian Chem. Soc.* **1980**, *57*, 744–745; *Chem. Abstr.* **1981**, *94*, 15159.
163. Lopez-Alvardo, P.; Avendano, C.; Menendez, J. C. *Synth. Commun.* **1992**, *22*, 2329–2333.
164. Barnikow, G.; Martin, A. A. *J. Prakt. Chem.* **1983**, *325*, 337–340.
165. Ponomarev, S. V.; Moskalenko, A. I.; Lutsenko, I. F. *Zh. Obshch. Khim.* **1978**, *48*, 296–301; *Engl. Transl.* **1978**, *48*, 263–267.
166. Kolodyazhnyi, I. I.; Kukhar, V. P. *Zh. Obshch. Khim.* **1981**, *51*, 2189–2194; *Engl. Transl.* **1981**, *51*, 1883–1887.
167. Kolodyazhnyi, O. I. *Zh. Obshch. Khim.* **1983**, *53*, 1226–1233; *Engl. Transl.* **1983**, *53*, 1093–1099.
168. Romanenko, V. D.; Shul'gin, V. F.; Scopenko, V. V.; Markovski, L. N. *J. Chem. Soc., Chem. Commun.* **1983**, 808–809.
169. Markovskii, L. N.; Romanenko, V. D.; Pidvarko, T. V. *Zh. Obshch. Khim.* **1983**, *53*, 1672–1673; *Engl. Transl.* **1983**, *53*, 1502–1503.
170. Well, M.; Schmutzler, R. *Phosphorous, Sulfur Silicon Relat. Elem.* **1992**, *72*, 189–199.
171. Halfon, S. E.; Fermin, M. C.; Bruno, J. W. *J. Am. Chem. Soc.* **1989**, *111*, 5490–5491.

5.5.3 Wittig Reactions

Wittig reactions of ketenes forming allenes involve nucleophilic attack by carbon on C_α of the ketenes. This reaction was first observed by Lüscher in the laboratory of Staudinger (equation 1),[1,2] and was latter developed by Wittig and Haag.[3] Bestmann and Hartung[4] and others[5] utilized the Horner–Emmons reaction of carboethoxyphosphonium ylides with ketenes and with acyl chlorides. The latter reaction was proposed to possibly involve in situ generation of ketenes (equation 2).

$$Ph_2C=C=O + Ph_3P=CPh_2 \longrightarrow Ph_2C=C=CPh_2 \quad (1)$$

$$PhCH_2COCl \xrightarrow{Et_3N} [PhCH=C=O] \xrightarrow{Ph_3P=CMeCO_2Et} PhCH=C=CMeCO_2Et \quad (2)$$

The use of phosphonates in this reaction has also been developed (equation 3),[6] as well as many other examples.[6–12] Ketene itself gives terminal allenes.[13,14]

$$\underset{Et}{\overset{Ph}{>}}C=C=O + Et_2O_3P-\overset{Ph}{\underset{CO_2Et}{\overset{|}{C}}}Na^+ \longrightarrow \underset{Et}{\overset{Ph}{>}}C=C=\underset{CO_2Et}{\overset{Ph}{<}} \quad (3)$$

Cinnamylidenetriphenylphosphorane (**1**) reacts with dimethyl- or diphenylketene to give normal Wittig reactions (equation 4) but the α-phenylphosphorane **2** adds two moles of $Me_2C=C=O$ to form **3** after hydrolysis (equation 5).[15]

$$R_2C=C=O + E-PhCH=CHCH=PPh_3 \longrightarrow E-PhCH=CHCH=C=CR_2 \quad (4)$$
$$\mathbf{1}$$

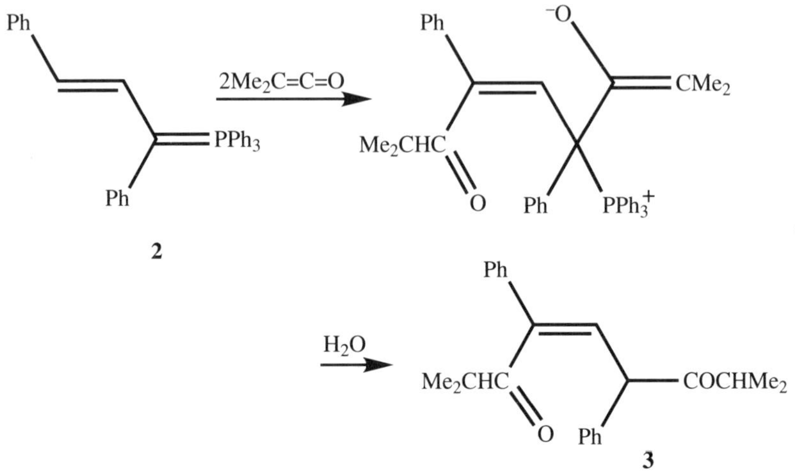

(5)

Reaction of unsymmetrical ketenes with optically active phosphinylacetate **4** gave enantiomerically enriched **5** (equation 6).[10]

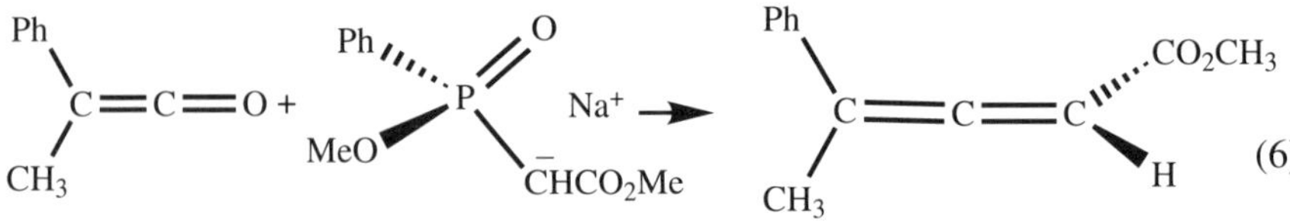

(6)

Photolysis of chromium–alkoxycarbene complexes produces reactive intermediates that undergo reactions characteristic of ketenes, and these intermediates also react with phosphoranes substituted with carboalkoxy or sulfone groups to give allenes (equation 7) which in some cases were not isolated but were hydrolyzed to ketones such as **7** or rearranged to 1,3-dienes.[16]

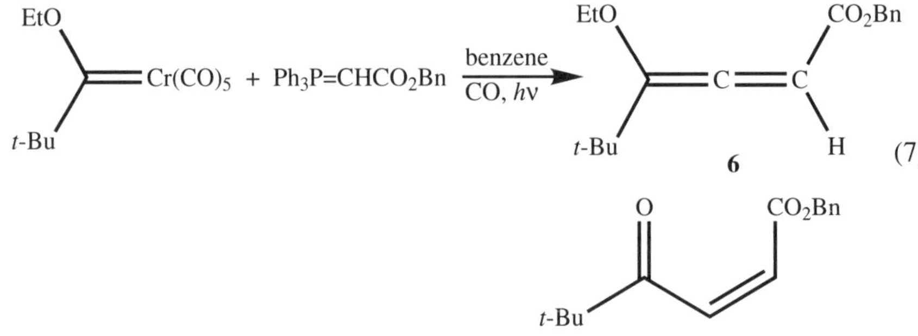

(7)

Ketenes also give aza-Wittig reactions with iminophosphoranes to form ketenimines in a reaction discovered by Staudinger.[17,18] When allyliminophosphoranes are utilized the product ketenimines give aza-Claisen rearrangements (equation 8).[19] N-Acylketenimines are also available by this route (equation 9).[20]

$$RR^1C{=}C{=}O \xrightarrow{Me_2C{=}CHCH_2N{=}PPh_3} \begin{array}{c} RR^1C{=}C{=}N \\ \diagup \\ Me_2C{=} \end{array} \longrightarrow \underset{\underset{Me_2CCH{=}CH_2}{|}}{RR^1CCN} \quad (8)$$

$$Ph_2C{=}C{=}O \xrightarrow{PhCON{=}PPh_3} Ph_2C{=}C{=}NCOPh \quad (9)$$

As discussed in Section 4.6, phosphaketene ylides such as $Ph_3P{=}C{=}C{=}O$ react with nucleophiles at the carbonyl carbon to form reactive acyl Wittig reagents.

References

1. Lüscher, G. Dissertation, ETH-Zürich, **1922,** as cited in ref. 2.
2. Lang, R. W.; Hansen, H. *Helv. Chim. Acta* **1980,** *63,* 438–455.
3. Wittig, G.; Haag, A. *Chem. Ber.* **1963,** *96,* 1535–1543.
4. Bestmann, H.; Hartung, H. *Chem. Ber.* **1966,** *99,* 1198–1207.
5. Nader, F. W.; Brecht, A.; Kreisz, S. *Chem. Ber.* **1986,** *119,* 1196–1207.
6. Runge, W.; Kresze, G.; Ruch, E. *Liebigs Ann. Chem.* **1975,** 1361–1378.
7. Kresze, G.; Runge, W.; Ruch, E. *Liebigs Ann. Chem.* **1972,** *756,* 112–127.
8. Lang, R. W.; Hansen, H.-J. *Org. Synth.* **1984,** *62,* 202–209.
9. Schweizer, E. E.; Hsueh, W.; Rheingold, A. L.; Durney, R. L. *J. Org. Chem.* **1983,** *48,* 3889–3894.
10. Musierowicz, S.; Wroblewski, A. E. *Tetrahedron* **1980,** *36,* 1375–1380.
11. Fillion, H.; Refouvelet, B.; Pera, M. H.; Dufaud, V.; Luche, J. L. *Synth. Commun.* **1989,** *19,* 3343–3348.
12. Aksnes, G.; Froyen, P. *Acta Chem. Scand.* **1968,** *22,* 2347–2352.
13. Rafizadeh, K.; Yates, K. *J. Org. Chem.* **1984,** *49,* 1500–1506.
14. Hamlet, Z.; Barker, W. D. *Synthesis,* **1970,** 543–544.
15. Capuano, L.; Wamprecht, C.; Willmes, A. *Chem. Ber.* **1982,** *115,* 3904–3907.
16. Sestrick, M. R.; Miller, M.; Hegedus, L. S. *J. Am. Chem. Soc.* **1992,** *114,* 4079–4088.
17. Staudinger, H.; Hauser, E. *Helv. Chem. Acta* **1921,** *4,* 887–896.
18. Lee, K.-W.; Singer, L. A. *J. Org. Chem.* **1974,** *39,* 3780–3781.
19. Molina, P.; Alajarin, M.; Lopez-Leonardo, C. *Tetrahedron Lett.* **1991,** *32,* 4041–4044.
20. Meier, S.; Würthwein, E.-U. *Chem. Ber.* **1990,** *123,* 2339–2347.

5.6 ELECTROPHILIC ADDITION TO KETENES

5.6.1 Protonation of Ketenes

The gas phase protonation of ketene and methylketene have been examined, and proton addition has been shown to occur on carbon, giving an acylium ion **1**[1-9] from ketene (equation 1).[2,3] The isomers **2-4** of **1** have also been examined experimentally and by theoretical methods, and **2** and **3** are also energy minimum structures that can be generated independently in the gas phase, and while **1-3** are distinct species that give different gas phase reactions their interconversion also occurs.[1] Molecular orbital calculations using ab initio methods indicate that **2** and **3** are 43.2 and 58.3 kcal/mol higher in energy than **1**, respectively, with barriers for unimolecular rearrangement to **1** of 68.6 and 20.3 kcal/mol, respectively.[9] Rearrangement of **3** to **1** proceeds through the twisted form of formylmethyl cation **4** as a transition state (Section 1.1.2).[9]

$$CH_2=C=O \xrightarrow{H^+} CH_3\overset{+}{C}=O \qquad (1)$$

1

$$CH_2=C=OH^+$$

2

(structure **3**: three-membered ring with CH_2—CH and O+ at apex)

3

(structure **4**: H-C(+)—C(=O)H with H substituents)

4

The gas phase proton affinities of ketene and methylketene are 196.2 and 202.2 kcal/mol, respectively.[6] As discussed in Section 1.1.3, a methyl group is destabilizing as a ketene substituent compared to hydrogen, and so the more favorable protonation of methylketene (equation 2) compared to ketene (equation 1) could arise from a higher ground state energy in methylketene combined with a stabilization of the cation **5** compared to **1**.

$$CH_3CH=C=O \xrightarrow{H^+} CH_3CH_2\overset{+}{C}=O \qquad (2)$$

5

Calculations of the preferred path of protonation of ketene thus are consistent with attack on the HOMO at C_β forming the acylium ion **1** which is stabilized by the resonance structure **1a**.[9,10] The reactivities of different substituted ketenes by this pathway are thus expected to be affected by the electronic effect of the substituent on the ground state stability of the ketene and on the acylium ion, and by any steric effects involving the substituents at C_β.

$$CH_3\overset{+}{C}{=}O \quad \longleftrightarrow \quad CH_3C{\equiv}O^+$$

$$\mathbf{1} \qquad\qquad\qquad \mathbf{1a}$$

For protonation of vinylketene, calculations indicate that protonation at C_δ is favored by 15.8 kcal/mol relative to protonation at C_β (equation 3).[10] The resulting ion **6** has an allylic resonance structure with acylium ion character.

$$CH_2{=}CH{-}CH{=}C{=}O \xrightarrow{H^+} CH_3\overset{+}{C}H{-}CH{=}C{=}O \longleftrightarrow CH_3CH{=}CH{-}\overset{+}{C}{=}O \quad (3)$$

$$\mathbf{6}$$

Acid-catalyzed hydration of ketenes is discussed in Section 5.5.1.3.

Protonation of diphenylketene and di-*tert*-butylketene in $(FSO_3H/SbF_5)/SO_2ClF$ at $-60\,°C$ give the corresponding acylium ions as detected by their ^{13}C NMR signals for CO^+ at $\delta\,154.7$ and 155.3, respectively.[11]

$$Ph_2C{=}C{=}O \xrightarrow{H^+} Ph_2CH\overset{+}{C}{=}O \qquad t{-}Bu_2C{=}C{=}O \xrightarrow{H^+} t{-}Bu_2CH\overset{+}{C}{=}O$$

5.6.2 Electrophilic Addition of Hydrogen Halides to Ketenes

The interconversion of ketenes and acyl halides via acylium ions as in equation 1 is a well-established process. In particular the generation of polyketone from the reaction of acyl halides with Lewis acids is ascribed to the addition of the acylium ion intermediates to ketene (Section 5.8).[12]

$$\underset{CH_3CCl}{\overset{O}{\|}} \underset{\longleftarrow}{\overset{-Cl^-}{\longrightarrow}} CH_3\overset{+}{C}{=}O \underset{\longleftarrow}{\overset{-H^+}{\longrightarrow}} CH_2{=}C{=}O \qquad (1)$$

However the addition of HCl and HBr to $Me_2C{=}C{=}O$ in ether at $-20\,°C$[13] and of HBr to $Ph_2C{=}C{=}O$ at $25\,°C$[14] has been attributed by Satchell and co-workers as involving formation of unobserved enol intermediates (equation 2).

$$Me_2C{=}C{=}O + HCl \underset{k_{-1}}{\overset{k_1}{\rightleftarrows}} Me_2C{=}C\overset{\displaystyle OH}{\underset{\displaystyle Cl}{\diagup\!\!\!\diagdown}} \xrightarrow[HCl]{k_2} Me_2CHCOCl \qquad (2)$$

As discussed in more detail elsewhere,[15] the evidence proposed for the scheme of equation 2 is scant and questionable. This evidence was an upward curvature from a first-order dependence of the rate on [HCl] below 10^{-2} M, which was interpreted as indicating a second-order dependence on [HCl] as in equation 2. However, as pointed out,[15] the reaction of HCl with alkenes frequently depends on the square

624   REACTIONS OF KETENES

of [HCl], but this indicates that protonation by HCl dimer occurs. Dimethylketene ($Me_2C=C=O$) is much more reactive than $Ph_2C=C=O$ towards HBr, and this is readily understood if protonation at C_β is rate limiting, but the mechanism of equation 2 provides no simple explanation of this fact. Thus protonation at dialkyl-substituted carbons may be slightly inhibited by the substituents, whereas diphenyl groups cause major decreases in protonation reactivity.[16] Thus rate-limiting protonation at C_β provides a consistent explanation for the addition of hydrogen halides to ketenes and of acid-catalyzed hydration.

Addition of HCl to γ-oxoketenes such as **7** leads to δ-chloro-δ-lactones such as **7** (equation 3).[17] These pseudo chlorides are unreactive toward MeOH, and may be envisaged as forming with initial protonation on either carbon or oxygen.[17] Trifluoroacetic acid reacts similarly.[18]

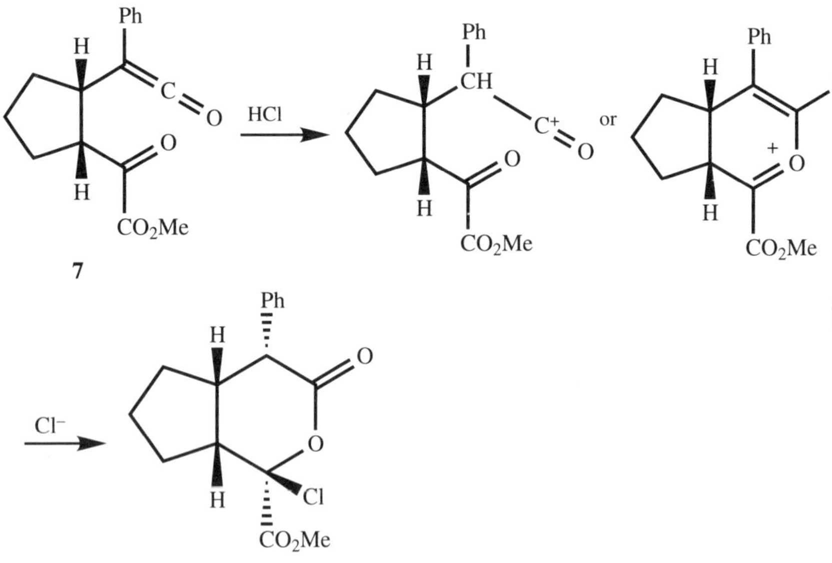

5.6.3   Electrophilic Additions of Other Reagents

The addition of acids to ketenes provides convenient preparations of mixed anhydrides, including acetyl dihydrogen phosphate (equation 1, see also equation 20, Section 5.5.2.2)[19] and those of cyclohexanecarboxylic acid with CH_3SO_3H, 4-TolSO$_3$H, and PhSO$_3$H (equation 2).[20] Carboxylic acids add to ketenes to give carboxylic acid anhydrides,[14,20a] but proposals that these proceed by concerted processes have been criticized.[15]

$$CH_2=C=O \xrightarrow{H_3PO_4} CH_3CO_2PO_3H_2 \qquad (1)$$

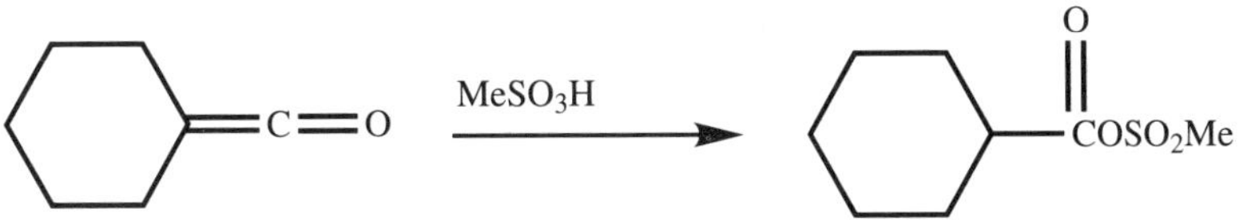

5.6 ELECTROPHILIC ADDITION TO KETENES

[Equation (2): cyclohexylidene ketene + MeSO₃H → cyclohexyl-C(=O)-COSO₂Me]

(2)

Ketenes react with stable carbocations as in equations 3 and 4.[21,22] The addition of triphenylmethyl cation to diphenylketene followed by decarbonylation was used to prepare the pentaphenylethyl cation which was not observed but was inferred as a reaction intermediate (equation 5).[11]

$$CH_2=C=O \xrightarrow[\text{2) MeOH}]{\text{1) R}\overset{+}{C}HOR^1} R^1OCHRCH_2CO_2Me \quad (3)^{21}$$

$$CH_2=C=O \xrightarrow{Ph_3C^+Cl^-} Ph_3CCH_2COCl \quad (4)^{22}$$

$$Ph_2C=C=O \xrightarrow{Ph_3C^+} Ph_3CCPh_2\overset{+}{C}=O \xrightarrow{-CO} Ph_3C\overset{+}{C}Ph_2 \quad (5)^{11}$$

The reaction of $(CF_3)_2C=O$ with $Me_3SiCH=C=O$ gave the 2-oxetanone **1** and the 2:1 adduct **2**.[23] These reactions were interpreted in terms of electrophilic attack on the ketene to give the zwitterion **3** as shown in equations 6 and 7.[23] The high electrophilicity of the carbonyl carbon of this ketone would favor this pathway, but the alternative of nucleophilic attack by the oxygen of the ketone on the carbonyl cation of the ketene is the normal path of oxetanone formation (Section 5.4.5) and would be less sterically hindered. There does not appear to be evidence which permits a choice between these alternatives.

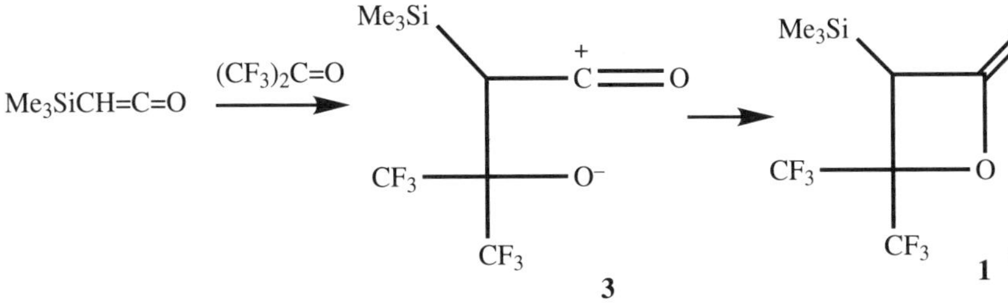

$$Me_3SiCH=C=O \xrightarrow{(CF_3)_2C=O} \mathbf{3} \longrightarrow \mathbf{1} \quad (6)$$

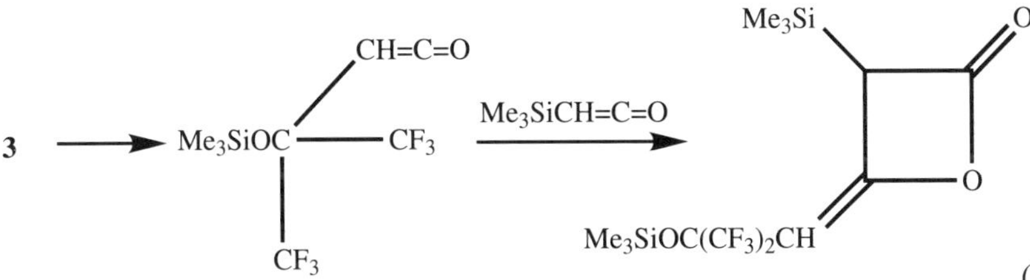

$$\mathbf{3} \longrightarrow Me_3SiOC(CF_3)_2CH=C=O \xrightarrow{Me_3SiCH=C=O} \mathbf{2} \quad (7)$$

Reaction of primary acyl chlorides with pyridine and trifluoroacetic anhydride followed by hydrolysis led to trifluoromethyl ketones (**4**).[24] This reaction was proposed to involved formation of ketene intermediates which underwent electrophilic attack by the trifluoroacetic anhydride to give β-trifluoroacetyl carboxylic acid derivatives **5** which upon hydrolysis formed β-trifluoroacetoacetic acids which spontaneously decarboxylated at room temperature (equation 8).[24]

$$RCH_2COCl \xrightarrow{\text{pyridine}} \underset{H}{\overset{R}{C}}=C=O \xrightarrow{(CF_3CO)_2O} \underset{CF_3}{\overset{O}{\|C}}\diagdown\underset{CHR}{\overset{O}{\|C}}\diagdown X$$

5 (X = Cl, CF$_3$CO$_2$ or pyridinium)   (8)

$$\mathbf{5} \xrightarrow[-CO_2]{H_2O} \underset{CF_3 \quad CH_2R}{\overset{O}{\|C}} \qquad \mathbf{5} \xrightarrow{MeOH} \underset{CF_3 \quad CHRCO_2Me}{\overset{O}{\|C}} \quad (9)$$

          4   **6**

Ketenes were not directly observed in these reactions and ketene–pyridine adducts or other enolate species may be the active nucleophile. Quenching of the reactions with methanol gave methyl 4,4,4-trifluoroacetoactates **6**, confirming the intermediacy of the adducts **5** (equation 9).[24] Trifluroacetic anhydride is known to undergo nucleophilic attack by electron-rich alkenes such as vinyl ethers in a process analogous to equation 8.[25,26]

The addition of carbonyl compounds to ketene is also enhanced by Ti(IV) alkoxides.[27] The reaction has been represented by the concerted process of equations 10 and 11,[27] and a stepwise pathway has also been envisaged.

$$(RO)_4Ti + R^1_2C=O \xrightarrow{CH_2=C=O} \begin{array}{c} R^1 \\ R^1 \diagdown \diagup \diagdown \diagup O \\ \vdots \diagup \diagdown \vdots \\ O \diagdown \quad \diagup OR \\ Ti \\ (OR)_3 \end{array} \longrightarrow (RO)_3TiOCR^1_2CH_2CO_2R \quad (10)$$

$$\xrightarrow{H_3O^+} R^1_2C(OH)CH_2CO_2R \quad (11)$$

A variety of metal alkoxides add to ketenes[27–39] and these reactions have been interpreted as possibly involving electrophilic attack by the metal (equation 12).[40] Organomercury compounds also add to ketenes.[41]

$$CH_2=C=O \xrightarrow{R_nMOR} R_nMCH_2\overset{+}{C}=O \longrightarrow R_nMCH_2CO_2R \quad (12)$$

Some boron and aluminum electrophiles react with ketenes with O-metal bond formation (equations 13 and 14).[42–46] The kinetics of the reaction of BH_3 with ketene in the gas phase were studied, and the formation of a 1:1 adduct was observed by mass spectrometry.[47] The structure of this gas phase product was not proven, but it was suggested to result from boron attack on the CH_2 group, and to have tetracoordinate boron, and so may be represented as **7**.

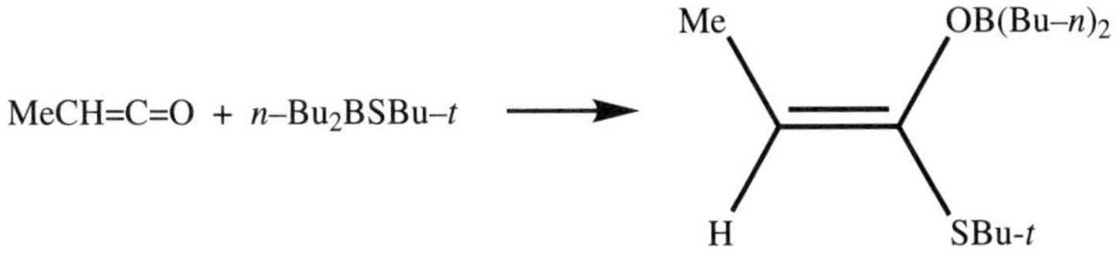

$$MeCH=C=O + n\text{-}Bu_2BSBu\text{-}t \longrightarrow \underset{H}{\overset{Me}{>}}\!=\!\underset{SBu\text{-}t}{\overset{OB(Bu-n)_2}{<}} \quad (13)^{45}$$

$$Ph_2C=C=O \xrightarrow{AlMe_3} Ph_2C=\underset{Me}{\overset{OAlMe_2}{<}} \quad (14)^{46}$$

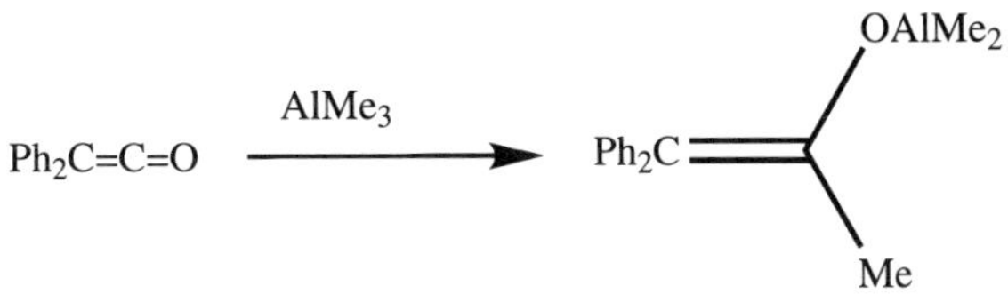

$$CH_2=C=O \xrightarrow{BH_3} H_3\overset{-}{B}\text{-}CH_2\overset{+}{C}=O$$

7

Additions of sulfur and phosphorous electrophiles are shown in equations 15–17.[48–53] In the reaction for equation 18 a solid intermediate was formed by the structure was not established.[53] The additions of Me_2AsNEt_2 (equation 19), Ph_2AsNMe_2, Ph_2AsNEt_2, Me_2SbNMe_2, and Me_2SbNEt_2 to the C=C bond of $CH_2=C=O$ have also been achieved.[54]

$$CH_2=C=O \xrightarrow{ArSCl} [ArSCH_2COCl] \xrightarrow{H_2O} ArSCH_2CO_2H \quad (15)^{48,49}$$

$$CH_2=C=O \xrightarrow{SCl_2} S(CH_2COCl)_2 \quad (16)^{50,51}$$

$$(CF_3)_2C=C=O \xrightarrow[CH_3CN]{PhSCl} PhSC(CF_3)_2COCl \quad (17)^{52}$$

$$Ph_2C=C=O \xrightarrow{P(OEt)_3} [Ph_2C=\overset{+}{C}O\overset{-}{P}(OEt)_3] \xrightarrow{EtOH} Ph_2CHCO_2Et \quad (18)^{53}$$

$$CH_2=C=O \xrightarrow{Me_2AsNEt_2} Me_2AsCH_2CONEt_2 \quad (19)^{54}$$

Additions of positive halogen are shown in equations 20–22,[56–61] and the reaction of $GeCl_4$ with ketene in equation 24.[62] The second-order rate constants in 1,2-dichloroethane for the iodination of $Me_2C=C=O$ (equation 21) at 20 °C and for the bromination of $Ph_2C=C=O$ at 30 °C were measured as 46 and 7 M^{-1} s^{-1}, respectively.[57] Preparative chlorine and bromine addition to $RCH=C=O$ (R = H, Me, Ph) have also been reported.[58,59]

$$Me_3SiCH=C=O \xrightarrow{Br_2} Me_3SiCHBrCOBr \quad (20)^{56}$$

$$Me_2C=C=O \xrightarrow{I_2} Me_2CICOI \xrightarrow[\text{2) } CH_2N_2]{\text{1) } H_2O} Me_2CICO_2Me \quad (21)^{57}$$

$$Ph_2C=C=O + ClNH_2 \longrightarrow Ph_2CClCONH_2 \quad (22)^{60}$$

$$Me_2C=C=O \xrightarrow{SOCl_2} Me_2C(SOCl)COCl \xrightarrow{-SO} Me_2CClCOCl \quad (23)^{61}$$

$$CH_2=C=O \xrightarrow{GeCl_4} Cl_3GeCH_2COCl \quad (24)^{62}$$

References

1. Eberline, M. N.; Majumdar, T. K.; Cooks, R. G. *J. Am. Chem. Soc.* **1992**, *114*, 2884–2896.
2. Ausloos, P.; Lias, S. G. *Chem. Phys. Lett.* **1977**, *51*, 53–56.
3. Vogt, J.; Williamson, A. D.; Beauchamp, J. L. *J. Am. Chem. Soc.* **1978**, *100*, 3478–3483.
4. Debrou, G. B.; Fulford, J. E.; Lewars, E. G.; March, R. E. *Int. J. Mass Spec. Ion Phys.* **1978**, *26*, 345–352.
5. Davidson, W. R.; Lau, Y. K.; Kebarle, P. *Can. J. Chem.* **1978**, *56*, 1016–1019.
6. Armitage, M. A.; Higgins, M. J.; Lewars, E. G.; March, R. E. *J. Am. Chem. Soc.* **1980**, *102*, 5064–5068.
7. Beach, D. B.; Eyermann, C. J.; Smit, S. P.; Xiang, S. F.; Jolly, W. L. *J. Am. Chem. Soc.* **1984**, *106*, 536–539.
8. Traeger, J. C.; McLoughlin, R. G.; Nicholson, A. J. C. *J. Am. Chem. Soc.* **1982**, *104*, 5318–5322.
9. Nobes, R. H.; Bouma, J. J.; Radom, L. *J. Am. Chem. Soc.* **1983**, *105*, 309–314.
10. Leung-Toung, R.; Peterson, M. R.; Tidwell, T. T. Csizmadia, I. G. *J. Mol. Struct. (Theochem)* **1989**, *183*, 319–330.

11. Olah, G. A.; Alemayehu, M.; Wu, A.; Farooq, O.; Prakash, G. K. S. *J. Am. Chem. Soc.* **1992**, *114*, 8042–8045.
12. Olah, G. A.; Zadok, E.; Edler, R.; Adamson, D. H.; Kasha, W.; Prakash, G. K. S. *J. Am. Chem. Soc.* **1989**, *111*, 9123–9124.
13. Lillford, P. J.; Satchell, D. P. N. *J. Chem. Soc. B* **1968**, 897–901.
14. Poon, N. L.; Satchell, D. P. N. *J. Chem. Res. Synop.* **1983**, 182–183.
15. Seikaly, H. R.; Tidwell, T. T. *Tetrahedron* **1986**, *42*, 2587–2613.
16. Chiang, Y.; Kresge, A. J.; Tidwell, T. T.; Walsh, P. A. *Can. J. Chem.* **1980**, *58*, 2203–2206.
17. Hegmann, J.; Ditterich, E.; Hüttner, G.; Christl, M.; Peters, E.; Peters, K.; von Schnering, H. G. *Chem. Ber.* **1992**, *125*, 1913–1918.
18. Christl, M.; Lanzendörfer, U.; Grötsch, M. M.; Hegmann, J.; Ditterich, E.; Hüttner, G.; Peters, E.; Peters, K.; von Schnering, H. G. *Chem. Ber.* **1993**, *126*, 797–902.
19. Bentley, R. *J. Am. Chem. Soc.* **1948**, *70*, 2183–2185.
20. Tempesti, E.; Giuffre, L.; Sioli, G.; Fornaroli, M.; Airoldi, G. *J. Chem. Soc., Perkin Trans. 1* **1974**, 771–773.
20a. Poon, N. L.; Satchell, D. P. N. *J. Chem. Res. Synop.* **1985**, 260–261.
21. Hurd, C. D.; Kimbrough, R. D. Jr. *J. Am. Chem. Soc.* **1961**, *83*, 236–240.
22. Blomquist, A. T.; Holley, R. W.; Sweeting, O. J. *J. Am. Chem. Soc.* **1947**, *69*, 2356–2358.
23. Zaitseva, G.-S.; Livantsova, L. I.; Bekker, R. A.; Baukov, Yu. I.; Lutsenko, I. F. *Zh. Obshch. Khim.* **1983**, *53*, 2068–2077; *Engl. Transl.* **1983**, *53*, 1867–1874.
24. Boivin, J.; El Kaim, L.; Zard, S. Z. *Tetrahedron Lett.* **1992**, *33*, 1285–1288.
25. Hojo, M.; Masuda, R.; Kokuryo, Y.; Shioda, H.; Matsuo, S. *Chem. Lett.* **1976**, 499–502.
26. Hojo, M.; Masuda, R.; Sakaguchi, S.; Takagawa, M. *Synthesis*, **1986**, 1016–1017.
27. Vuitel, L.; Jacot-Guillarmod, A. *Helv. Chim. Acta* **1974**, *57*, 1703–1713.
28. Lutsenko, I. F.; Foss, V. L.; Ivanova, N. L. *Dokl. Akad. Nauk SSSR*, **1961**, *141*, 1107–1108; *Chem. Abstr.* **1962**, *56*, 12920d.
29. Foss, V. L.; Zhadina, M. A.; Lutsenko, I. F.; Nesmeyanov, A. N. *Zh. Obshch. Khim.* **1963**, *33*, 1927–1933; *Engl. Transl.* **1963**, *33*, 1874–1879.
30. Lutsenko, I. F.; Ponomarev, S. V. *Zh. Obshch. Khim.* **1961**, *31*, 2025–2027; *Engl. Transl.* **1961**, *31*, 1894–1896.
31. Bloodworth, A. J.; Davies, A. G. *Proc. Chem. Soc.* **1963**, 315.
32. Ponomarev, S. V.; Baukov, Yu. I.; Lutsenko, I. F. *Zh. Obshch. Khim.* **1964**, *34*, 1938–1940; *Engl. Transl.* **1964**, *34*, 1951–1952.
33. Willemsens, L. C.; Van der Kerk, G. J. M. *J. Organomet. Chem.* **1965**, *4*, 241–244.
34. Burlachenko, G. S.; Avdeeva, V. I.; Baukov, Yu. I.; Lutsenko, I. F. *Zh. Obshch. Khim.* **1965**, *35*, 1881; *Chem. Abstr.* **1966**, *64*, 1607c.
35. Foss, V. L.; Besolova, E. A.; Lutsenko, I. F. *Zh. Obshch. Khim.* **1965**, *35*, 759–760; *Engl. Transl.* **1965**, *35*, 764.
36. Shibata, I.; Baba, A.; Matsuda, H. *Bull. Chem. Soc. Jpn.* **1986**, *59*, 4000–4002.
37. Blandy, C.; Hliwa, M. *C. R. Acad. Sci. Paris* **1983**, *296*, 51–52.

38. Blandy, C.; Gervais, D. *Inorg. Chim. Acta* **1981**, *47*, 197–199.
39. Akai, S.; Tsuzuki, Y.; Matsuda, S.; Kitagaki, S.; Kita, Y. *J. Chem. Soc., Perkin Trans. 2* **1992**, 2813–2820.
40. Ulrich, H. *Cycloaddition Reactions of Heterocumulenes*, Academic: New York, 1967; Chapter 2.
41. Gilman, H.; Woolley, B. L.; Wright, G. F. *J. Am. Chem. Soc.* **1933**, *55*, 2609.
42. Gutsche, C. D.; Kinoshita, K. *J. Org. Chem.* **1963**, *28*, 1762–1764.
43. Paetzold, P.; Kosma, S. *Chem. Ber.* **1979**, *112*, 654–662.
44. Hirama, M.; Masamune, S. *Tetrahedron Lett.* **1979**, 2225–2228.
45. Mukaiyama, T.; Inomata, K.; Muraki, M. *J. Am. Chem. Soc.* **1973**, *95*, 967–968.
46. Jeffery, E. A.; Meisters, A. *J. Organomet. Chem.* **1974**, *82*, 315–318.
47. Fehlner, T. P. *J. Phys. Chem.* **1972**, *76*, 3532–3538.
48. Roe, A.; McGeehee, J. W. *J. Am. Chem. Soc.* **1948**, *70*, 1662.
49. Kondrashov, N. V.; Sokol'skii, G. A.; Kolomiets, A. F.; Fokin, A. V. *Izv. Akad. Nauk SSSR, Ser. Khim.* **1987**, 2358–2360; *Chem. Abstr.* **1988**, *109*, 109845z.
50. Komatsu, M.; Harada, N.; Kashiwagi, H.; Oshiro, Y.; Agawa, T. *Phosphorous Sulfur* **1983**, *16*, 119–133.
51. N. V. de Bataafsche Petroleum Maatschappij *British Patent* 670, 130; *Chem. Abstr.* **1953**, *47*, 5430h.
52. Shkurak, S. N.; Ezhov, V. V.; Kolomiets, A. F.; Fokin, A. V. *Izv. Akad. Nauk SSSR, Ser. Chim.* **1984**, 1371–1378; *Engl. Transl.* **1984**, 1261–1264.
53. Mukaiyama, T.; Nambu, H.; Okamoto, M. *J. Org. Chem.* **1962**, *27*, 3651–3654.
54. Ando, F.; Kohmura, Y.; Koketsu, J. *Bull. Chem. Soc. Jpn.* **1987**, *60*, 1564–1566.
55. Böhme, H.; Bezzenberger, H.; Stachel, H.-D. *Liebigs Ann. Chem.* **1957**, *602*, 1–14.
56. Brady, W. T.; Owens, R. A. *Tetrahedron Lett.* **1976**, 1553–1556.
57. Ogata, Y.; Adachi, K. *J. Org. Chem.* **1982**, *47*, 1182–1184.
58. Gash, V. M.; Bissing, D. E. *Ger. Offen.* 2,247,764; *Chem. Abstr.* **1973**, *78*, 158969t.
59. Gash, V. M.; Bissing, D. E. *Ger. Offen.* 2,247,765; *Chem. Abstr.* **1973**, *79*, 31506e.
60. Coleman, G. H.; Petersen, R. L.; Goheen, G. E. *J. Am. Chem. Soc.* **1936**, *58*, 1874–1876.
61. Harpp, D. N.; Bao, L. Q.; Black, C. J.; Gleason, J. G.; Smith, R. A. *J. Org. Chem.* **1975**, *40*, 3420–3427.
62. Efrimova, I. V.; Kazankova, M. A.; Lutsenko, I. V. *Zh. Obshch. Khim.* **1983**, *53*, 950–951; *Engl. Transl.* **1983**, *53*, 838.

5.6.4 Oxygenation of Ketenes

In 1925 Staudinger and co-workers reacted diphenylketene (**1**) with O_2 and obtained the polyester **3** (equation 1).[1] They proposed that the reaction proceeded through the cyclic peroxyester **2**.

5.6 ELECTROPHILIC ADDITION TO KETENES

$$Ph_2C=C=O \xrightarrow{O_2} \underset{2}{\begin{array}{c}Ph\\Ph-C-C=O\\|\quad\quad|\\O—O\end{array}} \longrightarrow (CPh_2CO_2)_n \quad (1)$$

1   **2**   **3**

Bartlett and Gortler obtained polymeric material that appeared to be the same as **3** from the reaction of the bisperester **4**, and proposed that this arose via the α-lactone **5** (equation 2).[2] They suggested that the reaction of **1** with O_2 (equation 1) proceeded via a radical chain reaction also involving **5**, formed by radical attack at C_α of diphenylketene by intermediate oxygen-diphenyl ketene copolymer peroxy radicals **6**, which led to oxy radicals **6a**, which further fragmented to CO_2 and $Ph_2C=O$, which were observed from the reaction of **4** (equations 3 and 4).

$$Ph_2C(CO_3Bu\text{-}t)_2 \longrightarrow Ph_2\overset{\bullet}{C}CO_3Bu\text{-}t \longrightarrow \underset{Ph_2C——C\overset{\displaystyle O}{\diagdown}_O}{\triangle} \longrightarrow \mathbf{3} \quad (2)$$

4   **5**

$$\underset{\mathbf{6}}{\bullet OOCPh_2CO_3R} \xrightarrow{Ph_2C=C=O} Ph_2\overset{\bullet}{C}C\underset{OOCPh_2CO_3R}{\overset{\displaystyle O}{\diagup}} \xrightarrow{-OCPh_2CO_3R} \mathbf{5} \quad (3)$$

$$\underset{\mathbf{6a}}{\overset{\bullet}{O}CPh_2\overset{\displaystyle O}{\overset{\|}{C}}OOR} \longrightarrow Ph_2C=O + CO_2 + RO\bullet \quad (4)$$

The reaction of $Ph_2C=C=O$ (**1**) with O_3 at −78 °C also gave **3**, and this reaction was also interpreted as involving the α-lactone **5** (equation 5).[3] Further evidence for the intervention of **5** was obtained by carrying out the reaction in the presence of MeOH, which led to the formation of $Ph_2C(OMe)CO_2H$, the expected product of the capture of **5**.[3] Reaction of di-*tert*-butylketene (**7**) with O_3 gave the α-lactone **8** which could be identified by ^1H NMR at −60 °C and which on warming gave a polymer identified as **9** (equation 6).[3]

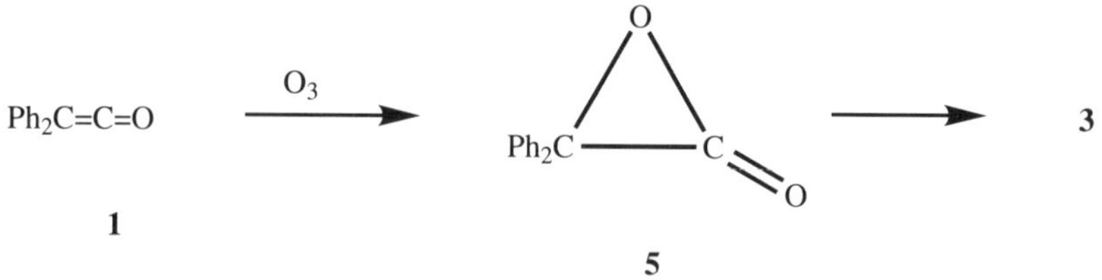

$$Ph_2C=C=O \xrightarrow{O_3} \underset{5}{\text{[Ph}_2C\text{—C(=O)—O (epoxide)]}} \longrightarrow 3 \quad (5)$$

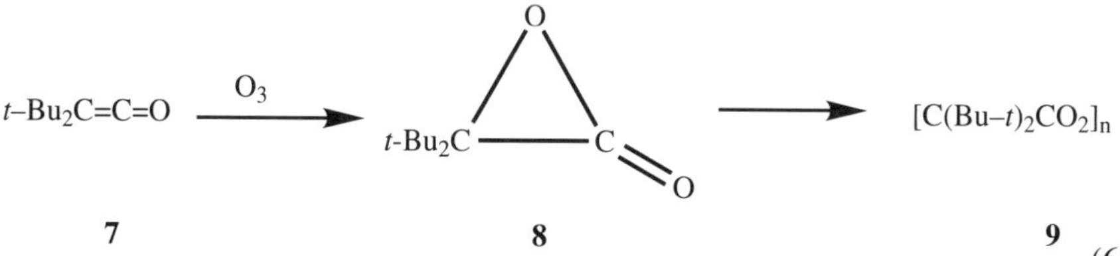

$$\mathbf{1}$$

$$t\text{-}Bu_2C=C=O \xrightarrow{O_3} \underset{8}{\text{[}t\text{-}Bu_2C\text{—C(=O)—O (epoxide)]}} \longrightarrow [C(Bu\text{-}t)_2CO_2]_n \quad (6)$$

$$\mathbf{7} \qquad \qquad \mathbf{9}$$

Singlet oxygen generated from $(PhO)_3PO_3$ reacted with $Ph_2C=C=O$ (**1**) and in the presence of a fluorescent material *Flr* [9,10-bis(phenylethynyl)anthracene] there was chemiluminescent emission from the fluorescer, with formation of the products $Ph_2C=O$ and CO_2.[4] This reaction was interpreted as proceeding through formation of Staudinger's peroxide **2**, which reacted with the fluorescer to give the excited fluorescer and the observed reaction products (equation 7).

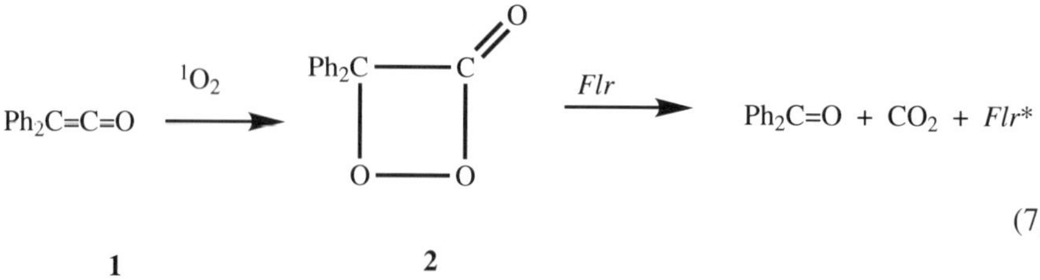

$$Ph_2C=C=O \xrightarrow{^1O_2} \underset{\mathbf{2}}{\text{[Ph}_2C\text{—C(=O), O—O four-membered ring]}} \xrightarrow{Flr} Ph_2C=O + CO_2 + Flr^* \quad (7)$$

$$\mathbf{1}$$

Other ketenes [$Me_2C=C=O$, $t\text{-}BuCH=C=O$, $n\text{-}PrCMe=C=O$, $n\text{-}BuCHPh=C=O$, and $(CF_3)C=C=O$] as well as **1** were similarly converted by $(PhO)_3PO_3$ to α-peroxylactones similar to **2**, and the structures of the lactones purified at −20 °C were confirmed by their ^1H NMR and IR spectra.[5,6] It was established that these processes involve reaction between the ketene and $^1O_2$ both by the reaction kinetics and by independent generation of $^1O_2$ by photooxidation.[5] When the reaction was carried out in the presence of MeOH, α-methoxyperacetic acids **10** were formed instead of α-peroxylactones, and so the reaction of ketenes with $^1O_2$ was interpreted as involving either a perepoxide (**11**) or a zwitterionic intermediate (**12**), and not direct formation of the α-peroxylactone (equation 8).[5]

$$R_2C=C=O \xrightarrow{^1O_2} R_2C\overset{\overset{O^-}{\underset{O^+}{|}}}{-\!\!\!-\!\!\!-}C\!\!=\!\!O \quad \text{or} \quad R_2C\overset{+}{-}C\overset{O-O^-}{\underset{\|}{}}O \xrightarrow{\text{MeOH}} \text{MeOCR}_2\text{CO}_3\text{H}$$

$$\qquad\qquad\qquad\qquad\qquad \mathbf{11} \qquad\qquad\qquad \mathbf{12} \qquad\qquad\qquad\qquad\qquad \mathbf{10}$$

(8)

Further study[7,8] of the reaction of ketenes with $^3O_2$ confirmed and extended the previous work of Staudinger.[1] Thus the reaction of $Me_2C=C=O$ with $^3O_2$ in ether at $-20\,°C$ gave the polyperester **13** (96%) and the polyester **14** (4%), whereas $Ph_2C=C=O$ (**1**) gave no polyperester, 70% polyester **3**, and 30% of $Ph_2C=O$ and CO_2. Reaction of $Me_2C=C=O$ with $^3O_2$ in MeOH gave $HOOCMe_2CO_2Me$ as the only major product, whereas $Ph_2C=C=O$ gave a significant amount of $Ph_2C(OMe)CO_3H$. These facts and other evidence were interpreted as showing that ketenes react with $^3O_2$ to form diradicals **15** which polymerize to **13** (equation 9), whereas $^1O_2$ gives perepoxides and/or zwitterions (**11, 12**) which are captured by MeOH to give α-methoxyperacid (equation 8). Reaction of **11** and **12** with ketenes gave α-lactones, which form polyester (equation 10), or **11/12** can form α-peroxylactones which lose CO_2 to form ketones (equation 11). The diradical **15** could undergo conversion to **11/12**, and the polyperester **13** undergoes thermal decomposition to ketone and CO_2 (equation 9). The reaction of $Ph_2C=C=O$ with $^3O_2$ in CCl_4 also gave $PhCO_2Ph$ in yields of 10–18%, and this was proposed to result from the carbonyl oxide $Ph_2C=O^+-O^-$.[8]

$$R_2C=C=O \xrightarrow{^3O_2} \underset{\mathbf{15}}{R_2\overset{\bullet}{C}-\overset{OO\bullet}{\underset{\|}{C}}\!\!=\!\!O} \xrightarrow[O_2]{R_2C=C=O} \underset{\mathbf{13}}{(CR_2COO)_n} \xrightarrow{-CO_2} R_2C=O$$

(9)

$$\mathbf{11/12} \xrightarrow{R_2C=C=O} \underset{\mathbf{14}}{R_2C\overset{O}{-\!\!\!-\!\!\!-}C\!\!=\!\!O} \longrightarrow (CR_2CO_2)_n \quad (10)$$

$$\mathbf{11/12} \longrightarrow \underset{\substack{| \quad | \\ O\!\!-\!\!\!-\!\!O}}{\overset{\substack{R_2C\!\!-\!\!\!-\!\!C=O}}{}} \longrightarrow R_2C=O + CO_2 \quad (11)$$

The reaction of the ketenes **16–19** generated by Wolff rearrangement in the presence of singlet oxygen gave the corresponding ketones by the route of equation 11.[9]

The ketene **20** was oxidized to **21** (equation 12),[10] and the reaction of bisketenes with oxygen is discussed in Section 4.9. Ketenes are formed by alkyne oxidation and in some cases the ketenes are oxidized further (Section 3.7).

16

17

18

19

(12)

20 **21**

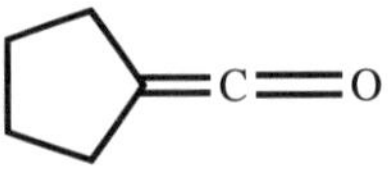

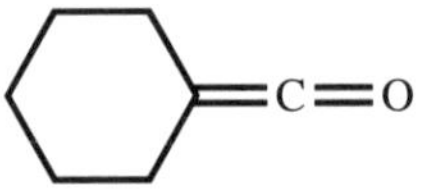

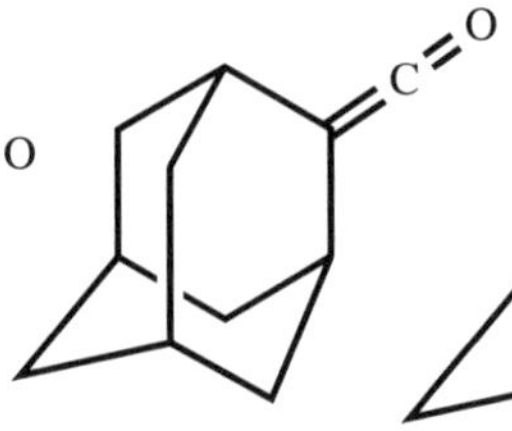

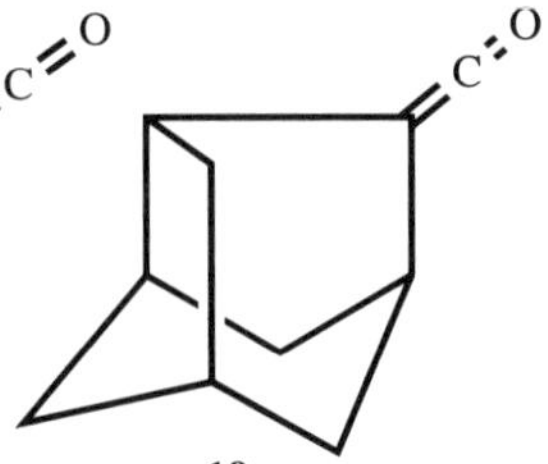

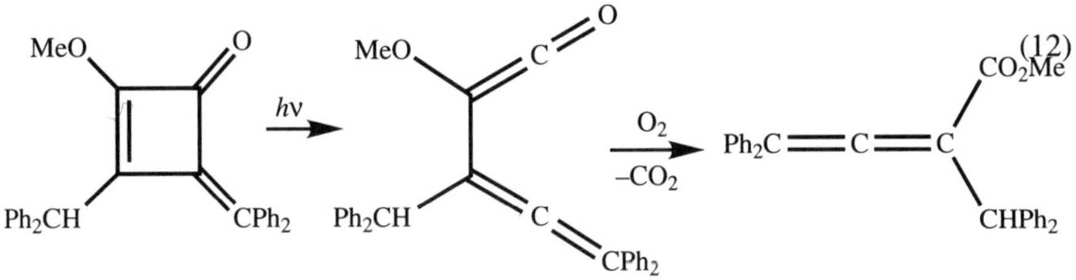

Further study[11–13] of the reaction of ketenes with O_3 showed that polyesters, presumably derived from α-lactones, and ketones were formed.[11,12] It was proposed[12] that α-lactones result from direct oxygen transfer from O_3 (equation 13), while ketones are formed from ozonides (equations 14 and 15).

(13)

(14)

(15)

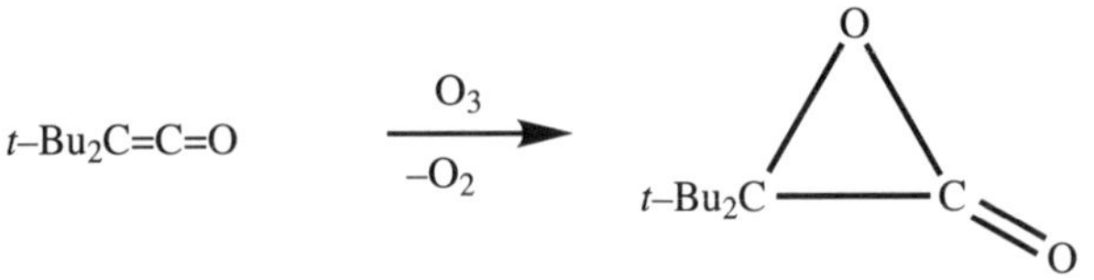

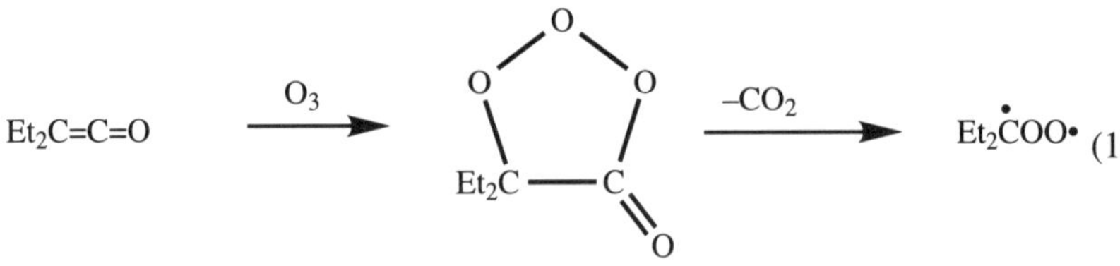

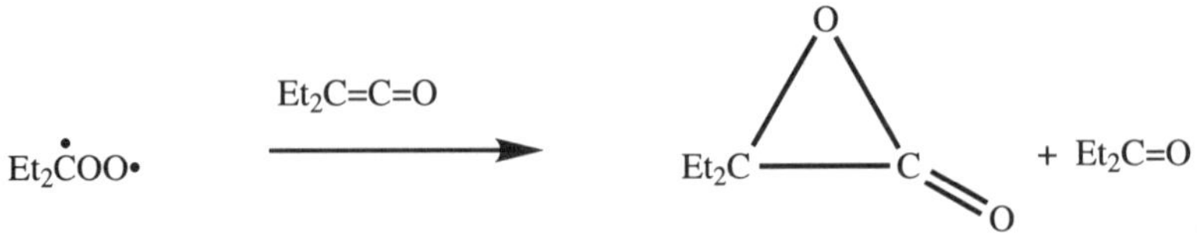

5.6 ELECTROPHILIC ADDITION TO KETENES

Oxidations of several ketenes with peracids were interpreted as involving α-lactone formation followed by decarbonylation to give ketones (equation 16).[11]

$$R_2C=C=O \xrightarrow{R'CO_3H} \left[\begin{array}{c} O \\ \triangle \\ R_2C\text{———}C \\ \phantom{R_2C\text{———}}\diagdown O \end{array} \right] \longrightarrow R_2C=O + \text{other products} \quad (16)$$

R = Ph, t–Bu, t–BuCH$_2$

Ketene **22** did not react with peracids but was cleaved by O_3 (equation 17).[11] The perfluorinated ketene **23** [$R_f = (C_2F_5)_2CCF_3$, $R_f^1 = C_2F_5CFCF_3$] reacted with NaOCl to give a stable α-lactone (equation 18).[14]

$$t\text{–BuC(CO}_2\text{Et)=C=O} \xrightarrow{O_3} t\text{–BuCOCO}_2\text{Et} + t\text{–BuO}_2\text{CCO}_2\text{Et} \quad (17)$$

22

$$R_fR_f^1C=C=O \xrightarrow{NaOCl} \begin{array}{c} O \\ \triangle \\ R_fR_f^1C\text{———}C \\ \phantom{R_fR_f^1C\text{———}}\diagdown O \end{array} \quad (18)^{14}$$

23

Reaction of $CH_2=C=O$, $CH_3CH=C=O$, and $Me_2C=C=O$ with atomic oxygen in the gas phase involves oxygen attack at the alkene carbons.[15] The kinetics of these reactions have been measured.[15] Iodosobenzene also serves as a ketene oxidant, and gives polyester product, presumably via an α-lactone.[16] It was suggested that the reaction occurs by nucleophilic attack of oxygen at C_β followed by formation of an α-lactone,[16] but as discussed in Section 5.4 nucleophilic attack is much more likely to occur at C_α (equation 19) and this might be testable experimentally using ^{18}O labeling. Oxidation of $Ph_2C=C=O$ by DMSO has been proposed to occur as in equation 20.[17] This process is similar to that of equation 19.

$$Ph_2C=C=O \xrightarrow{PhI=O} Ph_2C=C\begin{array}{c} \overset{+}{O}IPh \\ \diagdown \\ O^- \end{array} \xrightarrow{-PhI} \begin{array}{c} O \\ \triangle \\ Ph_2C\text{———}C \\ \phantom{Ph_2C\text{———}}\diagdown O \end{array} \longrightarrow \text{polymer} \quad (19)$$

5

$$Ph_2C=C=O \xrightarrow[Me_2S=O]{H^+} Ph_2C=C\begin{array}{c} \overset{+}{O}\text{———}SMe_2 \\ \diagdown \\ OH \end{array} \xrightarrow[-Me_2S]{-H^+} Ph_2C(OH)CO_2H \quad (20)$$

The reaction of an aldoketene generated from photolysis of a cyclohexanone derivative led to an aldehyde and carboxylic acid which had lost one carbon.[18] This result was attributed to air oxidation of the ketene.[18]

References

1. Staudinger, H.; Dyckerhoff, K.; Klever, H. W.; Ruzicka, L. *Chem. Ber.* **1925**, *58*, 1079–1087.
2. Bartlett, P. D.; Gortler, L. B. *J. Am. Chem. Soc.* **1963**, *85*, 1864–1869.
3. Wheland, R.; Bartlett, P. D. *J. Am. Chem. Soc.* **1970**, *92*, 6057–6058.
4. Bollyky, L. J. *J. Am. Chem. Soc.* **1970**, *92*, 3230–3232.
5. Turro, N. J.; Ito, Y.; Chow, M.; Adam, W.; Rodriquez, O.; Yany, F. *J. Am. Chem. Soc.* **1977**, *99*, 5836–5838.
6. Turro, N. J.; Chow, M. *J. Am. Chem. Soc.* **1980**, *102*, 5058–5064.
7. Turro, N. J.; Chow, M.-F.; Ito, Y. *J. Am. Chem. Soc.* **1978**, *100*, 5580–5582.
8. Bartlett, P. D.; McCluney, R. E. *J. Org. Chem.* **1983**, *48*, 4165–4168.
9. Majerski, Z.; Vinkovic, V. *Synthesis* **1989**, 559–560.
10. Toda, F.; Todo, Y.; Todo, E. *Bull. Chem. Soc. Jpn.* **1976**, *49*, 2645–2646.
11. Crandall, J. K.; Sojka, S. A.; Komin, J. B. *J. Org. Chem.* **1974**, *39*, 2172–2175.
12. Moriarty, R. M.; White, K. B.; Chin, A. *J. Am. Chem. Soc.* **1978**, *100*, 5582–5584.
13. Brady, W. T.; Saidi, K. *Tetrahedron Lett.* **1978**, 721–722.
14. Coe, P. L.; Sellars, A.; Tatlow, J. C.; Whittaker, G.; Fielding, H. C. *J. Chem. Soc., Chem. Commun.* **1982**, 362–363.
15. Washida, N.; Hatakeyama, S.; Takagi, M.; Kyogoku, T.; Sato, S. *J. Chem. Phys.* **1983**, *78*, 4533–4540.
16. Moriarty, R. M.; Gupta, S. C.; Hu, H.; Berenschot, D. R.; White, K. B. *J. Am. Chem. Soc.* **1981**, *103*, 686–688.
17. Lillien, I. *J. Org. Chem.* **1964**, *29*, 1631–1632.
18. Aoyagi, R.; Tsuyuki, T.; Takai, M.; Takahashi, T.; Kohen, F.; Stevenson, R. *Tetrahedron* **1973**, *29*, 4331–4340.

5.7 RADICAL REACTIONS OF KETENES

Radical additions to ketenes are observed to occur on either carbon, and this lack of regiochemical selectivity reflects the fact that the ketene HOMO lies perpendicular to the ketene plane with the highest coefficient at C_β, whereas the LUMO lies in the ketene plane with the highest coefficient at C_α. It may be expected that the tendency for radical attack to occur at C_β will increase with both the electrophilicity of the radical and the steric requirements of the radical, owing to the higher steric demands in the ketene plane. There is insufficient data, however, to test for these effects experimentally.

5.7 RADICAL REACTIONS OF KETENES

Hydrogen atom attack on ketene, methylketene, dimethylketene, and ethylketene results in formation of alkyl radicals and carbon monoxide, and evidently occurs with attack on C_β to give acyl radicals which undergo decarbonylation (equation 1).[1–3]

$$RCH=C=O \xrightarrow{H\cdot} RCH_2\dot{C}=O \xrightarrow{-CO} R\dot{C}H_2 \quad (1)$$

The addition of hydrogen atoms to t-Bu$_2$C=C=O (**1**) gave both the addition products **2** and **3**, which were observable by ESR (equation 2).[4] The acyl radical **2** was assigned the σ structure shown in which the unpaired electron is in an sp^2 orbital, and the *anti* conformation of the β-hydrogen was assigned on the basis of the large β-hydrogen splitting of 10.7 G.[4] This radical decays rapidly with first-order kinetics, with $\tau_{1/2}$ of 0.7 s at -90 °C, to give the t-Bu$_2$CH radical (equation 3). The radical t-Bu$_2\dot{C}$CHO (**3**) is much more stable, with a half-life of 220 s at 0 °C, and probably decays by attack on the solvent or some other species present.[4]

$$t\text{-Bu}_2C=C=O \xrightarrow[\text{Me}_6\text{Sn}_2]{\text{HI}, h\nu} t\text{-Bu}_2CH\dot{C}=O + t\text{-Bu}_2\dot{C}CH=O \quad (2)$$ 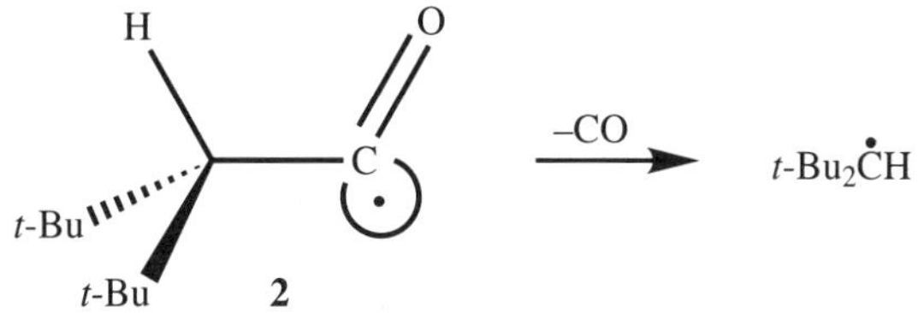
$$\textbf{1} \qquad\qquad\qquad \textbf{2} \qquad\qquad \textbf{3}$$

$$\xrightarrow{-CO} t\text{-Bu}_2\dot{C}H \quad (3)$$

The addition of C$_6$F$_5$ or CF$_3$ radicals to **1** occurred at C$_\alpha$ to give the radicals **4**, which were assigned perpendicular conformations based on their ESR spectra (equation 4).[4] These conformations of the acyl radicals are favored because of steric crowding in the planar conjugated structures, which are usually preferred.

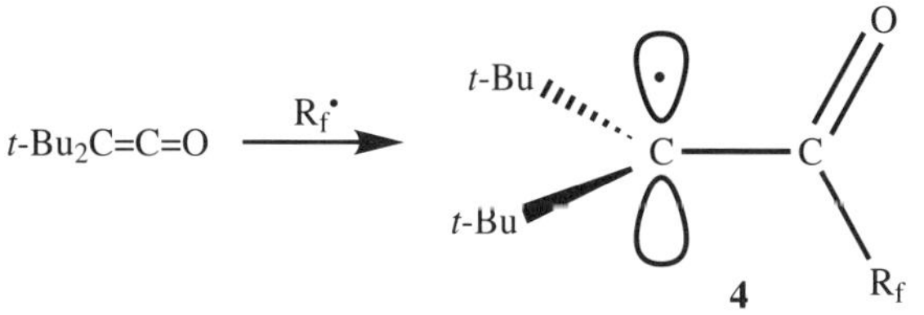

$$t\text{-Bu}_2C=C=O \xrightarrow{R_f\cdot} \textbf{4} \quad (4)$$

The regiochemistry of attack in equation 4 is not what would have been predicted if attack on the HOMO at C_β had been dictated by the expected electrophilic character of these radicals, and by the high steric requirements for attack in the plane. However the products **4** could result from attack on the HOMO perpendicular to the ketene plane at C_α, even though the coefficient of the orbital is less at this atom than at C_β. This pathway experiences minimal steric interference and leads directly to the observed products, but confirming evidence for this pathway is lacking.

Ethylthiol adds to $Me_2C=C=O$ in a reaction initiated by diisopropyl peroxydicarbonate.[5] A free-radical chain process with radical attack on C_α was proposed for this reaction (equation 5).[5]

$$Me_2C=C=O \xrightarrow{EtS^\bullet} Me_2\overset{\bullet}{C}C(=O)SEt \xrightarrow[-EtS^\bullet]{EtSH} Me_2CHC(=O)SEt \quad (5)$$

Thermolysis of azibenzil (**5**) produces an ESR signal attributed to α-acyl radical moiety **6**, which could result from radical addition to intermediate diphenylketene formed by Wolff rearrangement (equation 6).[6,7]

$$PhCOCN_2Ph \xrightarrow{\Delta} Ph_2C=C=O \xrightarrow{R^\bullet} Ph\overset{\bullet}{\underset{Ph}{C}}-C(=O)R \quad (6)$$

5   **6**

The gas phase reaction of oxygen atoms[8] and hydroxyl radicals[9-11] with ketenes have also been studied. The reactions are complex, but additions to both carbons followed by cleavage and rearrangement processes are implicated, as shown in equations 7 and 8.[10,11] Carbon radicals are also proposed to add to C_β of ketenes in mass spectrometric processes[12] (see also Section 2.5).

$$CH_2=C=O \xrightarrow{HO^\bullet} HOCH_2\overset{\bullet}{C}=O \longrightarrow HO\overset{\bullet}{C}H_2 + CO \quad (7)$$

$$CH_2=C=O \xrightarrow{HO^\bullet} \overset{\bullet}{C}H_2CO_2H \longrightarrow CH_3CO_2^\bullet \xrightarrow{-CO_2} \overset{\bullet}{C}H_3 \quad (8)$$

Photolysis of ketene gave the species CH by multiphoton dissociation, and the gas phase rate constant for the reaction of CH with $CH_2=C=O$ was measured.[13] The reaction was proposed to occur by insertion in a C—H bond to give transient formation of **7**, which underwent fragmentation as shown (equations 9 and 10).[13]

$$CH_2=C=O \xrightarrow{\overset{\bullet}{C}H} \overset{\bullet}{C}H_2CH=C=O \longrightarrow HC\equiv CH + H + CO \quad (9)$$

7

$$\mathbf{7} \longrightarrow C_2H_3 + CO \quad (10)$$

The rate constants for the reactions of F and Cl atoms with ketene in the gas phase have been reported.[14] The reactions with $F^\bullet$ were proposed to involve the steps shown in equations 11–13, and steps analogous to 11 and 12 were proposed for Cl•.[14]

$$CH_2=C=O \xrightarrow{F\bullet} \overset{\bullet}{C}H=C=O + HF \quad (11)$$

$$CH_2=C=O \xrightarrow{F\bullet} \overset{\bullet}{C}H_2F + CO \quad (12)$$

$$CH_2=C=O \xrightarrow{F\bullet} CHF=C=O + H\bullet \quad (13)$$

References

1. Michael, J. V.; Nava, D. F.; Payne, W. A.; Stief, L. J. *J. Chem. Phys.* **1979**, *70*, 5222–5227.
2. Umemoto, H.; Tsunashima, S.; Sato, S.; Washida, N.; Hatakeyama, S. *Bull. Chem. Soc. Jpn.* **1984**, *57*, 2578–2580.
3. Slemr, F.; Warneck, P. *Ber. Bunsenges. Phys. Chem.* **1975**, *79*, 152–156.
4. Malatesta, V.; Forrest, D.; Ingold, K. U. *J. Phys. Chem.* **1978**, *82*, 2370–2373.
5. Lillford, P. J.; Satchell, D. P. N. *J. Chem. Soc. B* **1970**, 1303–1305.
6. Singer, L. S.; Lewis, I. C. *J. Am. Chem. Soc.* **1968**, *90*, 4212–4218.
7. Contineanu, M. *Rev. Roum. Chim.* **1991**, *36*, 931–936.
8. Washida, N.; Hatakeyama, S.; Takagi, H.; Kyogoku, T.; Sato, S. *J. Chem. Phys.* **1983**, *78*, 4533–4540.
9. Hatakeyama, S.; Honda, S.; Washida, N.; Akimoto, H. *Bull. Chem. Soc. Jpn.* **1985**, *58*, 2157–2162.
10. Oehlers, C.; Temps, F.; Wagner, H. G.; Wolf, M. *Ber. Bunsenges. Phys. Chem.* **1992**, *96*, 171–175.
11. Brown, A. C.; Canosa-Mas, C. E.; Parr, A. D.; Wayne, R. P. *Chem. Phys. Lett.* **1989**, *161*, 491–496.
12. Smith, R. L.; Chyall, L. J.; Chou, P. K.; Kenttämaa, H. I. *J. Am. Chem. Soc.* **1994**, *116*, 781–782.
13. Hancock, G.; Heal, M. R. *J. Chem. Soc. Faraday Trans.* **1992**, *88*, 2121–2123.
14. Ebrecht, J.; Hack, W.; Wagner, H. G. *Ber. Bunsenges. Phys. Chem.* **1990**, *94*, 587–593.

5.8 POLYMERIZATION OF KETENES

Polymers based on ketenes[1–3] were discovered by Wedekind[4] and by Staudinger and co-workers.[5,6] Extensive studies of high polymers came later,[1–3,7] and three principal types have been identified. These are the 1,2-addition types involving the C=C and C=O bonds (**1** and **2**) which are ketones and vinyl acetals, respectively, and esters (**3**), which contain both types of linkages. All three can be prepared as crystalline materials from $Me_2C=C=O$ by appropriate choice of catalyst and solvent.[7]

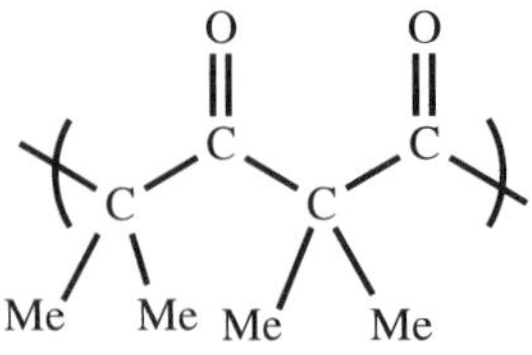

Thus **1–3** are formed with AlBr$_3$ in toluene, *n*-BuLi in ether, and AlEt$_3$ in toluene, respectively,[1,7–9] as shown in equations 1–3. Diphenylketene has also been polymerized by EtLi or BuLi.[10] Copolymers of ketenes with other carbonyl compounds[11,12] and with isocyanates[13,14] are also formed.

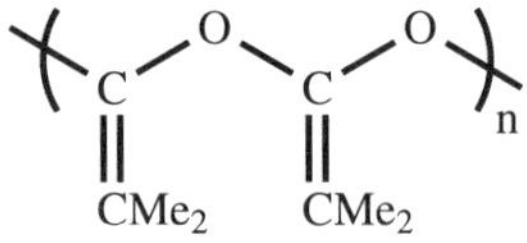

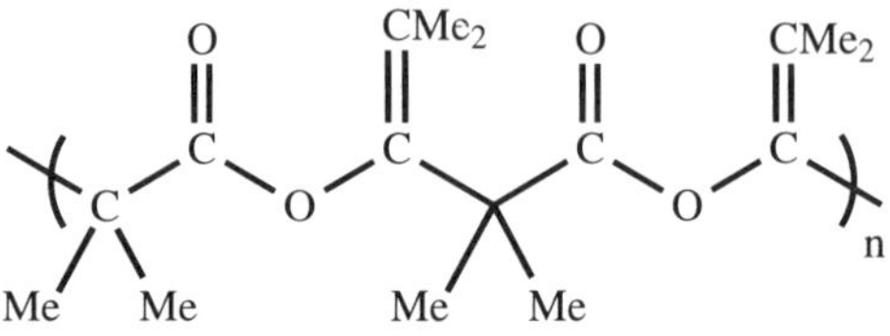

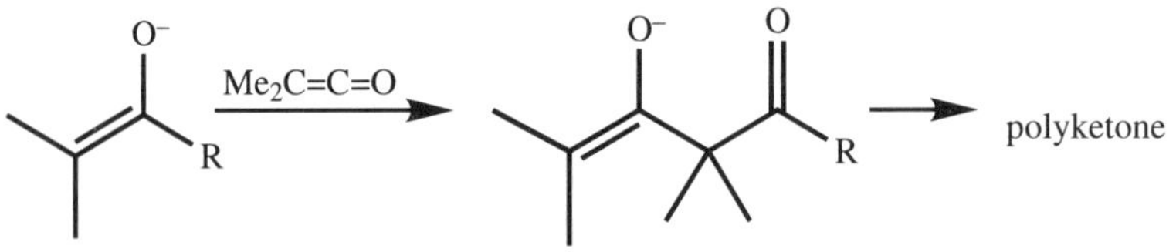

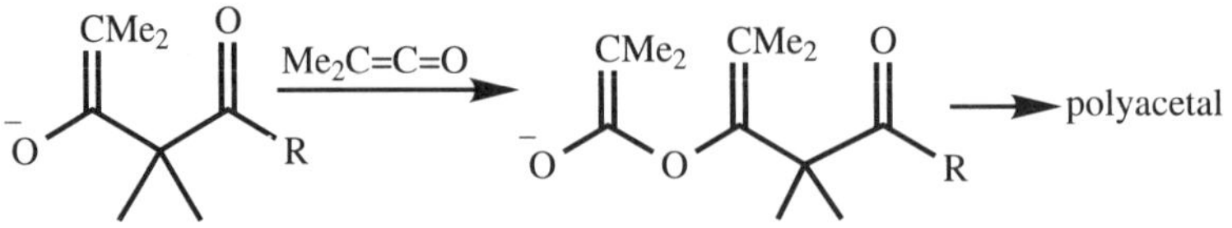

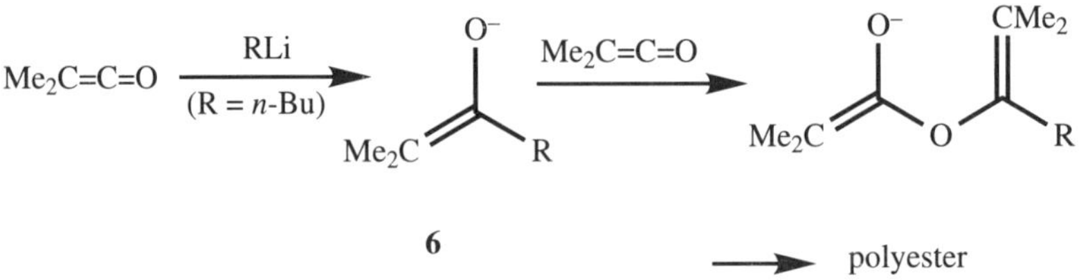

The preparation of polyketene by the reaction of acetyl chloride with Lewis acid catalysts has recently been examined and proposed to occur from addition of acylium ions to ketene (equation 4).[15] Polyketene has also been prepared by acidic polymerization of 1,1-di-(n-hexyloxy)ethene and mild hydrolysis of the resulting diether (equation 5).[16] This polyketene is extensively enolized, as represented in 7, and its conducting properties are of great interest.[17]

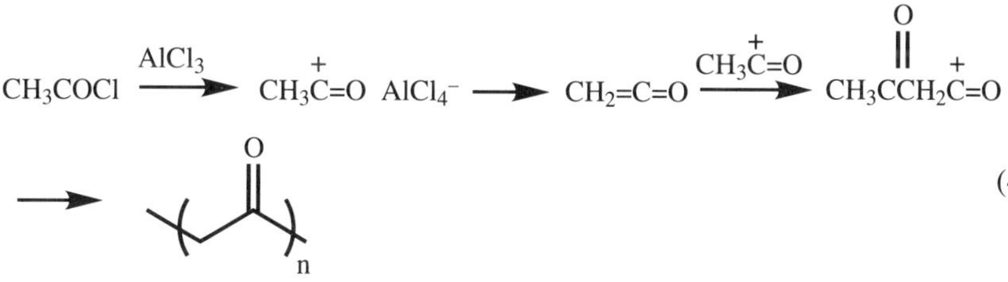

$$CH_3COCl \xrightarrow{AlCl_3} CH_3\overset{+}{C}=O \; AlCl_4^- \longrightarrow CH_2=C=O \xrightarrow{CH_3\overset{+}{C}=O} CH_3\overset{O}{\overset{\|}{C}}CH_2\overset{+}{C}=O$$

$$\longrightarrow \quad \text{—(CH}_2\text{—CO)}_n\text{—} \tag{4}$$

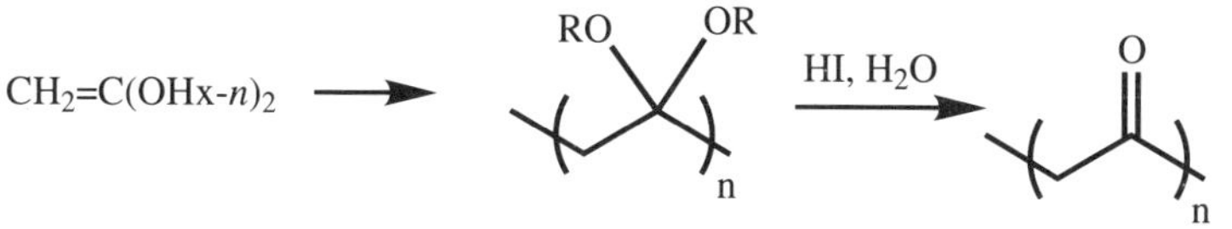

$$CH_2=C(OHx\text{-}n)_2 \longrightarrow \text{—[C(OR)}_2\text{CH}_2\text{]}_n\text{—} \xrightarrow{HI, H_2O} \text{—(CH}_2\text{CO)}_n\text{—} \tag{5}$$

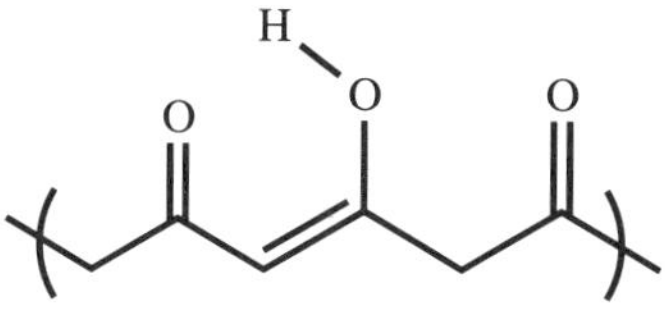

7

The photolysis of 2-diazo-1,3-diketones in diethyl ether leads to intermediate ketoketenes 8 which form novel oligomers 9 (equation 6).[18] A variety of polymers have also been formed from carbon suboxide (C_3O_2).[19,20] The formation of polymers from other bisketenes is noted in Section 4.9.

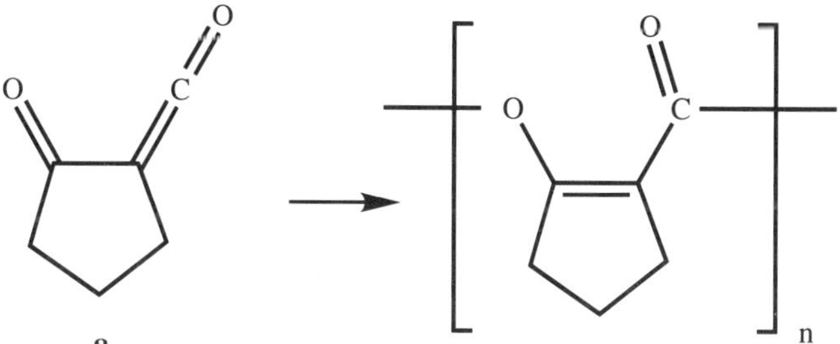

$$\tag{6}$$

8   9

References

1. Pregaglia, G. F.; Binaghi, M. In *Encyclopedia of Polymer Science and Technology*, Vol. 8; Wiley: New York, 1968; pp. 45–57.
2. Pasquon, I.; Porri, L.; Giannani, U. In *Encyclopedia of Polymer Science and Technology*, Vol. 15, 2nd ed.; Wiley: New York, 1989; pp. 713–715.
3. Pregaglia, G. F.; Binaghi, M. In *The Stereochemistry of Macromolecules*, Vol. 2; Ketley, A. D., Ed.; Dekker: New York, 1967; pp. 111–175.
3a. Zarras, P.; Vogl, D. *Prog. Polym. Sci.* **1991**, *16*, 173–201.
4. Wedekind, E. *Liebigs Ann. Chem.* **1902**, *323*, 246–257.
5. Staudinger, H.; Ruzicka, L. *Liebigs Ann. Chem.* **1911**, *380*, 278–303.
6. Staudinger, H. *Helv. Chim. Acta* **1925**, *8*, 306–332.
7. Natta, G.; Mazzanti, G.; Pregaglia, G.; Binaghi, M.; Peraldo, M. *J. Am. Chem. Soc.* **1960**, *82*, 4742–4743.
8. Yamashita, Y.; Miura, S.; Nakamura, M. *Makromol. Chem.* **1963**, *68*, 31–47.
9. Cash, G. O., Jr.; Martin, J. C. *U.S. Patent* 3,321,441; *Chem. Abstr.* **1968**, *68*, 13684e.
10. Nadzhimutdinov, Sh.; Cherneva, E. P.; Kargin, V. A. *Vysokomol. Soedin., Ser. B,* **1967**, *9*, 480–485; *Chem. Abstr.* **1967**, *67*, 100431t.
11. Natta, G.; Mazzanti, G.; Pregaglia, G. F.; Binaghi, M. *J. Am. Chem. Soc.* **1960**, *82*, 5511–5512.
12. Nunomoto, S.; Yamashita, Y. *Kogyo Kagaku Zasshi* **1968**, *71*, 2067–2072; *Chem. Abstr.* **1969**, *70*, 47947z.
13. Dyer, E.; Sincich, E. *J. Polym. Sci. Part A-1,* **1973**, *11*, 1249–1260.
14. Higashi, H.; Harada, H. *Kogyo Kagaku Zasshi* **1966**, *69*, 2344–2345; *Chem. Abstr.* **1967**, *66*, 105236x.
15. Olah, G. A.; Zadok, E.; Edler, R.; Adamson, D. H.; Kasha, W.; Prakash, G. K. S. *J. Am. Chem. Soc.* **1989**, *111*, 9123–9124.
16. Khemani, K. C.; Wudl, F. *J. Am. Chem. Soc.* **1989**, *111*, 9124–9125.
17. Cui, C. X.; Kertesz, M. *J. Am. Chem. Soc.* **1991**, *113*, 4404–4409.
18. Nikolaev, V. A.; Frenkh, Yu.; Korobitsyna, I. K. *Zh. Org. Khim.* **1978**, *14*, 1147–1160; *Engl. Transl.* **1978**, *14*, 1069–1080.
19. Porejko, S.; Makaruk, L.; Kepka, M. *Polimery* **1967**, *12*, 464–467; *Chem. Abstr.* **1968**, *69*, 3211e.
20. Sladkov, A. M.; Korshak, V. V.; Nepochatykh, V. P. *Izv. Akad. Nauk SSSR, Ser. Khim.* **1968**, 196–197; *Engl. Transl.* **1968**, 191–193.

5.9 STEREOSELECTIVITY IN KETENE REACTIONS

Reactions of ketenes frequently involve the formation of new chiral centers, and stereoselectivity in this process has been achieved in a variety of different ways, as discussed in several reviews.[1-3] The first study on this reaction was in 1919,[4] involving the reaction of menthol with phenyl-*p*-tolylketene, and while this initial effort did not lead to detectable diastereoselective reaction, later efforts have been more successful.

Thus additions of chiral alcohols to unsymmetrical ketenes such as arylmethylketenes (**1**),[5-7] alkylhaloketenes RCHal=C=O,[8] or phenyltrifluoromethylketene (**2**)[9] give α-arylpropionate esters with diastereoselectivities as high as 99% (equations 1 and 2). Reaction of sugars with **1** followed by hydrolysis of the resulting esters gives acids that have optical purities of up to 74%.[10]

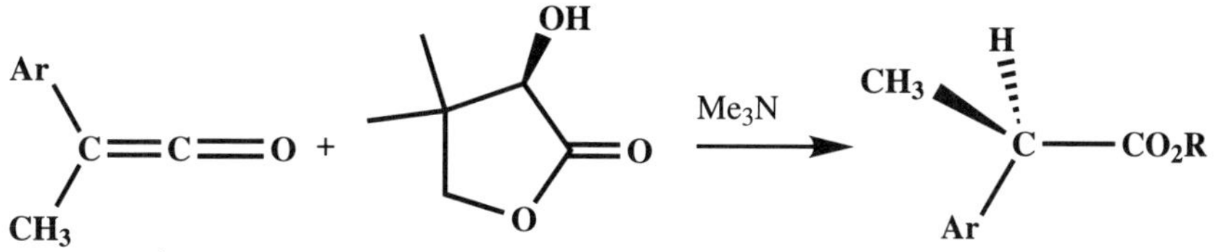

$$\underset{\underset{CH_3}{|}}{\overset{Ar}{|}}C=C=O \; + \; \text{(ROH)} \; \xrightarrow{Me_3N} \; \underset{Ar}{\overset{CH_3 \quad H}{\underset{|}{C}}}-CO_2R$$

1

90%, 99.5% R

(1)[5]

$$\underset{\underset{CF_3}{|}}{\overset{Ph}{|}}C=C=O \; + \; CH_3\text{-Ar-CH(OH)CH}_3 \; \longrightarrow \; \underset{Ph}{\overset{CF_3 \quad H}{\underset{|}{C}}}-CO_2R$$

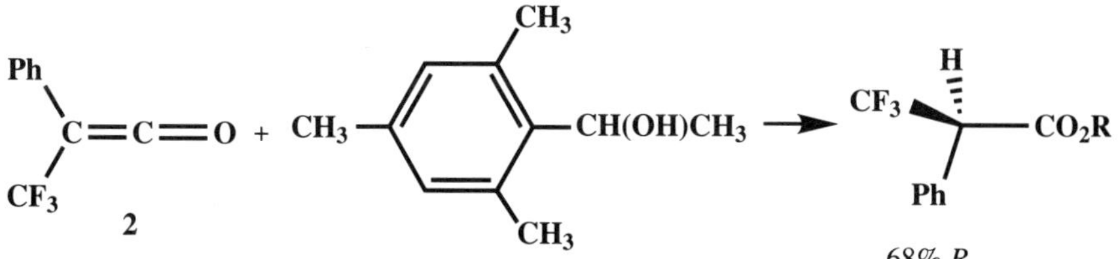

2 (S)

68% R

(2)[9]

(S)-(-)-1-Phenylethylamine reacts selectively with phenylmethylketene to product preferentially the S,S-amide (**3**), in 60% enantiomeric excess in toluene at −100 °C.[11] The proton transfer step to the β-carbon determines the stereochemistry of the product, and proton transfer to the face opposite the phenyl of the amine in the zwitterionic intermediate **4** would lead to this result (equation 3). However, the details of the proton transfer are not established. Other examples of this reaction are known.[1-3] Addition of the isopropyl ester of (R)–alanine to phthalimido-*tert*-butylketene (**5**) gives (R,R)-amide in up to 63% enantiomeric excess (equation 4).[12] Proton transfer opposite to the methyl group in **6** would explain this result.

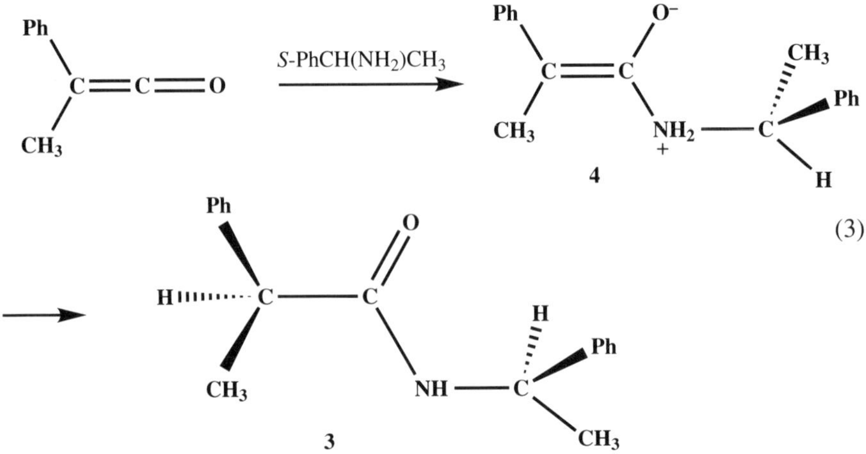

644 REACTIONS OF KETENES

$$\text{(3)}$$

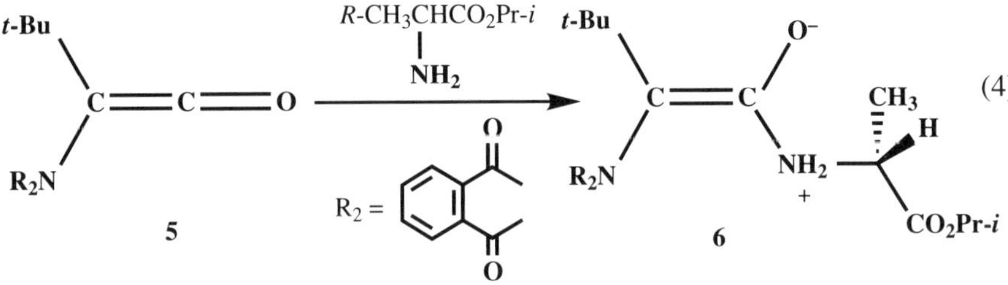

The reaction of **7** with chiral amines led to the formation of amides with stereoselectivity that did not depend upon the temperature but did change as a function of the amine and solvent.[13]

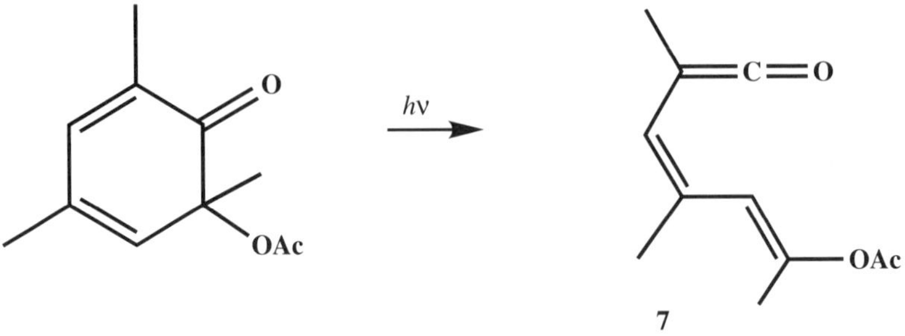

$$\text{(5)}$$

Photolyses of optically active chromium amine complexes such as **8** are suggested to give chromium complexed ketenes **9**, which react with esters of optically active amino acids such as **10** to give dipeptides **11** with a 98/2 diastereomeric ratio (equation 6).[14] This procedure has been adapted to solid state peptide synthesis.[14a]

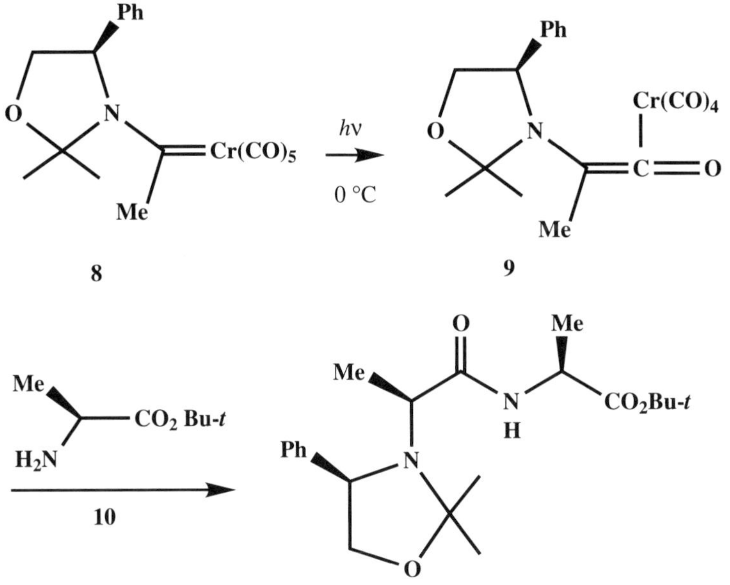

5.9 STEREOSELECTIVITY IN KETENE REACTIONS

[Structures 8, 9, 10, 11 shown with reaction scheme, equation (6)]

$$8 \xrightarrow{h\nu, 0°C} 9$$

$$10 \longrightarrow 11 \quad (6)$$

The addition of achiral alcohols to ketenes catalyzed by optically active bases also proceeds stereoselectivity.[15] Thus the reaction of phenylmethylketene with methanol catalyzed by brucine gives optically active methyl 2-phenylpropionate, with 25% excess S-(+)ester at -110 °C and a 10% excess of the R-(+)-ester at 80 °C. Selective addition to a zwitterionic intermediate 12 is proposed[15,16] to explain the results (equation 7), although the detailed reaction pathway is still not settled.[16a] Optically active amines attached to polymers were also effective catalysts in the reaction of equation 7.[17,18]

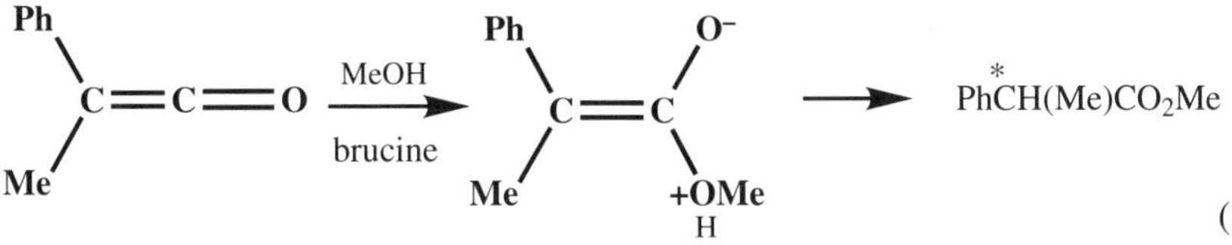

$$\text{Ph(Me)C=C=O} \xrightarrow[\text{brucine}]{\text{MeOH}} \mathbf{12} \longrightarrow \text{PhCH(Me)CO}_2\text{Me} \quad (7)$$

The reactions of carboxylic acids with unsymmetrical ketenes catalyzed by chiral amines give stereoselective formation of carboxylic acids and anhydrides, as in the example of equation 8.[19] Addition of lithium arylthiophenoxides to a ketene with protonation by a chiral proton donor gave the thiol ester in up to 97% *ee* (equation 9).[19a] Completely stereoselective addition of methanol to an aliphatic ketene generated by a Wolff rearrangement and bearing a chiral substituent has been observed.[19b]

REACTIONS OF KETENES

$$\underset{i\text{-Pr}}{\overset{\text{Ar}}{>}}C=C=O \xrightarrow[\text{chiral catalyst}]{HCO_2H} \underset{i\text{-Pr}}{\overset{\text{Ar}}{>}}\overset{H}{\underset{}{C}}-CO_2H \quad (8)$$

(equation 9 scheme with ArSLi, then HO-CH(Ph)-CH(Me)-N(i-Pr), giving thioester product)

(9)

Cycloaddition reactions of ketenes also proceeded with stereoselectivity. Thus the [2 + 2] cycloaddition of (menthyloxy)ketene (**13**), generated by dehydrochlorination, to Z-ethyl propenyl ether gave a 3:1 diastereoselectivity in the formation of each of the products **14** and **15** (equation 10).[20] Evidently **15** was formed by equilibration of **14**. The same menthyloxyketene, or its metal complex, was generated from chromium carbene complexes, and gave comparable reactivity in cycloaddition (equation 11).[21]

$$\underset{\text{Me}}{\overset{*\text{MenO}}{>}}C=C=O \xrightarrow{\underset{}{\text{EtO}}\overset{\text{Me}}{=}} \text{**14**} + \text{**15**} \quad (10)$$

13

$$\underset{\text{Me}}{\overset{*\text{MenO}}{>}}C=Cr(CO)_5 \xrightarrow[CO]{h\nu} \underset{\text{Me}}{\overset{*\text{MenO}}{>}}C=C=O \quad (11)$$

13

Enantiomerically enriched ketones were obtained when alkylarylketenes underwent SmI$_2$-mediated allylation or benzylation with allyl or benzyl halides followed by protonation with chiral acids (equation 4, Section 5.5.2.4).[22]

Chiral induction occurred in the [2 + 2] cycloaddition of the vinyl ether **16** with dichloroketene so that there was at least a 90% enantiomeric excess of the cyclobutanone **17**.[23,24] The high level of stereoselectivity was attributed to a preference of **16** for the conformation shown where only one face of the double bond was available for cycloaddition (equation 12). The product **17** was converted to a cyclopentanone by diazomethane treatment.[23]

16 17 (12)

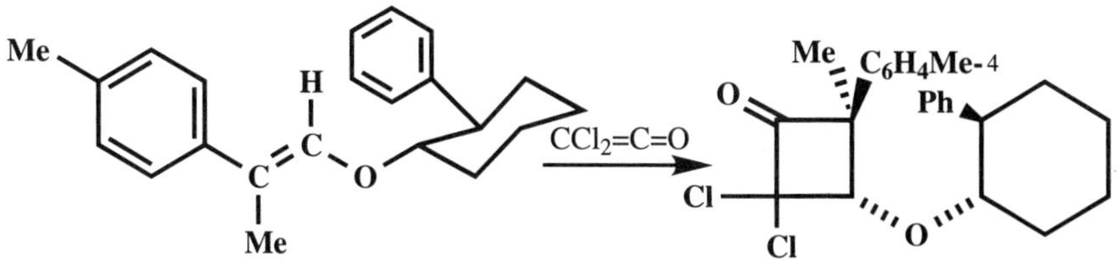

Reaction of dichloroketene with carbohydrate enol ethers **18** also forms optically active cyclobutanones **19** (equation 13).[25] Stepwise addition through a zwitterionic intermediate (equation 13) appears plausible. These were converted to lactones by Baeyer–Villiger reaction after removal of the chlorines by reduction.[25]

18 19 (13)

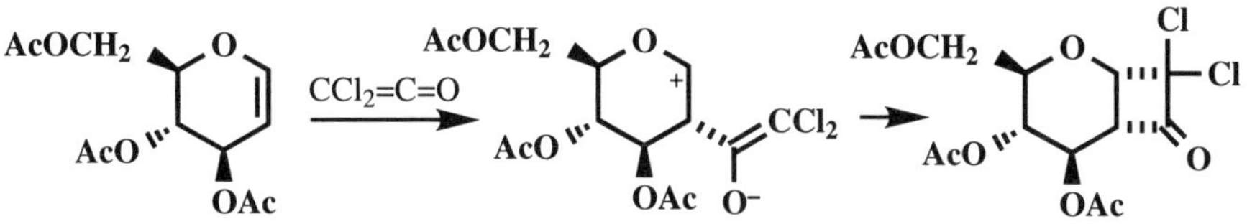

In the reaction of the furanose **20** it was thought unusual that attack from the α-side was favored, although only by a 1:1.2 ratio. For **21,** attack from the β-side was favored by a 4:1 ratio as expected (equations 14 and 15).[25]

20 (14)

21 (15)

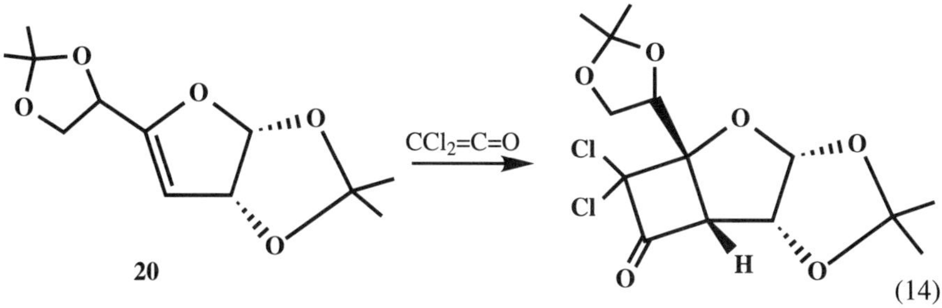

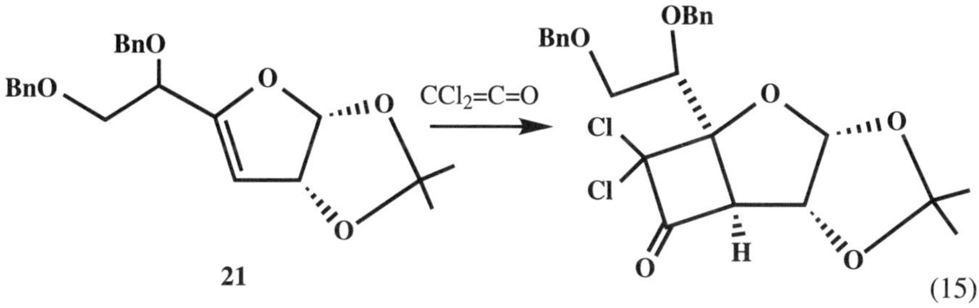

Reaction of **22E** (or **22Z**) occurred only from the α-face (equation 16).[25] It was suggested that this reaction would involve the ketene to be generated "fixed in the sphere of the carbohydrate derivative (**22Z**) such that the orientation of the addition is not controlled by normal steric effects. In particular the two halogen atoms of the dichloroketene must penetrate far into the concave region of the sugar molecule".[25] This explanation applies to a concerted process, but for an alternative stepwise process proceeding via a zwitterionic intermediate **23** the chlorines need not be as near the carbohydrate ring in the decisive first step. Attack from the side opposite methyl and below the benzyl ether function appears to be the major factor guiding the reaction.

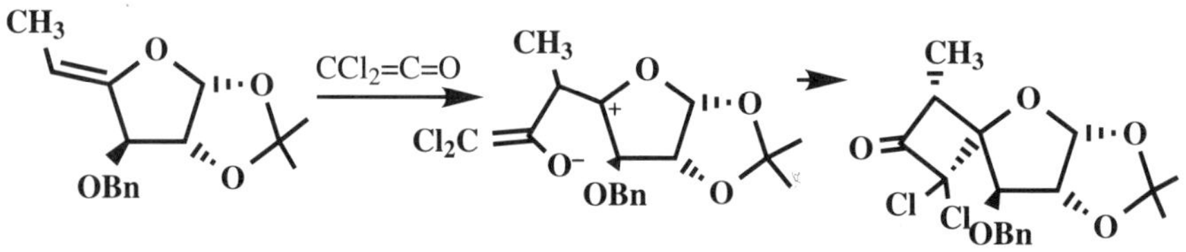

(16)

The reaction of chloral and other chlorinated aldehydes and ketones with ketene catalyzed by optically active tertiary amines gave optically active β-lactones such as **24** (equation 17).[26-28] This reaction was proposed to proceed through an amine–ketene complex which can be depicted as the zwitterion **25**.

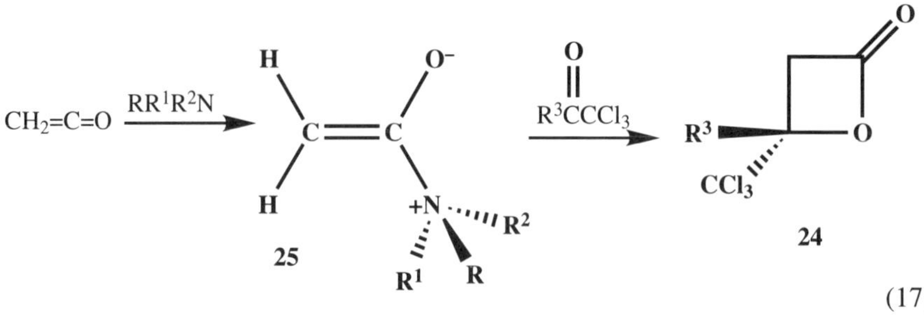

(17)

Chiral synthesis also occurs in addition–rearrangement sequences involving ketenes. In the example of equation 18 dichloroketene reacted with the optically active allyl thioether **26** to give the rearranged product **27** in essentially 99% ee, evidently via the process shown. Dechlorination of **27** was effected with Zn/HOAc.[29] This process is termed the "ketene Claisen rearrangement," and in the example of equation 19 proceeds with ≥94% diastereomeric selectivity through the proposed transition state **28**.[30]

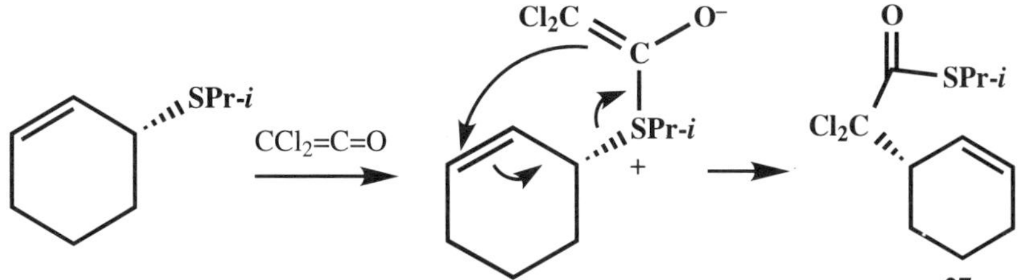

5.9 STEREOSELECTIVITY IN KETENE REACTIONS

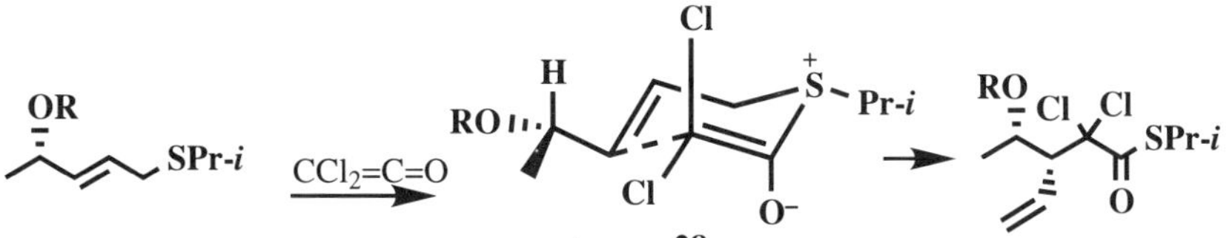

[Structures for equation 18: compound **26** with SPr-*i* substituent + CCl₂=C=O → intermediate with Cl₂C=C-O⁻ and SPr-*i* → compound **27** with Cl₂C, SPr-*i* groups]

(18)

[Structures for equation 19: starting material with OR and SPr-*i* groups + CCl₂=C=O → intermediate **28** with Cl, S⁺-Pr-*i*, O⁻ → product with RO, Cl, Cl, SPr-*i*]

R = *t*–BuSiMe₂

(19)

Chiral-*N*-(arylsulfonyl)indole sulfoxides react with CCl₂=C=O as shown in equation 20.[31]

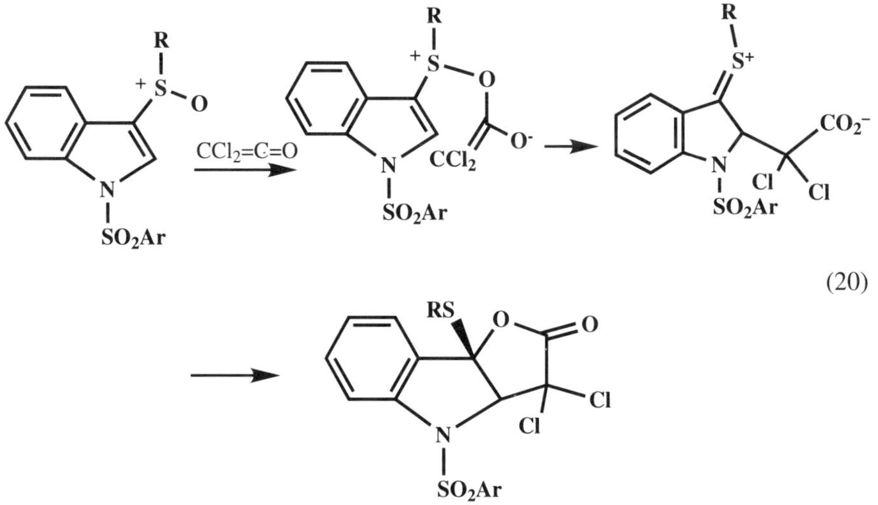

[Structures for equation 20: N-sulfonyl indole with +S–O and R + CCl₂=C=O → intermediate with CCl₂, O⁻ → indoline with S⁺, CO₂⁻, Cl, Cl, SO₂Ar → tricyclic product with RS, O, O, Cl, Cl, N–SO₂Ar]

(20)

Photolyses of optically active aminocarbene complexes **29** and **30** gave diastereoselective synthesis of amino acid derivatives (equations 21 and 22).[32] Reaction of **29**-OD (R = CH₃) gave **31** in 66% yield, with 95% deuterium incorporation and 90% *de*. However, with R = H the yield was only 21% with no stereoselectivity. Photolysis of **30** in the presence of CH₃OD gave **32** in 93% yield, with at least 97% incorporation of deuterium, and 86% diastereoselectivity.[32]

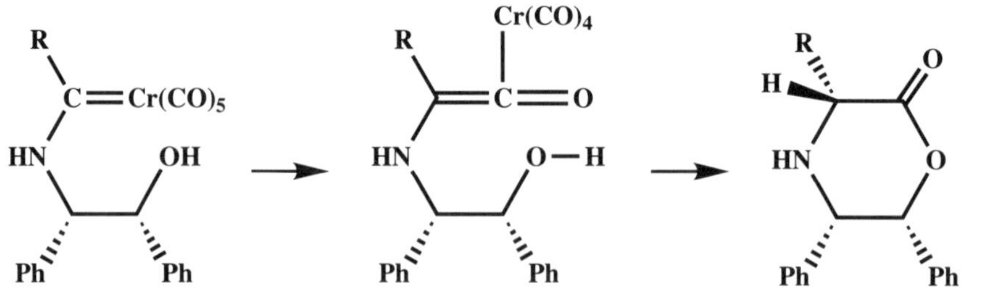

The reactions in equations 21 and 22 apparently proceed through chromium–ketene complexes as shown.[33,34] Further evidence for this supposition is provided by the reaction of **33** with Et$_3$N, followed by addition of CH$_3$OD.[32] This reaction is presumed[32] to generate the free ketene **34,** and the diastereoselectivity in the formation of **35** was 23%, as opposed to 74% when the ketene was generated as the chromium complex (equation 23). This sequence provides an efficient synthesis of the chiral deuteroglycine derivative **35.**

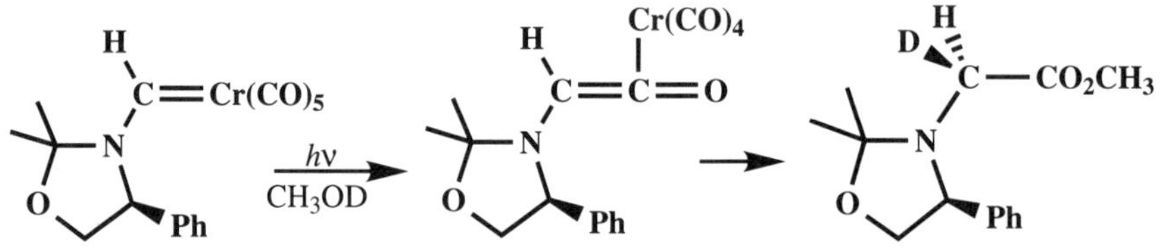

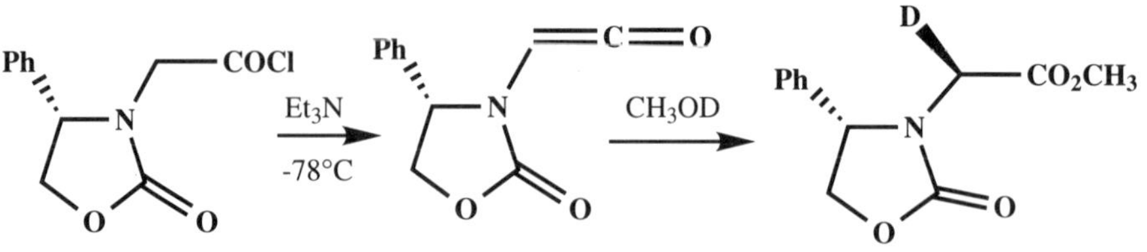

The preparation of β-lactams with control of the stereochemistry by [2 + 2] cycloadditions of ketenes or ketene equivalents with imines has been intensively studied and reviewed.[35–38] This reaction is discussed in more detail in Section 5.4.1.7.

Molecular modeling of the reaction between methylketene and *N*-methyl-2-methylimine predicted a strong preference for a perpendicular transition structure which favored formation of the *cis*-product (equation 24), in agreement with experiment.[36] Reaction of chiral imines with PhOCH=C=O also gave selectivity for formation of the *cis* product.[39] Calculations using AM1 gave good agreement with the experimentally observed diastereoselectivities.[40]

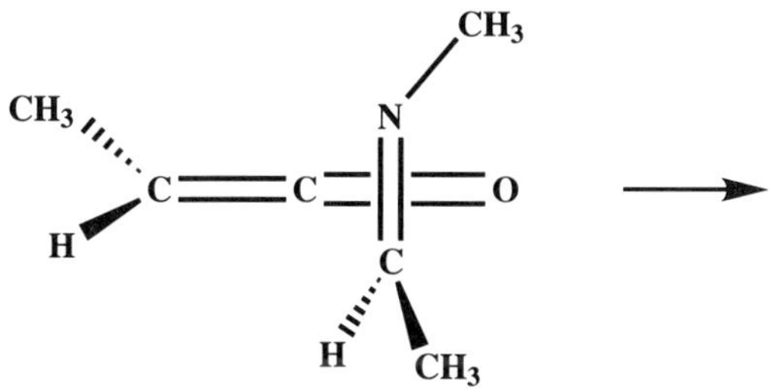

5.9 STEREOSELECTIVITY IN KETENE REACTIONS

(24)

A repetitive sequence in which imines are converted by ketene cycloaddition to β-lactams which are cleaved, deaminated, and converted to homologated imines which are again converted to β-lactams has been developed as a diastereoselective route to 1,3-polyols (equation 25).[37]

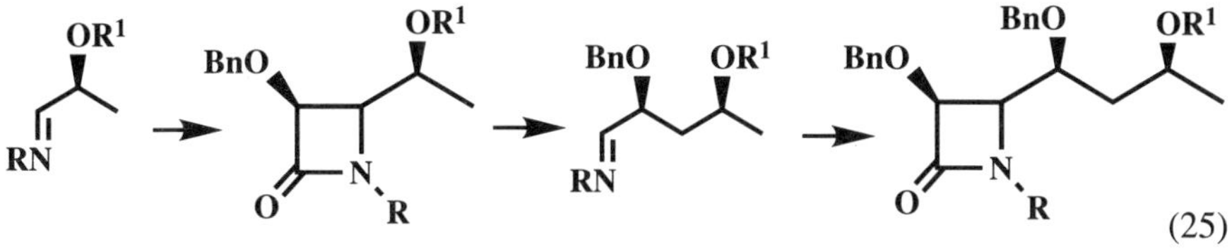

(25)

Reaction of dichloroketene with the steroidal alkene **36** gave only the cycloadduct **37** (equation 26).[41]

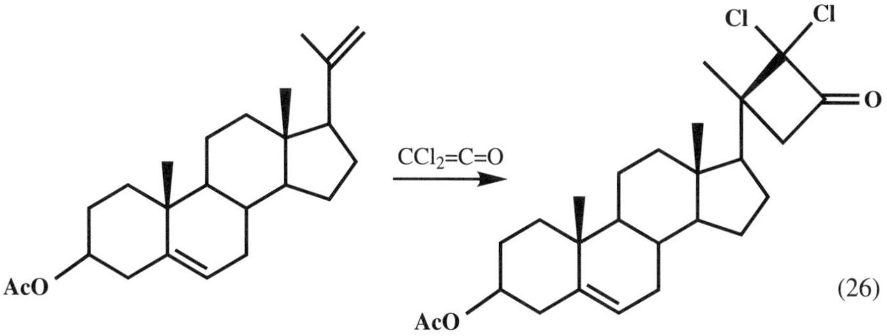

(26)

36

37

References

1. Morrison, J. D.; Mosher, H. S. *Asymmetric Organic Reactions*, Prentice Hall: New York, 1971.
2. Buschmann, H.; Scharf, H.; Hoffmann, N.; Esser, P. *Angew. Chem., Int. Ed. Engl.* **1991,** *30,* 447–515.
3. Pracejus, H. *Fortschr. Chem. Forsch.* **1967,** *8,* 493–553.
4. Weiss, R. *Monatsch. Chem.* **1919,** *40,* 391–402.

5. Larsen, R. D.; Corley, E. G.; Davis, P.; Reider, P. J.; Grabowski, E. J. J. *J. Am. Chem. Soc.* **1989**, *111*, 7650–7651.

5a. Senanayake, C. H.; Larsen, R. D.; Bill, T. J.; Liu, J.; Corley, E. G.; Reider, P. J. *Synlett* **1994**, 199–200.

6. Salz, U.; Rüchardt, C.; *Tetrahedron Lett.* **1982**, *23*, 4017–4020.

7. Jähme, J.; Rüchardt, C. *Angew. Chem., Int. Ed. Engl.* **1981**, *20*, 885–887.

8. Durst, T.; Koh, K. *Tetrahedron Lett.* **1992**, *33*, 6799–6802.

9. Anders, E.; Ruch, E.; Ugi, I. *Angew. Chem., Int. Ed. Engl.* **1973**, *12*, 25–29.

10. Bellucci, G.; Berti, G.; Bianchini, R.; Vecchiani, S. *Gazz. Chim. Ital.* **1988**, *118*, 451–456.

11. Pracejus, H.; Tille, A. *Chem. Ber.* **1963**, *96*, 854–865.

12. Winter, S.; Pracejus, H. *Chem. Ber.* **1966**, *99*, 151–159.

13. Schultz, A. G.; Kulkarni, Y. S. *J. Org. Chem.* **1984**, *49*, 5202–5206.

14. Miller, J. R.; Pulley, S. R.; Hegedus, L S.; DeLombaert, S. *J. Am. Chem. Soc.* **1992**, *114*, 5602–5607.

14a. Pulley, S. R.; Hegedus, L. S. *J. Am. Chem. Soc.* **1993**, *115*, 9037–9047.

15. Pracejus, H. *Liebigs Ann. Chem.* **1960**, *634*, 9–22.

16. Pracejus, H.; Kohl, G. *Liebigs Ann. Chem.* **1969**, *722*, 1–11.

16a. Weidert, P. J.; Geyer, E.; Horner, L. *Liebigs Ann. Chem.* **1989**, 533–538.

17. Yamashita, T.; Yasueda, H.; Nakamura, N. *Chem. Lett.* **1974**, 585–588.

18. Yamashita, T.; Mitsui, H.; Watanabe, H.; Nakamura, N. *Polym. J.* **1981**, *13*, 179–181.

19. Stoutamire, D. W. *U.S. Patent*, 4,570,017; *Chem. Abstr.* **1986**, *105*, 78661x.

19a. Fehr, C.; Stempf, I.; Galindo, J. *Angew. Chem., Int. Ed. Engl.* **1993**, *32*, 1044–1046.

19b. Lopez-Herrera, F. J.; Sarabia-Garcia, F. *Tetrahedron Lett.* **1994**, *35*, 2929–2932.

20. Fráter, G.; Müller, U.; Günther, W. *Helv. Chim. Acta* **1986**, *69*, 1858–1861.

21. Söderberg, B. J.; Hegedus, L. S.; Sierra, M. A. *J. Am. Chem. Soc.* **1990**, *112*, 4364–4374.

22. Takeuchi, S.; Miyoshi, N.; Ohgo, Y. *Chem. Lett.* **1992**, 551–554.

23. Greene, A. E.; Charbonnier, F.; Luche, M.; Moyano, A. *J. Am. Chem. Soc.* **1987**, *109*, 4752–4753.

24. Greene, A. E.; Charbonnier, F. *Tetrahedron Lett.* **1985**, *26*, 5525–5528.

25. Redlich, H.; Lenfers, J. B.; Kopf, J. *Angew. Chem., Int. Ed. Engl.* **1989**, *28*, 777–778.

26. Wynberg, H.; Staring, E. G. J. *J. Am. Chem. Soc.* **1982**, *104*, 166–168.

27. Wynberg, H.; Staring, E. G. J. *J. Org. Chem.* **1985**, *50*, 1977–1979.

28. Ketelaar, P. E. F.; Staring, E. G. J.; Wynberg, H. *Tetrahedron Lett.* **1985**, *26*, 4665–4668.

29. Öhrlein, R.; Jeschke, R.; Ernst, B.; Bellus, D. *Tetrahedron Lett.* **1989**, *30*, 3517–3520.

30. Nubbemeyer, U.; Öhrlein, R.; Gonda, J.; Ernst, B.; Bellus, D. *Angew. Chem., Int. Ed. Engl.* **1991**, *30*, 1465–1467.

31. Marino, J. P.; Kim, M.; Lawrence, R. *J. Org. Chem.* **1989**, *54*, 1782–1785.

32. Hedgedus, L. S.; Lastra, E.; Narukawa, Y.; Snustad, D. C. *J. Am. Chem. Soc.* **1992**, *114*, 2991–2994.

33. Hegedus, L. S.; de Weck, G.; D'Andrea, S. *J. Am. Chem. Soc.* **1988**, *110*, 2122–2126.

34. Hegedus, L. S.; Schwindt, M. A.; De Lombaert, S.; Imwinkelried, R. *J. Am. Chem. Soc.* **1990**, *112*, 2264–2273.
35. Cooper, R. D. G.; Daugherty, B. W.; Boyd, D. B. *Pure Appl. Chem.* **1987**, *59*, 485–492.
36. Palomo, C.; Cossio, F. P.; Cuevas, C.; Lecea, B.; Mielgo, A.; Roman, P.; Luque, A.; Martinez-Ripoll, M. *J. Am. Chem. Soc.* **1992**, *114*, 9360–9369.
37. Palomo, C.; Aizpurua, J. M.; Urchegui, R.; Garcia, J. M. *J. Org. Chem.* **1993**, *58*, 1646–1648.
38. Banik, B. K.; Manhas, M. S.; Bose, A. K. *J. Org. Chem.* **1993**, *58*, 307–309.
39. Georg, G. I.; Wu, Z. *Tetrahedron Lett.* **1994**, *35*, 381–384.
40. Cossio, F. P.; Arrieta, A.; Lecea, B.; Ugalde, J. M. *J. Am. Chem. Soc.* **1994**, *116*, 2085–2093.
41. Blaszczyk, K.; Tykarska, E.; Paryzek, Z. *J. Chem. Soc., Perkin Trans. 2* **1992**, 257–261.

5.10 OTHER ADDITIONS TO KETENES

5.10.1 Reaction with Diazomethanes

Staudinger and Reber[1] attempted to prepare tetraphenylcyclopropanone (**1**) from the reaction of diphenylketene with diphenyldiazomethane, and instead found that the intermediate underwent decarbonylation (equation 1).

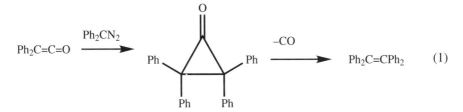

The reaction of ketenes with carbene precursors such as diazo compounds has been extensively applied to the synthesis of cyclopropanones (equations 2–6).[2–8] The reaction of silylketenes provides silylcyclopropanones (equations 7 and 8).[9–12] A carbene addition to a ketene is noted in Section 5.2 (equation 10).

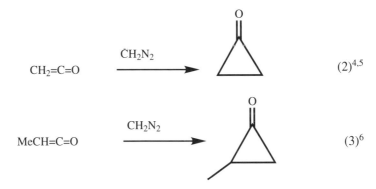

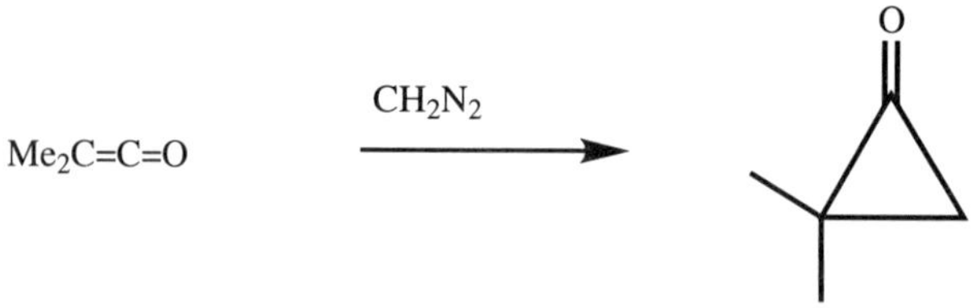

654 REACTIONS OF KETENES

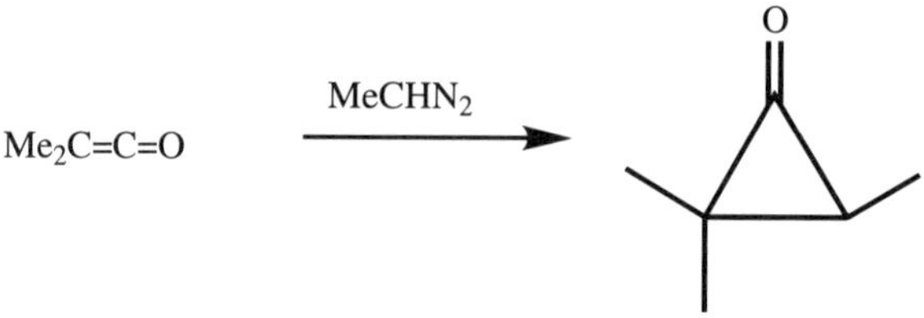

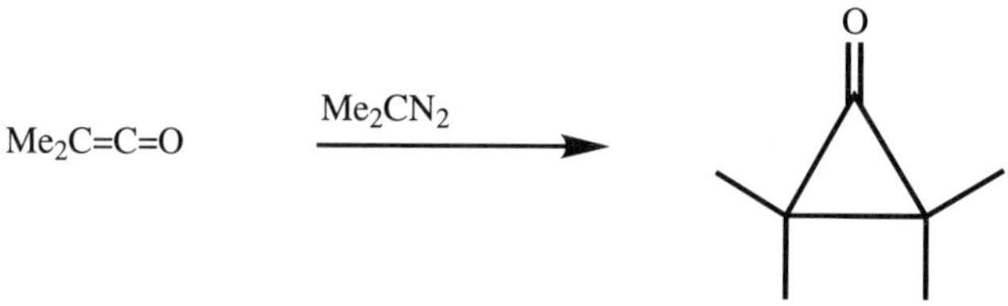

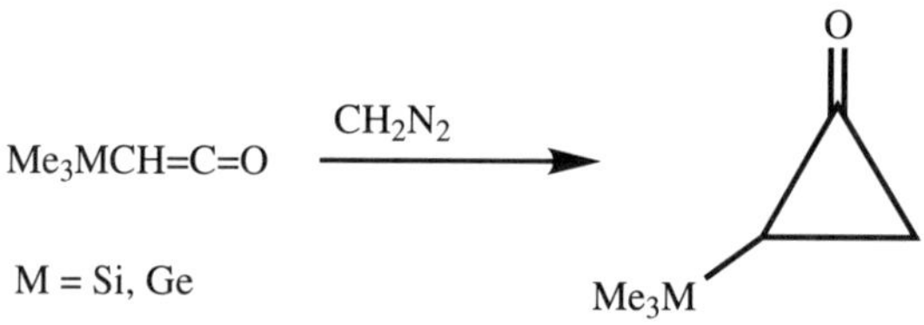

$$Me_2C=C=O \xrightarrow{CH_2N_2} \text{[cyclopropanone with 2 Me]} \quad (4)^7$$

$$Me_2C=C=O \xrightarrow{MeCHN_2} \text{[cyclopropanone with 3 Me]} \quad (5)^8$$

$$Me_2C=C=O \xrightarrow{Me_2CN_2} \text{[cyclopropanone with 4 Me]} \quad (6)^6$$

$$Me_3MCH=C=O \xrightarrow{CH_2N_2} \text{[cyclopropanone with Me}_3M\text{]} \quad (7)^9$$

M = Si, Ge

$$Me_3SiCH=C=O \xrightarrow{Me_3SiCHN_2} \text{[cyclopropanone with 2 SiMe}_3\text{]} \quad (8)^{12}$$

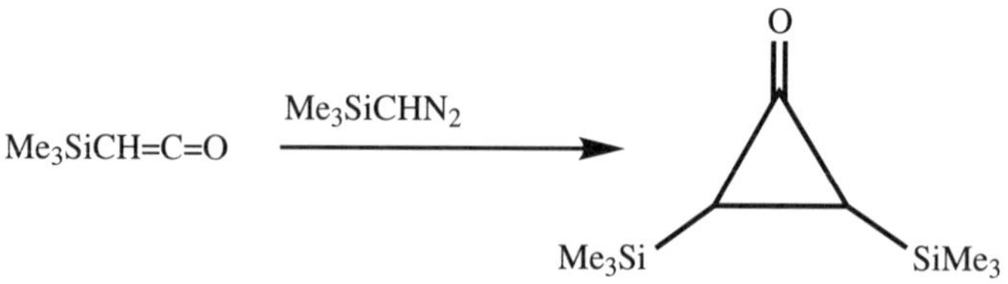

Reactions of other ketenes with diazomethanes gave cyclobutanones, presumably via cyclopropanones (equation 9).[8,13]

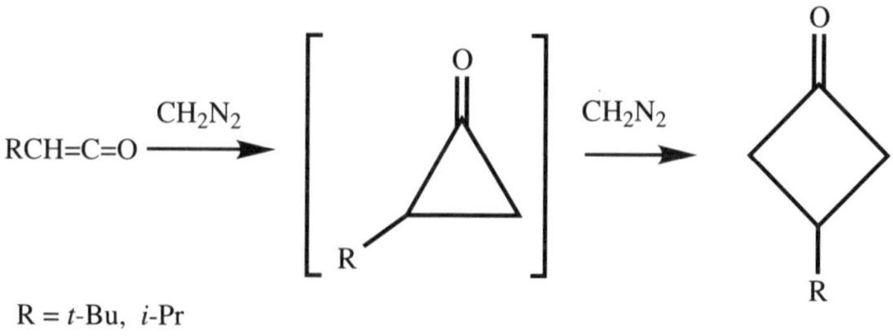

$$RCH=C=O \xrightarrow{CH_2N_2} [\text{cyclopropanone}] \xrightarrow{CH_2N_2} \text{[cyclobutanone]} \quad (9)^{13}$$

R = t-Bu, i-Pr

Thermal reactions of several bis(diazo)diketones gave enones, and these reactions were suggested to proceed by reaction of intermediate ketenes with the remaining diazoketone group to give intermediate cyclopropanones (equation 10).[14] Intermolecular reactions of ketenes with diazoketones are discussed in Section 5.4.2.

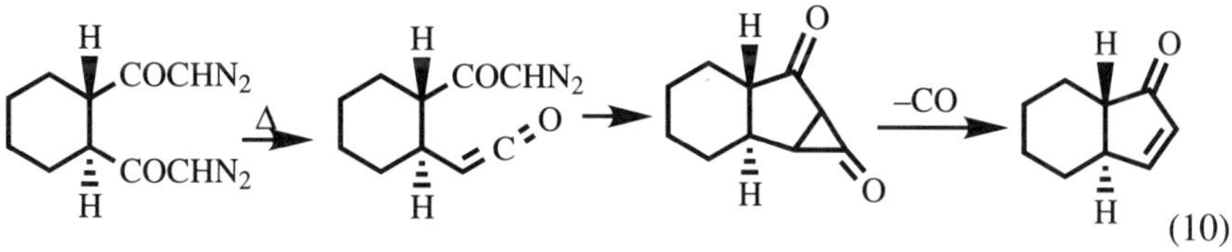

(10)[14]

5.10.2 Reaction with Sulfur Dioxide

Ketene reacts with sulfur dioxide at -78 °C to produce a white solid **1** which showed a single ^{1}H NMR signal at $\delta 2.30$ in SO_2 at -67 °C and reacted with aniline to produce **2**.[15] On the basis of this information **1** was assigned the structure shown. Other ketenes reacted with imines and SO_2 to produce adducts such as **3**.[16] The reaction of **1** with azines has also been reported.[17]

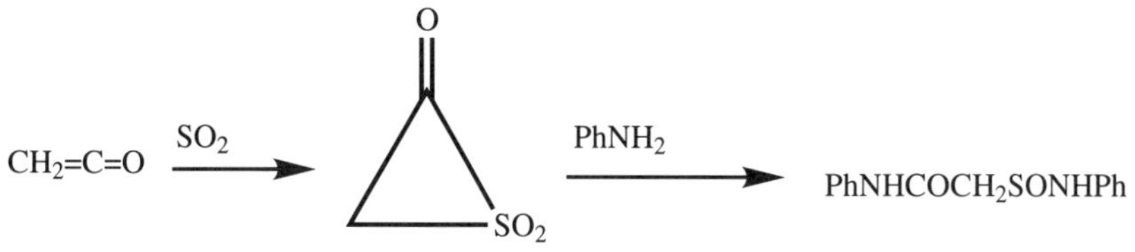

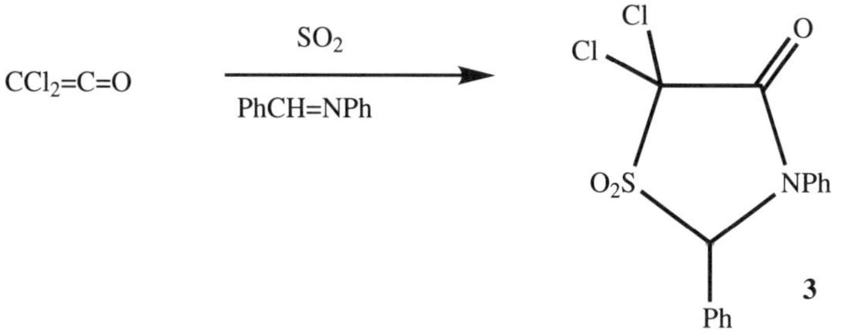

References

1. Staudinger, H.; Reber, Th. *Helv. Chem. Acta* **1921**, *4*, 3–23.
2. Wasserman, H. H.; Berdahl, D. R.; Lu, T. In *The Chemistry of the Cyclopropyl Group*; Rappoport, Z., Ed.; Wiley: New York, 1987; pp. 1455–1532.
3. Turro, N. J. *Acc. Chem. Res.* **1969**, *2*, 25–32.

4. Turro, N. J.; Hammond, W. B. *J. Am. Chem. Soc.* **1966,** *88,* 3672–3673.
5. Schaafsma, H.; Steinberg, H.; de Boer, Th. J. *Recl. Trav. Chim. Pays-Bas* **1966,** *85,* 1170–1172.
6. Turro, N. J.; Hammond, W. B. *Tetrahedron* **1968,** *24,* 6017–6028.
7. Hammond, W. B.; Turro, N. J. *J. Am. Chem. Soc.* **1966,** *88,* 2880–2881.
8. Turro, N. J.; Gagosian, R. B. *J. Am. Chem. Soc.* **1970,** *92,* 2036–2041.
9. Zaitseva, G. S.; Bogdanova, G. S.; Baukov, Yu. I.; Lutsenko, I. F. *Zh. Obshch. Chem.* **1978,** *48,* 131–137; *Engl. Transl.* **1978,** *48,* 111–117.
10. Zaitseva, G. S.; Novikova, O. P.; Livantsova, L. I.; Kisin, A. V.; Baukov, Yu. I. *Zh. Obshch. Khim.* **1990,** *60,* 1073–1077; *Engl. Transl.* **1990,** *60,* 947–951.
11. Zaitseva, G. S.; Krylova, G. S.; Perelygina, O. P.; Baukov, Yu. I.; Lutsenko, I. F.; *Zh. Obshch. Khim.* **1981,** *51,* 2252–2266; *Engl. Transl.* **1981,** *51,* 1935–1947.
12. Zaitseva, G. S.; Kisin, A. N.; Fedorenko, E. N.; Nosova, V. M.; Livantsova, L. I.; Baukov, Yu. I. *Zh. Obshch. Khim.* **1987,** *57,* 2049–2060; *Engl. Transl.* **1987,** *57,* 1836–1845.
13. Salaun, J. *Chem. Revs.* **1983,** 619–632.
14. Nakatani, K.; Takada, K.; Odagaki, Y.; Isoe, S. *J. Chem. Soc., Chem. Commun.* **1993,** 556–557.
15. Bohen, J. M.; Joullie, M. M. *J. Org. Chem.* **1973,** *38,* 2652–2657.
16. Bellus. D. *Helv. Chim. Acta* **1975,** *58,* 2509–2511.
17. Lysenko, Z.; Joullie, M. M. *J. Org. Chem.* **1976,** *41,* 3925–3927.

INDEX

Acetoacetate esters, as ketene precursors, 68, 71
Acetoxyketenes, 314–315
Acid anhydrides, as ketene precursors, 1, 53–55, 60–62, 322
  carboxylic-sulfonic, 61
  carboxylic-phosphoric, 62–63
Acid anhydrides, formation from ketenes, 597, 624–625
Acid chlorides, as ketene precursors, 57–65
Acidity of ketenes, 286, 389
Acyl radicals, 265–266
Acylaminoketenes, 300–301
Acylammonium ions, as ketene precursors, 57–59, 65
Acylium ions, 8–10, 46, 145, 148, 262–265, 330, 586, 622–625, 641
  ketenyl, 145
Acylketenes, 227–254, 418, 453, 456
  formation, 125–129, 392
  polymerization, 641
  stabilization, 228, 355
Adamantylideneketene, 153, 634
Agosta, W. C., 115
Aldoketenes, 151, 461
Alkenylketenes, 59, 103–107, 120–124, 132, 139–143, 176–193, 328, 352, 559, 608
  cycloaddition, 517, 550–551
  ene reactions, 132–133, 562
Alkylketenes, 150–176

Alkyne oxidation with ketene formation, 146
Alkynyl ethers:
  cycloaddition with ketenes, 106, 129, 204, 240, 516
  as ketene precursors, 129–130, 178, 195, 313, 349, 352, 360, 370–371, 396
Alkynylketenes, 193–203, 353, 581
Allenes:
  cycloaddition with ketenes, 329, 511–513
  from ketenes, 196, 301, 357, 392, 450, 455, 517, 604, 619–621
Allenylketenes, 259–260, 360, 634
Allyl ethers, reaction with ketenes, 595–597
Allylketenes, 131
Aminoketenes, 299–310
Arndt-Eistert reaction, 91, 92, 102
  double, 92, 420
Aromatic substitution by ketenes, 103, 105, 116, 134, 139, 142, 404, 561, 608
Aromaticity of ketenes and derivatives, 15, 270–299, 555
Arsenic substituted ketenes, 375
Asao, T., 275
Atomic carbon in ketene formation, 149, 239
Atomic charges of ketenes, 26, 29–30
Aza-Wittig reaction of ketenes, 621
Azides, as ketene precursors, 122, 195–199, 505
Azidoketenes, 61, 303
Azirines, reaction with ketenes, 601

Azoketenes, 303
Azulenones, from ketene cycloadditions, 204, 516
Azulenylketenes, 265, 422

Bader, Alfred, 125
Bartlett, P. D., 475, 631
Benzocyclobutenediones, 414–416
Benzocyclobutenones as ketene precursors, 276, 278
Bernardi, F., 498
Bestmann, H., 619
Biacetyl, cycloaddition with ketenes, 315
Biradicals, 7
   in ketene formation, 277
   from ketone photolysis, 108–115
   from cyclobutanones, 101
   from photolysis of alkynyl ethers, 130
Bis(carboethoxy)ketene, 240
Bisketenes, 405–428, 453
Bis(trifluoromethyl)ketene, 326–331, 472, 484–485, 632
Boron substituted ketenes, 13, 355, 395–399
Brady, W. T., 354, 524
Bromination of ketenes, 357, 628
Bromoketenes, 76, 198, 343
Butenolides, from ketenes, 132, 139, 222, 408, 410, 542, 567, 568
tert-Butylketene, 153, 542, 632

Calculations, see Theoretical studies
Captodative ketenes, 199
Carbanionic ketenes, 266–268, see also, Ynolates
Carbene complexes as ketene precursors, see Metal carbene complexes
Carbenic ketenes, 148, 262, 268, 330, 453
Carbon, atomic, 239
Carbon monoxide dimer, 434–435
Carbon suboxide, 434, 453, 641
Carbonylation of carbenes to ketenes, 137, 148, 206, 272, 274, 276, 311, 325–326, 455
Carbonylation of metal carbene complexes, 138–143, 220, 313, 352, 402–403, 644–645, 649–651
Carborane substituted ketenes, 397
Carboxylic acids, dehydration to ketenes, 53, 325, 348, 382
Carboxylic anhydrides as ketene precursors, 53–55, 322
Carroll, M. F., 125
Catalysis of ketene dimerization, 463–464
Cerium promoted ketene additions, 358, 607

Charged ketenes, 86, 262–268, 279, 330, 354, 392–394, 429–434, 464
Charges, atomic, 26, 29–30
Chiral ketenes, 59, 62, 141–142, 302, 645–646; see also, Stereoselectivity
Chlorocarbonylketenes, 235
Chlorocyclobutenones, ionization and rearrangement, 340–341
   dechlorination, 337–338, 342
Chloroketenes, 20, 336–346, 559
   cycloadditions, 552
Claisen rearrangement, 595, 648–649
Conformations:
   acylketenes, 126, 227–229, 231, 234
   alkenylketenes, 176–177, 180
   aminoketenes, 13, 15, 406, 417
   arylketenes, 20, 22, 37, 204, 582, 605
   bicyclo[3.2.0]hept-3-en-6-ones, 476–477
   bisketenes, 406, 417
   diazoamides, 303
   diazoketenes, 77–78, 81
   hydroxyketenes, 13, 15
   ketenes, 13, 42, 229, 300, 383
Copper ketenides, 399–401
Corey, E. J., 2
Cumulene substituted ketenes, 259–262, see also, Allenylketenes
Cumulenones, 330, 390, 434–444, 454, 563
Cyanide addition to ketenes, 607
Cyano N-oxides, see Nitrile oxides
Cyanoketenes, 20, 193–203
   cycloadditions, 470, 475, 483–485, 491, 495, 501, 503, 511, 566
Cyclization of ketenes, 6–7, 60, 554–564
   acyl, 207, 233, 234, 238–239, 555, 595
   with allyl ethers, 595
   amido, 301
   with aryl rings, 204, 516
   bisketenes, 419
   dienylketenes, 121–124, 181, 560
   hydroxyalkyl, 592–593
   imidoyl, 254–257, 279, 555
   vinyl, 7, 103, 106, 221, 554, 594
   with vinyl sulfoxides, 596
Cycloaddition of ketenes, 459–571, see also Cyclization, Dimerization
   [2 + 1], 653–656
   [2 + 2]: with alkynes, 338, 340, 360, 514–518
      with alkynyl ethers, 106
      with allenes, 329, 511–513
      of bis(trifluoromethyl)ketene, 327–330
      catalysis by $ZnCl_2$, 508

with diazo carbonyl compounds, 541–544, 656
with diazoalkanes, 536–537
with dienes, 220, 275, 314, 328, 336, 340, 372, 381, 475–477, 484, *see also* Dienes
with imines, 300, 303, 311, 314, 518–529
of ketene radical cations, 446
kinetics, 478, 492–495, 528
with nitrile oxides, 537–538
with nitrones, 539
with pyridine N-oxides, 539
theoretical studies, 6–7, 497–502, 518–519
[3 + 2], 82–83, 536–544, 601
[4 + 2], 374, 404, 407, 414, 422, 544–551, 560
[5 + 2], 551
[6 + 2], 551–552
[8 + 2], 552
Cycloalkanones, photolysis, 108–114, 116, 557
1,3-Cyclobutanediones, 53, 207, 314, 459, 461, 468–473
Cyclobutanones, 205
from cyclopropanones, 654
from ketene cycloaddition, 89, 154, 205, 336–343, 473–486, 558
as ketene precursors, 100–102, 106–107, 113, 131–132, 179, 323, 458, 495
kinetics of formation, 478, 492–496
ring expansion, 154, 342, 511, 646–647
stability, 475–477, 491, 504, 545
Cyclobutenediones, 199, 262, 388, 406–414, 514
Cyclobutenones, 204
from ketene cycloaddition, 106, 260, 338–340
from ketene cyclization, 179–180
as ketene precursors, 103–107, 182, 559–561
Cycloheptadienones from ketenes, 549
Cycloheptatriene, reaction with ketenes, 546, 552, 610
Cycloheptatrienylketene, 222
Cyclohexadienones, 120–121, 560, 644
Cyclohexylideneketene, 54, 61, 153, 634
Cyclooctatetraene, reaction with ketenes, 610
Cyclooctene, cycloaddition with ketenes, 491
Cyclopentadienones, 180, 260
Cyclopropanones from ketenes, 91, 260, 360, 450, 485, 653–655
Cyclopropenones:
from bis(diazo)ketones, 88, 261, 268
from bisketenes, 410, 453
ketene precursors, 133, 148, 257, 267, 388
reaction with ketenes, 566
Cyclopropenylketenes, 134

Cyclopropylketenes, 113, 218–227

Dane salt, 302
Dauben, W. G., 115
Decarbonylation of ketenes, 5, 37, 93, 109–110, 121, 180, 194, 219, 233, 239, 316, 322–323, 326, 383, 388, 410–411, 448–458
Decarbonylation, in ketene mass spectroscopy, 45
Dehalogenation in ketene formation, 1, 37, 74–76, 227, 323–324, 336, 340, 434, 464–466
Dehmlow, E. V., 471
Dehydration, of carboxylic acids, 53, 348, 382, 384, 434
Dehydrohalogenation in ketene formation, 36, 57–59, 151, 205, 241, 304, 312, 336–339, 383, 419, 626, 646
Deprotonation of ketenes, 10–11, 267, 358, 389, 392, 429
Diacylketenes, 230
Diazirines, formation from diazoketones, 78–79, 85
Diazoacetaldehyde, 5–6, 77, 80
Diazoalkanes:
reaction with ketenes, 360, 537, 653–654
Diazoalkylketenes, 88, 259, 261, 268
Diazoketene, 304
Diazomalonaldehyde, 227
Diazoniumketene, 86
Dibenzoylketene, 240
Dibromoketene, 34, 37, 75, 344
Di-*tert*-butylketene, 53, 78, 151–152, 446, 449, 586, 604, 611, 623, 632, 634, 637
Dichloroketene, 336–340, 481, 485–486, 489, 540, 648, 651
Dicyanoketene, 198, 566
Dieckmann reaction of acylketenes, 594–595
Diels, O., 2
Diels-Alder reaction, *see* Cycloaddition, Retro Diels-Alder
Dienes, reactivity with ketenes, 336, 340, 372, 410, 475–477, 544–546
Dimerization of ketenes, 6, 71, 75, 94, 116, 151, 204, 207, 218, 230–231, 234, 240, 314, 338, 348–349, 375, 459–473, 550, 603
catalyzed, 231, 603
stereochemistry, 468
Dimesitylketene, 20, 204, 450, 577, 603, 605, 611
Dimethylketene, 20, 32, 38, 52, 55, 57, 117, 150, 446, 449, 461, 465, 470, 490, 494, 503, 509, 513, 639

Dioxiranes, oxidation of alkynes, 147
Dioxinones, formation from ketenes, 70
Dioxinones, as ketene precursors, 125–128, 233, 324, 421
Diphenylketene: 203
  addition reactions, 402, 589, 608–612, 625–627
  bromination kinetics, 628
  cycloaddition reactions, 15, 473, 478–483, 492, 497, 528, 536–540, 543–548
  dimers, 207
  heat of formation, 32
  oxygenation, 445, 630–632, 635
  photolysis, 450
  preparation, 1, 82, 137, 203
  protonation, 623
  pyrolytic decarbonylation, 454
  reactivity in amine-catalyzed ethanolysis, 589
  reactivity toward phenol, 372
  structure, 23, 37
  UV and photoelectron spectra, 36, 203
1,3-Dipolar additions, 536
Dipole moments of ketenes, 20, 21, 44, 176
DNA cleavage by ketenes, 561
Dötz, K. H., 138

E1cb mechanism of ketene formation, 67–70, 196, 390, 590
Electron diffraction, 20, 462
Electron spin resonance spectroscopy, 265, 637–638
Electron transfer of ketenes, 445–447
Electronegativity and ketene stability, 13–14, 303, see also Stabilization energies
Electronic structure of ketenes, 4
Electrophilic addition to ketenes, 622–628, see also Ketene reactions, Protonation of ketenes
Enantioselective additions to ketenes, 206, 343, 599, 620, 643–651
Ene reactions, of ketenes, 179, 182, 277, 484–485, 562, 592
Eneamines, reaction with ketenes, 65, 509–511, 550, 608–609
Enediol, see Ketene hydrate
Energies of ketenes, 13–17
Enolates, additions to ketenes, 12, 575–576
Enols, 87, 236–238, 388, 590–591, 611, 623
Enols, ketenyl, 277
Enones, photolysis, 112–118
Entropies of activation of ketene cycloadditions, 472, 496
Enzymatic oxidation of alkynes to ketenes, 146

Enzyme catalyzed additions to ketenes, 356, 393–394
Esters as ketene precursors, 67–73, 198, 325, 382
Ethoxyketenes, 86, 89, 313, 314
Ethylenimines, reaction with ketenes, 596
Excitation energies of ketenes, 5

Facial diastereoselectivity, in cycloaddition, 485–486
Ferrocenylketenes, 206
Fluorenylideneketene, 69, 87, 205, 274, 520
Fluoroketenes, 20, 57, 58, 121, 126, 204, 321–333, 475, 524–525
Formylketene, 227–228, 453, 594
Fragmentation of ketenes in mass spectrometry, 44–47
Free radicals, see Radical
Friedel Crafts reactions of ketenes, 608, see also, Aromatic substitution
Frontier molecular orbitals of ketene, 5, 6, 581
Fukui, K., 2
Fulvenones, 15–17, 27–30, 205, 269–299
Furandiones, as ketene precursors, 135, 230, 234, 566
Furylketenes, 205–206

Gas phase reactions of ketenes, 7, 11, 44–47, 94, 238, 267, 389, 429, 576, 622, 627, 638
Geometry of ketenes, see Molecular structure
Germylketenes, 351, 360, 393, 395–396
Gold substituted ketenes, 399–401
Gomberg, M., 1
Gompper, R., 487
Grignard reagent additions to ketenes, 604

Halevi, A., 501
Haloacylketenes, from dioxinones, 126
Haloarylketenes, 204
Hammett correlations, 68, 373, 494, 577
Heat of formation of ketenes, 31–32
Heavy atom effects on ketene NMR, 34, 35
Heptafulvenones, 275–276
Hexadecylketene, 70
Hexaketene, 411
Hoffmann, R., 2, 486–487
HOMO of ketene, 4–6, 467, 489, 622, 636
Horner-Emmons reaction, 619
Houk, K. N., 497
Huisgen, R., 472
Hurd lamp, for ketene generation, 54

INDEX **661**

Hydration of ketenes:
  acid catalyzed, 585–587
  acylketenes, 238
  enzymatic, 393–394
  heat of reaction, 11
  kinetic studies, 68, 194, 203, 238, 272, 311, 325, 356, 576, 583
  mechanisms, 238, 576–585
  preparative, 155, 585
  reversibility, 585
  solvent effects, 577
  stereoselectivity, 584
  theoretical studies, 11–12
Hydroxyketenes, 311–312, 388, 407, 516

Imidoylketenes, 254–259
Imines, cycloaddition with ketenes, 300, 518–529
Iminoketenes, 304
Indenylideneketene, 273–274, 584
Infrared spectra of ketenes, 38–43, 57, 114, 120, 195, 198, 206, 219, 231, 232, 240, 260–262, 274, 301, 326, 400, 414, 519
Iron acetylides, reaction with ketenes, 610
Iron substituted ketenes, 400
Isocyanatoketene, 259, 304–305
Isocyanoketene, 304–305
Isodesmic reaction energies of ketene, 13–17, 271, 414, *see also,* Stabilization energies
Isomerization of ketene, 7
Isonitriles, addition to ketenes, 607
Isotope effects on ketene cycloadditions, 468, 496–497
  on ketene hydration, 581
  on ketene IR absorption, 180
Isotope labeled diketene, 466

Ketene acetals, as ketene precursors, 136–137, 198
  from ketenes, 594
  reaction with ketenes, 608
Ketene Claisen rearrangement, 595, 648–649
Ketene dimers, 52–53, 227, 230, 460–462, *see also,* Dimerization of ketenes
  ozonolysis, 55
Ketene equivalents, 459
Ketene formation from acetoacetate esters, kinetics, 70, 237
Ketene hydrate, 11, 271, 275, 572–573, 577, 581, 585, 588
Ketene-acetylene complex, microwave spectrum, 515

Ketene-ethylene complex, microwave spectrum, 502
Ketenes, reactions with:
  acetic anhydride, 609
  alcohols, 103, 229, 236–237, 357, 584, 587–589, 592, 643
  alkoxystannes, 357
  amides, 598, 602
  amines, 229, 237, 357, 589–590, 597–604, 643
  arsenic compounds, 628
  azides, 598
  azirines, 601
  azo compounds, kinetics, 527–528
  boron compounds, 396, 627
  bromine, 357, 628
  carbene metal complexes, 611
  carbenes, 449–450
  carbocations, 148, 625
  carboxylic acids, 624, 645–646
  cerium reagents, 358
  cyanides, 607
  diazirines, 602
  diazoalkanes, 360, 654
  dimethylsulfoxide, 635
  disilenes, 565
  enolates, 12, 575–576, 593
  ethylenimines, 596
  fluoride ion, 329–330
  formic acid, 206, 646
  free radicals, 636–639
  germanium hydrides, 611
  Grignard and organolithium reagents, 12, 358, 604–606
  hydrides, 12, 588, 591–592
  hydrogen atoms, 449, 637
  hydrogen halides, 623–624
  hypochlorite, 635
  imines, 518–527
  iodine, 628
  iodosobenzene, 635
  isocyanates, 528, 640
  isonitriles, 261–262, 404, 602
  metal salts and organometallics, 399–400, 402, 607, 609, 626–627
  methanol, kinetics, 587–588
  methanol and methoxide, 11, 596–597, 645
  niobium complexes, 613
  nitriles, 551, 607
  nitrones, 539
  nitroso compounds, 528
  nitrosyl chloride, 372
  oxiranes, 596–597

Ketenes, reactions with: *(Continued)*
  oxygen, 409, 630–636
  ozone, 631–632, 635
  peroxyacids, 146, 635
  phenoxides and phenols, 236, 357, 358, 372–373, 520
  phosphines and other phosphorous nucleophiles, 612
  phosphorous acids, 597
  pyridines, 12, 240, 574, 600
  pyridine N-oxides, 539
  silenes, 359
  silver ion, 399
  silyllithiums, 611
  sulfenyl halides, 327, 627
  sulfilimines, 329, 600
  sulfoxides, 596, 635
  sulfur dichloride, 627
  sulfur dioxide, 655
  thiols, 611–612, 638
  thiophenols, 645
  thionyl chloride, 628
  tin compounds, 612
  trifluoroacetic acid, 624
  trifluoroacetic anhydride, 626
  trimethylamine, 12
  water, 236, 572–573, 576–587
  ynolates, 392–393
Ketenides, 399–401
Keteniminylketenes, 259, 261–262
Ketenylcyclobutanones, 567–568
Ketoketenes, 461
Ketones, from ketene dimers, 419, 464

Lacey, R. N., 582
Lactams, from ketenes, 599, 601–602
β-Lactams, from ketenes, 124, 299–300, 314, 518–527, 650–651
Lactones, from ketenes, 195, 593, 595–597, 624, 650
α-Lactones, from ketenes, 631–635
β-Lactones, from ketenes, 60, 124, 239, 301, 314, 316, 359, 391, 565–569, 604, 625, 648
β-Lactone ketene dimers, 52, 419, 460–464
Leaving groups in ketene formation:
  aryloxy, 64, 69
  carboxylate, 60, 61, 559
  dienolate, 69
  halide, 57–59
  hydrogen, 266, 435, 592, *see also,* Ene reactions
  ketone, 70
  phenolate, 67–70, 590
  phosphoryl, 62–64
  pyridone, 62
  sulfonate, 60–61, 300
  thiol, 70
LUMO of ketenes, 4–6, 467, 474, 489, 581, 636
Lüscher, G., 619

Malonic anhydride, 55
Malonic anhydrides as ketene precursors, 54
Masochistic steric effect, 475
Mass spectrometry of ketenes, 44–47, 198, 411, 445
Matrix isolation studies, 180, 181, 198, 206, 219, 227, 231, 234, 236, 240, 254, 260, 261, 273, 279, 325, 369, 406, 414, 435, 450
Meier, H., 236
Meldrum's acid derivatives, 55, 93, 179, 241, 304–305, 316, 436, 457, 563, 594
Mercury substituted ketenes, 399–401
Metal carbene complexes, 138–143, 382, 402–403, 525–526, 620, 644–645, 649–650
Metal complexed bisketenes, 407
Metal complexed ketenes, 138–143, 402–405, 550–552, 568
Methoxyketenes, 313, 316, 416, 559
Methylenecyclobutenones, 260, 634
Microwave oven in ketene preparation, 59
Microwave spectroscopy of ketenes, 20, 462
Migration:
  alkoxy, in ketenes, 112
  in cyclopropylketenes, 218
  fluoride, 328
  hydrogen, in ketene formation, 133–134, 179–180, 206
  hydrogen, in ketene reactions, 195–196, 254, 277, 592
  methylthio, 136, 255, 380–381
  in oxiranylketene, 222
  oxygen, 279
  phenyl, 113, 117–118, 136, 148, 611
  phosphoryl, 369
  trialkylsilyl, 137, 456, 506
  in Wolff rearrangement, 94–95
Molecular mechanics calculations for ketenes, 15, 23
Molecular structure of ketenes, 20–30, 153, 355, 374
Moore, H. W., 194
Mukaiyama reagent, 62
Münchnones, 301, 384, 568

Naphthylketenes, 204, 516
Natta, G., 2
Nitrile oxides, cycloaddition with ketenes, 537–538
Nitrones, reaction with ketenes, 539
Nitroketenes, 304
Nitrosoketenes, 304–305
Norbornene, reaction with ketenes, 482–484
Norrish, R. G. W., 2
Norrish Type I cleavage, 108, 556–557
Norrish Type II cleavage, 54, 311
Nuclear magnetic resonance spectra of ketenes, 33–35, 177, 336, 356, 371, 435
Nucleophilic addition to ketenes, 571–619

Orbital symmetry, 486
Oxadiazinones, 134, 420–421, 595
Oxazinones, photolysis to ketenes, 123, 254
  from ketenes, 126, 255
Oxidation of ketenes, 445–447, 549
Oxiranes, reaction with ketenes, 596–597
  as ketene precursors, 111, 120, 132
Oxiranylketenes, 222
Oxirenes, 5–6, 79–80, 85, 137, 146, 273, 311, 448
α-Oxoketenes, see Acylketenes
Oxoquinonemethides, 15, 111, 207, 240, 257, 276–279, 587–588
Oxygen, reaction with ketenes, 409, 412, 630–636
Oxygen substituted ketenes, 93, 194, 220, 255, 310–321, 457
Ozonolysis of ketenes, 155, 632, 634–635

Pentafulvenones, 270–274, 583–584
Peracid oxidation of alkynes, 146, 635
Perkin conditions for ketene generation, 60, 312, 559
Peroxylactones:
  from ketene oxidation, 631–633
  as ketene precursors, 147–148
Persistent ketenes, 151–153, 178, 204, 220, 221, 228–230, 235, 240, 301, 325, 326, 348, 361, 371, 375, 381, 384, 396, 416, 591
Phenols, reactivity toward ketenes, 372–373
Phenoxyketenes, 313, 314, 316
Phenylketene, 15, 32, 37, 83, 149, 203, 391, 455, 537, 546–547, 619
Phosphines, in ketene formation, 75–76, 147–148, 336–337
Phosphorous substituted ketenes, 368–380, 556
Phosphorous pentoxide, in ketene formation, 53, 325, 382, 384

Photodissociation of ketene to CH_2 and CO, 5, 448, see also Decarbonylation of ketenes
Photoelectron spectroscopy of ketenes, 36–37, 445
  ketenyl anions, 429
Photolithography and photoresists, 94, 272
Photolysis:
  carboxylic anhydrides, 54
  cycloalkanones, 108–114, 117
  cyclobutanones, 101, 102
  cyclohexadienones, 120–124, 326
  ketenes, 410, 429, 448–455
  enones, 114–118, 123, 134, 219
Photoprotonation of ketenes, 451–452
Phthalimido-*tert*-butylketene, 301–302, 643
Polyketene, 145, 641
Polymerization of ketenes, 145, 418–420, 631, 639–641
Preparation of ketenes, 52–149
  from acetoacetate esters, kinetics, 237
  from acid anhydrides, 53–55, 60–64
  from acylammonium ions, 65
  from alkynyl ethers, 129–130
  by allene oxidation, 229
  from carbene complexes, 138–143
  from chloroacylhydrazines, 64
  from cyclobutanones and cyclobutenones, 100–107
  by dehalogenation of α-halo acid halides, 1, 74–76
  by dehydrochlorination, 44, 57–59, 229
    NMR study, 57
  from dioxinones, 125–128
  from ester enolates, 67–70
  by ester pyrolysis, 70–72, 229, 237
  from esters with P_2O_5, 325
  by flash vacuum thermolysis, 44, 46
  from ketene dimers, 52–53
  from malonic acids, 54–55, 145
  from vinylene carbonates, 447
  by Wolff-rearrangements, 77–96
  from ynol phosphates by enzymatic hydrolysis, 73
Pressure effects on ketene cycloadditions, 509
  on ketene formation, 67
Proton affinities of ketenes, 10, 622–623
Protonation of ketenes, 264, 356, 585–587, 597, 622–624
  stereoselective, 649–650
  theoretical studies, 7–10, 264
Proton abstraction, see Deprotonation
Pummerer rearrangement, with ketenes, 596

Pyrethroids, from ketenes, 341
Pyridine N-oxides, reaction with ketenes, 539
Pyridylketenes, 205, 206
Pyrimidones, photolysis to ketenes, 123, 254
Pyrolysis, see Thermolysis
Pyrones, photolysis to ketenes, 123, 133, 180, 206
  from ketene cycloaddition and dimerization, 234, 463, 517, 549, 570
Pyrolediones, in ketene formation, 304

Quinoketenes, 414–418
o-Quinones, precursors to bisketenes, 180, 417
  reactions with ketenes, 546
p-Quinones:
  as ketene precursors, 195, 197
  reaction with ketenes, 565
  synthesis via ketenes, 104–105, 182, 559–561

Racemization via ketenes, 103, 130
Radical ions of ketenes, 45–47, 445–447, 549
Radical reactions of ketenes, 449, 455, 631, 633, 636–639
Radical substituted ketenes, 265–266, 421, 429–433, 455
Rearrangements, see Migration
Reduction of ketenes, 445–447
  with hydride, 590–592
Regiochemistry of ketene cycloaddition, 489
Regitz, M., 368
Retro Diels–Alder reaction in ketene formation, 116, 117, 353, 458
Ritter reaction, 445
Roberts, J. D., 103
Runge, W., 354
Ruzicka, L., 2

Samarium iodide mediated addition to ketenes, 606–607, 646
Satchell, D. P. N., 582
Selenium substituted ketenes, 384, 399–400
Silver substituted ketenes, 399–401
Silylacetates:
  ketene precursors, 72
  from silylketenes, 356–357
Silylketenes, 260, 314, 348–368, 393
  dipole moments, 44
  stabilization, 33, 355–356
Silylmethylketenes, 342, 349–350, 526
Silyloxyketenes, 349, 352
Sodioketene, 355, 399
Solvent effects:
  on ketene cycloadditions, 472, 477, 493, 496, 545
  on ketene formation from cyclobutenones, 104
  on ketene hydration and alcoholysis, 238, 577, 583, 587
Spectroscopy of ketenes, see Infrared spectroscopy, Mass spectroscopy, Microwave spectroscopy, Nuclear magnetic resonance spectroscopy, Ultraviolet spectroscopy, and Photoelectron spectroscopy
Stabilization energies of ketene substituents, 10, 13, 150–151, 197, 203, 218, 228, 303, 310–311, 321, 326, 337, 349, 354–356, 368, 380, 395, 412–414
Stannylketenes, 351, 393
Staudinger, H., 1, 2, 74, 150, 240, 619, 630, 640
Stereoselectivity in additions to ketenes, 69, 313, 343, 372, 485–486, 519, 525–526, 575, 584, 588, 593, 597, 604–605, 611, 620, 642–652
Steric crowding in ketene hydration, 581
Steroids, 109–111, 154–155, 204, 651
Structures of ketenes, 20–30
Substituent effects on ketenes, 13–17, 414, 624, see also, Stabilization energies
Sulfoxides, reactions with ketenes, 597, 649
Sulfur substituted ketenes, 380–386
  sulfonium substituents, 68
Sulfur substituted bisketene, 411
Sydnones, 302–303

Theoretical studies, 4–32
  of bisketenes, 406
  of acylketenes, 228
  of ketene cyclization, 238
  of ketene cycloaddition, 488–489, 497–502, 515, 518–519, 547
  of ketene dication, 445
  of ketene dimerization, 461–462, 466
  ketene protonation, 622–623
  of nucleophilic additions to ketenes, 572–576
Thermochemistry of ketenes, 31–32, 429, 622–623
Thermolysis of ketenes, 410–411, 454–458
Thiele, J., 1
Thioketenes, 20, 327, 447
Thiols, additions to ketenes, 613
Triafulvenone, 267–270
Trifluoromethylketenes, 325–333
Trimers of ketenes, 464–465
Trimethyleneketene, 463
Triphenylphosphine, in ketene formation, 75

Tropone, adduct with diphenylketene, 101–102
Tropolones, from ketene cycloadducts, 338

Ultraviolet spectra of ketenes, 36, 68, 89, 311, 325, 383, 388, 414

Vinylketene:
  addition of alcohols, 592
  calculations, 177
  cycloaddition with alkynyl ethers, 516–517
  dimerization, 178, 550
  dipole moment and conformation, 44, 177
  heat of formation, 31
  formation, 53, 59, 116, 177, 179, 592
  NMR spectra, 34–35, 177
  photoelectron spectrum, 37, 177
  radical cation, 45
  protonation, 623
Vinylketenes, *see* Alkenylketenes
Vinylene carbonates, 239, 337, 389, 447

Wedekind, E., 2, 150, 640
Wilsmore, N. T. M., 1–2, 54, 459
Wittig reactions of ketenes, 357, 619–621
Wittig, G., 620
Witzeman, J. S., 237
Wolff-rearrangement, 5–6, 77–96, 106, 227, 313, 353, 380, 382, 420, 453, 537, 541–544, 562, 567, 584, 655
  catalyzed, 84, 218, 353, 397, 537, 543
  kinetics, 83–84, 368

migratory aptitudes, 94, 95
negative ion, 389–390
retro, 81, 455
theoretical studies, 5–7, 81
vinylogous, 95–96, 102, 112, 131
Wolff-type rearrangements:
  of acylketenes, 452–453
  of acylpyrazoles, 179
  of 2-bromo-1,3-diones, 241
  of dichlorovinylene carbonate, 337
  in mass spectroscopy, 46
  of sulfur ylides, 149
Woodward, R. B., 2, 460, 462–467, 479, 486–487

X-ray crystal structures, 20, 22–23, 143, 230, 311, 374, 407, 413, 460, 462, 470, 475

Yamabe, S., 502
Ynolate ions, 10, 360–361, 388–394
  reactions, gas phase, 238–239
Ynols, 179, 388, 449

Zinc, in ketene dehalogenation, 1, 74–75, 204, 323
Zinc salts, 74, 357, 463–464, 508, 514
Zwitterions:
  in ketene cycloadditions, 490, 506, 519
  from amine additions to ketenes, 143, 418, 520, 574, 600, 601, 603, 648